T0139042

The Math of Body, Soul, and the Universe

The Math of Body, Soul, and the Universe

Norbert Schwarzer

JENNY STANFORD
PUBLISHING

Published by

Jenny Stanford Publishing Pte. Ltd.
101 Thomson Road
#06-01, United Square
Singapore 307591

Email: editorial@jennystanford.com
Web: www.jennystanford.com

British Library Cataloguing-in-Publication Data
A catalogue record for this book is available from the British Library.

ISBN 978-981-4968-24-9 (Hardcover)
ISBN 978-1-003-33454-5 (eBook)

Dedication

To the victims of ignorant politicians.
We will not forget. We will not forgive.

Contents

THE FIFTH DAY: LET THERE BE A DIRAC EQUATION

THE SIXTH DAY: A MATH FOR BODY, SOUL, AND UNIVERSE

Acknowledgment

Thank you, reader, for your interest in my work!

About the Book

Dear reader,

Imagine your child being mortally ill—imagine your child asking you to explain the world before it will have to leave it for good—imagine yourself not being able to answer.

Now you know why I have written this book.

Norbert Schwarzer

In order to find the mathematical and physical fundament for the description of body, soul, and the whole universe, which is to say a "Theory of Everything," we think that we require "quantum gravity." That such a theory—in principle—already exists and was derived by the great German mathematician David Hilbert [α] was elaborated in our previous work [β]. Here now, we want to dig deeper and show that quantum gravity is more than just a physical theory, describing physical aspects, but that in fact it is covering "it all."

References

[α] D. Hilbert, Die Grundlagen der Physik, Teil 1, Göttinger Nachrichten, 395–407 (1915).

[β] N. Schwarzer, *The World Formula: A Late Recognition of David Hilbert's Stroke of Genius*, 2022, Jenny Stanford Publishing. ISBN: 9789814877206.

Technical Note

Even though it will mean quite some redundancy throughout the book, we will try to use as few as possible referencing within this work and rather repeat certain derivations than asking the reader to look up equations which are sometimes more than hundreds of pages apart or to seek for reference in a list of several hundred.

As we also intend to present our findings—roughly—in the chronological order as they appeared in the "space-time of the author," we will have to repeat some aspects several times in order to allow the reader to actually observe the evolution of our thoughts. This strategy has the inbuilt advantage and comfort of avoiding too much zigzagging in the search for certain sections and text parts and allows the individual adjustment of the topics in question to the field they are being applied to. This also has the advantage of chapters being roughly complete and self-consistent in case that a reader is only interested in one of them, respectively only wants to collect information about a certain topic (like socio-economy) without being bothered with physical basics on rather far off subjects like the 3-generation problem or the question of what is mass every time he simply wants to look up some of the more fundamental equations. With the redundancy used in this book, he will find most of what he needs to understand his topic inside the chapter where it is presented. Naturally, however, this could not have been done for all fundamental derivations, and thus a little bit of zigzagging will still be of need.

Personal Motivation

For my dearest father-in-law

It was on a Tuesday morning. The sun was shining on Tankow and no wind, as we usually have here, disturbed the wonderful silence. At the most, the birds were chirping in the bushes, which you had trimmed lovingly so often and with such pleasure. The ice glistened on the small fish pond. The first cranes hovered over the flat land and in the pasture in front of our house the geese chattered. Surely, they told each other where they had spent the winter or looked forward to the upcoming spring and breeding season.

Then you decided to leave this world.

You did not leave, as so many do, tearing a big void and leaving unfinished and untidy construction sites everywhere. No, your departure—though for us—unexpected and terribly sudden, was somehow organized and almost silent. Yes, our crying disturbed the silence and it was the mourning of so many people. People who had known you ... and loved you. Even our friend Andy, who rarely shows any feelings, couldn't help lighting a candle for you and writing "For a good friend" to it. But there was no open question like "what will be next" or "how are we going to do this without our grandpa, dad, husband or friend." No, and you had made sure of that: everyone would miss you, but no one could say you owed anyone anything.

You were (no: you ARE) someone who could live and you have lived indeed!

You could share and you have shared.

You could work and you have worked. Oh yes, and how! I almost believe that you could not have done without "digging, trimming, and mucking" and maybe that is why you said goodbye now, when your knees did not want to anymore and walking, and with it the "digging, trimming, and mucking," became more and more difficult, partly impossible.

Your four grandchildren loved you, your daughter adored you ... and me? Well, I was always there, in the middle of it or sometimes beside it, and I was so grateful for the freedom you gave us all because you were just the way you were ... or better: you are the way you are. If I needed time for myself—and that was often—or my dear wife and I needed time for us—that was seldom—then you were only too happy to be there and took over with dedication the tiresome tasks of a caring grandfather for a whole swarm of small and oh-so-lively grandchildren. I'm sure that's exactly why they became such wonderful people, because they had so much contact with you, grandma, and the other grandparents.

I didn't think it would make me so sad to see you go. Yes, I know you are not really gone, but it will be hard to play Yahtzee with you in the future, to have funny conversations, or to toast to the good marks of your grandchildren with that great beer from the Ore Mountains you enjoyed so much. I will also miss not having to look for the wheelbarrow or pruning shears, and I will hate having to take care of my lawn and the shrubs I mentioned earlier.

But I will know that you are there. When a twisted branch is suddenly no longer blocking the path, when the leaves have been "swept" off the paths as if by magic, or when a bent little flower stands upright again. Then I know that you have been "digging, trimming, and mucking" again and certainly had fun doing it.

See you!

Your son-in-law Norbert, who respects you so very much.

The goal of this book is to work out the mathematical fundamentals for the description of what often unscrupulously is called the "soul."

Some Fundamental Motivation

Why—apart from the philosophical question—would we need a quantum gravity or—as one also calls it—a Theory of Everything Approach for Everything?

From the Fermi-Paradox to a "Fermi-Prophecy"

In 1950, the famous physicist Enrico Fermi formulated a problem about the apparent absence of extraterrestrial intelligent life in our universe, where he expressed his astonishment about the fact that the earth is reached by the light of so many stars (about 10^{21}), but no signals from extraterrestrial civilizations.

Some scientists think that this is due to the simple fact that civilizations, having reached a state in development where they could send messages into the universe to reveal their presence, automatically also possess the technical ability to destroy themselves. This option for total obliteration then provides such a narrow bottleneck situation for all intelligent life-forms that—so far—no extraterrestrial civilization has made it though.

Seeing the action and nonaction of our political, social, media, and scientific "elite," this hypothesis appears quite reasonable. Just taking the anaphylactic shock reaction of almost all societies around the globe regarding the so-called COVID crisis proves our point. A holistic socioeconomic analysis (e.g., in our series "*Medical Socio-Economic Quantum Gravity*"), namely, clearly exposes the "elite" as an ever more tyrannic, selfish, and parasitic mass of absolutely incapable entities, which are securing their undeserved "leading" position only via a network of symbiotically entangled moochers.

With such "leadership" any progress permanently leads to the irresistible self-destruction of humanity and thus explains Fermi's paradox [1]. In this sense one may take the Fermi-Paradox and interpret it as a prophecy, which is to say, the "Fermi-Prophecy."

Figure 1   Spatial spinor as it occurs in 8 dimensions in a quantum gravity theory reverses sign if it is rotated 360°. Only a double rotation of 720° brings back the original object. There are tons of horrible applications for this little piece of science alone.

Would a Fundamental Theory of Everything Help Mankind to Avoid to Wipe Itself Out?

Einstein once said:

> **"Two things are infinite: the universe and human stupidity; and I'm not sure about the universe."**

and this author thinks he has had a good point there.

In [2], we discussed the example of the famous hockey stick curve, which was created by some Mr. M. Mann in order to show us that man-made CO_2

is responsible for a global temperature increase since man started to burn fossil fuels. After Mr. Mann lost his case against Prof. T. Ball, who had accused him of being a fraud (famous are "his words that Michael Mann *'belongs in the state pen, not Penn State'*, a comical reference to the fraudulent 'hockey stick' graph that knowledgeable scientists knew to be fakery," from [3]) many people slowly—very slowly indeed—started to realize how much this obviously wrong and made-up "man-made CO_2=global warming hypothesis" has cost humanity. Billions and billions of dollars, euros, yen, and whatever spent for nothing but a crude idea, a "scientist" having no scruples to fake data, and a sufficiently big number of opportunistic followers with no backbone but a big lust for good positions, totally pointless idle "science projects," and no conscience, inner compass, or at least logic orientation whatsoever. Thereby it was rather obvious that the climate community and their doing had very little to do with science ... any science at all, if truth be told [4]. This combined with the always omnipresent corrupt politicians jumping the bandwagon and some clever entrepreneurs smelling a nice chance for easy profit and personal gain on the cost of the society gave the right mix for the financial, political, and educational disaster of the century. But the worst of it was, and still is, that all these recourses spent on the nonsense fight against a harmless gas are lost for some reasonable measures against the process of man definitively destroying the planet. One simple glance on the trash, most especially the dangerous plastics, swimming around in the world's seas and another one on the forest destruction, wetland "cultivation" and disruption of the hydrological cycles, annihilation of natural water buffer systems, the wipeout of species, and so on should suffice to get the gist about the dimension of the true and important problems. In [2], we also have shown how an ideology-free, fundamental, and holistic approach, in other words TRUE SCIENCE in the sense of K. R. Popper, might have helped to prevent this disaster.

We also had discussed other examples from materials science, agriculture, and social science. We saw that wherever any form of dogma and ideology is allowed to enter scientific research, it will—over short or long—take over and govern this very research, because this is the only way—as an intrinsic and inconsistent lie or lying entity, which is to say as a parasite—the dogma or ideology can survive. Thereby it uses the one thing all scientific research—no matter how proper it is being performed and how well it is being monitored—contains and everything in this universe does possess, uncertainty. Parasites hide in the niches and the shadows of uncertainty and are being ignored or just not being detected there, and they can grow until they are big and powerful enough to seize control. Even the greatest rubbish like the gender nonsense or the most self-destructive hoaxes like the Nazi race theory or the COVID-19 lie [5] can become governing and society-defining rulers. This may not have been a threat to mankind as a whole as such mishaps and stupidities were only of local character, destroying "only" smaller societies. Even when this led to the destruction of empires like the Roman one, thereby plunging a whole continent into darkness for almost a millennium, the disaster was not global and therefore no threat fit to fulfill the "Fermi-Prophecy." Nowadays things are very much different and blunders, if done on a global scale, can easily lead to a global wipeout.

Quite an excellent example seems to be the current COVID-19 virus crisis (e.g., [5]), which we are going to make clear by a simple investigation of risk to benefit balance for the vaccination against this virus.

A Simple Calculation and a Terrible Result

If truth be told, my wife has begged me not to publish this text. The reason is the sheer terror which is being unleashed over everybody who says "the wrong things," in other words, the truth, about the COVID-19 virus. And this holds for almost all aspects of this crisis, be it the origin of the germ, the circumstances about how it got free, its pathogenic capabilities, treatments against the illness, and so on. We leave it to the reader to find out why this is, but obviously many people have many things to hide here and as we also see that many people are suffering, that there are deaths and forever disabled, that whole economic sectors are brought to extinction, we can easily understand the motivation for all the attempts to cover up bad happenings and suppress any honest information campaign. Under these circumstances, many scientists have decided to better keep quite rather than risking a well-paid job, their scientific and social credibility, the safety of their own family or—yes—even their own life. On the other side, there are always those sycophants among the scientists who give a damn about the truth, but only see their own forthcoming. The sudden appearance of those peculiar so-called "experts" in the mainstream media just speak volumes.

So, why meddling in this, if it is so dangerous and I'm not even, technically, an "expert" in this field? Wouldn't it better under these circumstances to just join the chorus of the silent?

May be?!

But then I tell myself again that I am a scientist and science is there to bring even uncomfortable truths out of the darkness of mendacity onto the brightly lit table. Science—real science, one must say—cannot be comfortable and a REAL scientist must not make himself comfortable. I thought of Martin Niemöller, a Christian priest who suffered under the Hitler regime, and his famous words:

First they came for the socialists, and I did not speak out—
Because I was not a socialist.

Then they came for the trade unionists, and I did not speak out—
Because I was not a trade unionist.

Then they came for the Jews, and I did not speak out—
Because I was not a Jew.

Then they came for me—and there was no one left to speak for me.

When would they come to us? I asked myself, and yet I 'stood there and couldn't help it', and I had to write down these facts. I had to name these uncomfortable truths, because just as with Niemöller, it was also about murder ... statistically secured murder, one should rather say, or, to keep it quite correct, about presumed statistical murder, which would go into the tens of thousands, if not even the tens of millions, or already went.

I am a scientist and not a lawyer and so I understand the word "presumably" here less as a legal quibble than as an expression of the fact that scientific findings should be expressed in a falsifiable way and thus everyone is enabled to verify these findings. In my case, these are fairly simple calculations that result from pooling the work of other scientists. Although the calculations seem simple to me, I cannot exclude having made a mistake ... because that is how science works. You have to start working somewhere and mistakes can always happen while working. However, since the results of my calculations seem so cruel, I had no other choice than to present them here together with the calculations—at least the simplest ones—myself. So, the reader and the potential critic can check everything and come to his own conclusion.

Should anyone be able to show me clearly and cleanly on the basis of these calculations that I am wrong, I am the last one who would oppose a correction.

So let us begin.

We start with the following scientific study by Walach et al. [6], which can be found at the following link or could be found until recently (more on this further below):

https://www.mdpi.com/2076-393X/9/7/693/htm (now: sciencefiles.org/ wp-content/uploads/2021/07/vaccines-09-00693-v2.pdf)

This study dealt with undesirable side effects of vaccination and the facts found there can now be summarized with those of other authors. Brown [7], for example, calculated the absolute efficacy of COVID-19 vaccination, while Prof. Luckhaus [8, 9] dealt with the question of how dangerous the virus really is.

Since in the Walach et al. study they found that "currently we see 16 serious adverse reactions per 100,000 vaccinations, and the number of fatal adverse reactions is 4.11/100,000 vaccinations," it is absolutely evident that vaccination is anything but harmless. This is even more true since:

1. so far, it has only been possible to record short-term damage, because to determine the medium- and long-term damage, it has not been long enough since the vaccinations were administered,

2. the dark figure might be extremely high, because, as already known from criminal statistics (see migrant criminality), there is a high reporting threshold for vaccination damages in an EXTREMELY pro-vaccination adjusted social psycho-pressure climate (as we have it at present, as every look into the mainstream media, and as the very vaccination fetishists and vaccination fascists themselves might confirm best), and

3. one could already find strong statistical evidence for large-scale cover-ups of vaccination damage (e.g., https://sciencefiles.org/2021/06/30/ indizien-dafur-dass-schwere-nebenwirkungen-von-covid-19-impfstoffen-vertuscht-werden-hochste-zeit-fur-eine-offentliche-diskussion/ and [11, 35]).

Now, Walach and co-authors in [6] already found that COVID-19 vaccinations kill about 4 people before any of the administered "shots" would have saved between 2 and 11 from COVID-19 death with just these vaccines. But here there is still considerable doubt as to whether we are really talking about "died from COVID-19" or just "died with COVID-19" deaths. Walach et al. did not distinguish by age but looked at the integral balance. Nevertheless, their calculation is already a disaster for vaccinators, and that is precisely why the authors' study is under attack. However, if the work of Luckhaus [8, 9] is taken into account, the situation becomes even worse. It has been known for a long time that COVID-19 is de facto only really dangerous for older people, while younger people mostly show only mild symptoms, if any at all. As an example, one may just take the number of active-duty members of the US army who had COVID-19, which is 202,567. Twenty-six died from or with the virus, which gives us a survival rate of 99.9872 percent. These numbers are from mid-July 2021 and they make clear that people younger than even the oldest employees of the US army have not much to fear from the COVID-19 virus. However, it is precisely the younger ones who increasingly have to complain about serious side effects of the vaccinations or even die from them, as the data from Israel shows (Fig. 2).

Benefits and risks after dose 2, by age group

For every **million** doses of mRNA vaccine given with current US exposure risk[1]

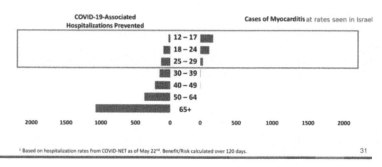

Figure 2 Cases of myocarditis after vaccination against COVID-19. *Source*: https://www.medpagetoday.com/opinion/second-opinions/93340.

Already in the rather upper age group of 50 to 64, instead of the ratio of 4 killed/2–11 saved found by Walach et al., the alarming number of 4 to 1 applies! We do not even want to talk about the 16 people who were seriously injured at the same time and will severely suffer for the rest of their lives. That is, four times more people die from the vaccination than from COVID and this with the above already listed points 1 to 3, which were not even taken into account. If doing so and applying the factor, which just came up in a law case in the USA [35], the number of deaths, caused by the COVID-19 vaccines, should be estimated to be at least 5 times higher. But, as said, we will not take this breaking news into account, because we do not want to speculate about dark figures, cover-ups, and medium and long-term damage.

However, what we could still look at is the issue of alternative treatments.

The simple rule about vaccination (any treatment) is or rather should be as follows:

Vaccination only makes sense if the disease it is intended to protect against is worse or more seriously dangerous than the vaccination itself. This criterion should also be used to select the people to be protected by the vaccination.

If, however, one uses the results from [7–9] and calculates a probability for under 65-year-olds (specifically 50-64; unvaccinated) who have already caught the infection, to also die from the infection and arrives at 0.1%, then the absolute protection from Coronavirus death caused by the vaccination would be just 0.0007 to at most 0.0011 percent. That is, according to Brown and Luckhaus, 88,000 to 142,000 vaccinated people WERE necessary to save even one person from Coronavirus death.

The data of Prof. Luckhaus, however, originate from a time (April 2020) when the German university hospitals still let people gargle with alcohol because they did not know how to help themselves otherwise, while alternatives to a directly aimed vaccination as well as the search for it were massively suppressed. That is why we used the word "WERE" in the paragraph above and had put it in capital letters.

Therefore, in order to find a comparison of vaccination with alternatives, weightings must be incorporated. These include reasonable treatment methods and the timely administration of active agents that have become known in the meantime, such as ivermectin, NAC, vitamins D and C, zinc, hydroxychloroquine, and so on (e.g., [10]). Such substances reduce the value of 0.1% found by Prof. Luckhaus by a factor of 1/10 to 1/50 and more (e.g., [10]).

Our motivation for this kind of weighting is simple: if you want to assess the sense of a vaccination, you have to compare it with the available alternatives and not just with the numbers that would come out without it (reminder: Luckhaus' death numbers referred to April 2020, that is, the "pre-ivermectin time," which is to say before the knowledge of its effect on COVID-19).

Only assuming the worst of the possible factors in the "calculation" for the rescue by the experimental vaccine injection, one now obtains that 880,000 to 1,420,000 vaccinated persons would (at least) be necessary to save even a single person in the age group under consideration from Coronavirus (age 50 to 65).

Already here now, the author certainly must have had—so he was hoping—a thinking error, because "the big wigs" could not be so stupid to want to vaccinate at all in this situation and even think about young people and children, to whom they want to administer the ultimately only emergency-approved and obviously very dangerous stuff (Vaccine Adverse Event Reporting System [VAERS], [6]) and with whom the COVID situation is still much more relaxed than with the elderly, to whom I apply my "calculation."

However, it seems that there was no error and it even gets worse. In fact, the following pinpoints about the COVID-19 crisis should be given. Everybody can/could easily extract them from official statistics and/or media (even mainstream), which is the reason why we only give very few references (things are just too obvious for the awaken—not "woke"—eye):

- COVID did not cause excess mortality at any time; that is, it was not and has never been anything more than a normal seasonal influenza [8, 9].
- Any excess mortality was only caused by lockdown measures, masks, and vaccination campaigns.
- Masks don't help, they do harm [12, 13].
- The average age of COVID deaths is consistently higher than the average life expectancy (so why vaccinate young healthy people?).
- Children and adolescents play a negligible role in the spread of COVID-19. Closing kindergartens and schools is therefore only counterproductive [12].
- The resulting record levels of suicides among children and young adults, as well as overcrowded child psychologies, speak volumes of regime-induced suffering.
- The economic damage caused by the nonsensical measures already exceeded by the end of 2020 what even the worst flu epidemic of recent years caused by at least 33 times. We are now at a stage where we can say that even the worst COVID events would have achieved at most 1/100th of the damage caused by the measures.
- The damage to humanity is so great that it already exceeds the dimensions of world wars.
- COVID vaccination has no positive outcome for those under 65 years of age (e.g., [6, 33]).
- For those under 50, per data, it is not an immunization program, but a statistically proven harm and kill program (see [6] and our own calculations above).

- PCR tests per se are not useful for COVID diagnosis and with CT values above 25, they are complete nonsense, but a welcome way to spread fear and panic. De facto, there would never have been a COVID "pandemic" without the PCR test (e.g., [12, 17]).

If all treatment optimizations, which were found in the meantime, were taken into account (and if one would concentrate on them in the treatment of patients) and if one would not assume the worst case, as I did, the numbers of people who would have to be vaccinated in order to save even one single person from the death of COVID-19 would be even higher. For younger people, it would be in the tens of millions that would have to be vaccinated to save even a single person from COVID. From this it seems—after only 6 months of vaccination campaign—almost inevitably to follow that the vaccination itself will produce many more deaths than it saves from COVID.

In the case of children, adolescents, and young adults, one can certainly assume murder will become part of the discussion as even the simplest estimates began to raise the alarm and is it inconceivable that the governmental vaccination authorities would not know this?

Now, coming back to our little evaluation, we want to combine our results with the ones of Walach et al. [6]. If we take only the possibilities of the drug ivermectin [10], which have been recognized by the WHO in the meantime, and the approx. 4/5 improved survival chances for COVID-19 patients, which are also confirmed by the WHO—apparently with gritted teeth, because there are also studies with even better figures [12]—when this cheap drug is used, the original Walach balance [6] deteriorates to 20 people killed or 80 seriously injured for one person saved. This, however, refers to the age group of 50–64 years. For younger people, the ratio is even more devastating.

Now, scientific honesty dictates us to mention that the paper by Walach et al. [6] has since been challenged [14]. If one searches the corresponding journal, one finds the following message:

If you take a closer look at this "statement," you will be surprised to see that it is not primarily the number of victims that is under discussion, but essentially the balance sheet that the courageous authors of the study determined from it. Of course, they also try to attack the connection between death and vaccination (see again points 1–3 in my text above). However, if one assumes that the link of deaths to vaccination was correctly determined by Walach et al. (and is probably worse, see [35]), a breakdown of the balance by age and the inclusion of effective and cheap alternative treatments result in a sheer grotesque skewing of benefit/harm.

Expression of Concern

Expression of Concern: Walach et al. The Safety of COVID-19 Vaccinations—We Should Rethink the Policy. *Vaccines* 2021, 9, 693

Vaccines Editorial Office

MDPI, St. Alban-Anlage 66, 4052 Basel, Switzerland; vaccines@mdpi.com

---

The journal is issuing this expression of concern to alert readers to significant concerns regarding the paper cited above [1].

Serious concerns have been raised about misinterpretation of the data and the conclusions.

The major concern is the misrepresentation of the COVID-19 vaccination efforts and misrepresentation of the data, e.g., Abstract: "For three deaths prevented by vaccination we have to accept two inflicted by vaccination". Stating that these deaths linked to vaccination efforts is incorrect and distorted.

We will provide an update following the conclusion of our investigation. The authors have been notified about this Expression of Concern.

Reference

1. Walach, H.; Klement, R.J.; Aukema, W. The Safety of COVID-19 Vaccinations—We Should Rethink the Policy. *Vaccines* **2021**, *9*, 693. [CrossRef]

Figure 3 "'Expression of concern' by the editors of *Vaccines*." *Source*: https://www.mdpi.com/2076-393X/9/7/705.

However, what all these sycophants of politicians who are now attacking the Walach paper seem to forget or even want to intentionally overlook is the following: It doesn't matter if we have 4 killed for every 2–11 saved, as Walach et al. concluded, if it's 4 to 1 or even if we get 20 to 1, the thing is just ethically unacceptable. Even much more favorable scenarios such as 1/10, 1/100, or even 1/1000 would still mean that one person would be killed by the COVID-19 injection, which was emergency approved without real need in order to potentially [1] save 10, 100, or even 1000 people from COVID-19 death. But that one person would not have died otherwise, not from COVID-19 or anything else.

He died solely because of the global experiment of COVID-19 vaccination!

The crucial point in Walach's and our calculation, however—and here, in the hype of COVID craziness, the critics apparently no longer see the forest for the trees—is the fact that there are large groups of people for whom COVID-19 poses no danger at all, neither the alpha, beta, nor the delta, nor even the DaDa variant. But for them the experimental vaccination is a danger and a very deadly danger it is. Who does not want to take note of this is not a scientist ... no matter how often he is invited by the system media as a "polithoric" twerp to so-called "interviews" and "expert rounds," no matter how much he may "research" at the expense of the taxpayer and give "his mustard" on twerps to the regime's dishes and the mainstream's one-size-fits-all pap. These persons are—scientifically seen—not worth to be taken seriously. For "scientific" journals that take part in such witch hunts and do not protect the true experts, like Walach and Co., this applies to the same extent.

So, now I still hope that I have made a huge calculation error and ask all readers to check my figures.

If they are even halfway correct, the following conclusion automatically applies:

Purely statistically seen, this COVID-19 vaccination campaign is not an immunization but a killing program!

Still, to no one's surprise, the editors of the journal *Vaccines* have meanwhile retracted the paper of Walach et al. with the following statement (https://www.mdpi.com/2076-393X/9/7/729/htm [see also Fig. 4]):

> The journal retracts the article, The Safety of COVID-19 Vaccinations—We Should Rethink the Policy [1], cited above.
> Serious concerns were brought to the attention of the publisher regarding misinterpretation of data, leading to incorrect and distorted conclusions.
> The article was evaluated by the Editor-in-Chief with the support of several Editorial Board Members. They found that the article contained several errors that fundamentally affect the interpretation of the findings.
> These include, but are not limited to:
> The data from the Lareb report (https://www.lareb.nl/coronameldingen) in The Netherlands were used to calculate the number of severe and fatal side effects per 100,000 vaccinations. Unfortunately, in the manuscript by Harald Walach et al. these data were incorrectly interpreted which led to erroneous conclusions. The data was presented as being causally related to adverse events by the authors. This is inaccurate. In The Netherlands, healthcare professionals and patients are invited to report suspicions of adverse events that may be associated with vaccination. For this type of reporting a causal relation between the event and the vaccine is not needed, therefore a reported event that occurred after vaccination is not necessarily attributable to vaccination. Thus, reporting of a death following vaccination does not imply that this is a vaccine-related event. There are several other inaccuracies in the paper by Harald Walach et al. one of which is that fatal cases were certified by medical specialists. It should be known that even this false claim does not imply causation, which the authors imply. Further, the authors have called the events 'effects' and 'reactions' when this is not established, and until causality is established they are 'events' that may or may not be caused by exposure to a vaccine. It does not matter what statistics one may apply, this is incorrect and misleading.
> The authors were asked to respond to the claims, but were not able to do so satisfactorily. The authors were notified of the retraction and did not agree.

Reference

1. Walach, H.; Klement, R.J.; Aukema, W. The Safety of COVID-19 Vaccinations—We Should Rethink the Policy. *Vaccines* **2021**, *9*, 693. [CrossRef]

Figure 4 Retraction comment by the editors of *Vaccines*. *Source*: https://www.mdpi.com/2076-393X/9/7/729/htm.

This time it is important to note that the editors claim that the authors "were asked to respond to the claims, but were not able to do so satisfactorily."

Well, this author directly asked the authors about this fact and received the original response they had sent to the editors. With permission from the authors, we here present their text.

"Response to 'Incorrect use of data …' by Prof. Dr. Eugène van Puijenbroek"

Harald Walach, Rainer J. Klement, and Wouter Aukema

We are grateful to Prof. van Puijenbroek for raising his concerns. This starts a long-overdue debate on how to gauge the safety of COVID-19 vaccines. We would like to remind Prof. van Puijenbroek and all readers: These vaccines have had an emergency approval without the necessary safety data. Although we would agree with Prof. van Puijenbroek that the self-reporting system of side effects for vaccines and other drugs is far from foolproof, it is the only data we have. So why should it not be put to use?

It is interesting to note that Prof. Puijenbroek, in his concern, describes the Lareb-ADR data as "spontaneous reporting." In a statement in Regulatory Science 2021 (https://www.regulatoryscience.nl/editions/2021/12/prof.-dr.- eugene-van-puijenbroek-on-the-nature-of-signals, accessed 29th June 2021), he says:

"The Netherlands Pharmacovigilance Center Lareb collected 34,000 reports of adverse drug reactions in 2019, of which 14,000 reports are submitted directly to Lareb by healthcare professionals and patients and more than 20,000 were forwarded by the marketing authorization holders. These reports are assessed and analyzed, which may lead to safety signals about adverse drug reactions. These are reported to and reviewed by the Medicines Evaluation Board (MEB), supporting the MEB in its decisions in pharmacovigilance in the Netherlands and Europe." (typos and grammatical errors removed, else identical with web quote at the end of the article)

So, what is really true and what should we go by: Is it true that roughly 60% of the adverse drug reaction (ADR) data come from market authorization holders, who, by law, are required to report, and is it true that the data are reviewed, as stated on the website and in this article, or are these data only true in all other cases but not in the case of COVID-19 vaccines? It would be good to have clarity on this point. We assumed that what Lareb says about all other ADR reports is also true of COVID-19 ADR reports. If we were mistaken in this assumption, perhaps Lareb should clearly state: "ADR reports are reviewed and evaluated in all cases of ADR reports, but not with COVID-19 vaccines." And, ideally, it should also give a reason, why this is so, if it is so.

Ideally the consequence of this debate is that someone sets up a systematic observational post-marketing surveillance study in a large number of vaccinated persons under public scrutiny to really document the side effects that can be causally related to the vaccine. Currently we only have association, we agree, and we never said anything else. But the same is true with fatalities as consequences of SARS-CoV-2 infections. The cases that are counted here as deaths are rarely vetted by autopsy or second opinion, but still counted as deaths due to COVID-19. And it is exactly this allegedly high number of COVID-19-related deaths that gave rise to an unprecedented sloppy regulation process that allowed new types of vaccines using a mechanism never before tested in humans to be widely distributed in the population. Prof. Puijenbroek basically argues that the largest vaccination experiment in the history of medicine cannot be assessed for safety and unforeseeable toxicities, because we should not use the ADR data for such inferences. In contrast, we argue that it is mandatory that those data which are at hand are used to gauge the safety, and this is what we have done.

We are happy to admit that these data are far from perfect. But we repeat: they are the only ones that are available. We quoted Lareb itself, which states on its website at the time we checked the data:

"All reports received are checked for completeness and possible ambiguities. If necessary, additional information is requested from the reporting party and/or the treating doctor. The report is entered into the database with all the necessary information. Side effects are coded according to the applicable (international) standards. Subsequently an individual assessment of the report is made. The reports are forwarded to the European database (Eudravigilance) and the database of the WHO Collaborating Center for International Drug Monitoring in Uppsala. The registration holders are informed about the reports concerning their product."

We took this statement to mean that those reports that are obviously without any foundation are taken out such that the final database is at least reliable to some degree. Would it not be like that, why else would one want to collect these data and make them public in the first place?

We are happy to concede that the data we used—the large Israeli field study to gauge the number needed to vaccinate and the Lareb data to estimate side effects and harms—are far from perfect, and we said so in our paper. But we did not use them incorrectly. We used imperfect data correctly. We are not responsible for the validity and correctness of the data, but for the correctness of the analysis. We contend that our analysis was correct. We agree with Lareb that their data are not good enough. But this is not our fault, nor can one deduce incorrect use of data or incorrect analysis.

And we hope that this stimulates governments or university consortia to collect valid data to prove us wrong. We would be the first to be happy about that. But the challenge is out: Prove that the vaccines are safe! No one has done so. We say they are not and we used the best data currently at hand. Our usage was correct. If the data were not, whose fault is this?

Ok, what was the second to last sentence in this statement of the editors again (c.f. Fig. 4):

"The authors were asked to respond to the claims, but were not able to do so satisfactorily."

No further comments are needed at this point, we think.

A Postscript for a Current Occasion

As it has since emerged there is now clear evidence that some countries are engaged in a massive cover-up of COVID-19 vaccine injuries (www. impfnebenwirkungen.net/report.pdf and [11, 35]), we want to further motivate the reader to perform his own evaluations. In the context of the facts and calculations presented above, the reader may conclude for himself how likely it can still be that this author could be significantly off in his results ... and if at all then certainly not in a direction that would somehow make the balance of damaged to saved look better. The opposite is more likely to be the case (see Fig. 5).

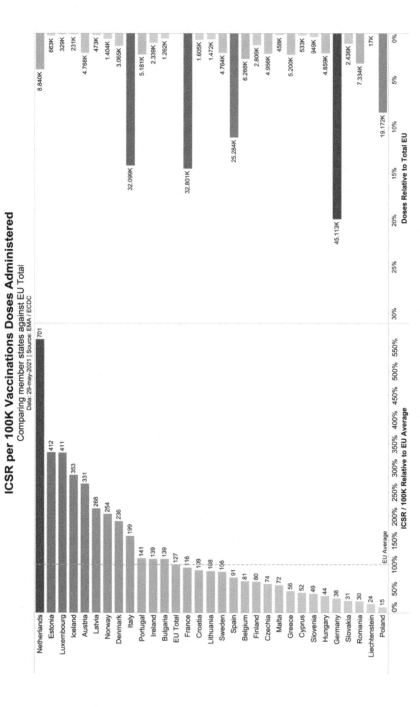

Figure 5 Overview of adverse effects of COVID-19 vaccinations for various countries (*source:* [6]). It is obvious that in Germany, with the highest vaccination rate, the reporting of adverse effects is massively suppressed (c.f. the Netherlands at the top).

Another Postscript

In the meantime, experts assume that the number of victims caused by vaccinations, which already exceeds any acceptable level, especially among young people and children, is still massively underestimated (up to a hundred times [16] and we are still talking ONLY about short-term damage). This also is supported by most recent reports about massive fraud and cover-up in the various vaccination damage reporting systems [35].

This is more than understandable for the author for many reasons:

First, it is easy to see that in a psychological environment that is more like a Coronavirus and vaccination terror regime than a democracy, reports that contradict the common and powerfully pushed propaganda are prevented. People simply do not dare to report their vaccination complaints or to indicate their own relatives as victims of these vaccinations.

Second, it is a long and difficult procedure to get from the detection of vaccine damage to the final entry in the corresponding database. There are many obstacles and—above all—stakeholders who have no interest in too many reports of vaccine damage being recorded.

Third, as has been recognized in the meantime (see Fig. 5), data sets on vaccination damage are subsequently removed to make the matter look not as bad as it actually is ... at least for a certain time to be able to persuade as many as possible to vaccinate.

Fourth, "ANAHEIM, California, July 22, 2021—A government related computer programmer, who works in health care data analytics, has made a declaration under penalty of perjury that according to medical claims submitted to the Centers for Medicare and Medicaid Services (CMS) there are 'at least 45,000' vaccine-related deaths due to experimental COVID-19 vaccine injections" (from [35]).

The author of this postscriptum does not have a very large circle of acquaintances or relatives. Nevertheless, only a few days ago a friend contacted me, whose aunt was forced by her own daughter and her husband to get vaccinated and who is now dying, although she was in perfect health before. The daughter and her family, who do not want to admit that they have possibly joint guilt in the meanwhile inevitable death of their own mother, will make naturally no report because of the clear suspicion on a vaccination damage. From the own and very close relationship, the author knows the case of a cousin of his wife, who, as a young firefighter, had to visit regularly the public health officer. While there was never the slightest problem for him during these examinations, now, immediately after the

COVID-19 vaccination, such a huge heart problem was detected that further examination was immediately stopped and the man was sent to a cardiologist (remember, myocarditis is a typical one of the adverse side effects the COVID-19 vaccination is known to cause; see Fig. 2). He, too, as a convinced vaccination advocate, will not file a damage report. The same applies to at least four other apparently lighter cases of vaccination side effects from the immediate acquaintance environment of the author. These persons are also convinced vaccination advocates (to put it politely) and will rather take their knowledge of suffering to the grave than admit that they have made a mistake. In psychology, this behavior falls under the term of "cognitive dissonance," but we rather hold it with Mark Twain, who once said:

"It's easier to fool people than to convince them that they have been fooled."

One may count now on all ten fingers how large the probability is for the fact that such an accumulation of inoculation damages and "reporting inhibitions" occurs with the author, who counts hardly 100 acquaintances or relatives to his closer environment. If there should be "only" 4 deaths per 100,000 vaccinations and 16 seriously injured persons according to official statistics, then it is already an almost unbelievable coincidence that the author "may" personally experience two such cases.

The reader may please make his own calculations and research in his "vaccinated" environment.

The Other Side of the Coin

Perhaps in only a few years from now, or perhaps even just a few months, one might state that COVID also has had its advantages:

- It separated the stupid, the clueless, and the faithful regime believers from the thinking ones.
- The mainstream media completely exposed itself as sources of lies and criminal instruments of power of the regimes.
- The "established science" exposed itself for everyone with enough brain mass and thinking ability as purely sycophantic entertainment entity and lie-supplying institution of the regimes.
- **The unvaccinated will soon be in great demand on the marriage markets of this world.**

The Other COVID Project for the Good Doctor Based on Honest Research and True Science

From our very simple and easily to reproduce evaluations, it clearly follows that even the dimmest and most brainwashed person must have meanwhile realized that there "seems" to be a huge gap between the rather alarmistic medial barrage about the danger the various variants of COVID-19 (with a new version or mutation whenever the topic seems to simmer down) are presenting and the number of actual REAL COVID-19 deaths one sees with one's own eyes. Just as a little reminder here again the most recent numbers (mid-July 2021) from the US army, where we find that there were 202,567 active-duty members who had COVID-19. Twenty-six died **from or with** the virus, which gives us a survival rate of 99.9872 percent. On the other hand, we are permanently bombarded with horror scenarios about the virus, which do not just come from manic journalists and lying politicians, but even from so-called "experts" and "scientists"... well, certain "scientists."

For the author of this proposal, this is not necessarily surprising, because natural processes have, besides an often strong parameter dependence (well known in chaos theory) in case of low dimensionality, often also course scenarios [18], which for the most part cannot show any tendency at all, because they, for example, in case of sigmoid curves, run much too flat in their initial and final stages, that is, do not show any decreases or increases exceeding the error and thus show a principal "simulation aversion." In order to discover such characteristics in systems like the Coronavirus spread, they not only have to be captured in as holistic a way as possible (i.e., with all their intrinsic and extrinsic degrees of freedom fully taken into account), but also have to be described in such a way that they can be observed along with their intrinsic uncertainties.

This can be achieved by a metric model [2] quantized by means of a volume- or sub-volume-scaled metric. The number of dimensions to be considered must be equal to the degrees of freedom (degree of freedom = dimension) of the system. The volume quantization of our model space now allows us to capture the uncertainties for all of these degrees of freedom over time. Time itself appears as a dimension and need not necessarily occupy only one dimension (distinguishing differentially relaxing or retarding effects, see [2] in conjunction with [19–23]). The advantages of this fundamental approach ("fundamental" because it combines the elements of general relativity [24, 25] and Quantum Theory (e.g., [26, 27])) are the following:

(a) The model controls itself. Since it knows its own uncertainty for all degrees of freedom at any time or times (even a multidimensional time with different "speeds of light" or relaxation or retardation parameters), it is not possible to overlook uncertain input values. That is, input values or measured data, which come from system areas, which have a larger uncertainty, than the statements of the model and/or the system, are recognized and separated.

(b) The model can deal with arbitrarily curved general spaces and space-times [28–30].

(c) The mathematical structure of the model by means of the volume quantization allows in principle to transfer the entire model into systems of multidimensional Klein–Gordon equations (see [1, 2, 31] and chapters 11 and 12 of this book). Thus, the nonlinearities otherwise known from metric theories [24, 25, 28–30] can be avoided.

(d) The principal option to incorporate different timescales is of especial interest in connection with crises like the COVID-19 because it allows the differentiated study of various evolutionary processes running on extremely different velocities. We point this out explicitly in connection with the experimental vaccination programs, where the already registered huge number of victims in the short term leads to quite some expectations with respect to the middle and long-term damage this program has done and will do to our societies [32].

(e) Since Klein–Gordon equations are known to be eigenvalue equations and thus have eigen solutions [27], it is possible to set such solutions as observable milestones or control points for the model to ensure a meaningful comparison with reality. For the project sponsor, investor, customer or funding provider (perhaps the taxpayer), this has the advantage of a strong and direct control of the project work.

(f) Even though the model can be kept linear, there is always also the potential of incorporating spinors (n-dimensional) and entanglements [33, 34] (or chapters 9 and 10 in this book) in order to connect the model with wider systems and potentially embed them in the socioeconomic space-time of a whole society for optimum decision-making.

Generalization for all types of project work

We here explicitly point out that the eigenvalue character of certain outputs and the control of all intrinsic and extrinsic uncertainties render the project a principal inherent honest one, because there is little until no chance to cheat with the milestones being defined on the basis of the eigenvalues and the uncertainties. Such a structure for scientific, political, economic, or financial project could easily assure a proper and fair use of the invested resources. Projects not fulfilling certain predefined "milestones" (defined via eigenvalues and uncertainties) could be recognized easily. They should be put under probation and may subsequently be terminated. Most of all publicly financed work should be put into such structures in order to assure that the taxpayer truly gets something for his "investment."

References

1. N. Schwarzer, *How will the Nearby Unification of Physics Change Future Warfare – a Brief Study, Part 10 of Medical Socio-Economic Quantum Gravity*, www.amazon. com, ASIN: B0971M85JQ.

2. N. Schwarzer, *The World Formula: A Late Recognition of David Hilbert's Stroke of Genius*, ISBN: 9789814877206 (hardcopy available in print in 2021 from T&F and Jenny Stanford Publishing).

3. www.quora.com/What-does-Michael-Mann-s-court-battle-loss-mean-to-the-notion-of-climate-change.

4. P. Frank, Propagation of error and reliability of global air temperature projection, *Front. Earth Sci.*, 06 September 2019, https://doi.org/10.3389/feart.2019.00223 (or www.frontiersin.org/articles/10.3389/feart.2019.00223/full).

5. N. Schwarzer, *Anaphylactic Shock Attack Strategy - The Assault You don't realize until it is TOO late, Part 10a of Medical Socio-Economic Quantum Gravity*, Self-published, Amazon Digital Services, 2021, Kindle.

6. H. Walach, R. J. Klement, W. Aukema, The safety of COVID-19 vaccinations—we should rethink the policy, *Vaccines*, 2021, **9**(7), 693; https://doi.org/10.3390/vaccines9070693 (now: https://newsvoice.se/wp-content/uploads/2021/07/vaccines-09-00693-v3.pdf).

7. R. B. Brown, Outcome Reporting Bias in COVID-19 mRNA Vaccine Clinical Trials, 2021.

8. St. Luckhaus, Mathematical epidemiology: SIR models and COVID-19, https://www.mis.mpg.de/publications/preprints/2020/prepr2020-60.html.

9. St. Luckhaus, Corona, mathematical epidemiology, herd immunity, and data, https://www.math.uni-leipzig.de/preprints/p2010.0010.pdf.

10. J. Merino, et al., Ivermectin and the odds of hospitalization due to COVID-19: evidence from a quasi-experimental analysis based on a public intervention in Mexico City, https://osf.io/preprints/socarxiv/r93g4/.

11. https://sciencefiles.org/2021/07/01/langsam-wird-es-kriminell-neue-indizien-dass-nebenwirkungen-von-covid-19-impfstoffen-unterschlagen-werden/.

12. D. Stock, https://dailyexpose.co.uk/2021/08/11/indiana-doctor-destroys-cdc-covid-narrative-at-school-board-meeting/ (see references given there).

13. H. Walach, R. Weikl, J. Prentice, et al., Experimental assessment of carbon dioxide content in inhaled air with or without face masks in healthy children – a randomized clinical trial, *JAMA Pediatr.*, published online on June 30, 2021, doi: 10.1001/jamapediatrics.2021.2659.

14. https://www.mdpi.com/2076-393X/9/7/705.

15. https://www.mdpi.com/2076-393X/9/7/729/htm.

16. R. Cole, A Pathologist Summary of What These Jabs Do to the Brain and Other Organs, https://rumble.com/vkopys-a-pathologist-summary-of-what-these-jabs-do-to-the-brain-and-other-organs.html.

17. P. Borger, B. R. Malhotra, M. Yeadon, C. Craig, K. McKernan, K. Steger, P. McSheehy, L. Angelova, F. Franchi, T. Binder, H. Ullrich, M. Ohashi, S. Scoglio, M. Doesburg-van Kleffens, D. Gilbert, R. Klement, R. Schruefer, B. W. Pieksma, J. Bonte, B. H. Dalle Carbonare, K. P. Corbett, U. Kämmerer, External peer review of the RTPCR test to detect SARS-CoV-2 reveals 10 major scientific flaws at the molecular and methodological level: consequences for false positive results, https://cormandrostenreview.com/report/.

18. A. Bejan, *The Physics of Life: The Evolution of Everything*, ISBN-10: 1250078822.

19. N. Schwarzer, About Holistic Optimization – Examples From In- and Outside the World of Coatings, 2015, Proceedings of the 58th Annual Technical SVC Conference.

20. N. Schwarzer, Scale invariant mechanical surface optimization applying analytical time dependent contact mechanics for layered structures, in A. Tiwari, ed., S. Natarajan, co-ed., *Applied Nanoindentation in Advanced Materials*, Chapter 22, ISBN: 978-1-119-08449-5, 2017, www.wiley.com/WileyCDA/WileyTitle/productCd-1119084490.html.

21. N. Schwarzer, From interatomic interaction potentials via Einstein field equation techniques to time dependent contact mechanics, *Mater. Res. Express*, 2014, **1**(1), http://dx.doi.org/10.1088/2053-1591/1/1/015042.

22. N. Schwarzer, Completely analytical tools for the next generation of surface and coating optimization, *Coatings*, 2014, **4**, 263–291, doi: 10.3390/coatings4020263.

23. A. E. Green, W. Zerna, *Theoretical Elasticity*, Oxford University Press, London, 1968.

24. A. Einstein, Grundlage der allgemeinen Relativitätstheorie, *Ann. Phys.*, 1916, **49** (ser. 4), 769–822.

25. D. Hilbert, Die Grundlagen der Physik, Teil 1, *Göttinger Nachrichten*, 1915, 395–407.

26. P. A. M. Dirac, The quantum theory of the electron, *Proc. R. Soc. A*, 1928, **117**(778), doi: 10.1098/rspa.1928.0023.

27. H. Haken, H. Chr. Wolf, *Atom- und Quantenphysik*, 4th ed. (in German), Springer Heidelberg, 1990, ISBN 0-387-52198-4.

28. Ch. W. Misner, K. S. Thorne, J. A. Wheeler, *Gravitation*, 20th ed., W. H. Freeman and Company, New York, 1997, ISBN 0-7167-0344.

29. C. A. Sporea, Notes on f(R) Theories of Gravity, arxiv.org/pdf/1403.3852.pdf.

30. H. Stephani, D. Kramer, M. MacCallum, C. Hoenselaers, E. Herlt, *Exact Solutions of Einstein's Field Equations*, Cambridge University Press, 2009, ISBN 978-0-521-46702-5.

31. N. Schwarzer, *My Horcruxes – A Curvy Math to Salvation, Part 9 of Medical Socio-Economic Quantum Gravity*, www.amazon.com, ASIN: B096SPB5MW.

32. St. Seneff, G. Nigh, Worse than the disease? Reviewing some possible unintended consequences of the mRNA vaccines against COVID-19, *Int. J. Vaccine Theory Pract. Res.*, 2021, **2**(1), https://ijvtpr.com/index.php/IJVTPR/article/view/23/51.

33. N. Schwarzer, *The Metric Dirac Equation Revisited and the Geometry of Spinors, Part 8 of Medical Socio-Economic Quantum Gravity*, www.amazon.com, ASIN: B08Y96TL3D.

34. N. Schwarzer, *The Dirac Miracle, Part 8a of Medical Socio-Economic Quantum Gravity*, www.amazon.com, ASIN: B0963Z1Z74.

35. Bombshell lawsuit: gov't whistleblower says coronavirus vaccine deaths at least 45,000, www.lifesitenews.com/news/bombshell-lawsuit-govt-whistleblower-says-coronavirus-vaccine-deaths-at-least-45000.

Seven Days

or
How to Explain the World to My Dying Child
(From T. Bodan, with Thanks)

Preface

There was a time when I thought this book would be difficult to read and certainly just as difficult to understand. This is hardly unexpected, as the world is not easy to explain after all. Learning and real understanding are never easy. There are tasks in life that we must endure and that are horrendous, complex, diverse, and as challenging as life itself. To live is to learn for a lifetime—to which dying is also part of. That, too, is learning and—YES—it is part of life. But then the only 12-year-old daughter of a good friend read my book and started—freely, completely according to her understanding—to draw some pictures. Pictures that would show how she saw the things she had read. We scanned the drawings and inserted them in the places of the book where the little reader wanted them to be.

... and lo and behold, all of a sudden, I no longer saw the book as a heavy read. A young girl had shown me how to approach and understand it. Only a few associative images had sufficed and reading became an easy and beautiful experience. My mind once again lived through the formerly tough pages, only this time it floated pleasantly along like a feather in the warm breeze of a summer evening by the sea and finally found what it had sought in vain for so long ... salvation.

(*Note*: For this scientific book, we do not consider the use of children's drawings appropriate, most of the readers will have no problem in imagining what is said anyway. But for those who are interested in the illustrated version, it can be found here:

T. Bodan, *7 Days – How to explain the world to my dying child*, Self-published, BoD Classic, ISBN 9783752639728.)

This is the story of a child, a most amazing child who mastered all these tasks. And so, her life, despite its brevity, was not meaningless. And so, her life was important and valuable.

With every bit of knowledge I manage to pass on to others, I do something in memory of this brave little child; I also think of all the many other children who have to leave much too early because we are not able to help them. None of these little characters, however, was unimportant. This book tells you why.

Our existence would be somewhat illogical if we did not have a task to fulfill. So far, we were probably not very good at recognizing this task as such, let alone to fulfill it. We will, this much is certain, leave this world without really having done anything—each of us.

But shouldn't we at least give our children a better start so that they can take stock of their lives differently later?

Why

A good friend has a small vacation home on the seaside. One day she had the idea to make this house available free of charge for a certain time of the year for families with seriously ill children.

Sometimes also for children who do not have long to live.

The Beginning in the End

"What are you reading?" I ask to begin this conversation as casually, gently, and carefully as possible.

My little one looks me straight in the eyes. "I'm writing something," she answers after some thought. Her voice is firm and yet, almost, dismissive.

I would have liked to ask now: "What is it?" or "What are you writing about?" but I feel that my girl does not want to talk about it. After all, I hadn't come for that either, and so I'm thinking about how to begin. But there my child starts before me. She has pushed the sheet of paper on which she was scribbling into the book that served as a base for her writing and then puts it on the bedside table. The action seems clumsy, like someone who feels infinite pain from every movement. I realize the book is "The Little Prince," the edition which my parents had given me when I passed my school leaving examination many years ago.

"Papa, I don't want these treatments anymore." Her little eyes stare at me from deep caves. "They hurt me and make me feel bad ... really bad." There's something hopeless in that look. Just as if the brain behind it knew what my answer would be. Too often this mouth had asked questions, these eyes had pleaded, and these ears had heard something soothing but meaningless.

However, this time I calmly respond, "They won't hurt you anymore, my love." After weeks/months, the burden leaves me, I no longer have to lie to my child. A decision has been made, a difficult, almost impudent decision.

*

This decision brought forth many more difficult conversations. "We can't give up hope," "we still have a few possibilities," and "just think about your child for once! She wants to live!"

"Yes, she wants to live, but she certainly doesn't want to suffer anymore." My wife and I had thought it over very well. We had studied all the information that was available. We had read, understood, and reread so much, in all kinds of languages. We had become experts, such specialists that many specialized doctors now seemed like a joke to us. They could no longer compete with our arguments, our knowledge, our treasure trove of quotations and literature, which we could simply rattle off without having to look it up. There is obviously nothing worse for a physician than having to have things explained to them by a layman. So, at some point they let us out of their clutches and declared the case of our child to be no longer treatable. The road to this was difficult. Medicine can be wonderful, but sometimes it is also terrible, an impersonal and inhumane paradigm, trimmed for efficiency. Unfortunately, this efficiency does not seem to leave much room for consideration of the patient, but one that is aimed at the optimal exploitation of a system, with many people involved who keep their hands in their pockets to appear important. Maybe it's not so bad if you have to work through 1000 appendixes, but it's bad for the seriously ill, it's terrible for little patients, and it's absolutely hell for dying children. I know by now why this is so, and perhaps also why there is no other way. Normal people within our civilization simply have only one possibility to exist or even work without harm in a system like the active core of our health care system: they have to make themselves completely insensitive to the suffering of others, especially their patients. They have to become like the machines they operate. Only in this way can they, the inner active core of the health care system, do anything at all for the patient, despite the fact that they are constantly clashing with the demands of an industry that is so excruciating large and leaching. A periphery that has wrapped itself around the core like an octopus, an all-encompassing drainage system, which now exceeds the core tenfold in size and financial output and which withdraws all the resources it can get, be it with open and hidden bureaucracy, cold-hearted regulatory frenzy or blatant cheeky charlatanism and inhuman cynicism. I also know now that this is only due to a deficit of knowledge. It is not money that the system lacks, it is reason. The system does not lack resources, but the reason for an optimized use of them. This lack of understanding and reason makes the system itself sick and a torture chamber for every truly sick person. A stupid man may well be kind, but he can never be truly understanding, because for understanding one must be able to understand and for that one needs knowledge.

The Task

My little girl looks at me in disbelief. She feels that something is different, that something has been decided. "Am I going to die?" she asks.

"Yes!" I answer and am amazed at how easily the answer comes. "I'm sorry that we have tortured you for so long, we still had some hope. It was wrong ... please forgive me," I add after a while.

My little child has tears rolling down her face. She just says, "What a pity ... that's a pity!" Then she starts to cry unrestrained.

I sit down with my little one and take her in my arms.

"I'm scared, Daddy."

"Me too, my love."

"I don't want to die."

I knew that these words would come. I just didn't expect it to happen so quickly. Our brave child probably knew far more than we, her parents, had thought. She had guessed, implied ... oh, if only we had listened to our little angle earlier. How much suffering we could have spared her!

"Nobody wants to die, sweetie. Some people wish for it, but this is never really the wish for death. Rather, some people do not want to suffer, not think, or not feel anymore. It is the desire for peace, and death seems to keep it ready for us, absolute peace. Death itself, however, is never the real wish, because you cannot truly wish what you don't know."

"I want nothing more than to live ... just a little bit more. And yet ..."

"Nature has given us some gifts, beautiful and terrible ones. One of them is our will to survive. This is even stronger with children than with us. That's why you fight stronger and suffer all the more, especially when the fight has become hopeless. I admire you, my child."

"Should I stop fighting ... not want to live anymore?" She does not speak the words like a question and continues immediately. "I cannot do that ... I see the people who come to me, I see their looks, and I know what everyone is thinking, but I can't ... I cannot believe it. I still want to live. Something in me wants to keep fighting ...." My child takes a break and almost apologetically adds: "But I don't want any more pain either."

By this she means the treatments, the operations ...

I see the little emaciated body; I see the places where once have been limbs. I can hardly stand it, but I know that I must bear it, because there is still much to do. There is still something very important to do. Before us lies the most difficult task that two people have ever had to face. And even though it may have happened a trillion times before, it will always be the most difficult task the universe has ever placed: To create hope out of pain and to receive solutions and salvation out of hope.

"You know, my love," I finally say, "this gift, your unconditional will to fight and survive, it was given to you for a very specific reason."

"A reason? For what reason?"

"Yes, my dear. Somewhere very deep inside of you and very deep inside of all of us is the desire not to give up. This drives us, even if it seems hopeless, when success is highly unlikely and this force resides throughout the whole universe and with every small step that evolution takes, it reinforces this urge, this force."

"What's the use if nobody can help me?"

"Into the structure and essence of all things has been woven a very important thing and that is indeterminacy. In general, this does not help us and it is good that we know this so that we really make an effort to solve our problems on our own. But sometimes exactly this uncertainty, which is present everywhere, holds surprises that help evolution to solve even the most hopeless problems. And it is the same with us: the indeterminacy also holds surprises for us and sometimes this helps us."

"But why can no one help me?" Sobs shake her little body.

"Because we're too stupid," I simply answer. "Because we just don't know enough to help you."

The sobbing gets stronger.

Finally, my child becomes calm. Her eyes look at me again. They show astonishment.

"Aren't you sad?"

"Yes," I answer, "but I'm also happy."

"Why?"

"Because it is easier now, because we don't have to torture you anymore and ...," involuntarily I take a little break because I suddenly realize how important this next sentence is to me,

"because I have to explain the world to you before you go."

Day One

"Will this help me?" Big eyes look out of my daughter's emaciated face with an incredible intensity, as if having a huge hunger and my lips could feed them.

"Yes," I answer. "It will help you a lot. It will make your way easier and your task more bearable."

"What task?"

"The universe created us for one reason alone."

"The universe?" My child is interrupting me. "Not God, not Creation?"

"It's all the same," I reply. "God is Creation, is the universe. It is absolutely the same as you call it. It is even completely the same as what you imagine it to be, because your imagination will never encompass the whole, but you should still try. Not to try would be foolish and a betrayal of our minds." After a short pause I add, "... and we are part of it, we are part of the universe, we are part of Creation, and we are also part of God, if you like."

"A part of almighty God as well?"

"Yes," I answer and ask back, "but where did this 'almighty' thing come from?"

"Well, that's what they always say, the 'almighty God' here and the 'Almighty' there, et cetera," my little one shrugs her shoulders.

"People talk a lot without thinking," I reply and ask myself at the same time whether I shouldn't leave the subject for now, but then I decide differently. By chance, my gaze falls on the book on the nightstand and I add:

"**Big people** talk a lot without thinking."

My daughter follows my gaze and smiles. She has understood and a mighty impatience sprays from her eyes.

"You know," I begin, "it is very easy to prove that God cannot be almighty."

There Is No Almighty God—and That's Just Fine

"How?"

"Well, if someone were almighty, would he allow himself to be imperfect?" I answer with a counterquestion.

"Of course, not," the answer comes promptly. And a little more hesitantly, she adds: "That would be as if I could heal myself and would not. Worse, it would be as if I had to heal myself and would not do it—a contradiction." The comparison surprises me, but I continue with my questions.

"And would you call someone who is not self-sufficient perfect?"

"What do you mean, Dad?"

"Well, I mean, would a perfect being need something like entertainment?"

"No, then it wouldn't be perfect. A perfect being doesn't need anything, it's enough for itself."

"That's what I think, too. Either the being is not perfect, or it is its own entertainment. But since being perfect also means perfection in all parts, the entertainment of the perfect being would also have to be perfect."

"A perfect entertainment?" my kid is frowning, "I can't imagine anything like that."

"You don't have to worry about it, because I can't imagine anything about it either. Nobody can, and this whole 'almighty' and 'perfect thing' is nothing that really exists anyway—at least not in this universe. And since this universe already exists and the rest can no longer be everything, perfection and omnipotence per se are excluded."

"Says who?" my dear little child asks.

"Our existence says so," I answer. "The fact that we exist with all our imperfections proves that there can be no perfect and therefore no almighty God, since he would not have created any imperfections."

"Why not?"

"Because it would be completely superfluous and illogical for a perfect being, and no perfect being does such a thing. Because for a perfect God, we would be less than for us an entertaining aquarium with a few fish in it ... infinitely much less. So, what would such a God do with it?"

"Maybe you see ...," my daughter falters briefly, "uh, maybe we just don't see perfection."

"Oh yes, I see it, but it is not just there. It arises only within the entire infinite existence of the universe and thus, of God and thus, of Creation and thus, ... fortunately also ... us."

"So, the universe is ..., God IS ... perfect?" my child asks somewhat confused.

"Yes, he is just as perfect as ideal Euclidean parallels intersect in infinity, namely, only in infinity, spatially and temporally seen. Actually, seen in the sense of all possible properties, that is, an infinity in an infinitely dimensional structure."

"In other words: never and nowhere." She smiles, almost indulging, and then adds with almost a grin, "Yes, but our real space is curved, and there parallels do not only intersect at infinity."

"An ingenious interjection," I admit. "We'll get to that, but it does indeed fit here too, because as far as perfection is concerned, there may well be effects in the universe that don't need infinite time to reach it. But nevertheless, the following is valid: Only the whole is perfect ... will be perfect ... well, or it is, if you take the time in its entirety and don't cut out 'time-frames', like we always do."

"So, we, here and now, are part of an imperfect universe and thus part of an imperfect, not-almighty God, because universe and God and Creation are one and the same thing?" my daughter summarizes somewhat hesitantly.

I nod.

Why Are We Here?

"Huh ...?" it sounds rather incredulous from the little bed, "And what is our task?"

"To explain to the universe why and for what it exists. To help it become perfect."

"What?" she asks sounding unbelieving and almost frightened. Oh, how long has it been since I've seen so much life in this body.

"Yes, you heard right," I continue unmoved. "Our task is to explain to the universe or God or Creation, whatever you want to call it, why we are at all."

An almost protesting look strikes me. "But God, universe, Creation, that's just there, isn't it?"

I smile. "It's good that you put the 'isn't it' in this sentence, my love. Because nothing is just there. Nothing was lifted onto the stage of being without having a history ... or being a history. The only thing that ever is, to say it once in this simple form with our mind trimmed to a continuous time, is a law."

"A law? What law?" those hungry eyes again.

"There is no rest."

Silence ... stillness, then even a little smile. Compassionate.

"It's a stupid law," comes the reply after a while from the little bed.

"You think so?" I ask back. "I think it's a wonderful law, because it guarantees you'll never fade."

"I will never pass away?" the pitiful smile gives way to a kind of hopeful incredulity. "But I will ..." my little one pauses and swallows.

"Die?" I ask.

"Yes, and ...," something heavy and oppressive is on the tip of her tongue, a deep fear, "... decay?" It almost sounds as if I am asked to deny this statement, which is emphasized as a question toward the end, no matter what logic says.

"No, you cannot decay," I answer reassuringly and recklessly at the same time, "but what you call your body will dissolve and return to the cycle of life."

"That is terrible," she almost sounds a bit reproachful that I don't want to find it terrible at all and speak of it as if it were the most normal thing in the world, that one day your own body will be taken away from you and recycled.

"You know, my dear child," I begin as empathetically as I can, "there would be none of us if the beings that were before us had not decomposed and what they were made of had not returned to the life cycle. Your body is made up of so many particles, it is very likely that there is a former trilobite, a tree fern, some dinosaurs—you love dinosaurs, don't you?—and there are certainly a few other people among them ... of course I mean that some of your atoms used to be the atoms of other inhabitants of this planet."

"So, I'm just borrowed," it sounds fascinated and a bit resigned at the same time.

"No, my child, you are only you, but your body is actually only borrowed. Indeed, to the borrowed body there are added many, many instincts and foreign thoughts, called memes, which we cannot really attribute to ourselves, but where we only store and reflect what we perceive."

"But that's not much of what I am, is it?"

"Oh yes, my love, it's a damn lot. Because everything that you take up in yourself is associated, processed, filed away, taken up again, reevaluated, and linked in its own way, so that it is just through and ONLY through you that it gains uniqueness and an unbelievable mass every day. Furthermore, it is so unique and valuable to have a piece of information that does not exist a second time in the universe in such a way that the universe cannot afford to give away this cornucopia of information ever again."

Why Am I Not Allowed to Live On?

"I am important to the universe?"

"Yes, my child. Very important."

"Then why won't it let me live?"

"Two reasons:

First: It has little or no direct influence on what happens to you. You have to imagine us like cells in the body of the universe, just like single cells in your body. If you did not have tools and hands, you would not be able to directly influence the fate of every single cell in your body. Second: It needs you as you are."

My child looks at me a little disappointed, "It's easy to say that."

'Easy when you know, as you do, that you will not die soon' she would like to say, but I don't think she can bear the words herself. I still hear them. They float in space and need an explanation. They need a consolation that is not an end in itself, but the truth. But how should I begin, and above all: WHERE should I begin?

Vibrations

"In the beginning, my child, there was perhaps a sound ..."

"A sound?"

"Yes, something in the universe led to the formation of a cellular structure, at least in our part of the universe. Let's call them Planck volumes after the great physicist Max Planck."

"I know him," says my little girl confidently. I nod approvingly.

"Actually, the term is not correct, because it is already used, namely for cubes with edges of the so-called Planck length. These are tiny 10 to the power of minus 35 meters and our cells are a little different, but that shouldn't bother us at the moment. To be on the safe side, we can say PlanckNvolumes, where the N stands for the number of dimensions. It is also important to know that these things are constantly wriggling and can't help it, because of the 'no-rest' law and so on. You know the experiment with the sawdust on a metal plate, don't you?"

"You mean, where such patterns do emerge when you let the plate swing?" my child asks back.

"Exactly! A sound, a vibration, could well have created such patterns, such structures in the universe. Of course, it must have been a very powerful vibration."

"Or a great bang," my little one smiles and ironically adds, "... a big bang maybe?"

We both laugh.

"Okay, anyway, in any case, these little mini things, these PlanckNvolumes, would then be left over."

"... be left over?" my child interrupts me, "At what? Or from what?"

I raise my hand and give her a thumbs up, nodding appreciatively. "Golly!" I say, "That's a wonderful question."

"It is?" my child asks in amazement. "I just thought I was too stupid to follow."

"No, not at all," I hurry to answer. "I must admit that I wanted to make it easy for myself and save the question about the essence of these small PlanckNvolumes for later. Perhaps I wanted to leave it out altogether, because for us humans, who as small children always tie our first understanding to grasping, that is touching things, it is infinitely difficult to accept the true 'substance' of all being."

"And that would be?" my girl asks.

"If you are lying on your bed or touching something, you surely believe that you are pressing a solid body, namely, your hands or fingers, onto another solid body; the object you are touching. And thus there is a physical connection between the two, isn't there?"

But my child thinks faster than I do and answers: "Yes, I may have the feeling that it is so, but of course I know that it is the electric fields of the electron shells of the atoms in both bodies that repel each other and that only simulate such physical contact for me."

"Excellent!" I nod appreciatively. "I can save myself a whole lot of explanation then. So, when we look closer and enlarge things, further and further, we always find that the physicality, strength, elasticity, or whatever you want to call it is not really there. It is only pretended and in reality, it is all just fields that are there and each of them needs its place and has its own characteristics. Some of these fields conspire to form structures that seem solid to us and others manifest themselves as rather fleeting things and so forth."

The Zero Is Prohibited

"And where do the fields come from?"

"They are there because 'the zero' is forbidden, because—as I said before—there can be no rest in the universe. They are there because there is a great universal law that does not allow nothingness."

"And why are they just as big as they are, the fields, I mean?"

"What do you mean?" I ask back with interest. I think I have understood the question, but nevertheless it seems to me a very far thought from my child and so I dig deeper.

"Well, they also could be smaller or larger, but instead ...," my child pauses and thinks, "the metal shavings on the swinging plate, your example from just now ... the shavings form patterns of a certain size, don't they?"

"Yes."

"Depending on the size of the plate and the height of the pitch, different 'cell sizes' are formed, right?"

"Yes."

"Ok, so if your comparison is right, the PlanckNvolumes must also have certain typical 'cell sizes'. Why?"

"This is due to the boundary conditions in our part of the universe. Here, the conditions were such that the Planck sizes emerged. They are about 10 to the power of minus 35 meters, which is really damn small."

"Could there have been other 'Planck sizes' elsewhere?"

I make a whistling sound. "You get us into the tough questions pretty damn fast. I am somewhat perplexed, because you have just broken through the boundaries of General Theory of Relativity."

Scale Invariance and Scale Relativity

"Pardon? Why?"

"In general relativity, there is something missing. A fact you just brought up: scale relativity."

"Sorry, Dad, but that's too high level, bring it back a notch ..."

"Okay, back to the sawdust on the metal plate. You know I can change the patterns or cells that form when the plate swings, for example, by making the plate bigger or smaller, right?"

"Yeah, sure."

"Well, what should have prevented the universe or our part of it from being bigger or smaller, more energetic or less energetic in its origin?"

"I don't know, but it will have had its reasons for becoming exactly what it is."

"You're probably right." I have to laugh, "but nevertheless there could have been other boundary conditions and then the PlanckNvolumes or primordial cells would have been different in size. And now comes the joke: Nothing would have changed for us anyway."

"Really? Why?"

"Because again the PlanckNvolumes would have formed the basis of our part of the universe and everything we perceive would have been based on it and for us the things would have been 10 to the power of minus 35 meters in diameter and everything would have been the same. You can even imagine the following: our part of the universe is only the size of a marble and the PlanckNvolumes, in absolute terms, that means seen from our scale, they are correspondingly smaller, say, 10 to the power of minus 63 meters and for us everything would be the same again. We would think we were sitting in a huge universe with a radius of more than 13 000 000 000 light years and from our point of view this would be true."

"Wow! A marble ... that's weird," my little one is impressed. "But wouldn't the other natural constants force a completely different world?"

"Good question, but I don't think so. I'm pretty sure that the size of the PlanckNvolumes also determines the size of the other natural constants, and this results in a scale invariance, which is a fundamental independence from the current absolute size of the universe. One could even say that the notion of absolute size is thus to some extent obsolete. But once the universe has decided on a size, the size of those inside the universe is no longer irrelevant; then the scale relativity will be valid."

"This is getting weirder and weirder, but I think I get it all the same." The fascination literally radiates from the small face. Nevertheless, my bright girl immediately has a new question:

"Does anyone know what shape these things have?"

"Oh!" I say, "another one of those big fish," and scratch my head to gain time. But my daughter just sits there and looks at me expectantly. So, I answer as best as I can:

"Yes, I have an idea. If you assume that Einstein's theory is still valid even in the smallest of cases—possibly with other constants—then all you would have to create are small Friedmann universes. These would be spheres, planes, or hyperfaces. I personally like the spheres best."

"Friedmann universes?" my little one repeats.

"Yes, after a Russian scientist who was a contemporary of Einstein."

"Huh?! Can't we call the little things Friedmen then?" my child asks, "PlanckNvolumes sounds terrible."

"Oh, I don't mind that at all," I quickly answer, but after a moment's thought I add: "But then the cosmos would be a bit too male dominated for my taste ... with all those Friedmen, don't you think?"

My child laughs out loud and finally suggests that one could well say Friedwomen and Friedmaids from time to time.

"I can live with that," I agree and laugh as well. Finally, I remember our topic before the Friedmann discussion and explain:

"It is getting even better with the little things, the grains of our universe: What we perceive as an expansion of our part of the universe may well be just a shrinking of the PlanckNvolumes, or Friedmen."

"Really?" I haven't seen such big eyes in my child for a long time.

"Yes, the Friedmen are shrinking and new volumes are forming. At the same time, however, no new matter is formed. Thus, the space, the matter-free space, becomes more everywhere and for us it seems to expand. For us, the number of Friedmen between two points counts as a measure of the distance between them and nothing else."

"Wicked!"

I take this as the highest possible acknowledgement from my child and continue.

"So, the entire big bang could be explained in a completely different way."

"Now I'm curious ..."

"Until now, we have always thought that our universe began in a point-like state and then expanded, right?"

"Yes, that's how I know it too ... with phases of varying speeds of expansion in between and so on."

"Oh yes, the phase of inflation," I agree, "we may come to that later. But just imagine that our part of the universe was at first just one single Friedmann-piece, perhaps in another universe, meaning part of this 'super-verse'."

"Ok?!" says my child to indicate that she can follow me.

"Suddenly this PlanckNvolumes or Friedmann began to divide. It split again and again with such speed and its parts split and so on."

"Why should it have started the dividing in the first place?" my child asked in between. "And besides," she adds in a hurry, "shouldn't every splitting also change the properties of space, so the wriggling of your volume-things becomes something else?"

I nod in recognition. "Both questions 100 points out of 100!" I think for a while and then answer: "Well, sharing is perhaps a stupid word, because you almost inevitably have to imagine bacteria constricting themselves in the middle and then one becomes two, two become four, and so on. After each division, the new bacteria have to grow again before they have reached the size of the original bacteria and are almost equal to it again. With the Friedmen this is somehow different, because here the growth phase would have to be so fast, at least in our present universe, that we do not notice anything. Or, new volumes simply emerge and the rest shrink so that the total size does not or hardly changes at all. Or there is a combination of both, shrinking and expanding during the increase of volumes OR ... the whole thing always takes place in the blink of an eye of the Planck time, unnoticed by us ..."

Planck Time and Constant Wriggling

"... of the what?" my little one asks in between.

"Of the Planck time, a tiny little period of time that we cannot dissolve any further. Everything that happens in such a time and beyond, we do not see as a process, but only as beginning and end."

"Aha."

"And now to the beginning of the 'dividing' or multiplying ... Perhaps it had been given a good blow, an enormously powerful burst of energy, and so it simply had to fall apart."

"So, was that blow the big bang?"

"Yes, that would be the big bang. During the decay, all the things we interpret as Creation would have unfolded: matter would have been created, stars, galaxies, planets, and the constantly progressing decay would have been a constant driving force, another ticking clock."

"ANOTHER clock? What else is there for a watch?"

"The wriggling itself is a clock too, but that comes later."

"And what about inflation?"

"You mean the phase of extremely rapid expansion in the first few moments after the 'Big Bang'?"

"Yes, exactly!" replies my child.

"Wow ... How do you know all that?! Well, have you ever heard of supercooled water?" I ask.

"I have even made some before," my girl answers with a wink at such a simple thing.

"So?" I ask suspiciously, "How did you do it?"

"I simply put a normal PE bottle with tap water outside the winter before last. We had around minus 10 degrees Celsius or even minus 15 ... hmm, I don't remember exactly ... and I wanted to see how the ice pushed the bottle apart. But when I looked at the bottle the other day, the water was still liquid. I was surprised and wanted to take the bottle inside. But as soon as I had the bottle in my hand, everything suddenly turned to ice. Well, actually, it was more like ice with water or water with ice, because when I unscrewed the bottle, quite a mixture came out."

"Aha, so you have even experienced the sudden phase jump from liquid to solid. Well, you could imagine the same thing with expansion: Just assume that the 'liquid' universe is already ripe for the solid state. It literally screams for interfaces that should cross it, but somehow the trigger is still missing. The temperature is already so low that the universe wants to become solid. In other words, it wants to disintegrate into many little Friedmen, but it can't. And then all of a sudden it starts: In a tiny moment the interfaces of the Friedmen form and because these Friedmen determine the size of the universe for us in their number, and their number suddenly increases tremendously, the inflation seems to us like a gigantic expansion of the whole universe, which takes place in a blink of an eye. But in reality, nothing expanded at all! The marble or the ball or the giant something remained in its size as it was. It was only crossed by interfaces, just like the surfaces of ice crystals cross the water when they freeze."

Matryoshka Game with Universes

"Can it happen again?" asks my child.

"I even think it happens all the time," I answer.

"How so?"

"Well, just imagine a single PlanckNvolume or a single Friedmann in our universe. For us who live in the super-verse of this PlanckNvolume, absolutely nothing would change if this volume suddenly became critical and decided to create more interfaces within it. These interfaces then form the Friedmen of this 'sub-verse'. There could also be an inflation and because of the world's invariance of scale—if the values are correct—stars could form and life could emerge, intelligence, ... and if they are very lucky in this sub-verse, they also get a Planck after which they could name the volumes in their I-don't-know-what-verse."

"Crazy!" my girl seems really amazed this time, "And how many of those sub-verses do you think there are?"

"Hmm ...," I say somewhat hesitantly, "maybe I should simply say that you, for example, meaning your body is full of such things."

Her eyes open wide, as big as tennis balls.

"And in order to round things off properly, I should add right away that there is no reason why there should not be further sub-verses in the sub-verse and there again some and so on. And with certainty we are also only a sub-verse in a super-verse, which itself is again a sub-verse and so on. Thus, I do not believe that there is a preferred universe or a preferred scale somewhere. There are innumerable nested universes. Maybe some of them can only 'crystallize out' when the temperature of the respective super-verse, that is, the surrounding temperature, has dropped sufficiently, but in principle they are there or will be there."

"It makes you dizzy," my child interjects.

"Yes, I agree, and we are not even finished yet ... because these infinitely many universes have an infinite number of possible properties that we can understand as dimensions. And some of these dimensions make up space for us, while other dimensions simply make up other properties for us."

"That is, if we were different minded, then we could certainly understand other properties as space dimensions, could we?"

"Yes, something like that."

"Wow, our universe is one of an infinite number of universes in infinite scales and what we perceive as space is only a collection of interfaces and, beyond that, only one of an infinite number of possible spaces that could form." My kid really looks dizzy now. After a short pause she adds, "Then we are even more insignificant than we thought?"

"Yes," I agree, "with regard to all beings, we are still quite a bit left off center than we have previously assumed. Yet we are infinitely important for the whole multiverse."

"Why?"

"Because, despite all our insignificance, we ask and try to answer the question that the multiverse constantly asks itself: 'Why?'"

"And that makes us important?"

"Not just us, but everything that surrounds us, just like nesting dolls, everything that is and was, everything that has ever existed from present time back to when it all began. That's why the multiverse does not discard any information, it stores everything and takes it up again and again and puts it together in new variations. That is the reason why we are immortal."

Silence, long, tangible silence. I feel a question, I feel that my child wants to dig deeper here but is apparently not sure if she wants to hear the answers ... if she wants to hear them now.

Suddenly and unexpectedly, my little one asks timidly: "And we can't figure out what our universe is now, marble or a huge structure?"

"Well, actually it doesn't matter, because seen from the outside the scale invariance applies, so everything is based on the Planck sizes anyway. But from the inside we are unfortunately limited in our access because of the scale relativity, I think. We can't get to the top because we cannot only not get out of our part of the universe, but we can neither get information from outside. And we are unfortunately also limited toward the bottom, meaning toward the smaller things."

"Why?"

"These small structures fill everything, they are the smallest structures we have access to, if you like. We cannot get below them from our level of existence, our scale. And we cannot investigate anything with greater accuracy than by resolving these 'particles'."

"Why not?"

"It's a little bit like Ping-Pong," I think. As soon as I have played a ball, it comes back again, always faster and more difficult to parry than before. But children are like that ... fortunately they are like that.

The Smallest Resolvable

"Just imagine a dark room that you want to explore. The only thing you have are your hands. You will not be able to feel any structures in the room that are smaller than what your fingertips can hold apart. In your fingertips are nerve cells, tiny sensors. You cannot feel anything smaller than the smallest distance between two of your fingertips."

"Yes, bacteria, tiny specks of dust, I will not be able to see them."

"Yes, very good, my child," I praise her, "but you could find some tools in the room, create a light source and build a microscope, and then you would see the dust and the bacteria. You would use the wavelength of the light, which is smaller than the sensor distance of your fingers. In our 'dark room' the Friedmen are the smallest things there are. I can't build a sensor that wouldn't have to consist of at least a few of these miniatures in order to be able to transmit information upward, that is, to our scale or size level. But this is definitely not the way to get to structures that are smaller than the Friedmen."

"Hmm ...," my child responds a little hesitantly, "I think I understand the example with the dark room, the bacteria, and my fingers better."

"Me too," I willingly admit, "but that hardly makes a difference, really."

"Good!" my child says somewhat laconically, "the world consists of nothing but small balls ..."

I raise my hand defensively and distort my face as if I had a toothache.

"... small things, volumes, okay? It doesn't matter. But ... where are we?"

"Aha!" I say meaningfully, "this is a really good question, because we are nothing but vibrations, distortions, or any other anomalies or information on the surfaces of these volumes."

"Huh?" my child responds.

Life Takes Place on Surfaces

"Well, all that is there, or what we perceive as BEING there are not the Friedmen themselves, but information, anomalies on their surfaces. Imagine a whole room full of perfect billiard balls, each of which has interesting patterns. You cannot look into the balls, but you can see the patterns on their surfaces. The universe for us is such a room full of balls with patterns on them and all we perceive and all we are these patterns. We cannot see into the volumes themselves, because we ourselves are only patterns on their surfaces."

Suddenly my child laughs out loud, "Ha, I always knew we were damn superficial beings."

I too have to laugh at this.

"Yes, we are really superficial and our whole existence is a reflection, a pattern on the little Friedmen."

"So, we are trapped in this room ... in this surface area, I mean. We cannot escape from it upward, toward the larger, or downward, toward the smaller, we cannot get out either."

"Yes, exactly because we ourselves are part of this space, we cannot get out of it. To escape from it would be like ...," I seek a suitable comparison, but notice that my initial ideas 'about a drop of water that cannot leave the sea' or 'a snail that has to stay in its house' are not strong enough. While I am still brooding, my child suddenly says, "... as if I wanted to leave this body without dying, right?"

She probably sees the shock in my face and I feel that she is looking for something comforting, but then she just shrugs her shoulders sadly.

"I am tired," she says after a long break.

I nod and lift her out of bed. My child is as light as a feather. I let water run into the tub and start to wash her. The stumps are still weeping and I have to be careful not to touch them too hard.

"Could you please wash my hair too, Daddy?" my child asks.

"I'm glad that you're demanding that, now … In the past, we always had to talk you into it, your mother and I."

My child laughs and then says: "Once you have been washed by 'Nurse Ruthless', then you know how good it was with you."

I blow dry her hair and comb it thoroughly. I know that this is good for my child. Then we brush teeth and I carry my little one back to her sickbed.

"Somehow I don't want to fall asleep yet, Dad," she whispers in my ear on the way.

"You don't have to, I have a surprise," I answer.

"A surprise?" asks my little one, "What kind of surprise?"

"I'll read you a story," I reply.

"What?" and behind this 'What?' I hear all the typical clichés, like 'Am I not too old for this' or 'this is something for babies' or just 'oh, stop it, you're crazy'. But I get ahead of these interjections by simply answering:

"It is a story that I once wrote myself. You will like it."

My child opens her mouth, but then closes it again. In the end, she simply says with unmistakable interest: "Okay, now I'm really curious."

"But before that, I have one more question: Which came first, the chicken or the egg?" I ask.

"Pardon? The question is nonsense," my child looks at me as if I am really crazy this time.

"You think so?" I ask back.

"Sure," my daughter says and recites it as if learned by heart: "J.K. Rowling has Loona Lovegood in 'Harry Potter and the deathly hallows' answer a similar question: 'A circle has no beginning' and that's it."

"Well, this answer is wrong, at least in our universe. You'll understand when we're done."

"Why don't you just tell me the answer?" my child asks almost demanding and a little offended.

"Anticipation," I answer briefly and smile.

Then I pull a bundle of folded sheets out of my pocket, make myself comfortable, and start reading. Here and there things falter a bit because I can't quite decipher my own writing, but my child is very indulgent with me and an excellent, patient listener.

*

Lilly

Little Lilly, although only five years old, already has three extremely unpleasant habits. The first one is: She loves to play the flute, the recorder to be exact. And to be even more precise: the soprano recorder in C.

For people who don't know what to do with the term soprano recorder: it is one of those small, elongated, wooden tubes in which a lot of holes have been drilled crosswise. If you cover all the holes with your fingers and blow into the instrument at the pointed end of the wooden tube, then, if you do everything correctly, you should hear exactly the tone that is on the piano in the middle, to the left of the two (the two, not the three, ok?) black keys, assuming that the piano is well tuned. So, now let's continue ... oh you do not know what a piano is .... My God, this is quite simple: it's the small form of that thing which is in almost every church and which is composed of nothing but flutes. That is called an organ. But at the piano, as a mini organ, if you want to make it easy, there are not flutes, but strings like on a guitar. So, a piano is an intermediate thing between an organ and a guitar. What? You want to know what a guitar is? Damn, it's a wooden box with wires on it and that's enough now! Back to Lilly.

The second unpleasant habit: Lilly plays this recorder in a way that is hardly bearable for the ear, namely, consistently and with great persistence wrong.

As if these two habits, especially in combination, weren't bad enough: there is a third, the most unpleasant of the three. Lilly easily gets flatulence. The slightest wrong bite and the most disgusting things happen.

Lilly McCoy, that is her full name, was on the train with her parents. The small family wanted to visit their grandparents in Des Moines. A long journey, from Denver via Mc Cool, Hartings, Lincoln, Omaha, and Council Bloffs. Father McCoy had, as always, taken care of the tickets and seat reservations while mother did the packing. Little Lilly played the famous song of the two flies in buttermilk on her soprano recorder until shortly before departure. She could not yet play another one. (If you don't know this song, just ask one of the big people for "Skip to my lou." Of course, it's best if you let them sing it for you right away, but I have to warn you, because most adults sing quite badly.) If you considered the matter in a sufficiently lenient way, then you could think that it actually sounded like "skip to my lou." Otherwise, Lilly preferred the great art of improvization (a terribly complicated word) because she had more freedom there. Improvization is just putting together the notes that seem to fit, and so you don't have to practice or memorize any songs. Unfortunately, even this form of music making does not give you all the freedom you need, because if you want it to sound pleasant to the ear, then you have to stick to certain tone sequences, keys, measures, and so on and of course play the notes as accurately as possible. Since Lilly McCoy did not master either one or the other, her improvizational playing was horrible. In the house where the McCoys lived, however, this was not a problem. Lilly had her own little room and the walls were very good for swallowing the high flute tones.

But unfortunately, there was a problem at the departure: The seat reservation was not correct. The McCoy family would have liked to have had tickets for only one compartment, but instead they got reservations for two people in one compartment and another person in a completely different compartment. Father McCoy was furious and called the whole thing a "mess." The lady behind the counter, who Lilly actually thought was quite nice, politely apologized for the, presumably, technical mishap and explained that it wasn't all that bad, since the train is usually very thinly occupied at this time. The angry Mr. McCoy did not care. "I want my right," he ranted, "I want …," he thought, with a bright red head, trying hard to think what he actually wanted, and since it is quite difficult to think with a bright red head, it took him a while to finish. "I want," he finally continued, "that you correct this now and we get the right reservations." With a friendly smile, the nice lady behind the counter replied, "Unfortunately, it is too late for that now. Reservations must be made at least five hours in advance." Her smile became even more friendly and conciliatory. "Of course, you can also ask for your money back for the wrong bookings. All you have to do is fill out this form and send it to the railroad administration."

The father was just about to start his lengthy thinking again when mother McCoy, also smiling friendly, bent over, accepted the offered form, and gently pulled her husband aside. "Come on honey, otherwise we'll miss our train," she said in a sweetly purring voice. Father McCoy reached for the luggage with a confused expression on his face and followed his wife. Lilly walked beside them and was proud of her father because he could change the color of his head so quickly. She couldn't do that, but she was eager to learn it.

Just as the nice lady at the counter had said, the train was quite empty and there was nobody in the McCoys' compartment but them.

Lilly loved that she was allowed to sit at the window. The landscape rushed past her. She saw animals in huge meadows, hills, trees, bushes and sometimes a few people. From her mother she had gotten the wrong seat reservation card.

"This is for your seat, so that you may sit there," she had said and added with a serious face, "you must not lose it." Then her father explained to her what the individual numbers on the card meant and that in her particular case not everything matched because the "knuckleheads of railroad employees had screwed up." Lilly listened attentively and nodded. In fact, the numbers above her head did not match the numbers on the card. So that was the screwing up that the "knuckleheads" had done. When her father came home from work, he often railed about such "screwing up" and his head got this interesting red color. So now she had personally experienced this kind of "screwing up" and also how her daddy always "fixed" it. She was proud of him.

For lunch they went to the train restaurant. Lilly ate a small steak and a children's portion of salad. Unfortunately, there were onions on the steak and some garlic in the salad. Nevertheless, it tasted delicious, and while chewing with relish, one could look out of the window and watch the passing landscape. The clouds in the sky looked like soft, white stuffed animals.

After the meal Lilly felt like playing the recorder. But her parents wanted to sleep, and so she sat in her window seat with her soprano recorder on her lap and was sad that she was not allowed to play music. Her stomach rumbled and she was urged to give in to a certain desire, but she knew that her parents didn't like it. She watched them. Her father was already fast asleep and had his mouth open. He snored a little. Her mother had her head snuggled against his strong shoulder and seemed to sleep just as soundly.

"Maybe I'll go and look for my right place," thought Lilly and let herself slide carefully from the seat. She took her recorder and the reservation ticket with her. It wasn't that hard to open the door of the compartment. She had thought she would not be able to do it alone.

In the corridor stood the train conductor. He looked out the window and smoked, although that was actually forbidden on the train. Lilly sneaked past him into the next wagon. When the door closed behind her, she suddenly felt relieved.

The train conductor sniffed the air in amazement. It looked like a wild animal suddenly sensing danger; then he looked around attentively, but couldn't see anything odd. Suspiciously he looked at his cigarette, then threw it disgustedly into the ashtray and pulled down the window. Eagerly he sucked in the fresh air. "That's horrible," he groaned.

Lilly compared the wagon number with the number on her ticket. '7, so it's here,' she thought. Now she just had to find the right place. The numbers were on the outside of the compartments, and little Lilly began to search. She had to walk back and forth a little because she could only count to ten and simply compared the numbers. But finally, she found it and opened the door.

Inside sat five giant young men. They belonged to a football team and obviously had just returned from an exhausting game. One of them was reading a book while the others were sleeping. The giant reading the book looked at Lilly in amazement when she entered, but Lilly confidently walked up to him and showed him her reservation card. The young lad nodded and, absurd as it seemed to him, a little girl alone in the train, took his bag from the seat next to him. Lilly was happy. It was the window seat again. The footballers had kept it free because it was railway-wise correctly marked as reserved outside, 'screwing up' or not.

Little Lilly was lifted to her seat. 'Wow, how strong he is,' she thought and looked at him with admiring, big children's eyes. The footballer smiled back, mildly embarrassed, and pointed to the recorder to do anything at all. "Can you play?" he asked very quietly so as not to wake his comrades. Lilly nodded proudly and with bright eyes led the instrument to her sweet mouth. The young man made a face as if he had just bitten into a lemon, but it was too late. Lilly started to play. She played her only song: "skip to my lou." You couldn't really say that she played that song badly, but a soprano recorder is a very penetrating noisemaker if you don't hit every note.

Lilly did her best. She played as well as she could. Her little fingers moved eagerly on the wooden tube and the upper part of her body swayed to the beat. One by one the other football players woke up and watched the little flute virtuoso at the window with sleepy amazed eyes.

Lilly played three verses, then repeated the last two bars "... skip to my lou, my darling" and finished her performance. The awakened boys gave a polite applause.

"That was great, kid. What's your name?"

"Lilly McCoy," the answer came proudly and she added in her eagerness, "I'm going to my grandparents in Des Moines."

"All by yourself, and such a long distance?"

"Yes!" answered Lilly and her head turned slightly red. It was really only a very light red, because it was only a little lie. There was some truth in it, because after all her parents were in one compartment and she, little Lilly, was in another. So, technically and in terms of scale size of railroad carriages, she was indeed traveling alone.

"Aha!" said the one with the book next to her, "Nice to meet you Lilly McCoy." He reached out his big, strong hand to her, "I'm Jeremy. That's Paddy in the corner and Tommy, Jack, and Edward over there."

"He's still sleeping," Lilly said.

"Yes, Edward is our sleepyhead, our big baby on defense. You might as well call him Teddy."

"Teddy, I like that," said Lilly and was happy that she could talk so well with the giant guys.

"Tell me Lilly," Paddy bent forward a little, "do you know any other songs on your recorder?"

His teammates gave him angry looks, but Paddy didn't understand what they wanted. But when Lilly keenly nodded "hmmm, hmmm," he realized and was looking for a way to prevent what he had conjured up. The tip of the flute moved inexorably toward the little mouth. Suddenly the movement stopped—was there still hope?—but no, Lilly just expertly turned the mouthpiece of her instrument, and then her concert began. She improvised as much as she could. At first, she played something sad, or what she thought was sad, to make the boys feel touched by her playing. In the end, she found her playing so heartrending that it started rumbling in her stomach again. The footballers looked at each other disgruntled, but smiled encouragingly as soon as Lilly looked up and glanced at one of them.

Suddenly it had happened, very quietly, the soft fabric upholstery swallowed the sound waves and also the other thing temporarily. At first Lilly faltered for a short moment in her audition, but then she thought she could avert the threatening danger by playing even more intensively. Changing hectically to dissonant, brisk passages (one could also call them fast, oblique tone sequences), she piped on.

But nothing helped. Jeremy was the first to turn up his nose. He looked reproachfully at Paddy, who understood at the same moment why this look was directed at him. Energetically he shook his head and pointed to the three opposite, who also, little by little, got a disgusted look on their faces, except for Teddy, who was still sleeping.

Lilly, who was busy with most violent puffing, knew this behavior only too well. It had happened to her once in a crowded movie theater, and it was interesting to watch how the clearing throats, coughing, staring in amazement, and loud "Yuck!" shrieks spread to an ever-growing circle of people. She then sat small and sunk together in her armchair, next to her, her big father, who was most excited by the terrible stench and finally even chanted "Throw the bastard out!"

In the small compartment it was now really hard to bear. "Booaaah," said Paddy, speaking from the heart of everyone else. Tommy and Jack had their noses hidden behind their T-shirts, so they looked like gangsters. Their eyes seemed to water. "What the hell, I don't believe it!" The four of them had agreed with their glances, and Jeremy finally kicked the giant Teddy in the shins.

Slowly he also livened up and rubbed his eyes. He had dreamed that he had fallen into a manure-barrel and now, still dawning, he had to realize that his dream had probably not been a dream after all. Panic-stricken, he rowed with his arms, and Jack and Tommy could hardly control him. When he finally understood the situation, he rushed wildly to the window and tore it down. Coughing and choking, he gasped for air.

"How can one produce such nasty farts in one's sleep?" Jeremy asked the others and pointed to poor Teddy. He still stood beside the window, looked guiltily into the compartment and shrugged his shoulders. When he noticed the little girl, he blushed. 'God, this is embarrassing,' he thought.

Lilly sat in her seat and continued playing the recorder now. She was very happy that no one suspected her and that the window was open because the next disaster was already looming. The strong wind would certainly drive the evil air away quickly, ... so off we go.

"All right now, Teddy. Close it again! Or the little one here will catch cold."

"Ok," the addressed giant answered meekly, closed the window, and sat down again on his seat. No sooner was he sitting than Jeremy gave him a reproachful look. Teddy thought that this was because of just now and raised his shoulders again.

Lilly prepared herself for the worst and tried desperately to suck up as much of the air as possible by breathing in and out fiercely. She blew in and out through the recorder and in her panic did not pay attention to what she was playing. She completely forgot that she actually had an instrument to play music in her hands and not a blowpipe. It sounded horrible—just like the rhythmic whistling of a steam whistle that had gone out of control.

"Teddy, look what you've done, the little girl can't breathe anymore, she's totally freaking out," Paddy said worriedly with teary eyes from under his T-shirt. At that moment the door opened and the conductor entered, meaning he wanted to enter, but it was as if he had run against an invisible wall. With eyes wide open, he stood there like paralyzed for two seconds, then turned on his heel and ran, shouting "Buooooaaa," into the corridor, to the nearest window.

Lilly's steam whistle performance now reached its climax. From outside one could hear disgusting choking sounds for a while; then the conductor left the wagon without checking the tickets.

"One more like that and you'll get tossed out, Teddy!" It was obvious that Jeremy was serious.

"That wasn't me," Teddy tried to justify himself, but the loud protest of the others quickly silenced him. Scratching his head in embarrassment, he looked at little Lilly, who was now improvising 'reasonably' again.

She played for another half an hour and was delighted with her attentive listeners. 'When I grow up, I want to marry one of them,' she thought. Otherwise, she was too busy with her performance to think much. It was sometimes loud, sometimes quiet, sometimes funny, sometimes sad, sometimes fast, sometimes slow, but always something special because she had made it all by herself. She ended with a fulminant trill that degenerated into a multi-octave change due to inadequate hole-covering and then put her instrument down. The boys applauded and smiled at her in a friendly manner.

"Thank you!" she said boldly and added with her chests swollen with pride, "This was my first real concert ever."

"Wow, that was great."

"Yeah, really great!"

The footballers praised her and Lilly basked in her glory for a while. 'It's time I went back to mom and dad or they'll worry,' she finally thought.

She slipped off her seat and took her soprano recorder and the ticket. "I'm going to play the recorder in other compartments now," she said confidently, and the boys hurried to reply, "aha, that's a wonderful idea!"

"Go ahead, I'm sure they're all happy."

Lilly turned to leave, but suddenly turned around again and put the ticket back to her seat. 'Let them have a souvenir of me and my concert,' she thought generously.

But when she took the first step toward the door, it happened again. With a quiet but clearly audible "Oooopps" it came out of her. Quickly she reached the door, opened it, slipped out, and pulled it shut behind her.

Through the glass door she once more glanced into the faces of her audience. They looked at her with an indescribable mixture of friendly smiles, bright astonishment, which was on the verge of horror and was slowly interspersed with disgust from below.

"Terrific guys," she thought and went back to her parents.

*

Outlook

My child had already been writhing with laughter throughout the story and it had taken me a lot of effort to stay 'serious' at least enough to keep reading.

Now, however, my little girl can no longer hold on to herself and laughs out loud unrestrainedly. It takes quite a while until she calms down again and since this laughter is contagious, I laugh too and cannot stop until I have tears in my eyes.

"Great! This is really from you?" my child finally asks, still giggling.

"Yes," I answer modestly and am a little bit proud, "I had actually written it in German and translated it at some point."

It takes quite a while until we are both back to 'normal' again. It is bedtime. I bend over my child and give her a kiss on the forehead.

"Are we going to continue tomorrow?" she asks.

"Absolutely!" I answer.

"With the wriggling on the surfaces of the Friedmen?" she keeps on digging.

"Not directly," I answer hesitantly and add, "rather we will be busy tomorrow with one exception."

"An exception?" the interest is obviously aroused. Actually, I didn't want to initiate this the night before, but now it had happened.

"Yes, there is an exception," I finally say thoughtfully. "One thing does not fit the pattern of superficialities."

"What is that?"

"**Gravity.**

And now go to sleep."

Day Two

"Where were we then, my child?" I ask.

"That we are all rather superficial," answers my little one cheekily and without hesitation. We both laugh again.

"Exactly, my love, you have grasped it. We live in a 3-dimensional space … at least we perceive it, but all information in this space is on the inner structural surfaces of the room, on the surfaces of the little Friedmen." I take a short break to have a sip of tea.

"Would you like some?" I ask my girl.

"Yes, I would love to have some of your lemon tea too, but please with a little more honey if you can."

"I'll be right back."

While I stir the honey into the cup I continue.

"If you look at it closely, our superficiality is even greater. I mean with respect to what we can recognize in terms of information at all, because for us, of a given volume of space—no matter how large it is—only those parts that are stored on the surface of the same volume are ever tangible."

"That ...," my little one answers hesitantly, "was now either a little weird or a little complicated or both."

"Okay, grab any piece of volume you want from the big wide universe. Take a close look at it and then look at the surface of exactly this room volume. You will find that all the information you can detect or sense is also stored on the surface. In a rather distorted form, similar to a hologram. That is why the principle is called the 'holographic principle'."

"Okay, it IS weird," my child now decides, "and why is that?"

"Well, the mathematician would say, because all information and its causes, no matter where it comes from and what it represents, in the end probably satisfy an equation of potential. But that is of course anything but illustrative. So, in simple words let me explain: On the edge of a volume, I see the average of what is going on in the volume and this is enough for me to be able to draw conclusions about what actually 'is going on' in general. To be able to draw conclusions within the scope of our observational accuracy, one should absolutely to add."

"So, we are not only quite superficial, but also extremely average," my child interjects.

"Exactly!" I reply enthusiastically due to the beautiful comparison. "We are average and superficial, and so is the universe we perceive, except ...," I pause for a theatrical pause, and indeed my child's eyes open a little further, "except for everything that has to do with mass and inertia."

"What's different?" my child immediately asks, "Aren't mass and inertia as pattern information on the **little things** ..." my little one grins again and probably wants to provoke me a little, but this time I only show the raised index finger and smile back, "... ok, ok imprinted on the little **Friedmen**? Where is this information?"

I can't help thinking how beautiful this little face had once been, how cute, open, and friendly. That was before the doctors hoped to control the disease with a few cruel cuts and amputations also in the front of the skull. I thought then that I would never be able to smile again when I looked at my child. Now I am amazed at myself and my ability to recognize and love the cute, open, and thoroughly friendly face far behind the scars and missing parts.

"In the volume change," I answer briefly and concisely.

"Ok?" replies my little one and lets the short reaction fade away in a thick question mark like a radio, which has been turned off, but still draws some energy from the capacitors, so that it doesn't die immediately in silence.

"Yes, this time it is actually quite simple. If the Friedmen are not so deformed that they do not experience a change in volume, then we perceive the room as massless ... gravity free, to be more precise."

"Good!" my child says, "and if the volume changes, then there is mass, right?"

"Yes, but let's rather say gravity, because mass only exists in isotropic and permanent volume changes."

"What is that supposed to mean?"

"Oh, it's not important," I wave aside, "just imagine a room full of flexible billiard balls again ..."

"Billiard balls are not flexible!" My kid interrupts me.

"Good point!" I answer with a smile, "but they are definitely elastically deformable. You can hardly perceive it, because even with great forces not much deformation happens. In any case, there are too many balls in this small room, and so they are somewhat under pressure. And for your sake, let's take very special billiard balls, which are made of a wonderfully elastic rubber. Suddenly we make the ball in the middle of the room shrink a little. What do the other balls do then?"

"They slide in as close to the ball as they can to fill the space, right?"

I nod approvingly and want to continue, but then my kid adds, "... And they're going to deform a little bit to fill the space really well, aren't they?"

"Really goooood!" I praise, "You're really on the ball. In fact, the rubber balls deform and the same happens to the little Friedmen who are near one of their own, which has a slightly smaller volume. The volumes deform around the strange volume element and we perceive this as curvature of space. The element in the middle, which causes the whole agitation, which creates the curvature and has a different volume, we perceive as a particle with mass. But we should honestly state that we perceive the whole agitation as mass and not just the small volume, because this is much too small for us."

"And how does attraction arise?"

"Imagine a second such deformed ball in your rubber billiard ball room. Also imagine that all balls are under pressure and actually want to have more space than they have at the moment. Now you shake the whole thing a little bit and allow the balls to move. What will happen?"

"I dunno!" it comes shrugging.

"I thought so," I laugh, "That's why I've prepared a little experiment. Wait here."

"Ha-ha," makes my kid not very amused.

I go to the bathroom and come back with a bowl of water. I put it on the bedside table, critically watched by my child.

"Let us now imagine that this is the universe," she jokes.

But I don't let myself be thrown off course and reply with a smile, "Well, not quite. But let's assume that the surface of the water here in the bowl is our space and this ...," I carefully place two thumbtacks at a proper distance from each other on the water, "... be two mass bodies within this space."

At first my child just watches with interest and is finally fascinated how both thumbtacks move toward each other without any apparent external reason and finally stand united in the center of the bowl.

"Wow!" my girl just says.

"That's not so hard to explain," I reply. "Each of the two thumbtacks deforms the water surface around itself and so the thumbtacks feel each other. This curvature causes a small increase in the surface energy of the water. But nature prefers to have less surface energy and is looking for a structure where the total surface of the two thumbtacks taken together is the smallest. This means that the two curvatures 'feel each other' and want to unite to a structure that needs less energy ..."

"You mean the one that uses the least energy, right?" my child interrupts me.

I smile. "That's correct, you say it ... The system searches for the structure with the least surface energy. Coincidentally, this happens just when the two pins meet. That's why they move toward each other."

"Uh-huh, that's obvious," my child says.

"Okay, so now back to the rubber billiard ball room with the two smaller balls in it."

"Well, if the other balls want to have more space and there is some kind of surface tension or even tension at all, then I guess they'll go best if they put the little ones together. The small balls would then move toward each other."

"Right! Very good!" I applaud, "But you must not confuse the effect. First, there is the volume element with the somewhat different volume content. This leads to the deformation of the Friedmen around it. The space is distorted. The same thing happens around other volume abnormal elements and they also distort the space around them and that brings the attraction in the end."

"How big must the change in volume actually be to get, say ..., a body like our Earth in terms of mass?"

"Oh, let me just do a little math on that." I fetch a writing pad, my calculator, and a book with some constants. "We need the mass of the Earth, the gravitational constant, radius of the Earth ... well, actually, its internal structure as well, but let's just assume it's a homogeneous sphere."

"Aha!" says my little one, "Let's **play** physics again and assume the cow is a sphere, okay?"

I laugh and answer, "Not quite, the cow is a point, but the Earth is a sphere, yes."

My child watches with interest as I put some equations on paper and play around with them. Finally, I type some numbers into the calculator.

"Well, in curved space the Earth seems 1.48 mm bigger in radius to us, because space is shrunk."

"What? Not more?" my child asks astonished, "the huge Earth with a radius of 6 378 km ..."

"Wow, you got that in your head?" I marvel.

"Sure! So, this large globe of 6 378 000 000 millimeters causes a spatial anomaly of not even 1.5 millimeters? And yet we feel such a strong attraction?"

"Yes," I answer meaningfully, "the effect of curvature is quite small, even with masses that are already very large by our standards ... that's why we usually don't notice it, I mean the curvature. But the attraction, we can feel it already."

"And the sun?"

"Hmm ...," I'm working on my calculator again, "Well, it's almost 500 meters there."

"Oh! Well, that's something!"

"Yes, fine," I answer, "but keep in mind that 500 meters is not exactly a lot with a total radius of almost 700 000 kilometers. Because the sun is damn big."

"Oh yes," my girl understands, "of course, ... 500 meters are not noticeable at all."

"Exactly, and moreover, where on the surface of the sun there are constantly powerful movements with turbulences of hundreds to thousands of kilometers of altitude difference."

"So, this is how mass is created?"

"Yes," I answer, "whenever space changes its properties in some way and bends, and so on, gravitational forces arise for us. If these forces are rotationally symmetric at a sufficient distance, they appear to us as mass ... resting mass."

"And not resting mass, how does that appear to us?" It was clear that my child would find a hitch again immediately.

"Oh yes, inertia ...," I begin.

"Why inertia?" my child asks immediately.

"Well, because the mass of an object that is not at rest or does not appear to us to be at rest has an inertia, and this inertia is strangely enough as great as the mass itself."

"But that sounded like saying the same thing twice over," criticizes my child.

"Okay, you're right, so let's try it another way." I'm thinking about it, stroking my chin. "If you try to throw a stone away, it won't be so easy because the stone resists the acceleration of your arm, right?"

My child nods, "Apart from the fact that throwing a stone won't work for me anymore, you are right, of course."

"Well, the strength of the resistance against your arm—or any other acceleration—is now exactly determined by the mass of the stone or whatever is to be thrown."

"And what's so special about that?"

"Well, it is special because the force with which a body is accelerated in a gravitational field is also precisely determined by its mass. There is no difference, no matter what the bodies consist of. Both the reaction to the attempt to set something in motion, meaning inertia, and force in a gravitational field are characterized by the same property: the mass of the object."

"What is so surprising in all this? In both cases, there is a reaction to the volume deformation of the Friedmen of the objects in the same way and since they determine the mass, the dependencies on the mass are the same." My child shrugs her shoulders.

"Super, you just explained the equivalence principle; hence, inertia and gravity are the same, because both couple to the same property."

"Then what about the Higgs particle?" my child suddenly surprises me with a completely unexpected question.

"Where did you hear about that?" I ask back in amazement.

"There were already reports everywhere," my little one is indignant about my astonishment. "They even called it the 'God particle'."

I laugh. Of course, I had heard about it and was amused by the typical exaggerations of the media. Even my child finds the expression 'God particle' ridiculous.

"Well," I answer, "eventually I can also formulate the effect of structural distortion as a field, a Higgs field. And like other fields, the Higgs field also has particles. If you want to put it this way, these are independent distortions or distortion solutions. Nevertheless, whatever they, the dear scientists I mean, will find in their accelerator experiments: Higgs is only a phenomenon, not the deeper structural cause of the mechanism of mass formation itself."

"Why was this particle invented in the first place?"

"Actually, because they just wanted it. They tried to get all the interactions, that is forces, under one roof and managed to do it pretty well. The only problem was gravity. It didn't fit into the picture."

"And with the Higgs would it have fit into the picture?"

"Yes, well, that's one way of putting it ... But it is just not true. It is a fact that I can't combine gravity with the other interactions, electromagnetism, weak and strong interaction, because they act on the surfaces of the Friedmen, while gravity comes through the volume."

"Ok, that's clear so far and electromagnetism also rings a bell ... magnets, electricity, and so on. But what are the other two forces?" asks my girl.

"That's quite simple: the strong force is holding together the atomic nuclei and the weak force is causing some forms of radioactive decay, the so-called beta decay, for example."

"Good!" my girl says like she just made a little discovery, "Gravity couples to the volume or change in volume of the little Planck or Friedmen things and the rest has something to do with a dog-and-pony show on their surface, right?"

"Right!" I confirm and laugh, "Instead of dog-and-pony show, you could also say wriggling, but both is exact enough ... highly scientific, indeed!"

"Well, but ...," she hesitates—aha, so she has discovered a flaw after all, "what about the constancy of the speed of light, entanglement, and ..." my child makes a significant pause, "ANTIMATTER?"

"Oha!" I burst out, "You've got me there pretty badly. How do you know all these weird things?"

My child smiles. "Star Trek," she replies.

I raise my right eyebrow in doubt and we both laugh.

"Ok, ok, there are these books about Star Trek technology on Johnny's shelf ..."

"Oh Johnny, that old freak ..."

"Yeah … He lent them to me from time to time and if I didn't understand something, he explained it to me … uh, I mean," my kid scratches her head a little embarrassed, "at least … he tried."

"Oh no," I quickly reassure, "Johnny obviously didn't do a poor job at all." And a little thoughtfully I add, "… and Star Trek apparently not either, who would have thought …"

"So, what about this stuff, speed of light, entanglement, and antimatter, hmm?" urges my girl.

"Oh, that's easy," I throw in succinctly and wink at her, "let's start with the speed of light and imagine that all light information resides as a surface structure on the Friedmen. From one Friedmann to the next, it travels in an incredibly short but finite time. This jump-time from one volume element to the next is always the same. Thus, the speed of such a light information depends ultimately only on the number of Friedmen along a stretch of line where the light signal is currently traveling. If this space is distorted, because for example a black hole is in the vicinity, then not the jump-time changes, but the number of Friedmen, that fit into a normal space area, changes. For the observer, however (that would be us), who does not live in the normal space, but can only see or perceive the space of the Friedmen (because we ourselves are a part of it), it looks as if light always has the same speed. In reality, however, it always has only the same jump-time from one volume element to the next."

"And what is it like when something moves?" my child continues to ask, "When we fly with our Earth through space, we always measure the same speed of light, no matter whether it comes from the front, where we are flying to, or from behind, where the light should actually run after us and catch up with us first."

"Again, the issue is not difficult to understand, my love. In front of us we compress the Friedmen a tiny little bit and behind us they are stretched. This does not change anything for the light, which only has to jump from one volume element to the next. Because the number of Friedmen it has to cross to reach us remains the same. But we measure the speed of light as the number of Friedmen divided by time, and because the jump-times remain the same, we 'see' the distortion of the volumes. That means, the constancy of the speed of light is nothing else but the constancy of the jump-times between the volume elements and the constancy of their number."

"And it doesn't matter what kind of light it is?"

"It doesn't matter at all," I answer. "Surface information like that of light is exchanged between volumes and all have the same jump-time from volume element to volume element. The other day I read an article somewhere about the observation of a stellar explosion, which happened ages ago. Because all light frequencies arrived at us at the same time, this was a good proof against a 'bubble structure' of space. Well, with the jump-time this would clarify itself, because there it does not matter which frequency the light signal has that is passed on."

"Entanglement?"

"Do you even know what this is?" I ask suspiciously and somewhat provocatively.

"The Einstein-Podolski-Rosen Paradox, of course, I know it. Two particles are created that are somehow connected to each other, entangled, as they say. Both move away from each other and are already tens of kilometers away when you make a certain measurement on both particles. They always prefer to measure the spin ... no idea why, but it doesn't matter. It is assumed that the particles, which flew in different directions, had nothing to do with each other from the moment they took off. Nevertheless, we find that the two particles seem to have been talking to each other all the time, and faster than the speed of light, because the results of the measurements are always 'as agreed'."

"Well, why do you think that is? Is there the spooky action at a distance that Einstein had spoken of in this context?"

"I don't know, you tell me!"

"The Friedmen can apparently also communicate with each other very directly. In certain situations, they succeed in bringing them into coherence, thus, into similar states of vibration. Like in a room full of dancers who all dance to the same music. In such a state, the volumes virtually merge with each other and can no longer be recognized as individual elements. One could say that large effective volumes are created, with their own super-wriggling on the common surface. Within such a structure the information is transported much faster than between individual Friedmen."

"How fast is that speed?" my child wants to know.

"No time passes at all, because on this level time does not really exist. This means that no time is measurable for **us**, no matter how large the respective effective volume is at the moment, because we are in the range of the smallest possible time intervals for **us**, you get it?"

My girl does not look very enthusiastic.

"Strictly speaking, the expansion time on such a super-Planck-thing is below the Planck time and therefore for us zero. Thus, Friedmen in coherence talk without any time passing, but effectively no distances are traveled either, because distance for us is always just the number of different Friedmen between two points. With coherence, however, the difference between Friedmen on the track is eliminated—sometimes only in one particular characteristic, while the difference for other characteristics is still manifest—and thus there is no distance at all, you see?"

"Hmm?!" my kid goes, "a little, maybe ... let's see. Now what about antimatter?"

"Oh, yes: antimatter. Well, you seem particularly interested in that, don't you?"

"Yes," my child replies briefly and honestly.

"You, I mean you and your siblings had a construction kit for a radio at some point, do I remember correctly?"

"A transistor radio, sure." My child nods, "We got it together all right, but then it fell to the floor and wouldn't work anymore."

"It's a pity, but things break, you can't help it ... especially with you guys."

"What does that have to do with antimatter?" my girl asks a little confused.

"In the transistor," I start without further ado, "there are two charge carriers, these are the particles that conduct electricity ..."

"I know, Daddy," my child interrupts me, annoyed.

"... just wanted to be sure. One is the electrons and the other the holes."

"The holes?"

"Yes, the holes," I answer unswervingly. "Actually, it is nonsense to differentiate that way, because in truth there are only the electrons, but the thing is: In one part of the transistor, the n-layer, where n stands for negative, there are a few free electrons that take over the charge transport. This is the normal charge transport, as it happens more or less in every conductor in the wall of a house."

"Sure, I know. The electrons scurry through the atomic lattice."

"Golly!" I acknowledge with astonishment, "Okay, but then there's another layer called the p-layer, and now the p is for positive. There is an incredible shortage of electrons here, so wherever an electron is missing, there is a p-charge, a positive charge, right?"

"Yeah, sure!" my child nods vigorously.

"However, these positive charges are firmly attached to the atomic lattice and thus cannot move. Nevertheless, when a voltage is applied to this p-layer, they seem to migrate. Do you have an idea how this works?"

My child ponders and then shakes her head. She seems a little disappointed with herself.

"It doesn't matter, you can't get at it that easily," I cheer her up. "Other, actually bound electrons simply jump into the missing parts of the electrons. And where they come from, there are the new p-positions. For an outsider it looks as if the p-positions are hopping or wandering, while in reality it is the surrounding electrons that take over the whole wandering."

"Ha, it's as simple as that?!" my child is happy, especially because she has understood immediately, "But I still don't know what that has to do with antimatter."

"Matter simply consists of Friedmen who have been given certain properties. You could imagine, for example, that a Planck sphere has become a Planck torus, with an axis of rotation around its center—that's the spin—and still a little wriggling on its torus surface and, of course, some volume distortion to 'simulate' mass."

"Oh, matter is degenerated …, I mean deformed, 'Planck spheres', right?"

"Exactly! Degenerated … good word," I praise, "and antimatter now is a flaw in the structure of the Friedmen, with the exactly contrary properties of the corresponding matter particles imprinted on the surrounding Friedmen."

"Huh?"

"Matter is degenerated Friedmen. Antimatter is defects or holes with the degeneration properties around them."

"Really, can it be that simple?"

"Yes, it's that easy. And we, as outsiders, cannot differentiate between them, because for us defects or real particles are one and the same thing … apparently, from our 'big' perspective."

"And that's why matter and antimatter always destroy each other in pairs, because a hole and the matching element want to merge in order to fill the space again." My child is miles ahead of me. I am impressed and can only nod approvingly. Nevertheless, she immediately discovers a catch, "Yes well," she starts and raises her weak hand, "but why is there no more antimatter in the universe?"

"Hm, you'll figure that out on your own," I answer encouraging.

My girl almost makes a little pout. It's a pity that you can't see it anymore under all the scars and the missing parts of her face. "Now come on, explain it to me!" she demands.

"There was a state in the universe where there must have been free surfaces. That is where the holes disappeared. Wherever they were near a surface or edge in the bunch of Friedmen, they couldn't survive there, because they need the surroundings of the other Friedmen **to BE** at all. The particles of matter, however, are really there; they don't need other volume elements around them to assure their existence. So, they are not getting lost, not even at the edges. Once again it is as simple as that."

"There is only matter in the universe today because there used to be edges where some of the hole particles, that is, the antimatter elements, disappeared, is it right?"

"Yes, exactly." I raise my thumb, "Whatever was left—and that was still most of all the antimatter—annihilated with matter. But because there wasn't as much antimatter as there was matter, some matter was left. Just a little, but enough for us."

"Wow!" says my child this time and I feel from her absent voice and the blurred look that she is flying in her mind with wondrous structures through an imaginary vacuum, sees particles forming and destroying themselves and I know that she understands ...

... probably better and more lasting than I ever could. Because I was simply too old to develop brain cells for such an imagination, when I finally realized how the universe we perceive works in the very small—possibly works ...

A long, long pause follows in which nobody says anything.

Day Three

"And what about the distortion of time? We didn't speak about that yet, did we?" My girl had obviously used the break to look for some more problems in the system I had built.

"No, that wasn't topic of our conversation yet, but since you're asking, we might as well get it over with. You have to imagine that the little Friedmen can't be still. They have to wriggle constantly, which means that their surfaces vibrate. This wriggling is what makes time for us. The smallest frequency in the whole hustle and bustle also determines the smallest pinch of time, the smallest time unit, so to speak."

"Time is grainy, then?"

"Quantized, yes," I speak the 'YES' somewhat hesitantly and after a pause I add, "you know, my love, time is only one effect of this wriggling, it is not there itself, not absolute ... it only appears to us because we are too big and too bumbling to be able to really look into this little clockwork. The truth is that the surfaces of the Friedmen change their shape and thus the positions of the surface elements in space. They simply are doing this and can't help it. Just like that, forwards, backwards, and again and again. This unspeakably small back and forth of elements in space makes the room receive a tiny inner period, let's call it elementary period. And this gives us a sense of time."

"And why does time only go in one direction?"

This time I have to laugh, "You've obviously overheard too much of these cosmological discussions, haven't you?"

"Well," says my little one somewhat defiantly, "that's the big mystery: Why does time always go only forward?"

"Big mystery? You'll see in a moment that it's quite simple. So, let's assume that time is a product of the wriggling of the Friedmen, okay?"

"Yes, okay."

"Well, then I can imagine a vast number of different time units by using different oscillation frequencies, meaning fast or slow wriggling, can't I?"

"Also clear, yes," my child nods.

"Damn small oscillation durations can come out of very nimble little wriggles."

"Okay?" my child wonders where this is all leading to.

"Also almost zero, that is, with incredibly short units for a wriggling period, right?"

"Yes!"

"Yes, but how the hell am I supposed to create a negative time? I can't even get below the point 'it doesn't wriggle anymore'. Or, the other way round, over the point where it wriggles infinitely fast."

Pause.

"Or do you see how I could make a time based on the wriggling of the Friedmen negative?"

Short break and then bright eyes and a big smile on my child's face. "No," she finally says, "that's not possible. If time is a product of the wriggling of space, then it can at most become zero, freeze, if nothing moves anymore. But it cannot go backwards."

"Exactly," I say, "but what do you mean by freezing?"

"Well, if the wriggling stops ...?" my child shrugs her shoulders as if my question was nonsense.

"I think," I say cautiously, "this does not fit into the one basic law that we had at the very beginning."

"Oh ... right ...," my girl nods, "'there can be no rest ...' The wriggling never really stops, does it? Well, then time can only ever run, like a watch that has an infinite battery and always in one and the same direction."

"Wow!" I say in amazement, "You put that damn well. If the wriggling is always there, so is the ticking of the universal clock, because the wriggling itself is the ticking."

My kid smiles proudly. Her gaze glides confidently through the window and into the distance, she sees the leaves gently swaying in the wind and the birds jumping around in the branches. It is springtime and the male songbirds are chirping what they can to impress the females. Then the gaze returns to the room, back to the sickbed, and finally gets stuck on the blankets in a certain place, the place where her legs would otherwise be. Almost too quiet to understand, my child says, "If time can only flow in one direction, then you can't undo anything, can you?"

It was probably formulated like a question, yet I knew that no answer was expected from me. My girl suddenly looks at me very firmly, "I want you to put my legs with me when I'm, ... when I'm dead, you understand?"

"Yes, I understand, and don't worry I've already thought of that."

"That's good!" my little one says very quietly again and almost whispering she adds, "Thank you."

Then there is silence for a long time.

"A little bit, however, I have to pour water into the wine with the negative time," I finally say with a slightly artificial sigh.

"Why?"

"Well, because I don't know exactly ... But let's take it one step at a time ... If we force the unification of the two leading and best tested theories of modern physics, quantum mechanics and General Theory of Relativity, we almost inevitably end up with tiny building blocks of which space must be made up."

"The Friedmen? We've already had that. So why going over it again?"

"Well, the Theory of Relativity says that space is elastic, that is, it can deform. Quantum mechanics, on the other hand, demands that all degrees of freedom, but also really ALL degrees of freedom, are subject to quantum fluctuations. This is simply demanded by the Heisenberg uncertainty principle. Simply put: everything that can wriggle is wriggling, and a believer might think that someone—the Creator—has imposed a few restrictions on us. But the physicist answers that such an interpretation would be too shallow and too brief, because without the wriggling we would not exist and we could not ask these questions. Regarding the possibility of space to bend, this simply means that it does this on a small scale all the time. In order that this mini-bending (we are talking about scales in the range of 10 to the power of −35 meters) does not violate the principle of energy conservation, an additional parameter is needed, thus, another degree of freedom, which we perceive as time."

"Time follows from Heisenberg uncertainty principle?" my child asks in between.

"Yes, Heisenberg's law of uncertainty applied to elastic space produces time."

"But such an uncertainty exists for time itself. How is that supposed to fit?"

"Well, that's another chicken-and-egg problem; so what is more fundamental: quantum mechanics and Heisenberg's uncertainty relation or time? I guess the former one. To put it simply: each space 'point' that we can perceive as such (in reality this is not a point, but a tiny little piece of volume—much, much smaller than a proton) is a clock with its very own shifting time. And this shifting time is insanely small, namely, around 10 to the power of –44 seconds. If we and everything that is, however, live (must live) in a space in which every 'point' is a clock, then we cannot escape the effect of 'time'. The space itself is the clock, or let us better say: the clocks."

"'To have no time' would therefore mean 'to have no space' and that would then mean that you are not at all," my girl philosophizes and I notice that she makes fun of me a little. But I ignore the point and continue.

"Well, the really funny thing is that if you measure the universe in these small units, thus, the mini-length and the mini-shifting time, then you get about 10 to the power of 60 of such lengths for the radius and 10 to the power of 60 of such mini-times for the age of the universe. So, the values are the same and no physicist can really explain this yet."

"Oh?!" my child only says and I feel that she takes me a little more seriously again.

"Yes, exactly, 'oh'!" I answer, "and there is something else that is funny: If you let space bend, as quantum mechanics demands, and assume that there are also fluctuations that lie below our time threshold, meaning below Planck time, then we would perceive these distortions as further degrees of freedom, thus as further dimensions."

"Err, ... so what?"

"Do you know how many distortions there are in 3 dimensions?"

"Well, three, of course!" my child answers without thinking about it for long.

"Well, that is a common misconception. Think again!"

My girl looks a little offended and defensively replies, "What's wrong with that? I have three axes in the 3-dimensional space: one in x-, one in y-, and one in z-direction. I can expand and stretch the x-axis, the y-axis, and then the z-axis. That makes three distortions ... or does it not?"

"Yes," I answer slowly, "but can I really move the x-axis (to take one direction out) only in x-direction, or can I do more?"

"Of course," my child smites her forehead, "I can also distort the x-axis in the y- and z-direction!"

"Excellent, kid! And how many distortions do I have then?"

In a low voice my child mumbles to herself, "... so the x-axis deforms in x-, y-, and z-direction, the y-axis in x, y, and z, and the z-axis of course just as well. That makes 9. Nine possible distortions," she exclaims, "or ...?"

"That's right," I confirm, "however, distortions x to y and y to x are exactly the same, because otherwise the space would rotate, which it obviously does not do; otherwise, we would have to notice it. Hence, x to y is coupled with y to x and thus not independent anymore. The same holds true for y to z with z to y, and last but not least, x to z with z to x. Therefore, six independent distortions remain."

"So that would be 9 dimensions then: 3 for our normal space plus 6 tiny ones that come from the distortions, is that right?"

"Absolutely correct!" I clap my hands and say, "And what else is missing?"

"Time," my little one answers without hesitation.

"Bingo! And you know what's funny again?"

A shrug of little shoulders and a tense look.

"There is a theory where physicists think they find the basic elements of quantum mechanics and general relativity combined ..."

"String theory!" interjects my girl.

"Wow, you really know a lot!

Well, in this theory, the people who work with it are always missing a few dimensions. And now, of all things, in the most hopeful approach of string theory, they are missing exactly ... how many spatial dimensions do you think?"

"Six? I mean 6 dimensions are missing?"

"Absolutely correct! Sit. A+!"

My child looks at me with astonishment and suspicion for a while. Finally, she says, "Freaky! That would be really phat if the dimensions the string people are missing were simply the fluctuating distortions of space!"

"Yes," I answer, "that would be phat, and it would also directly explain why we cannot see these additional degrees of freedom or dimensions, why they seem to be rolled up. It's just the distortion fluctuations at a magnitude level that is no longer accessible to us. An effect below our detection scale, if you will put it that way. We experience its effect as properties of space, but we cannot observe it directly."

My child touches her chin and broods for a while. Then she suddenly says:

"The curvature or distortion of time?"

"Excuse me?" I ask a startled question.

"Well, that's what I actually asked: How can time bend?"

"Oh, I see," I answer and am happy to be up-to-date again, "yes, I had completely lost sight of that. Let's not call it bending, but rather extending or stretching. One also speaks of time dilation."

"Anyway, how does it come about?"

"The guitar back there," I point to the instrument next to the children's desk, "when was the last time you actually played?"

"I don't know," it comes hesitantly and sadly at the same time, "it is not possible anymore ..."

"Do you still love the weird songs of this crazy guy? Of that Robbie Williams?"

"You shouldn't call him 'crazy'—he produces great music!" My girl knows that I'm not completely serious, "but what does that have to do with time ... what was that word again?"

"Dilation," I answer.

"Yes, with the time dila ...-thing?"

"What crazy Robbie has to do with time dilation?"

"Hey, stop calling him that!" my kid points at me with a raised finger.

"Okay, okay, you win." I grab my guitar, play a couple of chords of 'Angels' and hum along.

Promptly comes protest, "This thing is really out of tune."

"Exactly!" I just say and start tuning the guitar. Only on the last one, the E-string, I dawdle around a bit. "What happens when I turn this thing?" I ask, my fingers twiddling the last tuning key.

"You're tensioning the string, that's obvious." Shrugging.

"Okay, what if I pluck the string now?"

"It'll sound higher, of course," I realize that my kid starts to get annoyed with simple questions like that.

"Excellent, and that's the same with the Friedmen: When they get under tension because they get distorted, their movements, their wriggling, or fluctuations or oscillations, whatever we want to call it, also change."

"Okay?" my child is back and listening intently.

"Yes, but if the wriggling changes, then the smallest measure of time in the corresponding section of space changes, and thus time changes there. Got it?"

"When space is distorted, it comes under tension, and thus the wriggling of its smallest element changes, and that changes time?!"

"That's it, kid! And now a few more bars of music, right?" This time I play 'Angels' completely and I am not interrupted by my child, which must be the first time since I sang lullabies to her years ago.

"Does all the wriggling everywhere actually do more than generate time or 'fake' particles?"

"Yes, indeed!" I answer, "It stores information."

"It stores information … Why and how?"

"All these surfaces with their wriggling possibilities are nothing but giant information storages, my love."

My girl raises her eyebrows.

"Oh yes!" I hurry to say, "Everything in the universe we perceive is ultimately based on changes in the surfaces of the Friedmen. Even our information. While we are a few levels above the Friedmen in terms of size and timescales, everything is all just vibration. And even if our information storage in our layer should break down, the information on it is not really lost, because it only diffuses downward and eventually arrives at the lowest layer, the Planck surfaces."

"And that's where all the information goes?" asks my child.

"Yes," I answer confidently, "all information ends up there."

"Even those of us humans?"

"Yes, even those of us humans." I confirm.

"Of good and bad people alike?" my kid asks a little embarrassed.

I almost smile indulgently, but then I remember my own childhood and the seemingly foolish questions I had. And I also remember the silly answers I had been given by the big people back then. "You think of different places for good and evil, my dear, don't you?" I ask.

She just nods. Uncertain.

"Heaven and hell?" I continue to ask without any sign of evaluation or assessment of the question.

Again just a nod, this time even more insecure, almost uncomfortable.

"No, kid, there are no different levels for different information. They are all just information and all of them end up at some point on the lowest level of existence and there they cannot get lost."

"But …," my girl begins, still unsure whether one should ask such questions at all, "then in the end everyone is treated the same, no matter if they were good or bad?" Disappointment and deep unease sound from these simple words.

"You forget, my love, that on the lowest level there are no more potential differences and that only the amount of linear independent information decides. How does the amount of information of a bitterly evil individual 'down there' feel, for all times and eternities surrounded by his former victims … but this time without any possibility to keep them at a distance? What do you think? Forever and ever your own individuality would be trapped with the memories of your misdeeds, everything else of your former being would be small and unimportant."

My girl thinks for a while and then shakes, "This must be hell," she finally replies.

"Exactly!" I simply say.

"But then what is actually 'good'?" my kid asks quite softly and apparently more to herself than to me.

"That is the most difficult question of all, isn't it?"

My child just looks at me. She doesn't seem to expect me to have an answer.

"I would like to give you the worst example I can think of for this," I begin somewhat hesitantly.

"Why is it bad when the question is 'what is good?'" asks my little one.

"Because it is an example that has to do with us and it is always difficult 'to be good,'" I answer and continue, "suppose society had the resources to heal you …," I see all the muscles in the little body tightening, "… that would be good, wouldn't it?"

"Yes!" my kid replies without hesitation.

"The only problem is that your cure uses up so many resources that you could finance an annual world production of tetanus vaccines. You know what tetanus is, I suppose?"

"I know I have the shot, but I don't really know what it is."

"It's an infection you can get if you hurt yourself somewhere. It gradually paralyzes your entire body, you notice everything, but you are no longer able to move, talk, blink your eyes, and eventually you stop breathing."

"This is a terrible death," it breaks out of my little one.

"Yes, indeed, and a whole lot of people would be affected if we had to say goodbye to a full year's production of vaccine to …," I don't finish the sentence, but only make a hand movement that is supposed to say: 'Well, you know …'

"… to save me." My child ends the sentence after all.

"Yes, to save you … As I said, just a fictitious example," I add defensively and feel sort of bad.

"Then," my girl says very quietly and very slowly, "it wouldn't be **good** if I was rescued, would it?" Big round eyes look at me almost pleadingly from a disfigured little face that had once been so beautiful.

"Maybe not," I answer almost as slowly, "but now imagine that you, who has been saved, become a great scientist, who ... who, let's say, completely defeats cancer and saves many hundreds of times more people than ever died of tetanus ... what then? Wouldn't it then be **good** to save you, despite the few dying along the way?"

My child is just shrugging.

"But what if one of the people you saved from cancer became a second Hitler?" I continue mercilessly, "one who in turn destroys more people than you ever saved—wouldn't it then be **bad** if you were saved?"

Again, just a shrug of the shoulders and I begin to wonder how my child can stand it.

"We don't really know what is good, we can't know because we can't look into the future. To help just like that can be just as stupid and wrong and cruel as to do evil or cruel on purpose and at first sight seemingly evil can turn out to be good afterward.

We can only use our minds as best as possible, look ahead and into the broader picture, and try to do what is probably good. Ultimately, the question 'What is good?' is equal to the question of EVERYTHING, to the question 'Why are we and where is the journey of our existence and that of the universe heading ... of our universe?'"

My girl has tears in her eyes now and nods. I take her in my arms, I stroke her head and rock her. Finally, I add:

"But there is one thing that is most certainly always wrong."

"What?" it comes quietly from my arms.

"**Not to think about what is good ...,**" I answer.

Day Four

"You still look tired, how was your night?" I ask my little one.

"I don't really have any real nights anymore," she answers resignedly, "I can't sleep properly anymore and I'm not really awake either." She looks at me questioningly, just as if wanting to know from me what one could call this state she is in, but finally she says herself, "It is all somehow unreal ... I think I'm not really here anymore." And what she means by 'here' my girl doesn't have to explain to me.

But after a while my little one asks curiously, "What are we doing today?"

"Evolution," I reply.

"Evolution?"

"Yes," I answer, "today we learn why the world is evolving and why it can't help it."

"Why it can't help it?" replies my little kid in surprise. "Does that mean it's not a coincidence at all that we do exist?"

I have to laugh, "But yes," I answer quickly, "that was a damn coincidence. No, the world, our universe, cannot help but develop. Its inner structure forces it to do so, and in this process of development certain things inevitably occur at some point, one of them being intelligence. But how exactly this intelligence then is realized, there are many possibilities for it. Here on Earth, we were the product of this development by chance and we should be careful not to believe that we are an end in ourselves or an end product. That would be a boundless overestimation of our self and our existence."

"Ok, we are small and puny and insignificant," my child laughs, "but how did we come to be?"

"First of all: Whether we are insignificant is not yet clear. Even a failed attempt can be important, for example, to learn from it or to find new, better ways in the exclusion process. But how important our contribution will be in the end depends on us, on how far we are willing to overcome our, how did you call it ..., puniness ... to overcome ourselves, you could almost say."

"Hmm?!" my little one just makes and I see that she is actually more interested in her question than in my digression.

"What is a computer doing in standby?" I ask.

"Pardon me?"

"What does a computer do when it is idle?" I ask again, even though I know that my child didn't want me to repeat the question, but to explain its purpose, or rather its deeper meaning.

"What's the question?" it comes promptly, "It's not doing anything."

"Well, that's not quite right," I reply, "try again!"

My girl shrugs her shoulders and after a while she replies, "In idle state ... Well, a computer is probably just doing stupid stuff, pointlessly shifting zeros and ones back and forth."

"I can accept that," I say and smile. "Now imagine that the computer is huge and would work in idle mode for an infinite amount of time. What do you think would happen?"

"Well, since nothing is perfect and especially if the computer was built by people—who knows, maybe even the guys who put my laptop together which broke down a few days ago—then nothing is perfect anyway. At some point a little mistake would happen and the shifting of zeros and ones would turn into another shifting. It would change a bit, the pushing and shoving."

"And?" I dig deeper.

"And with time, these mistakes would become more and more, the changes would become bigger. Well, you know ..."

"Let's just assume that people hadn't made mistakes and the machine would work perfectly from our point of view, what then?"

"This is not possible because there is a natural blurring. At some point, a mistake has to happen somewhere," my child answers promptly.

"Very good! There we are. So, to the next question: What is the universe doing in idle?"

"Uh ...," but my girl is only a little bit off the mark for a short time and then answers, "It's wriggling!"

"Super!" I praise. "Exactly, it constantly carries out the small place changes or wriggles, because it can't help it; a trillion shifts are carried out in every cubic nanometer. Apparently quite senseless, but it goes back and forth."

"Just like in a computer," adds my child.

"Correct! Like in a computer, surfaces are shifted back and forth, and since moving a thing is identical to shifting information, information is constantly being shifted in the universe on an unimaginably large scale."

"And mistakes happen, don't they?"

"Aha, you know where all this is heading?" I show myself rather surprised.

"Yes, there are mistakes and the shifting, 'the program' is changing, but what does that have to do with evolution?"

"That, my love, is again much easier than you think and simpler than all clever people want to assume. There are a lot of stupid mistakes in the wriggling and they don't do anything. The senseless shifting remains a senseless shifting. But, with the huge number of shifts and mistakes that occur and with the presence of the right background, let's call it substrate, there are also some mistakes that lead to changes that can sustain themselves. Or let's say: that can slide up one level."

"Slide up one level?" asks my child.

"Yes, use the computer again as a visual aid! Instead of zeros and ones, suddenly whole packets of zeros and ones are moved. By chance, a pattern is created that is stable and with further errors this pattern gains the ability not only to maintain but to multiply. Eventually, several such patterns come together and form larger units. Maybe this is how the Friedmen, which build OUR recognizable universe, came into existence at all, who knows."

"And where does this ominous background come from, the substrate, as you called it?"

"Well, I think these are the initial fluctuations that were impressed on the space when it was created or when it took on its present structure," I reply.

"During the expansion?"

"Yeah, maybe then. In any case, there were some tiny, but apparently very important fluctuations frozen into the universe that was to be developed later; imprinted on it, so to speak. And evolution was able to 'trail its way', to grow along it like mussel beds on a substrate."

"With properties that led straight to more complex molecules?"

"Yes, exactly. After hell only knows how many different experiments, random structures, and accumulations of errors, there was somewhere ... someday ... a pattern that our evolving universe allowed. And, of course, it didn't end there: It is in the nature of the universe to change and new and more complex patterns can be found. Until ..."

"Calculating space creates intelligence," my child interrupts me.

"Calculating space?" I ask

"The first computer they named Z1 is in the Museum of Technology in Berlin ..."

"A replica of it," I correct, perhaps a bit too schoolmasterly.

"Yes, a replica then. Well, there was also a sign about the inventor, Zuse, and he had probably also already been of the opinion that space calculates."

"Uh-huh," I nod appreciatively, "you know a lot! Now anyway, there is still a long way to go until intelligence appears, but since the universal error-selection mechanism can use what it has already produced before, by moving it up in levels, the process itself accelerates. That is, the more already has been developed, the faster the development progresses. Of course, there are sometimes strange solutions, which may seem weird to us."

"Like sex, for example," my clever girl interjects.

"You find this a weird solution? Well, maybe you're right ... sex IS a really strange invention."

"Why did evolution actually invent sex?" my child looks at me expectantly.

"I don't know, really, but if I am to make an assumption, sex simply seems to be a powerful driving force for many, many more discoveries and endeavors. It seems to have a lot of accelerating side effects on evolution, if you want to persuade your partner, your sexual partner, to have sex and impress him with something." I take a break and then add, "But I mean that in general ... the impressing thing, you know?"

"Aha!" my child grins, "so you only put together the explanation of the world to impress mom?" Then she even giggles. Obviously, my sentence about generalization was completely lost.

I too have to laugh, "Exactly my love, you have understood it quite well: in order to impress mommy in such a way that she lures me with her charms, like a beautiful star, and to finally become a star for myself in an irresistible shower of lust that is brought down from heaven and manifested on her wonderful skin .... Do you mean it like this?"

Laughter.

"Well, in any case, evolution keeps making inventions with the help of which it can finally accelerate itself."

"Evolution is accelerating?" my child asks.

"Yes," I answer, "since nothing is really lost in the universal memory and the error mechanism cannot be stopped either, it is inevitably moving toward more complex systems and solutions, unless ...," I take a theatrical pause.

"Unless what?"

"Unless catastrophes occur that locally lead to the destruction of higher order levels. Or communication between parts of the universe is no longer possible because it expands too fast, expands at an accelerated rate, to be precise, or the universe runs out of energy."

"Can the universe run out of energy?" questions my kid further.

"The universe as a whole may not, but our part might ... Above all, it can run out of ordered energy, thus, energy that is not just heat."

"What is the difference between heat energy and ordered energy?"

"Well, thermal energy is simply general homogeneous wriggling in space. Imagine a room with particles flying around wildly and completely disordered and there you have nothing but heat."

"Sure!"

"Now imagine a charged capacitor. It won't be difficult for you to think of a way how to connect a radio to the latter to get it working ... even if only for a short time. But with the room full of flying particles, the room full of heat, it will be much harder for you."

"Yes, that makes sense." My girl sounds amused, "In a hot room, I can't just stick two wires in and attach them to the radio to make it work."

I also have to laugh, "In fact, that shouldn't help much," I agree.

"But if the ordered energy of the universe or our part of it is limited, then at some point evolution will not continue, will it?"

"Here again I agree with you. Our universe does not have infinite time to solve the riddle of its own existence, not infinite energy and resources, I was going to say."

"So, at some point only the lowest level of the general wriggling remains?" my child asks and she sounds worried.

"Not quite, because on the wriggling of the lowest level there will always remain a shadow or echo of everything that once was, is, and will be. But I rather believe that the universe as a whole has a different solution here."

"Well, what is it?"

"Self-organization! So, if a system is large enough, forced to wriggle, thus, to calculate, it automatically creates order."

"But does this not contradict the well-known fact that in a closed system disorder increases and order decreases?"

"You refer to the so-called 'second law of thermodynamics'."

My child is nodding.

"Well, you know, I don't think you can apply it to the universe as a whole. Exactly because the universe carries out an infinite number of computational processes, is itself infinite, quantum mechanics, or the requirement of restlessness in space, which forces calculations everywhere and at all times, and thus it is not a closed system."

"Restlessness and infinity provide for self-organization?"

"Yes," I simply answer.

"And what about black holes? If something is sucked into them, then its information is lost, isn't it?" even more concern.

"Wrong!" I answer as quick as a shot, "Fortunately!"

"Why fortunately?" asks my child.

"You don't find the idea that information that falls into black holes simply disappears into their so-called singularity very tolerable, do you?"

"No," my child admits, "that's a horrible thought."

"Not only you, but also some famous physicists like Leonard Susskind or Gerad d'Hooft found this idea unbearable and that's why they spent many years arguing with people like Steven Hawking, who believed that information is lost in the black hole. But even there I can reassure you: Black holes are formations in which indeed almost all the wriggling of higher order is frozen, but only inside. On the surface of the black hole, however, it wriggles happily on. This is why we find that black holes always have a surface size that is proportional to their entropy. One could perhaps better say that it is proportional to their information storage capacity."

"And what if that much information falls into a black hole whose surface is not large enough to store it there?"

"This, my child, fortunately cannot happen, because ordered information needs a minimum of space to be stored at all ... Well, actually it is better to say 'surface' than space, but that doesn't matter now. If this memory, a piece of space with information, falls into a black hole, then the mass it can bring into the black hole is always just large enough that the increase in mass of the hole increases the surface area just enough to store the information it brings with it. Any surplus must be radiated as energy, which also transports information away again."

"And what if two black holes that are already full of information were to merge?"

"Then the surface area of the resulting hole, thus, the information storage of the hole that remains in the end, is larger than the sum of the two initial holes. So even if the whole universe ends in a black hole, all information would remain on its surface, even if all wriggling inside the monster should freeze. However, this would only be the wriggling of our smallest level, the smallest we could still perceive. It is very likely that there it is not yet ... it will not stop for a long time yet."

"Okay, the information once determined cannot be lost, but if the resources of the universe are limited, how can it ever find all the answers?"

"Does it have to?" I ask back.

"I thought you had to have all the answers to be able to explain the question of 'why' as a whole?!"

"Well, I better make no bid, because we won't know until we're ready. It is the same as with the question 'What is good?' You can put it like that: Success is not guaranteed, but not to try would be a shame. And besides, our universe certainly goes on and on and on. It's just a question of what we contribute ..., what we leave for the universes after us, what starting advantages we give them so that they can solve the task or get even closer to their solution than our universe did."

"Is the universe or multiverse itself actually intelligent?"

I'm a little baffled and ask back, "What do you mean?"

"Well, if the space has so much computing power, then at some point it should start to think properly, shouldn't it?"

"That's an interesting thought," I admit. "The universe may really be thinking, but first of all I reckon that to develop intelligence with limited computing power, as with us, you need physicality and language, and since the universe does not have these things—at least not very obvious to us—its computing power must be much, much greater than ours before it develops intelligence. And secondly, its 'thoughts' may take place in timescales, complexities, and possibly also dimensions, which are completely foreign to us. It might not be that easy to connect or even to communicate. Completely different languages, media, spaces, you know?"

"Do you believe that the universe is thinking or intelligent despite the lack of physicality?"

"Yes, because the mere fact that we think and we are part of the universe shows that it thinks. But I also believe that there are many other levels of thinking besides our being intelligent and thinking; that is to say, space not only calculates, it obviously thinks, too, but it is certainly not thinking as we know it."

"Okay, but if we or the universe as a whole now come up against fundamental limits on the way to the final questions ... Such as those shown by this Gödel, for example?" my child continues to drill relentlessly.

"You mean the Gödel's incompleteness theorems. There we are lucky that our real world does not satisfy any strict, rigid formalism. Quantum mechanics applies everywhere in the universe and small fluctuations or general wriggling causes random jumps and movements everywhere. If we or the universe should really get stuck somewhere in front of a big mountain of ignorance, there is always a non-vanishing probability that a quantum jump leads over this potential wall or a tunnel effect leads through it to the other side of knowledge ... into the light, to put it somewhat prosaically."

"Into the light!" my little one repeats very quietly and her gaze disappears into the distance. I, too, get lost in a few thoughts and almost startle when I notice that my child looks directly at me. At first, I think I must have missed a question, but she hadn't said anything, she was just looking at me ... exploring. Finally, she asks, "Is there anything that you deeply regret?"

"What do you mean?" I ask back.

"Something in your life, I mean. Is there something you are truly sorry about?"

"Oh, there are many things I would do differently today ... things where I've probably hurt people, but I'm not sure that I could have done differently back then. Even with all or part of what I know today, many things would certainly not have been any different."

"And what was the worst of it?" my little one keeps on poking and I feel uneasy. I remember a little old woman looking out a tiny window. I try to repress the thought, but it keeps coming back like an ugly but very affectionate dog.

"Well?" that same look again.

"Why do you want to know?" I ask back defensively, but it is a weak attempt.

My little one thinks for a moment and finally says, "Maybe because I just want to know who my father is ... or was."

The Old Lady at the Window

At some point he had noticed her. Countless times he had walked along this road and had been in his thoughts, mostly staring at the pavement in front of him, whistling, or humming songs. He was thinking about the night before, his girls, what he would tell his boss, his work. He pondered about the weather, the latest news—this crazy world, life after death, and whether there is a God. He planned his plans, dreamed his dreams, chewed his lips, was hungry and thirsty, was full and satisfied, happy and sad. Sometimes he didn't feel like walking this way and sometimes it gave him pleasure. Almost always the traffic polluted the air when he walked up the mountain in the morning, occasionally stopping at the baker's to buy a loaf of bread, or sometimes a pancake, a piece of apple pie, if he had an appetite for it. And almost always people rushed past him who did not interest him and who did not care about him either. When it rained, large puddles quickly formed in the wide furrows, which were muddy in the places where the asphalt had already detached, and over which one had to jump. He liked such rainy weather, if he had an umbrella with him and did not get wet. The air was then not so thick and the other people did not get in his way, because they usually walked around the puddles instead of jumping over them. It was on such a rainy day when he felt like getting something warm and crispy and since he had just arrived there, he went to the baker. For sure it was again two pancakes and an apple pie he bought. His appetite always decided it for him at that time.

When he stepped out of the store, whose door pointed just enough to the corner that he could look up the hilly road, his pathway. That was when he saw an old woman at the window on the other side of the street, diagonally opposite, while he was opening the umbrella. A moment ago, he saw her, then she had disappeared behind the opening umbrella. "Funny," he thought, "exactly from this point I have to look at the tip of my umbrella, which is pointing slightly upward, to notice her," and the thought fascinated him. Perhaps he had seen her before, but had not consciously noticed her, neither had the people who rushed past him daily, nor the cars, the muddy and clear puddles, although they always formed in the same places. In any case, he could not remember having noticed her before.

So, there she was, the "old lady at the window." She was just like he had always imagined such an "old lady at the window," with always uniformly expressionless eyes and a white, dotted cushion under the elbows.

He took the umbrella down again and watched her. She observed the cars and the pedestrians on his side, for the pavement had ended on her side of the road. Only those who wanted to get to the house where she lived walked on her side, and that was probably very seldom.

A fat woman pushed past him and he realized that he was still standing at the entrance to the bakery. He had the feeling that the old lady was now looking at him and quickly he hid his face behind the big blue umbrella. "Am I stupid?" he asked himself and peeked out from under the umbrella again. Now the old lady looked away and stared with a blank expression over to the traffic and the people.

When he reached the drugstore at the next corner, the old lady disappeared behind the corner of her house. There were no windows on this side. You could clearly see that there used to be some, but they had been walled up for noise protection reasons. This was cheaper than soundproof glass. He continued on his way and soon stopped thinking about the old lady, ... the old lady at the window.

The next day, however, as he passed the bakery store, she came back to his mind and he looked up. And yes, indeed, there she was: the pillow, the gray cardigan, just like yesterday. And again, he had the feeling that she was looking at him. Their eyes had only met for a very short time, and it had been like a tiny flash, a breath of life, on her face, and then there were the cars and people again, the heavy air, and the puddles, which were now a little smaller, because it hadn't rained since yesterday.

He now looked up to her every day when he came up the hill. Every day he looked at her and was glad when he could see that little breath of life in her face again—then, he soon told himself, it would be a good day.

One day her eyes met a little longer, and he couldn't help but nod instinctively to her. She first looked to the side, but then she smiled. She smiled without moving. He nodded again and she smiled again, and then he had reached the drugstore and did not see her again.

On that day he received very good news. "The old lady brings me luck," he thought to himself and now greeted her every day with a friendly nod of his head. She thanked him with a tiny, motionless smile. Almost every morning, except on weekends, this ceremony was repeated. If he missed a day, her smile was even more intense the next morning.

Once he had to go to a conference and didn't come for a whole week. She seemed really excited when he nodded at her the following Monday, as usual. Her smile was much more vivid, her eyes awake. "Was she worried? Did she perhaps wonder where I have been so long?" he thought to himself and even greeted her briefly with his hand.

So, this is how it went every morning. He smiled and greeted by nodding or raising his hand, and she smiled back. Sometimes short, sometimes long, sometimes she nodded a little, sometimes she remained rigid, but always her eyes were shining.

He soon noticed that somehow she seemed more alive to him, if he was in a good mood himself. If he went grumpy and sleepy up the hill and greeting her was rather a tiresome duty, then her answer was also only short and cool. But when he came along cheerfully and exuberantly, literally bursting with energy and zest for action, her smile was almost enthusiastic.

One day, he dreamingly put one foot in front of the other and only very late noticed that he had forgotten something. He turned his head, nodded friendly, and walked on. Suddenly the old lady leaned far out of the window. Panic was in her face and she waved her arms so that he stood still in shock. At that moment a big BMW shot past him into the driveway at the drugstore. He would have walked right into it. White as chalk he looked at the speeding car and shook his head. Then he turned to the old lady and waved both hands gratefully. People around him looked at him somewhat uncomprehendingly, but he didn't care at that moment. The old lady waved back, then pointed to the street with a reproachful smile and raised her instructive index finger.

"Okay, I'll watch out next time," he said aloud, even though he knew that she couldn't hear him because of the traffic, and again there were a few people who wondered about him, and again he didn't care.

When he finally went on, he saw that the drugstore was called "Hilde's Drugstore." "Well," he thought, "then I'll call you Grandma Hilde from now on."

They now waved to each other daily, and he made an effort every morning to be well rested and cheerful. In fact, he went out much less in the evenings than usual. It was important to him not to disappoint her the next morning— she was his talisman, and if she was joyful, he was well. And for him to be well, he had nothing else to do but to be fit in the morning, for five minutes of open warmth and cordiality.

At All Souls' Day, he made a small poster, which he hid behind his back and took out only when he was close enough. He held it up and smiled at her.

Grandma Hilde cheered up, took her fists in front of her mouth with shining eyes, and then clapped her hands like a little child. The white paper only said: "Hello!"

The following day she, too, surprised him with a "Helo!," with one "l" only. He laughed loudly in the open street, but not because of the mistake, and waved.

For his birthday, someone had given him a bouquet of red roses. He took one of them and showed it to her. Incredulous, she pointed at herself, and he nodded. He tucked the rose into the fence so that she could see it. It would have been impossible to get across the road at that hour. The next morning the rose was gone, and he was angry about it, but all at once she held the rose in her hands and waved it. Overnight she had fetched the flower from the other side. He was pleased that she made such an effort for his little gift. With one hand he formed a well, just as they had always done when they were little children playing dice. With the other, he pointed to the flower, then made a motion to his "fountain." Grandma Hilde, smiling superiorly, waved him off. Proudly, she reached back and showed him the vase. Of course, she had thought of it! She had also carefully shortened the stem and cut it in a cross shape so that the rose would get water more easily. To prove her prudence, she poured some water from the vase onto the street, then she put the red rose into it.

The day when he was so lost in thought that he forgot to greet her, that day he will keep in his memory for the rest of his life. He could not succeed in anything that day and at night he could not sleep. With eyes rimmed with guilt, he crept up the hill the next morning, bent much farther than it would have been necessary by the slope's incline. He saw her reproachful look and tentatively tried to wave. Grandma Hilde, however, did not relent. Punishment had to be. She looked at him rigidly and reproachfully and did not move. Depressed, he crept on, feeling as if his arms were dragging on the floor. In the evening he got terribly drunk and now had a hangover as well.

It went on like that for a few days until he got up and said, "That's enough, I'll fix it!"

He had made a poster, with a handle and everything, a big one, DIN A3. Already at the foot of the hill, he took it out as if he were going to a demonstration. People around him secretly tapped their foreheads or shook their heads. But nobody dared to say anything, he looked too big and strong for that. Even from a distance he felt that his Grandma Hilde was smiling, and he walked upright, self-confidently raising his head. When he approached, he stopped and pointed with his hand to his poster. It said, "Excuse me." Grandma Hilde was touched. She had tears in her eyes and waved at him. Finally, she disappeared for a moment and came back with her old "Helo!" poster—she had saved it. He threw her a kiss on the hand, and like a shy little girl, she turned away and looked furtively through her gnarled hands. He felt how this gesture did her good.

For almost two years they greeted and waved to each other every morning. Sometimes there was a kiss on the hand, a flower or a nice word on a piece of paper, but only sometimes.

Then one morning there was no one at the window. He stopped and waited a while, finally he shrugged his shoulders and went on. The other morning, the same thing. Finally, on the third day, he noticed a mirror moving in the open window. He waved and the mirror waved too, in the same rhythm as he did. "She's fooling around," he thought and threw her a kiss. The mirror bobbed up and down gratefully, and he walked on, smiling.

Then, when no Grandma Hilde had appeared at the window for two weeks, he decided to make inquiries. He went into the drugstore, which he had never entered before, and asked for "his Grandma Hilde," the old lady across the street.

"She's dead," the young saleswoman told him, at first with a good deal of disinterest. But then she noticed his dismay and added that the old woman had fallen two weeks ago and had then been lying helplessly in her small apartment for three days with an open fracture.

They found her at the open window. In her stiff hands she had a broomstick, to which she had clamped a hand mirror with her hair clip. With her trembling old hands, it will have taken her a very long time to get it to hold properly.

When he went out, he had tears in his eyes and his walk was like that day when he forgot to greet her.

Sometimes there are fresh red roses stuck in the fence, across from Grandma Hilde's window, but only sometimes, and they are, almost always, still there the next day.

Absolution

"Daddy?" my little one almost whispers from her little bed. I have my hand on the light switch and turn around once more. I can only glimpse my child in the dim light, but I feel her gracious presence.

"Thank you, Daddy!" she says, and I feel as if God himself has forgiven me.

Day Five

"When did evolution finally create the first humans?" my child asks impatiently. Admittedly, the thing with the plants and animals had been very interesting, but sometime one would like to find oneself, nevertheless, in all the creation.

"And above all, why?" she adds even more urgently.

"That's almost a question in two, actually," I reply curtly.

My child just looks at me questioningly and goes, "Huh?"

"Yes, the evolution simply tries around with all possible things. It uses partly what it has already discovered and combines things anew, again and again. It was thus only a question of time, until it discovered also the intelligence and the first rational beings inevitably had to develop."

"Humans!" my little one is trying to sit up now, but only manages to lift her head a little. She had been given strong painkillers this morning and they had weakened her a lot. Nevertheless, her eyes are shining. Finally, the thing would come to the point, he would come to the point.

"Oh, I rather think not," I have to disappoint those eager eyes, though.

"No?" Oh, my poor little girl, if only I could make it easier for you.

"No!" I reply.

"What then? Or rather, who then?"

"Well, I have already told you, the experiment 'evolution' ran and runs not only here on Earth, but in probably hundreds of billions of other forms and at hundreds of billions of other places in the universe just as here. And 'just as here' does not mean of course that the things look like here on Earth, but only that there is also a kind of life and a pushing forward of the same."

"Forward? Where to?"

"Toward higher complexity, toward new variations, toward increased knowledge ... just like here."

"So, there are rational beings elsewhere, too." My child simply throws this sentence out as if it were not particularly surprising. I know many grown-up people who, in contrast, would find this statement quite unbearable. To me, who had already experienced so many things, the thought that we cannot be alone in the universe at all was not strange anymore. However, I had to work hard for its familiarity again, after I had spent my childhood in a quite religious, not to say geocentrically oriented environment. Children are probably not yet alien to the idea because it was not yet possible for them to want to see themselves as overly important ... except perhaps in their own family and at the dinner table.

"There has been, there is, and there will be," I reply, "the full range of possibilities for the existence of rational life has been and will be exploited by evolution."

"How is it born anyway, the *rational life*?"

"I'm sure you can come up with tons of possibilities for that again," I reply, "but I think here on Earth it was quite simple. At some point somewhere there was a slightly larger brain than would actually have been needed to control the body belonging to it. This was not necessarily an advantage at first, because such a brain consumes a lot of energy and the creature that was beaten with it had to get the additional food first. But for some reason the brain was an advantage. Perhaps because the external conditions changed and the brain-fortified creature could adjust to them more easily than its fellows. Or it could snatch itself with the daily food search not only the additional need for the brain, but still another surplus ..."

"Or it had the better ideas in flirting?" my kid interjects with a grin.

"Or something." I laugh. "I see you get it. Anyway, there was a little advantage and so the bigger brain not only persisted, but was followed up by evolution as a 'seemingly not-so-stupid idea'. One must imagine the concept of the 'brain' not at all Earth related, but as a quite general organ, with which information can be stored and calculated, and stored and/or calculated information can be called up again. Once this organ had developed and had brought advantages to its carrier, it was clear that the evolution would work sooner or later on a stronger variant of the same organ to see whether more advantages would arise then."

"Evolution doesn't see, it tries around and selects, that's its SEEING," my little one explains to me.

"Yeah right!" I agree. "Anyway, the brain-organ thingy could have evolved and ..."

"And so, there was, is, will be the rational being somewhere someday." My child mocks me, skillfully exaggerating my own attempts at generalization.

"Yes exactly!" I answer smiling.

"And then the brains became, become even bigger and the beings even more reasonable ..."

"Rational, my love! That doesn't necessarily mean reasonable," I interject.

I earn a reproachful look. My little one is not happy about the interruption and certainly not about the correction. Short and to the point, she replies, "Yes, but if there is enough gift for reason, then the beings in the sum should also be reasonable at some point, right?"

I hate to play teacher at this point, but what isn't right doesn't become true by coming out of the mouth of a little dying girl. And so, I answer, "Maybe you're right and it just has to happen sometime, but it's not guaranteed. Just look at what we are doing here on our planet, how we are destroying it ... how we are destroying our future and yours."

"Huh?" my child just goes.

"Many times it will have happened in the universe that rational beings arose, that thanks to their new abilities they began to dominate their home planet or wherever they were, and that in the end they destroyed themselves without producing anything to carry on. We're well on our way here to blowing it as well, and reasonable I don't think it is."

"Yes, that makes sense to me," my child now agrees, "but what do you mean by 'to carry on'?"

"Evolution always uses inventions that it has once made for its own acceleration. While it took almost a billion years to perfect cell division, the step from unicellular to multicellular organisms was so fruitful that in a few million years there were already vast numbers of different multicellular organisms. Once bisexuality was created, it took only a snap of the fingers for an almost unbelievable variety of sex-mad creatures to emerge."

My child giggles, but I do not let myself be distracted by this and continue unperturbed.

"Once the oversized brain was discovered, the planet was dominated and enslaved by the beings equipped with it in an even shorter time and in an even more serious manner. Should these beings eventually produce or bring forth artificial intelligence, gene design, and self-optimized beings or creations, these were and will take over the reins in yet another accelerating and broader manner."

"So, these artificial beings are the next step then?" my child asks.

"That only applies to the individual; something else usually happens before that."

"What is it?"

"Organization!" I reply. "Evolution had already made the 'experience' before, if you want to call it like that, that with simple structures, cells, or individuals, bees, ants, cleverly combined, it can produce completely new and significantly greater things. What only if one would do this with the new possibility of very complex and rational beings? Because of the self-accelerating character of evolution, this time it takes only fractions of the time it took to create the first true multicellular organism or large insect colony, and the optimization steps also run self-accelerating."

"What kind of optimization steps?" my child asks.

"Well, toward an optimized state system, global, efficient, holistic, and reasonable."

"And what should that look like?" I can see from my child's face that she is already making associations with the situation on Earth and simply cannot imagine how global efficiency and reason are ever supposed to come into it. "Socialism, capitalism, or a slave-owning system?" she adds almost contemptuously.

I have to laugh involuntarily, "You can forget about this system and left-right waffle, there's nothing really substantial about it. Those who operate with such terms usually only want to make themselves important, have no idea what they're actually talking about, and in their own verbal folly of system terms carry the proof of their inability to govern a society like a sign 'I'm stupid, please don't vote for me' large and visible for miles around."

My child laughs, "Yes, but most people can't read, can they, Dad?"

I also have to laugh, but I add, "No, most people could read, but the whole thing is organized like the psycho test with the gorilla in the basketball game. You know that, don't you?"

"Is that where they play people a film clip from a basketball game and all of a sudden there's a man in a gorilla suit running across the field and hardly any of the viewers notice him?"

"That's exactly what I mean, my girl. Very good!" I praise her good memory. "The fact that our politicians and other leaders are actually at most the intellectual gorillas—if anything so smart—in the great social basketball game cannot be denied, I'm afraid. They are all a bit too stupid for that. No, but they can try to make the basketball game around it as interesting as possible to minimize the number of people who see the monkeys, or whatever is crawling around."

"Bread and circuses!" my kid interjects.

I think of the ancient city-states of Greece, Rome, the Third Reich, and the European Union with their troublesome debt problems and I have to smile involuntarily. "You have associated that well, my child!"

"And they are helped in this by the media."

"Exactly," I agree.

"What would an optimized social system have to look like?" my girl comes back to the initial question.

"A social system or state structure is perfect when it optimally utilizes the performance of all individuals, ensures its continued existence, and releases innovation potential. In other words, it enables not only a certain stability but also optimized evolution. All this is always ensured in the long run, that is, at least 7 to 10 average life cycles of rational beings. In order to achieve this, one does not have to do actually much more than to place the brain of the state hard-core and mercilessly under a performance principle, whose evaluation criteria represent exactly the abovementioned optima. As you can see, the state itself here appears like an organism of its own, and ultimately that's what it is."

"Pha!" my little one exclaims scornfully and laughs.

"You're probably thinking of our political twerps and their stereotyped thinking in terms of votes and half-term election periods, aren't you?" I ask knowingly.

"Indeed!" she laughs back, "You can never make an optimal state with these dummies. 7 to 10 life cycles ..., that's absurd."

"Yes, but that wouldn't give us any problems today with the climate catastrophe that is already upon us and that will kill more people, primarily and secondarily, than all the wars of all time put together."

"But as it is now, this catastrophe will not be preventable, will it?"

"No," I admit bluntly. "Nevertheless, the structures are just such that the selection is poor. There are still too many parasitic, completely performance-free subsystems or groups of individuals who contribute nothing or only do harm without it hurting them. The do-gooders and dunderheads everywhere are still dancing around at the levers of power without realizing that what they are steering so miserably through the times—or pretending to—is not a paper boat, but a highly complex organism to which the laws of life apply just as they do to other organisms."

"What laws of life?"

"Actually, it can be summed up in one law:

Create order locally, accumulate and generate information, and use it while consuming as little energy or resources as possible," I say very slowly.

"But we are doing the complete opposite," my little one replies, sounding almost contemptuous.

"Exactly!" I reply curtly, "And that's why things will eventually go sour. Wrong-way drivers don't live long."

"What will happen?" asks my little girl, sounding worried now. 'Funny,' I think, 'she's worried about the future of humanity, even though she no longer has a future herself. What a great rational being!'

"Either external circumstances will destroy civilization as we know it, climate change being a good candidate, for example, or internal factors will force appropriate adjustments and generate a more optimal course."

"The next stage of rational beings, perhaps?" suggests my child.

"Exactly, for example, these artificial beings or hybrids or designed organisms, something along those lines, if you follow evolution and no local catastrophe stops the course of events, will be the next step, yes. And they will then put an end to the spooks, the stupidities of the previous generation," and a little quieter I add, almost muttering, "maybe just to do new stupidities again, but that will be a different story then."

"It's kind of a horrible idea. A designed being ..." My child shakes in thought, "Who would want something like that?"

"You think so?" I ask, "Just imagine if there were little nanorobots that could be put into our bodies. They'd nip your cancer in the bud. Wouldn't that perhaps be a reason to make yourself or your offspring hybrids after all?"

My child now looks thoughtful and I continue, "Or such tiny ones that would sit parallel to every one of your brain and nerve cells. Let's say 10 to every single one of your nerve cells. Your memory capacity would immediately be unrivaled, forgetfulness almost zero, your general brain capacity would increase not times 10, but almost 10 to the power of 10 ... minus a few losses for the management of so much computing power. Don't you think there would be rational beings who wouldn't want to use that immediately?"

"That is," my girl began timidly, "truly rational beings were and are created by evolution to improve themselves at some point?"

"Or rather, to create better themselves. That's how it is, my love. Sometime and somewhere this will happen. Whether we are already able to do this is completely unimportant," I answer.

"What if they, the rational beings that is, fail in this task?" my little one asks, looking concerned again.

"Then they have failed, and evolution, creation, God, the universe will follow the many, many backup systems in which things went better, and the fruitless process of useless 'rational existence' will become a dead branch in the family tree of all being and end someday."

"Dad, are you talking about us humans this time?"

"Oh, did I say 'humans' somewhere at some point today?" I ask, playing the artificially annoyed, "Shoot, that's exactly what I didn't mean to do just now."

"You very rarely said 'humans'," my little one good-naturedly reassures me, "but you talked about them all the time." My child shakes her head and smiles.

"You think so?" I pretend as if I must reflect again very intensively, "I think I have not spoken then, nevertheless, of humans, but alone of rational beings. That is probably something quite different, believe me, kid."

It was a wonderful evening. In the dense bushes in front of the house a thrush begins its song, although the sun has not even set yet.

"Are you tired already?" I ask my child. She shakes her head. "Then let's go for an excursion."

The Stream

Only a quarter of an hour's walk, a short distance out of our property, across a paddock where cows graze with relish, over the embankment, then along a narrow, overgrown field path, and we are at our destination. I can easily carry my child—she is so terribly light and delicate.

There it is in front of us, the stream with its wide reed belt, its willow- and birch-covered bank, the steep slope overgrown with brambles and broom, and the pronounced frogbite and weed zones.

There are some aisles in the reed beds. In one of them I find our boat, a catboat. I put my girl on the bank and make the boat ready. As I'm still thinking about whether to raise the foresail at all, my child, who probably suspects what I'm thinking about, calls out, "I want to hold the foresheet!"

"Okay!" I reply, "Then let's do it."

I climb back onto the shore, take my girl, and put her in the front part of the cockpit to portside. A few blankets help to support her battered little body. Then I loosen the lines, give us a little push, and sensitively take in the mainsail tight. A light southwest breeze pushes us gently out onto the stream. My girl has also pulled her sheet tight as best as she could, and now she looks ahead and seems to enjoy the almost silent glide. I give the boat a half-wind course to the northwest, and so we sail straight into the sunset. We head for the opposite shore, maybe 2 or 3 miles away. There, too, is a wide fringe of reeds with many aisles and coves.

One of these places is our destination this evening. We had often fished here and rarely returned home without a catch. But even when the fish didn't really want to come, we could experience many interesting things in this overgrown, shallow stretch of water. There were the coots wandering over the lake in search of food, or the warblers singing their peculiar song in the evening, similar to the chirping of crickets. We could never spot them, but my child recognized the birds by their song.

"You taught me that," she says, turning to me, "to recognize the bird calls, do you remember?"

Of course, I do remember.

"Some months we were out almost every weekend," my child continues, "together we cast the fishing rods and waited. While we were doing that, we talked a lot, and when night came, sometimes you'd fill a pipe."

"That was just because of the mosquitoes," I interrupted.

My child laughed, "I loved the spicy smell then, even though I got a terrible cough when I pulled on the pipe once. Do you remember?"

"Sure, I remember. You wouldn't believe me when I said I was just puffing and blowing at the mosquitoes."

"Yes, you then took the pipe from my hand, laughing, and slapped me on the back until the coughing stopped. Then you suddenly jumped up and grabbed your fishing rod. With a tense face, the pipe in your mouth, you waited for the right moment to strike."

"It's amazing what you remember!" I wonder aloud.

"When I used to think of you unconsciously, I always saw you like that in front of me: the fishing rod in your hand, the pipe in your mouth, and your gaze fixed on the pose." My child sighs softly, "It had always been nice with you, ... out here by the stream."

We see a muskrat with building material in its mouth swim by in front of our boat. It's time to haul in the mainsail. "Flap the jib!" I shout to my girl, and she responds immediately. She was watching a great reed warbler, whose busy activity catches her attention for a moment. I drop anchor and we slowly drift back a bit and sideways.

Otherwise, hardly anything seems to stir on this almost too warm spring evening. The wind continues to blow weakly from the southwest, rippling the water surface only slightly.

The bank is overgrown with high grass and young reed shoots protrude from the water across the entire width of the aisle. I point to it and my child nods. "A good sign," I say, "because it meant that no one could have been here for a long time." I quickly mounted the fishing rods, reel, float, and hook. We always used very narrow floats with signal red tips that were still easy to see in dim light—and today was no exception. Now come the baits. I'm sorry every time I have to pierce the poor worms, when they start to wriggle more violently because they feel the iron, but such a worm is just a damn good bait.

Another rod I bait with dough, which I had sneaked from my wife while baking a cake and made more palatable with vanilla powder.

My child watches me for a while during these preparations. I feel her desire to help, but she knows that she can no longer do this fine, delicate work and so she doesn't even ask.

Only when I am finished with the preparations do I slide forward in the boat and bring out the fishing rods. I then hand one rod to my little girl. She nods and whispers, "Still the good old sneak up technique." She smiles. I had taught her that. We could both only smile mildly when we watched shed game hunters acting like rioters, coming up to the fishing shore with a radio on board or even a car, and then noisily unpacking their odds and ends, stereo turned up fully under booming basses.

"These throws were good," my little one whispers. The poses stick out of the water about one to two meters in front of the reed edge. Satisfied, I now also sit down on a comfortable foam cushion in the cockpit of our boat and search my backpack for something to eat. A large piece of mother's delicious plum cake and a small water bottle full of cold cocoa are my finds.

I share the cake and hand my child a decent piece. She had hardly eaten anything in the last few days and so I am pleased that she is now biting into the cake with a tangible appetite. With a full mouth she asks for a sip of the cocoa and I hand her the water bottle. I have to help her a little bit to drink, because the bottle is still full and quite heavy. While chewing with pleasure we watch our poses, but nothing moves. They stick out of the water with the tips and move in the rhythm of the small waves.

So, we let the view wander over the stream. On the opposite side, between two short clumps of willow, sit more anglers, or is it just one? I can barely make them out, and so I take my binoculars, the good old Carl Zeiss, and scan the bank. It is a young fellow with his companion, apparently his girlfriend. With a light fishing rod they seem to be targeting rudd and roach. Right next to them begins the wide reed belt that reaches far into the water. At its edge facing the middle of the lake, coots could always be seen in the twilight hours, hiding in the dense reeds during the day. Now, when the sun is already touching the tree tops on the horizon, they are already sticking their bright beaks out of the reeds, just as if they wanted to see if the air is clear. Soon they would venture out completely and then swim across the stream individually or in small groups. I hand the glass to my child. However, she can hardly lift the heavy device and looks at me pleadingly. So, I just put the binoculars down on the gunwale and put the thick backpack underneath. Now the eyepieces are exactly at eye level of my child and she can sweep the stream in peace. When she sees the coots, she whispers, "You always mimicked their piercing, high-pitched call, and often they answered you from across the water. Do you remember?"

"Well," I reply modestly, "sometimes they answered." But my child doesn't seem to hear my interjection.

"Once they even came swimming over."

Then she lets her gaze wander further and hums softly to herself. I'm just trying to make out the melody when she suddenly says, "Too bad we can't do this with the whole family anymore."

"But maybe we can ..." I break off. My child gave me an indescribable look. It was something like 'Oh stop it Dad, you don't have to fool me anymore!' My smart girl knows that her time is not enough to get everyone together again to go on a family trip like in 'the old days'. With the sailboat and maybe the canoe or even with a campfire on the beach and staying outside overnight. My little one loved her brothers and sisters, but she also thought she understood that somehow it was no longer possible between them. They had a hard time around her, her siblings, with her disfigurements and terrible amputations, and certainly with the knowledge of her imminent death. I also know that my girl wants to protect her siblings, she feels especially sorry for her younger sister. It was like this at the beginning of the disease, even before we knew the fatal diagnosis. Her sister had been too little to understand anything at that time, but she had sensed that things were changing. She probably missed her big sister as a wonderful playmate, the best one anyone could ask for. She also missed her good-humored parents and she was always so whiny, a real crybaby, although my little one didn't like it when others called her that. "Adults are really stupid," she often mumbled to herself and made these small, quiet sighs again and again. "You have to be brave now! All of you have to be brave now!" I, their stupid father, had told them before we got into the car, with foreboding heaviness. It had been a very long journey, far, far away from home, a journey whose real distance could not be measured in kilometers or miles, but in destroyed hopes and dreams, in lost happiness, in endless agonies, and a disfigured little body. During the long wait I had written you a little poem at that time. Perhaps it had been more for me after all.

Helpless

From your father

Your little weak body
laid trembling in my arms
seconds melted away,
I felt you leaving
   ... so gently faded the warmth.

But I could not hold you,
I couldn't help you and time became a foe
your eyes pleaded wearily:
"Oh please, simply let me go!"

But what should I do here
   ... without you?

Fight, my little heart!
I can't lose you yet,
as pain would tear me apart

>*Please stay here, little heart*
>*here in our world*
>*Just one bit*
>*Perhaps one kiss*
>*As to me an eternity this is*

In a naked cold corridor,
they let you freeze,
after a hellish ride—blue light,
apathetic, heartless gaze … no ease.

Abandoned, lonely, deathly afraid,
like a small, sick birdie,
that can't go south,
Feel the winter coming on.
Some bitter juice to swallow,
even to cry you have no strength.
To them you are a number,
just a treatment, properly filed.
To me, an autumn leaf in the wind,
floating away

>… my child.

>*Please stay here, little heart*
>*here in our world*
>*Just one bit*
>*Perhaps one kiss*
>*As to me an eternity this is*

And finally, they push you on,
Away … your pleading, all in vain,
without words they work on you,
but you hardly feel the pain.

And then …
A warm soft hand
and you look at me,
so weak, so grateful,

how long
   a moment can be.

I hear myself
Words as if they came from far away,
as if nothing ever existed,
nothing was, except this final tear
My God we know!
  We are beyond the fear.

> *Please stay here, little heart*
> *here in our world*
> *Just one bit*
> *Perhaps one kiss*
> *As to me an eternity this is*

Bravery

The siblings didn't understand, but my incredibly smart girl did. I tried to explain to the others that it wasn't her, our girl, and it wasn't them, her siblings, and it wasn't us, the parents, that ... well, she simply had to go away.

Our little one had nodded with her head down and held the hand of her little sister. She must have realized that something was wrong, because she was crying softly. So did my wife and the rest of the siblings, and that hurt our little girl, too. She also felt like crying, but she clenched her teeth and finally reached out her hand to me. With a surprisingly firm voice she had said:

"Let's set off, Dad, it is time." Even today I wonder about the choice of words. She had said "setting off" not "going" or "leaving."

Once again, I had then taken my other children in my arms and kissed my wife. Then I turned around and walked to the car. Our child followed me very slowly, not because walking was already somewhat difficult for her, but rather because a walk to the scaffold is the hardest of all, but only my child knew that. The rest of us still had silly hopes. As she waved goodbye to the rest of the family, I saw in the rearview mirror that our girl had tears in her eyes ...

"Stupid!" I suddenly hear my little one mutter, "totally stupid."

I hand her the cocoa again and she takes a big gulp.

Suddenly, her float begins to twitch at short intervals. Quickly my child puts the cake on the backpack and takes the rod in her hand. The pose slowly dives toward the middle of the stream, the line begins to tighten. My little one waits a moment, then strikes with feeling. She notices the immediate, energetic resistance of the fish. The strike was perfect. The fish swims a bit parallel to the shore, shoots to the bottom there, and tries to get rid of the hook by shaking it violently. But my child makes sure that the line remains taut and the fish cannot throw out the hook. A few more cross-and-thrust escapes follow, to which my little one gives way with her soft hollow glass rod with a slight bounce. The attempts to free the fish soon weaken and my girl slowly brings it in. She pulls it over the landing net, which I have stretched far forward into the water in the meantime, and I lift the catch out. It's a nice perch, a good 14 inches long and certainly over a pound. It has the hook at the front of its mouth, so I get it out easily. Then I stow the fish in a roomy keep net.

"That must be your biggest perch yet," I praise. My child's eyes light up and she just nods.

The first one is followed by two more smaller perch, but we put them back.

Slowly it gets dusk. The wind dies down and the bloodred sun, only half above the horizon, is reflected on the smooth surface of the water. The frogs begin their evening concert and the coots finally come out of their hiding place. The cricket-like song of a reed warbler resounds from the reeds.

But also in the water it becomes vivid now. Here and there reed stalks are bent to the side, bubbles rise, and only a little away from the poses even a reasonably large carp jumps out of the water. This was the hour when I used to unpack and distribute some little snack. I remember the hazelnut chocolate bar I had left in my jacket pocket, and I dig it out. My child gratefully grabs it and bites into it with relish.

Then I haul in the fishing rods and unmount the poses. We can hardly see them now anyway. A small lead olive would keep the bait on the bottom and also serves as a casting weight. I now cast a little farther to the side, maybe 4 to 6 meters in front of the reed edge. We both know that there is a slight depression there, a sort of trough. To see if a fish was biting, I wrapped a small roll of aluminum foil around the line. In case of a bite, the line would rush through it, lifting the tube and making an audible rustling sound.

The sun has set in the meantime. The horizon shines in the light of the evening glow and the moon, which had just looked like a small cloud, now begins to glow dully. It is croaking, chirping, and splashing all around. The tips of the rods stand out clearly against the bright horizon.

Now that everything is done, I sit back on my cockpit stool, drink some cocoa, and watch the silvery bite indicators on our lines. My kid remembers the leftover cake on the backpack and reaches for it. She splits the piece and gives me one half. Chewing, we enjoy the nightfall, the soothing coolness after the hot day, and the diverse sounds all around.

A little further away, there was a rustling in the reeds. "Surely a bird looking for a place to sleep for the night," whispers my child. From the position where one of our poses had stood before comes the cry of a coot, and from everywhere the croaking of frogs can be heard.

But what is that? A sudden jerky rustle makes us both cringe—a bite. The silver foil had been pulled up to the first ring with the line, had snagged there, and was now causing a hell of a racket. So quickly that I could not have said how in the next moment, my child had dropped the cake and taken the rod in her hand. 'Where did she get this energy all of a sudden?' I think to myself and watch her, fascinated. The line is still whistling through the rings. My girl quickly pulls something from the reel, pinches the still loose line between the index and middle fingers of her right hand, and waits for it to tighten.

Hooked!

A force completely unexpected for both of us almost snatches the rod from my child's hands. A quick grip, the line brake is loosened. This is followed by a short escape toward the middle of the current, which my child simply has to give in to if she doesn't want to be pulled into the water. The boat rocks properly. Then suddenly nothing moves. The fish seems to have stopped on the spot, the line loosens. My child cranks back a little. "There you are again," she says. The fish comes straight at us. Just before the reed edge, however, it veers off to port and shoots into the clutch. With all the strength she can muster, my little girl tries to pull it out again. The rod bends almost to a semicircle, the line buzzes, but there's nothing to be done. "I can still feel it!" my little one exclaims, gasping, "But I can't get it out of there." Sensing panic and the onset of exhaustion, I ponder. I can't dare make an attempt to wade toward the fish. The water is shallow, but the bottom is completely muddy. With every step I would sink deeply. "I can't do it anymore," the words come almost with a groan, and I quickly reach for the rod.

"God!" I exclaim in dismay, "What a resistance! How could you hold it for so long?" All I hear is gasping, shallow rapid breathing. I struggle to maneuver the fish sideways out of the clutch, Nothing. Even tapping on the hand part of the rod, as my father had already shown me, or powerful strokes in quick succession show no effect. What now?

Seeking for help, I turn to my little one. She seems completely exhausted physically, but her eyes are shining and searching through the twilight for a solution. Her gaze finally falls on a weathered brick in the bow area of the boat, which had probably been forgotten there. I had always pushed it under the wheel of the trailer when we pulled the boat out of the water, and it should actually be on land with the trailer for that reason. But anyway, the stone gave us both an idea.

I quickly take the rod in my left hand, flip the line catching bow around, and lightly clamp the line between my thumb and the rod. So, I can give way to an escape sensitively at any time. With my right hand now free, I grab the stone and hurl it with all my might at the spot in the reeds where the fish had fled.

There is a loud clap, the water splashes up high, the thick drops glisten in the moonlight. An eerie jolt tears the line from my fingers. It rushes through the rings, but I quickly regain control and can sensitively stop the fish's escape. With a little effort I manage to get the rod back in my right hand and grab the crank of the reel. The line catching bow jumps around. Now I have our opponent back in my grasp, or so I think.

But what is this?

Again, the fish comes, this time a little slower, a little bit toward us. A few meters in front of us the water suddenly splits and a huge carp head looks out. The huge fish rolls slightly sideways so that we can see its full broadside for a moment. Its large scales shimmer golden in the moonlight. Then it dives. With its shovel-sized tail fin, it seems to be waving at us. It is a capital carp, certainly considerably heavier than thirty pounds, a good catch. I am completely overwhelmed, almost shocked by the sight of it. My child only exclaims, "Wow!" and is otherwise speechless with amazement. Only mechanically do I give in to the next flight. Uniformly, but inexorably, the carp pulls almost 30 meters of line from the reel, then the escape abruptly breaks off. Probably the fish has turned again and swam back a bit.

Farewell

Only for a tiny moment had the line been loose, but that had been enough.

As fast as I can, I turn the crank to get back in contact with our opponent, but the line is easily reeled in without resistance. With every centimeter, the hope of still having the fish on the hook fades.

When it finally splashes quietly directly in front of the boat and a moment later the bent-up hook swings in front of my face, I know that we have lost it, … this magnificent carp.

For minutes we sit there without moving. As if in a daze we stare at the dark water surface illuminated by the moon and at the bent-up hook. Only now do I notice that my knees are shaking. There is silence all around, only the wind, which in the meantime had awakened again, moves the reeds.

Suddenly I feel a hand on my shoulder, a gentle tap, "Never mind Dad, you'll get it some other time!"

I have to smile involuntarily. I put the rod aside and snuggle my child against me. What a wonderful person.

Slowly we disassemble our fishing rods, tidy up the things in the cockpit, and stow the backpack. Then I lift out the keep net and look at my little girl questioningly. She beams once again at her beautiful catch, then nods and I understand. We give the perch its freedom. It's a beautiful fish and I'm sure it would taste delicious, but we both don't want to kill it now.

When everything is stowed away, we take another look along the reed line. It is a beautiful night. The moon bathes everything around in a ghostly light. The grass on the shore, the reeds, the rose hip bush further up the slope, the birches and alders all around, everything appears in a strange, silvery gray. But we perceive all this beauty only unconsciously. Slowly we sail back. We use only the main, because my child is still too exhausted to hold the foresail.

When I finally lift my girl out of the cockpit and carry her away from the shore back to the house, we both feel incredibly fulfilled, as if we had been allowed to experience something very majestic. At the same time, there is a deep sadness. My girl's eyes shine as I lift her to my left side because my right arm is exhausted. They shine and look steadfastly over my shoulders in the direction of the stream from which we have just come. And there it is again, that soft sigh.

"Farewell!" my little girl whispers to the stream as we finally turn around a corner and dense shrubbery blocks the view to the water.

I try to cheer her up and so I reminisce about the adventure we just had together, "That was quite a fish, wasn't it?" I ask.

"Yes, it was!" my little one confirms somewhat absently. But then she adds a little more cheerfully, "But YOU lost that one, Dad." The slight reproach is only pretended. My girl knows that I myself have to bear the brunt of not having gotten this huge fish on board.

"Mea culpa!" I mumble. My child, who had put her arms around my neck, cuddles my back, "Oh, don't worry about it. I'm glad it got free and can go on living," and rather to herself she adds, "I wouldn't have been able to eat it anyway."

I carry my little one into the house, get her ready for bed. She is completely exhausted, almost apathetic as she endures the evening's routine. But in her face, there is something like a glow, a wonderful little smile, and I know that the excursion was not a mistake. I almost think she is already asleep when I finally put her gently into her bed. But then she suddenly opens her eyes and whispers, "You know, Dad, I would love to be that fish."

"Why, my little love?" I ask.

"Because then I would have come off the hook one more time."

Day Six

"What's with the humans, then?" Impatience lies in this question, as if it had been waiting to be spoken for a long time and now the time seems ripe. No, the time does not seem ripe, it is really pressing ... My little child was running out of time and so was she pushing as well. Fortunately, children are like that, in life as in death they have an unmistakable sense for time, for the things that are just right and important for them to experience and learn. We adults have merely forgotten how to read the signals they send us, our children. We have forgotten to interpret the little hints and to recognize the essentials behind all the superficial complaining. At some point they become like us and then the signs disappear. My child is lucky, she would leave this world without having lost this ability, without having lost her sense for what is essential.

"That's a difficult thing, certainly almost the most difficult of all," I answer very slowly after a while.

"Why only ***almost*** the most difficult one? What is the most difficult task then?" Again, my child surprises me. If I had just expected her not to want to let go of the topic 'humans', she immediately saw the weak point in my reaction and is now nailing me. Children don't want any doubts, they don't tolerate any excuses, and they don't want one thing above all: to be lied to. They don't want to hear 'you're too young for that' or 'you don't understand yet'. And they are right to do so, because their task in the world is to get further than we have been, and for that they must build on what we leave them. Poor children of the world, what will we be leaving you?

"Oh dear!" I answer somewhat timidly, "This question has more to do with humans than you think. The really hardest task for man is to detach himself from his own paltriness and to see himself in the world structure where he is, where he really stands. Namely not in the center of something and he is also nothing more important than the beautiful flowers here on your bedside table. No, we humans are a necessary product of a process that has occurred many, many times simultaneously in different places in the universe and has led to the same result." My child raises her eyebrows.

"To **the same** result, mind you, not to **the VERY same**," I hastily add.

"What is this result?"

"Intelligence."

"But what does that have to do with the most difficult task?"

"The most difficult task is to understand our smallness, to understand ourselves as part of a task and as a path, not as a goal and not as an end in itself."

"Yes, but if the goal of evolution was to produce intelligence, shouldn't this intelligence also help us to understand our position in the universe?" Apparently, it didn't seem so difficult to my child to grasp her own insignificance in all its unspeakably great significance. Perhaps children have much less problems with this than we grown-ups, who somehow always want to see ourselves as more important than we really are.

"You know, my dear, there is always a little problem here for evolution. It cannot create an intelligence that would be able to survive on its own and at the same time be of some use to the greater goal."

"I don't understand," oh those poor perplexed yet hungry eyes.

"In this way it is difficult to understand indeed, but let's make it simple. Let us imagine the first intelligent being as it may have begun to exist on this planet long ago. A being equipped with a large brain, a thirst for knowledge, and a fairly decent memory capacity, meaning memory, and so on."

"An Australopethicus?"

"Australopithecus," I correct casually and then reply, "yes, that one maybe, but surely it was even earlier in the evolutionary sequence." I smile. "What do you think this being feels when it first recognizes all the inexplicable things, like storms, lightning, earthquakes, huge predators around it, so it doesn't just subconsciously register them, but really **perceives** them? How do you think the being deals with the fact that it **recognizes** its fellow species as comrades, playmates, and lovemates for the first time and then at some point later has to experience how these familiar beings die?"

"… Or its children …" my child adds, somewhat lost in thought. Her gaze is far away, somewhere back then with those first beings with something like intelligence.

"Yes, or its children," I answer, delighted by this wonderful interjection. "The being would simply perish if it did not have, in addition to all its urge for knowledge and intelligence, a kind of buffer, a drawer, into which it could throw all the things that cannot be explained instantaneously, that cannot be grasped immediately, and at which a knowing and feeling being would become ill from all the apparent cruelty that reveals itself to it when it now consciously looks into the world around it."

"What is this buffer?"

"Spirituality, faith," I answer, adding thoughtfully, "one of the most beautiful and at the same time most terrible inventions of evolution ever. We have a separate brain area just for that."

"Why? What's so horrible about it, I believe you too, don't I?"

She looks thoughtfully at the water bottle on the bedside table. Just as if she was wondering whether a little sip could do her any more harm. I hand her the bottle, help her to drink, and she looks at me gratefully. Then she demands her answer. She says nothing, there is just that look, and that's enough.

"Yes, you believe me," I say, "but you certainly believe me not only because of your inherent spirituality, but also because the thoroughly rational part of your brain tells you that what we are putting together here is consistently logical and that it fits with all that is known and observed. Doesn't it?"

My child nods vigorously.

"I could even get you case studies of patients who had certain strange brain damage after accidents or surgeries and they were suddenly completely religious or even the complete opposite of that, that is, rather anti-religious, precisely because one had damaged their faith center or the counterpart, the ... let's call it the 'faith control part.'"

"But what's so bad about spirituality in our brains now?"

"Well, it helped us get over the dark times of ignorance. It was like a bridge over deep, terrifying canyons that we could not penetrate—for the time being. But at the same time, it has always been an all too comfortable refuge whenever we should have asked instead of simply blaming a higher being for a certain effect that could not be explained, a ghost or a god or even a wizard or a witch. Spirituality stood in our way when we would have wanted to recognize our own insignificance and yet also our importance and it prevents us more and more from recognizing and fulfilling our true task. The universe did not give us our brain, so that we hide behind stories, invented by other people and written down hundreds or even thousands of years ago, and cower there like frightened mice in front of the cat. It did not give us our ability to communicate to pass on nothing but fairy tales to our children. No, it wanted us to put you on a platform from which you could see further into the land than we could, and from which, in turn, you were to give an even greater gift to the next generation."

"But I don't fit into this cycle, do I?" I see them coming, those little tears, and I feel the unspeakable pain that is trapped here in this tiny body and that wants to come out in a sob that would shake the world. But the small weak body withstands this pain and only a few tear drops run down her cheeks as my girl softly says, "I won't have children!"

I stroke my little one.

"Yes," I finally say, "but it's precisely **YOUR** stepping out of line from the standard program that shows that, unlike the rest of us Earthlings, you must be something special, right?"

Behind the tears, a faint smile begins, then it grows stronger, and finally, with a "You other Earthlings, ha-ha!" it bursts out of her in a radiant, hearty laugh like I haven't seen in ages. The laughter is finally infectious and I, too, laugh and laugh until everything hurts and when I already can't take any more, I see through my tear-smeared gaze that my child is still giggling in her bed and her little leg stumps are wriggling under the covers in the wonderful rhythm of inner joy.

So, this is how my child laughs in the face of death. I am full of pride ...

When she finally calms down, she becomes very thoughtful again and at some point, she asks, "What about church and religion, and all that?"

I scratch my head. Actually, I wanted to have given up this silly habit long ago, because nothing itches there. Besides, I have the stupid imagination that it makes my hair fall out faster, but somehow it also helps me to concentrate a little bit. It's funny that sometimes we have to do one more completely superfluous thing to better grasp the main task. Finally, I answer, "It's not so easy to explain, but probably only because we are in the middle of it, and describing the outer facade of a house that you have hardly ever left or can leave is naturally a very difficult task.

Let me try anyway. First of all, you must know that we humans are herd animals. That means we need the company of other people. Only in the group can we ultimately survive and reproduce. Faith and spirituality help us to create and maintain such community structures, simply because individuals who believe a lot and don't question everything per se can more easily be talked into certain rules as let's say 'God-given'. Thus, for us 'rational beings' it has had a tremendous advantage for centuries, also in social terms, to be spiritually inclined or enabled. Unfortunately, with the rise of civilizations, this ability became more and more a disadvantage, because it enables single individuals, groups, media, institutions, and governments to mentally enslave the rest of us."

"They're taking us for fools!"

"Yes, something like that, but the web of lies consisting of theological fairy tales and the rules derived from them, of political mass stultification and opinion barriers, of media power and mental amputation is now so complex and pervades all areas that it is probably impossible to untangle this Gordian knot again. The easiest way is probably to always keep the following in mind: Consciously or unconsciously, directly or indirectly steered by own interests, each person uses the faith ability of the other. Especially there, where people want answers, there are always some, meaning other people, who give them 'answers' ... apparently give 'answers'. In truth and with deeper investigation, however, one then finds out incredibly often that the 'answers' are nothing but enormous structures of lies, which only serve the purpose of subordinating the 'seekers' to the own will and striving for power of the person giving the answer."

"But at least we can vote, right? I mean, we have a democracy and can decide what we want in the elections," my child points out.

"Yes, so it seems, but that is only a mirage, a show, or a huge circus that is put on for the masses to make them believe they can really have a say."

"But I do have a choice! I can decide which politician I want and which I'd rather not," my little girl protests.

"The question is not whether I have a choice, but which choice I have. A choice between programs that say nothing and do not differ or a choice between promises that are not kept is always worthless. More and more people are recognizing this. They have no more interest in this spectacle, in which so obviously all stultification instruments that the state has to offer, thus the media, always participate so enthusiastically. People understand that they have no real choice at all, that they are not really involved at all, but that it is all just a means of keeping them quiet in order to allow, in truth, a rotten, sworn-in, and well-organized caste in the most diverse fields to carry on as before, without having to perform reasonably and at the expense of everyone else in society."

"Yeah, the political babble in the media is equally stupid and boring everywhere, but I always thought that was because I didn't understand it," my child retorts.

"No, it is because it is not at all desired that you understand it. Because then you would immediately realize that 99% of the types and topics have no substance at all, but are just charlatanry and fraud. You know that society is a living system, an organism?"

"Yes, I have understood that and it seems absolutely logical to me."

"Then you would probably also think that a well-organized society, and thus a well-structured organism, always tries to occupy its control organ as optimally as possible, wouldn't you?"

"Sure, who wants a brain made of muscle cells," my little one laughs.

I, too, have to laugh, "Yes, that's a good comparison, but it's even worse with us. The individuals who get into the centers of power and control in our country, among other things, through the terrific election-show foolishness, cronyism, nepotism, and favoritism, are regarding their level unfortunately not even the muscle cells of society. Then one could at least say: ok, they are high achievers, let's see if they learn to think and control; their experiences as achievers, thus, muscles of society, cannot harm in any case. Oh no ..., what is messing around there at the levers of power and the tax instruments of the markets, the economy, the health policy, and so on are, if not parasites, but at most skin derivatives and that is really not good for us."

"Skin derivatives?" my little one asks.

"Yes, the derivatives of the skin ... sorry, I thought you had that in school ... Finger or toenails, for example."

"Toenails?" my child raises her eyebrows and I detect something like a laugh, "We are governed by toenails?" She is now rocking with laughter. Apparently, she is just developing a figurative imagination about it.

"Yeah, that's how it is. Toenails sit at the controls of our society and the dorky election shows allow us to choose between the nails of the big toes and the little toes, and those from the left and right feet. It's great, isn't it?"

Big laughter. Finally, an almost breathless, "And the ones from the left foot smell a bit stronger and have black moons ...! Finally, finally—I know—what—tee-that's all about—left and right—tee-hee-hee." Laughter again. I let myself be infected and we both laugh loud and resounding.

Suddenly, quite unexpectedly, my little one has to throw up. First, she coughs and then a gush of blackish, thick broth comes out of her mouth. With a practiced hand I grab the bowl next to the bed and support my little girl's head. The seizure lasts for a while and at its end, blood flows from my child's mouth. My little one is looking after the thick mucus. Her little chest rises and falls with difficulty.

Then she coughs and says unexpectedly, "And between toenails of clean and dirty feet, we can probably choose, right?" then she giggles until she has to cough again. She coughs, spits up blood, giggles. I hold her and giggle along. We both have tears in our eyes and most of them are from laughing and giggling.

When my girl finally sits up in her bed, clean and reasonably refreshed, she asks again, "And so that's why many people wouldn't like it at all if the masses found out that societies are actually living systems, correct?"

"Exactly. Because then it would also be clear immediately that the 'organism society' needs an adequate selection mechanism for its brain cells in order to really be able to exist in the long term. A selection mechanism that aims at performance and abilities, and not at promises, show qualities, dyed hair, intrigues, and false pretenses." I answer.

"But if there's so much going wrong ...," my little one begins, when I clarify, "Let's rather say: not optimal."

"Well, if so much is not optimal, why don't the experts, the economists, sociologists and political scientists, and all the rest of them, put their finger on the problem?"

"Ha!" I do at first, giving my child another raised thumb, "Why do you think that is?"

"Well ...," my child hunches her little shoulders, "why?"

"They're all in cahoots. I know, I know, that sounds delightfully conspiratorial and thus far too simplistic. But you know, it's not a prearranged conspiracy, but a kind of tacit agreement: You don't hurt me and I'll continue to finance your institutes and professorships and your actually superfluous 'research'. Every now and then, they put on a little show with the reports of some economic experts, or they fight on a talk show, but they don't really hurt each other."

"But wouldn't there have to be a public interest in being enlightened?"

"Good point well made! But here we have the problem of averageness again. That means: if our—at best—average politicians are criticized by likewise—at best—average 'experts', then everyone has fairly little work to do. A symbiotic system of togetherness has formed once more, which depends on both sides for its lousy quality and where both sides can only be so bad because none really demands anything of the other."

"But how could the 'experts' get any better? Surely, they're already doing what they can, right?"

"Oh, are they? Well, one thing is funny: in all sorts of disciplines, one has learned that one must always try to get beyond the human point of view in order to arrive at truly meaningful, neutral insights about systems. This is especially true when you are existentially linked to the systems, for example, the Earth or what we are interested in here now, a social system."

"Ok?"

"How was this solved on Earth? When was it possible to make a huge quantum leap of knowledge there?" I ask without expecting an answer and actually want to continue right away.

"Hmm," my girl interrupts me, however, and answers to my astonishment, "in school we were taught that you can learn umpteen times more about the Earth from space, thus, with satellites, than directly on the ground."

"Golly, look at that school ... In fact, by putting distance between oneself and the object of inquiry, one expanded one's horizons and gained additional, more *global* insights. Now how would you imagine that to happen with societies?"

"I really don't know."

"Ab inito!"

"What?"

"Model calculations from first principles. One must gain distance, get away from empirical or semi-empirical model approaches that contain at least as much individuality as many wishes and dreams and preferences of the creator of the model; just as the soup that someone cooks bears his signature, unique and unmistakable and not neutral from the outset. But if you step back a little bit and build societies out of small entities with well-defined properties and then let them evolve in a model, a simulation, then you can get a neutral, or let's say more neutral, view of what's really happening and what really matters."

"And that works?"

"Sure, it works, I've done it. It's not much different than molecular dynamics, except that the atoms and molecules are individuals, and the quantum mechanical interactions are the relationships we have with each other."

"Molecular dynamics with humans ... sounds sort of like science fiction."

"If those who have made themselves comfortable in the current substandard system have it their way, then it should not only sound like science fiction, but be immediately associated with the WORK OF THE DEVIL by the ordinary person on the street. That's how the dummies protect themselves."

"Hmm," my little one says, "I'm aware of that with churches, politicians, and the TV people, they have to bullshit the masses to keep themselves in power. But what about the scientists, aren't they committed to the truth?"

I have to laugh. "Committed perhaps yes, but they are also only humans," I answer. "The scientists are perhaps even the very best example to better understand the other 'cheaters'. Every human being, even the individual scientist, must constantly reassure himself of his importance in society. In the end, he must constantly prove that he is useful to society, that he has a unique selling point, yes, even that he is indispensable. My dear I am telling you: only one in a million people is really that good, the rest **have to pretend** to be."

"Are you that good, Dad?" my little one interrupts me and smiles cheekily.

I think, at first, I want to joke, but then I answer very seriously, "Only for you, my love. And there you must decide for yourself whether I am perhaps not just THE person among many others who means enough to you to be considered unique and irreplaceable."

My little girl suddenly has tears in her eyes. She makes an incredible effort to stretch out her little weak arms to me.

"For me, you are the most important person of all!"

I think I have never held my girl so tightly in my arms as I did in that moment ...

After a while, however, my child's curiosity gets the upper hand again, "Now what about the people who are 'not sooo good'?"

"They have to artificially create their niches, deepen them, make others believe that what they do is incredibly important to society, and make access to their field, their 'unique selling point', as difficult as possible for others."

"How do they do that?"

I chuckle. "Oh, there are almost as many tricks as there are people. But the simplest and most common are exaggerating, lying, and building language barriers."

"Building language barriers?"

"Yes, give your 'field of expertise' its own 'technical language' and you automatically make sure that not everyone can follow you anymore. In this way you can appear wise and important without ever accomplishing anything of significance. Hardly anyone will notice that you are only presenting platitudes in a new guise, and if someone notices, then the others will not be able to follow him in the inevitably arising dispute, a technical dispute held in technical language. And then they just say: 'Aha, a dispute among specialists, we can't discuss it anyway'."

"It's that easy, really?" my girl looks a little incredulous.

"Yes, I could show you some 'high profile publications' of really big shots here and in a few 'moves' shrink their essence down to less than two sentences in simple language."

"You could?"

I laugh. "All right, maybe it will be three sentences. But only for very few high-profile subjects. But then there are whole specialist areas that are complete humbug."

"Which ones?"

"I don't want to name individual black sheep, but I will give you a simple rule, how you can find especially many of such nonsensical fields: Look for the 'sciences' and 'special fields', where especially many foreign words circulate, where conspicuously little meaningful mathematics is used, where almost always the 'statistics' seems weak and often 'bent', where even the normal language is artificial and not suitable for everyday life. That is where you will find the most worthless 'researchers' there are."

"Hmm ... perhaps jurisprudence, political science, social science, and medicine?" my girl asks seemingly uncertainly and even gets a cheeky grin.

"YOU said that, but believe me: the charlatans are everywhere. In some disciplines, they just find it particularly easy to bullshit people and make themselves important without being able to do anything, or accomplish anything meaningful in any way."

"Maybe the weathermen, too?" adds my little one.

"Oh well," I answer timidly, "I would exclude those, because after all they are so honest and give stability data and probabilities for their predictions. But the example is good to show how wrong and superficial the opinion of the masses can be. Just because everybody seems to pick on a certain group, it doesn't necessarily have to be right, because our societies are so manipulated that probably about 80% of the people are domesticated and mentally amputated and follow the main opinion ... just like cattle behind the leading cow."

"And if the leading cow is called communism, socialism, or fascism, they run after it too, right?" my child interjects.

"Hmmm ...," I'm squirming a little bit, "initially, it may well be a running after, but what you absolutely have to understand here is the following: A stable society in the long term, all deceptive maneuvers included, can only ever come about if:

(a) the current boundary conditions are satisfied; thus, there must be a minimum flexibility that allows adaptation to external and internal conditions

(b) the basic principle of life, namely, the striving for local order, must always be achievable for every individual in society, or let's say for at least the majority of them.

In the social forms you mentioned, at least condition (b) is not fulfilled. In the case of (a), a temporary adaptation could still be achieved by force and trickery, but (b) in these systems is only possible by internal violence."

"That's why these systems always existed and still exist only as dictatorships, right?"

"Absolutely right," I answer.

"What would you have to do to prevent or at least minimize this charlatanism?" my child asks in a slightly resigned tone, as if not really expecting an answer.

"Astonishing as it may sound," I reply, "a quite sustainable solution would not be so difficult. One would simply have to always and regularly enforce a feedback to the masses, or let's better say: to the substance of society. That is, it must be clear that science does not belong to scientists, politics does not belong to politicians, art does not belong to artists, literature does not belong to writers and publishers, crafts do not belong to artisans, news do not belong to journalists, salvation does not belong to the church, truth does not belong to the 'good guys', medicine does not belong to doctors, and law does not belong to lawyers. This list can probably be continued infinitely, but it must be demanded that all these—often in addition self-proclaimed and highly conceited—'professionals' are obliged to share their work over and over again with the rest of society and in such a way that a participation of the others is possible. Everyone is not only obliged to do their job as well as possible, but also to explain it in an UNDERSTANDABLE way."

"And that's supposed to change something?"

"Yes, why do you think corruption is always lowest in places where, for example, tenders are completely public? That's exactly why Finland is always at the top of Transparency International's rankings.

Moreover, just imagine the judge who is forced to explain his judgments UNDERSTANDABLY and, above all, to justify them to the rest of society, perhaps even fearing a direct 'popular revision' if he makes complete nonsense. How much more possibilities would he have to slut now, if so many eyes look at his fingers or better lips? How much less should the probability of completely incomprehensible judgments be, if unintelligible reasons are simply not allowed?

Imagine a politician who is forbidden to dumb down the people under penalty of law, who is obliged to really explain what he is doing and not, as it is at present, to occasionally scatter a few crumbs of information in cryptic formulations among the common people, or who, in an effort to gain maximum votes and presence in a media-hungry environment, trumpets irresponsible stupidity into the world, promises that no one in the world will ever be able to keep, and bends and twists facts just so that no one out there can see how stupid he actually is. In addition, link his salary and pension to the well-being of the country, to indicators that reflect properly how the people he leads are doing, in the long run, via politicians' pensions, for example, and he will make an effort to perform well, you'll see!

Oh, and there's another wonderful proof of why this approach is the right one."

"So, which one?"

"The fact that all those lame ducks who benefit so wonderfully from the current system always present the possibility of such control as 'impossible' and work against it with every conceivable means is the best proof one can imagine. So, take a close look at the guys who don't want transparency and accountability and torpedo such solutions, and analyze what they do. You'll always find that they have incredible advantages in the current closed system, that they can grab resources free of charge, that they can parasitize unmolested.

Oh, and also imagine the doctor who is obligated to really explain to you what's happening to you ... I mean before they put you under the knife."

My child just looks at me with wide eyes. She thinks for a long time. Then she nods very slowly.

"I already realize that the established have no interest in establishing such a sustainable solution, but why does no one revolt and demand such a reorganization?" my child asks after a while.

"The magic word in this case is again discrediting," I reply. "You only have to explain often enough to the sufficiently stupid people that such a thing is not possible, then the people, when they hear it over and over again from enough corners, finally don't even ask anymore: 'Why actually not?' Only the stupidity of the masses secures power for the stupid and protects them from justified demands for accountability and proof of performance, even—God forbid—linking their incomes and pensions to their performance. Unfortunately, this will not last for long."

"Why not? Everyone's kind of happy ..." my child interjects.

"Yes, as happy as laying hens in their little cages." My girl chuckles. "But evolution didn't give us our brains to have at best mediocre dunderheads tell us what the world looks like and what to do in it. It gave us brains so that we might use them, so that we might exert ourselves, one by one and over and over again, to get behind the mystery of our own being and to contemplate and analyze the wonder of the universe over and over again. If we had been created to absorb and accept the truths of others, evolution could have been satisfied for us with eyes, ears, and a memory. But it gave us a damn powerful analysis apparatus and we have the duty to use it. Thereby it is ok if we do division of labor and some brains do Law, others Medicine, again others Physics, and so on in depth, but it is their damned duty and obligation to include all others and to take them along. Only then it is division of labor and not exploitation and parasitism.

A system that suppresses this progress cannot be stable in the long run. It will perish, be overrun by other more optimal forms of organization, meaning those which better train and exploit the resource 'spirit of society', and finally rot in the past of history as a useless, dead side arm of evolution."

"Are there no honest people at all, I mean those who want to tell the truth to others ... not even among scientists?"

"Yes, there are: me!" I answer and my child immediately understands that I'm just kidding, that I too can't get out of these constraints of omnipresent deceit.

"Yes, but they have already attacked you, too, in your subject. Why was that?"

"It's the same in small things as in big things. As soon as you step on others' toes, even if it's the right thing to do, and tell people they've been teaching and doing nonsense for years, they get mean," and smiling I add, "or try to be."

"What did they do to you?"

"The usual stuff, they arrange to meet at conferences, then go to your presentation together and try to discredit you there with stupid, damn stupid, to be exact, questions ... That is, to make a mockery of you."

"How did you fight back?" as weak as the little body seems, her eyes are wide awake and looking at me intently.

"With the best weapon the world of science has to offer."

"And what is that?"

"Mathematics!" I answer theatrically.

My girl, however, only looks at me questioningly and doubtfully, and so I quickly and laughingly add, "Well, when the first one of those guys asked me his stupid question, I just went to the flip chart, the blackboard, or whatever was there to 'scribble on' and wrote formulas on it. Large and visible for everyone in the room—especially for them, of course. I then explained in a bored tone, that 'as we all know' in this or that case, of course this or that 'would apply' and if one then still knows and from the other subject area the theorem of so and so would add, then one could make this transformation here, then integrate over everything and voila ... Quod erat demonstrandum, so as I said before ... The guy sat back down, looked at his other clown-friends in the room, shrugged his shoulders, and the whole thing was over."

"And that always worked?"

"Always!" I assure her.

"Didn't anyone ever find a mistake in your calculation?"

"You know, my love, people don't do the math. They don't even look at what you're doing up there, they're simply fascinated by the fact that someone has the nerve to leave the laptop with the prepared lecture, the lectern, the safe bulwark of the 'presenter', to stand in the most unpopular place in the whole universe, as is well known since his own examination time: namely, at the front of the blackboard in front of a room of perhaps 500 full-grown scientists, more than half of them veteran, mingy professors. You can paint any nonsense on the blackboard and claim that it would prove that everything in your lecture is correct (and be it that the Earth is a disk, which is also true in principle, because it is a discus but not a particularly flat one). They believe you and even if not, nobody will contradict, because the only possibility to do this correctly and sustainably or sincerely would be to come forward yourself—to the blackboard and in front of all the super smart people—and to PROVE that you and not the lecturer is right.

Once, at a conference in Colorado, I was so annoyed by the whole circus that I wrote your birth dates on the blackboard, those of you and your siblings, and just sprinkled a few integral and differential signs in between. It looked really pretty."

"But that must have stood out, because they were just numbers?!"

"Good thinking, kid!" I praise, "that's why I arranged them in a matrix and before that I specified that we are now moving in an n-dimensional space with n equal to the number of columns of the m matrices we need to prove my assertions. So, since there are six of us in the family, it was 6×6 matrices, and 6 birth dates fitted nicely in there."

"And that worked?" as much as her little body has to strain and no matter what it costs her, she is already on the verge of a laughing fit again.

"Yes," I answer, "I don't think I've ever gotten as much applause as I did then."

This time it takes a long time before we get serious again.

Still giggling, my child finally asks me, "Dad, is that the truth, too?"

I think a little while and answer, "For me it is, my dear. You, on the other hand, must seek your own truth."

"I don't have time for that anymore," my brave girl answers unexpectedly quick and slightly defensive.

"Finding one's own truth is not a question of time, but of courage," I answer without having to think, because here I know what I am talking about, or rather, I think I know.

My child nods. The movement seems incredibly slow and very quietly she suddenly says, "Yes, the truth can hurt, but I prefer it that way." Then my girl asks me for something to drink. She had probably laughed herself thirsty. With the glass in both hands and the oversized eyes directed at me, she asks, "And so there are no truly honest people ... I mean those who not only mean it sincerely, but who can be honest because they **know** instead of just guessing or parroting?"

"Oh, of course, there always have been. They used to burn them, as you know from history, and they don't do much better with them today."

"Do what?"

"There are many methods, too many to keep track of and to avoid all stumbling blocks. But in the end, if you have already managed to get other people to listen to you at all—because many prefer to stick to their fairy tales and the rules they have derived from them—then the 'experts' come along and talk everything down and ruin it in their incomprehensible technical language, so that the hoi polloi, unsettled, ultimately prefer to turn back to the ever larger crowd of fraudsters and the established or supposedly politically correct and so on ... Finally, also because it is easier socially."

"Terrible!" my little one shakes her head.

"Yes, my love, and this stupidity costs precious people their lives ... you, for example."

"How do you mean that, Dad?"

"Well, I told you a few days ago that we can't help you because we don't have enough knowledge, because we are too stupid, do you remember that?"

"Yes, I remember," my girl answers very hesitantly. Not so much because she has to search her memory for a long time, but because she is not sure where my question would lead. Some truths, unfortunately, are just too painful. Still, a child knows when she has no more time.

Eventually she asks timidly and in a very weak voice, "Could I be helped, for instance, if people were less stupid?"

"Yes!" I answer almost as quietly. Then I pick up my little girl and press her to my chest. Involuntarily, images appear before my eyes that I associate with that hug, when I held my little one as a baby, after her first swims, for the first, quite wonderful school report, the lost legs, the disfigured face.

Somehow, I feel that my little girl is now comforting me. And suddenly she says in an amazingly firm voice:

"Thank you, father! Thank you for everything!"

Only when she falls asleep and I turn off the light do I realize that my little one said 'father' and not 'dad'. That was the first time. She has grown up and my heart breaks when I realize that this will also be the end of her journey here on Earth.

Day Seven

"I'm very tired," comes weakly from her little mouth, "I want to rest."

"Yes!" I reply, "you have been quite wonderful and now you may rest."

"Will the universe ever manage to solve the fundamental mystery?" it's almost a whisper.

"I believe—and I hate to say 'I believe'—there are infinitely many answers to this question. And that is also the reason why the universe is infinite; infinitely wide, infinitely dimensional, infinitely nested, and infinitely existing. Infinitely transient and infinitely immortal at the same time, why it brings forth beings like us infinitely often and lets them pass away again. Nevertheless, every small solution found, indeed every microscopic step found toward it, even if it is only a quantum step further forward, is worth infinitely much."

Her small weak eyes search mine and there is so much knowledge in them that I feel we have succeeded. But then, for a brief moment, a tiny doubt, a last question.

"What will become of me when this body will no longer be mine?"

It seems that there is no longer fear in the question but rather something like need for a final reassurance.

"You mean you, that is your individuality, don't you?!" I am glad about this question and I am glad that I can give an answer of which I am absolutely and 100% sure. An answer of which I am sure, not because I BELIEVE, but because I KNOW! And so, I answer, "Information cannot get lost in this universe, my love. It cannot disappear, because there is a kind of net where it gets caught. A kind of bottom or a last level where it always arrives and continues to exist. As long as our universe remains, and it will remain until infinity, there will be this bottom; everything of individuality and uniqueness will be stored on this lowest level for all eternities ... Or should we call it the very first, the primal level? It doesn't matter! No matter what kind of information it is, as soon as it contains even a tiny bit of essence of uniqueness, it will remain, it will be stored. That, exactly *that* will also happen with you, thus your uniqueness and individuality."

"What is the universe doing with ... with my 'uniqueness'?" It almost seems as if my child wants to talk down her uniqueness, as in all her weakness she is adding a somewhat ironic note to this last word.

"It uses it, but you will have a say in that, I'm sure of it," I reply calmly.

Now, once again, something like fear gently twitches through the small, battered body, "But won't I get lost in this process?"

"No, my child," I reassure her, "it's just as if you had a thought, a memory, and combined it with another memory or a new thought to form a completely new idea. The old memories are not lost and yet they live on in something perhaps completely new. When you live, at every moment your self emerges from a previous self. The old self dies and the new one is there ... only in this way you can live. And that's exactly how it will go on only a little bit different than you knew it before."

My child begins to smile.

"And," I continue, also smiling, "because the old thought is not lost and there are still many trillions of other memories with which it could be linked, the one thought, the one unique thing, you and your individuality will be the starting point of infinitely many new and beautiful things. You are like a seed with inexhaustible fertility and immeasurable variety, my love."

"That which is easiest to create," my child whispers, and her look shows a cheeky little smile.

"Excuse me?" I ask.

"Chicken or egg, which came first? That which can be created most easily in this universe. That was first."

"You are utterly wonderful," and I am so proud and so grateful for this marvelous child. As proud as a father can be.

My girl smiles again. Her eyes want to tell me something else. A little imploringly, they ask me to listen again. I bend down to my child, touch her weak shoulders, and stroke her cheeks.

"I think I do understand now," my girl whispers in my ear.

"No!" she whispers with all the determination she can muster.

"I *know* I do understand now!" she adds with emphasis and all the strength that is still in her and then she closes her eyes forever and thinks her very last thought. A soupçon of a satisfied smile lies on her little face.

The thought, however, is like a puzzle piece that indelibly unites with everything that has ever happened, been thought, written, and said. It closes a huge gap, spans vast areas, and the universe begins to understand something entirely new.

It is done.

Obituary

I found it many weeks later while cleaning up. It was slipped into the book of 'the Little Prince'. An inconspicuous, rather worn sheet of paper.

For You

From my child

What does it mean
    to be all smart
to only wish
and get it all
to come and see and win
with just one call
and efforts so thin?

What does it mean
    to be admired and praised
for something too easy
and seen almost as waste?

What does it mean
    to be all known
to snip the fingers and get support
A fish in a swarm
no idea where it's heading
no gist, but who cares as long as the current it warm?

What does it mean
    to be hated
enemies wherever you go?
because so clear the thought
and too honest the mind
always enough to be everyone's foe.

Nothing does it mean
And nothing will it ever
I do not want to be
for all this rubbish THEY call life
not for all the riches
and the time is running
I see no need to strive.

> *But if you alone*
> *would mourn a bit*
> *when I'll stop to feel the pain*
> *then—so I think—*
> *my life was not in vain*

***

One could clearly see that the last stanza had been added later. It was squeezed to the side and written with a weak hand, barely legible. She must have mustered all her willpower to still bring them on paper, my little angle.

THE FIRST DAY:
LET THERE BE NOTHING

Chapter 1

Making Contact with Quantum Gravity

1.1 Brief Story about the Minimum Principle—Part I

I wasn't feeling so great and so I thought that maybe a little walk would do me good. Thus, I stopped working and asked our youngest, whom I affectionately called my Schabracke, if she would like to accompany me. She wanted to and so we dressed warmly because of the strong wind and headed toward the Bodden.

It was somehow beautiful despite the uncomfortable weather. At the edge of the field, a good 500 meters behind our house, two young foxes were playing. They were so engrossed in their little scuffle that we got very close, almost 40 meters, before they noticed us and ran away rather cautiously. At the Bodden we marveled at the high water that the storm had carried into the bay and Schabracke spotted a strange little bird on the paddock fence. I thought it was a wren, but only saw it for such a brief moment that I wasn't sure. My little daughter took it upon herself to look it up later in the nature guide.

On the way back, we tried to march with our eyes closed for a long distance and still not lose our direction. Because of the ruts, which you could feel quite well with your feet, it even worked.

The Math of Body, Soul, and the Universe
Norbert Schwarzer
Copyright © 2023 Jenny Stanford Publishing Pte. Ltd.
ISBN 978-981-4968-24-9 (Hardcover), 978-1-003-33454-5 (eBook)
www.jennystanford.com

"Dad," my youngest asked me abruptly, shortly after we had given up blind marching, "why does the world actually exist?"

I didn't know the answer to that right away and pondered. Schabracke, however, took my momentary silence for a failure to understand her question and added:

"I mean, if the world didn't exist, then there would be nothing ... so then nothing would be the world?"

I almost stopped, so fascinated was I by this statement, which was really "just" a question. But it was for me the answer to an ancient problem:

'Why does the universe follow laws that can be expressed or derived by minimal principles?'

My 9-year-old daughter had just solved this very problem and I answered her:

"Nothing is actually the world, or rather the world is the nothing, my little one, but the nothing wanted to know why it is there—just as you just asked why the world actually exists, the nothing wondered why it exists. And thus it built the world so it would get that answer."

"And how or with what did the Nothing build the world and thus answer its question ... because it is Nothing?"

"With nothing it built and it is still working on the question!" replied I, smiling to myself. I thought of the minimal principles that describe our universe and the many intelligent zeros that were actually behind it, and even though my health was not much better than before the walk, I suddenly felt much better.

Every other child would have surely protested about my answer and at least inquired further, but Schabracke suddenly stopped, bent over a trivial blade of grass, and told me the name of this kind of grass. I thought she had sufficiently worked off the topic with the nothing and the world, somehow—at least with this interlocutor—considered it unfruitful and had therefore switched to the determination of grass varieties, but there I should be badly mistaken. Because after studying intensively for a while the stalk together with its seeds, she suddenly said:

"It's hard to believe that this is all nothing, but surely it's like one and minus one. Together it also makes nothing, but when you see it separately, each is there by itself."

Then she looked up at me from the blade of grass and I nodded.

I could not and did not need to teach this child much more.

1.2 No Stress Without Contact and No Contact Without Stress

When this author started his work as a physicist in materials science, he had a lot of applications that came with mechanical contact problems. Often this was in connection with finding the reason for certain failure mechanisms such as fractures, plastic flow, and delamination of layers from surfaces. Thereby he learned that not the forces are determining the character and severity of a certain failure, but the stresses. So, one obvious way to avoid damage is to avoid stresses. On the other hand, most applications require certain forms of contact, which is to say surfaces connecting with other surfaces, thereby stresses just occur. In fact, there is no mechanical contact without stress.

Interestingly, what holds for materials and mechanical contacts also holds in general. Thermal or electrical connections are always also creating mechanical and other stresses inside the materials. Gravity produces distortions, leading to stresses. A good example is Jupiter's moon Titan, where intensive volcanic activity is caused by the gravitational forces between Jupiter and its moon and the subsequent shear stresses. We also find the concept of stresses in psychology [1] and socioeconomy [2]. This is not surprising. One should just try to have social contacts without also automatically producing at least some stress in oneself's social environment. One soon finds that this is impossible.

In my case now the solutions I had developed to treat mechanical contact problems [3–5] brought me into contact with ever more demanding and complex tasks, not only containing technical aspects but also sporting socioeconomic and even psychological dimensions [1, 6]. In addition to all these already very demanding aspects, there was always the need of knowing and perhaps even controlling the total uncertainty budget of a given problem. Thus, the contact with the mechanical contact problems created a lot of stress, which drove me into an ever more fundamental treatment of even the simplest (on first sight) tasks.

At some point then I had to realize that only the most holistic approach possible could deliver what was of need here and this meant a Theory of Everything or a world formula.

When starting this journey, however, I soon encountered a problem I could never have foreseen.

1.3 World Formula and Psychology ... What Does Not Want to Enter Our Heads

Very early in the world formula project, my partners, who wanted me to publish my results and were looking for a suitable publisher, asked me which explanation I would have for the fact that the so-called "world formula" was not discovered yet.

The answer I gave must have caused some confusion among the partners and with the publishing house, because ... but more on that later ...

> "I believe that finding a 'Theory of Everything' or a 'world formula' is not the problem," I had written. "On the contrary, I am convinced that the vast majority of people in this world should not find it too difficult to understand the fundamental laws of the universe. Rather they all have a problem to accept them."

Of course, my partners wanted to have this explained a little more closely, which I did as follows:

> *Years ago, I had occupied myself with the evolution of natural calculating machines, that is, brains, and came across a small problem again and again.*
>
> *I had used artificial calculating machines, so-called computers, to investigate the matter as holistically as possible and also thoroughly and found from a certain evolutionary stage that the whole development wanted to stop ... partly even regressed. Instead of further growing brains, my models preferred smaller thinking units and dumber carriers. This was somehow more favorable for the survival of the simulated species. The reason was not that there were problems with the improving brains because they consumed more energy than they brought in, no, this balance was always ok, but the natural thinking apparatuses developed an intrinsic, that is, an inner problem child. As soon as they could have crossed the threshold of self-knowledge and environment mirroring, the brains began to strike. There had to be something what made the whole Darwinian superiority compass swing in connection with the self-knowledge to finally reverse its direction. Like in a resonance catastrophe, the evolution vector, formerly set on brain growth, went haywire and finally switched into the opposite direction.*
>
> *The reason was found quickly—after all, I had my artificial calculators to help me find it—and when I did, I just thought:*
>
> *"Yeah, right!"*

You only have to look at people with all their everyday and even existential worries, how they perhaps look enviously at their pets, who are happy when they get food, a place to sleep, and a few strokes, to understand where the problem lies.

The ability to be self-aware is anything but a blessing for many people: it is a burden, because with self-awareness comes self-responsibility and even responsibility for others ... at least one's own offspring. There's a lot of hard stuff to endure ... to endure consciously, and that didn't always make the smarter brains really fitter than the dumber ones.

But there are inhabitants on this planet capable of self-knowledge (5 or so, though the author is not quite sure about himself) and the species to which these entities are belonging must have managed—at least in parts—to overcome this threshold somehow. Evolution must have found a way to slip through that bottleneck.

So, what was the solution that evolution found here?

Evolution is minimalistic in the end; it always tries to solve problems with the least effort and so it was—perhaps—obvious to slay the occurred problem with its own cause. If the bigger brains caused the problem, then they should get rid of it themselves. The evolution thus tried around and then also once a brain came out that was still a tad bigger than all the others. This time, however, this more at neuronal mass was not immediately put into survival strategies (well, indirectly in fact it was), but in something completely different, something apparently secondary.

*The evolution had obviously found the spirituality ... ATTENTION: It has not found **to** spirituality, but it has invented **the** spirituality. There is a huge difference!*

Now the thinking apparatuses equipped in such a way had the possibility to push recognized or self-recognized things, which came up in the daily life, but were of no use, but on the contrary only brought damage, into a drawer, where it neither disturbed, nor caused pain, nor distracted otherwise somehow too much from the survival fight. If, for example, one of the own descendants was so seriously ill that that he/she was on his/her deathbed, then the spiritual drawer helped the affected "parent" to keep the grounding and just not to neglect the rest of the brood. With sufficiently stupid brains, however, this dissonance did not occur at all. They had no problem with it and needed consequently also no "drawer for faith questions."

Why am I telling you this?

Well, almost exactly 2 weeks ago I received mail from a member of the EU Parliament, MEP. In the e-mail a link was given, which referred to a nice lecture of the MEP. There, in the lecture, it was shown in a wonderful shortness what has gone wrong in our country—GERMANY—under the Merkel regime. The enumeration was conclusive and so vivid that one could already clearly hear, see, and feel the impending collapse of all socioeconomic structures, not just in Germany, but in whole Europe. Every word in the lecture was so coherent that I would have signed it immediately.

Now, politicians often tend to name the problems created by others with relish in order to then announce that they themselves would do everything better. This did not do said delegate, however. On the contrary, he reported the request, which would be frequently brought up to him and his party and which read there:

"You've got to do something about that!"

"No," the congressman said in his message, "YOU have to do something!"

*And then he referred to the elections coming up, the many small battles you could fight yourself and the resistance you could offer ... you would just have to get out of your feel-good bubble once in a while and face what is commonly called **life**.*

We all have the spiritual drawer. It helps us in many, otherwise unbearable, life situations.

But we must be careful not to get too comfortable in this drawer.

Some awake people ask themselves what "for God's sake" may have ridden or still rides the maniacs who were at the German train stations in the fall of 2015 to applaud to the hordes of false refugee, who conceal or gloss over the dramatic increase of knife murders we since then observe in Europe in the regime media, or who rage as modern Mengeles in vaccination fascism against a rather harmless flu through the country to "inoculate" us with dangerous cytotoxins. Well, believe me, these people (most of them anyway) are also only caught in their spiritual drawer, in which they have made themselves comfortable with catchy and fitting theses, ideologies, or even religious dogmas. From there they see through the slit to the outside and seen from their limited view, what they do is abysmally good. The only thing that prevents them from recognizing the stupidity of their actions is their narrowness ... the vast majority of them are not really stupid (I think).

And most of these people—even though very dangerous in their ideologic mania— are not the true enemies of reason, logic, and free thought. No, those are the brainwashers and "masterminds" behind the dumbing down.

The station applauders do not see the contradictions of migration, not even the damage and paralysis that migration causes in the countries of origin.

The system docile scribbler actually means to serve the social peace, the climate hopper the climate and all the small and large mask, and polymerase testing and "vaccination" fascists serve the 'public health'.

But are these cognitively dissonant people so much worse than the rest of us? Aren't we also often only too happy to crawl into our spiritual bubble and think that some benevolent power will finally save us?

"So, what was the story with the publishing?" you may now ask.

Oh yes, I almost forgot.

Well, they would like me to write a book on this as well.

But I really don't know whether I should make that effort.

I don't want to exclude the possibility that the message is understood here and there, but the whole thing only makes sense if it is also accepted and applied, doesn't it?

Still, my partners didn't stop nudging.

"Norbert, you are right, but there is only one way to find out, isn't there?"

"Good point, my friends!" I said eventually and thus

Shall we begin ...

References

1. N. Schwarzer, *The World Formula: A Late Recognition of David Hilbert's Stroke of Genius*, 2022, Jenny Stanford Publishing. ISBN: 9789814877206.
2. T. Bodan, N. Schwarzer, *Quantum Economy*, www.amazon.com/dp/B01N80I0NG.
3. N. Schwarzer, Scale invariant mechanical surface optimization applying analytical time dependent contact mechanics for layered structures, in A. Tiwari, ed., S. Natarajan, co-ed., *Applied Nanoindentation in Advanced Materials*, Chapter 22, 2017, ISBN: 978-1-119-08449-5, www.wiley.com/WileyCDA/WileyTitle/productCd-1119084490.html.

4. N. Schwarzer, From interatomic interaction potentials via Einstein field equation techniques to time dependent contact mechanics, *Mater. Res. Express*, 2014, **1**(1), http://dx.doi.org/10.1088/2053-1591/1/1/015042.

5. N. Schwarzer, Completely analytical tools for the next generation of surface and coating optimization, *Coatings*, 2014, **4**, 263–291, doi: 10.3390/coatings4020263.

6. N. Schwarzer, About Holistic Optimization – Examples From In- and Outside the World of Coatings, 2015, Proceedings of the 58th Annual Technical SVC Conference.

THE SECOND DAY:
LET THERE BE PHOTONS

Chapter 2

Societons and Ecotons

Figure 2.1 "Societons at Work."

The Math of Body, Soul, and the Universe
Norbert Schwarzer
Copyright © 2023 Jenny Stanford Publishing Pte. Ltd.
ISBN 978-981-4968-24-9 (Hardcover), 978-1-003-33454-5 (eBook)
www.jennystanford.com

2.1 Why Considering Socioeconomic Interaction via Photons?

In this section, we will derive that human societies, if also following principal natural laws, must have their very own photonic background, time, and consequently also a fundamental Heisenberg uncertainty.

We will base our considerations on a world formula approach originated from David Hilbert and Albert Einstein. We will give proof that this approach truly considers "it all" by deriving a few principle—classical—gravity and quantum equations. Thereby we are not aiming for completeness but only present examples. Further proof for the world formula character of the approach will come in the later chapters of this book.

2.1.1 Motivation

Understanding the working of a human society usually, which is to say if not performed by an extraterrestrial intelligence, has the principal inbuilt flaw that the understanding is done by a human or a group of humans. This automatically renders the process of the understanding kind of impartial.

As the author has not the option to ask an extraterrestrial entity to do the analysis for him, but still is determined not to make the mistake of many other society-understanders, the author resorts to the simple method of starting so fundamental that further digging would be futile, because he already has started from the lowest ground ... at least if taken the current scientific understanding.

Our starting point here shall be the Einstein–Hilbert action in arbitrary space-times and with arbitrary numbers of dimensions. We consider this to be fundamental enough—if applied and investigated thoroughly enough—to provide us with a world formula [1].

2.2 A Most Fundamental Starting Point and How to Proceed from There

Before introducing math, we want to give a very brief verbal explanation of the principle structure and appearance of a world formula, which—as being a world formula—also should be able to describe a human society. The latter may be extremely complex and of very high dimensional order, but this does not mean that a mathematical description would not mirror it (not in principle anyway). A more thorough verbal explanation can be found in [1].

1. Let there be a set of properties.

2. Such properties could also be seen as degrees of freedom of a system, which mathematically would allow us to interpret and treat them as the system's dimensions, whereas the system itself is a space, formed by such dimensions (properties). Thus, we consider space as an ensemble of properties.

3. These properties, which—as said—could just be seen as degrees of freedom and thus, dimensions, are subjected to a Hamilton extremal principle.

4. This leads, if taking the Ricci scalar as the Lagrange density, to the Einstein–Hilbert action [2] and subsequently to the Einstein field equations [2, 3], when following just Hilbert's mathematical path of variating the action with respect to the metric tensor [2] or Einstein's argumentation [3].

5. However, the simplest way to satisfy the Einstein–Hilbert action could also be achieved by just demanding the curvature R (the so-called scalar Ricci curvature) of the space or space-time of properties to be a constant, which leads to the well-known quantum equations such as Klein–Gordon, Schrödinger, and Dirac [1, 4, 5]. It was shown in [1] and it will be evaluated later in this book how this is still completely consistent with the Hamilton minimum principle approach.

6. Thus, and most interestingly, this $R = 0$ condition, as it can be shown [32], is just a by-product of the same Hamilton minimum condition that also gave us the Einstein field equations.

2.3   An Equation for Everything

In 1915, David Hilbert [2] was able to show that a mathematical structure, very similar to a volume equation (a volume integral to be precise), apparently contained Einstein's famous General Theory of Relativity [3], which, as we all know, is a theory about gravity. Thereby the fascinating aspect was that something so very much physical, like gravity, came out of a completely mathematical source, namely, Hilbert's "volume integral."[1] In fact, it is a bit more than just a "volume integral," but an integral that actually looks for an extremum, which means maximum or minimum, of the volume result.

---

[1]Please note that this "volume integral," in the literature, is usually known under the expression "Einstein–Hilbert action."

More than one hundred years after these groundbreaking works of Hilbert and Einstein, we were able to show that not only gravity resided inside the Hilbert equation, but obviously just everything [1].

But why, with the Hilbert equation already being there for so long, wasn't this fact discovered much earlier?

In order to grasp the implications here, we need to understand that, even though the Hilbert equation looks quite simple on first sight, it has many degrees of freedom and a fairly complicated intrinsic structure. Therefore, this author suspects that Hilbert, Einstein, and many others simply have not seen all the possibilities the apparently so simple "volume integral" offered. For one thing, as this is going to be the one we are most interested in here, obviously nobody ever bothered about, was investigating the Hilbert equation with respect to the number of dimensions in which a certain problem is considered[2]. Almost everybody usually just concentrated on our "classical" 4-dimensional space-time. Also, the option of a function-scaled metric (a scaled volume or sub-volume, then) was not properly investigated earlier on (c.f. [1]).

2.4 From Hilbert via Klein–Gordon to an Approximated Dirac Equation

Our starting point shall be the classical Einstein–Hilbert action [2] with the Ricci scalar R^* as kernel or Lagrange density:

$$\delta W = 0 = \delta \int_V d^n x \left(\sqrt{-G} \cdot R^* \right) \tag{1}$$

and, in contrast to the classical form in [2], with a somewhat adapted metric tensor $G_{\delta\gamma} = F[f] \cdot g_{\delta\gamma}$. Thereby G shall denote the determinant of the metric tensor $G_{\alpha\beta}$, while g will later stand for the corresponding determinant of the metric tensor $g_{\alpha\beta}$. In order to distinguish our Ricci scalar, being based on $G_{\delta\gamma} = F[f] \cdot g_{\delta\gamma}$ from the usual one, being based on the metric tensor $g_{\alpha\beta}$, we marked it with the * superscript. Please note that (1) should allow us to just say that W is a constant. Then, of course, with W being a constant, it would automatically variate to zero. As a sharper (less general) condition one may even assume that:

---

[2]Yes, there are many n-dimensional solutions to the Einstein field equations, but this is not what is meant here. We are looking for variations of the Einstein–Hilbert action with respect to the number of dimensions (e.g., [9]).

$$\sqrt{-G}\cdot R^* = \text{const},\tag{2}$$

or we even may demand:

$$\sqrt{-G}\cdot R^* = 0\tag{3}$$

with either $G = 0$ or $R^* = 0$, whereby the latter case definitively is less trivial and thus probably also makes more sense.

Please note that with $G_{\delta\gamma} = F[f]\cdot g_{\delta\gamma}$ we have used the simplest form of metric adaptation, which we could construct as a simplification from a generalization of the typical form of tensor transformations, namely:

$$G_{\alpha\beta} = F\Big[f\big[t,x,y,z,\ldots,\xi_k,\ldots,\xi_n\big]\Big]^{ij}_{\alpha\beta}\,g_{ij},\tag{4}$$

$$\rightarrow G_{\alpha\beta} = F\Big[f\big[t,x,y,z,\ldots,\xi_k,\ldots,\xi_n\big]\Big]\cdot\delta^i_\alpha\delta^j_\beta g_{ij}$$

$$\rightarrow G_{\alpha\beta} = F\Big[f\big[t,x,y,z,\ldots,\xi_k,\ldots,\xi_n\big]\Big]\cdot g_{\alpha\beta} = F[f]\cdot g_{\alpha\beta}.\tag{5}$$

The generalization is elaborated in [1].

It should be pointed out that there will always be a possibility to choose a $g_{\alpha\beta}$ allowing us for the simple transformation version (5) instead of (4). Thus, for the time being, we can reduce our considerations here to this simple option.

Now we evaluate the resulting Ricci scalar for the metric $G_{\alpha\beta}$ at first in n and then, subsequently, simplify to 4 dimensions (the complete derivation is given later in this book):

$$R^* = \frac{1}{F[f]^3}\cdot\left(\left(C_{N1}\cdot\left(\frac{\partial F[f]}{\partial f}\right)^2 - C_{N2}\cdot F[f]\cdot\frac{\partial^2 F[f]}{\partial f^2}\right)\cdot\overbrace{\left(\tilde{\nabla}_g f\right)^2}^{=f_{,\alpha}g^{\alpha\beta}f_{,\beta}} - C_{N2}\cdot F[f]\cdot\frac{\partial F[f]}{\partial f}\cdot\Delta_g f\right)$$

$$\xrightarrow{n=4} = \frac{1}{F[f]^3}\cdot\left(\left(\frac{3}{2}\cdot\left(\frac{\partial F[f]}{\partial f}\right)^2 - 3\cdot F[f]\cdot\frac{\partial^2 F[f]}{\partial f^2}\right)\cdot\left(\tilde{\nabla}_g f\right)^2 - 3\cdot F[f]\cdot\frac{\partial F[f]}{\partial f}\cdot\Delta_g f\right)$$

$$\xrightarrow{\frac{3}{2}\left(\frac{\partial F[f]}{\partial f}\right)^2 - 3\cdot F[f]\cdot\frac{\partial^2 F[f]}{\partial f^2}=0} = -\frac{3}{F[f]^2}\cdot\left(\frac{\partial F[f]}{\partial f}\cdot\Delta_g f\right).\tag{6}$$

With respect to the constants C_{N1} and C_{N2} in arbitrary n dimensions, the reader is referred to [1]. In chapter 8, we will derive these constants for arbitrary n-dimensional spaces and space-times.

One may assume that with $R^* = 0$ one would immediately satisfy the Einstein–Hilbert action. Taking the general n-dimensional case in the first line of (6) and always demanding that:

$$C_{N1} \cdot \left(\frac{\partial F[f]}{\partial f} \right)^2 - C_{N2} \cdot F[f] \cdot \frac{\partial^2 F[f]}{\partial f^2} = 0, \tag{7}$$

which simply is a fixing condition for the wrapper function $F[f]$ but leaves f completely arbitrary, and we find:

$$R^* = -\frac{C_{N2}}{F[f]^2} \cdot \left(\frac{\partial F[f]}{\partial f} \cdot \Delta_g f \right) = 0$$

$$\Rightarrow \qquad\qquad \Delta_g f = 0. \tag{8}$$

This is the Laplace equation, but here we are often referring to it as a "homogeneous" Klein–Gordon equation, respectively, the mass- and potential-less Klein–Gordon equation. The reason for this peculiar renaming becomes clear when we ask:

Where to get the mass from?

This question will be answered in the next sub-section 2.4.1, where we will show that a simple entanglement of dimensions leads to the necessary mass terms and in fact gives the usual, which is to say the classical, Klein–Gordon equation.

Before we go there, however, we should point out that by applying an inner function $f \rightarrow H(f, f_1)$, we should always be able to obtain a constant as follows:

$$R^* = R^*_{\text{const}} + 0, \tag{9}$$

with the function f then satisfying condition (6) in the simpler manner:

$$0 = \frac{1}{F[f]^3} \cdot \left(\begin{array}{c} \left(C_{N1} \cdot \left(\frac{\partial F[f]}{\partial f} \right)^2 - C_{N2} \cdot F[f] \cdot \frac{\partial^2 F[f]}{\partial f^2} \right) \cdot \overbrace{\left(\tilde{\nabla}_g f \right)^2}^{= f_{,\alpha} g^{\alpha\beta} f_{,\beta}} \\ -C_{N2} \cdot F[f] \cdot \frac{\partial F[f]}{\partial f} \cdot \Delta_g f \end{array} \right). \tag{10}$$

Things are immediately and more directly clear when choosing condition (7). In this case, we have from (6):

$$R^* = -\frac{C_{N2}}{F\left[H(f,f_1)\right]^2} \cdot \left(\frac{\partial F[f]}{\partial H(f,f_1)} \cdot \Delta_g H(f,f_1)\right)$$

$$\xrightarrow{H(f,f_1)=f+f_1} = -\frac{C_{N2}}{F\left[f+f_1\right]^2} \cdot \left(\frac{\partial F[f+f_1]}{\partial(f+f_1)} \cdot \Delta_g(f+f_1)\right)$$

$$= -\frac{C_{N2}}{F\left[f+f_1\right]^2} \cdot \frac{\partial F[f+f_1]}{\partial(f+f_1)} \cdot \left(\Delta_g f + \Delta_g f_1\right)$$

$$\xrightarrow{-\frac{C_{N2}}{F\left[f+f_1\right]^2} \cdot \frac{\partial F[f+f_1]}{\partial(f+f_1)} \cdot \Delta_g f_1 \equiv R^*_{\text{const}}} = -\frac{C_{N2}}{F\left[f+f_1\right]^2} \cdot \frac{\partial F[f+f_1]}{\partial(f+f_1)} \cdot \overbrace{\Delta_g f}^{=0} + R^*_{\text{const}}$$

$$\Rightarrow \qquad\qquad \Delta_g f = 0 \qquad\qquad\qquad (11)$$

and would be back with the mass-free Klein–Gordon equation.

As the classical Dirac equation is just the quaternion square root of the Klein–Gordon equation, we consider to have derived this, too, but it will be shown further below that there is a better way. In fact, there exists a truly metrical-based Dirac equation (see book chapters 9 and 10 in the part "The fifth day ..." and sub-section "2.5.2 Toward a Dirac Equation in the Metric Picture and Its Connection to the Classical Quaternion Form").

2.4.1 Entanglement as Origin of Mass (and Potential) and Thus Possibly Also Any Kind of Inertia and Interaction (Including Interaction in Socioeconomic Space-Times)

It was shown in [6–10] (or more comprehensively, see also [1] and chapter 6 of this book) that mass automatically results from certain forms of entanglements of dimensions. Here, we only give one example in 6 dimensions, but it should be pointed out that the production (creation) of mass via entanglement works in any number of dimensions. There are other forms of entanglement that cause mass to occur, but apparently also spin and more forms of inertia as matter and energy [1, 7, 10]. This statement also holds for the entanglement of more than just 2 dimensions (e.g., [10]). But as said, here we consider only the creation of mass in its simplest form, namely, via the entanglement of 2 dimensions.

The reader may ask why we need such a principle derivation of the origin of inertia if in the end we intend "only" to discuss the results with respect to a human society. Well, masses (or inertia in general), resulting from the entanglement of dimensions, can occur in any space-time (system) consisting of any kind of ensemble of dimensions/properties and as we know how important masses and inertia are in the "real" world, it is obvious that finding the driving forces for the creation of such inertia in any system is of great importance. How can we otherwise understand economic cycles without understanding what—within such cycles—the inertia and the driving potentials are? After all, cycles are oscillations, respectively, processes with periodic behavior and inertia are a key factor in such processes. **Even wars may just be seen as resonance catastrophes of such cyclic oscillations and in truly knowing how they can occur, one might also be able to actually hinder them to happen at all.**

Introducing a metric for the coordinates $t, r, \vartheta, \varphi, u, v$ of the form:

$$g_{\alpha\beta}^6 = \begin{pmatrix} -c^2 & 0 & 0 & 0 & 0 & 0 \\ 0 & 1 & 0 & 0 & 0 & 0 \\ 0 & 0 & r^2 & 0 & 0 & 0 \\ 0 & 0 & 0 & r^2 \cdot \sin[\vartheta]^2 & 0 & 0 \\ 0 & 0 & 0 & 0 & g[v] & 0 \\ 0 & 0 & 0 & 0 & 0 & g[v] \end{pmatrix} \cdot f[t,r,\vartheta,\varphi], \qquad (12)$$

with the entanglement solution for u and v as follows:

$$g[v] = \frac{C_{v1}^2}{C_{v0}^2 \cdot \left(1 + \cosh\left[C_{v1} \cdot \left(v + C_{v2}\right)\right]\right)}, \qquad (13)$$

gives us the following Ricci scalar from (6) with condition (7):

$$R^* = \frac{1}{F[f]^3} \cdot \left(-C_{N2} \cdot F[f] \cdot \frac{\partial F[f]}{\partial f} \cdot \Delta_g f\right); \quad F[f] = C_f + f[t,r,\vartheta,\varphi]$$

$$\xrightarrow{6D} R^* = \frac{5}{f^2} \cdot \left(\frac{C_{v0}^2}{5} \cdot f + \frac{\partial^2 f}{c^2 \cdot \partial t^2} - \Delta_{3D-sphere} f\right); \quad f = f[t,r,\vartheta,\varphi]; \quad C_f = 0.$$

$$(14)$$

Here $\Delta_{3D-sphere} f$ denotes the Laplace operator in spherical coordinates.

As for $R^* = 0$ and $\dfrac{C_{v0}^2}{5} = \dfrac{m^2 \cdot c^2}{\hbar^2}$ (m = mass, c = speed of light in vacuum, $\hbar$ = reduced Planck constant), we have obtained the classical quantum Klein–

Gordon equation with mass and therefore conclude that mass can obviously be constructed by a suitable set of additional (meaning in addition to our 4-dimensional space-time), entangled coordinates.

2.5   A Little Bit Proof and Reassurance

2.5.1   Deriving the Electromagnetic Field

In order to show that our approach (in fact it is not ours, but the approach of David Hilbert from over 100 years ago) represents a world formula, we need a convincing demonstrator. This was found in just deriving the electromagnetic interaction out of the equation for the constant or zero Ricci scalar [1]. For convenience, we repeat the evaluation in the appendix of this chapter.

But there is more proof and we want to give it here.

2.5.2   Toward a Dirac Equation in the Metric Picture and Its Connection to the Classical Quaternion Form

It should be pointed out that with the condition for f to be a scalar, there are still many more options to have vector and tensor functions included, namely, for example, via:

$$f = h + h_\alpha q^\alpha + h_{\alpha\beta} q^{\alpha\beta} + \ldots . \tag{15}$$

As $F[f]$ could still be freely chosen, we take the simple option that we seek for a solution for f within the minimum or maximum region of F, giving us $\dfrac{\partial F[f]}{\partial f} \approx 0$ and simplifying (6) to:

$$R^* = -\frac{C_{N2} \cdot \dfrac{\partial^2 F[f]}{\partial f^2} \cdot \overbrace{\left(\tilde{\nabla}_g f\right)^2}^{= f_{,\alpha} g^{\alpha\beta} f_{,\beta}}}{F[f]^2} \xrightarrow{f = h_\lambda q^\lambda} = -\frac{C_{N2} \cdot \dfrac{\partial^2 F[f]}{\partial f^2} \cdot \left(h_\lambda q^\lambda\right)_{,\alpha} g^{\alpha\beta} \left(h_\lambda q^\lambda\right)_{,\beta}}{F[f]^2}$$

$$\Rightarrow \qquad \frac{F[f]^2 \cdot R^*}{C_{N2} \cdot \dfrac{\partial^2 F[f]}{\partial f^2}} = -\left(h_\lambda q^\lambda\right)_{,\alpha} g^{\alpha\beta} \left(h_\lambda q^\lambda\right)_{,\beta}$$

$$= -\left(h_\lambda q^\lambda_{,\alpha} + h_{\lambda,\alpha} q^\lambda\right) g^{\alpha\beta} \left(h_\lambda q^\lambda_{,\beta} + h_{\lambda,\beta} q^\lambda\right)$$

$$= -h_\lambda q^\lambda{}_{,\alpha} g^{\alpha\beta} h_\lambda q^\lambda{}_{,\beta} - h_{\lambda,\alpha} q^\lambda g^{\alpha\beta} h_\lambda q^\lambda{}_{,\beta}$$
$$- h_\lambda q^\lambda{}_{,\alpha} g^{\alpha\beta} h_{\lambda,\beta} q^\lambda - h_{\lambda,\alpha} q^\lambda g^{\alpha\beta} h_{\lambda,\beta} q^\lambda. \tag{16}$$

As we haven't had any need to fix both vectors h_λ or q^λ yet, we assume that one is a vector of constants and by making q^λ this constant vector, we obtain:

$$\frac{F[f]^2 \cdot R^*}{C_{N2} \cdot \dfrac{\partial^2 F[f]}{\partial f^2}} = -h_{\lambda,\alpha} q^\lambda g^{\alpha\beta} h_{\lambda,\beta} q^\lambda \xrightarrow{g^{\alpha\beta} = \mathbf{e}^\alpha \cdot \mathbf{e}^\beta} = -h_{\lambda,\alpha} q^\lambda \mathbf{e}^\alpha \cdot \mathbf{e}^\beta h_{\lambda,\beta} q^\lambda$$

$$\xrightarrow{q^\lambda \mathbf{e}^\alpha = \mathbf{Q}^{\lambda\alpha}} = -h_{\lambda,\alpha} \mathbf{Q}^{\lambda\alpha} \cdot \mathbf{Q}^{\lambda\beta} h_{\lambda,\beta}. \tag{17}$$

Now we approximate the curvature term on the left-hand side as a square of the linear function of the scalar product f, and therefore we can write:

$$\frac{F[f]^2 \cdot R^*}{C_{N2} \cdot \dfrac{\partial^2 F[f]}{\partial f^2}} \simeq h_\lambda q^\lambda \mathbf{M} \cdot \mathbf{M} h_\lambda q^\lambda = -h_{\lambda,\alpha} q^\lambda g^{\alpha\beta} h_{\lambda,\beta} q^\lambda$$

$$\xrightarrow{g^{\alpha\beta} = \mathbf{e}^\alpha \cdot \mathbf{e}^\beta} = -h_{\lambda,\alpha} q^\lambda \mathbf{e}^\alpha \cdot \mathbf{e}^\beta h_{\lambda,\beta} q^\lambda$$

$$\Rightarrow \qquad \mp h_\lambda q^\lambda \mathbf{M} = i \cdot h_{\lambda,\alpha} q^\lambda \mathbf{e}^\alpha, \tag{18}$$

$$\Rightarrow \qquad 0 = i \cdot h_{\lambda,\alpha} \mathbf{e}^\alpha \pm h_\lambda \mathbf{M}. \tag{19}$$

There, when reducing our general and arbitrarily n-dimensional equation to 4 dimensions, we almost directly recognize the Dirac equation, only that in our—the metric—case the mass terms have to be at least a vector, while the Dirac matrices for the derivatives of the function vector $\mathbf{h}_\lambda$ (c.f. [11]) are becoming the metric base vector components. In order to obtain the classical Dirac equation in the natural units form:

$$0 = i \cdot \gamma^\alpha h_{\lambda,\alpha} \pm h_\lambda \cdot m, \tag{20}$$

which we can also write as:

$$0 = i \cdot \gamma^\alpha h_{\lambda,\alpha} \pm I \cdot h_\lambda \cdot m, \tag{21}$$

and with the Dirac gamma matrices γ^α and the unit matrix I given as follows:

$$\gamma^0 = \begin{pmatrix} 1 & & & \\ & 1 & & \\ & & -1 & \\ & & & -1 \end{pmatrix}; \quad \gamma^1 = \begin{pmatrix} & & & 1 \\ & & 1 & \\ & -1 & & \\ -1 & & & \end{pmatrix}$$

$$\gamma^2 = \begin{pmatrix} & & -i \\ & i & \\ & i & \\ -i & & \end{pmatrix} ; \quad \gamma^3 = \begin{pmatrix} & 1 & \\ & & -1 \\ -1 & & \\ & 1 & \end{pmatrix} ; \quad I = \begin{pmatrix} 1 & & \\ & 1 & \\ & & 1 \\ & & & 1 \end{pmatrix}, \quad (22)$$

we simply need to multiply (19) with a suitable object **C**. This gives us the identities[3]:

$$\mathbf{C} \cdot \mathbf{e}^\alpha = \gamma^\alpha ; \quad \mathbf{C} \cdot \mathbf{M} = I \cdot m \tag{23}$$

and provides a connection between the classical quaternion and the metric Dirac equation. As the classical Dirac equation was given in Minkowski coordinates, with base vectors of the kind:

$$\mathbf{e}^0 = \begin{pmatrix} i \\ 0 \\ 0 \\ 0 \end{pmatrix} ; \quad \mathbf{e}^1 = \begin{pmatrix} 0 \\ 1 \\ 0 \\ 0 \end{pmatrix} ; \quad \mathbf{e}^2 = \begin{pmatrix} 0 \\ 0 \\ 1 \\ 0 \end{pmatrix} ; \quad \mathbf{e}^3 = \begin{pmatrix} 0 \\ 0 \\ 0 \\ 1 \end{pmatrix}, \tag{24}$$

we could find the corresponding structure of the object **C** via the first equation in (23) and it becomes clear that we need something of the kind:

$$C^\alpha_\beta \cdot \mathbf{e}^\beta = \gamma^\alpha, \tag{25}$$

$$\Rightarrow \quad \left\{ \begin{aligned} C^0_\alpha &= \begin{pmatrix} -i & & \\ & 1 & \\ & & -1 \\ & & & -1 \end{pmatrix} ; \quad C^1_\alpha = \begin{pmatrix} & & & 1 \\ & & 1 & \\ & -1 & & \\ i & & & \end{pmatrix} ; \\ C^2_\alpha &= \begin{pmatrix} & & & -i \\ & & i & \\ & i & & \\ -1 & & & \end{pmatrix} ; \quad C^3_\alpha = \begin{pmatrix} & & 1 & \\ & & & -1 \\ i & & & \\ & 1 & & \end{pmatrix} . \end{aligned} \right. \tag{26}$$

---

[3]It should be noted, that classically one does not use the identity matrix for the second equation

in (23), but the so-called β-matrix instead. It reads $\beta = \begin{pmatrix} 1 & & \\ & 1 & \\ & & -1 \\ & & & -1 \end{pmatrix}$, but will not be

considered here, as it makes no difference for the results of this chapter whether we use I or β in (23).

From here we deduce that **M** must be something more complex than just a vector as only the following approach gives us what we need:

$$C_\beta^\alpha \cdot \mathbf{M} = I \cdot m \quad \Rightarrow \quad C_\beta^\alpha \cdot M_\lambda^\beta = \delta_\lambda^\alpha \cdot m = I \cdot m. \tag{27}$$

Thereby we need to point out that $h_\lambda q^\lambda \mathbf{M} \cdot \mathbf{M} h_\lambda q^\lambda$ in (18) only is a linear approximation of the complex expression $\dfrac{F[f]^2 \cdot R^*}{C_{N2} \cdot \dfrac{\partial^2 F[f]}{\partial f^2}}$ and that the classical

Dirac equation was just postulated [5, 11], while our new Dirac equation (19) has a fundamental metric and extremal principle origin. The result after solving (23) with respect to the object M_λ^β reads:

$$\Rightarrow \quad \left\{ \begin{array}{ll} M_0^\alpha = \begin{pmatrix} i & & & \\ & 1 & & \\ & & -1 & \\ & & & -1 \end{pmatrix}; & M_1^\alpha = \begin{pmatrix} & & & -i \\ & & -1 & \\ & 1 & & \\ 1 & & & \end{pmatrix} \\[30pt] M_2^\alpha = \begin{pmatrix} & & -1 & \\ & -i & & \\ & & & -i \\ i & & & \end{pmatrix}; & M_3^\alpha = \begin{pmatrix} & & & -i \\ & & 1 & \\ 1 & & & \\ & -1 & & \end{pmatrix} \end{array} \right. \tag{28}$$

and gives us the connection of the classical Dirac equation (21) from [11] with our new metric Dirac equation (18) in its linear approximated form (19). We realize that in the metric picture the corresponding inertia parameters are not just a scalar.

It should be pointed out that a more rigorous derivation of the Dirac equation from the Einstein–Hilbert action of a scaled metric is given in chapters 9 and 10 of this book.

2.5.3 About a Small Extension and Generalization

Please note that when demanding the possible solutions to our metric Dirac equation (18) to be a harmonic function, meaning that we always have:

$$\Delta f = \Delta(h + h_\alpha q^\alpha + h_{\alpha\beta} q^{\alpha\beta} + \cdots) = 0, \tag{29}$$

we would not need the condition $\dfrac{\partial F[f]}{\partial f} \simeq 0$ any longer and (6) simplifies to:

$$R^* = \frac{1}{F[f]^3} \cdot \left(\left(C_{N1} \cdot \left(\frac{\partial F[f]}{\partial f} \right)^2 - C_{N2} \cdot F[f] \cdot \frac{\partial^2 F[f]}{\partial f^2} \right) \cdot \overbrace{\left(\tilde{\nabla}_g f \right)^2}^{= f_{,\alpha} g^{\alpha\beta} f_{,\beta}} \right). \quad (30)$$

This generalizes and dis-approximates (18) to:

$$\frac{F[f]^3 \cdot R^*}{C_{N1} \cdot \left(\frac{\partial F[f]}{\partial f} \right)^2 - C_{N2} \cdot F[f] \cdot \frac{\partial^2 F[f]}{\partial f^2}} = h_{\lambda,\alpha} q^\lambda g^{\alpha\beta} h_{\lambda,\beta} q^\lambda$$

$$\xrightarrow{g^{\alpha\beta} = \mathbf{e}^\alpha \cdot \mathbf{e}^\beta} = h_{\lambda,\alpha} q^\lambda \mathbf{e}^\alpha \cdot \mathbf{e}^\beta h_{\lambda,\beta} q^\lambda$$

$$\xrightarrow{q^\lambda \mathbf{e}^\alpha = \mathbf{Q}^{\lambda\alpha}} = -h_{\lambda,\alpha} \mathbf{Q}^{\lambda\alpha} \cdot \mathbf{Q}^{\lambda\beta} h_{\lambda,\beta}. \quad (31)$$

Now we could just again linearly approximate the curvature or gravity term via:

$$\frac{F[f]^3 \cdot R^*}{C_{N1} \cdot \left(\frac{\partial F[f]}{\partial f} \right)^2 - C_{N2} \cdot F[f] \cdot \frac{\partial^2 F[f]}{\partial f^2}} \simeq h_\lambda q^\lambda \mathbf{M} \cdot \mathbf{M} h_\lambda q^\lambda = h_{\lambda,\alpha} q^\lambda g^{\alpha\beta} h_{\lambda,\beta} q^\lambda$$

$$\xrightarrow{g^{\alpha\beta} = \mathbf{e}^\alpha \cdot \mathbf{e}^\beta} = h_{\lambda,\alpha} q^\lambda \mathbf{e}^\alpha \cdot \mathbf{e}^\beta h_{\lambda,\beta} q^\lambda$$

$$\Rightarrow \qquad \mp h_\lambda q^\lambda \mathbf{M} = i \cdot h_{\lambda,\alpha} q^\lambda \mathbf{e}^\alpha$$

$$\Rightarrow \qquad 0 = i \cdot h_{\lambda,\alpha} \mathbf{e}^\alpha \pm h_\lambda \mathbf{M} \quad (32)$$

and regarding the connection to the classical Dirac equation [11], we would have the same calculation as before ... only that this time it is even more general.

2.5.4 Elastic Space-Times

Now we apply the following approach for the scalar f:

$$f = h^\alpha_{\ ,\beta} q^\beta_\alpha \quad (33)$$

and again demand f to be a Laplace function. This leads us—via Eq. (6)—to the following equation:

$$\frac{F[f]^3 \cdot R^*}{C_{N1} \cdot \left(\frac{\partial F[f]}{\partial f} \right)^2 - C_{N2} \cdot F[f] \cdot \frac{\partial^2 F[f]}{\partial f^2}} = \xrightarrow{f = h^\alpha_{\ ,\beta} q^\beta_\alpha} = \left(h^\lambda_{\ ,\kappa} q^\kappa_\lambda \right)_{,\alpha} g^{\alpha\beta} \left(h^\lambda_{\ ,\kappa} q^\kappa_\lambda \right)_{,\beta}$$

$\Rightarrow$

$$\frac{F[f]^3 \cdot R^*}{C_{N1} \cdot \left(\frac{\partial F[f]}{\partial f}\right)^2 - C_{N2} \cdot F[f] \cdot \frac{\partial^2 F[f]}{\partial f^2}}$$

$$= \left(h^\lambda{}_{,\kappa}q^\kappa_{\lambda,\alpha} + h_{\lambda,\kappa,\alpha}q^{\lambda\kappa}\right)g^{\alpha\beta}\left(h^\lambda{}_{,\kappa}q^\kappa_{\lambda,\beta} + h_{\lambda,\kappa,\beta}q^{\lambda\kappa}\right)$$

$$= h^\lambda{}_{,\kappa}q^\kappa_{\lambda,\alpha}g^{\alpha\beta}h^\lambda{}_{,\kappa}q^\kappa_{\lambda,\beta} + h^\lambda{}_{,\kappa}q^\kappa_{\lambda,\alpha}q^{\lambda\kappa}g^{\alpha\beta}h^\lambda{}_{,\kappa,\beta}q^\kappa_\lambda$$

$$+ h^\lambda{}_{,\kappa,\alpha}q^\kappa_\lambda g^{\alpha\beta}h^\lambda{}_{,\kappa}q^\kappa_{\lambda,\beta} + h^\lambda{}_{,\kappa,\alpha}q^\kappa_\lambda g^{\alpha\beta}h^\lambda{}_{,\kappa,\beta}q^\kappa_\lambda. \tag{34}$$

Assuming the object q^κ_λ just to consists of constants (c.f. sub-section above), we can dramatically simplify to:

$$\frac{F[f]^3 \cdot R^*}{C_{N1} \cdot \left(\frac{\partial F[f]}{\partial f}\right)^2 - C_{N2} \cdot F[f] \cdot \frac{\partial^2 F[f]}{\partial f^2}} = h^\lambda{}_{,\kappa,\alpha}q^\kappa_\lambda g^{\alpha\beta}h^\lambda{}_{,\kappa,\beta}q^\kappa_\lambda$$

$$\xrightarrow{g^{\alpha\beta} = e^\alpha \cdot e^\beta} = h^\lambda{}_{,\kappa,\alpha}q^\kappa_\lambda e^\alpha \cdot e^\beta h^\lambda{}_{,\kappa,\beta}q^\kappa_\lambda$$

$$\xrightarrow{q^{\lambda\kappa}e^\alpha = Q^{\lambda\kappa\alpha}} = h^\lambda{}_{,\kappa,\alpha}Q^{\kappa\alpha}_\lambda \cdot Q^{\kappa\beta}_\lambda h^\lambda{}_{,\kappa,\beta}. \tag{35}$$

We recognize the terms $h^\lambda{}_{,\kappa,\alpha}q^\kappa_\lambda e^\alpha, h^\lambda{}_{,\kappa,\alpha}Q^{\kappa\alpha}_\lambda$ as being also parts of the fundamental equations of elasticity [12] and remember that solutions to the case $R^* = 0$ can be found by the means of gradients and spin-like approaches [1] as follows:

$$h^j = G^j[x_k] = g^{jl}\partial_l G[x_k]; \quad \Delta G[x_k] = 0$$

$$G^j[x_k] = x_\xi \cdot g^{jl}\partial_l G[x_k] + \theta \cdot G[x_k]; \quad \xi \text{ any of } 0,1,\dots,n-1, \tag{36}$$

$$G^j[x_k] = \left\{ \dots, \overbrace{\frac{\partial G[x_k]}{\partial x_\zeta}}^{pos\ \xi}, \dots, -\overbrace{\frac{\partial G[x_k]}{\partial x_\xi}}^{pos\ \zeta}, \dots \right\}; \quad \forall(\dots,\dots) = 0$$

$$G^j[x_k] = \left\{ \dots, -\overbrace{\frac{\partial G[x_k]}{\partial x_\zeta}}^{pos\ \xi}, \dots, \overbrace{\frac{\partial G[x_k]}{\partial x_\xi}}^{pos\ \zeta}, \dots \right\}; \quad \forall(\dots,\dots) = 0. \tag{37}$$

Any combination of the basic solutions (36) and (37) is also a solution. We see that we can construct nontrivial solutions to (6) in the case of $R^* = 0$, which are automatically also solutions to the completely flat space case:

$$\frac{F[f]^3 \cdot R^*}{C_{N1} \cdot \left(\frac{\partial F[f]}{\partial f}\right)^2 - C_{N2} \cdot F[f] \cdot \frac{\partial^2 F[f]}{\partial f^2}} = 0 = h^\lambda_{,\kappa\alpha} q^\kappa_\lambda g^{\alpha\beta} h^\lambda_{,\kappa\beta} q^\kappa_\lambda$$

$$\xrightarrow{g^{\alpha\beta} = e^\alpha \cdot e^\beta} 0 = h^\lambda_{,\kappa\alpha} q^\kappa_\lambda e^\alpha \cdot e^\beta h^\lambda_{,\kappa\beta} q^\kappa_\lambda$$

$$\xrightarrow{q^{\lambda\kappa} e^\alpha = Q^{\lambda\kappa\alpha}} 0 = h^\lambda_{,\kappa\alpha} Q^{\kappa\alpha}_\lambda \cdot Q^{\kappa\beta}_\lambda h^\lambda_{,\kappa\beta}. \qquad (38)$$

Interestingly and in accordance with the classical Quantum Theory, we have to construct these solutions out of combinations of harmonic functions and their derivatives. Thereby, in order to satisfy (6) in its currently given form (35), we have to set $\theta = -1$ (c.f. [13] three-function approach, which here was extended to a *n*-function approach).

We may consider the second type of solutions (37) spin-like. More such forms can be found easily via the following recipe:

$$G^j[x_k] = \left\{ \ldots, A_1 \overbrace{\frac{\partial^2 G}{\partial\xi_2 \partial\xi_3}}^{pos\ \xi_1}, \ldots, A_2 \overbrace{\frac{\partial^2 G}{\partial\xi_1 \partial\xi_3}}^{pos\ \xi_2}, \ldots, A_3 \overbrace{\frac{\partial^2 G}{\partial\xi_1 \partial\xi_2}}^{pos\ \xi_3}, \ldots \right\}$$

$$G^j[x_k] = \left\{ \begin{array}{cc} \ldots, A_1 \overbrace{\frac{\partial^3 G}{\partial\xi_2 \partial\xi_3 \partial\xi_4}}^{pos\ \xi_1}, \ldots, A_2 \overbrace{\frac{\partial^3 G}{\partial\xi_1 \partial\xi_3 \partial\xi_4}}^{pos\ \xi_2}, \\[3ex] \ldots, A_3 \underbrace{\frac{\partial^3 G}{\partial\xi_1 \partial\xi_2 \partial\xi_4}}_{pos\ \xi_3}, \ldots, A_4 \underbrace{\frac{\partial^3 G}{\partial\xi_1 \partial\xi_2 \partial\xi_3}}_{pos\ \xi_4}, \ldots \end{array} \right\}$$

$$\ldots$$

$$\forall(\ldots,\ldots) = 0; \quad \sum_{\forall k} A_k = 0. \qquad (39)$$

In a Minkowski space-time with $c = 1$ and the coordinates t, x, y, and z, we could have the following of such solutions:

$$G^j[x_k] = \pm \left\{ \frac{\partial G}{\partial z}, 0, 0, -\frac{\partial G}{\partial t} \right\}; \quad G^j[x_k]$$

$$= \pm \left\{ \frac{\partial G}{\partial y}, 0, -\frac{\partial G}{\partial t}, 0 \right\}; \quad G^j[x_k] = \pm \left\{ \frac{\partial G}{\partial x}, -\frac{\partial G}{\partial t}, 0, 0 \right\}$$

$$G^j\left[x_k\right]=\pm\left\{0,\frac{\partial G}{\partial z},0,-\frac{\partial G}{\partial x}\right\}; \quad G^j\left[x_k\right]$$

$$=\pm\left\{0,\frac{\partial G}{\partial y},-\frac{\partial G}{\partial x},0\right\}; \quad G^j\left[x_k\right]=\pm\left\{0,0,\frac{\partial G}{\partial z},-\frac{\partial G}{\partial y}\right\}, \tag{40}$$

$$G^j\left[x_k\right]=\left\{A_1\frac{\partial^2 G}{\partial x\partial y},A_2\frac{\partial^2 G}{\partial t\partial y},A_3\frac{\partial^2 G}{\partial x\partial t},0\right\};$$

$$G^j\left[x_k\right]=\left\{A_1\frac{\partial^2 G}{\partial x\partial z},A_2\frac{\partial^2 G}{\partial t\partial z},0,A_3\frac{\partial^2 G}{\partial x\partial t}\right\}$$

$$G^j\left[x_k\right]=\left\{A_1\frac{\partial^2 G}{\partial z\partial y},0,A_2\frac{\partial^2 G}{\partial t\partial z},A_3\frac{\partial^2 G}{\partial z\partial t}\right\};$$

$$G^j\left[x_k\right]=\left\{0,A_1\frac{\partial^2 G}{\partial z\partial y},A_2\frac{\partial^2 G}{\partial x\partial z},A_3\frac{\partial^2 G}{\partial x\partial y}\right\}. \tag{41}$$

$$G^j\left[x_k\right]=\left\{A_1\frac{\partial^3 G}{\partial x\partial y\partial z},A_1\frac{\partial^3 G}{\partial t\partial y\partial z},A_1\frac{\partial^3 G}{\partial x\partial t\partial z},A_4\frac{\partial^3 G}{\partial x\partial y\partial t}\right\}; \quad \sum_{\forall k}A_k=0. \tag{42}$$

Thereby we find that the last solution in (41) with a setting of the kind $A_1=1/3$, $A_2=1/3$, and $A_3=-2/3$ shows a peculiar closeness to the charges of quarks, the building blocks of hadrons and baryons.

It should be noted that an approach of the form:

$$f=h^\lambda{}_{,\kappa}q^\kappa_\lambda+\theta\cdot h^\chi{}_{,\beta}g^{\beta\alpha}p_{\alpha\chi}, \tag{43}$$

with the object $p_{\alpha\chi}$ also only consisting of constants, automatically leads to the fundamental linear elastic equation in the isotropic case with the Poisson's ratio $\sigma\neq0$ [12, 13] and $\theta=-(1-2\cdot\sigma)$. Setting (43) into (6), thereby still assuming that f is a Laplace function, gives us:

$$\frac{F[f]^3\cdot R^*}{C_{N1}\cdot\left(\frac{\partial F[f]}{\partial f}\right)^2-C_{N2}\cdot F[f]\cdot\frac{\partial^2 F[f]}{\partial f^2}}$$

$$=\left(h^\lambda{}_{,\kappa\alpha}q^\kappa_\lambda+\theta\cdot\left(h^\lambda{}_{,\rho}g^{\rho\chi}\right)_{,\chi}\delta^\chi_\alpha p_{\lambda\chi}\right)g^{\alpha\beta}\left(h^\lambda{}_{,\kappa\beta}q^\kappa_\lambda+\theta\cdot\left(h^\lambda{}_{,\rho}g^{\rho\chi}\right)_{,\beta}\delta^\chi_\alpha p_{\lambda\chi}\right)$$

$$=\left(h^\lambda{}_{,\kappa\alpha}q^\kappa_\lambda+\theta\cdot p_{\lambda\alpha}\Delta h^\lambda\right)g^{\alpha\beta}\left(h^\lambda{}_{,\kappa\beta}q^\kappa_\lambda+\theta\cdot p_{\lambda\beta}\Delta h^\lambda\right)$$

$$\xrightarrow{g^{\alpha\beta}=e^\alpha\cdot e^\beta}=\left(h^\lambda{}_{,\kappa\alpha}q^\kappa_\lambda+\theta\cdot p_{\lambda\alpha}\Delta h^\lambda\right)e^\alpha\cdot e^\beta\left(h^\lambda{}_{,\kappa\beta}q^\kappa_\lambda+\theta\cdot p_{\lambda\beta}\Delta h^\lambda\right). \tag{44}$$

Any of the two factors made to zero would immediately fulfill the whole equation in the flat space case and it was shown in [14–17] that interesting elementary particle solutions can be derived this way.

The importance of the derivation of elastic equations out of our fundamental metric approach for the most neutral and deepest originated modeling and understanding also of human societies will become clear in the next section.

2.6   Modeling Socioeconomic Space-Times

Now, with some very major equations being traced back to the Einstein–Hilbert action, respectively the Ricci scalar and assuming that also a human society (or socioeconomic space-time) just follows the principal laws of nature, we may conclude that models describing societies should also be based on the metric equations presented and derived above (naturally, in potentially very high numbers of dimensions and with many centers of gravity and/or quantum centers, c.f. chapter 8).

The complexity of any such metrical very first principal socioeconomic system grows rapidly with its size (number of entities and their properties). In these cases, we may resort to the option of simplification via the introduction of an "elastic" society as already outlined in previous publications (e.g., [18]). The corresponding math has been given in a variety of publications, but with respect to a comprehensive introduction and for convenience, we suggest the section "14.1 What is Good?" and chapter "10 An Elastic World Formula" in [1].

But why do we think that—from all possible simplifications—the elastic approach should be a good starting point if it comes to describe human societies?

The answer to this question is almost obvious as we will see in the next sub-section.

2.6.1   The Space-Time of Socioeconomy

The attentive reader will note that the text in this sub-section is almost identical to an abstract this author entitled "The New Space-Time of Psychology," which was first published on www.worldformulaapps.com (see also [1]), and the attentive reader will also immediately understand why there is absolutely nothing to wonder about.

(a) It obviously is no accident that our notion of socioeconomic spaces or space-times (or just societies) and their aspects is expressible in terms that we usually relate to things like the stability and reliability of objects or—in other words—to materials science. Obviously, it is not just an association or a metaphorical description when, talking about societies, we speak about friction, stress and strain, social defects, mismatches, dislocations, or—rather general—elastic and inelastic deformations. Taking the many forms of socioeconomic dissonances alone and considering their deforming effects regarding various time and spatial levels and scales, we cannot ignore the striking similarity to the same or mirroring situations in the world of energy and matter. But can this really come as a surprise in a society made out of entities, which are themselves evolved from matter and energy ... and—what is more—are consisting of it?

So we ask:

Is it possible that there is a much deeper connection between the understanding of socioeconomy and a most general theory of matter than this similarity of words and expressions?

Before falling under the suspicion of esoteric waffle, the author wants to point out that—after all—the origin of all our social activities is based on the interaction of electromagnetism, while the latter—of course—is based on the fundamental laws of matter. Thus, even though we are here talking about the interaction of intelligent entities in the field of socioeconomy, there may well be/has to be such a deeper connection, because the evolution of bigger and bigger brains [19] has had its starting point within the same fundamental laws that are creating, forming, and permanently reshaping our matter surrounding. It is, therefore, most likely that the matter-based interactions, which bring about our psychological self (our feelings, self-awareness, consciousness, and our whole cognitive being) and the subsequent social interaction, are following the same rules. Then, however, it should also be possible to apply these rules to the field of socioeconomy.

(b) Assuming the existence of a true "Theory of Everything" [1, 20], one should, of course, be able to extract from this theory all physical interactions that are effectively creating the complex pattern of a dynamic society. Additionally, incorporating what was said under point A, it has to be possible to formulate a socioeconomic theory in consistent mathematical form (similarly to [21] in classical medicine or as in [22] in philosophy) and taking its fundamental origin, one might name it a quantum gravity socioeconomy.

(c) And how to incorporate the omnipresent substructuring and formation of social and economic clusters in such a theory? Well, here we have to understand that substructures (e.g., the family as the potentially smallest cell, groups, clusters of individuals of certain interests, and so on) are equivalent to the formation of gravity centers in a quantum gravity theory. It was shown in [1, 23, 24] that this automatically brings about the necessary internal interactions and thermodynamic effects. Just like the formation of matter, we thereby also need to understand the omnipresent Russian doll structure of such cluster effects, where smaller clusters form bigger ones, the bigger ones again can cluster together, and so on. Some boundaries can dissolve while principal ones have to remain intact as their dissolution would be equivalent to the biological cell death. Permanent formation and reformation of clusters provide one important aspect within the dynamic society and give it the character of a living "thing."

(d) The new approach should be seen as an essential part for the realization of a "Virtual Society" just like the holistic "Virtual Patient" concept in medicine as proposed by Leuenberger [25]. It is clear from the holistic and top-down character of this approach that the simulation and understanding of incidents like the recent COVID-19 or Coronavirus event would be very straightforward and also allows for the consideration of a wide range of uncertainties as these are automatically contained in a quantum gravity-based approach.

2.6.2 Photons

It can easily be shown that in the case of $R^* = 0$ and Cartesian coordinates, Eq. (6) can always be solved by a simple wave function of the kind:

$$f = C_f \cdot e^{\sum_{\forall j} c_j \cdot x_j} \ ; \quad \sum_{\forall j} c_j^2 = 0. \tag{45}$$

This requires at least one coordinate to become time-like. We consider these solutions as general photons or photonic fields and in order to allow for various energies, rewrite (45) as follows:

$$f = C_f \cdot e^{v \cdot \sum_{\forall j} c_j \cdot x_j} \; ; \; \sum_{\forall j} c_j^2 = 0, \tag{46}$$

with v denotes the frequency. After linearization of (6) via condition (7), we can even localize our plane wave solutions via superposition. This was evaluated and extensively discussed in [20, 26].

The important conclusions we extract from the simple solution (45) are as follows:

(a) A space being flat can develop photonic solutions of the kind (46).

(b) It automatically then also develops a time-coordinate (or several time-coordinates) due to the condition $\sum_{\forall j} c_j^2 = 0$.

(c) With the photonic solutions providing the space's background, the time-coordinate must be "felt" as omnipresent and general inside the corresponding space. Only spatial regions without any photonic background could be "time-less," which is to say free of a general time-coordinate.

(d) As demonstrated by the author, for example, in [1], such a photonic background also provides the necessary conditions for an omnipresent Heisenberg uncertainty principle (see section 3.7 and chapter 11 in [1]).

In other words: A flat space can have time, while a space with an omnipresent dominant time-coordinate must globally (here meaning on a cosmologic scale) be flat. A space with time must also sprout an omnipresent uncertainty principle.

2.7 Societons and Ecotons—The Photons of a Human Society

As said earlier in this chapter, we assume a society or socioeconomic space to obey the same principal laws as nature does, which simply follows from the fact that societies are a part of nature, respectively, its derivatives ... even if many human societies have a neck to destroy mother nature or consider themselves as "superior" to nature.

Assuming that a society cannot exist without time, at least not a living and evolving one, it is clear from the sections above that such societies must also have a photonic background. Of what nature this background is does not matter at the moment, but we know (c.f. section "2.2 A Most Fundamental Starting Point and How to Proceed from There") that any set of properties can form spaces in which the properties are the degrees of freedom and thus dimensions. With these properties forming the dimensions of the human socioeconomic spaces (societies), however, they should also be able (just as in the ordinary cosmologic space) to form the abovementioned photonic background, only that here the photons are not made up by space and time but by other "spatial" properties of the society. Thereby the math is by no means and at any kind restricted to the type of properties. It is for this reason that we want to distinguish the ordinary photons from the ones circulating inside a society by naming the latter "societons" and "ecotons," thereby pointing out that we have social and economic properties and thus also the corresponding socio- and economic photons and of course also combinations that we may name "socio-ecotons."

Now, from the sections above we can also extract the following:

When forming this photonic background, automatically at least one such property becomes time-like.

AND: When forming this photonic background, automatically a fundamental uncertainty principle, being of global scale to the whole society (respectively the part in which the corresponding photonic background is active), appears and no observable (measurable parameter within the society) can be given with an arbitrary certainty or precision. Meaning: the certainty in which any statement, regarding the society, can be made is limited by the omnipresent uncertainty and thus the society's own photonic background. This holds for social and economical statements alike and it is clear that this finding has a great effect to all aspects having to do with socioeconomic predictions and decision-making.

Thus, understanding and controlling the photonic background results in understanding and controlling the time-vectors and the uncertainties of a society.

2.7.1 How to Prove the Existence of Societons and Ecotons and Measure the Strength of the Effect?

The photonic background field can be revealed by the famous Casimir experiment [31] and thus, we think, that just something similar should be designed for societies.

2.8 Appendix to "Societons and Ecotons"

In order to make things convenient for the reader, we here repeat all equations being of need, even though some may already have been brought up earlier in the chapter.

Our starting point shall be the classical Einstein–Hilbert action [2, 3]:

$$\delta_g W = 0 = \delta_g \int_V d^n x \left(\sqrt{-g} \left[R - 2\kappa L_M - 2\Lambda \right] \right). \tag{47}$$

Thereby g denotes the determinant of the metric tensor, W and V are giving the action and the volume of the n-dimensional space, respectively, R is denoting the scalar curvature or Ricci scalar, and $f[R]$ shall be an arbitrary function of R in case we intend to consider the following generalization of (47):

$$\delta_g W = 0 = \delta_g \int_V d^n x \left(\sqrt{-g} \left[f(R) - 2\kappa L_M - 2\Lambda \right] \right), \tag{48}$$

and the last term $2\kappa L_M$ describes the Lagrange density of matter.

Over one hundred years ago, Einstein [3] and Hilbert [2] have shown that, as a result of the variation (47), a space of a given metric $g^{\alpha\beta}$ must curve and that the curvature described by the Ricci tensor $R^{\alpha\beta}$ must satisfy certain conditions. The **classical** result for the metric tensor $g^{\alpha\beta}$ of the curved space can be given as follows:

$$R^{\alpha\beta} - \frac{1}{2} R g^{\alpha\beta} + \Lambda g^{\alpha\beta} = -\kappa T^{\alpha\beta}. \tag{49}$$

Here, we have $R^{\alpha\beta}$, $T^{\alpha\beta}$ the Ricci and the energy momentum tensor, respectively, while the parameters Λ and κ are constants (usually called cosmological and coupling constant, respectively). These are the well-known Einstein field equations in n dimensions with the indices α and β running from 1 to n. The theory behind is called "General Theory of Relativity."

But even though most people connect the "General Theory of Relativity" only with big scales like cosmology, galaxies, planetary movements, and black holes, we will demonstrate that apparently also all other aspects and laws of natural science actually reside within the action given above, be it in the classical linear form (47) or the generalization (48).

2.8.1 Derivation of Electromagnetic Interaction (and Matter) via a Set of Creative Transformations

In order to allow for an easy entrance into the theoretical apparatus, we shall start with the Minkowski flat space metric tensor in Cartesian coordinates t, x, y, z as follows:

$$g_{\alpha\beta}^{\text{flat}} = \begin{pmatrix} -c^2 & 0 & 0 & 0 \\ 0 & 1 & 0 & 0 \\ 0 & 0 & 1 & 0 \\ 0 & 0 & 0 & 1 \end{pmatrix}. \tag{50}$$

This metric solves the Einstein field equations (49) (see [3]) and corresponds to a Ricci scalar of $R = 0$. Now we introduce the following special transformation:

$$G_{\alpha\beta} = g_{\alpha\beta}^{\text{flat}} \cdot F\big[f[t,x,y,z]\big] = \begin{pmatrix} -c^2 & 0 & 0 & 0 \\ 0 & 1 & 0 & 0 \\ 0 & 0 & 1 & 0 \\ 0 & 0 & 0 & 1 \end{pmatrix} \cdot F\big[f[t,x,y,z]\big] \tag{51}$$

and evaluate the resulting new Ricci scalar R^*:

$$R^* = \frac{R}{F[f]} + \left(\begin{array}{c} -\Gamma_{\sigma\alpha}^{\mu}\Gamma_{\beta\mu}^{**\sigma} + \Gamma_{\alpha\beta}^{\sigma}\Gamma_{\sigma\mu}^{**\mu} - \Gamma_{\sigma\alpha}^{**\mu}\Gamma_{\beta\mu}^{\sigma} + \Gamma_{\alpha\beta}^{**\sigma}\Gamma_{\sigma\mu}^{\mu} \\ +\Gamma_{\alpha\beta,\sigma}^{**\sigma} - \Gamma_{\beta\sigma,\alpha}^{**\sigma} - \Gamma_{\sigma\alpha}^{**\mu}\Gamma_{\beta\mu}^{**\sigma} + \Gamma_{\alpha\beta}^{**\sigma}\Gamma_{\sigma\mu}^{**\mu} \end{array} \right) \frac{g^{\alpha\beta}}{F[f]}$$

$$\xrightarrow{R=0;\, g_{\alpha\beta}^{\text{flat}}} = \frac{1}{F[f]^3} \cdot \left(\begin{array}{c} \left(C_{N1} \cdot \left(\frac{\partial F[f]}{\partial f}\right)^2 - C_{N2} \cdot F[f] \cdot \frac{\partial^2 F[f]}{\partial f^2} \right) \cdot \left(\tilde{\nabla}_g f\right)^2 \\ -C_{N2} \cdot F[f] \cdot \frac{\partial F[f]}{\partial f} \cdot \Delta_g f \end{array} \right)$$

$$= \frac{1}{F[f]^3} \cdot \left(\left(\frac{3}{2} \cdot \left(\frac{\partial F[f]}{\partial f}\right)^2 - 3 \cdot F[f] \cdot \frac{\partial^2 F[f]}{\partial f^2} \right) \cdot \left(\tilde{\nabla}_g f\right)^2 - 3 \cdot F[f] \cdot \frac{\partial F[f]}{\partial f} \cdot \Delta_g f \right)$$

$$= \frac{3}{2 \cdot F[f]^3} \cdot \left(\begin{array}{c} \left(\left(\frac{\partial F[f]}{\partial f}\right)^2 - 2 \cdot F[f] \cdot \frac{\partial^2 F[f]}{\partial f^2} \right) \cdot \left[-\frac{(\partial_t f)^2}{c^2} + (\partial_x f)^2 + (\partial_y f)^2 + (\partial_z f)^2 \right] \\ -2 \cdot F[f] \cdot \frac{\partial F[f]}{\partial f} \cdot \left[-\frac{\partial_t^2}{c^2} + \partial_x^2 + \partial_y^2 + \partial_z^2 \right] f \end{array} \right). \tag{52}$$

Thereby we have used:

$$\Gamma^{*\gamma}_{\alpha\beta} = \frac{g^{\gamma\sigma}}{2\cdot F[f]}\left(\left[F[f]\cdot g_{\sigma\alpha}\right]_{,\beta} + \left[F[f]\cdot g_{\sigma\beta}\right]_{,\alpha} - \left[F[f]\cdot g_{\alpha\beta}\right]_{,\sigma}\right)$$

$$= \frac{g^{\gamma\sigma}}{2\cdot F[f]}\left(\left[F[f]\cdot g_{\sigma\alpha}\right]_{,\beta} + \left[F[f]\cdot g_{\sigma\beta}\right]_{,\alpha} - \left[F[f]\cdot g_{\alpha\beta}\right]_{,\sigma}\right)$$

$$= \frac{g^{\gamma\sigma}}{2}(g_{\sigma\alpha,\beta} + g_{\sigma\beta,\alpha} - g_{\alpha\beta,\sigma}) + \frac{g^{\gamma\sigma}}{2\cdot F[f]}\left(F[f]_{,\beta}\cdot g_{\sigma\alpha} + F[f]_{,\alpha}\cdot g_{\sigma\beta} - F[f]_{,\sigma}\cdot g_{\alpha\beta}\right)$$

$$= \Gamma^{\gamma}_{\alpha\beta} + \frac{g^{\gamma\sigma}}{2\cdot F[f]}\left(F[f]_{,\beta}\cdot g_{\sigma\alpha} + F[f]_{,\alpha}\cdot g_{\sigma\beta} - F[f]_{,\sigma}\cdot g_{\alpha\beta}\right)$$

$$\equiv \Gamma^{\gamma}_{\alpha\beta} + \Gamma^{**\gamma}_{\alpha\beta}. \tag{53}$$

The reader will find a more comprehensive elaboration about the motivation of our special transformation in the book [1] in section "3.3 The Ricci Scalar Quantization." However, it is quite entertaining to follow this simple trial and observe its amazing evolution into something rather unexpected (at least if taking its origin as a metric solution to the Einstein field equations) out of our "wrapper-transformation" (51).

The symbol $\tilde{\nabla}_g$ in (52) denotes a first-order differential operator similar to the Nabla operator in the metric $g_{\alpha\beta}$. The symbols C_{Ni} are standing for constants, which depend only on the number of dimensions. Their derivation will be presented in chapter 8 of this book. Please note that the transformation (5) is just the simplest form of a general approach like:

$$G_{\alpha\beta} = F\left[f[t,x,y,z]\right]^{ij}_{\alpha\beta} g_{ij} \rightarrow G_{\alpha\beta} = F\left[f[t,x,y,z]\right]\cdot\delta^i_\alpha\delta^j_\beta g_{ij}. \tag{54}$$

Without loss of generality, we can now demand that the first term in parenthesis in (52) would be zero, which is to say we choose our arbitrary function F such that we have:

$$C_{N1}\cdot\left(\frac{\partial F[f]}{\partial f}\right)^2 - C_{N2}\cdot F[f]\cdot\frac{\partial^2 F[f]}{\partial f^2} = 0. \tag{55}$$

For instance, in 4 dimensions this would always be the case for $F[f] = f^2$, giving us the Klein–Gordon-like equation:

$$\left[\Delta_g + \frac{R^*\cdot f^4}{f\cdot C_{N2}\cdot 2\cdot f}\right]f = \left[\Delta_g + \frac{R^*\cdot f^2}{2}\right]f = 0; \quad \Delta_{g-\text{Coordinates}} = \Delta_g. \tag{56}$$

Setting the solution $F[f] = f^2$ into our transformation starting point (51), however, one would directly obtain an all-scale quantum dominated metric solution for $G_{\alpha\beta}$. This is in total contrast to the daily experiences that quantum effects are only important in smaller scales. In order to overcome this problem, we evaluate the general solution of condition (55) in 4 dimensions and find:

$$F[f] = C_1 \cdot f + \frac{C_1^2 \cdot f^2}{4 \cdot C_2} + C_2. \tag{57}$$

Therefore, the constants C_i are arbitrary.

For entertainment and further motivation, it should be pointed out here that the functional wrapper $F[f]$ from (57) could also be adapted as follows:

$$F[f] = C_1 \cdot f^2 + \frac{C_1^2 \cdot f^4}{4 \cdot C_2} + C_2. \tag{58}$$

With arbitrary constants C_i, this assures the appearance of linear Laplace operator terms according to the condition (55) for the resulting Ricci scalar of the transformed metric in (6) as follows:

$$R^* = \frac{1}{F[f]^3} \cdot \left(-C_{N2} \cdot F[f] \cdot \frac{\partial F[f]}{\partial f} \cdot \Delta_g f^2 \right) \xrightarrow{4D} R^* \cdot \frac{\left(2 \cdot C_2 + C_1 \cdot f^2 \right)^3}{8 \cdot C_2 \cdot C_1} = -3 \cdot \Delta_g f^2. \tag{59}$$

Now we assume f to be a constant, which automatically leads to the simple equation:

$$R^* \cdot \frac{\left(2 \cdot C_2 + C_1 \cdot f^2 \right)^3}{8 \cdot C_2 \cdot C_1} = 0. \tag{60}$$

Also, assuming that the Ricci scalar curvature R^* shall be proportional to f^n, with an arbitrary exponent $n \geq 0$, we obtain the familiar trivial solution of $f_0 = 0$. However, from (60) we also obtain the nontrivial ground states:

$$f^2 = -\frac{2 \cdot C_2}{C_1}. \tag{61}$$

As the constants C_i are arbitrary, we could simply set them as follows:

$$2 \cdot C_2 = -\mu^2; \quad C_1 = 2 \cdot \lambda. \tag{62}$$

This gives us the additional ground state solutions directly in the classical Higgs field style [27], namely:

$$\left(f_{1,2}\right)^2 = \frac{\mu^2}{2\cdot\lambda} \quad \Rightarrow \quad f_{1,2} = \pm\frac{\mu}{\sqrt{2\cdot\lambda}}. \tag{63}$$

The measured value for the $f_{1,2}$ is known to be [28]:

$$\left|f_{1,2}\right| = \frac{\mu}{\sqrt{2\cdot\lambda}} = \frac{246\,\text{GeV}}{\sqrt{2}\cdot c^2}. \tag{64}$$

Rewriting (60) with the use of (62) as:

$$R^* \cdot \frac{\left(-\mu^2 + 2\cdot\lambda\cdot f^2\right)^3}{8\cdot\mu^2\cdot 2\cdot\lambda} = f^n \cdot \frac{\left(2\cdot\lambda\cdot f^2 - \mu^2\right)^3}{16\cdot\mu^2\cdot\lambda} = 3\cdot\Delta_g f^2 \tag{65}$$

gives us the total Higgs–Ricci-curvature connection. It also, automatically, gives us a curvature value R^* at the ground state, which is:

$$R^* \sim \left|f_{1,2}\right|^n = \left(\frac{\mu}{\sqrt{2\cdot\lambda}}\right)^n = \left(\frac{246\,\text{GeV}}{\sqrt{2}\cdot c^2}\right)^n. \tag{66}$$

Now, with the Higgs mass m_H known to be 125 GeV/c^2 and the fact that we have [28]:

$$m_H = \sqrt{2\cdot\mu^2} = \sqrt{4\cdot\lambda\cdot\left(f_{1,2}\right)^2}, \tag{67}$$

we can obtain:

$$\lambda \approx 0.13; \quad \mu \approx 88.8\,\text{GeV}/c^2. \tag{68}$$

Please note that we obtain the simple case $F[f] = f^2$ with the condition:

$$C_1^2 = 2\cdot C_2 \tag{69}$$

and the subsequent limit procedure $C_2 \to 0$ as follows:

$$\Rightarrow \qquad F[f] = \sqrt{2\cdot C_2}\cdot f + f^2 + C_2 \xrightarrow[C_2\to 0]{\lim} F[f] = f^2. \tag{70}$$

Using (57) instead of $F = f^2$ in (6) gives us:

$$R^* = \frac{1}{F[f]^3}\cdot\left(\begin{array}{c}\left(\left(C_{N1}\cdot\left(\frac{\partial F[f]}{\partial f}\right)^2 - C_{N2}\cdot F[f]\cdot\frac{\partial^2 F[f]}{\partial f^2}\right)\cdot\left(\tilde{\nabla}_g f\right)^2\right)\\[2mm] -C_{N2}\cdot F[f]\cdot\frac{\partial F[f]}{\partial f}\cdot\Delta_g f\end{array}\right)$$

$$= -C_{N2}\cdot\frac{\frac{\partial F[f]}{\partial f}\cdot\Delta_g f}{F[f]^2}\xrightarrow{4D} = -3\cdot\frac{\left(C_1 + \frac{C_1^2\cdot f}{2\cdot C_2}\right)}{\left(C_1\cdot f + \frac{C_1^2\cdot f^2}{4\cdot C_2} + C_2\right)^2}\cdot\Delta_g f. \tag{71}$$

Immediately we see that in the many cases with $R^* = 0$, we would obtain classical quantum Klein–Gordon equations. Later in this book we will learn that the condition $R^* = 0$ also directly follows from the Hamilton minimum principle and is not just a simple definition. Taylor expansion for small f (which is to say at $f = 0$) for the general case in the second line in (71) also gives us equations leading to these classical equations. For instance, if only considering up to linear f as follows:

$$\frac{R^*}{3} \cdot \frac{\left(C_1 \cdot f + \frac{C_1^2 \cdot f^2}{4 \cdot C_2} + C_2 \right)^2}{C_1 + \frac{C_1^2 \cdot f}{2 \cdot C_2}} + \Delta_g f = 0 \xrightarrow{\text{small } f} \frac{R^*}{3} \cdot \left(\frac{C_2^2}{C_1} + \frac{3}{2} \cdot C_2 \cdot f + \dots \right) + \Delta_g f = 0,$$

$$(72)$$

and assuming constant R^*, we already have the typical structure of the Klein–Gordon equation plus a constant term $\dfrac{R^*}{3} \cdot \dfrac{C_2^2}{C_1}$. Such a constant would still not influence the principal structural character of our resulting quantum equations because a simple transformation of:

$$f^* = \frac{C_2^2}{C_1} + \frac{3}{2} \cdot C_2 \cdot f; \quad \Delta_g f^* = \Delta_g f \tag{73}$$

does give us back the ordinary Klein–Gordon equation with:

$$\frac{R^*}{3} \cdot f^* + \Delta_g f^* = 0. \tag{74}$$

The resulting transformation in 4 dimensions would give the following quantum metric:

$$G_{\alpha\beta} = g_{\alpha\beta} \cdot F[f] = g_{\alpha\beta} \cdot \left(C_1 \cdot f + \frac{C_1^2 \cdot f^2}{4 \cdot C_2} + C_2 \right). \tag{75}$$

This clearly allows for the classical physics or Einstein situation with quantum effects becoming dominant only at small scales and thus a dominating part C_2 at bigger scales. In the result, one might separate as follows in 4 dimensions:

$$G_{\alpha\beta} = \overbrace{g_{\alpha\beta} \cdot C_2}^{\text{class. GTR}} + \overbrace{g_{\alpha\beta} \cdot \left(C_1 \cdot f \left(1 + \frac{C_1 \cdot f}{4 \cdot C_2} \right) \right)}^{\substack{\text{Quantum Gravity in } 4D \\ \text{quantum metric}}}. \tag{76}$$

Thereby it does not come as a surprise that we have already found so many cases where f has to be determined by classical quantum equations (see [1] and references therein), respectively where our equations give similar or even equal solutions. Normalization of the quantum gravity metric $G_{\alpha\beta}$ requires division by C_2 and gives:

$$
G_{\alpha\beta}^{norm} = \frac{G_{\alpha\beta}}{C_2} = \overbrace{g_{\alpha\beta}}^{\text{class. GTR}} + \overbrace{g_{\alpha\beta} \cdot \left(\frac{C_1}{C_2} \cdot f \cdot \left(1 + \frac{C_1 \cdot f}{4 \cdot C_2} \right) \right)}^{\text{quantum metric}}.
\tag{77}
$$

Quantum Gravity in 4D

Now we apply the recipe outlined above to our simple flat space example in order to see how the transformation can bring about matter.

We saw that by taking the metric (5), using rule (55) with the subsequent solution (57), we obtain the new Ricci scalar curvature:

$$
R^* = -3 \cdot \frac{\left(C_1 + \dfrac{C_1^2 \cdot f}{2 \cdot C_2} \right)}{\left(C_1 \cdot f + \dfrac{C_1^2 \cdot f^2}{4 \cdot C_2} + C_2 \right)^2} \cdot \Delta_g f.
\tag{78}
$$

Now we demand that the Ricci scalar should be a conserved quantity and as we have $R = 0$ with the flat space metric (50), we shall also demand $R^* = 0$. This gives us the simplest equation:

$$
0 = \Delta_g f = \left[\partial_x^2 + \partial_y^2 + \partial_z^2 - \frac{\partial_t^2}{c^2} \right] f.
\tag{79}
$$

Just for an increase of generality, we extend (50) as follows (A, B, D are constants):

$$
g_{\alpha\beta}^{flat} = \begin{pmatrix} -c^2 & 0 & 0 & 0 \\ 0 & A^2 & 0 & 0 \\ 0 & 0 & B^2 & 0 \\ 0 & 0 & 0 & D^2 \end{pmatrix},
\tag{80}
$$

which makes (79) to:

$$
0 = \Delta_g f = \left[\frac{\partial_x^2}{A^2} + \frac{\partial_y^2}{B^2} + \frac{\partial_z^2}{D^2} - \frac{\partial_t^2}{c^2} \right] f.
\tag{81}
$$

Applying the separation approach $f[t, x, y, z] = T[t]*X[x]*Y[y]*Z[z]$ gives us the solutions:

$$T[t] = C_{t1} \cdot \cos\left[c \cdot C_t \cdot t\right] + C_{t2} \cdot \sin\left[c \cdot C_t \cdot t\right],$$ (82)

$$X[x] = C_{x1} \cdot \cos\left[A \cdot C_x \cdot x\right] + C_{x2} \cdot \sin\left[A \cdot C_x \cdot x\right],$$ (83)

$$Y[y] = C_{y1} \cdot \cos\left[B \cdot C_y \cdot y\right] + C_{y2} \cdot \sin\left[B \cdot C_y \cdot y\right],$$ (84)

$$Z[z] = C_{z1} \cdot \cos\left[D \cdot C_z \cdot z\right] + C_{z2} \cdot \sin\left[D \cdot C_z \cdot z\right]$$ (85)

and the following characteristic equation:

$$0 = C_x^2 + C_y^2 + C_z^2 - C_t^2.$$ (86)

The matter coded with this solution for R^* can now directly be obtained via the Einstein field equations with matter, which reads:

$$R^{*\alpha\beta} - \frac{1}{2} R^* \cdot g^{\alpha\beta} + \Lambda \cdot g^{\alpha\beta} = -\kappa \cdot T^{\alpha\beta}.$$ (87)

As we had per demand $R^* = R = 0$ and as we also assume the cosmological constant to be equal to zero, we obtain the following identity:

$$R^{*\alpha\beta} = -\kappa \cdot T^{\alpha\beta}.$$ (88)

Thus, in order to find out what kind of matter our transformation:

$$G_{\alpha\beta} = g_{\alpha\beta}^{\text{flat}} \cdot F\left[f[t,x,y,z]\right] = \begin{pmatrix} -c^2 & 0 & 0 & 0 \\ 0 & A^2 & 0 & 0 \\ 0 & 0 & B^2 & 0 \\ 0 & 0 & 0 & D^2 \end{pmatrix} \cdot F\left[f[t,x,y,z]\right]$$ (89)

has created, we simply need to evaluate the corresponding Ricci tensor of the transformed metric $G_{\alpha\beta}$. In order to keep the presentation general[4] and as the evaluation of the derivatives is simple, we give the Ricci tensor with the function f. But for simplicity we simplified $F[f]$ to $F[f] = (f + C_1)^2$:

$$R^{*00} = \frac{(A \cdot B \cdot c)^2 \left(f_{,z}\right)^2 + D^2 \left(A^2 c^2 \left(f_{,y}\right)^2 + B^2 \left(\begin{array}{c} c^2 \left(f_{,x}\right)^2 + 3A^2 \left(f_{,t}\right)^2 \\ -2A^2 \left(C_1 + f\right) f_{,t,t} \end{array} \right) \right)}{(A \cdot B \cdot D)^2 c^4 \left(C_1 + f\right)^6}.$$

(90)

---

[4]This comes in handy the moment we want to exploit the additive character of our metric solutions. Thereby the additivity is assured by the means of condition (55). See further below in the main text!

$$R^{*11} = \frac{-(A \cdot B \cdot c)^2 (f_{,z})^2 + D^2 \left(-A^2 c^2 (f_{,y})^2 + B^2 \left(\begin{array}{c} 3c^2 (f_{,x})^2 \\ -2c^2 (C_1 + f) f_{,x,x} \end{array} \right) + A^2 (f_{,t})^2 \right)}{(c \cdot B \cdot D)^2 A^4 (C_1 + f)^6}.$$

(91)

$$R^{*22} = \frac{-(A \cdot B \cdot c)^2 (f_{,z})^2 + D^2 \left(\begin{array}{c} 3A^2 c^2 (f_{,y})^2 - 2A^2 c^2 (C_1 + f) f_{,y,y} \\ -B^2 c^2 (f_{,x})^2 + B^2 A^2 (f_{,t})^2 \end{array} \right)}{(c \cdot A \cdot D)^2 B^4 (C_1 + f)^6}.$$

(92)

$$R^{*33} = \frac{3(A \cdot B \cdot c)^2 (f_{,z})^2 + D^2 \left(\begin{array}{c} -A^2 c^2 (f_{,y})^2 - B^2 c^2 (f_{,x})^2 + B^2 A^2 (f_{,t})^2 \\ +2(C_1 + f) \left(A^2 c^2 f_{,y,y} + B^2 \left(c^2 f_{,x,x} - A^2 f_{,t,t} \right) \right) \end{array} \right)}{(c \cdot A \cdot D)^2 B^4 (C_1 + f)^6}.$$

(93)

$$R^{*01} = \frac{2(-2 f_{,x} f_{,t} + (C_1 + f) f_{,t,x})}{A^2 c^2 (C_1 + f)^6}; \quad R^{*02} = \frac{2(-2 f_{,y} f_{,t} + (C_1 + f) f_{,t,y})}{B^2 c^2 (C_1 + f)^6}$$

$$R^{*03} = \frac{2(-2 f_{,z} f_{,t} + (C_1 + f) f_{,t,z})}{D^2 c^2 (C_1 + f)^6}; \quad R^{*12} = \frac{2(2 f_{,y} f_{,x} - (C_1 + f) f_{,x,y})}{A^2 B^2 (C_1 + f)^6}$$

$$R^{*13} = \frac{2(2 f_{,z} f_{,x} - (C_1 + f) f_{,x,z})}{A^2 D^2 (C_1 + f)^6}; \quad R^{*23} = \frac{2(2 f_{,y} f_{,z} - (C_1 + f) f_{,z,y})}{B^2 D^2 (C_1 + f)^6}. \quad (94)$$

Previously we had discussed corresponding solutions in spherical [29] and cylindrical [30] geometries and it is clear that we have obtained photons or photonic matter forms. What is more, with condition (55) we have automatically achieved additive character for all our solutions (82) to (86), allowing to apply simple integral transform methods in order to construct almost arbitrary photonic matter forms.

Without applying the technique of superposition of our solution (82) to (86), however, the subsequent photonic solution would be a plane wave structure with no changes along the x-axis. It was shown in [26] how such a structure can be localized also in lateral directions via standard integral transform methods and that the subsequent solutions do fulfill the Maxwell equations. The connection with the energy momentum tensor for the electromagnetic field is therefore achieved via (88). The electrostatic and magnetic vector fields **E** and **B** with components E_i and B_i are forming the energy momentum tensor in contravariant form as follows:

$$\left(T^{\alpha\beta}\right) = \begin{pmatrix} \dfrac{1}{2} \cdot (\mathbf{E} \cdot \mathbf{E} + \mathbf{B} \cdot \mathbf{B}) & (\mathbf{E} \times \mathbf{B})^T \\ \mathbf{E} \times \mathbf{B} & \dfrac{1}{2} \cdot (\mathbf{E} \cdot \mathbf{E} + \mathbf{B} \cdot \mathbf{B}) \cdot \delta_{ik} - E_i E_k - B_i B_k \end{pmatrix}$$

$$\delta_{ik} = \begin{pmatrix} 1 & 0 & 0 \\ 0 & 1 & 0 \\ 0 & 0 & 1 \end{pmatrix}. \tag{95}$$

Please note that in SI units this tensor reads:

$$\left(T^{\alpha\beta}\right) = \begin{pmatrix} \dfrac{\varepsilon_0 \cdot \mathbf{E} \cdot \mathbf{E} + \dfrac{\mathbf{B} \cdot \mathbf{B}}{\mu_0}}{2} & c \cdot \varepsilon_0 \cdot (\mathbf{E} \times \mathbf{B})^T \\ c \cdot \varepsilon_0 \cdot \mathbf{E} \times \mathbf{B} & \dfrac{\varepsilon_0 \cdot \mathbf{E} \cdot \mathbf{E} + \dfrac{\mathbf{B} \cdot \mathbf{B}}{\mu_0}}{2} \cdot \delta_{ik} - \varepsilon_0 \cdot E_i E_k - \dfrac{B_i B_k}{\mu_0} \end{pmatrix}. \tag{96}$$

The constants ε_0, μ_0 are the electric and the magnetic field constants, respectively.

For brevity, we shall leave it to the mathematically skilled and interested reader to work out the detailed connection of the two classical fields with our quantum metric distortion. For illustration, however, we just consider the energy density of the electromagnetic field given via the component T^{00} and thus obtain:

$$R^{*00} = \frac{(A \cdot B \cdot c)^2 \left(f_{,z}\right)^2 + D^2 \left[A^2 c^2 \left(f_{,y}\right)^2 + B^2 \begin{pmatrix} c^2 \left(f_{,x}\right)^2 + 3A^2 \left(f_{,t}\right)^2 \\ -2A^2 \left(C_1 + f\right) f_{,t,t} \end{pmatrix} \right]}{(A \cdot B \cdot D)^2 c^4 \left(C_1 + f\right)^6}$$

$$= -\kappa \cdot T^{00} = -\kappa \cdot \frac{\varepsilon_0 \cdot \mathbf{E} \cdot \mathbf{E} + \dfrac{\mathbf{B} \cdot \mathbf{B}}{\mu_0}}{2} = -\frac{\kappa}{2} \cdot \left(\varepsilon_0 \cdot \left(E_1^2 + E_2^2 + E_3^2\right) + \frac{B_1^2 + B_2^2 + B_3^2}{\mu_0} \right). \tag{97}$$

In fact, we find quadratic terms on both sides of the equation and it appears kind of attractive to seek the connections between the classical E and B fields and the quantum metric distortion in the following way:

$$\frac{\left(f_{,x}\right)^2 + C_{tx}\left(3\left(f_{,t}\right)^2 - 2\left(C_1 + f\right)f_{,t,t}\right)}{\left(C_1 + f\right)^6} = -\frac{\kappa_x}{2}\cdot\left(\varepsilon_0 \cdot E_1^2 + \frac{B_1^2}{\mu_0}\right)$$

$$\frac{\left(f_{,y}\right)^2 + C_{ty}\left(3\left(f_{,t}\right)^2 - 2\left(C_1 + f\right)f_{,t,t}\right)}{\left(C_1 + f\right)^6} = -\frac{\kappa_y}{2}\cdot\left(\varepsilon_0 \cdot E_2^2 + \frac{B_2^2}{\mu_0}\right)$$

$$\frac{\left(f_{,z}\right)^2 + C_{tz}\left(3\left(f_{,t}\right)^2 - 2\left(C_1 + f\right)f_{,t,t}\right)}{\left(C_1 + f\right)^6} = -\frac{\kappa_z}{2}\cdot\left(\varepsilon_0 \cdot E_3^2 + \frac{B_3^2}{\mu_0}\right). \tag{98}$$

Thus, obviously we have created photonic matter from a flat space vacuum solution to the Einstein field equations by the means of a simple metric transformation. The scalar Ricci curvature was thereby kept unchanged, which is to say $R = R^*$. The equations necessary to solve this were the classical quantum Klein–Gordon equations, which we directly obtain from the $R = R^*$ condition.

References

1. N. Schwarzer, *The World Formula: A Late Recognition of David Hilbert's Stroke of Genius*, 2022, Jenny Stanford Publishing. ISBN: 9789814877206.

2. D. Hilbert, Die Grundlagen der Physik, Teil 1, *Göttinger Nachrichten*, 1915, 395–407.

3. A. Einstein, Grundlage der allgemeinen Relativitätstheorie, *Ann. Phys.*, 1916, **49** (ser. 4), 769–822.

4. N. Schwarzer, *Brief Proof of Hilbert's World Formula – Dirac, Klein-Gordon, Schrödinger, Einstein, Evolution and the 2nd Law of Thermodynamics all from one origin*, www.amazon.com, ASIN: B08585TRB8.

5. N. Schwarzer, *The "New" Dirac Equation – Originally Postulated, Now Geometrically Derived and – along the way – Generalized to Arbitrary Dimensions and Arbitrary Metric Space-Times*, www.amazon.com, ASIN: B085636Q88.

6. N. Schwarzer, *Science Riddles – Riddle No. 11: What is Mass?*, www.amazon.com, ASIN: B07SSF1DFP.

7. N. Schwarzer, *Science Riddles – Riddle No. 12: Is There a Cosmological Spin?*, www.amazon.com, ASIN: B07T3WS7XK.

8. N. Schwarzer, *Science Riddles – Riddle No. 13: How to Solve the Flatness Problem?*, www.amazon.com, ASIN: B07T9WXZVH.

9. N. Schwarzer, *Science Riddles – Riddle No. 14: And what if Time...? A Paradigm Shift*, www.amazon.com, ASIN: B07VTMP2M8.

10. N. Schwarzer, *Science Riddles – Riddle No. 15: Is there an Absolute Scale?*, www.amazon.com, ASIN: B07V9F2124.

11. P. A. M. Dirac, The quantum theory of the electron, *Proc. R. Soc. A*, 1928, **117**(778), doi: 10.1098/rspa.1928.0023.

12. A. E. Green, W. Zerna, *Theoretical Elasticity*, Oxford University Press, London, 1968.

13. H. Neuber, *Kerspannungslehre*, in German, 3rd ed., Springer-Verlag, Berlin, Heidelberg, New York, Tokyo, 1985, ISBN 3-540-13558-8.

14. N. Schwarzer, *Epistle to Elementary Particle Physicists: A Chance You Might not Want to Miss*, www.amazon.com, ASIN: B07XDMLDQQ.

15. N. Schwarzer, *2nd Epistle to Elementary Particle Physicists: Brief Study about Gravity Field Solutions for Confined Particles*, www.amazon.com, ASIN: B07XN8GLXD.

16. N. Schwarzer, *3rd Epistle to Elementary Particle Physicists: Beyond the Standard Model – Metric Solutions for Neutrino, Electron, Quark*, www.amazon.com, ASIN: B07XJJ535T.

17. N. Schwarzer, *Science Riddles – Riddle No. 26: Is there an Elastic Quantum Gravity?*, www.amazon.com, ASIN: B081K5LZLD.

18. N. Schwarzer, *Social Distancing: COVID-19 may give one reason, but why not discussing this a bit more general and fundamentally?*, www.amazon.com, ASIN: B086RP4S6W.

19. N. Schwarzer, *The Relativistic Quantum Bible: Genesis and Revelation*, www.amazon.com, ASIN: B01M1CJH1B.

20. N. Schwarzer, *The Theory of Everything: Quantum and Relativity Is Everywhere – A Fermat Universe*, Pan Stanford Publishing, 2020, ISBN-10: 9814774472.

21. N. Schwarzer, *Einstein had it, but he did not see it, Part LXIX: The Hippocratic Oath in Mathematical Form and why – so often – it will be of no Use*, www.amazon.com, ASIN: B07KDSMNSK.

22. N. Schwarzer, *Philosophical Engineering, Part 1: The Honest Non-Parasitic Philosopher and the Universal "GOOD" Derived from a Theory of Everything*, www.amazon.com, ASIN: B07KNWRDYW.

23. N. Schwarzer, *Quantum Gravity Thermodynamics: And it May Get Hotter*, www.amazon.com, ASIN: B07XC2JW7F.

24. N. Schwarzer, *Quantum Gravity Thermodynamics II: Derivation of the Second Law of Thermodynamics and the Metric Driving Force of Evolution*, www.amazon.com, ASIN: B07XWPXF3G.

25. H. Leuenberger, What is life?, a new human model of life, disease and death – a challenge for artificial intelligence and bioelectric medicine specialists, SWISS PHARMA, 2019, **41**(1), www.ifiip.ch.

26. N. Schwarzer, *The Photon*, www.amazon.com/dp/B06XGC4NDM.

27. Peter W. Higgs, Broken symmetries and the masses of gauge bosons. *Phys. Rev. Lett.*, 1964, **13**, 508.

28. en.wikipedia.org/wiki/Higgs_boson.

29. N. Schwarzer, *Einstein had it, but he did not see it, Part LXXVI: Quantum Universes – We don't need no... an Inflation*, www.amazon.com, ASIN: B07NLH3JJV.

30. N. Schwarzer, *Einstein had it, but he did not see it, Part LXXVII: Matter is Nothing and so Nothing Matters*, www.amazon.com, ASIN: B07NQKKC31.

31. www.britannica.com/science/Casimir-effect.

32. N. Schwarzer, *My Horcruxes – A Curvy Math to Salvation, Part 9 of Medical Socio-Economic Quantum Gravity*, www.amazon.com, ASIN: B096SPB5MW.

Chapter 3

Humanitons—The Intrinsic Photons of the Human Body Understand Them and Cure Yourself

Figure 3.1 "Humanitons at Work" (by Livia Schwarzer).

The Math of Body, Soul, and the Universe
Norbert Schwarzer
Copyright © 2023 Jenny Stanford Publishing Pte. Ltd.
ISBN 978-981-4968-24-9 (Hardcover), 978-1-003-33454-5 (eBook)
www.jennystanford.com

3.1 Why Should We Model the Intrinsic Communication of a Biological Entity Like Our Own Organism via Generalized Photons?

Intrinsic communication of the human body—including the human psychology—is the key for the understanding of immune reactions, immune strength, autoimmune diseases—naturally—the psycho-neuro-immunologic complex, and ... well, in essence, everything the human being makes out. But how, when only just considering the psycho-neuro-immunologic complex, with more alien cells than the human body has own cells and in certain aspects these aliens taking up to 90% of the immune system and one hundred billion brain cells permanently producing trillions of connections, communication being on a biochemical and biophysical level so complex, multidimensional, and thus currently only rudimentary understood, can such an understanding be ever built up?

How about a top-down approach, starting from the fundamental quantum gravity simulation of a very general organic system and then just refining as needed to the desired complexity?

We will base our considerations on a world formula approach originated from David Hilbert and Albert Einstein. Proof that this approach truly considers "it all" was already given in the first issues of our series "*Medical Socio-Economic Quantum Gravity*" and [6]. We also gave prove here in chapter 2 by deriving a few principal—classical—gravity and quantum equations as demonstrators.

In this chapter, we will now show how this approach may also be of use in providing a suitable toolbox for a holistic description of complex medical aspects like the whole human mind–body system.

3.1.1 A Simple Motivation

Only taking the human body cells with core, the ratio of the "others" (mainly bacteria) to our cells would be about 10:1. Also, including the simple body cells without core, the ratio reduces to about 1.3:1, but still shows a significant dominance of the "aliens." The new field of neuroimmunology shows us that the nerve system, the other—own—body cells, and the aliens are working together. It is even assumed that the intestines, which host most of the "aliens," make up to about 80% of the immune system.

So how do these partners communicate?

By electro-biochemical means we may answer and this would principally be correct. But would this also be of any use ... meaning does this help in understanding the complex communication patterns among such often so very different and often even more distant interacting partners such as human nerve and body cells and bacteria?

Not really, because the biochemical reactor "human being" is by far too huge and complex to ever completely mirror it without the help of a quantum gravity supercomputer [1–4]. After all, we know from the field of thermodynamics that systems with long range and multiple scale interactions are extremely difficult to handle and are usually dealt with by renormalization group theoretical approaches.

But what about starting one level below? After all, all biochemical processes and interactions are based on electromagnetism. In finding and using just this fundamental starting point, perhaps, we are able to model the internal communication patterns and thus also better understand important immune/autoimmune reactions. The latter may then—naturally—also hold for the interaction of the neuro system, be it among itself, with other body cells, the aliens or the environment.

Our starting point here shall be the Einstein–Hilbert action in arbitrary space-times and with arbitrary numbers of dimensions. We consider this to be fundamental enough—if applied and investigated thoroughly enough—to provide us with a world formula. From there we are able to derive the most fundamental theory of electromagnetism. The derivation was performed in the appendix of [5] (also see [6]), respectively the appendix of chapter 2 in this book. The necessary complexity in order to take care about thermodynamic and evolutionary processes was elaborated in [7].

3.2   Fundamental Top-Down Approach for Any System

Here we repeat the recipe from section 2.2 but add an essential point.

In order to avoid complex math, we want to give a very brief verbal explanation of the principal structure and appearance of a world formula, which—as being a world formula—also should be able to describe complex organic systems like a human body. The latter may be extremely complex and of a very high-dimensional order, but this does not mean that a mathematical description would not mirror it. A more thorough verbal explanation can be found in [6].

1. Let there be a set of properties.

2. Such properties could also be seen as degrees of freedom of a system, which mathematically would allow us to see them as the system's dimensions, whereas the system itself is a space, formed by such dimensions (properties). Thus, we consider space as an ensemble of properties.

3. These properties, which—as said—could just be seen as degrees of freedom and thus dimensions, are subjected to a Hamilton extremal principle.

4. This leads to the Einstein–Hilbert action [8] and subsequently to the Einstein field equations [8, 9] when following just Hilbert's mathematical path of variating the action with respect to the metric tensor [8] or Einstein's argumentation [9].

5. However, the simplest way to satisfy the Einstein–Hilbert action could also be achieved by just demanding the curvature R (the so-called scalar Ricci curvature) of the space or space-time of properties to be a constant, which leads to the well-known quantum equations such as Klein–Gordon, Schrödinger, and Dirac [5, 6, 10, 11].

6. Most interestingly, this $R = 0$ condition, as it can be shown [32], is just a by-product of the same Hamilton minimum condition that also gave us the Einstein field equations.

7. Inertia, evolution forces, and the laws of thermodynamics appear via entanglement of dimensions, the variation with respect to the number of dimensions of sub-systems, and the position of centers of gravity [5–7, 12–16]. This way also the required complexity of a system can be driven up arbitrarily without leaving the path of Hilbert's fundamentality.

3.3 The Human Body as a VERY Complex Communication System

In one way or the other, all parts of the human body including its billions of alien cells are connected with each other. Without taking all these connections into account, thereby also including the interaction of the system "human body" with its surrounding and other human beings, the individual human entity can never be fully understood and subsequently never be properly treated.

On the other hand, it is impossible to simulate such a system in its full complexity (except with a gravity quantum computer already at hand [1–4]).

Thus, we require a model that allows us to keep the task identity without the need of also keeping the full complicacy, which is to say we would like to have a model or model toolbox allowing us to just extract the complexity we need. Thereby we should not fall for the usual flaw in linearizing a potentially nonlinear system ... at least we should not do this too early (meaning before we are sure about its true nonlinearity). Especially when it comes to simulating complex systems, humans tend to linearize their understanding of the model about the system by far too early a stage during the process of abstraction.

Hilbert's action formula [8], being in fact a world formula approach [6], shows a method that retains full 'complexity and nonlinearity' for any complex system, while remaining mathematically manageable [6]. This leads to a fundamental quantum gravity description of the system and thus allows for an extremely general consideration of any problem in mathematical form (e.g., [5, 7]).

As complexity is equal to the number of degrees of freedom and thus to the number of dimensions—the dimensions form space, entangle, and subsequently give rise to masses, spin, inertia, potentials, and so on ... Then, there we are with the world formula to have the best description for any complex system one can ever have.

Thus, seeing the "Human Body as a VERY Complex Communication System," we have all we need to try for a most holistic and convergingly ever better description of it. This also includes the mathematically rather complicated treatment of psychological problems as they occur in the so-called "affect logic" (e.g., [17–19]).

The necessary toolbox was already given in a variety of previous papers by the author (e.g., [5–7]), but for convenience we will here consider a few examples in Cartesian coordinates.

3.4   From Hilbert's World Formula

As elaborated in chapter 2, our starting point shall be the classical Einstein–Hilbert action [8] with the Ricci scalar R^* as kernel or Lagrange density:

$$\delta W = 0 = \delta \int_V d^n x \left(\sqrt{-G} \cdot R^* \right) \tag{99}$$

which, in contrast to the classical form in [8], contains a somewhat adapted metric tensor $G_{\delta\gamma} = F[f] \cdot g_{\delta\gamma}$. Thereby G shall denote the determinant of the metric tensor $G_{\alpha\beta}$, while g will later stand for the corresponding determinant of the metric tensor $g_{\alpha\beta}$. In order to distinguish our Ricci scalar, being based on $G_{\delta\gamma} = F[f] \cdot g_{\delta\gamma}$ from the usual one, being based on the metric tensor $g_{\alpha\beta}$, we marked it with the * superscript.

Now we evaluate the resulting Ricci scalar for the metric $G_{\alpha\beta}$ at first in n and then, subsequently, simplify to 4 dimensions:

$$R^* = \frac{1}{F[f]^3} \cdot \left(\begin{array}{c} \left(C_{N1} \cdot \left(\frac{\partial F[f]}{\partial f} \right)^2 - C_{N2} \cdot F[f] \cdot \frac{\partial^2 F[f]}{\partial f^2} \right) \cdot \overbrace{\left(\tilde{\nabla}_g f \right)^2}^{= f_{,\alpha} g^{\alpha\beta} f_{,\beta}} \\ - C_{N2} \cdot F[f] \cdot \frac{\partial F[f]}{\partial f} \cdot \Delta_g f \end{array} \right)$$

$$\xrightarrow{n=4} = \frac{1}{F[f]^3} \cdot \left(\left(\frac{3}{2} \cdot \left(\frac{\partial F[f]}{\partial f} \right)^2 - 3 \cdot F[f] \cdot \frac{\partial^2 F[f]}{\partial f^2} \right) \cdot \left(\tilde{\nabla}_g f \right)^2 - 3 \cdot F[f] \cdot \frac{\partial F[f]}{\partial f} \cdot \Delta_g f \right)$$

$$\xrightarrow{\frac{3}{2}\left(\frac{\partial F[f]}{\partial f} \right)^2 - 3 \cdot F[f] \cdot \frac{\partial^2 F[f]}{\partial f^2} = 0} = -\frac{3}{F[f]^2} \cdot \left(\frac{\partial F[f]}{\partial f} \cdot \Delta_g f \right). \tag{100}$$

With respect to the constants C_{N1} and C_{N2} in arbitrary n dimensions, the reader is referred to [6], respectively the part "The fourth day ..." in this book (mainly chapter 8).

Taking the general n-dimensional case in the first line of (100) and always demanding that:

$$C_{N1} \cdot \left(\frac{\partial F[f]}{\partial f} \right)^2 - C_{N2} \cdot F[f] \cdot \frac{\partial^2 F[f]}{\partial f^2} = 0, \tag{101}$$

which simply is a fixing condition for the wrapper function $F[f]$, we obtain:

$$R^* = -\frac{C_{N2}}{F[f]^2} \cdot \left(\frac{\partial F[f]}{\partial f} \cdot \Delta_g f \right) = 0$$

$$\Rightarrow \qquad \Delta_g f = 0. \tag{102}$$

This is the "homogeneous" Laplace equation, respectively the mass- and potential-free Klein–Gordon equation.

So, the question would be where to get the mass from?

This question was answered in earlier papers by the author (e.g., [5]) and represented here in section "2.4.1 Entanglement as Origin of Mass (and Potential) and Thus Possibly Also Any Kind of Inertia and Interaction (Including Interaction in Socioeconomic Space-Times)."

3.5   The Situation in 4 Dimensions

In [22], we considered a variety of examples, mainly starting with the Schwarzschild metric [23] and some of its derivatives. Thereby we automatically found the classical quantum equations for the central field problem and derived, among other things, also the typical Schrödinger hydrogen solutions [24]. The fact that we so clearly obtained quantum solutions (without any approximation) in connection with Schwarzschild or flat space metrics of spherical symmetry also brings in a problem with our transformation rule (5). In 4 dimensions, we applied $F[f] = f^2$ in order to fulfill the condition (7), (101) within our derivations in [22]. This, however, taking our transformation rule from 2.4 automatically leads to the following new—quantized—metric:

$$G_{\alpha\beta} = g_{\alpha\beta} \cdot F[f] = g_{\alpha\beta} \cdot f^2. \tag{103}$$

Taking the Minkowski metric in generalized spherical coordinates as an example:

$$g_{\alpha\beta} = \begin{pmatrix} c^2 \cdot g_t'[t]^2 \cdot F[f] & 0 & 0 & 0 \\ 0 & -g_r'[r]^2 \cdot F[f] & 0 & 0 \\ 0 & 0 & -r^2 \cdot g_\vartheta'[\vartheta]^2 \cdot F[f] & 0 \\ 0 & 0 & 0 & g_{33} \end{pmatrix}$$

$$f = f\left[g_t[t], g_r[r], g_\vartheta[\vartheta], g_\varphi[\varphi]\right]; \quad g_{33} = -r^2 \cdot g_\varphi'[\varphi]^2 \cdot \sin[\vartheta]^2 \cdot F[f], \tag{104}$$

one would directly obtain result in an all-scale quantum dominated metric solution for $G_{\alpha\beta}$. This, however, is in total contrast to the daily experiences that quantum effects are only important in smaller scales.

In order to solve this problem, we evaluate the general solution of condition (101) in 4 dimensions and find [21]:

$$\Rightarrow \qquad F[f] = C_1 \cdot f + \frac{C_1^2 \cdot f^2}{4 \cdot C_2} + C_2. \tag{105}$$

Thereby the constants C_i in (105) are arbitrary. Please note that we obtain the simple case $F[f] = f^2$, which was applied in [22], with the condition:

$$C_1^2 = 2 \cdot C_2 \tag{106}$$

and the subsequent limit procedure $C_2 \to 0$ as follows:

$$F[f] = 2 \cdot C_2 \cdot f + f^2 + C_2 \xrightarrow[C_2 \to 0]{\lim} F[f] = f^2. \tag{107}$$

Using (105) instead of $F = f^2$ in (100) for general metrics gives us:

$$R^* = \frac{1}{F[f]^3} \cdot \left(\left(\left(C_{N1} \cdot \left(\frac{\partial F[f]}{\partial f} \right)^2 - C_{N2} \cdot F[f] \cdot \frac{\partial^2 F[f]}{\partial f^2} \right) \cdot \left(\tilde{\nabla}_g f \right)^2 \right. \right.$$
$$\left. \left. - C_{N2} \cdot F[f] \cdot \frac{\partial F[f]}{\partial f} \cdot \Delta_g f \right) \right)$$

$$= -C_{N2} \cdot \frac{\frac{\partial F[f]}{\partial f} \cdot \Delta_g f}{F[f]^2} \xrightarrow{4D} = -3 \cdot \frac{\left(C_1 + \frac{C_1^2 \cdot f}{2 \cdot C_2} \right)}{\left(C_1 \cdot f + \frac{C_1^2 \cdot f^2}{4 \cdot C_2} + C_2 \right)^2} \cdot \Delta_g f. \quad (108)$$

Immediately we see that in the many cases with $R^* = 0$, we have considered in [22], nothing would change for the result of f. Taylor expansion for small f (which is to say at $f = 0$) also gives us equations leading to the results obtained and discussed previously in our series, especially [22]. For instance, if only considering up to linear f as follows:

$$\frac{R^*}{3} \cdot \frac{\left(C_1 \cdot f + \frac{C_1^2 \cdot f^2}{4 \cdot C_2} + C_2 \right)^2}{C_1 + \frac{C_1^2 \cdot f}{2 \cdot C_2}}$$

$$+ \Delta_g f = 0 \xrightarrow{small \ f} \frac{R^*}{3} \cdot \left(\frac{C_2^2}{C_1} + \frac{3}{2} \cdot C_2 \cdot f + \ldots \right) + \Delta_g f = 0, \quad (109)$$

and assuming constant R^*, we already have the typical structure of the Klein–Gordon equation plus a constant term $\dfrac{R^*}{3} \cdot \dfrac{C_2^2}{C_1}$. Such a constant would still not influence the principal structural character of our previous solution because a simple transformation of:

$$f^* = \frac{C_2^2}{C_1} + \frac{3}{2} \cdot C_2 \cdot f; \quad \Delta_g f^* = \Delta_g f \quad (110)$$

does give us back the ordinary Klein–Gordon equation with:

$$\frac{R^*}{3} \cdot f^* + \Delta_g f^* = 0. \quad (111)$$

The resulting transformation in 4 dimensions would give the following quantum metric:

$$G_{\alpha\beta} = g_{\alpha\beta} \cdot F[f] = g_{\alpha\beta} \cdot \left(C_1 \cdot f + \frac{C_1^2 \cdot f^2}{4 \cdot C_2} + C_2 \right). \tag{112}$$

This clearly allows for the classical Einstein situation with a dominating part C_2 at bigger scales and the quantum behavior terms with all the f. In result, one might separate as follows in 4 dimensions:

$$G_{\alpha\beta} = \overbrace{g_{\alpha\beta} \cdot C_2}^{\text{class. GTR}} + \overbrace{g_{\alpha\beta} \cdot \left(C_1 \cdot f \cdot \left(1 + \frac{C_1 \cdot f}{4 \cdot C_2} \right) \right)}^{\text{quantum metric}}. \tag{113}$$

(Quantum Gravity in 4D)

Thereby it does not come as a surprise that we have already found so many cases where f has to be determined by classical quantum equations, where our equations give similar or even equal solutions. Normalization of the quantum gravity metric $G_{\alpha\beta}$ requires division by C_2 and gives:

$$G_{\alpha\beta}^{\text{norm}} = \frac{G_{\alpha\beta}}{C_2} = \overbrace{g_{\alpha\beta}}^{\text{class. GTR}} + \overbrace{g_{\alpha\beta} \cdot \left(\frac{C_1}{C_2} \cdot f \cdot \left(1 + \frac{C_1 \cdot f}{4 \cdot C_2} \right) \right)}^{\text{quantum metric}}. \tag{114}$$

(Quantum Gravity in 4D)

Now we apply the recipe outlined above to a few flat space examples in a variety of dimensions and investigate the quantized solutions.

3.6 The Situation in *n* Dimensions

At first, however, we need to generalize our results from the section above to the situation in *n* dimensions. Generally structured metrics (c.f. the part "The fourth day …" of this book) will always give a Ricci scalar R^* as presented in the first line in (100). Applying condition (101) and thereby fixing $F[f]$ then always leads to structures as follows:

$$R^* + C_{N2} \cdot \frac{\dfrac{\partial F[f]}{\partial f} \cdot \Delta_g f}{F[f]^2} = 0. \tag{115}$$

Here we give the subsequent equations for the dimensions $n = 2$ to $n = 10$. Partially simpler, respectively, simplified forms are given in [21].

The following table gives the constants as used in (100) for $n = 2$ to 10:

n = number of dimensions	C_{N1}	C_{N2}
2	1	1
3	3/2	2
4	3/2	3
5	1	4
6	0	5
7	-3/2	6
8	-7/2	7
9	-6	8
10	-9	9

As a sidenote we point out that $C_{N2} = n - 1$, while there is a perfect parabolic behavior found for C_{N1} (see Fig. 3.2). Most interestingly, we find the maxima for C_{N1} at $n = 3$ and $n = 4$. We also find that the sum of the coefficients $C_{N1} + C_{N2}$ has its maximum at $n = 5$, 6, and 7 with $C_{N1} + C_{N2} = 5$ and give $C_{N1} + C_{N2} = 0$ for $n = 10$, which peculiarly is the number of dimensions most string- and brane-theoreticians consider as the right number of dimensions for our universe (see Fig. 3.3).

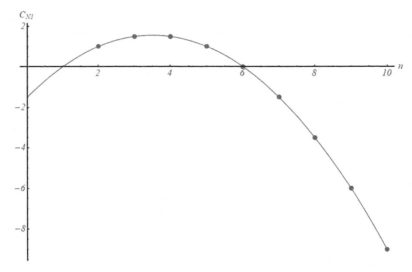

Figure 3.2 Coefficient C_{N1} from the transformed Ricci scalar as given in (6) and (100) (first line) as function of the number of dimensions n. While the dots give the exact values (c.f. table above), the line shows a parabolic fit with an approach: $C_{N1} = c_0 + c_1 * n + c_2 * n^2$ (with c_0, c_1, c_2 as fitting constants).

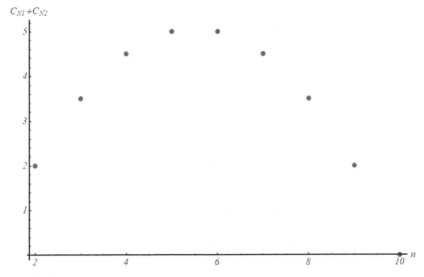

Figure 3.3 Sum of coefficients $C_{N1} + C_{N2}$ from the transformed Ricci scalar as given in (6) and (100) (first line) as function of the number of dimensions n.

3.6.1 The 2-Dimensional Space

Equation (101) gives the following solution for the function $F[f]$:

$$F[f] = C_1 \cdot \exp[f \cdot C_2] \tag{116}$$

and results in the General Theory of Relativity (GTR)–Klein–Gordon equation as follows:

$$\left[\Delta_g + \frac{R^* \cdot \left(C_1 \cdot \exp[f \cdot C_2] \right)^2}{f \cdot C_{N2} \cdot C_1 \cdot C_2 \cdot \exp[f \cdot C_2]} \right] f = \left[\Delta_g + \frac{R^* \cdot C_1 \cdot \exp[f \cdot C_2]}{f \cdot C_2} \right] f = 0. \tag{117}$$

3.6.2 The 3-Dimensional Space

Equation (101) gives the following solution for the function $F[f]$:

$$F[f] = C_2 \cdot \left(f - C_1 \right)^4 \tag{118}$$

and results in the GTR–Klein–Gordon equation as follows:

$$C_2 \cdot \frac{(f-C_1)^5}{8} \cdot R^* + \Delta_g f = \left[C_2 \cdot \frac{(f-C_1)^5}{f \cdot 8} \cdot R^* + \Delta_g \right] f$$

$$= \left[C_2 \cdot \frac{(f-C_1)^4}{8} \cdot R^* + \Delta_g \right] (f-C_1) = 0. \tag{119}$$

3.6.3 The 4-Dimensional Space

See section "3.5 The Situation in 4 Dimensions."

3.6.4 The 5-Dimensional Space

Equation (101) gives the following solution for the function $F[f]$:

$$F[f] = C_2 \cdot (f-C_1)^{4/3} \tag{120}$$

and results in the GTR–Klein–Gordon equation as follows:

$$\left[\Delta_g + \frac{3 \cdot R^* \cdot C_2 \cdot (f-C_1)^{7/3}}{f \cdot 16} \right] f = \left[\Delta_g + \frac{3 \cdot R^* \cdot C_2 \cdot (f-C_1)^{4/3}}{16} \right] (f-C_1) = 0. \tag{121}$$

3.6.5 The 6-Dimensional Space

Equation (101) gives the following solution for the function $F[f]$:

$$F[f] = f + C_1 \tag{122}$$

and results in the GTR–Klein–Gordon equation as follows:

$$R^* + \frac{5}{F[f]^2} \cdot \frac{\partial F[f]}{\partial f} \cdot \Delta_g f = R^* + \frac{5}{(f+C_1)^2} \cdot \Delta_g f$$

$$\Rightarrow \quad \frac{(f+C_1)^2}{5} \cdot R^* + \Delta_g f = \left(\frac{(f+C_1)^2}{f \cdot 5} \cdot R^* + \Delta_g \right) f$$

$$= \left(\frac{f+C_1}{5} \cdot R^* + \Delta_g \right) (f+C_1) = 0. \tag{123}$$

3.6.6 The 7-Dimensional Space

Equation (101) gives the following solution for the function $F[f]$:

$$F[f] = C_2 \cdot (f + C_1)^{4/5} \tag{124}$$

and results in the GTR–Klein–Gordon equation as follows:

$$R^* + \frac{6}{F[f]^2} \cdot \frac{\partial F[f]}{\partial f} \cdot \Delta_g f = 0$$

$$\Rightarrow \quad \frac{5 \cdot C_2 (f + C_1)^{9/5}}{24} \cdot R^* + \Delta_g f = \left(\frac{5 \cdot C_2 (f + C_1)^{9/5}}{24 \cdot f} \cdot R^* + \Delta_g \right) f$$

$$= \left(\frac{5 \cdot C_2 (f + C_1)^{4/5}}{24} \cdot R^* + \Delta_g \right) (f + C_1) = 0. \tag{125}$$

3.6.7 The 8-Dimensional Space

Equation (101) gives the following solution for the function $F[f]$:

$$F[f] = C_2 \cdot (3 \cdot f - C_1)^{2/3} \tag{126}$$

and results in the GTR–Klein–Gordon equation as follows:

$$R^* + \frac{7}{F[f]^2} \cdot \frac{\partial F[f]}{\partial f} \cdot \Delta_g f = 0$$

$$\Rightarrow \quad \frac{C_2 \cdot 7 \cdot (3 \cdot f - C_1)^{5/3}}{2} \cdot R^* + \Delta_g f = \left(\frac{C_2 \cdot 7 \cdot (3 \cdot f - C_1)^{5/3}}{f \cdot 2} \cdot R^* + \Delta_g \right) f$$

$$= \left(\frac{C_2 \cdot 7 \cdot (3 \cdot f - C_1)^{2/3}}{2} \cdot R^* + \Delta_g \right) (9 \cdot f - C_1) = 0. \tag{127}$$

3.6.8 The 9-Dimensional Space

Equation (101) gives the following solution for the function $F[f]$:

$$F[f] = C_2 \cdot (f + C_1)^{4/7} \tag{128}$$

and results in the GTR–Klein–Gordon equation as follows:

$$R^* + \frac{8}{F[f]^2} \cdot \frac{\partial F[f]}{\partial f} \cdot \Delta_g f = 0$$

$$\Rightarrow \quad \frac{C_2 \cdot 7 \cdot (f+C_1)^{11/7}}{32} \cdot R^* + \Delta_g f = \left(\frac{C_2 \cdot 7 \cdot (f+C_1)^{11/7}}{32 \cdot f} \cdot R^* + \Delta_g \right) f$$

$$= \left(\frac{C_2 \cdot 7 \cdot (f+C_1)^{4/7}}{32} \cdot R^* + \Delta_g \right) (f+C_1) = 0. \tag{129}$$

3.6.9 The 10-Dimensional Space

Equation (101) gives the following solution for the function $F[f]$:

$$F[f] = C_2 \cdot (2 \cdot f - C_1)^{1/2} \tag{130}$$

and results in the GTR–Klein–Gordon equation as follows:

$$R^* + \frac{9}{F[f]^2} \cdot \frac{\partial F[f]}{\partial f} \cdot \Delta_g f = 0$$

$$\Rightarrow \quad \frac{C_2 \cdot 2 \cdot (2 \cdot f - C_1)^{3/2}}{9} \cdot R^* + \Delta_g f = \left(\frac{C_2 \cdot 2 \cdot (2 \cdot f - C_1)^{3/2}}{f \cdot 9} \cdot R^* + \Delta_g \right) f$$

$$= \left(\frac{C_2 \cdot 2 \cdot (2 \cdot f - C_1)^{1/2}}{9} \cdot R^* + \Delta_g \right) (2 \cdot f - C_1) = 0. \tag{131}$$

3.7 Periodic Space-Time Solutions

One of the most important solutions for the human body's internal working and communication are oscillations and wave-like solutions. As the waves (photonic solutions) are derived and discussed in chapter 2 and [5], we here concentrate on oscillations.

Thus, we are interested in learning how the simplest oscillating solutions could be obtained for a variety of settings to R^*. Only assuming Minkowski-like flat spaces, the Ricci scalar R of the original metric $g_{\alpha\beta}$ before the transformation would be zero, of course. Picking the simplest example from [6] (see also [5, 7]), which was found to be the 6-dimensional space-time, we assume a flat space of 6 dimensions with the coordinates t, x, y, z, u, w. With the following ansatz for the Ricci scalar and the function f in 6 dimensions:

$$R^* = \frac{5}{(f+C_1)} \cdot \sum_{k=1}^{6} R_k^*; \quad f + C_1 = \prod_{k=1}^{6} \left(f_k[\xi_k] + C_{1k} \right); \quad \xi_k = \{t, x, y, z, u, w\} \tag{132}$$

and the trace of the metric $g_{\alpha\beta}$ being {1, 1, 1, 1, 1, 1} (please note that, for simplicity, we ignore the special character of the time coordinate by setting c = 1 and do not apply any sign convention as such does not matter here), we obtain the following sets of ordinary differential equations from our quantum gravity Klein–Gordon equation following from the adapted Einstein–Hilbert action (99) (regarding the evaluation, see [5–7] and sections above):

$$\left(R_k^* + \frac{\partial^2}{\partial \xi_k^2} \right) \left(f_k \left[\xi_k \right] + C_{1k} \right) = 0. \tag{133}$$

The subsequent solutions for all these equations in (133) read:

$$f_k \left[\xi_k \right] + C_{1k} = C_{k1} \cdot \sin \left[\xi_k \cdot \sqrt{R_k^*} \right] + C_{k2} \cdot \cos \left[\xi_k \cdot \sqrt{R_k^*} \right]. \tag{134}$$

Applying the rule of generalization via coordinate transformation (e.g., see [20]), which is valid for any arbitrary solution of the Einstein field equations, we can generalize our solutions (134) to:

$$f_k \left[\xi_k \right] + C_{1k} = C_{k1} \cdot \sin \left[g_k \left[\xi_k \right] \cdot \sqrt{R_k^*} \right] + C_{k2} \cdot \cos \left[g_k \left[\xi_k \right] \cdot \sqrt{R_k^*} \right]. \tag{135}$$

Please note that by using this transformation rule as follows:

$$g_{ij} \left[x_0, x_1, \ldots, x_n \right] \Rightarrow G_{ij} = g'_{x_i} \left[x_i \right] \cdot g'_{x_j} \left[x_j \right] \cdot g_{ij} \left[g_{x_0} \left[x_0 \right], g_{x_1} \left[x_1 \right], \ldots, g_{x_n} \left[x_n \right] \right]$$

$$\text{with}: \quad g'_{x_i} \left[x_i \right] \equiv \frac{\partial g_{x_i} \left[x_i \right]}{\partial x_i}, \tag{136}$$

this will only change the Ricci scalar R^* as follows:

$$R^* \left[f_k \left[\xi_k \right] + C_{1k} \right] \Rightarrow R^* \left[f_k \left[g_k \left[\xi_k \right] \right] + C_{1k} \right], \tag{137}$$

which means that there is no principal structural change, as the g_k stay intrinsic and the special choice of R^* (132) in our 6-dimensional space becomes:

$$R^* = \frac{5}{\prod_{k=1}^{6} \left(f_k \left[g_k \left[\xi_k \right] \right] + C_{1k} \right)} \cdot \sum_{k=1}^{6} R_k^*. \tag{138}$$

Thus, in order to obtain R^* free of singularities, we chose our constants C_{1k} such that we always have:

$$f_k \left[\xi_k \right] + C_{1k} \neq 0. \tag{139}$$

Obviously, there are infinitely many options fulfilling the conditions above. Applying the transformation rule from [20] and using the periodic solutions we have obtained yields:

$$G_{\alpha\beta} = g_{\alpha\beta} \cdot F[f] = g_{\alpha\beta} \cdot (f + C_1) = g_{\alpha\beta} \cdot \prod_{k=1}^{6} \left(f_k \left[g_k \left[\xi_k \right] \right] + C_{1k} \right)$$

$$= g_{\alpha\beta} \cdot \prod_{k=1}^{6} C_{1k} + g_{\alpha\beta} \cdot \left[\prod_{k=1}^{6} \left(f_k \left[g_k \left[\xi_k \right] \right] + C_{1k} \right) - \prod_{k=1}^{6} C_{1k} \right]. \qquad (140)$$

Normalizing the result by dividing it by $\displaystyle\prod_{k=1}^{6} C_{1k}$ results in the following split up of classical General Theory of Relativity metric $g_{\alpha\beta}$ and its quantized part:

$$\underbrace{\frac{G_{\alpha\beta}}{\displaystyle\prod_{k=1}^{6} C_{1k}}}_{} = \overbrace{g_{\alpha\beta}}^{\text{classical GTR}} + \overbrace{g_{\alpha\beta} \cdot \left[\prod_{k=1}^{6} \left(f_k \left[g_k \left[\xi_k \right] \right] + C_{1k} \right) - \prod_{k=1}^{6} C_{1k} \right]}^{\text{Quantum Gravity}}. \qquad (141)$$

Thus, we ended up with a quantized, but still completely Einstein field equations compatible metric solution in 6 dimensions. In [21], it was demonstrated how similar (periodic) solutions in other numbers of dimensions can be obtained. Here now we intend to investigate a variety of options in connection with our quantum gravity approach. In order not to overload the paper, we thereby concentrate only on cases in 4 dimensions.

Definitively, so it can easily be seen, our transformed, respectively, quantized metric $G_{\alpha\beta}$ describes states with matter, which is to say the corresponding Einstein–Hilbert action—effectively—contains a nonzero Lagrange density term L_M for the matter. As our starting point always was a flat space without anything in it, we have to conclude that the quantum transformation $F[f]$ can be used to create matter from Einstein–Hilbert vacua, which is to say from nothing. And as, regarding the human body, we are just talking about communication patterns, we have to conclude that matter is communication and communication is matter, whereby here the matter type could just be seen as energy.

3.8 Cartesian Coordinates

Similar to the 6-dimensional case we have considered in the section "3.7 Periodic Space-Time Solutions," we can also construct a periodic solution in 4 dimensions.

3.8.1 A Somewhat More General Case

Thereby we assume a flat space of 4 dimensions with the coordinates t, x, y, z. The equation we have to solve would be [21]:

$$\frac{R^*}{3} \cdot \frac{\left(C_1 \cdot f + \dfrac{C_1^2 \cdot f^2}{4 \cdot C_2} + C_2\right)^2}{C_1 + \dfrac{C_1^2 \cdot f}{2 \cdot C_2}} + \Delta_g f = 0. \tag{142}$$

As demonstrated in [21], we assume f to be small against the constant C_2 and Taylor expansion gives us the simple equation:

$$\frac{R^*}{3} \cdot f^* + \Delta_g f^* = 0. \tag{143}$$

With the following ansatz for the Ricci scalar and the function f in 4 dimensions:

$$\frac{R^*}{3} = \sum_{k=1}^{4} R_k^*; \quad f^* = \prod_{k=1}^{4} f_k\left[\xi_k\right]; \quad \xi_k = \{t, x, y, z\}, \tag{144}$$

and the trace of the metric $g_{\alpha\beta}$ being $\{-c^2, 1, 1, 1\}$, we obtain the following sets of ordinary differential equations from our quantum gravity Klein–Gordon equation (143):

$$\left(R_1^* - \frac{\partial^2}{c^2 \cdot \partial \xi_1^2}\right) f_1\left[\xi_1\right] = 0; \quad \left(R_k^* + \frac{\partial^2}{\partial \xi_k^2}\right) f_k\left[\xi_k\right] = 0 \;\forall\; k > 1. \tag{145}$$

The subsequent solutions for all these equations in (133) read:

$$k = 1: \quad f_k\left[\xi_k\right] = C_{k1} \cdot \sin\left[\xi_k \cdot c \cdot \sqrt{-R_k^*}\right] + C_{k2} \cdot \cos\left[\xi_k \cdot c \cdot \sqrt{-R_k^*}\right]$$

$$k > 1: \quad f_k\left[\xi_k\right] = C_{k1} \cdot \sin\left[\xi_k \cdot \sqrt{R_k^*}\right] + C_{k2} \cdot \cos\left[\xi_k \cdot \sqrt{R_k^*}\right]. \tag{146}$$

Applying the rule of generalization allows us to generalize our solutions to:

$$k = 1: \quad f_k\left[\xi_k\right] = C_{k1} \cdot \sin\left[g_k\left[\xi_k\right] \cdot c \cdot \sqrt{-R_k^*}\right] + C_{k2} \cdot \cos\left[g_k\left[\xi_k\right] \cdot c \cdot \sqrt{-R_k^*}\right]$$

$$k > 1: \quad f_k\left[\xi_k\right] = C_{k1} \cdot \sin\left[g_k\left[\xi_k\right] \cdot \sqrt{R_k^*}\right] + C_{k2} \cdot \cos\left[g_k\left[\xi_k\right] \cdot \sqrt{R_k^*}\right].$$

$$\tag{147}$$

As before, note that using this transformation rule as follows:

$$g_{ij}\left[x_0, x_1, \ldots, x_n\right] \Rightarrow G_{ij} = g'_{x_i}\left[x_i\right] \cdot g'_{x_j}\left[x_j\right] \cdot g_{ij}\left[g_{x_0}\left[x_0\right], g_{x_1}\left[x_1\right], \ldots, g_{x_n}\left[x_n\right]\right]$$

with
$$g'_{x_i}\left[x_i\right] \equiv \frac{\partial g_{x_i}\left[x_i\right]}{\partial x_i} \tag{148}$$

will only change the Ricci scalar R^* as follows:

$$R^*\left[f_k\left[\xi_k\right]\right] \Rightarrow R^*\left[f_k\left[g_k\left[\xi_k\right]\right]\right], \tag{149}$$

which means that there is no principal structural change, as the g_k stay intrinsic and the special choice of R^* is maintained.

Again, we note that there are infinitely many options fulfilling the conditions above. One might consider these solutions as quantum gravity waves or just photons of any kind as we have not specified that character of the dimensions, yet. These dimensions could be any degree of freedom and not just the classical space and time axes. It could be the special position and character of just one single alien cell inside our body determining in its being the communication within our immune system.

Applying the general transformation rule and using the periodic solutions we have obtained yields the following generalizations:

$$G_{\alpha\beta}^{\text{norm}} = \frac{G_{\alpha\beta}}{C_2} = \overbrace{g_{\alpha\beta}}^{\text{class. GTR}} + g_{\alpha\beta} \cdot \overbrace{\left(\frac{C_1}{C_2} \cdot f \cdot \left(1 + \frac{C_1 \cdot f}{4 \cdot C_2}\right)\right)}^{\substack{\text{Quantum Gravity in } 4D \\ \text{quantum metric}}}$$

$$= \overbrace{g_{\alpha\beta}}^{\text{class. GTR}} + \frac{g_{\alpha\beta}}{9} \cdot \overbrace{\left(\frac{C_1 \cdot f^* \cdot \left(4 \cdot C_2^2 + C_1 \cdot f^*\right)}{C_2^4} - 5\right)}^{\substack{\text{Quantum Gravity in } 4D \\ \text{quantum metric}}}$$

$$= \overbrace{g_{\alpha\beta}}^{\text{class. GTR}} + \frac{g_{\alpha\beta}}{9} \cdot \overbrace{\left(\frac{C_1 \cdot \prod_{k=1}^{4} f_k\left[\xi_k\right] \cdot \left(4 \cdot C_2^2 + C_1 \cdot \prod_{k=1}^{4} f_k\left[\xi_k\right]\right)}{C_2^4} - 5\right)}^{\substack{\text{Quantum Gravity in } 4D \\ \text{quantum metric}}}. \tag{150}$$

3.8.2   The Total Vacuum Case → Photonic Solutions

Now we assume $R^* = 0$ again and solve the subsequent equation resulting from (142) as a simple:

$$\frac{R^*}{3} \cdot \frac{\left(C_1 \cdot f + \frac{C_1^2 \cdot f^2}{4 \cdot C_2} + C_2\right)^2}{C_1 + \frac{C_1^2 \cdot f}{2 \cdot C_2}} + \Delta_g f = 0 \xrightarrow{R^* = 0} \Delta_g f = 0. \qquad (151)$$

Applying the separation approach as before:

$$f = \prod_{k=1}^{4} f_k[\xi_k]; \quad \xi_k = \{t, x, y, z\}, \qquad (152)$$

this time results in:

$$\frac{1}{f}\left(\sum_{k=2}^{4} \frac{\partial^2}{\partial \xi_k^2} - \frac{\partial^2}{c^2 \cdot \partial \xi_1^2}\right)f = \sum_{k=2}^{4} \frac{\partial^2 f_k}{f_k \cdot \partial \xi_k^2} - \frac{\partial^2 f_1}{f_1 \cdot c^2 \cdot \partial \xi_1^2} = 0$$

$$\Rightarrow \qquad \sum_{k=2}^{4} \frac{\partial^2 f_k}{f_k \cdot \partial \xi_k^2} = -C_t^2 = \frac{\partial^2 f_1}{f_1 \cdot c^2 \cdot \partial \xi_1^2}. \qquad (153)$$

As before, the time function $f_1 = f_1[\xi_1] = f_1[t]$ gives:

$$f_1[t] = C_{11} \cdot \cos[c \cdot C_t \cdot t] + C_{12} \cdot \sin[c \cdot C_t \cdot t], \qquad (154)$$

while for convenience, we split up the constant C_t with respect to the spatial coordinates into:

$$-C_t^2 = -C_x^2 - C_y^2 - C_z^2 = -C_2^2 - C_3^2 - C_4^2 \qquad (155)$$

and allocate as follows:

$$\frac{\partial^2 f_k}{f_k \cdot \partial \xi_k^2} = -C_k^2. \qquad (156)$$

This gives the solutions:

$$f_k[\xi_k] = C_{k1} \cdot \cos[C_k \cdot \xi_k] + C_{k2} \cdot \sin[C_k \cdot \xi_k]. \qquad (157)$$

As discussed in chapter 2 and [5], this forces at least one coordinate to become time-like and results in photonic solutions. In the case of the human body, we may just say that we are talking about the means of interaction and thus photons of the human body or—as named in the title of this chapter—"humanitons."

As shown in the previous chapter 2, we can have this type of humaniton solutions in any number of dimensions.

3.9 Higher Numbers of Dimensions

But taking the association with the position of just one alien cell as used above at the end of sub-section "3.8.1 A Somewhat More General Case," we see that we would need a general solution to $F[f]$ for a very high number of dimensions.

In order to find this F function, we simply use the parabolic approach from the caption of Fig. 3.2 for the constant C_{1N} and obtain $c_0 = -3/2$, $c_1 = 7/4$, $c_2 = -1/4$. The corresponding function $F[f]$ for any number $n > 2$ would then be:

$$F[f] = C_2 \cdot \left(f \cdot (n-2) + C_1 \right)^{\frac{4}{n-2}}. \tag{158}$$

A direct derivation from the scale-transformed Ricci scalar will be given in chapter 8. We note that for $n \to \infty$, we obtain just a constant, namely:

$$\lim_{n \to \infty} F[f] = \lim_{n \to \infty} C_2 \cdot \left(f \cdot (n-2) + C_1 \right)^{\frac{4}{n-2}} = C_2. \tag{159}$$

As this would make the right-hand side of (100) (first line) to vanish, we conclude that in infinite dimensions the Ricci scalar must be zero. This might explain why in a universe with extremely many degrees of freedom and thus dimensions, the space appears to be cosmologically flat [25–32]. However, as there is a more elegant explanation for the so-called flatness problem [14], we will not propose such a dimensional reason here. Because when applying condition (101) and performing the limiting procedure after the derivation in (100), the result would not be a zero outcome, but:

$$\lim_{n \to \infty} R^* = \lim_{n \to \infty} \frac{1}{F[f]^3} \cdot \left(-C_{N2} \cdot F[f] \cdot \frac{\partial F[f]}{\partial f} \cdot \Delta_g f \right) = -\frac{4}{f \cdot C_2} \cdot \Delta_g f$$

$$\Rightarrow \qquad \left(\frac{\lim_{n \to \infty} R^*}{4} \cdot C_2 + \Delta_g \right) f = 0. \tag{160}$$

This gives us a linear differential equation in higher numbers of dimensions the moment we can assume the Ricci scalar R^* to be independent on f. Thereby the constant C_2 is yet unknown.

3.10 Conclusions about Our "Photonic Math" for the Description of the Human "Body System"

We found that the system "human body" (just as every complex and thus multidimensional system) including its psychological aspect would be described by the classical Klein–Gordon quantum equations of a quantum function f and its derivatives (Schrödinger and Dirac) as long as the Ricci scalar R^* would be independent on the quantum function f, respectively, could be approximated by a constant. In cases where the Ricci scalar is influenced by the quantum function/quantum effects, however, the resulting equations become highly nonlinear.

This result was not postulated as the classical quantum equations were, but it was derived from the Einstein–Hilbert action [8].

Thus, we conclude that our quantum gravity-based apparatus, motivated by Hilbert and Einstein and introduced here, is the most holistic approach there can be and may perhaps provide a suitable toolbox for better models of complex problems as they occur in medical fields such as immunology and affect logic (e.g., [17–19]).

References

1. N. Schwarzer, *Einstein had it, but he did not see it, Part XXXIX: EQ or The Einstein Quantum Computer*, www.amazon.com, ASIN: B07D9MBRS3.

2. N. Schwarzer, *The Einstein Quantum Computer: Mathematical Principle and Transition to the Classical Discrete and Quantum Computer Design*, www.amazon.com, ASIN: B07D9J5VLV.

3. N. Schwarzer, *Is there an ultimate, truly fundamental and universal Computer Machine?*, www.amazon.com, ASIN: B07V52RB2F.

4. H. Leuenberger, What is life?, a new human model of life, disease and death – a challenge for artificial intelligence and bioelectric medicine specialists, SWISS PHARMA, 2019, **41**(1), www.ifiip.ch.

5. N. Schwarzer, *Societons and Ecotons – The Photons of the Human Society: Control them and Rule the World, Part 1 of Medical Socio-Economic Quantum Gravity*, www.amazon.com, ASIN: B0876CLT7C.

6. N. Schwarzer, *The World Formula: A Late Recognition of David Hilbert's Stroke of Genius*, 2022, Jenny Stanford Publishing. ISBN: 9789814877206.

7. N. Schwarzer, *Mastering Human Crises with Quantum-Gravity-based but still Practicable Models – First Measure: SEEING and UNDERSTANDING the WHOLE,*

Part 3 of Medical Socio-Economic Quantum Gravity, www.amazon.com, ASIN: B087M321W9.

8. D. Hilbert, Die Grundlagen der Physik, Teil 1, *Göttinger Nachrichten*, 1915, 395–407.

9. A. Einstein, Grundlage der allgemeinen Relativitätstheorie, *Ann. Phys.*, 1916, **49** (ser. 4), 769–822.

10. N. Schwarzer, *Brief Proof of Hilbert's World Formula – Dirac, Klein-Gordon, Schrödinger, Einstein, Evolution and the 2nd Law of Thermodynamics all from one origin*, www.amazon.com, ASIN: B08585TRB8.

11. N. Schwarzer, *The "New" Dirac Equation – Originally Postulated, Now Geometrically Derived and – along the way – Generalized to Arbitrary Dimensions and Arbitrary Metric Space-Times*, www.amazon.com, ASIN: B085636Q88.

12. N. Schwarzer, *Science Riddles – Riddle No. 11: What is Mass?*, www.amazon.com, ASIN: B07SSF1DFP.

13. N. Schwarzer, *Science Riddles – Riddle No. 12: Is There a Cosmological Spin?*, www.amazon.com, ASIN: B07T3WS7XK.

14. N. Schwarzer, *Science Riddles – Riddle No. 13: How to Solve the Flatness Problem?*, www.amazon.com, ASIN: B07T9WXZVH.

15. N. Schwarzer, *Science Riddles – Riddle No. 14: And what if Time...? A Paradigm Shift*, www.amazon.com, ASIN: B07VTMP2M8.

16. N. Schwarzer, *Science Riddles – Riddle No. 15: Is there an Absolute Scale?*, www.amazon.com, ASIN: B07V9F2124.

17. L. Ciompi, *The Psyche and Schizophrenia – The Bond between Affect and Logic*, Harvard University Press, Cambridge/Mass. (USA) and London (GB), 1988, ISBN-10: 0674719905, ISBN-13: 978-0674719903.

18. L. Ciompi, *Die emotionalen Grundlagen des Denkens – Entwurf einer fraktalen Affektlogik*, (in German), Vandenhoeck & Ruprecht, Göttingen, 1997, ISBN 3-525-01437-6.

19. L. Ciompi, Non-linear dynamics of complex systems. The chaos-theoretical approach to schizophrenia, in: H.-D. Brenner, W. Böker, R. Genner, eds, *Toward a Comprehensive Therapy of Schizophrenia*, Hogrefe & Huber, Seattle-Toronto-Bern-Göttingen, 1997, pp. 18–31.

20. N. Schwarzer, *Einstein had it, but he did not see it, Part LXXVII: Matter is Nothing and so Nothing Matters*, www.amazon.com, ASIN: B07NQKKC31.

21. N. Schwarzer, *Einstein Already had it, But He Did not See it, Part LXV: Swing When You're Singing*, www.amazon.com, ASIN: B07JFMMVKQ.

22. N. Schwarzer, *Einstein had it, but he did not see it, Part LXXV: The Metric Creation of Matter*, www.amazon.com, ASIN: B07ND3LWZJ.

23. K. Schwarzschild, On the gravitational field of a mass point according to Einstein's theory (translation and foreword by S. Antoci and A. Loinger), arXiv:physics/9905030v1.

24. H. Haken, H. Chr. Wolf, *Atom- und Quantenphysik*, 4th ed. (in German), Springer Heidelberg, 1990, ISBN 0-387-52198-4.

25. www.astro.umd.edu/~richard/ASTRO340/class23_RM_2015.pdf.

26. en.wikipedia.org/wiki/Flatness_problem.

27. A. Friedman, Über die Krümmung des Raumes, *Z. Phys.* (in German), 1922, **10**(1), 377–386, Bibcode:1922ZPhy...10..377F, doi: 10.1007/BF01332580 (English translation: A. Friedman, On the curvature of space, *Gen. Relativ. Gravitation*, 1999, **31**(12), 1991–2000, Bibcode:1999GReGr..31.1991F, doi:10.1023/A:1026751225741).

28. A. Friedmann, Über die Möglichkeit einer Welt mit konstanter negativer Krümmung des Raumes, *Z. Phys.* (in German), 1924, **21**(1), 326–332, Bibcode:1924ZPhy...21..326F, doi:10.1007/BF01328280 (English translation: A. Friedmann, On the possibility of a world with constant negative curvature of space, *Gen. Relativ. Gravitation*, 1999, **31**(12): 2001–2008, Bibcode:1999GReGr..31.2001F, doi:10.1023/A:1026755309811).

29. A. H. Guth, Fluctuations in the new inflationary universe, *Phys. Rev. Lett.*, 1982, **49**(15), 1110–1113, Bibcode:1982PhRvL..49.1110G, doi: 10.1103/PhysRevLett.49.1110.

30. A. Linde, A new inflationary universe scenario: a possible solution of the horizon, flatness, homogeneity, isotropy and primordial monopole problems, *Phys. Lett. B*, 1982, **108**(6), 389–393, Bibcode:1982PhLB..108..389L, doi: 10.1016/0370-2693(82)91219-9.

31. St. Hawking, Th. Hertog, A smooth exit from eternal inflation, *J. High Energy Phys.*, 2018, arXiv:1707.07702, Bibcode:2018JHEP...04..147H, doi: 10.1007/JHEP04(2018)147.

32. S. F. Bramberger, A. Coates, J. Magueijo, S. Mukohyama, R. Namba, Y. Watanabe, Solving the flatness problem with an anisotropic instanton in Hořava-Lifshitz gravity, *Phys. Rev. D*, 2018, **97**, 043512, arXiv:1709.07084.

33. N. Schwarzer, *My Horcruxes – A Curvy Math to Salvation, Part 9 of Medical Socio-Economic Quantum Gravity*, www.amazon.com, ASIN: B096SPB5MW.

Chapter 4

Social Distancing

COVID-19 may give one reason, but why not discuss this a bit more generally and fundamentally?

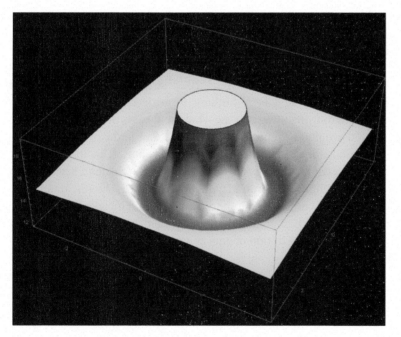

Figure 4.1 A social distancing potential.

The Math of Body, Soul, and the Universe
Norbert Schwarzer
Copyright © 2023 Jenny Stanford Publishing Pte. Ltd.
ISBN 978-981-4968-24-9 (Hardcover), 978-1-003-33454-5 (eBook)
www.jennystanford.com

4.1 Motivation

Social distancing is one of the suggested key recipes to deal with the COVID-19 crisis in order to keep the numbers of those patients in check, which need intensive health care in connection with this Coronavirus "pandemic" (c.f. section "Motivation by an Example" in this book).

However, a first principal socioeconomic investigation of societies shows hints of social distancing (see Fig. 4.1) or separation of a semipermeable character as a much more general concept to reduce social stresses, intrinsic society friction, and economic mismatches. If our assumption (being more a deduction from the solutions rather than an assumption) is correct, then this would have quite some consequences for business managers, financial analysists, and—naturally—politicians.

4.2 Open Letter to Ray Dalio

Dear Mr. Dalio,

With great interest I read your "2020 Strategic Report: Our Economic Outlook." In principle I have nothing to criticize, but I'm afraid I'm not completely wrong in noting that the whole analysis is pretty much "financially lopsided." I wouldn't even dare to point that out and contacting you in this way if it wasn't for the fact that in this report you used an analogy to physics by applying a comparison of the socio-economy with a ball having potential and (when falling to the ground) kinetic energy. On the one hand, in using this analogy one could assume that you see the financial markets in a much more holistic way than many others in the field, but, on the other hand, the report shows that the notion of energy, especially of the potential one, is not truly understood.

This is a drawback ... not just with respect to the report, but also to the conclusions you draw and the—little—advice you can give in the end. And what is more, it may also be seen as a disadvantage for your own company as—in not seeing the whole picture with all possibilities and POTENTIAL options—your decision-making might not be as perfect as it could be.

I'd like to give you an example of a way to describe the state in a certain (perhaps even global) socio-economy allowing a more direct and—hopefully—easier to achieve solution finding. Thereby I explicitly avoid the direct confrontation with the problems you have addressed and give a completely different one, but one which—as you will see—contains many of the others.

As this is an open letter and not an open paper; however, I refrain from presenting any math here. The abstract below just gives the essentials including the necessary references.

We keep things as simple as possible, which in our case just means that we want to build our example on as few numbers of dimensions as possible. Let a number of entities form a socioeconomic space-time (society). Let them have positions inside the society and let them have a socioeconomic parameter, summing up all their skills, law-abiding performance, innovativeness, and so on. One may say that the socioeconomic parameter sums up the ability of the entity to contribute to the society it lives in. Now we have the interaction of those entities within the society and this automatically leads to strain and stresses. Please note, stress is not bad in principle. After all, without stress, there is no contact and without contact, there is no interaction. But without interaction, there can also be no economy, no society. This holds in materials science as in human interaction alike. Just prove it by pressing your hand on the table or by trying to have contact with somebody without also having the slightest bit of stress. It is impossible ... you cannot have the one without the other.

Knowing this and being able to mathematically understand and model it, you now have the means to avoid bad stresses, respectively the evolution of such.

Let's simply assume that in your model society of entities from above, there are fairly big differences in the socioeconomic parameters among the various entities. Immediately your society results in high internal shearing stresses, subsequent frictional forces and potential yield, which is to say disintegration of the society itself. Politicians and most finance people would just look at such a constellation and call it a "failed state," with little hope to find a solution avoiding the big crash. But seen holistically, one can easily show that not all ensembles of entities provide the same amount of intrinsic (and dangerous) shear. In other words, even without the immediate ability to raise the low socio-economy parameter for the lower entities, but starting with a suitable arrangement in placing them, the overall friction can be dramatically reduced and resources can be freed to subsequently raise the necessary parameter minima, which are endangering the society via potentially causing the yielding problems.

Sincerely,

Norbert Schwarzer

4.3 The New Space-Time of Socio-Economy

The attentive reader will note that the following text is almost identical to an abstract this author entitled "The new Space-Time of Psychology," which was published on www.worldformulaapps.com and—after reading this text—the even more attentive reader will also immediately understand why there is absolutely nothing to wonder about.

(a) It obviously is no accident that our notion of socioeconomic spaces or space-times (or just societies) and their aspects is expressible in terms that we usually relate to things like the stability and reliability of objects or—in other words—to materials science. Obviously, it is not just an association or a metaphorical description when, talking about societies, we speak about friction, stress and strain, social defects, mismatches, dislocations, or—rather general—elastic and inelastic deformations. Taking the many forms of socioeconomic dissonances alone and considering their deforming effects regarding various time and spatial levels and scales, we cannot ignore the striking similarity to the same or mirroring situations in the world of energy and matter. But can this really come as a surprise in a society made out of entities, which are themselves evolved from matter and energy … and—what is more—are consisting of it?

So we ask:

Is it possible that there is a much deeper connection between the understanding of socio-economy and a most general theory of matter than this similarity of words and expressions?

Before falling under the suspicion of esoteric waffle, the author wants to point out that—after all—the origin of all our social activities is based on the interaction of electromagnetism, while the latter—of course—is based on the fundamental laws of matter. Thus, even though we are here talking about the interaction of intelligent entities in the field of socio-economy, there may well be/has to be such a deeper connection, because the evolution of bigger and bigger brains [1] has had its starting point within the same fundamental laws that are creating, forming, and permanently reshaping our matter surrounding. It is, therefore, most likely that the matter-based interactions, which bring about our psychological self (our feelings, self-awareness, consciousness, and our whole cognitive being) and the subsequent social interaction, are following the same rules. Then, however, it should also be possible to apply these rules to the field of socio-economy.

(b) Assuming the existence of a true "Theory of Everything" [2, 3], one should, of course, be able to extract from this theory all physical interactions that are effectively creating the complex pattern of a dynamic society. Additionally incorporating what was said under point A, it has to be possible to formulate a socioeconomic theory in consistent mathematical form (similarly to [4] in classical medicine or [5] in philosophy) and taking its fundamental origin, one might name it a quantum gravity socio-economy.

(c) And how to incorporate the omnipresent substructuring and formation of social and economic clusters in such a theory? Well, here we have to understand that substructures (like the family as the potentially smallest cell, groups, clusters of individuals of certain interests, and so on) are equivalent to the formation of gravity centers in a quantum gravity theory. It was shown in [2, 6, 7] that this automatically brings about the necessary internal interactions and thermodynamic effects. Just like the formation of matter, we thereby also need to understand the omnipresent Russian doll structure of such cluster effects, where smaller clusters form bigger ones, the bigger ones again can cluster together, and so on. Some boundaries can dissolve, while principal ones have to remain intact as their dissolution would be equivalent to the biological cell death. Permanent formation and reformation of clusters provide one important aspect within the dynamic society.

(d) The new approach should be seen as an essential part for the realization of a "Virtual Society" just like the holistic "Virtual Patient" concept in medicine as proposed by Leuenberger [8]. It is clear from the holistic and top-down character of this approach that the simulation and understanding of incidents like the recent COVID-19 or Corona event would be very straightforward and also allows for the consideration of a wide range of uncertainties as these are automatically contained in a quantum gravity-based approach.

4.4 The Fundamental Equation for Everything

In 1915, David Hilbert [9] was able to show that a mathematical structure, very similar to a volume equation (a volume integral to be precise), apparently contained Einstein's famous General Theory of Relativity [10], which, as we all know, is a theory about gravity. Thereby the fascinating aspect was that something so very much physical, like gravity, came out of a completely mathematical source, namely, Hilbert's "volume integral"[1]. In fact, it is a bit more than just a "volume integral," but an integral that actually looks for an extremum, which means maximum or minimum, of the volume result.

More than one hundred years after these groundbreaking works of Hilbert and Einstein, we were able to show that not only gravity resided inside the Hilbert equation, but obviously just everything [2].

But why, with the Hilbert equation already being there for so long, wasn't this fact discovered much earlier?

In order to grasp the implications here, we need to understand that even though the Hilbert equation looks quite simple on first sight, it has many degrees of freedom and a fairly complicated intrinsic structure. Therefore, this author suspects that Hilbert, Einstein, and many others simply have not seen all the possibilities the apparently so simple "volume integral" offered. For one thing, as this is going to be the one, we are most interested in here, obviously nobody ever bothered about, was investigating the Hilbert equation with respect to the number of dimensions in which a certain problem is considered[2]. Almost everybody usually just concentrated on our "classical" 4-dimensional space-time.

4.5 From Hilbert via Klein–Gordon to Dirac

For convenience we now repeat our evaluation from the most fundamental starting point, namely, the classical Einstein–Hilbert action [9] with the "special" Ricci scalar R^* as kernel or Lagrange density:

$$\delta W = 0 = \delta \int_V d^n x \left(\sqrt{-G} \cdot R^* \right). \tag{161}$$

In contrast to the classical form in [9], we consider a somewhat adapted metric tensor $G_{\delta\gamma} = F[f] \cdot g_{\delta\gamma}$. Thereby G shall denote the determinant of the metric tensor $G_{\alpha\beta}$, while g will later stand for the corresponding determinant of the metric tensor $g_{\alpha\beta}$. In order to distinguish our Ricci scalar, being based on $G_{\delta\gamma} = F[f] \cdot g_{\delta\gamma}$ from the usual one, being based on the metric tensor $g_{\alpha\beta}$, we

---

[1]Please note that this "volume integral," in the literature, is usually known under the expression "Einstein-Hilbert action."

[2]Yes, there are many n-dimensional solutions to the Einstein field equations, but this is not what is meant here. We are looking for variations of the Einstein–Hilbert action with respect to the number of dimensions (e.g., [9]).

marked it with the * superscript. Please note that (1) should allow us to just say that W is a constant. Then, of course, with W being a constant, it would automatically variate to zero. As a sharper (less general) condition, one may even assume that:

$$\sqrt{-G} \cdot R^* = \text{const}, \tag{162}$$

or we even may demand:

$$\sqrt{-G} \cdot R^* = 0 \tag{163}$$

with either $G = 0$ or $R^* = 0$, whereby the latter case definitively is less trivial and thus probably also makes more sense.

Please note that with $G_{\delta\gamma} = F[f] \cdot g_{\delta\gamma}$ we have used the simplest form of metric adaptation that we could construct as a simplification from a generalization of the typical form of tensor transformations, namely:

$$G_{\alpha\beta} = F\Big[f\big[t,x,y,z,\ldots,\xi_k,\ldots,\xi_n\big]\Big]_{\alpha\beta}^{ij} g_{ij}$$

$$\rightarrow G_{\alpha\beta} = F\Big[f\big[t,x,y,z,\ldots,\xi_k,\ldots,\xi_n\big]\Big] \cdot \delta_\alpha^i \delta_\beta^j g_{ij}$$

$$\rightarrow G_{\alpha\beta} = F\Big[f\big[t,x,y,z,\ldots,\xi_k,\ldots,\xi_n\big]\Big] \cdot g_{\alpha\beta} = F[f] \cdot g_{\alpha\beta}. \tag{164}$$

The generalization is elaborated in [2].

Now we evaluate the resulting Ricci scalar for the metric $G_{\alpha\beta}$ at first in n and then, subsequently, simplify to 4 dimensions:

$$R^* = \frac{1}{F[f]^3} \cdot \left(\begin{array}{c} \left(C_{N1} \cdot \left(\frac{\partial F[f]}{\partial f}\right)^2 - C_{N2} \cdot F[f] \cdot \frac{\partial^2 F[f]}{\partial f^2} \right) \cdot \overbrace{\left(\tilde{\nabla}_g f\right)^2}^{= f_{,\alpha} g^{\alpha\beta} f_{,\beta}} \\ - C_{N2} \cdot F[f] \cdot \frac{\partial F[f]}{\partial f} \cdot \Delta_g f \end{array} \right)$$

$$\xrightarrow{n=4} = \frac{1}{F[f]^3} \cdot \left(\begin{array}{c} \left(\frac{3}{2} \cdot \left(\frac{\partial F[f]}{\partial f}\right)^2 - 3 \cdot F[f] \cdot \frac{\partial^2 F[f]}{\partial f^2} \right) \\ \cdot \left(\tilde{\nabla}_g f\right)^2 - 3 \cdot F[f] \cdot \frac{\partial F[f]}{\partial f} \cdot \Delta_g f \end{array} \right)$$

$$\xrightarrow{\frac{3}{2}\left(\frac{\partial F[f]}{\partial f}\right)^2 - 3 \cdot F[f] \cdot \frac{\partial^2 F[f]}{\partial f^2} = 0} = -\frac{3}{F[f]^2} \cdot \left(\frac{\partial F[f]}{\partial f} \cdot \Delta_g f\right). \tag{165}$$

With respect to the constants C_{N1} and C_{N2} in arbitrary n dimensions, the reader is referred to [2] or chapter 8 in this book.

With $R^* = 0$, one would immediately satisfy the Einstein–Hilbert action. It will be shown later in this book that this simple scalar condition can also directly be derived from the Einstein–Hilbert action with a scaled metric. Taking the general n-dimensional case in the first line of (165) and always demanding that:

$$C_{N1} \cdot \left(\frac{\partial F[f]}{\partial f} \right)^2 - C_{N2} \cdot F[f] \cdot \frac{\partial^2 F[f]}{\partial f^2} = 0, \tag{166}$$

which simply is a fixing condition for the wrapper function $F[f]$, we obtain:

$$R^* = -\frac{C_{N2}}{F[f]^2} \cdot \left(\frac{\partial F[f]}{\partial f} \cdot \Delta_g f \right) = 0$$

$$\Rightarrow \qquad\qquad \Delta_g f = 0. \tag{167}$$

This is the "homogeneous" Klein–Gordon equation, respectively the mass- and potential-less Klein–Gordon equation.

So, the question would be where to get the mass from?

4.5.1 Entanglement as Origin of Mass (and Potential) and Thus Possible Also Any Kind of Inertia and Interaction (Including Interaction in Socioeconomic Space-Times)

It was shown in [11] and [12] (or more comprehensively, see also [2] and the part "The third day …" in this book) that mass automatically results from certain forms of entanglements of dimensions.

Introducing a metric for the coordinates $t, r, \vartheta, \varphi, u, v$ of the form:

$$g_{\alpha\beta}^6 = \begin{pmatrix} -c^2 & 0 & 0 & 0 & 0 & 0 \\ 0 & 1 & 0 & 0 & 0 & 0 \\ 0 & 0 & r^2 & 0 & 0 & 0 \\ 0 & 0 & 0 & r^2 \cdot \sin[\vartheta] & 0 & 0 \\ 0 & 0 & 0 & 0 & g[v] & 0 \\ 0 & 0 & 0 & 0 & 0 & g[v] \end{pmatrix} \cdot f[t,r,\vartheta,\varphi], \tag{168}$$

with the entanglement solution for u and v as follows:

$$g[v] = \frac{C_{v1}^2}{C_{v0}^2 \cdot \left(1 + \cosh\left[C_{v1} \cdot (v + C_{v2}) \right] \right)}, \tag{169}$$

gives us the following Ricci scalar from (165) with condition (7):

$$R^* = \frac{1}{F[f]^3} \cdot \left(-C_{N2} \cdot F[f] \cdot \frac{\partial F[f]}{\partial f} \cdot \Delta_g f \right); \quad F[f] = C_f + f[t, r, \vartheta, \varphi]$$

$$\xrightarrow{6D} R^* = \frac{5}{f^2} \cdot \left(\frac{C_{v0}^2}{5} \cdot f + \frac{\partial^2 f}{c^2 \cdot \partial t^2} - \Delta_{3D-\text{sphere}} f \right); \quad f = f[t, r, \vartheta, \varphi]; \quad C_f = 0.$$

$$(170)$$

Here, $\Delta_{3D-\text{sphere}} f$ denotes the Laplace operator in spherical coordinates.

As for $R^* = 0$ and $\dfrac{C_{v0}^2}{5} = \dfrac{m^2 \cdot c^2}{\hbar^2}$ (m = mass, c = speed of light in vacuum,

$\hbar$ = reduced Planck constant), we have obtained the classical quantum Klein–Gordon equation with mass and therefore conclude that mass can obviously be constructed by a suitable set of additional (meaning in addition to our 4-dimensional space-time) entangled coordinates.

Having achieved this, we only need to follow the classical Dirac path as elaborated in [13] for the extraction of the Dirac equation as square root of the Klein–Gordon equation by the introduction of quaternions. As this is dull transcription work, which adds no knowledge whatsoever, we refrain from representing it here, but simply refer to Dirac's original work [13]. Alternatively, the reader could also use [2]. Regarding an alternative, direct evaluation of the Dirac equation from the Ricci scalar of a scaled metric, the reader is referred to [14] (first trials) and chapters 9 and 10 of this book.

4.6 Modeling Socioeconomic Space-Times

Now with all major equations being traced back to the Einstein–Hilbert action respectively the Ricci scalar and assuming that also a human society (or socioeconomic space-time) just follows the principal laws of nature, we may conclude that models describing societies should also be based on the metric equations presented and derived above (naturally, in potentially very high numbers of dimensions and with many centers of gravity).

The complexity of any such metrical very first principal socioeconomic system grows rapidly with its size (number of entities and their properties). We, therefore, resort to the option of simplification via the introduction of an "elastic" society as already outlined in the abstract. The corresponding math has been given in a variety of papers (e.g., [4, 5, 15–20]), but with respect to a comprehensive introduction and convenience, we suggest the section "14.1 What is Good?" and chapter "10 An Elastic World formula" in [2]. As for a real-sized simulation probably only supercomputers or even super quantum computers will allow for sufficiently complex socioeconomic space-times, we here apply an extremely simplified model where we consider the socioeconomic performance of groups of entities as a kind of summed-up parameter. This performance can be positive (contributes to the society in a synergetic and symbiotic manner) or negative (foe-like or parasitic behavior). As we have seen in the chapter above that all sorts of inertia appear in compactifiable shape within their own space-time (e.g., see solution $g[v]$ from (169)), we apply such "gaussian-like" distributions as results of our quantum-gravity equations [2, 20] to the various groups of entities we intend to consider. Figure 4.2 shows a selection of such distributions of equivalent Gauss shapes for our inertia (entities) as we will apply them further in this chapter.

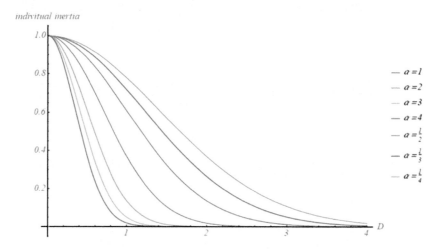

Figure 4.2 Equivalent Gauss shapes for individual gravity centers as considered here with the approximate (one-dimensions) dependency $f(x) = e^{-\alpha \cdot D^2}$, where the symbol D stands for "distance." Please note the D can stand for a whole set of dimensions, meaning $D^2 = x^2 + y^2 + z^2 + \dots$

Now we apply the math for the elastic space-times and evaluate the second invariant of the deviator strain tensor. As known from the theory of elasticity (e.g., [21]), this is just formed as difference of the strain tensor minus the unit tensor times the first strain-tensor invariant (usually divided by the number of dimensions) and therefore nothing special, which is to say materials science textbook material. The result gives us a scalar field of all shear as being distributed over the whole society. Shear, however, can destroy every society as it can also destroy material. The process is called "yielding" or "plastic yielding." Considering a society or any other socioeconomic system, it is therefore of great interest to find states of interaction strains with the minimum total shear as this would give the socioeconomic state the best performance. Integration over the whole socioeconomic space-time results in the total shear of the whole society.

Now we want to investigate this total integral shear as a function of the distance of two different groups of entities of only slightly different socioeconomic performance. Thereby this total performance not only determines the widths of the group's individual inertia, but also its height and sign. For simplicity we consider almost equal heights, but different widths (see Fig. 4.2).

The results are given in Figs. 4.3 and 4.4, which show that there is always a minimum of total integral shear for a certain distance between the entities, respectively groups of entities.

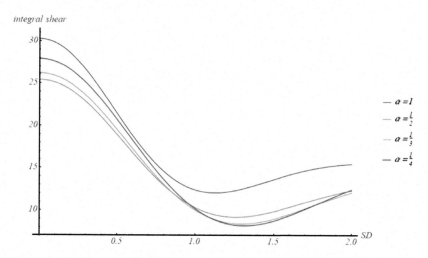

Figure 4.3 Total integral shear for a variety of entity pairings (one always having an equivalent $\alpha = 1$, while the other α are given in the legend) as a function of the social distancing, which is here abbreviated by SD.

We also find that there is extreme shear increase for closer packaging beyond the minimum, which is so for SD values (SD = "social distancing") below the various optima for our entity ensembles. Interestingly, we also find an integral shear increase for growing social distancing beyond the point of optima, which is to say no matter what social entities we are talking about, be it corporation, individuals, assets of any kind, ethnic groups, and so on, there is always not just a minimum distance these entities have to be kept away from each other, but it also holds that the total performance of the whole system decreases with such distances growing too big.

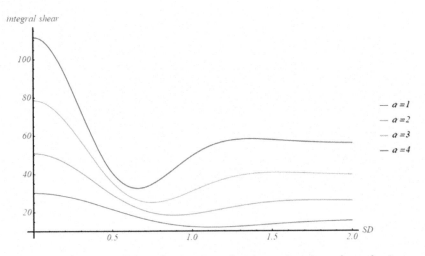

Figure 4.4 Total integral shear for a variety of entity parings (one always having an equivalent $\alpha = 1$, while the other α are given in the legend) as a function of the social distancing, which is here abbreviated by SD.

4.7 Interpretation

Knowing that social shear is always better to be avoided and finding here that there obviously are optimum distances among the individual cells of the society with all their degrees of freedom (not just the spatial distance), the author is aware of the conflict potential behind this finding and therefore prefers an interpretation in a more figurative rather than scientific manner. Thereby we will see that our finding could be seen as some kind of "universal wall principle," which, motivated by a biological analog, we also named "the Lipid principle" elsewhere [22]. This name was created by T. Bodan, who had also written the 7-days introduction part of this book and from him are also the original examples or illustrating associations we will reuse here.

4.7.1 The Universal Wall Principle—Why Everything in Existence Needs Boundaries

"Who believes that a liberal society needs open borders has not understood that freedom requires the protection of property." [23]

When German star economist Hans-Werner Sinn said this in connection with the BREXIT, probably only very few did realize that Prof. Sinn had just talked about a very principal, if not to say, universal law. Perhaps not even Prof. Sinn himself did know at that time that there wouldn't even be our universe if it weren't for boundaries leading to an internal structure of the whole space and only this way allowing it to BE in the first place. The slogan "property needs borders" is so universal that we consider it worthwhile to look a bit closer at this principal law.

We will see that we can generalize Prof. Sinn's sentence to:

"Creation needs boundaries"

Most people know of course that life would never have had the slightest chance to come into existence if it weren't for the formation of membranes. Only with membranes the first cells could become stable and evolve after they had learned to establish a border between their "inner self" and the "outer world." Only after the introduction of semipermeable walls, the genesis of so-called Coacervates (which means heaped or agglomerated, and thus the first quasi cells) and the ability to defend such boundaries, life was able to become true life. Interestingly, the physical effect of self-organization helped a lot in this process. If truth be told, this effect was so crucial for the genesis of life that it seems to be important to look out for more deep down, meaning even more principal forms of self-organization even before life. The question comes up whether perhaps the whole universe was created by such a self-organizing process?

With respect to life—the first cells—the process could be described as follows:

Stage 1: Certain molecules (lipids) orientate themselves along phase boundaries (in the primordial soup, this could just have been the water–air boundary).

Stage 2: Micelles are forming, which is to say small conglomerates and liposomes are coming into being.

Stage 3: This way also the so-called double membrane is formed.

Stage 4: This membrane is stabilized by the attachment of proteins.

As said before, without such primitive borders, life would never have been possible and the same holds for its permanent evolution. Yes, with respect to the maintenance of optimum distances between individual biological entities, there was an evolution of "walls" and ever "cleverer" ways of defense and—what is more—without it, there would not have been any other, which is to say further evolution at all. Life would not have come over the threshold of its first beginning and evolve into such manifolds of forms and shapes as we can see today. It is no wonder that exactly at places somehow confined and distinguished from a surrounding area, life almost exploded in various structural solutions ... all of which properly separated and are distinguishable from each other. This alone should clearly demonstrate the principal importance and universality of distances and boundaries for everything having to do with forward development and evolution, with creation, social-economic structuring, technological, medical and scientific progress, and—of course—property.

However, studying Figs 4.2 to 4.4 again, we see that it strongly depends on the interaction whether there is a minimum (optimum distance) at all. Surely, there are also interactions where bosonic behavior is possible and everything can come together to just one mathematical point (at least at extremely low temperatures). This, however, does not hold for interactions allowing for the formation of life.

Because of this universality regarding the existence for optimum distances for certain interactions and thus boundaries for everything having to do with evolution and development, we want to give this principle a name. In the German language and within a scientific article, we chose the rather boring name "Lipid-principle," which was mainly due to the omnipresent censorship meanwhile haunting almost the whole world except for a very few countries. In American English, however, we think we better give it a name with which everybody immediately can associate something. We name it:

The universal WALL principle.

Before we want to show that this principle, most people only connect with living forms and life in general, is even more basic, which is to say it created our whole universe, we want to consider a few examples. These examples shall give us a bit of insight into the severity of consequences if the principle is being ignored and boundaries or borders are not being respected and defended.

For a very good reason, our journey starts in the animal kingdom and from there we chose something up-to-date. In Australia lives a rather cute creature that belongs to the marsupials. Its name is Wombat. It lives underground and can reach up to 70 pounds of weight. This harmless herbivore is—as so many other species, too—at the brink of extinction. Not so much holistic thinkers like to blame the climate change, but that's not even half of the whole picture. The truth is that the introduction of foreign species, of alien plants, and those cute, big-eyed little rabbits had laid the foundation for the Wombat's extinction. The Wombat can't eat the strange plants; they are like poison for it; however, the rabbits not only eat all the remaining vegetation, but also reproduce many, many times faster than the Wombat can. While for certain types of Wombats, like the northern hairy-nosed Wombat, the population went down to just one hundred exemplars, the number of rabbits in the same area, however, went up to a few millions and there is still no holding it up.

We see that caused by the disregarding of boundaries a whole population (or several of such) was crushed out. It might well be that some readers think that the original author, Bodan, has used this example because of the big-eyed, fast reproducing, aggressive, and invasive rabbit aliens and some similarities in other fields (like the mitten crab invasion that almost completely wiped out European noble crayfish). Well, perhaps they are right, but if it was so, then it was a rather unconscious decision, which brings us right to the next example.

The Roman Empire, as most historians agree, would never have fallen apart in the way it did in the end, if it had not started to neglect its borders. If the Romans not had started to uncontrollably flood or allow to flood their territory with primitive and uncultivated hordes from the north, their Empire might still exist today. Of course, the Barbarian Invasion within the so-called "Migration Period" might have been triggered by external, perhaps climatic changes or extremes, but this did not cause the collapse of the Roman Empire. With properly defended boundaries, so reasonable historians claim, the Empire could still have maintained its integrity and thus survived also the difficult times of those Barbarian attacks. It should be pointed out that "properly defended" does not necessarily mean "to close." No, as with semipermeable cell walls, it—of course—stands for a well-controlled, which is to say selective, in- and outlet of people, goods, and—necessarily—troops. Taking the example of the Roman Empire again, it might have sufficed to disarm the barbarian hordes before letting them in.

In the next example we consider a virus. Thereby we need to point out that we do not consider the current form of "detection" and "characterization" of viruses even nearly scientifically acceptable. Thus, the classical way of "verifying the existence" of viruses [30] or the modern way of "virus assembly via computer" may suffice in virology to distinguish a certain "viral" disease from another, but this is by no means a scientific proof for the classical virus concept. This needs to be emphasized before introducing our virus example. This assumed to be non- or half-alive parasite needs to smuggle its own genetic code into a living cell in order to force it to start to produce virus codes and virus proteins that form thousands of new little virus parasites until the cell is blasted apart and the virus is set free. Immediately it starts to seek more cells to capture and to bring them also under its control and this way to reproduce further and further ... until, yes, being stopped by the immune system and properly defended cell walls. It is not a coincidence that the disintegration of the cell wall is often used as the definition for the moment of cell death.

It is interesting that the original author, who has first published this set of examples in German, merely to start a reasonable and open discussion about these findings, immediately was attacked by certain left-greenish do-gooders and no-border fetishists because of these very examples ... especially the one with the wombat. But after there was nothing factually wrong about this association, they started concentrating on the virus example. After all, so it seems, somebody who scientifically finds that boundaries, walls, even borders are obviously essential for the existence of all living structures, including societies, simply has to be attacked. Such people, who, according to their own motto, "only want to make the world a better place" obviously have big problems in recognizing fine differences. For instance, they permanently mix up true charity, meaning to "give one's own property to a few in need," with giving property they never owned and they never had acquired any right to. In their inability to see things holistically, they slaughter whole civilizations in their attempt to equalize everything and everyone (except their very own property, of course). It was this kind of people who now attacked the original author about the illustrative examples above and most of all the virus one. They argued that the author has only used this example to illustrate the fact that northern African and Arabic migrants usually (which means in average) do not contribute to their Western host countries but cost by far more than they bring in. What is more, statistically, they contribute above average with the number of crimes (e.g., [24]) even in the third or fourth generation. Proper and honest evaluations show that migrants of black-

African or Arabic origin commit about three to four times more crimes than natives. The statistic is similar to the American one with—of course—a few substitutions regarding the types of "aliens" we are talking about. Neither the original nor the present author could help these facts, but in just using examples, which draw associations in the minds of the oh so "woke," they were immediately attacked as "racists" or "Nazis." This, however, is typical, because the "woke" were not interested in the facts, but only in suppressing the truth, which obviously clashed rather hard with their ideology. Thereby it did not help of course that in addition to the parallels the "woke" already worked out themselves, the reproduction rates of such aliens is usually much higher than the one of the native population. The funny thing now is that those do-gooders accused the author to have used the example of the virus in order to "demonstrate" that the migrants behave like viruses while the natives are the cells being abused by those viruses to help them reproduce, thereby being annihilated themselves.

Well, once again, the original author gives credit to his critics in this point, because they do have found an interesting parallel here. Still, Bodan would never have come up with such a bold comparison by himself. But a great difference needs to be pointed out: Namely, viruses cannot reproduce at all without other cells, at least not if following the classical virus concept [30]. Thus, the parallel the "woke" wanted and still want to see does not exist. This does not mean, however, that the concepts of optimum social distancing and semipermeable walls should not be seen as of great importance in societies also with respect to viral and viral-parasitic entities in such very societies. Suppressing a scientific discussion will never help but only lead to disaster. History is full of suitable examples, but for here and now "the great leap forward" and others of Mao's ridiculous ideas may help to get the gist about the danger of an ideologically restricted science.

And again: This whole association with the migration problems wasn't even part of the German version of this little text, but had been brought up by the red-greenish do-gooders, equalizer fetishists, "woke," and aggressive-invasive alien friends and this is racism at its best. Such a comparison, namely, is both scientifically incorrect (with many aspects) and ethically unworthy.

All these examples show one thing more than clearly: It is not only property that requires distancing, boundaries, borders, frontiers, but also life itself cannot be without a well-defined and properly marked, respected, and—if need be—defended periphery. The fact that these boundaries are—usually—, respectively, even have to be semipermeable does not change the principle of their need for life in general. In the contrary, it does point to other extremely important facts, namely:

- A healthy cell only lets pass through its own cell walls what it needs.
- The passage of toxic or harmful substances, parasites, or any kind of vermin over the boundaries of an organism poisons, captures, damages, or kills that organism.
- Boundaries need to be defended and at this point life does not know about mercy or limits.

4.7.2 The Generalization of the WALL Principle

Now, the most fascinating aspect about the Wall principle resides in the fact that it did not came into this world after the genesis of life, but that it was there long before. It was used by the universe right from its beginning. De facto, the universe could not have been without it and we are going to prove this sentence here and now.

It should be clear that a universe with the ability to create life must be able to work with and store information.

Thus, our first question has to be: What is the simplest and general form a universe must have in order to be able to store information?

In order to answer this question—most generally answer it, so that nobody would argue with us about excluding certain universes and call us universophob or so—we simply state that such a universe has properties. These properties form a thing we name space and in dependence on the number of properties the space has a certain number of dimensions. There are many possibilities for such a space of properties. It could range from tastes, smells, colors, or just length in certain directions we name x, y, and z. Thus, we could call such properties the coordinates of our space. When now measuring a certain small distance in our space, we find that we could do this measurement in two different ways:

$$ds = \sqrt{g_{ij} dx^i dx^j}$$

$$ds = \sqrt{g^{nm} dx_n dx_m}. \tag{171}$$

This is extremely interesting, no matter in what kind of universe we would be in, any smooth property's length could be measured in such two ways. The interested reader can check the evaluation in [1, 3, 22], but for here and now it totally suffices to know that these two ways of measuring and describing a piece of length in the space of our arbitrary universe are giving the same thing. Thus, when subtracting one from the other, the result—of course—should be zero. Scientists call such a thing, such a difference, an intelligent naught:

$$ds^2 - ds^2 = 0 = g_{ij}dx^i dx^j - g^{nm}dx_n dx_m. \tag{172}$$

Now we simply ask ourselves what kind of structures our space can have in order to fulfill the condition the intelligent naught requires in order to give, respectively, to be naught. We perform the following derivation:

$$0 = ds^2 - ds^2 = g_{ij}dx^i dx^j - g^{nm}dx_n dx_m$$

$$= g_{ij}dx^i dx^j - g^{nm}d(p_n f_n)d(p_m f_m)$$

$$= g_{ij}e^i e^j dy^2 - g^{nm}dy(p_n f_n)dy(p_m f_m) \quad / : dy^2$$

$$\frac{0}{dy^2} = g_{ij}g^{ik}\frac{\partial f}{\partial x^k}g^{jk}\frac{\partial f}{\partial x^k} - g^{nm}(p_n f_n)(p_m f_m)$$

$$= g^{ik}\frac{\partial f}{\partial x^k}e_i g^{jk}\frac{\partial f}{\partial x^k}e_j - g^{nm}(p_n f_n)(p_m f_m)$$

$$= g^{ik}\frac{\partial f_i}{\partial x^k}g^{jk}\frac{\partial f_j}{\partial x^k} - g^{nm}(p_n f_n)(p_m f_m)$$

$$= g^{ik}\frac{\partial f_i}{\partial x^k}g^{jk}\frac{\partial f_j}{\partial x^k} - g^{ij}(p_i f_i)(p_j f_j) = 0. \tag{173}$$

The results are oscillations and we here refer to [22] for the derivation, because we have meanwhile learned that also the much more elaborate intelligent naught coming from the Hamilton minimum principle (161) just gives oscillations (c.f. sections from 3.6 to 3.8). This means our space, in order to fulfill this intelligent naught condition, no matter whether it comes from a primitive zero like (172) or the Hamilton minimum principle, which is to say in order to be at all or—even more drastically put—to be allowed to be at all, must JITTER. As this holds for any arbitrary universe with any arbitrary number of dimensions or properties, our question from above comes down to the following one:

In what kind of permanently oscillating space can information be stored?

The answer is pretty simple. When our space jitters, one could store information within the jitter, but in order to do this properly, which is to say in a lasting manner, the modes of such oscillations must be properly distinguishable. In order to understand this, we simply imagine a piano.

There the modes of oscillations are clearly distinguishable notes one can let sound by pushing a certain key. I can easily imagine to code any information just within a set of such notes. Now imagine a string on a guitar. Imagine you can only continuously produce various tunes by increasing or decreasing the tension of this string. Without further definition or rules, for example, by defining discrete tensions or introducing the so-called frets, it is not possible for you to clearly code information with this string. Thus, we would like to have keys and clearly distinguishable tunes (notes) for your universal space and its oscillations. This, however, can only be achieved by putting boundaries into our space or space-time.

Here we have a problem, because, as space is smooth and continuous, there are no such boundaries!

Well, of course, when considering the space at scales of everyday experiences, space and time are smooth, but when going to smaller and smaller scales one suddenly experiences funny properties leading to the conclusion that at such smaller scales space is not as homogeneous as we thought. Over 100 years ago, science was pretty flabbergasted when discovering the funny behavior at smaller scales. This was at the beginning of last century and to be correct, it wasn't science that was flabbergasted, because science is a thing, but the scientists of course, who were astounded and some of them were less dumbfounded than others. It was interesting to see that the more taken aback were those who also were the ones with rather dogmatic ideologies in their heads. This old scientific elite with its almost religious believes and convictions had to be washed away to allow the truth to be given a chance to come forth. So, it took a while until intellect won over ideology and dogma. After realizing that these effects somehow had to do with small portions of things such as energy, mass, momentum, and so on, the scientists named the theory describing such effects Quantum Theory. What the scientists did not realize—even though some of them suspected it all along—was that all this quantum stuff comes from a structured space. In fact, if using our intelligent naught from above, where we were putting together the two ways to measure distances in space, we derive exactly at those equations and solutions already known from Quantum Theory. Interestingly, the structured space not only brings about the Quantum effects we see, but also makes the permanent jitter of space discrete, which means it makes space fit for the storage of information and thus, consequently, life.

But why should space or space-time have started to form such structures in the first place?

The answer to this question is a little bit demanding and we are going to consider quite some possible reasons for such "quantizations" in the next chapters of this book. In order to have something figurative (non-mathematical) to start with, however, we require the results of the work of a man who probably was one of the greatest human geniuses this universe has brought about. We are talking about Albert Einstein and his famous field theory. We have already discussed the fact that the intelligent naught forces the space to jitter. The jitter however, is nothing else but energy and a space containing energy must curve. This is what Einstein's field theory demands. The curvature of space thereby depends on the distribution of the jitter (energy). As there is no reason to assume any inhomogeneities (after all, the space starts in a completely smooth way), we will obtain result in a fairly homogeneous curvature. Thus, we either have (multidimensional) planes, spheres, or hyperboloids (see chapter 12). Taking this result from the Einstein field equations, one simply needs to imagine the space curled into equally sized spheres in order to see that each sphere is a separate little universe of its own. The boundaries of such spheres are the universal WALLs we would need for our space to be able to store information.

There is one more thing to learn, namely, that the energy stored or confined within the little spheres also must be finite. Otherwise the curvature would go to infinity and the spheres would simply be crushed to mathematical points, which is to say to nothingness. Perhaps the whole even came from there, which makes nothingness equal to infinity, but let's not bother about this aspect here and now. Anyway, with the jitter energy being finite, however, so must be the curvature and thus also the boundaries of the spheres (or hyperboloids) cannot be infinitely high potential walls. From this directly follows that these boundaries must be semipermeable with respect to oscillations (standing waves) being transmitted from one such sphere or hyperboloid to another. This automatically provides a kind of "decision-making process" for the universe being structured like this. One simply imagines an oscillation being confined to a certain number of substructures (spheres). There could be states where it would be completely coincidental whether such an oscillation concerns only one sphere or two or three and so on. When the "decision" is made, however, over how many spheres the oscillation should stretch, the state of energy may manifest the "conclusion" of the spatial "decision-making" process and the result is stored as a permanent oscillation ... or standing wave. This process could be considered as a very basic way of calculation, and thus the calculating space might be considered a very natural, if not to say UNIVERSAL, computer machine [25–27]. The results of its evaluations are complex forms of matter, stars, planets, and yes also plants, animals, and, of course, us humans. Even something like a typical politician or—even worse—polit-sycophant scientist is nothing but a result of such spatial calculations (not a very good result, if you ask me, but after all the spatial computer uses trial and error and so naturally ... well, we all know).

As said before, one can perform this kind of derivation with almost any kind of smooth space and always obtain results in the same jittery stuff producing substructures, bringing about boundaries allowing the space to perform evaluations and store information. Thus, Quantum Theory and Einstein's General Theory of Relativity (nothing else are these Einstein field equations) are providing the ingredients, which is to say the driving forces, for the creation of everything, when there is only a bit of space or something like a collection of properties we could call space.

The recipe seems to be as follows:

Properties form spaces, respectively, space-times. Then Quantum Theory brings about jitter, the jitter is nothing but energy, and the General Theory of Relativity immediately forces the jittery space to curl into sub-spaces leaving boundaries and thus WALLS. These boundaries require what is needed to make the space a basic computer and then the divine—spatial—calculation, which is to say creation can start. The interesting point we already have worked out here (sections 2.2, 2.4, 3.4, and 4.5) is that we just need the Einstein–Hilbert action to already have it all. By just applying scaled metrics, we automatically have the Einstein field equations AND quantum equations and the two together give us curvature and jitter.

One could even say that the jitter and the subsequent substructures are a kind of Coacervates of first order. Then stars, galaxies, and planets would be such things of higher orders and those Coacervates in the primordial soup appear already the x-ordered ones. Thus, Coacervates had to form many times in the universe and on many scales, too, before life was possible at all. Just in order to name the most important steps:

(a) In order to create a "calculating space" (also being able to store information), substructures with boundaries are of need.

(b) Then many spatially limited structures such as stars, galaxies, and planets were necessary in order to allow the creation of the ingredients of the life as we know it (elements of higher order, for instance).

(c) Finally, those Coacervates of the "real" primordial soup came into being.

We see that the WALL principle was activated many times and so was the need for boundaries. Without it no evolution would have been possible neither with respect to non-life nor life-forms.

It appears evident, especially when now knowing how omnipresent and universal this WALL principle really is, that also culture, societies, and— naturally—property require boundaries that have to be protected and—if needed—defended BY ALL MEANS.

That is why we can easily conclude that the very principal laws of physics demand that:

> The well-ordered world does not need any "open-border" or "no-border" fascists, but the understanding of properly adjusted social distances and semipermeable boundaries.

Yes, this is the very principal physical law of creation ... of any kind of creation, to be precise.

In fact, one might even see these "no-border" types as by far the most effective destroyers of everything evolution has created, be it of natural or human (cultural) origin. As a reminder, one simply needs to consider the definition of the death of a cell again.

It seems that every open door policy as being performed by the likes of some currently leading politicians and globalists is so principally in contrast to the universal laws of evolution and progress that even the title "enemy of humanity," as given to such folks by knowledgeable scientists and well-informed people, does not even remotely grasp the true complexity of their crime and dogmatic stupidity.

4.7.3 Information and Progress Require Boundaries

So, we have learned that the sentence of the famous economist Prof. Sinn that property requires borders in fact is much more omnipresent and much more universal.

It does not matter whether I say:

"Property needs borders"

"Life requires boundaries"

"Space needs boundaries"

"A civilization even more so"

"Culture needs protection"

"A society needs a borderline"

"The Wombat needs boundaries"

"Our children need protected borders"

"Some countries need a great WALL"

It always comes down to the same simple message:

Everything that wants to be and remain an entity needs to circumvent itself from the surrounding world and everything needs to protect and defend its boundaries. The moment you have created something special and good, the moment there are differences, which is to say thresholds or stronger gradients, to the "outside" and you are interested to keep such differences, you need walls to protect your creation. Without this protection your creation, your little cell, simply dies, because there is no such thing as a cell without its cell wall. There is no such thing as "tearing down the wall and everything will rise to the beauty of your creation." No, the result will always be the lowest level, meaning your creation will be torn apart and it will make no difference whatsoever or it will not be different to the surrounding you have not been able to protect it from. This is just the second law of thermodynamics and it cannot be betrayed except by using symmetry and energy to maintain the order of difference, because order is the acceptance of differences. Of course, there are red-greenish do-gooders like those globalists who will tell you that things are "a bit different with us humans," but this is nothing else but saying "we are above nature," "we are above creation," and I severely doubt that even the dimmest dunderheads would fall for this one.

Thus, in case you have a creation or property you are interested to keep, then find the optimum (lowest shearing force) distance to its environment, circumvent, and protect it ... and protect it properly. Those who want to convince you that such a protection is not of need are only interested in depriving you from what is rightfully yours.

4.8 Another Open Letter to the Poor Wombats of This World

Dear Wombat!

It is time to make a decision!

In fact, it is already almost too late for you!

The laws of the universe, of life, economy, and society are all telling you the same thing. If truth be told, mathematically it all REALLY IS the same thing, but we leave this for the theoreticians. Of importance for you only is the following:

Either you are going to die and your children, grandchildren, even your whole species are going down with you or you start to fight and defend your habitat. Either you start to restore and maintain the integrity of your living environment, your little ecosystem, or the game is over for you and your loved ones. Then you will be the one who has given up what generations of your kind had managed to protect for you and your—never to come— descendants.

Oh yes, it might well be that along the way, I mean along this process of restoring some order, you might be forced to shot at one or two of those alien big-eyed rabbits (perhaps you start with rubber bullets to show some humanity, sorry, I meant some Wombaty), but you better believe me:

- First: Their cuteness is nothing but pretense.
- Second: There are more than enough of them elsewhere, which is to say outside your territory.
- Third: In fact, they have their own ecosystem and space elsewhere.
- Fourth: Their reproduction rate is the one of—naturally—rabbits.
- Fifth: The alternative is your species to get extinct. The alternative is the annihilation of your children, your grandchildren, your culture, your people.

4.9 Epilogue

The usual reaction of the do-gooders, red-greenish career politicians, equalizer fetishists, alien friends, and all those other "woke" people now of course is to discredit the author as a Nazi. I'm not going to argue with these folks, but I'm asking you, the reader, to have a look at reference [28] or the introduction part of this book and tell me (or those do-gooders) how well this fits to your picture of a German Nazi.

And there is another point that needs to be taken care about. Not all walls are good for progress, positive evolution, respectively, creation. We are talking about walls only to protect parasitic behavior within a society or such forms of life. The imprisonment of people in failed states like the former socialistic ones is such an example. But the more important example OF GLOBAL CHARACTER are those walls cementing the parasitic life of those few extremely rich elements completely dependent on an unfair finance system and the "business model of invented money." Such parasitic forms of existence are of no good to anyone except for the parasites themselves (see [2, 5] and [29]). Tearing these walls down, which only protect cadgers, of course, is good for the evolution of a society, but the reader should be warned, may all those parasitic and globalistic elements in the world preach about an open door policy with respect to those bad (poor) migrants and rabbit migration in general, they will find no mercy for you the moment you are going to even try to scratch such parasitic walls of theirs.

And why is that? Well, the reader only needs to analyze the network those "intrinsic parasites" have woven and then he probably will be able to give the answer himself [2, 5, 29].

4.10 Conclusions

By building up socioeconomic models from metric first principles, we automatically obtained scale and size parameters as important structural ingredients to any such model. It was found that "social distancing" is an important parameter in any socioeconomic space-time. This holds for all kinds of entities forming the socioeconomic space or society and thus should be of great importance to finance market simulations, asset evaluation, any kind of economic/entrepreneurial decision-making, and—naturally—also political planning.

4.11 Outlook

With the models quickly increasing in size and dimensions when trying to move toward real-sized societies, it requires sophisticated and powerful computer technology to perform such simulations in bigger (and consequently more realistic) scales.

References

1. N. Schwarzer, *The Relativistic Quantum Bible: Genesis and Revelation*, www. amazon.com, ASIN: B01M1CJH1B.
2. N. Schwarzer, *The World Formula: A Late Recognition of David Hilbert's Stroke of Genius*, 2022, Jenny Stanford Publishing. ISBN: 9789814877206.

3. N. Schwarzer, *The Theory of Everything: Quantum and Relativity Is Everywhere – A Fermat Universe*, Pan Stanford Publishing, 2020, ISBN-10: 9814774472.

4. N. Schwarzer, *Einstein had it, but he did not see it, Part LXIX: The Hippocratic Oath in Mathematical Form and why – so often – it will be of no Use*, www.amazon.com, ASIN: B07KDSMNSK.

5. N. Schwarzer, *Philosophical Engineering, Part 1: The Honest Non-Parasitic Philosopher and the Universal "GOOD" Derived from a Theory of Everything*, www.amazon.com, ASIN: B07KNWRDYW.

6. N. Schwarzer, *Quantum Gravity Thermodynamics: And it May Get Hotter*, www.amazon.com, ASIN: B07XC2JW7F.

7. N. Schwarzer, *Quantum Gravity Thermodynamics II: Derivation of the Second Law of Thermodynamics and the Metric Driving Force of Evolution*, www.amazon.com, ASIN: B07XWPXF3G.

8. H. Leuenberger, What is life?, a new human model of life, disease and death – a challenge for artificial intelligence and bioelectric medicine specialists, SWISS PHARMA, 2019, **41**(1), www.ifiip.ch.

9. D. Hilbert, Die Grundlagen der Physik, Teil 1, *Göttinger Nachrichten*, 1915, 395–407.

10. A. Einstein, Grundlage der allgemeinen Relativitätstheorie, *Ann. Phys.*, 1916, **49** (ser. 4), 769–822.

11. N. Schwarzer, *Science Riddles – Riddle No. 12: Is There a Cosmological Spin?*, www.amazon.com, ASIN: B07T3WS7XK.

12. N. Schwarzer, *Science Riddles – Riddle No. 13: How to Solve the Flatness Problem?*, www.amazon.com, ASIN: B07T9WXZVH.

13. P. A. M. Dirac, The quantum theory of the electron, *Proc. R. Soc. A*, 1928, **117**(778), doi: 10.1098/rspa.1928.0023.

14. N. Schwarzer, *The "New" Dirac Equation – Originally Postulated, Now Geometrically Derived and – along the way – Generalized to Arbitrary Dimensions and Arbitrary Metric Space-Times*, www.amazon.com, ASIN: B085636Q88.

15. N. Schwarzer, *Einstein had it, but he did not see it, Part LIV: Mathematical Philosophy & Quantum Gravity Ethic*, www.amazon.com, ASIN: B07GR994HT.

16. N. Schwarzer, *Einstein had it, but he did not see it, Part LXVIII: Most fundamental Tools for Optimum Decision-Making based on Quantum Gravity*, www.amazon.com, ASIN: B07KDFDZVZ.

17. N. Schwarzer, *Epistle to Elementary Particle Physicists: A Chance You Might not Want to Miss*, www.amazon.com, ASIN: B07XDMLDQQ.

18. N. Schwarzer, *2nd Epistle to Elementary Particle Physicists: Brief Study about Gravity Field Solutions for Confined Particles*, www.amazon.com, ASIN: B07XN8GLXD.

19. N. Schwarzer, 3rd *Epistle to Elementary Particle Physicists: Beyond the Standard Model – Metric Solutions for Neutrino, Electron, Quark*, www.amazon.com, ASIN: B07XJJ535T.

20. N. Schwarzer, *Science Riddles – Riddle No. 26: Is there an Elastic Quantum Gravity?*, www.amazon.com, ASIN: B081K5LZLD.

21. A. E. Green, W. Zerna, *Theoretical Elasticity*, Oxford University Press, London, 1968.

22. N. Schwarzer, *Quantized Relativized Theology – Where is God?*, www.amazon.com/dp/B01M0XPXTT.

23. H.-W. Sinn, in German, www.fondsprofessionell.de/news/maerkte/headline/hans-werner-sinn-die-wichtigste-lehre-aus-dem-brexit-126024/

24. F. Menzel, Ausländergewalt, das Verschweigen & Vertuschen der Medien, in German, https://www.youtube.com/watch?v=kjUWz7rTfjI.

25. N. Schwarzer, *Einstein had it, but he did not see it, Part XXXIX: EQ or The Einstein Quantum Computer*, www.amazon.com, ASIN: B07D9MBRS3.

26. N. Schwarzer, *The Einstein Quantum Computer: Mathematical Principle and Transition to the Classical Discrete and Quantum Computer Design*, www.amazon.com, ASIN: B07D9J5VLV.

27. N. Schwarzer, *Is there an ultimate, truly fundamental and universal Computer Machine?*, www.amazon.com, ASIN: B07V52RB2F.

28. T. Bodan, *The Eighth Day – Two Jews against The Third Reich*, illustrated version, Self-published, BoD Classic, 2021, ISBN 9783753417257.

29. T. Bodan, N. Schwarzer, *Quantum Economy*, www.amazon.com/dp/B01N80I0NG.

30. J. F. Enders, T. C. Peebles, Propagation in tissue cultures of cytopathogenic agents from patients with measles, *Proc. Soc. Exp. Biol. Med.*, 1954, **86**(2), 277–286, https://pubmedinfo.files.wordpress.com/2017/01/propagation-in-tissue-cultures-of-cytopathogenic-agents-from-patients-with-measles.pdf.

Chapter 5

Mastering Human Crises with Quantum Gravity-Based but Still Practicable Models

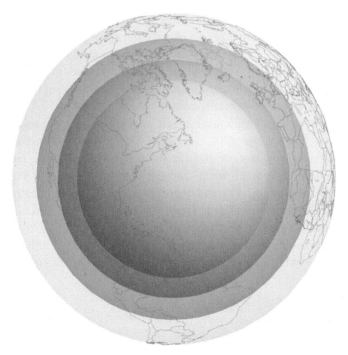

Figure 5.1 "Seeing the whole and avoiding the collapse."

The Math of Body, Soul, and the Universe
Norbert Schwarzer
Copyright © 2023 Jenny Stanford Publishing Pte. Ltd.
ISBN 978-981-4968-24-9 (Hardcover), 978-1-003-33454-5 (eBook)
www.jennystanford.com

5.1 First Measure: SEEING and UNDERSTANDING the WHOLE

Those who expect that this chapter will give advices like "government should now spend in order to revive the economy" will not find what they seek.

The truly good doctor always sees a patient as a whole before giving any advice about how to treat this or that. The truly good doctor always seeks to understand a certain illness of a certain patient as the result of a chain of events and an environment that has led to the patient's problem. In other, slightly more mathematical words, the good doctor tries to consider the whole space-time system of the patient before even trying to form an opinion, not to speak about uttering a suggestion.

The patient is the society and the diseases are of multiple character, which are brought into a state of multiple organ failure due to external and internal influences, one of which being the COVID virus. The cure will not come by the means of the usual recipes as these have brought the patient into this state in the first place. Thus, the first measure is understanding ... understanding of the patient and their environment as a whole.

We will base our considerations on a world formula approach originated from David Hilbert and Albert Einstein. Proof that this approach truly considers "it all" was already given in the first issues of our series "*Medical Socio-Economic Quantum Gravity*" and the chapters above by deriving a few principal—classical—gravity and quantum equations as demonstrators.

In this chapter, we will only present a most practical and simple path forward in dealing with—mainly nonlinear and classically unaccountable[1]— complex modeling situations as they especially are of need in human crises situations. We will not give examples or applications here, but only provide the math.

5.2 Motivation

Understanding the working of a human society usually, which is to say if not performed by an extraterrestrial intelligence, has the principal inbuilt flaw that the understanding is done by a human or a group of humans. This automatically renders the process of the understanding kind of impartial.

---

[1]That is, the problem with a crisis (even a global one as illustrated in Fig. 5.1), namely, that it usually occurs, because no one has seen it coming and this often happens when classical understandings and models are failing spectacularly, simply because they are too simple or even just wrong.

In ordinary circumstances, this usually does not matter, because even the hocus pocus models of the socioeconomic "specialists" of the worst kind or those polit-sycophants of virologists currently dealing with the COVID crisis do not immediately reveal themselves as useless and—too often—dangerous ideologies rather than scientific theories. This holds because the deviations to the reality are not necessarily very obvious, that is, obvious enough to allow the ordinary human being to see the flaws. This especially holds when the corresponding models are sexed up by loads of data-mined parameters (permanently adjusted), bamboozled in such a way that in essence they can describe everything (and thus nothing) or simply often consisting just of "Yes, we can!"—or "Wir schaffen das!"—wordings without any real substance. Only in situations like the recent COVID crisis, states of war, or severe economic breakdowns (nonequilibrium states and phase changes), the nonsense theories reveal themselves as what they are, namely, rubbish.

As the author has not the option to ask an extraterrestrial entity to do the analysis for him, but still is determined not to make the mistake of many other society understanders, the author resorts to the simple method of starting so fundamental that further digging would be futile, because he already has started from the lowest ground ... at least if taking the current state of science.

Our starting point here shall be the Einstein–Hilbert action in arbitrary space-times and with arbitrary numbers of dimensions. We consider this to be fundamental enough—if applied and investigated thoroughly enough—to provide us with a world formula.

5.3 How to Achieve the Necessary Complexity of Socioeconomic Space-Times?

Even though the mathematical basics were already given in the previous sections of this book [1], we here repeat the essential parts for convenience.

With respect to our most general starting point, however, we refer to sections 3.2 and 4.5 and only point out certain extensions we intend to use here.

Our starting point shall be the classical Einstein–Hilbert action [3] with the Ricci scalar R^* as kernel or Lagrange density:

$$\delta W = 0 = \delta \int_V d^n x \left(\sqrt{-G} \cdot R^* \right). \tag{174}$$

Please note that with $G_{\delta\gamma} = F[f] \cdot g_{\delta\gamma}$ we have used the simplest form of metric adaptation, which we could construct as a simplification from a generalization of the typical form of tensor transformations (5), respectively:

$$G_{\alpha\beta} = F\left[f\left[t,x,y,z,\ldots,\xi_k,\ldots,\xi_n\right]\right]^{ij}_{\alpha\beta} g_{ij}, \tag{175}$$

$$\rightarrow G_{\alpha\beta} = F\left[f\left[t,x,y,z,\ldots,\xi_k,\ldots,\xi_n\right]\right] \cdot \delta^i_\alpha \delta^j_\beta g_{ij}$$

$$\rightarrow G_{\alpha\beta} = F\left[f\left[t,x,y,z,\ldots,\xi_k,\ldots,\xi_n\right]\right] \cdot g_{\alpha\beta} = F[f] \cdot g_{\alpha\beta}. \tag{176}$$

The generalization is elaborated in [2].

Now we evaluate the resulting Ricci scalar for the metric $G_{\alpha\beta}$ at first in n and then, subsequently, simplify to 4 dimensions:

$$R^* = \frac{1}{F[f]^3} \cdot \left(\left(C_{N1} \cdot \left(\frac{\partial F[f]}{\partial f}\right)^2 - C_{N2} \cdot F[f] \cdot \frac{\partial^2 F[f]}{\partial f^2}\right) \cdot \overbrace{\left(\tilde{\nabla}_g f\right)^2}^{= f_{,\alpha} g^{\alpha\beta} f_{,\beta}} \right.$$
$$\left. - C_{N2} \cdot F[f] \cdot \frac{\partial F[f]}{\partial f} \cdot \Delta_g f \right)$$

$$\xrightarrow{n=4} = \frac{1}{F[f]^3} \cdot \left(\left(\frac{3}{2}\left(\frac{\partial F[f]}{\partial f}\right)^2 - 3 \cdot F[f] \cdot \frac{\partial^2 F[f]}{\partial f^2}\right) \cdot (\tilde{\nabla}_g f)^2 - 3 \cdot F[f] \cdot \frac{\partial F[f]}{\partial f} \cdot \Delta_g f \right)$$

$$\xrightarrow{\frac{3}{2}\left(\frac{\partial F[f]}{\partial f}\right)^2 - 3 \cdot F[f] \cdot \frac{\partial^2 F[f]}{\partial f^2} = 0} = -\frac{3}{F[f]^2} \cdot \left(\frac{\partial F[f]}{\partial f} \cdot \Delta_g f\right). \tag{177}$$

With respect to the constants C_{N1} and C_{N2} in arbitrary n dimensions, the reader is referred to [2] or chapter 8 in this book.

With $R^* = 0$ one would immediately satisfy the Einstein–Hilbert action. Taking the general n-dimensional case in the first line of (177) and always demanding that:

$$C_{N1} \cdot \left(\frac{\partial F[f]}{\partial f}\right)^2 - C_{N2} \cdot F[f] \cdot \frac{\partial^2 F[f]}{\partial f^2} = 0, \tag{178}$$

which simply is a fixing condition for the wrapper function $F[f]$, we obtain:

$$R^* = -\frac{C_{N2}}{F[f]^2} \cdot \left(\frac{\partial F[f]}{\partial f} \cdot \Delta_g f\right) = 0$$

$$\Rightarrow \qquad\qquad \Delta_g f = 0. \tag{179}$$

This is the "homogeneous" Klein–Gordon equation, respectively, the mass- and potential-less Klein–Gordon equation.

So, the question would be where to get the mass from?

This question will be answered in the next chapter.

We should point out that by applying an inner function $f \to H(f, f_1)$, we should always be able to obtain a constant as follows:

$$R^* = R^*_{const} + 0, \tag{180}$$

with the function f then satisfying condition (177) in the simpler manner:

$$0 = \frac{1}{F[f]^3} \cdot \left(\overbrace{\left(C_{N1} \cdot \left(\frac{\partial F[f]}{\partial f} \right)^2 - C_{N2} \cdot F[f] \cdot \frac{\partial^2 F[f]}{\partial f^2} \right) \cdot \left(\tilde{\nabla}_g f \right)^2}^{= f_{,\alpha} g^{\alpha\beta} f_{,\beta}} - C_{N2} \cdot F[f] \cdot \frac{\partial F[f]}{\partial f} \cdot \Delta_g f \right). \tag{181}$$

Things are immediately and more directly clear when choosing condition (178). In this case, we have from (177):

$$R^* = -\frac{C_{N2}}{F\left[H(f, f_1)\right]^2} \cdot \left(\frac{\partial F[f]}{\partial H(f, f_1)} \cdot \Delta_g H(f, f_1) \right)$$

$$\xrightarrow{H(f,f_1)=f+f_1} = -\frac{C_{N2}}{F[f + f_1]^2} \cdot \left(\frac{\partial F[f + f_1]}{\partial (f + f_1)} \cdot \Delta_g (f + f_1) \right)$$

$$= -\frac{C_{N2}}{F[f + f_1]^2} \cdot \frac{\partial F[f + f_1]}{\partial (f + f_1)} \cdot \left(\Delta_g f + \Delta_g f_1 \right)$$

$$\xrightarrow{-\frac{C_{N2}}{F[f+f_1]^2} \cdot \frac{\partial F[f+f_1]}{\partial(f+f_1)} \Delta_g f_1 = R^*_{const}} = -\frac{C_{N2}}{F[f + f_1]^2} \cdot \overbrace{\frac{\partial F[f + f_1]}{\partial (f + f_1)} \cdot \Delta_g f}^{=0} + R^*_{const}$$

$$\Rightarrow \qquad \Delta_g f = 0 \tag{182}$$

and would be back with the mass-free Klein–Gordon equation.

It should be pointed out that condition (182) could also be generalized to more Hilbert- and action-like conditions as follows:

$$\delta W = 0 = \delta \int_V d^n x \left(\sqrt{-G} \cdot R^* \right) = -\delta \int_V d^n x \left(\frac{\sqrt{-G} \cdot C_{N2}}{F[H(f, f_1)]^2} \cdot \left(\frac{\partial F[f]}{\partial H(f, f_1)} \cdot \Delta_g H(f, f_1) \right) \right)$$

$$\underset{H(f,f_1)=f+f_1}{\underline{\hspace{3cm}}} \rightarrow = -\delta \int_V d^n x \left(\frac{\sqrt{-G} \cdot C_{N2}}{F[f+f_1]^2} \cdot \left(\frac{\partial F[f+f_1]}{\partial (f+f_1)} \cdot \Delta_g (f+f_1) \right) \right)$$

$$= -\delta \int_V d^n x \left(\frac{\sqrt{-G} \cdot C_{N2}}{F[f+f_1]^2} \cdot \left(\frac{\partial F[f+f_1]}{\partial (f+f_1)} \cdot (\Delta_g f + \Delta_g f_1) \right) \right)$$

$$= -\delta \int_V d^n x \left(\frac{\sqrt{-G} \cdot C_{N2}}{F[f+f_1]^2} \cdot \left(\frac{\partial F[f+f_1]}{\partial (f+f_1)} \cdot \Delta_g f \right) \right)$$

$$\overbrace{- \delta \int_V d^n x \left(\frac{\sqrt{-G} \cdot C_{N2}}{F[f+f_1]^2} \cdot \left(\underbrace{\frac{\partial F[f+f_1]}{\partial (f+f_1)} \cdot \Delta_g f_1}_{\frac{\sqrt{-G} \cdot C_{N2}}{F[f+f_1]^2} \frac{\partial F[f+f_1]}{\partial (f+f_1)} \Delta_g f_1 \equiv \sqrt{-G} \cdot R^{**} \equiv w^*_{\text{const}} \longrightarrow} \right) \right)}^{=0}$$

$$= -\delta \int_V d^n x \left(\frac{\sqrt{-G} \cdot C_{N2}}{F[f+f_1]^2} \cdot \left(\underbrace{\frac{\partial F[f+f_1]}{\partial (f+f_1)} \cdot (\Delta_g f)}_{=0} \right) \right)$$

$$+ \int_V d^n x \cdot \delta w^*_{\text{const}}$$

$$\Rightarrow \qquad\qquad\qquad \Delta_g f = 0. \qquad\qquad\qquad (183)$$

Now the complete quantum gravity solution consists of a function f, which has to be a Laplace function, and the following equation for the function f_1 and the metric:

$$-\frac{\sqrt{-G} \cdot C_{N2}}{F[f+f_1]^2} \cdot \frac{\partial F[f+f_1]}{\partial (f+f_1)} \cdot \Delta_g f_1 \equiv \sqrt{-G} \cdot R^{**} \equiv w^*_{\text{const}}$$

$$-\frac{\sqrt{-g \cdot F[f+f_1]^n} \cdot C_{N2}}{F[f+f_1]^2} \cdot \frac{\partial F[f+f_1]}{\partial (f+f_1)} \cdot \Delta_g f_1 = \sqrt{-g \cdot F[f+f_1]^n} \cdot R^{**} \equiv w^*_{\text{const}} = \text{const.}$$

$$(184)$$

For those who prefer the classical separation in physics, one may see the last line in (183) as the quantum and (184) as the gravity part of our quantum gravity approach.

Assuming that there always exists such a solution for f_1 in (184) for any suitable metric, we could just concentrate on finding the right set of solutions for f in (183).

As already said in the introduction part above, the modeling of real societies requires many degrees of freedom and thus dimensions. This should also hold with respect to the options for the function f, which should allow for a greater flexibility, especially when we want to observe our societies in the vicinity of extreme events or phase transitions.

It should be pointed out that with the condition for f to be a scalar, there is still a great flexibility for options to have vector and tensor functions included, for example, via:

$$f + f_1 = \theta_0 \cdot h + \theta_1 \cdot h_\alpha Q^\alpha + \theta_2 \cdot H^\alpha P_\alpha + \theta_3 \cdot h_{\alpha\beta} q^{\alpha\beta}$$
$$+ \theta_4 \cdot H^\lambda{}_\kappa Q^\kappa_\lambda + \theta_5 \cdot S^{\alpha\chi} p_{\alpha\chi} + \ldots + f_1. \tag{185}$$

In order to quickly come to a solution or a set of solutions, we use the following degrees of freedom:

(a) We demand the function f to be a Laplace function $\Delta_g f = 0$.
(b) We demand condition (7), (178) to be fulfilled.
(c) We chose function f_1 according to (165) and (182).

This way we end up with the simple total "world formula" condition $\Delta_g f = 0$ plus the choice for a suitable metric $g_{\alpha\beta}$. However, as it was shown in the previous chapters and some papers (for convenience we suggest [2, 5]), there are also other degrees of freedom under the variational integral. We learned that the number of dimensions and the variation of the number and positions of centers of gravity automatically bring in many other physical effects such as thermodynamic, interaction, and evolution. In order to account for these effects, we can still use our comfortable f structure as linear (or nearly linear) separator. Let there be an f like:

$$f \to \Phi + \Theta + \Psi$$

$$= \theta_0 \cdot h + \theta_1 \cdot h_\alpha Q^\alpha + \theta_2 \cdot H^\alpha P_\alpha + \theta_3 \cdot h_{\alpha\beta} q^{\alpha\beta}$$
$$+ \theta_4 \cdot H^\lambda{}_\kappa Q^\kappa_\lambda + \theta_5 \cdot S^{\alpha\chi} p_{\alpha\chi} + \ldots + \Theta + \Psi. \tag{186}$$

Now we go back to (183) and extend as follows:

$$\delta W = 0 = \delta \int_V d^n x \left(\sqrt{-G} \cdot R^* \right) = \sum_{\forall g_{ij} |n| \xi_k} \delta_{g_{ij} |n| \xi_k} \int_V d^n x \left(\sqrt{-G} \cdot R^* \right)$$

$$= -\sum_{\forall g_{ij} |n| \xi_k} \delta_{g_{ij} |n| \xi_k} \int_V d^n x \left(\frac{\sqrt{-G} \cdot C_{N2}}{F^2} \cdot \left(\frac{\partial F}{\partial f} \cdot \Delta_g f_{g_{ij} |n| \xi_k} \right) \right)$$

$$= -\sum_{\forall g_{ij}|n|\xi_k} \delta_{g_{ij}|n|\xi_k} \int_V d^n x \left(\frac{\sqrt{-G}\cdot C_{N2}}{F^2} \cdot \left(\frac{\partial F}{\partial(\Phi+\Theta+\Psi)} \cdot \left(\Delta_g \Phi_{g_{ij}} + \Delta_g \Theta_n + \Delta_g \Psi_{\xi_k} \right) \right) \right).$$

(187)

Here we have assumed that we may be allowed to separate the Laplace functions $f_{g_{ij}|n|\xi_k}$ into those we can allocate to the various metrics g_{ij}, various dimensions n, and various gravity centers ξ_k. Further assuming that the functional choice with respect to the variation of the metric could always be chosen as in the fifth line of (183), respectively, condition (184), which is to say:

$$= -\sum_{\forall g_{ij}|n|\xi_k} \delta_{g_{ij}|n|\xi_k} \int_V d^n x \left(\frac{\sqrt{-G}\cdot C_{N2}}{F^2} \cdot \left(\frac{\partial F}{\partial f} \cdot \left(\Delta_g \Phi + \Delta_g \Theta + \Delta_g \Psi \right) \right) \right)$$

$$= -\sum_{\forall g_{ij}|n|\xi_k} \delta_{n|\xi_k} \overbrace{\int_V d^n x \left(\frac{\sqrt{-G}\cdot C_{N2}}{F^2} \cdot \left(\frac{\partial F}{\partial f} \cdot \Delta_g f_{g_{ij}|n|\xi_k} \right) \right)}^{=0} + \overbrace{\delta_{g_{ij}} \int_V d^n x \cdot \overset{*}{w}_{\mathrm{const}}\big|_{g_{ij}|n|\xi_k}}^{=0},$$

(188)

we can get rid of the metric variation and obtain condition (184) instead. The condition (184) would now become more general and reads:

$$-\sum_{\forall g_{ij}|n|\xi_k} \left(\frac{\sqrt{-G}\cdot C_{N2}}{F^2} \frac{\partial F}{\partial f} \cdot \Delta_g f_{g_{ij}|n|\xi_k} \right) \equiv \sum_{\forall g_{ij}|n|\xi_k} \sqrt{-G}_{g_{ij}|n|\xi_k} \cdot \overset{**}{R}_{g_{ij}|n|\xi_k}$$

$$\Rightarrow \quad -\sum_{\forall g_{ij}|n|\xi_k} \left(\frac{\sqrt{-G}\cdot C_{N2}}{F^2} \frac{\partial F}{\partial f} \cdot \Delta_g f_{g_{ij}|n|\xi_k} \right) = \sum_{\forall g_{ij}|n|\xi_k} \sqrt{-g}_{g_{ij}|n|\xi_k} \cdot F^n \cdot \overset{**}{R}_{g_{ij}|n|\xi_k}$$

$$\equiv \sum_{\forall g_{ij}|n|\xi_k} \overset{*}{w}_{\mathrm{const}}\big|_{g_{ij}|n|\xi_k} = \sum_{\forall g_{ij}|n|\xi_k} \mathrm{const}_{g_{ij}|n|\xi_k}. \qquad (189)$$

Thus, when constructing a set of functions $f_{g_{ij}|n|\xi_k}$ in such a way that the expression:

$$-\sum_{\forall g_{ij}|n|\xi_k} \left(\frac{\sqrt{-g}_{g_{ij}|n|\xi_k} \cdot F^n \cdot C_{N2}}{F^2} \frac{\partial F}{\partial f} \cdot \Delta_g f_{g_{ij}|n|\xi_k} \right) = \sum_{\forall g_{ij}|n|\xi_k} \mathrm{const}_{g_{ij}|n|\xi_k} \quad (190)$$

remains constant under any changes of the metric tensor, we would just be left with the task:

$$0 = \sum_{\forall g_{ij}|n|\xi_k} \delta_{n|\xi_k} \int_V d^n x \left(\frac{\sqrt{-G} \cdot C_{N2}}{F^2} \cdot \left(\frac{\partial F}{\partial f} \cdot \Delta_g f_{g_{ij}|n|\xi_k} \right) \right)$$

$$= \sum_{\forall g_{ij}|n|\xi_k} \delta_{n|\xi_k} \int_V d^n x \left(\frac{\sqrt{-g_{g_{ij}|n|\xi_k}} \cdot F^n \cdot C_{N2}}{F^2} \cdot \left(\frac{\partial F}{\partial f} \cdot \Delta_g f_{g_{ij}|n|\xi_k} \right) \right). \quad (191)$$

Obviously, the simplest solution to this equation would be:

$$0 = \Delta_g f_{g_{ij}|n|\xi_k}, \quad (192)$$

but it needs to be pointed out that it is not the only one.

However, even if heading for the simplest solution with just suitable selections of Laplace functions additionally fulfilling condition (190), we already have an infinitely huge toolbox to describe any system in a Hilbert-conform manner, thereby taking all its physical aspects into account. Thereby "physical" here stands for all interactions, inertia, and states and can—depending on the task in question—stand for "economic," "social," "financial," "health," "viral," and so on.

In other words, in principle we have all we need to comprehensively model and simulate societies in a most neutral manner. Such models could help us understand and potentially hinder or just mitigate the effects of bigger crises, like the current Corona virus "epidemic," which more and more shows to be nothing else but an anaphylactic shock caused by incompetent political leaders.

5.3.1 Self-Similarity, Bi-harmonics, and Other Options

As we have seen, the only condition to the function f is that it has to be a scalar. Thus, we also have options to demand that f would just be a curvature (Ricci scalar) itself, leading us to:

$$f \to R_1 \Rightarrow F[f] = F[R_1]. \quad (193)$$

This does allow us to define some self-similarity like:

$$F[f] = F[R_1] = F\left[R_1\left[F_1[R_2]\right]\right] = F\left[R_1\left[F_1\left[R_2\left[F_2\left[R_3[\ldots]\right]\right]\right]\right]\right], \quad (194)$$

which we will consider in later publications.

It may also be useful to apply higher harmonic orders like:

$$f \to \Delta\varphi \Rightarrow F[\Delta\varphi] = F[\Delta\varphi]. \quad (195)$$

5.4 Simple Symmetries

For illustration and in order to give some examples, we will consider space-times in spherical and cylindrical spatial coordinates in the next sections.

Thereby we do not suggest to model a human society as simple as a hydrogen atom or a cylindrically symmetric ensemble of just 4 dimensions. However, using these examples as potential starting points—perhaps even sticking to the perfect spherical or cylindrical symmetry in generalized, higher dimensional form—may give some nice fundamental solutions from which one can draw conclusions toward more practical (and thus higher dimensional) problems.

5.5 Spherical Coordinates

The case with spatial spherical coordinates was partially investigated in connection with the Schwarzschild metric in [12] and with the Robertson–Walker metric in [13]. Here we shall directly start with the vacuum case:

$$
g_{\alpha\beta}^{\text{flat}} = \begin{pmatrix} -c^2 & 0 & 0 & 0 \\ 0 & 1 & 0 & 0 \\ 0 & 0 & r^2 & 0 \\ 0 & 0 & 0 & r^2 \cdot \sin^2\vartheta \end{pmatrix}.
\tag{196}
$$

For convenience we here want to repeat the investigation with condition R^* = 0, which was also investigated in [12]. From (182) and condition (7), we obtain the simple 4D Laplacian:

$$
R^* = -3 \cdot \frac{\left(C_1 + \dfrac{C_1^2 \cdot f}{2 \cdot C_2}\right)}{\left(C_1 \cdot f + \dfrac{C_1^2 \cdot f^2}{4 \cdot C_2} + C_2\right)^2} \cdot \Delta_g f = 0 = \Delta_g f = \Delta_{3D-\text{sphere}} f - \frac{f^{(2,0,0,0)}}{c^2}.
\tag{197}
$$

Applying the separation approach $f[t, r, \vartheta, \varphi] = g[t]*h[r]*Y[\vartheta, \varphi]$ gives us the solution (197) for $g[t]$:

$$
g[t] = C_1 \cdot \cos\left[c \cdot C_t \cdot t\right] + C_2 \cdot \sin\left[c \cdot C_t \cdot t\right],
\tag{198}
$$

and the spherical Bessel functions $j_n[z]$ and $y_n[z]$ for $h[r]$ as already derived in [12], only that this time the solution reads as follows:

$$h[r] = C_j \cdot j_k \left[C_t \cdot r \right] + C_y \cdot y_k \left[C_t \cdot r \right]; \quad k = \frac{\sqrt{1 + 4\omega^2} - 1}{2}. \tag{199}$$

Solutions, not producing singularities or growing infinitely with respect to r, are only found for the spherical Bessel function of the first kind $j_k[z]$ with the usual $\omega^2 = l*(l+1)$ (or $\omega^2 = l*(l-1)$) and $l = 1, 2, 3, \ldots$ Also, as discussed in [12] (there see sub-section "The $r_s = 0$-Case Now Without Feeding from Classical Quantum Theory") and very different from the Schrödinger hydrogen solution, we have no main quantum number.

This has quite some effect on the possible selection for the quantum numbers l and m for the spherical harmonics $Y_l^m[\vartheta, \varphi]$, be it for the usual spherical metric or our hyperbolical spheres, which can be given as follows (simply set $r_s = \Lambda = 0$ in order to have the Einstein–Hilbert vacuum situation there, too):

$$g_{\alpha\beta}^{\text{hyp}} = \begin{pmatrix} h \cdot g_t'[t]^2 \cdot \alpha[r] & 0 & 0 & 0 \\ 0 & a \cdot G_r'[r]^2 \alpha[r]^{-1} & 0 & 0 \\ 0 & 0 & g_{22} & 0 \\ 0 & 0 & 0 & g_{33} \end{pmatrix}$$

$$g_{22} = b \cdot g_\vartheta'[c_\vartheta \vartheta]^2 G_r[r]^2 \cdot \begin{Bmatrix} \sinh[g_\vartheta[c_\vartheta \vartheta]]^{-2} \\ \cosh[g_\vartheta[c_\vartheta \vartheta]]^{-2} \end{Bmatrix}$$

$$g_{33} = d \cdot g_\varphi'[c_\varphi \varphi]^2 \cdot G_r[r]^2 \cdot \begin{Bmatrix} \sinh[g_\vartheta[c_\vartheta \vartheta]]^{-2} \\ \cosh[g_\vartheta[c_\vartheta \vartheta]]^{-2} \end{Bmatrix}$$

$$\alpha[r] = \left(C_1 + \frac{r_s}{G_r[r]} + \frac{\Lambda}{3 \cdot c^2} \cdot G_r[r]^2 \right). \tag{200}$$

For one thing we have the condition that $-|l - 1| \leq m \leq |l - 1|$ and for a second we find that $l = 0$ and $l = 1$ are providing the same trivial result, namely, $Y_{l=0,1}^0[\vartheta, \varphi] = 1$. Thereby we note that only for $l = 0$ also $m = l$ gives a nonvanishing solution. We also note that, in contrast to the Schwarzschild metric, we have considered in [12], it does not matter which spherical harmonics we chose. Both forms:

$$Y_l^m[\vartheta, \varphi] = e^{i \cdot m \cdot \varphi} \cdot \left(C_{jP} \cdot P_{l_j}^{m_j}[\cos \vartheta] + C_{jQ} \cdot Q_{l_j}^{m_j}[\cos \vartheta] \right), \tag{201}$$

$$Y_l^m[\vartheta,\varphi] = e^{i\cdot m\cdot\varphi}\cdot\left(C_{jP}\cdot P_{l_j}^{m_j}[\tanh\vartheta] + C_{jQ}\cdot Q_{l_j}^{m_j}[\tanh\vartheta]\right), \qquad (202)$$

being the spherical and the spherical-hyperbolical one, respectively, give reasonable distributions with the conditions for the quantum numbers m and l as demanded in the paragraph above. However, if we insist on vanishing values on the poles, we should also exclude the quantum number $m = 0$ (except in the case $l = 0$, where the result is a constant anyway).

It should be noted that the quantum solutions of our transformation are principally additive in the case of $R^* = 0$. It was demonstrated in [13] that while the Einstein field equations are not additive with respect to the metric, we find that our differential equations for gravity quantum fields (197) are perfectly linear operations on f. This gives us the option of superposition. Thus, this time applying the separation approach in form of a sum of functions

$$f = \sum_{j=1}^{\Omega} f_j \text{, with all } f_j\,[t, r, \vartheta, \varphi] = g_j[t]*h_j\,[r]*Y_j\,[\vartheta, \varphi]\text{, and solving the angular}$$

part in the usual way (c.f. [14]) makes (197):

$$\Delta_g f = \Delta_f \sum_{j=1}^{\Omega} f_j = 0. \qquad (203)$$

Now we simply demand that each of the summands gives zero and have:

$$\Delta_g f = \Delta_g \sum_{j=1}^{\Omega} f_j = \sum_{j=1}^{\Omega} \Delta_g f_j = 0 \Rightarrow \Delta_g f_j = 0. \qquad (204)$$

The separation calculation with our form $f = \sum_{j=1}^{\Omega} f_j$ from above has now simply to be performed for each function separately f_j. Subsequently, we obtain result in a set of solutions $Y_{jl}^m[\vartheta,\varphi]$, $h_j[r]$, and $g_j[t]$ with parameters C_{jt} and ω_j, respectively, l_j and m_j, as follows:

$$Y_{jl}^m[\vartheta,\varphi] = e^{i\cdot m_j\cdot\varphi}\cdot\left(C_{jP}\cdot P_{l_j}^{m_j}[\cos\vartheta] + C_{jQ}\cdot Q_{l_j}^{m_j}[\cos\vartheta]\right), \qquad (205)$$

$$Y_{jl}^m[\vartheta,\varphi] = e^{i\cdot m_j\cdot\varphi}\cdot\left(C_{jP}\cdot P_{l_j}^{m_j}[\tanh\vartheta] + C_{jQ}\cdot Q_{l_j}^{m_j}[\tanh\vartheta]\right), \qquad (206)$$

for the usual spherical metric and the spherical hyperbolical metric (200), respectively, plus:

$$g_j[t] = C_{j1}\cdot\cos\left[c\cdot C_{jt}\cdot t\right] + C_{j2}\cdot\sin\left[c\cdot C_{jt}\cdot t\right] \qquad (207)$$

for the time functions and:

$$h_{jl}[r] = C_{jj} \cdot j_{jk}[C_{jt} \cdot r] + C_{jy} \cdot y_{jk}[C_{jt} \cdot r]; \quad k_j = \frac{\sqrt{1 + 4\omega_j^2} - 1}{2} \quad (208)$$

for the radius functions.

An important simplification is obtained in the case $l = 0$, where, as an interesting side effect of our superposition approach, we find that the radial distribution in the case of $l = 0$ and the limit of the flat space can be made a Dirac delta distribution. From (208), setting $C_{jj} = 1$ and $C_{jy} = 0$, we derive the limit $l = 0$ as follows:

$$h_{j0}[r] = \frac{\sin[C_{jt} \cdot r]}{C_{jt} \cdot r}. \quad (209)$$

The following integrals could be seen as superpositions of solutions $h_{j0}[r]$ for certain distributions of C_{jt} according to (208):

$$h_{\Sigma\infty 0}[r] = \int_{-\infty}^{\infty} \frac{\sin[C_t \cdot r]}{C_t \cdot r} \cdot dC_t = \frac{\pi}{r}$$

$$h_{\Sigma 0}[r] = \int_{-A}^{A} \frac{\sin[C_t \cdot r]}{C_t \cdot r} \cdot dC_t = 2 \cdot \frac{Si[A \cdot r]}{r}; \quad Si[z] = \int_{0}^{z} \frac{\sin[u]}{u} \cdot du. \quad (210)$$

Thereby it needs to be pointed out that the integration should be performed over the whole function $f[t, r, \vartheta, \varphi]$ and with $g[t]$ also depending on C_t, this needs to be taken into account. Here, for simplicity, however, we pick a $t = t_0$ for which we would have $g[t_0]$ = const and concentrate solely on the radial dependency. It requires a skilled mathematician to work out the situation in the general cases. The limits for $r = 0$ at $t = t_0$ are:

$$\lim_{r \to 0} h_{\Sigma\infty 0}[r] = \lim_{r \to 0} \int_{-\infty}^{\infty} \frac{\sin[C_t \cdot r]}{C_t \cdot r} \cdot dC_t = \lim_{r \to 0} \frac{\pi}{r} = \infty$$

$$\lim_{r \to 0} h_{\Sigma 0}[r] = \lim_{r \to 0} \int_{-A}^{A} \frac{\sin[C_t \cdot r]}{C_t \cdot r} \cdot dC_t = \lim_{r \to 0} 2 \cdot \frac{Si[A \cdot r]}{r} = 2 \cdot A. \quad (211)$$

The interesting fact here is that the second superposition, one might see as finite, gives finite values at $r = 0$, while the first—infinite—superposition results in a singularity at $r = 0$. This not only leaves some interesting options for the construction of space out of miniscule FLRW universes with $k = 0$

as suggested in [13][2], but also opens up a completely new possibility to explain gravity in the limit to the Newton cases without resorting to the Schwarzschild singularity.

For the reason of generality, we should point out that the superposition above can be extended as follows:

$$h_{\Sigma\infty 0}[r] = \int_{-\infty}^{\infty} C_j[C_t] \cdot \frac{\sin[C_t \cdot r]}{C_t \cdot r} \cdot dC_t; \quad h_{\Sigma 0}[r] = \int_{-A}^{A} C_j[C_t] \cdot \frac{\sin[C_t \cdot r]}{C_t \cdot r} \cdot dC_t.$$

(212)

Further discussion for our special vacuum quantum gravity solutions to the spatial spherical symmetry is given in [12].

Things are getting more difficult (but also more interesting) in the case of $R^* \neq 0$. From (197) we then have:

$$\frac{R^*}{3} \cdot \frac{\left(C_1 \cdot f + \dfrac{C_1^2 \cdot f^2}{4 \cdot C_2} + C_2\right)^2}{C_1 + \dfrac{C_1^2 \cdot f}{2 \cdot C_2}} + \Delta_g f = 0$$

(213)

and Taylor expansion with respect to f (at $f = 0$) we get (c.f. [18] section "3.5 The Situation in 4 Dimensions"):

$$\xrightarrow{\text{Taylor at } f=0} \frac{R^*}{3} \cdot \left(\frac{C_2^2}{C_1} + \frac{3}{2} \cdot C_2 \cdot f + \ldots\right) + \Delta_g f = 0.$$

(214)

Now we introduce a simple transformation of the following form:

$$f^* = \frac{C_2^2}{C_1} + \frac{3}{2} \cdot C_2 \cdot f; \quad \Delta_g f^* = \Delta_g f$$

(215)

and obtain the ordinary Klein–Gordon equation with:

$$\frac{R^*}{3} \cdot f^* + \Delta_g f^* = 0.$$

(216)

In [12] we investigated a variety of assumptions regarding the curvature term R^*, of which we here shall only represent the basic results.

With the separation approach $f^*[t, r, \vartheta, \varphi] = g[t] * h[r] * Y[\vartheta, \varphi]$, we are able to solve this equation for quite some assumptions regarding R^*. Solving the angular part in the usual way (c.f. [14]) gives us:

---

[2] ... and thereby itself resulting from an idea in a fictive story [15, 16, 17].

$$0 = R^* + \frac{3}{f^*} \cdot \left(\frac{2 \cdot f^{*(0,1,0,0)} + r \cdot f^{*(0,2,0,0)}}{r} - \frac{\omega^2}{r^2} \cdot f^* - \frac{f^{*(2,0,0,0)}}{c^2} \right); \quad R_6 \equiv \frac{R^*}{3}$$

$$0 = \left(R_6 - \frac{\omega^2}{r^2} \right) \cdot f^*[t,r,\vartheta,\varphi] + \left(g[t] \cdot \left(\frac{2}{r} \frac{\partial h[r]}{\partial r} + \frac{\partial^2 h[r]}{\partial r^2} \right) - h[r] \cdot \frac{\partial^2 g[t]}{c^2 \cdot \partial t^2} \right) \cdot Y[\vartheta,\varphi].$$

(217)

Thereby $Y[\vartheta, \varphi]$ denotes the spherical harmonics, which are usually given with the quantum numbers l and m and written as $Y_l^m[\vartheta,\varphi]$. The quantum numbers l and m are integers with the conditions $l = 1, 2, 3,...$ and $-l \le m \le +l$. Division by $f^*[...]$ on both sides of (217) yields:

$$0 = \left(R_6 - \frac{\omega^2}{r^2} \right) + \left(\frac{1}{h[r]} \cdot \left(\frac{2}{r} \frac{\partial h[r]}{\partial r} + \frac{\partial^2 h[r]}{\partial r^2} \right) - \frac{1}{g[t]} \cdot \frac{\partial^2 g[t]}{c^2 \cdot \partial t^2} \right)$$

$$\frac{1}{g[t]} \cdot \frac{\partial^2 g[t]}{c^2 \cdot \partial t^2} = \left(R_6 - \frac{\omega^2}{r^2} \right) + \frac{1}{h[r]} \cdot \left(\frac{2}{r} \frac{\partial h[r]}{\partial r} + \frac{\partial^2 h[r]}{\partial r^2} \right). \quad (218)$$

Assuming R_6 not to be a function of t or the angles, meaning that we just have $R_6 = R_6[r]$, demands the terms on both sides to be a constant. Thus, we can write:

$$\frac{1}{g[t]} \cdot \frac{\partial^2 g[t]}{c^2 \cdot \partial t^2} \equiv -C_t^2 \equiv \left(R_6 - \frac{\omega^2}{r^2} \right) + \frac{1}{h[r]} \cdot \left(\frac{2}{r} \frac{\partial h[r]}{\partial r} + \frac{\partial^2 h[r]}{\partial r^2} \right). \quad (219)$$

We shall start with the assumption of R_6 just being a constant. While the solution for $g[t]$ is a simple periodic function:

$$g[t] = C_1 \cdot \cos[c \cdot C_t \cdot t] + C_2 \cdot \sin[c \cdot C_t \cdot t], \quad (220)$$

we obtain the spherical Bessel functions $j_n[z]$ and $y_n[z]$ for $h[r]$ as follows:

$$h[r] = C_j \cdot j_k\left[-i \cdot r \cdot \sqrt{-C_t^2 - R_6} \right] + C_y \cdot y_k\left[-i \cdot r \cdot \sqrt{-C_t^2 - R_6} \right]; \quad k = \frac{\sqrt{1 + 4\omega^2} - 1}{2}.$$

(221)

As before, solutions, which are not producing singularities or growing infinitely with respect to r, are only found for the spherical Bessel function of the first kind $j_k[z]$. Please note that in contrast to the Schrödinger hydrogen solution, we have no main quantum number.

We get much closer to the classical Schrödinger hydrogen by choosing more suitable Ricci curvature terms, which automatically bring matter into our quantum gravity picture (c.f. [18] section "Intermediate Sum-up"). Thus, the character of the curvature determines inertia and potentials within our Klein–Gordon equations. Later in this book we will learn that the same effect can also be obtained via the entanglement of higher dimensions.

In effect, we obtain the classical hydrogen case completely out of a metric with scalar wrapper.

However, in order to simplify the comparison with the Schrödinger case, we divide (217), (218) by $\dfrac{12 \cdot M}{\hbar^2}$ and obtain:

$$
\begin{aligned}
0 &= \frac{\hbar^2}{2 \cdot M} \cdot R_6 + \frac{\hbar^2}{2 \cdot f^* \cdot M} \cdot \left(\Delta_{3D-\text{sphere}} f^* - \frac{f^{*(2,0,0,0)}}{c^2} \right) \\
&= R_6^* + \frac{\hbar^2}{2 \cdot f^* \cdot M} \cdot \left(\Delta_{3D-\text{sphere}} f^* - \frac{f^{*(2,0,0,0)}}{c^2} \right).
\end{aligned}
\tag{222}
$$

A very interesting setting for R_6^* would be as follows:

$$
R_6^* = R_6^*[r] = R_{60} + \frac{R_{61}}{r} + \frac{R_{62}}{r^2}.
\tag{223}
$$

This completely reproduces the Schrödinger hydrogen problem with (218) being changed to:

$$
0 = \left(R_{60} + \frac{R_{61}}{r} + \frac{R_{62}}{r^2} - \frac{\omega^2}{r^2} \right) + \frac{\hbar^2}{2 \cdot M} \cdot \left(\frac{1}{h[r]} \cdot \left(\frac{2}{r} \cdot \frac{\partial h[r]}{\partial r} + \frac{\partial^2 h[r]}{\partial r^2} \right) - \frac{1}{g[t]} \cdot \frac{\partial^2 g[t]}{c^2 \cdot \partial t^2} \right)
$$

$$
\frac{\hbar^2}{2 \cdot M} \frac{1}{g[t]} \cdot \frac{\partial^2 g[t]}{c^2 \cdot \partial t^2} = \left(R_{60} + \frac{R_{61}}{r} + \frac{R_{62}}{r^2} - \frac{\omega^2}{r^2} \right) + \frac{\hbar^2}{2 \cdot M} \frac{1}{h[r]} \cdot \left(\frac{2}{r} \cdot \frac{\partial h[r]}{\partial r} + \frac{\partial^2 h[r]}{\partial r^2} \right).
\tag{224}
$$

Comparing with the Schrödinger derivation (e.g., [14], p. 155) and assuming that R_{60} is a constant gives us:

$$
\overbrace{\frac{\hbar^2}{2 \cdot M} \cdot \frac{1}{g[t]} \cdot \frac{\partial^2 g[t]}{c^2 \cdot \partial t^2}}^{-E_{n-\text{Schrödinger}}} = -R_{60} - C_t^2
$$

$$
= \left(\overbrace{\frac{R_{61}}{r}}^{-V/r} + \overbrace{\frac{R_{62}}{r^2}}^{-\omega^2_{\text{Schrödinger}}/r^2} - \frac{\omega^2}{r^2} \right) + \frac{\hbar^2}{2 \cdot M} \frac{1}{h[r]} \cdot \left(\frac{2}{r} \cdot \frac{\partial h[r]}{\partial r} + \frac{\partial^2 h[r]}{\partial r^2} \right).
\tag{225}
$$

Using the result for $g[t]$ from above, which is to say (219) and (220) with the following adaptation:

$$g[t] = C_1 \cdot \cos\left[c \cdot C_{tt} \cdot t\right] + C_2 \cdot \sin\left[c \cdot C_{tt} \cdot t\right]; \quad C_{tt} = \frac{\sqrt{2 \cdot M \cdot \left(R_{60} + C_t^2\right)}}{\hbar},$$

(226)

we can now rewrite (225) as follows:

$$E_{n-\text{Schrödinger}} = R_{60} + C_t^2 = -\left(\overbrace{\frac{R_{61}}{r}}^{-V/r} + \overbrace{\frac{R_{62}}{r^2}}^{-\omega_{\text{Schrödinger}}^2/r^2} - \frac{\omega^2}{r^2}\right) - \frac{\hbar^2}{2 \cdot M} \cdot \frac{1}{h[r]} \cdot \left(\frac{2}{r} \cdot \frac{\partial h[r]}{\partial r} + \frac{\partial^2 h[r]}{\partial r^2}\right)$$

(227)

and thus get the classical radial Schrödinger solution for the radial part of (222).

By the way, the classical wave function solution to the hydrogen atom can be given as follows [14]:

$$\Psi_{n,l,m}[r,\vartheta,\varphi] = e^{i \cdot m \cdot \varphi} \cdot P_l^m[\cos\vartheta] \cdot R_{n,l}[r]$$

$$= \sqrt{\left(\frac{2}{n \cdot a_0}\right)^3 \frac{(n-l-1)!}{2 \cdot n \cdot (n+l)!}} \cdot e^{-\rho/2} \cdot \rho^l \cdot L_{n-l-1}^{2l+1}[\rho] \cdot Y_l^m[\vartheta,\varphi]; \quad \rho = \frac{2 \cdot r}{n \cdot a_0}.$$

(228)

The constant a_0 is denoting the Bohr radius with (m_e = electron rest mass, ε_0 = permittivity of free space, Q_e = elementary charge):

$$a_0 = \frac{4 \cdot \pi \cdot \varepsilon_0 \cdot \hbar^2}{m_e \cdot Q_e^2} = 5.292 \cdot 10^{-11} \text{ meter.}$$

(229)

The functions P, L, and Y denote the associated Legendre function, the Laguerre polynomials, and the spherical harmonics, respectively. The wave function (228) directly solves our flat space limit (222) for static cases where f does not depend on t. The extension to time-dependent f^* according to (222) thereby is simple. By using the results from above, we obtain:

$$f^*_{n,l,m}[t,r,\vartheta,\varphi] = g[t] \cdot \Psi_{n,l,m}[r,\vartheta,\varphi] = g[t] \cdot e^{i \cdot m \cdot \varphi} \cdot P_l^m[\cos\vartheta] \cdot R_{n,l}[r]$$

$$= \left(C_1 \cdot \cos\left[c \cdot C_{tt} \cdot t\right] + C_2 \cdot \sin\left[c \cdot C_{tt} \cdot t\right]\right) \cdot N \cdot e^{-\rho/2} \cdot \rho^l \cdot L_{n-l-1}^{2l+1}[\rho] \cdot Y_l^m[\vartheta,\varphi]$$

$$\rho = \frac{2 \cdot r}{n \cdot a_{00}}; \quad N = \sqrt{\left(\frac{2}{n \cdot a_0}\right)^3 \frac{(n-l-1)!}{2 \cdot n \cdot (n+l)!}}; \quad C_{tt} = \frac{\sqrt{2 \cdot M \cdot \left(R_{60} + C_t^2\right)}}{\hbar}$$

$$C_1 = \pm C_2 = 1; \quad a_{00} = \frac{\hbar^2}{M} \cdot \frac{1}{R_{61}}; \quad E_n = -\left[R_{60} + C_t^2\right]_n = -\frac{R_{61}^2}{n^2} \cdot \frac{M}{2 \cdot \hbar^2}; \quad \omega^2 - R_{62} = l^2 + l$$

$$\Rightarrow \quad f_{n,l,m}\left[t,r,\vartheta,\varphi\right] = e^{\pm i \cdot c \cdot C_{tt} \cdot t} \cdot N \cdot e^{-\rho/2} \cdot \rho^l \cdot L_{n-l-1}^{2l+1}\left[\rho\right] \cdot Y_l^m\left[\vartheta,\varphi\right]. \tag{230}$$

Regarding the conditions for the quantum numbers n, l, and m, we have the usual:

$$\{n,l,m\} \in \mathbb{Z}; \quad n \geq 0; \quad l < n; \quad -l \leq m \leq +l. \tag{231}$$

Assuming an omnipresent constant curvature R_{60} does not seem to make much sense in our current system, and thus we should set $R_{60} = 0$. Doing the same with the curvature parameter R_{62} just gives us the usual condition for the spherical harmonics, namely, $\omega^2 = l \cdot (l+1) = l^2 + l$. Thus, we have metrically derived a "hydrogen atom." In addition to the Schrödinger structure, our form also sports a time-dependent factor clearly showing the options for matter and antimatter via the ± sign in the $g[t]$ function.

Please note that according to [19] we should not expect to have modeled the true hydrogen atom here. What we have considered was a 4-dimensional space with a 3D central symmetry and not a true 2-body problem. According to [18], a fully described problem consisting of two 3-dimensional objects would require 6 spatial dimensions plus time. The reason for this lies in the fact that in the metric theory not the apparent spatial dimensions are the dimensions for the metric, but that in the latter we have to consider all degrees of freedom of the system we intent to investigate. Thus, the true metric hydrogen atom would require 7 dimensions instead of 4. Things are getting simpler, of course, if one could assume the objects as point-like and without inner properties. Regarding the true hydrogen atom, it also helps a lot that the electron is so much lighter than the proton.

We therefore conclude that the metric transformation (5) applied on a spherically symmetric flat space given via (196) produces energy states that are similar to the ones of the Schrödinger hydrogen problem if we assume certain (rather simple) r dependencies (223) for the Ricci-like curvature R^*. Thereby the connection to the Einstein field equations with matter and the corresponding Einstein–Hilbert action (c.f. (1)) is given as follows:

$$\frac{\hbar^2}{12 \cdot M} \cdot F[f] \cdot R^* = \frac{\hbar^2}{2 \cdot M} \cdot \frac{R^{**}}{6} = \frac{\hbar^2}{2 \cdot M} \cdot R_6 = R_6^* = R_6^*[r] = R_{60} + \frac{R_{61}}{r} + \frac{R_{62}}{r^2}.$$

$$(232)$$

Now we can also give the metric of the hydrogen atom (matter and antimatter cases). Applying the results from [18] and combining it with (215) yields the normalized quantum gravity metric, including our—metric—wave function f^*:

$$G_{\alpha\beta}^{\text{norm}} = \frac{G_{\alpha\beta}}{C_2} = \overbrace{g_{\alpha\beta}}^{\text{class. GTR}} + g_{\alpha\beta} \cdot \left(\frac{C_1}{C_2} \cdot f \cdot \left(1 + \frac{C_1 \cdot f}{4 \cdot C_2} \right) \right)$$

(Quantum Gravity in 4D / quantum metric)

$$= \overbrace{g_{\alpha\beta}}^{\text{class. GTR}} + \frac{g_{\alpha\beta}}{9} \cdot \left(\frac{C_1 \cdot f^* \cdot \left(4 \cdot C_2^2 + C_1 \cdot f^* \right)}{C_2^4} - 5 \right).$$

$$(233)$$

(Quantum Gravity in 4D / quantum metric)

Please note that no quantum equation was postulated or constructed. Instead, we directly derived our "quantum hydrogen" solution from the scalar Ricci curvature of a scale-transformed metric.

5.6 Half Spin Hydrogen

We always learned that it requires the Dirac equation in order to obtain half spin solutions to the hydrogen problem, while the Schrödinger equation for the hydrogen atom does give singularity-free solutions only for integer quantum numbers.

It was found earlier by this author that this assumption is not correct. In fact, there are half spin hydrogen solutions that can directly be obtained from the classical Schrödinger equation [20, 21]. This aspect will be considered further in later chapters of this book (e.g., see chapter 9).

5.7 Cylindrical Coordinates

In cylindrical coordinates with the metric:

$$
g_{\alpha\beta}^{\text{flat}} = \begin{pmatrix} -c^2 & 0 & 0 & 0 \\ 0 & 1 & 0 & 0 \\ 0 & 0 & 1 & 0 \\ 0 & 0 & 0 & r^2 \end{pmatrix}; \quad \xi_k = \{t, x, r, \varphi\}, \tag{234}
$$

we will only concentrate on the case $R^* = 0$. As this leads to the simple 4D Laplacian:

$$
\frac{R^*}{3} \cdot \frac{\left(C_1 \cdot f + \dfrac{C_1^2 \cdot f^2}{4 \cdot C_2} + C_2\right)^2}{C_1 + \dfrac{C_1^2 \cdot f}{2 \cdot C_2}} + \Delta_g f = 0 \xrightarrow{R^* = 0} \Delta_g f = 0
$$

$$
\Rightarrow \qquad \Delta_{\text{3D-Cylindrical}} f - \frac{\partial^2 f}{c^2 \cdot \partial t^2} = 0 \tag{235}
$$

again and gives us, apart from the time derivative, just the well-known cylindrical Laplacian in 3D, we will not show the evaluation in detail but only present the result. A solution can be found via a separation approach $f[t, x, r, \varphi] = g[t]*f_x[x]*h[r]*f_\varphi[\varphi]$ with:

$$
g[t] = C_1 \cdot \cos\left[c \cdot C_t \cdot t\right] + C_2 \cdot \sin\left[c \cdot C_t \cdot t\right], \tag{236}
$$

$$
f_x[x] = C_{x1} \cdot e^{x \cdot i \cdot \sqrt{C_t^2 + \omega_{r\varphi}^2}} + C_{x2} \cdot e^{-x \cdot i \cdot \sqrt{C_t^2 + \omega_{r\varphi}^2}}, \tag{237}
$$

$$
f_\varphi[\varphi] = C_{\varphi 1} \cdot \cos\left[\omega_\varphi \cdot \varphi\right] + C_{\varphi 2} \cdot \sin\left[\omega_\varphi \cdot \varphi\right], \tag{238}
$$

$$
h[r] = C_J \cdot J_k\left[\omega_{r\varphi} \cdot r\right] + C_Y \cdot Y_k\left[\omega_{r\varphi} \cdot r\right]; \quad k = \omega_\varphi. \tag{239}
$$

Thereby we have the Bessel functions $J_n[z]$ and $Y_n[z]$ as well as the separation parameters C_t, ω_φ, and $\omega_{r\varphi}$. We immediately realize that we have derived a rather fundamental photonic solution. As due to the linearity of (235) (last line) our solution (236) to (239) is additive (c.f. text between (202) and (211)), many photon forms could be constructed via simple superposition.

Again, we see that out of pure Einstein–Hilbert vacuum states, we have been able to obtain states of photonic energy and thus matter. A few illustrative examples are given in [18].

5.8   Cartesian Coordinates

Examples in Cartesian coordinates have been quite extensively considered in [22] and section 3.8 here.

References

1. N. Schwarzer, *Societons and Ecotons – The Photons of the Human Society: Control them and Rule the World, Part 1 of Medical Socio-Economic Quantum Gravity*, www.amazon.com, ASIN: B0876CLT7C.

2. N. Schwarzer, *The World Formula: A Late Recognition of David Hilbert's Stroke of Genius*, 2022, Jenny Stanford Publishing. ISBN: 9789814877206.

3. D. Hilbert, Die Grundlagen der Physik, Teil 1, *Göttinger Nachrichten*, 1915, 395–407.

4. A. Einstein, Grundlage der allgemeinen Relativitätstheorie, *Ann. Phys.*, 1916, **49** (ser. 4), 769–822.

5. N. Schwarzer, *Brief Proof of Hilbert's World Formula – Dirac, Klein-Gordon, Schrödinger, Einstein, Evolution and the 2nd Law of Thermodynamics all from one origin*, www.amazon.com, ASIN: B08585TRB8.

6. N. Schwarzer, *The "New" Dirac Equation – Originally Postulated, Now Geometrically Derived and – along the way – Generalized to Arbitrary Dimensions and Arbitrary Metric Space-Times*, www.amazon.com, ASIN: B085636Q88.

7. N. Schwarzer, *Science Riddles – Riddle No. 11: What is Mass?*, www.amazon.com, ASIN: B07SSF1DFP.

8. N. Schwarzer, *Science Riddles – Riddle No. 12: Is There a Cosmological Spin?*, www.amazon.com, ASIN: B07T3WS7XK.

9. N. Schwarzer, *Science Riddles – Riddle No. 13: How to Solve the Flatness Problem?*, www.amazon.com, ASIN: B07T9WXZVH.

10. N. Schwarzer, *Science Riddles – Riddle No. 14: And what if Time...? A Paradigm Shift*, www.amazon.com, ASIN: B07VTMP2M8.

11. N. Schwarzer, *Science Riddles – Riddle No. 15: Is there an Absolute Scale?*, www.amazon.com, ASIN: B07V9F2124.

12. N. Schwarzer, *Einstein had it, but he did not see it, Part LXXV: The Metric Creation of Matter*, www.amazon.com, ASIN: B07ND3LWZJ.

13. N. Schwarzer, *Einstein had it, but he did not see it, Part LXXVI: Quantum Universes – We don't need no... an Inflation*, www.amazon.com, ASIN: B07NLH3JJV.

14. H. Haken, H. Chr. Wolf, *Atom- und Quantenphysik*, 4th ed. (in German), Springer Heidelberg, 1990, ISBN 0-387-52198-4.

15. T. Bodan, *7 Days – How to explain the world to my dying child, in German – 7 Tage: Wie erkläre ich meinem sterbenden Kind die Welt*, www.amazon.com, ASIN: 1520917562.

16. T. Bodan, *The Eighth Day – Holocaust and the World's Biggest Mysteries: The other Final Solution – with mathematical elaborations* (English ed.), www.amazon.com, ASIN: B019M9ZHIE.

17. T. Bodan, *The Eighth Day – Two Jews against The Third Reich*, illustrated version, Self-published, BoD Classic, 2021, ISBN 9783753417257.

18. N. Schwarzer, *Einstein had it, but he did not see it, Part LXXVII: Matter is Nothing and so Nothing Matters*, www.amazon.com, ASIN: B07NQKKC31.

19. N. Schwarzer, *Einstein had it... Part XXXVI: The Classical and Principle Misinterpretation of the Einstein-Field-Equations AND How it Might be Done Correctly*, www.amzon.com, ASIN: B07D68G9M9.

20. N. Schwarzer, *Einstein had it, but he did not see it – Part LXXIX: Dark Matter Options*, www.amazon.com, ASIN: B07PDMH2JB.

21. N. Schwarzer, *Einstein had it, but he did not see it – Part LXXXII: Half Spin Hydrogen*, www.amazon.com, ASIN: B07Q3NFB39.

22. N. Schwarzer, *Humanitons – The Intrinsic Photons of the Human Body – Understand them and Cure Yourself, Part 2 of Medical Socio-Economic Quantum Gravity*, www.amazon.com, ASIN: B088QGK6ST.

The Third Day:
Let There Be Mass and Inertia

Chapter 6

Masses and the Infinity Options Principle

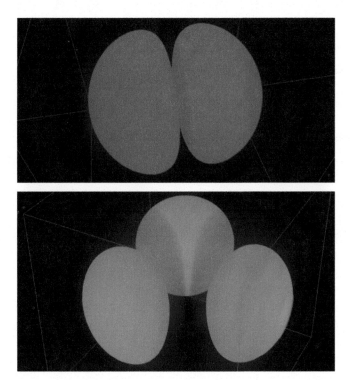

Figure 6.1 Half spin solutions and question: Can we explain the 3-generation and the quantized mass problem?

The Math of Body, Soul, and the Universe
Norbert Schwarzer
Copyright © 2023 Jenny Stanford Publishing Pte. Ltd.
ISBN 978-981-4968-24-9 (Hardcover), 978-1-003-33454-5 (eBook)
www.jennystanford.com

6.1 Abstract: Brief Description of Goals of This Section

After having found that in spaces or space-times with an infinite number of options the quantized Einstein field equations take on their simplest (linear) form [A1] (see also chapter 8 of this book), we here investigated the problem of how mass could be quantized.

In fact, we will see that, with mass appearing as entanglement of dimensions [A2–A8], its quantization can be easily achieved.

Along the way, we also tried to answer the question why there are just three generations of elementary particles (Fig. 6.1). It was derived that in order to satisfy the necessary outcome for Newton's law of gravitation in the far distance, the Ricci curvature has to contain a certain term with a power law r-dependency of the type $1/r^3$ (please note that the exponent is just the number of our ordinary space' spatial dimensions). Inserting this into the quantum gravity equation, which is resulting from the scaling of any metric solution to the Einstein field equations, we found three possible ways for the extraction of the three generations:

(a) A simple algebraic equation of third order for the Schwarzschild radius allowing for three real solutions.

(b) A quantum solution for the wave function f only allowing for certain Schwarzschild radii.

(c) Extraction of a wave solution to the metric Dirac equation.

Unfortunately, by sticking to the classical Klein–Gordon or Dirac approaches, we were not able to complete the task, but found many interesting facts along the way. With respect to the three generations of elementary particles, we might say that in fact we have found a fairly potent way to explain an X-generation problem, only that we do not really see why X should be restricted to three. However, in demanding the space-time to be restricted to four big-scale dimensions, we found that the product of mass and the quantum gravity wave function occurs in a polynomial of third order, which, as it is well known, can have three solutions.

In the end, a solution to the 3-generation problem, so it seems, can be found by combining the Klein–Gordon approach and the Dirac approach, and excepting the two, respectively their solutions, to exist simultaneously. This then automatically leaves no other choice for the product of mass and quantum gravity wave function, but to algebraically fall apart into three different solutions. Most interestingly, it was found that this combined approach favors octonions as a tool (not the only one as it will be shown here) to solve the subsequent differential equations. Thus, so our current conclusion, the way to solve the 3-generation problem could be outlined here but—so far—this path was not treated in a manner one could call complete.

6.1.1   References for the Abstract

A1. N. Schwarzer, *Humanitons—The Intrinsic Photons of the Human Body—Understand them and Cure Yourself*, Part 2 of "Medical Socio-Economic Quantum Gravity," Self-published, Amazon Digital Services, 2020, Kindle.

A2. N. Schwarzer, *The World Formula: A Late Recognition of David Hilbert's Stroke of Genius*, 2022, Jenny Stanford Publishing. ISBN: 9789814877206.

A3. N. Schwarzer, *Societons and Ecotons—The Photons of the Human Society—Control them and Rule the World*, Part 1 of "Medical Socio-Economic Quantum Gravity," Self-published, Amazon Digital Services, 2020, Kindle.

A4. N. Schwarzer, *Science Riddles—Riddle No. 11: What is Mass?*, Self-published, Amazon Digital Services, 2019, Kindle.

A5. N. Schwarzer, *Science Riddles—Riddle No. 12: Is There a Cosmological Spin?*, Self-published, Amazon Digital Services, 2019, Kindle.

A6. N. Schwarzer, *Science Riddles—Riddle No. 13: How to Solve the Flatness Problem?*, Self-published, Amazon Digital Services, 2019, Kindle.

A7. N. Schwarzer, *Science Riddles—Riddle No. 14: And what if Time … ? A Paradigm Shift*, Self-published, Amazon Digital Services, 2019, Kindle.

A8. N. Schwarzer, *Science Riddles—Riddle No. 15: Is there an Absolute Scale?*, Self-published, Amazon Digital Services, 2019, Kindle.

6.2   Repetition: Fundamental Starting Point and How to Proceed from There

Before introducing math, we want to give a very brief verbal explanation of the principle structure and appearance of a world formula, which—as being a world formula—should be able to describe any system, including a human society [1], organisms [2], and more [3, 4]. Thereby the system may be extremely complex and of very high dimensional order, but this does not mean that a mathematical description would not mirror it (not in principle, anyway). A more thorough verbal explanation can be found in [4].

1. Let there be a set of properties.

2. Such properties could also be seen as degrees of freedom of a system, which mathematically would allow us to see them as the system's dimensions, whereas the system itself is a space, formed by such dimensions (properties). Thus, we consider space as an ensemble of properties.

3. These properties, which, as said, could just be seen as degrees of freedom, and thus, dimensions are subjected to a Hamilton extremal principle.

4. This leads to the Einstein–Hilbert action [5] and subsequently to the Einstein field equations [5, 6], when following just Hilbert's mathematical path of variating the action with respect to the metric tensor [5] or Einstein's argumentation [6].

5. However, the simplest way to satisfy the Einstein–Hilbert action could also be achieved by just demanding the curvature R (the so-called scalar Ricci curvature) of the space or space-time of properties to be a constant, which leads to the well-known quantum equations like Klein–Gordon, Schrödinger, and Dirac [1–6]. In chapter 10, it will be shown how this condition directly follows from the usual variation of the Einstein–Hilbert action.

6. Inertia, evolution forces, and the laws of thermodynamics appear via entanglement of dimensions, variation with respect to the number of dimensions of sub-systems, and position of centers of gravity [5–12]. This way also the required complexity of a system can be driven up arbitrarily without leaving the path of Hilbert's fundamentality.

6.3 Repetition: Hilbert's World Formula

Our starting point shall be the classical Einstein–Hilbert action [5] with the Ricci scalar R^* as kernel or Lagrange density:

$$\delta W = 0 = \delta \int_V d^n x \left(\sqrt{-G} \cdot R^* \right) \tag{240}$$

and, in contrast to the classical form in [8], with a somewhat adapted metric tensor $G_{\delta\gamma} = F[f] \cdot g_{\delta\gamma}$. Thereby G shall denote the determinant of the metric tensor $G_{\alpha\beta}$, while g will later stand for the corresponding determinant of the metric tensor $g_{\alpha\beta}$. In order to distinguish our Ricci scalar, being based on $G_{\delta\gamma} = F[f] \cdot g_{\delta\gamma}$ from the usual one, being based on the metric tensor $g_{\alpha\beta}$, we marked it with the * (superscript). Please note that Eq. (240) should allow us to just say that W is a constant. Then, of course, with W being a constant, it would automatically variate to zero. As a sharper (less general) condition, one may even assume that:

$$\sqrt{-G} \cdot R^* = \text{const,} \tag{241}$$

or we even may demand:

$$\sqrt{-G} \cdot R^* = 0 \tag{242}$$

with either $G = 0$ or $R^* = 0$, whereby the latter case definitively is less trivial and thus, probably also makes more sense.

Please note that with $G_{\delta\gamma} = F[f] \cdot g_{\delta\gamma}$, we have used the simplest form of metric adaptation, which we could construct as a simplification from a generalization of the typical form of tensor transformations, namely:

$$G_{\alpha\beta} = F\left[f\left[t,x,y,z,\ldots,\xi_k,\ldots,\xi_n\right]\right]^{ij}_{\alpha\beta} g_{ij}, \tag{243}$$

$$\rightarrow G_{\alpha\beta} = F\left[f\left[t,x,y,z,\ldots,\xi_k,\ldots,\xi_n\right]\right] \cdot \delta^i_\alpha \delta^j_\beta g_{ij}$$

$$\rightarrow G_{\alpha\beta} = F\left[f\left[t,x,y,z,\ldots,\xi_k,\ldots,\xi_n\right]\right] \cdot g_{\alpha\beta} = F[f] \cdot g_{\alpha\beta}. \tag{244}$$

The generalization is been elaborated in [4].

It should be pointed out that there will always be a possibility to choose a $g_{\alpha\beta}$ allowing us for the simple transformation version (244) instead of (243). Thus, for the time being, we can reduce our considerations here to this simple option. In this case the function $F[f]$ just acts as a scaling factor.

Now, we evaluate the resulting Ricci scalar for the metric $G_{\alpha\beta}$ at the first in n and then, subsequently, simplify to 4 dimensions:

$$R^* = \frac{1}{F[f]^3} \cdot \left(\left(C_{N1} \cdot \left(\frac{\partial F[f]}{\partial f}\right)^2 - C_{N2} \cdot F[f] \cdot \frac{\partial^2 F[f]}{\partial f^2} \right) \cdot \overbrace{\left(\tilde{\nabla}_g f\right)^2}^{= f_{,\alpha} g^{\alpha\beta} f_{,\beta}} - C_{N2} \cdot F[f] \cdot \frac{\partial F[f]}{\partial f} \cdot \Delta_g f \right) \Bigg|_{n=4} \rightarrow$$

$$= \frac{1}{F[f]^3} \cdot \left(\left(\frac{3}{2} \cdot \left(\frac{\partial F[f]}{\partial f}\right)^2 - 3 \cdot F[f] \cdot \frac{\partial^2 F[f]}{\partial f^2} \right) \cdot (\tilde{\nabla}_g f)^2 - 3 \cdot F[f] \cdot \frac{\partial F[f]}{\partial f} \cdot \Delta_g f \right)$$

$$\xrightarrow{\frac{3}{2}\left(\frac{\partial F[f]}{\partial f}\right)^2 - 3 \cdot F[f] \cdot \frac{\partial^2 F[f]}{\partial f^2} = 0} = -\frac{3}{F[f]^2} \cdot \left(\frac{\partial F[f]}{\partial f} \cdot \Delta_g f \right). \tag{245}$$

With respect to the constants C_{N1} and C_{N2} in arbitrary n dimensions, the reader is referred to [6] or Eq. (503) in section "6.12 Toward an Explanation for the 3-Generation Problem [28]" of this book.

With $R^* = 0$, one would immediately satisfy the Einstein–Hilbert action. Taking the general n-dimensional case in the first line of Eq. (245) and always demanding that:

$$C_{N1} \cdot \left(\frac{\partial F[f]}{\partial f} \right)^2 - C_{N2} \cdot F[f] \cdot \frac{\partial^2 F[f]}{\partial f^2} = 0, \qquad (246)$$

which simply is a fixing-condition for the wrapper function $F[f]$, we get:

$$R^* = -\frac{C_{N2}}{F[f]^2} \cdot \left(\frac{\partial F[f]}{\partial f} \cdot \Delta_g f \right) = 0$$

$$\Rightarrow \qquad \qquad \Delta_g f = 0. \qquad (247)$$

This is the "homogenous" Klein–Gordon equation, respectively the mass- and potential-less Klein–Gordon equation.

The question where to get the mass from will be discussed in section 6.6.

This question was also answered in earlier chapters of this book and papers of the author. Therefore, the reader can refer to [1, 4]. Alternatively, also chapters 2 and 9 should be recommended.

We point out again that the function $F[f]$ is just a scale-parameter to the metric $g_{\alpha\beta}$.

6.4   The Infinity Options Principle

It could be shown in [2] that in spaces or space-times with infinite numbers of dimensions, the function $F[f]$ satisfying condition (246) would converge to a constant, namely:

$$\lim_{n\to\infty} F[f] = \lim_{n\to\infty} C_2 \cdot \left(f \cdot (n-2) + C_1 \right)^{\frac{4}{n-2}} = C_2. \qquad (248)$$

Now applying condition (246) and performing the limiting procedure after the derivation in Eq. (245), the result would be:

$$\lim_{n\to\infty} R^* = \lim_{n\to\infty} \frac{1}{F[f]^3} \cdot \left(-C_{N2} \cdot F[f] \cdot \frac{\partial F[f]}{\partial f} \cdot \Delta_g f \right) = -\frac{4}{f \cdot C_2} \cdot \Delta_g f$$

$$\Rightarrow \qquad \left(\frac{\lim_{n\to\infty} R^*}{4} \cdot C_2 + \Delta_g \right) f = 0. \qquad (249)$$

This gives us a linear differential equation in higher (infinite) numbers of dimensions the moment we can assume the Ricci scalar R^* to be independent on f. Thereby the constant C_2 is yet unknown.

We conclude that an infinity of options/dimensions leads to linearity.

We also conclude that in space-times with truly infinite options any functional quantum effect disappears, because the function f then converges toward a constant (Eq. (248)).

As the postulated classical quantum equations being linear and work so well, we have to further conclude that, in fact, there is a fairly high number of dimensions allowing for a good linearity approximation, but that—at least for smaller scales—there cannot be a true dimensional infinity, because this would completely rule out any quantum effects from the corresponding space-times. There have to be at least sub-spaces with finite numbers of dimensions.

6.5 Considering the Vacuum State Leads to Broken Symmetry and Higgs Field

Directly combining Eq. (245) and Eq. (246), the resulting Ricci scalar of the transformed metric in Eq. (244) reads as follows:

$$R^* = \frac{1}{F[f]^3} \cdot \left(-C_{N2} \cdot F[f] \cdot \frac{\partial F[f]}{\partial f} \cdot \Delta_g f \right) = \frac{1}{F[f]^2} \cdot \left(-C_{N2} \cdot \frac{\partial F[f]}{\partial f} \cdot \Delta_g f \right). \quad (250)$$

From [2], we know that in n-dimensions, the function $F[f]$ satisfies condition (246), which could be given as:

$$F[f] = C_2 \cdot \left(f \cdot (n-2) + C_1 \right)^{\frac{4}{n-2}} = \left(c_1 \cdot f \cdot (n-2) + c_2 \right)^{\frac{4}{n-2}}. \quad (251)$$

Thereby the constants C_1, C_2, c_1 and c_2 are arbitrary.

Now, we assume f to be a constant, which automatically leads to the simple equation:

$$R^* \cdot \left[\left(c_1 \cdot f \cdot (n-2) + c_2 \right)^{\frac{4}{n-2}} \right]^2 = 0. \quad (252)$$

Also assuming that the Ricci scalar curvature R^* shall either be a constant or proportional to f^n, with an arbitrary exponent $n \geq 0$, we obtain the familiar trivial solution of $f_0 = 0$ in the proportionality case, while the $R^* =$ constant, situation only leaves the second factor to become zero. From Eq. (252), we here obtain the non-trivial ground state, which also holds in any other case of R^*:

$$f = -\frac{c_2}{c_1 \cdot (n-2)}. \tag{253}$$

As the constants c_1 and c_2 are arbitrary, we could simply set them as follows:

$$c_2 = -\mu^2; \quad c_1 \times (n-2) = 2 \cdot \lambda. \tag{254}$$

This gives us the additional ground state solutions directly in the classical Higgs field style [14], namely:

$$f = \frac{\mu}{\sqrt{2 \cdot \lambda}}. \tag{255}$$

The measured value assumed to be constant f, which is known to be [15]:

$$f = \frac{\mu}{\sqrt{2 \cdot \lambda}} = \frac{246\,\text{GeV}}{\sqrt{2} \cdot c^2}. \tag{256}$$

Now, with the Higgs mass m_H known to be 125 GeV/c² and the fact that we have [15]:

$$m_H = \sqrt{2 \cdot \mu^2} = \sqrt{4 \cdot \lambda \cdot f} = \sqrt{2 \cdot c_1 \cdot (n-2) \cdot f}, \tag{257}$$

we can obtain:

$$\frac{c_1 \cdot (n-2)}{2} = \lambda \approx 0.13; \quad \mu \approx 88.8\,\text{GeV/c}^2. \tag{258}$$

6.6 The Entanglement of Dimensions and the Production of Mass

It was shown in [9–13] that mass automatically results from certain forms of entanglements of dimensions. For a start we first only give one example in 6 dimensions, but it should be pointed out that the production (creation) of mass via entanglement works in any number of dimensions. There are other forms of entanglement, which cause mass to occur, but apparently also spin and more forms of inertia as matter and energy [4, 10, 13], can be created via dimensional entanglement. This statement also holds for the entanglement of more than just 2 dimensions [12, 13]. But as said, here we only consider the creation of mass in its simplest form, namely via the entanglement of 2 dimensions.

The reader may ask why we need such a principle derivation of the origin of inertia if in the end we intend "only" to discuss the results with respect to certain—and apparently completely different—problems in various fields. Well, masses (or inertia in general) resulting from the entanglement of dimensions can occur in any space-time (system) consisting of any kind of ensemble of dimensions/properties and as we know how important masses and inertia are in the "real" world, it is obvious that finding the driving forces for the creation of such inertia in any system is of great importance. This also holds for fields far away from physics like socio economics, for instance. How can we otherwise understand economic cycles if not understanding what—within such cycles—are the inertia and what are the driving potentials? Even wars may just be seen as resonance catastrophes of such cyclic oscillations and in truly knowing how they can occur one might also be able to actually hinder them to happen at all.

Introducing a metric for the coordinates t, r, ϑ, φ, u, and v of the form:

$$g^6_{\alpha\beta} = \begin{pmatrix} -c^2 & 0 & 0 & 0 & 0 & 0 \\ 0 & 1 & 0 & 0 & 0 & 0 \\ 0 & 0 & r^2 & 0 & 0 & 0 \\ 0 & 0 & 0 & r^2 \cdot \sin[\vartheta]^2 & 0 & 0 \\ 0 & 0 & 0 & 0 & g[v] & 0 \\ 0 & 0 & 0 & 0 & 0 & g[v] \end{pmatrix} \cdot f[t,r,\vartheta,\varphi], \qquad (259)$$

with the entanglement solution for u and v as:

$$g[v] = \frac{C_{v1}^2}{C_{v0}^2 \cdot \left(1 + \cosh\left[C_{v1} \cdot (v + C_{v2})\right]\right)}, \qquad (260)$$

which gives us the following Ricci scalar from (245) with condition (246):

$$R^* = \frac{1}{F[f]^3} \cdot \left(-C_{N2} \cdot F[f] \cdot \frac{\partial F[f]}{\partial f} \cdot \Delta_g f\right); \quad F[f] = C_f + f[t,r,\vartheta,\varphi]$$

$$\xrightarrow{6D} R^* = \frac{5}{f^2} \cdot \left(\frac{C_{v0}^2}{5} \cdot f + \frac{\partial^2 f}{c^2 \cdot \partial t^2} - \Delta_{3\text{D-sphere}} f\right); \quad f = f[t,r,\vartheta,\varphi]; \quad C_f = 0.$$

$$(261)$$

Here $\Delta_{3\text{D-sphere}} f$ denotes the Laplace operator in spherical coordinates.

As for $R^* = 0$ and $\dfrac{C_{v0}^2}{5} = \dfrac{m^2 \cdot c^2}{\hbar^2}$ (m = mass, c = speed of light in vacuum, $\hbar$ = reduced Planck constant), we have obtained the classical quantum Klein–Gordon equation with mass and therefore conclude that mass can obviously be constructed by a suitable set of additional (meaning in addition to our 4-dimensional space-time), entangled coordinates.

In the case considered in Eq. (259), the mass does not seem to be quantized. In order to also consider an example, where we have reason to assume the quantization of mass, we introduce a slightly different metric via:

$$g_{\alpha\beta}^6 = \begin{pmatrix} -c^2 & 0 & 0 & 0 & 0 & 0 \\ 0 & 1 & 0 & 0 & 0 & 0 \\ 0 & 0 & r^2 & 0 & 0 & 0 \\ 0 & 0 & 0 & r^2 \cdot \sin[\vartheta]^2 & 0 & 0 \\ 0 & 0 & 0 & 0 & 1 & 0 \\ 0 & 0 & 0 & 0 & 0 & g[v] \end{pmatrix} \cdot f[t,r,\vartheta,\varphi]. \qquad (262)$$

Now, we apply $g[v]$ as follows:

$$g[v] = C_{v1} \cdot \cos^2 \left[\sqrt{\frac{5}{2}} \cdot C_{v0} \cdot (v + C_{v2}) \right], \qquad (263)$$

which gives us the following Ricci scalar from (245) with condition (246):

$$R^* = \frac{1}{F[f]^3} \cdot \left(-C_{N2} \cdot F[f] \cdot \frac{\partial F[f]}{\partial f} \cdot \Delta_g f \right); \quad F[f] = C_f + f[t,r,\vartheta,\varphi]$$

$$\xrightarrow{6D} R^* = \frac{5}{f^2} \cdot \left(C_{v0}^2 \cdot f + \frac{\partial^2 f}{c^2 \cdot \partial t^2} - \Delta_{\text{3D-sphere}} f \right); \quad f = f[t,r,\vartheta,\varphi]; \quad C_f = 0.$$

$$(264)$$

And again, for $R^* = 0$ and $C_{v0}^2 = \dfrac{m^2 \cdot c^2}{\hbar^2}$ (m = mass, c = speed of light in vacuum, $\hbar$ = reduced Planck constant), we have obtained the classical quantum Klein–Gordon equation with mass.

As for $C_f \neq 0$, we would obtain:

$$R^* = \frac{1}{F[f]^3} \cdot \left(-C_{N2} \cdot F[f] \cdot \frac{\partial F[f]}{\partial f} \cdot \Delta_g f \right); \quad F[f] = C_f + f[t,r,\vartheta,\varphi]$$

$$\xrightarrow{6D} R^* = \frac{5}{f^2} \cdot \left(C_{v0}^2 \cdot C_f + C_{v0}^2 \cdot f + \frac{\partial^2 f}{c^2 \cdot \partial t^2} - \Delta_{\text{3D-sphere}} f \right); \quad f = f[t, r, \vartheta, \varphi],$$

(265)

we realize that the multiplicator function $f[\dots]$, which we see as a quantum solution, always acts on the whole metric and does not allow for a combined solution of the kind:

$$F[f] = C_f + f[t, r, \vartheta, \varphi],$$

(266)

because the solution to the $R^* = 0$ case leads to:

$$C_{v0}^2 \cdot C_f + C_{v0}^2 \cdot f + \frac{\partial^2 f}{c^2 \cdot \partial t^2} - \Delta_{\text{3D-sphere}} f = 0$$

(267)

and thus,

$$F[f] = C_f + G[t, r, \vartheta, \varphi] - C_f,$$

(268)

with:

$$C_{v0}^2 \cdot G + \frac{\partial^2 G}{c^2 \cdot \partial t^2} - \Delta_{\text{3D-sphere}} G = 0.$$

(269)

In other words, we have harmonics as solutions and in order to obtain a flat space solution for the far distance, we either have to assume many sources spread over, respectively placed within the flat space-time (many centers of gravity) or we assume a series of solutions $G[\dots]$ to the linear Eq. (269) in such a way that it guarantees the nearly flat outcome. The latter—naturally— leads to a wiggling quantum background of space and time. However, as the function for $g[v]$ has a nonlinear source, we cannot add-up masses there as easy as for the linear equation on the function f. This—if insisting on a convergence to a flat space solution—either favors the "many centers of gravity" approach or requires a completely different approach for the entanglement introduction of masses via add-on dimensions. In chapter 11, we will see how such problems can be overcome in 8 dimensions.

Things are getting easier with a Cartesian approach as follows:

$$g_{\alpha\beta}^6 = \begin{pmatrix} -c^2 & 0 & 0 & 0 & 0 & 0 \\ 0 & 1 & 0 & 0 & 0 & 0 \\ 0 & 0 & 1 & 0 & 0 & 0 \\ 0 & 0 & 0 & 1 & 0 & 0 \\ 0 & 0 & 0 & 0 & 1 & 0 \\ 0 & 0 & 0 & 0 & 0 & 1 \end{pmatrix} \cdot \left(C_f + f[t, x, y, z] \cdot g[v] \right).$$

(270)

Now also the differential equation for $g[v]$ becomes linear as we can demand:

$$\frac{\partial^2 g}{\partial v^2} + C_{v0}^2 \cdot f = 0 \quad \Rightarrow \quad g[v] = C_{v1} \cdot \cos\left[C_{v0} \cdot v\right] + C_{v2} \cdot \sin\left[C_{v0} \cdot v\right]. \quad (271)$$

This leads us to (267) again, but allows for arbitrary series of parameters $C_{v0}^2 = \dfrac{m^2 \cdot c^2}{\hbar^2}$ and thus, masses, when still assuming $R^* = 0$. In addition, we can choose the constant C_f arbitrarily and thus, automatically can have the asymptotic flat space behavior with asymptotically vanishing functions $g[v]$ and $f[\,\dots\,]$.

In the next section, we will consider spherical base coordinates and a somewhat more general approach.

6.7 Obtaining Quantized Masses in Situations of $R^* = 0$

6.7.1 Several Time-Like Dimensions

Assuming curled up or compactified time-like dimensions, one might take a metric as follows:

$$g_{\alpha\beta}^6 = \begin{pmatrix} -c^2 & 0 & 0 & 0 & 0 & 0 \\ 0 & 1 & 0 & 0 & 0 & 0 \\ 0 & 0 & r^2 & 0 & 0 & 0 \\ 0 & 0 & 0 & r^2 \cdot \sin[\vartheta]^2 & 0 & 0 \\ 0 & 0 & 0 & 0 & -c_u^2 & 0 \\ 0 & 0 & 0 & 0 & 0 & -c_v^2 \end{pmatrix} \cdot \left(C_f + g[u,v] \cdot f[t,r,\vartheta,\varphi]\right).$$

$$(272)$$

Further assuming that t, r, ϑ, φ are the usual 4-dimensional space-time coordinates, while u and v are compactified and applying the approach $g[u,v] = Gu[u]*Gv[v]$ plus the usual separation approach for the function $f[\,\dots\,] = f_t[t] \cdot f_r[r] \cdot f_\vartheta[\vartheta] \cdot f_\varphi[\varphi]$, we obtain the following solution for the case $R^* = 0$:

$$Gv[v] = C_{v1} \cdot \cos[c_v \cdot E_v \cdot v] + [C_{v2} \cdot \sin[c_v \cdot E_v \cdot v], \quad (273)$$

$$Gu[u] = C_{u1} \cdot \cos[c_u \cdot E_u \cdot u] + C_{u2} \cdot \sin[c_u \cdot E_u \cdot u], \quad (274)$$

$$f_t[t] = C_{t1} \cdot \cos[c \cdot E \cdot t] + C_{t2} \cdot \sin[c \cdot E \cdot t], \quad (275)$$

$$f_\varphi[\varphi] = C_{\varphi 1} \cdot \cos[m \cdot \varphi] + C_{\varphi 2} \cdot \sin[m \cdot \varphi], \qquad (276)$$

$$f_\vartheta[\vartheta] = C_{P\vartheta} \cdot P_\ell^m[\cos[\vartheta]] + C_{Q\vartheta} \cdot Q_\ell^m[\cos[\vartheta]], \qquad (277)$$

with the associated Legendre polynomials P_ℓ^m. In the sub-section below, we are going to discuss the option for half spin solutions.

The intermediate result above lets us end up with the following differential equation for $f_r[r]$:

$$R^* = \frac{1}{F[g \cdot f]^3} \cdot \left(-C_{N2} \cdot F[g \cdot f] \cdot \frac{\partial F[g \cdot f]}{\partial(g \cdot f)} \cdot \Delta_g f \right);$$

$$F[g \cdot f] = C_f + f[t,r,\vartheta,\varphi] \cdot g[u,v]$$

$$C_f \neq 0$$

$$\xrightarrow{\ 6D\ } R^* = 0$$

$$\Rightarrow \qquad 0 = \left(\ell(1 + \ell) - \left(E^2 + E_u^2 + E_v^2 \right) r^2 \right) f_r[r] - r \left(2 f_r[r]' + r \cdot f_r[r]'' \right). \qquad (278)$$

This time we do not have to demand $C_f = 0$ and thus, can assure the asymptotic flat space-time behavior. The solution for Eq. (278) (last line) consists of the spherical Bessel functions j_n and y_n:

$$f_r[r] = C_{r1} \cdot j_\ell\left[\sqrt{E^2 + E_u^2 + E_v^2} \cdot r \right] + C_{r2} \cdot y_\ell[\sqrt{E^2 + E_u^2 + E_v^2} \cdot r]. \qquad (279)$$

Changing the metric (272) to space-like coordinates u and v:

$$g_{\alpha\beta}^6 = \begin{pmatrix} -c^2 & 0 & 0 & 0 & 0 & 0 \\ 0 & 1 & 0 & 0 & 0 & 0 \\ 0 & 0 & r^2 & 0 & 0 & 0 \\ 0 & 0 & 0 & r^2 \cdot \sin[\vartheta]^2 & 0 & 0 \\ 0 & 0 & 0 & 0 & c_u^2 & 0 \\ 0 & 0 & 0 & 0 & 0 & c_v^2 \end{pmatrix} \cdot \left(C_f + g[u,v] \cdot f[t,r,\vartheta,\varphi] \right), \qquad (280)$$

but leaving the approach (273) to (277) results in the following differential equation for f_r:

$$0 = \left(\ell(1 + \ell) - \left(E^2 - E_u^2 - E_v^2 \right) r^2 \right) f_r[r] - r \left(2 f_r[r]' + r \cdot f_r[r]'' \right), \qquad (281)$$

which results in:

$$f_r[r] = C_{r1} \cdot j_{-1-\ell}\left[\sqrt{E^2 - E_u^2 - E_v^2} \cdot r \right] + C_{r2} \cdot y_{-1-\ell}[\sqrt{E^2 - E_u^2 - E_v^2} \cdot r]. \qquad (282)$$

While with the solution (279), the function j should be favored as it does not lead to a singularity for $r = 0$, we have to choose the function y in the case of (282).

However, when also changing the approach for the function Gu and Gv to:

$$Gv[v] = C_{v1} \cdot e^{-c_v \cdot E_v \cdot v} + C_{v2} \cdot e^{c_v \cdot E_v \cdot v}, \tag{283}$$

$$Gu[u] = C_{u1} \cdot e^{-c_u \cdot E_u \cdot u} + C_{u2} \cdot e^{c_u \cdot E_u \cdot u}, \tag{284}$$

we are back with Eq. (278) and the corresponding solution (279) for f_r. Now, together with the choice $C_{u2} = C_{v2} = 0$ for positive u and v, and $C_{u1} = C_{v1} = 0$ for negative, we can even assure asymptotically vanishing (thus, compactified) add-on coordinates u and v.

More such dimensions ω_i could be added easily in the same way and seeing the parameters $E_{\omega i}$ either as energies or masses $m_{\omega i}$ via $E_{\omega i} = c^{2*} m_{\omega i}$, we do not only have the equivalency of mass and energy, but also a great variety of options to add mass and energy as such entangled (entangled via the function $g[u,v, \dots] = g[\omega_0, \omega_1, \omega_2, \dots]$) dimensions.

Thus, this way we can add arbitrary masses and energies just as additional dimensions either time- or space-like. Together with the add-on coordinates, respectively dimensions the additional masses and energies will automatically come as quanta.

6.7.2   Wave-Like Entanglement with Higher Dimensions

Does this also solve the so-called 3-generation problem, where elementary particles appear to exist in three types of differently heavy, but otherwise similar particles?

This would not only force us to accept that, at least with respect to the creation of elementary particles, the number of dimensions we could add is restricted (it obviously ends with the third generation), but also requires us to find a way to connect the additional masses with the function f_t for t (275).

Once again choosing metric (272), we seek to solve the $R^* = 0$ task as follows:

$$g_{\alpha\beta}^6 = \begin{pmatrix} -c^2 & 0 & 0 & 0 & 0 & 0 \\ 0 & 1 & 0 & 0 & 0 & 0 \\ 0 & 0 & r^2 & 0 & 0 & 0 \\ 0 & 0 & 0 & r^2 \cdot \sin[\vartheta]^2 & 0 & 0 \\ 0 & 0 & 0 & 0 & -c_u^2 & 0 \\ 0 & 0 & 0 & 0 & 0 & -c_v^2 \end{pmatrix}$$

$$\times\left(C_f + g\left[C_t \cdot t + C_u \cdot u + C_v \cdot v\right] \cdot f_r[r] \cdot f_\vartheta[\vartheta] \cdot f_\varphi[\varphi]\right), \tag{285}$$

and find:

$$g\left[C_t \cdot t + C_u \cdot u + C_v \cdot v\right]$$

$$= C_1 \cdot \cos\left[E \cdot \left(C_t \cdot t + C_u \cdot u + C_v \cdot v\right)\right] + C_2 \cdot \sin\left[E \cdot \left(C_t \cdot t + C_u \cdot u + C_v \cdot v\right)\right], \tag{286}$$

$$f_\varphi[\varphi] = C_{\varphi 1} \cdot \cos[m \cdot \varphi] + C_{\varphi 2} \cdot \sin[m \cdot \varphi], \tag{287}$$

$$f_\vartheta[\vartheta] = C_{P\vartheta} \cdot P_\ell^m[\cos[\vartheta]] + C_{Q\vartheta} \cdot Q_\ell^m[\cos[\vartheta]], \tag{288}$$

$$f_r[r] = \left(\begin{array}{c} C_{r1} \cdot j_\ell \left[E \cdot \dfrac{\sqrt{C_t^2 c_u^2 c_v^2 + c^2\left(C_u^2 c_v^2 + c_u^2 C_v^2\right)}}{c \cdot c_u \cdot c_v} \cdot r \right] \\ \\ + C_{r2} \cdot y_\ell \left[E \cdot \dfrac{\sqrt{C_t^2 c_u^2 c_v^2 + c^2\left(C_u^2 c_v^2 + c_u^2 C_v^2\right)}}{c \cdot c_u \cdot c_v} \cdot r \right] \end{array} \right). \tag{289}$$

Seeing the function g now as the function of just one effective time-coordinate τ:

$$g\left[C_t \cdot t + C_u \cdot u + C_v \cdot v\right] = C_1 \cdot \cos[E \cdot \tau] + C_2 \cdot \sin[E \cdot \tau], \tag{290}$$

with the auxiliary times u and v being unstable, we may not only have a recipe for the construction of additional particle masses, but also a comfortable explanation for the fact that none of the higher order generation particles is stable. On the other hand, we see no reason why the number of generations and thus, in our case, the number of additional time dimensions should be limited to three or two, respectively.

One possible reason might be that there can be no more time-like dimensions than there are already—"visible"—spatial ones. Thereby "visible" means our ability to detect only three spatial dimensions and as we only detect those, we may not be able to detect masses (particle generations) being connected with other—probably rolled-up or compactified—dimensions.

6.7.3 Periodic Coordinate Quantization

From the section above it became evident that there is a huge amount of possibilities to extract inertia and masses of any kind. So far, however, these masses are not quantized. This changes the moment we apply certain symmetries.

For instance, using the 6-dimensional case again and changing Eq. (259) to:

$$
g^6_{\alpha\beta} = \begin{pmatrix}
-c^2 & 0 & 0 & 0 & 0 & 0 \\
0 & 1 & 0 & 0 & 0 & 0 \\
0 & 0 & r^2 & 0 & 0 & 0 \\
0 & 0 & 0 & r^2 \cdot \sin[\vartheta]^2 & 0 & 0 \\
0 & 0 & 0 & 0 & 1 & 0 \\
0 & 0 & 0 & 0 & 0 & \sin[u]^2
\end{pmatrix} \cdot \left(C_f + g[u,v] \cdot f[t,r,\vartheta,\varphi] \right),
$$

(291)

thereby applying the approach $g[u,v] = Gu[u]*Gv[v]$ plus the usual separation approach for the function $f[\ldots] = f_t[t] \cdot f_r[r] \cdot f_\vartheta[\vartheta] \cdot f_\varphi[\varphi]$ and obtaining the solutions:

$$
Gv[v] = C_{v1} \cdot \cos[\mu \cdot v] + C_{v2} \cdot \sin[\mu \cdot v],
$$

(292)

$$
Gu^\mu_\ell[u] = C_{Pu} \cdot P^\mu_\ell[\cos[u]] + C_{Qu} \cdot Q^\mu_\ell[\cos[u]],
$$

(293)

$$
f_t[t] = C_{t1} \cdot \cos[c \cdot E \cdot t] + C_{t2} \cdot \sin[c \cdot E \cdot t],
$$

(294)

$$
f_\varphi[\varphi] = C_{\varphi 1} \cdot \cos[m \cdot \varphi] + C_{\varphi 2} \cdot \sin[m \cdot \varphi],
$$

(295)

$$
f_\vartheta[\vartheta] = C_{P\vartheta} \cdot P^m_L[\cos[\vartheta]] + C_{Q\vartheta} \cdot Q^m_L[\cos[\vartheta]],
$$

(296)

with the associated Legendre polynomials P^m_ℓ, we end up with the following differential equation for $f_r[r]$:

$$
R^* = \frac{1}{F[f]^3} \cdot \left(-C_{N2} \cdot F[f] \cdot \frac{\partial F[f]}{\partial f} \cdot \Delta_g f \right);
$$

$$
F[f] = C_f + f[t,r,\vartheta,\varphi] \cdot g[u,v]; \quad C_f = 0 \xrightarrow{6D} R^* = 0
$$

$$
\Rightarrow \quad 0 = \left(\left(L(1+L) + \left(\frac{2}{5} + \ell(1+\ell) \right) r^2 \right) - r^2 E^2 \right) f_r[r] - r \left(2 f_r[r]' + r \cdot f_r[r]'' \right).
$$

(297)

Once again, we had to demand $C_f = 0$. Other metric approaches, also treated with the separation ansatz, give similar results. Thereby it is found that metric components treated as angles and therefore multiplied with r^2 result in spin-like quantum properties (see parameter L), while metric components not being multiplied with r^2 give energy or mass-like quantum behavior.

The solution for Eq. (297) (second line) consists of the spherical Bessel functions j_n and y_n:

$$f_r[r] = C_{r1} \cdot j_{-1-L}\left[-i\sqrt{\left(\frac{2}{5} + L(1+L)\right) - E^2} \cdot r\right]$$

$$+ C_{r2} \cdot y_{-1-L}\left[-i\sqrt{\left(\frac{2}{5} + L(1+L)\right) - E^2} \cdot r\right]. \tag{298}$$

From classical Quantum Theory [16], we know that the numbers μ, m, l, and L are quantized. However, it should be pointed out here that in addition to the classical quantization where it was assumed that μ and L have to be integers with the boundary $-\mu \leq L \leq \mu+$, it was found meanwhile that there are also half spin solutions [17, 18]. As this finding is very important, we are going to present it in the appendix of this chapter.

Again, for $R^* = 0$ and $\left(\frac{2}{5} + L + L^2\right) = \frac{m^2 \cdot c^2}{\hbar^2}$ (m = mass, c = speed of light in vacuum, $\hbar$ = reduced Planck constant), we have obtained the classical quantum Klein–Gordon equation with mass. This time, however, the mass would automatically be quantized.

Most interestingly, there would be the option for a minimum mass for $L = 0$. Unfortunately, however, there is no way that for the ordinary subsequent quantum numbers of $L = 1/2, 1, 3/2, \dots$, we would be able to construct the mass distribution of the known elementary particle generations.

Of great interest should be the case of the 5-dimensional spatial sphere with the space-time metric:

$$g_{\alpha\beta}^6 = \begin{pmatrix} -c^2 & 0 & 0 & 0 & 0 & 0 \\ 0 & 1 & 0 & 0 & 0 & 0 \\ 0 & 0 & r^2 & 0 & 0 & 0 \\ 0 & 0 & 0 & r^2 \cdot \sin[\vartheta]^2 & 0 & 0 \\ 0 & 0 & 0 & 0 & r^2 \cdot \sin[\vartheta]^2 \sin[\varphi]^2 & 0 \\ 0 & 0 & 0 & 0 & 0 & r^2 \cdot \sin[\vartheta]^2 \sin[\varphi]^2 \sin[u]^2 \end{pmatrix}$$

$$\times \left(C_f + f[t, r, \vartheta, \varphi, u, v]\right). \tag{299}$$

This time the solutions of the separation approach read:

$$Gv[v] = C_{v1} \cdot \cos[\mu \cdot v] + C_{v2} \cdot \sin[\mu \cdot v], \tag{300}$$

$$Gu_\lambda^\mu[u] = C_{Pu} \cdot P_\lambda^\mu[\cos[u]] + C_{Qu} \cdot Q_\lambda^\mu[\cos[u]], \tag{301}$$

$$f_t[t] = C_{t1} \cdot \cos[c \cdot E \cdot t] + C_{t2} \cdot \sin[c \cdot E \cdot t], \tag{302}$$

$$f_\varphi[\varphi] = \frac{C_{P\varphi} \cdot P_\ell^{\lambda + \frac{1}{2}}[\cos[\varphi]] + C_{Q\varphi} \cdot Q_\ell^{\lambda + \frac{1}{2}}[\cos[\varphi]]}{\left(-\sin[\varphi]^2\right)^{\frac{1}{4}}}, \tag{303}$$

$$f_\vartheta[\vartheta] = \frac{C_{P\vartheta} \cdot P_L^{\ell + \frac{1}{2}}[\cos[\vartheta]] + C_{Q\vartheta} \cdot Q_L^{\ell + \frac{1}{2}}[\cos[\vartheta]]}{\left(-\sin[\vartheta]^2\right)^{\frac{1}{2}}}. \tag{304}$$

We end up with the following differential equation for $f_r[r]$ in the case of $R^* = 0$:

$$R^* = \frac{1}{F[f]^3} \cdot \left(-C_{N2} \cdot F[f] \cdot \frac{\partial F[f]}{\partial f} \cdot \Delta_g f\right); \quad F[f] = C_f + f[t,r,\vartheta,\varphi] \cdot g[u,v];$$

$$C_f \neq 0 \xrightarrow{6D} R^* = 0$$

$$\Rightarrow \quad 0 = \left(\left(L(1+L) - 2\right) - r^2 E^2\right) f_r[r] - r\left(4 f_r[r]' + r \cdot f_r[r]''\right). \tag{305}$$

The solution consists of Bessel functions J_n and Y_n:

$$f_r[r] = \frac{C_{r1} \cdot J_{\frac{1}{2} + L}[E \cdot r] + C_{r2} \cdot Y_{\frac{1}{2} + L}[E \cdot r]}{r^{\frac{3}{2}}}. \tag{306}$$

In order to avoid singularities with the Legendre polynomials, we can only have the parameter set: $\mu = \pm 1/2$, $\lambda = 1/2$, $\ell = 1$ and $L = 3/2$. This automatically leads to two "centers of gravity" for the distribution if we assume the ordinary space to be span out by the coordinates r, ϑ, φ.

In 5 dimensions, the corresponding metric and the $R^* = 0$ solution would be:

$$g_{\alpha\beta}^5 = \begin{pmatrix} -c^2 & 0 & 0 & 0 & 0 \\ 0 & 1 & 0 & 0 & 0 \\ 0 & 0 & r^2 & 0 & 0 \\ 0 & 0 & 0 & r^2 \cdot \sin[\vartheta]^2 & 0 \\ 0 & 0 & 0 & 0 & r^2 \cdot \sin[\vartheta]^2 \sin[\varphi]^2 \end{pmatrix}$$

$$\times F\big[f[t,r,\vartheta,\varphi,u]\big]. \tag{307}$$

The solution of the separation approach reads:

$$g[u] = C_{u1} \cdot \cos[\mu \cdot u] + C_{u2} \cdot \sin[\mu \cdot u], \tag{308}$$

$$f_t[t] = C_{t1} \cdot \cos[c \cdot E \cdot t] + C_{t2} \cdot \sin[c \cdot E \cdot t], \tag{309}$$

$$f_\varphi[\varphi] = C_{P\varphi} \cdot P_\lambda^\mu\big[\cos[\varphi]\big] + C_{Qu} \cdot Q_\lambda^\mu\big[\cos[\varphi]\big], \tag{310}$$

$$f_\vartheta[\vartheta] = \frac{C_{P\vartheta} \cdot P_\ell^{\lambda+\frac{1}{2}}\big[\cos[\vartheta]\big] + C_{Q\vartheta} \cdot Q_\ell^{\lambda+\frac{1}{2}}\big[\cos[\vartheta]\big]}{\left(-\sin[\vartheta]^2\right)^{\frac{1}{4}}}. \tag{311}$$

We end up with the following differential equation for $f_r[r]$ in the case of R^* = 0:

$$R^* = \frac{1}{F[f]^3} \cdot \left(-C_{N2} \cdot F[f] \cdot \frac{\partial F[f]}{\partial f} \cdot \Delta_g f\right); \quad F[f] = \big(C_f + f[t,r,\vartheta,\varphi] \cdot g[u]\big)^{\frac{4}{3}};$$

$$C_f \neq 0 \xrightarrow{\text{5D}} R^* = 0$$

$$\Rightarrow \qquad 0 = \left(\left(\ell(1+\ell) - \frac{3}{4}\right) - r^2 E^2\right) f_r[r] - r\big(3 f_r[r]' + r \cdot f_r[r]''\big). \tag{312}$$

The solution consists of Bessel functions J_n and Y_n:

$$f_r[r] = \frac{C_{r1} \cdot J_{\frac{1}{2}+\ell}[E \cdot r] + C_{r2} \cdot Y_{\frac{1}{2}+\ell}[E \cdot r]}{r}. \tag{313}$$

This time, in order to avoid singularities with the Legendre polynomials, we can only have the parameter set: $\mu = \pm 1/2$, $\lambda = 1/2$, and $\ell = 1$.

In both cases, 5 and 6 dimensions, we find two spin states as an intrinsic property in connection with additional angular dimensions (in addition to the usual angular coordinates ϑ, φ. From there, we may deduce that particles with spin 1/2 are in fact higher dimensional objects of spherical symmetry. The two centers of gravity are found for all other possible singularity-free solutions with starting values $\mu = \pm k/2$ (k = 1,2,3, ...) (see Fig. 6.2) as long as we choose ϑ, φ as our ordinary spatial dimensions. The situation changes, however, when assuming that (taking the 6-dimensional case) the coordinate v becomes one of the usual spatial dimensions. Then taking $\mu = \pm 3/2$, we end up with three centers of gravity (Fig. 6.3) and we are strongly reminded of the spatial appearance of the three quarks inside hadrons like proton and neutron.

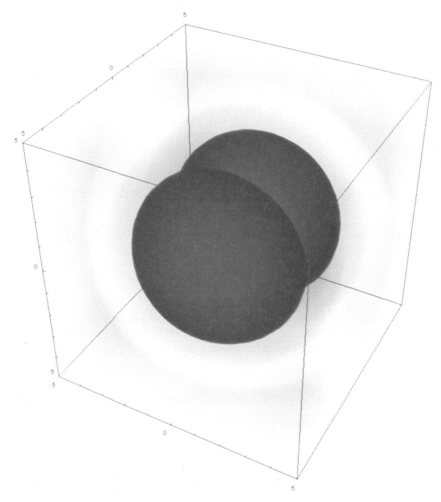

Figure 6.2 Two gravity centers solution in 6-dimensional space-times with coordinates t, r, and the angles ϑ, φ as our ordinary spatial dimensions. The metric trace for these "visual" coordinates looks classical and reads:

$$\text{trace}\left\{g_{\alpha\beta}^{\text{visual part of 6}}\right\} = \left\{-c^2, 1, r^2, r^2 \cdot \sin[\vartheta]^2, \underbrace{\ldots,\ldots}_{\text{hidden}}\right\} \times \left(C_f + g[u,v] \cdot f[t,r,\vartheta,\varphi]\right).$$

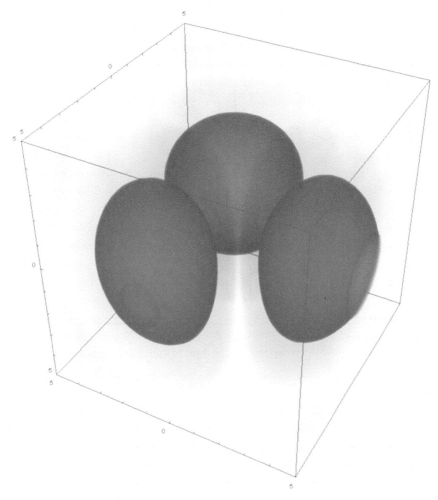

Figure 6.3 Three gravity centers solution in 6-dimensional space-times with coordinates t, r, and the angles ϑ, v as our ordinary spatial dimensions. The metric trace for these "visual" coordinates looks non-classical and reads as follows:

$$\text{trace}\left\{g_{\alpha\beta}^{\text{visual part of 6}}\right\} = \left\{-c^2, 1, r^2, \underbrace{\ldots,\ldots,}_{\text{hidden}}, r^2 \cdot \sin[\vartheta]^2 \sin[\varphi]^2 \sin[u]^2\right\} \times (C_f + g[\varphi,u] \cdot f[t,r,$$

$\vartheta,v]).$

6.7.4 Quantum Well Quantization

A more or less classical idea to explain the 3-generation problem of elementary particles was the introduction of suitable quantum wells for the different generations. Figure 6.4 shows such wells for the three leptons: electron, muon, and tauon. The length scale has been chosen in such a way that the rather narrow structure still is visible, even though the difference between the generations is fairly huge.

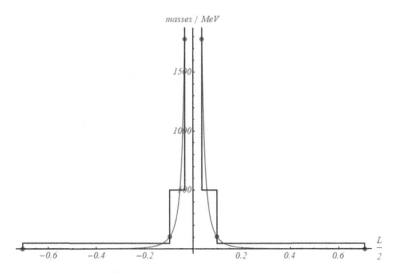

Figure 6.4 Quantum well structures for the three leptons electron, muon, and tauon.

The idea behind the quantum well is the enclosure of one and the same particle in an ever-smaller box with increasing generation. The smaller the box, the higher the frequency, and thus, the mass of the subsequent particle.

The reason for the existence of such boxes, however, remained unclear.

The interesting fact about this approach, however, may be seen in the fact that trials to fit this mass or box-size behavior result in exponential laws with powers equal or close to three, which is the number of spatial dimensions (at least if one concentrates on the easily "visible" ones).

6.8 The Schwarzschild Case and the Meaning of the Schwarzschild Radius

Having already established the fact that the Einstein–Hilbert action also contains the principle means for the extraction of the fundamental quantum equations (including the Dirac equation as shown in [1] and—more clearly— later in chapters 9 and 10 of this book), we saw that mass can be introduced via dimensional entanglement, respectively additional dimensions, being "somehow" connected (entangled) with our ordinary space-time. Thereby even quantized masses and the asymptotic flat space behavior can be achieved rather easily. This was shown in a variety of example in the previous section.

Here, now we want to apply our recipe on the classical Schwarzschild solution with the metric (317), which leads to the following differential equation:

$$F[f] \cdot R^* = R^{**} = -\frac{6}{f} \cdot \Delta_g f = -\frac{6}{f} \cdot \Delta_{SS} f$$

$$= -\frac{6}{f} \cdot \left(\Delta_{\text{sphere}} f - r_s \cdot \left(\frac{f^{(0,1,0,0)} + r \cdot f^{(0,2,0,0)}}{r^2} + \frac{f^{(2,0,0,0)}}{c^2(r-r_s)} \right) \right)$$

$$= -\frac{6}{f} \cdot \left(\Delta_{\text{3D-sphere}} f - \frac{f^{(2,0,0,0)}}{c^2} - r_s \cdot \left(\frac{f^{(0,1,0,0)} + r \cdot f^{(0,2,0,0)}}{r^2} + \frac{f^{(2,0,0,0)}}{c^2(r-r_s)} \right) \right)$$

$$f^{(0,1,0,0)} = \frac{\partial f[t,r,\vartheta,\varphi]}{\partial r}; \quad f^{(0,2,0,0)} = \frac{\partial^2 f[t,r,\vartheta,\varphi]}{\partial r^2}; \quad f^{(2,0,0,0)} = \frac{\partial^2 f[t,r,\vartheta,\varphi]}{\partial t^2}.$$

$$(314)$$

Thereby we denoted the Laplace operator in spherical polar coordinates with $\Delta_{\text{3D-sphere}}$. It is connected with the Laplace operator in spherical Minkowski coordinates as follows:

$$\Delta_{\text{sphere}} f = \Delta_{\text{3D-sphere}} f - \frac{f^{(2,0,0,0)}}{c^2}. \tag{315}$$

Please note that the functional wrapper $F[f]$ reading:

$$G_{\alpha\beta} = g_{\alpha\beta} \cdot F[f] = g_{\alpha\beta} \cdot (C_1 + f)^2 = g_{\alpha\beta} \cdot (C_1 + f[t,r,\vartheta,\varphi])^2 \tag{316}$$

fulfills the necessary condition (246) in 4 dimensions and was applied in (314).

In order to construct the complete Schwarzschild Laplacian, we took the Schwarzschild metric from [19] as follows:

$$g_{\alpha\beta}^{\text{Schwarzschild}} = g_{\alpha\beta}^{SS} = \begin{pmatrix} -c^2 \left(1 - \frac{2 \cdot G \cdot m}{c^2 \cdot r} \right) & 0 & 0 & 0 \\ 0 & \left(1 - \frac{2 \cdot G \cdot m}{c^2 \cdot r} \right)^{-1} & 0 & 0 \\ 0 & 0 & r^2 & 0 \\ 0 & 0 & 0 & r^2 \cdot \sin^2 \vartheta \end{pmatrix}.$$

$$(317)$$

Here r gives the spherical radius, M stands for the gravitational mass of the Schwarzschild object, and constants c and G are giving the speed of light in vacuum and the gravitational or Newton constant, respectively. One easily sees that the solution above has two singularities, namely one at $r = r_s = \dfrac{2 \cdot m \cdot G}{c^2}$, which is a result of our choice of coordinates and could be transformed away (see next sub-section and Eq. (319)), and at $r = 0$. The parameter r_s is called the Schwarzschild radius. In n dimensions, the metric solution looks as follows:

$$
g_{\alpha\beta}^{SS} = \begin{pmatrix}
-c^2 \left(1 - \left(\dfrac{r_s}{r}\right)^{n-3}\right) & 0 & 0 & 0 & 0 & 0 \\
0 & \left(1 - \left(\dfrac{r_s}{r}\right)^{n-3}\right)^{-1} & 0 & 0 & 0 & 0 \\
0 & 0 & r^2 & 0 & 0 & 0 \\
0 & 0 & 0 & r^2 \cdot \sin[\vartheta]^2 & 0 & 0 \\
0 & 0 & 0 & 0 & r^2 \cdot \sin[\vartheta]^2 \sin[\varphi]^2 & 0 \\
0 & 0 & 0 & 0 & 0 & \ddots
\end{pmatrix}.
$$

$$(318)$$

6.8.1 Brief Discussion of Singularity Problem

More or less for completeness, we here point out that the famous $1/r^{n-3}$ singularity may be avoided in constructing a metric of the kind:

$$
g_{\alpha\beta}^{SS} = \begin{pmatrix}
-c^2 \left(1 - \left(\dfrac{r_s}{R[r]}\right)^{n-3}\right) & 0 & 0 & 0 & 0 & 0 \\
0 & g_{11}^{SS} & 0 & 0 & 0 & 0 \\
0 & 0 & R[r]^2 & 0 & 0 & 0 \\
0 & 0 & 0 & R[r]^2 \cdot \sin[\vartheta]^2 & 0 & 0 \\
0 & 0 & 0 & 0 & g_{44}^{SS} & 0 \\
0 & 0 & 0 & 0 & 0 & \ddots
\end{pmatrix}
$$

$$
g_{11}^{SS} = \left(1 - \left(\frac{r_s}{R[r]}\right)^{n-3}\right)^{-1} \cdot \left(\frac{\partial R[r]}{\partial r}\right)^2 \; ; \; R[r] = r_s + r \cdot e^{-\frac{r_s}{r}} \; ; \; \frac{\partial R[r]}{\partial r} = e^{-\frac{r_s}{r}} \cdot \left(1 + \frac{r_s}{r}\right)
$$

$$g_{44}^{SS} = R[r]^2 \cdot \sin[\vartheta]^2 \sin[\varphi]^2, \tag{319}$$

which is just the same metric as given in Eq. (318) only that it has a transformed radius coordinate. Evaluation of the metric components for time and radius shows us that all singularities have now disappeared (see Fig. 6.5 and Fig. 6.6). Thereby the transformation function $R[r]$ just assures that there is the smallest piece of space the radius coordinate cannot reach, which is to say, it cannot penetrate the area $r < r_s$.

Considering a possible generalization of $R[r]$ with respect to the "grain" size of space, one may even introduce a radius transformation of the kind:

$$R[r] = r_b + r \cdot e^{-\frac{r_s}{r}}, \tag{320}$$

or

$$R[r] = r_{b1} + r \cdot e^{-\frac{r_{b2}}{r}}, \tag{321}$$

with r_{b1} and r_{b2} giving some kind of boundary or typical grain size radii, which have to be suitably chosen or—whenever this may be possible—experimentally determined.

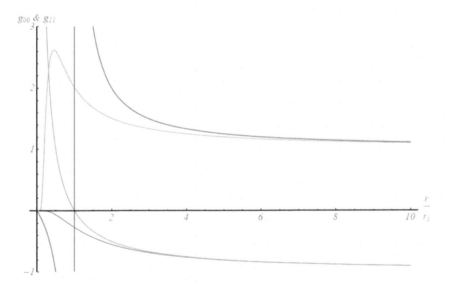

Figure 6.5 Metric components for time g_{00} and radius g_{11} in a 4-dimensional space-time for the new metric (319) (blue and green) and the classical Schwarzschild solution (318) (yellow and red), respectively.

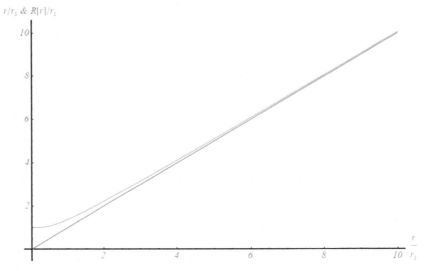

r/r_s & $R[r]/r_s$

Figure 6.6 Ordinary spherical radius (blue) and transformed radius $R[r]$ (yellow) as used in Eq. (319).

6.8.2 Back to the Case $r \gg r_s$

However, as being manly interested in the situation for bigger r (meaning $r > r_s$ or even $r \gg r_s$), we can always evaluate:

$$\lim_{r\to\infty} R[r] = \lim_{r\to\infty}\left(r_s + r \cdot e^{-\frac{r_s}{r}} \right) = r, \tag{322}$$

and thus, by sticking to the classical form (318), the corresponding R^* (245) results in the following equation:

$$R^* = -\frac{C_{N2}\cdot F[f]\cdot \dfrac{\partial F[f]}{\partial f}}{F[f]^3}\cdot \Delta_g f = -\frac{C_{N2}\cdot F[f]\cdot \dfrac{\partial F[f]}{\partial f}}{F[f]^3}\cdot \Delta_{SS} f$$

$$= -\frac{C_{N2}\cdot F[f]\cdot \dfrac{\partial F[f]}{\partial f}}{F[f]^3}\cdot \left(\Delta_{n-1\text{-sphere}} f - \frac{f^{(2,0,0,0)}}{c^2} - \left(\frac{r_s}{r}\right)^{n-3}\cdot \left(\frac{\dfrac{f^{(0,1,0,0)}}{r} + f^{(0,2,0,0)}}{} + \frac{f^{(2,0,0,0)}}{c^2\left(1-\left(\dfrac{r_s}{r}\right)^{n-3}\right)} \right) \right)$$

$$f^{(0,1,0,0)} = \frac{\partial f[t,r,\vartheta,\varphi]}{\partial r}; \quad f^{(0,2,0,0)} = \frac{\partial^2 f[t,r,\vartheta,\varphi]}{\partial r^2}; \quad f^{(2,0,0,0)} = \frac{\partial^2 f[t,r,\vartheta,\varphi]}{\partial t^2}.$$

(323)

Please note that for all n, we always took care to apply the function $F[f]$ in the form such that it fulfills condition of Eq. (246), no matter what number of space-time dimensions we are considering.

Closely observing Eq. (314), however, we find that for the Schwarzschild metric, the resulting Klein–Gordon equation does not contain the mass-term as linear factor to the linear form of the quantum function f as we were able to construct it with the dimensional entanglement in the section above and as it was—quite successfully—postulated with the classical quantum equations.

How can we explain the discrepancy between the obviously not too incorrect linear appearance of mass in the quantum equations (including those being directly extracted from the Einstein–Hilbert action, respectively the Ricci scalar with the introduction of a variated metric (our functional or quantum wrapper function $F[f]$)) and its totally different form in the case of the Schwarzschild metric?

Is heavy or gravity mass, as described best as Schwarzschild radius within the Schwarzschild approach, so much different than the quantum mass?

After all, while in the Schwarzschild case, we have something rather complicated, namely:

$$R^* \Rightarrow \Delta_{n\text{-space-time}} f - \left(\frac{r_s}{r}\right)^{n-3} \cdot \left(\frac{f^{(0,1,0,0)}}{r} + f^{(0,2,0,0)} + \frac{f^{(2,0,0,0)}}{c^2\left(1-\left(\frac{r_s}{r}\right)^{n-3}\right)} \right)$$

$$= \Delta_{n-1\text{-sphere}} f - \frac{f^{(2,0,0,0)}}{c^2} - \left(\frac{r_s}{r}\right)^{n-3} \cdot \left(\frac{f^{(0,1,0,0)}}{r} + f^{(0,2,0,0)} + \frac{f^{(2,0,0,0)}}{c^2\left(1-\left(\frac{r_s}{r}\right)^{n-3}\right)} \right),$$

(324)

we are always able to construct a linear form for the flat space with our entanglement approach, namely:

$$R^* \Rightarrow \Delta_{n\text{-space-time}} f - M \cdot f.$$

(325)

Something must be wrong with our approach and/or the classical theory.

Or are we missing something else here?

Before answering the latter question, we need to remind ourselves about the way black holes and thus, Schwarzschild objects store information.

6.8.3 A Brief Reminder about Bekenstein Problem

In [20], where we wanted to investigate the problem of finding the most fundamental Turing machine, we also had to take care about the question of how the universe stores information. One of the most famous problems in this context is been seen in the Bekenstein Bit-problem, where it was found that black holes can store information, but it is been seen as a mystery how these objects actually do this. In [20, 21], we have shown that bit-like information is been stored as dimensions and that each bit becomes 1 dimension. For convenience we here will repeat parts of the evaluation.

In the early seventies, J. Bekenstein [22] investigated the connection between black hole surface area and information (see also [23]). Thereby he simply considered the surfaces change of a black hole, which would be hit by a photon just of the same size as the black hole. His idea was that with such a geometric constellation, the outcome of the information would just consist of the information whether the photon fell into the black hole or whether it did not. Thus, it would be a 1-bit information. His calculations led him to the funny proportionality of area and information. He found that the number of bits, coded by a certain black hole, is proportional to the surface area of this very black hole if measured in Planck area ℓ_P^2. In fact, the dependency how one bit of information changes the area of the black hole (ΔA) reads:

$$\Delta A = 32 \cdot \pi^2 \cdot \ell_P^2 + 64 \cdot \pi^3 \cdot \frac{\ell_P^4}{r_s^2}. \tag{326}$$

Ignoring the extremely small second term, one could just assume our black hole to be constructed out of many such bit surface pieces. Thus, we could write:

$$q \cdot \Delta A = q \cdot 32 \cdot \pi^2 \cdot \ell_P^2 = 4 \cdot \pi \cdot r_s^2 \quad \Rightarrow \quad r_s^2 = q \cdot 8 \cdot \pi \cdot \ell_P^2, \tag{327}$$

where r_s gives the radius of the black hole. We see that our black hole radius is proportional to the square root of the bits q thrown into it.

Now, we want to compare the dependency $r_s[q]$ with the radii $r_{max}[n]$ resulting in maximum surface of n-spheres for a certain number of space-time dimensions n.

Note: The correct solution for the evaluation of the Schwarzschild radius r_s as function of the bits thrown into a black hole object (if using Eq. (326)) would be:

$$r_s = 2 \cdot \ell_p \cdot \sqrt{\cdot \left(q + \sqrt{q \cdot (1+q)} \right)}!$$

(328)

We find a perfect fit to the *n*-sphere with maximized surface to a given radius (dots in Fig. 6.7) with a Planck length of $\ell_p = 0.07878133250775303$[1] (Fig. 6.7).

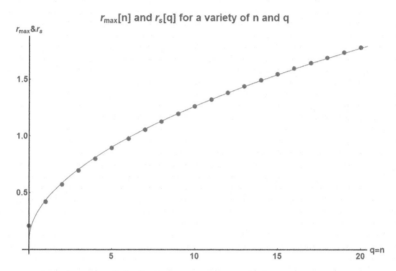

Figure 6.7 Radius r_{max} for which at a certain number of dimensions the *n*-sphere has maximum surface in dependence on *n* compared with the increase of the Schwarzschild radius r_s of a black hole in dependence on the number of bits *q* thrown into it by using Eq. (328). We find that $q = n$.

Our finding does not only connect the intrinsic dimension of a black hole with its mass respectively its surface, but also gives an explanation to the hitherto unsolved problem of "what are the micro states of a black hole giving it temperature and allowing it to store information." According to the evaluation in this section, these microstates are just various states of dimensions realized within the black hole in dependence on the number of bits it contains (and thus its mass). The bigger the number of bits, the higher the intrinsic dimensions the black hole has. In fact, the connection even is a direct one and only seems to deviate from the simple direct proportionality

---

[1]Please note that this value came out slightly different when using the volume [14, 15, 17], where we obtained $\ell_p = 0.07881256452824544$.

for very low numbers of masses[2], respectively Schwarzschild radii r_s, and numbers of bits q the black hole has swallowed.

This finding also gives us a direct connection between a principle mathematical law (the maximum volume as function of the dimension for a given radius of an n-sphere) to the number of bits a black hole contains, to the mass or Schwarzschild radius of this very black hole and the number and character of microstates the black hole actually uses to internally code the bits.

It has to be pointed out that the expression "intrinsic dimension" truly stands for the part of space for $r < r_s$. For more information about this aspect and its handling, the reader can refer to [20, 21].

Assuming that the number of dimensions and thus, information, a black hole can contain, increases from the center of the black hole to its periphery, we see that right in the center at a dimension of $n = 3$ the singularity disappears (one may check this easily with the classical Schwarzschild solution (318)).

6.8.4 Consequences with Respect to the Dimension of Time

Even though we currently do not have as suitable metric solution for the black hole with pairwise entangled dimensions, we will still try to discuss the problem of time for these objects.

We start with the metric of an n-dimensional spherical object (potentially a black hole):

$$
g^n_{\alpha\beta} =
\begin{pmatrix}
-c^2 \cdot g[r] & 0 & 0 & 0 & \cdots & 0 \\
0 & \dfrac{1}{g[r]} & 0 & 0 & \cdots & 0 \\
0 & 0 & r^2 & 0 & \cdots & 0 \\
0 & 0 & 0 & r^2 \cdot \sin^2 \varphi_1 & \cdots & 0 \\
\cdots & \cdots & \cdots & \cdots & \ddots & 0 \\
0 & 0 & 0 & 0 & 0 & g_{nn}
\end{pmatrix}
$$

$$
g_{44} = r^2 \cdot \sin^2 \varphi_1 \cdot \sin^2 \varphi_2; \quad g_{nn} = r^2 \cdot \prod_{j=1}^{n-2} \sin^2 \varphi_j; \quad g[r] = 1 - \frac{r_s^{n-3}}{r^{n-3}}. \tag{329}
$$

---

[2]Besides, this deviation is also suggested by the Bekenstein finding summed up in Eq. (326), where we could assume the second term to become of importance at lower numbers of r_s.

We see that the classical time component $g_{00} = -c^2 \times g[r]$ vanishes at $r = r_s$ and becomes space-like for $r < r_s$, which is inside the black hole. There, however, we find that r-component g_{11} would become time-like and thus, we would still have a "classical" time dimension behind the event horizon.

What about our entangled time? Could it still be obtained inside a black hole?

In order to investigate this question, we introduce a p-dimensional black hole ($p \geq 4$) within a space of $n > 6$ dimensions as follows:

$$g_{\alpha\beta}^n = \begin{pmatrix} -c^2 \cdot g[r] & 0 & 0 & 0 & \cdots & 0 \\ 0 & \dfrac{1}{g[r]} & 0 & 0 & \cdots & 0 \\ 0 & 0 & r^2 & 0 & \cdots & 0 \\ 0 & 0 & 0 & r^2 \cdot \sin^2 \varphi_1 & \cdots & 0 \\ \cdots & \cdots & \cdots & \cdots & \ddots & 0 \\ 0 & 0 & 0 & 0 & 0 & g_{nn} \end{pmatrix}$$

$$g_{44} = r^2 \cdot \sin^2 \varphi_1 \cdot \sin^2 \varphi_2; \quad g[r] = 1 - \frac{r_s^{p-3}}{r^{p-3}}$$

$$g_{aa} = g_{nn} = F\left[f\left[x_a, x_n \right] \right] \quad a = n - 1. \tag{330}$$

As before [12], we have the general solution:

$$F[f] = C_1 \cdot e^{C_2 \cdot f}; \quad f\left[x_a, x_n \right] = C_+ \left[x_n + i \cdot x_a \right] + C_- \left[x_n - i \cdot x_a \right]. \tag{331}$$

Thus, as long as our black hole has additional degrees of freedom, independent on its spherical geometry, we can easily have entangled time coordinates as given in [12].

How could it be understood that a black hole of a certain mass or radius should contain a corresponding number of dimensions? Obviously the 4-dimensional black hole would not be a very massive and therefore not a very stable object (c.f. [21] and see Schwarzschild radius for $n = q = 4$ in Fig. 6.7).

The way to overcome that particular problem would be to say goodbye to the idea of black holes of a certain radius existing independent on the number of dimensions they contain. Thereby especially the number of dimensions making out the Schwarzschild n-sphere is of importance. The simplest way to construct objects being compatible with the Bekenstein-observation and our finding of the $r_s[q = n]$-dependency would be a radius-dependent change of dimensions.

6.8.4.1 How can a *q*-dimensional black hole object reside inside an only 4-dimensional space-time? Answer: Our supposedly 4-dimensional space-time is not 4- but *n*-dimensional!

In [12], we assumed that there must be a transition region that adjusts the number of dimensions corresponding to the mass/size and thus, dimensions of the black hole to our ordinary 4-dimensional space-time. Back then we suggested:

"When moving toward the black hole, the dimension near the event horizon increases until it reaches the necessary n at the corresponding $r_s[q = n]$. While it appears logic to assume an n[r]-dependency being simply inverse to Eq. (328) (note that we have q = n) for $r < r_s$, we can only guess[3] what might be a suitable approach for the outside region $r > r_s$. For simplicity, we here assume an exponential behavior. Let the inverse function to Eq. (328) be denoted as n[r_s], we approach the total and n[r]-variable metric via the Schwarzschild function:

$$g[r] = 1 - \frac{r_s^{N[r]-3}}{r^{N[r]-3}}; \quad n[r] = \begin{cases} n[r] \text{ for } r < r_s \\ \left(C \cdot \left(n[r_s] \cdot e^{-\left(\frac{r}{r_s}\right)^p} + 4 \right) \text{ for } r \geq r_s \right. \end{cases}$$

$$\text{with:} \quad \left[C \cdot \left(n[r_s] \cdot e^{-\left(\frac{r}{r_s}\right)^p} + 4 \right) \right]_{r=r_s} = n[r_s]; \quad p > 1. \tag{332}$$

This way we would end up with the usual outside behavior for spherical massive objects, but avoid the singularity for r = 0. In addition, we obtain a solution being able to store the Bekenstein-bits q by the means of dimensions n via the simple relation n = q."

Is our ordinary space-time truly 4-dimensional or should we rather consider all options it is containing as degrees of freedom and thus, however, small dimensions?

Are we not living in a universe of almost infinite options?

Shouldn't we therefore see our space-time as effective—almost— infinitely dimensional with only four of the dimensions being a bit "bigger" or "obvious" than all the others (c.f. [24])?

One immediate consequence of such an assumption is linearity for the quantum function *f* allocated to the Ricci curvature R^* as derived

---

[3]Respectively experimentally determine.

in section "**6.4 The Infinity Options Principle**" and given in Eq. (249) (second line). The second consequence is the absolute equivalency of entanglement and the Ricci curvature *R** (with respect to the linear appearance of the function *f*). Thereby the latter consequence is probably just giving a relatively fundamental explanation for energy, mass, and heavy and inertial masses.

In order to still have the correct transition to the Einstein theory, respectively Newton theory in the outside region around the central symmetric gravity object, we may just assume that for the consideration of bigger scales at $r = r_s$ and for $r > r = r_s$ the n-dimensional term $(r_s/r)^n$ becomes r_s/r, which gives us the classical Schwarzschild solution for the outside space-time. The fact that this does not automatically provide us a smooth transition shall not bother us at the moment, because we would not expect such a behavior anyway as—just as said—there is a leap in dimensions.

What do we mean by "the consideration of bigger scales"?

Well, we think that when observing the space-time around a gravity object of symmetry of revolution under various resolutions, we will obtain different results.

At bigger scales, we will always see the Schwarzschild metric behavior (as long as the object is not rotating, which then leads to the Kerr metric, of course). It may even have a quantum halo as evaluated in [25] with the assumption of $R^* = 0$ and no time-dependency:

$$G_{\alpha\beta}^{Q-SS} = \left(1+C_2 \cdot \ln\left[\frac{r-r_s}{r}\right]\right)^2 \cdot \left(Z_m \cdot Y_0^m\right)^2 \cdot \begin{pmatrix} -c^2\left(\frac{r-r_s}{r}\right) & 0 & 0 & 0 \\ 0 & \left(\frac{r-r_s}{r}\right)^{-1} & 0 & 0 \\ 0 & 0 & r^2 & 0 \\ 0 & 0 & 0 & r^2 \cdot \sin^2\vartheta \end{pmatrix}.$$

$$(333)$$

There we have:

$$Z_m[\varphi] = C_{\varphi 1} \cdot \cos[m \cdot \varphi] + C_{\varphi 2} \cdot \sin[m \cdot \varphi], \qquad (334)$$

$$Y_0^m = Y[\vartheta] = C_{\vartheta 1} \cdot \cosh\left[B \cdot m\left(\ln\left[\cot\left[\frac{\vartheta}{2}\right]\right]\right)\right] - i \cdot C_{\vartheta 2} \cdot \sinh\left[B \cdot m\left(\ln\left[\cot\left[\frac{\vartheta}{2}\right]\right]\right)\right]. \qquad (335)$$

We also were able to show in [25] that for small C_2 (an arbitrary constant) the functional wrapper $\left(1 + C_2 \cdot \ln\left[\dfrac{r - r_s}{r}\right]\right)^2 \cdot \left(Z_m \cdot Y_0^m\right)^2$ will not do any harm to the Schwarzschild metric in such a way that it destroys its perfect agreement with the experimental findings as for r only slightly bigger than the Schwarzschild radius, we asymptotically get back the classical solution. For more information the reader is kindly referred to [25].

At very small scales, however, when we will be able to see the intrinsic and probably curled up (compactified) dimensions, we would immediately see our gravity object in the corresponding higher dimensional Schwarzschild form (in correspondence to its mass, of course).

How to represent such a constellation metrically?

Well, instead of the usual, which is to say, classical metric Schwarzschild solution as given in Eq. (329), we simply imagine an approach as follows:

$$g_{\alpha\beta}^n = \begin{pmatrix} -c^2 \cdot g[r] & 0 & 0 & 0 & \cdots & 0 \\ 0 & \dfrac{1}{g[r]} & 0 & 0 & \cdots & 0 \\ 0 & 0 & r^2 & 0 & \cdots & 0 \\ 0 & 0 & 0 & r^2 \cdot \sin^2 \varphi_1 & \cdots & 0 \\ \cdots & \cdots & \cdots & \cdots & \ddots & 0 \\ 0 & 0 & 0 & 0 & 0 & g_{nn} \end{pmatrix}$$

$$g_{44} = a \cdot r^2 \cdot \sin^2 \varphi_1 \cdot \sin^2 \varphi_2; \quad g_{ii} = a \cdot r^2 \cdot \prod_{j=1}^{i-2} \sin^2 \varphi_j;$$

$$g_{nn} = a \cdot r^2 \cdot \prod_{j=1}^{n-2} \sin^2 \varphi_j; \quad g[r] = 1 - \frac{r_s}{r}; \quad a \ll 1. \tag{336}$$

Now, we evaluate the Einstein tensor and thus, obtain the equivalent momentum energy results for the metric given above. With a result fading away sufficiently quickly with increasing distance to the object, we can be quite sure that the classical descriptions of a gravity object using the 4-dimensional space-time Schwarzschild or Kerr approach do a very good job if it comes to compare with experimental results. In fact, we find that, no matter in which number of add-on dimensions we want to consider the problem, only the energy component G^{00} fades with $1/r^2$, while all other components fade with higher negative powers of r. Here we only present the result for $n = 4, 5, 6,$ and 8 dimensions and just give the energy component of the Einstein tensor:

$$G^{00} = \left\{ \underset{n=4}{\underset{\sim}{0}} , \underbrace{\frac{3 \cdot r_s}{2 \cdot c^2 r^2 (r - r_s)}}_{n=5} , -\underbrace{\frac{(a - 1) r \csc[\vartheta]^2 \csc[\varphi]^2}{a \cdot c^2 (r^3 - r_s)}}_{n=6} , \underbrace{\frac{15 a \cdot r_s - 6 (a - 1) r \csc[\vartheta]^2 \csc[\varphi]^2}{a \cdot c^2 r^2 (r - r_s)}}_{n=8} \right\}. \tag{337}$$

It should be noted that Eq. (336) is neither a solution of the Einstein field equations nor a quantum gravity form of the latter. It only is a suggested metric with the usual approach of the classic Schwarzschild solution applied on a space-time with much more dimensions than our assumed 4-dimensional space-time, thereby having all of the add-on dimensions scaled down to almost "non-existence" by a scaling factor a. This way we are able to estimate the error, the classical 4-dimensional-Schwarzschild approach makes on real (as assumed here) gravity objects residing in spaces with higher dimensions similar to Eq. (336). The error found this way is going to tell us how far away our assumptions are from the experimental observations.

We realize that for ordinary massive objects, like planets and stars, their "energy-halo" already fades and probably disappears inside the objects. For black holes or perhaps even neutron stars, we may see some problematic aspects in connection with the terms:

$$G^{00}_{problem} = \left\{ \underset{n=4}{\underset{\sim}{0}} , \underset{n=5}{\underset{\sim}{0}} , \underbrace{\frac{r \csc[\vartheta]^2 \csc[\varphi]^2}{a \cdot c^2 (r^3 - r_s)}}_{n=6} , \underbrace{\frac{6 \cdot \csc[\vartheta]^2 \csc[\varphi]^2}{a \cdot c^2 r (r - r_s)}}_{n=8} \right\}, \tag{338}$$

where especially the angular functions are causing poles also for bigger radii. However, we should remember at this point that—after all—Eq. (336) is not a solution to the scaled n-D-Schwarzschild object, but only a way to estimate the potential deviation of the r-behavior. Finding this difference to disappear with the square of r and thus, probably not even being detectable for non-degenerated objects (stars and planets), we saw all we wanted to see.

For the skilled mathematicians, it might be an interesting quest to find correct solutions for the partially compactified space-times. A good approach may be the assumption of a functionally dependent scale-function or—perhaps even better—a whole set of such scale functions (one for each metric component).

The author on the other side prefers to aim for a simple solution of the following kind:

$$g^n_{\alpha\beta} = \begin{pmatrix} -c^2 \cdot g[r] & 0 & 0 & 0 & \cdots & 0 \\ 0 & \dfrac{1}{g[r]} & 0 & 0 & \cdots & 0 \\ 0 & 0 & r^2 & 0 & \cdots & 0 \\ 0 & 0 & 0 & r^2 \cdot \sin^2 \varphi_1 & \cdots & 0 \\ \cdots & \cdots & \cdots & \cdots & \ddots & 0 \\ 0 & 0 & 0 & 0 & 0 & g_{nn} \end{pmatrix}$$

$$g_{44} = \left(g_{\varphi_3}\left[\varphi_3\right]'\right)^2 \cdot r^2 \cdot \sin^2 \varphi_1 \cdot \sin^2 \varphi_2; \quad g_{ii} = \left(g_{\varphi_{i-1}}\left[\varphi_{i-1}\right]'\right)^2 \cdot r^2 \cdot \prod_{j=1}^{i-2} \sin^2 g_{\varphi_j}\left[\varphi_j\right];$$

$$g_{nn} = \left(g_{\varphi_{n-1}}\left[\varphi_{n-1}\right]'\right)^2 \cdot r^2 \cdot \prod_{j=1}^{n-2} \sin^2 g_{\varphi_j}\left[\varphi_j\right]; \quad g[r] = 1 - \frac{r_s}{r}; \quad g_{\varphi_k}\left[\varphi_k\right] = \varphi_k, \, k = 1,2.$$

$$(339)$$

Now, we simply need to assume that all add-on coordinates are trapped inside a minimum, which makes their first derivative $g_{\varphi_{i-1}}\left[\varphi_{i-1}\right]'$ and thus, the scale factor to the corresponding metric component—almost—zero. Reevaluating the energy component G^{00} for Eq. (339) gives us the corresponding error with respect to the correct solution of the vacuum Einstein field equations as follows:

$$G^{00} = \left\{ \underset{n=4}{0}, \underbrace{\frac{3 \cdot r_s}{2c^2 r^2 \left(r - r_s\right)}}_{n=5}, \underbrace{\frac{4 \cdot r_s}{c^2 r^2 \left(r - r_s\right)}}_{n=6}, \underbrace{\frac{12 \cdot r_s}{c^2 r^2 \left(r - r_s\right)}}_{n=8} \right\}. \qquad (340)$$

We see that by applying the approach (339) to our "wrong" classical Schwarzschild solution to our assumed 4+x-dimensional space-time holding an *n*-spherical gravity object, the error is fading even faster than with r^{-2} as obtained in Eq. (337) for the approach with constant scale factors (336) and the problematic angular terms have disappeared (at least for the energy component, because there now appears one with the component G^{33}, but it fades with power 5 in *r*).

Please note, however, that we chose the special 4-dimensional Schwarzschild approach for *g[r]* with $g[r] = 1 - \dfrac{r_s}{r}$ **only in order to**

see the error of this "classical" assumption of a 4-dimensional space-time within an in-reality n-dimensional (n fairly big) space-time with compactified add-on dimensions (added on top of the classical 4 dimensions). Of course, we know that the correct solution to the Einstein field equations would be:

$$g_{\alpha\beta}^n = \begin{pmatrix} -c^2 \cdot g[r] & 0 & 0 & 0 & \cdots & 0 \\ 0 & \dfrac{1}{g[r]} & 0 & 0 & \cdots & 0 \\ 0 & 0 & r^2 & 0 & \cdots & 0 \\ 0 & 0 & 0 & r^2 \cdot \sin^2 \varphi_1 & \cdots & 0 \\ \cdots & \cdots & \cdots & \cdots & \ddots & 0 \\ 0 & 0 & 0 & 0 & 0 & g_{nn} \end{pmatrix}$$

$$g_{44} = \left(g_{\varphi_3} \left[\varphi_3 \right]' \right)^2 \cdot r^2 \cdot \sin^2 \varphi_1 \cdot \sin^2 \varphi_2; \quad g_{ii} = \left(g_{\varphi_{i-1}} \left[\varphi_{i-1} \right]' \right)^2 \cdot r^2 \cdot \prod_{j=1}^{i-2} \sin^2 g_{\varphi_j} \left[\varphi_j \right];$$

$$g_{nn} = \left(g_{\varphi_{n-1}} \left[\varphi_{n-1} \right]' \right)^2 \cdot r^2 \cdot \prod_{j=1}^{n-2} \sin^2 g_{\varphi_j} \left[\varphi_j \right]$$

$$g[r] = 1 - \left(\frac{r_s}{r} \right)^{n-3}; \quad g_{\varphi_k} \left[\varphi_k \right] = \varphi_k, \, k = 1,2. \tag{341}$$

Most interestingly, however, we found that the error to the assumed 4-dimensional Schwarzschild solution is relatively small if it comes to compare the various solutions for cases of $r \gg r_s$. Thus, it appears to be extremely difficult to experimentally prove that, in fact, we live in an n-dimensional space-time ($n > 4$) rather than a 4-dimensional one.

6.8.5 The Observer and the Problem of the Transition to Newton's Gravity Law

It may well be that, just as shown above, the error, been made in assuming a 4-dimensional space-time and applying such a metric solution in a 4++-dimensional space-time, is fading quickly with bigger radii.

What about the transition to Newton's gravity law?

Here the reader may prove easily that the classical way of introducing a sample mass and expanding the square of the radius derivative with respect to proper time does not yield the correct Newton's limits[4] if not the 4-dimensional Schwarzschild solution (plus a test mass), but a 4++-dimensional metric is been used instead.

Why is that and how can we overcome that particular problem?

Well, in principle, the answer was already given in [26, 27]. We simply cannot assume to have a correct metric description of a reality with two bodies within a given space-time system when choosing to incorporate only one of them into the metric and degrading the other to a "test mass."

We think that a one-body problem in principle does not exist as there is always an observer to be counted into a system, which contains the "one body" in the first place. Without the observer, the whole mathematical description would not make much sense. This, however, automatically, renders any one-body problem, in fact, a two-body problem. The space-time spent out by such a system, thereby considering both, observer and mass as mathematical points, would have 3 dimensions, namely one in time and two in space and thus—applying (341)—contains no singularity. A system consisting of two masses plus observer has one time-dimension and three spatial dimensions. The classical 4-dimensional Schwarzschild metric should be applied to describe this system, thereby seeing the spherical coordinates as relative coordinates from one center of gravity to the other. In other words, the classical coordinates are co-moving with one of the two bodies the observer considers.

Possible solutions to the Einstein field equations in such a case are— among others—the flat space, the Kerr, and the Schwarzschild metric in 4 dimensions. Now the author is aware of the fact that usually authors explain the transition to Newton's law of gravity in a different manner (see last footnote), but here we seek for a way being consistent to our quantum gravity approach with a wrapper function $F[f]$ as introduced in (243). Thereby we remember that the Ricci scalar of such a "wrapped metric" is been given via (c.f. (314)):

$$-\frac{f \cdot F[f]}{6} \cdot R^* = \Delta_g f = \Delta_{SS} f = \Delta_{\text{sphere}} f - r_s \cdot \left(\frac{f^{(0,1,0,0)} + r \cdot f^{(0,2,0,0)}}{r^2} + \frac{f^{(2,0,0,0)}}{c^2 (r - r_s)} \right)$$

$$= \Delta_{\text{3D-sphere}} f - \frac{f^{(2,0,0,0)}}{c^2} - r_s \cdot \left(\frac{f^{(0,1,0,0)} + r \cdot f^{(0,2,0,0)}}{r^2} + \frac{f^{(2,0,0,0)}}{c^2 (r - r_s)} \right). \quad (342)$$

---

[4]A nice presentation was given by Sam Dietterich in "The General Relativity Correction to Newton's Gravity Law" (see also: https://www.youtube.com/watch?v=N5qTCpQf4nw).

With a bit of rearrangement:

$$-\frac{f \cdot F[f]}{6} \cdot R^* = \Delta_{\text{3D-sphere}} f - \frac{f^{(2,0,0,0)}}{c^2} - r_s \cdot \left(\frac{f^{(0,1,0,0)} + r \cdot f^{(0,2,0,0)}}{r^2} + \frac{f^{(2,0,0,0)}}{c^2 (r - r_s)} \right)$$

$$= \Delta_{\text{3D-sphere}} f - \frac{f^{(2,0,0,0)}}{c^2} - \frac{r_s}{r} \cdot \left(\frac{f^{(0,1,0,0)}}{r} + f^{(0,2,0,0)} + \frac{r \cdot f^{(2,0,0,0)}}{c^2 (r - r_s)} \right)$$

$$= \Delta_{\text{3D-sphere}} f - \frac{f^{(2,0,0,0)}}{c^2} \cdot \left(1 + \frac{r_s}{r - r_s} \right) - \frac{r_s}{r} \cdot \left(\frac{f^{(0,1,0,0)}}{r} + f^{(0,2,0,0)} \right), \quad (343)$$

approximation:

$$-\frac{f \cdot F[f]}{6} \cdot R^* \xrightarrow{r \gg r_s} \simeq \Delta_{\text{3D-sphere}} f - \frac{f^{(2,0,0,0)}}{c^2} \cdot \left(1 + \frac{r_s}{r} \right) - \frac{r_s}{r} \cdot \left(\frac{f^{(0,1,0,0)}}{r} + f^{(0,2,0,0)} \right)$$

$$(344)$$

and plugging in the Schwarzschild radius:

$$-\frac{f \cdot F[f]}{6} \cdot R^* \xrightarrow{r \gg r_s} \simeq \Delta_{\text{3D-sphere}} f - \frac{f^{(2,0,0,0)}}{c^2} \cdot \left(1 + \frac{r_s}{r} \right) - \frac{r_s}{r} \cdot \left(\frac{f^{(0,1,0,0)}}{r} + f^{(0,2,0,0)} \right)$$

$$= \Delta_{\text{3D-sphere}} f - \frac{f^{(2,0,0,0)}}{c^2} \cdot \left(1 + \frac{2 \cdot G \cdot m}{r \cdot c^2} \right) - \frac{2 \cdot G \cdot m}{r \cdot c^2} \cdot \left(\frac{f^{(0,1,0,0)}}{r} + f^{(0,2,0,0)} \right)$$

$$\xrightarrow{R^* = 0} 0 = \Delta_{\text{3D-sphere}} f \cdot \frac{c^2}{2} - f^{(2,0,0,0)} \cdot \left(\frac{1}{2} + \frac{G \cdot m}{r} \right) - \frac{G \cdot m}{r} \cdot \left(\frac{f^{(0,1,0,0)}}{r} + f^{(0,2,0,0)} \right),$$

$$(345)$$

we are able to extract Newton's law of gravity in the desired way. Namely, by demanding that the operator $\left(\dfrac{f^{(0,1,0,0)}}{r} + f^{(0,2,0,0)} \right)$ with respect to the radius coordinate times the term $\dfrac{G \cdot m}{r}$ would mirror the classical potential, we obtain the condition:

$$\left(\frac{f^{(0,1,0,0)}}{r} + f^{(0,2,0,0)} \right) = -f \cdot B^2 \qquad (346)$$

and the corresponding solution:

$$f[r] = C_{rJ} \cdot J_0[r \cdot B] + C_{rY} \cdot Y_0[r \cdot B], \qquad (347)$$

we are able to rewrite the last line in Eq. (345) as follows:

$$-\frac{f \cdot F[f]}{6} \cdot R^* \xrightarrow{R^*=0 \ \& \ r \gg r_s} 0 = \Delta_{\text{3D-sphere}} f \cdot \frac{c^2}{2} - f^{(2,0,0,0)} \cdot \left(\frac{1}{2} + \frac{G \cdot m}{r}\right) + f \cdot \frac{G \cdot m}{r} \cdot B^2$$

$$\Rightarrow \quad 0 = \Delta_{\text{3D-sphere}} f \cdot \frac{c^2}{2} - f^{(2,0,0,0)} \cdot \frac{G \cdot m}{r} - \frac{f^{(2,0,0,0)}}{2} + f \cdot \frac{G \cdot m}{r} \cdot B^2. \qquad (348)$$

Now choosing a suitable constant B may bring us toward the Newton-type of gravity, only that we have a rather "wiggly" space in r and this might be seen as inappropriate. It has to be pointed out, however, that, in accordance with Eq. (316), in a 4-dimensional space-time the wrapper function F[f] is of the type $(C_1+f)^2$ and that therefore the radial wiggles can be infinitively small in comparison to the constant C_1. This automatically leaves us with the question whether the Newton "approximation" lesser represents a simplified version of Einstein's metric concept rather than already a simple version of quantum gravity. After all, in its simplest version, realized via a wrapper function F[f] subjected to condition (246), the quantum gravity—here just to be seen as a functional scale factor to a given metric—consists of just one Laplace-dominated equation and not 10 field equations. As also Newton's gravity law consists of just one equation, we think it is kind of obvious that the quantum gravity (containing Einstein's theory) is closer to Newton than the Einstein concept.

Could there be an even simpler path, one where we see the transition to Newton gravity more directly?

We think there is! Here it goes:

By assuming that R^* would not give zero, but could be asymptotically (for $r \gg r_s$) approximated by something like:

$$-\frac{F[f]}{6} \cdot R^* \simeq \left(\sum_{i=1}^{\infty} c_i \left(\frac{r_s}{r}\right)^i\right) \cdot F[f], \qquad (349)$$

which is to say by a radial curvature fading with the distance to the mass's center and choosing an approximated quantum gravity function as $f[t,r]$) $= f_t[t]^* f_r[r]$ with the f_r-term obtained from the time-independent vacuum solution as given in Eq. (333), where we had (with respect to the derivation we refer to [25]):

$$F[f] = \left(1 + C_2 \cdot \ln\left[1 - \frac{r_s}{r}\right]\right)^2 \cdot \left(Z_m \cdot Y_0^m\right)^2$$

$$Z_m[\varphi] = C_{\varphi 1} \cdot \cos[m \cdot \varphi] + C_{\varphi 2} \cdot \sin[m \cdot \varphi]$$

$$Y_0^m = Y[\vartheta] = C_{\vartheta 1} \cdot \cosh\left[B \cdot m\left(\ln\left[\cot\left[\frac{\vartheta}{2}\right]\right]\right)\right] - i \cdot C_{\vartheta 2} \cdot \sinh\left[B \cdot m\left(\ln\left[\cot\left[\frac{\vartheta}{2}\right]\right]\right)\right]$$

$$(350)$$

$$\Rightarrow \qquad G_{\alpha\beta}^{Q-SS} = F[f] \cdot \begin{pmatrix} -c^2\left(1 - \dfrac{r_s}{r}\right) & 0 & 0 & 0 \\ 0 & \left(1 - \dfrac{r_s}{r}\right)^{-1} & 0 & 0 \\ 0 & 0 & r^2 & 0 \\ 0 & 0 & 0 & r^2 \cdot \sin^2 \vartheta \end{pmatrix}$$

$$= \left(1 + C_2 \cdot \ln\left[1 - \frac{r_s}{r}\right]\right)^2 \cdot \left(Z_m \cdot Y_0^m\right)^2 \cdot g_{\alpha\beta}^{SS}$$

$$\xrightarrow{\quad f[t,r,\vartheta,\varphi] \to f[t,r] \quad} f_r[r] = 1 + C_2 \cdot \ln\left[1 - \frac{r_s}{r}\right]$$

$$\Rightarrow \qquad G_{\alpha\beta}^{Q-SS} = \left(1 + C_2 \cdot \ln\left[1 - \frac{r_s}{r}\right]\right)^2 \cdot g_{\alpha\beta}^{SS}, \qquad (351)$$

the last line of Eq. (343) (reminder, there were no further approximations until that point (!) except for our constant-assumption with respect to the quantum gravity Ricci scalar in Eq. (349)) gives us:

$$-\frac{f \cdot F[f]}{6} \cdot R^* \simeq f \cdot F[f] \cdot \left(\sum_{i=1}^{\infty} c_i \left(\frac{r_s}{r}\right)^i\right) \cdot \mathrm{const}$$

$$= \Delta_{3D\text{-sphere}} f - \frac{f^{(2,0)}}{c^2}\left(1 + \frac{r_s}{r - r_s}\right) - \frac{r_s}{r} \cdot \left(\frac{f^{(0,1)}}{r} + f^{(0,2)}\right)$$

$$\xrightarrow{\quad f = f[t,r] = f_t[t] \cdot f_r[r] = f_t[t] \cdot \left(1 + C_2 \cdot \ln\left[1 - \frac{r_s}{r}\right]\right) \quad},$$

$$(352)$$

$$f \cdot F[f] \cdot \left(\sum_{i=1}^{\infty} c_i \left(\frac{r_s}{r} \right)^i \right) = \overbrace{\Delta_{\text{3D-sphere}} f}^{=0} - \frac{f^{(2,0)}}{c^2} \cdot \left(1 + \frac{r_s}{r - r_s} \right) - \frac{r_s}{r} \cdot \left(\overbrace{\frac{f^{(0,1)}}{r}}^{=0} + f^{(0,2)} \right)$$

$$= -\frac{\partial^2 f_t[t]}{\partial t^2} \cdot \frac{\left[1 + C_2 \cdot \ln \left[1 - \frac{r_s}{r} \right] \right]}{c^2} \cdot \left(1 + \frac{r_s}{r - r_s} \right)$$

$$\Rightarrow \quad f_t[t]^3 \cdot \left(1 + C_2 \cdot \ln \left[1 - \frac{r_s}{r} \right] \right)^2 \cdot \left(\sum_{i=1}^{\infty} c_i \left(\frac{r_s}{r} \right)^i \right) = -\frac{\partial^2 f_t[t]}{c^2 \cdot \partial t^2} \cdot \left(1 + \frac{r_s}{r - r_s} \right)$$

$$\Rightarrow \quad \left(1 + C_2 \cdot \ln \left[1 - \frac{r_s}{r} \right] \right)^2 \cdot c^2 \cdot \left(\sum_{i=1}^{\infty} c_i \left(\frac{r_s}{r} \right)^i \right) = -\frac{\partial^2 f_t[t]}{f_t[t]^3 \cdot \partial t^2} \cdot \left(1 + \frac{r_s}{r - r_s} \right).$$

(353)

Now, we expand the left-hand side with respect to small r_s:

$$\Rightarrow \left\{ \begin{array}{l} \left[1 - C_2 \cdot \frac{r_s}{r} + (C_2^2 - C_2) \cdot \left(\frac{r_s}{r} \right)^2 + \left(C_2^2 - \frac{2}{3} \cdot C_2 \right) \cdot \left(\frac{r_s}{r} \right)^3 \ldots \right] \cdot c^2 \cdot \text{const} \cdot \left(\sum_{i=1}^{\infty} c_i \left(\frac{r_s}{r} \right)^i \right) \\ \\ = -\frac{\partial^2 f_t[t]}{f_t[t]^3 \cdot \partial t^2} \cdot \left(1 + \frac{r_s}{r - r_s} \right), \end{array} \right.$$

(354)

avoid all terms being small of higher order (thereby simply assuming $r \gg r_s$):

$$\Rightarrow \quad \left(1 - C_2 \cdot \frac{r_s}{r} \ldots \right) \cdot c_1 \frac{r_s}{r} \cdot c^2 = \left(1 - C_2 \cdot \frac{r_s}{r} \ldots \right) \cdot c_1 \frac{r_s}{r} \cdot c^2 =$$

$$-\frac{\partial^2 f_t[t]}{f_t[t]^3 \cdot \partial t^2} \cdot \left(1 + \frac{r_s}{r - r_s} \right) \approx -\frac{\partial^2 f_t[t]}{f_t[t]^3 \cdot \partial t^2} \cdot \left(1 + \frac{r_s}{r} \right)$$

$$\Rightarrow \quad c^2 \cdot c_1 \frac{r_s}{r} \ldots \approx -\frac{\partial^2 f_t[t]}{f_t[t]^3 \cdot \partial t^2} \cdot \left(1 + \frac{r_s}{r} \right) \qquad (355)$$

and substitute the time-dependent term $\dfrac{\partial^2 f_t[t]}{f_t[t]^3 \cdot \partial t^2}$ by something we may consider an acceleration field A. This gives us:

$$\Rightarrow \qquad c^2 \cdot \frac{r_s}{r} \cdot c_1 \ldots \simeq -A \cdot \left(1+\frac{r_s}{r}\right). \qquad (356)$$

Plugging in the Schwarzschild radius:

$$\Rightarrow \qquad c^2 \cdot \frac{2 \cdot G \cdot m}{c^2 \cdot r} \cdot c_1 \ldots = \frac{2 \cdot G \cdot m}{r} \cdot c_1 \ldots \simeq -A \cdot \left(1+\frac{r_s}{r}\right), \qquad (357)$$

we already recognize something similar to the Newton potential $V[r] \sim 1/r$ on the left-hand side and in order to obtain the classical form of Newton's gravity law, we apply the spatial Nabla-operator, which in our case is just the derivative with respect to r, onto both sides of the last line of Eq. (353) and have to evaluate the following:

$$\nabla\left(1+C_2 \cdot \ln\left[1-\frac{r_s}{r}\right]\right)^2 \cdot c^2 \cdot \left(\sum_{i=1}^{\infty} c_i \left(\frac{r_s}{r}\right)^i\right) = -\nabla A \cdot \left(1+\frac{r_s}{r}\right) = -A \cdot \frac{\partial}{\partial r}\left(1+\frac{r_s}{r}\right),$$
$$(358)$$

giving us:

$$\Rightarrow \qquad -A \cdot \frac{r_s}{c^2 \cdot (r-r_s)^2} =$$

$$\sum_{i=1}^{\infty} c_i \frac{\left(\frac{r_s}{r}\right)^i \left(1+C_2 \cdot \log\left[1-\frac{r_s}{r}\right]\right)\left(i \cdot r - (2C_2+i)r_s + C_2 \cdot i\left(1-\frac{r_s}{r}\right)\log\left[1-\frac{r_s}{r}\right]\right)}{r-r_s}.$$
$$(359)$$

Taylor expansion on the left-hand side of Eq. (359):

$$-A \cdot \frac{r_s}{c^2 \cdot (r-r_s)^2} = -A \cdot \frac{r_s}{c^2 \cdot r^2}\left(1+\frac{2}{r}+\frac{3}{r^2}+\ldots\right) \qquad (360)$$

tells us that, in order to obtain Newton's law, the Ricci-curvature should converge to zero for increasing r with $1/r^3$. Thus, we set all $c_i = 0$ except for $i = 3$ and Taylor expand the right-hand side of Eq. (359) thereby using the results from Eq. (360), leading us to:

$$\Rightarrow \qquad -A \cdot \frac{r_s}{c^2 \cdot r^2}\left(1+\frac{2}{r}+\frac{3}{r^2}+\ldots\right) = \frac{c_3}{r}\left(\frac{r_s}{r}\right)^3\left(3-8\cdot C_2 \cdot \frac{r_s}{r}\ldots\right)$$

$$\Rightarrow \qquad -A \cdot \frac{1}{c^2 \cdot r^2}\left(1+\frac{2}{r}+\frac{3}{r^2}+\ldots\right) = \frac{c_3}{r^2}\left(\frac{r_s}{r}\right)^2\left(3-8\cdot C_2 \cdot \frac{r_s}{r}\ldots\right)$$

$$\Rightarrow \qquad -A \cdot \frac{1}{c^2}\left(1 + \frac{2}{r} + \frac{3}{r^2} + \ldots\right) = c_3\left(\frac{r_s}{r}\right)^2\left(3 - 8 \cdot C_2 \cdot \frac{r_s}{r} \ldots\right)$$

$$\Rightarrow \qquad A \cdot (1 + \ldots) = -c^2 c_3\left(\frac{r_s}{r}\right)^2 (3 + \ldots) \quad \Rightarrow \quad A \simeq -3 \cdot c^2 c_3 \left(\frac{r_s}{r}\right)^2. \qquad (361)$$

Plugging in the Schwarzschild radius and choosing a suitable constant c_3 results in:

$$A \simeq -\frac{12 \cdot c_3 \cdot G^2 \cdot m^2}{r^2 \cdot c^2}. \qquad (362)$$

Now, we apply the Newton's second law and multiply our acceleration A by the mass of the second body M, which finally gives us the classical form of Newton's law of gravity:

$$F_{\text{Newton}} = M \cdot A \simeq -M \cdot \frac{12 \cdot c_3 \cdot G^2 \cdot m^2}{r^2 \cdot c^2} \xrightarrow{c_3 = \frac{c^2}{12 \cdot G \cdot m}} = -G \cdot \frac{M \cdot m}{r^2}. \qquad (363)$$

The negative sign just points out that the direction of the force is toward the mass and antiparallel to the radius vector.

Note: **We have three interesting results here.**

(a) **The Classical Gravity Law of Newton was extracted not from the Einstein General Theory of Relativity via the Schwarzschild solution, but from its quantized form by the means of a scalar wrapper function applied onto the metric of a Schwarzschild object. This automatically puts Newton's law closer to quantum gravity than its non-quantized classical version, which is to say the Einstein field equations and their solution (here Schwarzschild). This result is amazing, as it explains the gravitational force emanating from a massive object by the means of its quantum field and not just the metric distortion it has created.**

(b) **In order to obtain the classical outcome of Newton's law, we had to assume a spatial Ricci curvature with the functionality $1/r^3$, which is to say, with a power-law exactly having the number of ordinary (our scale) spatial dimensions in the negative exponent.**

(c) **With now also gravity being extracted from a quantum equation (just like the other interactions, electromagnetic, weak and strong), we may have the means to bring all fundamental interactions together.**

6.8.6 Back to Quantum-Mass-Schwarzschild Radius Discrepancy

Taking the Bekenstein results into account, we immediately realize that there is no ordinary sized black hole with only 4 dimensions. Taking our standards (our scales) and Eq. (328) (also see Fig. 6.7), the number of dimensions for every "visible" black hole immediately becomes very large. The same holds for any other heavy (gravity) object, classically being described by the Schwarzschild solution. Thus, taking Eq. (249), we can almost always assume that:

$$\lim_{n \to \text{big}} R^* = \lim_{n \to \text{big}} \frac{1}{F[f]^3} \cdot \left(-C_{N2} \cdot F[f] \cdot \frac{\partial F[f]}{\partial f} \cdot \Delta_g f \right) \simeq -\frac{4}{f \cdot C_2} \cdot \Delta_g f$$

$$\Rightarrow \quad \left(\frac{\lim\limits_{n \to \text{big}} R^*}{4} \cdot C_2 + \Delta_g \right) f = 0. \tag{364}$$

Now using the result already established for the n-dimensional Schwarzschild metric as presented in Eq. (324), gives us:

$$\lim_{n \to \text{big}} R^* = \lim_{n \to \text{big}} \frac{1}{F[f]^3} \cdot \left(-C_{N2} \cdot F[f] \cdot \frac{\partial F[f]}{\partial f} \cdot \Delta_g f \right)$$

$$= \frac{1}{F[f]^3} \cdot \left(-C_{N2} \cdot F[f] \cdot \frac{\partial F[f]}{\partial f} \cdot \left(\Delta_{n\text{-space-time}} f - \left(\frac{r_s}{r} \right)^{n-3} \cdot \left(\frac{f^{(0,1,0,0)}}{r} + f^{(0,2,0,0)} + \frac{f^{(2,0,0,0)}}{c^2 \left(1 - \left(\frac{r_s}{r} \right)^{n-3} \right)} \right) \right) \right)$$

$$\simeq -\frac{4}{f \cdot C_2} \cdot \left(\Delta_{n\text{-space-time}} f - \left(\frac{r_s}{r} \right)^{n-3} \cdot \left(\frac{f^{(0,1,0,0)}}{r} + f^{(0,2,0,0)} + \frac{f^{(2,0,0,0)}}{c^2 \left(1 - \left(\frac{r_s}{r} \right)^{n-3} \right)} \right) \right)$$

$$\xrightarrow{r \gg r_s} \simeq -\frac{4}{f \cdot C_2} \cdot \left(\Delta_{n\text{-space-time}} f \right)$$

$$\Rightarrow \quad \left(\frac{\lim\limits_{n \to \text{big}} R^*}{4} \cdot C_2 + \Delta_{n\text{-space-time}} \right) f = 0. \tag{365}$$

Thereby we have used the fact that:

$$\lim_{r \to \infty} \left(\frac{r_s}{r} \right)^{n-3} = 0. \tag{366}$$

Thus, with $\dfrac{\lim\limits_{n \to \text{big}} R^*}{4} \cdot C_2$ assuming to give the classical mass term plus a

potential V via $\dfrac{\lim\limits_{n \to \text{big}} R^*}{4} \cdot C_2 = \dfrac{m^2 \cdot c^2}{\hbar^2} + V$, we have obtained the classical

linear form and thereby found a potentially deeper reason for the equivalency of gravitational (heavy) and inertial masses.

Assuming that there is no potential V and thus, $\dfrac{\lim\limits_{n \to \text{big}} R^*}{4} \cdot C_2 = \dfrac{m^2 \cdot c^2}{\hbar^2}$ and

applying the definition of the Schwarzschild radius $r_s = \dfrac{2 \cdot m \cdot G}{c^2}$, we obtain

the relation:

$$\frac{\lim\limits_{n \to \text{big}} R^*}{4} \cdot C_2 \cdot \frac{\hbar^2}{c^2} = m^2 = \frac{r_s^2 \cdot c^4}{4 \cdot G^2} \quad \Rightarrow \quad \lim_{n \to \text{big}} R^* = \frac{r_s^2 \cdot c^6}{\hbar^2 \cdot C_2 \cdot G^2}. \tag{367}$$

This result, however, requires the limit $r \gg r_s$ (c.f. (366)). Thus, gravity forces/interactions, which, as shown in the last sub-section, require a $R^* \sim 1/r^3$-behavior in order to mirror the Newton gravity law, would easily be overshadowed by other interactions whose curvature terms follow smaller negative exponents or are even constants as obtained in Eq. (367).

On the other side, we have some conceptual problems in imagining a constant curvature, even though it may only be caused by a curvature in higher dimensions. We, therefore, remind ourselves at this point that each constant curvature can be produced by an entanglement of dimensions, just as demonstrated in sections "6.6 The Entanglement of Dimensions and the Production of Mass" and "6.7 Obtaining Quantized Masses in Situations of $R^* = 0$." This leaves us free to choose suitable dimensional entanglements for all inertia, but keep the gravity masses by the means of a global Ricci scalar curvatures of the $R^* \sim 1/r^3$-type. Thereby we also remember that this $R^* \sim 1/r^3$-behavior had to be introduced for the limit $r \gg r_s$ in order to obtain the correct asymptotic distance functionality of gravity in the Newton limit. There are no conditions set, not yet anyway, for the behavior in the vicinity of the origin $r = 0$. In order to avoid any singularity there, we might just assume a "more realistic" (realistic in the sense of avoiding any singularities) radius dependency like:

$$R^* = C_R \cdot e^{-\left(\frac{r_s}{r}\right)^3} \cdot \left(\frac{r_s}{r}\right)^3.$$ (368)

Summing it all up, we found that quantum gravity equations in the far distance limit can be constructed as follows:

$$\left(\frac{\lim\limits_{n \to \text{big}} \overbrace{R^*_{\text{gravity}} + R^*_{\text{quantum}} + \overbrace{R^*_{\text{other}}}^{\text{here } = 0}}^{R^*}}{4} \cdot C_2 + \Delta_{n\text{-space-time}} \right) f$$

$$= \left(\overbrace{C_R \cdot e^{-\left(\frac{r_s}{r}\right)^3} \cdot \left(\frac{r_s}{r}\right)^3}^{R^*_{\text{gravity}}} + R^*_{\text{quantum}} + \Delta_{n\text{-space-time}} \right) f = 0.$$ (369)

Now, we can further distinguish between our ordinary 4-dimensional space-time and the entangled and compactified additional dimensions leading to inertia and masses as demonstrated in sections "6.6 The Entanglement of Dimensions and the Production of Mass" and "6.7 Obtaining Quantized Masses in Situations of $R^* = 0$." This may be written as:

$$\left(\overbrace{C_R \cdot e^{-\left(\frac{r_s}{r}\right)^3} \cdot \left(\frac{r_s}{r}\right)^3}^{R^*_{\text{gravity}}} + R^*_{\text{quantum}} + \Delta_{n\text{-space-time}} \right) f$$

$$\left(\overbrace{C_R \cdot e^{-\left(\frac{r_s}{r}\right)^3} \cdot \left(\frac{r_s}{r}\right)^3}^{R^*_{\text{gravity}}} + R^*_{\text{quantum}} + \Delta_{4D} + \overbrace{\Delta_{\text{compact}}}^{\text{hidden}} + \overbrace{\Delta_{\text{entangled}}}^{\text{masses/inertia}} \right) f = 0.$$ (370)

It should explicitly be pointed out that Eq. (370) is only an extremely simple suggestion. In section "6.12 Toward an Explanation for the 3-Generation Problem," we are going to consider generalizations and more complex dependencies in connection with the fundamental interactions detectable in this universe.

In general, we conclude at this point that it might be worthwhile to consider a few examples for Eq. (370) in higher dimensional space-times with partially compactified dimensions.

6.9 More Metric Options and Subsequent Quantum Equations Plus Their Solutions

In the previous section, we saw that metric wrapper functions, which are just scaling functions to a given metric, do not only give quantum equations and thus, a quantization option to every metric solution of the Einstein field equations, but also can produce masses, spins, and all sorts of inertia and—more generally seen—matter. Thereby, as a most convenient aspect if it comes to make our approach compatible to classical theories, all add-on dimensions can be metrically introduced as compactified. In the following, more or less out of interest and for completeness, we will just derive a few more solutions in the cases of 5, 6, and 8 dimensions in the case of spherical symmetry, respectively spherically motivated coordinate systems.

6.9.1 Five Dimensions

Starting with just one additional dimension (in addition to the ordinary 4-dimensional space-time we mean), we want to more or less systemically investigate the case of spherical coordinates with a scaled 5th coordinate … thereby leaving the question open which one of the angles we actually make this very 5th coordinate. The latter point is motivated by the fact that certain interchanges here lead to different symmetries and—obviously—particle appearances as it was shown in section "6.7.3 Periodic Coordinate Quantization" (c.f. Figs. 6.2 and 6.3, there obtained for the case of 6 dimensions).

Assuming coordinates t, r, ϑ, φ, u, we start with the following metric approach:

$$g_{\alpha\beta}^{5} = \begin{pmatrix} -c^2 & 0 & 0 & 0 & 0 \\ 0 & 1 & 0 & 0 & 0 \\ 0 & 0 & r^2 & 0 & 0 \\ 0 & 0 & 0 & r^2 \cdot \sin[\vartheta]^2 & 0 \\ 0 & 0 & 0 & 0 & \left(g_u[u]'\right)^2 \cdot r^2 \cdot \sin[\vartheta]^2 \sin[\varphi]^2 \end{pmatrix}$$

$$\times F\big[g[u] \cdot f[t,r,\vartheta,\varphi]\big]. \tag{371}$$

There is a variety of options regarding the choice of the functions $g[u]$ and $g_u[u]$. We begin with the apparently simple choice of $g[u] = 1$, thereby leaving $g_u[u]$ arbitrary and, with the separation approach, obtain $f[t, r, \vartheta, \varphi] = f_t[t] \cdot f_r[r] \cdot f_\vartheta[\vartheta] \cdot f_\varphi[\varphi]$ the following solution:

$$f_t[t] = C_{t1} \cdot \cos[c \cdot E \cdot t] + C_{t2} \cdot \sin[c \cdot E \cdot t], \tag{372}$$

$$f_\varphi[\varphi] = C_{\varphi 1} \cdot \cos[m \cdot \varphi] + C_{\varphi 2} \cdot \sin[m \cdot \varphi], \tag{373}$$

$$f_\vartheta[\vartheta] = \csc[\vartheta] \begin{pmatrix} \left(C_{\vartheta-} \cdot e^{i(1-\ell)\vartheta} + C_{\vartheta+} \cdot e^{i(1+\ell)\vartheta}\right) \\ \begin{pmatrix} \dfrac{{}_2F_1\left[1, \dfrac{1-\ell}{2}, \dfrac{3-\ell}{2}, e^{2i\vartheta}\right]}{(\ell-1)} \\ i \cdot e^{i\vartheta} m(1+m) \\ \dfrac{}{\ell} \\ + (\ell-1)\dfrac{{}_2F_1\left[1, \dfrac{1+\ell}{2}, \dfrac{3+\ell}{2}, e^{2i\vartheta}\right]}{(\ell^2+1)} \end{pmatrix} \end{pmatrix}. \tag{374}$$

Here ${}_2F_1$ denotes the corresponding hyper geometric function. We end up with the following differential equation for $f_r[r]$ in the case of $R^* = 0$:

$$R^* = \frac{1}{F[f \cdot g]^3} \cdot \left(-C_{N2} \cdot F[f \cdot g] \cdot \frac{\partial F[f \cdot g]}{\partial (f \cdot g)} \cdot \Delta_g (f \cdot g) \right)$$

$$F[f \cdot g] = \left(C_f + f[t,r,\vartheta,\varphi] \cdot g[u]\right)^{\frac{4}{3}}; \quad C_f \neq 0$$

$$\xrightarrow{5D} \quad 0 = \left(1 - \ell^2 + r^2 E^2\right) f_r[r] + r\left(3 f_r[r]' + r \cdot f_r[r]''\right) - \frac{R^* \cdot F[f \cdot g]^2}{4 \cdot \dfrac{\partial F[f \cdot g]}{\partial (f \cdot g)}}. \tag{375}$$

The solution in the case $R^* = 0$ consists of Bessel functions J_n and Y_n:

$$f_r[r] = \frac{C_{r1} \cdot J_\ell[E \cdot r] + C_{r2} \cdot Y_\ell[E \cdot r]}{r}. \tag{376}$$

This time, in difference to the solution (313), we have integer values for the Bessel functions and no integers $+1/2$.

We obtain an amazing simplification when setting $g[u] = g_u[u]$ ($g_u[u]$ still arbitrary), which leads us to the following solution (this time with the ordinary Legendre polynomials for the functions $f_\varphi[\varphi]$ instead of the associated ones):

$$f_t[t] = C_{t1} \cdot \cos[c \cdot E \cdot t] + C_{t2} \cdot \sin[c \cdot E \cdot t], \tag{377}$$

$$f_\varphi[\varphi] = C_{P\varphi} \cdot P_\lambda \big[\cos[\varphi]\big] + C_{Qu} \cdot Q_\lambda \big[\cos[\varphi]\big], \tag{378}$$

$$f_\vartheta[\vartheta] = \frac{C_{P\vartheta} \cdot P_\ell^{\lambda + \frac{1}{2}} \big[\cos[\vartheta]\big] + C_{Q\vartheta} \cdot Q_\ell^{\lambda + \frac{1}{2}} \big[\cos[\vartheta]\big]}{\left(-\sin[\vartheta]^2\right)^{\frac{1}{4}}}. \tag{379}$$

We end up with the following differential equation for $f_r[r]$:

$$R^* = \frac{1}{F[f \cdot g]^3} \cdot \left(-C_{N2} \cdot F[f \cdot g] \cdot \frac{\partial F[f \cdot g]}{\partial (f \cdot g)} \cdot \Delta_g (f \cdot g)\right)$$

$$F[f \cdot g] = \left(C_f + f[t, r, \vartheta, \varphi] \cdot g[u]\right)^{\frac{4}{3}}; \quad C_f \neq 0$$

$$\xrightarrow{\quad 5D \quad}$$

$$\Rightarrow \quad 0 = \left(\left(\ell(1+\ell) - \frac{3}{4}\right) - r^2 E^2\right) f_r[r] - r\left(3 f_r[r]' + r \cdot f_r[r]''\right) + \frac{R^* \cdot F[f \cdot g]^2}{4 \cdot \dfrac{\partial F[f \cdot g]}{\partial (f \cdot g)}}. \tag{380}$$

The setting $R^* = 0$ reproduces the solution (313), consisting of the Bessel functions J_n and Y_n:

$$f_r[r] = \frac{C_{r1} \cdot J_{\frac{1}{2} + \ell}[E \cdot r] + C_{r2} \cdot Y_{\frac{1}{2} + \ell}[E \cdot r]}{r}. \tag{381}$$

This time, in order to avoid singularities with the Legendre polynomials, we can only have the parameter settings: $\lambda = k$ (k ... integer), $\ell = k \pm 1/2$.

Now, we interchange the compactified coordinate with a "visible" one and apply the metric:

$$g^5_{\alpha\beta} = \begin{pmatrix} -c^2 & 0 & 0 & 0 & 0 \\ 0 & 1 & 0 & 0 & 0 \\ 0 & 0 & r^2 & 0 & 0 \\ 0 & 0 & 0 & \left(g_\varphi[\varphi]'\right)^2 \cdot r^2 \cdot \sin[\vartheta]^2 & 0 \\ 0 & 0 & 0 & 0 & r^2 \cdot \sin[\vartheta]^2 \sin[\varphi]^2 \end{pmatrix}$$

$$\times F\left[f[t,r,\vartheta,\varphi,u]\right]. \tag{382}$$

Just as before we apply the separation approach: $f[t, r, \vartheta, \varphi, u] = f_t[t] \cdot f_r[r] \cdot f_\vartheta[\vartheta] \cdot f_\varphi[\varphi] \cdot f_u[u]$ together with the simplification setting for the compactified coordinate $g_\varphi[\varphi] = f_\varphi[\varphi]$ ($f_\varphi[\varphi]$ at the moment arbitrary), which leads us to the following solution:

$$f_t[t] = C_{t1} \cdot \cos[c \cdot E \cdot t] + C_{t2} \cdot \sin[c \cdot E \cdot t], \tag{383}$$

$$f_u[u] = C_{u1} \cdot \cos[\mu \cdot u] + C_{u2} \cdot \sin[\mu \cdot u], \tag{384}$$

$$f_\vartheta[\vartheta] = \frac{C_{P\vartheta} \cdot P_\ell^{\frac{1}{2}\sqrt{1-K_\vartheta}} [\cos[\vartheta]] + C_{Q\vartheta} \cdot Q_\ell^{\frac{1}{2}\sqrt{1-K_\vartheta}} [\cos[\vartheta]]}{\left(-\sin[\vartheta]^2\right)^{\frac{1}{4}}}$$

$$K_\vartheta = \frac{4 \cdot \left(\cot[f_\varphi[\varphi]] - \mu^2 \cdot \csc[f_\varphi[\varphi]]^2 f_\varphi[\varphi]\right)}{f_\varphi[\varphi]} = 4 \cdot \left(\frac{\cot[f_\varphi[\varphi]]}{f_\varphi[\varphi]} - \mu^2 \cdot \csc[f_\varphi[\varphi]]^2\right). \tag{385}$$

Again, we end up with the following differential equation for $f_r[r]$, which we can solve in the case of $R^* = 0$:

$$R^* = \frac{1}{F[f]^3} \cdot \left(-C_{N2} \cdot F[f] \cdot \frac{\partial F[f]}{\partial f} \cdot \Delta_g f\right)$$

$$F[f] = \left(C_f + f[t,r,\vartheta,\varphi,u]\right)^{\frac{4}{3}} ; \quad C_f \neq 0 \xrightarrow{\text{5D}}$$

$$\Rightarrow \quad 0 = \left(\left(\ell(1+\ell) - \frac{3}{4}\right) - r^2 E^2\right) f_r[r] - r\left(3 f_r[r]' + r \cdot f_r[r]''\right) + \frac{R^* \cdot F[f]^2}{4 \cdot \frac{\partial F[f]}{\partial f}}. \tag{386}$$

The setting $R^* = 0$ reproduces the solution (313), consisting of the Bessel functions J_n and Y_n:

$$f_r[r] = \frac{C_{r1} \cdot J_{\frac{1}{2}+\ell}[E \cdot r] + C_{r2} \cdot Y_{\frac{1}{2}+\ell}[E \cdot r]}{r}. \tag{387}$$

We realize that a parameter setting of the kind $1 - K_\vartheta = 25 \;\rightarrow\; K_\vartheta = -24; \; \ell = \frac{5}{2}; \; \mu = \pm\frac{3}{2}$ would just give us the 3-gravity center solutions again (c.f. Fig. 6.3). When assuming that there are different realizations with respect to the sin and cos options for the function f_u in Eq. (384), we see that there exists a possibility of combining the two solutions in such a way that the one system could just fill the spatial gaps in the "plane of existence" (here it would be the $z = 0$-plane) the other one leaves open. This might appear as attraction, while the combination of two equal solutions (with the same spatial distribution) most likely creates repulsion. We may just see this as another realization of the Pauli Exclusion Principle for spin-$k/2$ particles (see Fig. 6.8).

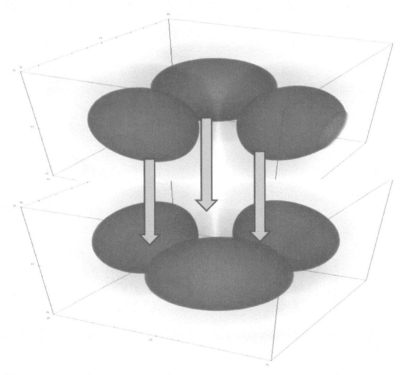

Figure 6.8 Illustration of the spatial distribution of $R^* = 0$-solutions to the metric (382) with two different settings for the function f_u (see text) and the parameters: $1 - K_\vartheta = 25 \;\rightarrow\; K_\vartheta = -24; \; \ell = \frac{5}{2}; \; \mu = \pm\frac{3}{2}.$

6.9.2 Six Dimensions

As shown in the sub-section above, there is quite a variety of options already in 5 dimensions. We here, therefore, restrict ourselves to the simplest case in 6 dimensions, where we choose the metric as follows:

$$
g_{\alpha\beta}^6 = \begin{pmatrix}
-c^2 & 0 & 0 & 0 & 0 & 0 \\
0 & 1 & 0 & 0 & 0 & 0 \\
0 & 0 & r^2 & 0 & 0 & 0 \\
0 & 0 & 0 & r^2 \cdot \sin[\vartheta]^2 & 0 & 0 \\
0 & 0 & 0 & 0 & \left(g_u[u]'\right)^2 \cdot r^2 \cdot \sin[\vartheta]^2 \sin[\varphi]^2 & 0 \\
0 & 0 & 0 & 0 & 0 & g_{55}
\end{pmatrix}
$$

$$
\times \left(C_f + f[t,r,\vartheta,\varphi,u,v]\right); \quad f[t,r,\vartheta,\varphi,u,v] = f_t[t] \cdot f_r[r] \cdot f_\vartheta[\vartheta] \cdot f_\varphi[\varphi] \cdot f_u[u] \cdot f_v[v]
$$

$$
g_{55} = \left(g_v[v]'\right)^2 \cdot r^2 \cdot \sin[\vartheta]^2 \sin[\varphi]^2 \sin[u]^2, \tag{388}
$$

and partially "fix" the functions g_v and g_u. With this the solution reads:

$$
g_v[v] = f_v[v], \tag{389}
$$

$$
g_u[u] = f_u[u], \tag{390}
$$

$$
f_t[t] = C_{t1} \cdot \cos[c \cdot E \cdot t] + C_{t2} \cdot \sin[c \cdot E \cdot t], \tag{391}
$$

$$
f_\varphi[\varphi] = \frac{C_{P\varphi} \cdot P_\ell^{\frac{1}{2}\sqrt{1-4 \cdot \frac{\cot[g_u[u]]}{g_u[u]}}}\left[\cos[\varphi]\right] + C_{Q\varphi} \cdot Q_\ell^{\frac{1}{2}\sqrt{1-4 \cdot \frac{\cot[g_u[u]]}{g_u[u]}}}\left[\cos[\varphi]\right]}{\left(-\sin[\varphi]^2\right)^{\frac{1}{4}}}, \tag{392}
$$

$$
f_\vartheta[\vartheta] = \frac{C_{P\vartheta} \cdot P_L^{\ell+\frac{1}{2}}\left[\cos[\vartheta]\right] + C_{Q\vartheta} \cdot Q_L^{\ell+\frac{1}{2}}\left[\cos[\vartheta]\right]}{\left(-\sin[\vartheta]^2\right)^{\frac{1}{2}}}. \tag{393}
$$

We end up with the following differential equation for $f_r[r]$:

$$
R^* = \frac{1}{F[f]^3} \cdot \left(-C_{N2} \cdot F[f] \cdot \frac{\partial F[f]}{\partial f} \cdot \Delta_g f\right); \quad F[f] = C_f + f[t,r,\vartheta,\varphi,u,v]; \quad C_f \neq 0
$$

$$\xrightarrow{\quad 6D \quad}$$

$$\Rightarrow \quad 0 = \left(\left(L(1+L) - 2 \right) - r^2 E^2 \right) f_r[r] - r \left(4 f_r[r]' + r \cdot f_r[r]'' \right) + \frac{R^* \cdot F[f]^2}{5 \cdot \dfrac{\partial F[f]}{\partial f}},$$

(394)

which, in the case of $R^* = 0$, gives the solution:

$$f_r[r] = \frac{C_{r1} \cdot J_{\frac{1}{2}+L}[E \cdot r] + C_{r2} \cdot Y_{\frac{1}{2}+L}[E \cdot r]}{r^{\frac{3}{2}}}.$$

(395)

We shall leave the discussion to the interested reader.

6.9.3   Eight Dimensions

In 8 dimensions and—apart from the compactification of certain dimensions—8-sphere symmetry (compressed in the down-scaled dimensions), the corresponding metric could be given as follows:

$$g_{\alpha\beta}^{8} = \begin{pmatrix} -c^2 & 0 & 0 & 0 & \cdots & 0 \\ 0 & 1 & 0 & 0 & \cdots & 0 \\ 0 & 0 & r^2 & 0 & \cdots & 0 \\ 0 & 0 & 0 & r^2 \cdot \sin^2 \vartheta & \cdots & 0 \\ \cdots & \cdots & \cdots & \cdots & \ddots & 0 \\ 0 & 0 & 0 & 0 & 0 & g_{77} \end{pmatrix}$$

$$\times F \left[f \left[t, r, \vartheta, \varphi_1, \varphi_2, \varphi_3, \varphi_4, \varphi_5 \right] \right];$$

$$f \left[t, r, \vartheta, \varphi_1, \varphi_2, \varphi_3, \varphi_4, \varphi_5 \right] = f_t[t] \cdot f_r[r] \cdot f_\vartheta[\vartheta] \cdot \prod_{i=1}^{5} f_{\varphi_i} \left[\varphi_i \right];$$

$$g_{44} = \left(g_{\varphi_2} \left[\varphi_2 \right]' \right)^2 \cdot r^2 \cdot \sin^2 \vartheta \cdot \sin^2 \varphi_1;$$

$$g_{55} = \left(g_{\varphi_3} \left[\varphi_3 \right]' \right)^2 \cdot r^2 \cdot \sin^2 \vartheta \cdot \sin^2 \varphi_1 \cdot \sin^2 \varphi_2;$$

$$g_{66} = \left(g_{\varphi_4} \left[\varphi_4 \right]' \right)^2 \cdot r^2 \cdot \sin^2 \vartheta \cdot \sin^2 \varphi_1 \cdot \sin^2 \varphi_2 \cdot \sin^2 \varphi_3;$$

$$g_{77} = \left(g_{\varphi_5} \left[\varphi_5 \right]' \right)^2 \cdot r^2 \cdot \sin^2 \vartheta \cdot \sin^2 \varphi_1 \cdot \sin^2 \varphi_2 \cdot \sin^2 \varphi_3 \cdot \sin^2 \varphi_4; \quad (396)$$

again, we partially "fix" the functions g as follows (for $i = 1, \ldots, 5$):

$$g_{\varphi_i}\left[\varphi_i\right] = f_{\varphi_i}\left[\varphi_i\right]. \tag{397}$$

With this the solution reads:

$$f_t[t] = C_{t1} \cdot \cos[c \cdot E \cdot t] + C_{t2} \cdot \sin[c \cdot E \cdot t], \tag{398}$$

$$f_{\varphi_1}\left[\varphi_2\right] = \frac{C_{P\varphi} \cdot P_\ell^{\frac{1}{2}\sqrt{9-4\cdot\frac{AA1}{At\cdot Az\cdot Azz}}}\left[\cos[\varphi]\right] + C_{Q\varphi} \cdot Q_\ell^{\frac{1}{2}\sqrt{9-4\cdot\frac{AA1}{At\cdot Az\cdot Azz}}}\left[\cos[\varphi]\right]}{\left(-\sin[\varphi]^2\right)^{\frac{3}{4}}}$$

$$Az = f_{\varphi_2}\left[\varphi_2\right]; \quad Azz = f_{\varphi_3}\left[\varphi_3\right]; \quad At = f_{\varphi_4}\left[\varphi_4\right]; \quad Att = f_{\varphi_5}\left[\varphi_5\right];$$

$$\Rightarrow AA1 = 3 \cdot At \cdot Azz \cdot \cot[Az] + Az \cdot \csc[Az]^2 \left(2 \cdot At \cdot \cot[Azz] + Azz \cdot \cot[At] \cdot \csc[Azz]^2\right), \tag{399}$$

$$f_\vartheta[\vartheta] = \csc[\vartheta]^2 \left(C_{P\vartheta} \cdot P_L^{\ell+\frac{1}{2}}\left[\cos[\vartheta]\right] + C_{Q\vartheta} \cdot Q_L^{\ell+\frac{1}{2}}\left[\cos[\vartheta]\right]\right). \tag{400}$$

We end up with the following differential equation for $f_r[r]$:

$$R^* = \frac{1}{F[f]^3} \cdot \left(-C_{N2} \cdot F[f] \cdot \frac{\partial F[f]}{\partial f} \cdot \Delta_g f\right)$$

$$F[f] = \left(C_f + f\left[t, r, \vartheta, \varphi_1, \varphi_2, \varphi_3, \varphi_4, \varphi_5\right]\right)^{\frac{2}{3}}; \quad C_f \neq 0$$

$$\xrightarrow{8D}$$

$$\Rightarrow \quad 0 = \left(\left(L(1+L)-6\right)-r^2 E^2\right) f_r[r] - r\left(6 \cdot f_r[r]' + r \cdot f_r[r]''\right) + \frac{R^* \cdot F[f]^2}{7 \cdot \frac{\partial F[f]}{\partial f}}, \tag{401}$$

which, in the case of $R^* = 0$, gives the solution:

$$f_r[r] = \frac{C_{r1} \cdot J_{\frac{1}{2}+L}\left[E \cdot r\right] + C_{r2} \cdot Y_{\frac{1}{2}+L}\left[E \cdot r\right]}{r^{\frac{5}{2}}}. \tag{402}$$

6.10 A Few Words about *n* Dimensions and *n*–1 Spheres

Comparing the results for the dimensions 5, 6, and 8, we realize that with increasing numbers of dimensions ever higher quantum numbers L are required for the function f_r in order to avoid $r = 0$ singularities. Thus, the number of dimensions itself could be seen as a quantum number. In fact, comparing the types of solutions resulting from 4 (c.f. [17, 18]), 5, 6, 7 (not shown here), and 8 dimensions, one may deduce that what is been seen as the potential in classical physics, in reality is the add-on of various dimensions. Thereby the number of dimensions plays the role of the main quantum number. The general radius function $f_r[r]$ for the n-sphere symmetry with potentially compactified add-on coordinates (added to our ordinary 4-dimensional space-time) would look as follows:

$$f_r[r] = \frac{C_{r1} \cdot J_{\frac{1}{2}+L}[E \cdot r] + C_{r2} \cdot Y_{\frac{1}{2}+L}[E \cdot r]}{r^{\frac{n-3}{2}}}. \tag{403}$$

Now, not only does the number of dimensions play the role of a quantum number, but we also find a behavior similar to the case of the Schrödinger hydrogen solutions with a potential $V[r] \sim 1/r$. Thus, we ask: Is it possible that the dimensional-$V[n] \sim V[r]$-model is just some kind of analog to the classical potential?

Of course, this idea is compromised by the fact that the parameter L needs to have a minimum value ($L \geq n-6$) in order to assure that there is no singularity at $r = 0$.

Further discussion will be left to the interested reader or for later publications (whatever comes first).

6.11 Intermediate Sum-Up: Direct Extraction of Klein–Gordon and Schrödinger Equations from Metric Tensor

With the means at hand now, we are able to directly derive classical quantum equations directly from the metric tensor via a functional wrapper approach … just as shown above. For convenience we are partially going to repeat the approach in compact form, before showing how to add in mass and potential.

In addition, we demonstrate the derivation of the Schrödinger equation out of the Klein–Gordon equation.

6.11.1   Repetition: Metric Klein–Gordon Equation without Mass or Potential

In the previous sections, it was demonstrated how we can find a direct extraction of the Klein–Gordon equation from the Ricci scalar R^* of a modified metric of the kind $G_{\alpha\beta} = F[f]^* g_{\alpha\beta}$. Namely, with $F[f]$ satisfying the condition:

$$C_{N1} \cdot \left(\frac{\partial F[f]}{\partial f} \right)^2 - C_{N2} \cdot F[f] \cdot \frac{\partial^2 F[f]}{\partial f^2} = 0, \tag{404}$$

the corresponding modified Ricci scalar R^* reads:

$$R^* = -\frac{C_{N2} \cdot \dfrac{\partial F[f]}{\partial f}}{F[f]^2} \cdot \Delta_g f \xrightarrow{n=4} = -\frac{3}{F[f]^2} \cdot \left(\frac{\partial F[f]}{\partial f} \cdot \Delta_g f \right). \tag{405}$$

Thereby the C_{N1} and C_{N2} are constants depending on the number of dimensions n of the space-time being considered. It was shown in [2] that we can write:

$$C_{N1} = -\frac{3}{2} + \frac{7}{4} \cdot n - \frac{1}{4} \cdot n^2$$

$$C_{N2} = n - 1. \tag{406}$$

In chapter 8, we are going to derive these parameters directly from the Ricci scalar of a scale-transformed metric.

The corresponding function $F[f]$, which satisfies condition (404) for every dimension n, could be given as:

$$F[f] = C_2 \cdot \left(f \cdot (n-2) + C_1 \right)^{\frac{4}{n-2}} = \left(c_1 \cdot f \cdot (n-2) + c_2 \right)^{\frac{4}{n-2}}. \tag{407}$$

Thereby the constants C_1, C_2, c_1, and c_2 are arbitrary.

For the case $R^* = 0$, we directly obtain the mass-free Klein–Gordon equation:

$$R^* = 0 = -\frac{C_{N2} \cdot \dfrac{\partial F[f]}{\partial f}}{F[f]^2} \cdot \Delta_g f = \Delta_g f. \tag{408}$$

Mass and other forms of matter or inertia could be included via suitable dependencies for R^*, but this would lead to nonlinear equations in general space-times of arbitrary numbers of dimensions as one can directly see in Eq. (405) with the plugged-in $F[f]$ from (407). Classical Klein–Gordon equations, however, do not show such non-linearities and still provide a good description of the quantum reality when compared with experimental findings.

Thus, so we ask, is there a way to derive the complete Klein–Gordon equation (with mass and potential) in its classical form, thereby avoiding the non-linearities?

6.11.2   Metric Klein–Gordon Equation with Mass

When using dimensional entanglement, we can easily incorporate mass in a linear manner. This was already shown above, which is why here we give one additional example. Introducing a 5-dimensional space-time $t, r, \vartheta, \varphi, u$ with a 3-dimensional spherical symmetry (this we choose in order to provide the metric origin and quantum gravity transition to the important Schrödinger hydrogen problem later on):

$$g^5_{\alpha\beta} = \begin{pmatrix} -c^2 & 0 & 0 & 0 & 0 \\ 0 & 1 & 0 & 0 & 0 \\ 0 & 0 & r^2 & 0 & 0 \\ 0 & 0 & 0 & r^2 \cdot \sin[\vartheta]^2 & 0 \\ 0 & 0 & 0 & 0 & \left(g_u[u]'\right)^2 \end{pmatrix} \times F\Big[f\big[t,r,\vartheta,\varphi,g_u[u]\big]\Big], \quad (409)$$

$$F[f] = \Big(C_f + f\big[t,r,\vartheta,\varphi,g_u[u]\big]\Big)^{\frac{4}{3}}; \quad C_f \neq 0$$

$$f\big[t,r,\vartheta,\varphi,g_u[u]\big] = f_t[t] \cdot f_r[r] \cdot f_\vartheta[\vartheta] \cdot f_\varphi[\varphi] \cdot f_u[u]$$

$$f_u[u] = C_{u1} \cdot \cos\big[A_u \cdot g_u[u]\big] + C_{u2} \cdot \sin\big[A_u \cdot g_u[u]\big], \quad (410)$$

will bring us what we require. We obtain the following equation for the adapted Ricci scalar R^* in the case of $R^* = 0$:

$$R^* = 0 = \left(\Delta_{\text{3D-sphere}} - \frac{\partial^2}{c^2 \cdot \partial t^2} - A_u^2 \right) f. \quad (411)$$

Please note that for the reason of synergy (see section "6.11.5 More General r-Dependent Potentials") we in fact used an 8-dimensional metric and simply avoided dependencies with respect to the coordinates 5, 6, and 7, thereby $t =$ time = coordinate zero.

With the setting $A_u^2 = \dfrac{m^2 \cdot c^2}{\hbar^2}$ (c = speed of light in vacuum, m = rest mass,

$\hbar$ = reduced Planck constant), we would just have obtained the classical Klein–Gordon equation (completely metrically derived (!)) with mass.

Please note that the function g_u is still arbitrary and that we do not harm our solution above when demanding that the function $g_u[u]$ should be minimal near the u we consider while dealing with the system described by the metric (409). This automatically makes the 5th dimension to appear "invisible," "curled up," or just—as they say—"compactified." It should be pointed out that the compactification has not to be demanded, but is simply an organic part of the metric approach and no approximation is needed to assure the near-invisibility of the u-component (apart from the fact that it provides mass).

We conclude, with Eq. (411) we have obtained the classical Klein–Gordon equation in a 4-dimensional space-time with spatial spherical symmetry with mass. In contrast to the classical equation, which was postulated, we have derived this equation directly out of the condition of the Ricci scalar R^* of a variated metric of the kind $G_{\alpha\beta} = F[f]^*g_{\alpha\beta}$ being zero. In choosing any other symmetry in Eq. (409), we would obtain the classical Klein–Gordon equation in this very metric.

6.11.3 Metric Klein–Gordon Equation with Mass and Potential

This time we choose a metric in 6 dimensions of the following kind:

$$g_{\alpha\beta}^6 = \begin{pmatrix} -c^2 & 0 & 0 & 0 & 0 & 0 \\ 0 & 1 & 0 & 0 & 0 & 0 \\ 0 & 0 & r^2 & 0 & 0 & 0 \\ 0 & 0 & 0 & r^2 \cdot \sin[\vartheta]^2 & 0 & 0 \\ 0 & 0 & 0 & 0 & \left(g_u[u]'\right)^2 & 0 \\ 0 & 0 & 0 & 0 & 0 & r \cdot \left(g_v[v]'\right)^2 \end{pmatrix}$$

$$\times \overbrace{\left(C_f + f[t,r,\vartheta,\varphi,u,v]\right)}^{=F[f]\xrightarrow{6D} C_f + f}$$

$$f[t,r,\vartheta,\varphi,u,v] = f_t[t] \cdot f_r[r] \cdot f_\vartheta[\vartheta] \cdot f_\varphi[\varphi] \cdot f_u[u] \cdot f_v[v]. \tag{412}$$

Please note that for the reason of synergy (see section "6.11.5 More General r-Dependent Potentials") we in fact used an 8-dimensional metric and simply avoided dependencies with respect to the coordinates 6 and 7, thereby $t =$ time = coordinate zero).

Applying the partial solution:

$$F[f] = \left(C_f + f\left[t,r,\vartheta,\varphi,g_u[u],g_v[v]\right]\right); \quad C_f = 0$$

$$f\left[t,r,\vartheta,\varphi,g_u[u],g_v[v]\right] = f_t[t] \cdot f_r[r] \cdot f_\vartheta[\vartheta] \cdot f_\varphi[\varphi] \cdot f_u[u] \cdot f_v[v] \equiv \Psi \cdot f_u[u] \cdot f_v[v]$$

$$f_u[u] = C_{u1} \cdot e^{-A_u \cdot g_u[u]} + C_{u2} \cdot e^{A_u \cdot g_u[u]}$$

$$f_v[v] = C_{v1} \cdot e^{-A_v \cdot g_v[v]} + C_{v2} \cdot e^{A_v \cdot g_v[v]}, \tag{413}$$

gives us the following differential equation:

$$R^* = 0 = \left(\frac{1}{2} \cdot \frac{\partial}{r \cdot \partial r} + \left(\frac{A_v^2}{r} + A_u^2\right) + \Delta_{\text{3D-sphere}} - \frac{\partial^2}{c^2 \cdot \partial t^2} + \frac{9}{28 \cdot r^2}\right) \Psi$$

$$\Psi = \Psi[t,r,\vartheta,\varphi]. \tag{414}$$

We recognize the mass term $A_u^2 = \dfrac{m^2 \cdot c^2}{\hbar^2}$ and the classical Coulomb potential $\dfrac{A_v^2}{r}$, but also obtained the additional terms $\dfrac{1}{2} \cdot \dfrac{\partial}{r \cdot \partial r} \Psi$ and $\dfrac{9}{28 \cdot r^2} \Psi$.

A solution to (414) can be found via the classical hydrogen approach with the Laguerre series for f_r in a slightly adapted form than being used for the Schrödinger hydrogen, where we only had the $\dfrac{A_v^2}{r}$ for the potential part.

Please note that with additional coordinates and settings for the metric components, it is easily possible to come ever closer to the classical Schrödinger hydrogen problem, but this shall not be our concern here in this chapter, where we only intended to show briefly how the Klein–Gordon equation with mass and potential can be obtained from a modified metric of the kind $G_{\alpha\beta} = F[f]^* g_{\alpha\beta}$.

It would probably be an interesting task to find the metric structure, giving the classical Coulomb potential. For here and now, however, we leave this task to the interested reader.

6.11.4   Derivation of the Schrödinger Equation

We start from our result of the last section and use Eq. (411), only that, for the reason of recognition, instead of the symbol f we apply the classical Greek symbol Ψ. We write (411) as follows:

$$0 = A_u^2 \cdot f + \frac{\partial^2 f}{c^2 \cdot \partial t^2} - \Delta_{\text{3D-sphere}} f; \quad f = f[t, r, \vartheta, \varphi]$$

$$\Rightarrow \qquad -M^2 \cdot f + \Delta_{4D} f = -M^2 \cdot f + \Delta f = 0. \tag{415}$$

Then we can separate the time derivative as follows:

$$0 = \left[-M^2 + \Delta \right] \Psi = \left[-M^2 + \left(C1 \cdot \frac{\partial_t^2}{c^2} + \underbrace{\frac{1}{\sqrt{g}} \partial_\alpha \sqrt{g} \cdot g^{\alpha\beta} \partial_\beta}_{\text{3D-}\Delta\text{-Operator}} \right) \right] \Psi. \tag{416}$$

Please note that our Greek indices α and β are running only from 1 to n, because the time or 0-component has been separated. Now, we introduce a function $\Psi = \Phi + X$ and demand the following additional condition:

$$\partial_t \Psi = c^2 \cdot C2 \cdot (\Phi - X). \tag{417}$$

Together with (416) we obtain:

$$0 = -M^2 (\Phi + X) + \left(C1 \cdot C2 \cdot \partial_t (\Phi - X) + \frac{1}{\sqrt{g}} \partial_\alpha \sqrt{g} \cdot g^{\alpha\beta} \partial_\beta (\Phi + X) \right). \tag{418}$$

The following two equations summed up would result in Eq. (418):

$$0 = -M^2 \Phi + \left(C1 \cdot C2 \cdot \partial_t \Phi + \frac{1}{\sqrt{g}} \partial_\alpha \sqrt{g} \cdot g^{\alpha\beta} \partial_\beta \Phi \right)$$

$$0 = -M^2 X + \left(\frac{1}{\sqrt{g}} \partial_\alpha \sqrt{g} \cdot g^{\alpha\beta} \partial_\beta X - C1 \cdot C2 \cdot \partial_t X \right). \tag{419}$$

Comparing with the original Schrödinger equation as given in the form below:

$$\left[-i \cdot \hbar \cdot \partial_t - \frac{\hbar^2}{2m} \Delta_{\text{Schrödinger}} + V_{\text{Schrödinger}} \right] \Psi = 0, \tag{420}$$

it not only gives us the matter and antimatter solutions (c.f. [4, 37]), but also shows us—as seen and discussed before [40]—that mass m and the potential $V_{\text{Schrödinger}}$ is now been taken on by the metric and the results in hidden (potentially compactified) dimensions, potentially due to entanglement. Thus, a distorted metric and additional dimensions act like an effective potential and/or mass in the Schrödinger approximation and vice versa, which is to say: what in classical physics is been described as a potential would now become a set of potentially compactified dimensions, which is providing the necessary interaction. Similarly, we have to formulate for the mass M: what appears as (rest) mass to us is just permanently and locally curved space, whereby the curvature is the one of additional dimensions, being potentially entangled. Disregarding the antimatter solution here, setting the constants $C1$, $C2$, and reshaping the classical Schrödinger equation as:

$$C1 \cdot C2 = \frac{i \cdot 2 \cdot m}{\hbar}; \quad X = 0; \quad 0 = \left[\frac{i \cdot 2 \cdot m}{\hbar} \cdot \partial_t + \Delta_{\text{Schrödinger}} - 2 \cdot V_{\text{Schrödinger}} \right] \Psi,$$

(421)

gives us the proportionality:

$$\Rightarrow \quad \left[\Delta_{3D} - 2 \cdot V_{\text{Schrödinger}} \right] \Psi \triangleq -M^2 \Phi + \frac{1}{\sqrt{g}} \partial_\alpha \sqrt{g} \cdot g^{\alpha\beta} \partial_\beta \Phi$$

$$\Rightarrow \quad \Delta_{3D} \Psi \triangleq \Delta_{3D} \Phi; \quad \Rightarrow V_{\text{Schrödinger}} \triangleq \frac{M^2 \Phi}{2 \cdot \Psi}.$$

(422)

This way we can now link classical Schrödinger solutions and the corresponding Schrödinger potentials with their metric analog.

6.11.5 More General *r*-Dependent Potentials

The attentive reader may have already deduced from our approach (412) that there can be more general constructions of r-dependent potentials simply via different exponents of r at the metric components. Just in order to give an example we go to 8 dimensions and set:

$$g^8_{\alpha\beta} = \begin{pmatrix} -c^2 & 0 & 0 & 0 & \cdots & 0 \\ 0 & 1 & 0 & 0 & \cdots & 0 \\ 0 & 0 & r^2 & 0 & \cdots & 0 \\ 0 & 0 & 0 & r^2 \cdot \sin^2 \varphi_1 & \cdots & 0 \\ \cdots & \cdots & \cdots & \cdots & \ddots & 0 \\ 0 & 0 & 0 & 0 & 0 & g_{77} \end{pmatrix}$$

$$\times F\left[f\left[t, r, \vartheta, \varphi_1, \varphi_2, \varphi_3, \varphi_4, \varphi_5, \varphi_6 \right] \right]; \quad F[f] = f^{\frac{2}{3}}.$$

$$f\left[t, r, \vartheta, \varphi_1, \varphi_2, \varphi_3, \varphi_4, \varphi_5, \varphi_6 \right] = f_t[t] \cdot f_r[r] \cdot f_\vartheta[\vartheta] \cdot \prod_{i=1}^{6} f_{\varphi_i}[\varphi_i];$$

$$g_{44} = \left(g_{\varphi_3}[\varphi_3]' \right)^2 \cdot r^0;$$

$$g_{55} = \left(g_{\varphi_4}\left[\varphi_4\right]'\right)^2 \cdot r^1;$$

$$g_{66} = \left(g_{\varphi_5}\left[\varphi_5\right]'\right)^2 \cdot r^2;$$

$$g_{77} = \left(g_{\varphi_6}\left[\varphi_6\right]'\right)^2 \cdot r^3; \tag{423}$$

We partially "fix" some of the functions as follows (for $i = 3, 4, 5, 6$):

$$f_{\varphi_i}\left[\varphi_i\right] = f_{\varphi_i}\left[g_{\varphi_i}\left[\varphi_i\right]\right] \equiv f_{\varphi_i}\left[g_{\varphi_i}\right]; \quad f_{\varphi_i}\left[g_{\varphi_i}\right] = C_{-i} \cdot e^{-A_i \cdot g_{\varphi_i}} + C_{+i} \cdot e^{+A_i \cdot g_{\varphi_i}}$$

$$\tag{424}$$

$$f_t[t] = C_{t1} \cdot \cos[c \cdot E \cdot t] + C_{t2} \cdot \sin[c \cdot E \cdot t]. \tag{425}$$

This gives us the following differential equation:

$$R^* = 0 = \left(\frac{3}{2} \cdot \frac{\partial}{r \cdot \partial r} + \left(\frac{A_6^2}{r^3} + \frac{A_5^2}{r^2} + \frac{A_4^2}{r} + A_3^2\right) + \Delta_{\text{3D-sphere}} + E^2 + \frac{111}{28 \cdot r^2}\right)\Psi$$

$$\Psi = \Psi[r, \vartheta, \varphi]. \tag{426}$$

Also negative exponents can be constructed, which are leading to infinitely growing potential terms. So, for example:

$$g_{\alpha\beta}^8 = \begin{pmatrix} -c^2 & 0 & 0 & 0 & \cdots & 0 \\ 0 & 1 & 0 & 0 & \cdots & 0 \\ 0 & 0 & r^2 & 0 & \cdots & 0 \\ 0 & 0 & 0 & r^2 \cdot \sin^2 \varphi_1 & \cdots & 0 \\ \cdots & \cdots & \cdots & \cdots & \ddots & 0 \\ 0 & 0 & 0 & 0 & 0 & g_{77} \end{pmatrix}$$

$$\times F\left[f\left[t, r, \vartheta, \varphi_1, \varphi_2, \varphi_3, \varphi_4, \varphi_5, \varphi_6\right]\right]; \quad F[f] = f^{\frac{2}{3}}$$

$$f\left[t, r, \vartheta, \varphi_1, \varphi_2, \varphi_3, \varphi_4, \varphi_5, \varphi_6\right] = f_t[t] \cdot f_r[r] \cdot f_\vartheta[\vartheta] \cdot \prod_{i=1}^{6} f_{\varphi_i}\left[\varphi_i\right];$$

$$g_{44} = \left(g_{\varphi_3}\left[\varphi_3\right]'\right)^2 \cdot r^0;$$

$$g_{55} = \left(g_{\varphi_4}\left[\varphi_4\right]'\right)^2 \cdot r^1;$$

$$g_{66} = \left(g_{\varphi_5}\left[\varphi_5\right]'\right)^2 \cdot r^2;$$

$$g_{77} = \left(g_{\varphi_6}\left[\varphi_6\right]'\right)^2 \cdot r^{-1}. \tag{427}$$

This gives us the following differential equation for the spatial coordinates:

$$R^* = 0 = \left(\frac{1}{2} \cdot \frac{\partial}{r \cdot \partial r} + \left(r \cdot A_6^2 + \frac{A_5^2}{r^2} + \frac{A_4^2}{r} + A_3^2\right) + \Delta_{\text{3D-sphere}} + E^2 + \frac{27}{28 \cdot r^2}\right)\Psi$$

$$\Psi = \Psi\left[r, \vartheta, \varphi\right]. \tag{428}$$

Note: Further generalization of potentials also with respect to other coordinates is easily possible via structural "playing" with the additional—compactified—coordinate components. For instance, the metrics already considered in section "6.9 More Metric Options and Subsequent Quantum Equations Plus Their Solutions" are giving angular dependencies.

More options for radius settings and quite some discussion will be presented in chapter 11.

6.12   Toward an Explanation for the 3-Generation Problem [28]

In order to discuss possible ways to solve the so-called 3-generation problem, we first want to derive the Dirac equation [29] from the Einstein–Hilbert action in order to see where the mass in this—probably most important of the quantum equations—comes in. The derivation was already shown in [1], but nevertheless we here want to repeat it in order to point out certain aspects with respect to heavy and inertial mass and curvature. A more fundamental and comprehensive consideration about the Dirac equation will be given in chapters 9 and 10 of this book.

6.12.1   Repetition: Dirac in the Metric Picture and Its Connection to the Classical Quaternion Form

As already pointed out earlier in this book, in connection with our scaling or quantum wrapping approach F[f] (see section "6.3 Repetition: Hilbert's World Formula"), it should be pointed out that with the condition for f to be a scalar there are still many more options to have vector and tensor functions included via:

$$f = h + h_\alpha q^\alpha + h_{\alpha\beta} q^{\alpha\beta} + \ldots. \tag{429}$$

As $F[f]$ could still be freely chosen, we take the two simple options that we either seek for a solution for f within the minimum or maximum region of F, giving us $\dfrac{\partial F[f]}{\partial f} \simeq 0$ and simplifying (245) to:

$$R^* = -\frac{C_{N2} \cdot \dfrac{\partial^2 F[f]}{\partial f^2} \cdot \overbrace{\left(\tilde{\nabla}_g f\right)^2}^{= f_{,\alpha} g^{\alpha\beta} f_{,\beta}}}{F[f]^2} \xrightarrow{f = h_\lambda q^\lambda} = -\frac{C_{N2} \cdot \dfrac{\partial^2 F[f]}{\partial f^2} \cdot (h_\lambda q^\lambda)_{,\alpha} \, g^{\alpha\beta} (h_\lambda q^\lambda)_{,\beta}}{F[f]^2}$$

$$\Rightarrow \qquad \frac{F[f]^2 \cdot R^*}{C_{N2} \cdot \dfrac{\partial^2 F[f]}{\partial f^2}} = -\left(h_\lambda q^\lambda\right)_{,\alpha} g^{\alpha\beta} \left(h_\lambda q^\lambda\right)_{,\beta}$$

$$= -\left(h_\lambda q^\lambda_{,\alpha} + h_{\lambda,\alpha} q^\lambda\right) g^{\alpha\beta} \left(h_\lambda q^\lambda_{,\beta} + h_{\lambda,\beta} q^\lambda\right)$$

$$= -h_\lambda q^\lambda_{,\alpha} g^{\alpha\beta} h_\lambda q^\lambda_{,\beta} - h_{\lambda,\alpha} q^\lambda g^{\alpha\beta} h_\lambda q^\lambda_{,\beta} - h_\lambda q^\lambda_{,\alpha} g^{\alpha\beta} h_{\lambda,\beta} q^\lambda - h_{\lambda,\alpha} q^\lambda g^{\alpha\beta} h_{\lambda,\beta} q^\lambda ,$$

$$\tag{430}$$

or we just demand:

$$\Delta f = \Delta\left(h + h_\alpha q^\alpha + h_{\alpha\beta} q^{\alpha\beta} + \ldots\right) = 0, \tag{431}$$

longer and consequently simplify (245) to:

$$R^* = \frac{1}{F[f]^3} \cdot \left(\left(C_{N1} \cdot \left(\frac{\partial F[f]}{\partial f}\right)^2 - C_{N2} \cdot F[f] \cdot \frac{\partial^2 F[f]}{\partial f^2}\right) \cdot \overbrace{\left(\tilde{\nabla}_g f\right)^2}^{= f_{,\alpha} g^{\alpha\beta} f_{,\beta}}\right). \tag{432}$$

A first, for brevity and simplicity, we proceed with the $\dfrac{\partial F[f]}{\partial f} \simeq 0$-option.

As we haven't had any need to fix both vectors h_λ and q^λ, yet, we assume that one is a vector of constants and by making q^λ this constant vector we obtain:

$$\frac{F[f]^2 \cdot R^*}{C_{N2} \cdot \dfrac{\partial^2 F[f]}{\partial f^2}} = -h_{\lambda,\alpha} q^\lambda g^{\alpha\beta} h_{\lambda,\beta} q^\lambda \xrightarrow{g^{\alpha\beta} = \mathbf{e}^\alpha \cdot \mathbf{e}^\beta} = -h_{\lambda,\alpha} q^\lambda \mathbf{e}^\alpha \cdot \mathbf{e}^\beta h_{\lambda,\beta} q^\lambda$$

$$\xrightarrow{q^\lambda \mathbf{e}^\alpha = \mathbf{Q}^{\lambda\alpha}} = -h_{\lambda,\alpha} \mathbf{Q}^{\lambda\alpha} \cdot \mathbf{Q}^{\lambda\beta} h_{\lambda,\beta} . \tag{433}$$

Now, only in order to obtain the classical Dirac equation, we approximate the curvature term on the left-hand side as a square of the linear function of the scalar product f and therefore we can write[5]:

$$\frac{F[f]^2 \cdot R^*}{C_{N2} \cdot \dfrac{\partial^2 F[f]}{\partial f^2}} \simeq h_\lambda q^\lambda \mathbf{M} \cdot \mathbf{M} h_\lambda q^\lambda = -h_{\lambda,\alpha} q^\lambda g^{\alpha\beta} h_{\lambda,\beta} q^\lambda$$

$$\xrightarrow{g^{\alpha\beta} = \mathbf{e}^\alpha \cdot \mathbf{e}^\beta} = -h_{\lambda,\alpha} q^\lambda \mathbf{e}^\alpha \cdot \mathbf{e}^\beta h_{\lambda,\beta} q^\lambda$$

$$\Rightarrow \qquad \mp h_\lambda q^\lambda \mathbf{M} = i \cdot h_{\lambda,\alpha} q^\lambda \mathbf{e}^\alpha$$

$$\Rightarrow \qquad 0 = i \cdot h_{\lambda,\alpha} \mathbf{e}^\alpha \pm h_\lambda \mathbf{M}, \qquad (434)$$

$$\Rightarrow \qquad 0 = i \cdot h_{\lambda,\alpha} \mathbf{e}^\alpha \pm h_\lambda \mathbf{M}. \qquad (435)$$

Please note that Eq. (435) represents a generalized Dirac equation in an arbitrary number of dimensions. By reducing this general and arbitrary n-dimensional equation to 4 dimensions, we almost directly recognize the Dirac equation, only that in our—the metric—case the mass-terms have to be at least a vector, while the Dirac matrices for the derivatives of the function vector $\mathbf{h}_\lambda$ (c.f. [29]) are becoming the metric base vector components. In order to obtain the classical Dirac equation in the natural units form:

$$0 = i \cdot \gamma^\alpha h_{\lambda,\alpha} \pm h_\lambda \cdot m, \qquad (436)$$

---

[5]Please note that the approximation $\dfrac{F[f]^2 \cdot R^*}{C_{N2} \cdot \dfrac{\partial^2 F[f]}{\partial f^2}} = h_\lambda q^\lambda \mathbf{M} \cdot \mathbf{M} h_\lambda q^\lambda$ as introduced in Eq. (434)

may just be the reason for the restriction of the rest masses for elementary particles to only three generations. Knowing namely that $\dfrac{F[f]^2 \cdot R^*}{C_{N2} \cdot \dfrac{\partial^2 F[f]}{\partial f^2}}$ has to be proportional to $\left(\dfrac{r_s}{r}\right)^3$ in order to

provide the right limit to Newton's law of gravity for $r \gg r_s$ (c.f. sub-section "6.8.5. The Observer and the Problem of Transition to Newton's Gravity Law"), but also realizing that the squared mass approximation in Eq. (434) does such an excellent job in praxis, one might be tempted to conclude that the possibility of such an approximation has a rather fundamental meaning. In other words, perhaps there is a deeper reason for the existence of a relation of the kind

$\left(\dfrac{r_s}{r}\right)^3 \Bigg|_{r=?} = r_s^2 + c_1 \cdot r_s + c_0$. We will come across such a possibility further below in this section.

which we can also write as:

$$0 = i \cdot \gamma^{\alpha} h_{\lambda,\alpha} \pm I \cdot h_{\lambda} \cdot m, \tag{437}$$

and with the Dirac Gamma matrices γ^{α} and the unit matrix I given as follows:

$$\gamma^0 = \begin{pmatrix} 1 & & & \\ & 1 & & \\ & & -1 & \\ & & & -1 \end{pmatrix} ; \quad \gamma^1 = \begin{pmatrix} & & & 1 \\ & & 1 & \\ & -1 & & \\ -1 & & & \end{pmatrix}$$

$$\gamma^2 = \begin{pmatrix} & & & -i \\ & & i & \\ & i & & \\ -i & & & \end{pmatrix} ; \quad \gamma^3 = \begin{pmatrix} & & 1 & \\ & & & -1 \\ -1 & & & \\ & 1 & & \end{pmatrix} ; \quad I = \begin{pmatrix} 1 & & & \\ & 1 & & \\ & & 1 & \\ & & & 1 \end{pmatrix}, \tag{438}$$

we simply need to multiply Eq. (435) with a suitable object **C**. This gives us the identities[6]:

$$\mathbf{C} \cdot \mathbf{e}^{\alpha} = \gamma^{\alpha} ; \quad \mathbf{C} \cdot \mathbf{M} = I \cdot m, \tag{439}$$

and provides a connection between the classical quaternion and the metric Dirac equation. As the classical Dirac equation was given in Minkowski coordinates, with base vectors of the kind:

$$\mathbf{e}^0 = \begin{pmatrix} i \\ 0 \\ 0 \\ 0 \end{pmatrix} ; \quad \mathbf{e}^1 = \begin{pmatrix} 0 \\ 1 \\ 0 \\ 0 \end{pmatrix} ; \quad \mathbf{e}^2 = \begin{pmatrix} 0 \\ 0 \\ 1 \\ 0 \end{pmatrix} ; \quad \mathbf{e}^3 = \begin{pmatrix} 0 \\ 0 \\ 0 \\ 1 \end{pmatrix}, \tag{440}$$

we could find the corresponding structure of the object **C** via the first equation in Eq. (439) and it becomes clear that we need something of the kind:

$$C_{\beta}^{\alpha} \cdot \mathbf{e}^{\beta} = \gamma^{\alpha}, \tag{441}$$

---

[6]It should be noted that classically one does not use the identity matrix for the second equation in Eq. (439), but the so called β-matrix instead. It reads $\beta = \begin{pmatrix} 1 & & & \\ & 1 & & \\ & & -1 & \\ & & & -1 \end{pmatrix}$, but will not be considered here, as it makes no difference for the results of this chapter whether we use I or β in Eq. (439).

$$\Rightarrow \quad \begin{cases} C_\alpha^0 = \begin{pmatrix} -i & & & \\ & 1 & & \\ & & -1 & \\ & & & -1 \end{pmatrix}; \quad C_\alpha^1 = \begin{pmatrix} & & & 1 \\ & & 1 & \\ & & -1 & \\ & i & & \end{pmatrix} \\ \\ C_\alpha^2 = \begin{pmatrix} & & -i & \\ & i & & \\ & i & & \\ -1 & & & \end{pmatrix}; \quad C_\alpha^3 = \begin{pmatrix} & & & 1 \\ & & -1 & \\ & i & & \\ 1 & & & \end{pmatrix}. \end{cases} \quad (442)$$

From here we deduce that **M** must be something more complex than just a vector as only the following approach gives us what we need:

$$C_\beta^\alpha \cdot \mathbf{M} = I \cdot m \quad \Rightarrow \quad C_\beta^\alpha \cdot M_\lambda^\beta = \delta_\lambda^\alpha \cdot m = I \cdot m. \quad (443)$$

Thereby we need to point out that $h_\lambda q^\lambda \mathbf{M} \cdot \mathbf{M} h_\lambda q^\lambda$ in Eq. (434) only is a linear approximation of the complex expression $\dfrac{F[f]^2 \cdot R^*}{C_{N2} \cdot \dfrac{\partial^2 F[f]}{\partial f^2}}$ and that the classical Dirac equation was just postulated [29], while our new Dirac equation (Eq. (433)) has a fundamental metric and extremal-principle origin. The result after solving Eq. (443) with respect to the object M_λ^β reads:

$$\Rightarrow \quad \begin{cases} M_0^\alpha = \begin{pmatrix} i & & & \\ & 1 & & \\ & & -1 & \\ & & & -1 \end{pmatrix}; \quad M_1^\alpha = \begin{pmatrix} & & & -i \\ & & -1 & \\ & 1 & & \\ 1 & & & \end{pmatrix} \\ \\ M_2^\alpha = \begin{pmatrix} & & -1 & \\ & -i & & \\ & -i & & \\ i & & & \end{pmatrix}; \quad M_3^\alpha = \begin{pmatrix} & & & -i \\ & & 1 & \\ 1 & & & \\ & & & -1 \end{pmatrix} \end{cases} \quad (444)$$

and gives us the connection of the classical Dirac equation (Eq. (436)) from [29] with our new metric Dirac equation (Eq. (433)) in its linear approximated form Eq. (434). We realize that in the metric picture, the corresponding inertia parameters are not just a scalar.

6.12.1.1 Restriction to Laplace functions

It needs to be pointed out that the classical Dirac equation explicitly only considers Laplace functions satisfying condition (Eq. (431)). In fact, when deriving his equation as a quaternion square root from the Klein–Gordon equation, the whole derivation of Dirac was founded onto this point. This generalizes and dis-approximates Eq. (433) to:

$$\frac{F[f]^3 \cdot R^*}{C_{N1} \cdot \left(\frac{\partial F[f]}{\partial f}\right)^2 - C_{N2} \cdot F[f] \cdot \frac{\partial^2 F[f]}{\partial f^2}} = h_{\lambda,\alpha} q^\lambda g^{\alpha\beta} h_{\lambda,\beta} q^\lambda$$

$$\xrightarrow{g^{\alpha\beta} = e^\alpha \cdot e^\beta} = h_{\lambda,\alpha} q^\lambda e^\alpha \cdot e^\beta h_{\lambda,\beta} q^\lambda$$

$$\xrightarrow{q^\lambda e^\alpha = Q^{\lambda\alpha}} = -h_{\lambda,\alpha} Q^{\lambda\alpha} \cdot Q^{\lambda\beta} h_{\lambda,\beta}. \tag{445}$$

Now, we could just again linearly approximate the curvature or gravity term via:

$$\frac{F[f]^3 \cdot R^*}{C_{N1} \cdot \left(\frac{\partial F[f]}{\partial f}\right)^2 - C_{N2} \cdot F[f] \cdot \frac{\partial^2 F[f]}{\partial f^2}} \approx h_\lambda q^\lambda \mathbf{M} \cdot \mathbf{M} h_\lambda q^\lambda = h_{\lambda,\alpha} q^\lambda g^{\alpha\beta} h_{\lambda,\beta} q^\lambda$$

$$\xrightarrow{g^{\alpha\beta} = e^\alpha \cdot e^\beta} = h_{\lambda,\alpha} q^\lambda e^\alpha \cdot e^\beta h_{\lambda,\beta} q^\lambda$$

$$\Rightarrow \qquad \mp h_\lambda q^\lambda \mathbf{M} = i \cdot h_{\lambda,\alpha} q^\lambda e^\alpha$$

$$\Rightarrow \qquad 0 = i \cdot h_{\lambda,\alpha} e^\alpha \pm h_\lambda \mathbf{M} \tag{446}$$

and, regarding the connection to the classical Dirac equation [29], we would have the same calculation as before ... only that this time it is even more general and no approximation or restriction of the kind $\frac{\partial F[f]}{\partial f} \approx 0$ was needed.

6.12.2 Getting Rid of Quaternions

Knowing now that the fundamental metric origin of the quaternions, which Dirac had to apply in order to extract the root of the Klein–Gordon equation, to be the metric and subsequently the base vectors (see transition from the first to the second line in Eq. (434)), we also have the means to potentially avoid these mathematical structures. We are keen to do so, because:

(a) They are kind of nasty ... in a purely practical-mathematical sense, of course.

(b) It is difficult to change the number of dimensions (e.g., the classical Dirac equation is only defined in a 4-dimensional space-time).

In order to achieve this goal, we have to find a different way to separate, respectively decompose the metric tensor. Instead of applying base vectors for the decomposition, we intend to use vectors. This "vectorial root extraction" or "vector decomposition" has been introduced by the author a long time ago. It was motivated by an adventurous story of two Jews in Nazi Germany, who had left a mathematical message inside the gas chamber where they had been killed in 1944 [41, 42].

Apparently, as demonstrated above, there also seems to be an approach that extracts the Dirac operator directly out of the Einstein field equations or the Einstein–Hilbert action and thus, the General Theory of Relativity. With respect to the connection with the classical approach, we refer to the subsection above and present here only the new technique that does not require the use of quaternions as Dirac needed them.

It is easy to prove that the scalar product of the following vector:

$$V_\Omega = \begin{cases} a+b+c+d+e, a+b+c+d-e, a+b+c-d+e, a+b+c-d-e, \\ a+b-c+d+e, a+b-c+d-e, a+b-c-d+e, a+b-c-d-e, \\ a-b+c+d+e, a-b+c+d-e, a-b+c-d+e, a-b+c-d-e, \\ a-b-c+d+e, a-b-c+d-e, a-b-c-d+e, a-b-c-d-e \end{cases}$$

$$\equiv \left[a \pm b \pm c \pm d \pm e\right]_\Omega \qquad (447)$$

gives:

$$V_\Omega \cdot V_\Omega = a^2 + b^2 + c^2 + d^2 + e^2. \qquad (448)$$

Even the introduction of "virtual" parameters ε can be incorporated into the vector $\mathbf{V_\Omega}$ as follows:

$$V_\Omega = \left\{a+b+c, a+b-c, a-b+i\cdot c, a-b-i\cdot c\right\} \equiv \left[a \pm b \pm I \cdot c\right]_\Omega$$

$$\text{with}: \quad V_\Omega \cdot V_\Omega = a^2 + b^2. \qquad (449)$$

We also have the possibility to explicitly demand mixed terms, which is to say, we can give any metric tensor in form of a metric equivalent as $\mathbf{V_\Omega}$-vector. The "peculiar" thing about the new form only is to be seen in the fact that our vector has too many components. Taking, for example, the possibility of the purely diagonal metric case with the subsequent scalar product (448) we already end up with 2^{n-1} components, which consequently give more "Dirac" equations than we have metric conditions. Thus, the form (447), as simple and attractive it may appear, cannot be the irreducible form for the vector $\mathbf{V_\Omega}$. In order to find the latter, we simply have to compose a general vector of the type:

$$V_\Omega = \begin{cases} a_0 \cdot a + b_0 \cdot b + c_0 \cdot c + d_0 \cdot d + e_0 \cdot e + \ldots, \\ a_0 \cdot a + b_1 \cdot b + c_1 \cdot c + d_1 \cdot d + e_1 \cdot e + \ldots, \\ a_0 \cdot a + b_1 \cdot b + c_2 \cdot c + d_2 \cdot d + e_2 \cdot e + \ldots, \\ a_0 \cdot a + b_1 \cdot b + c_2 \cdot c + d_3 \cdot d + e_3 \cdot e + \ldots, \\ a_0 \cdot a + b_1 \cdot b + c_2 \cdot c + d_3 \cdot d + e_4 \cdot e + \ldots, \\ a_0 \cdot a + b_1 \cdot b + c_2 \cdot c + d_3 \cdot d + e_4 \cdot e + \ldots, \\ \ldots \end{cases}$$

$$\equiv \left[a_j \cdot a + b_j \cdot b + c_j \cdot c + d_j \cdot d + e_j \cdot e + \ldots \right]_\Omega. \tag{450}$$

The constants a_j, b_j, c_j, and so on have to be determined from the line element equation, we here assume to have the structure:

$$ds^2 = a^2 \cdot dx_0^2 + b^2 \cdot dx_1^2 + c^2 \cdot dx_2^2 + d^2 \cdot dx_3^2 + e^2 \cdot dx_4^2 + \ldots$$

$$+ a \cdot b \cdot dx_0 dx_1 + a \cdot c \cdot dx_0 dx_2 + a \cdot d \cdot dx_0 dx_3 + a \cdot e \cdot dx_0 dx_4 + \ldots$$

$$+ b \cdot c \cdot dx_1 dx_2 + b \cdot d \cdot dx_1 dx_3 + b \cdot e \cdot dx_1 dx_4 + \ldots$$

$$+ c \cdot d \cdot dx_2 dx_3 + c \cdot e \cdot dx_2 dx_4 + \ldots + d \cdot e \cdot dx_3 dx_4 + \ldots \tag{451}$$

with the coordinates x_i, in comparison with the scalar product:

$$V_\Omega \cdot V_\Omega$$

$$= \left[a_j \cdot a + b_j \cdot b + c_j \cdot c + d_j \cdot d + e_j \cdot e + \ldots \right]_\Omega \cdot \left[a_j \cdot a + b_j \cdot b + c_j \cdot c + d_j \cdot d + e_j \cdot e + \ldots \right]_\Omega. \tag{452}$$

The resulting equations for the determination of the constants a_j, b_j, c_j, and so on are to be taken from:

$$V_\Omega \cdot V_\Omega$$

$$= \left[a_j \cdot a + b_j \cdot b + c_j \cdot c + d_j \cdot d + e_j \cdot e + \ldots \right]_\Omega \cdot \left[a_j \cdot a + b_j \cdot b + c_j \cdot c + d_j \cdot d + e_j \cdot e + \ldots \right]_\Omega$$

$$= a^2 + b^2 + c^2 + d^2 + e^2 + \ldots + a \cdot b + a \cdot c + a \cdot d + a \cdot e + \ldots$$

$$+ b \cdot c + b \cdot d + b \cdot e + \ldots + c \cdot d + c \cdot e + \ldots + d \cdot e + \ldots \tag{453}$$

via comparing of coefficients.

The number of coefficients and thus, the number of "Dirac" equations will now be equal to the number of dimensions.

Let us choose an example in 3 dimensions, with the line element been given as:

$$ds^2 = a^2 \cdot dx_0^2 + b^2 \cdot dx_1^2 + c^2 \cdot dx_2^2 + a \cdot b \cdot dx_0 dx_1 + a \cdot c \cdot dx_0 dx_2 + b \cdot c \cdot dx_1 dx_2.$$

$$(454)$$

The irreducible V_Ω-vector would read:

$$V_\Omega = \left\{ a_0 \cdot a + b_0 \cdot b + c_0 \cdot c, a_0 \cdot a + b_1 \cdot b + c_1 \cdot c, a_0 \cdot a + b_1 \cdot b + c_2 \cdot c \right\}$$

$$\equiv \left[a_j \cdot a + b_j \cdot b + c_j \cdot c \right]_\Omega \qquad (455)$$

and the subsequent equations for the determination of the constants a_j, b_j, and c_j can be written as:

$$a^2 = 3 \cdot a_0^2 \cdot a^2, \quad b^2 = b^2 \cdot \left(2 \cdot b_0^2 + b_1^2 \right),$$

$$c^2 = c^2 \cdot \left(b_0^2 + b_1^2 + b_2^2 \right), \quad a \cdot b = a \cdot b \cdot 2 \cdot a_0 \cdot \left(2 \cdot b_0 + b_1 \right),$$

$$b \cdot c = b \cdot c \cdot 2 \cdot \left(b_0 \cdot c_0 + b_0 \cdot c_2 + b_1 \cdot c_1 \right), \quad a \cdot c = a \cdot c \cdot 2 \cdot a_0 \cdot \left(c_0 + c_1 + c_2 \right). \quad (456)$$

There are eight different solutions to the equations above and one of them would just be:

$$a_0 = -\frac{1}{\sqrt{3}}, \quad b_0 = -\frac{1}{2}\sqrt{\frac{5}{6} + \sqrt{\frac{2}{3}}}, \quad b_1 = \frac{1}{\sqrt{2}} - \frac{1}{2\sqrt{3}},$$

$$c_0 = -\frac{18 + \sqrt{6}}{12\sqrt{3}}, \quad c_1 = \frac{1}{6}\left(\sqrt{2} - \sqrt{3} \right), \quad c_2 = \frac{1}{6}\sqrt{\frac{7}{2} - \sqrt{6}}. \quad (457)$$

For practical purposes shorter settings for the V_Ω-vector could be useful:

$$V_\Omega = \left\{ a_0 \cdot a + b_0 \cdot b + c_0 \cdot c, b_1 \cdot b + c_1 \cdot c, c_2 \cdot c \right\}$$

$$\equiv \left[a_j \cdot a + b_j \cdot b + c_j \cdot c \right]_\Omega. \qquad (458)$$

The corresponding equations for the determination of the constants a_j, b_j, and c_j then read:

$$a^2 = a_0^2 \cdot a^2, \quad b^2 = b^2 \cdot \left(b_0^2 + b_1^2 \right), \quad c^2 = c^2 \cdot \left(c_0^2 + c_1^2 + c_2^2 \right),$$

$$a \cdot b = a \cdot b \cdot 2 \cdot a_0 \cdot b_0, \quad b \cdot c = b \cdot c \cdot 2 \cdot \left(b_0 \cdot c_0 + b_1 \cdot c_1 \right), \quad a \cdot c = a \cdot c \cdot 2 \cdot a_0 \cdot c_0 \quad (459)$$

and we obtain somewhat simpler results for the subsequent eight solutions (here only one example):

$$a_0 = -1, \quad b_0 = -\frac{1}{2}, \quad b_1 = -\frac{\sqrt{3}}{2}$$

$$c_0 = -\frac{1}{2}, \quad c_1 = -\frac{1}{2\sqrt{3}}, \quad c_2 = -\sqrt{\frac{2}{3}}. \qquad (460)$$

In 4 dimensions, we have ten equations for the a_j, b_j, and c_j of the form:

$$a^2 = a_0^2 \cdot a^2, \quad b^2 = b^2 \cdot \left(b_0^2 + b_1^2\right), \quad c^2 = c^2 \cdot \left(c_0^2 + c_1^2 + c_2^2\right),$$

$$d^2 = d^2 \cdot \left(d_0^2 + d_1^2 + d_2^2 + d_3^2\right),$$

$$a \cdot b = a \cdot b \cdot 2 \cdot a_0 \cdot b_0, \quad b \cdot c = b \cdot c \cdot 2 \cdot \left(b_0 \cdot c_0 + b_1 \cdot c_1\right),$$

$$b \cdot d = b \cdot d \cdot 2 \cdot \left(b_0 \cdot d_0 + b_1 \cdot d_1\right), \quad a \cdot c = a \cdot c \cdot 2 \cdot a_0 \cdot c_0,$$

$$a \cdot d = a \cdot d \cdot 2 \cdot a_0 \cdot d_0, \quad c \cdot d = c \cdot d \cdot 2 \cdot \left(c_0 \cdot d_0 + c_1 \cdot d_1 + c_2 \cdot d_2\right), \quad (461)$$

when applying the short $\mathbf{V_\Omega}$-vector as:

$$V_\Omega = \left\{a_0 \cdot a + b_0 \cdot b + c_0 \cdot c + d_0 \cdot d, b_1 \cdot b + c_1 \cdot c + d_1 \cdot d, c_2 \cdot c + d_2 \cdot d, d_3 \cdot d\right\}$$

$$\equiv \left[a_j \cdot a + b_j \cdot b + c_j \cdot c + d_j \cdot d\right]_\Omega. \quad (462)$$

One solution out of the 16 shall be given here:

$$a_0 = -1, \quad b_0 = -\frac{1}{2}, \quad b_1 = -\frac{\sqrt{3}}{2}$$

$$c_0 = -\frac{1}{2}, \quad c_1 = -\frac{1}{2\sqrt{3}}, \quad c_2 = -\sqrt{\frac{2}{3}}$$

$$d_0 = -\frac{1}{2}, \quad d_1 = -\frac{1}{2\sqrt{3}}, \quad d_2 = -\frac{1}{2\sqrt{6}}, \quad d_3 = -\frac{\sqrt{\frac{5}{2}}}{2}. \quad (463)$$

The simplification in the case of purely diagonal metrics is dramatic as the solutions then consist only of 1 and -1. Here just the simplest case:

$$a_0 = 1, \quad b_0 = 0, \quad b_1 = 1, \quad c_0 = 0, \quad c_1 = 0, \quad c_2 = 1$$

$$d_0 = 0, \quad d_1 = 0, \quad d_2 = 0, \quad d_3 = 1. \quad (464)$$

Now, we go back to the Ricci scalar of our scale-wrapped metric tensor $G_{\alpha\beta} = F[f]^* g_{\alpha\beta}$ and the resulting Eq. (430) in the case of $\Delta f = 0$. Taking what we have just derived about the $\mathbf{V_\Omega}$-vector, we can rewrite the right-hand side of (430) as follows:

$$h_{\lambda,\alpha} q^\lambda g^{\alpha\beta} h_{\lambda,\beta} q^\lambda \xrightarrow{g^{\alpha\beta} \to V_\Omega \cdot V_\Omega} = V\left[q^\lambda h_{\lambda,\alpha}\right]_\Omega \cdot V\left[h_{\lambda,\beta} q^\lambda\right]_\Omega$$

$$= V\left[q^\lambda h_{\lambda,\alpha}\right]_\Omega \cdot V\left[h_{\lambda,\beta} q^\lambda\right]_\Omega$$

$$\left[q^\lambda \left(a_j \cdot \partial_0 + b_j \cdot \partial_1 + c_j \cdot \partial_2 + d_j \cdot \partial_3 + \ldots \right) h_\lambda \right]_\Omega \cdot$$
$$\left[q^\lambda \left(a_j \cdot \partial_0 + b_j \cdot \partial_1 + c_j \cdot \partial_2 + d_j \cdot \partial_3 + \ldots \right) h_\lambda \right]_\Omega, \tag{465}$$

which in 4-dimensional space-times and the $\mathbf{V_\Omega}$-vector from Eq. (462) results in:

$$h_{\lambda,\alpha} q^\lambda g^{\alpha\beta} h_{\lambda,\beta} q^\lambda \xrightarrow{g^{\alpha\beta} \rightarrow V_\Omega \cdot V_\Omega}$$

$$= \left\{ \begin{array}{c} q^\lambda \left(a_0 \cdot \partial_0 + b_0 \cdot \partial_1 + c_0 \cdot \partial_2 + d_0 \cdot \partial_3 \right) h_\lambda \\ q^\lambda \left(b_1 \cdot \partial_1 + c_1 \cdot \partial_2 + d_1 \cdot \partial_3 \right) h_\lambda \\ q^\lambda \left(c_2 \cdot \partial_2 + d_2 \cdot \partial_3 \right) h_\lambda \\ q^\lambda \left(d_3 \cdot \partial_3 \right) h_\lambda \end{array} \right\}$$

$$\cdot \left\{ \begin{array}{c} q^\lambda \left(a_0 \cdot \partial_0 + b_0 \cdot \partial_1 + c_0 \cdot \partial_2 + d_0 \cdot \partial_3 \right) h_\lambda \\ q^\lambda \left(b_1 \cdot \partial_1 + c_1 \cdot \partial_2 + d_1 \cdot \partial_3 \right) h_\lambda \\ q^\lambda \left(c_2 \cdot \partial_2 + d_2 \cdot \partial_3 \right) h_\lambda \\ q^\lambda \left(d_3 \cdot \partial_3 \right) h_\lambda \end{array} \right\}. \tag{466}$$

Subsequently, the complete Eq. (446) would read:

$$\frac{F[f]^3 \cdot R^*}{C_{N1} \cdot \left(\dfrac{\partial F[f]}{\partial f} \right)^2 - C_{N2} \cdot F[f] \cdot \dfrac{\partial^2 F[f]}{\partial f^2}}$$

$$\xrightarrow{4D\,(\text{see:}[1])} \frac{2 \cdot F[f]^3 \cdot R^*}{3 \cdot \left(\left(\dfrac{\partial F[f]}{\partial f} \right)^2 - 2 \cdot F[f] \cdot \dfrac{\partial^2 F[f]}{\partial f^2} \right)}$$

$$= h_{\lambda,\alpha} q^\lambda g^{\alpha\beta} h_{\lambda,\beta} q^\lambda \xrightarrow{g^{\alpha\beta} \rightarrow V_\Omega \cdot V_\Omega}$$

$$= \left\{ \begin{array}{c} q^\lambda \left(a_0 \cdot \partial_0 + b_0 \cdot \partial_1 + c_0 \cdot \partial_2 + d_0 \cdot \partial_3 \right) h_\lambda \\ q^\lambda \left(b_1 \cdot \partial_1 + c_1 \cdot \partial_2 + d_1 \cdot \partial_3 \right) h_\lambda \\ q^\lambda \left(c_2 \cdot \partial_2 + d_2 \cdot \partial_3 \right) h_\lambda \\ q^\lambda \left(d_3 \cdot \partial_3 \right) h_\lambda \end{array} \right\}$$

$$\left. \begin{cases} q^\lambda \left(a_0 \cdot \partial_0 + b_0 \cdot \partial_1 + c_0 \cdot \partial_2 + d_0 \cdot \partial_3 \right) h_\lambda \\ q^\lambda \left(b_1 \cdot \partial_1 + c_1 \cdot \partial_2 + d_1 \cdot \partial_3 \right) h_\lambda \\ q^\lambda \left(c_2 \cdot \partial_2 + d_2 \cdot \partial_3 \right) h_\lambda \\ q^\lambda \left(d_3 \cdot \partial_3 \right) h_\lambda \end{cases} \right\}. \tag{467}$$

Applying the linearization from the postulated Klein–Gordon equation, respectively the subsequent classical Dirac theory as also used in Eq. (434) in order to show the transition from the metric Dirac equation to its classical form, we obtain:

$$\frac{2 \cdot F[f]^3 \cdot R^*}{3 \cdot \left(\left(\frac{\partial F[f]}{\partial f} \right)^2 - 2 \cdot F[f] \cdot \frac{\partial^2 F[f]}{\partial f^2} \right)} \simeq h_\lambda q^\lambda \mathbf{M} \cdot \mathbf{M} h_\lambda q^\lambda = h_{\lambda,\alpha} q^\lambda g^{\alpha\beta} h_{\lambda,\beta} q^\lambda$$

$$\xrightarrow{g^{\alpha\beta} \to V_\Omega \cdot V_\Omega}$$

$$= \left. \begin{cases} q^\lambda \left(a_0 \cdot \partial_0 + b_0 \cdot \partial_1 + c_0 \cdot \partial_2 + d_0 \cdot \partial_3 \right) h_\lambda \\ q^\lambda \left(b_1 \cdot \partial_1 + c_1 \cdot \partial_2 + d_1 \cdot \partial_3 \right) h_\lambda \\ q^\lambda \left(c_2 \cdot \partial_2 + d_2 \cdot \partial_3 \right) h_\lambda \\ q^\lambda \left(d_3 \cdot \partial_3 \right) h_\lambda \end{cases} \right\} \cdot \left\{ \begin{cases} q^\lambda \left(a_0 \cdot \partial_0 + b_0 \cdot \partial_1 + c_0 \cdot \partial_2 + d_0 \cdot \partial_3 \right) h_\lambda \\ q^\lambda \left(b_1 \cdot \partial_1 + c_1 \cdot \partial_2 + d_1 \cdot \partial_3 \right) h_\lambda \\ q^\lambda \left(c_2 \cdot \partial_2 + d_2 \cdot \partial_3 \right) h_\lambda \\ q^\lambda \left(d_3 \cdot \partial_3 \right) h_\lambda \end{cases} \right\}$$

$$\Rightarrow \qquad h_\lambda q^\lambda \mathbf{M} = \left\{ \begin{cases} q^\lambda \left(a_0 \cdot \partial_0 + b_0 \cdot \partial_1 + c_0 \cdot \partial_2 + d_0 \cdot \partial_3 \right) h_\lambda \\ q^\lambda \left(b_1 \cdot \partial_1 + c_1 \cdot \partial_2 + d_1 \cdot \partial_3 \right) h_\lambda \\ q^\lambda \left(c_2 \cdot \partial_2 + d_2 \cdot \partial_3 \right) h_\lambda \\ q^\lambda \left(d_3 \cdot \partial_3 \right) h_\lambda \end{cases} \right\}$$

$$\Rightarrow \qquad 0 = \left\{ \begin{cases} q^\lambda \left(a_0 \cdot \partial_0 + b_0 \cdot \partial_1 + c_0 \cdot \partial_2 + d_0 \cdot \partial_3 \right) h_\lambda \\ q^\lambda \left(b_1 \cdot \partial_1 + c_1 \cdot \partial_2 + d_1 \cdot \partial_3 \right) h_\lambda \\ q^\lambda \left(c_2 \cdot \partial_2 + d_2 \cdot \partial_3 \right) h_\lambda \\ q^\lambda \left(d_3 \cdot \partial_3 \right) h_\lambda \end{cases} \right\} \pm h_\lambda q^\lambda \mathbf{M} \tag{468}$$

and thus:

$$\Rightarrow \qquad 0 = \left\{ \begin{cases} \left(a_0 \cdot \partial_0 + b_0 \cdot \partial_1 + c_0 \cdot \partial_2 + d_0 \cdot \partial_3 \right) h_\lambda \\ \left(b_1 \cdot \partial_1 + c_1 \cdot \partial_2 + d_1 \cdot \partial_3 \right) h_\lambda \\ \left(c_2 \cdot \partial_2 + d_2 \cdot \partial_3 \right) h_\lambda \\ \left(d_3 \cdot \partial_3 \right) h_\lambda \end{cases} \right\} \pm h_\lambda \mathbf{M}$$

$$\Rightarrow \qquad 0 = \left[\left\{ \begin{pmatrix} a_0 \cdot \partial_0 + b_0 \cdot \partial_1 + c_0 \cdot \partial_2 + d_0 \cdot \partial_3 \\ b_1 \cdot \partial_1 + c_1 \cdot \partial_2 + d_1 \cdot \partial_3 \\ c_2 \cdot \partial_2 + d_2 \cdot \partial_3 \\ d_3 \cdot \partial_3 \end{pmatrix} \pm \begin{Bmatrix} M_0 \\ M_1 \\ M_2 \\ M_3 \end{Bmatrix} \right\} h_\lambda . \right. \qquad (469)$$

The latter could be seen as a curvature-linearized metric Dirac equation in vector form.

There are quite some differences to the classical Dirac case:

(a) The functions $\mathbf{h}_\lambda$ need all to be Laplace functions, while in the postulated Dirac equation, the resulting solutions and thus, the outcome function f also fulfills the Klein–Gordon equation (for more see the sub-sections above), which, as we know, gives a 4-dimensional Laplace equation in the case of vanishing masses. Thus, we always have to guarantee the condition $\Delta f = \Delta h_\lambda q^\lambda = 0$ in addition to the metric Dirac equation (Eq. (469)) (or its non-linearized equivalent (467)). An exception from this rule can be made in the case where $F[f]$ is chosen such that we have $\dfrac{\partial F}{\partial f} \approx 0$ around the f being considered.

(b) The masses term $\mathbf{M}$ is a vector and not just a scalar as in the classical Dirac case.

(c) We obtain extremely simple equations in the case of purely diagonal metrics, namely:

$$\Rightarrow \qquad 0 = \left[\left\{ \begin{pmatrix} a_0 \cdot \partial_0 \\ b_1 \cdot \partial_1 \\ c_2 \cdot \partial_2 \\ d_3 \cdot \partial_3 \end{pmatrix} \pm \begin{pmatrix} M_0 \\ M_1 \\ M_2 \\ M_3 \end{pmatrix} \right\} h_\lambda . \right. \qquad (470)$$

(d) With the time t-dependency of a great variety of metrics and subsequent Laplace solutions for these very metrics usually being found of the type $\exp[\pm c^* C_t^* x_0] = \exp[\pm c^* C_t^* t]$ (c.f. the many examples given above), we have no problem in recognizing the rest mass term $M_t^2 = M_0^2 = \dfrac{m^2 \cdot c^2}{\hbar^2}$ in the case of the "particle at rest," where we have:

$$\Rightarrow \qquad 0 = \left[\left\{ \begin{pmatrix} a_0 \cdot \partial_0 \\ b_1 \cdot \partial_1 \\ c_2 \cdot \partial_2 \\ d_3 \cdot \partial_3 \end{pmatrix} \pm \begin{pmatrix} M_0 \\ M_1 = 0 \\ M_2 = 0 \\ M_3 = 0 \end{pmatrix} \right\} h_\lambda . \right. \qquad (471)$$

Thus, we are also back with the Dirac matter and antimatter solutions.

(e) Still there are four functions $\mathbf{h}_{\lambda=0,1,2,3}$ with the classical matter $\exp[-c^*M_0{}^*t]$ and antimatter $\exp[c^*M_0{}^*t]$ solution, plus two more functions, which should code the two classical spin states. Thereby the origin of the latter could either be a residual from the linearization of the curvature term in (468) (first line) or, as demonstrated on many examples in the upper sections, the dimensional entanglement with other degrees of freedom and thus, dimensions of higher order. We also saw in connection with elastic approaches that the spin comes in quite naturally in the case of shear-approaches ([1, 4] and the subsection below).

(f) It should be pointed out, however, that in order to obtain non-vanishing solutions in Cartesian (Minkowski) coordinates—even for the particle at rest $f[\ \dots\ ] = f[t]$-, we either to have curvature of space-time, which is to say $R^* \neq 0$ or dimensional entanglement with higher order dimensions (see the sections above). Otherwise, all M_i would vanish and with it (as one can easily deduce from Eq. (471)) all $\mathbf{h}_\lambda$. Of course, one might just demand the derivatives of the scalar product $\mathbf{h}_\lambda q^\lambda$ or the scalar itself to become zero, but here we are not interested in such trivial zero-sum solutions.

This, meaning the fact that the metric Dirac equation requires curvature or entanglement in order to "have something to work with," is in severe contrast to the classical, postulated Dirac theory, which is completely developed on the basis of the flat Minkowski space-time and where no curvature is of need in order to "obtain something." This, however, is a result of the postulation, while, obviously, the metrically derived Dirac equation does not allow for "something to be" without this something also to produce curvature or entanglement, along the way creating inertia, spin, energy and so on (c.f. the sections above).

Of course, things are getting much more interesting when moving to certain symmetries like the spatial spherical one, but for here and now, we shall leave this to the interested reader or own later publications (whatever comes first).

So, apparently, there is a way to avoid Dirac's quaternions. It should be pointed out, however, that the introduction of the $\mathbf{V}_\Omega$-vector alone does not do the trick. Looking carefully at Eq. (468), we realize the combination of the $\mathbf{V}_\Omega$-vector with the vector of constants $\mathbf{q}^\lambda$ residing in there. We could easily assume the two forming one mathematical object like:

$$\left\{ \begin{matrix} q^\lambda \left(a_0 \cdot \partial_0 + b_0 \cdot \partial_1 + c_0 \cdot \partial_2 + d_0 \cdot \partial_3 \right) h_\lambda \\ q^\lambda \left(b_1 \cdot \partial_1 + c_1 \cdot \partial_2 + d_1 \cdot \partial_3 \right) h_\lambda \\ q^\lambda \left(c_2 \cdot \partial_2 + d_2 \cdot \partial_3 \right) h_\lambda \\ q^\lambda \left(d_3 \cdot \partial_3 \right) h_\lambda \end{matrix} \right\} = \left\{ \begin{matrix} q^\lambda \left(a_0 \cdot \partial_0 + b_0 \cdot \partial_1 + c_0 \cdot \partial_2 + d_0 \cdot \partial_3 \right) \\ q^\lambda \left(b_1 \cdot \partial_1 + c_1 \cdot \partial_2 + d_1 \cdot \partial_3 \right) \\ q^\lambda \left(c_2 \cdot \partial_2 + d_2 \cdot \partial_3 \right) \\ q^\lambda \left(d_3 \cdot \partial_3 \right) \end{matrix} \right\} h_\lambda$$

$$= q^\lambda \left\{ \begin{matrix} \left(a_0 \cdot \partial_0 + b_0 \cdot \partial_1 + c_0 \cdot \partial_2 + d_0 \cdot \partial_3 \right) \\ \left(b_1 \cdot \partial_1 + c_1 \cdot \partial_2 + d_1 \cdot \partial_3 \right) \\ \left(c_2 \cdot \partial_2 + d_2 \cdot \partial_3 \right) \\ \left(d_3 \cdot \partial_3 \right) \end{matrix} \right\} h_\lambda$$

$$= q^\lambda \left\{ \begin{matrix} a_0 \cdot \partial_0 \\ 0 \\ 0 \\ 0 \end{matrix} \right\} h_\lambda + q^\lambda \left\{ \begin{matrix} b_0 \cdot \partial_1 \\ b_1 \cdot \partial_1 \\ 0 \\ 0 \end{matrix} \right\} h_\lambda + q^\lambda \left\{ \begin{matrix} c_0 \cdot \partial_2 \\ b_1 \cdot \partial_2 \\ c_2 \cdot \partial_2 \\ 0 \end{matrix} \right\} h_\lambda + q^\lambda \left\{ \begin{matrix} d_0 \cdot \partial_3 \\ d_1 \cdot \partial_3 \\ d_2 \cdot \partial_3 \\ d_3 \cdot \partial_3 \end{matrix} \right\} h_\lambda \equiv \mathbf{Q}^{\lambda\alpha} h_{\lambda,\alpha},$$

$$\tag{472}$$

where we are indeed back with some quaternion-like objects. This time, however, we also have a way to decompose these objects and separate the vector of constants $\mathbf{q}^\lambda$ from it. Some special care needs to be taken about curvature and mass terms resulting from entanglement (c.f. [1, 4] and sections above), but even though some of this will be dealt with in chapters 9 and 10 of this book, we shall leave this for later or the interested and mathematically skilled readers.

6.12.2.1 Treating classical Dirac approach with the new method

In principle, we should also be able to apply the $\mathbf{V_\Omega}$-vector method onto the classical Klein–Gordon equation, thereby following the Dirac path only that we—this time—avoid the quaternions. At first, we bring the Minkowski Laplace operator in the $\mathbf{V_\Omega}$-vector form:

$$\Delta f = \frac{\left(\sqrt{g} \cdot g^{\alpha\beta} f_{,\beta} \right)_{,\alpha}}{\sqrt{g}} \xrightarrow{g^{\alpha\beta} \to \text{Minkowski}} = \left(g^{\alpha\beta} f_{,\beta} \right)_{,\alpha}$$

$$= \left\{ \begin{matrix} a_0 \cdot \partial_0 + b_0 \cdot \partial_1 + c_0 \cdot \partial_2 + d_0 \cdot \partial_3 \\ b_1 \cdot \partial_1 + c_1 \cdot \partial_2 + d_1 \cdot \partial_3 \\ c_2 \cdot \partial_2 + d_2 \cdot \partial_3 \\ d_3 \cdot \partial_3 \end{matrix} \right\} \cdot \left\{ \begin{matrix} a_0 \cdot \partial_0 + b_0 \cdot \partial_1 + c_0 \cdot \partial_2 + d_0 \cdot \partial_3 \\ b_1 \cdot \partial_1 + c_1 \cdot \partial_2 + d_1 \cdot \partial_3 \\ c_2 \cdot \partial_2 + d_2 \cdot \partial_3 \\ d_3 \cdot \partial_3 \end{matrix} \right\} f.$$

$$\tag{473}$$

Now, we add the mass-vector to our decomposed Klein–Gordon equation:

$$\Rightarrow \qquad 0 = \left\{ \begin{matrix} a_0 \cdot \partial_0 + b_0 \cdot \partial_1 + c_0 \cdot \partial_2 + d_0 \cdot \partial_3 \\ b_1 \cdot \partial_1 + c_1 \cdot \partial_2 + d_1 \cdot \partial_3 \\ c_2 \cdot \partial_2 + d_2 \cdot \partial_3 \\ d_3 \cdot \partial_3 \end{matrix} \right\} f \pm f \cdot \mathbf{M} \qquad (474)$$

and, remembering that we can give the scalar function f any arbitrary intrinsic structure, we obtain the $\mathbf{V}_\Omega$-vector equivalent of the Dirac equation:

$$0 = \left[\left\{ \begin{matrix} a_0 \cdot \partial_0 + b_0 \cdot \partial_1 + c_0 \cdot \partial_2 + d_0 \cdot \partial_3 \\ b_1 \cdot \partial_1 + c_1 \cdot \partial_2 + d_1 \cdot \partial_3 \\ c_2 \cdot \partial_2 + d_2 \cdot \partial_3 \\ d_3 \cdot \partial_3 \end{matrix} \right\} \pm \begin{pmatrix} M \\ M \\ M \\ M \end{pmatrix} \right] q^\lambda h_\lambda$$

$$\Rightarrow \qquad 0 = \left[q^\lambda \left\{ \begin{matrix} a_0 \cdot \partial_0 + b_0 \cdot \partial_1 + c_0 \cdot \partial_2 + d_0 \cdot \partial_3 \\ b_1 \cdot \partial_1 + c_1 \cdot \partial_2 + d_1 \cdot \partial_3 \\ c_2 \cdot \partial_2 + d_2 \cdot \partial_3 \\ d_3 \cdot \partial_3 \end{matrix} \right\} \pm q^\lambda \begin{pmatrix} M \\ M \\ M \\ M \end{pmatrix} \right] h_\lambda. \qquad (475)$$

6.12.3 Repetition: Elastic Space-Times

Now, we apply the following approach for the scalar f:

$$f = h^\alpha{}_{,\beta} q^\beta_\alpha \qquad (476)$$

and again demand f to be a Laplace function. This leads us—via Eq. (245)—to the following equation:

$$\frac{F[f]^3 \cdot R^*}{C_{N1} \cdot \left(\frac{\partial F[f]}{\partial f} \right)^2 - C_{N2} \cdot F[f] \cdot \frac{\partial^2 F[f]}{\partial f^2}} \xrightarrow{f = h^\alpha{}_{,\beta} q^\beta_\alpha} = \left(h^\lambda{}_{,\kappa} q^\kappa_\lambda \right)_{,\alpha} g^{\alpha\beta} \left(h^\lambda{}_{,\kappa} q^\kappa_\lambda \right)_{,\beta}$$

$$\Rightarrow \qquad \frac{F[f]^3 \cdot R^*}{C_{N1} \cdot \left(\frac{\partial F[f]}{\partial f} \right)^2 - C_{N2} \cdot F[f] \cdot \frac{\partial^2 F[f]}{\partial f^2}}$$

$$= \left(h^\lambda{}_{,\kappa} q^\kappa_{\lambda,\alpha} + h_{\lambda,\kappa,\alpha} q^{\lambda\kappa} \right) g^{\alpha\beta} \left(h^\lambda{}_{,\kappa} q^\kappa_{\lambda,\beta} + h_{\lambda,\kappa,\beta} q^{\lambda\kappa} \right)$$

$$= h^\lambda{}_{,\kappa} q^\kappa_{\lambda,\alpha} g^{\alpha\beta} h^\lambda{}_{,\kappa} q^\kappa_{\lambda,\beta} + h^\lambda{}_{,\kappa} q^\kappa_{\lambda,\alpha} q^{\lambda\kappa} g^{\alpha\beta} h^\lambda{}_{,\kappa,\beta} q^\kappa_\lambda$$

$$+ h^\lambda{}_{,\kappa,\alpha} q^\kappa_\lambda g^{\alpha\beta} h^\lambda{}_{,\kappa} q^\kappa_{\lambda,\beta} + h^\lambda{}_{,\kappa,\alpha} q^\kappa_\lambda g^{\alpha\beta} h^\lambda{}_{,\kappa,\beta} q^\kappa_\lambda. \qquad (477)$$

Please note that by plugging in the dimension-dependent constants C_{N1} and C_{N2} from [2] (see Eq. (503) further below) and applying the suitable assumption for the behavior of $F[f]$ satisfying:

$$C_{N1} \cdot \left(\frac{\partial F[f]}{\partial f}\right)^2 - C_{N2} \cdot F[f] \cdot \frac{\partial^2 F[f]}{\partial f^2} = F[f]^2, \tag{478}$$

we may also write (477) as follows (no approximation!):

$$\frac{F[f]^3 \cdot R^*}{C_{N1} \cdot \left(\dfrac{\partial F[f]}{\partial f}\right)^2 - C_{N2} \cdot F[f] \cdot \dfrac{\partial^2 F[f]}{\partial f^2}} = F[f] \cdot R^*$$

$$= C_{f1} \cdot \cosh\left[\frac{1}{2} \cdot \sqrt{\frac{1}{n-1} - 1} \cdot (f - C_{f2})\right]^{\frac{4}{n-2}} \cdot R^*$$

$$\xrightarrow{f = h^\alpha{}_{,\beta} q^\beta_\alpha} = \left(h^\lambda{}_{,\kappa} q^\kappa_\lambda\right)_{,\alpha} g^{\alpha\beta} \left(h^\lambda{}_{,\kappa} q^\kappa_\lambda\right)_{,\beta}$$

$$\Rightarrow \qquad F[f] \cdot R^* = C_{f1} \cdot \cosh\left[\frac{1}{2} \cdot \sqrt{\frac{1}{n-1} - 1} \cdot (f - C_{f2})\right]^{\frac{4}{n-2}} \cdot R^*$$

$$= \left(h^\lambda{}_{,\kappa} q^\kappa_{\lambda,\alpha} + h_{\lambda,\kappa,\alpha} q^{\lambda\kappa}\right) g^{\alpha\beta} \left(h^\lambda{}_{,\kappa} q^\kappa_{\lambda,\beta} + h_{\lambda,\kappa,\beta} q^{\lambda\kappa}\right)$$

$$= h^\lambda{}_{,\kappa} q^\kappa_{\lambda,\alpha} g^{\alpha\beta} h^\lambda{}_{,\kappa} q^\kappa_{\lambda,\beta} + h^\lambda{}_{,\kappa} q^\kappa_{\lambda,\alpha} q^{\lambda\kappa} g^{\alpha\beta} h^\lambda{}_{,\kappa,\beta} q^\kappa_\lambda$$

$$+ h^\lambda{}_{,\kappa,\alpha} q^\kappa_\lambda g^{\alpha\beta} h^\lambda{}_{,\kappa} q^\kappa_{\lambda,\beta} + h^\lambda{}_{,\kappa,\alpha} q^\kappa_\lambda g^{\alpha\beta} h^\lambda{}_{,\kappa,\beta} q^\kappa_\lambda. \tag{479}$$

Assuming that the object q^κ_λ just consists of constants (c.f. the sub-section above), we can dramatically simplify to:

$$\frac{F[f]^3 \cdot R^*}{C_{N1} \cdot \left(\dfrac{\partial F[f]}{\partial f}\right)^2 - C_{N2} \cdot F[f] \cdot \dfrac{\partial^2 F[f]}{\partial f^2}} = h^\lambda{}_{,\kappa\alpha} q^\kappa_\lambda g^{\alpha\beta} h^\lambda{}_{,\kappa\beta} q^\kappa_\lambda$$

$$\xrightarrow{g^{\alpha\beta} = e^\alpha \cdot e^\beta} = h^\lambda{}_{,\kappa\alpha} q^\kappa_\lambda e^\alpha \cdot e^\beta h^\lambda{}_{,\kappa\beta} q^\kappa_\lambda$$

$$\xrightarrow{q^{\lambda\kappa} e^\alpha = Q^{\kappa\alpha}} = h^\lambda{}_{,\kappa\alpha} Q^{\kappa\alpha}_\lambda \cdot Q^{\kappa\beta}_\lambda h^\lambda{}_{,\kappa\beta}. \tag{480}$$

We recognize the terms $h^\lambda{}_{,\kappa\alpha} q^\kappa_\lambda e^\alpha$, $h^\lambda{}_{,\kappa\alpha} Q^{\kappa\alpha}_\lambda$ as being also parts of the fundamental equations of elasticity [30] and remember that solutions to the case $R^* = 0$ can be found by the means of gradients and shear- or spin-like approaches [4] as follows:

$$h^j = G^j\left[x_k\right] = g^{jl}\partial_l G\left[x_k\right]; \quad \Delta G\left[x_k\right] = 0$$

$$G^j\left[x_k\right] = x_\xi \cdot g^{jl}\partial_l G\left[x_k\right] + \theta \cdot G\left[x_k\right]; \quad \xi \text{ any of } 0,1,\dots,n-1, \quad (481)$$

$$G^j\left[x_k\right] = \left\{\dots,\overbrace{\frac{\partial G\left[x_k\right]}{\partial x_\zeta}}^{\text{pos }\xi},\dots,\overbrace{-\frac{\partial G\left[x_k\right]}{\partial x_\xi}}^{\text{pos }\zeta},\dots\right\}; \quad \forall(\dots,\dots) = 0$$

$$G^j\left[x_k\right] = \left\{\dots,\overbrace{-\frac{\partial G\left[x_k\right]}{\partial x_\zeta}}^{\text{pos }\xi},\dots,\overbrace{\frac{\partial G\left[x_k\right]}{\partial x_\xi}}^{\text{pos }\zeta},\dots\right\}; \quad \forall(\dots,\dots) = 0. \quad (482)$$

Any combination of the basic solutions (481) and (482) is also a solution. We see that we can construct non-trivial solutions to Eq. (480) in the case of $R^* = 0$, which are automatically also solutions to the completely flat space case:

$$\frac{F[f]^3 \cdot R^*}{C_{N1} \cdot \left(\frac{\partial F[f]}{\partial f}\right)^2 - C_{N2} \cdot F[f] \cdot \frac{\partial^2 F[f]}{\partial f^2}} = 0 = h^\lambda_{,\kappa\alpha}q^\kappa_\lambda g^{\alpha\beta}h^\lambda_{,\kappa\beta}q^\kappa_\lambda$$

$$\xrightarrow{g^{\alpha\beta} = \mathbf{e}^\alpha \cdot \mathbf{e}^\beta} 0 = h^\lambda_{,\kappa\alpha}q^\kappa_\lambda \mathbf{e}^\alpha \cdot \mathbf{e}^\beta h^\lambda_{,\kappa\beta}q^\kappa_\lambda$$

$$\xrightarrow{q^{\lambda\kappa}\mathbf{e}^\alpha = \mathbf{Q}^{\lambda\kappa\alpha}} 0 = h^\lambda_{,\kappa\alpha}\mathbf{Q}^{\kappa\alpha}_\lambda \cdot \mathbf{Q}^{\kappa\beta}_\lambda h^\lambda_{,\kappa\beta}. \quad (483)$$

Interestingly and in accordance with the classical Quantum Theory, we have to construct these solutions out of combinations of harmonic functions and their derivatives. Thereby, in order to satisfy Eq. (483) in its currently given form, we have to set $\theta = -1$ (c.f. [31] and Neuber's three-function approach, which here was extended to an n-function approach).

We may consider the second type of solution (482) spin-like. More such forms can be found easily via the following recipe:

$$G^j\left[x_k\right] = \left\{\dots,A_1\overbrace{\frac{\partial^2 G}{\partial\xi_2\partial\xi_3}}^{\text{pos }\xi_1},\dots,A_2\overbrace{\frac{\partial^2 G}{\partial\xi_1\partial\xi_3}}^{\text{pos }\xi_2},\dots,A_3\overbrace{\frac{\partial^2 G}{\partial\xi_1\partial\xi_2}}^{\text{pos }\xi_3},\dots\right\}$$

$$G^j\left[x_k\right] = \left\{\dots,A_1\overbrace{\frac{\partial^3 G}{\partial\xi_2\partial\xi_3\partial\xi_4}}^{\text{pos }\xi_1},\dots,A_2\overbrace{\frac{\partial^3 G}{\partial\xi_1\partial\xi_3\partial\xi_4}}^{\text{pos }\xi_2},\dots,A_3\overbrace{\frac{\partial^3 G}{\partial\xi_1\partial\xi_2\partial\xi_4}}^{\text{pos }\xi_3},\dots,A_4\overbrace{\frac{\partial^3 G}{\partial\xi_1\partial\xi_2\partial\xi_3}}^{\text{pos }\xi_4}\dots\right\}$$

$$\dots$$

$$\forall(\ldots,\ldots)=0; \quad \sum_{\forall k} A_k = 0. \tag{484}$$

In a Minkowski space-time with $c = 1$ and the coordinates t, x, y, and z, we could have the following of such solutions:

$$G^j\left[x_k\right]=\pm\left\{\frac{\partial G}{\partial z},0,0,-\frac{\partial G}{\partial t}\right\}; \quad G^j\left[x_k\right]=\pm\left\{\frac{\partial G}{\partial y},0,-\frac{\partial G}{\partial t},0\right\};$$

$$G^j\left[x_k\right]=\pm\left\{\frac{\partial G}{\partial x},-\frac{\partial G}{\partial t},0,0\right\}$$

$$G^j\left[x_k\right]=\pm\left\{0,\frac{\partial G}{\partial z},0,-\frac{\partial G}{\partial x}\right\}; \quad G^j\left[x_k\right]=\pm\left\{0,\frac{\partial G}{\partial y},-\frac{\partial G}{\partial x},0\right\};$$

$$G^j\left[x_k\right]=\pm\left\{0,0,\frac{\partial G}{\partial z},-\frac{\partial G}{\partial y}\right\}, \tag{485}$$

$$G^j\left[x_k\right]=\left\{A_1\frac{\partial^2 G}{\partial x\partial y},A_2\frac{\partial^2 G}{\partial t\partial y},A_3\frac{\partial^2 G}{\partial x\partial t},0\right\}$$

$$G^j\left[x_k\right]=\left\{A_1\frac{\partial^2 G}{\partial x\partial z},A_2\frac{\partial^2 G}{\partial t\partial z},0,A_3\frac{\partial^2 G}{\partial x\partial t}\right\}$$

$$G^j\left[x_k\right]=\left\{A_1\frac{\partial^2 G}{\partial z\partial y},0,A_2\frac{\partial^2 G}{\partial t\partial z},A_3\frac{\partial^2 G}{\partial z\partial t}\right\}$$

$$G^j\left[x_k\right]=\left\{0,A_1\frac{\partial^2 G}{\partial z\partial y},A_2\frac{\partial^2 G}{\partial x\partial z},A_3\frac{\partial^2 G}{\partial x\partial y}\right\}. \tag{486}$$

$$G^j\left[x_k\right]=\left\{A_1\frac{\partial^3 G}{\partial x\partial y\partial z},A_1\frac{\partial^3 G}{\partial t\partial y\partial z},A_1\frac{\partial^3 G}{\partial x\partial t\partial z},A_4\frac{\partial^3 G}{\partial x\partial y\partial t}\right\}; \quad \sum_{\forall k} A_k = 0. \tag{487}$$

Thereby we find that the last solution in Eq. (486) with a setting of the kind $A_1 = 1/3$, $A_2 = 1/3$, and $A_3 = -2/3$ shows a peculiar closeness to the charges of quarks and the building blocks of hadrons and baryons.

It should be noted that an approach of the form:

$$f = h^\lambda_{\ ,\kappa}q^\kappa_\lambda + \theta\cdot h^\chi_{\ ,\beta}g^{\beta\alpha}p_{\alpha\chi}, \tag{488}$$

with the object $p_{\alpha\chi}$ also only consisting of constants, automatically leads to the fundamental linear elastic equation in the isotropic case with the Poisson's ratio $\sigma \neq 0$ [30, 31] and $\theta = -(1 - 2\cdot\sigma)$. Setting Eq. (488) into Eq. (245), thereby still assuming that f is a Laplace function, gives us:

$$\frac{F[f]^3 \cdot R^*}{C_{N1} \cdot \left(\dfrac{\partial F[f]}{\partial f}\right)^2 - C_{N2} \cdot F[f] \cdot \dfrac{\partial^2 F[f]}{\partial f^2}}$$

$$= \left(h^\lambda{}_{,\kappa\alpha} q^\kappa_\lambda + \theta \cdot \left(h^\lambda{}_{,\rho} g^{\rho\chi}\right)_{,\chi} \delta^\chi_\alpha p_{\lambda\chi}\right) g^{\alpha\beta} \left(h^\lambda{}_{,\kappa\beta} q^\kappa_\lambda + \theta \cdot \left(h^\lambda{}_{,\rho} g^{\rho\chi}\right)_{,\beta} \delta^\chi_\alpha p_{\lambda\chi}\right)$$

$$= \left(h^\lambda{}_{,\kappa\alpha} q^\kappa_\lambda + \theta \cdot p_{\lambda\alpha} \Delta h^\lambda\right) g^{\alpha\beta} \left(h^\lambda{}_{,\kappa\beta} q^\kappa_\lambda + \theta \cdot p_{\lambda\beta} \Delta h^\lambda\right)$$

$$\xrightarrow{g^{\alpha\beta} = \mathbf{e}^\alpha \cdot \mathbf{e}^\beta} = \left(h^\lambda{}_{,\kappa\alpha} q^\kappa_\lambda + \theta \cdot p_{\lambda\alpha} \Delta h^\lambda\right) \mathbf{e}^\alpha \cdot \mathbf{e}^\beta \left(h^\lambda{}_{,\kappa\beta} q^\kappa_\lambda + \theta \cdot p_{\lambda\beta} \Delta h^\lambda\right). \quad (489)$$

Any of the two factors made to zero would immediately fulfill the whole equation in the flat space case and it was show in [32–35] that interesting elementary particle solutions can be derived this way.

6.12.4 Back to Main Section

For historical reasons, we here concentrate on the Dirac approach (after all the elastic approach is quite new in particle physics). As the scalar wrapper function $F[f]$ is not fixed yet, we try to find it in such a way that the metric Dirac equation (Eq. (446)) becomes solvable.

Unfortunately, we see no way in obtaining something of the kind:

$$\frac{F[f]^3 \cdot R^*}{C_{N1} \cdot \left(\dfrac{\partial F[f]}{\partial f}\right)^2 - C_{N2} \cdot F[f] \cdot \dfrac{\partial^2 F[f]}{\partial f^2}} = h_{\lambda,\alpha} q^\lambda \mathbf{e}^\alpha \cdot \mathbf{e}^\beta h_{\lambda,\beta} q^\lambda$$

$$\Rightarrow \qquad f^2 \cdot R^* = h_{\lambda,\alpha} q^\lambda \mathbf{e}^\alpha \cdot \mathbf{e}^\beta h_{\lambda,\beta} q^\lambda. \quad (490)$$

So instead we choose to seek for a solution to:

$$C_{N1} \cdot \left(\frac{\partial F[f]}{\partial f}\right)^2 - C_{N2} \cdot F[f] \cdot \frac{\partial^2 F[f]}{\partial f^2} = F[f]^2. \quad (491)$$

We obtain the following for $F[f]$:

$$F[f] = C_{f1} \cdot \cos\left[\frac{1}{C_{N2}} \cdot \sqrt{C_{N2} - C_{N1}} \cdot \left(f - C_{f2}\right)\right]^{-\frac{C_{N2}}{C_{N1} - C_{N2}}}$$

$$= C_{f1} \cdot \cosh\left[\frac{1}{C_{N2}} \cdot \sqrt{C_{N1} - C_{N2}} \cdot \left(f - C_{f2}\right)\right]^{-\frac{C_{N2}}{C_{N1} - C_{N2}}} \quad (492)$$

and subsequently for Eq. (489):

$$F[f] \cdot R^* = C_{f1} \cdot \cosh\left[\frac{1}{C_{N2}} \cdot \sqrt{C_{N1} - C_{N2}} \cdot \left(f - C_{f2}\right)\right]^{-\frac{C_{N2}}{C_{N1} - C_{N2}}} \cdot R^*$$

$$= h_{\lambda,\alpha} q^{\lambda} e^{\alpha} \cdot e^{\beta} h_{\lambda,\beta} q^{\lambda}. \tag{493}$$

Please note, a more general choice for Eq. (491) would be:

$$C_{N1} \cdot \left(\frac{\partial F[f]}{\partial f}\right)^2 - C_{N2} \cdot F[f] \cdot \frac{\partial^2 F[f]}{\partial f^2} = C_F^2 \cdot F[f]^2, \tag{494}$$

which gives us:

$$F[f] = C_{f1} \cdot \cos\left[\frac{C_F}{C_{N2}} \cdot \sqrt{C_{N2} - C_{N1}} \cdot \left(f - C_{f2}\right)\right]^{-\frac{C_{N2}}{C_{N1} - C_{N2}}}$$

$$= C_{f1} \cdot \cosh\left[\frac{C_F}{C_{N2}} \cdot \sqrt{C_{N1} - C_{N2}} \cdot \left(f - C_{f2}\right)\right]^{-\frac{C_{N2}}{C_{N1} - C_{N2}}} \tag{495}$$

and subsequently for Eq. (489):

$$F[f] \cdot R^* = C_{f1} \cdot \cosh\left[\frac{C_F}{C_{N2}} \cdot \sqrt{C_{N1} - C_{N2}} \cdot \left(f - C_{f2}\right)\right]^{-\frac{C_{N2}}{C_{N1} - C_{N2}}} \cdot R^*$$

$$= h_{\lambda,\alpha} q^{\lambda} e^{\alpha} \cdot e^{\beta} h_{\lambda,\beta} q^{\lambda}. \tag{496}$$

By plugging in the dimension-dependent constants C_{N1} and C_{N2} from [2] (see Eq. (503)), we may also write:

$$F[f] \cdot R^* = C_{f1} \cdot \cosh\left[\frac{1}{2} \cdot \sqrt{\frac{1}{n-1} - 1} \cdot \left(f - C_{f2}\right)\right]^{\frac{4}{n-2}} \cdot R^* = h_{\lambda,\alpha} q^{\lambda} e^{\alpha} \cdot e^{\beta} h_{\lambda,\beta} q^{\lambda}, \tag{497}$$

respectively, on the more general setting (494):

$$F[f] \cdot R^* = C_{f1} \cdot \cosh\left[\frac{C_F}{2} \cdot \sqrt{\frac{1}{n-1} - 1} \cdot \left(f - C_{f2}\right)\right]^{\frac{4}{n-2}} \cdot R^* = h_{\lambda,\alpha} q^{\lambda} e^{\alpha} \cdot e^{\beta} h_{\lambda,\beta} q^{\lambda}. \tag{498}$$

Series expansion of $F[f]$ to the second order for small f plus setting $C_{f2} = 0$ in Eq. (492) and plugging this into Eq. (489) gives us:

$$F[f] \cdot R^* = C_{f1} \cdot \left(1 - \frac{f^2}{2 \cdot C_{N2}} + \ldots\right) \cdot R^* = h_{\lambda,\alpha} q^{\lambda} e^{\alpha} \cdot e^{\beta} h_{\lambda,\beta} q^{\lambda}$$

$$\Rightarrow \quad F[f] \cdot R^* \simeq C_{f1} \cdot \left(1 - \frac{f}{\sqrt{2 \cdot C_{N2}}}\right) \cdot \left(1 + \frac{f}{\sqrt{2 \cdot C_{N2}}}\right) \cdot R^* = h_{\lambda,\alpha} q^{\lambda} e^{\alpha} \cdot e^{\beta} h_{\lambda,\beta} q^{\lambda}.$$

(499)

The corresponding result for the elastic approach would look as follows:

$$F[f] \cdot R^* = C_{f1} \cdot \left(1 - \frac{f^2}{2 \cdot C_{N2}} + \ldots\right) \cdot R^*$$

$$= h^{\lambda}{}_{,\kappa\alpha} q^{\kappa}_{\lambda} g^{\alpha\beta} h^{\lambda}{}_{,\kappa\beta} q^{\kappa}_{\lambda} \xrightarrow{g^{\alpha\beta} = e^{\alpha} \cdot e^{\beta}} = h^{\lambda}{}_{,\kappa\alpha} q^{\kappa}_{\lambda} e^{\alpha} \cdot e^{\beta} h^{\lambda}{}_{,\kappa\beta} q^{\kappa}_{\lambda}$$

$$\Rightarrow \quad F[f] \cdot R^* \simeq C_{f1} \cdot \left(1 - \frac{f}{\sqrt{2 \cdot C_{N2}}}\right) \cdot \left(1 + \frac{f}{\sqrt{2 \cdot C_{N2}}}\right) \cdot R^* = h^{\lambda}{}_{,\kappa\alpha} q^{\kappa}_{\lambda} e^{\alpha} \cdot e^{\beta} h^{\lambda}{}_{,\kappa\beta} q^{\kappa}_{\lambda}.$$

(500)

Now, we could assume the Ricci scalar to be itself dominated by the scaling function $F[f]$, respectively f and choose something like:

$$R^* = f \cdot R_{\alpha}^{**} e^{\alpha} \cdot e^{\beta} R_{\beta}^{**} \cdot f$$

$$\Rightarrow \quad f \cdot R_{\alpha}^{**} e^{\alpha} \cdot e^{\beta} R_{\beta}^{**} \cdot f = C_{f1} \cdot \left(1 - \frac{f^2}{2 \cdot C_{N2}} + \ldots\right) \cdot f \cdot R_{\alpha}^{**} e^{\alpha} \cdot e^{\beta} R_{\beta}^{**} \cdot f$$

$$= h_{\lambda,\alpha} q^{\lambda} e^{\alpha} \cdot e^{\beta} h_{\lambda,\beta} q^{\lambda}$$

$$\simeq C_{f1} \cdot \left(1 - \frac{f}{\sqrt{2 \cdot C_{N2}}}\right) \cdot f \cdot R_{\alpha}^{**} e^{\alpha} \cdot e^{\beta} R_{\beta}^{**} \cdot f \cdot \left(1 + \frac{f}{\sqrt{2 \cdot C_{N2}}}\right) = h_{\lambda,\alpha} q^{\lambda} e^{\alpha} \cdot e^{\beta} h_{\lambda,\beta} q^{\lambda}$$

$$\Rightarrow \quad \sqrt{C_{f1}} \cdot \left(1 - \frac{f}{\sqrt{2 \cdot C_{N2}}}\right) \cdot f \cdot R_{\alpha}^{**} e^{\alpha} = h_{\lambda,\alpha} q^{\lambda} e^{\alpha}$$

$$\Rightarrow \quad \sqrt{C_{f1}} \cdot \left(1 - \frac{f}{\sqrt{2 \cdot C_{N2}}}\right) \cdot f \cdot R_{\alpha}^{**} = h_{\lambda,\alpha} q^{\lambda} \tag{501}$$

for the Dirac form, and consequently:

$$R^* = f \cdot R_{\alpha}^{**} e^{\alpha} \cdot e^{\beta} R_{\beta}^{**} \cdot f$$

$$\Rightarrow \quad f \cdot R_{\alpha}^{**} e^{\alpha} \cdot e^{\beta} R_{\beta}^{**} \cdot f = C_{f1} \cdot \left(1 - \frac{f^2}{2 \cdot C_{N2}} + \ldots\right) \cdot f \cdot R_{\alpha}^{**} e^{\alpha} \cdot e^{\beta} R_{\beta}^{**} \cdot f$$

$$= h^\lambda_{,\kappa\alpha} q^\kappa_\lambda \mathbf{e}^\alpha \cdot \mathbf{e}^\beta h^\lambda_{,\kappa\beta} q^\kappa_\lambda$$

$$\simeq C_{f1} \cdot \left(1 - \frac{f}{\sqrt{2 \cdot C_{N2}}}\right) \cdot f \cdot R^{**}_\alpha \mathbf{e}^\alpha \cdot \mathbf{e}^\beta R^{**}_\beta \cdot f \cdot \left(1 + \frac{f}{\sqrt{2 \cdot C_{N2}}}\right) = h^\lambda_{,\kappa\alpha} q^\kappa_\lambda \mathbf{e}^\alpha \cdot \mathbf{e}^\beta h^\lambda_{,\kappa\beta} q^\kappa_\lambda$$

$$\Rightarrow \qquad \sqrt{C_{f1}} \cdot \left(1 - \frac{f}{\sqrt{2 \cdot C_{N2}}}\right) \cdot f \cdot R^{**}_\alpha = h^\lambda_{,\kappa\alpha} q^\kappa_\lambda \tag{502}$$

for the elastic form.

Thereby it needs to be emphasized that our choice for the separation of R^* is only motivated by the original simple linear form of Dirac (see step (446) in our derivation of the metric origin of the original Dirac equation above) and our desire to obtain simple equations in the end. There are other options, of course, and our choice here may not be the cleverest one.

Inserting what we know about the constants C_{N1} and C_{N2} (see [2]), which is to say (for $n \geq 2$):

$$C_{N1} = -\frac{3}{2} + \frac{7}{4} \cdot n - \frac{1}{4} \cdot n^2$$

$$C_{N2} = n - 1, \tag{503}$$

we see that for either high numbers of dimensions n or small functions f, we could further simplify our equations above to:

$$\Rightarrow \sqrt{C_{f1}} \cdot f \cdot R^{**}_\alpha = \xrightarrow{f = h_\lambda q^\lambda} = \sqrt{C_{f1}} \cdot h_\lambda q^\lambda \cdot R^{**}_\alpha = h_{\lambda,\alpha} q^\lambda \Rightarrow \sqrt{C_{f1}} \cdot R^{**}_\alpha \cdot h_\lambda = h_{\lambda,\alpha}$$

$$\Rightarrow \qquad \boxed{\sqrt{C_{f1}} \cdot R^{**}_\alpha \cdot h_\lambda - h_{\lambda,\alpha} \equiv \left(R^{***}_\alpha - \partial_\alpha\right) h_\lambda = 0} \tag{504}$$

for the Dirac form, and:

$$\Rightarrow \sqrt{C_{f1}} \cdot f \cdot R^{**}_\alpha = \sqrt{C_{f1}} \cdot h^\lambda_{,\kappa} q^\kappa_\lambda \cdot R^{**}_\alpha = h^\lambda_{,\kappa\alpha} q^\kappa_\lambda \Rightarrow \sqrt{C_{f1}} \cdot R^{**}_\alpha \cdot h^\lambda_{,\kappa} = h^\lambda_{,\kappa\alpha}$$

$$\Rightarrow \qquad \boxed{\sqrt{C_{f1}} \cdot R^{**}_\alpha \cdot h^\lambda_{,\kappa} - h^\lambda_{,\kappa\alpha} \equiv \left(R^{***}_\alpha - \partial_\alpha\right) \partial_\kappa h^\lambda = 0} \tag{505}$$

for the elastic form.

So, we have obtained very simple metric Dirac or elastic space-time equations in arbitrary numbers of dimensions n with either small scaling functions f or big numbers of dimensions with then arbitrary f.

From section "6.8.5 The Observer and the Problem of Transition to Newton's Gravity Law," we learned that, in order to obtain the right limit for the Newton gravity, the Ricci scalar R^* has to fade with $1/r^3$ (for bigger r). Taking the Dirac form (504) and the singularity-free R^* approach from Eq. (368), this would lead to the condition:

$$\sqrt{C_{f1}} \cdot f \cdot R_{*\alpha}^{**} = \sqrt{C_{f1}} \cdot f \cdot \sqrt{C_R \cdot e^{-\left(\frac{r_s}{r}\right)^3} \cdot \left(\frac{r_s}{r}\right)^3} \cdot R_\alpha^{****} \equiv f \cdot \sqrt{e^{-\left(\frac{r_s}{r}\right)^3} \cdot \left(\frac{r_s}{r}\right)^3} \cdot C_\alpha^*$$

$$\Rightarrow \xrightarrow{f = h_\lambda q^\lambda} = \sqrt{e^{-\left(\frac{r_s}{r}\right)^3} \cdot \left(\frac{r_s}{r}\right)^3} \cdot C_\alpha^* \cdot h_\lambda q^\lambda = h_{\lambda,\alpha} q^\lambda \Rightarrow \sqrt{e^{-\left(\frac{r_s}{r}\right)^3} \cdot \left(\frac{r_s}{r}\right)^3} \cdot C_\alpha^* \cdot h_\lambda = h_{\lambda,\alpha}$$

$$\Rightarrow \boxed{\sqrt{e^{-\left(\frac{r_s}{r}\right)^3} \cdot \left(\frac{r_s}{r}\right)^3} \cdot C_\alpha^* \cdot h_\lambda - h_{\lambda,\alpha} \equiv \left(\sqrt{e^{-\left(\frac{r_s}{r}\right)^3} \cdot \left(\frac{r_s}{r}\right)^3} \cdot C_\alpha^* - \partial_\alpha\right) h_\lambda = 0}$$

$$(506)$$

Thereby we have the potentially problematic intermediate step with the usage of:

$$C_R \cdot e^{-\left(\frac{r_s}{r}\right)^3} \cdot \left(\frac{r_s}{r}\right)^3 = R^* = f \cdot R_\alpha^{**} e^\alpha \cdot e^\beta R_\beta^{**} \cdot f$$

$$= f \cdot \underbrace{\sqrt{e^{-\left(\frac{r_s}{r}\right)^3} \cdot \left(\frac{r_s}{r}\right)^3} \cdot C_\alpha^* e^\alpha}_{=\frac{R_{*\alpha}^{**}}{\sqrt{C_{f1}}}} \cdot e^\beta \underbrace{\sqrt{e^{-\left(\frac{r_s}{r}\right)^3} \cdot \left(\frac{r_s}{r}\right)^3} \cdot C_\beta^* \cdot f}_{=\frac{R_{*\beta}^{**}}{\sqrt{C_{f1}}}}, \qquad (507)$$

which we introduced in order to keep the eigencharacter of the metric Dirac equation.

To avoid any trouble arising from there, we move back to Eq. (501) and introduce the Newton-limit (368) for R^* as follows:

$$R^* = \underbrace{\sqrt{e^{-\left(\frac{r_s}{r}\right)^3} \cdot \left(\frac{r_s}{r}\right)^3} \cdot C_\alpha^* e^\alpha}_{=R_{*\alpha}^{**}} \cdot e^\beta \underbrace{\sqrt{e^{-\left(\frac{r_s}{r}\right)^3} \cdot \left(\frac{r_s}{r}\right)^3} \cdot C_\beta^*}_{=R_{*\beta}^{**}}$$

$$\Rightarrow \quad \left(1 - \frac{f^2}{2 \cdot C_{N2}} + \dots\right) \cdot R_{*\alpha}^{**} e^\alpha \cdot e^\beta R_{*\beta}^{**} = h_{\lambda,\alpha} q^\lambda e^\alpha \cdot e^\beta h_{\lambda,\beta} q^\lambda$$

$$\approx \left(1 - \frac{f}{\sqrt{2 \cdot C_{N2}}}\right) \cdot R_{*\alpha}^{**} e^\alpha \cdot e^\beta R_{*\beta}^{**} \cdot \left(1 + \frac{f}{\sqrt{2 \cdot C_{N2}}}\right) = h_{\lambda,\alpha} q^\lambda e^\alpha \cdot e^\beta h_{\lambda,\beta} q^\lambda$$

$$\Rightarrow \qquad \left(1 - \frac{f}{\sqrt{2 \cdot C_{N2}}}\right) \cdot \sqrt{e^{-\left(\frac{r_s}{r}\right)^3} \cdot \left(\frac{r_s}{r}\right)^3} \cdot C_\alpha^* = h_{\lambda,\alpha} q^\lambda$$

or

$$\Rightarrow \qquad \left(1 + \frac{f}{\sqrt{2 \cdot C_{N2}}}\right) \cdot \sqrt{e^{-\left(\frac{r_s}{r}\right)^3} \cdot \left(\frac{r_s}{r}\right)^3} \cdot C_\beta^* = h_{\lambda,\beta} q^\lambda. \qquad (508)$$

This time we choose the Ricci curvature R^* not to contain the function f and consequently also obtain no eigenequation.

The corresponding elastic results would take the form:

$$\Rightarrow \qquad \left(1 - \frac{f}{\sqrt{2 \cdot C_{N2}}}\right) \cdot \sqrt{e^{-\left(\frac{r_s}{r}\right)^3} \cdot \left(\frac{r_s}{r}\right)^3} \cdot C_\alpha^* = h^\lambda_{,\kappa\alpha} q^\kappa_\lambda$$

or

$$\Rightarrow \qquad \left(1 + \frac{f}{\sqrt{2 \cdot C_{N2}}}\right) \cdot \sqrt{e^{-\left(\frac{r_s}{r}\right)^3} \cdot \left(\frac{r_s}{r}\right)^3} \cdot C_\beta^* = h^\lambda_{,\kappa\beta} q^\kappa_\lambda. \qquad (509)$$

However, when adding a constant to the function f and reevaluating everything from scratch, we would end up with the following result:

$$\frac{f}{\sqrt{2 \cdot C_{N2}}} = \Phi \pm 1 \quad \Rightarrow \quad \Phi = \frac{f}{\sqrt{2 \cdot C_{N2}}} \mp 1 = \frac{h_\lambda q^\lambda}{\sqrt{2 \cdot C_{N2}}} \mp 1$$

$$R^* = \underbrace{\sqrt{e^{-\left(\frac{r_s}{r}\right)^3} \cdot \left(\frac{r_s}{r}\right)^3} \cdot C_\alpha^*} e^\alpha \cdot e^\beta \underbrace{\sqrt{e^{-\left(\frac{r_s}{r}\right)^3} \cdot \left(\frac{r_s}{r}\right)^3} \cdot C_\beta^*}$$

$$\underbrace{}_{=R_{*\alpha}^{**}} \qquad \underbrace{}_{=R_{*\beta}^{**}}$$

$$\Rightarrow \qquad \left(1 - \frac{f^2}{2 \cdot C_{N2}} + \dots\right) \cdot R_{*\alpha}^{**} e^\alpha \cdot e^\beta R_{*\beta}^{**} = h_{\lambda,\alpha} q^\lambda e^\alpha \cdot e^\beta h_{\lambda,\beta} q^\lambda$$

$$\approx \Phi \cdot R_{*\alpha}^{**} e^\alpha \cdot e^\beta R_{*\beta}^{**} \cdot \left(1 + \frac{f}{\sqrt{2 \cdot C_{N2}}}\right) = h_{\lambda,\alpha} q^\lambda e^\alpha \cdot e^\beta h_{\lambda,\beta} q^\lambda \qquad (510)$$

or

$$\approx \left(1 - \frac{f}{\sqrt{2 \cdot C_{N2}}}\right) \cdot R_{*\alpha}^{**} e^\alpha \cdot e^\beta R_{*\beta}^{**} \cdot \Phi = h_{\lambda,\alpha} q^\lambda e^\alpha \cdot e^\beta h_{\lambda,\beta} q^\lambda,$$

$$\Phi \cdot \sqrt{e^{-\left(\frac{r_s}{r}\right)^3} \cdot \left(\frac{r_s}{r}\right)^3} \cdot C_\alpha^* = h_\lambda q^\lambda \cdot \sqrt{e^{-\left(\frac{r_s}{r}\right)^3} \cdot \left(\frac{r_s}{r}\right)^3} \cdot C_\alpha^* = h_{\lambda,\alpha} q^\lambda$$

$$\Rightarrow \qquad h_\lambda \cdot \sqrt{e^{-\left(\frac{r_s}{r}\right)^3} \cdot \left(\frac{r_s}{r}\right)^3} \cdot C_\alpha^* = h_{\lambda,\alpha}$$

or

$$\Phi \cdot \sqrt{e^{-\left(\frac{r_s}{r}\right)^3} \cdot \left(\frac{r_s}{r}\right)^3} \cdot C_\beta^* = h_\lambda q^\lambda \cdot \sqrt{e^{-\left(\frac{r_s}{r}\right)^3} \cdot \left(\frac{r_s}{r}\right)^3} \cdot C_\beta^* = h_{\lambda,\beta} q^\lambda$$

$$\Rightarrow \qquad h_\lambda \cdot \sqrt{e^{-\left(\frac{r_s}{r}\right)^3} \cdot \left(\frac{r_s}{r}\right)^3} \cdot C_\beta^* = h_{\lambda,\beta}. \qquad (511)$$

For completeness, we also give the elastic form, which would read:

$$\Phi \cdot \sqrt{e^{-\left(\frac{r_s}{r}\right)^3} \cdot \left(\frac{r_s}{r}\right)^3} \cdot C_\alpha^* = h^\lambda_{,\kappa} q^\kappa_\lambda \cdot \sqrt{e^{-\left(\frac{r_s}{r}\right)^3} \cdot \left(\frac{r_s}{r}\right)^3} \cdot C_\alpha^* = h^\lambda_{,\kappa\alpha} q^\kappa_\lambda$$

$$\Rightarrow \qquad h^\lambda_{,\kappa} \cdot \sqrt{e^{-\left(\frac{r_s}{r}\right)^3} \cdot \left(\frac{r_s}{r}\right)^3} \cdot C_\alpha^* = h^\lambda_{,\kappa\alpha}$$

or

$$\Phi \cdot \sqrt{e^{-\left(\frac{r_s}{r}\right)^3} \cdot \left(\frac{r_s}{r}\right)^3} \cdot C_\beta^* = h^\lambda_{,\kappa} q^\kappa_\lambda \cdot \sqrt{e^{-\left(\frac{r_s}{r}\right)^3} \cdot \left(\frac{r_s}{r}\right)^3} \cdot C_\beta^* = h^\lambda_{,\kappa\beta} q^\kappa_\lambda$$

$$\Rightarrow \qquad h^\lambda_{,\kappa} \cdot \sqrt{e^{-\left(\frac{r_s}{r}\right)^3} \cdot \left(\frac{r_s}{r}\right)^3} \cdot C_\beta^* = h^\lambda_{,\kappa\beta}. \qquad (512)$$

Now, we are back with an eigenequation for the vector function $\mathbf{h}_\lambda$. It should be pointed out that the special r-dependent Ricci curvature (368) was introduced to assure the Newton limit for the gravity interaction. Remembering Eq. (370), we also have to consider the possibility of other terms of curvature, which acts in addition to the gravity-part. Using our hypothesis from Eq. (370) and incorporating it into Eq. (510) leads us to:

$$R^* = \overbrace{C_R \cdot e^{-\left(\frac{r_s}{r}\right)^3} \cdot \left(\frac{r_s}{r}\right)^3}^{R^*_{\text{gravity}}} + R^*_{\text{quantum}}$$

$$\equiv \left(R^{**}_{\text{gravity}|\alpha} + R^{**}_{\text{quantum}|\alpha}\right) \mathbf{e}^{\alpha} \cdot \mathbf{e}^{\beta} \left(R^{**}_{\text{gravity}|\beta} + R^{**}_{\text{quantum}|\beta}\right)$$

$$\Rightarrow \quad C_{f1} \cdot \left(1 - \frac{f^2}{2 \cdot C_{N2}} + \ldots\right) \cdot \left(R^{**}_{\text{gravity}|\alpha} + R^{**}_{\text{quantum}|\alpha}\right) \mathbf{e}^{\alpha} \cdot \mathbf{e}^{\beta} \left(R^{**}_{\text{gravity}|\beta} + R^{**}_{\text{quantum}|\beta}\right)$$

$$= h_{\lambda,\alpha} q^{\lambda} \mathbf{e}^{\alpha} \cdot \mathbf{e}^{\beta} h_{\lambda,\beta} q^{\lambda}$$

$$\simeq \Phi \cdot \left(R^{**}_{\text{gravity}|\alpha} + R^{**}_{\text{quantum}|\alpha}\right) \mathbf{e}^{\alpha} \cdot \mathbf{e}^{\beta} \left(R^{**}_{\text{gravity}|\beta} + R^{**}_{\text{quantum}|\beta}\right) \cdot \left(1 + \frac{f}{\sqrt{2 \cdot C_{N2}}}\right)$$

$$= h_{\lambda,\alpha} q^{\lambda} \mathbf{e}^{\alpha} \cdot \mathbf{e}^{\beta} h_{\lambda,\beta} q^{\lambda}$$

or

$$\simeq \left(1 - \frac{f}{\sqrt{2 \cdot C_{N2}}}\right) \cdot \left(R^{**}_{\text{gravity}|\alpha} + R^{**}_{\text{quantum}|\alpha}\right) \mathbf{e}^{\alpha} \cdot \mathbf{e}^{\beta} \left(R^{**}_{\text{gravity}|\beta} + R^{**}_{\text{quantum}|\beta}\right) \cdot \Phi$$

$$= h_{\lambda,\alpha} q^{\lambda} \mathbf{e}^{\alpha} \cdot \mathbf{e}^{\beta} h_{\lambda,\beta} q^{\lambda}, \tag{513}$$

respectively:

$$\simeq h_{\lambda} \cdot \left(R^{**}_{\text{gravity}|\alpha} + R^{**}_{\text{quantum}|\alpha}\right) = h_{\lambda,\alpha}$$

or

$$\simeq \left(R^{**}_{\text{gravity}|\beta} + R^{**}_{\text{quantum}|\beta}\right) \cdot h_{\lambda} \cdot = h_{\lambda,\beta}. \tag{514}$$

Similar equations are obtained for the elastic approach:

$$\simeq h^{\lambda}_{,\kappa} \cdot \left(R^{**}_{\text{gravity}|\alpha} + R^{**}_{\text{quantum}|\alpha}\right) = h^{\lambda}_{,\kappa\alpha}$$

or

$$\simeq \left(R^{**}_{\text{gravity}|\beta} + R^{**}_{\text{quantum}|\beta}\right) \cdot h^{\lambda}_{,\kappa} \cdot = h^{\lambda}_{,\kappa\beta}. \tag{515}$$

Please note that the simple additive connection of various aspects for the Ricci curvature as introduced above is only to be seen as a suggestion. Surely, other options are possible and may even be more suitable in certain circumstances and applications.

It needs to be pointed out that the function vector $\mathbf{h}_{\lambda}$ has to satisfy not just the conditions of the equations given here (e.g., the last one (514)), but that also its components have to be Laplace functions.

6.12.5   Intermediate Section: Consequences of Infinite Options Principle

Taking our result for the function $F[f]$ from Eq. (492), thereby taking into account our knowledge about the dimension-dependent constants C_{N1} and C_{N2} from [2] (see Eq. (503)), we may now investigate the situation in space-times with—nearly—infinitely many options: with infinitely many dimensions (hidden, open, or compactified does not matter with respect to the C_{N1} and C_{N2}). We, therefore, evaluate the following limit:

$$
\lim_{n\to\infty} F[f] = \lim_{n\to\infty} C_{f1} \cdot \cos\left[\frac{1}{C_{N2}} \cdot \sqrt{C_{N2} - C_{N1}} \cdot \left(f - C_{f2}\right)\right]^{-\frac{C_{N2}}{C_{N1} - C_{N2}}}
$$

$$
= \lim_{n\to\infty} C_{f1} \cdot \cosh\left[\frac{1}{C_{N2}} \cdot \sqrt{C_{N1} - C_{N2}} \cdot \left(f - C_{f2}\right)\right]^{-\frac{C_{N2}}{C_{N1} - C_{N2}}}
$$

$$
= \lim_{n\to\infty} C_{f1} \cdot \cosh\left[\frac{1}{2} \cdot \sqrt{\frac{1}{n-1} - 1} \cdot \left(f - C_{f2}\right)\right]^{\frac{4}{n-2}} = C_{f1} \tag{516}
$$

and subsequently obtain for the metric Dirac equation (Eq. (493)):

$$
\lim_{n\to\infty} F[f] \cdot R^* = C_{f1} \cdot R^* = \lim_{n\to\infty} h_{\lambda,\alpha} q^{\lambda}_{\alpha} e^{\alpha} \cdot e^{\beta} h_{\lambda,\beta} q^{\lambda}_{\beta}. \tag{517}
$$

The corresponding result for the elastic approach would look as follows:

$$
\lim_{n\to\infty} F[f] \cdot R^* = C_{f1} \cdot R^*
$$

$$
= \lim_{n\to\infty} h^{\lambda}_{,\kappa\alpha} q^{\kappa}_{\lambda} g^{\alpha\beta} h^{\lambda}_{,\kappa\beta} q^{\kappa}_{\lambda} \xrightarrow{g^{\alpha\beta} = e^{\alpha} \cdot e^{\beta}} \lim_{n\to\infty} h^{\lambda}_{,\kappa\alpha} q^{\kappa}_{\lambda} e^{\alpha} \cdot e^{\beta} h^{\lambda}_{,\kappa\beta} q^{\kappa}_{\lambda}. \tag{518}
$$

We realize that insisting on the eigencharacter of the metric Dirac equation (and any similar approach like the elastic one in the chapter here, for instance) in the limit of n to infinity requires the f^2-dependency of R^*.

This is in difference to the Laplace-like Eq. (370) as outcome for the application of a scalar wrapper $F[f]$ as multiplicator for a given metric when demanding condition (246). Then—also in connection with the Infinity Options Principle—our quantum equation for the determination of f reads (c.f. section "6.4 The Infinity Options Principle"):

$$\left(\frac{\lim\limits_{n \to \infty} R^*}{4} \cdot C_2 + \Delta_g \right) f = 0 \tag{519}$$

and automatically is an eigenequation with respect to f even in the case that R^* does not contain any explicit f-dependencies. As our choice for F determines the outcome of the type of the final equation (Dirac- or Laplace-like), we consequently end up with different solutions for the scalar wrapper's kernel f.

6.12.6 Back to the Sub-Section and the 3-Generation Problem of Elementary Particles

Now, as we have established a consistency between the metric Dirac (or elastic) approach and our Laplace-like Eq. (370), with both containing gravity terms, we still wonder why we have a $1/r^3$-dependency for gravity (when keen on obtaining the Newton limit for bigger r) within the Ricci scalar R^*, but face constants when dealing with masses as results of dimensional entanglement. Could there be an equivalency between the curvature and the inertia resulting from such entanglement? And if so, would this equivalency be a possible reason to explain the three generations of elementary particles?

In order to try and answer these questions, we want to reconsider Eq. (370).

If there was a principle, demanding that the gravity dependency for $R^* \sim (r_s/r)^3$ (in the far distance limit) in connection with heavy masses has to provide equivalency to the masses resulting from dimensional entanglement (always leading to quadratic terms of the masses), we obviously end up with polynomial expressions for the masses, respectively the Schwarzschild radii r_s, containing third order and quadratic terms (if not more). For such algebraic expressions of third order in r_s, three solutions can exist.

Let us simply assume that the approach (370) could be generalized as follows:

$$\left(\overbrace{C_R \cdot e^{-\left(\frac{r_s}{r}\right)^3} \cdot \left(\frac{r_s}{r}\right)^3}^{R^*_{\text{gravity}}} + R^*_{\text{quantum}} + \Delta_{n\text{-space-time}} \right) f$$

$$
\left(
\overbrace{C_R \cdot e^{-\left(\frac{r_s}{r}\right)^3} \cdot \left(\frac{r_s}{r}\right)^3}^{R^*_{\text{gravity}}}
+ R^*_{\text{quantum}} + \Delta_{4D} + \overbrace{\Delta_{\text{compact}}}^{\text{hidden}} + \overbrace{\Delta_{\text{entangled}}}^{\text{masses/inertia}}
\right) f
$$

$$
\left(
\overbrace{C_{R3} \cdot e^{-\left(\frac{r_s}{r}\right)^3} \cdot \left(\frac{r_s}{r}\right)^3}^{R^*_{\text{gravity}}}
+ C_{R2} \cdot e^{-\left(\frac{r_s}{r}\right)^2} \cdot \left(\frac{r_s}{r}\right)^2
+ C_{R1} \cdot e^{-\frac{r_s}{r}} \cdot \frac{r_s}{r}
+ C_{R0} \cdot r_s + C_c \ldots + \Delta_{4D}
\right) f = 0.
$$

$$
\tag{520}
$$

In addition, we demand that for certain positions r, the Schwarzschild radius of the system under consideration has to be the same for all terms, which leads to something like:

$$
C_{R03} \cdot r_s^3 + C_{R02} \cdot r_s^2 + \left(C_{R01} + C_{R0} \right) \cdot r_s + C_c = 0
$$

$$
\Rightarrow \qquad \left(r_s - r_{s3} \right) \cdot \left(r_s - r_{s2} \right) \cdot \left(r_s - r_{s1} \right) = 0. \tag{521}
$$

Here now the three solutions might just be the three masses of one family of elementary particles consisting of three generations. As some additional (weak, though) circumstantial evidence supporting our suggestion, we may take the fact that in the right scales the masses of the three generations can always be fitted by power laws with exponents 3 (or very close to three as mentioned in sub-section "6.7.4 Quantum Well Quantization").

Of course, another option may be that—just as in the case of the Schrödinger hydrogen atom—the gravity term requires certain quantum numbers r_s, which assure a singularity-free outcome to the quantum solution of Eq. (520). Unfortunately, we are unable to derive such a solution in closed form at the moment and have to leave this task to skilled mathematicians.

6.12.6.1 Applying metric Dirac or elastic approach

Using the infinity options principle, thereby assuming that the space-time has a very high number of degrees of freedom and thus, dimensions (one can easily evaluate from Eq. (516) that this is already the case for fairly low numbers of n ... even below 10), we could just concentrate on the simple Dirac and/or elastic equation or similar forms (c.f. (429) or (476) and the more general form shown in [1]) as given in Eqs. (517) and (518). If further assuming that we only have to deal with gravity Ricci curvature, the conditions would read:

$$\lim_{n\to\infty} F[f]\cdot R^* = C_{f1}\cdot R^* = C_{f1}\cdot \overbrace{C_R \cdot e^{-\left(\frac{r_s}{r}\right)^3}\cdot\left(\frac{r_s}{r}\right)^3}^{R^*_{\text{gravity}}} = \lim_{n\to\infty} h_{\lambda,\alpha}q^\lambda e^\alpha \cdot e^\beta h_{\lambda,\beta}q^\lambda.$$

(522)

For those who insist on the general form of arbitrary n, it has to be Eq. (489) or Eq. (493). The corresponding result for the elastic approach would look as follows:

$$\lim_{n\to\infty} F[f]\cdot R^* = C_{f1}\cdot R^* = \lim_{n\to\infty} h^\lambda_{,\kappa\alpha}q^\kappa_\lambda g^{\alpha\beta}h^\lambda_{,\kappa\beta}q^\kappa_\lambda$$

$$\xrightarrow{g^{\alpha\beta}=e^\alpha\cdot e^\beta} = C_{f1}\cdot \overbrace{C_R \cdot e^{-\left(\frac{r_s}{r}\right)^3}\cdot\left(\frac{r_s}{r}\right)^3}^{R^*_{\text{gravity}}} = \lim_{n\to\infty} h^\lambda_{,\kappa\alpha}q^\kappa_\lambda e^\alpha \cdot e^\beta h^\lambda_{,\kappa\beta}q^\kappa_\lambda. \quad (523)$$

Also a combination of the two equations above could do or—yes—an extension of the approach to arbitrary complex h-tensor forms of higher ranks, leading to:

$$\lim_{n\to\infty} F[f]\cdot R^* = C_{f1}\cdot R^* = C_{f1}\cdot \overbrace{C_R \cdot e^{-\left(\frac{r_s}{r}\right)^3}\cdot\left(\frac{r_s}{r}\right)^3}^{R^*_{\text{gravity}}}$$

$$= \lim_{n\to\infty}\left(h_{\lambda,\alpha}q^\lambda + h^\lambda_{,\kappa\alpha}q^\kappa_\lambda +...\right)e^\alpha \cdot e^\beta \left(h_{\lambda,\beta}q^\lambda + h^\lambda_{,\kappa\beta}q^\kappa_\lambda +...\right). \quad (524)$$

Remember, the only condition we have to watch is f being a scalar and fulfilling the Laplace equation.

In order to just have an example, we choose Eq. (522) for here and now and assume the function f to be solely dependent on the radius r. As before we apply a base vector separation for R^* and result in the following:

$$\overbrace{C_{f1}\cdot C_R \cdot e^{-\left(\frac{r_s}{r}\right)^3}\cdot\left(\frac{r_s}{r}\right)^3}^{R^*_{\text{gravity}}} = C^2_{fR}\cdot \overbrace{e^{-\left(\frac{r_s}{r}\right)^3}\cdot\left(\frac{r_s}{r}\right)^3}^{\equiv G^2_R} = C^2_{fR}G^2_R e^1 \cdot e^1 = h_{\lambda,1}q^\lambda e^1 \cdot e^1 h_{\lambda,1}q^\lambda$$

$$\Rightarrow \qquad C_{fR} \cdot G_R = C_{fR} \cdot e^{-\frac{1}{2}\left(\frac{r_s}{r}\right)^3} \cdot \left(\frac{r_s}{r}\right)^{\frac{3}{2}} = q^\lambda \partial_r h_\lambda = q^\lambda \frac{\partial h_\lambda}{\partial r}$$

$$\Rightarrow \qquad C_{fR} \cdot \int e^{-\frac{1}{2}\left(\frac{r_s}{r}\right)^3} \cdot \left(\frac{r_s}{r}\right)^{\frac{3}{2}} dr = q^\lambda h_\lambda. \tag{525}$$

As our space has a high number of dimensions, the sum on the right-hand side could be arbitrarily long and we should be able to construct solutions to the last line of Eq. (525) with components of $\mathbf{h}_\lambda$ all being Laplace functions and thus, all satisfying the condition:

$$\frac{(n-2)}{r} \cdot \frac{\partial h_\lambda}{\partial r} + \frac{\partial^2 h_\lambda}{\partial r^2} = 0. \tag{526}$$

If we want to avoid the $\mathbf{q}$-vector, we could either assume $\mathbf{h}_\lambda$ to only consist of one component, or apply a different R^*-separation as follows:

$$C_{f1} \cdot C_R \cdot \overbrace{e^{-\left(\frac{r_s}{r}\right)^3} \cdot \left(\frac{r_s}{r}\right)^3}^{R^*_{\text{gravity}}} = \underbrace{C_{fR\lambda} C_{fR\lambda} q^\lambda \mathbf{e}^1 \cdot \mathbf{e}^1 q^\lambda}_{\equiv C_{f1} \cdot C_R} \cdot \overbrace{e^{-\left(\frac{r_s}{r}\right)^3} \cdot \left(\frac{r_s}{r}\right)^3}^{\equiv G_R^2} = G_R^2 C_{fR\lambda} q^\lambda \mathbf{e}^1 \cdot \mathbf{e}^1 C_{fR\lambda} q^\lambda$$

$$= h_{\lambda,1} q^\lambda \mathbf{e}^1 \cdot \mathbf{e}^1 h_{\lambda,1} q^\lambda$$

$$\Rightarrow \qquad C_{fR\lambda} \cdot G_R q^\lambda = C_{fR\lambda} q^\lambda \cdot e^{-\frac{1}{2}\left(\frac{r_s}{r}\right)^3} \cdot \left(\frac{r_s}{r}\right)^{\frac{3}{2}} = q^\lambda \partial_r h_\lambda = q^\lambda \frac{\partial h_\lambda}{\partial r}$$

$$\Rightarrow \qquad C_{fR\lambda} \cdot e^{-\frac{1}{2}\left(\frac{r_s}{r}\right)^3} \cdot \left(\frac{r_s}{r}\right)^{\frac{3}{2}} = \partial_r h_\lambda = \frac{\partial h_\lambda}{\partial r}$$

$$\Rightarrow \qquad C_{fR\lambda} \cdot \int e^{-\frac{1}{2}\left(\frac{r_s}{r}\right)^3} \cdot \left(\frac{r_s}{r}\right)^{\frac{3}{2}} dr = h_\lambda. \tag{527}$$

Now, instead of a sum of derivatives to the $\mathbf{h}_\lambda$ components (second last line in Eq. (525)) and consequently a scalar sum $q^\lambda h_\lambda$ over all components of $\mathbf{h}_\lambda$ (last line in Eq. (525)), we have a list of integrals, namely, one to each component. Sticking to Eq. (525), assuming $\mathbf{h}_\lambda$ to only consist of one component and incorporating the Laplace condition (526), we could write:

$$\frac{\partial h_\lambda}{\partial r} = -r \cdot \frac{q^\lambda}{n-2} \cdot \frac{\partial^2 h_\lambda}{\partial r^2} = 0$$

$$\Rightarrow \quad C_{fR} \cdot G_R = C_{fR} \cdot e^{-\frac{1}{2}\left(\frac{r_s}{r}\right)^3} \cdot \left(\frac{r_s}{r}\right)^{\frac{3}{2}} = q^\lambda \frac{\partial h_\lambda}{\partial r} = -r \cdot \frac{q^\lambda}{n-2} \cdot \frac{\partial^2 h_\lambda}{\partial r^2}$$

$$\Rightarrow \quad C_{fR} \cdot e^{-\frac{1}{2}\left(\frac{r_s}{r}\right)^3} \cdot \left(\frac{r_s}{r}\right)^{\frac{3}{2}} + r \cdot \frac{q^\lambda}{n-2} \cdot \frac{\partial^2 h_\lambda}{\partial r^2} = 0. \tag{528}$$

Solving the differential equation in the last line gives us:

$$q^\lambda h_\lambda = q^\lambda h_\lambda[r] = C_{r0} + r \cdot C_{r1} + \frac{(n-2)}{C_{fR}}\left(\sqrt{2 \cdot \pi}(r - r_s) erf\left[\sqrt{\frac{r_s}{2 \cdot r}}\right] - 2 \cdot e^{-\frac{1}{2}\left(\frac{r_s}{r}\right)}\sqrt{r \cdot r_s}\right)$$

$$\text{with: } erf(z) = \frac{2}{\sqrt{\pi}}\int_0^z e^{-t^2}\,dt. \tag{529}$$

Similarly, we would obtain the following for the approach (527):

$$h_\lambda = h_\lambda[r] = C_{r0\lambda} + r \cdot C_{r1\lambda} + \frac{(n-2)}{C_{fR\lambda}}\left(\sqrt{2 \cdot \pi}(r - r_s) erf\left[\sqrt{\frac{r_s}{2 \cdot r}}\right] - 2 \cdot e^{-\frac{1}{2}\left(\frac{r_s}{r}\right)}\sqrt{r \cdot r_s}\right). \tag{530}$$

Reasonable solutions are being found for settings where the arbitrary constants $C_{r0\lambda}$, $C_{r1\lambda}$ and C_{r0}, C_{r1} are chosen to be zero. Figures 6.9 and 6.10 show that for all dimensions $n > 2$, we obtain quantum gravity solutions being finite at the origin $r = 0$ and vanishing for $r \to \infty$. Thereby higher numbers of dimensions lead to lesser fading of the quantum gravity field. We also find (Figs. 6.11 and 6.12) that for increasing Schwarzschild radii, the field shows bigger absolute values at the origin $r = 0$ and slower fading for increasing r.

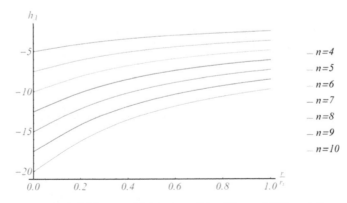

Figure 6.9 Function $h_\lambda[r]$ for equal Schwarzschild radii r_s and different dimensions n with arbitrary scaling near $r = 0$.

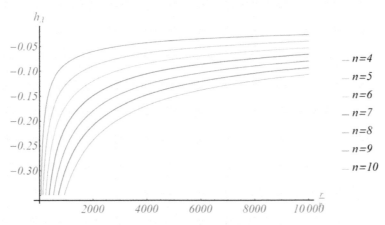

Figure 6.10 Function $h_\lambda[r]$ for equal Schwarzschild radii r_s and different dimensions n with arbitrary scaling for $r \gg r_s$.

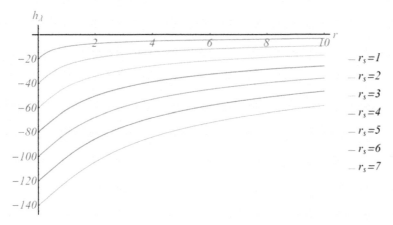

Figure 6.11 Function $h_\lambda[r]$ for equal dimensions $n = 10$ and different Schwarzschild radii r_s with arbitrary scaling near $r = 0$.

As our original wrapper approach for the metric (244) together with the infinity options principle, and thus, Eq. (516) would result in just a constant multiplicator for the original metric, which is to say:

$$\lim_{n\to\infty} G_{\alpha\beta} = \lim_{n\to\infty} F\left[f\left[t,x,y,z,\ldots,\xi_k,\ldots,\xi_n\right]\right] \cdot \delta_\alpha^i \delta_\beta^j g_{ij}$$

$$\lim_{n\to\infty} G_{\alpha\beta} = \lim_{n\to\infty} F\left[f\left[t,x,y,z,\ldots,\xi_k,\ldots,\xi_n\right]\right] \cdot g_{\alpha\beta} = C_{f1} \cdot g_{\alpha\beta}, \qquad (531)$$

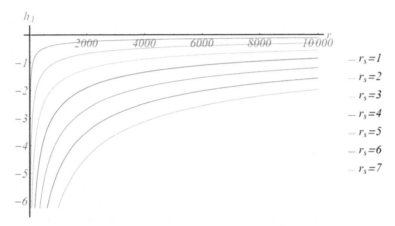

Figure 6.12 Function $h_\lambda[r]$ for equal dimensions n and different Schwarzschild radii r_s with arbitrary scaling for $r \gg r_s$.

we have to remind ourselves that the infinite options principle does not really mean that we assume the space-time to be infinitely dimensional, but only that the number of dimensions is sufficiently high to allow for the simplification (522) with respect to the derivation of the linearized metric Dirac equation (or its elastic equivalent or any other—similar—multiple tensor approach with inner f-structures). Thereby the simplification or linearization (522) is still justified because (as already said above) the function $F[f]$ converges very fast against a constant for either increasing n (see and evaluate directly from Eq. (492)) or small values of f itself. With respect to the interpretation of our result regarding the metric $G_{\alpha\beta}$, however, we have to resort once more to our series expansion (e.g., Eq. (499) for the metric Dirac vector approach) and find that instead of a trivial Eq. (531) the complete solution would be rather:

$$\lim_{n\to\text{big}} G_{\alpha\beta} = \lim_{n\to\text{big}} F\left[f\left[t,x,y,z,\ldots,\xi_k,\ldots,\xi_n\right]\right] \cdot g_{\alpha\beta}$$

$$= C_{f1} \cdot \lim_{n\to\text{big}} \cosh\left[\frac{1}{C_{N2}} \cdot \sqrt{C_{N1} - C_{N2}} \cdot \left(f - C_{f2}\right)\right]^{-\frac{C_{N2}}{C_{N1}-C_{N2}}} \cdot g_{\alpha\beta}$$

$$= C_{f1} \cdot \left(1 - \frac{f^2}{2 \cdot C_{N2}} + \ldots\right) \cdot g_{\alpha\beta} \simeq C_{f1} \cdot \left(1 - \frac{f^2}{2 \cdot C_{N2}}\right) \cdot g_{\alpha\beta}. \qquad (532)$$

Choosing $C_{f1} = 1$, the metric Dirac quantum gravity field would be finite at $r = 0$, which is to say the gravitational origin, and vanish in the infinity. Thus, the metric $G_{\alpha\beta}$ does show a distortion around the gravitational origin and then asymptotically takes on the non-disturbed metric $g_{\alpha\beta}$ in the far distance (Figs. 6.13 and 6.14).

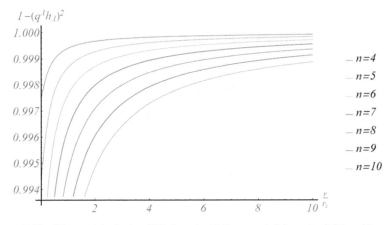

Figure 6.13 The metric factor $F[f]$ (see text) for equal Schwarzschild radii r_s and different dimensions n with arbitrary scaling for $r \gg r_s$.

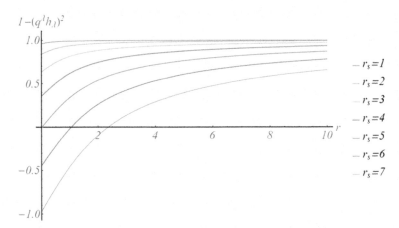

Figure 6.14 The metric factor $F[f]$ (see text) for a fixed number of dimensions $n = 10$ and different Schwarzschild radii r_s with arbitrary scaling for $r \gg r_s$.

It needs to be pointed out that, in Figs. 6.13 and 6.14, we did apply an arbitrary factor to the function f, which was constant to all n and r_s, because we only wanted to show the differences between various n and r_s settings. In reality, the factor should assure that at least in the region of $r \gg r_s$, f will be small against one also in cases of higher dimensions n or bigger Schwarzschild radii r_s. As we also see the functional wrapper $F[f]$ as quantum gravity effect, the multiplicator should probably be somehow dependent on the Planck constant $\hbar$.

6.12.7 Incorporation of Our Bekenstein Information Solution with *n*-Dependent Schwarzschild Radius

The moment we accept the dependency of the Schwarzschild radius on the dimension of the object in question (c.f. section "6.8 The Schwarzschild Case and the Meaning of the Schwarzschild Radius"), we find that we face a problem when moving from 4 dimensions to three for smaller r_s. We easily see that the general solution to the Schwarzschild metric in *n*-dimensional space-time (318) does not allow for any coding of the mass as the term $\left(\dfrac{r_s}{r}\right)^{n-3}$ simply disappears in 3-dimensional spaces. We meanwhile know, however, that we also have the option to metrically code mass via dimensional entanglement (c.f. the previous sections, especially section "6.6 The Entanglement of Dimensions and the Production of Mass"). It is, therefore, necessary to demand that in 4-dimensional space-time, there exists a—potentially smooth—transition between the corresponding quantized Schwarzschild solution (333) and the flat metric equivalent with entangled mass creation. The latter might look like:

$$
g_{\alpha\beta}^{\text{4-flat}} = \begin{pmatrix} -c^2 \cdot \left(g[t]'\right)^2 & 0 & 0 & 0 \\ 0 & 1 & 0 & 0 \\ 0 & 0 & r^2 & 0 \\ 0 & 0 & 0 & r^2 \cdot \sin^2 \vartheta \end{pmatrix} F\left[f\left[g[t],r,\vartheta,\varphi\right]\right]
$$

$$
f\left[g[t],r,\vartheta,\varphi\right] = f_t\left[g[t]\right] \cdot f_r[r] \cdot f_\vartheta[\vartheta] \cdot f_\varphi[\varphi]. \tag{533}
$$

With the usual approach for the matter-antimatter time functions $f_t[t]$:

$$
f_t[t] = f_t\left[g_t[t]\right] = C_{t-} \cdot e^{-E \cdot c \cdot g_t[t]} + C_{t+} \cdot e^{E \cdot c \cdot g_t[t]}, \tag{534}
$$

we obtain the classical Klein–Gordon or Schrödinger equation in spherical coordinates with mass (and/or energy):

$$
R^* = 0 = \left(\Delta_{\text{3D-sphere}} - E^2\right)\Psi; \quad \Psi = \Psi[r,\vartheta,\varphi]. \tag{535}
$$

With $E^2 = \dfrac{m^2 \cdot c^2}{\hbar^2}$, we would just have the mass-only option.

Further reduction of dimensions would then just lead to equations of the type:

$$R^* = 0 = \left(\Delta_{\text{2D-sphere}} - E^2 \right) \Psi; \quad \Psi = \Psi \left[\vartheta, \varphi \right] \tag{536}$$

and

$$R^* = 0 = \left(\frac{\partial^2}{\partial \varphi^2} - E^2 \right) \Psi; \quad \Psi = \Psi \left[\varphi \right]. \tag{537}$$

It may well be that there are no solutions for suitable transitions of the metrics and that therefore further degrees of freedom are of need. In this case, the Kerr solution of a rotating or even the Kerr–Newman metric of a charged rotating massive object instead of the simple Schwarzschild solution should provide the necessary additional parameters. This, however, shall not bother us within this chapter. We, therefore, leave it to interested readers to consider such options.

Here, we are interested in the possible occurrence of conditions helping us to explain the "funny" 3-generation problem. Comparing the quantum gravity differential equation for a Schwarzschild object (314) with (535) in the time-independent case gives us:

$$R^* = 0 = \left(\overbrace{\Delta_{\text{3D-sphere}} - E^2}^{B} \right) \Psi$$

$$= \underbrace{-\frac{6}{f} \cdot \left(\Delta_{\text{3D-sphere}} f - \frac{f^{(2,0,0,0)}}{c^2} - r_s \cdot \left(\frac{f^{(0,1,0,0)} + r \cdot f^{(0,2,0,0)}}{r^2} + \frac{f^{(2,0,0,0)}}{c^2 \left(r - r_s \right)} \right) \right)}_{A}$$

$$f^{(0,1,0,0)} = \frac{\partial f \left[t, r, \vartheta, \varphi \right]}{\partial r}; \quad f^{(0,2,0,0)} = \frac{\partial^2 f \left[t, r, \vartheta, \varphi \right]}{\partial r^2};$$

$$f^{(2,0,0,0)} = \frac{\partial^2 f \left[t, r, \vartheta, \varphi \right]}{\partial t^2}; \quad \Psi = \Psi \left[r, \vartheta, \varphi \right]. \tag{538}$$

We see that the Schwarzschild radius, which is linear in the mass m, needs to be somehow equal to the quadratic mass in the flat metric quantum gravity equation. It may appear as a problem that we have further derivatives with respect to r in equation A, while there is only the constant term in equation B, but as we saw in section "6.11 Intermediate Sum-up: Direct Extraction of the Klein–Gordon and the Schrödinger Equation from the Metric Tensor" that we can easily obtain such additional terms in connection with dimensional entanglement with hidden or compactified dimensions, we assume that there is always a suitable structure of such entanglements, which could be applied in order to provide a perfect fit.

Now, we may easily assume more complicated settings for E, giving higher potentials of m and thus, providing conditions leading to algebraic expressions of the type (521) and thus, explaining the 3-generation problem. The same could be achieved via the assumption that higher dimensional Schwarzschild solutions with r-dependencies in the quantum gravity regime of the form:

$$A: \quad \Rightarrow 0 = \left((n-2) \cdot \frac{\partial}{r \cdot \partial r} + \frac{\partial^2}{r \cdot \partial r^2} - \left(\frac{r_s}{r} \right)^{n-3} \cdot \left(\frac{\partial}{r \cdot \partial r} + \frac{\partial^2}{\partial r^2} \right) \right) f[r] \quad (539)$$

requires an equivalent flat space quantum gravity description with entangled masses instead of Schwarzschild-coded ones (Schwarzschild-coded means via terms of the form $\left(\dfrac{r_s}{r} \right)^{n-3}$ as we find them in the n-dimensional Schwarzschild solution (318)). Once again, however, we are slightly puzzled about the fact that there are only connections up to a certain number of dimensions, because—obviously—only algebraic equations of the kind (521) for r_s, respectively m, of the third order are of need.

Thus, we conclude that in fact have found a fairly potent way to explain an X-generation elementary particle problem, only that we do not really see why X should be restricted to three.

6.12.7.1 Other inner solutions applying rigid spheres and elastic shells

It should only be mentioned here that the flat space metric, we have used above, is not the only option. We only took it for the reason of simplicity. There are also options for inner solutions of the type:

$$g_{\alpha\beta}^{4-\text{flat}} = \begin{pmatrix} -c^2 \cdot G[r] & 0 & 0 & 0 \\ 0 & G[r]^{-1} & 0 & 0 \\ 0 & 0 & A[r]^2 & 0 \\ 0 & 0 & 0 & B[r]^2 \cdot \sin^2 \vartheta \end{pmatrix} F\big[f[t,r,\vartheta,\varphi] \big]$$

$$f[t,r,\vartheta,\varphi] = f_t[t] \cdot f_r[r] \cdot f_\vartheta[\vartheta] \cdot f_\varphi[\varphi]. \quad (540)$$

Already with a simple setting of $G[r] = C_{r0}+C_{r1}*x$ and $B[r] = A[r]$, we are not only able to find smooth transitions with an outside Schwarzschild solution for any predefined radius, but also find nice approximated solutions for the corresponding quantum gravity equations under certain assumptions for the function $A[r]$. Thereby prominent among the $A[r]$ choices are those of the rigid sphere and the elastic shell types.

6.12.8 Summing It Up and Going for the Simplest Explanation of the Question: "Why Are There Three Masses for Charged Leptons, Neutrons, and Quarks?"

6.12.8.1 Brief repetition: Ricci scalar for a scaled metric

In the sections above (and quite some other papers by this author), it is demonstrated how we can find a direct extraction of the Klein–Gordon, the Schrödinger, and the Dirac equation from the Ricci scalar R^* of a modified metric of the kind $G_{\alpha\beta} = F[f]^*g_{\alpha\beta}$. With an arbitrary scalar function $F[f]$, the corresponding modified Ricci scalar R^* reads:

$$R^* = \frac{1}{F[f]^3}\cdot\left(\left(C_{N1}\cdot\left(\frac{\partial F[f]}{\partial f}\right)^2 - C_{N2}\cdot F[f]\cdot\frac{\partial^2 F[f]}{\partial f^2}\right)\cdot\overbrace{\left(\tilde{\nabla}_g f\right)^2}^{=f_{,\alpha}g^{\alpha\beta}f_{,\beta}} - C_{N2}\cdot F[f]\cdot\frac{\partial F[f]}{\partial f}\cdot\Delta_g f\right)\xrightarrow{\ n=4\ }$$

$$= \frac{1}{F[f]^3}\cdot\left(\left(\frac{3}{2}\cdot\left(\frac{\partial F[f]}{\partial f}\right)^2 - 3\cdot F[f]\cdot\frac{\partial^2 F[f]}{\partial f^2}\right)\cdot\left(\tilde{\nabla}_g f\right)^2 - 3\cdot F[f]\cdot\frac{\partial F[f]}{\partial f}\cdot\Delta_g f\right).$$

(541)

Thereby the C_{N1} and C_{N2} are constants depending on the number of dimensions n of the space-time being considered. For illustration, we also gave the case $n = 4$ (n = number of dimensions). Thereby f is to be understood as a function of the coordinates $f = f[x_0, x_1, x_2, \ldots]$. It was shown above that we can write:

$$C_{N1} = -\frac{3}{2}+\frac{7}{4}\cdot n-\frac{1}{4}\cdot n^2$$

$$C_{N2} = n-1.$$

(542)

Demanding certain conditions for the function $F[f]$ then gives us Dirac or Klein–Gordon equations (see the sections above).

So, for instance, by setting $F[f]$ to fulfill the following condition:

$$C_{N1} \cdot \left(\frac{\partial F[f]}{\partial f}\right)^2 - C_{N2} \cdot F[f] \cdot \frac{\partial^2 F[f]}{\partial f^2} = 0, \tag{543}$$

we obtain:

$$R^* = -\frac{1}{F[f]^3} \cdot C_{N2} \cdot F[f] \cdot \frac{\partial F[f]}{\partial f} \cdot \Delta_g f$$

$$\xrightarrow{n=4 \ \& \ F[f]=\left(f+\frac{C_f}{M^2}\right)^2} = -\frac{1}{F[f]^2} \cdot 3 \cdot \frac{\partial F[f]}{\partial f} \cdot \Delta_g f \xrightarrow{F[f]=\left(f+\frac{C_f}{M^2}\right)^2}$$

$$= -\frac{6}{\left(f+\dfrac{C_f}{M^2}\right)^3} \cdot \Delta_g f$$

$$\Rightarrow \qquad \frac{R^*}{6} \cdot \left(f+\frac{C_f}{M^2}\right)^3 + \Delta_g f = 0. \tag{544}$$

6.12.8.2 Intermediate remark

Now, we assume that the two differential operators in Eqs. (245) (or (541)) give constants A and B, which result to:

$$R^* = \frac{1}{F[f]^3} \cdot \left(\left(C_{N1} \cdot \left(\frac{\partial F[f]}{\partial f}\right)^2 - C_{N2} \cdot F[f] \cdot \frac{\partial^2 F[f]}{\partial f^2}\right) \cdot A - C_{N2} \cdot F[f] \cdot \frac{\partial F[f]}{\partial f} \cdot B\right). \tag{545}$$

In the case of $R^* = 0$, the equation could be solved for $F[f]$ and we obtain:

$$F[f] = C_2 \cdot e^{\dfrac{C_{N2}\left(B \cdot f - A \cdot \log\left[e^{\frac{B \cdot f}{A}} + A \cdot (C_{N1} - C_{N2}) e^{B C_{N2} C_1}\right]\right)}{A \cdot (C_{N1} - C_{N2})}}. \tag{546}$$

Evaluation for a variety of settings tells us an interesting story about the special case of $n = 3$ and $n = 4$ (see Fig. 6.15).

A similar result is obtained when investigating the case of the differential operators in Eq. (245) (or Eq. (541)) giving eigenvalues, which result to:

$$R^* = \frac{f}{F[f]^3} \cdot \left(\left(C_{N1} \cdot \left(\frac{\partial F[f]}{\partial f}\right)^2 - C_{N2} \cdot F[f] \cdot \frac{\partial^2 F[f]}{\partial f^2}\right) \cdot f \cdot A - C_{N2} \cdot F[f] \cdot \frac{\partial F[f]}{\partial f} \cdot B\right). \tag{547}$$

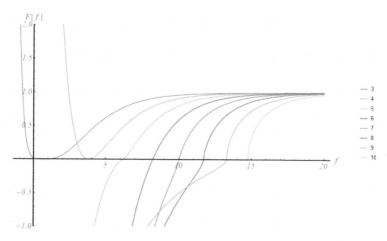

Figure 6.15 $F[f]$ (only real part is plotted) for the setting $B = 1/2$, $A = 1$, $C_1 = C_2 = 1$. The legend gives the corresponding dimensions. We see interesting behavior for the cases $n = 3$ and $n = 4$ where there are minima for certain values of f.

In the case of $R^* = 0$, the equation above could be solved for $F[f]$ and we obtain:

$$F[f] = C_2 \cdot f^{\frac{C_{N2} \cdot B}{A \cdot (C_{N1} - C_{N2})}} \left(A \cdot (C_{N2} - C_{N1}) \cdot f + (A - B) \cdot C_{N2} \cdot f^{\frac{B}{A}} \cdot C_1 \right)^{\frac{C_{N2}}{C_{N2} - C_{N1}}}.$$

(548)

As before with constant outcomes, evaluation for a variety of settings seems to set the case of $n = 3$ and $n = 4$ apart from the others (see Fig. 6.16).

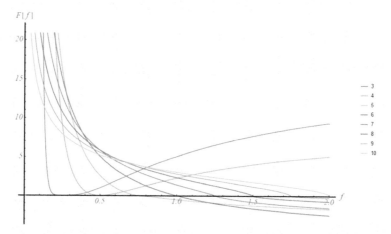

Figure 6.16 $F[f]$ (only real part is plotted) for the setting $B = 2$, $A = 1$, $C_1 = C_2 = 1$. The legend gives the corresponding dimensions. We see interesting behavior for the cases $n = 3$ and $n = 4$ where there are minima for certain values of f.

Quite automatically we ask, whether in a universe, apparently governed by a minimum principle, the preferred choice of numbers of dimensions should be one where there is a minimum when demanding certain conditions for the curvature. In this case, this condition was just $R^* = 0$ and we should remind ourselves that on bigger scales, our universe appears to be perfectly flat.

6.12.8.3 Finally, Klein–Gordon: Toward simplest explanation for the 3-generation problem

Now, we know that the classical Klein–Gordon equation always is of the kind:

$$-M^2 f + \Delta_g f = \left(-M^2 + \Delta_g\right) f = 0$$

$$\text{with}: \quad M^2 \equiv \frac{m^2 \cdot c^2}{\hbar^2}, \tag{549}$$

where c = speed of light in vacuum, m = rest mass, and $\hbar$ = reduced Planck constant.

Comparing the last line of Eq. (544) with the first line of Eq. (549), tells us that finding a linearization for the expression $\dfrac{R^*}{6} \cdot \left(f + \dfrac{C_f}{M^2}\right)^3$ with respect to the function f, would immediately give us the classical Klein–Gordon quantum equation from a completely metric origin. Thus, we have to look for possible—potentially approximated—solutions to the equation:

$$M^2 \cdot f \equiv \frac{m^2 \cdot c^2}{\hbar^2} \cdot f = \frac{R^*}{6} \cdot \left(f + \frac{C_f}{M^2}\right)^3. \tag{550}$$

This, however, is a polynomial of the third order and it can have three solutions.

A bit of reshaping the last equation is going to help us realize how this will possibly also solve the 3-generation mass problem:

$$\Psi \equiv M^2 \cdot f \equiv \frac{m^2 \cdot c^2}{\hbar^2} \cdot f = \frac{R^*}{6} \cdot \left(f + \frac{C_f}{M^2}\right)^3 = \frac{R^*}{M^6 \cdot 6} \cdot \left(\Psi + C_f\right)^3$$

$$\xrightarrow[\quad R^{**} = \frac{R^*}{M^6 \cdot 6}; \ R^{***} = C_f \quad]{} R^{**} \cdot \left(\Psi + R^{***}\right)^3 - \Psi = 0$$

$$= \left(\Psi + R^{***}\right)^3 - \frac{\Psi}{R^{**}} = \left(R^{***}\right)^3 - \frac{\Psi}{R^{**}} + 3 \cdot \left(R^{***}\right)^2 \cdot \Psi + 3 \cdot R^{***} \cdot \Psi^2 + \Psi^3. \tag{551}$$

The general solution to a three-order polynomial could be given via the following product form:

$$\left(\Psi - \Psi_1\right)\cdot\left(\Psi - \Psi_2\right)\cdot\left(\Psi - \Psi_3\right) =$$

$$\Psi^3 - \Psi^2 \cdot \left(\Psi_1 + \Psi_2 + \Psi_3\right) + \Psi \cdot \left(\Psi_1\Psi_2 + \Psi_1\Psi_3 + \Psi_2\Psi_3\right) - \Psi_1\Psi_2\Psi_3. \quad (552)$$

Comparing the latter with the last line in Eq. (551):

$$\left(R^{***}\right)^3 - \frac{\Psi}{R^{**}} + 3\cdot\left(R^{***}\right)^2 \cdot \Psi + 3\cdot R^{***} \cdot \Psi^2 + \Psi^3 \qquad (553)$$

gives us:

$$3\cdot R^{***} = 3\cdot C_f = -\left(\Psi_1 + \Psi_2 + \Psi_3\right)$$

$$3\cdot\left(R^{***}\right)^2 - \frac{1}{R^{**}} = 3\cdot\left(C_f\right)^2 - \frac{6\cdot M^6}{R^*} = \Psi_1\Psi_2 + \Psi_1\Psi_3 + \Psi_2\Psi_3$$

$$\left(C_f\right)^3 = -\Psi_1\Psi_2\Psi_3. \qquad (554)$$

Thus, we have obtained the three generations of quantum gravity solutions to the combined mass²-times-f-function, given via $\Psi \equiv M^2 \cdot f \equiv \dfrac{m^2 \cdot c^2}{\hbar^2} \cdot f$, as functions or dependencies of the Ricci scalar R^* of the quantum-gravity variated (with respect to its scale only) metric $G_{\alpha\beta} = F[f]^* g_{\alpha\beta}$, the mass-values $M^2 \equiv \dfrac{m^2 \cdot c^2}{\hbar^2}$ and a constant C_f.

Please note: As the expression $\Psi_1\Psi_2 + \Psi_1\Psi_3 + \Psi_2\Psi_3$ on the right-hand side in the second line in Eq. (554) should be just a constant and thus, could not depend on the various masses M_i, we have to demand the Ricci scalar R^* to be directly connected with the M_i via a constant const:

$$\frac{6\cdot M^6}{R^*} = \text{const} \quad \Rightarrow \quad R^* = \frac{6\cdot M^6}{\text{const}}. \qquad (555)$$

This gives two equations for the extraction of two of the three states Ψ_i out of the other:

$$R^{***} = C_f = -\frac{\Psi_1 + \Psi_2 + \Psi_3}{3}$$

$$\Rightarrow \quad \frac{1}{3}\cdot\left(\Psi_1 + \Psi_2 + \Psi_3\right)^2 - \frac{6\cdot M^6}{R^*} = \frac{1}{3}\cdot\left(\Psi_1 + \Psi_2 + \Psi_3\right)^2 - \text{const}$$

$$= \Psi_1\Psi_2 + \Psi_1\Psi_3 + \Psi_2\Psi_3$$

$$\Rightarrow \quad \left(\frac{\Psi_1 + \Psi_2 + \Psi_3}{3}\right)^3 = \Psi_1\Psi_2\Psi_3. \qquad (556)$$

We find that with a vanishing constant const (the curvature term), we would obtain one solution, namely:

$$\Psi_1 = \Psi_2 = \Psi_3. \tag{557}$$

We also find that with a vanishing constant C_f (the shift of the function f with the value $F[f] = f = 0$ moving away from the origin), we would only obtain the solution:

$$\Psi_2 = \Psi_3 = \frac{1}{2}\left(-\Psi_1 \pm \sqrt{4 \cdot \text{const} - 3 \cdot \Psi_1^2}\right). \tag{558}$$

We saw that with the mass M_i also the corresponding quantum gravity function f_i should change, which is only logic. We also find that for different masses the curvature term R^* must change, which is also easily understood as the curvature of space-time actually represents mass. We only have three generations apparently as a consequence of the number of bigger scale dimensions, i.e., 4. In space-times of other numbers of dimensions, no polynomial of the third order would occur and thus, also no three generations of elementary particles. We are, obviously, living in a universe with exactly four big-scale dimensions of which three are of spatial character. Coincidence or a possible explanation for the 3-generation problem! Well, at least quite a few things already seem to fit here.

Before truly accepting this explanation, however, we should check the Dirac approach, too.

Please note that one has to see it as a major restriction that the gravity wave function, which after all should be a function of the coordinates, is bound by so many harsh conditions (554). This aspect will be discussed—and potentially mitigated—further below.

6.12.8.4 Finally, Dirac I: Toward simplest explanation for the 3-generation problem

We saw that by demanding certain conditions for the function $F[f]$ and/or f then gives us Dirac or Klein–Gordon equations. Thus, when demanding f to be a Laplace function, we obtain from Eq. (245):

$$R^* = -\frac{1}{F[f]^3} \cdot \left(\left(C_{N1} \cdot \left(\frac{\partial F[f]}{\partial f}\right)^2 - C_{N2} \cdot F[f] \cdot \frac{\partial^2 F[f]}{\partial f^2} \right) \cdot \overbrace{\left(\tilde{\nabla}_g f\right)^2}^{= f_{,\alpha} g^{\alpha\beta} f_{,\beta}} \right)$$

$$\xrightarrow{n=4} = \frac{1}{F[f]^3} \cdot \left(\left(\frac{3}{2} \cdot \left(\frac{\partial F[f]}{\partial f}\right)^2 - 3 \cdot F[f] \cdot \frac{\partial^2 F[f]}{\partial f^2} \right) \cdot \left(\tilde{\nabla}_g f\right)^2 \right), \tag{559}$$

which—so it was shown in [1, 4]—gives the metric equivalent to the Dirac equation. It was also shown in [1] how this gives the classical Dirac equation in flat space Minkowski metrics.

There is a variety of options to derive discrete masses for only three generations in the Dirac case:

(a) Via the classical Dirac path by extracting the square root from the Klein–Gordon operator, thereby applying quaternions. Having previously shown that there is a 3-generation solution for the Klein–Gordon equation, the same then also follows for the subsequent Dirac equation. Thus, we can consider the job of following this option already done.

(b) Via an $\Delta_g f = 0$ approach as used here for the derivation of Eq. (559) and a suitable choice for the function $F[f]$.

(c) Via a suitable assumption for $\dfrac{\partial F[f]}{\partial f} \approx 0$ leading to an approach with a scalar field somewhat similar to the Higgs mechanics.

(d) ...

All these options split up into more possibilities when distinguishing the way of root extraction (Dirac-like or via vectors as recently shown here (see section: "6.12.2 Getting Rid of Quaternions")).

Here, in order to have an easy start, we begin with the fairly simple option b.

We apply the following exponential form for $F[f]$ via $F[f] = (f + C_f/M)^q$. This gives us, from Eq. (245) in the case of $q = 4$, the following:

$$R^* = \frac{4 \cdot (n-3)(1-n)}{\left(f + \dfrac{C_f}{M}\right)^6} \cdot \overbrace{\left(\tilde{\nabla}_g f\right)^2}^{= f_{,\alpha} g^{\alpha\beta} f_{,\beta}}. \tag{560}$$

We have learned from [1] that Eq. (559) or—just simpler (560)—is just the square of the metric Dirac equation from which the classical Dirac equation can easily be derived when moving toward Minkowski coordinates (again readers can refer to [1] or the sections above with respect to the mathematical details—see section "6.12.1 Repetition: Dirac in the Metric Picture and its Connection to the Classical Quaternion Form"). Thus, regarding the 3-generation problem of elementary particles, we simply can proceed with the square form (560), in order to show that also the metric Dirac equation could result in the three generations. Subsequent extraction of the square root (via Quaternions or vectors as also given above) will not change the results. This can easily be shown by remembering that f definitively is a scalar (and thus, has to be treated as one in the following calculation), however, when moving toward the metric Dirac equation, we simply have to assume this scalar to be decomposable as $f = h_\alpha q^\alpha$ (see section "6.12.1 Repetition: Dirac in the Metric Picture and its Connection to the Classical Quaternion Form").

We know that the operator square of the classical Dirac equation gives the classical Klein–Gordon equation. It always is of the kind:

$$-M^2 f + \Delta_g f = \left(-M^2 + \Delta_g\right) f = 0$$

$$\text{with:} \quad M^2 \equiv \frac{m^2 \cdot c^2}{\hbar^2}, \tag{561}$$

where c = speed of light in vacuum, m = rest mass, and $\hbar$ = reduced Planck constant.

Because these linear forms work so well, we conclude that—in this universe—with respect to the mass term M^2, there exists a linear form to Eq. (560), and solutions have to be at least temporarily stable. It should read:

$$M^2 \cdot f^2 = \left(\tilde{\nabla}_g f\right)^2 \Rightarrow M \cdot f = \left(\tilde{\nabla}_g f\right). \tag{562}$$

Please note that for here and now (for the reason of simplicity and brevity mainly), we did not care about potential inner vector characters of the mass, the function f, and the operator terms. For the moment we just assume that this does not have any influence on the 3-generation problem, we want to consider here. The proof for this can easily be obtained by applying $f = h_\alpha q^\alpha$ in all derivations below. It will not change the results.

Comparing the last Eq. (562) with the square of the metric Dirac equation of our choice for $F[f]] = (f + C_f/M)^q$ tells us that finding a squared linearization for the expression $\dfrac{R^*}{4 \cdot (n-3)(1-n)} \cdot \left(f + \dfrac{C_f}{M}\right)^6$ with respect to the function f, would immediately give us the classical Dirac quantum equation from a completely metric origin. Thus, we have to look for possible—potentially approximated—solutions to the equation:

$$M^2 \cdot f^2 \equiv \frac{m^2 \cdot c^2}{\hbar^2} \cdot f^2 = \frac{R^*}{4 \cdot (n-3)(1-n)} \cdot \left(f + \frac{C_f}{M}\right)^6$$

$$\Rightarrow \qquad M \cdot f = \frac{\sqrt{R^*}}{2 \cdot \sqrt{(n-3)(1-n)}} \cdot \left(f + \frac{C_f}{M} \right)^3. \qquad (563)$$

This, however, is a polynomial of the third order and it can have three solutions.

A bit of reshaping the last equation is going to help us realize how this will possibly also solve the 3-generation mass problem:

$$\Psi \equiv M \cdot f \equiv \frac{m \cdot c}{\hbar} \cdot f = \frac{\sqrt{R^*}}{2 \cdot \sqrt{(n-3)(1-n)}} \cdot \left(f + \frac{C_f}{M} \right)^3$$

$$= \frac{\sqrt{R^*}}{2 \cdot M^3 \cdot \sqrt{(n-3)(1-n)}} \cdot \left(\Psi + C_f \right)^3$$

$$R^{**} = \frac{\sqrt{R^*}}{2 \cdot M^3 \cdot \sqrt{(n-3)(1-n)}}; \quad R^{***} = C_f \longrightarrow R^{**} \cdot \left(\Psi + R^{***} \right)^3 - \Psi = 0$$

$$= \left(\Psi + R^{***} \right)^3 - \frac{\Psi}{R^{**}} = \left(R^{***} \right)^3 - \frac{\Psi}{R^{**}} + 3 \cdot \left(R^{***} \right)^2 \cdot \Psi + 3 \cdot R^{***} \cdot \Psi^2 + \Psi^3. \qquad (564)$$

The general solution to a three-order polynomial could be given via the following product form:

$$\left(\Psi - \Psi_1 \right) \cdot \left(\Psi - \Psi_2 \right) \cdot \left(\Psi - \Psi_3 \right) =$$

$$\Psi^3 - \Psi^2 \cdot \left(\Psi_1 + \Psi_2 + \Psi_3 \right) + \Psi \cdot \left(\Psi_1 \Psi_2 + \Psi_1 \Psi_3 + \Psi_2 \Psi_3 \right) - \Psi_1 \Psi_2 \Psi_3. \qquad (565)$$

Comparing the latter with the last line in Eq. (564):

$$\left(R^{***} \right)^3 - \frac{\Psi}{R^{**}} + 3 \cdot \left(R^{***} \right)^2 \cdot \Psi + 3 \cdot R^{***} \cdot \Psi^2 + \Psi^3 \qquad (566)$$

gives us:

$$3 \cdot R^{***} = 3 \cdot C_f = -\left(\Psi_1 + \Psi_2 + \Psi_3 \right)$$

$$3 \cdot \left(R^{***} \right)^2 - \frac{1}{R^{**}} = 3 \cdot (C_f)^2 - \frac{2 \cdot M^3 \cdot \sqrt{(n-3)(1-n)}}{\sqrt{R^*}} = \Psi_1 \Psi_2 + \Psi_1 \Psi_3 + \Psi_2 \Psi_3$$

$$(C_f)^3 = -\Psi_1 \Psi_2 \Psi_3. \qquad (567)$$

Thus, we have obtained the three generations of quantum gravity solutions to the combined mass-times-f-function, given via $\Psi \equiv M \cdot f \equiv \frac{m \cdot c}{\hbar} \cdot f$, as functions or dependencies of the Ricci scalar R^* of the quantum-gravity (just scalar) variated metric $G_{\alpha\beta} = F[f]^{*}g_{\alpha\beta}$, the mass-values $M \equiv \dfrac{m \cdot c}{h}$ and a constant C_f.

Please note: As the expression $\Psi_1\Psi_2 + \Psi_1\Psi_3 + \Psi_2\Psi_3$ on the right-hand side in the second line in Eq. (554) should be just a constant and thus, could not depend on the various masses M_i, we have to demand the Ricci scalar R^{*} to be directly connected with the M_i via a constant const:

$$\frac{2 \cdot M^3 \cdot \sqrt{(n-3)(1-n)}}{\sqrt{R^{*}}} = \sqrt{\text{const}} \quad \Rightarrow \quad R^{*} = 4 \cdot \frac{(n-3)(1-n) \cdot M^3}{\text{const}}. \quad (568)$$

This gives two equations for the extraction of two of the three states Ψ_i out of the other:

$$R^{***} = C_f = -\frac{\Psi_1 + \Psi_2 + \Psi_3}{3}$$

$$\Rightarrow \qquad \frac{1}{3} \cdot \left(\Psi_1 + \Psi_2 + \Psi_3\right)^2 - \frac{2 \cdot M^3 \cdot \sqrt{(n-3)(1-n)}}{\sqrt{R^{*}}}$$

$$= \frac{1}{3} \cdot \left(\Psi_1 + \Psi_2 + \Psi_3\right)^2 - \text{const} = \Psi_1\Psi_2 + \Psi_1\Psi_3 + \Psi_2\Psi_3$$

$$\left(\frac{\Psi_1 + \Psi_2 + \Psi_3}{3}\right)^3 = \Psi_1\Psi_2\Psi_3. \quad (569)$$

As before, we find that with a vanishing constant const (the curvature term), we would obtain one solution, namely:

$$\Psi_1 = \Psi_2 = \Psi_3. \quad (570)$$

We also find that with a vanishing constant C_f (the shift of the function f with the value $F[f] = f = 0$ moving away from the origin), we would only obtain the solution:

$$\Psi_2 = \Psi_3 = \frac{1}{2}\left(-\Psi_1 \pm \sqrt{4 \cdot \text{const} - 3 \cdot \Psi_1^2}\right). \quad (571)$$

Please note that, as said in the sub-section above, one has to see it as a major restriction that the gravity wave function, which after all should be a function of the coordinates, is bound by so many harsh conditions (567). This aspect will be discussed—and potentially mitigated—further below.

More discussion and the other options are presented in chapter 7 of this book or in [43].

6.12.8.5 Finally, Dirac II: Toward simplest explanation for the 3-generation problem

Now, we apply the following simplest linear form for $F[f]$ via $F[f] = (f + C_f/M)$. From Eq. (245), together with the assumption of f being a Laplace solution, we get the following result:

$$R^* = \frac{(n-6)(1-n)}{4 \cdot \left(f + \dfrac{C_f}{M}\right)^3} \cdot \overbrace{\left(\tilde{\nabla}_g f\right)^2}^{= f_{,\alpha} g^{\alpha\beta} f_{,\beta}}. \tag{572}$$

We have learned from [1] that Eq. (559) or—even simpler here (572)—is just the square of the metric Dirac equation of which the classical Dirac equation can easily be derived when moving toward Minkowski coordinates (readers can refer to [1] or the sections above with respect to the mathematical details—especially see section "6.12.1 Repetition: Dirac in the Metric Picture and its Connection to the Classical Quaternion Form"). Thus, regarding the 3-generation problem of elementary particles, we simply can proceed with the square form (572) in order to show that also the metric Dirac equation could result in three generations. Subsequent extraction of the square root (via Quaternions or vectors as also given above in here) will not change the results. With respect to the "square character" of Eqs. (559), (560), and (572), we need to remember that this only holds in the case of the classical approximation (c.f. section "6.12.1 Repetition: Dirac in the Metric Picture and its Connection to the Classical Quaternion Form"), where we assumed the left-hand side of the metric Dirac equation with the curvature term in Eq. (434) (first line) to be of square character in f. We already saw above, that this way the possible explanation for the 3-generation problem disappears (simply because the power-3 term of f has been approximated/classically postulated away). Thus, here now, we want to avoid this approximation and as long as we stick to the scalar f, things should stay simple and mathematically feasible.

Instead now of starting with the assumption that the linear appearance of f in the classical Klein–Gordon equation:

$$-M^2 f + \Delta_g f = \left(-M^2 + \Delta_g\right) f = 0$$

$$\text{with}: \quad M^2 \equiv \frac{m^2 \cdot c^2}{\hbar^2}, \tag{573}$$

(c = speed of light in vacuum, m = rest mass, and $\hbar$ = reduced Planck constant) is kind of a "natural law," we now ignore this classical postulation of linearity, completely stick to the metric results we derived so far, and only assume—for simplicity—that we can find f^2-type eigensolutions to the operator term $f_{,\alpha}g^{\alpha\beta}f_{,\beta}$ in Eqs. (559), (560), and (572). For the latter the equation would then read:

$$R^* = \frac{(n-6)(1-n)}{4\cdot\left(f+\dfrac{C_f}{M}\right)^3}\cdot\overbrace{\left(\tilde{\nabla}_g f\right)^2}^{=f_{,\alpha}g^{\alpha\beta}f_{,\beta}} = \frac{(n-6)(1-n)}{4\cdot\left(f+\dfrac{C_f}{M}\right)^3}\cdot C_D^2\cdot M^2\cdot f^2$$

$$\Rightarrow R^*\cdot\frac{4\cdot\left(f+\dfrac{C_f}{M}\right)^3}{(n-6)(1-n)} = C_D^2\cdot M^2\cdot f^2 \xrightarrow{\Psi^2\equiv M^2\cdot f^2} R^*\cdot\frac{4\cdot\left(\dfrac{\Psi}{M}+\dfrac{C_f}{M}\right)^3}{(n-6)(1-n)} = C_D^2\cdot\Psi^2.$$
(574)

Again, we point out that for here and now (for the reason of simplicity and brevity mainly), we need not care about potential inner vector characters of the mass, the function f, and the operator terms. For the moment we just assume that this does not have any influence on the 3-generation problem we want to consider here. The proof for this can easily be obtained by applying $f = h_\alpha q^\alpha$ in all derivations below. It will not change the results with regards to the 3-generation mass problem.

Reshaping of Eq. (574) leads to:

$$\frac{R^*}{M^3}\cdot\frac{4\cdot\left(\Psi+C_f\right)^3}{(n-6)(1-n)} \xrightarrow{R^{**}\equiv\frac{R^*}{M^3}\frac{4}{(n-6)(1-n)}} = R^{**}\cdot\left(\Psi+C_f\right)^3 = C_D^2\cdot\Psi^2. \quad (575)$$

This, again, is a polynomial of third order and it can have three solutions. Expansion helps us to realize how this will possibly also solve the 3-generation mass problem:

$$R^{**}\cdot\left(\Psi+C_f\right)^3 - C_D^2\cdot\Psi^2 = 0$$

$$\Rightarrow \quad \left(C_f\right)^3 - C_D^2\cdot\frac{\Psi^2}{R^{**}} + 3\cdot\left(C_f\right)^2\cdot\Psi + 3\cdot C_f\cdot\Psi^2 + \Psi^3 = 0. \quad (576)$$

The general solution to a three-order polynomial could be given via the following product form:

$$\left(\Psi-\Psi_1\right)\cdot\left(\Psi-\Psi_2\right)\cdot\left(\Psi-\Psi_3\right) =$$

$$\Psi^3 - \Psi^2\cdot\left(\Psi_1+\Psi_2+\Psi_3\right) + \Psi\cdot\left(\Psi_1\Psi_2+\Psi_1\Psi_3+\Psi_2\Psi_3\right) - \Psi_1\Psi_2\Psi_3. \quad (577)$$

Comparing the latter with the last line in Eq. (576):

$$(C_f)^3 - C_D^2 \cdot \frac{\Psi^2}{R^{**}} + 3 \cdot (C_f)^2 \cdot \Psi + 3 \cdot C_f \cdot \Psi^2 + \Psi^3 = 0 \qquad (578)$$

gives us:

$$3 \cdot C_f - \frac{C_D^2}{R^{**}} = -\left(\Psi_1 + \Psi_2 + \Psi_3\right)$$

$$3 \cdot (C_f)^2 = \Psi_1 \Psi_2 + \Psi_1 \Psi_3 + \Psi_2 \Psi_3$$

$$(C_f)^3 = -\Psi_1 \Psi_2 \Psi_3. \qquad (579)$$

Thus, we have obtained the three generations of quantum gravity solutions to the combined mass-times-f-function, given via $\Psi \equiv M \cdot f \equiv \dfrac{m \cdot c}{\hbar} \cdot f$, as functions or dependencies of the Ricci scalar R^* of the quantum-gravity (just scalar) variated metric $G_{\alpha\beta} = F[f]^* g_{\alpha\beta}$, the mass-values $M \equiv \dfrac{m \cdot c}{\hbar}$ and a constant C_f.

Please note: As the expression $\Psi_1 + \Psi_2 + \Psi_3$ on the right-hand side in the first line in Eq. (579) should be just a constant and thus, could not depend on the various masses M_i, we have to demand the Ricci scalar R^* to be directly connected with the M_i via a constant const:

$$\frac{C_D^2}{R^{**}} = \text{const} = \frac{C_D^2}{\dfrac{R^*}{M^3} \cdot \dfrac{4}{(n-6)(1-n)}} \quad \Rightarrow \quad R^* = \frac{C_D^2 \cdot M^3 \cdot (n-6)(1-n)}{4 \cdot \text{const}}. \qquad (580)$$

This gives two equations for the extraction of two of the three states Ψ_i out of the other:

$$C_f = -\sqrt[3]{\Psi_1 \Psi_2 \Psi_3}$$

$$\Rightarrow \qquad \frac{C_D^2}{\dfrac{R^*}{M^3} \cdot \dfrac{4}{(n-6)(1-n)}} = \Psi_1 + \Psi_2 + \Psi_3 - 3 \cdot \sqrt[3]{\Psi_1 \Psi_2 \Psi_3}$$

$$\Rightarrow \qquad 0 = \Psi_1 \Psi_2 + \Psi_1 \Psi_3 + \Psi_2 \Psi_3 - 3 \cdot \left(\sqrt[3]{\Psi_1 \Psi_2 \Psi_3}\right)^2. \qquad (581)$$

We find that with a vanishing constant C_f (the shift of the function f with the value $f = 0$ moving away from the origin), we would only obtain one non-trivial solution, namely:

$$\Psi_1 = \text{const}, \Psi_2 = \Psi_3 = 0; \ \Psi_2 = \text{const}, \Psi_1 = \Psi_3 = 0; \ \Psi_3 = \text{const}, \Psi_2 = \Psi_1 = 0.$$

For a vanishing constant, we find equality of all Ψ_i with $\Psi_i = \dfrac{1}{3} \cdot C_f$.

More discussion and the other options shall be presented elsewhere. The fact that we do not consider this a full solution to the 3-generation mass problem lays in the fact that we had to assume the existence of eigensolutions to the operator $f_{,\alpha} g^{\alpha\beta} f_{,\beta}$ in addition to the fact that f also has to be a Laplace function. It needs to be demonstrated that such solutions exist and that they have the right behavior. We will leave this task for later.

Please note that, as said in the sub-sections above, one has to see it as a major restriction that the gravity wave function, which after all should be a function of the coordinates, is bound by so many harsh conditions (579). This aspect will be discussed—and potentially mitigated—further below.

More discussion and the other options are presented in chapter 7 of this book or in [43].

6.12.8.6 Finally, Klein–Gordon + Dirac: The simplest explanation for the 3-generation problem

This time, we will not try to get rid of one of the two differential operators in Eq. (245) or Eq. (541), but assume them to have eigensolutions with eigenvalues A and B, as shown in Eq. (547). Now, just as in the last sub-section, we apply the following simplest linear form for $F[f]$ via $F[f] = (f + C_f/M)$. This gives us, taking Eq. (245) and this time without the assumption of f being a Laplace solution, the following result:

$$
\begin{aligned}
R^* &= \frac{(1-n)}{4 \cdot \left(f + \dfrac{C_f}{M}\right)^3} \cdot \left((6-n) \cdot f_{,\alpha} g^{\alpha\beta} f_{,\beta} + 4 \cdot \left(f + \frac{C_f}{M}\right) \cdot \Delta f \right) \\
&= f \cdot \frac{(1-n)}{4 \cdot \left(f + \dfrac{C_f}{M}\right)^3} \cdot \left(A \cdot (6-n) \cdot f + 4 \cdot B \cdot \left(f + \frac{C_f}{M}\right) \right).
\end{aligned}
\tag{582}
$$

Now, we follow the path outlined in the sub-sections above and algebraically solve the equation above. In order to make things easier with respect to the 3-generation problem, we substitute as follows:

$$\Psi = M \cdot f; \ R^{**} = 4 \cdot \frac{R^*}{M}; \ B_s = 4 \cdot (1-n) \cdot B; \ A_s = A \cdot (1-n) \cdot (n-6)$$

$$\Rightarrow \qquad R^* = \frac{\Psi}{M} \cdot \frac{(1-n)}{4 \cdot \left(\dfrac{\Psi}{M} + \dfrac{C_f}{M}\right)^3} \cdot \left(A \cdot (6-n) \cdot \frac{\Psi}{M} + 4 \cdot B \cdot \left(\frac{\Psi}{M} + \frac{C_f}{M}\right)\right)$$

$$\Rightarrow \qquad 4 \cdot R^* \cdot \left(\frac{\Psi}{M} + \frac{C_f}{M}\right)^3 = \frac{\Psi}{M} \cdot (1-n) \cdot \left(A \cdot (6-n) \cdot \frac{\Psi}{M} + 4 \cdot B \cdot \left(\frac{\Psi}{M} + \frac{C_f}{M}\right)\right)$$

$$\Rightarrow \qquad R^{**} \cdot \left(\Psi + C_f\right)^3 = 4 \cdot \frac{R^*}{M} \cdot \left(\Psi + C_f\right)^3 = \Psi \cdot \left(A_s \cdot \Psi + B_s \cdot \left(\Psi + C_f\right)\right), \quad (583)$$

which gives—after expansion and division by R^{**}—from Eq. (582):

$$(C_f)^3 - \frac{B_s \cdot C_f}{R^{**}} + \left(3 \cdot (C_f)^2 - \frac{A_s}{R^{**}} - \frac{B_s}{R^{**}}\right) \cdot \Psi + 3 \cdot C_f \cdot \Psi^2 + \Psi^3 = 0. \qquad (584)$$

Again, we point out that for here and now (for the reason of simplicity and brevity mainly), we need not care about potential inner vector characters of the mass, the function f, and the operator terms. For the moment we just assume that this does not have any influence on the 3-generation problem we want to consider here. The proof for this can easily be obtained by applying $f = h_\alpha q^\alpha$ in all derivations below. It will not change the results with regards to the 3-generation mass problem.

Equation (584) is a polynomial of third order and it can have three solutions. The general solution to a three-order polynomial could be given via the following product form:

$$\left(\Psi - \Psi_1\right) \cdot \left(\Psi - \Psi_2\right) \cdot \left(\Psi - \Psi_3\right) =$$

$$\Psi^3 - \Psi^2 \cdot \left(\Psi_1 + \Psi_2 + \Psi_3\right) + \Psi \cdot \left(\Psi_1 \Psi_2 + \Psi_1 \Psi_3 + \Psi_2 \Psi_3\right) - \Psi_1 \Psi_2 \Psi_3. \qquad (585)$$

Comparing the latter with Eq. (584) gives us:

$$3 \cdot C_f = -\left(\Psi_1 + \Psi_2 + \Psi_3\right)$$

$$3 \cdot (C_f)^2 - \frac{A_s}{R^{**}} - \frac{B_s}{R^{**}} = \Psi_1 \Psi_2 + \Psi_1 \Psi_3 + \Psi_2 \Psi_3$$

$$(C_f)^3 - \frac{B_s \cdot C_f}{R^{**}} = -\Psi_1 \Psi_2 \Psi_3. \qquad (586)$$

Thus, we have obtained the three generations of quantum gravity solutions to the combined mass-times-f-function, given via $\Psi \equiv M \cdot f \equiv \dfrac{m \cdot c}{h} \cdot f$, as functions or dependencies of the Ricci scalar $R*$ of the quantum-gravity (just scalar) variated metric $G_{\alpha\beta} = F[f]*g_{\alpha\beta}$, the mass-values $M \equiv \dfrac{m \cdot c}{h}$ and a constant C_f.

We see it as major restriction to the approach that certain combinations of the functions Ψ_1, Ψ_2, Ψ_3 have to combine to constants as given in Eq. (586). In this context it has to be noted that, as the expressions $\Psi_1 + \Psi_2 + \Psi_3$, $\Psi_1\Psi_2 + \Psi_1\Psi_3 + \Psi_2\Psi_3$, $\Psi_1\Psi_2\Psi_3$ on the right-hand side in (586) should be just constants and thus, could not depend on the various masses M_j, we have to demand the Ricci scalar R^* to be directly connected with the M_i via the two constants $const_1$ and $const_2$:

$$
\text{(I)}\left\{
\begin{aligned}
&\frac{A_s}{R^{**}} + \frac{B_s}{R^{**}} = const_1 = \frac{(1-n)\cdot M}{4\cdot R^*}\cdot\big(A\cdot(n-6)+4\cdot B\big)\\[2mm]
&\Rightarrow\ R^* = \frac{(1-n)}{4}\cdot M\cdot\frac{\big(A\cdot(n-6)+4\cdot B\big)}{const_1}
\end{aligned}
\right.
$$

$$
\text{(II)}\left\{
\begin{aligned}
&\frac{B_s\cdot C_f}{R^{**}} = const_2 = \frac{(1-n)\cdot B\cdot C_f}{R^*}\cdot M\\[2mm]
&\Rightarrow\ R^* = \frac{(1-n)\cdot B\cdot C_f}{const_2}\cdot M.
\end{aligned}
\right.
\tag{587}
$$

$$
\Leftrightarrow
$$

$$
\Rightarrow\qquad \frac{\big(A\cdot(n-6)+4\cdot B\big)}{const_1} = 4\cdot\frac{B\cdot C_f}{const_2}.
\tag{588}
$$

Assuming that the parameters C_f, A, and B do not directly depend on the various masses M_i within the solutions for the Ψ_1, Ψ_2, Ψ_3, we have a connection of the eigenvalues A and B of the differential operators in Eq. (582) (first line). Most interestingly, in 6 dimensions, Eq. (588) significantly simplifies and gives us:

$$
\Rightarrow\qquad C_f = \frac{const_2}{const_1} = \frac{\dfrac{(1-n)\cdot B\cdot C_f}{R^*}\cdot M}{\dfrac{(1-n)}{4\cdot R^*}\cdot\big(A\cdot(n-6)+4\cdot B\big)\cdot M} \xrightarrow{n=6} = C_f.
\tag{589}
$$

As before we can extract two equations for the determination of two of the three states Ψ_i out of the other:

$$C_f = -\frac{\left(\Psi_1 + \Psi_2 + \Psi_3\right)}{3}$$

$$\Rightarrow \qquad \text{const}_1 = \frac{1}{3}\cdot\left(\Psi_1 + \Psi_2 + \Psi_3\right)^2 - \Psi_1\Psi_2 - \Psi_1\Psi_3 - \Psi_2\Psi_3$$

$$\Rightarrow \qquad \text{const}_2 = \Psi_1\Psi_2\Psi_3 + \left(\frac{\left(\Psi_1 + \Psi_2 + \Psi_3\right)}{3}\right)^3. \qquad (590)$$

More discussion and the other options shall be presented elsewhere.

In contrast to the above sub-sections, this time we have no need for an additional condition $\Delta f = 0$, as we had to demand it in the two previous sub-sections about the Dirac case. We only need to demand that the function f has eigenvalue solutions A and B to the differential operators in Eq. (582) (first line) as follows $\Delta f = B \cdot f$ and $f_{,\alpha}g^{\alpha\beta}f_{,\beta} = A \cdot f^2$. This means, the moment these eigenvalues exist, we automatically have an explanation for the 3-generation problem of the elementary particles, namely simply through the fact that the Ricci scalar of a scale variated metric with the simple variation $G_{\alpha\beta} = F[f]^*g_{\alpha\beta} = (f + C_f/M)^*g_{\alpha\beta}$ forms an algebraic equation of the third order and by solving this equation with respect to f and M, we obtain three solutions for the combination of mass M and the quantum gravity function f.

It also should be pointed out that the combined demand for the existence of eigenvalue solutions to both operators $\Delta f = B \cdot f$ and $f_{,\alpha}g^{\alpha\beta}f_{,\beta} = A \cdot f^2$ in 4 dimensions would automatically favor octonion, which is to say the bigger brothers of Dirac's quaternions, as just the obvious tool for the solution of the subsequent complete differential equation in Eq. (582) (first line).

However, as demonstrated in section "6.12.2 Getting Rid of Quaternions," there may be a way around these complex features and the subsequent cumbersome evaluations with the many restrictions they automatically bring with themselves. Thus, in the next section, we will try to "Avoid the Octonions" and still find first order differential equations to eigenvalue equations with mixed differential operators $\Delta f = B \cdot f$ and $f_{,\alpha}g^{\alpha\beta}f_{,\beta} = A \cdot f^2$ leading to:

$$R^* = \frac{(1-n)}{4\cdot\left(f+\dfrac{C_f}{M}\right)^3}\cdot\left((6-n)\cdot f_{,\alpha}g^{\alpha\beta}f_{,\beta} + 4\cdot\left(f+\frac{C_f}{M}\right)\cdot\Delta f\right)$$

$$= f \cdot \frac{(1-n)}{4 \cdot \left(f + \dfrac{C_f}{M} \right)^3} \cdot \left(A \cdot (6-n) \cdot f + 4 \cdot B \cdot \left(f + \frac{C_f}{M} \right) \right)$$

$$\xrightarrow{\text{via octonions ?}}$$

$$\Rightarrow \quad \boxed{f_{,\alpha} g^{\alpha\beta} f_{,\beta} + \frac{4}{(6-n)} \cdot \left(f + \frac{C_f}{M} \right) \cdot \Delta f = A \cdot f^2 + \frac{4 \cdot B}{(6-n)} \cdot f \cdot \left(f + \frac{C_f}{M} \right)}. \quad (591)$$

Please note that, as said in the sub-section above, one has to see it as a major restriction that the gravity wave function, which after all should be a function of the coordinates, is bound by so many harsh conditions (586). This aspect will be discussed—and potentially mitigated—further below.

More discussion and the other options are presented in chapter 7 of this book or in [43].

6.12.8.7 How to avoid the octonions?

Our goal is to decompose Eq. (591) (last line) into a set of equations of first order differential equations. We try to do so via a vector approach as demonstrated in section "6.12.2 Getting Rid of Quaternions." At first we try to bring Eq. (591) into a more suitable form:

$$f_{,\alpha} g^{\alpha\beta} f_{,\beta} + \frac{4}{(6-n)} \cdot \left(f + \frac{C_f}{M} \right) \cdot \Delta f - A \cdot f^2 - \frac{4 \cdot B}{(6-n)} \cdot f \cdot \left(f + \frac{C_f}{M} \right) = 0$$

$$\xrightarrow{g = \det[g_{\alpha\beta}]}$$

$$f_{,\alpha} g^{\alpha\beta} f_{,\beta} + \frac{4}{(6-n)} \cdot \left(f + \frac{C_f}{M} \right) \cdot \frac{\left(\sqrt{g} \cdot g^{\alpha\beta} f_{,\beta} \right)_{,\alpha}}{\sqrt{g}} - A \cdot f^2 - \frac{4 \cdot B}{(6-n)} \cdot f \cdot \left(f + \frac{C_f}{M} \right) = 0.$$

$$(592)$$

For brevity, we introduce a few new symbols and rewrite the last line of Eq. (592):

$$f_{,\alpha} g^{\alpha\beta} f_{,\beta} + \frac{4}{(6-n)} \cdot \left(f + \frac{C_f}{M} \right) \cdot \frac{\left(\sqrt{g} \cdot g^{\alpha\beta} f_{,\beta} \right)_{,\alpha}}{\sqrt{g}} - A \cdot f^2 - \frac{4 \cdot B}{(6-n)} \cdot f \cdot \left(f + \frac{C_f}{M} \right) = 0$$

$$= f_{,\alpha} g^{\alpha\beta} f_{,\beta} + a \cdot (f + \mu) \cdot \frac{\left(\sqrt{g} \cdot g^{\alpha\beta} f_{,\beta} \right)_{,\alpha}}{\sqrt{g}} - A \cdot f^2 - b \cdot f \cdot (f + \mu)$$

$$= f_{,\alpha} g^{\alpha\beta} f_{,\beta} - (A-b) \cdot f^2 - b \cdot \mu \cdot f + a \cdot (f+\mu) \cdot \frac{\left(\sqrt{g} \cdot g^{\alpha\beta} f_{,\beta}\right)_{,\alpha}}{\sqrt{g}}$$

$$= f_{,\alpha} g^{\alpha\beta} f_{,\beta} - X \cdot f^2 - Y \cdot f + Z \cdot (f+\mu) \cdot \frac{\left(\sqrt{g} \cdot g^{\alpha\beta} f_{,\beta}\right)_{,\alpha}}{\sqrt{g}}$$

$$= f_{,\alpha} g^{\alpha\beta} f_{,\beta} - X \cdot f^2 - Y \cdot f + \Theta \cdot \left(\sqrt{g} \cdot g^{\alpha\beta} f_{,\beta}\right)_{,\alpha} = 0. \tag{593}$$

Now, just as shown in section "6.12.2 Getting Rid of Quaternions," we decompose (factorize) (593) with the help of two—this time different— $\mathbf{V_\Omega}$-vectors and—for illustration—assume a 4-dimensional space-time:

$$0 = \left\{ \begin{array}{l} \begin{pmatrix} a_0 \cdot f_{,0} + b_0 \cdot f_{,1} + c_0 \cdot f_{,2} + d_0 \cdot f_{,3} + r_0 \cdot f + s_0 \\ +\Theta \cdot e_0 \cdot \partial_0 + \Theta \cdot g_0 \cdot \partial_1 + \Theta \cdot o_0 \cdot \partial_2 + \Theta \cdot p_0 \cdot \partial_3 \end{pmatrix} \\ \begin{pmatrix} b_1 \cdot f_{,1} + c_1 \cdot f_{,2} + d_1 \cdot f_{,3} + r_1 \cdot f + s_1 + \Theta \cdot e_1 \cdot \partial_0 \\ +\Theta \cdot g_1 \cdot \partial_1 + \Theta \cdot o_1 \cdot \partial_2 + \Theta \cdot p_1 \cdot \partial_3 \end{pmatrix} \\ \begin{pmatrix} c_2 \cdot f_{,2} + d_2 \cdot f_{,3} + r_2 \cdot f + s_2 + \Theta \cdot e_2 \cdot \partial_0 \\ +\Theta \cdot g_2 \cdot \partial_1 + \Theta \cdot o_2 \cdot \partial_2 + \Theta \cdot p_2 \cdot \partial_3 \end{pmatrix} \\ \vdots \\ \Theta \cdot p_9 \cdot \partial_3 \end{array} \right.$$

$$\cdot \begin{bmatrix} \begin{pmatrix} a_0 \cdot f_{,0} + b_0 \cdot f_{,1} + c_0 \cdot f_{,2} + d_0 \cdot f_{,3} + r_0 \cdot f + s_0 \\ +e_0 \cdot E \cdot f_{,0} + G \cdot g_0 \cdot f_{,1} + O \cdot o_0 \cdot f_{,2} + P \cdot p_0 \cdot f_{,3} \end{pmatrix} \\ \begin{pmatrix} b_1 \cdot f_{,1} + c_1 \cdot f_{,2} + d_1 \cdot f_{,3} + r_1 \cdot f + s_1 \\ +E \cdot e_1 \cdot f_{,0} + G \cdot g_1 \cdot f_{,1} + O \cdot o_1 \cdot f_{,2} + P \cdot p_1 \cdot f_{,3} \end{pmatrix} \\ \begin{pmatrix} c_2 \cdot f_{,2} + d_2 \cdot f_{,3} + r_2 \cdot f + s_2 \\ +E \cdot e_2 \cdot f_{,0} + G \cdot g_2 \cdot f_{,1} + O \cdot o_2 \cdot f_{,2} + P \cdot p_2 \cdot f_{,3} \end{pmatrix} \\ \vdots \\ P \cdot p_9 \cdot f_{,3} \end{bmatrix}. \tag{594}$$

Thereby the symbols with the capital letters E, G, O, and P are to be understood as functions to assure the correct outcome for the Laplace operator $\dfrac{\left(\sqrt{g} \cdot g^{\alpha\beta} f_{,\beta}\right)_{,\alpha}}{\sqrt{g}}$ after the scalar product. As we have 55 components, we have to determine via comparison of the scalar product outcome with Eq. (593), we here refrain from any further consideration of this still very cumbersome approach, but leave it for later.

Note: Only the second—the inner—factor is of "ordinary" character, while the first factor—the outer—is an operator. As we see no way—not now anyway—to decompose the general Laplace operator in arbitrary coordinates in a completely symmetric or at least non-operational way (meaning that none of the factors has to operate on the other one), we can only demand the inner factor to satisfy the whole equation via the inner or intrinsic solution:

$$0 = \begin{cases} \begin{pmatrix} a_0 \cdot f_{,0} + b_0 \cdot f_{,1} + c_0 \cdot f_{,2} + d_0 \cdot f_{,3} + r_0 \cdot f + s_0 \\ + e_0 \cdot E \cdot f_{,0} + G \cdot g_0 \cdot f_{,1} + O \cdot o_0 \cdot f_{,2} + P \cdot p_0 \cdot f_{,3} \end{pmatrix} \\ \begin{pmatrix} b_1 \cdot f_{,1} + c_1 \cdot f_{,2} + d_1 \cdot f_{,3} + r_1 \cdot f + s_1 \\ + E \cdot e_1 \cdot f_{,0} + G \cdot g_1 \cdot f_{,1} + O \cdot o_1 \cdot f_{,2} + P \cdot p_1 \cdot f_{,3} \end{pmatrix} \\ \begin{pmatrix} c_2 \cdot f_{,2} + d_2 \cdot f_{,3} + r_2 \cdot f + s_2 \\ + E \cdot e_2 \cdot f_{,0} + G \cdot g_2 \cdot f_{,1} + O \cdot o_2 \cdot f_{,2} + P \cdot p_2 \cdot f_{,3} \end{pmatrix} \\ \vdots \\ P \cdot p_9 \cdot f_{,3} \end{cases} . \tag{595}$$

This is quite similar to the classical Dirac approach, where we also have operators as factors (only that they are built of quaternions). Please note: Just as in section "6.12.2 Getting Rid of Quaternions" the scalar function f has to be decomposed into a 10-vector $f = q^\lambda h_\lambda$ in order to satisfy all 10 equations.

6.12.8.8 Interpretation and particle at rest

Finding a general solution to Eq. (592) seems to be rather difficult even in the simplest cases. So, we were not able to find a solution for the so-called particle at rest problem with $f = f[t]$. Only in the case of $C_f = 0$, we found the solution:

$$f[t] = C_{t1} \cdot \cos\left[\frac{\sqrt{(n-2) \cdot (A \cdot (6-n) - 4B)}}{4} \cdot (t + C_{t2}) \right]^{\frac{4}{-2+n}} . \tag{596}$$

These are the usual oscillating solutions for positive values under the square root. Thus, the solution, even though we did not resort to the Dirac factorization, already is not very different from the classical result (c.f. [29] and "particle at rest solution" in the text books). The oscillations are completely symmetric in positive and negative directions of time and thus, there is no asymmetry between matter and antimatter. We guess, however, that things are changing here the moment we allow for $C_f \neq 0$.

Could this lead then to some asymmetry, potentially also explaining the matter antimatter discrepancy of our universe?

Summing up now what we have found, we can state the following:

(a) When introducing a scale-variated metric of the kind $G_{\alpha\beta} = F[f]^* g_{\alpha\beta}$, we obtain a Ricci scalar R^* which contains the function f in such a form, that we can extract both, Dirac and Klein–Gordon equations. As the origin is completely geometry, respectively of 100% metric character, we consider our approach as a quantum gravity one and, thus, consequently, the function $F[f]$, respectively f as a quantum gravity wave function.

(b) Adding dimensions with special dependencies to the "main (space-time) dimensions," which we here saw AND treated as entanglement, brings all forms of inertia, spin and stuff, even when demanding the total scalar curvature R^* to vanish.

(c) Not demanding the total scalar curvature to vanish and assuming that there are eigensolutions to the differential equation of the Ricci scalar R^* for the function f, leads to algebraic equations allowing for three different solutions to the combined product of mass and the quantum gravity wave function f. We think that this is the answer to the riddle of the three generations of elementary particles.

(d) The interesting point is here, that all classical quantum effects, including an obviously rather potent solution to the 3-generaion problem, arise from a simple scale factor $F[f]$ to the metric $g_{\alpha\beta}$.

(e) We also saw, even though we were not able—so far anyway—to completely solve the subsequent complex eigenequation for the function f resulting from the adapted Ricci scalar, that the so-called "particle at rest" sports the usual oscillations in time as we already know them from the classical Dirac theory [29].

(f) From this automatically also follows (among other things, which should be discussed later on) that particle masses are apparently just static (or quasi-static) scale oscillations of space and time. In other words, where space-time vibrates, there is mass; and where

space-time vibrates in a stable manner, there are stable particles with mass.

(g) **Still we see it as rather problematic that our combined mass-times-gravity-wave-functions shall form constants in a certain manner as given in Eqs. (586) and (590). This extremely restricts the possibility for such solutions and may be seen as a major drawback.**

6.12.8.9 The further path forward

We already saw that with the assumption for eigenfunction solutions to the differential operators in Eq. (245) (or see Eq. (582) first to second line), we could end up with three solutions to the product of mass times the wave function f, but as the problem of the three generations of elementary particles (Fig. 6.17) also requires the calculation of the exact masses of these particles, we cannot consider our task done yet. In order to move forward, we start with the assumption of the existence of eigenvalue solutions to Eq. (245) as incorporated in Eq. (582), which leads us to:

$$\Delta f = B \cdot f. \tag{597}$$

$$f_{,\alpha} g^{\alpha\beta} f_{,\beta} = A \cdot f^2. \tag{598}$$

Figure 6.17   The three generations of particles, by Livia Schwarzer.

Thereby the two eigenequations (597) and (598) could be brought together, forming:

$$(\Delta f)_{,\alpha} g^{\alpha\beta} (\Delta f)_{,\beta} = B^2 \cdot A \cdot f^2. \tag{599}$$

We find a particle at rest solution in six forms, three for each matter and antimatter. Only two (one matter and one antimatter particle) of the six solutions are stable oscillations.

Please note that in the case of the particle of rest, it would be much more convenient to seek the solution in the following way:

$$\Delta f = b^2 \cdot f, \tag{600}$$

$$(\Delta f)_{,\alpha} g^{\alpha\beta} (\Delta f)_{,\beta} = b^4 \cdot a^6 \cdot f^2. \tag{601}$$

Thereby the condition (600) gives separate solutions for b for each of the six solutions to Eq. (601). With the outcomes from Eqs. (601) and (600), we should now go into the Einstein–Hilbert action:

$$\delta W = 0 = \delta \int_V d^n x \left(\sqrt{-G} \cdot \left(R^* - 2\Lambda + L_M \right) \right), \tag{602}$$

where, in difference to the introduction part and Eq. (240), we took into account the cosmological constant Λ and the Lagrange matter density L_M. Setting Eq. (591) into Eq. (602) results in the variation:

$$\delta W = 0 = \delta \int_V d^n x \left(\begin{array}{c} \sqrt{-g \cdot \left(f + \dfrac{C_f}{M} \right)^n} \\ \times \left(f \cdot \dfrac{(1-n)}{4 \cdot \left(f + \dfrac{C_f}{M} \right)^3} \left(A \cdot (6-n) \cdot f + 4 \cdot B \cdot \left(f + \dfrac{C_f}{M} \right) \right) - 2\Lambda + L_M \right) \end{array} \right). \tag{603}$$

Ignoring the cosmological constant and not worrying about the kind of variation (which is to say with respect to which quantity it will be varied), we just obtain:

$$\delta W \Rightarrow \delta_? W = 0$$

$$= \int_V d^n x \left| \begin{array}{c} \left(\left(f \cdot \dfrac{(1-n)}{4 \cdot \left(f + \dfrac{C_f}{M} \right)^3} \left(A \cdot (6-n) \cdot f + 4 \cdot B \cdot \left(f + \dfrac{C_f}{M} \right) \right) + L_M \right) \right. \\[2em] \times \delta_? \sqrt{-g \cdot \left(f + \dfrac{C_f}{M} \right)^n} + \sqrt{-g \cdot \left(f + \dfrac{C_f}{M} \right)^n} \\[2em] \left. \times \delta_? \left(f \cdot \dfrac{(1-n)}{4 \cdot \left(f + \dfrac{C_f}{M} \right)^3} \left(A \cdot (6-n) \cdot f + 4 \cdot B \cdot \left(f + \dfrac{C_f}{M} \right) \right) + L_M \right) \right) \end{array} \right| . \quad (604)$$

Now, we assume that demanding the total Lagrange density term:

$$f \cdot \frac{(1-n)}{4 \cdot \left(f + \dfrac{C_f}{M} \right)^3} \left(A \cdot (6-n) \cdot f + 4 \cdot B \cdot \left(f + \dfrac{C_f}{M} \right) \right) + L_M \quad (605)$$

to give zero only leaves the variation of a potentially "intelligent" zero, which hopefully we can consider to be small:

$$\delta W \Rightarrow \delta_? W = 0$$

$$= \int_V d^n x \left| \begin{array}{c} \overbrace{\left(f \cdot \dfrac{(1-n)}{4 \cdot \left(f + \dfrac{C_f}{M} \right)^3} \left(A \cdot (6-n) \cdot f + 4 \cdot B \cdot \left(f + \dfrac{C_f}{M} \right) \right) + L_M \right)}^{=0} \\[2em] \times \delta_? \sqrt{-g \cdot \left(f + \dfrac{C_f}{M} \right)^n} + \sqrt{-g \cdot \left(f + \dfrac{C_f}{M} \right)^n} \\[2em] \times \delta_? \left(f \cdot \dfrac{(1-n)}{4 \cdot \left(f + \dfrac{C_f}{M} \right)^3} \left(A \cdot (6-n) \cdot f + 4 \cdot B \cdot \left(f + \dfrac{C_f}{M} \right) \right) + L_M \right) \end{array} \right| \Rightarrow$$

$$= \int_V d^n x \left(\times \delta_? \left[f \cdot \underbrace{\frac{(1-n)}{4 \cdot \left(f + \frac{C_f}{M}\right)^3} \left(A \cdot (6-n) \cdot f + 4 \cdot B \cdot \left(f + \frac{C_f}{M}\right) \right) + L_M}_{\approx 0} \right] \right) \cdot$$

$$\overbrace{\sqrt{-g \cdot \left(f + \frac{C_f}{M}\right)^n}}^{=0}$$

$$(606)$$

This, however, gives us exactly the starting point for our evaluation from section "6.12.8.6 Finally, Klein–Gordon + Dirac: The simplest explanation for the 3-generation problem" only that this time we have to deal with the Lagrange density for matter L_M rather than an assumed curvature R^*.

This just changes (582) to:

$$R^* = -L_M = \frac{(1-n)}{4 \cdot \left(f + \frac{C_f}{M}\right)^3} \cdot \left((6-n) \cdot f_{,\alpha} g^{\alpha\beta} f_{,\beta} + 4 \cdot \left(f + \frac{C_f}{M}\right) \cdot \Delta f \right)$$

$$= f \cdot \frac{(1-n)}{4 \cdot \left(f + \frac{C_f}{M}\right)^3} \cdot \left(A \cdot (6-n) \cdot f + 4 \cdot B \cdot \left(f + \frac{C_f}{M}\right) \right), \qquad (607)$$

and via:

$$\Psi = M \cdot f; R^{**} = 4 \cdot \frac{R^*}{M} = -4 \cdot \frac{L_M}{M}; B_s = 4 \cdot (1-n) \cdot B; A_s = A \cdot (1-n) \cdot (n-6)$$

$$\Rightarrow \qquad R^* = \frac{\Psi}{M} \cdot \frac{(1-n)}{4 \cdot \left(\frac{\Psi}{M} + \frac{C_f}{M}\right)^3} \cdot \left(A \cdot (6-n) \cdot \frac{\Psi}{M} + 4 \cdot B \cdot \left(\frac{\Psi}{M} + \frac{C_f}{M}\right) \right)$$

$$\Rightarrow \qquad 4 \cdot R^* \cdot \left(\frac{\Psi}{M} + \frac{C_f}{M}\right)^3 = \frac{\Psi}{M} \cdot (1-n) \cdot \left(A \cdot (6-n) \cdot \frac{\Psi}{M} + 4 \cdot B \cdot \left(\frac{\Psi}{M} + \frac{C_f}{M}\right) \right)$$

$$\Rightarrow \qquad R^{**} \cdot \left(\Psi + C_f\right)^3 = 4 \cdot \frac{R^*}{M} \cdot \left(\Psi + C_f\right)^3 = \Psi \cdot \left(A_s \cdot \Psi + B_s \cdot \left(\Psi + C_f\right)\right), \qquad (608)$$

subsequently gives us:

$$\left(C_f\right)^3 + M \cdot \frac{B_s \cdot C_f}{4 \cdot L_M} + \left(3 \cdot \left(C_f\right)^2 + \frac{M \cdot A_s}{4 \cdot L_M} + \frac{M \cdot B_s}{4 \cdot L_M}\right) \cdot \Psi + 3 \cdot C_f \cdot \Psi^2 + \Psi^3 = 0.$$

$$(609)$$

This way, we have connected our approach for the solutions of the 3-generation problem with the Einstein–Hilbert action. However, it became clear that this is neither the only way nor is it a perfect one, due to the conditions (586), we have to demand rather harsh boundaries for the functions Ψ_k. This dramatically restricts the options for massive particle solutions, suitable metrics, and matter forms in general. On the other hand, we know that the so far unfixed variation in Eq. (604) opens up quite a variety of new doors to move on from here. The same holds for the fact that the eigenvalue approach from Eqs. (597) and (598), leading to polynomial conditions for the constants A and B.

6.12.8.10 Moving on

Thus, in order to move on, we go back to a more general form for the function $F[f]$ and obtain for the Ricci scalar R^*:

$$R^* = \frac{1}{F[f]^3} \cdot \left(\begin{array}{c} \left(C_{N1} \cdot \left(\dfrac{\partial F[f]}{\partial f}\right)^2 - C_{N2} \cdot F[f] \cdot \dfrac{\partial^2 F[f]}{\partial f^2} \right) \cdot f_{,\alpha} g^{\alpha\beta} f_{,\beta} \\[4mm] - C_{N2} \cdot F[f] \cdot \dfrac{\partial F[f]}{\partial f} \cdot \Delta_g f \end{array} \right)$$

$$\xrightarrow[\substack{C_{N1} = -\frac{3}{2} + \frac{7}{4} \cdot n - \frac{1}{4} \cdot n^2 \\ C_{N2} = n - 1}]{} = \frac{1-n}{4 \cdot F[f]^3} \cdot \left(\begin{array}{c} \left((n-6) \cdot F'[f]^2 + 4 \cdot F[f] \cdot F''[f]\right) \cdot f_{,\alpha} g^{\alpha\beta} f_{,\beta} \\[2mm] + 4 \cdot F[f] \cdot F'[f] \cdot \Delta_g f \end{array} \right)$$

$$= \frac{1-n}{4 \cdot F[f]^3} \cdot \left(\left((n-6) \cdot F'[f]^2 + 4 \cdot F[f] \cdot F''[f]\right) \cdot A \cdot f^2 + 4 \cdot F[f] \cdot F'[f] \cdot B \cdot f\right)$$

$$\text{with}: \quad F''[f] = \frac{\partial^2 F[f]}{\partial f^2}; \quad F'[f] = \frac{\partial F[f]}{\partial f}. \qquad (610)$$

Setting Eq. (610) into Eq. (240), respectively Eq. (602) results in the variation:

$$\delta W = 0 = \delta \int_V d^n x \left(\sqrt{-g \cdot F[f]^n} \cdot \left(f \cdot \frac{(1-n)}{4 \cdot F[f]^3} \cdot \left(A \cdot \left(\begin{array}{c} (n-6) \cdot F'[f]^2 \\ +4 \cdot F[f] \cdot F''[f] \end{array} \right) \cdot f \\ +4 \cdot B \cdot F[f] \cdot F'[f] \end{array} \right) - 2\Lambda + L_M \right) \right).$$

$$(611)$$

At this point, we do not fix the variation, which is to say, it can be a variation with respect to anything. Now, we go back to the eigenequations (597) and (598), and see easily from the combined form (599) that in the case of the particle at rest, it would be convenient to start with eigenvalues a and b instead of A and B in the form:

$$\Delta f = b^2 \cdot f, \qquad (612)$$

$$(\Delta f)_{,\alpha} g^{\alpha\beta} (\Delta f)_{,\beta} = b^4 \cdot a^6 \cdot f^2. \qquad (613)$$

Thereby the condition (600) gives separate solutions for b for each of the six solutions to Eq. (601). This changes the variational task (611) to:

$$\delta W = 0 = \delta \int_V d^n x \left(\sqrt{-g \cdot F[f]^n} \cdot \left(f \cdot \frac{(1-n)}{4 \cdot F[f]^3} \cdot \left(a^6 \cdot \left(\begin{array}{c} (n-6) \cdot F'[f]^2 \\ +4 \cdot F[f] \cdot F''[f] \end{array} \right) \cdot f \\ +4 \cdot b^2 \cdot F[f] \cdot F'[f] \end{array} \right) - 2\Lambda + L_M \right) \right)$$

$$(614)$$

and, also considering the dependency of $b(a)$ as discussed above, already shows the algebraic dependency we need to—hopefully—explain the 3-generation problem. This is because the particle at rest approaches $f = f[t]$, set into Eq. (601) and assuming a simple Minkowski space-time results in:

$$f[t] = e^{\pm(-1)^{5/6} a \cdot b^{2/3} c^{1/3} t} C_{1\pm} + e^{\pm i \cdot a \cdot b^{2/3} c^{1/3} t} C_{2\pm} + e^{\pm(-1)^{1/6} a \cdot b^{2/3} c^{1/3} t} C_{3\pm}. \quad (615)$$

In addition, we have the condition (600), which results in dependencies of the type $b \sim a^3$. The task left is to find a suitable variation and a function $F[f]$ to give us an algebraic equation of (effectively) the third order with respect to a (or a^2). Thereby we take into account the similarity of the Dirac solutions [7] and our solution (615) and recognize the $M \sim a^2$-dependency. We note that still the metric is not fixed, but—for the reason of simplicity—here was assumed to be a Minkowski metric. Still we already have (without the introduction of quaternions) three matter and three antimatter solutions plus the typical mass-exponent just like in the Dirac approach for the particle at rest.

We, therefore, think that the path forward to solve the 3-generation riddle is rather clear now and we leave it to skilled readers to solve the problem completely.

6.13 Toward an Understanding of Fundamental Interactions and Their Origin

Please note that Eq. (520) could also be extended toward higher and even positive exponents:

$$
\left(
\begin{array}{l}
\ldots + C_{R4} \cdot e^{-\left(\frac{r_s}{r}\right)^4} \cdot \left(\frac{r_s}{r}\right)^4 + C_{R3} \cdot e^{-\left(\frac{r_s}{r}\right)^3} \cdot \left(\frac{r_s}{r}\right)^3 + C_{R2} \cdot e^{-\left(\frac{r_s}{r}\right)^2} \cdot \left(\frac{r_s}{r}\right)^2 \\[2em]
+ C_{R1} \cdot e^{-\frac{r_s}{r}} \cdot \frac{r_s}{r} + C_{R0} \cdot r_s + C_c + C_{R(-1)} \cdot \left(\frac{r_s}{r}\right)^{-1} + C_{R(-2)} \cdot \left(\frac{r_s}{r}\right)^{-2} \\[2em]
+ C_{R(-3)} \cdot \left(\frac{r_s}{r}\right)^{-3} + C_{R(-4)} \cdot \left(\frac{r_s}{r}\right)^{-4} \ldots + \Delta_{\text{other}} + \Delta_{4D}
\end{array}
\right) f = 0,
$$

$$(616)$$

or—in order to avoid monotonic increase for $r \to$ infinity—something like:

$$
\left(
\begin{array}{l}
\ldots + C_{R4} \cdot e^{-\left(\frac{r_s}{r}\right)^4} \cdot \left(\frac{r_s}{r}\right)^4 + C_{R3} \cdot e^{-\left(\frac{r_s}{r}\right)^3} \cdot \left(\frac{r_s}{r}\right)^3 + C_{R2} \cdot e^{-\left(\frac{r_s}{r}\right)^2} \cdot \left(\frac{r_s}{r}\right)^2 \\[2em]
+ C_{R1} \cdot e^{-\frac{r_s}{r}} \cdot \frac{r_s}{r} + C_{R0} \cdot r_s + C_c + C_{R(-1)} \cdot e^{-\left(\frac{r}{r_s}\right)^1} \cdot \left(\frac{r_s}{r}\right)^{-1} + C_{R(-2)} \cdot e^{-\left(\frac{r}{r_s}\right)^2} \cdot \left(\frac{r_s}{r}\right)^{-2} \\[2em]
+ C_{R(-3)} \cdot e^{-\left(\frac{r}{r_s}\right)^3} \cdot \left(\frac{r_s}{r}\right)^{-3} + C_{R(-4)} \cdot e^{-\left(\frac{r}{r_s}\right)^4} \cdot \left(\frac{r_s}{r}\right)^{-4} \ldots + \Delta_{\text{other}} + \Delta_{4D}
\end{array}
\right) f = 0,
$$

$$(617)$$

or even more general with different parameters for the exponents, different buffer functions to avoid singularities (here we only used simple exponential functions), and possible functionalities with respect to the other dimensions (not just the radius as applied here).

As long as concentrating on symmetry of revolution (n-spherical symmetry even), however, we see no reason to restrict the possible Ricci curvature approaches to—potentially very complex—r-dependencies.

We boldly suspect that the various terms in Eq. (520), (616), or (617) have something to do with our fundamental interactions. As there are four of them (strong, weak, gravity, and electromagnetic), we may further assume that either not all terms can be "activated" or that some of them just lead to too compactified or too weak interactions to be experimentally detectable. For instance, with the term $C_{R3} \cdot e^{-\left(\frac{r_s}{r}\right)^3} \cdot \left(\frac{r_s}{r}\right)^3$ already been allocated to gravity (the weakest interaction known), we might deduce that an interaction of the type $C_{R4} \cdot e^{-\left(\frac{r_s}{r}\right)^4} \cdot \left(\frac{r_s}{r}\right)^4$, in fact does exist, but that it is too weak to be seen in practical experiments.

6.13.1 But What Could Be the Origin for such Interaction Dependencies and from Where Do the Differences in Their Strength Come?

In previous papers and books, we often came across an interesting concept of nature, namely that it uses intelligent zeros and ones for its own amazing creation [4, 36, 37].

Could there also be such an explanation for the interactions we see within our portion or scale of the universe?

In order to answer this question, we shall start with the assumption of a global Ricci curvature being zero and then try to end up with a list of interactions as seen above. Here we go:

$$0 = R^* = 1 - 1 = 1 - e^0 = 1 - e^{\frac{r_s}{r} - \frac{r_s}{r}} = 1 - e^{-\frac{r_s}{r}} \cdot e^{\frac{r_s}{r}} = 1 - e^{-\frac{r_s}{r}} \cdot \sum_{k=0}^{\infty} \frac{1}{k!}\left(\frac{r_s}{r}\right)^k$$

$$= 1 - e^{-\frac{r_s}{r}} \cdot \left(\begin{array}{l} \cdots + \frac{1}{5040}\left(\frac{r_s}{r}\right)^7 + \frac{1}{720}\left(\frac{r_s}{r}\right)^6 + \frac{1}{120}\left(\frac{r_s}{r}\right)^5 \\ + \frac{1}{24}\left(\frac{r_s}{r}\right)^4 + \frac{1}{6}\left(\frac{r_s}{r}\right)^3 + \frac{1}{2}\left(\frac{r_s}{r}\right)^2 + \frac{r_s}{r} + 1 \end{array} \right)$$

$$= 1 - e^{-\frac{r_s}{r}} \cdot \left(\frac{r_s}{r}\right)^X \left(\begin{array}{l} \cdots + \frac{1}{5040}\left(\frac{r_s}{r}\right)^{7-X} + \frac{1}{720}\left(\frac{r_s}{r}\right)^{6-X} + \frac{1}{120}\left(\frac{r_s}{r}\right)^{5-X} \\ + \frac{1}{24}\left(\frac{r_s}{r}\right)^{4-X} + \frac{1}{6}\left(\frac{r_s}{r}\right)^{3-X} + \frac{1}{2}\left(\frac{r_s}{r}\right)^{2-X} + \left(\frac{r_s}{r}\right)^{1-X} + \left(\frac{r_s}{r}\right)^{-X} \end{array} \right)$$

$$= 1 - e^{-\frac{r_s}{r}} \cdot \left(\frac{r_s}{r}\right)^X \sum_{k=0}^{\infty} \frac{1}{k!} \cdot \left(\frac{r_s}{r}\right)^{k-X}. \tag{618}$$

Now, we assume that our experimental ability to detect certain of the $1/r^{q=k-X}$ is limited to a stretch of $q = \ldots, -1, 0, 1, 2, 3, \ldots$ or so. Then, in order to find the right differences between the various terms above in order to make them represent our fundamental interactions, we only needed to find a suitable value for X. Not knowing about this X, however, for us, the list of interactions we are able to detect would just appear as:

$$0 = R^* = 1 - e^{-\frac{r_s}{r}} \cdot \left(\frac{r_s}{r}\right)^X \sum_{k=0}^{\infty} \frac{1}{k!} \cdot \left(\frac{r_s}{r}\right)^{k-X}$$

$$= 1 - e^{-\frac{r_s}{r}} \cdot \left(\frac{r_s}{r}\right)^X \left(\underbrace{\ldots + C_{R4} \cdot \left(\frac{r_s}{r}\right)^4 + C_{R3} \cdot \left(\frac{r_s}{r}\right)^3 + C_{R2} \cdot \left(\frac{r_s}{r}\right)^2 + C_{R1} \cdot \frac{r_s}{r} + C_{R0} \cdot r_s \ldots}_{\text{what we might "see"}} \right), \tag{619}$$

where we simply do not know (because we have no means to find out) at what position within the series "we exist."

Please note that our choice for the intelligent zero and the subsequent construction of the intelligent 1 here, namely: $\left(1 - e^{\frac{r_s}{r} - \frac{r_s}{r}} = 1 - e^{-\frac{r_s}{r}} \cdot e^{\frac{r_s}{r}}\right)$, was completely arbitrary. Any other functional split up may also do the job. The only condition is that—in the end—we obtain the right series explaining the differences of the various r-dependencies (or more general functionalities) and thus, our fundamental interactions.

With R^* being a constant, on the other way, Eq. (619) would lose its non-functional term and look as follows:

$$C_R = R^* = C_R \cdot e^{-\frac{r_s}{r}} \cdot \left(\frac{r_s}{r}\right)^X \sum_{k=0}^{\infty} \frac{1}{k!} \cdot \left(\frac{r_s}{r}\right)^{k-X}$$

$$= C_R \cdot e^{-\frac{r_s}{r}} \cdot \left(\frac{r_s}{r}\right)^X \left(\ldots + C_{R4} \cdot \left(\frac{r_s}{r}\right)^4 + C_{R3} \cdot \left(\frac{r_s}{r}\right)^3 + C_{R2} \cdot \left(\frac{r_s}{r}\right)^2 + C_{R1} \cdot \frac{r_s}{r} + C_{R0} \cdot r_s \ldots \right). \tag{620}$$

We may now assume that we only "feel" a certain part of interactions, which is to say, "see" only a few of the terms above and experimentally realize something like:

$$C_R = R^* \simeq C_R \cdot e^{-\frac{r_s}{r}} \cdot \left(\frac{r_s}{r}\right)^X \left(C_{R3} \cdot \left(\frac{r_s}{r}\right)^3 + C_{R2} \cdot \left(\frac{r_s}{r}\right)^2 + C_{R1} \cdot \frac{r_s}{r} + C_{R0} \cdot r_s\right)$$

$$\simeq C_{R3}^* \cdot \left(\frac{r_s}{r}\right)^3 + C_{R3}^* \cdot \left(\frac{r_s}{r}\right)^2 + C_{R3}^* \cdot \frac{r_s}{r} + C_{R3}^* \cdot r_s. \tag{621}$$

Setting $C_{R3}^* = 0$, the corresponding differential equation becomes the classical Schrödinger hydrogen equation:

$$\left(C_{R2}^* \cdot \left(\frac{r_s}{r}\right)^2 + C_{R1}^* \cdot \frac{r_s}{r} + C_{R0}^* \cdot r_s + \Delta_{4D}\right) f = 0. \tag{622}$$

Its solution (including spin $k/2$ forms with k = 1,3,5,7, ...) is given in the appendix together with a few illustrations for the half spin cases.

6.14 Consequences of the Infinity Options Principle and the Disappearance of Quantum Theory with $n \to \infty$

6.14.1 Repetition

It is shown that in a space of infinite number of dimensions (degrees of freedom/options), Quantum Theory does not exist. In fact, and in a bit more mathematical phrasing, with increasing number of dimensions n, the strength of quantum effects decreases and converges toward zero for the limit $n \to \infty$.

6.14.2 Repetition—Part I: Metric Klein–Gordon, Schrödinger, and Dirac Equations Potential

In this chapter (and a few other papers by this author), it is demonstrated how we can find a direct extraction of the most important classical quantum equations from the Ricci scalar R^* of a modified metric of the kind $G_{\alpha\beta}$ = $F[f]^* g_{\alpha\beta}$. Here, for demonstration, the example for the massless Klein–Gordon equation shall be repeated. With $F[f]$ satisfying the condition:

$$C_{N1} \cdot \left(\frac{\partial F[f]}{\partial f}\right)^2 - C_{N2} \cdot F[f] \cdot \frac{\partial^2 F[f]}{\partial f^2} = 0, \tag{623}$$

the corresponding modified Ricci scalar R^* reads:

$$R^* = -\frac{C_{N2} \cdot \dfrac{\partial F[f]}{\partial f}}{F[f]^2} \cdot \Delta_g f \xrightarrow{\;n=4\;} = -\frac{3}{F[f]^2} \cdot \left(\frac{\partial F[f]}{\partial f} \cdot \Delta_g f \right). \quad (624)$$

Thereby the C_{N1} and C_{N2} are constants depending on the number of dimensions n of the space-time being considered. It was shown above that we can write:

$$C_{N1} = -\frac{3}{2} + \frac{7}{4} \cdot n - \frac{1}{4} \cdot n^2$$

$$C_{N2} = n - 1. \quad (625)$$

The corresponding function $F[f]$, which satisfies condition (623) for every dimension n, could be given as:

$$F[f] = C_2 \cdot \left(f \cdot (n-2) + C_1 \right)^{\frac{4}{n-2}} = \left(c_1 \cdot f \cdot (n-2) + c_2 \right)^{\frac{4}{n-2}}. \quad (626)$$

Thereby the constants C_1, C_2, c_1, and c_2 are arbitrary and we can simplify to:

$$F[f] = C_2 \cdot \left(f + C_1 \right)^{\frac{4}{n-2}}. \quad (627)$$

For the case $R^* = 0$, we directly obtain the mass-free Klein–Gordon equation:

$$R^* = 0 = -\frac{C_{N2} \cdot \dfrac{\partial F[f]}{\partial f}}{F[f]^2} \cdot \Delta_g f = \Delta_g f. \quad (628)$$

Regarding the other equations (Schrödinger and Dirac), also including masses, spin, and inertia in general, potential readers can refer to section "6.11 Intermediate Sum-up: Direct Extraction of the Klein–Gordon and the Schrödinger Equation from the Metric Tensor" and—with respect to the Dirac equation—to the next sub-section of this chapter and chapters 9 and 10, where we will present the whole derivation in very comprehensive manner.

6.14.3 Repetition—Part II: Direct Extraction of the Dirac Equation from the Metric Tensor

In section "6.12.1 Repetition: Dirac in the Metric Picture and its Connection to the Classical Quaternion Form," it was demonstrated how we can obtain the Dirac equation from the Ricci scalar R^* of a modified metric of the kind $G_{\alpha\beta} = F[f]^* g_{\alpha\beta}$. With arbitrary $F[f]$ and f being a Laplace function, we get:

$$\frac{F[f]^3 \cdot R^*}{C_{N1} \cdot \left(\dfrac{\partial F[f]}{\partial f}\right)^2 - C_{N2} \cdot F[f] \cdot \dfrac{\partial^2 F[f]}{\partial f^2}} = h_{\lambda,\alpha} q^\lambda g^{\alpha\beta} h_{\lambda,\beta} q^\lambda$$

$$\xrightarrow{g^{\alpha\beta} = \mathbf{e}^\alpha \cdot \mathbf{e}^\beta} = h_{\lambda,\alpha} q^\lambda \mathbf{e}^\alpha \cdot \mathbf{e}^\beta h_{\lambda,\beta} q^\lambda. \tag{629}$$

Thereby the C_{Ni} are constants depending on the number of dimensions (eq. (503)) of the space-time and $\mathbf{q}^\lambda$ is a vector of constants. Please note that we also used the base vector separation of the metric $g^{\alpha\beta} = \mathbf{e}^\alpha \cdot \mathbf{e}^\beta$. With this, the derivation of Dirac equations in space-times of arbitrary dimensions and in metric form is straight forward and has been presented in this chapter already. Also, it presented the transition to the classical Dirac form in 4-dimensional Minkowski space-times. We, therefore, do not repeat any of this here, but only point out a few connections or extensions.

A suitable choice for the determination of the function $F[f]$ seems to be:

$$C_{N1} \cdot \left(\frac{\partial F[f]}{\partial f}\right)^2 - C_{N2} \cdot F[f] \cdot \frac{\partial^2 F[f]}{\partial f^2} = F[f]^2. \tag{630}$$

We obtain the following for $F[f]$:

$$F[f] = C_{f1} \cdot \cos\left[\frac{1}{C_{N2}} \cdot \sqrt{C_{N2} - C_{N1}} \cdot \left(f - C_{f2}\right)\right]^{-\frac{C_{N2}}{C_{N1} - C_{N2}}}$$

$$= C_{f1} \cdot \cosh\left[\frac{1}{C_{N2}} \cdot \sqrt{C_{N1} - C_{N2}} \cdot \left(f - C_{f2}\right)\right]^{-\frac{C_{N2}}{C_{N1} - C_{N2}}} \tag{631}$$

and subsequently for the metric Dirac equation (Eq. (629)):

$$F[f] \cdot R^* = C_{f1} \cdot \cosh\left[\frac{1}{C_{N2}} \cdot \sqrt{C_{N1} - C_{N2}} \cdot \left(f - C_{f2}\right)\right]^{-\frac{C_{N2}}{C_{N1} - C_{N2}}} \cdot R^*$$

$$= h_{\lambda,\alpha} q^\lambda \mathbf{e}^\alpha \cdot \mathbf{e}^\beta h_{\lambda,\beta} q^\lambda. \tag{632}$$

6.14.4 Repetition—Part III: The Infinity Options Principle

Now, we want to evaluate the limits of the quantum wrapper functions $F[f]$ for $n \to \infty$, which gives us for the second order quantum or Laplace-dominated equations (Klein–Gordon, Schrödinger, ...) (407):

$$\lim_{n\to\infty} F[f] = \lim_{n\to\infty} C_2 \cdot \left(f \cdot (n-2) + C_1\right)^{\frac{4}{n-2}} = C_2 \qquad (633)$$

and for the Dirac (or elastic or similar ... see section "6.12.3 Repetition: Elastic Space-Times" or [4]) case:

$$\lim_{n\to\infty} F[f] = \lim_{n\to\infty} C_{f1} \cdot \cos\left[\frac{1}{C_{N2}} \cdot \sqrt{C_{N2} - C_{N1}} \cdot \left(f - C_{f2}\right)\right]^{-\frac{C_{N2}}{C_{N1} - C_{N2}}}$$

$$= \lim_{n\to\infty} C_{f1} \cdot \cosh\left[\frac{1}{C_{N2}} \cdot \sqrt{C_{N1} - C_{N2}} \cdot \left(f - C_{f2}\right)\right]^{-\frac{C_{N2}}{C_{N1} - C_{N2}}}$$

$$= \lim_{n\to\infty} C_{f1} \cdot \cosh\left[\frac{1}{2} \cdot \sqrt{\frac{1}{n-1} - 1} \cdot \left(f - C_{f2}\right)\right]^{\frac{4}{n-2}} = C_{f1}. \qquad (634)$$

We see that in both cases, we obtain constants for the limit $\lim_{n\to\infty}$.

Thus, the resulting quantum transformation or scalar quantum variation of any metric would read $G_{\alpha\beta} = F[f]*g_{\alpha\beta} = \text{const}*g_{\alpha\beta}$. In other words, the functional $F[f]$ would become a constant scaling and no quantum effects could be visible any more. This, however, only holds for the $F[f]$-factor to the whole metric with $n\to\infty$. We still have the possibility of intrinsic scaling factors, which we will discuss later in this book (see chapter 11, especially section "11.12 The Math for Body, Soul, and Universe").

This means, in spaces or space-times of infinitely many options/ degrees of freedom and thus, consequently, dimensions, there can be no global (meaning with respect to the whole metric) quantum effects ... at least not in the classical sense of Klein–Gordon, Schrödinger, Dirac, elastic, and similar!

Intrinsic quantum effects, based on limited numbers of dimensions of inner metrics and mathematically being brought to light via intrinsic scaling factors, would still be possible (c.f. chapter 11).

6.14.5 Consequences

We were able to derive that global quantum effects vanish in systems with an infinite number of dimensions. This means that also the principle uncertainty (quantum originated, therefore fundamental and inevitable) disappears ... at least on a global level.

There is no uncertainty and consequently no Heisenberg uncertainty principle in space or space-times with infinitely many dimensions, when considering the global metric with $n\to\infty$. Intrinsic quantum effects, on the other hand, are still possible (c.f. chapter 11).

In other words, an entity being able to see the whole infinite space-time, feels no uncertainty with respect to the whole, but still realizes uncertainty with respect to the details.

This finding has dramatic consequences with respect to practical problems regarding big system analysis, system science, decision making, and thus, naturally, socio-economics, politics, medicine, military, and so on. It also means that any ideology-containing or perhaps even ideology-dominated process of data handling, analysis, and subsequent decision-making automatically increases its own—intrinsic and fundamental—uncertainty by the simple fact that all ideologies can only exist under the condition of ideology driven information selection ... which is to say, an ideology dominated system always tends to (has to tend to) degeneration and thus, dimensional system reduction. In this sense, the old work of Konrad Lorenz (Nobel Prize in Physiology or Medicine in 1973) about the "Verhausschweinung des Menschen" [C1] (loosely translated one might put it as "the domestication of man") appears in a completely new light. Whatever Lorenz' original intention when writing his paper during the Nazi terror in the Third Reich and his apparent attempt to support some Nazi race theories, it obviously is a fundamental fact that the reduction of information an entity (any entity, including a human being) considers before making decisions, dramatically increases the uncertainty in an inevitable and very principle manner. The number of information a given system can hold—as we have been able to show (e.g., section "6.8 The Schwarzschild Case and the Meaning of the Schwarzschild Radius" and references given there)—is equal to the number of dimensions of this very system. The exclusion of information, be it via degeneration, domestication, ideology-governed "political correctness," false "tolerance" or "politeness," dogmatic narratives as in the current Covid-crisis or simply due to system immanent and entity-typical boundaries, reduces the number of information the entity can hold and digest and thus increases the uncertainty. This uncertainty increase is on the lowest quantum and thus, most fundamental, level of the system in question. Therefore:

(a) Any system-inherent attempt of uncertainty reduction is futile.

(b) However, so it was shown here, increasing the system by incorporating more information (degrees of freedom, properties, and dimensions) and thus, thinking outside the box (meaning the system) is going to lead to success regarding the goal of a reduced uncertainty (see also [1–4]).

C1. K. Lorenz, Durch Domestikation verursachte Störungen arteigenen Verhaltens, *Zeitschrift für angewandte Psychologie und Charakterkunde 59*, 1940, **1** (2), 2–81.

6.15 A Few Intermediate Conclusions

After deriving fundamental quantum equations from the Einstein–Hilbert action and thus, combining the General Theory of Relativity with Quantum Theory, we found that infinity of possibilities/degrees of freedom/ dimensions leads to linearity of the fundamental quantum-gravity equations of the system … at least with respect to the derivative operators.

We also found options to metrically derive quantized masses, spins, energies, and inertia via dimensional entanglement of various kinds.

Along the way we came across a variety of options to solve the so-called 3-generation problem, but did not succeed (not truly anyway) in solving the corresponding and rather complicated differential equations by the means of either the Klein–Gordon or the Dirac approach.

However, in demanding the space-time to be restricted to four big-scale dimensions, we found that the product of mass and the quantum gravity wave function occur in a polynomial of third order, which, as it is well known, can have three solutions.

In the end, an apparently complete solution to the 3-generation problem was found by combining the Klein–Gordon and the Dirac approaches and excepting the two, respectively their solutions, to exist simultaneously. This then automatically leaves no other choice for the product of mass and quantum gravity wave function, but to algebraically fall apart into three different solutions. However, more investigation and discussion are of need (see chapter 7).

Most interestingly, it was found that this combined approach favors octonions as a tool (not the only one as it was shown here) to solve the subsequent differential equations.

6.16 Appendix: A Few Words about Spin 1/2, 3/2, 5/2, and So On

In the following, we want to investigate potential half spin solutions for the associated Legendre polynomials, which we needed in various forms within this chapter.

Here, as an example, we intend to consider the simple form:

$$Y_l^m[v] = C_{Pv} \cdot P_l^m[\cos[v]] + C_{Qv} \cdot Q_l^m[\cos[v]]. \tag{635}$$

Observing the solution $Y[v]$ more closely, we find that there exist singularity-free spin $l = n/2$-solutions for $m = -n/2$ with $n = 1,3,5,7 \ldots$ in the case of $C_{Pv} \neq 0$, $C_{Qv} = 0$ and for $m = +n/2$ in the case of $C_{Pv} = 0$, $C_{Qv} \neq 0$. Figures 6.18 and 6.19 illustrate the corresponding distribution within the definition range for the angle v from 0 to π.

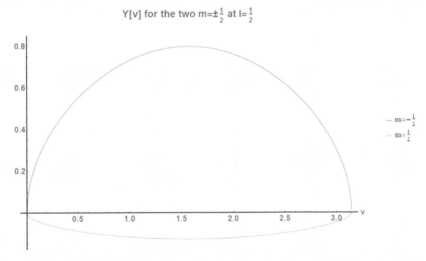

Figure 6.18 Spin $l = 1/2$ situation with the two possible spin states $m = \pm1/2$ according to our angular quantum gravity solution (635). For better illustration and comparability, we divided the "Q-Legendres" by 10.

Y[v] for a variety of m at l=$\frac{5}{2}$

Figure 6.19 Spin $l = 5/2$ situation with the five possible spin states $m = \pm\{1/2,3/2,5/2\}$ according to our angular quantum gravity solution (635). Regarding the evaluation, see text. For better illustration we divided the "Q-Legendres" by 10.

The general solution to the partial differential Eq. (622) can be given as follows:

$$f_{n,l,m}\left[t,r,\vartheta,\varphi\right] = g[t]\cdot\Psi_{n,l,m}\left[r,\vartheta,\varphi\right] = g[t]\cdot e^{i\cdot m\cdot\varphi}\cdot P_l^m\left[\cos\vartheta\right]\cdot R_{n,l}[r]$$

$$= \left(C_1\cdot\cos\left[c\cdot C_{tt}\cdot t\right] + C_2\cdot\sin\left[c\cdot C_{tt}\cdot t\right]\right)\cdot N\cdot e^{-\rho/2}\cdot\rho^l\cdot L_{n-l-1}^{2l+1}[\rho]\cdot Y_l^m[\vartheta,\varphi]$$

$$\rho = \frac{2\cdot r}{n\cdot a_0};\quad N = \sqrt{\left(\frac{2}{n\cdot a_0}\right)^3\frac{(n-l-1)!}{2\cdot n\cdot(n+l)!}}. \tag{636}$$

For nostalgic reasons, we stuck to the symbol n (here not the dimension) as the so-called main quantum number. All quantum numbers n and l and the parameter a_0 depend on the constants $C_{R2}^*, C_{R1}^*, C_{R0}^*$ in Eq. (622) and have to satisfy certain quantum conditions in order to result in singularity-free solution for $f[\,...\,]$.

Regarding the conditions for the quantum numbers n, l, and m, we not only have the usual:

$$\{n,l,m\}\in\mathbb{Z};\quad n\geq 0;\quad l < n;\quad -l\leq m\leq +l, \tag{637}$$

but also found the suitable solutions for the half spin forms as discussed above and derived in Figs. 6.18 and 6.19. The corresponding main quantum numbers for half spin *l*-numbers with *l* = 1/2, 3/2, ... are simply (just as before with the integers) $n = l + 1 = 3/2, 5/2, 7/2, ...$.

It should explicitly be noted, however, that the usual spherical harmonics are inapplicable in cases of half spin. For $\{n, l, m\} = \{1/2, 3/2, 5/2, 7/2, ... \}$, the wave function (636) has to be adapted as follows:

$$f_{n,l,m}[t,r,\vartheta,\varphi] = e^{\pm i \cdot c \cdot C_t \cdot t} \cdot N \cdot e^{-\rho/2} \cdot \rho^l \cdot L_{n-l-1}^{2l+1}[\rho] \cdot Z_m[\varphi] \cdot \begin{Bmatrix} P_l^{m<0}[\cos\vartheta] \\ Q_l^{m>0}[\cos\vartheta] \end{Bmatrix}$$

$$= e^{\pm i \cdot c \cdot C_t \cdot t} \cdot N \cdot e^{-\rho/2} \cdot \rho^l \cdot L_{n-l-1}^{2l+1}[\rho] \cdot \begin{Bmatrix} \cos[m \cdot \varphi] \\ \sin[m \cdot \varphi] \end{Bmatrix} \cdot \begin{Bmatrix} P_l^{m<0}[\cos\vartheta] \\ Q_l^{m>0}[\cos\vartheta] \end{Bmatrix}. \quad (638)$$

Thereby we will elaborated elsewhere [18] that in fact the sin- and the cos-functions seem to make the Pauli exclusion and not the "+" and "−" of the *m*. However, in order to have the usual Fermionic statistic, we can simply define as follows:

$$f_{n,l,m}[t,r,\vartheta,\varphi] = e^{\pm i \cdot c \cdot C_t \cdot t} \cdot N \cdot e^{-\rho/2} \cdot \rho^l \cdot L_{n-l-1}^{2l+1}[\rho] \cdot \begin{Bmatrix} \sin[m \cdot \varphi]_{m<0} \\ \cos[m \cdot \varphi]_{m>0} \end{Bmatrix} \cdot \begin{Bmatrix} P_l^{m<0}[\cos\vartheta] \\ Q_l^{m>0}[\cos\vartheta] \end{Bmatrix}.$$

$$(639)$$

As derived in [38], the resolution of the degeneration with respect to half spin requires a break of the symmetry, which we achieved by introducing elliptical geometry instead of the spherical one in [38].

Thus, we have metrically derived a fairly general "hydrogen atom." In addition to the Schrödinger structure, our form also spots a time-dependent factor clearly showing the options for matter and antimatter via the ±sign in the *g*[*t*]-function. We also found the half spin states were able to resolve the spin-degeneration via a simple symmetry break by switching from spherical to elliptical coordinates (see [38] regarding the evaluation).

6.16.1 A Few Illustrations

Instead of interpreting the solutions obtained above as electron orbitals of a simple hydrogen atom, we might also see options to use them as descriptions for a variety of particles. However, as such a consideration, if performed holistically, would by far exceed the intentions of this chapter, which after all was thought to just treat the problem of the origin and appearance of masses from a fundamental minimum condition like the Einstein–Hilbert action, we will only take one example.

As the classical integer quantum number solutions are already well documented and known, we here concentrate on the illustration of half spin fields.

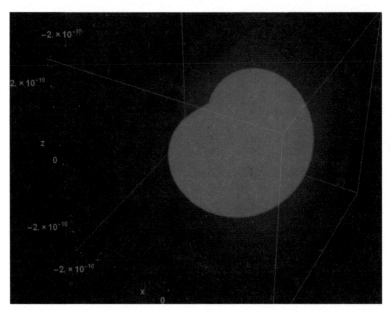

Figure 6.20 Absolute space geometry of distortion (or wave) function in the case of quantum gravity solution (638) for the state $n = 3/2$, $l = 1/2$, and $m = 1/2$. Please note that in spherical coordinates, the sign of $m = \pm 1/2$ does not matter with respect to the resulting spatial deformation. Resolving this degeneration requires a change of the φ-function with the sign of the m-value (639) or a symmetry break (c.f. [38]).

Thereby, only in order to keep things familiar and potentially compare with the classical integer hydrogen states, we keep the normalization and the scale of the Bohr radius a_0. One might perhaps call the resulting objects "half spin hydrogen atoms." With only one exception (Fig. 6.22) we start with the general setting of $m > 0$ ($\rightarrow Z[\varphi] = \cos[m^*\varphi]$). It can easily be seen with the simplest half spin states ($n = 3/3$, $l = 1/2$, and $|m| = 1/2$) as presented in Figs. 6.20 and 6.21 that the gross of deformed space-time is to be found on the right-hand side of the $x = 0$-plane. Now choosing $n = 3/3$, $l = 1/2$, and $m = -1/2$ and applying the statistic rule defined in Eq. (639), leading to $Z[\varphi] = \sin[m^*\varphi]$, gives us a concentration of deformed space-time on the other side of the $x = 0$-plane (Fig. 6.22). It appears intuitively logic to assume that the combination of objects with deformation maxima on left and right (anti-parallel spin) is easier than the combination of objects having the maximum deformations on the same sides of the $x = 0$-plane (parallel spin combination). We might see this as the geometric manifestation of the Pauli Exclusion Principle [39].

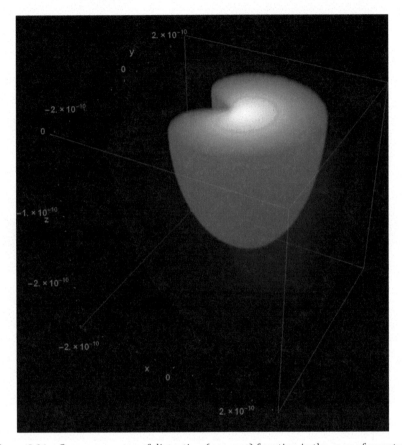

Figure 6.21 Space geometry of distortion (or wave) function in the case of quantum gravity solution (638) for the state $n = 3/2$, $l = 1/2$, and $m = 1/2$. Please note that in spherical coordinates, the sign of $m = \pm1/2$ does not matter with respect to the resulting spatial deformation. Resolving this degeneration requires a change of the φ-function with the sign of the m-value (639) or a symmetry break (c.f. [38]). This time the inside of the distribution is shown.

As negative m would only bring in a mirror effect at the $x = 0$-plane to the deformation fields in our graphics, we will present further examples of half spin states only with positive m-quantum numbers according to our statistic rule as defined in Eq. (639) (see the following Figs. 6.23 to 6.34).

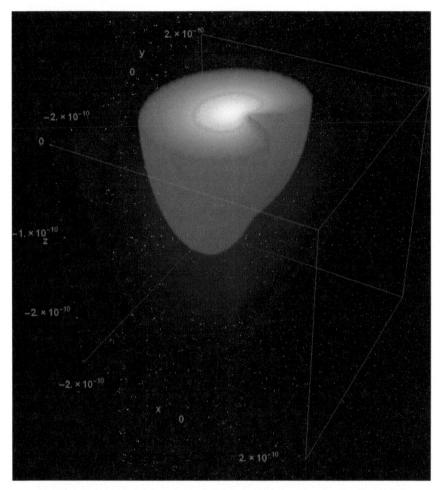

Figure 6.22 Space geometry of distortion (or wave) function in the case of quantum gravity solution (639) for the state $n = 3/2$, $l = 1/2$, and $m = -1/2$. Please see text and figure captions of Figs. 6.20 and 6.21.

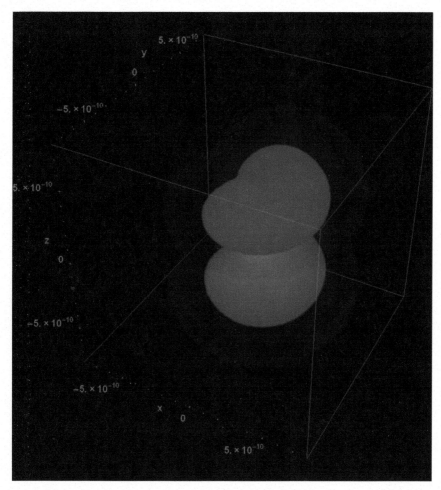

Figure 6.23 Space geometry of distortion (or wave) function in the case of quantum gravity solution (638) for the state $n = 5/2$, $l = 3/2$, and $m = 1/2$.

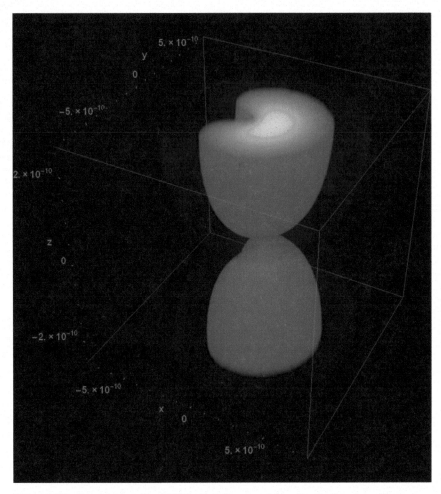

Figure 6.24 Space geometry of distortion (or wave) function in the case of quantum gravity solution (638) for the state $n = 5/2$, $l = 3/2$, and $m = 1/2$. Upper distribution was cut open to show inside metric distribution.

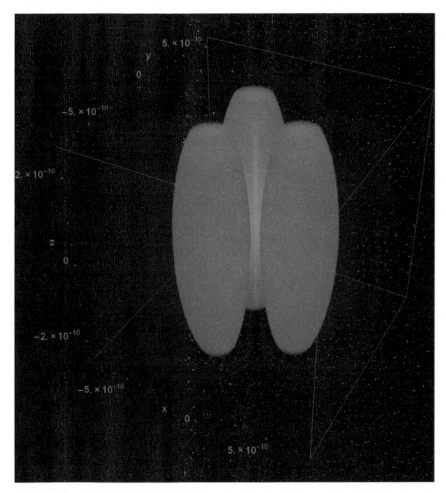

Figure 6.25   Space geometry of distortion (or wave) function in the case of quantum gravity solution (638) for the state $n = 5/2$, $l = 3/2$, and $m = 3/2$.

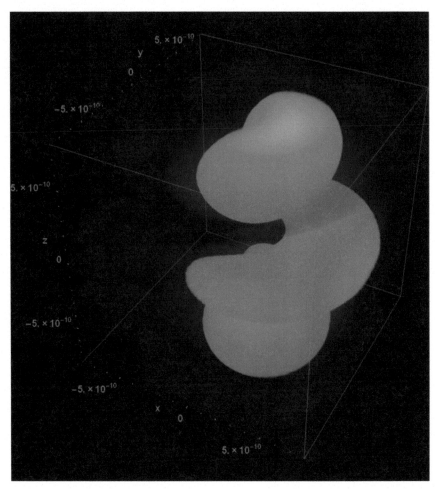

Figure 6.26 Space geometry of distortion (or wave) function in the case of quantum gravity solution (638) for the state $n = 7/2$, $l = 5/2$, and $m = 1/2$.

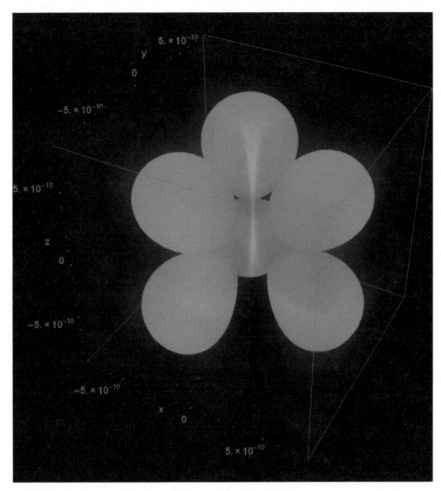

Figure 6.27 Space geometry of distortion (or wave) function in the case of quantum gravity solution (638) for the state $n = 7/2$, $l = 5/2$, and $m = 3/2$.

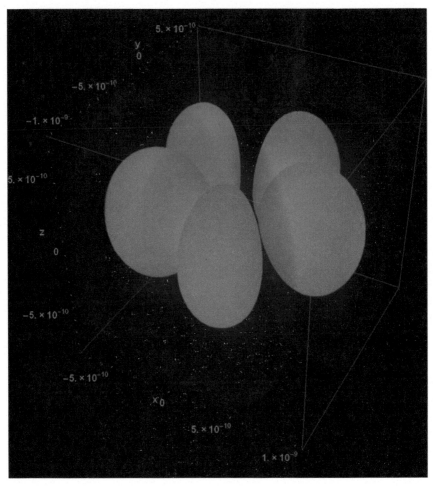

Figure 6.28 Space geometry of distortion (or wave) function in the case of quantum gravity solution (638) for the state $n = 7/2$, $l = 5/2$, and $m = 5/2$.

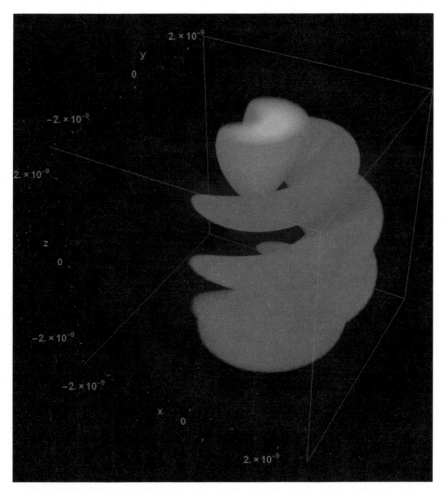

Figure 6.29 Space geometry of distortion (or wave) function in the case of quantum gravity solution (638) for the state $n = 13/2$, $l = 11/2$, and $m = 1/2$.

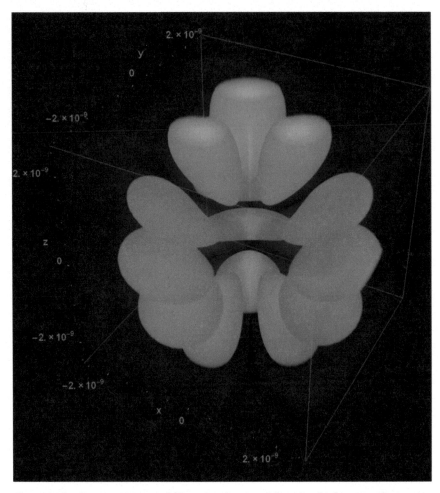

Figure 6.30 Space geometry of distortion (or wave) function in the case of quantum gravity solution (638) for the state $n = 13/2$, $l = 11/2$, and $m = 3/2$.

Figure 6.31 Space geometry of distortion (or wave) function in the case of quantum gravity solution (638) for the state $n = 13/2$, $l = 11/2$, and $m = 5/2$.

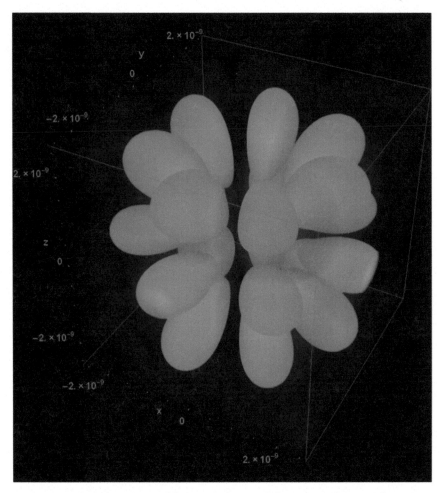

Figure 6.32 Space geometry of distortion (or wave) function in the case of quantum gravity solution (638) for the state $n = 13/2$, $l = 11/2$, and $m = 7/2$.

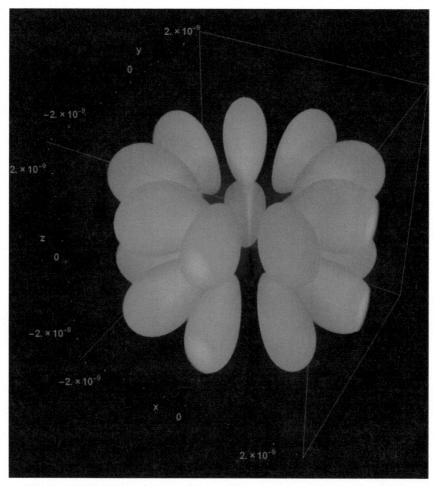

Figure 6.33 Space geometry of distortion (or wave) function in the case of quantum gravity solution (638) for the state $n = 13/2$, $l = 11/2$, and $m = 9/2$.

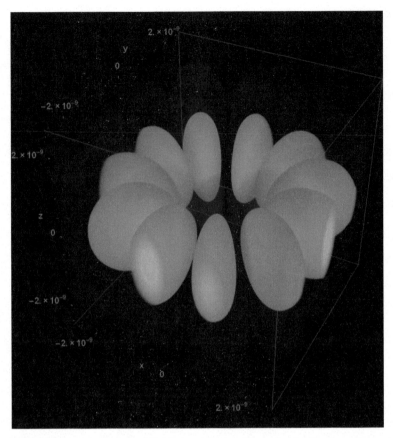

Figure 6.34 Space geometry of distortion (or wave) function in the case of quantum gravity solution (638) for the state $n = 13/2$, $l = 11/2$, and $m = 11/2$.

References

1. N. Schwarzer, *Societons and Ecotons—The Photons of the Human Society—Control them and Rule the World*, Part 1 of "Medical Socio-Economic Quantum Gravity," Self-published, Amazon Digital Services, 2020, Kindle.

2. N. Schwarzer, *Humanitons —The Intrinsic Photons of the Human Body—Understand them and Cure Yourself*, Part 2 of "Medical Socio-Economic Quantum Gravity," Self-published, Amazon Digital Services, 2020, Kindle.

3. N. Schwarzer, *Mastering Human Crises with Quantum-Gravity-based but still Practicable Models—First Measure: SEEING and UNDERSTANDING the WHOLE,*

Part 3 of "Medical Socio-Economic Quantum Gravity," Self-published, Amazon Digital Services, 2020, Kindle.

4. N. Schwarzer, *The World Formula: A Late Recognition of David Hilbert's Stroke of Genius*, 2022, Jenny Stanford Publishing. ISBN: 9789814877206.

5. D. Hilbert, Die Grundlagen der Physik, Teil 1, *Göttinger Nachrichten*, 1915, 395–407.

6. A. Einstein, Grundlage der allgemeinen Relativitätstheorie, *Ann. Phys.*, 1916, **49** (ser. 4), 769–822.

7. N. Schwarzer, *Brief Proof of Hilbert's World Formula—Dirac, Klein–Gordon, Schrödinger, Einstein, Evolution and the Second Law of Thermodynamics all from One Origin*, Self-published, Amazon Digital Services, 2020, Kindle.

8. N. Schwarzer, *The New Dirac Equation—Originally Postulated, Now Geometrically Derived and—along the way—Generalized to Arbitrary Dimensions and Arbitrary Metric Space-Times*, Self-published, Amazon Digital Services, 2020, Kindle.

9. N. Schwarzer, *Science Riddles—Riddle No. 11: What is Mass?*, Self-published, Amazon Digital Services, 2019, Kindle.

10. N. Schwarzer, *Science Riddles—Riddle No. 12: Is There a Cosmological Spin?*, Self-published, Amazon Digital Services, 2019, Kindle.

11. N. Schwarzer, *Science Riddles—Riddle No. 13: How to Solve the Flatness Problem?*, Self-published, Amazon Digital Services, 2019, Kindle.

12. N. Schwarzer, *Science Riddles—Riddle No. 14: And what if Time … ? A Paradigm Shift*, Self-published, Amazon Digital Services, 2019, Kindle.

13. N. Schwarzer, *Science Riddles—Riddle No. 15: Is there an Absolute Scale?*, Self-published, Amazon Digital Services, 2019, Kindle.

14. Peter W. Higgs, Broken Symmetries and the Masses of Gauge Bosons, *Phys. Rev. Lett.*, 1964, **13**, 508.

15. en.wikipedia.org/wiki/Higgs_boson

16. H. Haken and H. Chr. Wolf, *Atom- und Quantenphysik* (in German), 4th Ed., 1990, Springer Heidelberg, ISBN: 0-387-52198-4.

17. N. Schwarzer, *Einstein had it, but he did not see it—Part LXXIX: Dark Matter Options*, Self-published, Amazon Digital Services, 2019, Kindle.

18. N. Schwarzer, *Einstein had it, but he did not see it—Part LXXXII: Half Spin Hydrogen*, Self-published, Amazon Digital Services, 2019, Kindle.

19. K. Schwarzschild, Über das Gravitationsfeld einer Kugel aus inkompressibler Flüssigkeit nach der Einsteinschen Theorie *[On the Gravitational Field of a Ball of Incompressible Fluid Following Einstein's Theory]*, Sitzungsberichte der Königlich-Preussischen Akademie der Wissenschaften (in German), Berlin, 1916, 424–434.

20. N. Schwarzer, *Is There an Ultimate, Truly Fundamental and Universal Computer Machine?*, Self-published, Amazon Digital Services, 2019, Kindle.

21. N. Schwarzer, *Einstein had it, but he did not see it—Part LXXXIII: Quantum Relativity*, Self-published, Amazon Digital Services, 2019, Kindle.

22. J. D. Bekenstein, Information in the Holographic Universe, *Scientific American*, 2003, **289** (2), 61.

23. J. D. Bekenstein, Black Holes and Entropy, *Phys. Rev.*, 1973, **D 7**, 2333–2346.

24. N. Schwarzer, *Einstein had it ... Part XXXVI: The Classical and Principle Misinterpretation of the Einstein Field Equations and How it Might be Done Correctly*, Self-published, Amazon Digital Services, 2018, Kindle.

25. N. Schwarzer, *Science Riddles—Riddle No. 8: Could the Schwarzschild Metric Contain its own Quantum Solution?*, Self-published, Amazon Digital Services, 2019, Kindle.

26. N. Schwarzer, *Einstein had it ... Part XXXV: The 2-Body Problem and the GTR-Origin of Mass*, Self-published, Amazon Digital Services, 2018, Kindle.

27. N. Schwarzer, *Einstein had it ... Part XXXVI: The Classical and Principle Misinterpretation of the Einstein Field Equations and How it Might be Done Correctly*, Self-published, Amazon Digital Services, 2018, Kindle.

28. https://en.wikipedia.org/wiki/Generation_(particle_physics)

29. P. A. M. Dirac, *The Quantum Theory of the Electron*, 1928. *Proceedings of the Royal Society A*. DOI: 10.1098/rspa.1928.0023.

30. A. E. Green and W. Zerna, *Theoretical Elasticity*, 1968, Oxford University Press, London.

31. H. Neuber, *Kerspannungslehre*, in German, 3rd Ed., 1985, Springer-Verlag, Berlin, Heidelberg, New York, Tokyo, ISBN: 3-540-13558-8.

32. N. Schwarzer, *Epistle to Elementary Particle Physicists—A Chance You Might not Want to Miss*, Self-published, Amazon Digital Services, 2019, Kindle.

33. N. Schwarzer, *Second Epistle to Elementary Particle Physicists—Brief Study about Gravity Field Solutions for Confined Particles*, Self-published, Amazon Digital Services, 2019, Kindle.

34. N. Schwarzer, *3rd Epistle to Elementary Particle Physicists—Beyond the Standard Model—Metric Solutions for Neutrino, Electron, Quark*, Self-published, Amazon Digital Services, 2019, Kindle.

35. N. Schwarzer, *Science Riddles—Riddle No. 26: Is there an Elastic Quantum Gravity?*, Self-published, Amazon Digital Services, 2019, Kindle.

36. N. Schwarzer, *Our Universe, Nothing but an Intelligent Zero? The Dark Lord's Zero-Sum- & God's Non-Zero-Sum-Game*, Self-published, Amazon Digital Services, 2016, Kindle.

37. N. Schwarzer, *The Theory of Everything—Quantum and Relativity is Everywhere—A Fermat Universe*, 2020, Pan Stanford Publishing, ISBN: 10: 9814774472.

38. N. Schwarzer, *Einstein had it, but he did not see it—Part LXXIX: Dark Matter Options*, Self-published, Amazon Digital Services, 2016, Kindle.

39. W. Pauli, Über den Zusammenhang des Abschlusses der Elektronengruppen im Atom mit der Komplexstruktur der Spektren, 1925, Zeitschrift für Physik 31: 765–783, Bibcode:1925Zphy ... 31 ... 765P. doi:10.1007/BF02980631

40. N. Schwarzer, *General Quantum Relativity*, Self-published, Amazon Digital Services, 2016, Kindle.

41. T. Bodan, *The Eighth Day—Holocaust and the World's Biggest Mysteries: The other Final Solution—with Mathematical Elaborations,* English Edition, Self-published, Amazon Digital Services, 2015, Kindle.

42. T. Bodan, *The Eighth Day: Two Jews against The Third Reich*, 2021, Illustrated Version, Self-published, BoD Classic, ISBN: 9783753417257.

43. N. Schwarzer, *The 3 Generations of Elementary Particles*, Part 6 of "Medical Socio-Economic Quantum Gravity," Self-published, Amazon Digital Services, 2020, Kindle.

Chapter 7

The Three Generations of Elementary Particles

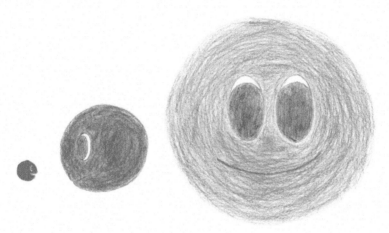

Figure 7.1   The three generations of elementary particles, by Livia Schwarzer.

7.1   Abstract: From Quantum Gravity to the 3-Generation Problem

In our earlier paper [A1], we saw that Quantum Theory emerges when we construct a scaled metric by simply multiplying the metric with a factor $F[f]$,

The Math of Body, Soul, and the Universe
Norbert Schwarzer
Copyright © 2023 Jenny Stanford Publishing Pte. Ltd.
ISBN 978-981-4968-24-9 (Hardcover), 978-1-003-33454-5 (eBook)
www.jennystanford.com

thereby assuming the inner part of the scale factor to be an arbitrary function *f*. The *f* itself then—so it was derived—becomes the wave function[1]**. Subjecting this scaled metric in the usual way to the Einstein–Hilbert action [A2] does not only present us with quantum Einstein field equations [A3] and thus, a quantized version of Einstein's General Theory of Relativity [A4], but—along the way (e.g., [A5–A10])—gives us all classical and non-classical quantum equations, including Schrödinger, Klein–Gordon, and Dirac. Also, thermodynamics can be derived from, respectively found inside this simple scaled metric approach [A3, A10].

Now in our previous chapter of this book (also see [A1]), we saw that there are apparently several paths to come up with an explanation for the existence of the generations of elementary particles (Fig. 7.1). None of these explanations, however, was truly satisfying and so we concluded the abstract to chapter 6 and [A1] with the words:

> "Thus, so our current conclusion, the way to solve the 3-generation problem [A11] could be outlined here but—so far—this path was not treated in a manner one could call complete."

Here, now we want to move on by applying the scaled metric approach and combining it with the Hilbert concept [A2] of the variated Ricci scalar, only that this time, we do not only look for the variation with respect to the metric tensor but go a bit broader, we think that we have found the answer.

7.1.1 References for the Abstract

A1. N. Schwarzer, *Masses and the Infinity Options Principle—Can We Explain the 3-Generation and the Quantized Mass Problem?*, Part 5 of "Medical Socio-Economic Quantum Gravity," Self-published, Amazon Digital Services, 2020, Kindle.

A2. D. Hilbert, Die Grundlagen der Physik, Teil 1, *Göttinger Nachrichten*, 1915, 395–407.

A3. N. Schwarzer, *Toward Quantum Einstein Field Equations*, Part 7 of "Medical Socio-Economic Quantum Gravity," Self-published, Amazon Digital Services, 2020, Kindle.

A4. A. Einstein, Grundlage der allgemeinen Relativitätstheorie, *Annalen der Physik (ser. 4)*, 1916, **49**, 769–822.

A5. N. Schwarzer, *Brief Proof of Hilbert's World Formula—Dirac, Klein–Gordon, Schrödinger, Einstein, Evolution and the Second Law of*

---

[1]c.f. section 6.14 in this book.

Thermodynamics all from One Origin, Self-published, Amazon Digital Services, 2020, Kindle.

A6. N. Schwarzer, *Societies and Ecotons—The Photons of the Human Society—Control them and Rule the World*, Part 1 of "Medical Socio-Economic Quantum Gravity," Self-published, Amazon Digital Services, 2020, Kindle.

A7. N. Schwarzer, *Humanitons—The Intrinsic Photons of the Human Body— Understand them and Cure Yourself*, Part 2 of "Medical Socio-Economic Quantum Gravity," Self-published, Amazon Digital Services, 2020, Kindle.

A8. N. Schwarzer, *Mastering Human Crises with Quantum-Gravity-based but still Practicable Models—First Measure: SEEING and UNDERSTANDING the WHOLE*, Part 3 of "Medical Socio-Economic Quantum Gravity," Self-published, Amazon Digital Services, 2020, Kindle.

A9. N. Schwarzer, *Self-Similar Quantum Gravity—How Infinity Brings Simplicity?*, Part 4 of "Medical Socio-Economic Quantum Gravity," Self-published, Amazon Digital Services, 2020, Kindle.

A10. N. Schwarzer, *The World Formula: A Late Recognition of David Hilbert's Stroke of Genius*, 2022, Jenny Stanford Publishing. ISBN: 9789814877206.

A11. https://en.wikipedia.org/wiki/Generation_(particle_physics).

7.2 Extended (Quantum Gravity) Einstein Field Equations ... of Third Order?

In order to find an answer to the 3-generation problem in particle physics [1], it is clear that we should somehow look for an equation of third order.

In the previous section $[2-7]^2$, and quite some other papers by this author, it was demonstrated how we can find a direct extraction of the Klein–Gordon, the Schrödinger, and the Dirac equations from the Ricci scalar R^* of a modified metric of the kind $G_{\alpha\beta} = F[f]^* g_{\alpha\beta}$.^{3}

With a yet arbitrary scalar function $F[f]$, the corresponding modified Ricci scalar R^* reads^4:

---

^{2}Please note than in most papers, we thereby considered metrics g_{ij} with $R = 0$. Here, now we need to introduce the general form.

^{3}For convenience, we here repeat and partially extend these derivations in the appendices of this paper.

^{4}The complete evaluation was presented in the appendix A of reference [8] and is repeated in chapter 8 of this book.

$$R^* = R^*_{\alpha\beta} G^{\alpha\beta} = \begin{pmatrix} \Gamma^{\sigma}_{\alpha\beta,\sigma} g^{\alpha\beta} - \Gamma^{\sigma}_{\beta\sigma,\alpha} g^{\alpha\beta} - \Gamma^{\mu}_{\sigma\alpha} \Gamma^{\sigma}_{\beta\mu} g^{\alpha\beta} + \Gamma^{\sigma}_{\alpha\beta} \Gamma^{\mu}_{\sigma\mu} g^{\alpha\beta} \\ + \dfrac{F'}{F}(1-n)\Delta f + \dfrac{f_{,\alpha} f_{,\beta} g^{\alpha\beta}(1-n)}{4F^2}\left(4FF'' + (F')^2(n-6)\right) \end{pmatrix} \frac{1}{F}$$

$$= \left(R + \frac{F'}{F}(1-n)\Delta f + \frac{f_{,\alpha} f_{,\beta} g^{\alpha\beta}(1-n)}{4F^2}\left(4FF'' + (F')^2(n-6)\right) \right) \frac{1}{F}$$

$$\text{with}: \quad F = F[f]; \quad F' = \frac{\partial F[f]}{\partial f}; \quad F'' = \frac{\partial^2 F[f]}{\partial f^2}. \tag{640}$$

Demanding certain conditions for the function $F[f]$ and/or f then gives us Dirac or Klein–Gordon equations [2–4]. Thus, when demanding f to be a Laplace function, we obtain from Eq. (640):

$$R^* = \frac{R}{F[f]} + \frac{(1-n)}{F[f]^3} \cdot \left(\left(\frac{(n-6)}{4} \cdot \left(\frac{\partial F[f]}{\partial f} \right)^2 + F[f] \cdot \frac{\partial^2 F[f]}{\partial f^2} \right) \cdot \overbrace{\left(\tilde{\nabla}_g f \right)^2}^{= f_{,\alpha} g^{\alpha\beta} f_{,\beta}} \right)$$

$$\xrightarrow{n=4; R=0} = \frac{1}{F[f]^3} \cdot \left(\left(3 \cdot F[f] \cdot \frac{\partial^2 F[f]}{\partial f^2} - \frac{3}{2} \cdot \left(\frac{\partial F[f]}{\partial f} \right)^2 \right) \cdot \left(\tilde{\nabla}_g f \right)^2 \right), \tag{641}$$

which—so it was shown above and in [2–4]—gives the metric equivalent to the Dirac equation. It was also shown in the previous chapter and [2, 3] how this gives the classical Dirac equation in flat space Minkowski metrics.

Does this scaled metric approach also give us quantum Einstein field equations?

Yes, it seems so!

As demonstrated in [8] and chapter 8 of this book, when starting with the classical Einstein–Hilbert action [9] with the Ricci scalar R^* as kernel or Lagrange density:

$$\delta W = 0 = \delta \int_V d^n x \left(\sqrt{-G} \cdot \left(R^* - 2\Lambda + L_M \right) \right), \tag{642}$$

we were able to derive new field equations, not only containing the classical part from Einstein's General Theory of Relativity [10], but also a quantum part [8], which is governed by the wave-function (and metric scale) f.

The reader will have realized that in Eq. (642), in contrast to the classical form, we used a scaled metric tensor $G_{\delta\gamma} = F[f] \cdot g_{\delta\gamma}$. Thereby G shall denote the determinant of the metric tensor $G_{\delta\gamma}$, while g will later stand for the corresponding determinant of the metric tensor $g_{\alpha\beta}$. In order to distinguish our new Ricci scalar R^*, being based on $G_{\delta\gamma} = F[f] \cdot g_{\delta\lambda}$ from the usual one R, being based on the metric tensor $g_{\alpha\beta}$, we marked it with the *-superscript. We also have the matter density L_M and the cosmological constant Λ.

Please note that with $G_{\delta\gamma} = F[f] \cdot g_{\delta\lambda}$ we have used the simplest form of metric adaptation, which we could construct as a simplification from a generalization of the typical form of tensor transformations, namely:

$$G_{\alpha\beta} = F\left[f\left[t,x,y,z,\ldots,\xi_k,\ldots,\xi_n\right]\right]_{\alpha\beta}^{ij} g_{ij}, \tag{643}$$

$$\rightarrow G_{\alpha\beta} = F\left[f\left[t,x,y,z,\ldots,\xi_k,\ldots,\xi_n\right]\right] \cdot \delta_\alpha^i \delta_\beta^j g_{ij}$$

$$\rightarrow G_{\alpha\beta} = F\left[f\left[t,x,y,z,\ldots,\xi_k,\ldots,\xi_n\right]\right] \cdot g_{\alpha\beta} = F[f] \cdot g_{\alpha\beta}. \tag{644}$$

The generalization is been elaborated in [4] and its need (and/or redundancy) has been discussed in [8] and chapter 11 of this book.

Setting Eq. (640) into Eq. (642) results in the variation:

$$\delta W = 0 = \delta \int_V d^n x \sqrt{-g \cdot F^n} \times \left(\left(\left(\begin{array}{c} R + \dfrac{F'}{F}(1-n)\Delta f \\ + \dfrac{f_{,\alpha} f_{,\beta} g^{\alpha\beta}(1-n)}{4F^2}\left(4FF'' + (F')^2(n-6)\right) \\ -2\Lambda + L_M \end{array}\right) \dfrac{1}{F}\right)\right). \tag{645}$$

Please note that the cosmological constant term as given in Eq. (645) in the current form requires the variation with respect to the metric $G_{\alpha\beta}$. When insisting on the variation with respect to the metric $g_{\alpha\beta}$ instead, we better rewrite Eq. (645) as follows:

$$\delta_{g_{\alpha\beta}} W = 0 = \left\{\begin{array}{c} \delta_{g_{\alpha\beta}} \int_V d^n x \dfrac{\sqrt{-g \cdot F^n}}{F} \times \left(\begin{array}{c} R + \dfrac{F'}{F}(1-n)\Delta f \\ + \dfrac{f_{,\alpha} f_{,\beta} g^{\alpha\beta}(1-n)}{4F^2}\left(4FF'' + (F')^2(n-6)\right) \end{array}\right) \\ + \delta_{g_{\alpha\beta}} \int_V d^n x \sqrt{-g}\left(L_M - 2\Lambda\right) \end{array}\right. \tag{646}$$

in order to make it clear that we do not intend to also scale-adapt the cosmological constant or the Hilbert matter term.

However, when ignoring the cosmological constant and PERHAPS assuming that we would not need any postulated matter terms L_M, simply because our scale-adaptation $F[f]$ and similar "tricks" or "add-ons" automatically provides matter, we just obtain:

$$\delta W = 0 = \delta \int_V d^n x \left| \sqrt{-g \cdot F^n} \times \left(\left(\begin{array}{c} R + \dfrac{F'}{F}(1-n)\Delta f \\[2ex] + \dfrac{f_{,\alpha} f_{,\beta} g^{\alpha\beta}(1-n)}{4F^2}\left(4FF'' + (F')^2(n-6)\right) \end{array} \right)^{\frac{1}{F}} \right) \right)$$

$$= \delta \int_V d^n x \left| \sqrt{-g \cdot F^n} \times \left(\left(\begin{array}{c} R + \dfrac{F'}{\sqrt{-g} \cdot F}(1-n)\partial_\beta \sqrt{-g} \cdot g^{\alpha\beta} f_{,\alpha} \\[2ex] + \dfrac{f_{,\alpha} f_{,\beta} g^{\alpha\beta}(1-n)}{4F^2}\left(4FF'' + (F')^2(n-6)\right) \end{array} \right)^{\frac{1}{F}} \right) \right).$$

(647)

Performing the usual Hilbert variation with respect to the metric tensor $g_{\alpha\beta}$ now leads us to (regarding the full evaluation see [8] or chapter 8):

$$\Rightarrow \quad 0 = \left\{ \begin{array}{l} R^{\kappa\lambda} - \dfrac{1}{2}Rg^{\kappa\lambda} + \left(\left(\underbrace{\dfrac{8\pi G T^{\kappa\lambda}}{\kappa} = \text{matter}}_{} \atop 0 = \text{vacuum} \right) + \Lambda \cdot g^{\kappa\lambda} \right) \\[4ex] - \dfrac{1}{\sqrt{F^n}} \cdot \left(\partial^{;\kappa}\partial^{;\lambda} - g^{\kappa\lambda}\Delta_g \right) F^{\frac{n}{2}-1} \\[3ex] + \left(g^{\kappa\lambda}g^{\alpha\beta} - g^{\kappa\alpha}g^{\lambda\beta} - g^{\lambda\alpha}g^{\kappa\beta} \right)(1-n) \\[2ex] \quad \cdot \left[\dfrac{f_{,\alpha} f_{,\beta}}{F^2}\left(FF'' + (F')^2 \cdot \dfrac{3 \cdot n - 14}{8} \right) + f_{,\alpha\beta}\dfrac{F'}{F} \right] \\[3ex] - \dfrac{F' \cdot (1-n)}{2 \cdot F}\left(g^{\kappa\lambda}g^{\alpha\beta} - g^{\kappa\alpha}g^{\lambda\beta} - g^{\lambda\alpha}g^{\kappa\beta} \right)\Gamma^{\gamma}_{\alpha\beta} f_{,\gamma} \\[3ex] + \dfrac{F'}{F} \cdot (1-n) g^{\alpha\beta} g^{\gamma\lambda}\Gamma^{\kappa}_{\alpha\beta} f_{,\gamma} \\[3ex] + \dfrac{F'}{F} \cdot \dfrac{(1-n)}{2}\left(g^{\kappa\lambda}g^{\alpha\beta} - g^{\kappa\alpha}g^{\lambda\beta} - g^{\lambda\alpha}g^{\kappa\beta} \right)_{,\beta} f_{,\alpha} \\[3ex] + \dfrac{g^{\xi\zeta}g_{\xi\zeta,\beta}}{2} \cdot \left(\dfrac{F'}{F} \cdot \dfrac{(1-n)}{2}\left(g^{\kappa\lambda}g^{\alpha\beta} - g^{\kappa\alpha}g^{\lambda\beta} - g^{\lambda\alpha}g^{\kappa\beta} \right) f_{,\alpha} \right). \end{array} \right.$$

(648)

Quite some discussion to this new quantum Einstein field equation is presented in chapter 8.

It has to be pointed out that Hilbert's choice of the Ricci scalar R acting as the Lagrange density in Eq. (642), which was motivated by his attempt (with success) to reproduce the Einstein field equations in a completely mathematical manner. As his scale factor F was set to one, which was to say as Hilbert did not consider a $F[f]$-scaled metric tensor, we now, when aiming for the Hilbert approach in the $F \rightarrow 1$ limit, have the principle degree of freedom of adding a factor F^q to the integrand. This action, reading (when setting L_M and Λ to zero and choosing an additional scale factor F^q):

$$\delta W = 0 = \delta \int_V d^n x \left(\sqrt{-g \cdot F^n} \times F^q \cdot R^* \right)$$

$$= \delta \int_V d^n x \left[\sqrt{-g \cdot F^n} \times \left(F^q \cdot \left(\left(R + \frac{F'}{F}(1-n)\Delta f \atop + \frac{f_{,\alpha} f_{,\beta} g^{\alpha\beta}(1-n)}{4F^2}\left(4FF'' + (F')^2(n-6)\right) \right) \frac{1}{F} \right) \right) \right]$$

$$= \delta \int_V d^n x \left[\sqrt{-g \cdot F^{n+2q}} \times \left(\left(R + \frac{F'}{F}(1-n)\Delta f \atop + \frac{f_{,\alpha} f_{,\beta} g^{\alpha\beta}(1-n)}{4F^2}\left(4FF'' + (F')^2(n-6)\right) \right) \frac{1}{F} \right) \right] \quad (649)$$

would then lead (for the full derivation see chapter 8) to:

$$\Rightarrow \quad 0 = \left\{ \begin{array}{l} R^{\kappa\lambda} - R^* g^{\kappa\lambda} + \left(\left(\underbrace{\frac{8\pi G}{\kappa} T^{\kappa\lambda}}_{} = \text{matter} \atop 0 = \text{vacuum} \right) + \Lambda \cdot g^{\kappa\lambda} \right) \\[2em] - \frac{1}{\sqrt{F^p}} \cdot \left(\partial^{:\kappa} \partial^{:\lambda} - g^{\kappa\lambda} \Delta_g \right) F^{\frac{p}{2}-1} \\[1em] + \left(g^{\kappa\lambda} g^{\alpha\beta} - g^{\kappa\alpha} g^{\lambda\beta} - g^{\lambda\alpha} g^{\kappa\beta} \right)(1-n) \\[1em] \cdot \left[\frac{f_{,\alpha} f_{,\beta}}{F^2}\left(FF'' + (F')^2 \cdot \frac{2 \cdot p + n - 14}{8} \right) + f_{,\alpha\beta} \frac{F'}{F} \right] \\[1em] - \frac{F' \cdot (1-n)}{2 \cdot F}\left(g^{\kappa\lambda} g^{\alpha\beta} - g^{\kappa\alpha} g^{\lambda\beta} - g^{\lambda\alpha} g^{\kappa\beta} \right) \Gamma^{\gamma}_{\alpha\beta} f_{,\gamma} \\[1em] + \frac{F'}{F} \cdot (1-n) g^{\alpha\beta} g^{\gamma\lambda} \Gamma^{\kappa}_{\alpha\beta} f_{,\gamma} \\[1em] + \frac{F'}{F} \cdot \frac{(1-n)}{2}\left(g^{\kappa\lambda} g^{\alpha\beta} - g^{\kappa\alpha} g^{\lambda\beta} - g^{\lambda\alpha} g^{\kappa\beta} \right)_{,\beta} f_{,\alpha} \\[1em] + \frac{g^{\xi\zeta} g_{\xi\zeta,\beta}}{2} \cdot \left(\frac{F'}{F} \cdot \frac{(1-n)}{2}\left(g^{\kappa\lambda} g^{\alpha\beta} - g^{\kappa\alpha} g^{\lambda\beta} - g^{\lambda\alpha} g^{\kappa\beta} \right) f_{,\alpha} \right). \end{array} \right. \quad (650)$$

Note that we defined $p = n + 2q$.

7.3 Centers of Gravity and Quantum Centers

In a variety of previous books and papers [4, 11–14], the author also discussed the question of centers of gravity, which is to say ensembles of origins to the metric base vectors, and was able to show that the principles of thermodynamics and evolution lurk behind this additional degree of freedom (especially see [4]).

For convenience, we here repeat some of the essentials of the idea.

7.3.1 Repetition: The Base-Vector Variation (from [4])

Our starting point shall be the usual tensor transformation rule for the covariant metric tensor:

$$g_{\delta\gamma} = \mathbf{g}_\delta \cdot \mathbf{g}_\gamma = \frac{\partial G^\alpha \left[x_k \right]}{\partial x^\delta} \frac{\partial G^\beta \left[x_k \right]}{\partial x^\gamma} g_{\alpha\beta}. \tag{651}$$

This already leads to quite some varieties with respect to the apparently simple term $\delta g_{\mu\nu}$ [4, 15]. The base vectors $\mathbf{g}_\delta$ to a certain metric are given as:

$$\mathbf{g}_\delta = \frac{\partial G^\alpha \left[x_k \right]}{\partial x^\delta} \mathbf{e}_\alpha, \tag{652}$$

where the functions $G^\alpha[\ \dots \ ]$ denote arbitrary functions of the coordinates x_k. Here the vectors $\mathbf{e}_\alpha$ shall denote the base vectors of a fundamental coordinate system of the right (in principle arbitrary) number of dimensions. Thus, we have the variation for $\delta g_{\delta\lambda}$ as follows:

$$\delta g_{\delta\gamma} = \delta\left(\mathbf{g}_\delta \cdot \mathbf{g}_\gamma \right) = \mathbf{g}_\delta \cdot \delta\mathbf{g}_\gamma + \delta\mathbf{g}_\delta \cdot \mathbf{g}_\gamma$$

$$= \frac{\partial G^\alpha \left[x_k \right]}{\partial x^\delta} \mathbf{e}_\alpha \cdot \delta\left(\frac{\partial G^\beta \left[x_k \right]}{\partial x^\gamma} \mathbf{e}_\beta \right) + \delta\left(\frac{\partial G^\alpha \left[x_k \right]}{\partial x^\delta} \mathbf{e}_\alpha \right) \cdot \frac{\partial G^\beta \left[x_k \right]}{\partial x^\gamma} \mathbf{e}_\beta. \tag{653}$$

Now, we introduce two additional degrees of freedom, namely:

(a) neither the number of dimensions in which the base vectors exist and form a complete transformation (652) needs to be the same as the metric space they define,

(b) nor that the variation δ is defined or fixed in any way.

In order to properly account for point (a), we shall rewrite Eqs. (651) and (652) as:

$$\delta g_{\delta\gamma} = \delta\left(\mathbf{g}_\delta \cdot \mathbf{g}_\gamma \right) = \mathbf{g}_\delta \cdot \delta\mathbf{g}_\gamma + \delta\mathbf{g}_\delta \cdot \mathbf{g}_\gamma$$

$$= \frac{\partial G^i \left[x_k \right]}{\partial x^\delta} \mathbf{e}_i \cdot \delta \left(\frac{\partial G^j \left[x_k \right]}{\partial x^\gamma} \mathbf{e}_j \right) + \delta \left(\frac{\partial G^i \left[x_k \right]}{\partial x^\delta} \mathbf{e}_i \right) \cdot \frac{\partial G^j \left[x_k \right]}{\partial x^\gamma} \mathbf{e}_j$$

$$\mathbf{g}_\delta = \frac{\partial G^j \left[x_k \right]}{\partial x^\delta} \mathbf{e}_j. \tag{654}$$

Thereby the Latin indices are running to a different (potentially higher) number of dimensions N than the Greek indices, which shall be defined for a space or space-time of n dimensions.

Now, having extended the options for the metric and its base vectors in such a way, which is to say by allowing them to have some kind of arbitrary intrinsic structure, we should ask ourselves whether this should not also be possible for our quantum scale factors $F[f]$. As already shown in connection with the solutions to the elastic equations in [4, 8], we may assume intrinsic structures and various quantum centers (see section "8.8 Centers of Gravity and Quantum Centers" in the next chapter). The latter then automatically leads to ensembles of $F[f]$s and thus, to statistics (quantum statistics) and—necessarily— thermodynamics (quantum thermodynamics, which we may distinguish from the metric ensemble thermodynamics, because the quantum thermodynamics is completely based on the wave function $F[f]$ [the scale factor to the metric]), while the metric thermodynamics is based on the base vectors only.

In order to better understand this, we here briefly repeat the corresponding considerations with respect to the metric or base vector ensemble variation from [4]. A somewhat more comprehensive summary is been given in [8] and the next chapter.

At first, we assume to have an ensemble of gravity centers as follows:

$$\frac{\partial G^j \left[x_k \right]}{\partial x^\gamma} \mathbf{e}_j = \frac{\partial G^j \left[x_k - \xi_{kj} \right]}{\partial x^\gamma} \mathbf{e}_j, \tag{655}$$

with the ξ_{kj} denoting the various centers of gravity. Now one could perform the variation with respect to exactly these coordinates and find (thereby simplifying via the condition of the closed system, which leaves the number of dimensions a constant and thus, $\frac{\partial g_{\alpha\beta}}{\partial n} \delta n = 0$, see [4] for more information):

$$X_k \equiv x_k - \xi_k; \quad \delta_g W = \int_V d^n x \left(\overbrace{R^{\delta\gamma} - \frac{1}{2} R g^{\delta\gamma} + \Lambda g^{\delta\gamma} + \kappa T^{\delta\gamma}}^{\text{Relativity}} \right)$$

$$
\times \left(
\begin{array}{c}
\underbrace{\dfrac{\partial G^i\left[X_k\right]}{\partial x^\delta}\mathbf{e}_i \cdot \delta_x\left(\dfrac{\partial G^j\left[X_k\right]}{\partial x^\gamma}\mathbf{e}_j\right) + \delta_x\left(\dfrac{\partial G^i\left[X_k\right]}{\partial x^\delta}\mathbf{e}_i\right)\cdot\dfrac{\partial G^j\left[X_k\right]}{\partial x^\gamma}\mathbf{e}_j}_{\text{Quantum}} \\[2em]
\underbrace{+\dfrac{\partial G^i\left[X_k\right]}{\partial x^\delta}\mathbf{e}_i \cdot \delta_\xi\left(\dfrac{\partial G^j\left[X_k\right]}{\partial x^\gamma}\mathbf{e}_j\right) + \delta_\xi\left(\dfrac{\partial G^i\left[X_k\right]}{\partial x^\delta}\mathbf{e}_i\right)\cdot\dfrac{\partial G^j\left[X_k\right]}{\partial x^\gamma}\mathbf{e}_j}_{\text{"?"}\,\Rightarrow\,2^{\text{nd}}\text{ law of Thermodynamics}}
\end{array}
\right).
$$

$$(656)$$

It is evident that with the introduction of centers of gravity, we should also consider the option for interaction among those various centers. This would simply be achieved by allowing the G^j to not only depend on the differential coordinates X_k as given in Eq. (656), but also to have dependencies as:

$$
G^j\left[\ldots\right] = G^j\left[X_{kj}\right] \equiv G^j\left[X_{k0},X_{k1},\ldots,X_{ki},\ldots,X_{k(N-1)}\right]; \quad X_{ki} \equiv x_k - \xi_{ki}.
$$

$$(657)$$

This immediately renders the variation with respect to the relative or position coordinates extremely complex even in the smallest ensembles. But this is not so different from the usual complexity known from statistical mechanics. Here, now we just face quantum gravity interactive statistics, where the complete variation should read:

$$
\delta_g W = \int_V d^n x \left(\overbrace{R^{\delta\gamma} - \frac{1}{2}Rg^{\delta\gamma} + \Lambda g^{\delta\gamma}}^{\text{Relativity}} + \kappa T^{\delta\gamma} \right)
$$

$$
\times \left(
\begin{array}{c}
\underbrace{\dfrac{\partial G^i\left[X_{ki}\right]}{\partial x^\delta}\mathbf{e}_i \cdot \delta_x\left(\dfrac{\partial G^j\left[X_{kj}\right]}{\partial x^\gamma}\mathbf{e}_j\right) + \delta_x\left(\dfrac{\partial G^i\left[X_{ki}\right]}{\partial x^\delta}\mathbf{e}_i\right)\cdot\dfrac{\partial G^j\left[X_{kj}\right]}{\partial x^\gamma}\mathbf{e}_j}_{\text{Quantum}} \\[2em]
\underbrace{+\sum_{i,j=0}^{N-1}\left(\dfrac{\partial G^i\left[X_{ki}\right]}{\partial x^\delta}\mathbf{e}_i \cdot \delta_{\xi j}\left(\dfrac{\partial G^j\left[X_{kj}\right]}{\partial x^\gamma}\mathbf{e}_j\right) + \delta_{\xi i}\left(\dfrac{\partial G^i\left[X_{ki}\right]}{\partial x^\delta}\mathbf{e}_i\right)\cdot\dfrac{\partial G^j\left[X_{kj}\right]}{\partial x^\gamma}\mathbf{e}_j\right)}_{2^{\text{nd}}\text{ law of Thermodynamics \& Interaction}}
\end{array}
\right).
$$

$$(658)$$

Summing up all variations considered here (meaning without the scale factor) in just one scale-factor free, but otherwise "world formula like" also for N- and n-variable (perhaps open) systems would now give us the following two forms [4]:

$$\delta_g W = \left(\int_V d^n x \left[\overbrace{\left[R - 2\kappa L_M - 2\Lambda \right] \cdot \frac{\partial \sqrt{-g}}{\partial n} + \sqrt{-g} \cdot \frac{\partial \left[R - 2\kappa L_M - 2\Lambda \right]}{\partial n}}^{\text{Thermodynamics / Exchange / Open Systems}} \right] \right) \delta n$$

$$+ \int_V d^n x \left(\overbrace{R^{\delta\gamma} - \frac{1}{2} R g^{\delta\gamma} + \Lambda g^{\delta\gamma} + \kappa T^{\delta\gamma}}^{\text{Relativity}} \right)$$

$$\times \left(\underbrace{\frac{\partial G^i [X_{ki}]}{\partial x^\delta} \mathbf{e}_i \cdot \delta_x \left(\frac{\partial G^j [X_{kj}]}{\partial x^\gamma} \mathbf{e}_j \right) + \delta_x \left(\frac{\partial G^i [X_{ki}]}{\partial x^\delta} \mathbf{e}_i \right) \cdot \frac{\partial G^j [X_{kj}]}{\partial x^\gamma} \mathbf{e}_j}_{\text{Quantum}} + \underbrace{\frac{\partial g_{\delta\gamma}}{\partial n} \delta n}_{\text{ThD}} \right.$$

$$\left. \underbrace{+ \delta_N \left(\sum_{i,j=0}^{N-1} \frac{\partial G^i [X_{ki}]}{\partial x^\delta} \mathbf{e}_i \cdot \delta_{\xi j} \left(\frac{\partial G^j [X_{kj}]}{\partial x^\gamma} \mathbf{e}_j \right) + \delta_{\xi i} \left(\frac{\partial G^i [X_{ki}]}{\partial x^\delta} \mathbf{e}_i \right) \cdot \frac{\partial G^j [X_{kj}]}{\partial x^\gamma} \mathbf{e}_j \right)}_{2^{\text{nd}} \text{ law of Thermodynamics \& Interaction}} \right)$$

$$(659)$$

or somewhat simpler with the *N*-variation as separate addend:

$$\delta_g W = \left(\int_V d^n x \left[\overbrace{\left[R - 2\kappa L_M - 2\Lambda \right] \cdot \frac{\partial \sqrt{-g}}{\partial n} + \sqrt{-g} \cdot \frac{\partial \left[R - 2\kappa L_M - 2\Lambda \right]}{\partial n}}^{\text{Thermodynamics / Exchange / Open Systems}} \right] \right) \delta n$$

$$+ \int_V d^n x \left(\overbrace{R^{\delta\gamma} - \frac{1}{2} R g^{\delta\gamma} + \Lambda g^{\delta\gamma} + \kappa T^{\delta\gamma}}^{\text{Relativity}} \right)$$

$$\times \left(\underbrace{\frac{\partial G^i [X_{ki}]}{\partial x^\delta} \mathbf{e}_i \cdot \delta_x \left(\frac{\partial G^j [X_{kj}]}{\partial x^\gamma} \mathbf{e}_j \right) + \delta_x \left(\frac{\partial G^i [X_{ki}]}{\partial x^\delta} \mathbf{e}_i \right) \cdot \frac{\partial G^j [X_{kj}]}{\partial x^\gamma} \mathbf{e}_j}_{\text{Quantum}} + \underbrace{\frac{\partial g_{\delta\gamma}}{\partial n} \delta n}_{\text{ThD}} \right.$$

$$\left. \underbrace{+ \frac{\partial g_{\alpha\beta}}{\partial N} \delta_N + \sum_{i,j=0}^{N-1} \left(\begin{array}{c} \frac{\partial G^i [X_{ki}]}{\partial x^\delta} \mathbf{e}_i \cdot \delta_{\xi j} \left(\frac{\partial G^j [X_{kj}]}{\partial x^\gamma} \mathbf{e}_j \right) \\ + \delta_{\xi i} \left(\frac{\partial G^i [X_{ki}]}{\partial x^\delta} \mathbf{e}_i \right) \cdot \frac{\partial G^j [X_{kj}]}{\partial x^\gamma} \mathbf{e}_j \end{array} \right)}_{2^{\text{nd}} \text{ law of Thermodynamics \& Interaction}} \right)$$

$$(660)$$

7.3.2 Quantum Centers

As it was shown in [4] that there are obviously quite some interesting things behind the idea of extending the variation of the Einstein–Hilbert action with respect to ensemble parameters like the position of various centers of gravity, we now want to briefly investigate the question of what happens when also allowing for different centers of scaling or quantum functions $F[f]$.

Thus, similar to Eq. (657), we could also introduce quantum or scaling centers of the following kind:

$$f_k[\ldots] = f\left[Y_{kj}\right] \equiv f\left[Y_{k0}, Y_{k1}, \ldots, Y_{ki}, \ldots, Y_{k(N-1)}\right]; \quad Y_{ki} \equiv x_k - \zeta_{ki}. \quad (661)$$

The attentive reader may have realized that, just as with the centers of gravity approach (657), we also have the possibility of interaction. This renders the variability of the general approach rather enormous and we have to move further and carry out deeper investigations and discussions to follow-up publications.

The question we here want to discuss, however, should be the one about how these quantum centers could actually be "activated" during the variation. Within the last section it became clear that the centers of gravity are becoming "active" simply by elongating the usual metric variation beyond the classical point and also varying the metric itself with respect to its possible intrinsic structure, which is the base vector (generalized according to Eq. (657)) and the various positions ξ_{ki} of the centers of gravity. Also having to deal with quantum centers in accordance with Eq. (661) requires an additional variation with respect to the parameters ζ_{ki}. This, however, leads automatically to an additional derivation of the function f with respect to the coordinates, because the task reads:

$$\delta W = 0 = \delta_\zeta \int_V d^n x \left(\sqrt{-G} \cdot \left(R^* - 2\Lambda + L_M\right)\right)$$

$$= \delta_\zeta \int_V d^n x \left(\sum_{\forall f_k} \sqrt{-G} \cdot \left(R^* - 2\Lambda + L_M\right)\right) = \sum_{\forall f_k} \delta_\zeta \int_V d^n x \left(\sqrt{-G} \cdot \left(R^* - 2\Lambda + L_M\right)\right).$$

$$(662)$$

Inserting our result for R^* from above (Eq. (640)), we obtain a third-order differential equation for f, which in fact becomes f_k now. Please note that during the evaluation, it has to be assured that the derivative ∂_σ only acts on the function f. After simplification we have (for the full evaluation see next chapter):

$$\delta W = 0 =$$

$$= \sum_{\forall f_k} \int_V d^n x \left(\sqrt{-g} \cdot F^{\frac{n}{2}-1} \left(\begin{array}{c} \left(\dfrac{n}{2}-1\right)\dfrac{f_{,\sigma}}{F} \\ \cdot \left(R + \dfrac{f_{,\alpha}f_{,\beta}g^{\alpha\beta}(1-n)}{4F^2}\left(4FF'' + \left(F'\right)^2(n-6)\right)\right) \\ + \dfrac{f_{,\sigma}}{F}\left(\dfrac{FF'' + \left(\dfrac{n}{2}-2\right)\left(F'\right)^2}{F}\right)(1-n)\Delta f \\ g^{\alpha\beta}\left(\begin{array}{c} +\dfrac{F'}{F}(1-n)\left(f_{,\alpha\beta\sigma} - \Gamma^{\gamma}_{\alpha\beta}f_{,\gamma\sigma}\right) \\ + \left(\dfrac{f_{,\alpha}f_{,\beta}(1-n)}{4F^2}\left(4FF'' + \left(F'\right)^2(n-6)\right)\right)_{,\sigma} \end{array}\right) \\ + f_{,\sigma}\dfrac{n}{2}\cdot\left(L_M - 2\Lambda\right) \end{array}\right) \right) Y^{\sigma}_{,k}\delta\zeta^k.$$

(663)

We realize that even without the sum over the whole k-ensemble, it should be possible to find solutions of the form:

$$0 = \left\{ \begin{array}{c} \left(\dfrac{n}{2}-1\right)\dfrac{f_{,\sigma}}{F}\cdot\left(R + \dfrac{f_{,\alpha}f_{,\beta}g^{\alpha\beta}(1-n)}{4F^2}\left(4FF'' + \left(F'\right)^2(n-6)\right)\right) \\ + \dfrac{f_{,\sigma}}{F^2}\left(FF'' + \left(\dfrac{n}{2}-2\right)\left(F'\right)^2\right)(1-n)\Delta f \\ + g^{\alpha\beta}\left(\begin{array}{c} \dfrac{F'}{F}(1-n)\left(f_{,\alpha\beta\sigma} - \Gamma^{\gamma}_{\alpha\beta}f_{,\gamma\sigma}\right) \\ + \left(\dfrac{f_{,\alpha}f_{,\beta}(1-n)}{4F^2}\left(4FF'' + \left(F'\right)^2(n-6)\right)\right)_{,\sigma} \end{array}\right) \\ + f_{,\sigma}\dfrac{n}{2}\cdot\left(L_M - 2\Lambda\right). \end{array} \right.$$

(664)

For completeness we also give the results in the case of the generalized (F^q-scale-factor generalized) action (649) with the total integral form in the case of non-zero cosmological constant and matter density:

$$\delta W = 0 = \delta_{\zeta}\int_V d^n x\left(\sqrt{-G}\cdot F^q\cdot\left(R^* - 2\Lambda + L_M\right)\right)$$

$$= \delta_\zeta \int_V d^n x \left(\sum_{\forall f_k} \sqrt{-G} \cdot F^q \cdot \left(R^* - 2\Lambda + L_M \right) \right) = \sum_{\forall f_k} \delta_\zeta \int_V d^n x \left(\sqrt{-G} \cdot F^q \cdot \left(R^* - 2\Lambda + L_M \right) \right)$$

(665)

(with respect to the derivation we refer to chapter 8 of this book),

$$0 = \sum_{\forall f_k} \int_V d^n x \left(\sqrt{-g} \cdot F^{\frac{p}{2}-1} \left(\left(\begin{array}{c} \left(\frac{p}{2} - 1 \right) \cdot F' \frac{f_{,\sigma}}{F} \\ \\ \left(R + \frac{F'}{F}(1-n)\Delta f \\ + \frac{f_{,\alpha} f_{,\beta} g^{\alpha\beta}(1-n)}{4F^2} \left(4FF'' + (F')^2 (n-6) \right) \right) \\ \\ + \left(\frac{F''F + \left(\frac{p}{2} - 2 \right)(F')^2}{F^2} \right) \cdot f_{,\sigma}(1-n)\Delta f \\ \\ + g^{\alpha\beta} \left(\begin{array}{c} + \frac{F'}{F}(1-n)\left(f_{,\alpha\beta\sigma} - \Gamma^\gamma_{\alpha\beta} f_{,\gamma\sigma} \right) \\ + \left(\frac{f_{,\alpha} f_{,\beta}(1-n)}{4F^2} \left(4FF'' + (F')^2 (n-6) \right) \right)_{,\sigma} \end{array} \right) \\ \\ + f_{,\sigma} \frac{p}{2} F' \cdot \left(L_M - 2\Lambda \right) \end{array} \right) \right) Y^\sigma_{,k} \delta\zeta^k$$

(666)

and the corresponding intrinsic equation:

$$0 = \left\{ \begin{array}{c} \left[\left(\frac{p}{2} - 1 \right) F' \frac{f_{,\sigma}}{F} \cdot \left(R + \frac{f_{,\alpha} f_{,\beta} g^{\alpha\beta}(1-n)}{4F^2} \left(4FF'' + (F')^2 (n-6) \right) \right) \right. \\ \\ \left. + \frac{f_{,\sigma}}{F^2} \left(FF'' \left(\frac{n}{2} - 2 \right)(F')^2 \right)(1-n)\Delta f \right] \\ \\ + g^{\alpha\beta} \left(\begin{array}{c} \frac{F'}{F}(1-n)\left(f_{,\alpha\beta\sigma} - \Gamma^\gamma_{\alpha\beta} f_{,\gamma\sigma} \right) \\ + \left(\frac{f_{,\alpha} f_{,\beta}(1-n)}{4F^2} \left(4FF'' + (F')^2 (n-6) \right) \right)_{,\sigma} \end{array} \right) \\ \\ + f_{,\sigma} \frac{p}{2} F' \cdot \left(L_M - 2\Lambda \right). \end{array} \right.$$

(667)

7.3.2.1 A possible connection to the 3-generation problem?

Solutions to Eq. (664) would be intrinsic ones and we here want to investigate some simpler types for certain settings of $F[f]$. Thus, in the case of:

$$0 = 4FF'' + \left(F'\right)^2 (n-6) \quad \Rightarrow \quad F = \left(f - C_{f0}\right)^{\frac{4}{n-2}} C_{f1}, \tag{668}$$

Eq. (664) simplifies to:

$$0 = \begin{cases} \left(\dfrac{n}{2} - 1\right)\dfrac{f_{,\sigma}}{F} \cdot R + f_{,\sigma}\dfrac{n}{2}\cdot\left(L_M - 2\Lambda\right) \\[2ex] + \dfrac{f_{,\sigma}}{F^2}\left(FF'' + \left(\dfrac{n}{2} - 2\right)(F')^2\right)(1-n)\Delta f + g^{\alpha\beta}\left(\dfrac{F'}{F}(1-n)\left(f_{,\alpha\beta\sigma} - \Gamma^{\gamma}_{\alpha\beta}f_{,\gamma\sigma}\right)\right) \end{cases}$$

$$= \begin{cases} \left(\dfrac{n}{2} - 1\right)\dfrac{f_{,\sigma}}{F} \cdot R + f_{,\sigma}\dfrac{n}{2}\cdot\left(L_M - 2\Lambda\right) \\[2ex] + \dfrac{f_{,\sigma}}{F^2}\left(FF'' + \left(\dfrac{n}{2} - 2\right)(F')^2\right)(1-n)\Delta f + g^{\alpha\beta}\left(\dfrac{F'}{F}(1-n)\left(f_{,\sigma\alpha\beta} - \Gamma^{\gamma}_{\alpha\beta}f_{,\sigma\gamma}\right)\right) \end{cases}$$

$$= \begin{cases} \left(\dfrac{n}{2} - 1\right)\dfrac{f_{,\sigma}}{F} \cdot R + f_{,\sigma}\dfrac{n}{2}\cdot\left(L_M - 2\Lambda\right) \\[2ex] + \dfrac{(1-n)}{F}\left(\dfrac{f_{,\sigma}}{F}\left(FF'' + \left(\dfrac{n}{2} - 2\right)(F')^2\right)\Delta f + F'\Delta\left(f_{,\sigma}\right)\right). \end{cases} \tag{669}$$

Setting $n = 4$ and inserting $F[f]$ from Eq. (668) yields:

$$0 = 4\cdot f_{,\sigma}\cdot\left(L_M - 2\Lambda\right)\cdot\left(f - C_{f0}\right)\cdot C_{f1} + \frac{2\cdot\left(\left(f - C_{f0}\right)\cdot\left(f_{,\sigma}\cdot R - 3\cdot\Delta\left(f_{,\sigma}\right)\right) - 3\cdot f_{,\sigma}\Delta f\right)}{\left(f - C_{f0}\right)^2}$$

$$\Rightarrow 0 = 2\cdot f_{,\sigma}\cdot\left(L_M - 2\Lambda\right)\cdot\left(f - C_{f0}\right)\cdot C_{f1} + \frac{\left(f - C_{f0}\right)\cdot\left(f_{,\sigma}\cdot R - 3\cdot\Delta\left(f_{,\sigma}\right)\right) - 3\cdot f_{,\sigma}\Delta f}{\left(f - C_{f0}\right)^2}. \tag{670}$$

With the assumption of the existence of Eigenvalue solutions to the Laplace equation of the kind $\Delta f = B \cdot f$ and the additional demand that we can interchange the Laplace operator and the derivation with respect to σ, we obtain from Eq. (670):

$$0 = 2 \cdot f_{,\sigma} \cdot \left(L_M - 2\Lambda\right) \cdot \left(f - C_{f0}\right) \cdot C_{f1} + \frac{\left(f - C_{f0}\right) \cdot \left(f_{,\sigma} \cdot R - 3 \cdot \overbrace{\Delta\left(f_{,\sigma}\right)}^{=B \cdot f_{,\sigma}}\right) - 3 \cdot f_{,\sigma} \overbrace{\Delta f}^{=B \cdot f}}{\left(f - C_{f0}\right)^2}$$

$$0 = f_{,\sigma} \cdot \left[2 \cdot \left(L_M - 2\Lambda\right) \cdot \left(f - C_{f0}\right) \cdot C_{f1} + \frac{\left(f - C_{f0}\right) \cdot \left(R - 3 \cdot B\right) - 3 \cdot B \cdot f}{\left(f - C_{f0}\right)^2}\right]$$

$$\Rightarrow f_{1,\sigma} = 0 \quad \& : 0 = 2 \cdot \left(L_M - 2\Lambda\right) \cdot \left(f - C_{f0}\right)^3 \cdot C_{f1} + \left(f - C_{f0}\right) \cdot \left(R - 3 \cdot B\right) - 3 \cdot B \cdot f.$$

$$(671)$$

Thereby $f_{1,\sigma} = 0$ just means that we have one solution with:

$$f_1 = \text{const}. \tag{672}$$

Setting $L_M - 2\Lambda = 0$ we can further simplify to:

$$0 = \left(f - C_{f0}\right) \cdot R + 3 \cdot B \cdot \left(C_{f0} - 2 \cdot f\right). \tag{673}$$

This is a simple linear equation with the following solution:

$$0 = \left(f - C_{f0}\right) \cdot R + 3 \cdot B \cdot \left(C_{f0} - 2 \cdot f\right)$$

$$\Rightarrow \qquad f_2 = \frac{C_{f0}\left(R - 3 \cdot B\right)}{\left(R - 6 \cdot B\right)}. \tag{674}$$

Please note that in the case of $L_M - 2\Lambda \neq 0$ we even end up with three solutions instead of only one solution (not counting the f = const) as the equation we have to solve would now be:

$$0 = 2 \cdot \left(L_M - 2\Lambda\right) \cdot \left(f - C_{f0}\right)^3 \cdot C_{f1} + \left(f - C_{f0}\right) \cdot \left(R - 3 \cdot B\right) - 3 \cdot B \cdot f \tag{675}$$

and it is of third order with respect to f.

Could this already give or at least lead to the answer to the 3-generation riddle in elementary particle physics?

A bit of reshaping of the last equation is going to help us to realize how this could possibly be connected with the 3-generation mass problem:

$$0 = \begin{cases} 3 \cdot B \cdot C_{f0} - C_{f0}{}^3 \cdot C_{f1} \cdot 2 \cdot \left(L_M - 2\Lambda\right) - C_{f0} \cdot R \\ + f \cdot \left(6 \cdot B + 6 \cdot C_{f0}{}^2 \cdot \left(L_M - 2\Lambda\right) + R\right) \\ + f^2 \cdot 6 \cdot C_{f0} \cdot \left(L_M - 2\Lambda\right) \\ + f^3 \cdot 2 \cdot \left(L_M - 2\Lambda\right) \end{cases} . \tag{676}$$

The general solution to a three-order polynomial could be given via the following product form:

$$(f - f_2) \cdot (f - f_3) \cdot (f - f_4) =$$

$$f^3 - f^2 \cdot (f_4 + f_2 + f_3) + f \cdot (f_4 f_2 + f_4 f_3 + f_2 f_3) - f_4 f_2 f_3. \qquad (677)$$

Comparing the latter with Eq. (675), divided by $2 \cdot (L_M - 2\Lambda)$ gives us:

$$f_4 f_2 f_3 = -\frac{3 \cdot B \cdot C_{f0} - C_{f0}{}^3 \cdot C_{f1} \cdot 2 \cdot (L_M - 2\Lambda) - C_{f0} \cdot R}{2 \cdot (L_M - 2\Lambda)}$$

$$f_4 f_2 + f_4 f_3 + f_2 f_3 = \frac{6 \cdot B + 6 \cdot C_{f0}{}^2 \cdot (L_M - 2\Lambda) + R}{2 \cdot (L_M - 2\Lambda)}$$

$$f_4 + f_2 + f_3 = -3 \cdot C_{f0}. \qquad (678)$$

Thus, we have obtained three solutions to the function f, which satisfy Eq. (675) and together with (668) fix the scale factor $F[f]$ to the metric tensor. As this means nothing else than the determination of the volume for a given metric, we might indeed see some connection to the 3-generation problem.

But before drawing a final conclusion here, we first want to investigate other options.

7.3.3 Centers of Gravity or Metric Centers

Within our repetition above we have only considered the situation with metric centers in connection with the classical (which is to say non-$F[f]$-scaled) metrics and the subsequent Einstein–Hilbert action derivation. Extension to the situation with the scaled metric is straight forward. Using the result from Eq. (648) and putting this back into the Hilbert integral, thereby assuming that we have an ensemble of metric or gravity centers, we obtain:

$$0 = \sum_{\forall G_k ; \xi^k} \int_V d^n x \cdot \frac{\sqrt{-g \cdot F^n}}{F}$$

$$
\begin{pmatrix}
R^{\kappa\lambda} - \dfrac{1}{2} R g^{\kappa\lambda} + \left(\begin{pmatrix} \underbrace{\dfrac{8\pi G T^{\kappa\lambda}}{\kappa}} = \text{matter} \\ 0 = \text{vacuum} \end{pmatrix} + \Lambda \cdot g^{\kappa\lambda} \right) \\[2em]
- \dfrac{1}{\sqrt{F^n}} \cdot \left(\partial^{;\kappa} \partial^{;\lambda} - g^{\kappa\lambda} \Delta_g \right) F^{\frac{n}{2}-1} \\[1em]
+ \left(g^{\kappa\lambda} g^{\alpha\beta} - g^{\kappa\alpha} g^{\lambda\beta} - g^{\lambda\alpha} g^{\kappa\beta} \right)(1-n) \\[1em]
\cdot \left[\dfrac{f_{,\alpha} f_{,\beta}}{F^2} \left(FF'' + \left(F' \right)^2 \cdot \dfrac{3\cdot n - 14}{8} \right) + f_{,\alpha\beta} \dfrac{F'}{F} \right] \\[1.5em]
- \dfrac{F' \cdot (1-n)}{2\cdot F} \left(g^{\kappa\lambda} g^{\alpha\beta} - g^{\kappa\alpha} g^{\lambda\beta} - g^{\lambda\alpha} g^{\kappa\beta} \right) \Gamma^{\gamma}_{\alpha\beta} f_{,\gamma} \\[1.5em]
+ \dfrac{F'}{F} \cdot (1-n) g^{\alpha\beta} g^{\gamma\lambda} \Gamma^{\kappa}_{\alpha\beta} f_{,\gamma} \\[1.5em]
+ \dfrac{F'}{F} \cdot \dfrac{(1-n)}{2} \left(g^{\kappa\lambda} g^{\alpha\beta} - g^{\kappa\alpha} g^{\lambda\beta} - g^{\lambda\alpha} g^{\kappa\beta} \right)_{,\beta} f_{,\alpha} \\[1.5em]
+ \dfrac{g^{\xi\zeta} g_{\xi\zeta,\beta}}{2} \cdot \left(\dfrac{F'}{F} \cdot \dfrac{(1-n)}{2} \left(g^{\kappa\lambda} g^{\alpha\beta} - g^{\kappa\alpha} g^{\lambda\beta} - g^{\lambda\alpha} g^{\kappa\beta} \right) f_{,\alpha} \right)
\end{pmatrix} X^{\sigma}_{\;,k} \delta \xi^{k}. \qquad (679)
$$

7.3.4 Summing Up: Centers of Gravity and Quantum Centers $\rightarrow$ Thermodynamics

We realize that in addition to any intrinsic solution to Eq. (648) or (664) (or—just simpler due to a pre-setting of $F[f]$—Eq. (669) and—in the case of $n = 4$—also Eq. (670)), for a given volume, we can always also aim for a statistic satisfaction of the extended Einstein–Hilbert actions via Eqs. (663) and (679). This kind of statistics, however, is just thermodynamics.

7.4 The Infinity Options Principle [3] or What Happens in Infinitely Many Dimensions

In order to investigate the situation in spaces or space-times with infinitely many dimensions, we simply take the results from above (Eqs. (663) and

(679)) divide by n^3, respectively n^2, and then look for the limit of $n \to \infty$. Thus, for Eqs. (663) and (679), respectively, we obtain:

$$\delta W = 0 =$$

$$= \sum_{\forall f_k} \int_V \lim_{n \to \infty} d^n x \left(\begin{array}{c} \dfrac{\sqrt{-g} \cdot F^{\frac{n}{2}-1}}{n^3} \times \\[2mm] \left(\begin{array}{c} \left(\dfrac{n}{2}-1\right)\dfrac{f_{,\sigma}}{F} \\[3mm] \cdot\left(R + \dfrac{f_{,\alpha}f_{,\beta}g^{\alpha\beta}(1-n)}{4F^2}\left(4FF'' + (F')^2(n-6)\right)\right) \\[3mm] +\dfrac{f_{,\sigma}}{F}\left(\dfrac{FF'' + \left(\dfrac{n}{2}-2\right)(F')^2}{F}\right)(1-n)\Delta f \\[3mm] g^{\alpha\beta}\left(\begin{array}{c} +\dfrac{F'}{F}(1-n)\left(f_{,\alpha\beta\sigma} - \Gamma^{\gamma}_{\alpha\beta}f_{,\gamma\sigma}\right) \\[3mm] +\left(\dfrac{f_{,\alpha}f_{,\beta}(1-n)}{4F^2}\left(4FF'' + (F')^2(n-6)\right)\right)_{,\sigma} \end{array}\right) \\[3mm] +f_{,\sigma}\dfrac{n}{2}\cdot\left(L_M - 2\Lambda\right) \end{array}\right) \end{array}\right) Y^{\sigma}_{,k}\delta\zeta^k$$

$$\delta W = 0 = \sum_{\forall f_k} \int_V \lim_{n \to \infty} d^n x \left(\sqrt{-g}\cdot F^{\frac{n}{2}-1}\cdot\left(f_{,\sigma}\dfrac{f_{,\alpha}f_{,\beta}g^{\alpha\beta}}{F^3}(F')^2\right)\right) Y^{\sigma}_{,k}\delta\zeta^k$$

$$(680)$$

and:

$$0 = \sum_{\forall G_k : \xi^k} \int_V \lim_{n \to \infty} d^n x \dfrac{\sqrt{-g}\cdot F^n}{F\cdot n^2}$$

$$\left(\cdot \begin{pmatrix} R^{\kappa\lambda} - \dfrac{1}{2} R \cdot g^{\kappa\lambda} + \left(\begin{pmatrix} \underbrace{\dfrac{8\pi G T^{\kappa\lambda}}{\kappa}}_{} = \text{matter} \\ 0 = \text{vacuum} \end{pmatrix} + \Lambda \cdot g^{\kappa\lambda} \right) \\[2em] - \dfrac{1}{\sqrt{F^n}} \cdot \left(\partial^{;\kappa}\partial^{;\lambda} - g^{\kappa\lambda}\Delta_g \right) F^{\frac{n}{2}-1} \\[1em] + \left(g^{\kappa\lambda} g^{\alpha\beta} - g^{\kappa\alpha} g^{\lambda\beta} - g^{\lambda\alpha} g^{\kappa\beta} \right)(1-n) \\[1em] \cdot \left[\dfrac{f_{,\alpha} f_{,\beta}}{F^2} \left(FF'' + (F')^2 \cdot \dfrac{3 \cdot n - 14}{8} \right) + f_{,\alpha\beta} \dfrac{F'}{F} \right] \\[1em] - \dfrac{F' \cdot (1-n)}{2 \cdot F} \left(g^{\kappa\lambda} g^{\alpha\beta} - g^{\kappa\alpha} g^{\lambda\beta} - g^{\lambda\alpha} g^{\kappa\beta} \right) \Gamma^{\gamma}_{\alpha\beta} f_{,\gamma} \\[1em] + \dfrac{F'}{F} \cdot (1-n) g^{\alpha\beta} g^{\gamma\lambda} \Gamma^{\kappa}_{\alpha\beta} f_{,\gamma} \\[1em] + \dfrac{F'}{F} \cdot \dfrac{(1-n)}{2} \left(g^{\kappa\lambda} g^{\alpha\beta} - g^{\kappa\alpha} g^{\lambda\beta} - g^{\lambda\alpha} g^{\kappa\beta} \right)_{,\beta} f_{,\alpha} \\[1em] + \dfrac{g^{\xi\zeta} g_{\xi\zeta,\beta}}{2} \cdot \left(\dfrac{F'}{F} \cdot \dfrac{(1-n)}{2} \left(g^{\kappa\lambda} g^{\alpha\beta} - g^{\kappa\alpha} g^{\lambda\beta} - g^{\lambda\alpha} g^{\kappa\beta} \right) f_{,\alpha} \right) \end{pmatrix} \right) X^{\sigma}{}_{,k} \delta\xi^k$$

$$= \sum_{\forall G_k;\xi^k} \int_V \lim_{n\to\infty} d^n x \frac{\sqrt{-g \cdot F^n}}{F} \cdot \left(-\left(g^{\kappa\lambda} g^{\alpha\beta} - g^{\kappa\alpha} g^{\lambda\beta} - g^{\lambda\alpha} g^{\kappa\beta} \right) \cdot \frac{f_{,\alpha} f_{,\beta}}{F^2} (F')^2 \cdot \frac{3}{8} \right) X^{\sigma}{}_{,k} \delta\xi^k$$

$$= \sum_{\forall G_k;\xi^k} \int_V \lim_{n\to\infty} d^n x \frac{\sqrt{-g \cdot F^n}}{F} \cdot \left(2 \cdot f^{,\kappa} f^{,\beta} - g^{\kappa\lambda} f^{,\beta} f_{,\beta} \right) \cdot \frac{(F')^2}{F^2} \cdot X^{\sigma}{}_{,k} \delta\xi^k . \tag{681}$$

Thereby, this we want to point out explicitly, we do not necessarily need to assume truly infinite numbers of dimensions, but simply big enough n so that all terms with n^p and $p < 3$ in Eq. (663) and $p < 2$ in Eq. (679) are exceedingly small against the n^3 in Eq. (663) and n^2 in Eq. (679).

We are going to discuss these last results in the next section ... at least with respect to Eq. (679), which—surely this is obvious—is, because of the power-3 term for the derivative of the quantum scale function f in there.

7.5 Toward a Solution to the 3-Generation Problem

7.5.1 Pre-Variation Results: The Simplest Path

Slightly intrigued by the fact that we already derived most classical quantum equations directly from the variated (scaled) Ricci scalar, we wonder whether it might be possible to already get an answer to the 3-generation riddle directly from the Ricci scalar equation itself rather than any of its variated forms as given in the sections above. In the next chapters (especially chapters 10 and 11), we will see that the condition of the vanishing or constant Ricci scalar is itself just a result of the variation of the generalized Einstein–Hilbert action.

Thus, we intend to derive equations for the masses of elementary or fundamental objects directly from the Ricci scalar of the scaled metric without any additional variation. Setting $F[f] = f + C_f$, we obtain from Eq. (640):

$$R^* = R^*_{\alpha\beta} G^{\alpha\beta} = \begin{pmatrix} \Gamma^\sigma_{\alpha\beta,\sigma} g^{\alpha\beta} - \Gamma^\sigma_{\beta\sigma,\alpha} g^{\alpha\beta} - \Gamma^\mu_{\sigma\alpha} \Gamma^\sigma_{\beta\mu} g^{\alpha\beta} + \Gamma^\sigma_{\alpha\beta} \Gamma^\mu_{\sigma\mu} g^{\alpha\beta} \\ + \dfrac{F'}{F}(1-n)\Delta f + \dfrac{f_{,\alpha} f_{,\beta} g^{\alpha\beta}(1-n)}{4F^2}\left(4FF'' + (F')^2(n-6)\right) \end{pmatrix} \dfrac{1}{F}$$

$$= \left(R + \frac{F'}{F}(1-n)\Delta f + \frac{f_{,\alpha} f_{,\beta} g^{\alpha\beta}(1-n)}{4F^2}\left(4FF'' + (F')^2(n-6)\right) \right) \frac{1}{F}$$

$$= \left(R + \frac{1}{f+C_f}(1-n)\Delta f + \frac{f_{,\alpha} f_{,\beta} g^{\alpha\beta}(1-n)}{4(f+C_f)^2}(n-6) \right) \frac{1}{f+C_f}. \qquad (682)$$

Now, we evaluate as follows:

$$\Xi = R^* \cdot F^3 = R^* (f+C_f)^3 = R \cdot (f+C_f)^2 + (f+C_f)(1-n)\Delta f + \frac{f_{,\alpha} f_{,\beta} g^{\alpha\beta}(1-n)}{4}(n-6)$$

$$(683)$$

and demand the integrated absolute value of this term to give zero, which is to say:

$$\int_0^{t_0} \int_{V_3} \sqrt{-g \cdot \Xi \cdot \Xi^*}\, dx^{n-1} dt = 0. \qquad (684)$$

Thereby we have to choose a suitable t_0. The superscript star as Ξ^* in Eq. (684) denotes the conjugate complex of the term Ξ (this is in difference to the Ricci scalar R^* of a $F[f]$-scaled metric tensor, where the $*$ is only been used to point out the difference to the unscaled R). The derivation with respect of the changes to the Hilbert evaluation of the Einstein field equations out of a $\sqrt{-g \cdot R \cdot R^*}$ -kernel instead of the usual $\sqrt{-g} \cdot R$ is been given in the appendix of this chapter (section 7.8).

It should be pointed out that currently it is not completely clear why we cannot simply use R^* in Eq. (684), but needed to resort to the form (683), which is to say $R* \cdot F^3$, in order to obtain suitable results. Here we try for a first explanation, where—just to have a start—we assume that the Ricci scalars, determinants, and so on, are all real and we do not need to bother about the evaluation of absolute values:

(a) The classical Einstein–Hilbert action does not deal with a scale factor, respectively there the scale factor $F[f]$ is one and so it is not clear what exponent one has to put onto it in order to obtain the correct asymptotic non-quantum outcome for the Lagrange density under the integral of the Einstein–Hilbert action. Thus, ignoring the artificially introduced and only postulated matter density L_M and the cosmological constant from Eq. (642), the exponent to this scale factor function F could be arbitrary within this integral making it to:

$$\delta W = 0 = \delta \int_V d^n x \sqrt{-G} \cdot F^q \cdot R^*. \tag{685}$$

(b) The setting $p = 3$ in Eq. (683) was motivated by the fact that one factor F was necessary anyway to give the term R in R^* (c.f. (682)) its usual appearance in the integral, namely just as R and not at R/F. Further two multiplications with F, making it $F*F^2 = F^3$ then, result from the determinant of the metric tensor in the case of $n = 4$, where we have:

$$\delta W = 0 = \delta \int_V d^{n=4} x \sqrt{-G} \cdot F \cdot R^* = \delta \int_V d^{n=4} x \sqrt{-g \cdot F^{n=4}} \cdot F \cdot R^* = \delta \int_V d^{n=4} x \sqrt{-g} \cdot F^3 \cdot R^*. \tag{686}$$

From this it follows that while the power-3 obviously gives no completely unreasonable results for $n = 4$, we simply do not know whether $p = 3$ should always be applied in Eq. (683) or rather $p = n/2 + 1$ or $p = n/2+?$. Also weak—at least at the moment—is our assumption that, while the integral with the square root in Eq. (684) is usually very difficult to handle, we can simply switch to the squared integrand and demanding this to be zero, which is to say:

$$\int_{0}^{t_0} \int_{V_3} g \cdot \Xi \cdot \Xi^* dx^{n-1} dt = 0. \tag{687}$$

The idea behind this extension of the usual Einstein–Hilbert action integral is that when scale variating the metric and allowing for potentially complex scale functions $F[f]$, here also playing the role of the quantum theoretical wave function, we shall also use the quantum theoretical way to extract observables. And as we here want to extract masses, being our observables of interest at the moment and thereby apply the particle at rest in the classical (Dirac-like) and thus, complex manner, we have to adjust the Einstein–Hilbert action accordingly. There might come up some criticism about the fact that our Ricci scalar R^* is not varied, which is true, but only if one excludes the scale variation as a proper variation and only considers the classical (derivative-like) variation as suitable. This author, however, is convinced that every form of variation should be considered as it is not in the nature of the universe (or the nature of nature) to leave out options for its own development and evolution. In addition, it will be demonstrated in the next chapters of this book (especially chapters 10 and 11) that the vanishing Ricci scalar conditions do indeed also result from the variation of the Einstein–Hilbert action.

The simplest explanation—at least at the moment—seems to be that we stick to the classical form without any scale factor used in the Einstein–Hilbert action, which is to say $q = 0$ in Eq. (685). Then, in order to obtain $p = 3$, the number of dimensions has to be $n = 6$. However, as we only "see" 4 dimensions, we can only assume that 2 further dimensions are compactified.

The rest, which is to say in order to come to Eq. (687), is just the usual quantum normalization.

Further discussion will be needed here and we are planning to present it in upcoming issues to one of our series.

Now, we assume a particle at rest with the following approach:

$$f = h_{,t} = \pm C_{h1} \cdot m \cdot i \cdot e^{\pm t \cdot m \cdot i} \tag{688}$$

and obtain ($t_0 = 2\pi/m$):

$$0 = \begin{cases} C_{h1}{}^2 m^6 \left(16 C_f{}^2 + C_{h1}{}^2 m^2 (n-2)^2\right)(n-1)^2 \\ + 8 C_{h1}{}^2 m^4 \left(8 C_f{}^2 + C_{h1}{}^2 m^2 (n-2)\right)(n-1) R \cdot \\ + 16\left(C_f{}^4 + 4 C_f{}^2 C_{h1}{}^2 m^2 + C_{h1}{}^4 m^4\right) R^2 \end{cases} \tag{689}$$

We substitute $\mu = m^2$:

$$0 = \begin{cases} 16 \cdot C_f{}^4 R^2 + \mu \cdot 64 \cdot C_f{}^2 C_{h1}{}^2 R^2 + \mu^2 \cdot 16 \cdot C_{h1}{}^2 R\left(4 \cdot C_f{}^2 (n-1) + C_{h1}{}^2 R\right) \\ + \mu^3 \cdot 8 \cdot C_{h1}{}^2 (n-1)\left(2 \cdot C_f{}^2 (n-1) + C_{h1}{}^2 (n-2) R\right) + \mu^4 \cdot C_{h1}{}^4 \left(2 - 3n + n^2\right)^2 \end{cases} \tag{690}$$

and using the general solution to a four-order polynomial via the following product form:

$$\left(\mu - \mu_1\right) \cdot \left(\mu - \mu_2\right) \cdot \left(\mu - \mu_3\right) \cdot \left(\mu - \mu_4\right) =$$
$$\mu^4 - \mu^3 \cdot \left(\mu_1 + \mu_2 + \mu_3 + \mu_4\right) + \mu^2 \cdot \left(\mu_1 \mu_2 + \mu_1 \mu_3 + \mu_2 \mu_3 + \mu_1 \mu_4 + \mu_3 \mu_4 + \mu_2 \mu_4\right)$$
$$- \mu \cdot \left(\mu_1 \mu_2 \mu_3 + \mu_2 \mu_3 \mu_4 + \mu_1 \mu_2 \mu_4 + \mu_1 \mu_3 \mu_4\right) + \mu_1 \mu_2 \mu_3 \mu_4 \tag{691}$$

and comparing the latter with Eq. (713), respectively (690) divided by $C_{h1}{}^4 \left(2 - 3n + n^2\right)^2$, gives us:

$$\mu_1 \mu_2 \mu_3 \mu_4 = \frac{16 \cdot C_f{}^4 R^2}{C_{h1}{}^4 \left(2 - 3n + n^2\right)^2}$$

$$\mu_1 \mu_2 \mu_3 + \mu_2 \mu_3 \mu_4 + \mu_1 \mu_2 \mu_4 + \mu_1 \mu_3 \mu_4 = -\frac{64 \cdot C_f{}^2 C_{h1}{}^2 R^2}{C_{h1}{}^4 \left(2 - 3n + n^2\right)^2}$$

$$\mu_1 \mu_2 + \mu_1 \mu_3 + \mu_2 \mu_3 + \mu_1 \mu_4 + \mu_3 \mu_4 + \mu_2 \mu_4 = \frac{16 \cdot C_{h1}{}^2 R\left(4 \cdot C_f{}^2 (n-1) + C_{h1}{}^2 R\right)}{C_{h1}{}^4 \left(2 - 3n + n^2\right)^2}$$

$$\mu_1 + \mu_2 + \mu_3 + \mu_4 = -\frac{8 \cdot C_{h1}{}^2 (n-1)\left(2 \cdot C_f{}^2 (n-1) + C_{h1}{}^2 (n-2) R\right)}{C_{h1}{}^4 \left(2 - 3n + n^2\right)^2}$$

$$\Rightarrow \qquad m_{1,5} = \pm\sqrt{\mu_1}; \quad m_{2,6} = \pm\sqrt{\mu_2}; \quad m_{3,7} = \pm\sqrt{\mu_3}; \quad m_{4,8} = \pm\sqrt{\mu_4}. \qquad (692)$$

We see that we have four pairs of matter and antimatter solutions instead of the experimentally known three. Far from considering this a solution, we just name the objects after the author's children, Julius, Livia, Felix, and Filia, and ask ourselves where—the heck—is Filia (Fig. 7.2)? As the latter (the child we mean) also has a habit of sudden disappearance, we find this a rather suitable analogy. Nevertheless, we should try to discuss why Filia is not been seen, neither in high energy accelerator experiments nor in cosmic radiation nor anywhere else. Observing the term of fourth order in Eq. (690), we see that it is vanishing in the following circumstances with respect to the number of dimensions *n*:

$$C_{h1}{}^4 \left(2 - 3n + n^2\right)^2 = 0 \quad \Rightarrow \quad n = 1, 2. \qquad (693)$$

Thus, one might conclude that elementary particles are all based on surfaces (constructed out of surfaces) and so they can only have three different masses, because then Eq. (690) reduces to an algebraic equation of third order. The other possible explanation is that the coefficient $C_{h1}{}^4 \left(2 - 3n + n^2\right)^2$ is always small against the other coefficients, which is to say:

$$C_{h1}{}^4 \left(2 - 3n + n^2\right)^2 \ll 16 \cdot C_f{}^4 R^2$$

$$C_{h1}{}^4 \left(2 - 3n + n^2\right)^2 \ll 64 \cdot C_f{}^2 C_{h1}{}^2 R^2$$

$$C_{h1}{}^4 \left(2 - 3n + n^2\right)^2 \ll 16 \cdot C_{h1}{}^2 R \left(4 \cdot C_f{}^2 (n-1) + C_{h1}{}^2 R\right)$$

$$C_{h1}{}^4 \left(2 - 3n + n^2\right)^2 \ll 8 \cdot C_{h1}{}^2 (n-1) \left(2 \cdot C_f{}^2 (n-1) + C_{h1}{}^2 (n-2) R\right)$$

$$\Rightarrow$$

$$C_{h1}{}^2 \left(2 - 3n + n^2\right)^2 \ll 64 \cdot C_f{}^2 R^2$$

$$C_{h1}{}^2 \left(2 - 3n + n^2\right)^2 \ll 16 \cdot R \left(4 \cdot C_f{}^2 (n-1) + C_{h1}{}^2 R\right)$$

$$C_{h1}{}^2 \left(2 - 3n + n^2\right)^2 \ll 8 \cdot (n-1) \left(2 \cdot C_f{}^2 (n-1) + C_{h1}{}^2 (n-2) R\right). \qquad (694)$$

We find that this is always fulfilled if we assume:

$$C_f{}^2 \gg C_{h1}{}^2. \qquad (695)$$

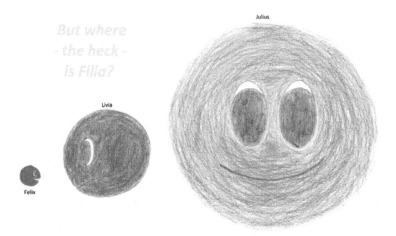

Figure 7.2 Where—the heck—is Filia (the fourth particle)?

This being the case, however, we can even formulate the conditions resulting from Eq. (695) with respect to Eq. (690) as being realized in our universe in a partially stronger manner, namely:

$$C_{h1}{}^4\left(2-3n+n^2\right)^2 \lll 16\cdot C_f{}^4 R^2$$

$$C_{h1}{}^2\left(2-3n+n^2\right)^2 \ll 64\cdot C_f{}^2 R^2$$

$$C_{h1}{}^2\left(2-3n+n^2\right)^2 \ll 16\cdot R\left(4\cdot C_f{}^2\left(n-1\right)+\varepsilon_1\right)$$

$$C_{h1}{}^2\left(2-3n+n^2\right)^2 \ll 8\cdot\left(n-1\right)\left(2\cdot C_f{}^2\left(n-1\right)+\varepsilon_2\right). \tag{696}$$

Thereby ε_i stands for very small terms. Ignoring all small contributions, we can rewrite Eq. (690) as follows:

$$0 = \begin{cases} 16\cdot C_f{}^4 R^2 + \mu\cdot 64\cdot C_f{}^2 C_{h1}{}^2 R^2 + \mu^2\cdot 64\cdot C_{h1}{}^2 R\cdot C_f{}^2\left(n-1\right) \\ \quad +\mu^3\cdot 16\cdot C_{h1}{}^2\left(n-1\right)^2\cdot C_f{}^2 \end{cases}$$

$$= 16\cdot C_f{}^2\left(C_f{}^2 R^2 + C_{h1}{}^2\cdot\left(\mu\cdot 4\cdot R^2 + \left(\mu^2\cdot 4\cdot R + \mu^3\cdot\left(n-1\right)\right)\cdot\left(n-1\right)\right)\right)$$

$$\Rightarrow \qquad 0 = C_f{}^2 R^2 + C_{h1}{}^2\cdot\left(\mu\cdot 4\cdot R^2 + \left(\mu^2\cdot 4\cdot R + \mu^3\cdot\left(n-1\right)\right)\cdot\left(n-1\right)\right). \tag{697}$$

From the last line, we cannot only extract the fact that the three solutions to the parameter μ are algebraically appearing in the same strength ($C_{h1}{}^2$), we also might deduce from the structure of the last equation why gravity, which is to say μ- or m-related interaction, is so weak compared to other interactions. This weakness just seems to follow from the fact that we have the factor C_{h1}^2 combined with all μ-terms, while the term $C_f^2 R^2$ usually dominates the scene. This fact would then also explain why the classical quantum equations can so easily be derived from the scaled Ricci scalar with just the setting $R = 0$ (c.f. [2–7] and chapters 10 and 11 of this book).

The condition (695), however, is just the basic assumption of our scaling approach, namely that the scale factor $F[f]$ should have a structure where it guaranties the classical non-quantum outcome for our ordinary experimental surrounding and just adds a wave-function (of quantum character), being usually rather small compared to the absolute values to the classical observables. In other words, the usual experimental finding and every day observation that quantum effects are small also prevent us from producing (and observing) the Filia-object.

Now let us see: We have $F[f] = f + C_f$ and f being proportional to C_{h1}. We realize that the condition (695), which hinders us to see Filia anywhere in the experimental reality we can create, also is the guarantee for quantum effects being small. Thus, it would be no surprise to have the theoretical option for a Filia, which is to say a fourth particle, a fourth mass, or a fourth generation, but its practical realization is almost impossible within our universe, because only under conditions where quantum effects would become equal to classical ones $C_f^2 \approx C_{h1}^2$, we could hope to find it.

But!

Is this already the answer or are there also other possibilities?

Well, at first, we should investigate the possibility that our outcome is just the result of a special choice for the particle at rest function approach Eq. (688). We find that we obtain the same solution for approaches like:

$$F[f] = f + C_f; \quad f = h_{,t} = \pm C_{h1} \cdot m \cdot e^{\pm t \cdot m \cdot i}$$

$$F[f] = f + \frac{C_f}{m}; \quad f = h_{,t} = \pm C_{h1} \cdot e^{\pm t \cdot m \cdot i}. \tag{698}$$

Polynomials with even higher powers of m can easily be obtained when choosing F of the kind ($q > 1$):

$$F[f] = (f + C_f)^q ; \quad f = h_{,t} = \pm C_{h1} \cdot m \cdot e^{\pm t \cdot m \cdot i}$$

$$F[f] = \left(f + \frac{C_f}{m} \right)^q ; \quad f = h_{,t} = \pm C_{h1} \cdot e^{\pm t \cdot m \cdot i}. \tag{699}$$

Then, however, these higher powers for m, respectively μ are all connected with ever higher powers of the constant C_{h1}, which as stated above, is small against C_f and therefore, those higher powers for m will fall even further behind the lower powers than we already have realized for the terms μ^4, respectively the m^8 and as discussed above.

Admittedly, the structures (698) look a bit arbitrary and we can give no explanation why the scale factor should take such a form. We should also point out that, due to lack of time and recourses; we have not been able to check other options for $F[f]$-f-settings so far.

So, we shall see what our post-variation equations will bring about.

7.5.2   In Connection with Our Simplified Inner Equation (669)

We already saw that in section "7.3.2.1 A possible connection to the 3-generation problem," the variation of the Einstein–Hilbert action for a scaled metric with respect to quantum centers (centers of the function f, we already made out as the classical wave-function) can lead to equations of third order. Thereby we tweaked the situation in such a way that the resulting equation was algebraic and did not contain any derivatives anymore (c.f. Eq. (675)). Here now, our starting point should be a bit more general and we use Eq. (670) without any further simplifications, but the assumption of a particle at rest, which is to say $f = f[t]$. The equation becomes:

$$0 = 2 \cdot f_{,t} \cdot (L_M - 2\Lambda) \cdot (f - C_{f0}) \cdot C_{f1} + \frac{(f - C_{f0}) \cdot (f_{,t} \cdot R - 3 \cdot f_{,ttt}) - 3 \cdot f_{,t} \cdot f_{,tt}}{(f - C_{f0})^2}.$$

$$\tag{700}$$

Unfortunately, we are unable to find any suitable solution, but will keep this approach in mind and revisit it later again.

7.5.2.1   A foam of surfaces as by-product

Thus, we go back to Eq. (669) and taking condition (668) in only 2 dimensions:

$$0 = 4FF'' + (F')^2 (2 - 6) = 0 = FF'' - (F')^2 \quad \Rightarrow \quad F = e^{f \cdot C_{f0}} \cdot C_{f1}, \tag{701}$$

leads us to:

$$0 = \left\{ -\frac{1}{F}\left(\frac{f_{,\sigma}}{F}\left(FF'' + \left(\frac{n}{2} - 2\right)(F')^2 \right) \Delta f + F'\Delta\left(f_{,\sigma}\right) \right) \overbrace{f_{,\sigma}\cdot\left(L_M - 2\Lambda\right)}^{=0} \right.$$

$$= f_{,\sigma}\cdot\left(L_M - 2\Lambda\right) - \frac{F'}{F}\Delta\left(f_{,\sigma}\right)$$

$$\Rightarrow \qquad 0 = f_{,\sigma}\cdot\frac{L_M - 2\Lambda}{C_{f0}} - \Delta\left(f_{,\sigma}\right). \tag{702}$$

The subsequent three solutions to the last line in Eq. (702) in the case of a particle at rest would be:

$$f_{1,2,3} = C_{f2}\cdot e^{-t\cdot\sqrt{\frac{L_M - 2\Lambda}{C_{f0}}}} + C_{f3}\cdot e^{+t\cdot\sqrt{\frac{L_M - 2\Lambda}{C_{f0}}}} + C_{f4}. \tag{703}$$

Extraction of f_0 = constant as one solution to Eq. (702), we can substitute $h = f_{,\sigma}$ and get:

$$\Rightarrow \qquad 0 = h\cdot\frac{L_M - 2\Lambda}{C_{f0}} - \Delta h, \tag{704}$$

which gives us an Eigenvalue equation to the function h.

Here we give a small selection of simple (if not to say simplest and obvious) solutions.

We start with functions consisting of the wave-solutions in the case of $0 = \frac{L_M - 2\Lambda}{C_{f0}}$ and 2-dimensional Minkowski coordinates t and x:

$$0 = \Delta h = h_{,xx} - \frac{h_{,tt}}{c^2} \quad\Rightarrow\quad h_{1,2} = \Psi[x \pm c\cdot t], \tag{705}$$

with arbitrary functions Ψ.

For $\frac{L_M - 2\Lambda}{C_{f0}} \neq 0$ and applying the separation approach $h = h_x[x]\cdot h_t[t]$, we obtain:

$$0 = \frac{L_M - 2\Lambda}{C_{f0}}\cdot h - \left(h_{,xx} - \frac{h_{,tt}}{c^2} \right) \quad\Rightarrow\quad \begin{cases} h_t = e^{m\cdot c\cdot t}\cdot C_{t1} + e^{-m\cdot c\cdot t}\cdot C_{t2} \\[2mm] h_x = e^{x\sqrt{\frac{L_M - 2\Lambda}{C_{f0}} - m^2}}\cdot C_{x1} + e^{-x\sqrt{\frac{L_M - 2\Lambda}{C_{f0}} - m^2}}\cdot C_{t2} \end{cases}. \tag{706}$$

An interesting solution is obtained in spherical polar coordinates with the approach $h = h_\varphi[\varphi] \cdot h_\vartheta[\vartheta]$ for the unit sphere ϑ and φ, where we get:

$$0 = \frac{L_M - 2\Lambda}{C_{f0}} \cdot h - \left(\frac{h_{,\varphi\varphi}}{\sin(\vartheta)^2} + \frac{(\sin(\vartheta) \cdot h_{,\vartheta})_{,\vartheta}}{\sin(\vartheta)} \right)$$

$$\Rightarrow \begin{cases} h_\varphi = e^{m \cdot \varphi} \cdot C_{\varphi 1} + e^{-m \cdot \varphi} \cdot C_{\varphi 2} \\ h_\vartheta = P^m_{\frac{1}{2}\left(\sqrt{1-4 \cdot \frac{L_M - 2\Lambda}{C_{f0}}} - 1\right)} [\cos(\vartheta)] \cdot C_{\vartheta 1} + Q^m_{\frac{1}{2}\left(\sqrt{1-4 \cdot \frac{L_M - 2\Lambda}{C_{f0}}} - 1\right)} [\cos(\vartheta)] \cdot C_{\vartheta 2} \end{cases}$$

$$(707)$$

Here the capital letters P and Q stand for the associated Legendre functions of the first and second kind, respectively.

Note: The solution allows some kind of 2-sphere ensemble. As the same can also be obtained for other closed or open surfaces, we end up with the option of a structure work of surfaces or a foam, which was also proposed by other authors with different starting points than ours (e.g., [16−18]). The foam, however, automatically forms a 3-dimensional structure and as, due to the variation of the ensemble parameters according to Eq. (663), this structure has the ability to reorder (in whatever manner) we also automatically have time. Thereby time would just be measured or felt via the period of shifts (the circuit time) between two sets of ensemble positionings. Such a positioning or repositioning should have a maximum velocity which would be c, the speed of light in vacuum.

7.5.2.2 Back to arbitrary *n*

It is clear from the structure of Eq. (663) that there are more types of solutions with three options, possibly suitable for discussion in connection with the 3-generation problem, but we shall leave it to interested readers to try more things out.

7.5.3 In Connection with the Infinity Options Principle

Here, now we just want to see the outcome of particle at rest solutions to Eq. (663) in connection with the infinity options principle, leading us to Eq. (680). In the case of the particle at rest this would read:

$$\delta W = 0 = \sum_{\forall f_k} \int_V \lim_{n \to \infty} d^n x \left(\sqrt{-g} \cdot F^{\frac{n}{2}-1} \cdot \left(f_{,t} \frac{f_{,t} f_{,t} g^{00}}{F^3} (F')^2 \right) \right) Y^t_{,k} \delta \zeta^k. \quad (708)$$

Approaching this equation with a $f[t] \sim \exp(m^* t)$, one obtains a characteristic equation of third order in m. With the function $F[f]$, the metric, its determinant, and the $Y^t{}_{,k}\delta\zeta^k$ still being arbitrary, we could always construct an algebraic form for this characteristic equation in such a way that the parameter m would give three solutions. These three solutions are the masses of the three generations.

Why should an elementary particle have an underlying equation, based on the infinity options principle ... or in other words, what on the earth makes us even dare to assume something so obviously ridiculous like infinitely many dimensions in connection with elementary particles?

Well, perhaps elementary particles are just universes of their very own.

7.5.4 Back Again to the Simplified Inner Equation and *n* = 4

In section "7.3.2.1 A possible connection to the 3-generation problem," we had found a very simple algebraic equation (Eq. (675)) in the case of $n = 4$. Once again assuming a particle at rest with $L_M - 2\Lambda = 0$ and setting:

$$f = h_{,t} = \pm C_{h1} \cdot m \cdot i \cdot e^{\pm t \cdot m \cdot i}, \tag{709}$$

leading us to $B = -m^2$, we obtain the integrand:

$$0 = 2 \cdot \left(L_M - 2\Lambda\right) \cdot \left(f - C_{f0}\right)^3 \cdot C_{f1} + \left(f - C_{f0}\right) \cdot (R - 3 \cdot B) - 3 \cdot B \cdot f. \tag{710}$$

Now, we perform the same evaluation as we had introduced in section "7.5.1 Pre-Variation Results: The Simplest Path" in order to obtain observable parameters, which gives us:

$$\left\{ \begin{array}{l} \left[2 \cdot i \cdot C_{h1} \cdot e^{i \cdot m \cdot t} \cdot m^2 \cdot \left(C_{h1} \cdot e^{i \cdot m \cdot t} \cdot m \cdot \left(6 \cdot m^2 + R\right) - C_{f0}\left(3 \cdot m^2 + R\right)\right)\right] \\ \times \left[2 \cdot i \cdot C_{h1} \cdot e^{i \cdot m \cdot t} \cdot m^2 \cdot \left(C_{h1} \cdot e^{i \cdot m \cdot t} \cdot m \cdot \left(6 \cdot m^2 + R\right) - C_{f0}\left(3 \cdot m^2 + R\right)\right)\right]^* \end{array} \right\} = \Xi\Xi^*. \tag{711}$$

As usual in Quantum Theory when evaluating observables, we integrate the absolute value as follows (here we refer to our corresponding discussion in section "7.5.1 Pre-Variation Results: The Simplest Path"[5]):

---

[5]Note: The derivation with respect of the changes to the Hilbert evaluation of the Einstein field equations out of a $\sqrt{-g \cdot R \cdot R}^*$ -kernel instead of the usual $\sqrt{-g} \cdot R$ is been given in the appendix of this chapter.

$$\int_{0}^{\frac{2\cdot\pi}{m}} \sqrt{\Xi\cdot\Xi^{*}}\,dt \tag{712}$$

and demanding this to give zero, we obtain:

$$0 = \begin{cases} 36\cdot C_{h1}{}^{2}\cdot m^{6} \\ +\left(9\cdot C_{f0}{}^{2}+12C_{h1}{}^{2}R\right)\cdot m^{4} \\ +\left(6\cdot C_{f0}{}^{2}R+C_{h1}{}^{2}R^{2}\right)\cdot m^{2}+C_{f0}{}^{2}R^{2} \end{cases}. \tag{713}$$

Now, we substitute $\mu = m^2$ and using the general solution to a three-order polynomial via the following product form:

$$\begin{aligned}\left(\mu-\mu_1\right)\cdot\left(\mu-\mu_2\right)\cdot\left(\mu-\mu_3\right)= \\ \mu^3 - \mu^2\cdot\left(\mu_1+\mu_2+\mu_3\right)+\mu\cdot\left(\mu_1\mu_2+\mu_1\mu_3+\mu_2\mu_3\right)-\mu_1\mu_2\mu_3\end{aligned} \tag{714}$$

and comparing the latter with Eq. (713), divided by $\left(36\cdot C_{h1}{}^{2}\right)$, gives us:

$$\mu_1\mu_2\mu_3 = -\frac{C_{f0}{}^{2}R^{2}}{36\cdot C_{h1}{}^{2}}$$

$$\mu_1\mu_2+\mu_1\mu_3+\mu_2\mu_3 = \frac{6\cdot C_{f0}{}^{2}R+C_{h1}{}^{2}R^{2}}{36\cdot C_{h1}{}^{2}}$$

$$\mu_1+\mu_2+\mu_3 = -\frac{3\cdot C_{f0}{}^{2}+4\cdot C_{h1}{}^{2}R}{12\cdot C_{h1}{}^{2}}$$

$$\Rightarrow \qquad m_{1,4}=\pm\sqrt{\mu_1}\,;\quad m_{2,5}=\pm\sqrt{\mu_2}\,;\quad m_{3,6}=\pm\sqrt{\mu_3}\,. \tag{715}$$

Thus, we have obtained three solutions to the parameter μ, resulting in three matter and antimatter values for m. Taking this m as the Dirac mass to an elementary particle family, we clearly have obtained three generations with the masses $m_{1,4}=\pm\sqrt{\mu_1}\,;\ m_{2,5}=\pm\sqrt{\mu_2}\,;\ m_{3,6}=\pm\sqrt{\mu_3}$.

We realize that while we do not need Einstein's and/or Hilbert's postulated matter or constant add-ons, neither in form of a cosmological constant Λ nor a Lagrange matter density L_M, in order to end up with the three different masses, we definitely require a non-zero Ricci scalar curvature R. With the character of this curvature being unclear, we still cannot consider the 3-generation problem as being solved. Especially as this curvature should not simply be a constant as it has been treated

here, but is expected to fade with increasing distance to the source (the mass of the elementary particle) and thus, $R = R[r]$, we still see a lot of work ahead of us. This follows that the "particle at rest," meaning $f = f[t]$ can only be an approximation as it should at least be something like $f = f[t,r]$. For the stable elementary particles, we could clearly expect $R = R[r]$, but this should become $R = R[t,r]$ in the case of the unstable ones. Taking the charged Leptons, for instance, we expect an $R[r, ...]$ (no t-dependency) for the electron and the positron while we should have $R[t,r, ...]$ for the muon, the tauon and their antimatter partners.

7.5.4.1 Generalization of the approach

In order to overcome the problems mentioned above, we remember that one essential move in our approach with the simplified inner equation (Eq. (675)) was the demand of Eigenvalue solutions to the Laplace term in Eq. (664) as $\Delta f = B \cdot f$. Keeping the function f general (not purely t-dependent as in our particle at rest approach) and using the Minkowski space-time in spatial spherical coordinates t, r, ϑ, φ as follows:

$$
g_{\alpha\beta}^{\text{Spatial\_Sphere}} = g_{\alpha\beta}^{SS} = \begin{pmatrix} -c^2 & 0 & 0 & 0 \\ 0 & 1 & 0 & 0 \\ 0 & 0 & r^2 & 0 \\ 0 & 0 & 0 & r^2 \cdot \sin^2\vartheta \end{pmatrix}, \tag{716}
$$

we now intend to give our approach from the sub-section above a bit more substance.

Applying the usual separation approach for the function $f[...] = f_t[t] \cdot f_r[r] \times f_\vartheta[\vartheta] \cdot f_\varphi[\varphi]$, we obtain the following solutions:

$$
f_t[t] = C_{t1} \cdot \cos[c \cdot E \cdot t] + C_{t2} \cdot \sin[c \cdot E \cdot t], \tag{717}
$$

$$
f_\varphi[\varphi] = C_{\varphi1} \cdot \cos[\lambda \cdot \varphi] + C_{\varphi2} \cdot \sin[\lambda \cdot \varphi], \tag{718}
$$

$$
f_\vartheta[\vartheta] = C_{P\vartheta} \cdot P_\ell^\lambda[\cos[\vartheta]] + C_{Q\vartheta} \cdot Q_\ell^\lambda[\cos[\vartheta]], \tag{719}
$$

with the associated Legendre functions P_ℓ^λ and Q_ℓ^λ. In the appendix of our previous chapter 6 (section 6.16, also see [3]), we have also discussed the option for half spin solutions.

The intermediate result above makes us end up with the following differential equation for $f_r[r]$:

$$
\Rightarrow \qquad 0 = \left(\left(E^2 - B \right) r^2 - \ell(1+\ell) \right) f_r[r] + r \left(2 f_r[r]' + r \cdot f_r[r]'' \right). \tag{720}
$$

The solution for Eq. (278) consists of the spherical Bessel functions j_n and y_n:

$$f_r[r] = C_{r1} \cdot j_\ell\left[\sqrt{E^2 - B} \cdot r\right] + C_{r2} \cdot y_\ell[\sqrt{E^2 - B} \cdot r]. \qquad (721)$$

Now, similar to the situation above, where we dealt with the pure $f[t]$-case, we assume to have $B = -m^2 {*} b^2$ and $E^2 = -m^2 {*} a^2$. We also take it that—just as before—the first derivative of $f_t[t]$ with respect to t has to be chosen instead of the function itself, leaving us with:

$$f[t, r, \vartheta, \varphi] = \partial_t f_t[t] \cdot f_r[r] \cdot f_\vartheta[\vartheta] \cdot f_\varphi[\varphi]. \qquad (722)$$

This results in the total solution as given in Eqs. (718) and (277) plus[6]:

$$\partial_t f_t[t] = m \cdot C_{h1} \cdot e^{\pm i \cdot m \cdot a \cdot t}, \qquad (723)$$

$$f_r[r] = C_{r1} \cdot j_\ell\left[m \cdot \sqrt{b^2 - a^2} \cdot r\right] + C_{r2} \cdot y_\ell[m \cdot \sqrt{b^2 - a^2} \cdot r], \qquad (724)$$

where only for the reason of simplicity and recognition with the classical Dirac particle at rest approach [19], we have set $C_{t2} = \pm i \cdot C_{t1}$. Instead of the simple time-integral (712), we have now to integrate over the spatial coordinates, due to the fact that our function is fading in r, which gives a finite result and—along the way—also allows us to normalize the various product terms of our solution, thereby fixing the corresponding constants. As before we obtain six solutions for m, which divide into pairs of matter and antimatter values and can be evaluated from the following equations:

$$\mu_1 \mu_2 \mu_3 = -\frac{C_{f0}^2 R^2}{36 \cdot a^4 b^4 C_{h1}^2}$$

$$\mu_1 \mu_2 + \mu_1 \mu_3 + \mu_2 \mu_3 = \frac{6 a^2 b^2 C_{f0}^2 R + C_{h1}^2 R^2}{36 \cdot a^4 b^4 C_{h1}^2}$$

$$\mu_1 + \mu_2 + \mu_3 = -\frac{9 a^4 b^4 C_{f0}^2 + 12 a^2 b^2 C_{h1}^2 R}{36 \cdot a^4 b^4 C_{h1}^2}$$

$$\Rightarrow \qquad m_{1,4} = \pm\sqrt{\mu_1}; \quad m_{2,5} = \pm\sqrt{\mu_2}; \quad m_{3,6} = \pm\sqrt{\mu_3}. \qquad (725)$$

7.5.4.2 Brief discussion of the fading Ricci curvature

Regarding our previous discussion about the Ricci scalar not being a constant, we go back to the original equation (Eq. (675)), where we have assumed the

---

[6]This time we omitted the explicit appearance of the first derivative with respect to time but only multiplied with m, but it should be noted that the results remain with the derivative-approach (709).

Hilbert matter term and the cosmological constant to be zero:

$$0 = \left(f - C_{f0}\right) \cdot R + 3 \cdot B \cdot \left(C_{f0} - 2 \cdot f\right). \tag{726}$$

Simply assuming that R could be split into a functional part, depending on the coordinates $R[\dots]$ and a constant R_c, we may rewrite Eq. (726) as follows:

$$0 = R_c \cdot \overbrace{R[t,r,\vartheta,\varphi]}^{\equiv R_f} + 3 \cdot \frac{B \cdot \left(C_{f0} - 2 \cdot f\right)}{\left(f - C_{f0}\right)}. \tag{727}$$

Now, just as before, we have to set:

$$0 = R_c \cdot \overbrace{R[t,r,\vartheta,\varphi]}^{\equiv R_f} + 3 \cdot \frac{B \cdot \left(C_{f0} - 2 \cdot f\right)}{\left(f - C_{f0}\right)} = \Xi \tag{728}$$

and evaluate the volume integral:

$$\int\limits_{0}^{\frac{2 \cdot \pi}{m \cdot a}} \int\limits_{V_3} \Xi \cdot \Xi^* dx^3 dt. \tag{729}$$

Assuming both functional terms f and R_f to fade, the finite outcome is guaranteed even for infinite boundaries of the integral and we will result in equations of the kind (725) once again.

It should be noted that the same result is been obtained in section "7.5.4 Back Again to the Simplified Inner Equation and $n = 4$" with the approach:

$$f = \pm C_{h1} \cdot m \cdot e^{\pm t \cdot m \cdot i}, \tag{730}$$

while there appears asymmetric odd powers of m in the case of approaches of the kind:

$$F = \left(f - \frac{C_{f0}}{m}\right)^{\frac{4}{n-2}} C_{f1}. \tag{731}$$

Could this lead to an explanation for the observed matter antimatter asymmetry in our universe?

7.6   Discussion with Respect to F^q

In connection with a generalized action with an additional scale factor $F[f]^q$ attached to the Ricci scalar (649) or (665), we need to discuss the origin of such a factor. At first, we note that the classical variation result has no such factor, respectively this factor would be $F[f] = 1$ or consequently $F[f]^q = 1$. Thus, we have no possibility of extracting any hint for q from the classical theory. However, instead of choosing an arbitrary q, we may also consider the possibility of $q = 0$, but an arbitrary n instead. Thereby, only a certain number of dimensions (perhaps 4) may be present, while others are compactified or otherwise hidden. Then, for instance, in the case $n = 6$, a wave function f could depend only on 4 dimensions, but the Einstein–Hilbert action integral would still provide a scale factor $F[f]^3$ due to the spare root of the determinant of the metric. We may also point out the opportunity of f depending on all 6 dimensions, but—as shown in [3, 4] or the previous chapter of this book— two of these are producing certain kinds of inertia and do not show up as degrees of freedom like t, x, y, and z (if choosing Cartesian coordinates for the 4-dimensional space-time we observe). Then again, the determinant of the metric tensor would provide the scale factor $F[f]^3$ as needed during the evaluation of the mass-generations in the sections above.

7.7 Conclusions about Our Attempts to Find an Explanation to the 3-Generation Problem

Yes, we think that, perhaps, we have found a way to explain the 3-generation problem of elementary particles. In fact, there are at least two possible ways.

The first and simpler path is to variate the metric tensor with a scale factor $F[f]$, where f reveals itself as the quantum wave function. Subsequent evaluation of the Ricci scalar of this scaled metric and inserting the result into a classical observable integral, thereby demanding this integral to give zero, results in pairs of matter and antimatter masses. Assuming that quantum effects caused by the wave function f are small in comparison to the otherwise static (stable) metric, one automatically obtains only three pairs of such masses, while other masses of higher generations can only appear where the quantum effects become at least equal to the classical physics ... meaning where the quantum fields become equal in strength to the classical interactions and subsequent forces.

The second path requires the variation of the scaled metric Ricci scalar with respect to quantum centers. Evaluation of the intrinsic solution of the resulting equation in 4 dimensions gives, after some simplifications, exactly three pairs of masses for each matter and antimatter particles. As an add-on, this approach also produces asymmetries between matter and antimatter for certain settings of the function F.

Still, we are of the opinion that much more discussion is needed in order to come to truly satisfying conclusions.

7.8 Appendix: Taking Care of Complex Curvatures R or R^*

It was discussed in section "7.5 Toward a Solution to the 3-Generation Problem," sections "7.5.1 Pre-Variation Results: The Simplest Path" and "7.5.4 Back Again to the Simplified Inner Equation and $n = 4$" of this book that as we can have complex results for the scaled Ricci curvature R^*, we require a slightly different Einstein–Hilbert action, which should follow the typical quantum way of extracting observables.

In this sense the action (240) should be adapted as follows:

$$\delta W = 0 = \delta \int_V d^n x \left(\sqrt{-G} \cdot \sqrt{\left(R^* - 2\Lambda + L_M\right)\left(\bar{R}^* - 2\bar{\Lambda} + \bar{L}_M\right)} \right). \qquad (732)$$

Thereby the bar denotes the conjugate complex of the symbol underneath. In the classical case, the corresponding adapted action should read:

$$\delta W = 0 = \delta \int_V d^n x \left(\sqrt{-g} \cdot \sqrt{\left(R - 2\Lambda + L_M\right)\left(\bar{R} - 2\bar{\Lambda} + \bar{L}_M\right)} \right). \qquad (733)$$

Applying the results from [22], we can obtain the variation results of Eq. (733) as follows:

$$0 = \left\{ \begin{array}{l} \dfrac{\left(R^{\kappa\lambda} - Rg^{\kappa\lambda} + \left(\left(\underbrace{\dfrac{8\pi G}{\kappa} T^{\kappa\lambda}}_{} = \text{matter} \atop 0 = \text{vacuum} \right) + \Lambda \cdot g^{\kappa\lambda} \right) \right)}{2\sqrt{\left(R - 2\Lambda + L_M\right)\left(\bar{R} - 2\bar{\Lambda} + \bar{L}_M\right)}} \cdot \left(\bar{R} - 2\bar{\Lambda} + \bar{L}_M \right) \\[3em] + \dfrac{\left(\bar{R}^{\kappa\lambda} - \bar{R}g^{\kappa\lambda} + \left(\left(\underbrace{\dfrac{8\pi G}{\kappa} \bar{T}^{\kappa\lambda}}_{} = \text{matter} \atop 0 = \text{vacuum} \right) + \bar{\Lambda} \cdot g^{\kappa\lambda} \right) \right)}{2\sqrt{\left(R - 2\Lambda + L_M\right)\left(\bar{R} - 2\bar{\Lambda} + \bar{L}_M\right)}} \cdot \left(R - 2\Lambda + L_M \right) \\[3em] -\left(\partial^{;\kappa}\partial^{;\lambda} - g^{\kappa\lambda}\Delta_g \right) \dfrac{\left(R + \bar{R}\right)}{2\sqrt{\left(R - 2\Lambda + L_M\right)\left(\bar{R} - 2\bar{\Lambda} + \bar{L}_M\right)}}. \end{array} \right.$$

$$(734)$$

We see that in the case of a purely real R we obtain the usual classical Einstein field equations [10]. The corresponding equations in the scaled metric case are much more complicated and—as we did not need them in this paper—will not be presented here.

References

1. https://en.wikipedia.org/wiki/Generation_(particle_physics)

2. N. Schwarzer, *Societies and Ecotons—The Photons of the Human Society—Control them and Rule the World*, Part 1 of "Medical Socio-Economic Quantum Gravity," Self-published, Amazon Digital Services, 2020, Kindle.

3. N. Schwarzer, *Masses and the Infinity Options Principle—Can We Explain the 3-Generation and the Quantized Mass Problem?*, Part 5 of "Medical Socio-Economic Quantum Gravity," Self-published, Amazon Digital Services, 2020, Kindle.

4. N. Schwarzer, *The World Formula: A Late Recognition of David Hilbert's Stroke of Genius*, 2022, Jenny Stanford Publishing. ISBN: 9789814877206.

5. N. Schwarzer, *Humanitons —The Intrinsic Photons of the Human Body—Understand them and Cure Yourself*, Part 2 of "Medical Socio-Economic Quantum Gravity," Self-published, Amazon Digital Services, 2020, Kindle.

6. N. Schwarzer, *Mastering Human Crises with Quantum-Gravity-based but still Practicable Models—First Measure: SEEING and UNDERSTANDING the WHOLE*, Part 3 of "Medical Socio-Economic Quantum Gravity," Self-published, Amazon Digital Services, 2020, Kindle.

7. N. Schwarzer, *Self-Similar Quantum Gravity—How Infinity Brings Simplicity*, Part 4 of "Medical Socio-Economic Quantum Gravity," Self-published, Amazon Digital Services, 2020, Kindle.

8. N. Schwarzer, *Toward Quantum Einstein Field Equations*, Part 7 of "Medical Socio-Economic Quantum Gravity," Self-published, Amazon Digital Services, 2020, Kindle.

9. D. Hilbert, Die Grundlagen der Physik, Teil 1, *Göttinger Nachrichten*, 1915, 395–407.

10. A. Einstein, Grundlage der allgemeinen Relativitätstheorie, *Annalen der Physik (ser. 4)*, 1916, **49**, 769–822.

11. N. Schwarzer, *Quantum Gravity Thermodynamics II—Derivation of the Second Law of Thermodynamics and the Metric Driving Force of Evolution*, Self-published, Amazon Digital Services, 2019, Kindle.

12. N. Schwarzer, *Einstein had it, but he did not see it—Part LXXXV: In Conclusion*, Self-published, Amazon Digital Services, 2019, Kindle.

13. N. Schwarzer, *Science Riddles—Riddle No. 20: Second Law of Thermodynamics—Where is its Fundamental Origin?*, Self-published, Amazon Digital Services, 2019, Kindle.

14. N. Schwarzer, *Quantum Gravity Thermodynamics—And it May Get Hotter*, Self-published, Amazon Digital Services, 2019, Kindle.

15. N. Schwarzer, *Einstein had it, but he did not see it—Part LXXXI: More Dirac Killing*, Self-published, Amazon Digital Services, 2019, Kindle.

16. A. A. Kirillova and D. Turaev, Foam-like structure of the Universe, *Physics Letters B.*, 2007, **656** (1–3), 1–8. doi.org/10.1016/j.physletb.2007.09.025

17. A. Linde, Inflation in Supergravity and String Theory: Brief History of the Multiverse, 2012. ctc.cam.ac.uk. Archived from the original on 14 July 2014. http://www.ctc.cam.ac.uk/stephen70/talks/swh70_linde.pdf

18. A. Linde, Inflation and String Cosmology, 2005. arxiv.org/pdf/hep-th/0503195v1.pdf

19. P. A. M. Dirac, The Quantum Theory of the Electron, 1928. DOI: 10.1098/rspa.1928.0023.

20. N. Schwarzer, *The Theory of Everything—Quantum and Relativity is Everywhere—A Fermat Universe*, 2020, Pan Stanford Publishing, ISBN: 10: 9814774472.

21. N. Schwarzer, *General Quantum Relativity*, Self-published, Amazon Digital Services, 2016, Kindle.

22. C. A. Sporea, Notes on f(R) Theories of Gravity, 1914. arxiv.org/pdf/1403.3852.pdf.

THE FOURTH DAY:
LET THERE BE QUANTUM GRAVITY

Chapter 8

Toward Quantum Einstein Field Equations

A Full Derivation of Some Fundamental Equations to *Quantum Gravity* and a Few of Its Applications in Space-Times of Arbitrary Numbers of Dimensions

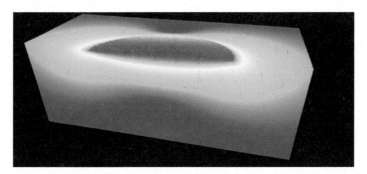

Figure 8.1 A spin field in space-time according to the elastic solution of the quantum Einstein field equations, evaluated with the software FilmDoctor [19] using Eq. (830) with solution (831).

8.1 Abstract: About the Source

It will be shown how quantum Einstein field equations can directly be derived from the classical Einstein–Hilbert action [A1] by a simple generalization of

The Math of Body, Soul, and the Universe
Norbert Schwarzer
Copyright © 2023 Jenny Stanford Publishing Pte. Ltd.
ISBN 978-981-4968-24-9 (Hardcover), 978-1-003-33454-5 (eBook)
www.jennystanford.com

the kernel of the latter. The kernel Hilbert had chosen in his famous work [A1], leading to the Einstein field equations [A2] in an elegant and purely mathematical manner, was the Ricci scalar R and this author is aware of similar trials by other scientists who had introduced arbitrary functions of that classical kernel like $f(R)$. Our approach is different, as it leaves the classical kernel, namely, the Ricci scalar, unharmed, but allows a scale factor to the metric tensor. This changes the Ricci scalar in such a way that immediately—as already shown by this author in the previous chapters and a variety of papers (e.g., [A3–A8])—the most important classical quantum equations appear. Here in this chapter, we will now evaluate the result of the Einstein–Hilbert action for such a generalized (scaled) Ricci kernel and thus, derive new Einstein field equations, clearly showing the characteristics of their classical analog plus quantum properties in connection with the scale factor.

We, therefore, conclude that we have found quantum Einstein field equations.

It needs to be pointed out that the classical Hilbert approach already contains our scaling concept, but it appears that this degree of freedom was neither seen nor used by other researchers.

As our calculation is performed in an arbitrary number of dimensions, we result in very general forms of these new equations.

It seems most intriguing that a surprising and rather dramatic simplification appears in a variety of dimensions and certain approaches or boundary conditions. In this case, we suddenly find the quantum Einstein field equations to be almost equivalent to the equations of linear elasticity ... allowing for many interesting solutions, like:

(a) Shear stress fields as spin fields

(b) Spherical stress or strain fields as particles

(c) Sub-space-time solutions

(d) Time-planes and time-layer solutions, which could explain many features of our universe—including the thing (universe) itself—as impacts, imprints, or indents

(e) Quark-like 1/3 and 2/3 charges

(f) Many other things

For illustration, we are going to present some of these applications and try—here and there—for a brief discussion.

We also find, even though it will not be discussed in this paper, that the scaled metric approach, which is—so at least it seems—so nicely giving us the connection between Einstein's General Theory of Relativity and Quantum Theory, apparently also delivers the necessary ingredients to *R*. Penrose's "before the big bang hypothesis" [A9], which requires a scaled metric.

8.1.1 References for the Abstract

A1. D. Hilbert, Die Grundlagen der Physik, Teil 1, *Göttinger Nachrichten*, 1915, 395–407.

A2. A. Einstein, Grundlage der allgemeinen Relativitätstheorie, *Annalen der Physik (ser. 4)*, 1916, **49**, 769–822.

A3. N. Schwarzer, *Brief Proof of Hilbert's World Formula: Dirac, Klein–Gordon, Schrödinger, Einstein, Evolution and the Second Law of Thermodynamics all from One Origin*, Self-published, Amazon Digital Services, 2020, Kindle.

A4. N. Schwarzer, *Societies and Ecotons—The Photons of the Human Society—Control them and Rule the World*, Part 1 of "Medical Socio-Economic Quantum Gravity," Self-published, Amazon Digital Services, 2020, Kindle.

A5. N. Schwarzer, *Humanitons—The Intrinsic Photons of the Human Body—Understand them and Cure Yourself*, Part 2 of "Medical Socio-Economic Quantum Gravity," Self-published, Amazon Digital Services, 2020, Kindle.

A6. N. Schwarzer, *Mastering Human Crises with Quantum-Gravity-based but still Practicable Models—First Measure: SEEING and UNDERSTANDING the WHOLE*, Part 3 of "Medical Socio-Economic Quantum Gravity," Self-published, Amazon Digital Services, 2020, Kindle.

A7. N. Schwarzer, *Self-Similar Quantum Gravity: How Infinity Brings Simplicity?*, Part 4 of "Medical Socio-Economic Quantum Gravity," Self-published, Amazon Digital Services, 2020, Kindle.

A8. N. Schwarzer, *Masses and the Infinity Options Principle: Can We Explain the 3-Generations and the Quantized Mass Problem?*, Part 5 of "Medical Socio-Economic Quantum Gravity," Self-published, Amazon Digital Services, 2020, Kindle.

A9. https://www.youtube.com/watch?v=ypjZF6Pdrws

8.2 Extended (Quantum Gravity) Einstein Field Equations

In the previous chapters [1–6][1] (and quite some other papers by this author), it was demonstrated how we can find a direct extraction of the Klein–Gordon, the Schrödinger, and the Dirac equation from the Ricci scalar R^* of a modified metric of the kind $G_{\alpha\beta} = F[f]^* g_{\alpha\beta}$. With a yet arbitrary scalar function $F[f]$, the corresponding modified Ricci scalar R^* reads[2]:

$$R^* = R^*_{\alpha\beta} G^{\alpha\beta} = \left(\begin{array}{c} \Gamma^{\sigma}_{\alpha\beta,\sigma} g^{\alpha\beta} - \Gamma^{\sigma}_{\beta\sigma,\alpha} g^{\alpha\beta} - \Gamma^{\mu}_{\sigma\alpha} \Gamma^{\sigma}_{\beta\mu} g^{\alpha\beta} + \Gamma^{\sigma}_{\alpha\beta} \Gamma^{\mu}_{\sigma\mu} g^{\alpha\beta} \\ + \dfrac{F'}{F}(1-n)\Delta f + \dfrac{f_{,\alpha} f_{,\beta} g^{\alpha\beta}(1-n)}{4F^2} \left(4FF'' + (F')^2(n-6) \right) \end{array} \right) \dfrac{1}{F}$$

$$= \left(R + \frac{F'}{F}(1-n)\Delta f + \frac{f_{,\alpha} f_{,\beta} g^{\alpha\beta}(1-n)}{4F^2} \left(4FF'' + (F')^2(n-6) \right) \right) \frac{1}{F}$$

$$\text{with}: \quad F = F[f]; \quad F' = \frac{\partial F[f]}{\partial f}; \quad F'' = \frac{\partial^2 F[f]}{\partial f^2}. \tag{735}$$

As the non-scaled Ricci scalar R also provides the kernel to the classical Einstein–Hilbert action [7], we here intend to investigate the effect of the scaling with respect to the metric variation of the latter.

Note: Demanding certain conditions for the function $F[f]$ and/or f then gives us Dirac or Klein–Gordon equations [1–3]. Thus, when demanding f to be a Laplace function, we obtain from Eq. (735):

$$R^* = \frac{R}{F[f]} + \frac{(1-n)}{F[f]^3} \cdot \left(\left(\frac{(n-6)}{4} \cdot \left(\frac{\partial F[f]}{\partial f} \right)^2 + F[f] \cdot \frac{\partial^2 F[f]}{\partial f^2} \right) \cdot \overbrace{\left(\tilde{\nabla}_g f \right)^2}^{= f_{,\alpha} g^{\alpha\beta} f_{,\beta}} \right)$$

$$\xrightarrow{n=4; R=0} = \frac{1}{F[f]^3} \cdot \left(\left(3 \cdot F[f] \cdot \frac{\partial^2 F[f]}{\partial f^2} - \frac{3}{2} \cdot \left(\frac{\partial F[f]}{\partial f} \right)^2 \right) \cdot \left(\tilde{\nabla}_g f \right)^2 \right), \tag{736}$$

which—so it was shown in [1–3] (see also chapters 9 and 10)—gives the metric equivalent to the Dirac equation. It was also shown in [1, 2] how this gives the classical Dirac equation in flat space Minkowski metrics. A nice sum-up of the derivation of classical quantum equations from Eq. (735), which also includes the case $R \neq 0$, was presented in chapter 7 and [8].

---

[1]Please note than in most papers we thereby considered metrics g_{ij} with $R = 0$. Here now we need to introduce the general form.

[2]The complete evaluation will be presented in the appendix A.

Our starting point shall now be the classical Einstein–Hilbert action [7] with the Ricci scalar R^* as kernel or Lagrange density:

$$\delta W = 0 = \delta \int_V d^n x \left(\sqrt{-G} \cdot \left(R^* - 2\Lambda + L_M \right) \right) \tag{737}$$

and, in contrast to the classical form, with a somewhat adapted metric tensor $G_{\delta\gamma} = F[f] \cdot g_{\delta\gamma}$. Thereby G shall denote the determinant of the metric tensor $G_{\delta\gamma}$, while g will later stand for the corresponding determinant of the metric tensor $g_{\alpha\beta}$. In order to distinguish our new Ricci scalar R^*, being based on $G_{\delta\gamma} = F[f] \cdot g_{\delta\gamma}$ from the usual one, R, being based on the metric tensor $g_{\alpha\beta}$, we marked it with the *-superscript. We also have the matter density L_M and the cosmological constant Λ.

Please note that with $G_{\delta\gamma} = F[f] \cdot g_{\delta\gamma}$, we have used the simplest form of metric adaptation, which we could construct as a simplification from a generalization of the typical form of tensor transformations, namely:

$$G_{\alpha\beta} = F \left[f \left[t, x, y, z, \ldots, \xi_k, \ldots, \xi_n \right] \right]_{\alpha\beta}^{ij} g_{ij}, \tag{738}$$

$$\rightarrow G_{\alpha\beta} = F \left[f \left[t, x, y, z, \ldots, \xi_k, \ldots, \xi_n \right] \right] \cdot \delta_\alpha^i \delta_\beta^j g_{ij}$$

$$\rightarrow G_{\alpha\beta} = F \left[f \left[t, x, y, z, \ldots, \xi_k, \ldots, \xi_n \right] \right] \cdot g_{\alpha\beta} = F[f] \cdot g_{\alpha\beta}. \tag{739}$$

The generalization is been elaborated in [3] and in chapter 11.

It should be pointed out that we chose a simple form (737) only because it is the closest to Hilbert's classical action integral. As in his case F was 1, any generalization of the following shape:

$$\delta W = 0 = \delta \int_V d^n x \left(\sqrt{-G} \cdot F^q \cdot \left(R^* - 2\Lambda + L_M \right) \right) \tag{740}$$

could also be possible and still converges to the classical form for $F \rightarrow 1$. Here, which is to say in this paper, we will mainly consider examples with $q = 0$, but for completeness and later investigation we here also evaluate the variational integrals for the cases of general q (see appendices D and E).

Setting Eq. (735) into Eq. (737) results in the variation:

$$\delta W = 0 = \delta \int_V d^n x \left(\sqrt{-g \cdot F^n} \times \left(\left(\begin{array}{c} R + \dfrac{F'}{F}(1-n)\Delta f \\[2mm] + \dfrac{f_{,\alpha} f_{,\beta} g^{\alpha\beta}(1-n)}{4F^2} \left(4FF'' + \left(F' \right)^2 (n-6) \right) \\[2mm] -2\Lambda + L_M \end{array} \right)^{\frac{1}{F}} \right) \right) \tag{741}$$

Please note that the cosmological constant term requires the variation with respect to the metric $G_{\alpha\beta}$. When insisting on the variation with respect to the metric instead $g_{\alpha\beta}$, we better rewrite Eq. (741) as follows:

$$\delta_{g_{\alpha\beta}} W = 0 = \left\{ \begin{array}{l} \delta_{g_{\alpha\beta}} \int_V d^n x \dfrac{\sqrt{-g \cdot F^n}}{F} \times \left(\begin{array}{l} R + \dfrac{F'}{F}(1-n)\Delta f \\ + \dfrac{f_{,\alpha} f_{,\beta} g^{\alpha\beta}(1-n)}{4F^2}\left(4FF'' + (F')^2(n-6)\right) \end{array} \right) \\ + \delta_{g_{\alpha\beta}} \int_V d^n x \sqrt{-g}\left(L_M - 2\Lambda\right) \end{array} \right.$$

(742)

in order to make it clear that we do not intend to also scale-adapt the cosmological constant or the Hilbert matter term.

However, when ignoring the cosmological constant and—perhaps—assuming that we would not need any postulated matter terms L_M, simply because our scale-adaptation $F[f]$ and similar "tricks" or "add-ons" automatically provide matter, we just obtain:

$$\delta W = 0 = \delta \int_V d^n x \left(\sqrt{-g \cdot F^n} \times \left(\left(\begin{array}{l} R + \dfrac{F'}{F}(1-n)\Delta f \\ + \dfrac{f_{,\alpha} f_{,\beta} g^{\alpha\beta}(1-n)}{4F^2}\left(4FF'' + (F')^2(n-6)\right) \end{array} \right) \dfrac{1}{F} \right) \right)$$

$$= \delta \int_V d^n x \left(\sqrt{-g \cdot F^n} \times \left(\left(\begin{array}{l} R + \dfrac{F'}{\sqrt{-g} \cdot F}(1-n)\partial_\beta \sqrt{-g} \cdot g^{\alpha\beta} f_{,\alpha} \\ + \dfrac{f_{,\alpha} f_{,\beta} g^{\alpha\beta}(1-n)}{4F^2}\left(4FF'' + (F')^2(n-6)\right) \end{array} \right) \dfrac{1}{F} \right) \right).$$

(743)

Performing the usual Hilbert variation with respect to the metric tensor $g_{\alpha\beta}$ now leads us to:

$$\delta_{g_{\alpha\beta}} W = 0 = \delta_{g_{\alpha\beta}} \int_V d^n x \left(\sqrt{-g \cdot F^n} \times \left(\left(\begin{array}{l} R + \dfrac{F'}{\sqrt{-g} \cdot F}(1-n)\partial_\beta \sqrt{-g} \cdot g^{\alpha\beta} f_{,\alpha} \\ + \dfrac{f_{,\alpha} f_{,\beta} g^{\alpha\beta}(1-n)}{4F^2}\left(4FF'' + (F')^2(n-6)\right) \end{array} \right) \dfrac{1}{F} \right) \right)$$

$$= \int_V d^n x \frac{\sqrt{-g \cdot F^n}}{F} \times \left(\begin{array}{c} R^{\kappa\lambda} - \dfrac{R}{2} g^{\kappa\lambda} \\[2mm] f_{,\alpha} f_{,\beta} \left(\dfrac{g^{\kappa\lambda} g^{\alpha\beta}}{2} - \dfrac{g^{\kappa\alpha} g^{\lambda\beta} + g^{\lambda\alpha} g^{\kappa\beta}}{2} \right) (1-n) \\[2mm] + \dfrac{}{4F^2} \left[\begin{array}{c} (F')^2 (n-6) \\ +4FF'' \end{array} \right] \end{array} \right) \delta g_{\kappa\lambda}$$

$$+\delta_{g_{\alpha\beta}} \int_V d^n x \frac{F^{n/2}}{F^2} \cdot \left(F' \cdot (1-n) \partial_\beta \sqrt{-g} \cdot g^{\alpha\beta} f_{,\alpha} \right) + \int_V d^n x \frac{\sqrt{-g \cdot F^n}}{F} \times g^{\alpha\beta} \delta_{g_{\alpha\beta}} R_{\alpha\beta}.$$

$$(744)$$

Please note that we did not—as Hilbert has done in [7]— automatically[3] set the variation of the Ricci tensor equal to zero but keep it as $g^{\alpha\beta} \delta_{g_{\alpha\beta}} R_{\alpha\beta}$. We will deal later with this term.

Please also note that we separated the Laplace term, because it cannot be variated as easily as the other terms. This is because we have the metric within this operator. So just like with the Hilbert variation of the Ricci tensor [7] we have to apply caution. We see this, when taking the first term from the last line of Eq. (744) and rewriting the Laplace operator in a different form:

$$\delta_{g_{\alpha\beta}} \int_V d^n x \sqrt{-g} \cdot \frac{F^{n/2}}{F^2} \cdot \left(F' \cdot (1-n) \frac{1}{\sqrt{-g}} \partial_\beta \sqrt{-g} \cdot g^{\alpha\beta} f_{,\alpha} \right)$$

$$= \delta_{g_{\alpha\beta}} \int_V d^n x \sqrt{-g} \cdot \frac{F^{n/2}}{F^2} \cdot \left(F' \cdot (1-n) \left(g^{\alpha\beta} f_{,\alpha\beta} - g^{\mu\nu} \Gamma^\gamma_{\mu\nu} f_{,\gamma} \right) \right)$$

$$= \int_V d^n x \sqrt{-g} \cdot \frac{F^{n/2}}{F^2} \cdot \left(F' \cdot (1-n) \left(\frac{g^{\kappa\lambda} g^{\alpha\beta}}{2} - \frac{g^{\kappa\alpha} g^{\lambda\beta} + g^{\lambda\alpha} g^{\kappa\beta}}{2} \right) (f_{,\alpha\beta} - g^{\mu\nu} \Gamma^\gamma_{\mu\nu} f_{,\gamma}) \right) \delta g_{\kappa\lambda}$$

$$- \int_V d^n x \sqrt{-g} \cdot \frac{F^{n/2}}{F^2} \cdot \left(F' \cdot (1-n) \frac{g^{\alpha\beta}}{n} \delta_{g_{\alpha\beta}} g^{\mu\nu} \Gamma^\gamma_{\mu\nu} f_{,\gamma} \right). \qquad (745)$$

While the first integral makes no problems, we find that the second integral (last line of Eq. (745)) gives us:

$$\int_V d^n x \sqrt{-g} \cdot \frac{F^{n/2}}{F^2} \cdot \left(F' \cdot (1-n) \frac{g^{\alpha\beta}}{n} \delta_{g_{\alpha\beta}} g^{\mu\nu} \Gamma^\gamma_{\mu\nu} f_{,\gamma} \right)$$

$$= \int_V d^n x \sqrt{-g} \cdot \frac{F^{n/2}}{F^2} \cdot \left(F' \cdot (1-n) \frac{g^{\alpha\beta}}{n} \left(\Gamma^\gamma_{\mu\nu} f_{,\gamma} \delta_{g_{\alpha\beta}} g^{\mu\nu} + g^{\mu\nu} \delta_{g_{\alpha\beta}} \Gamma^\gamma_{\mu\nu} f_{,\gamma} \right) \right)$$

---

[3]After proving that it delivers a surface term. However, the Ricci tensor in the current integral with the function $F[f]$ does not give such a simple solution.

$$= \int_V d^n x \sqrt{-g} \cdot \frac{F^{n/2}}{F^2} \cdot \left(F' \cdot (1-n) \frac{g^{\alpha\beta}}{n} f_{,\gamma} \left(\Gamma^\gamma_{\mu\nu} \delta_{g_{\alpha\beta}} g^{\mu\nu} + g^{\mu\nu} \frac{1}{2} g^{\gamma\sigma} (\delta g_{\sigma\mu;\nu} + \delta g_{\sigma\nu;\mu} - \delta g_{\mu\nu;\sigma}) \right) \right).$$

(746)

Thus, we have covariant derivatives of metric variations $\delta g_{\sigma\mu;\nu} + \delta g_{\sigma\nu;\mu} - \delta g_{\mu\nu;\sigma}$ instead of the variation itself $\delta g_{\kappa\lambda}$ as obtained for the other integrals. The total result looks as follows:

$$\delta_{g_{\alpha\beta}} W = 0 = \int_V d^n x \frac{\sqrt{-g \cdot F^n}}{F} \times g^{\alpha\beta} \delta_{g_{\alpha\beta}} R_{\alpha\beta}$$

$$+ \int_V d^n x \frac{\sqrt{-g \cdot F^n}}{F} \times \left(\begin{array}{c} R^{\kappa\lambda} - \dfrac{R}{2} g^{\kappa\lambda} \\[2mm] + \dfrac{f_{,\alpha} f_{,\beta} \left(\dfrac{g^{\kappa\lambda} g^{\alpha\beta}}{2} - \dfrac{g^{\kappa\alpha} g^{\lambda\beta} + g^{\lambda\alpha} g^{\kappa\beta}}{2} \right) (1-n)}{4F^2} \left(\begin{array}{c} (F')^2 (n-6) \\ +4FF'' \end{array} \right) \end{array} \right) \delta g_{\kappa\lambda}$$

$$+ \int_V d^n x \sqrt{-g} \cdot \frac{F^{n/2}}{F^2} \cdot \left(F' \cdot (1-n) \left(\frac{g^{\kappa\lambda} g^{\alpha\beta}}{2} - \frac{g^{\kappa\alpha} g^{\lambda\beta} + g^{\lambda\alpha} g^{\kappa\beta}}{2} \right) \left(f_{,\alpha\beta} - \frac{1}{n} g^{\mu\nu} \Gamma^\gamma_{\mu\nu} f_{,\gamma} \right) \right) \delta g_{\kappa\lambda}$$

$$- \int_V d^n x \sqrt{-g} \cdot \frac{F^{n/2}}{F^2} \cdot \left(F' \cdot (1-n) \frac{g^{\alpha\beta}}{n} f_{,\gamma} \left(\Gamma^\gamma_{\mu\nu} \delta_{g_{\alpha\beta}} g^{\mu\nu} + g^{\mu\nu} \frac{1}{2} g^{\gamma\sigma} (\delta g_{\sigma\mu;\nu} + \delta g_{\sigma\nu;\mu} - \delta g_{\mu\nu;\sigma}) \right) \right)$$

$$= \int_V d^n x \frac{\sqrt{-g \cdot F^n}}{F} \times g^{\alpha\beta} \delta_{g_{\alpha\beta}} R_{\alpha\beta}$$

$$+ \int_V d^n x \frac{\sqrt{-g \cdot F^n}}{F} \times \left(\begin{array}{c} R^{\kappa\lambda} - \dfrac{R}{2} g^{\kappa\lambda} \\[2mm] + \dfrac{f_{,\alpha} f_{,\beta} \left(\dfrac{g^{\kappa\lambda} g^{\alpha\beta}}{2} - \dfrac{g^{\kappa\alpha} g^{\lambda\beta} + g^{\lambda\alpha} g^{\kappa\beta}}{2} \right) (1-n)}{4F^2} \left(\begin{array}{c} (F')^2 (n-6) \\ +4FF'' \end{array} \right) \\[4mm] + \dfrac{F' \cdot (1-n)}{F} \left(\dfrac{g^{\kappa\lambda} g^{\alpha\beta}}{2} - \dfrac{g^{\kappa\alpha} g^{\lambda\beta} + g^{\lambda\alpha} g^{\kappa\beta}}{2} \right) \left(f_{,\alpha\beta} - \dfrac{1}{n} g^{\mu\nu} \Gamma^\gamma_{\mu\nu} f_{,\gamma} \right) \end{array} \right) \delta g_{\kappa\lambda}$$

$$- \int_V d^n x \frac{\sqrt{-g \cdot F^n}}{F} \cdot \left(\frac{F' \cdot (1-n)}{F} \frac{g^{\alpha\beta}}{n} f_{,\gamma} \left(\Gamma^\gamma_{\mu\nu} \delta_{g_{\alpha\beta}} g^{\mu\nu} + g^{\mu\nu} \frac{1}{2} g^{\gamma\sigma} (\delta g_{\sigma\mu;\nu} + \delta g_{\sigma\nu;\mu} - \delta g_{\mu\nu;\sigma}) \right) \right).$$

(747)

Not being able to place everything under one integral, it seems that we are stuck with our attempt to derive a quantum Einstein field equation ... or are we not?

8.2.1 Variation of the Laplace Operator

$$\delta_{g_{\alpha\beta}} \int_V d^n x \sqrt{-g} \cdot \frac{F^{n/2}}{F^2} \cdot \left(F' \cdot (1-n) \frac{1}{\sqrt{-g}} \partial_\beta \sqrt{-g} \cdot g^{\alpha\beta} f_{,\alpha} \right)$$

$$= \delta_{g_{\alpha\beta}} \int_V d^n x \sqrt{-g} \cdot \frac{F^{n/2}}{F^2} \cdot \left(F' \cdot (1-n) \left(g^{\alpha\beta} f_{,\alpha\beta} - g^{\alpha\beta} \Gamma^\gamma_{\alpha\beta} f_{,\gamma} \right) \right)$$

$$= \int_V d^n x \sqrt{-g} \cdot \frac{F^{n/2}}{F^2} \cdot \left(F' \cdot (1-n) \left(\frac{g^{\kappa\lambda} g^{\alpha\beta}}{2} - \frac{g^{\kappa\alpha} g^{\lambda\beta} + g^{\lambda\alpha} g^{\kappa\beta}}{2} \right) \left(f_{,\alpha\beta} - g^{\alpha\beta} \Gamma^\gamma_{\alpha\beta} f_{,\gamma} \right) \right) \delta g_{\kappa\lambda}$$

$$+ \int_V d^n x \sqrt{-g} \cdot \frac{F^{n/2}}{F^2} \cdot \left(F' \cdot (1-n) g^{\alpha\beta} f_{,\gamma} \delta_{g_{\alpha\beta}} \Gamma^\gamma_{\alpha\beta} \right). \tag{748}$$

Further evaluation of the last integral yields:

$$\int_V d^n x \sqrt{-g} \cdot \frac{F^{n/2}}{F^2} \cdot \left(F' \cdot (1-n) g^{\alpha\beta} f_{,\gamma} \delta_{g_{\alpha\beta}} \Gamma^\gamma_{\alpha\beta} \right)$$

$$= \int_V d^n x \sqrt{-g} \cdot \frac{F^{n/2}}{F^2} \cdot \left(F' \cdot (1-n) \frac{g^{\alpha\beta}}{n} g^{\mu\nu} f_{,\gamma} \left[\frac{g^{\gamma\sigma}}{2} GX^{\kappa\lambda\rho}_{\mu\nu\sigma} \delta g_{\kappa\lambda,\rho} - g^{\gamma\lambda} \Gamma^\kappa_{\mu\nu} \delta g_{\kappa\lambda} \right] \right)$$

with: $\delta_g \left(\Gamma^\gamma_{\mu\nu} \right) = \frac{1}{2} g^{\gamma\sigma} \left(\delta g_{\sigma\mu;\nu} + \delta g_{\sigma\nu;\mu} - \delta g_{\mu\nu;\sigma} \right) = \frac{g^{\gamma\sigma}}{2} GX^{\kappa\lambda\rho}_{\mu\nu\sigma} \delta g_{\kappa\lambda,\rho} - g^{\gamma\lambda} \Gamma^\kappa_{\mu\nu} \delta g_{\kappa\lambda}$

$$\left(g^\kappa_\mu g^\lambda_\sigma g^\rho_\nu + g^\kappa_\sigma g^\lambda_\nu g^\rho_\mu - g^\kappa_\mu g^\lambda_\nu g^\rho_\sigma \right) \equiv GX^{\kappa\lambda\rho}_{\mu\nu\sigma}. \tag{749}$$

Thereby:

$$\delta_g \left(\Gamma^\gamma_{\alpha\beta} \right) = \frac{1}{2} g^{\gamma\sigma} \left(\delta g_{\sigma\alpha;\beta} + \delta g_{\sigma\beta;\alpha} - \delta g_{\alpha\beta;\sigma} \right) = \frac{g^{\gamma\sigma}}{2} \left(\delta g_{\sigma\alpha,\beta} + \delta g_{\sigma\beta,\alpha} - \delta g_{\alpha\beta,\sigma} \right)$$

$$- \frac{g^{\gamma\sigma}}{2} \left(\Gamma^\xi_{\beta\sigma} \delta g_{\xi\alpha} + \Gamma^\xi_{\beta\alpha} \delta g_{\sigma\xi} + \Gamma^\xi_{\alpha\sigma} \delta g_{\xi\beta} + \Gamma^\xi_{\alpha\beta} \delta g_{\sigma\xi} - \Gamma^\xi_{\sigma\alpha} \delta g_{\xi\beta} - \Gamma^\xi_{\sigma\beta} \delta g_{\alpha\xi} \right)$$

$$= \frac{g^{\gamma\sigma}}{2} \left(\delta g_{\sigma\alpha,\beta} + \delta g_{\sigma\beta,\alpha} - \delta g_{\alpha\beta,\sigma} \right)$$

$$- \frac{g^{\gamma\sigma}}{2} \left(\Gamma^\kappa_{\beta\sigma} \delta g_{\kappa\alpha} + \Gamma^\kappa_{\beta\alpha} \delta g_{\sigma\kappa} + \Gamma^\kappa_{\alpha\sigma} \delta g_{\kappa\beta} + \Gamma^\kappa_{\alpha\beta} \delta g_{\sigma\kappa} - \Gamma^\kappa_{\sigma\alpha} \delta g_{\kappa\beta} - \Gamma^\kappa_{\sigma\beta} \delta g_{\alpha\kappa} \right)$$

$$\xrightarrow{\delta g_{ij}=\delta g_{ji}} = \frac{g^{\gamma\sigma}}{2}\left(\delta g_{\sigma\alpha,\beta}+\delta g_{\sigma\beta,\alpha}-\delta g_{\alpha\beta,\sigma}\right)$$

$$-\frac{g^{\gamma\sigma}}{2}\left(\Gamma^{\kappa}_{\beta\sigma}g^{\lambda}_{\alpha}+\Gamma^{\kappa}_{\beta\alpha}g^{\lambda}_{\sigma}+\Gamma^{\kappa}_{\alpha\sigma}g^{\lambda}_{\beta}+\Gamma^{\kappa}_{\alpha\beta}g^{\lambda}_{\sigma}-\Gamma^{\kappa}_{\sigma\alpha}g^{\lambda}_{\beta}-\Gamma^{\kappa}_{\sigma\beta}g^{\lambda}_{\alpha}\right)\delta g_{\kappa\lambda}$$

$$\xrightarrow{\Gamma^{\kappa}_{ij}=\Gamma^{\kappa}_{ji}} = \frac{g^{\gamma\sigma}}{2}\left(\delta g_{\sigma\alpha,\beta}+\delta g_{\sigma\beta,\alpha}-\delta g_{\alpha\beta,\sigma}\right)$$

$$-\frac{g^{\gamma\sigma}}{2}\left(\overbrace{\Gamma^{\kappa}_{\beta\sigma}g^{\lambda}_{\alpha}-\Gamma^{\kappa}_{\sigma\beta}g^{\lambda}_{\alpha}}^{=0}+\overbrace{\Gamma^{\kappa}_{\alpha\beta}g^{\lambda}_{\sigma}+\Gamma^{\kappa}_{\beta\alpha}g^{\lambda}_{\sigma}}^{=2\Gamma^{\kappa}_{\alpha\beta}g^{\lambda}_{\sigma}}+\overbrace{\Gamma^{\kappa}_{\alpha\sigma}g^{\lambda}_{\beta}-\Gamma^{\kappa}_{\sigma\alpha}g^{\lambda}_{\beta}}^{=0}\right)\delta g_{\kappa\lambda}$$

$$= \frac{g^{\gamma\sigma}}{2}\left(\delta g_{\sigma\alpha,\beta}+\delta g_{\sigma\beta,\alpha}-\delta g_{\alpha\beta,\sigma}\right)-g^{\gamma\lambda}\Gamma^{\kappa}_{\alpha\beta}\delta g_{\kappa\lambda}. \tag{750}$$

Now we proceed with the result of Eq. (749) (second line) and by applying integration by parts and setting all surface terms equal to zero, we end up with:

$$=\int_{V}d^{n}x\sqrt{-g}\cdot\frac{F^{n/2}}{F^{2}}\cdot\left(F'\cdot(1-n)g^{\alpha\beta}\frac{g^{\gamma\sigma}}{2}GX^{\kappa\lambda\rho}_{\alpha\beta\sigma}f_{,\gamma}\right)\delta g_{\kappa\lambda,\rho}$$

$$=\int_{V}d^{n}x\left[\sqrt{-g}\cdot\frac{F^{n/2}}{F^{2}}\cdot\left(F'\cdot(1-n)g^{\alpha\beta}\frac{g^{\gamma\sigma}}{2}GX^{\kappa\lambda\rho}_{\alpha\beta\sigma}f_{,\gamma}\right)\right]_{,\rho}\delta g_{\kappa\lambda}$$

$$-\int_{V}d^{n}x\left[\sqrt{-g}\cdot\frac{F^{n/2}}{F^{2}}\cdot\left(F'\cdot(1-n)g^{\alpha\beta}\frac{g^{\gamma\sigma}}{2}GX^{\kappa\lambda\rho}_{\alpha\beta\sigma}f_{,\gamma}\right)\delta g_{\kappa\lambda}\right]_{,\rho}$$

$$=\int_{V}d^{n}x\left[\sqrt{-g}\cdot\frac{F^{n/2}}{F^{2}}\cdot\left(F'\cdot(1-n)g^{\alpha\beta}\frac{g^{\gamma\sigma}}{2}GX^{\kappa\lambda\rho}_{\alpha\beta\sigma}f_{,\gamma}\right)\right]_{,\rho}\delta g_{\kappa\lambda}$$

$$\overbrace{-\int_{\partial V}d^{n}x\cdot n_{\rho}\sqrt{-g}\cdot\frac{F^{n/2}}{F^{2}}\cdot\left(F'\cdot(1-n)g^{\alpha\beta}\frac{g^{\gamma\sigma}}{2}GX^{\kappa\lambda\rho}_{\alpha\beta\sigma}f_{,\gamma}\right)\delta g_{\kappa\lambda}}^{=0\;\;?\;\;\text{Gibbons-Hawking-York}}$$

$$=\int_{V}d^{n}x\left[\sqrt{-g}\cdot\frac{F^{n/2}}{F^{2}}\cdot\left(F'\cdot(1-n)g^{\alpha\beta}\frac{g^{\gamma\sigma}}{2}GX^{\kappa\lambda\rho}_{\alpha\beta\sigma}f_{,\gamma}\right)\right]_{,\rho}\delta g_{\kappa\lambda}. \tag{751}$$

8.2.2 Back to the Total Variation of Our Scaled Metric Ricci Scalar

With the help of the results of the last sub-section, we obtain from Eq. (747):

$$\delta_{g_{\alpha\beta}} W = 0 = \int_V d^n x \frac{\sqrt{-g}\cdot F^n}{F}\times g^{\alpha\beta}\delta_{g_{\alpha\beta}}R_{\alpha\beta} + \int_V d^n x$$

$$\frac{\sqrt{-g}\cdot F^n}{F}\cdot\Bigg\{ R^{\kappa\lambda} - \frac{R}{2}g^{\kappa\lambda}$$

$$+ f_{,\alpha}f_{,\beta}\frac{\left(\dfrac{g^{\kappa\lambda}g^{\alpha\beta}}{2} - \dfrac{g^{\kappa\alpha}g^{\lambda\beta}+g^{\lambda\alpha}g^{\kappa\beta}}{2}\right)(1-n)\begin{pmatrix}(F')^2(n-6)\\ +4FF''\end{pmatrix}}{4F^2}$$

$$+ \frac{F'\cdot(1-n)}{F}\left(\dfrac{g^{\kappa\lambda}g^{\alpha\beta}}{2} - \dfrac{g^{\kappa\alpha}g^{\lambda\beta}+g^{\lambda\alpha}g^{\kappa\beta}}{2}\right)\left(f_{,\alpha\beta}-\Gamma^\gamma_{\alpha\beta}f_{,\gamma}\right)$$

$$+\frac{F'}{F}\cdot(1-n)g^{\alpha\beta}g^{\gamma\lambda}\Gamma^\kappa_{\alpha\beta}f_{,\gamma}$$

$$-\left(\frac{F'}{F}\cdot(1-n)g^{\alpha\beta}\frac{g^{\gamma\sigma}}{2}GX^{\kappa\lambda\rho}_{\alpha\beta\sigma}f_{,\gamma}\right)_{,\rho}$$

$$-\left[\frac{\sqrt{-g}\cdot F^n}{F}\right]_{,\rho}\cdot\left(\frac{F'}{F}\cdot(1-n)g^{\alpha\beta}\frac{g^{\gamma\sigma}}{2}GX^{\kappa\lambda\rho}_{\alpha\beta\sigma}f_{,\gamma}\right)_{,\rho}\Bigg\}\delta g_{\kappa\lambda}.$$

$$(752)$$

For simplification, we further consider the integral above without further treatment of the variation term for the Ricci tensor $\dfrac{\sqrt{-g}\cdot F^n}{F}\cdot\left(g^{\alpha\beta}\delta_{g_{\alpha\beta}}R_{\alpha\beta}\right)$. The rest gives us:

$$\text{Integrand} - \frac{\sqrt{-g \cdot F^n}}{F} \cdot \left(g^{\alpha\beta} \delta_{g_{\alpha\beta}} R_{\alpha\beta} \right)$$

$$\Rightarrow \frac{\sqrt{-g \cdot F^n}}{F} \cdot \left(\begin{array}{l} R^{\kappa\lambda} - \dfrac{R}{2} g^{\kappa\lambda} \\[2ex] + \dfrac{f_{,\alpha} f_{,\beta} \left(\dfrac{g^{\kappa\lambda} g^{\alpha\beta}}{2} - \dfrac{g^{\kappa\alpha} g^{\lambda\beta} + g^{\lambda\alpha} g^{\kappa\beta}}{2} \right)(1-n)}{4F^2} \left(\begin{array}{c} (F')^2 (n-6) \\ +4FF'' \end{array} \right) \\[3ex] + \dfrac{F' \cdot (1-n)}{F} \left(\dfrac{g^{\kappa\lambda} g^{\alpha\beta}}{2} - \dfrac{g^{\kappa\alpha} g^{\lambda\beta} + g^{\lambda\alpha} g^{\kappa\beta}}{2} \right) \left(f_{,\alpha\beta} - \Gamma^{\gamma}_{\alpha\beta} f_{,\gamma} \right) \\[3ex] + \dfrac{F'}{F} \cdot (1-n) g^{\alpha\beta} g^{\gamma\lambda} \Gamma^{\kappa}_{\alpha\beta} f_{,\gamma} \\[3ex] - \left(\dfrac{F'}{F} \cdot (1-n) g^{\alpha\beta} \dfrac{g^{\gamma\sigma}}{2} GX^{\kappa\lambda\rho}_{\alpha\beta\sigma} f_{,\gamma} \right)_{,\rho} \end{array} \right)$$

$$- \left[\frac{\sqrt{-g \cdot F^n}}{F} \right]_{,\rho} \cdot \left(\frac{F'}{F} \cdot (1-n) g^{\alpha\beta} \frac{g^{\gamma\sigma}}{2} GX^{\kappa\lambda\rho}_{\alpha\beta\sigma} f_{,\gamma} \right), \tag{753}$$

which then gives (after equalizing some of the dummy indices):

$$0 = \left\{ \begin{array}{l} R^{\kappa\lambda} - \dfrac{R}{2} g^{\kappa\lambda} \\[2ex] + \left(g^{\kappa\lambda} g^{\alpha\beta} - g^{\kappa\alpha} g^{\lambda\beta} - g^{\lambda\alpha} g^{\kappa\beta} \right)(1-n) \times \\[2ex] \left[\begin{array}{l} \dfrac{f_{,\alpha} f_{,\beta}}{2 \cdot F^2} \left(\dfrac{1}{4} \left((F')^2 (n-6) + 4FF'' \right) + FF'' - (F')^2 + (F')^2 \left(\dfrac{n}{2} - 1 \right) \right) \\[3ex] + f_{,\alpha\beta} \left(\dfrac{F'}{2 \cdot F} + \dfrac{F'}{2 \cdot F} \right) \end{array} \right] \\[5ex] - \dfrac{F' \cdot (1-n)}{2 \cdot F} \left(g^{\kappa\lambda} g^{\alpha\beta} - g^{\kappa\alpha} g^{\lambda\beta} - g^{\lambda\alpha} g^{\kappa\beta} \right) \Gamma^{\gamma}_{\alpha\beta} f_{,\gamma} \\[3ex] + \dfrac{F'}{F} \cdot (1-n) g^{\alpha\beta} g^{\gamma\lambda} \Gamma^{\kappa}_{\alpha\beta} f_{,\gamma} \\[3ex] + \dfrac{F'}{F} \cdot \dfrac{(1-n)}{2} \left(g^{\kappa\lambda} g^{\alpha\beta} - g^{\kappa\alpha} g^{\lambda\beta} - g^{\lambda\alpha} g^{\kappa\beta} \right)_{,\beta} f_{,\alpha} \\[3ex] + \dfrac{g^{\xi\zeta} g_{\xi\zeta,\beta}}{2} \cdot \left(\dfrac{F'}{F} \cdot \dfrac{(1-n)}{2} \left(g^{\kappa\lambda} g^{\alpha\beta} - g^{\kappa\alpha} g^{\lambda\beta} - g^{\lambda\alpha} g^{\kappa\beta} \right) f_{,\alpha} \right). \end{array} \right. \tag{754}$$

On first sight, we consider it quite surprising that the product of the two Nabla-kind operator terms for f are giving such a strange sum with $FF'' + (F')^2 \cdot \dfrac{3 \cdot n - 14}{8}$. On the other side, when inserting $n = 4$, one gets $FF'' - \dfrac{(F')^2}{4}$, which does not look so very strange at all. We obtain:

$$0 = \left\{ \begin{array}{l} R^{\kappa\lambda} - \dfrac{R}{2} g^{\kappa\lambda} \\[2mm] + \left(g^{\kappa\lambda} g^{\alpha\beta} - g^{\kappa\alpha} g^{\lambda\beta} - g^{\lambda\alpha} g^{\kappa\beta} \right)(1-n) \cdot \left[\dfrac{f_{,\alpha} f_{,\beta}}{F^2} \left(FF'' + (F')^2 \cdot \dfrac{3 \cdot n - 14}{8} \right) + f_{,\alpha\beta} \dfrac{F'}{F} \right] \\[4mm] - \dfrac{F' \cdot (1-n)}{2 \cdot F} \left(g^{\kappa\lambda} g^{\alpha\beta} - g^{\kappa\alpha} g^{\lambda\beta} - g^{\lambda\alpha} g^{\kappa\beta} \right) \Gamma^{\gamma}_{\alpha\beta} f_{,\gamma} \\[4mm] + \dfrac{F'}{F} \cdot (1-n) g^{\alpha\beta} g^{\gamma\lambda} \Gamma^{\kappa}_{\alpha\beta} f_{,\gamma} \\[4mm] + \dfrac{F'}{F} \cdot \dfrac{(1-n)}{2} \left(g^{\kappa\lambda} g^{\alpha\beta} - g^{\kappa\alpha} g^{\lambda\beta} - g^{\lambda\alpha} g^{\kappa\beta} \right)_{,\beta} f_{,\alpha} \\[4mm] + \dfrac{g^{\xi\zeta} g_{\xi\zeta,\beta}}{2} \cdot \left(\dfrac{F'}{F} \cdot \dfrac{(1-n)}{2} \left(g^{\kappa\lambda} g^{\alpha\beta} - g^{\kappa\alpha} g^{\lambda\beta} - g^{\lambda\alpha} g^{\kappa\beta} \right) f_{,\alpha} \right). \end{array} \right.$$

(755)

Now we have to incorporate the variation of the Ricci tensor, respectively its new form, which is now connected with the scale function $F[f]$.

8.2.3 Variation of the Scaled Ricci Tensor Term

It becomes immediately clear from the structure of the Ricci tensor integral term from Eq. (744):

$$\int_V d^n x \sqrt{-g} \cdot \dfrac{F^{n/2}}{F} \cdot g^{\alpha\beta} \delta_{g_{\alpha\beta}} R_{\alpha\beta} = \int_V d^n x \sqrt{-g} \cdot F^{n/2-1} \cdot g^{\alpha\beta} \delta R_{\alpha\beta} \qquad (756)$$

that this cannot give the same surface term (and thus, vanishing [7]) result as the classical form, reading:

$$\int_V d^n x \sqrt{-g} \cdot g^{\alpha\beta} \delta_{g_{\alpha\beta}} R_{\alpha\beta} = \int_V d^n x \sqrt{-g} \cdot g^{\alpha\beta} \delta R_{\alpha\beta}. \qquad (757)$$

The reason is that while Eq. (757) can be made a complete divergence, reading:

$$\int_V d^n x \sqrt{-g} \cdot g^{\alpha\beta} \delta R_{\alpha\beta} = \int_V d^n x \left(\sqrt{-g} \cdot g^{\alpha\beta} \delta \Gamma^\gamma_{\alpha\beta} \right)_{,\gamma} - \int_V d^n x \left(\sqrt{-g} \cdot g^{\alpha\beta} \delta \Gamma^\sigma_{\beta\sigma} \right)_{,\alpha},$$

$$(758)$$

we will have the following from (756):

$$\int_V d^n x \sqrt{-g} \cdot F^{n/2-1} \cdot g^{\alpha\beta} \delta R_{\alpha\beta}$$

$$= \int_V d^n x \cdot F^{n/2-1} \cdot \left(\sqrt{-g} \cdot g^{\alpha\beta} \delta \Gamma^\gamma_{\alpha\beta} \right)_{,\gamma} - \int_V d^n x \cdot F^{n/2-1} \cdot \left(\sqrt{-g} \cdot g^{\alpha\beta} \delta \Gamma^\sigma_{\beta\sigma} \right)_{,\alpha}.$$

$$(759)$$

Only after integration by parts:

$$\int_V d^n x \sqrt{-g} \cdot F^{n/2-1} \cdot g^{\alpha\beta} \delta R_{\alpha\beta}$$

$$= \int_V d^n x \cdot F^{n/2-1} \cdot \left(\sqrt{-g} \cdot g^{\alpha\beta} \delta \Gamma^\gamma_{\alpha\beta} \right)_{,\gamma} - \int_V d^n x \cdot F^{n/2-1} \cdot \left(\sqrt{-g} \cdot g^{\alpha\beta} \delta \Gamma^\sigma_{\beta\sigma} \right)_{,\alpha}$$

$$= \int_V d^n x \cdot F^{n/2-1}{}_{,\gamma} \cdot \sqrt{-g} \cdot g^{\alpha\beta} \delta \Gamma^\gamma_{\alpha\beta} - \int_V d^n x \cdot \left(F^{n/2-1} \cdot \sqrt{-g} \cdot g^{\alpha\beta} \delta \Gamma^\gamma_{\alpha\beta} \right)_{,\gamma}$$

$$- \int_V d^n x \cdot F^{n/2-1}{}_{,\alpha} \cdot \sqrt{-g} \cdot g^{\alpha\beta} \delta \Gamma^\sigma_{\beta\sigma} + \int_V d^n x \cdot \left(F^{n/2-1} \cdot \sqrt{-g} \cdot g^{\alpha\beta} \delta \Gamma^\sigma_{\beta\sigma} \right)_{,\alpha}$$

$$= \int_V d^n x \cdot F^{n/2-1}{}_{,\gamma} \cdot \sqrt{-g} \cdot g^{\alpha\beta} \delta \Gamma^\gamma_{\alpha\beta} - \int_{\partial V} d^n x \cdot n_\gamma \left(F^{n/2-1} \cdot \sqrt{-g} \cdot g^{\alpha\beta} \delta \Gamma^\gamma_{\alpha\beta} \right)$$

$$- \int_V d^n x \cdot F^{n/2-1}{}_{,\alpha} \cdot \sqrt{-g} \cdot g^{\alpha\beta} \delta \Gamma^\sigma_{\beta\sigma} + \int_{\partial V} d^n x \cdot n_\alpha \left(F^{n/2-1} \cdot \sqrt{-g} \cdot g^{\alpha\beta} \delta \Gamma^\sigma_{\beta\sigma} \right),$$

$$(760)$$

we have the desired surface integrals, which we can assume to vanish (!) using Hilbert's assumption of a boundary free space-time (caution here (!)). The remains are the two integrals:

$$\int_V d^n x \sqrt{-g} \cdot F^{n/2-1} \cdot g^{\alpha\beta} \delta R_{\alpha\beta}$$

$$= \int_V d^n x \cdot F^{n/2-1}{}_{,\gamma} \cdot \sqrt{-g} \cdot g^{\alpha\beta} \delta \Gamma^\gamma_{\alpha\beta} - \int_V d^n x \cdot F^{n/2-1}{}_{,\alpha} \cdot \sqrt{-g} \cdot g^{\alpha\beta} \delta \Gamma^\sigma_{\beta\sigma}, \quad (761)$$

where we once again find the variated Levi–Civita connections $\delta\Gamma^{\gamma}_{\alpha\beta}, \delta\Gamma^{\sigma}_{\beta\sigma}$, which this time result in:

$$\int_V d^n x \cdot F^{n/2-1}_{,\gamma} \cdot \sqrt{-g} \cdot g^{\alpha\beta} \delta\Gamma^{\gamma}_{\alpha\beta} - \int_V d^n x \cdot F^{n/2-1}_{,\alpha} \cdot \sqrt{-g} \cdot g^{\alpha\beta} \delta\Gamma^{\sigma}_{\beta\sigma}$$

$$= \int_V d^n x \cdot F^{n/2-1}_{,\gamma} \cdot \sqrt{-g} \cdot g^{\alpha\beta} \left(\frac{g^{\gamma\sigma}}{2} \left(\delta g_{\sigma\alpha,\beta} + \delta g_{\sigma\beta,\alpha} - \delta g_{\alpha\beta,\sigma} \right) - g^{\gamma\lambda} \Gamma^{\kappa}_{\alpha\beta} \delta g_{\kappa\lambda} \right)$$

$$- \int_V d^n x \cdot F^{n/2-1}_{,\alpha} \cdot \sqrt{-g} \cdot g^{\alpha\beta} \frac{1}{2} \cdot g^{\sigma\lambda} \delta g_{\sigma\lambda;\beta}$$

$$= \int_V d^n x \cdot F^{n/2-1}_{,\gamma} \cdot \sqrt{-g} \cdot g^{\alpha\beta} \left(\frac{g^{\gamma\sigma}}{2} GX^{\kappa\lambda\rho}_{\alpha\beta\sigma} \delta g_{\kappa\lambda,\rho} - g^{\gamma\lambda} \Gamma^{\kappa}_{\alpha\beta} \delta g_{\kappa\lambda} \right)$$

$$- \int_V d^n x \cdot F^{n/2-1}_{,\alpha} \cdot \sqrt{-g} \cdot g^{\alpha\beta} \frac{1}{2} \cdot g^{\kappa\lambda} \delta g_{\kappa\lambda;\beta}$$

$$= \int_V d^n x \cdot F^{n/2-1}_{,\gamma} \cdot \sqrt{-g} \cdot g^{\alpha\beta} \left(\frac{g^{\gamma\sigma}}{2} GX^{\kappa\lambda\rho}_{\alpha\beta\sigma} \delta g_{\kappa\lambda,\rho} - g^{\gamma\lambda} \Gamma^{\kappa}_{\alpha\beta} \delta g_{\kappa\lambda} \right)$$

$$- \int_V d^n x \cdot F^{n/2-1}_{,\alpha} \cdot \sqrt{-g} \cdot g^{\alpha\beta} \frac{1}{2} \cdot g^{\kappa\lambda} \left(\delta g_{\kappa\lambda,\beta} - \Gamma^{\xi}_{\beta\kappa} \delta g_{\xi\lambda} - \Gamma^{\xi}_{\beta\lambda} \delta g_{\xi\kappa} \right)$$

$$= \int_V d^n x \cdot F^{n/2-1}_{,\gamma} \cdot \sqrt{-g} \cdot g^{\alpha\beta} \left(\frac{g^{\gamma\sigma}}{2} GX^{\kappa\lambda\rho}_{\alpha\beta\sigma} \delta g_{\kappa\lambda,\rho} - g^{\gamma\lambda} \Gamma^{\kappa}_{\alpha\beta} \delta g_{\kappa\lambda} \right)$$

$$- \int_V d^n x \cdot F^{n/2-1}_{,\alpha} \cdot \sqrt{-g} \cdot g^{\alpha\beta} \frac{1}{2} \left(g^{\kappa\lambda} \delta g_{\kappa\lambda,\beta} - g^{\kappa\lambda} \Gamma^{\xi}_{\beta\kappa} \delta g_{\xi\lambda} - g^{\kappa\lambda} \Gamma^{\xi}_{\beta\lambda} \delta g_{\xi\kappa} \right)$$

$$= \int_V d^n x \cdot F^{n/2-1}_{,\gamma} \cdot \sqrt{-g} \cdot g^{\alpha\beta} \left(\frac{g^{\gamma\sigma}}{2} GX^{\kappa\lambda\rho}_{\alpha\beta\sigma} \delta g_{\kappa\lambda,\rho} - g^{\gamma\lambda} \Gamma^{\kappa}_{\alpha\beta} \delta g_{\kappa\lambda} \right)$$

$$- \int_V d^n x \cdot F^{n/2-1}_{,\alpha} \cdot \sqrt{-g} \cdot g^{\alpha\beta} \frac{1}{2} \left(g^{\kappa\lambda} \delta g_{\kappa\lambda,\beta} - g^{\xi\lambda} \Gamma^{\kappa}_{\beta\xi} \delta g_{\kappa\lambda} - g^{\kappa\xi} \Gamma^{\lambda}_{\beta\xi} \delta g_{\lambda\kappa} \right),$$

$$(762)$$

$$= \int_V d^n x \cdot F^{n/2-1}{}_{,\gamma} \cdot \sqrt{-g} \cdot g^{\alpha\beta} \left(\frac{g^{\gamma\sigma}}{2} GX^{\kappa\lambda\rho}_{\alpha\beta\sigma} \delta g_{\kappa\lambda,\rho} - g^{\gamma\lambda} \Gamma^{\kappa}_{\alpha\beta} \delta g_{\kappa\lambda} \right)$$

$$- \int_V d^n x \cdot F^{n/2-1}{}_{,\alpha} \cdot \sqrt{-g} \cdot g^{\alpha\beta} \frac{1}{2} \left(g^{\kappa\lambda} \delta g_{\kappa\lambda,\beta} - g^{\xi\lambda} \Gamma^{\kappa}_{\beta\xi} \delta g_{\kappa\lambda} - g^{\kappa\xi} \Gamma^{\lambda}_{\beta\xi} \delta g_{\lambda\kappa} \right)$$

$$= \int_V d^n x \cdot \sqrt{-g} \cdot g^{\alpha\beta} \left(F^{n/2-1}{}_{,\gamma} \cdot \frac{g^{\gamma\sigma}}{2} GX^{\kappa\lambda\rho}_{\alpha\beta\sigma} \delta g_{\kappa\lambda,\rho} - F^{n/2-1}{}_{,\gamma} \cdot g^{\gamma\lambda} \Gamma^{\kappa}_{\alpha\beta} \delta g_{\kappa\lambda} \right)$$

$$- \int_V d^n x \cdot \sqrt{-g} \cdot g^{\alpha\beta} \frac{1}{2} \left(F^{n/2-1}{}_{,\alpha} \cdot g^{\kappa\lambda} \delta g_{\kappa\lambda,\beta} - F^{n/2-1}{}_{,\alpha} \cdot g^{\xi\lambda} \Gamma^{\kappa}_{\beta\xi} \delta g_{\kappa\lambda} - F^{n/2-1}{}_{,\alpha} \cdot g^{\kappa\xi} \Gamma^{\lambda}_{\beta\xi} \delta g_{\lambda\kappa} \right)$$

$$= \int_V d^n x \cdot \sqrt{-g} \cdot g^{\alpha\beta} F^{n/2-1}{}_{,\gamma} \cdot \frac{g^{\gamma\sigma}}{2} GX^{\kappa\lambda\rho}_{\alpha\beta\sigma} \delta g_{\kappa\lambda,\rho} - \int_V d^n x \cdot \sqrt{-g} \cdot g^{\alpha\beta} \frac{1}{2} F^{n/2-1}{}_{,\alpha} \cdot g^{\kappa\lambda} \delta g_{\kappa\lambda,\beta}$$

$$\int_V d^n x \cdot \sqrt{-g} \cdot g^{\alpha\beta} F^{n/2-1}{}_{,\gamma} \cdot g^{\gamma\lambda} \Gamma^{\kappa}_{\alpha\beta} \delta g_{\kappa\lambda}$$

$$+ \int_V d^n x \cdot \sqrt{-g} \cdot g^{\alpha\beta} \frac{1}{2} \left(F^{n/2-1}{}_{,\alpha} \cdot g^{\xi\lambda} \Gamma^{\kappa}_{\beta\xi} + F^{n/2-1}{}_{,\alpha} \cdot g^{\kappa\xi} \Gamma^{\lambda}_{\beta\xi} \right) \delta g_{\kappa\lambda}. \quad (763)$$

While the integrals with the metric variation terms (second last and last line in Eq. (763)) are no problem, as they directly become a part of the whole integral in Eq. (744), we need to find a way to also treat the derivatives of the metric variation. Thus, we further consider the two integrals:

$$\int_V d^n x \cdot \sqrt{-g} \cdot g^{\alpha\beta} F^{n/2-1}{}_{,\gamma} \cdot \frac{g^{\gamma\sigma}}{2} GX^{\kappa\lambda\rho}_{\alpha\beta\sigma} \delta g_{\kappa\lambda,\rho} - \int_V d^n x \cdot \sqrt{-g} \cdot g^{\alpha\beta} \frac{1}{2} F^{n/2-1}{}_{,\alpha} \cdot g^{\kappa\lambda} \delta g_{\kappa\lambda,\beta}$$

$$= \int_V d^n x \cdot \left(\sqrt{-g} \cdot g^{\alpha\beta} F^{n/2-1}{}_{,\gamma} \cdot \frac{g^{\gamma\sigma}}{2} GX^{\kappa\lambda\rho}_{\alpha\beta\sigma} \right)_{,\rho} \delta g_{\kappa\lambda}$$

$$- \int_V d^n x \cdot \left(\sqrt{-g} \cdot g^{\alpha\beta} F^{n/2-1}{}_{,\gamma} \cdot \frac{g^{\gamma\sigma}}{2} GX^{\kappa\lambda\rho}_{\alpha\beta\sigma} \delta g_{\kappa\lambda} \right)_{,\rho}$$

$$- \int_V d^n x \cdot \left(\sqrt{-g} \cdot g^{\alpha\beta} \frac{1}{2} F^{n/2-1}{}_{,\alpha} \cdot g^{\kappa\lambda} \right)_{,\beta} \delta g_{\kappa\lambda}$$

$$+ \int_V d^n x \cdot \left(\sqrt{-g} \cdot g^{\alpha\beta} \frac{1}{2} F^{n/2-1}{}_{,\alpha} \cdot g^{\kappa\lambda} \delta g_{\kappa\lambda} \right)_{,\beta}$$

$$= \int_V d^n x \cdot \left(\sqrt{-g} \cdot g^{\alpha\beta} F^{n/2-1}{}_{,\gamma} \cdot \frac{g^{\gamma\sigma}}{2} GX^{\kappa\lambda\rho}_{\alpha\beta\sigma} \right)_{,\rho} \delta g_{\kappa\lambda}$$

$$\overbrace{- \int_{\partial V} d^n x \cdot n_\rho \sqrt{-g} \cdot g^{\alpha\beta} F^{n/2-1}{}_{,\gamma} \cdot \frac{g^{\gamma\sigma}}{2} GX^{\kappa\lambda\rho}_{\alpha\beta\sigma} \delta g_{\kappa\lambda}}^{=0}$$

$$- \int_V d^n x \cdot \left(\sqrt{-g} \cdot g^{\alpha\beta} \frac{1}{2} F^{n/2-1}{}_{,\alpha} \cdot g^{\kappa\lambda} \right)_{,\beta} \delta g_{\kappa\lambda}$$

$$\overbrace{+ \int_{\partial V} d^n x \cdot n_\beta \sqrt{-g} \cdot g^{\alpha\beta} \frac{1}{2} F^{n/2-1}{}_{,\alpha} \cdot g^{\kappa\lambda} \delta g_{\kappa\lambda}}^{=0}. \tag{764}$$

This leads us to:

$$\int_V d^n x \cdot \sqrt{-g} \cdot g^{\alpha\beta} F^{n/2-1}{}_{,\gamma} \cdot \frac{g^{\gamma\sigma}}{2} GX^{\kappa\lambda\rho}_{\alpha\beta\sigma} \delta g_{\kappa\lambda,\rho} - \int_V d^n x \cdot \sqrt{-g} \cdot g^{\alpha\beta} \frac{1}{2} F^{n/2-1}{}_{,\alpha} \cdot g^{\kappa\lambda} \delta g_{\kappa\lambda,\beta}$$

$$= \int_V d^n x \cdot \left[\left(\sqrt{-g} \cdot g^{\alpha\beta} F^{n/2-1}{}_{,\gamma} \cdot \frac{g^{\gamma\sigma}}{2} GX^{\kappa\lambda\rho}_{\alpha\beta\sigma} \right)_{,\rho} - \left(\sqrt{-g} \cdot g^{\alpha\beta} \frac{1}{2} F^{n/2-1}{}_{,\alpha} \cdot g^{\kappa\lambda} \right)_{,\beta} \right] \delta g_{\kappa\lambda}. \tag{765}$$

Thus, in total we have from (761):

$$\int_V d^n x \sqrt{-g} \cdot F^{n/2-1} \cdot g^{\alpha\beta} \delta R_{\alpha\beta}$$

$$= \int_V d^n x \cdot \left(\sqrt{-g} \cdot g^{\alpha\beta} \left(\frac{F^{n/2-1}{}_{,\alpha} \cdot g^{\xi\lambda} \Gamma^\kappa_{\beta\xi} + F^{n/2-1}{}_{,\alpha} \cdot g^{\kappa\xi} \Gamma^\lambda_{\beta\xi}}{2} + F^{n/2-1}{}_{,\gamma} \cdot g^{\gamma\lambda} \Gamma^\kappa_{\alpha\beta} \right) \\ + \left(\sqrt{-g} \cdot g^{\alpha\beta} F^{n/2-1}{}_{,\gamma} \cdot \frac{g^{\gamma\sigma}}{2} GX^{\kappa\lambda\rho}_{\alpha\beta\sigma} \right)_{,\rho} - \left(\sqrt{-g} \cdot g^{\alpha\beta} \frac{1}{2} F^{n/2-1}{}_{,\alpha} \cdot g^{\kappa\lambda} \right)_{,\beta} \right) \delta g_{\kappa\lambda}$$

$$= \int_V d^n x \cdot \left(\begin{array}{l} \sqrt{-g} \cdot g^{\alpha\beta} \left(\dfrac{F^{n/2-1}{}_{,\alpha} \cdot g^{\xi\lambda} \Gamma^\kappa_{\beta\xi} + F^{n/2-1}{}_{,\alpha} \cdot g^{\kappa\xi} \Gamma^\lambda_{\beta\xi}}{2} + F^{n/2-1}{}_{,\gamma} \cdot g^{\gamma\lambda} \Gamma^\kappa_{\alpha\beta} \right) \\[4mm] + \left(\sqrt{-g} \cdot g^{\alpha\beta} F^{n/2-1}{}_{,\gamma} \dfrac{\left(g^{\kappa\rho} g^{\gamma\lambda} + g^{\rho\lambda} g^{\gamma\kappa} - g^{\kappa\lambda} g^{\gamma\rho} \right)}{2} \right)_{,\rho} \\[4mm] - \left(\sqrt{-g} \cdot g^{\alpha\beta} \dfrac{1}{2} F^{n/2-1}{}_{,\alpha} \cdot g^{\kappa\lambda} \right)_{,\beta} \end{array} \right) \delta g_{\kappa\lambda}.$$

$$(766)$$

8.2.4 Back to the Total Variation of Our Scaled Metric Ricci Scalar

Now incorporating the results for the variated Ricci tensor from section "8.2.3 Variation of the Scaled Ricci Tensor Term" gives us:

$$0 = \frac{\sqrt{-g} \cdot F^n}{F} \cdot \left(\begin{array}{l} R^{\kappa\lambda} - \dfrac{R}{2} g^{\kappa\lambda} \\[3mm] + \left(g^{\kappa\lambda} g^{\alpha\beta} - g^{\kappa\alpha} g^{\lambda\beta} - g^{\lambda\alpha} g^{\kappa\beta} \right)(1-n) \\[3mm] \cdot \left[\dfrac{f_{,\alpha} f_{,\beta}}{F^2} \left(FF'' + (F')^2 \cdot \dfrac{3 \cdot n - 14}{8} \right) + f_{,\alpha\beta} \dfrac{F'}{F} \right] \\[3mm] - \dfrac{F' \cdot (1-n)}{2 \cdot F} \left(g^{\kappa\lambda} g^{\alpha\beta} - g^{\kappa\alpha} g^{\lambda\beta} - g^{\lambda\alpha} g^{\kappa\beta} \right) \Gamma^\gamma_{\alpha\beta} f_{,\gamma} \\[3mm] + \dfrac{F'}{F} \cdot (1-n) g^{\alpha\beta} g^{\gamma\lambda} \Gamma^\kappa_{\alpha\beta} f_{,\gamma} \\[3mm] + \dfrac{F'}{F} \cdot \dfrac{(1-n)}{2} \left(g^{\kappa\lambda} g^{\alpha\beta} - g^{\kappa\alpha} g^{\lambda\beta} - g^{\lambda\alpha} g^{\kappa\beta} \right)_{,\beta} f_{,\alpha} \\[3mm] + \dfrac{g^{\xi\zeta} g_{\xi\zeta,\beta}}{2} \cdot \left(\dfrac{F'}{F} \cdot \dfrac{(1-n)}{2} \left(g^{\kappa\lambda} g^{\alpha\beta} - g^{\kappa\alpha} g^{\lambda\beta} - g^{\lambda\alpha} g^{\kappa\beta} \right) f_{,\alpha} \right) \end{array} \right)$$

$$+ \sqrt{-g} \cdot g^{\alpha\beta} \left(\frac{F^{n/2-1}{}_{,\alpha} \cdot g^{\xi\lambda} \Gamma^\kappa_{\beta\xi} + F^{n/2-1}{}_{,\alpha} \cdot g^{\kappa\xi} \Gamma^\lambda_{\beta\xi}}{2} + F^{n/2-1}{}_{,\gamma} \cdot g^{\gamma\lambda} \Gamma^\kappa_{\alpha\beta} \right)$$

$$+ \left(\sqrt{-g} \cdot g^{\alpha\beta} F^{n/2-1}{}_{,\gamma} \frac{g^{\gamma\sigma}}{2} GX^{\kappa\lambda\rho}_{\alpha\beta\sigma} \right)_{,\rho} - \left(\sqrt{-g} \cdot \frac{g^{\alpha\beta}}{2} F^{n/2-1}{}_{,\alpha} \cdot g^{\kappa\lambda} \right)_{,\beta}. \quad (767)$$

Inserting the extension for $GX^{\kappa\lambda\rho}_{\alpha\beta\sigma}$ yields:

$$
0 = \frac{\sqrt{-g} \cdot F^n}{F} \cdot \left(
\begin{array}{c}
R^{\kappa\lambda} - \dfrac{R}{2} g^{\kappa\lambda} \\[2mm]
+ \left(g^{\kappa\lambda} g^{\alpha\beta} - g^{\kappa\alpha} g^{\lambda\beta} - g^{\lambda\alpha} g^{\kappa\beta} \right) (1-n) \\[2mm]
\cdot \left[\dfrac{f_{,\alpha} f_{,\beta}}{F^2} \left(FF'' + \left(F'\right)^2 \cdot \dfrac{3 \cdot n - 14}{8} \right) + f_{,\alpha\beta} \dfrac{F'}{F} \right] \\[2mm]
- \dfrac{F' \cdot (1-n)}{2 \cdot F} \left(g^{\kappa\lambda} g^{\alpha\beta} - g^{\kappa\alpha} g^{\lambda\beta} - g^{\lambda\alpha} g^{\kappa\beta} \right) \Gamma^{\gamma}_{\alpha\beta} f_{,\gamma} \\[2mm]
+ \dfrac{F'}{F} \cdot (1-n) g^{\alpha\beta} g^{\gamma\lambda} \Gamma^{\kappa}_{\alpha\beta} f_{,\gamma} \\[2mm]
+ \dfrac{F'}{F} \cdot \dfrac{(1-n)}{2} \left(g^{\kappa\lambda} g^{\alpha\beta} - g^{\kappa\alpha} g^{\lambda\beta} - g^{\lambda\alpha} g^{\kappa\beta} \right)_{,\beta} f_{,\alpha} \\[2mm]
+ \dfrac{g^{\xi\zeta} g_{\xi\zeta,\beta}}{2} \cdot \left(\dfrac{F'}{F} \cdot \dfrac{(1-n)}{2} \left(g^{\kappa\lambda} g^{\alpha\beta} - g^{\kappa\alpha} g^{\lambda\beta} - g^{\lambda\alpha} g^{\kappa\beta} \right) f_{,\alpha} \right)
\end{array}
\right)
$$

$$
+ \sqrt{-g} \cdot g^{\alpha\beta} \left(\frac{F^{n/2-1}_{,\alpha} \cdot g^{\xi\lambda} \Gamma^{\kappa}_{\beta\xi} + F^{n/2-1}_{,\alpha} \cdot g^{\kappa\xi} \Gamma^{\lambda}_{\beta\xi}}{2} + F^{n/2-1}_{,\gamma} \cdot g^{\gamma\lambda} \Gamma^{\kappa}_{\alpha\beta} \right)
$$

$$
+ \left(\sqrt{-g} \cdot F^{n/2-1}_{,\gamma} \cdot \frac{\left(g^{\kappa\rho} g^{\gamma\lambda} + g^{\rho\lambda} g^{\gamma\kappa} - g^{\kappa\lambda} g^{\gamma\rho} \right)}{2} \right)_{,\rho} - \left(\sqrt{-g} \cdot \frac{g^{\alpha\beta}}{2} F^{n/2-1}_{,\alpha} \cdot g^{\kappa\lambda} \right)_{,\beta} \cdot
$$

$$(768)$$

From there we can evaluate:

$$
0 = \frac{\sqrt{-g} \cdot F^n}{F} \cdot \left(
\begin{array}{c}
R^{\kappa\lambda} - \dfrac{R}{2} g^{\kappa\lambda} \\[2mm]
+ \left(g^{\kappa\lambda} g^{\alpha\beta} - g^{\kappa\alpha} g^{\lambda\beta} - g^{\lambda\alpha} g^{\kappa\beta} \right) (1-n) \\[2mm]
\cdot \left[\dfrac{f_{,\alpha} f_{,\beta}}{F^2} \left(FF'' + \left(F'\right)^2 \cdot \dfrac{3 \cdot n - 14}{8} \right) + f_{,\alpha\beta} \dfrac{F'}{F} \right] \\[2mm]
- \dfrac{F' \cdot (1-n)}{2 \cdot F} \left(g^{\kappa\lambda} g^{\alpha\beta} - g^{\kappa\alpha} g^{\lambda\beta} - g^{\lambda\alpha} g^{\kappa\beta} \right) \Gamma^{\gamma}_{\alpha\beta} f_{,\gamma} \\[2mm]
+ \dfrac{F'}{F} \cdot (1-n) g^{\alpha\beta} g^{\gamma\lambda} \Gamma^{\kappa}_{\alpha\beta} f_{,\gamma} \\[2mm]
+ \dfrac{F'}{F} \cdot \dfrac{(1-n)}{2} \left(g^{\kappa\lambda} g^{\alpha\beta} - g^{\kappa\alpha} g^{\lambda\beta} - g^{\lambda\alpha} g^{\kappa\beta} \right)_{,\beta} f_{,\alpha} \\[2mm]
+ \dfrac{g^{\xi\zeta} g_{\xi\zeta,\beta}}{2} \cdot \left(\dfrac{F'}{F} \cdot \dfrac{(1-n)}{2} \left(g^{\kappa\lambda} g^{\alpha\beta} - g^{\kappa\alpha} g^{\lambda\beta} - g^{\lambda\alpha} g^{\kappa\beta} \right) f_{,\alpha} \right)
\end{array}
\right)
$$

$$+\sqrt{-g} \cdot g^{\alpha\beta} \left(\frac{n}{2} - 1 \right) F^{n/2-2} F' \left(\frac{f_{,\alpha} \cdot g^{\xi\lambda} \Gamma^{\kappa}_{\beta\xi} + f_{,\alpha} \cdot g^{\kappa\xi} \Gamma^{\lambda}_{\beta\xi}}{2} + f_{,\gamma} \cdot g^{\gamma\lambda} \Gamma^{\kappa}_{\alpha\beta} \right)$$

$$+ \left(\frac{n}{2} - 1 \right) \left(\begin{pmatrix} \sqrt{-g} \cdot F^{n/2-2} F' f_{,\gamma} \cdot \dfrac{\left(g^{\kappa\rho} g^{\gamma\lambda} + g^{\rho\lambda} g^{\gamma\kappa} - g^{\kappa\lambda} g^{\gamma\rho} \right)_{,\rho}}{2} \\[2em] + \sqrt{-g} \cdot F^{n/2-2} F' f_{,\gamma\rho} \cdot \dfrac{\left(g^{\kappa\rho} g^{\gamma\lambda} + g^{\rho\lambda} g^{\gamma\kappa} - g^{\kappa\lambda} g^{\gamma\rho} \right)}{2} \\[2em] + \sqrt{-g} \cdot F^{n/2-2} F'' f_{,\rho} f_{,\gamma} \cdot \dfrac{\left(g^{\kappa\rho} g^{\gamma\lambda} + g^{\rho\lambda} g^{\gamma\kappa} - g^{\kappa\lambda} g^{\gamma\rho} \right)}{2} \\[2em] + \sqrt{-g} \cdot \left(\frac{n}{2} - 2 \right) F^{n/2-3} f_{,\rho} F'^2 f_{,\gamma} \cdot \dfrac{\left(g^{\kappa\rho} g^{\gamma\lambda} + g^{\rho\lambda} g^{\gamma\kappa} - g^{\kappa\lambda} g^{\gamma\rho} \right)}{2} \\[2em] + \sqrt{-g} \cdot \dfrac{g^{\xi\zeta} g_{\xi\zeta,\rho}}{2} \cdot F^{n/2-2} F' f_{,\gamma} \cdot \dfrac{\left(g^{\kappa\rho} g^{\gamma\lambda} + g^{\rho\lambda} g^{\gamma\kappa} - g^{\kappa\lambda} g^{\gamma\rho} \right)}{2} \end{pmatrix} \\[6em] - \begin{pmatrix} \sqrt{-g} \cdot \dfrac{g^{\alpha\beta}}{2} \cdot F^{n/2-2} F' f_{,\alpha} \cdot g^{\kappa\lambda}{}_{,\beta} + \sqrt{-g} \cdot \dfrac{g^{\alpha\beta}}{2} \cdot F^{n/2-2} F' f_{,\alpha\beta} \cdot g^{\kappa\lambda} \\[2em] + \sqrt{-g} \cdot \dfrac{g^{\alpha\beta}}{2} \cdot F^{n/2-2} F'' f_{,\beta} f_{,\alpha} \cdot g^{\kappa\lambda} \\[2em] + \sqrt{-g} \cdot \dfrac{g^{\alpha\beta}}{2} \cdot \left(\frac{n}{2} - 2 \right) F^{n/2-3} f_{,\beta} F'^2 f_{,\alpha} \cdot g^{\kappa\lambda} \\[2em] + \sqrt{-g} \cdot \dfrac{g^{\alpha\beta}{}_{,\beta}}{2} \cdot F^{n/2-2} F' f_{,\alpha} \cdot g^{\kappa\lambda} \\[2em] + \sqrt{-g} \cdot \dfrac{g^{\xi\zeta} g_{\xi\zeta,\beta}}{2} \cdot \dfrac{g^{\alpha\beta}}{2} \cdot F^{n/2-2} F' f_{,\alpha} \cdot g^{\kappa\lambda} \end{pmatrix} \right) .$$

$$(769)$$

This could be a bit simplified:

$$0 = \frac{\sqrt{-g \cdot F^n}}{F} \cdot \left(\begin{array}{c} R^{\kappa\lambda} - \dfrac{R}{2} g^{\kappa\lambda} \\[2mm] + \left(g^{\kappa\lambda} g^{\alpha\beta} - g^{\kappa\alpha} g^{\lambda\beta} - g^{\lambda\alpha} g^{\kappa\beta} \right)(1-n) \\[2mm] \left[\dfrac{f_{,\alpha} f_{,\beta}}{F^2} \left(FF'' + (F')^2 \cdot \dfrac{3 \cdot n - 14}{8} \right) + f_{,\alpha\beta} \dfrac{F'}{F} \right] \\[2mm] - \dfrac{F' \cdot (1-n)}{2 \cdot F} \left(g^{\kappa\lambda} g^{\alpha\beta} - g^{\kappa\alpha} g^{\lambda\beta} - g^{\lambda\alpha} g^{\kappa\beta} \right) \Gamma^{\gamma}_{\alpha\beta} f_{,\gamma} \\[2mm] + \dfrac{F'}{F} \cdot (1-n) g^{\alpha\beta} g^{\gamma\lambda} \Gamma^{\kappa}_{\alpha\beta} f_{,\gamma} \\[2mm] + \dfrac{F'}{F} \cdot \dfrac{(1-n)}{2} \left(g^{\kappa\lambda} g^{\alpha\beta} - g^{\kappa\alpha} g^{\lambda\beta} - g^{\lambda\alpha} g^{\kappa\beta} \right)_{,\beta} f_{,\alpha} \\[2mm] + \dfrac{g^{\xi\zeta} g_{\xi\zeta,\beta}}{2} \cdot \left(\dfrac{F'}{F} \cdot \dfrac{(1-n)}{2} \left(g^{\kappa\lambda} g^{\alpha\beta} - g^{\kappa\alpha} g^{\lambda\beta} - g^{\lambda\alpha} g^{\kappa\beta} \right) f_{,\alpha} \right) \\[2mm] + \left(\dfrac{n}{2} - 1 \right) \left(\begin{array}{c} + g^{\alpha\beta} \dfrac{F'}{F} \left(\dfrac{f_{,\alpha} \cdot g^{\xi\lambda} \Gamma^{\kappa}_{\beta\xi} + f_{,\alpha} \cdot g^{\kappa\xi} \Gamma^{\lambda}_{\beta\xi}}{2} + f_{,\gamma} \cdot g^{\gamma\lambda} \Gamma^{\kappa}_{\alpha\beta} \right) \\[2mm] + \dfrac{F'}{F} f_{,\gamma} \cdot \dfrac{\left(g^{\kappa\rho} g^{\gamma\lambda} + g^{\rho\lambda} g^{\gamma\kappa} - g^{\kappa\lambda} g^{\gamma\rho} \right)_{,\rho}}{2} \\[2mm] + \dfrac{F'}{F} f_{,\gamma\rho} \cdot \dfrac{\left(g^{\kappa\rho} g^{\gamma\lambda} + g^{\rho\lambda} g^{\gamma\kappa} - g^{\kappa\lambda} g^{\gamma\rho} \right)}{2} \\[2mm] + \dfrac{F''}{F} f_{,\rho} f_{,\gamma} \cdot \dfrac{\left(g^{\kappa\rho} g^{\gamma\lambda} + g^{\rho\lambda} g^{\gamma\kappa} - g^{\kappa\lambda} g^{\gamma\rho} \right)}{2} \\[2mm] + \left(\dfrac{n}{2} - 2 \right) \dfrac{F'^2}{F^2} f_{,\rho} f_{,\gamma} \cdot \dfrac{\left(g^{\kappa\rho} g^{\gamma\lambda} + g^{\rho\lambda} g^{\gamma\kappa} - g^{\kappa\lambda} g^{\gamma\rho} \right)}{2} \\[2mm] + \dfrac{g^{\xi\zeta} g_{\xi\zeta,\rho}}{2} \dfrac{F'}{F} f_{,\gamma} \cdot \dfrac{\left(g^{\kappa\rho} g^{\gamma\lambda} + g^{\rho\lambda} g^{\gamma\kappa} - g^{\kappa\lambda} g^{\gamma\rho} \right)}{2} \\[2mm] - \dfrac{F'}{F} \dfrac{g^{\alpha\beta}}{2} \cdot f_{,\alpha} \cdot g^{\kappa\lambda}_{,\beta} - \dfrac{F'}{F} \dfrac{g^{\alpha\beta}}{2} \cdot f_{,\alpha\beta} \cdot g^{\kappa\lambda} - \dfrac{F''}{F} \dfrac{g^{\alpha\beta}}{2} \cdot f_{,\beta} f_{,\alpha} \cdot g^{\kappa\lambda} \\[2mm] - \dfrac{g^{\alpha\beta}}{2} \cdot \left(\dfrac{n}{2} - 2 \right) \dfrac{F'^2}{F^2} f_{,\beta} f_{,\alpha} \cdot g^{\kappa\lambda} - \dfrac{F'}{F} \dfrac{g^{\alpha\beta}_{,\beta}}{2} \cdot f_{,\alpha} \cdot g^{\kappa\lambda} \\[2mm] - \dfrac{F'}{F} \dfrac{g^{\xi\zeta} g_{\xi\zeta,\beta}}{2} \dfrac{g^{\alpha\beta}}{2} \cdot f_{,\alpha} \cdot g^{\kappa\lambda} \end{array} \right) \end{array} \right).$$

(770)

Now, with the help of the work of other authors (c.f. section "8.6.2 About a More General Kernel within the Einstein–Hilbert Action"), we can derive some further simplifications:

$$
\Rightarrow \quad 0 = \left\{
\begin{aligned}
& R^{\kappa\lambda} - \frac{1}{2}Rg^{\kappa\lambda} + \left(\left(\begin{array}{c} \underbrace{\frac{8\pi G}{\kappa}T^{\kappa\lambda}} = \text{matter} \\ 0 = \text{vacuum} \end{array}\right) + \Lambda \cdot g^{\kappa\lambda}\right) \\[6pt]
& \quad -\frac{1}{\sqrt{F^n}} \cdot \left(\partial^{;\kappa}\partial^{;\lambda} - g^{\kappa\lambda}\Delta_g\right)F^{\frac{n}{2}-1} \\[6pt]
& \quad + \left(g^{\kappa\lambda}g^{\alpha\beta} - g^{\kappa\alpha}g^{\lambda\beta} - g^{\lambda\alpha}g^{\kappa\beta}\right)(1-n) \\[6pt]
& \quad \cdot \left[\frac{f_{,\alpha}f_{,\beta}}{F^2}\left(FF'' + (F')^2 \cdot \frac{3\cdot n - 14}{8}\right) + f_{,\alpha\beta}\frac{F'}{F}\right] \\[6pt]
& \quad - \frac{F'\cdot(1-n)}{2\cdot F}\left(g^{\kappa\lambda}g^{\alpha\beta} - g^{\kappa\alpha}g^{\lambda\beta} - g^{\lambda\alpha}g^{\kappa\beta}\right)\Gamma^{\gamma}_{\alpha\beta}f_{,\gamma} \\[6pt]
& \quad + \frac{F'}{F}\cdot(1-n)g^{\alpha\beta}g^{\gamma\lambda}\Gamma^{\kappa}_{\alpha\beta}f_{,\gamma} \\[6pt]
& \quad + \frac{F'}{F}\cdot\frac{(1-n)}{2}\left(g^{\kappa\lambda}g^{\alpha\beta} - g^{\kappa\alpha}g^{\lambda\beta} - g^{\lambda\alpha}g^{\kappa\beta}\right)_{,\beta}f_{,\alpha} \\[6pt]
& \quad + \frac{g^{\xi\zeta}g_{\xi\zeta,\beta}}{2}\cdot\left(\frac{F'}{F}\cdot\frac{(1-n)}{2}\left(g^{\kappa\lambda}g^{\alpha\beta} - g^{\kappa\alpha}g^{\lambda\beta} - g^{\lambda\alpha}g^{\kappa\beta}\right)f_{,\alpha}\right).
\end{aligned}
\right. \tag{771}
$$

However, this author is sure that there are still shorter and simpler forms, only we'd rather like to consider an illustrative example than further stressing the point of structurally shaping Eq. (770) or (771) (for this c.f. chapter 11). Therefore, we will choose a metric of constants (like the Minkowski metric perhaps) and consider this in the next section(s).

8.3 Metric of Constants as an Example

Assuming metrics consisting only of constants, which make all terms with derivatives on the metric tensor to disappear (including all Levi–Civita connections), give us from Eq. (770):

$$0 = \frac{\sqrt{-g} \cdot F^n}{F} \cdot \begin{pmatrix} 0 \\ +\left(g^{\kappa\lambda} g^{\alpha\beta} - g^{\kappa\alpha} g^{\lambda\beta} - g^{\lambda\alpha} g^{\kappa\beta}\right)(1-n) \\ \cdot\left[\frac{f_{,\alpha} f_{,\beta}}{F^2}\left(FF'' + (F')^2 \cdot \frac{3 \cdot n - 14}{8}\right) + f_{,\alpha\beta} \frac{F'}{F}\right] \\ +0 \end{pmatrix}$$

$$+0$$

$$+\left(\sqrt{-g} \cdot F^{n/2-1}_{,\gamma,\rho} \cdot \frac{\left(g^{\kappa\rho} g^{\gamma\lambda} + g^{\rho\lambda} g^{\gamma\kappa} - g^{\kappa\lambda} g^{\gamma\rho}\right)}{2}\right) - \left(\sqrt{-g} \cdot \frac{g^{\alpha\beta}}{2} F^{n/2-1}_{,\alpha,\beta} \cdot g^{\kappa\lambda}\right),$$

$$(772)$$

$$0 = \frac{\sqrt{-g} \cdot F^n}{F} \cdot \left(g^{\kappa\lambda} g^{\alpha\beta} - g^{\kappa\alpha} g^{\lambda\beta} - g^{\lambda\alpha} g^{\kappa\beta}\right)(1-n)$$

$$\cdot\left[\frac{f_{,\alpha} f_{,\beta}}{F^2}\left(FF'' + (F')^2 \cdot \frac{3 \cdot n - 14}{8}\right) + f_{,\alpha\beta} \frac{F'}{F}\right]$$

$$+\begin{pmatrix} \sqrt{-g} \cdot \left(\frac{n}{2}-1\right)\left(F' F^{n/2-2} \cdot f_{,\gamma\rho} + \left(F'' F^{n/2-2} + \left(\frac{n}{2}-2\right)F'^2 F^{n/2-3}\right) \cdot f_{,\rho} f_{,\gamma}\right) \\ \cdot\frac{\left(g^{\kappa\rho} g^{\gamma\lambda} + g^{\rho\lambda} g^{\gamma\kappa} - g^{\kappa\lambda} g^{\gamma\rho}\right)}{2} \end{pmatrix}$$

$$-\left(\sqrt{-g} \cdot \frac{g^{\alpha\beta}}{2}\left(\frac{n}{2}-1\right)\left(F' F^{n/2-2} \cdot f_{,\alpha\beta} + \left(F'' F^{n/2-2} + \left(\frac{n}{2}-2\right)F'^2 F^{n/2-3}\right) \cdot f_{,\alpha} f_{,\beta}\right) \cdot g^{\kappa\lambda}\right)$$

$$\xrightarrow{\text{cleaning up dummies}}$$

$$= \frac{\sqrt{-g} \cdot F^n}{F} \cdot \left(g^{\kappa\lambda} g^{\alpha\beta} - g^{\kappa\alpha} g^{\lambda\beta} - g^{\lambda\alpha} g^{\kappa\beta}\right)(1-n)$$

$$\cdot\left[\frac{f_{,\alpha} f_{,\beta}}{F^2}\left(FF'' + (F')^2 \cdot \frac{3 \cdot n - 14}{8}\right) + f_{,\alpha\beta} \frac{F'}{F}\right]$$

$$-\left(\sqrt{-g} \cdot \left(\frac{n}{2}-1\right)\left(F' F^{n/2-2} \cdot f_{,\alpha\beta} + \left(\frac{n}{2}-2\right)F^{n/2-3} F'^2 \cdot f_{,\alpha} f_{,\beta}\right) \cdot \frac{\left(g^{\kappa\lambda} g^{\alpha\beta} - g^{\kappa\beta} g^{\alpha\lambda} - g^{\beta\lambda} g^{\alpha\kappa}\right)}{2}\right)$$

$$-\left(\sqrt{-g} \cdot \frac{g^{\alpha\beta}}{2}\left(\frac{n}{2}-1\right)\left(F' F^{n/2-2} \cdot f_{,\alpha\beta} + \left(\frac{n}{2}-2\right)F^{n/2-3} F'^2 \cdot f_{,\alpha} f_{,\beta}\right) \cdot g^{\kappa\lambda}\right), \quad (773)$$

$$= \frac{\sqrt{-g \cdot F^n}}{F} \cdot \left(g^{\kappa\lambda} g^{\alpha\beta} - g^{\kappa\alpha} g^{\lambda\beta} - g^{\lambda\alpha} g^{\kappa\beta} \right) (1-n)$$

$$\cdot \left[\frac{f_{,\alpha} f_{,\beta}}{F^2} \left(FF'' + \left(F' \right)^2 \cdot \frac{3 \cdot n - 14}{8} \right) + f_{,\alpha\beta} \frac{F'}{F} \right]$$

$$- \left(\begin{array}{c} \sqrt{-g} \cdot \left(\frac{n}{2} - 1 \right) \left(F' F^{n/2-2} \cdot f_{,\alpha\beta} + \left(F'' F^{n/2-2} + \left(\frac{n}{2} - 2 \right) F'^2 F^{n/2-3} \right) \cdot f_{,\alpha} f_{,\beta} \right) \\ \cdot \dfrac{\left(2g^{\kappa\lambda} g^{\alpha\beta} - g^{\kappa\beta} g^{\alpha\lambda} - g^{\beta\lambda} g^{\alpha\kappa} \right)}{2} \end{array} \right),$$

$$(774)$$

$$\Rightarrow 0 = \left\{ \begin{array}{l} - \left(g^{\kappa\lambda} g^{\alpha\beta} - g^{\kappa\alpha} g^{\lambda\beta} - g^{\lambda\alpha} g^{\kappa\beta} \right)(1-n) \cdot \left[\dfrac{f_{,\alpha} f_{,\beta}}{F^2} \left(FF'' + \left(F' \right)^2 \cdot \dfrac{3 \cdot n - 14}{8} \right) \right] \\[2ex] + (n-2) \left(2 \cdot F'' F + (n-4) \left(F' \right)^2 \right) \cdot \dfrac{f_{,\alpha} f_{,\beta}}{8F^2} \left(2g^{\kappa\lambda} g^{\alpha\beta} - g^{\kappa\beta} g^{\alpha\lambda} - g^{\beta\lambda} g^{\alpha\kappa} \right) \\[2ex] + \dfrac{F' \cdot f_{,\alpha\beta}}{4F} \left((n-2) g^{\kappa\lambda} g^{\alpha\beta} + \left(g^{\kappa\lambda} g^{\alpha\beta} - g^{\kappa\beta} g^{\alpha\lambda} - g^{\beta\lambda} g^{\alpha\kappa} \right) (5 \cdot n - 6) \right) \end{array} \right.$$

$$= \left\{ \begin{array}{l} - \left(g^{\kappa\lambda} g^{\alpha\beta} - g^{\kappa\alpha} g^{\lambda\beta} - g^{\lambda\alpha} g^{\kappa\beta} \right)(1-n) \cdot \dfrac{f_{,\alpha} f_{,\beta}}{F^2} FF'' \\[2ex] + (n-2) F'' F \cdot \dfrac{f_{,\alpha} f_{,\beta}}{4F^2} \left(2g^{\kappa\lambda} g^{\alpha\beta} - g^{\kappa\beta} g^{\alpha\lambda} - g^{\beta\lambda} g^{\alpha\kappa} \right) \\[2ex] + \dfrac{\left(F' \right)^2}{8F^2} \cdot f_{,\alpha} f_{,\beta} \left(\begin{array}{c} (n-2)(n-4) g^{\kappa\lambda} g^{\alpha\beta} + \left(g^{\kappa\lambda} g^{\alpha\beta} - g^{\kappa\beta} g^{\alpha\lambda} - g^{\beta\lambda} g^{\alpha\kappa} \right) \\ \cdot \left[(n-2)(n-4) - (1-n) \cdot (3 \cdot n - 14) \right] \end{array} \right) \\[2ex] + \dfrac{F' \cdot f_{,\alpha\beta}}{4F} \left((n-2) g^{\kappa\lambda} g^{\alpha\beta} + \left(g^{\kappa\lambda} g^{\alpha\beta} - g^{\kappa\beta} g^{\alpha\lambda} - g^{\beta\lambda} g^{\alpha\kappa} \right) (5 \cdot n - 6) \right) \end{array} \right.$$

$$= \left\{ \begin{array}{l} \dfrac{f_{,\alpha} f_{,\beta}}{4 \cdot F^2} FF'' \left(g^{\kappa\lambda} g^{\alpha\beta} \cdot (6 \cdot n - 8) - \left(g^{\kappa\beta} g^{\alpha\lambda} + g^{\beta\lambda} g^{\alpha\kappa} \right) (5 \cdot n - 6) \right) \\[2ex] + \dfrac{\left(F' \right)^2}{8F^2} \cdot f_{,\alpha} f_{,\beta} \left(\begin{array}{c} (n-2)(n-4) g^{\kappa\lambda} g^{\alpha\beta} + \left(g^{\kappa\lambda} g^{\alpha\beta} - g^{\kappa\beta} g^{\alpha\lambda} - g^{\beta\lambda} g^{\alpha\kappa} \right) \\ \times [22 + n \cdot (4 \cdot n - 23)] \end{array} \right) \\[2ex] + \dfrac{F' \cdot f_{,\alpha\beta}}{4F} \left((n-2) g^{\kappa\lambda} g^{\alpha\beta} + \left(g^{\kappa\lambda} g^{\alpha\beta} - g^{\kappa\beta} g^{\alpha\lambda} - g^{\beta\lambda} g^{\alpha\kappa} \right) (5 \cdot n - 6) \right) \end{array} \right.$$

$$
= \begin{cases}
\dfrac{f_{,\alpha} f_{,\beta}}{4 \cdot F^2} F F'' \left(g^{\kappa\lambda} g^{\alpha\beta} \cdot (6 \cdot n - 8) - \left(g^{\kappa\beta} g^{\alpha\lambda} + g^{\beta\lambda} g^{\alpha\kappa} \right) (5 \cdot n - 6) \right) \\[2mm]
+ \dfrac{(F')^2}{8F^2} \cdot f_{,\alpha} f_{,\beta} \left(\begin{array}{l} ((n-2)(n-4) + 22 + n \cdot (4 \cdot n - 23)) g^{\kappa\lambda} g^{\alpha\beta} \\ - \left(g^{\kappa\beta} g^{\alpha\lambda} + g^{\beta\lambda} g^{\alpha\kappa} \right) \times (22 + n \cdot (4 \cdot n - 23)) \end{array} \right) \\[3mm]
+ \dfrac{F' \cdot f_{,\alpha\beta}}{4F} \left((n-2) g^{\kappa\lambda} g^{\alpha\beta} + \left(g^{\kappa\lambda} g^{\alpha\beta} - g^{\kappa\beta} g^{\alpha\lambda} - g^{\beta\lambda} g^{\alpha\kappa} \right) (5 \cdot n - 6) \right).
\end{cases}
$$

$$(775)$$

While we obtain nothing very exciting in the case of $F[f] = f$:

$$\xrightarrow{\quad F[f]=f \quad}$$

$$
0 = \begin{cases}
\dfrac{f_{,\alpha} f_{,\beta}}{2 \cdot f} \cdot \left((n-2)(n-4) g^{\kappa\lambda} g^{\alpha\beta} + \left(g^{\kappa\lambda} g^{\alpha\beta} - g^{\kappa\beta} g^{\alpha\lambda} - g^{\beta\lambda} g^{\alpha\kappa} \right) \times [22 + n \cdot (4 \cdot n - 23)] \right) \\[2mm]
+ f_{,\alpha\beta} \left((n-2) g^{\kappa\lambda} g^{\alpha\beta} + \left(g^{\kappa\lambda} g^{\alpha\beta} - g^{\kappa\beta} g^{\alpha\lambda} - g^{\beta\lambda} g^{\alpha\kappa} \right) (5 \cdot n - 6) \right),
\end{cases}
$$

$$(776)$$

we find some interesting options in 2 dimensions:

$$\xrightarrow{\quad n=2 \quad}$$

$$
\Rightarrow \qquad 0 = \begin{cases}
\left(g^{\kappa\lambda} g^{\alpha\beta} - g^{\kappa\alpha} g^{\lambda\beta} - g^{\lambda\alpha} g^{\kappa\beta} \right) \cdot \left[\dfrac{f_{,\alpha} f_{,\beta}}{F^2} \left(F F'' - (F')^2 \right) \right] \\[3mm]
+ \dfrac{F' \cdot f_{,\alpha\beta}}{F} \left(g^{\kappa\lambda} g^{\alpha\beta} - g^{\kappa\beta} g^{\alpha\lambda} - g^{\beta\lambda} g^{\alpha\kappa} \right).
\end{cases}
$$

$$(777)$$

As we could also demand the Nabla-kind operator terms to vanish via a suitable $F[f]$:

$$\xrightarrow{\quad n=2; \;\; FF'' - (F')^2 = 0 \;\; \Rightarrow \;\; F[f] = C_{f1} \cdot e^{f \cdot C_{f0}} \quad}$$

$$
\Rightarrow \qquad \begin{cases}
0 = F' \cdot f_{,\alpha\beta} \cdot \left(g^{\kappa\lambda} g^{\alpha\beta} - g^{\kappa\beta} g^{\alpha\lambda} - g^{\beta\lambda} g^{\alpha\kappa} \right) \\[2mm]
\to 0 = C_{f0} \cdot f_{,\alpha\beta} \cdot \left(g^{\kappa\lambda} g^{\alpha\beta} - g^{\kappa\beta} g^{\alpha\lambda} - g^{\beta\lambda} g^{\alpha\kappa} \right),
\end{cases}
$$

$$(778)$$

we end up with three possible "options for the solution," namely:

$$
\Rightarrow \qquad \left\{ C_{f0} = 0; \quad f_{,\alpha\beta} = 0; \quad g^{\kappa\lambda} g^{\alpha\beta} - \left(g^{\kappa\beta} g^{\alpha\lambda} + g^{\beta\lambda} g^{\alpha\kappa} \right) = 0^{\kappa\beta\alpha\lambda} \right., \qquad (779)
$$

where we explicitly used the "..."-signs in order to point out that—of course—setting the metric ensemble to zero isn't really a solution.

Ignoring the metric option (which appears rather strange anyway) and remembering that in a metric of constants, when also having $f_{,\alpha\beta} = 0$, we end up—so it seems—with apparently meaningless linear functions as solutions in the case of $n = 2$ and $F[f] = C_{f1} \cdot e^{f \cdot C_{f0}}$.

Thus, overall, it seems that our equation does not allow for sufficiently interesting cases for the function f.

The way out—so our current conclusion—would be the assumption of an internal structure for the function f as it is was introduced in the previous chapters 5 and 6, but before discussing this possibility here, we want to get a somewhat clearer picture about the whole structure in general.

8.3.1 A Few Options to Think About

In order to develop the idea of an internally structured function f, we start with a few general—almost brain-storming like—approaches.

Perhaps, we should look for a solution where there is a connection between the derivatives of f and the metric. Maybe something as follows (here in the case of $n = 4$):

$$C \cdot g^{\kappa\lambda} f^{,\beta} f_{,\beta} = f^{,\kappa} f^{,\lambda} \quad \& \quad D \cdot g^{\kappa\lambda} f_{,\beta}{}^{\beta} = f^{,\kappa\lambda}$$

$$\Rightarrow \quad 0 = \begin{cases} 3 \cdot \left(g^{\kappa\lambda} f^{,\beta} f_{,\beta} - 2 \cdot C \cdot g^{\kappa\lambda} f^{,\beta} f_{,\beta} \right) \cdot \dfrac{1}{F^2} \left(FF'' - \dfrac{(F')^2}{4} \right) \\ + F'' \cdot \dfrac{f^{,\beta} f_{,\beta}}{F} (1 - C) \cdot g^{\kappa\lambda} + \dfrac{F'}{2F} \left(g^{\kappa\lambda} f_{,\beta}{}^{\beta} + 7 \cdot \left(g^{\kappa\lambda} f_{,\beta}{}^{\beta} - 2 \cdot D \cdot g^{\kappa\lambda} f_{,\beta}{}^{\beta} \right) \right) \end{cases}$$

$$0 = \frac{g^{\kappa\lambda} f^{,\beta} f_{,\beta}}{F^2} \left((4 - 7 \cdot C) \cdot FF'' - 3 \cdot (1 - 2 \cdot C) \cdot \frac{(F')^2}{4} \right) + \frac{F'}{2F} g^{\kappa\lambda} f_{,\beta}{}^{\beta} (8 - 14 \cdot D)$$

$$0 = \frac{f^{,\beta} f_{,\beta}}{F^2} \left((4 - 7 \cdot C) \cdot FF'' - 3 \cdot (1 - 2 \cdot C) \cdot \frac{(F')^2}{4} \right) + \frac{F'}{2F} f_{,\beta}{}^{\beta} (8 - 14 \cdot D).$$

$$(780)$$

In the general *n* case, this would look like:

$$0 = g^{\kappa\lambda} \left(f^{,\beta} f_{,\beta} \begin{bmatrix} (1-2\cdot C)(n-1)\cdot\dfrac{F''}{F}+(n-2)\dfrac{F''}{2F^2}(1-C) \\[2mm] +\dfrac{(F')^2}{8F^2}\left((n-2)(n-4)+(1-2\cdot C)\cdot[22+n\cdot(4\cdot n-23)]\right) \\[2mm] +\dfrac{F'\cdot f_{,\beta}{}^{\beta}}{4F}\left((n-2)+(1-2\cdot D)(5\cdot n-6)\right) \end{bmatrix} \right)$$

$$\Rightarrow \quad 0 = \left\{ f^{,\beta} f_{,\beta} \begin{bmatrix} \left((1-2\cdot C)(n-1)+\dfrac{(n-2)}{2}(1-C)\right)\cdot\dfrac{F''}{F} \\[2mm] +\dfrac{(F')^2}{8F^2}\left((n-2)(n-4)+(1-2\cdot C)\cdot[22+n\cdot(4\cdot n-23)]\right) \\[2mm] +\dfrac{F'\cdot f_{,\beta}{}^{\beta}}{4F}\left((n-2)+(1-2\cdot D)(5\cdot n-6)\right) \end{bmatrix} \right.$$

$$= \left\{ f^{,\beta} f_{,\beta} \begin{bmatrix} \left(C\cdot(6-5\cdot n)+3\cdot n-4\right)\cdot\dfrac{F''}{2\cdot F} \\[2mm] +\dfrac{(F')^2}{8F^2}\left((n-2)(n-4)+(1-2\cdot C)\cdot[22+n\cdot(4\cdot n-23)]\right) \\[2mm] +\dfrac{F'\cdot f_{,\beta}{}^{\beta}}{4F}\left((n-2)+(1-2\cdot D)(5\cdot n-6)\right) \end{bmatrix} \right. \quad (781)$$

Driving the idea a bit further leads to:

$$f_{,\alpha} f_{,\beta} = C[x_i]^2 \cdot g_{\alpha\beta} = C[x_i]\cdot e_\alpha \cdot C[x_i]\cdot e_\beta$$

$$\Rightarrow \qquad f_{,\alpha} = C[x_i]\cdot e_\alpha$$

$$\Rightarrow \qquad f_{,\alpha\beta} = \left(C[x_i]\cdot e_\alpha\right)_{,\beta}$$

$$\Rightarrow \quad 0 = \left\{ \begin{array}{l} \dfrac{C[x_i]^2\cdot g_{\alpha\beta}}{4\cdot F^2}FF''\left(g^{\kappa\lambda}g^{\alpha\beta}\cdot(6\cdot n-8)-\left(g^{\kappa\beta}g^{\alpha\lambda}+g^{\beta\lambda}g^{\alpha\kappa}\right)(5\cdot n-6)\right) \\[3mm] +\dfrac{(F')^2}{8F^2}\cdot C[x_i]^2\cdot g_{\alpha\beta}\left(\begin{array}{l}((n-2)(n-4)+22+n\cdot(4\cdot n-23))g^{\kappa\lambda}g^{\alpha\beta} \\ -\left(g^{\kappa\beta}g^{\alpha\lambda}+g^{\beta\lambda}g^{\alpha\kappa}\right)\times(22+n\cdot(4\cdot n-23))\end{array}\right) \\[5mm] +\dfrac{F'\cdot f_{,\alpha\beta}}{4F}\left((n-2)g^{\kappa\lambda}g^{\alpha\beta}+\left(g^{\kappa\lambda}g^{\alpha\beta}-g^{\kappa\beta}g^{\alpha\lambda}-g^{\beta\lambda}g^{\alpha\kappa}\right)(5\cdot n-6)\right) \end{array} \right.$$

$$
=\left\{
\begin{array}{l}
\dfrac{C\left[x_i\right]^2 \cdot g^{\kappa\lambda}}{2\cdot F^2} FF''\left(2\cdot(6\cdot n-8)-(5\cdot n-6)\right) \\[3mm]
+\dfrac{\left(F'\right)^2}{4F^2}\cdot C\left[x_i\right]^2\cdot g^{\kappa\lambda}\left(
\begin{array}{l}
2\cdot\left((n-2)(n-4)+22+n\cdot(4\cdot n-23)\right) \\
-\left(22+n\cdot(4\cdot n-23)\right)
\end{array}
\right) \\[3mm]
+\dfrac{F'\cdot f_{,\alpha\beta}}{4F}\left((n-2)g^{\kappa\lambda}g^{\alpha\beta}+\left(g^{\kappa\lambda}g^{\alpha\beta}-g^{\kappa\beta}g^{\alpha\lambda}-g^{\beta\lambda}g^{\alpha\kappa}\right)(5\cdot n-6)\right).
\end{array}
\right.
$$

$$(782)$$

This gives us two equations, namely one for $F[f]$ and the other one for f:

$$
0=\frac{\left(F'\right)^2}{4F^2}\cdot\left(38+n\cdot(6\cdot n-35)\right)+\left(7\cdot n-10\right)\cdot\frac{F''}{F}
$$

$$
0=\frac{F'\cdot f_{,\alpha\beta}}{4F}\left((6\cdot n-8)g^{\kappa\lambda}g^{\alpha\beta}-\left(g^{\kappa\beta}g^{\alpha\lambda}+g^{\beta\lambda}g^{\alpha\kappa}\right)(5\cdot n-6)\right). \quad (783)
$$

The first equation can easily be solved and gives:

$$
F[f]=C_{f1}\cdot\left(f-C_{f0}\right)^{\frac{40-28\cdot n}{2+(7-6\cdot n)\cdot n}}. \quad (784)
$$

Looking closely at the second equation and assuming an internally (one may call this Dirac like) structured function f, one recognizes the features of the fundamental equation of linear elasticity [9]. We can rewrite the last line in Eq. (783) as follows:

$$
0=\frac{F'}{4F}\left((6\cdot n-8)g^{\kappa\lambda}f_{,\beta}{}^{\beta}-\left(f^{,\kappa\lambda}+f^{,\lambda\kappa}\right)(5\cdot n-6)\right). \quad (785)
$$

Giving f now an internal structure like:

$$
f=Q^{\gamma}\phi_{\gamma}, \quad (786)
$$

where we assume Q^{γ} to be a vector of constants, leads us to:

$$
\begin{aligned}
0&=\frac{F'}{4F}\left((6\cdot n-8)g^{\kappa\lambda}Q^{\gamma}\phi_{\gamma,\beta}{}^{\beta}-\left(Q^{\gamma}\phi_{\gamma}{}^{,\kappa\lambda}+Q^{\gamma}\phi_{\gamma}{}^{,\lambda\kappa}\right)(5\cdot n-6)\right) \\
&=\frac{F'}{4F}\left((6\cdot n-8)g^{\kappa\lambda}Q^{\gamma}\phi_{\gamma,\beta}{}^{\beta}-\left(q^{\kappa}h_{\gamma}{}^{,\gamma\lambda}+q^{\lambda}h_{\gamma}{}^{,\gamma\kappa}\right)(5\cdot n-6)\right). \quad (787)
\end{aligned}
$$

Comparison with the fundamental (load free) equation of linear elasticity, which in the isotropic case with Poisson's ratio μ and the displacement vector $\mathbf{h}_{\gamma}$ reads[4] (c.f. [9], p. 166):

---

[4]It should be noted that in the original equation [9] the non-Laplacian term has a positive sign, but this is not of interest to the shear solution we are discussing here.

$$0 = 2 \cdot (1 - 2 \cdot \mu) h_{\gamma,\beta}{}^{\beta} - \left(h_{\gamma}{}^{,\gamma\lambda} + h_{\gamma}{}^{,\gamma\kappa} \right), \tag{788}$$

directly does not only lead us to a great variety of possible solutions, but also to some rather interesting analogies. One of these solutions is most interesting for our case here, namely the pure shear stress solution with the example (in Cartesian coordinates):

$$h_{\gamma} = \left\{ -\partial_y, \partial_x, 0 \right\} \psi \quad \text{with:} \quad \psi_{,\beta}{}^{\beta} = 0. \tag{789}$$

Applied onto Eq. (787), taking $n = 4$ or just the function f to depend on 4 dimensions, we may set:

$$h_{\gamma} = D_{\gamma}^{\sigma} \phi_{\sigma} = \left\{ 0, -\partial_y, \partial_x, 0 \right\} \psi \quad \text{with:} \quad \psi_{,\beta}{}^{\beta} = 0. \tag{790}$$

The charm of this type of solution to Eq. (787) could be seen in the fact that it provides the characteristics of a spin field. Thus, a spin in elementary particle physics would become a shear field in the quantum Einstein field equations.

It appears entertaining to evaluate the corresponding Poisson's ratio with the ones we'd effectively obtain from Eq. (788). At first, we ignore the sign-difference between the original elastic equations and our form (787). The result would be a rather glass-like behavior for all dimensions, because we find:

$$\mu = \frac{n-1}{5 \cdot n - 6}. \tag{791}$$

As we will see in other derivations, presented below and leading to elastic equations, too, the case $n = 1$, where a Poisson's ratio apparently makes no sense, we also result in $\mu = 0$, while for $n \rightarrow \infty$ we get $\mu = 0.2$, which would be a glass-like behavior, as we have already mentioned. Not ignoring the sign-difference, however, the Poisson's ratio varies between 0.75 for $n = 2$ and 0.8 for $n \rightarrow \infty$, while we obtain a strange $\mu = 1$ in the case of $n = 1$. At this point it should be noted, however, that this peculiar result could just be a product of our choice for $F[f]$. So, we check other paths, also leading to elastic equations. Going back to Eq. (775) and setting $n = 1$, we obtain:

$$\Rightarrow \quad 0 = \begin{cases} -\left(2 \cdot F''F - 3 \cdot \left(F' \right)^2 \right) \cdot \dfrac{f_{,\alpha} f_{,\beta}}{8F^2} \left(2 g^{\kappa\lambda} g^{\alpha\beta} - g^{\kappa\beta} g^{\alpha\lambda} - g^{\beta\lambda} g^{\alpha\kappa} \right) \\[4mm] + \dfrac{F' \cdot f_{,\alpha\beta}}{4F} \left(-g^{\kappa\lambda} g^{\alpha\beta} - \left(g^{\kappa\lambda} g^{\alpha\beta} - g^{\kappa\beta} g^{\alpha\lambda} - g^{\beta\lambda} g^{\alpha\kappa} \right) \right) \end{cases}$$

$$\xrightarrow{\quad 2 \cdot F''F - 3 \cdot \left(F' \right)^2 = 0 \;\; \Rightarrow \;\; F[f] = C_{f1} \left(f - C_{f0} \right)^{-2} \quad}$$

$$0 = \frac{F' \cdot f_{,\alpha\beta}}{4F} \left(-g^{\kappa\lambda} g^{\alpha\beta} - \left(g^{\kappa\lambda} g^{\alpha\beta} - g^{\kappa\beta} g^{\alpha\lambda} - g^{\beta\lambda} g^{\alpha\kappa} \right) \right)$$

$$\Rightarrow \qquad 0 = f_{,\alpha\beta}\left(2 \cdot g^{\kappa\lambda}g^{\alpha\beta} - \left(g^{\kappa\beta}g^{\alpha\lambda} + g^{\beta\lambda}g^{\alpha\kappa}\right)\right)$$

$$= f_{,\alpha\beta}\left(2 \cdot (1 - 2 \cdot \mu)g^{\kappa\lambda}g^{\alpha\beta} + \left(g^{\kappa\beta}g^{\alpha\lambda} + g^{\beta\lambda}g^{\alpha\kappa}\right)\right)$$

$$\Rightarrow \qquad\qquad \mu = 1. \qquad\qquad (792)$$

Thus, the peculiar result would still be the same.

8.3.2 Separating the Metric from the Quantum Part: First Simple Trials

As with $F[f]$ and f itself, we have the possibility to adjust two functions, we may also separate the metric terms in such a way that only equations with either F or f in it have to be solved. This gives us the following approach:

$$\Rightarrow \quad 0 = \left\{ \begin{array}{l} \dfrac{f_{,\alpha}f_{,\beta}}{4 \cdot F^2}FF''\left(g^{\kappa\lambda}g^{\alpha\beta} \cdot (6 \cdot n - 8) - \left(g^{\kappa\beta}g^{\alpha\lambda} + g^{\beta\lambda}g^{\alpha\kappa}\right)(5 \cdot n - 6)\right) \\[3mm] +\dfrac{\left(F'\right)^2}{8F^2}\cdot f_{,\alpha}f_{,\beta}\left(\begin{array}{l}\left((n-2)(n-4)+22+n\cdot(4\cdot n-23)\right)g^{\kappa\lambda}g^{\alpha\beta} \\ -\left(g^{\kappa\beta}g^{\alpha\lambda}+g^{\beta\lambda}g^{\alpha\kappa}\right)\times(22+n\cdot(4\cdot n-23))\end{array}\right) \\[5mm] +\dfrac{F' \cdot f_{,\alpha\beta}}{4F}\left((n-2)g^{\kappa\lambda}g^{\alpha\beta} + \left(g^{\kappa\lambda}g^{\alpha\beta} - g^{\kappa\beta}g^{\alpha\lambda} - g^{\beta\lambda}g^{\alpha\kappa}\right)(5\cdot n-6)\right) \end{array} \right.$$

$$= \left\{ \begin{array}{l} g^{\kappa\lambda}g^{\alpha\beta}\left(\begin{array}{l}\dfrac{\left(F'\right)^2}{8F^2}\cdot f_{,\alpha}f_{,\beta}\left((n-2)(n-4)+[22+n\cdot(4\cdot n-23)]\right) \\[3mm] +(6\cdot n-8)\cdot\dfrac{f_{,\alpha}f_{,\beta}}{4\cdot F^2}FF''+\dfrac{F'\cdot f_{,\alpha\beta}}{4F}\left((n-2)+(5\cdot n-6)\right)\end{array}\right) \\[10mm] -\left(g^{\kappa\alpha}g^{\lambda\beta}+g^{\lambda\alpha}g^{\kappa\beta}\right)\left(\begin{array}{l}(5\cdot n-6)\cdot\dfrac{f_{,\alpha}f_{,\beta}}{4\cdot F^2}FF''+\dfrac{\left(F'\right)^2}{8F^2}\cdot f_{,\alpha}f_{,\beta}\left(22+n\cdot(4\cdot n-23)\right) \\[3mm] +\dfrac{F'\cdot f_{,\alpha\beta}}{4F}(5\cdot n-6)\end{array}\right) \end{array} \right\}.$$

$$(793)$$

Now we boldly want to kick the metrics out completely. In result, we obtain two equations:

$$0 = \begin{cases} \dfrac{(F')^2}{8F^2} \cdot f_{,\alpha} f_{,\beta} \big((n-2)(n-4) + [22 + n \cdot (4 \cdot n - 23)] \big) \\[4mm] + (6 \cdot n - 8) \cdot \dfrac{f_{,\alpha} f_{,\beta}}{4 \cdot F^2} FF'' + \dfrac{F' \cdot f_{,\alpha\beta}}{4F} \big((n-2) + (5 \cdot n - 6) \big) \end{cases}$$

$$0 = \begin{cases} (5 \cdot n - 6) \cdot \dfrac{f_{,\alpha} f_{,\beta}}{4 \cdot F^2} FF'' + \dfrac{(F')^2}{8F^2} \cdot f_{,\alpha} f_{,\beta} \big(22 + n \cdot (4 \cdot n - 23) \big) \\[4mm] + \dfrac{F' \cdot f_{,\alpha\beta}}{4F} (5 \cdot n - 6). \end{cases} \tag{794}$$

Subtracting the second from the first equation gives us:

$$0 = \begin{cases} (6 \cdot n - 8) \cdot \dfrac{f_{,\alpha} f_{,\beta}}{4 \cdot F^2} FF'' + \dfrac{(F')^2}{8F^2} \cdot f_{,\alpha} f_{,\beta} \big((n-2)(n-4) + 22 + n \cdot (4 \cdot n - 23) \big) \\[4mm] + \dfrac{F' \cdot f_{,\alpha\beta}}{4F} \big((n-2) + (5 \cdot n - 6) \big) \end{cases}$$

$$-0 = \begin{cases} (5 \cdot n - 6) \cdot \dfrac{f_{,\alpha} f_{,\beta}}{4 \cdot F^2} FF'' + \dfrac{(F')^2}{8F^2} \cdot f_{,\alpha} f_{,\beta} \big(22 + n \cdot (4 \cdot n - 23) \big) \\[4mm] + \dfrac{F' \cdot f_{,\alpha\beta}}{4F} (5 \cdot n - 6) \end{cases}$$

$$\Rightarrow \quad 0 = (n-2) \cdot \frac{f_{,\alpha} f_{,\beta}}{4 \cdot F^2} FF'' + f_{,\alpha} f_{,\beta} \left(\frac{(F')^2}{8 \cdot F^2} \cdot (n-4)(n-2) \right) + \frac{F' \cdot f_{,\alpha\beta}}{4 \cdot F} (n-2)$$

$$\Rightarrow \qquad 0 = \frac{f_{,\alpha} f_{,\beta}}{4 \cdot F} FF'' + f_{,\alpha} f_{,\beta} \frac{(F')^2}{8 \cdot F} \cdot (n-4) + \frac{F' \cdot f_{,\alpha\beta}}{4}$$

$$\Rightarrow \qquad F' \cdot f_{,\alpha\beta} = \frac{f_{,\alpha} f_{,\beta}}{F} \cdot \left((F')^2 \cdot \frac{(4-n)}{2} - F \cdot F'' \right), \tag{795}$$

and by putting this result back into one of the two equations in Eq. (794), we subsequently have the simple equation:

$$0 = (22 + n \cdot (4 \cdot n - 23)) + (4 - n)(5 \cdot n - 6)$$

$$\Rightarrow \qquad\qquad n = 1, 2. \tag{796}$$

This automatically makes the last line in Eq. (795):

$$n = 2: \quad F' \cdot f_{,\alpha\beta} = \frac{f_{,\alpha} f_{,\beta}}{F} \cdot \left((F')^2 - F \cdot F'' \right)$$

$$n=1: \quad F' \cdot f_{,\alpha\beta} = \frac{f_{,\alpha}f_{,\beta}}{F} \cdot \left(\left(F'\right)^2 \cdot \frac{3}{2} - F \cdot F'' \right). \tag{797}$$

One may consider the two last equations, the quantum part of our quantum Einstein field equations in the simplest case of metrics with constant components (like Minkowski) and the lowest possible number of dimensions, namely $n = 1$ and $n = 2$ (at least as long as we not also allow for fractal space-times).

One point of criticism immediately arises due to the fact that in Eq. (794) the sum over the indices α and β had been completely ignored, which dramatically narrows down the number of options. In order to have a somewhat less rigorous condition than (794), one may also just demand:

$$0 = \left\{ \begin{array}{l} g^{\kappa\lambda} g^{\alpha\beta} \left(\begin{array}{l} \dfrac{\left(F'\right)^2}{8F^2} \cdot f_{,\alpha}f_{,\beta} \left((n-2)(n-4) + 22 + n \cdot (4 \cdot n - 23) \right) \\[3mm] + (6 \cdot n - 8) \cdot \dfrac{f_{,\alpha}f_{,\beta}}{4 \cdot F^2} FF'' + \dfrac{F' \cdot f_{,\alpha\beta}}{4F} \left((n-2) + (5 \cdot n - 6) \right) \end{array} \right) \\[8mm] - (5 \cdot n - 6) \cdot \dfrac{f^{,\lambda}f^{,\kappa}}{2 \cdot F^2} FF'' - \dfrac{\left(F'\right)^2}{4F^2} \cdot f^{,\lambda}f^{,\kappa} \left(22 + n \cdot (4 \cdot n - 23) \right) \\[5mm] \qquad\qquad - \dfrac{F' \cdot f^{,\lambda\kappa}}{2F} (5 \cdot n - 6) \end{array} \right. \tag{798}$$

and try to simplify this via:

$$0 = \frac{f^{,\lambda}f^{,\kappa}}{2 \cdot F^2} \left((5 \cdot n - 6) \cdot FF'' + \frac{\left(F'\right)^2}{2} \cdot (22 + n \cdot (4 \cdot n - 23)) \right)$$

$$\Rightarrow \qquad F[f] = \left(f - C_{f0} \right)^{\frac{2(5n-6)}{(n-2)(4 \cdot n - 5)}} C_{f1}, \tag{799}$$

which gives:

$$0 = \left\{ \begin{array}{l} g^{\kappa\lambda} g^{\alpha\beta} \left(\begin{array}{l} \dfrac{\left(F'\right)^2}{8F^2} \cdot f_{,\alpha}f_{,\beta} \left((n-2)(n-4) + 22 + n \cdot (4 \cdot n - 23) \right) \\[3mm] + (6 \cdot n - 8) \cdot \dfrac{f_{,\alpha}f_{,\beta}}{4 \cdot F^2} FF'' + \dfrac{F' \cdot f_{,\alpha\beta}}{4F} \left((n-2) + (5 \cdot n - 6) \right) \end{array} \right) \\[8mm] \qquad\qquad - \dfrac{F' \cdot f^{,\lambda\kappa}}{2F} (5 \cdot n - 6) \end{array} \right.$$

$$= g^{\kappa\lambda} g^{\alpha\beta} \left(\begin{array}{c} \dfrac{(F')^2}{8F} \cdot f_{,\alpha} f_{,\beta} \big((n-2)(n-4)+22+n\cdot(4\cdot n-23)\big) \\[2mm] +(6\cdot n-8)\cdot \dfrac{f_{,\alpha} f_{,\beta}}{4} F'' + \dfrac{F'\cdot f_{,\alpha\beta}}{4}\big((n-2)+(5\cdot n-6)\big) \end{array} \right) - \dfrac{F'\cdot f^{,\lambda\kappa}}{2}(5\cdot n-6).$$

$$(800)$$

Inserting $F[f]$ from Eq. (799), we find that in $n = 2,1$ the terms

$$\frac{(F')^2}{8F^2}\cdot f_{,\alpha} f_{,\beta}\big((n-2)(n-4)+22+n\cdot(4\cdot n-23)\big)+(6\cdot n-8)\cdot\frac{f_{,\alpha} f_{,\beta}}{4\cdot F^2}FF'' \quad \text{in}$$

Eq. (800) above gives zero due to our solution for $F[f]$ from Eq. (799), leaving us with the total equation:

$$0 = g^{\kappa\lambda} \frac{f_{,\beta}{}^{\beta}}{2}(6\cdot n-8) - f^{,\lambda\kappa}(5\cdot n-6); \quad \text{for} \quad n=2,1. \qquad (801)$$

In the more interesting case of 2 dimensions, this simplifies to:

$$0 = g^{\kappa\lambda} f_{,\beta}{}^{\beta} - 2\cdot f^{,\lambda\kappa}. \qquad (802)$$

As before we recognize the similarity to the equations in linear elastic theory [9] and expect to find suitable solutions with functions f of an internal structure as introduced in the section above (see Eq. (786)). Comparison with Eq. (788) also gives us the Poisson's ratio of the 2-dimensional space (or space-time) as (when ignoring the sign-difference of the non-Laplace term to the classical equation of linear elasticity):

$$1-2\cdot\mu = \frac{1}{2} \quad \Rightarrow \quad \mu = \frac{1}{4} = 0.25, \qquad (803)$$

which makes our 2D "space-time" rather glass-like.

While in the case of $n = 2$ we would exactly have $\mu = \frac{1}{4} = 0.25$, in 1 dimension (quite understandably), the Poisson's ration would be zero. This last result is not surprising as with no lateral dimensions the lateral contraction would be quite meaningless. Obviously, the fundamental quantum Einstein field equations already take care about this and save us from the need of discussing a non-zero Poisson's ratio in a space of only 1 dimension.

Not ignoring the sign difference to the classical elastic equations give a very unusual Poisson's ratio of:

$$1-2\cdot\mu = -\frac{1}{2} \quad \Rightarrow \quad \mu = \frac{3}{4} = 0.75. \qquad (804)$$

From Eq. (797), the clever reader will make out that by assuming Eigen solutions for the Laplace operator (here degenerated as we only have a metric of constants) and investigating the particle at rest, we end up in having six solutions (three for each, matter and antimatter) and thus, may have found an elegant way to solve the problem of the three generations of elementary particles. On top, so it was already shown (either [2, 8] or chapter 7), we always find that two of the three corresponding solutions for matter and antimatter are unstable, while only one can exist infinitely. This, however, shall not be discussed here as we already dealt with the matter in chapters 6 and—more intensely—7.

8.3.2.1 Elastic equations in *n* > 2 dimensions

At the beginning of this paper, we mentioned that in order to obtain a quantum gravity theory, based on an extended form of the Einstein–Hilbert action and falling back to the Einstein–Hilbert action in its classical form, we principally have the degree of freedom of choosing an arbitrary scale factor $F[f]$ with exponent q in connection with the classical integrand. The corresponding generalized action (see Eq. (740) and appendices D and E with respect to the subsequent derivation) would then be:

$$\delta W = 0 = \delta \int_V d^n x \left(\sqrt{-G} \cdot F^q \cdot \left(R^* - 2\Lambda + L_M \right) \right). \tag{805}$$

Now we ask whether there is a way in using this additional degree of freedom and obtain elastic equations in higher numbers of dimensions.

Using the results from appendix D and assuming a metric of constants, we have:

$$0 = \frac{\sqrt{-g} \cdot F^p}{F} \cdot \begin{pmatrix} 0 \\ +\left(g^{\kappa\lambda} g^{\alpha\beta} - g^{\kappa\alpha} g^{\lambda\beta} - g^{\lambda\alpha} g^{\kappa\beta} \right)(1-n) \\ \cdot \left[\frac{f_{,\alpha} f_{,\beta}}{F^2} \left(FF'' + (F')^2 \cdot \frac{2 \cdot p + n - 14}{8} \right) + f_{,\alpha\beta} \frac{F'}{F} \right] \\ +0 \\ +\left(\frac{p}{2} - 1 \right) \begin{pmatrix} +0 \\ + \frac{\left(g^{\kappa\beta} g^{\alpha\lambda} + g^{\beta\lambda} g^{\alpha\kappa} - 2 \cdot g^{\kappa\lambda} g^{\alpha\beta} \right)}{2} \\ \cdot \left(f_{,\beta} f_{,\alpha} \left(\frac{F''}{F} + \left(\frac{p}{2} - 2 \right) \frac{F'^2}{F^2} \right) + \frac{F'}{F} f_{,\alpha\beta} \right) \\ +0 \end{pmatrix} \end{pmatrix}. \tag{806}$$

Summing up leads to:

$$0 = \begin{pmatrix} +g^{\kappa\lambda}g^{\alpha\beta} \cdot \begin{bmatrix} \frac{f_{,\alpha}f_{,\beta}}{F^2}\left(FF''\left(2-n-\frac{p}{2}\right) + \left(F'\right)^2 \cdot \left((1-n)\frac{2\cdot p+n-14}{8} - \left(\frac{p}{2}-1\right)\cdot\left(\frac{p}{2}-2\right)\right)\right) \\ +f_{,\alpha\beta}\frac{F'}{F}\left(2-n-\frac{p}{2}\right) \end{bmatrix} \\ -\left(g^{\kappa\alpha}g^{\lambda\beta}+g^{\lambda\alpha}g^{\kappa\beta}\right) \\ \cdot \begin{bmatrix} \frac{f_{,\alpha}f_{,\beta}}{F^2}\left(FF''\left(\frac{3}{2}-n-\frac{p}{4}\right) + \left(F'\right)^2 \cdot \left((1-n)\frac{2\cdot p+n-14}{8} - \frac{1}{2}\left(\frac{p}{2}-1\right)\cdot\left(\frac{p}{2}-2\right)\right)\right) \\ +f_{,\alpha\beta}\frac{F'}{F}\left(\frac{3}{2}-n-\frac{p}{4}\right) \end{bmatrix} \end{pmatrix}.$$

$$(807)$$

Now we set $p = 4$ and assume a linear function for $F[f]$, reading $F = C_{f1}{}^*(f - C_{f0})$. This gives us:

$$0 = \begin{pmatrix} +g^{\kappa\lambda}g^{\alpha\beta} \cdot \left[\frac{f_{,\alpha}f_{,\beta}}{\left(f-C_{f0}\right)^2}\cdot(1-n)\frac{n-6}{8} - f_{,\alpha\beta}\frac{1}{f-C_{f0}}(n+1)\right] \\ -\left(g^{\kappa\alpha}g^{\lambda\beta}+g^{\lambda\alpha}g^{\kappa\beta}\right)\cdot\left[\frac{f_{,\alpha}f_{,\beta}}{\left(f-C_{f0}\right)^2}\cdot(1-n)\frac{n-6}{8} + f_{,\alpha\beta}\frac{1}{f-C_{f0}}\left(\frac{1}{2}-n\right)\right] \end{pmatrix}.$$

$$(808)$$

We find that in the case of $n = 6$, the Nabla-terms vanish and we are left with:

$$0 = \left(11\cdot\left(g^{\kappa\alpha}g^{\lambda\beta}+g^{\lambda\alpha}g^{\kappa\beta}\right) - 14\cdot g^{\kappa\lambda}g^{\alpha\beta}\right)f_{,\alpha\beta}. \qquad (809)$$

As long as we just go for "solutions" of the type $0 = f_{,\alpha\beta}$, it seems that our equation does not allow for sufficiently interesting cases for the function f.

The way out—so we already hinted above—would be the assumption of an internal structure for the function f, but before discussing this possibility, we—once again—want to get a somewhat clearer picture about the whole structure in general.

As an "appetizer" it should only be mentioned that with an internal structure for f allowed, Eq. (809) becomes similar to the equations of elasticity (e.g., [9], pp. 166) and we are going to discuss certain solutions in the sub-section "8.4.2 The Linear Elastic Space-Time" further below. These equations can be given for an isotropic material with the Poisson's ratio ν as:

$$(1-2\cdot\mu)\cdot\Delta G^{j}\left[x_{k}\right]+\left(\frac{\partial^{2}G^{j}\left[x_{k}\right]}{\partial x^{\gamma}\partial x^{\delta}}\right)=0. \tag{810}$$

Comparing this with Eq. (809) gives us the peculiar Poisson's ratio:

$$1-2\cdot\mu=-\frac{14}{22} \quad\Rightarrow\quad \mu=\frac{9}{11}\cong0.8182. \tag{811}$$

8.3.3 Separating the Metric from the Quantum Part: A Bit More General

This time we apply the following approach:

$$D\cdot f\cdot g^{\alpha\beta}f_{,\alpha\beta}=g^{\alpha\beta}f_{,\alpha}f_{,\beta} \tag{812}$$

and subsequently obtain from Eq. (772), respectively (793):

$$\Rightarrow$$

$$0=\left\{\begin{array}{l}g^{\kappa\lambda}g^{\alpha\beta}f_{,\alpha\beta}\left(\begin{array}{l}\dfrac{\left(F'\right)^{2}}{8F^{2}}\cdot D\cdot f\left((n-2)(n-4)+22+n\cdot(4\cdot n-23)\right)\\[2mm]+(6\cdot n-8)\cdot\dfrac{D\cdot f}{4\cdot F^{2}}FF''+\dfrac{F'}{4F}\left((n-2)+(5\cdot n-6)\right)\end{array}\right)\\[8mm]-\left(g^{\kappa\alpha}g^{\lambda\beta}+g^{\lambda\alpha}g^{\kappa\beta}\right)\left(\begin{array}{l}(5\cdot n-6)\cdot\dfrac{f_{,\alpha}f_{,\beta}}{4\cdot F^{2}}FF''+\dfrac{\left(F'\right)^{2}}{8F^{2}}\cdot f_{,\alpha}f_{,\beta}\left(22+n\cdot(4\cdot n-23)\right)\\[2mm]+\dfrac{F'\cdot f_{,\alpha\beta}}{4F}(5\cdot n-6)\end{array}\right)\end{array}\right\}. \tag{813}$$

Also demanding:

$$0=\frac{\left(F'\right)^{2}}{8F^{2}}\cdot D\cdot f\left((n-2)(n-4)+22+n\cdot(4\cdot n-23)\right)+(6\cdot n-8)\cdot\frac{D\cdot f}{4\cdot F^{2}}FF''$$

$$\Rightarrow \qquad F[f]=\left(f-C_{f0}\right)^{\frac{4\cdot(3\cdot n-4)}{(n-2)(5\cdot n-7)}}C_{f1}, \tag{814}$$

leaves us with:

$$\Rightarrow$$
$$0 = \left\{ \begin{array}{l} g^{\kappa\lambda} f_{,\beta}{}^{\beta} \dfrac{F'}{4F} \big((n-2) + (5 \cdot n - 6)\big) - \Big(f^{,\kappa\lambda} + f^{,\lambda\kappa}\Big) \dfrac{F'}{4F} (5 \cdot n - 6) \\[4mm] - \Big(g^{\kappa\alpha} g^{\lambda\beta} + g^{\lambda\alpha} g^{\kappa\beta}\Big) \left((5 \cdot n - 6) \cdot \dfrac{f_{,\alpha} f_{,\beta}}{4 \cdot F^2} FF'' + \dfrac{(F')^2}{8F^2} \cdot f_{,\alpha} f_{,\beta} \big(22 + n \cdot (4 \cdot n - 23)\big) \right), \end{array} \right.$$

$$(815)$$

where—in the first line—we have the essential derivative terms of the elastic equation again. As, however, we also have the second line (acting as gravitational fields, if sticking to the elastic picture) with the Nabla-terms plus the fact that F depends on the function f, our equation renders rather non-linear and thus, difficult in a general number of dimensions.

Only with an additional setting similar to Eq. (812) like:

$$C \cdot f \cdot \Big(f^{,\kappa\lambda} + f^{,\lambda\kappa}\Big) = \Big(g^{\kappa\alpha} g^{\lambda\beta} + g^{\lambda\alpha} g^{\kappa\beta}\Big) f_{,\alpha} f_{,\beta} = f^{,\kappa} f^{,\lambda} + f^{,\lambda} f^{,\kappa}$$

$$\Rightarrow$$
$$0 = \left\{ \begin{array}{l} g^{\kappa\lambda} f_{,\beta}{}^{\beta} \dfrac{F'}{4F} \big((n-2) + (5 \cdot n - 6)\big) - \Big(f^{,\kappa\lambda} + f^{,\lambda\kappa}\Big) \dfrac{F'}{4F} (5 \cdot n - 6) \\[4mm] - \Big(g^{\kappa\alpha} g^{\lambda\beta} + g^{\lambda\alpha} g^{\kappa\beta}\Big) \left((5 \cdot n - 6) \cdot \dfrac{f_{,\alpha} f_{,\beta}}{4 \cdot F^2} FF'' + \dfrac{(F')^2}{8F^2} \cdot f_{,\alpha} f_{,\beta} \big(22 + n \cdot (4 \cdot n - 23)\big) \right), \end{array} \right.$$

$$(816)$$

the situation clears up fairly nicely and gives us the following (and quite elastic) differential equation:

$$0 = \left\{ \begin{array}{l} g^{\kappa\lambda} f_{,\beta}{}^{\beta} F' \big((n-2) + (5 \cdot n - 6)\big) - \Big(f^{,\kappa\lambda} + f^{,\lambda\kappa}\Big) F' (5 \cdot n - 6) \\[4mm] - C \cdot f \cdot \Big(f^{,\kappa\lambda} + f^{,\lambda\kappa}\Big) \left((5 \cdot n - 6) \cdot F'' + \dfrac{(F')^2}{2F} \cdot \big(22 + n \cdot (4 \cdot n - 23)\big) \right). \end{array} \right.$$

$$(817)$$

8.3.4 Brief Consideration of the Situation with Eigenvalue Solutions

Again assuming a metric of constants, we now take it that we can find Eigenvalue solutions to the two differential operators on f and obtain from Eq. (770) via the reordering of Eq. (798):

$$
0 = \begin{cases} g^{\kappa\lambda} \left(\begin{array}{l} \dfrac{(F')^2}{8F^2} \cdot f^{,\beta} f_{,\beta} \left((n-2)(n-4) + 22 + n \cdot (4 \cdot n - 23) \right) \\[2ex] + (6 \cdot n - 8) \cdot \dfrac{f^{,\beta} f_{,\beta}}{4 \cdot F^2} FF'' + \dfrac{F' \cdot f_{,\beta}{}^{\beta}}{4F} \left((n-2) + (5 \cdot n - 6) \right) \end{array} \right) \\[5ex] -(5 \cdot n - 6) \cdot \dfrac{f^{,\lambda} f^{,\kappa}}{2 \cdot F^2} FF'' - \dfrac{(F')^2}{4F^2} \cdot f^{,\lambda} f^{,\kappa} \left(22 + n \cdot (4 \cdot n - 23) \right) \\[4ex] \qquad\qquad - \dfrac{F' \cdot f^{,\lambda\kappa}}{2F} (5 \cdot n - 6) \end{cases}
$$

$$\xrightarrow{\quad f^{,\beta} f_{,\beta} = A \cdot f^2; \quad f_{,\beta}{}^{\beta} = B \cdot f \quad}$$

$$
\Rightarrow \quad 0 = \left\{ \begin{array}{l} g^{\kappa\lambda} \left(\begin{array}{l} \dfrac{(F')^2}{8F^2} \cdot A \cdot f \cdot \left((n-2)(n-4) + 22 + n \cdot (4 \cdot n - 23) \right) \\[2ex] + (6 \cdot n - 8) \cdot \dfrac{A \cdot f}{4 \cdot F^2} FF'' + \dfrac{F' \cdot B}{4F} \left((n-2) + (5 \cdot n - 6) \right) \end{array} \right) \\[5ex] + \left(\begin{array}{l} -(5 \cdot n - 6) \cdot \dfrac{f^{,\lambda} f^{,\kappa}}{2 \cdot F^2} FF'' - \dfrac{(F')^2}{4F^2} \cdot f^{,\lambda} f^{,\kappa} \left(22 + n \cdot (4 \cdot n - 23) \right) \\[4ex] \qquad\qquad - \dfrac{F' \cdot f^{,\lambda\kappa}}{2F} (5 \cdot n - 6) \end{array} \right) \end{array} \right.
$$

$$(818)$$

Now we try to find a solution for $F[f]$ such that we have:

$$
0 = \begin{cases} \dfrac{(F')^2}{8F^2} \cdot A \cdot f \cdot \left((n-2)(n-4) + 22 + n \cdot (4 \cdot n - 23) \right) \\[2ex] + (6 \cdot n - 8) \cdot \dfrac{A \cdot f}{4 \cdot F^2} FF'' + \dfrac{F' \cdot B}{4F} \left((n-2) + (5 \cdot n - 6) \right), \end{cases}
$$

$$(819)$$

leading us to:

$$
0 = \dfrac{f^{,\lambda} f^{,\kappa}}{2 \cdot F^2} \left((5 \cdot n - 6) \cdot FF'' + \dfrac{(F')^2}{2} \left(22 + n \cdot (4 \cdot n - 23) \right) \right) + \dfrac{F' \cdot f^{,\lambda\kappa}}{2F} (5 \cdot n - 6).
$$

$$(820)$$

From here, for the time being, we leave it to the interested reader to find out whether this approach does bring any advantage.

8.4 Intermediate Sum-Up and Repetition of a Few Important Results

8.4.1 The Special Case of 2 Dimensions

Assuming metrics consisting only of constants, which make all terms with derivatives on the metric tensor to disappear (including all Levi–Civita connections), give us from Eq. (770) the following simplification:

$$
0 = \left\{ \begin{array}{l} \dfrac{\sqrt{-g} \cdot F^n}{F} \cdot \left(\begin{array}{l} + \left(g^{\kappa\lambda} g^{\alpha\beta} - g^{\kappa\alpha} g^{\lambda\beta} - g^{\lambda\alpha} g^{\kappa\beta} \right) (1-n) \\[2mm] \cdot \left[\dfrac{f_{,\alpha} f_{,\beta}}{F^2} \left(FF'' + \left(F' \right)^2 \cdot \dfrac{3 \cdot n - 14}{8} \right) + f_{,\alpha\beta} \dfrac{F'}{F} \right] \end{array} \right) \\[8mm] + \left(\sqrt{-g} \cdot F^{n/2-1} \right)_{,\gamma,\rho} \cdot \dfrac{\left(g^{\kappa\rho} g^{\gamma\lambda} + g^{\rho\lambda} g^{\gamma\kappa} - g^{\kappa\lambda} g^{\gamma\rho} \right)}{2} \right) - \left(\sqrt{-g} \cdot \dfrac{g^{\alpha\beta}}{2} F^{n/2-1} \right)_{,\alpha,\beta} \cdot g^{\kappa\lambda} . \end{array} \right.
$$

$$(821)$$

We find a most intriguing simplification in 2 dimensions:

$$\xrightarrow{\;n=2\;} \quad 0 = g^{\kappa\lambda} f_{,\beta}{}^{\beta} - \left(f^{,\kappa\lambda} + f^{,\lambda\kappa} \right). \tag{822}$$

Thereby the only condition needed would be:

$$FF'' - (F')^2 = 0, \tag{823}$$

which fixes the function $F[f]$ as follows:

$$F[f] = C_{f1} \cdot e^{f \cdot C_{f0}}. \tag{824}$$

8.4.2 The Linear Elastic Space-Time

Comparing our peculiarly simple result (822) in 2 dimensions, Eq. (809) in 6 dimensions and Eq. (787) in n dimensions (peculiar if one considers the complexity of the general starting point (744) and its even more complicated result) with the fundamental equation of equilibrium for the free of body forces linear elastic space [9, 10], one finds a great similarity.

Why is this?

Does it mean that our space-time is linearly elastic and what do all those many possibilities of various stress, strain, and displacement fields then mean with respect to—let's say—elementary particle physics?

Well, let us see.

We concentrate on the case $p = 4$, $n = 6$ where just as elaborated in [2], we assume 2 dimensions to be curled up or compactified and f only depends on the remaining 4 dimensions. It was also shown in [2] that the compactified dimensions can then bring about inertia.

Giving f an internal structure like:

$$f = Q^\gamma \phi_\gamma, \tag{825}$$

where we assume Q^γ to be a vector of constants and inserting this into Eq. (822), leads us to:

$$0 = \left(11 \cdot \left(g^{\kappa\alpha} g^{\lambda\beta} + g^{\lambda\alpha} g^{\kappa\beta} \right) - 14 \cdot g^{\kappa\lambda} g^{\alpha\beta} \right) f_{,\alpha\beta}$$

$$\Rightarrow \quad 0 = \left(14 \cdot g^{\kappa\lambda} g^{\alpha\beta} - 11 \cdot \left(g^{\kappa\alpha} g^{\lambda\beta} + g^{\lambda\alpha} g^{\kappa\beta} \right) \right) Q^\gamma \phi_{\gamma,\alpha\beta}$$

$$= Q^\gamma \left(14 \cdot g^{\kappa\lambda} g^{\alpha\beta} - 11 \cdot \left(g^{\kappa\alpha} g^{\lambda\beta} + g^{\lambda\alpha} g^{\kappa\beta} \right) \right) \phi_{\gamma,\alpha\beta}$$

$$\Rightarrow \quad 0 = 14 \cdot g^{\kappa\lambda} Q^\gamma \phi_{\gamma,\beta}{}^\beta - 11 \cdot \left(Q^\gamma \phi_\gamma{}^{,\kappa\lambda} + Q^\gamma \phi_\gamma{}^{,\lambda\kappa} \right). \tag{826}$$

Comparison with the fundamental (load free) equation of linear elasticity, which in the isotropic case with Poisson's ratio μ and the displacement vector $\mathbf{h}_\gamma$ reads[5]:

$$0 = 2 \cdot (1 - 2 \cdot \mu) h_{\gamma,\beta}{}^\beta - \left(h_\gamma{}^{,\gamma\lambda} + h_\gamma{}^{,\gamma\kappa} \right), \tag{827}$$

directly does not only lead us to a great variety of possible solutions, but also to some rather interesting analogies. One of these solutions, which was already discussed here, is the pure shear stress solution with the following example (in Cartesian coordinates[6]):

$$h_\gamma = \left\{ -\partial_y, \partial_x, 0 \right\} \psi \quad \text{with:} \quad \psi_{,\beta}{}^\beta = 0. \tag{828}$$

Applied onto Eq. (826), we may set:

$$h_\gamma = D_\gamma^\sigma \phi_\sigma = \left\{ 0, -\partial_y, \partial_x, 0 \right\} \psi \quad \text{with:} \quad \psi_{,\beta}{}^\beta = 0. \tag{829}$$

8.4.2.1 Shear fields = spin fields

It seems that this type of solution to Eq. (826) provides the characteristics of a spin field (Fig. 8.1). Thus, a spin in elementary particle physics would

---

[5]It should be noted that in the original equation [9], the non-Laplacian term has a positive sign, but, as already stated above, this is not of interest to the shear solutions we are going to discuss here.

[6]Please note that in Cartesian coordinates, the difference between co- and contra-variant tensor forms vanishes.

become a shear field in the quantum Einstein field equations. We apply this approach as follows:

$$0 = 14 \cdot g^{\kappa\lambda} Q^{\gamma} \phi_{\gamma,\beta}{}^{\beta} - 11 \cdot \left(Q^{\gamma} \phi_{\gamma}{}^{,\kappa\lambda} + Q^{\gamma} \phi_{\gamma}{}^{,\lambda\kappa} \right)$$

$$= 14 \cdot g^{\kappa\lambda} Q^{\gamma} \phi_{\gamma,\beta}{}^{\beta} - 11 \cdot \left(q^{\kappa} h_{\gamma}{}^{,\gamma\lambda} + q^{\lambda} h_{\gamma}{}^{,\gamma\kappa} \right) \tag{830}$$

and directly see that because of the structure of our solution:

$$h_{\gamma} = \pm \left\{ \begin{array}{l} \{-\partial_x, \partial_t, 0, 0\}, \{-\partial_y, 0, \partial_t, 0\}, \{-\partial_z, 0, 0, \partial_t\} \\ \{0, -\partial_y, \partial_x, 0\}, \{0, -\partial_z, 0, \partial_x\}, \{0, 0, -\partial_z, \partial_y\} \end{array} \right\} \psi; \quad Q^{\gamma} \phi_{\gamma,\beta}{}^{\beta} = 0, \tag{831}$$

we automatically get:

$$14 \cdot g^{\kappa\lambda} \overbrace{Q^{\gamma} \phi_{\gamma,\beta}{}^{\beta}}^{=0} - 11 \cdot \left(\overbrace{q^{\kappa} h_{\gamma}{}^{,\gamma\lambda}}^{=0} + \overbrace{q^{\lambda} h_{\gamma}{}^{,\gamma\kappa}}^{=0} \right) = \quad \Rightarrow \quad = 0. \tag{832}$$

8.4.2.2 Quark-like 1/3 and 2/3 charges

Now we extend this approach a bit and apply it as follows:

$$0 = 14 \cdot g^{\kappa\lambda} Q^{\gamma\sigma} \phi_{\gamma\sigma,\beta}{}^{\beta} - 11 \cdot \left(Q^{\gamma\sigma} \phi_{\gamma\sigma}{}^{,\kappa\lambda} + Q^{\gamma\sigma} \phi_{\gamma\sigma}{}^{,\lambda\kappa} \right)$$

$$= 14 \cdot g^{\kappa\lambda} Q^{\gamma\sigma} \phi_{\gamma\sigma,\beta}{}^{\beta} - 11 \cdot \left(q^{\kappa\sigma} h_{\gamma\sigma}{}^{,\gamma\lambda} + q^{\lambda\sigma} h_{\gamma\sigma}{}^{,\gamma\kappa} \right) \tag{833}$$

or even:

$$0 = 14 \cdot g^{\kappa\lambda} Q^{\gamma\sigma\mu} \phi_{\gamma\sigma\mu,\beta}{}^{\beta} - 11 \cdot \left(Q^{\gamma\sigma\mu} \phi_{\gamma\sigma\mu}{}^{,\kappa\lambda} + Q^{\gamma\sigma\mu} \phi_{\gamma\sigma\mu}{}^{,\lambda\kappa} \right)$$

$$= 14 \cdot g^{\kappa\lambda} Q^{\gamma\sigma\mu} \phi_{\gamma\sigma\mu,\beta}{}^{\beta} - 11 \cdot \left(q^{\kappa\sigma\mu} h_{\gamma\sigma\mu}{}^{,\gamma\lambda} + q^{\lambda\sigma\mu} h_{\gamma\sigma\mu}{}^{,\gamma\kappa} \right). \tag{834}$$

Then, as shown in [3], more such forms of shear-like solutions can be found easily via the following recipe:

$$h_{\gamma}[x_k] = \left\{ \ldots, A_1 \overbrace{\frac{\partial^2 \psi}{\partial \xi_2 \partial \xi_3}}^{\text{pos } \xi_1}, \ldots, A_2 \overbrace{\frac{\partial^2 \psi}{\partial \xi_1 \partial \xi_3}}^{\text{pos } \xi_2}, \ldots, A_3 \overbrace{\frac{\partial^2 \psi}{\partial \xi_1 \partial \xi_2}}^{\text{pos } \xi_3}, \ldots \right\}$$

$$h_{\gamma}[x_k] = \left\{ \ldots, A_1 \overbrace{\frac{\partial^3 \psi}{\partial \xi_2 \partial \xi_3 \partial \xi_4}}^{\text{pos } \xi_1}, \ldots, A_2 \overbrace{\frac{\partial^3 \psi}{\partial \xi_1 \partial \xi_3 \partial \xi_4}}^{\text{pos } \xi_2}, \ldots, A_3 \overbrace{\frac{\partial^3 \psi}{\partial \xi_1 \partial \xi_2 \partial \xi_4}}^{\text{pos } \xi_3}, \ldots, A_4 \overbrace{\frac{\partial^3 \psi}{\partial \xi_1 \partial \xi_2 \partial \xi_3}}^{\text{pos } \xi_4} \ldots \right\}$$

$$\ldots$$

$$\forall(\ldots,\ldots) = 0; \quad \sum_{\forall k} A_k = 0. \tag{835}$$

In a Minkowski space-time with $c = 1$ and the coordinates t, x, y, and z, we could have the following types of such solutions:

$$h_\gamma = \pm\left\{\frac{\partial}{\partial z},0,0,-\frac{\partial}{\partial t}\right\}\psi; \quad h_\gamma = \pm\left\{\frac{\partial}{\partial y},0,-\frac{\partial}{\partial t},0\right\}\psi; \quad h_\gamma = \pm\left\{\frac{\partial\psi}{\partial x},-\frac{\partial}{\partial t},0,0\right\}\psi$$

$$h_\gamma = \pm\left\{0,\frac{\partial}{\partial z},0,-\frac{\partial}{\partial x}\right\}\psi; \quad h_\gamma = \pm\left\{0,\frac{\partial}{\partial y},-\frac{\partial}{\partial x},0\right\}\psi; \quad h_\gamma = \pm\left\{0,0,\frac{\partial}{\partial z},-\frac{\partial}{\partial y}\right\}\psi,$$

$$(836)$$

$$h_\gamma = \left\{A_1\frac{\partial^2}{\partial x\partial y},A_2\frac{\partial^2}{\partial t\partial y},A_3\frac{\partial^2}{\partial x\partial t},0\right\}\psi; \quad h_\gamma = \left\{A_1\frac{\partial^2}{\partial x\partial z},A_2\frac{\partial^2}{\partial t\partial z},0,A_3\frac{\partial^2}{\partial x\partial t}\right\}\psi$$

$$h_\gamma = \left\{A_1\frac{\partial^2}{\partial z\partial y},0,A_2\frac{\partial^2}{\partial t\partial z},A_3\frac{\partial^2}{\partial z\partial t}\right\}\psi; \quad h_\gamma = \left\{0,A_1\frac{\partial^2}{\partial z\partial y},A_2\frac{\partial^2}{\partial x\partial z},A_3\frac{\partial^2}{\partial x\partial y}\right\}\psi,$$

$$(837)$$

$$h_\gamma = \left\{A_1\frac{\partial^3}{\partial x\partial y\partial z},A_1\frac{\partial^3}{\partial t\partial y\partial z},A_1\frac{\partial^3}{\partial x\partial t\partial z},A_4\frac{\partial^3}{\partial x\partial y\partial t}\right\}\psi; \quad \sum_{\forall k}A_k = 0. \quad (838)$$

Thereby we find that the last solution in Eq. (837) with a setting of the kind $A_1 = 1/3$, $A_2 = 1/3$, and $A_3 = -2/3$ shows a peculiar closeness to the charges of quarks, the building blocks of hadrons and baryons.

8.4.2.3 Hydrostatic particle fields

As said above, comparison of Eq. (826) with the fundamental (load free) equation of linear elasticity, which in the isotropic case with Poisson's ratio μ and the displacement vector $\mathbf{h}_\gamma$ reads:

$$0 = 2\cdot(1-2\cdot\mu)h_{\gamma,\beta}{}^\beta - \left(h_\gamma{}^{,\gamma\lambda} + h_\gamma{}^{,\gamma\kappa}\right), \quad (839)$$

directly does not only lead us to a great variety of possible solutions, but also to some rather interesting analogies. Another interesting solution could be the pure normal stress (or hydrostatic) solution with the example (in Cartesian coordinates):

$$h_\gamma = \left\{\partial_x,\partial_y,\partial_z\right\}\psi \quad \text{with:} \quad \psi_{,\beta}{}^\beta = 0. \quad (840)$$

Applied onto Eq. (826), we may set:

$$h_\gamma = D_\gamma^\sigma \phi_\sigma = \begin{cases} \{\partial_t, \partial_x, \partial_y, \partial_z\}\psi & \text{with:} \quad \psi_{,k}{}^k = 0; \quad k = 0,1,2,3 \\ \{\partial_t, \partial_x, \partial_y, 0\}\psi & \text{with:} \quad \psi_{,k}{}^k = 0; \quad k = 0,1,2 \\ \{\partial_t, \partial_x, 0, \partial_z\}\psi & \text{with:} \quad \psi_{,k}{}^k = 0; \quad k = 0,1,3 \\ \{\partial_t, 0, \partial_y, \partial_z\}\psi & \text{with:} \quad \psi_{,k}{}^k = 0; \quad k = 0,2,3 \\ \{\partial_t, \partial_x, 0, 0\}\psi & \text{with:} \quad \psi_{,k}{}^k = 0; \quad k = 0,1 \\ \{\partial_t, 0, 0, \partial_z\}\psi & \text{with:} \quad \psi_{,k}{}^k = 0; \quad k = 0,3 \\ \{\partial_t, 0, \partial_z, 0\}\psi & \text{with:} \quad \psi_{,k}{}^k = 0; \quad k = 0,2 \\ \{0, \partial_x, \partial_y, \partial_z\}\psi & \text{with:} \quad \psi_{,k}{}^k = 0; \quad k = 1,2,3 \\ \{0, \partial_x, \partial_y, 0\}\psi & \text{with:} \quad \psi_{,k}{}^k = 0; \quad k = 1,2 \\ \{0, \partial_x, 0, \partial_z\}\psi & \text{with:} \quad \psi_{,k}{}^k = 0; \quad k = 1,3 \\ \{0, 0, \partial_y, \partial_z\}\psi & \text{with:} \quad \psi_{,k}{}^k = 0; \quad k = 2,3. \end{cases} \tag{841}$$

The charm of this type of solution to Eq. (826) could be seen in the fact that it provides the characteristics of a spherical pressure field (spherical in four, three, or 2 dimensions). This could mean a charge, mass, or something. Thus, a certain charge in elementary particle physics would become a symmetric normal pressure field in the quantum Einstein field equations. Thus, picking—for instance—the spatial solution:

$$h_\gamma = D_\gamma^\sigma \phi_\sigma = \{0, \partial_x, \partial_y, \partial_z\}\psi \quad \text{with:} \quad \psi_{,k}{}^k = 0; \quad k = 1,2,3, \tag{842}$$

we'd result in a stable (time-independent) homogeneous pressure field around the spatial point {0,0,0}. The stresses and strains it eradiates into its neighborhood might be registered there as some kind of force fields and their origin would then be a charged particle, which sends out the field.

8.4.2.4 Time-planes and time-layers

Another elastic solution, which appears to be of cosmological character, is the plane boundary solution with the example (in Cartesian coordinates):

$$h_\gamma = z \cdot \{\partial_x, \partial_y, \partial_z\}\psi + (1 - 4 \cdot \mu)\{0, 0, \partial_z\}\psi \quad \text{with:} \quad \psi_{,\beta}{}^\beta = 0. \tag{843}$$

Please note that in the classical literature of elasticity (e.g., [9]), where we have Eq. (788) with a positive sign as:

$$0 = 2 \cdot (1 - 2 \cdot \mu) h_{\gamma,\beta}{}^{\beta} + \left(h_{\gamma}{}^{,\gamma\lambda} + h_{\gamma}{}^{,\gamma\kappa} \right), \tag{844}$$

solution (843) does read:

$$h_{\gamma} = z \cdot \{\partial_x, \partial_y, \partial_z\} \psi - (3 - 4 \cdot \mu)\{0, 0, \partial_z\} \psi \quad \text{with}: \quad \psi_{,\beta}{}^{\beta} = 0. \tag{845}$$

Applied onto Eq. (826), we may set:

$$f = Q^{\gamma} \phi_{\gamma} \Leftrightarrow h_{\gamma} = D_{\gamma}^{\sigma} \phi_{\sigma} = t \cdot \{\partial_x, \partial_x, \partial_y, \partial_z\} \psi + \left(1 - 4 \cdot \frac{9}{11} \right) \{\partial_t, 0, 0, 0\} \psi$$

with: $\psi_{,\beta}{}^{\beta} = 0$

$$\tag{846}$$

and obtain the solution to Eq. (826) via:

$$\Rightarrow \qquad 0 = 14 \cdot g^{\kappa\lambda} Q^{\gamma} \phi_{\gamma,\beta}{}^{\beta} - 11 \cdot \left(Q^{\gamma} \phi_{\gamma}{}^{,\kappa\lambda} + Q^{\gamma} \phi_{\gamma}{}^{,\lambda\kappa} \right)$$

$$= \left(1 - 2 \cdot \frac{9}{11} \right) \cdot g^{\kappa\lambda} q^{\beta} \phi_{\gamma,\beta}{}^{\gamma} - Q^{\kappa} \phi_{\gamma}{}^{,\gamma\lambda}. \tag{847}$$

The importance of this type of solution to Eq. (826), here could be seen in the fact that it provides the characteristics of the fundamental elastic solution to an impact or indent on the plane $t = 0$. Especially in connection with the other solutions, we already presented here, we would obtain a great variety of indent solutions. Particles (and other stuff, usually been named matter) may just be seen as defects [11], which are the remainders of the hypothetic primordial impact.

As shown elsewhere [12–16], this type of solution also allows the introduction of layered structures, which are in this case layers of time. However, as the type of solution (846) can be easily generalized to layers within all dimensions:

$$f = Q^{\gamma} \phi_{\gamma} \Leftrightarrow h_{\gamma} = \begin{cases} t \cdot \{\partial_x, \partial_x, \partial_y, \partial_z\} \psi + \left(1 - 4 \cdot \frac{9}{11} \right) \{\partial_t, 0, 0, 0\} \psi \\[2mm] x \cdot \{\partial_x, \partial_x, \partial_y, \partial_z\} \psi + \left(1 - 4 \cdot \frac{9}{11} \right) \{0, \partial_x, 0, 0\} \psi \\[2mm] y \cdot \{\partial_x, \partial_x, \partial_y, \partial_z\} \psi + \left(1 - 4 \cdot \frac{9}{11} \right) \{0, 0, \partial_y, 0\} \psi \\[2mm] z \cdot \{\partial_x, \partial_x, \partial_y, \partial_z\} \psi + \left(1 - 4 \cdot \frac{9}{11} \right) \{0, 0, 0, \partial_z\} \psi \end{cases}$$

with: $\psi_{,\beta}{}^{\beta} = 0$

$$\tag{848}$$

or even be used for the introduction of contacting and/or layered manifolds via (here we only give one example in 3D and 2D. For more complex solutions, readers can refer to the books of Fabrikant [17, 18], for instance):

$$
h_\gamma = \begin{cases} x \cdot \left\{0, \partial_x, \partial_y, \partial_z\right\}\psi + C_\psi \cdot \left\{0, \partial_x, 0, 0\right\}\psi \\ y \cdot \left\{0, \partial_x, \partial_y, \partial_z\right\}\psi + C_\psi \cdot \left\{0, 0, \partial_y, 0\right\}\psi \\ z \cdot \left\{0, \partial_x, \partial_y, \partial_z\right\}\psi + C_\psi \cdot \left\{0, 0, 0, \partial_z\right\}\psi \end{cases} \quad \text{with:} \quad \psi_{,k}{}^k = 0; \quad k = 1,2,3
$$

$$
h_\gamma = \begin{cases} x \cdot \left\{0, \partial_x, \partial_y, 0\right\}\psi + C_\psi \cdot \left\{0, \partial_x, 0, 0\right\}\psi \\ y \cdot \left\{0, \partial_x, \partial_y, 0\right\}\psi + C_\psi \cdot \left\{0, 0, \partial_y, 0\right\}\psi \end{cases} \quad \text{with:} \quad \psi_{,k}{}^k = 0; \quad k = 1,2,
$$

(849)

we end up with a space-time brim-full with "elastic options."

One may even use the analogies found here for an easy way to actually illustrate features of our universe as stress, strain, displacement, and deformation energy fields [19] (c.f. cover picture).

Thus: Our Universe—Not an Accident, Not an Event, but Perhaps an Indent.

There have been hypothesizes about the big bang just being an impact event. Thereby it was assumed that some colliding superbranes started what we realize as our universe.

Now, by reevaluating the Einstein–Hilbert action for a scaled metric, thereby deriving a set of quantum Einstein field equations, we found solutions which seem to support this idea.

Our universe could indeed just be the deformation field of some kind of "mechanical contact," impact or indent ... and the surface at which the contact "was, is, and will be" originated "was, is, and will be" *t* = 0, the beginning of time.

The charm of this idea is not just the fact that it provides the features the original hypothesizes were aiming for anyway, but that it also explains time as the propagation of the shock wave, which is still pushing through our universe, if not to say, which is making our universe.

Yes, if this hypothesis takes shape and gets more supporting evidence, it may be a shock to some people to find the result of an impact and its subsequent shockwave, but at least this scenario would assure everyone and everything to create an impact as it, he, or she actually is impact material.

So, let us just move on and try to make an impression then!

8.5 Brief Discussion Regarding Our Peculiar Results for the "Poisson's Ratio"

In the two sections above, we found a setting for the quantum Einstein field equations, which resulted in equations similar to the governing equations in elasticity, only that here our equations sported a funny Poisson's ratio μ > 0.5. Such a Poisson's ratio would be equivalent to a negative compression module and means that space-time, if put under pressure, would expand and not compress.

We also find some strange behavior when putting the space under certain loading situations. In Figs. 8.2–8.4 below, we illustrated this in the case of some pulling forces being active on planar surface parts within the space. Such loading situations are completely hypothetical and only shown here in order to demonstrate the strange effect of the high μ. The colors code the total shear of von Mises stress of the spatial space.

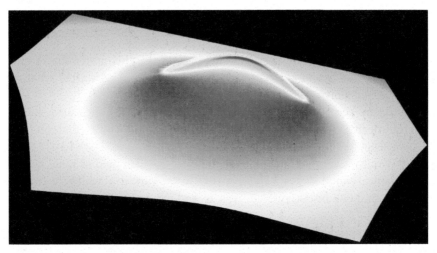

Figure 8.2 A stress field from a pulling force in space-time according to the elastic solution of the quantum Einstein field equations, evaluated with the software FilmDoctor [19] using Eq. (830) with solution (849).

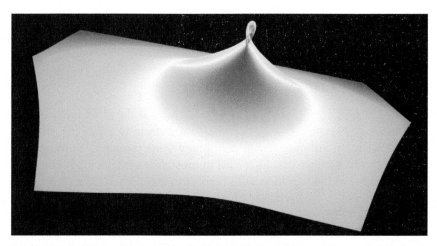

Figure 8.3 A stress field from a pulling force in space-time according to the elastic solution of the quantum Einstein field equations, evaluated with the software FilmDoctor [19] using Eq. (830) with solution (849).

Figure 8.4 A stress field from a pulling force in space-time according to the elastic solution of the quantum Einstein field equations, evaluated with the software FilmDoctor [19] using Eq. (830) with solution (849).

8.6 Discussion with Respect to More General Scale Factors

8.6.1 About a Tensor Scaled Metric

It was discussed at the beginning of this paper that the scale factor $F[f]$ as used in Eq. (739) could be seen as the simplest form to Eq. (738), which is just similar to an ordinary tensor coordinate transformation. However, this author has the opinion that Eq. (738) in most cases (c.f. chapter 11) is not of need, because the variation with respect to a general metric (leaving all its components to be variated) plus the scalar scale factor as performed via approach (739) and put into the Einstein–Hilbert action; and makes an explicit tensor scaling according to Eq. (738) obsolete, because in essence we already have a scaling of the components and the volume. Thereby the latter is realized via the $F[f]$-factor, while the first is been taken care about via the usual metric variation.

Nevertheless, it may provide some insights or technical advantages, when performing the complete variation with a generalized approach (738). Therefore, we add to the starting point to this chapter and put it in the appendix C. Later in this book, in chapter 11, we will see how such a general approach could be of use.

8.6.2 About a More General Kernel within the Einstein–Hilbert Action

Even though there is no experimental evidence that the kernel in the integral of the Einstein–Hilbert action should be anything else than just the Ricci scalar, there has been a lot of discussion about alternatives. So, in [20], it is shown that the generalization of the kernel for the Einstein–Hilbert action to a function $\varphi(R)$ instead of just R, leads to the following extension of the Einstein field equations:

$$\varphi'(R)R_{\mu\nu} - \frac{1}{2}\varphi(R)g_{\mu\nu} + \Lambda g_{\mu\nu} + \varphi''(R)\left[R_{;\mu\nu} - \Delta_g R \cdot g_{\mu\nu}\right]$$

$$+\varphi'''(R)\left[R_{;\mu}R_{;\nu} - R^{;\sigma}R_{;\sigma} \cdot g_{\mu\nu}\right] = \begin{cases} 0 & \ldots\text{``vacuum''} \\ -8\pi G T_{\mu\nu} & \ldots\text{postulated matter.} \end{cases} \quad (850)$$

The ";" denotes covariant derivatives and the symbol Δ_g gives the Laplace–Beltrami operator in the metric $g_{\alpha\beta}$. Einstein and Hilbert had used $\varphi[R] = R$, which avoids higher orders of curvature respectively higher orders for the Ricci scalar and/or the metric [7, 21, 22].

In connection with the structure of our previous calculations, we might prefer the contra-variant form of Eq. (850), which reads:

$$0 = \begin{cases} \varphi'[R]R^{\delta\gamma} - \dfrac{1}{2}\varphi[R]g^{\delta\gamma} + \Lambda \cdot g^{\delta\gamma} + \varphi''[R]\left(R^{;\delta\gamma} - \Delta_g R \cdot g^{\delta\gamma}\right) \\[2mm] + \varphi'''[R]\left(R^{;\delta}R^{;\gamma} - R^{;\sigma}R_{;\sigma} \cdot g^{\delta\gamma}\right) + \begin{cases} \dfrac{8\pi G T^{\delta\gamma}}{\kappa} = \text{matter} \\[2mm] 0 = \text{vacuum.} \end{cases} \end{cases} \qquad (851)$$

For our purposes, it is sometimes better to have the compacter—non-expanded—form (see [26]):

$$\varphi'(R)R_{\mu\nu} - \frac{1}{2}\varphi(R)g_{\mu\nu} + \Lambda g_{\mu\nu} - \left(\partial_{;\mu}\partial_{;\nu} - g_{\mu\nu}\Delta_g\right)\varphi'(R)$$

$$= \begin{cases} 0 & \dots\text{``vacuum''} \\ -8\pi G T_{\mu\nu} & \dots\text{postulated matter.} \end{cases} \qquad (852)$$

The evaluation in the case of R^* instead of R is quite lengthy and will therefore not be performed here. We just give the starting point, a few hints and the subsequent result. At first, we need to write down the $\varphi(R^*)$-extended Einstein–Hilbert action, reading:

$$\delta W = 0 = \delta \int_V d^n x \left(\sqrt{-G} \cdot \left(\varphi(R^*) - 2\Lambda + L_M\right)\right)$$

$$= \delta \int_V d^n x \left(\sqrt{-g} \cdot F^n \cdot \left(\varphi\left(\begin{array}{c}\left(\begin{array}{c}\Gamma^\sigma_{\alpha\beta,\sigma}g^{\alpha\beta} - \Gamma^\sigma_{\beta\sigma,\alpha}g^{\alpha\beta} - \Gamma^\mu_{\sigma\alpha}\Gamma^\sigma_{\beta\mu}g^{\alpha\beta} + \Gamma^\sigma_{\alpha\beta}\Gamma^\mu_{\sigma\mu}g^{\alpha\beta} \\[2mm] + \dfrac{F'}{F}(1-n)\Delta f \\[2mm] + \dfrac{f_{,\alpha}f_{,\beta}g^{\alpha\beta}(1-n)}{4F^2}\left(4FF'' + (F')^2(n-6)\right)\end{array}\right)\dfrac{1}{F} \\[4mm] -2\Lambda + L_M\end{array}\right)\right)\right)$$

$$= \delta \int_V d^n x \left(\sqrt{-g} \cdot F^n \cdot \left(\varphi\left(\begin{array}{c}\left(R + \dfrac{F'}{F}(1-n)\Delta f \\[2mm] + \dfrac{f_{,\alpha}f_{,\beta}g^{\alpha\beta}(1-n)}{4F^2}\left(4FF'' + (F')^2(n-6)\right)\right)\dfrac{1}{F} \\[4mm] -2\Lambda + L_M\end{array}\right)\right)\right)$$

with:
$$F = F[f]; \quad F' = \frac{\partial F[f]}{\partial f}; \quad F'' = \frac{\partial^2 F[f]}{\partial f^2}. \qquad (853)$$

From there we can deduce the $\varphi(R^*)$-generalization for a scaled metric according to Eq. (739) as follows:

$$= \int_V d^n x \left(\begin{array}{c} \varphi\left(\begin{array}{c} \dfrac{R}{F} + \dfrac{F'}{F^2}(1-n)\Delta f \\[2mm] + \dfrac{f_{,\alpha} f_{,\beta} g^{\alpha\beta}(1-n)}{4F^3}\left(4FF'' + (F')^2(n-6)\right) \end{array}\right) \delta\sqrt{-g \cdot F^n} \\[6mm] + \sqrt{-g \cdot F^n} \cdot \left(\left(\begin{array}{c} \underbrace{\dfrac{8\pi G}{\kappa}T^{\kappa\lambda}}_{} = \text{matter} \\ 0 = \text{vacuum} \end{array}\right) + \Lambda \cdot g^{\kappa\lambda}\right)\delta g_{\kappa\lambda} \\[6mm] + \sqrt{-g \cdot F^n} \cdot \varphi' \left(\begin{array}{c} \dfrac{R}{F} + \dfrac{F'}{F^2}(1-n)\Delta f \\[2mm] + \dfrac{f_{,\alpha} f_{,\beta} g^{\alpha\beta}(1-n)}{4F^3}\left(4FF'' + (F')^2(n-6)\right) \end{array}\right) \\[6mm] \times \delta\left(\begin{array}{c} \dfrac{R}{F} + \dfrac{F'}{F^2}(1-n)\Delta f \\[2mm] + \dfrac{f_{,\alpha} f_{,\beta} g^{\alpha\beta}(1-n)}{4F^3}\left(4FF'' + (F')^2(n-6)\right) \end{array}\right) \end{array}\right)$$

with: $\quad \varphi'(R) = \dfrac{\partial\varphi[R]}{\partial R}; \quad \varphi'(R^*) = \dfrac{\partial\varphi[R^*]}{\partial R^*},$ (854)

$$= \int_V d^n x \left(\begin{array}{c} -\dfrac{\sqrt{-g \cdot F^n}}{2}\varphi \left(\begin{array}{c} \dfrac{R}{F} + \dfrac{F'}{F^2}(1-n)\Delta f \\[2mm] + \dfrac{f_{,\alpha} f_{,\beta} g^{\alpha\beta}(1-n)}{4F^3}\left(4FF'' + (F')^2(n-6)\right) \end{array}\right) g^{\kappa\lambda}\delta g_{\kappa\lambda} \\[6mm] + \sqrt{-g \cdot F^n} \cdot \left(\left(\begin{array}{c} \underbrace{\dfrac{8\pi G}{\kappa}T^{\kappa\lambda}}_{} = \text{matter} \\ 0 = \text{vacuum} \end{array}\right) + \Lambda \cdot g^{\kappa\lambda}\right)\delta g_{\kappa\lambda} \\[6mm] + \dfrac{\sqrt{-g \cdot F^n}}{F} \cdot \varphi' \left(\begin{array}{c} \dfrac{R}{F} + \dfrac{F'}{F^2}(1-n)\Delta f \\[2mm] + \dfrac{f_{,\alpha} f_{,\beta} g^{\alpha\beta}(1-n)}{4F^3}\left(4FF'' + (F')^2(n-6)\right) \end{array}\right) \\[6mm] \times \delta\left(\begin{array}{c} R + \dfrac{F'}{F}(1-n)\Delta f \\[2mm] + \dfrac{f_{,\alpha} f_{,\beta} g^{\alpha\beta}(1-n)}{4F^2}\left(4FF'' + (F')^2(n-6)\right) \end{array}\right) \end{array}\right),$$

(855)

$$
= \int\limits_V d^n x \left(
\begin{array}{l}
-\dfrac{\sqrt{-g \cdot F^n}}{2} \varphi \left(
\begin{array}{l}
\dfrac{R}{F} + \dfrac{F'}{F^2}(1-n)\Delta f \\[2mm]
+\dfrac{f_{,\alpha} f_{,\beta} g^{\alpha\beta}(1-n)}{4F^3}\left(4FF'' + (F')^2(n-6)\right)
\end{array}
\right) g^{\kappa\lambda}\delta g_{\kappa\lambda} \\[6mm]
+\sqrt{-g \cdot F^n} \cdot \left(\left(
\begin{array}{l}
\underbrace{\dfrac{8\pi G}{\kappa} T^{\kappa\lambda}}_{} = \text{matter} \\[2mm]
0 = \text{vacuum}
\end{array}
\right) + \Lambda \cdot g^{\kappa\lambda} \right)\delta g_{\kappa\lambda} \\[8mm]
+\dfrac{\sqrt{-g \cdot F^n}}{F} \cdot \varphi' \left(
\begin{array}{l}
\dfrac{R}{F} + \dfrac{F'}{F^2}(1-n)\Delta f \\[2mm]
+\dfrac{f_{,\alpha} f_{,\beta} g^{\alpha\beta}(1-n)}{4F^3}\left(4FF'' + (F')^2(n-6)\right)
\end{array}
\right) \\[6mm]
\times \left(
\begin{array}{l}
R^{\delta\gamma}\delta g_{\kappa\lambda} + g^{\alpha\beta}\delta R_{\alpha\beta} + \\[2mm]
\delta\left(\dfrac{F'}{F}(1-n)\Delta f + \dfrac{f_{,\alpha} f_{,\beta} g^{\alpha\beta}(1-n)}{4F^2}\left(4FF'' + (F')^2(n-6)\right)\right)
\end{array}
\right)
\end{array}
\right), \quad (856)
$$

$$
= \int\limits_V d^n x \left(
\begin{array}{l}
-\dfrac{\sqrt{-g \cdot F^n}}{2} \varphi \left(
\begin{array}{l}
\dfrac{R}{F} + \dfrac{F'}{F^2}(1-n)\Delta f \\[2mm]
+\dfrac{f_{,\alpha} f_{,\beta} g^{\alpha\beta}(1-n)}{4F^3}\left(4FF'' + (F')^2(n-6)\right)
\end{array}
\right) g^{\kappa\lambda}\delta g_{\kappa\lambda} \\[6mm]
+\sqrt{-g \cdot F^n} \cdot \left(\left(
\begin{array}{l}
\underbrace{\dfrac{8\pi G}{\kappa} T^{\kappa\lambda}}_{} = \text{matter} \\[2mm]
0 = \text{vacuum}
\end{array}
\right) + \Lambda \cdot g^{\kappa\lambda} \right)\delta g_{\kappa\lambda} \\[8mm]
+\dfrac{\sqrt{-g \cdot F^n}}{F} \cdot \left[
\begin{array}{l}
\varphi' \overbrace{\left(
\begin{array}{l}
\dfrac{R}{F} + \dfrac{F'}{F^2}(1-n)\Delta f \\[2mm]
+\dfrac{f_{,\alpha} f_{,\beta} g^{\alpha\beta}(1-n)}{4F^3}\left(4FF'' + (F')^2(n-6)\right)
\end{array}
\right)}^{=R^*}\delta g_{\kappa\lambda} \\[4mm]
+\varphi''(R^*)\left(R^{;\kappa\lambda} - \Delta_g R \cdot g^{\kappa\lambda}\right) \\[2mm]
+\varphi'''(R^*)\left(R^{;\kappa}R^{;\lambda} - R^{;\sigma}R_{;\sigma} \cdot g^{\kappa\lambda}\right) \\[2mm]
+FSTs + \varphi'(R^*)R^{\kappa\lambda}
\end{array}
\right] \\[8mm]
+\dfrac{\sqrt{-g \cdot F^n}}{F} \cdot \varphi'(R^*) \\[4mm]
\times\delta\left(\dfrac{F'}{F}(1-n)\Delta f + \dfrac{f_{,\alpha} f_{,\beta} g^{\alpha\beta}(1-n)}{4F^2}\left(4FF'' + (F')^2(n-6)\right)\right)
\end{array}
\right). \quad (857)
$$

Thereby the expression FSTs stands for the former surface terms which now (just as elaborated in sub-section "8.2.3 Variation of the Scaled Ricci Tensor Term"), due to $\dfrac{\sqrt{-g}\cdot F^n}{F}$, do not give surface terms anymore (at least not that easily). Thus, we only have the terms FSTs and

$$\delta\left(\frac{F'}{F}(1-n)\Delta f + \frac{f_{,\alpha}f_{,\beta}g^{\alpha\beta}(1-n)}{4F^2}\left(4FF'' + (F')^2(n-6)\right) \right)$$ left to variate and the

latter was already done in the previous sections of this paper. The total result now looks as follows:

$$= \int_V d^n x \frac{\sqrt{-g}\cdot F^n}{F} \cdot \left(\begin{array}{c} \left(\begin{array}{c} -\varphi\left(\begin{array}{c} \dfrac{R}{F} + \dfrac{F'}{F^2}(1-n)\Delta f \\[2mm] + \dfrac{f_{,\alpha}f_{,\beta}g^{\alpha\beta}(1-n)}{4F^3}\left(4FF'' + (F')^2(n-6)\right) \end{array}\right) \Bigg|\dfrac{F}{2}\cdot g^{\delta\gamma} \\[6mm] + \left(\left(\begin{array}{c} \dfrac{8\pi G\,T^{\kappa\lambda}}{\kappa} = \text{matter} \\[2mm] 0 = \text{vacuum} \end{array}\right) + \Lambda\cdot g^{\kappa\lambda}\right)\cdot F \\[6mm] +\varphi'(R^*) + \varphi''(R^*)\left(R^{;\kappa\lambda} - \Delta_g R\cdot g^{\kappa\lambda}\right) \\[2mm] +\varphi'''(R^*)\left(R^{;\kappa}R^{;\lambda} - R^{;\sigma}R_{;\sigma}\cdot g^{\kappa\lambda}\right) \\[2mm] +\text{FSTs} + \varphi'(R^*)R^{\delta\gamma} + \varphi'(R^*) \\[2mm] +\left(g^{\kappa\lambda}g^{\alpha\beta} - g^{\kappa\alpha}g^{\lambda\beta} - g^{\lambda\alpha}g^{\kappa\beta}\right)(1-n) \\[2mm] \left[\dfrac{f_{,\alpha}f_{,\beta}}{F^2}\left(FF'' + (F')^2\cdot\dfrac{3\cdot n-14}{8}\right) + f_{,\alpha\beta}\dfrac{F'}{F}\right] \\[4mm] -\dfrac{F'\cdot(1-n)}{2\cdot F}\left(g^{\kappa\lambda}g^{\alpha\beta} - g^{\kappa\alpha}g^{\lambda\beta} - g^{\lambda\alpha}g^{\kappa\beta}\right)\Gamma^\gamma_{\alpha\beta}f_{,\gamma} \\[4mm] +\dfrac{F'}{F}\cdot(1-n)g^{\alpha\beta}g^{\gamma\lambda}\Gamma^\kappa_{\alpha\beta}f_{,\gamma} \\[4mm] +\dfrac{F'}{F}\cdot\dfrac{(1-n)}{2}\left(g^{\kappa\lambda}g^{\alpha\beta} - g^{\kappa\alpha}g^{\lambda\beta} - g^{\lambda\alpha}g^{\kappa\beta}\right)_{,\beta}f_{,\alpha} \\[4mm] +\dfrac{g^{\xi\zeta}g_{\xi\zeta,\beta}}{2}\cdot\left(\dfrac{F'}{F}\cdot\dfrac{(1-n)}{2}\left(g^{\kappa\lambda}g^{\alpha\beta} - g^{\kappa\alpha}g^{\lambda\beta} - g^{\lambda\alpha}g^{\kappa\beta}\right)f_{,\alpha}\right) \end{array}\right)\dfrac{\delta g_{\kappa\lambda}}{F}\end{array}\right),$$

$$(858)$$

where we can conclude—in the usual way—that for arbitrary integrals, the whole integrand must vanish in order to fulfill the $\delta W = 0$-condition and we obtain:

$$
0 = \left\{
\begin{array}{l}
\left(\left(\begin{array}{c} \underbrace{8\pi G\, T^{\kappa\lambda}}_{\kappa} = \text{matter} \\ 0 = \text{vacuum} \end{array} \right) + \Lambda \cdot g^{\kappa\lambda} \right) \cdot F \\[3em]
-\varphi \left(\begin{array}{c} \dfrac{R}{F} + \dfrac{F'}{F^2}(1-n)\Delta f \\[1em] + \dfrac{f_{,\alpha} f_{,\beta}\, g^{\alpha\beta}(1-n)}{4F^3}\left(4FF'' + \left(F'\right)^2 (n-6) \right) \end{array} \right) \cdot \dfrac{F}{2} \cdot g^{\kappa\lambda} \\[3em]
+\varphi'\left(R^*\right) + \varphi''\left(R^*\right)\left(R^{;\kappa\lambda} - \Delta_g R \cdot g^{\kappa\lambda} \right) \\[1em]
+\varphi'''\left(R^*\right)\left(R^{;\kappa} R^{;\lambda} - R^{;\sigma} R_{;\sigma} \cdot g^{\kappa\lambda} \right) \\[1em]
+FSTs + \varphi'\left(R^*\right) R^{\kappa\lambda} + \varphi'\left(R^*\right) \\[1em]
+\left(g^{\kappa\lambda} g^{\alpha\beta} - g^{\kappa\alpha} g^{\lambda\beta} - g^{\lambda\alpha} g^{\kappa\beta} \right)(1-n) \\[1em]
\cdot \left[\dfrac{f_{,\alpha} f_{,\beta}}{F^2}\left(FF'' + \left(F'\right)^2 \cdot \dfrac{3 \cdot n - 14}{8} \right) + f_{,\alpha\beta}\dfrac{F'}{F} \right] \\[2em]
-\dfrac{F' \cdot (1-n)}{2 \cdot F}\left(g^{\kappa\lambda} g^{\alpha\beta} - g^{\kappa\alpha} g^{\lambda\beta} - g^{\lambda\alpha} g^{\kappa\beta} \right)\Gamma^{\gamma}_{\alpha\beta} f_{,\gamma} \\[2em]
+\dfrac{F'}{F}\cdot(1-n) g^{\alpha\beta} g^{\gamma\lambda}\Gamma^{\kappa}_{\alpha\beta} f_{,\gamma} \\[2em]
+\dfrac{F'}{F}\cdot\dfrac{(1-n)}{2}\left(g^{\kappa\lambda} g^{\alpha\beta} - g^{\kappa\alpha} g^{\lambda\beta} - g^{\lambda\alpha} g^{\kappa\beta} \right)_{,\beta} f_{,\alpha} \\[2em]
+\dfrac{g^{\xi\zeta} g_{\xi\zeta,\beta}}{2}\cdot\left(\dfrac{F'}{F}\cdot\dfrac{(1-n)}{2}\left(g^{\kappa\lambda} g^{\alpha\beta} - g^{\kappa\alpha} g^{\lambda\beta} - g^{\lambda\alpha} g^{\kappa\beta} \right) f_{,\alpha} \right)
\end{array}
\right.
$$

(859)

The interested reader may also sort out the situation with the FSTs, which should be possible with the recipe given in sub-section "8.2.3 Variation of the Scaled Ricci Tensor Term." Here we do not bother with this rather lengthy calculation, because—as said at the beginning of this section—there is no experimental evidence for a deviation from the linear R- or R^*-dependency as kernel for the Einstein–Hilbert action. Instead, we here only give the result:

$$0 = \left\{ \left[-\varphi \left(\begin{array}{c} \dfrac{R}{F} + \dfrac{F'}{F^2}(1-n)\Delta f \\ + \dfrac{f_{,\alpha} f_{,\beta} g^{\alpha\beta}(1-n)}{4F^3}\left(4FF'' + (F')^2(n-6)\right) \end{array} \right) \cdot \dfrac{F}{2} \cdot g^{\kappa\lambda} \right. \right.$$

$$+\varphi'(R^*) + \varphi''(R^*)\left(R^{;\kappa\lambda} - \Delta_g R \cdot g^{\kappa\lambda}\right)$$

$$+\varphi'''(R^*)\left(R^{;\kappa}R^{;\lambda} - R^{;\sigma}R_{,\sigma} \cdot g^{\kappa\lambda}\right)$$

$$+\varphi'(R^*)R^{\kappa\lambda} + \varphi'(R^*)$$

$$+\left(g^{\kappa\lambda}g^{\alpha\beta} - g^{\kappa\alpha}g^{\lambda\beta} - g^{\lambda\alpha}g^{\kappa\beta}\right)(1-n)$$

$$\cdot\left[\dfrac{f_{,\alpha}f_{,\beta}}{F^2}\left(FF'' + (F')^2 \cdot \dfrac{3\cdot n - 14}{8}\right) + f_{,\alpha\beta}\dfrac{F'}{F}\right]$$

$$-\dfrac{F'\cdot(1-n)}{2\cdot F}\left(g^{\kappa\lambda}g^{\alpha\beta} - g^{\kappa\alpha}g^{\lambda\beta} - g^{\lambda\alpha}g^{\kappa\beta}\right)\Gamma^{\gamma}_{\alpha\beta}f_{,\gamma}$$

$$+\dfrac{F'}{F}\cdot(1-n)g^{\alpha\beta}g^{\gamma\lambda}\Gamma^{\kappa}_{\alpha\beta}f_{,\gamma}$$

$$+\dfrac{F'}{F}\cdot\dfrac{(1-n)}{2}\left(g^{\kappa\lambda}g^{\alpha\beta} - g^{\kappa\alpha}g^{\lambda\beta} - g^{\lambda\alpha}g^{\kappa\beta}\right)_{,\beta}f_{,\alpha}$$

$$\left. +\dfrac{g^{\xi\zeta}g_{\xi\zeta,\beta}}{2}\cdot\left(\dfrac{F'}{F}\cdot\dfrac{(1-n)}{2}\left(g^{\kappa\lambda}g^{\alpha\beta} - g^{\kappa\alpha}g^{\lambda\beta} - g^{\lambda\alpha}g^{\kappa\beta}\right)f_{,\alpha}\right) \right]$$

$$+\left(\dfrac{n}{2}-1\right)\left[+g^{\alpha\beta}\dfrac{F'}{F}\left(\dfrac{f_{,\alpha}\cdot g^{\xi\lambda}\Gamma^{\kappa}_{\beta\xi} + f_{,\alpha}\cdot g^{\kappa\xi}\Gamma^{\lambda}_{\beta\xi}}{2} + f_{,\gamma}\cdot g^{\gamma\lambda}\Gamma^{\kappa}_{\alpha\beta}\right) \right.$$

$$+\dfrac{F'}{F}f_{,\gamma}\cdot\dfrac{\left(g^{\kappa\rho}g^{\gamma\lambda} + g^{\rho\lambda}g^{\gamma\kappa} - g^{\kappa\lambda}g^{\gamma\rho}\right)_{,\rho}}{2}$$

$$+\dfrac{F'}{F}f_{,\gamma\rho}\cdot\dfrac{\left(g^{\kappa\rho}g^{\gamma\lambda} + g^{\rho\lambda}g^{\gamma\kappa} - g^{\kappa\lambda}g^{\gamma\rho}\right)}{2}$$

$$+\dfrac{F''}{F}f_{,\rho}f_{,\gamma}\cdot\dfrac{\left(g^{\kappa\rho}g^{\gamma\lambda} + g^{\rho\lambda}g^{\gamma\kappa} - g^{\kappa\lambda}g^{\gamma\rho}\right)}{2}$$

$$+\left(\dfrac{n}{2}-2\right)\dfrac{F'^2}{F^2}f_{,\rho}f_{,\gamma}\cdot\dfrac{\left(g^{\kappa\rho}g^{\gamma\lambda} + g^{\rho\lambda}g^{\gamma\kappa} - g^{\kappa\lambda}g^{\gamma\rho}\right)}{2}$$

$$+\dfrac{g^{\xi\zeta}g_{\xi\zeta,\rho}}{2}\dfrac{F'}{F}f_{,\gamma}\cdot\dfrac{\left(g^{\kappa\rho}g^{\gamma\lambda} + g^{\rho\lambda}g^{\gamma\kappa} - g^{\kappa\lambda}g^{\gamma\rho}\right)}{2}$$

$$-\dfrac{F'}{F}\dfrac{g^{\alpha\beta}}{2}\cdot f_{,\alpha}\cdot g^{\kappa\lambda}{}_{,\beta} - \dfrac{F'}{F}\dfrac{g^{\alpha\beta}}{2}\cdot f_{,\alpha\beta}\cdot g^{\kappa\lambda} - \dfrac{F''}{F}\dfrac{g^{\alpha\beta}}{2}\cdot f_{,\beta}f_{,\alpha}\cdot g^{\kappa\lambda}$$

$$-\dfrac{g^{\alpha\beta}}{2}\cdot\left(\dfrac{n}{2}-2\right)\dfrac{F'^2}{F^2}f_{,\beta}f_{,\alpha}\cdot g^{\kappa\lambda} - \dfrac{F'}{F}\dfrac{g^{\alpha\beta}{}_{,\beta}}{2}\cdot f_{,\alpha}\cdot g^{\kappa\lambda}$$

$$\left.\left.\left. -\dfrac{F'}{F}\dfrac{g^{\xi\zeta}g_{\xi\zeta,\beta}}{2}\dfrac{g^{\alpha\beta}}{2}\cdot f_{,\alpha}\cdot g^{\kappa\lambda} \right]\right]\right\}.$$

(860)

Hopefully somebody would be willing to find some suitable simplifications for this. Here we can aim for an indirect simplification by going back to Eq. (856) and considering the term:

$$
= \int_V d^n x \left(\begin{array}{c} \cdots \\ \left(+\dfrac{\sqrt{-g} \cdot F^n}{F} \cdot \varphi' \left(\begin{array}{c} \dfrac{R}{F} + \dfrac{F'}{F^2}(1-n)\Delta f \\ + \dfrac{f_{,\alpha} f_{,\beta} g^{\alpha\beta}(1-n)}{4F^3} \left(4FF'' + (F')^2 (n-6) \right) \end{array} \right) \right) \\ \times \left(\cdots + g^{\alpha\beta} \delta R_{\alpha\beta} + \delta \left(\begin{array}{c} \dfrac{F'}{F}(1-n)\Delta f \\ + \dfrac{f_{,\alpha} f_{,\beta} g^{\alpha\beta}(1-n)}{4F^2} \left(4FF'' + (F')^2 (n-6) \right) \end{array} \right) \right) \end{array} \right),
$$

$$(861)$$

in connection with the evaluations of Eq. [26]. The variation result for the part shown in Eq. (861) is:

$$
= \int_V d^n x \left(\begin{array}{c} \cdots \\ \left(+\dfrac{\sqrt{-g} \cdot F^n}{F} \cdot \varphi' \left(\begin{array}{c} \dfrac{R}{F} + \dfrac{F'}{F^2}(1-n)\Delta f \\ + \dfrac{f_{,\alpha} f_{,\beta} g^{\alpha\beta}(1-n)}{4F^3} \left(4FF'' + (F')^2 (n-6) \right) \end{array} \right) \right) \\ \times \left(\cdots + g^{\alpha\beta} \delta R_{\alpha\beta} + \delta \left(\begin{array}{c} \dfrac{F'}{F}(1-n)\Delta f \\ + \dfrac{f_{,\alpha} f_{,\beta} g^{\alpha\beta}(1-n)}{4F^2} \left(4FF'' + (F')^2 (n-6) \right) \end{array} \right) \right) \end{array} \right)
$$

$$
= \int_V d^n x \left(\begin{array}{c} \cdots \\ -\left(\partial_{;\mu} \partial_{;\nu} - g_{\mu\nu} \Delta_g \right) \left[\varphi'\left(R^*\right) \cdot \dfrac{\sqrt{F^n}}{F} \right] \\ +\dfrac{\sqrt{-g} \cdot F^n}{F} \cdot \varphi'\left(R^*\right) \delta \left(\begin{array}{c} \dfrac{F'}{F}(1-n)\Delta f \\ + \dfrac{f_{,\alpha} f_{,\beta} g^{\alpha\beta}(1-n)}{4F^2} \left(4FF'' + (F')^2 (n-6) \right) \end{array} \right) \\ +\cdots \end{array} \right),
$$

$$(862)$$

which significantly abbreviates our Eq. (860), namely:

$$0 = \int_V d^n x \frac{\sqrt{-g \cdot F^n}}{F} \cdot \left(\begin{array}{c} -F \cdot \dfrac{\varphi(R^*)}{2} g^{\kappa\lambda} \\[2ex] +F \cdot \left(\left(\begin{array}{c} \underbrace{\dfrac{8\pi G T^{\kappa\lambda}}{\kappa} = \text{matter}} \\ 0 = \text{vacuum} \end{array} \right) + \Lambda \cdot g^{\kappa\lambda} \right) \\[4ex] -F^{1-\frac{n}{2}} \left(\partial^{;\kappa} \partial^{;\lambda} - g^{\kappa\lambda} \Delta_g \right) \left[\varphi'(R^*) \cdot F^{\frac{n}{2}-1} \right] + \varphi'(R^*) R^{\kappa\lambda} \\[2ex] + \cdots \end{array} \right) \delta g_{\kappa\lambda}$$

$$\Rightarrow \quad 0 = \left\{ \begin{array}{c} \varphi'(R^*) R^{\kappa\lambda} - \dfrac{1}{2} \varphi(R^*) g^{\kappa\lambda} + \left(\left(\begin{array}{c} \underbrace{\dfrac{8\pi G T^{\kappa\lambda}}{\kappa} = \text{matter}} \\ 0 = \text{vacuum} \end{array} \right) + \Lambda \cdot g^{\kappa\lambda} \right) \\[4ex] -\dfrac{1}{\sqrt{F^n}} \cdot \left(\partial^{;\kappa} \partial^{;\lambda} - g^{\kappa\lambda} \Delta_g \right) \left[\varphi'(R^*) \cdot F^{\frac{n}{2}-1} \right] + \varphi'(R^*) \\[4ex] \times \left(\begin{array}{c} + \left(g^{\kappa\lambda} g^{\alpha\beta} - g^{\kappa\alpha} g^{\lambda\beta} - g^{\lambda\alpha} g^{\kappa\beta} \right)(1-n) \\[1ex] \cdot \left[\dfrac{f_{,\alpha} f_{,\beta}}{F^2} \left(FF'' + (F')^2 \cdot \dfrac{3 \cdot n - 14}{8} \right) + f_{,\alpha\beta} \dfrac{F'}{F} \right] \\[2ex] - \dfrac{F' \cdot (1-n)}{2 \cdot F} \left(g^{\kappa\lambda} g^{\alpha\beta} - g^{\kappa\alpha} g^{\lambda\beta} - g^{\lambda\alpha} g^{\kappa\beta} \right) \Gamma^\gamma_{\alpha\beta} f_{,\gamma} \\[2ex] + \dfrac{F'}{F} \cdot (1-n) g^{\alpha\beta} g^{\gamma\lambda} \Gamma^\kappa_{\alpha\beta} f_{,\gamma} \\[2ex] + \dfrac{F'}{F} \cdot \dfrac{(1-n)}{2} \left(g^{\kappa\lambda} g^{\alpha\beta} - g^{\kappa\alpha} g^{\lambda\beta} - g^{\lambda\alpha} g^{\kappa\beta} \right)_{,\beta} f_{,\alpha} \\[2ex] + \dfrac{g^{\xi\zeta} g_{\xi\zeta,\beta}}{2} \cdot \left(\dfrac{F'}{F} \cdot \dfrac{(1-n)}{2} \left(g^{\kappa\lambda} g^{\alpha\beta} - g^{\kappa\alpha} g^{\lambda\beta} - g^{\lambda\alpha} g^{\kappa\beta} \right) f_{,\alpha} \right) \end{array} \right) \end{array} \right.$$

(863)

8.7 Discussion with Respect to Alternatives and Potentially Missing Features

It was shown in [3] that the typical equation of elasticity can also be derived when extending the classical variation in the following way:

$$\delta g_{\delta\gamma} = \delta\left(\mathbf{g}_\delta \cdot \mathbf{g}_\gamma\right) = \mathbf{g}_\delta \cdot \delta\mathbf{g}_\gamma + \delta\mathbf{g}_\delta \cdot \mathbf{g}_\gamma$$

$$= \frac{\partial G^i\left[x_k\right]}{\partial x^\delta}\mathbf{e}_i \cdot \left(\frac{\partial^2 G^j\left[x_k\right]}{\partial x^\beta \partial x^\gamma}\mathbf{e}_j\right)\delta x^\beta + \left(\frac{\partial^2 G^i\left[x_k\right]}{\partial x^\chi \partial x^\delta}\mathbf{e}_i\right)\delta x^\chi \cdot \frac{\partial G^j\left[x_k\right]}{\partial x^\gamma}\mathbf{e}_j$$

$$= \frac{\partial G^i\left[x_k\right]}{\partial x^\delta}\mathbf{e}_i \cdot \left(\frac{\partial^2 G^j\left[x_k\right]}{\partial x^\gamma \partial x^\gamma}\mathbf{e}_j\right)C_\beta^\gamma \delta x^\beta + \left(\frac{\partial^2 G^i\left[x_k\right]}{\partial x^\chi \partial x^\delta}\mathbf{e}_i\right)\delta x^\chi \cdot \frac{\partial G^j\left[x_k\right]}{\partial x^\gamma}\mathbf{e}_j$$

$$= \frac{\partial G^i\left[x_k\right]}{\partial x^\delta}\mathbf{e}_i \cdot \left(\frac{\partial^2 G^j\left[x_k\right]}{\left(\partial x^\gamma\right)^2}\mathbf{e}_j\right)C_\beta^\gamma \delta x^\beta + \left(\frac{\partial^2 G^i\left[x_k\right]}{\partial x^\chi \partial x^\delta}\mathbf{e}_i\right)\delta x^\beta C_\beta^\chi \cdot C_\gamma^\delta \frac{\partial G^j\left[x_k\right]}{\partial x^\delta}\mathbf{e}_j.$$

$$(864)$$

So far it is not clear to the author how this could be connected to our scaled metric derivation from above, but it appears intriguing that elastic equations seem to pop up in connection with the base vector of the metric (general relativity) and the scale factor (quantum part). Before moving on and investigating this connection in this chapter, however, we first need to repeat some essentials and try to work out what might be missing.

8.7.1 Alternative Way for Deriving the Equations of Elasticity from the Metric Origin out of the Einstein–Hilbert Action

The experienced reader recognizes the structural elements of the basic equation of elasticity (e.g., [9], pp. 166), which can be given for an isotropic material with the Poisson's ratio μ as:

$$\left(\overset{\overset{\triangleq a}{}}{\left(1 - 2\cdot\mu\right)}\cdot \Delta G^j\left[x_k\right] + \left(\frac{\partial^2 G^j\left[x_k\right]}{\partial x^\gamma \partial x^\delta}\right)\right)\mathbf{e}_j = 0. \qquad (865)$$

A simple exchange of the dummy indices in Eq. (864) leads us to:

$$\delta g_{\delta\gamma} = \delta\left(\mathbf{g}_\delta \cdot \mathbf{g}_\gamma\right) = \mathbf{g}_\delta \cdot \delta\mathbf{g}_\gamma + \delta\mathbf{g}_\delta \cdot \mathbf{g}_\gamma$$

$$= \frac{\partial G^i\left[x_k\right]}{\partial x^\delta}\mathbf{e}_i \cdot \left(\frac{\partial^2 G^j\left[x_k\right]}{\left(\partial x^\gamma\right)^2}\mathbf{e}_j\right)C_\beta^\gamma \delta x^\beta + \left(\frac{\partial^2 G^i\left[x_k\right]}{\partial x^\chi \partial x^\delta}\mathbf{e}_i\right)\delta x^\beta C_\beta^\chi \cdot C_\gamma^\delta \frac{\partial G^j\left[x_k\right]}{\partial x^\delta}\mathbf{e}_j$$

$$= \frac{\partial G^j\left[x_k\right]}{\partial x^\delta}\mathbf{e}_j \cdot \left(\frac{\partial^2 G^i\left[x_k\right]}{\left(\partial x^\gamma\right)^2}\mathbf{e}_i\right)C_\beta^\gamma \delta x^\beta + \left(\frac{\partial^2 G^i\left[x_k\right]}{\partial x^\chi \partial x^\delta}\mathbf{e}_i\right)\delta x^\beta C_\beta^\chi \cdot C_\gamma^\delta \frac{\partial G^j\left[x_k\right]}{\partial x^\delta}\mathbf{e}_j$$

$$= \left(\left(\frac{\partial^2 G^i\left[x_k\right]}{\left(\partial x^\gamma\right)^2}\mathbf{e}_i\right)C_\beta^\gamma + \left(\frac{\partial^2 G^i\left[x_k\right]}{\partial x^\chi \partial x^\delta}\mathbf{e}_i\right)C_\beta^\chi C_\gamma^\delta\right)\cdot \delta x^\beta \frac{\partial G^j\left[x_k\right]}{\partial x^\delta}\mathbf{e}_j, \qquad (866)$$

and assuming $C_\beta^\gamma = b \cdot \delta_\beta^\gamma, C_\beta^\chi C_\gamma^\delta = b^2 \cdot \delta_\beta^\chi \delta_\gamma^\delta, a = 1/b$ yields the isotropic equation:

$$\delta g_{\delta\gamma} = \delta\left(\mathbf{g}_\delta \cdot \mathbf{g}_\gamma\right) = \mathbf{g}_\delta \cdot \delta\mathbf{g}_\gamma + \delta\mathbf{g}_\delta \cdot \mathbf{g}_\gamma$$

$$= \left(\left(\frac{\partial^2 G^i\left[x_k\right]}{\left(\partial x^\gamma\right)^2}\mathbf{e}_i\right)C_\beta^\gamma + \left(\frac{\partial^2 G^i\left[x_k\right]}{\partial x^\chi \partial x^\delta}\mathbf{e}_i\right)C_\beta^\chi C_\gamma^\delta\right) \cdot \delta x^\beta \frac{\partial G^j\left[x_k\right]}{\partial x^\delta}\mathbf{e}_j$$

$$= b^2 \cdot \left(\frac{\Delta G^i\left[x_k\right]\mathbf{e}_i}{b} + \left(\frac{\partial^2 G^i\left[x_k\right]}{\partial x^\gamma \partial x^\delta}\mathbf{e}_i\right)\right) \cdot \delta x^\beta \frac{\partial G^j\left[x_k\right]}{\partial x^\delta}\mathbf{e}_j$$

$$= b^2 \cdot \left(a \cdot \Delta G^i\left[x_k\right]\mathbf{e}_i + \left(\frac{\partial^2 G^i\left[x_k\right]}{\partial x^\gamma \partial x^\delta}\mathbf{e}_i\right)\right) \cdot \delta x^\beta \frac{\partial G^j\left[x_k\right]}{\partial x^\delta}\mathbf{e}_j. \tag{867}$$

With the n-function-ansatz from [3], we can solve:

$$a \cdot \Delta G^j\left[x_k\right]\mathbf{e}_j + \left(\frac{\partial^2 G^j\left[x_k\right]}{\partial x^\gamma \partial x^\delta}\mathbf{e}_j\right) = \left(\overbrace{\left(1 - 2 \cdot \mu\right)}^{\hat{=}a} \cdot \Delta G^j\left[x_k\right] + \left(\frac{\partial^2 G^j\left[x_k\right]}{\partial x^\gamma \partial x^\delta}\right)\right)\mathbf{e}_j = 0 \tag{868}$$

with $\alpha = -1 - 2*a$.

As in connection with the general equation:

$$\delta g_{\delta\gamma} = \delta\left(\mathbf{g}_\delta \cdot \mathbf{g}_\gamma\right) = \mathbf{g}_\delta \cdot \delta\mathbf{g}_\gamma + \delta\mathbf{g}_\delta \cdot \mathbf{g}_\gamma$$

$$= \frac{\partial G^i\left[x_k\right]}{\partial x^\delta}\mathbf{e}_i \cdot \left(\frac{\partial^2 G^j\left[x_k\right]}{\partial x^\alpha \partial x^\gamma}\mathbf{e}_j\right)\delta x^\alpha + \left(\frac{\partial^2 G^i\left[x_k\right]}{\partial x^\alpha \partial x^\delta}\mathbf{e}_i\right)\delta x^\alpha \cdot \frac{\partial G^j\left[x_k\right]}{\partial x^\gamma}\mathbf{e}_j$$

$$= \frac{\partial G^i\left[x_k\right]}{\partial x^\delta}\mathbf{e}_i \cdot \sum_{\alpha=0}^{n-1}\left(\frac{\partial^2 G^j\left[x_k\right]}{\partial x^\alpha \partial x^\gamma}\mathbf{e}_j\right)\delta x^\alpha + \sum_{\alpha=0}^{n-1}\left(\frac{\partial^2 G^i\left[x_k\right]}{\partial x^\alpha \partial x^\delta}\mathbf{e}_i\right)\delta x^\alpha \cdot \frac{\partial G^j\left[x_k\right]}{\partial x^\gamma}\mathbf{e}_j$$

$$= \frac{\partial G^i\left[x_k\right]}{\partial x^\delta}\mathbf{e}_i \cdot \sum_{\alpha=0}^{n-1}\left(\frac{\partial^2 G^j\left[x_k\right]}{\partial x^\alpha \partial x^\gamma}\mathbf{e}_j\right)\delta^\alpha + \sum_{\alpha=0}^{n-1}\left(\frac{\partial^2 G^i\left[x_k\right]}{\partial x^\alpha \partial x^\delta}\mathbf{e}_i\right)\delta^\alpha \cdot \frac{\partial G^j\left[x_k\right]}{\partial x^\gamma}\mathbf{e}_j$$

$$\Rightarrow \quad \sum_{\alpha=0}^{n-1}\left(\frac{\partial^2 G^j\left[x_k\right]}{\partial x^\alpha \partial x^\gamma}\mathbf{e}_j\right)\delta^\alpha = 0 \tag{869}$$

from [3], it should be pointed out that with non-equal δx^α or δ^α for the various α, we could even derive anisotropic equations, respectively equations for anisotropic space-times.

It should be noted that, as shown before in [3] (see sub-section "4.1 Matrix Option and Classical Dirac Form"), one could also extract a Dirac-like equation from Eq. (867). This could be obtained via $dx^\beta \to dx^\delta$ leading to:

$$\delta g_{\delta\gamma} = \delta\left(\mathbf{g}_\delta \cdot \mathbf{g}_\gamma\right) = \mathbf{g}_\delta \cdot \delta\mathbf{g}_\gamma + \delta\mathbf{g}_\delta \cdot \mathbf{g}_\gamma$$

$$= b^2 \cdot \left(a \cdot \Delta G^i\left[x_k\right]\mathbf{e}_i + \left(\frac{\partial^2 G^i\left[x_k\right]}{\partial x^\gamma \partial x^\delta}\mathbf{e}_i\right)\right) \cdot \delta x^\beta \frac{\partial G^j\left[x_k\right]}{\partial x^\delta}\mathbf{e}_j$$

$$= b^2 \cdot \left(a \cdot \Delta G^i\left[x_k\right]\mathbf{e}_i + \left(\frac{\partial^2 G^i\left[x_k\right]}{\partial x^\gamma \partial x^\delta}\mathbf{e}_i\right)\right) \cdot \delta x^\delta \frac{\partial G^j\left[x_k\right]}{\partial x^\delta}\mathbf{e}_j$$

$$\Rightarrow \qquad \sum_{\delta=0}^{n-1} \delta x^\delta \frac{\partial G^j\left[x_k\right]}{\partial x^\delta}\mathbf{e}_j = 0. \qquad (870)$$

We have to point out that, even though these Dirac-like equations are anything but close to classical elasticity equations, they still give displacement distributions of spaces and might automatically include defects (like dislocations), which require quite some mathematical construction work in the classical technical mechanics. Thus, our purely metric approach here may even find some use in more down to earth applications of materials science.

8.7.2 Where Is Thermodynamics?

Having worked out the similarity of quantum gravity and the theory of elasticity has also allowed us to try and find other fundamental fields inside the already derived general equation(s). Such general considerations have been performed in [3]. Here we repeat the question about the quantum gravity origin of thermodynamics.

It is well known from the thermodynamics of deformed bodies that the work of deformation can be described as the inner product of the stress tensor σ^{jk} with the deformation tensor u_{jk}:

$$W_D = \sigma^{jk}u_{jk}. \qquad (871)$$

Knowing that the functions G^j from the sub-section above are displacement fields and closely investigating them, we conclude that the variation of the metric tensor (870) is nothing else than the holistic equivalent of a deformation tensor and thus, that the integrand in the classical and the extended Einstein–Hilbert action is giving us nothing else but a very general form of deformation work.

If this is the case, however, and especially as we assume the Einstein–Hilbert action to "contain it all" (at least in its extended form), we have to ask where we could find the equivalent to thermal energy? After all, sticking to the simpler classical action form, which reads:

$$\left(R^{\delta\gamma} - \frac{1}{2} R g^{\delta\gamma} + \Lambda g^{\delta\gamma} + \kappa T^{\delta\gamma} \right) \delta g_{\delta\gamma}, \tag{872}$$

none of the addends in parenthesis has the makings of a term T^*dS or S^*dT with S denoting the entropy and T denoting temperature.

Obviously, we are missing something ... at least as long as we consider thermodynamics as so fundamental that it has to show up in a truly fundamental theory.

Thus, with thermodynamics missing inside the classical metric and the newer base-vector variation of the Einstein–Hilbert action, we have to look for a different way to obtain what we need.

8.7.3   Repetition: The Base-Vector Variation

Our starting point shall be the usual tensor transformation rule for the covariant metric tensor:

$$g_{\delta\gamma} = \mathbf{g}_\delta \cdot \mathbf{g}_\gamma = \frac{\partial G^\alpha \left[x_k \right]}{\partial x^\delta} \frac{\partial G^\beta \left[x_k \right]}{\partial x^\gamma} g_{\alpha\beta}. \tag{873}$$

This already leads to quite some varieties with respect to the apparently simple term $\delta g_{\mu\nu}$ (e.g., see [3, 21]). The base vectors $\mathbf{g}_\delta$ to a certain metric are given as:

$$\mathbf{g}_\delta = \frac{\partial G^\alpha \left[x_k \right]}{\partial x^\delta} \mathbf{e}_\alpha, \tag{874}$$

where the functions $G^\alpha[\; ... \;]$ denote arbitrary functions of the coordinates x_k. Here the vectors $\mathbf{e}_\alpha$ shall denote the base vectors of a fundamental coordinate system of the right (in principle arbitrary) number of dimensions. Thus, we have the variation for $\delta g_{\delta\gamma}$ as follows:

$$\delta g_{\delta\gamma} = \delta \left(\mathbf{g}_\delta \cdot \mathbf{g}_\gamma \right) = \mathbf{g}_\delta \cdot \delta \mathbf{g}_\gamma + \delta \mathbf{g}_\delta \cdot \mathbf{g}_\gamma$$

$$= \frac{\partial G^\alpha \left[x_k \right]}{\partial x^\delta} \mathbf{e}_\alpha \cdot \delta \left(\frac{\partial G^\beta \left[x_k \right]}{\partial x^\gamma} \mathbf{e}_\beta \right) + \delta \left(\frac{\partial G^\alpha \left[x_k \right]}{\partial x^\delta} \mathbf{e}_\alpha \right) \cdot \frac{\partial G^\beta \left[x_k \right]}{\partial x^\gamma} \mathbf{e}_\beta. \tag{875}$$

Now we introduce two additional degrees of freedom, namely:

(a) neither the number of dimensions in which the base vectors exist and form a complete transformation (874) needs to be the same as the metric space they define,

(b) nor the variation δ is defined or fixed in any way.

In order to properly account for point (a), we shall rewrite Eqs. (873) and (874) as:

$$\delta g_{\delta\gamma} = \delta\left(\mathbf{g}_\delta \cdot \mathbf{g}_\gamma\right) = \mathbf{g}_\delta \cdot \delta\mathbf{g}_\gamma + \delta\mathbf{g}_\delta \cdot \mathbf{g}_\gamma$$

$$= \frac{\partial G^i\left[x_k\right]}{\partial x^\delta}\mathbf{e}_i \cdot \delta\left(\frac{\partial G^j\left[x_k\right]}{\partial x^\gamma}\mathbf{e}_j\right) + \delta\left(\frac{\partial G^i\left[x_k\right]}{\partial x^\delta}\mathbf{e}_i\right) \cdot \frac{\partial G^j\left[x_k\right]}{\partial x^\gamma}\mathbf{e}_j$$

$$\mathbf{g}_\delta = \frac{\partial G^j\left[x_k\right]}{\partial x^\delta}\mathbf{e}_j. \tag{876}$$

Thereby the Latin indices are running to a different (potentially higher) number of dimensions N than the Greek indices, which shall be defined for a space or space-time of n dimensions.

Now, having extended the options for the metric and its base vectors in such a way, which is to say by allowing them to have some kind of arbitrary intrinsic structure, we should ask ourselves whether this should not also be possible for our quantum scale factors $F[f]$. As already shown in connection with the solutions to the elastic equations, we may assume intrinsic structures and various quantum centers (see section "8.8 Centers of Gravity and Quantum Centers" further below). The latter then automatically leads to ensembles of $F[f]$s and thus, to statistics and—necessarily—thermodynamics. In order to better understand this, we here repeat the corresponding considerations with respect to the metric or base vector variation from [3].

8.7.4 The Variation with Respect to Ensemble Parameters

Reconsideration of the classical Einstein–Hilbert action according to [3], thereby incorporating our "new" degrees of freedom (876) leads to:

$$\delta_g W = 0 = \int_V d^n x \overbrace{\left(R^{\delta\gamma} - \frac{1}{2}Rg^{\delta\gamma} + \Lambda g^{\delta\gamma} + \kappa T^{\delta\gamma}\right)}^{\text{Relativity}}$$

$$\times \underbrace{\left(\frac{\partial G^i\left[x_k\right]}{\partial x^\delta}\mathbf{e}_i \cdot \delta\left(\frac{\partial G^j\left[x_k\right]}{\partial x^\gamma}\mathbf{e}_j\right) + \delta\left(\frac{\partial G^i\left[x_k\right]}{\partial x^\delta}\mathbf{e}_i\right) \cdot \frac{\partial G^j\left[x_k\right]}{\partial x^\gamma}\mathbf{e}_j}_{\text{Quantum}} + \underbrace{\frac{\partial g_{\delta\gamma}}{\partial n}\delta n}_{\text{ThD}}\right).$$

$$\tag{877}$$

We have assumed that we have a closed system and that therefore the amount of information, respectively the number of dimensions should not change, meaning $\dfrac{\partial g_{\alpha\beta}}{\partial n}\delta n = 0$ (c.f. [23, 24]). What still could change, respectively be variated, however, would be the number of *i-j*-ensembles and the corresponding base-vectors. For instance, we could understand these ensembles as objects with different centers of gravity, like:

$$\frac{\partial G^{j}\left[x_{k}\right]}{\partial x^{\gamma}}\mathbf{e}_{j} = \frac{\partial G^{j}\left[x_{k}-\xi_{kj}\right]}{\partial x^{\gamma}}\mathbf{e}_{j}, \tag{878}$$

with the ξ_{kj} denoting the various centers of gravity. Now one could perform the variation with respect to exactly these coordinates and find (thereby simplifying via our condition of the closed system $\dfrac{\partial g_{\alpha\beta}}{\partial n}\delta n = 0$):

$$X_{k} \equiv x_{k}-\xi_{k}; \quad \delta_{g}W = \int_{V} d^{n}x \overbrace{\left(R^{\delta\gamma}-\frac{1}{2}Rg^{\delta\gamma}+\Lambda g^{\delta\gamma}+\kappa T^{\delta\gamma}\right)}^{\text{Relativity}}$$

$$\times \left(\underbrace{\frac{\partial G^{i}\left[X_{k}\right]}{\partial x^{\delta}}\mathbf{e}_{i}\cdot\delta_{x}\left(\frac{\partial G^{j}\left[X_{k}\right]}{\partial x^{\gamma}}\mathbf{e}_{j}\right)+\delta_{x}\left(\frac{\partial G^{i}\left[X_{k}\right]}{\partial x^{\delta}}\mathbf{e}_{i}\right)\cdot\frac{\partial G^{j}\left[X_{k}\right]}{\partial x^{\gamma}}\mathbf{e}_{j}}_{\text{Quantum}} \right.$$
$$\left. \underbrace{+\frac{\partial G^{i}\left[X_{k}\right]}{\partial x^{\delta}}\mathbf{e}_{i}\cdot\delta_{\xi}\left(\frac{\partial G^{j}\left[X_{k}\right]}{\partial x^{\gamma}}\mathbf{e}_{j}\right)+\delta_{\xi}\left(\frac{\partial G^{i}\left[X_{k}\right]}{\partial x^{\delta}}\mathbf{e}_{i}\right)\cdot\frac{\partial G^{j}\left[X_{k}\right]}{\partial x^{\gamma}}\mathbf{e}_{j}}_{\text{"?"}\,\Rightarrow\,2^{\text{nd}}\text{ law of Thermodynamics}} \right).$$

$$(879)$$

We realize that the additional variation is just the positioning of the first derivatives of metric displacements and their vectors. In other words, this variation is about positioning the individual centers of gravity within the metric space-time and their vectors of changes against the various coordinates *x*. In our "Science Riddle 20" [25], we discussed the effect of chaotic distributions for huge ensembles. We came to the conclusion that states are statistically favored where the various displacements cancel each other out. This then leads to a disappearance of the term "?". Interestingly, however, such equal distributions of vector orientations and magnitudes also coincide with states with maximum classical entropy, where we exactly obtain

? = 0 (c.f. Fig. 8.5). Thus, it is clear that the "?" must stand for thermodynamics and that it is a driving force for the second law of thermodynamics.

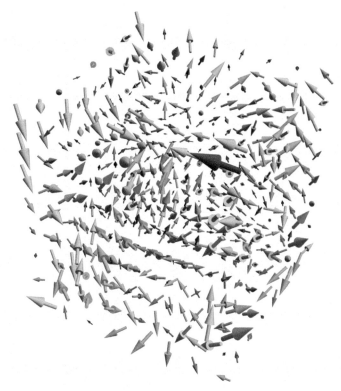

Figure 8.5 Random distribution of a displacement vector field for G^j in order to achieve a structural solution to Eq. (879). The solution also results in maximum entropy.

However, there is also the chance to make this variational term to zero with suitable symmetrical structures of the ensembles also canceling each other out (Figs. 8.6 and 8.7). Then the term "?" would stand for a force of "self-organization" and/or self-structuring of the very system (or parts of it). The whole could also be achieved as a proper combination of the "?" and the "quantum"-term which then has to give a resulting zero in sum.

In this whole context, we also make out the parameter N (the number to which the indices i and j can run) as yet another quantity defining upon the form of enclosure of a given system. With n and N being fixed, we should speak about a truly closed system in the classical meaning.

Figure 8.6 Symmetric distribution of a displacement vector field for G^j in order to achieve a structural solution to Eq. (879). The solution corresponds to a non-extremal entropy.

From all what has been elaborated above we conclude that the term "?" does not only present the metric manifestation of the second law of thermodynamics, but also gives the reason for the self-organization vector and evolution we observe within our universe as a rather omnipresent property.

With this, we might want to rewrite Eq. (879) as follows:

$$X_k \equiv x_k - \xi_k; \quad \delta_g W = \int_V d^n x \overbrace{\left(R^{\delta\gamma} - \frac{1}{2} R g^{\delta\gamma} + \Lambda g^{\delta\gamma} + \kappa T^{\delta\gamma} \right)}^{\text{Relativity}}$$

$$\times \left(\underbrace{\frac{\partial G^{i}[X_{k}]}{\partial x^{\delta}}\mathbf{e}_{i} \cdot \delta_{x}\left(\frac{\partial G^{j}[X_{k}]}{\partial x^{\gamma}}\mathbf{e}_{j}\right) + \delta_{x}\left(\frac{\partial G^{i}[X_{k}]}{\partial x^{\delta}}\mathbf{e}_{i}\right) \cdot \frac{\partial G^{j}[X_{k}]}{\partial x^{\gamma}}\mathbf{e}_{j}}_{\text{Quantum}} \right.$$
$$\left. \underbrace{+ \frac{\partial G^{i}[X_{k}]}{\partial x^{\delta}}\mathbf{e}_{i} \cdot \delta_{\xi}\left(\frac{\partial G^{j}[X_{k}]}{\partial x^{\gamma}}\mathbf{e}_{j}\right) + \delta_{\xi}\left(\frac{\partial G^{i}[X_{k}]}{\partial x^{\delta}}\mathbf{e}_{i}\right) \cdot \frac{\partial G^{j}[X_{k}]}{\partial x^{\gamma}}\mathbf{e}_{j}}_{\text{Thermodynamics \& Evolution}} \right).$$

$$(880)$$

It should be pointed out that the structure also allows an easy incorporation of interaction among the various centers of gravity (e.g., [25]).

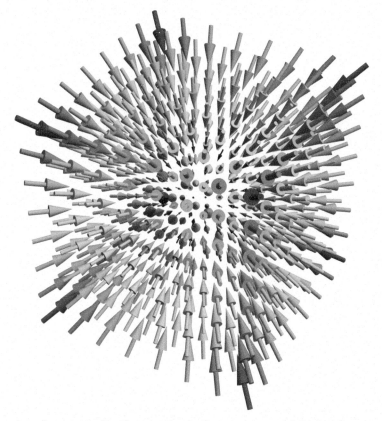

Figure 8.7 Symmetric distribution of a displacement vector field for G^j in order to achieve a structural solution to Eq. (879). The solution corresponds to a non-extremal entropy.

In [3], we considered a variety of options with respect to the variation of ensemble parameters. In order to cover at least a few of these options before moving on to distinguish between centers of gravity and quantum centers (see below), we here repeat some of our earlier results.

8.7.4.1 Ordinary derivative variation and the ideal gas

It should be noted that performing the variation as ordinary derivation and with the simple setting for X_k as set in Eq. (880), the variation of the quantum and the thermo-evolution term can cancel each other out, because we could perform it as follows:

$$\delta_x \left(\frac{\partial G^j [X_k]}{\partial x^\gamma} \mathbf{e}_j \right) = \frac{\partial}{\partial x^\beta} \left(\frac{\partial G^j [X_k]}{\partial x^\gamma} \mathbf{e}_j \right) \delta x^\beta$$

$$\delta_\xi \left(\frac{\partial G^j [X_k]}{\partial x^\gamma} \mathbf{e}_j \right) = \frac{\partial}{\partial \xi^\beta} \left(\frac{\partial G^j [X_k]}{\partial x^\gamma} \mathbf{e}_j \right) \delta \xi^\beta = \frac{\partial}{\partial x^\gamma} \left(\frac{\partial G^j [X_k]}{\partial \xi^\beta} \mathbf{e}_j \right) \delta \xi^\beta. \quad (881)$$

Now we assume that we can perform the following simplification in the second line:

$$\frac{\partial}{\partial \xi^\beta} \left(\frac{\partial G^j [X_k]}{\partial x^\gamma} \mathbf{e}_j \right) \delta \xi^\beta = \left(\frac{\partial^2 G^j [X_k]}{\partial x^\gamma \partial \xi^\beta} \mathbf{e}_j \right) \delta \xi^\beta = \left(-\frac{\partial^2 G^j [X_k]}{\partial x^\gamma \partial x^\beta} \mathbf{e}_j \right) \delta x^\beta. \quad (882)$$

It has to be pointed out that this is not possible in cases of position dependent base vectors, for instance. However, in Cartesian coordinates, we could also simplify the first line of Eq. (881) and then obtain in sum:

$$\left(\frac{\partial^2 G^j [X_k]}{\partial x^\beta \partial x^\gamma} \mathbf{e}_j \right) \delta x^\beta$$

$$\Rightarrow \qquad \delta_x \left(\frac{\partial G^j [X_k]}{\partial x^\gamma} \mathbf{e}_j \right) + \delta_\xi \left(\frac{\partial G^j [X_k]}{\partial x^\gamma} \mathbf{e}_j \right) = 0, \quad (883)$$

which automatically is giving us:

$$0 = \begin{pmatrix} \underbrace{\frac{\partial G^i [X_k]}{\partial x^\delta} \mathbf{e}_i \cdot \delta_x \left(\frac{\partial G^j [X_k]}{\partial x^\gamma} \mathbf{e}_j \right) + \delta_x \left(\frac{\partial G^i [X_k]}{\partial x^\delta} \mathbf{e}_i \right) \cdot \frac{\partial G^j [X_k]}{\partial x^\gamma} \mathbf{e}_j}_{\text{Quantum}} \\ + \underbrace{\frac{\partial G^i [X_k]}{\partial x^\delta} \mathbf{e}_i \cdot \delta_\xi \left(\frac{\partial G^j [X_k]}{\partial x^\gamma} \mathbf{e}_j \right) + \delta_\xi \left(\frac{\partial G^i [X_k]}{\partial x^\delta} \mathbf{e}_i \right) \cdot \frac{\partial G^j [X_k]}{\partial x^\gamma} \mathbf{e}_j}_{\text{Thermodynamics \& Evolution}} \end{pmatrix}. \quad (884)$$

In order to avoid such a triviality, we either shall demand the variation with respect to ξ_k to be different or we set X_k as follows:

$$X_k \equiv x_k - f_k \left[\xi_k \right]. \tag{885}$$

Obviously already the simplest linear form like $f[\xi_k] = m_k * \xi_k$ suffices to force both terms in Eq. (881) to give zero independently, which, having assumed the same structural variation with respect to ξ_k and x_k, only requires the solution of one. This only leaves statistical laws to take care about the distribution of the various i-j-gravity centers. There is no interaction.

As example we want to consider solutions to the following variational outcome (from [27]):

$$\sum_{\alpha=0}^{n-1} \frac{\partial^2 G^j \left[x_k \right]}{\partial x^\alpha \partial x^\gamma} \mathbf{e}_j = 0 = \sum_{\alpha=0}^{n-1} \frac{\partial^2 G^i \left[x_k \right]}{\partial x^\alpha \partial x^\delta} \mathbf{e}_i. \tag{886}$$

Even when introducing point solutions from [27], reading for a single center at $\xi_k = 0$:

$$G^j \left[x_k \right] = \pm C_s \cdot \left(\frac{t}{\left(x^2 + y^2 + z^2 \right)^2 \left(1 + \frac{t^2}{x^2 + y^2 + z^2} \right)} + \frac{\arctan\left[\frac{t}{\sqrt{x^2 + y^2 + z^2}} \right]}{\left(x^2 + y^2 + z^2 \right)^{3/2}} \right) \cdot \left\{ t, x, y, z \right\}, \tag{887}$$

we still do not get anything but an identical solution already to the quantum term (regarding the evaluation in connection with the variation we have to refer to [27]). Thereby C_s stands for a suitable constant. We recognize the potential of a point charge as limit for $t \to \infty$. The resulting $t \to \infty$-limit would then read:

$$\lim_{t \to \infty} G^j \left[x_k \right] = \mp C_s \cdot \frac{\pi}{2} \cdot \frac{1}{\left(x^2 + y^2 + z^2 \right)^{3/2}} \cdot \left\{ 0, x, y, z \right\}. \tag{888}$$

Assuming two equal centers placed at $\pm x_0$ would give us:

$$\lim_{t \to \infty} G^j \left[x_k \right] = \mp C_s \cdot \frac{\pi}{2} \cdot \frac{1}{\left(\left(x \pm x_0 \right)^2 + y^2 + z^2 \right)^{3/2}} \cdot \left\{ 0, x \pm x_0, y, z \right\}, \tag{889}$$

but still already the fulfillment of the quantum term, which is guaranteed due to our solution (887) to (886) and thus, both terms in Eq. (881). This means that even the presence of "point charges" would not force our system to follow anything else but statistics, which leads to the second law of thermodynamics and does neither provide any interaction nor evolutionary driving forces.

We might see this approach above therefore as a quantum gravity model for an ideal gas.

8.7.4.2 Combined successive variation

Things are getting significantly different the moment we allow the variation to be performed somehow simultaneously or in a successive manner, but then we have variations of the second order and these are considered small of the second order. So, here is an example:

$$
\delta_\xi \left(\frac{\partial G^i [X_k]}{\partial x^\delta} \mathbf{e}_i \cdot \delta_x \left(\frac{\partial G^j [X_k]}{\partial x^\gamma} \mathbf{e}_j \right) + \delta_x \left(\frac{\partial G^i [X_k]}{\partial x^\delta} \mathbf{e}_i \right) \cdot \frac{\partial G^j [X_k]}{\partial x^\gamma} \mathbf{e}_j \right)
$$

$$
= \delta_\xi \left(\frac{\partial G^i [X_k]}{\partial x^\delta} \mathbf{e}_i \right) \cdot \delta_x \left(\frac{\partial G^j [X_k]}{\partial x^\gamma} \mathbf{e}_j \right) + \delta_\xi \delta_x \left(\frac{\partial G^i [X_k]}{\partial x^\delta} \mathbf{e}_i \right) \cdot \frac{\partial G^j [X_k]}{\partial x^\gamma} \mathbf{e}_j
$$

$$
+ \frac{\partial G^i [X_k]}{\partial x^\delta} \mathbf{e}_i \cdot \delta_\xi \delta_x \left(\frac{\partial G^j [X_k]}{\partial x^\gamma} \mathbf{e}_j \right) + \delta_x \left(\frac{\partial G^i [X_k]}{\partial x^\delta} \mathbf{e}_i \right) \cdot \delta_\xi \left(\frac{\partial G^j [X_k]}{\partial x^\gamma} \mathbf{e}_j \right).
$$

$$(890)$$

Assuming all double variations to be small of higher order, we could simplify:

$$
\delta_\xi \left(\frac{\partial G^i [X_k]}{\partial x^\delta} \mathbf{e}_i \cdot \delta_x \left(\frac{\partial G^j [X_k]}{\partial x^\gamma} \mathbf{e}_j \right) + \delta_x \left(\frac{\partial G^i [X_k]}{\partial x^\delta} \mathbf{e}_i \right) \cdot \frac{\partial G^j [X_k]}{\partial x^\gamma} \mathbf{e}_j \right)
$$

$$
= \delta_\xi \left(\frac{\partial G^i [X_k]}{\partial x^\delta} \mathbf{e}_i \right) \cdot \delta_x \left(\frac{\partial G^j [X_k]}{\partial x^\gamma} \mathbf{e}_j \right) + \delta_x \left(\frac{\partial G^i [X_k]}{\partial x^\delta} \mathbf{e}_i \right) \cdot \delta_\xi \left(\frac{\partial G^j [X_k]}{\partial x^\gamma} \mathbf{e}_j \right).
$$

$$(891)$$

Still we have obtained products of variated terms that, we might also consider small of the second order and thus, for here and now we are not going to consider this path any further.

Thus, we conclude that our form of rather simple (even though extended) variation by the means of ordinary derivatives does not do the job and we have to investigate more general options.

8.7.4.3 Incorporating interaction

It is evident that with the introduction of centers of gravity, we should also consider the option for interactions among those various centers. This would simply be achieved by allowing the G^j to not only depend on the differential coordinates X_k as given in Eq. (879), but also to have dependencies as:

$$G^j[...] = G^j[X_{kj}] \equiv G^j[X_{k0}, X_{k1}, ..., X_{ki}, ..., X_{k(N-1)}]; \quad X_{ki} \equiv x_k - \xi_{ki}.$$

$$(892)$$

This immediately renders the variation with respect to the relative or position coordinates extremely complex even in the smallest ensembles, which are not so very much different from the usual complexity known from statistical mechanics. Here now we just face a quantum gravity interactive statistics, where the complete variation should read:

$$\delta_g W = \int_V d^n x \overbrace{\left(R^{\delta\gamma} - \frac{1}{2} R g^{\delta\gamma} + \Lambda g^{\delta\gamma} + \kappa T^{\delta\gamma} \right)}^{\text{Relativity}}$$

$$\times \left(\underbrace{\left(\frac{\partial G^i[X_{ki}]}{\partial x^\delta} \mathbf{e}_i \cdot \delta_x \left(\frac{\partial G^j[X_{kj}]}{\partial x^\gamma} \mathbf{e}_j \right) + \delta_x \left(\frac{\partial G^i[X_{ki}]}{\partial x^\delta} \mathbf{e}_i \right) \cdot \frac{\partial G^j[X_{kj}]}{\partial x^\gamma} \mathbf{e}_j \right)}_{\text{Quantum}} \right.$$

$$\left. + \underbrace{\sum_{i,j=0}^{N-1} \left(\frac{\partial G^i[X_{ki}]}{\partial x^\delta} \mathbf{e}_i \cdot \delta_{\xi j} \left(\frac{\partial G^j[X_{kj}]}{\partial x^\gamma} \mathbf{e}_j \right) + \delta_{\xi i} \left(\frac{\partial G^i[X_{ki}]}{\partial x^\delta} \mathbf{e}_i \right) \cdot \frac{\partial G^j[X_{kj}]}{\partial x^\gamma} \mathbf{e}_j \right)}_{\text{2}^{nd}\text{ law of Thermodynamics \& Interaction}} \right).$$

$$(893)$$

Summing up all variations considered here (meaning without the scale factor) in just one world formula also for N- and n-variable (perhaps open) systems would now give us the following two forms [3]:

$$\delta_g W = \left(\int_V d^n x \overbrace{\left[[R - 2\kappa L_M - 2\Lambda] \cdot \frac{\partial \sqrt{-g}}{\partial n} + \sqrt{-g} \cdot \frac{\partial [R - 2\kappa L_M - 2\Lambda]}{\partial n} \right]}^{\text{Thermodynamics / Exchange / Open Systems}} \right) \delta n$$

$$+ \int_V d^n x \overbrace{\left(R^{\delta\gamma} - \frac{1}{2} R g^{\delta\gamma} + \Lambda g^{\delta\gamma} + \kappa T^{\delta\gamma} \right)}^{\text{Relativity}}$$

$$
\times \left|
\begin{array}{c}
\underbrace{\frac{\partial G^i\left[X_{ki}\right]}{\partial x^\delta}\mathbf{e}_i \cdot \delta_x\left(\frac{\partial G^j\left[X_{kj}\right]}{\partial x^\gamma}\mathbf{e}_j\right) + \delta_x\left(\frac{\partial G^i\left[X_{ki}\right]}{\partial x^\delta}\mathbf{e}_i\right)\cdot\frac{\partial G^j\left[X_{kj}\right]}{\partial x^\gamma}\mathbf{e}_j}_{\text{Quantum}} + \underbrace{\frac{\partial g_{\delta\gamma}}{\partial n}\delta n}_{\text{ThD}} \\[2em]
\underbrace{+\delta_N\left(\sum_{i,j=0}^{N-1}\frac{\partial G^i\left[X_{ki}\right]}{\partial x^\delta}\mathbf{e}_i\cdot\delta_{\xi j}\left(\frac{\partial G^j\left[X_{kj}\right]}{\partial x^\gamma}\mathbf{e}_j\right) + \delta_{\xi i}\left(\frac{\partial G^i\left[X_{ki}\right]}{\partial x^\delta}\mathbf{e}_i\right)\cdot\frac{\partial G^j\left[X_{kj}\right]}{\partial x^\gamma}\mathbf{e}_j\right)}_{2^{nd}\text{ law of Thermodynamics \& Interaction}}
\end{array}
\right)
$$

$$(894)$$

or somewhat simpler with the *N*-variation as separate addend:

$$
\delta_g W = \left(\int_V d^n x \overbrace{\left[\left[R - 2\kappa L_M - 2\Lambda\right]\cdot\frac{\partial\sqrt{-g}}{\partial n} + \sqrt{-g}\cdot\frac{\partial\left[R - 2\kappa L_M - 2\Lambda\right]}{\partial n}\right]}^{\text{Thermodynamics / Exchange / Open Systems}}\right)\delta n
$$

$$
+ \int_V d^n x \overbrace{\left(R^{\delta\gamma} - \frac{1}{2}Rg^{\delta\gamma} + \Lambda g^{\delta\gamma} + \kappa T^{\delta\gamma}\right)}^{\text{Relativity}}
$$

$$
\times \left|
\begin{array}{c}
\underbrace{\frac{\partial G^i\left[X_{ki}\right]}{\partial x^\delta}\mathbf{e}_i \cdot \delta_x\left(\frac{\partial G^j\left[X_{kj}\right]}{\partial x^\gamma}\mathbf{e}_j\right) + \delta_x\left(\frac{\partial G^i\left[X_{ki}\right]}{\partial x^\delta}\mathbf{e}_i\right)\cdot\frac{\partial G^j\left[X_{kj}\right]}{\partial x^\gamma}\mathbf{e}_j}_{\text{Quantum}} + \underbrace{\frac{\partial g_{\delta\gamma}}{\partial n}\delta n}_{\text{ThD}} \\[2em]
\underbrace{+\frac{\partial g_{\alpha\beta}}{\partial N}\delta_N + \sum_{i,j=0}^{N-1}\left(\begin{array}{c}\frac{\partial G^i\left[X_{ki}\right]}{\partial x^\delta}\mathbf{e}_i\cdot\delta_{\xi j}\left(\frac{\partial G^j\left[X_{kj}\right]}{\partial x^\gamma}\mathbf{e}_j\right) \\[1em] +\delta_{\xi i}\left(\frac{\partial G^i\left[X_{ki}\right]}{\partial x^\delta}\mathbf{e}_i\right)\cdot\frac{\partial G^j\left[X_{kj}\right]}{\partial x^\gamma}\mathbf{e}_j\end{array}\right)}_{2^{nd}\text{ law of Thermodynamics \& Interaction}}
\end{array}
\right).
$$

$$(895)$$

8.7.4.4   Repetition: Derivation of the diffusion equation

As a very simple example, we here want to repeat the essentials of the derivation of the diffusion equation from the fundamental Eq. (895) and demand the following boundary conditions to be fulfilled:

$$
\int_V d^n x \left[\left[R - 2\kappa L_M - 2\Lambda\right]\cdot\frac{\partial\sqrt{-g}}{\partial n} + \sqrt{-g}\cdot\frac{\partial\left[R - 2\kappa L_M - 2\Lambda\right]}{\partial n}\right]\cdot\delta n = 0
$$

$$\frac{\partial g_{\alpha\beta}}{\partial n}\delta n = 0$$

$$\frac{\partial G^i\left[X_{ki}\right]}{\partial x^\delta}\mathbf{e}_i\cdot\delta_x\left(\frac{\partial G^j\left[X_{kj}\right]}{\partial x^\gamma}\mathbf{e}_j\right)+\delta_x\left(\frac{\partial G^i\left[X_{ki}\right]}{\partial x^\delta}\mathbf{e}_i\right)\cdot\frac{\partial G^j\left[X_{kj}\right]}{\partial x^\gamma}\mathbf{e}_j = 0, \quad (896)$$

which results in the remaining equation:

$$\frac{\partial g_{\delta\gamma}}{\partial N}\delta_N + \sum_{i,j=0}^{N-1}\left(\frac{\partial G^i\left[X_{ki}\right]}{\partial x^\delta}\mathbf{e}_i\cdot\delta_{\xi j}\left(\frac{\partial G^j\left[X_{kj}\right]}{\partial x^\gamma}\mathbf{e}_j\right)+\delta_{\xi i}\left(\frac{\partial G^i\left[X_{ki}\right]}{\partial x^\delta}\mathbf{e}_i\right)\cdot\frac{\partial G^j\left[X_{kj}\right]}{\partial x^\gamma}\mathbf{e}_j\right) = 0.$$
$$(897)$$

For the reason of simplicity, we ignore any particle interaction and reduce Eq. (897) to:

$$\frac{\partial g_{\delta\gamma}}{\partial N}\delta_N + \frac{\partial G^i\left[X_k\right]}{\partial x^\delta}\mathbf{e}_i\cdot\delta_\xi\left(\frac{\partial G^j\left[X_k\right]}{\partial x^\gamma}\mathbf{e}_j\right)+\delta_\xi\left(\frac{\partial G^i\left[X_k\right]}{\partial x^\delta}\mathbf{e}_i\right)\cdot\frac{\partial G^j\left[X_k\right]}{\partial x^\gamma}\mathbf{e}_j = 0,$$
$$(898)$$

with the definition for X_k as given in Eq. (879). Similar to our evaluation from the section "Motivation" in [28] (see also [3], sub-section "2.2 Intelligent Zero Approaches: Just one Example"), we perform the variation of the second and third addend in Eq. (898) as follows:

$$G^j\left[X_k\right]\equiv G^j;\quad \frac{\partial G^i}{\partial x^\delta}\mathbf{e}_i\cdot\delta_\xi\left(\frac{\partial G^j}{\partial x^\gamma}\mathbf{e}_j\right)+\delta_\xi\left(\frac{\partial G^i}{\partial x^\delta}\mathbf{e}_i\right)\cdot\frac{\partial G^j}{\partial x^\gamma}\mathbf{e}_j$$

$$= -C_\delta^\gamma\frac{\partial G^i}{\partial x^\gamma}\mathbf{e}_i\cdot\left(\frac{\partial^2 G^j}{\partial x^\sigma\partial x^\gamma}\mathbf{e}_j\right)\delta^\sigma - \left(\frac{\partial^2 G^i}{\partial x^\sigma\partial x^\delta}\mathbf{e}_i\right)\delta^\sigma\cdot\frac{\partial G^j}{\partial x^\gamma}\mathbf{e}_j. \quad (899)$$

Exchanging the dummy indices and assuming $C_\delta^\gamma = b\cdot\delta_\delta^\gamma$ leads to:

$$= -\left(C_\delta^\gamma\left(\frac{\partial^2 G^i}{\partial x^\sigma\partial x^\gamma}\mathbf{e}_i\right)+\left(\frac{\partial^2 G^i}{\partial x^\sigma\partial x^\delta}\mathbf{e}_i\right)\right)\delta^\sigma\cdot\frac{\partial G^j}{\partial x^\gamma}\mathbf{e}_j$$

$$= -(b+1)\left(\frac{\partial^2 G^i}{\partial x^\sigma\partial x^\delta}\mathbf{e}_i\right)\delta^\sigma\cdot\frac{\partial G^j}{\partial x^\gamma}\mathbf{e}_j = -(b+1)C_\sigma^\delta\left(\frac{\partial^2 G^i}{\partial x^\sigma\partial x^\delta}\mathbf{e}_i\right)\delta^\sigma\cdot\frac{\partial G^j}{\partial x^\gamma}\mathbf{e}_j$$

$$= -(b+1)\cdot b\cdot\delta_\sigma^\delta\left(\frac{\partial^2 G^i}{\partial x^\sigma\partial x^\delta}\mathbf{e}_i\right)\delta^\sigma\cdot\frac{\partial G^j}{\partial x^\gamma}\mathbf{e}_j = -(b+1)\cdot b\left(\frac{\partial^2 G^i}{\left(\partial x^\sigma\right)^2}\mathbf{e}_i\right)\delta^\sigma\cdot\frac{\partial G^j}{\partial x^\gamma}\mathbf{e}_j$$

$$= -(b+1) \cdot b \left(\frac{\partial^2 G^i \mathbf{e}_i}{(\partial x^\sigma)^2} \right) \delta^\sigma \cdot \frac{\partial G^j}{\partial x^\gamma} \mathbf{e}_j = -(b+1) \cdot b \cdot \Delta\left(G^i \mathbf{e}_i\right) \delta^\sigma \cdot \frac{\partial G^j}{\partial x^\gamma} \mathbf{e}_j. \quad (900)$$

Now we assume the variation regarding N to result in zeros with respect to all spatial derivatives. The justification for this is the typical continuity equation approach where we demand that under the volume considered in our variational integral (895) any change of N does not change the derivatives with respect to the spatial coordinates, meaning:

$$\frac{\partial g_{\delta\gamma}}{\partial N} \delta_N = \frac{\partial\left(\frac{\partial G^j}{\partial x^\gamma} \mathbf{e}_j \frac{\partial G^i}{\partial x^\delta} \mathbf{e}_i \right)}{\partial N} \delta_N = 0; \quad \delta, \gamma \neq 0. \quad (901)$$

With respect to the time-coordinate, however, we can have very complex dependencies. The simplest would be a linear one like:

$$\frac{\partial\left(\frac{\partial G^j}{\partial x^0} \mathbf{e}_j \right)}{\partial N} \delta_N = D^{-1} \cdot \frac{\partial G^j}{\partial x^0} \mathbf{e}_j \cdot \delta^\delta. \quad (902)$$

This gives us the following for the first addend:

$$\frac{\partial g_{\delta\gamma}}{\partial N} \delta_N = D^{-1} \cdot \left(\frac{\partial G^i}{\partial x^\delta} \mathbf{e}_i \cdot \frac{\partial G^j}{\partial x^0} \mathbf{e}_j + \frac{\partial G^i}{\partial x^0} \mathbf{e}_i \cdot \frac{\partial G^j}{\partial x^\gamma} \mathbf{e}_j \right) \cdot \delta^\delta. \quad (903)$$

In connection with the result from Eq. (900), again using $C_\delta^\gamma = b \cdot \delta_\delta^\gamma$ and exchanging the dummy indices, we result in:

$$0 = D^{-1} \cdot \left(\frac{\partial G^i}{\partial x^\delta} \mathbf{e}_i \cdot \frac{\partial G^j}{\partial x^0} \mathbf{e}_j + \frac{\partial G^i}{\partial x^0} \mathbf{e}_i \cdot \frac{\partial G^j}{\partial x^\gamma} \mathbf{e}_j \right) \cdot \delta^\sigma - (b+1) \cdot b \cdot \Delta\left(G^i \mathbf{e}_i\right) \delta^\sigma \cdot \frac{\partial G^j}{\partial x^\gamma} \mathbf{e}_j$$

$$= D^{-1} \cdot (b+1) \cdot \left(\frac{\partial G^i}{\partial x^0} \mathbf{e}_i \right) \cdot \frac{\partial G^j}{\partial x^\gamma} \mathbf{e}_j \cdot \delta^\sigma - (b+1) \cdot b \cdot \Delta\left(G^i \mathbf{e}_i\right) \delta^\sigma \cdot \frac{\partial G^j}{\partial x^\gamma} \mathbf{e}_j$$

$$= (b+1) \cdot \left(\left(\frac{\partial G^i}{\partial x^0} \mathbf{e}_i \right) - D \cdot b \cdot \Delta\left(G^i \mathbf{e}_i\right) \right) \delta^\sigma \cdot \frac{\partial G^j}{\partial x^\gamma} \mathbf{e}_j$$

$$\Rightarrow \qquad \left(\frac{\partial G^i}{\partial x^0} \mathbf{e}_i \right) = D \cdot b \cdot \Delta\left(G^i \mathbf{e}_i\right), \quad (904)$$

where we recognize the diffusion equation in its simplest (homogeneous) form. Along the way of our derivation, we can easily make out passages where we can incorporate anisotropy and inhomogeneity in very general manners.

8.8   Centers of Gravity and Quantum Centers

8.8.1   Quantum Centers

As we have seen that there are obviously quite some interesting things behind the idea of extending the variation of the Einstein–Hilbert action with respect to ensemble parameters like the position of various centers of gravity (see the section above and [3]), we now want to briefly investigate the question of what happens when also allowing for different centers of scaling or quantum functions $F[f]$.

Thus, similar to Eq. (892), we could also introduce quantum or scaling centers of the following kind:

$$f_k [\ldots] = f\left[Y_{kj}\right] \equiv f\left[Y_{k0}, Y_{k1}, \ldots, Y_{ki}, \ldots, Y_{k(N-1)}\right]; \quad Y_{ki} \equiv x_k - \zeta_{ki}. \quad (905)$$

The attentive reader may have realized that, just as with the centers of gravity approach Eq. (892), we also have the possibility of interaction. This renders the variability of the general approach rather enormous and we have to move further and carry out deeper investigations and discussions to follow-up publications.

The question we here want to discuss, however, should be the one about how these quantum centers could actually be "activated" during the variation. Within the last section, it became clear that the centers of gravity are becoming "active" simply by elongating the usual metric variation beyond the classical point and also varying the metric itself with respect to its possible intrinsic structure, which is the base vector (generalized according to Eq. (892)) and the various positions ξ_{ki} of the centers of gravity. Also having to deal with quantum centers in accordance to Eq. (905) requires an additional variation with respect to the parameters ζ_{ki}. This, however, leads automatically to an additional derivation of the function f with respect to the coordinates, because the task reads:

$$\delta W = 0 = \delta_\zeta \int_V d^n x \left(\sqrt{-G} \cdot \left(R^* - 2\Lambda + L_M\right)\right)$$

$$= \delta_\zeta \int_V d^n x \left(\sum_{\forall f_k} \sqrt{-G} \cdot \left(R^* - 2\Lambda + L_M\right)\right) = \sum_{\forall f_k} \delta_\zeta \int_V d^n x \left(\sqrt{-G} \cdot \left(R^* - 2\Lambda + L_M\right)\right).$$

$$(906)$$

Inserting our result for R^* from the Appendix A, we have:

$$\delta W = 0 = \delta_\zeta \int_V d^n x \left(\sqrt{-G} \cdot \left(R^* - 2\Lambda + L_M \right) \right)$$

$$= \delta_\zeta \int_V d^n x \left(\sum_{\forall f_k} \sqrt{-G} \cdot \left(R^* - 2\Lambda + L_M \right) \right) = \sum_{\forall f_k} \delta_\zeta \int_V d^n x \left(\sqrt{-G} \cdot \left(R^* - 2\Lambda + L_M \right) \right)$$

$$0 = \delta_\zeta \int_V d^n x \left(\sqrt{-G} \cdot \left(R^* - 2\Lambda + L_M \right) \right)$$

$$= \sum_{\forall f_k} \delta_\zeta \int_V d^n x \left[\sqrt{-G} \cdot \left(\left(\begin{array}{c} R + \dfrac{F'}{F}(1-n)\Delta f \\ + \dfrac{f_{,\alpha} f_{,\beta} g^{\alpha\beta}(1-n)}{4F^2} \left(4FF'' + (F')^2(n-6) \right) \end{array} \right) \dfrac{1}{F} - 2\Lambda + L_M \right) \right]$$

$$= \sum_{\forall f_k} \delta_\zeta \int_V d^n x \left[\sqrt{-G} \cdot \left(\left(\begin{array}{c} R + \dfrac{F'}{F}(1-n)g^{\alpha\beta}\left(f_{,\alpha\beta} - \Gamma^\gamma_{\alpha\beta} f_{,\gamma} \right) \\ + \dfrac{f_{,\alpha} f_{,\beta} g^{\alpha\beta}(1-n)}{4F^2} \left(4FF'' + (F')^2(n-6) \right) \end{array} \right) \dfrac{1}{F} - 2\Lambda + L_M \right) \right]$$

$$= \sum_{\forall f_k} \delta_\zeta \int_V d^n x \left[\sqrt{-g} \cdot F^{\frac{n}{2}} \cdot \left(\left(\begin{array}{c} R + \dfrac{F'}{F}(1-n)g^{\alpha\beta}\left(f_{,\alpha\beta} - \Gamma^\gamma_{\alpha\beta} f_{,\gamma} \right) \\ + \dfrac{f_{,\alpha} f_{,\beta} g^{\alpha\beta}(1-n)}{4F^2} \left(4FF'' + (F')^2(n-6) \right) \end{array} \right) \dfrac{1}{F} - 2\Lambda + L_M \right) \right],$$

$$\tag{907}$$

$$= \sum_{\forall f_k} \int_V d^n x \left[\sqrt{-g} \cdot \left(\begin{array}{c} \left(\dfrac{n}{2} - 1 \right) F' \cdot F^{\frac{n}{2}-2} f_{,\sigma} \\ \cdot \left(\begin{array}{c} R + \dfrac{F'}{F}(1-n)g^{\alpha\beta}\left(f_{,\alpha\beta} - \Gamma^\gamma_{\alpha\beta} f_{,\gamma} \right) \\ + \dfrac{f_{,\alpha} f_{,\beta} g^{\alpha\beta}(1-n)}{4F^2} \left(4FF'' + (F')^2(n-6) \right) \end{array} \right) \\ + F^{\frac{n}{2}-1} \delta_\zeta \left(\begin{array}{c} \dfrac{F'}{F}(1-n)g^{\alpha\beta}\left(f_{,\alpha\beta} - \Gamma^\gamma_{\alpha\beta} f_{,\gamma} \right) \\ + \dfrac{f_{,\alpha} f_{,\beta} g^{\alpha\beta}(1-n)}{4F^2} \left(4FF'' + (F')^2(n-6) \right) \end{array} \right) \\ + f_{,\sigma} F^{\frac{n}{2}-1} \dfrac{n}{2} \cdot \left(L_M - 2\Lambda \right) \end{array} \right) Y^\sigma_{,k} \delta\zeta^k \right]$$

with: $Z_{,k} = \dfrac{\partial Z}{\partial \zeta^k}; \quad Y^{\sigma} = Y^{\sigma} - \zeta^{\sigma k},$ (908)

$$
= \sum_{\forall f_k} \int_V d^n x \left| \sqrt{-g} \cdot F^{\frac{n}{2}-1} \left(\begin{array}{c} \left(\left(\dfrac{n}{2}-1\right)F'\dfrac{f_{,\sigma}}{F} \right. \\[2mm] \left(\begin{array}{c} R + \dfrac{F'}{F}(1-n)\Delta f \\[2mm] + \dfrac{f_{,\alpha}f_{,\beta}g^{\alpha\beta}(1-n)}{4F^2}\left(4FF'' + (F')^2(n-6)\right) \end{array} \right) \\[6mm] + g^{\alpha\beta} \left(\begin{array}{c} \left(\dfrac{F'}{F}\right)_{,\sigma}(1-n)\left(f_{,\alpha\beta}-\Gamma^{\gamma}_{\alpha\beta}f_{,\gamma}\right) \\[2mm] + \dfrac{F'}{F}(1-n)\left(f_{,\alpha\beta\sigma}-\Gamma^{\gamma}_{\alpha\beta}f_{,\gamma\sigma}\right) \\[2mm] + \left(\dfrac{f_{,\alpha}f_{,\beta}(1-n)}{4F^2}\left(4FF''+(F')^2(n-6)\right)\right)_{,\sigma} \end{array} \right) \\[8mm] + f_{,\sigma}\dfrac{n}{2}\cdot(L_M - 2\Lambda) \end{array} \right) \right| Y^{\sigma}_{,k}\delta\zeta^k,
$$

(909)

and subsequently obtain the third-order differential equation for f, which in fact becomes f_k now. Please note that during the evaluation, it has to be assured that the derivative ∂_{σ} only acts on the function f. With a bit of simplification:

$$\delta W = 0 =$$

$$
= \sum_{\forall f_k} \int_V d^n x \left| \sqrt{-g} \cdot F^{\frac{n}{2}-1} \left(\begin{array}{c} \left(\left(\dfrac{n}{2}-1\right)F'\dfrac{f_{,\sigma}}{F} \right. \\[2mm] \cdot \left(R + \dfrac{f_{,\alpha}f_{,\beta}g^{\alpha\beta}(1-n)}{4F^2}\left(4FF''+(F')^2(n-6)\right) \right) \\[4mm] + \dfrac{f_{,\sigma}}{F}\left(\dfrac{FF''-(F')^2}{F} + \left(\dfrac{n}{2}-1\right)\dfrac{(F')^2}{F} \right)(1-n)\Delta f \\[4mm] g^{\alpha\beta}\left(\begin{array}{c} + \dfrac{F'}{F}(1-n)\left(f_{,\alpha\beta\sigma}-\Gamma^{\gamma}_{\alpha\beta}f_{,\gamma\sigma}\right) \\[2mm] + \left(\dfrac{f_{,\alpha}f_{,\beta}(1-n)}{4F^2}\left(4FF''+(F')^2(n-6)\right)\right)_{,\sigma} \end{array} \right) \\[8mm] + f_{,\sigma}\dfrac{n}{2}\cdot(L_M - 2\Lambda) \end{array} \right) \right| Y^{\sigma}_{,k}\delta\zeta^k,
$$

(910)

$$\delta W = 0 =$$

$$= \sum_{\forall f_k} \int_V d^n x \left| \sqrt{-g} \cdot F^{\frac{n}{2}-1} \left| \begin{pmatrix} \left(\frac{n}{2}-1\right)F'\frac{f_{,\sigma}}{F} \\ \cdot\left(R + \frac{f_{,\alpha}f_{,\beta}g^{\alpha\beta}(1-n)}{4F^2}\left(4FF'' + (F')^2(n-6)\right)\right) \\ + \frac{f_{,\sigma}}{F}\left(\frac{FF'' + \left(\frac{n}{2}-2\right)(F')^2}{F}\right)(1-n)\Delta f \\ g^{\alpha\beta}\begin{pmatrix} +\frac{F'}{F}(1-n)\left(f_{,\alpha\beta\sigma} - \Gamma^{\gamma}_{\alpha\beta}f_{,\gamma\sigma}\right) \\ +\left(\frac{f_{,\alpha}f_{,\beta}(1-n)}{4F^2}\left(4FF'' + (F')^2(n-6)\right)\right)_{,\sigma} \end{pmatrix} \\ + f_{,\sigma}\frac{n}{2}\cdot(L_M - 2\Lambda) \end{pmatrix} \right| Y^{\sigma}_{,k}\delta\zeta^k, \right.$$

$$(911)$$

we realize that even without the sum over the whole k-ensemble, it should be possible to find solutions of the form:

$$0 = \left\{ \begin{array}{l} \left(\frac{n}{2}-1\right)F'\frac{f_{,\sigma}}{F}\cdot\left(R + \frac{f_{,\alpha}f_{,\beta}g^{\alpha\beta}(1-n)}{4F^2}\left(4FF'' + (F')^2(n-6)\right)\right) \\ + \frac{f_{,\sigma}}{F^2}\left(FF'' + \left(\frac{n}{2}-2\right)(F')^2\right)(1-n)\Delta f \\ + g^{\alpha\beta}\begin{pmatrix} \frac{F'}{F}(1-n)\left(f_{,\alpha\beta\sigma} - \Gamma^{\gamma}_{\alpha\beta}f_{,\gamma\sigma}\right) \\ +\left(\frac{f_{,\alpha}f_{,\beta}(1-n)}{4F^2}\left(4FF'' + (F')^2(n-6)\right)\right)_{,\sigma} \end{pmatrix} \\ + f_{,\sigma}\frac{n}{2}\cdot(L_M - 2\Lambda). \end{array} \right. \qquad (912)$$

Such solutions would be intrinsic ones and we here want to investigate some simpler types for certain settings of $F[f]$. Thus, in the case of:

$$0 = 4FF'' + (F')^2(n-6) \quad \Rightarrow \quad F = (f - C_{f0})^{\frac{4}{n-2}}C_{f1}, \qquad (913)$$

Eq. (912) simplifies to:

$$
0 = \begin{cases} \left(\dfrac{n}{2}-1\right)F'\dfrac{f_{,\sigma}}{F}\cdot R + f_{,\sigma}\dfrac{n}{2}\cdot(L_M - 2\Lambda) \\[2mm] +\dfrac{f_{,\sigma}}{F^2}\left(FF'' + \left(\dfrac{n}{2}-2\right)(F')^2\right)(1-n)\Delta f + g^{\alpha\beta}\left(\dfrac{F'}{F}(1-n)\left(f_{,\alpha\beta\sigma} - \Gamma^{\gamma}_{\alpha\beta}f_{,\gamma\sigma}\right)\right) \end{cases}
$$

$$
= \begin{cases} \left(\dfrac{n}{2}-1\right)F'\dfrac{f_{,\sigma}}{F}\cdot R + f_{,\sigma}\dfrac{n}{2}\cdot(L_M - 2\Lambda) \\[2mm] +\dfrac{f_{,\sigma}}{F^2}\left(FF'' + \left(\dfrac{n}{2}-2\right)(F')^2\right)(1-n)\Delta f + g^{\alpha\beta}\left(\dfrac{F'}{F}(1-n)\left(f_{,\sigma\alpha\beta} - \Gamma^{\gamma}_{\alpha\beta}f_{,\sigma\gamma}\right)\right) \end{cases}
$$

$$
= \begin{cases} \left(\dfrac{n}{2}-1\right)F'\dfrac{f_{,\sigma}}{F}\cdot R + f_{,\sigma}\dfrac{n}{2}\cdot(L_M - 2\Lambda) \\[2mm] +\dfrac{(1-n)}{F}\left(\dfrac{f_{,\sigma}}{F}\left(FF'' + \left(\dfrac{n}{2}-2\right)(F')^2\right)\Delta f + F'\Delta\left(f_{,\sigma}\right)\right). \end{cases}
\tag{914}
$$

Setting $n = 4$ and inserting $F[f]$ from Eq. (913) yields:

$$
0 = 4\cdot f_{,\sigma}\cdot(L_M - 2\Lambda)\cdot(f - C_{f0})\cdot C_{f1} + \dfrac{2\cdot\left((f - C_{f0})\cdot\left(f_{,\sigma}\cdot R - 3\cdot\Delta\left(f_{,\sigma}\right)\right) - 3\cdot f_{,\sigma}\Delta f\right)}{\left(f - C_{f0}\right)^2}
$$

$$
\Rightarrow\ 0 = 2\cdot f_{,\sigma}\cdot(L_M - 2\Lambda)\cdot(f - C_{f0})\cdot C_{f1} + \dfrac{\left(f - C_{f0}\right)\cdot\left(f_{,\sigma}\cdot R - 3\cdot\Delta\left(f_{,\sigma}\right)\right) - 3\cdot f_{,\sigma}\Delta f}{\left(f - C_{f0}\right)^2}.
$$

$$\tag{915}$$

Another potentially simplified intrinsic equation can be obtained for $n = 4$ with the approach:

$$
F = \left(f - C_{f0}\right)C_{f1} \quad\Rightarrow\quad FF'' = 0; \quad F' = C_{f1}; \quad \left(\dfrac{n}{2}-2\right) = 0.
\tag{916}
$$

This will lead to:

$$
0 = \begin{cases} \dfrac{f_{,\sigma}}{f - C_{f0}}\cdot\left(R + \dfrac{3\cdot f_{,\alpha}f_{,\beta}g^{\alpha\beta}}{\left(f - C_{f0}\right)^2}\right) \\[3mm] +g^{\alpha\beta}\left(3\cdot\left(\dfrac{\left(f_{,\alpha\sigma}f_{,\beta} + f_{,\alpha}f_{,\beta\sigma}\right)\cdot\left(f - C_{f0}\right) - 2\cdot f_{,\alpha}f_{,\beta}f_{,\sigma}}{\left(f - C_{f0}\right)^3}\right) - \dfrac{3\cdot\left(f_{,\alpha\beta\sigma} - \Gamma^{\gamma}_{\alpha\beta}f_{,\gamma\sigma}\right)}{f - C_{f0}}\right) \\[3mm] +f_{,\sigma}2\cdot(L_M - 2\Lambda) \end{cases}
$$

$$= \left\{ \frac{f_{,\sigma}}{f - C_{f0}} \cdot \left(R - \frac{3 \cdot f_{,\alpha} f_{,\beta} g^{\alpha\beta}}{\left(f - C_{f0}\right)^2} \right) + 3 \cdot g^{\alpha\beta} \left(\frac{\left(f_{,\alpha\sigma} f_{,\beta} + f_{,\alpha} f_{,\beta\sigma}\right)}{\left(f - C_{f0}\right)^2} - \frac{\left(f_{,\alpha\beta\sigma} - \Gamma^{\gamma}_{\alpha\beta} f_{,\gamma\sigma}\right)}{f - C_{f0}} \right) \right.$$

$$\left. + f_{,\sigma} 2 \cdot \left(L_M - 2\Lambda\right). \right.$$

(917)

8.8.2 Centers of Gravity or Metric Centers

Within our repetition above we have only considered the situation with metric centers in connection with the classical (which is to say non-$F[f]$-scaled) metrics and subsequent Einstein–Hilbert action derivation. Extension to the situation with the scaled metrics is straight forward. Using the result from Eq. (771) and putting this back into the Hilbert integral, thereby assuming that we have an ensemble of metric or gravity centers, we obtain:

$$0 = \sum_{\forall G_k ; \xi^k} \int_V d^n x \frac{\sqrt{-g \cdot F^n}}{F}$$

$$\left(R^{\kappa\lambda} - \frac{1}{2} R \cdot g^{\kappa\lambda} + \left(\begin{array}{c} \frac{8\pi G T^{\kappa\lambda}}{\kappa} = \text{matter} \\ 0 = \text{vacuum} \end{array} \right) + \Lambda \cdot g^{\kappa\lambda} \right)$$

$$- \frac{1}{\sqrt{F^n}} \cdot \left(\partial^{;\kappa}\partial^{;\lambda} - g^{\kappa\lambda}\Delta_g\right) F^{\frac{n}{2}-1}$$

$$+ \left(g^{\kappa\lambda} g^{\alpha\beta} - g^{\kappa\alpha} g^{\lambda\beta} - g^{\lambda\alpha} g^{\kappa\beta}\right)(1 - n)$$

$$\cdot \left[\frac{f_{,\alpha} f_{,\beta}}{F^2}\left(FF'' + \left(F'\right)^2 \cdot \frac{3 \cdot n - 14}{8}\right) + f_{,\alpha\beta} \frac{F'}{F}\right]$$

$$- \frac{F' \cdot (1 - n)}{2 \cdot F}\left(g^{\kappa\lambda} g^{\alpha\beta} - g^{\kappa\alpha} g^{\lambda\beta} - g^{\lambda\alpha} g^{\kappa\beta}\right)\Gamma^{\gamma}_{\alpha\beta} f_{,\gamma}$$

$$+ \frac{F'}{F} \cdot (1 - n) g^{\alpha\beta} g^{\gamma\lambda} \Gamma^{\kappa}_{\alpha\beta} f_{,\gamma}$$

$$+ \frac{F'}{F} \cdot \frac{(1 - n)}{2}\left(g^{\kappa\lambda} g^{\alpha\beta} - g^{\kappa\alpha} g^{\lambda\beta} - g^{\lambda\alpha} g^{\kappa\beta}\right)_{,\beta} f_{,\alpha}$$

$$\left. + \frac{g^{\xi\zeta} g_{\xi\zeta,\beta}}{2} \cdot \left(\frac{F'}{F} \cdot \frac{(1 - n)}{2}\left(g^{\kappa\lambda} g^{\alpha\beta} - g^{\kappa\alpha} g^{\lambda\beta} - g^{\lambda\alpha} g^{\kappa\beta}\right) f_{,\alpha}\right)\right)$$

$$X^{\sigma}_{,k} \delta \xi^k. \quad (918)$$

8.8.3 Summing up: Centers of Gravity and Quantum Centers → Thermodynamics

We realize that in addition to any intrinsic solution to Eq. (771) or (912) (or just simpler due to a pre-setting of $F[f]$—(914) and in the case of $n = 4$—also (915)), for a given volume, we can always aim for a statistic satisfaction of the extended Einstein–Hilbert actions via Eqs. (910) and (918). This kind of quantum gravity statistics, however, is just thermodynamics.

8.9 Repetition: The Infinity Options Principle or What Happens in Infinitely Many Dimensions

In order to investigate the situation in spaces or space-times with infinitely many dimensions, we simply take the results from above (Eqs. (910) and (918)), divide by n^3, respectively n^2, and then look for the limit of $n \to \infty$. Thus, for Eqs. (910) and (918), respectively, we obtain:

$$\delta W = 0 =$$

$$= \sum_{\forall f_k} \int_V \lim_{n \to \infty} d^n x \left(\begin{array}{c} \dfrac{\sqrt{-g} \cdot F^{\frac{n}{2}-1}}{n^3} \times \\[2mm] \left(\begin{array}{c} \left(\dfrac{n}{2}-1\right) F' \dfrac{f_{,\sigma}}{F} \\[3mm] \cdot \left(R + \dfrac{f_{,\alpha} f_{,\beta} g^{\alpha\beta}(1-n)}{4F^2}\left(4FF'' + (F')^2(n-6) \right) \right) \\[3mm] + \dfrac{f_{,\sigma}}{F}\left(\dfrac{FF'' + \left(\dfrac{n}{2}-2\right)(F')^2}{F} \right)(1-n)\Delta f \\[4mm] g^{\alpha\beta}\left(\begin{array}{c} +\dfrac{F'}{F}(1-n)\left(f_{,\alpha\beta\sigma} - \Gamma^{\gamma}_{\alpha\beta} f_{,\gamma\sigma} \right) \\[3mm] +\left(\dfrac{f_{,\alpha} f_{,\beta}(1-n)}{4F^2}\left(4FF'' + (F')^2(n-6) \right) \right)_{,\sigma} \end{array} \right) \\[4mm] + f_{,\sigma}\dfrac{n}{2}\cdot (L_M - 2\Lambda) \end{array} \right) \end{array} \right) Y^{\sigma}_{,k} \delta\zeta^k$$

$$\delta W = 0 = \sum_{\forall f_k} \int_V \lim_{n\to\infty} d^n x \left(\sqrt{-g} \cdot F^{\frac{n}{2}-1} \cdot \left(f_{,\sigma} \frac{f_{,\alpha} f_{,\beta} g^{\alpha\beta}}{F^3} (F')^2 \right) \right) Y^\sigma_{,k} \delta\zeta^k \quad (919)$$

and:

$$0 = \sum_{\forall G_k ; \xi^k} \int_V \lim_{n\to\infty} d^n x \frac{\sqrt{-g \cdot F^n}}{F \cdot n^2}$$

$$\left(\begin{array}{c} R^{\kappa\lambda} - \dfrac{1}{2} R \cdot g^{\kappa\lambda} + \left(\left(\begin{array}{c} \underbrace{\dfrac{8\pi G}{\kappa} T^{\kappa\lambda}}_{} = \text{matter} \\[4pt] 0 = \text{vacuum} \end{array} \right) + \Lambda \cdot g^{\kappa\lambda} \right) \\[20pt] - \dfrac{1}{\sqrt{F^n}} \cdot \left(\partial^{:\kappa}\partial^{:\lambda} - g^{\kappa\lambda}\Delta_g \right) F^{\frac{n}{2}-1} \\[12pt] + \left(g^{\kappa\lambda} g^{\alpha\beta} - g^{\kappa\alpha} g^{\lambda\beta} - g^{\lambda\alpha} g^{\kappa\beta} \right)(1-n) \\[8pt] \cdot \left[\dfrac{f_{,\alpha} f_{,\beta}}{F^2} \left(FF'' + (F')^2 \cdot \dfrac{3\cdot n - 14}{8} \right) + f_{,\alpha\beta} \dfrac{F'}{F} \right] \\[12pt] - \dfrac{F'\cdot(1-n)}{2\cdot F} \left(g^{\kappa\lambda} g^{\alpha\beta} - g^{\kappa\alpha} g^{\lambda\beta} - g^{\lambda\alpha} g^{\kappa\beta} \right)\Gamma^\gamma_{\alpha\beta} f_{,\gamma} \\[12pt] + \dfrac{F'}{F} \cdot (1-n) g^{\alpha\beta} g^{\gamma\lambda} \Gamma^\kappa_{\alpha\beta} f_{,\gamma} \\[12pt] + \dfrac{F'}{F} \cdot \dfrac{(1-n)}{2} \left(g^{\kappa\lambda} g^{\alpha\beta} - g^{\kappa\alpha} g^{\lambda\beta} - g^{\lambda\alpha} g^{\kappa\beta} \right)_{,\beta} f_{,\alpha} \\[12pt] + \dfrac{g^{\xi\zeta} g_{\xi\zeta,\beta}}{2} \cdot \left(\dfrac{F'}{F} \cdot \dfrac{(1-n)}{2} \left(g^{\kappa\lambda} g^{\alpha\beta} - g^{\kappa\alpha} g^{\lambda\beta} - g^{\lambda\alpha} g^{\kappa\beta} \right) f_{,\alpha} \right) \end{array} \right) \cdot X^\sigma_{,k} \delta\xi^k$$

$$= \sum_{\forall G_k ; \xi^k} \int_V \lim_{n\to\infty} d^n x \frac{\sqrt{-g \cdot F^n}}{F} \cdot \left(-\left(g^{\kappa\lambda} g^{\alpha\beta} - g^{\kappa\alpha} g^{\lambda\beta} - g^{\lambda\alpha} g^{\kappa\beta} \right) \cdot \frac{f_{,\alpha} f_{,\beta}}{F^2} (F')^2 \cdot \frac{3}{8} \right) X^\sigma_{,k} \delta\xi^k$$

$$= \sum_{\forall G_k ; \xi^k} \int_V \lim_{n\to\infty} d^n x \frac{\sqrt{-g \cdot F^n}}{F} \cdot \left(2 \cdot f^{:\kappa} f^{,\beta} - g^{\kappa\lambda} f^{,\beta} f_{,\beta} \right) \cdot \frac{(F')^2}{F^2} \cdot X^\sigma_{,k} \delta\xi^k. \quad (920)$$

Thereby, this we want to point out explicitly, we do not necessarily need to assume truly infinite numbers of dimensions, but simply big enough n so that all terms with n^p and $p < 3$ in Eq. (910) and $p < 2$ in Eq. (918) are exceedingly small against the n^3 in Eq. (910) and n^2 in Eq. (918).

We already have discussed the results of this section in chapter 7 (also in [8] … at least with respect to Eq. (918), which surely is obvious, because of the power-3 term for the derivative of the quantum scale function f and its potential connection with the so-called 3-generation problem in elementary particle physics).

8.10 Appendix A: The Derivation of the Scaled Ricci Scalar R^*

In order to evaluate the Ricci scalar of the scaled metric $G_{\alpha\beta} = F[f]^* g_{\alpha\beta} = F^* g_{\alpha\beta}$, we need a variety of identities. At first, we need some principle definitions:

$$G_{\alpha\gamma} G^{\alpha\beta} = \delta_\gamma^\beta; \quad G_{\alpha\beta} = F \cdot g_{\alpha\beta} = F\big[f[x_k]\big] \cdot g_{\alpha\beta}; \quad G^{\alpha\beta} = \frac{g^{\alpha\beta}}{F} = \frac{g^{\alpha\beta}}{F\big[f[x_k]\big]},$$

$$(921)$$

$$
\begin{aligned}
\Gamma_{\alpha\beta}^{*\gamma} &= \frac{g^{\gamma\sigma}}{2 \cdot F[f]}\left(\big[F[f] \cdot g_{\sigma\alpha}\big]_{,\beta} + \big[F[f] \cdot g_{\sigma\beta}\big]_{,\alpha} - \big[F[f] \cdot g_{\alpha\beta}\big]_{,\sigma}\right) \\[2mm]
&= \frac{g^{\gamma\sigma}}{2 \cdot F[f]}\left(\big[F[f] \cdot g_{\sigma\alpha}\big]_{,\beta} + \big[F[f] \cdot g_{\sigma\beta}\big]_{,\alpha} - \big[F[f] \cdot g_{\alpha\beta}\big]_{,\sigma}\right) \\[2mm]
&= \frac{g^{\gamma\sigma}}{2}\left(g_{\sigma\alpha,\beta} + g_{\sigma\beta,\alpha} - g_{\alpha\beta,\sigma}\right) + \frac{g^{\gamma\sigma} F'}{2 \cdot F[f]}\left(f_{,\beta} \cdot g_{\sigma\alpha} + f_{,\alpha} \cdot g_{\sigma\beta} - f_{,\sigma} \cdot g_{\alpha\beta}\right) \\[2mm]
&= \Gamma_{\alpha\beta}^\gamma + \frac{F'}{2 \cdot F[f]}\left(f_{,\beta} \cdot \delta_\alpha^\gamma + f_{,\alpha} \cdot \delta_\beta^\gamma - f_{,\sigma} \cdot g^{\gamma\sigma} g_{\alpha\beta}\right) \\[2mm]
&\equiv \Gamma_{\alpha\beta}^\gamma + \Gamma_{\alpha\beta}^{**\gamma},
\end{aligned}
$$

$$(922)$$

and then the necessary terms for the evaluation for the Ricci scalar:

$$\Gamma_{\alpha\beta}^{*\gamma} = \Gamma_{\alpha\beta}^\gamma + \frac{F'}{2 \cdot F[f]}\left(f_{,\beta} \cdot \delta_\alpha^\gamma + f_{,\alpha} \cdot \delta_\beta^\gamma - f_{,\sigma} \cdot g^{\gamma\sigma} g_{\alpha\beta}\right)$$

$$R^* = R^*{}_{\alpha\beta} G^{\alpha\beta} = \left(\Gamma_{\alpha\beta,\sigma}^{*\sigma} - \Gamma_{\beta\sigma,\alpha}^{*\sigma} - \Gamma_{\sigma\alpha}^{*\mu}\Gamma_{\beta\mu}^{*\sigma} + \Gamma_{\alpha\beta}^{*\sigma}\Gamma_{\sigma\mu}^{*\mu}\right) G^{\alpha\beta},$$

$$(923)$$

$$\Gamma_{\alpha\beta}^{*\sigma} = \Gamma_{\alpha\beta}^\sigma + \frac{F'}{2 \cdot F[f]}\left(f_{,\beta} \cdot \delta_\alpha^\sigma + f_{,\alpha} \cdot \delta_\beta^\sigma - f_{,\gamma} \cdot g^{\gamma\sigma} g_{\alpha\beta}\right)$$

$$\Gamma^{*\mu}_{\sigma\mu} = \Gamma^{\mu}_{\sigma\mu} + \frac{F'}{2 \cdot F[f]}\left(f_{,\sigma} \cdot \delta^{\mu}_{\mu} + f_{,\mu} \cdot \delta^{\mu}_{\sigma} - f_{,\gamma} \cdot g^{\gamma\mu}g_{\sigma\mu}\right)$$

$$= \Gamma^{\mu}_{\sigma\mu} + \frac{F'}{2 \cdot F[f]}\left(f_{,\sigma} \cdot n + f_{,\sigma} - f_{,\sigma}\right) = \Gamma^{\mu}_{\sigma\mu} + \frac{F' f_{,\sigma} \cdot n}{2 \cdot F[f]}, \tag{924}$$

$$\Gamma^{*\sigma}_{\beta\mu} = \Gamma^{\sigma}_{\beta\mu} + \frac{1}{2 \cdot F[f]}\left(F[f]_{,\beta} \cdot \delta^{\sigma}_{\mu} + F[f]_{,\mu} \cdot \delta^{\sigma}_{\beta} - F[f]_{,\gamma} \cdot g^{\gamma\sigma}g_{\mu\beta}\right)$$

$$\Gamma^{*\mu}_{\sigma\alpha} = \Gamma^{\mu}_{\sigma\alpha} + \frac{1}{2 \cdot F[f]}\left(F[f]_{,\sigma} \cdot \delta^{\mu}_{\alpha} + F[f]_{,\alpha} \cdot \delta^{\mu}_{\sigma} - F[f]_{,\gamma} \cdot g^{\gamma\mu}g_{\alpha\sigma}\right), \tag{925}$$

$$\Gamma^{*\sigma}_{\beta\sigma} = \Gamma^{\sigma}_{\beta\sigma} + \frac{1}{2 \cdot F[f]}\left(F[f]_{,\sigma} \cdot \delta^{\sigma}_{\beta} + F[f]_{,\beta} \cdot n - F[f]_{,\beta}\right)$$

$$= \Gamma^{\sigma}_{\beta\sigma} + \frac{1}{2 \cdot F[f]}\left(F[f]_{,\beta} + F[f]_{,\beta} \cdot n - F[f]_{,\beta}\right) = \Gamma^{\sigma}_{\beta\sigma} + \frac{F' f_{,\beta} \cdot n}{2 \cdot F[f]}. \tag{926}$$

Without much fuss we now evaluate R^* as follows:

$$R^* = R^*_{\alpha\beta}G^{\alpha\beta}$$

$$= \begin{pmatrix} \left(\Gamma^{\sigma}_{\alpha\beta} + \dfrac{F'}{2 \cdot F[f]}\left(f_{,\beta} \cdot \delta^{\sigma}_{\alpha} + f_{,\alpha} \cdot \delta^{\sigma}_{\beta} - f_{,\gamma} \cdot g^{\gamma\sigma}g_{\alpha\beta}\right)\right)_{,\sigma} \\[2ex] -\left(\Gamma^{\sigma}_{\beta\sigma} + \dfrac{F' f_{,\beta} \cdot n}{2 \cdot F[f]}\right)_{,\alpha} \\[2ex] -\left(\Gamma^{\mu}_{\sigma\alpha} + \dfrac{1}{2 \cdot F[f]}\left(F[f]_{,\sigma} \cdot \delta^{\mu}_{\alpha} + F[f]_{,\alpha} \cdot \delta^{\mu}_{\sigma} - F[f]_{,\gamma} \cdot g^{\gamma\mu}g_{\alpha\sigma}\right)\right) \\[2ex] \cdot\left(\Gamma^{\sigma}_{\beta\mu} + \dfrac{1}{2 \cdot F[f]}\left(F[f]_{,\beta} \cdot \delta^{\sigma}_{\mu} + F[f]_{,\mu} \cdot \delta^{\sigma}_{\beta} - F[f]_{,\gamma} \cdot g^{\gamma\sigma}g_{\mu\beta}\right)\right) \\[2ex] +\left(\Gamma^{\sigma}_{\alpha\beta} + \dfrac{F'}{2 \cdot F[f]}\left(f_{,\beta} \cdot \delta^{\sigma}_{\alpha} + f_{,\alpha} \cdot \delta^{\sigma}_{\beta} - f_{,\gamma} \cdot g^{\gamma\sigma}g_{\alpha\beta}\right)\right)\left(\Gamma^{\mu}_{\sigma\mu} + \dfrac{F' f_{,\sigma} \cdot n}{2 \cdot F[f]}\right) \end{pmatrix} \frac{g^{\alpha\beta}}{F}$$

$$\tag{927}$$

$$R^* = R^*_{\alpha\beta}G^{\alpha\beta}$$

$$
\left(
\begin{array}{l}
\left(
\begin{array}{l}
\Gamma^{\sigma}_{\alpha\beta,\sigma} + \dfrac{FF'' - \left(F'\right)^2}{2\cdot F[f]^2} f_{,\sigma}\left(f_{,\beta}\cdot\delta^{\sigma}_{\alpha} + f_{,\alpha}\cdot\delta^{\sigma}_{\beta} - f_{,\gamma}\cdot g^{\gamma\sigma}g_{\alpha\beta}\right) \\
+\dfrac{F'}{2\cdot F[f]}\left(f_{,\beta\sigma}\cdot\delta^{\sigma}_{\alpha} + f_{,\alpha\sigma}\cdot\delta^{\sigma}_{\beta} - f_{,\gamma\sigma}\cdot g^{\gamma\sigma}g_{\alpha\beta} - f_{,\gamma}\cdot\left(g^{\gamma\sigma}g_{\alpha\beta}\right)_{,\sigma}\right)
\end{array}
\right) \\
-\left(\Gamma^{\sigma}_{\beta\sigma,\alpha} - \dfrac{\left(F'\right)^2 f_{,\alpha}f_{,\beta}\cdot n}{2\cdot F[f]^2} + \dfrac{\left(F'' f_{,\alpha}f_{,\beta} + F' f_{,\alpha\beta}\right)\cdot n}{2\cdot F[f]}\right) \\
-\left(\Gamma^{\mu}_{\sigma\alpha} + \dfrac{F'}{2\cdot F[f]}\left(f_{,\sigma}\cdot\delta^{\mu}_{\alpha} + f_{,\alpha}\cdot\delta^{\mu}_{\sigma} - f_{,\gamma}\cdot g^{\gamma\mu}g_{\alpha\sigma}\right)\right) \\
\cdot\left(\Gamma^{\sigma}_{\beta\mu} + \dfrac{F'}{2\cdot F[f]}\left(f_{,\beta}\cdot\delta^{\sigma}_{\mu} + f_{,\mu}\cdot\delta^{\sigma}_{\beta} - f_{,\gamma}\cdot g^{\gamma\sigma}g_{\mu\beta}\right)\right) \\
+\left(\Gamma^{\sigma}_{\alpha\beta} + \dfrac{F'}{2\cdot F[f]}\left(f_{,\beta}\cdot\delta^{\sigma}_{\alpha} + f_{,\alpha}\cdot\delta^{\sigma}_{\beta} - f_{,\gamma}\cdot g^{\gamma\sigma}g_{\alpha\beta}\right)\right)\left(\Gamma^{\mu}_{\sigma\mu} + \dfrac{F' f_{,\sigma}\cdot n}{2\cdot F[f]}\right)
\end{array}
\right)\dfrac{g^{\alpha\beta}}{F} \quad (928)
$$

$$
R^{*} = R^{*}_{\alpha\beta}G^{\alpha\beta}
$$

$$
\left(
\begin{array}{l}
\left(
\begin{array}{l}
\Gamma^{\sigma}_{\alpha\beta,\sigma} + \dfrac{FF'' - \left(F'\right)^2}{2\cdot F[f]^2} f_{,\sigma}\left(f_{,\beta}\cdot\delta^{\sigma}_{\alpha} + f_{,\alpha}\cdot\delta^{\sigma}_{\beta} - f_{,\gamma}\cdot g^{\gamma\sigma}g_{\alpha\beta}\right) \\
+\dfrac{F'}{2\cdot F[f]}\left(f_{,\beta\sigma}\cdot\delta^{\sigma}_{\alpha} + f_{,\alpha\sigma}\cdot\delta^{\sigma}_{\beta} - f_{,\gamma\sigma}\cdot g^{\gamma\sigma}g_{\alpha\beta} - f_{,\gamma}\cdot\left(g^{\gamma\sigma}g_{\alpha\beta}\right)_{,\sigma}\right)
\end{array}
\right) \\
-\left(\Gamma^{\sigma}_{\beta\sigma,\alpha} - \dfrac{\left(F'\right)^2 f_{,\alpha}f_{,\beta}\cdot n}{2\cdot F[f]^2} + \dfrac{\left(F'' f_{,\alpha}f_{,\beta} + F' f_{,\alpha\beta}\right)\cdot n}{2\cdot F[f]}\right) \\
-\left(
\begin{array}{l}
\Gamma^{\mu}_{\sigma\alpha}\Gamma^{\sigma}_{\beta\mu} + \dfrac{\Gamma^{\sigma}_{\beta\mu}F'}{2\cdot F[f]}\left(f_{,\sigma}\cdot\delta^{\mu}_{\alpha} + f_{,\alpha}\cdot\delta^{\mu}_{\sigma} - f_{,\gamma}\cdot g^{\gamma\mu}g_{\alpha\sigma}\right) \\
+\Gamma^{\mu}_{\sigma\alpha}\dfrac{F'}{2\cdot F[f]}\left(f_{,\beta}\cdot\delta^{\sigma}_{\mu} + f_{,\mu}\cdot\delta^{\sigma}_{\beta} - f_{,\gamma}\cdot g^{\gamma\sigma}g_{\mu\beta}\right) \\
+\left(\dfrac{F'}{2\cdot F[f]}\right)^2\left(
\begin{array}{l}
f_{,\beta}\cdot\delta^{\sigma}_{\mu}f_{,\sigma}\cdot\delta^{\mu}_{\alpha} + f_{,\mu}\cdot\delta^{\sigma}_{\beta}f_{,\sigma}\cdot\delta^{\mu}_{\alpha} - f_{,\gamma}\cdot g^{\gamma\sigma}g_{\mu\beta}f_{,\sigma}\cdot\delta^{\mu}_{\alpha} \\
+f_{,\beta}\cdot\delta^{\sigma}_{\mu}f_{,\alpha}\cdot\delta^{\mu}_{\sigma} + f_{,\mu}\cdot\delta^{\sigma}_{\beta}f_{,\alpha}\cdot\delta^{\mu}_{\sigma} - f_{,\gamma}\cdot g^{\gamma\sigma}g_{\mu\beta}f_{,\alpha}\cdot\delta^{\mu}_{\sigma} \\
-f_{,\gamma}\cdot g^{\gamma\mu}g_{\alpha\sigma}f_{,\beta}\cdot\delta^{\sigma}_{\mu} - f_{,\gamma}\cdot g^{\gamma\mu}g_{\alpha\sigma}f_{,\mu}\cdot\delta^{\sigma}_{\beta} \\
+f_{,\gamma}\cdot g^{\gamma\mu}g_{\alpha\sigma}f_{,\gamma}\cdot g^{\gamma\sigma}g_{\mu\beta}
\end{array}
\right)
\end{array}
\right) \\
+\left(
\begin{array}{l}
\Gamma^{\sigma}_{\alpha\beta}\Gamma^{\mu}_{\sigma\mu} + \dfrac{\Gamma^{\mu}_{\sigma\mu}F'}{2\cdot F[f]}\left(f_{,\beta}\cdot\delta^{\sigma}_{\alpha} + f_{,\alpha}\cdot\delta^{\sigma}_{\beta} - f_{,\gamma}\cdot g^{\gamma\sigma}g_{\alpha\beta}\right) \\
+\dfrac{F' f_{,\sigma}\cdot n}{2\cdot F[f]}\Gamma^{\sigma}_{\alpha\beta} + \left(\dfrac{F'}{2\cdot F[f]}\right)^2 n\left(f_{,\sigma}f_{,\beta}\cdot\delta^{\sigma}_{\alpha} + f_{,\sigma}f_{,\alpha}\cdot\delta^{\sigma}_{\beta} - f_{,\sigma}f_{,\gamma}\cdot g^{\gamma\sigma}g_{\alpha\beta}\right)
\end{array}
\right)
\end{array}
\right)\dfrac{g^{\alpha\beta}}{F} \quad (929)
$$

$$
\begin{aligned}
R^* = R^*_{\alpha\beta} G^{\alpha\beta} = \frac{1}{F}\Bigg\{
&\overbrace{\Gamma^\sigma_{\alpha\beta,\sigma} g^{\alpha\beta} + \frac{FF'' - (F')^2}{2\cdot F[f]^2}\, f_{,\sigma}\left(f_{,\beta}\cdot\delta^\sigma_\alpha + f_{,\alpha}\cdot\delta^\sigma_\beta - f_{,\gamma}\cdot g^{\gamma\sigma} g_{\alpha\beta}\right) g^{\alpha\beta}}^{\frac{1}{F}} \\
&+ \frac{F'}{2\cdot F[f]}\left(f_{,\beta\sigma}\cdot\delta^\sigma_\alpha + f_{,\alpha\sigma}\cdot\delta^\sigma_\beta - f_{,\gamma\sigma}\cdot g^{\gamma\sigma} g_{\alpha\beta} - f_{,\gamma}\cdot\left(g^{\gamma\sigma} g_{\alpha\beta}\right)_{,\sigma}\right) g^{\alpha\beta} \\[4pt]
-\; & \Bigg(\Gamma^\sigma_{\beta\sigma,\alpha} g^{\alpha\beta} - \frac{(F')^2 f_{,\alpha} f_{,\beta}\cdot n}{2\cdot F[f]^2}\, g^{\alpha\beta} + \frac{\left(F'' f_{,\alpha} f_{,\beta} + F' f_{,\alpha\beta}\right)\cdot n}{2\cdot F[f]}\, g^{\alpha\beta} \Bigg) \\[4pt]
+\; & \Gamma^\mu_{\sigma\alpha}\Gamma^\sigma_{\beta\mu} g^{\alpha\beta} + \frac{\Gamma^\sigma_{\beta\mu} F'}{2\cdot F[f]}\, g^{\alpha\beta}\left(f_{,\sigma}\cdot\delta^\mu_\alpha + f_{,\alpha}\cdot\delta^\mu_\sigma - f_{,\gamma}\cdot g^{\gamma\mu} g_{\alpha\sigma}\right) \\
&+ \Gamma^\mu_{\sigma\alpha}\,\frac{F'}{2\cdot F[f]}\left(f_{,\beta}\cdot\delta^\sigma_\mu + f_{,\mu}\cdot\delta^\sigma_\beta - f_{,\gamma}\cdot g^{\gamma\sigma} g_{\mu\beta}\right) g^{\alpha\beta} \\
&+ \left(\frac{F'}{2\cdot F[f]}\right)^2 \overbrace{\begin{pmatrix}
f_{,\beta}\cdot\delta^\sigma_\mu f_{,\sigma}\cdot\delta^\mu_\alpha + f_{,\mu}\cdot\delta^\sigma_\beta f_{,\sigma}\cdot\delta^\mu_\alpha - f_{,\gamma}\cdot g^{\gamma\sigma} g_{\mu\beta} f_{,\sigma}\cdot\delta^\mu_\alpha \\
+\, f_{,\beta}\cdot\delta^\sigma_\mu f_{,\alpha}\cdot\delta^\mu_\sigma + f_{,\beta}\cdot\delta^\sigma_\mu f_{,\alpha}\cdot\delta^\mu_\sigma - f_{,\gamma}\cdot g^{\gamma\sigma} g_{\mu\beta} f_{,\alpha}\cdot\delta^\mu_\sigma \\
-\, f_{,\gamma}\cdot g^{\gamma\mu} g_{\alpha\sigma} f_{,\beta}\cdot\delta^\sigma_\mu - f_{,\gamma}\cdot g^{\gamma\mu} g_{\alpha\sigma} f_{,\mu}\cdot\delta^\sigma_\beta \\
+\, f_{,\gamma}\cdot g^{\gamma\mu} g_{\alpha\sigma} f_{,\gamma}\cdot g^{\gamma\sigma} g_{\mu\beta}
\end{pmatrix}}^{g^{\alpha\beta}} \\[4pt]
+\; & \Gamma^\sigma_{\alpha\beta}\Gamma^\mu_{\sigma\mu} g^{\alpha\beta} + \frac{\Gamma^\mu_{\sigma\mu} F'}{2\cdot F[f]}\left(f_{,\beta}\cdot\delta^\sigma_\alpha + f_{,\alpha}\cdot\delta^\sigma_\beta - f_{,\gamma}\cdot g^{\gamma\sigma} g_{\alpha\beta}\right) g^{\alpha\beta} + \frac{F' f_{,\sigma}\cdot n}{2\cdot F[f]}\Gamma^\sigma_{\alpha\beta} g^{\alpha\beta} \\
&+ \left(\frac{F'}{2\cdot F[f]}\right)^2 n\left(f_{,\sigma} f_{,\beta}\cdot\delta^\sigma_\alpha + f_{,\sigma} f_{,\alpha}\cdot\delta^\sigma_\beta - f_{,\sigma} f_{,\gamma}\cdot g^{\gamma\sigma} g_{\alpha\beta}\right) g^{\alpha\beta}
\Bigg\}
\end{aligned}
\tag{930}
$$

$$R^* = R^*_{\alpha\beta}G^{\alpha\beta} =$$

$$\frac{1}{F}\left(\begin{array}{l}
\Gamma^\sigma_{\alpha\beta,\sigma}g^{\alpha\beta} + \dfrac{FF'' - (F')^2}{2\cdot F[f]^2}\,f_{,\sigma}\left(f_{,\beta\gamma}g^{\sigma\beta} + f_{,\alpha}\cdot g^{\alpha\sigma} - f_{,\gamma}\cdot g^{\gamma\sigma}n\right)\\[2mm]
+\dfrac{F'}{2\cdot F[f]}\left(f_{,\beta\sigma}g^{\sigma\beta} + f_{,\alpha\sigma}g^{\alpha\sigma} - f_{,\gamma\sigma}\cdot g^{\gamma\sigma}n - f_{,\gamma}\cdot\left(g^{\gamma\sigma}g_{\alpha\beta}\right)_{,\sigma}g^{\alpha\beta}\right)\\[4mm]
-\Gamma^\sigma_{\beta\sigma,\alpha}g^{\alpha\beta} - \dfrac{(F')^2}{2\cdot F[f]^2}f_{,\alpha}f_{,\beta}\cdot n\,g^{\alpha\beta} + \dfrac{\left(F''f_{,\alpha}f_{,\beta} + F'f_{,\alpha\beta}\right)\cdot n}{2\cdot F[f]^2}g^{\alpha\beta}\\[4mm]
-\left(\begin{array}{l}
\Gamma^\mu_{\alpha\sigma}\Gamma^\sigma_{\beta\mu}g^{\alpha\beta} + \dfrac{\Gamma^\sigma_{\beta\mu}F'}{2\cdot F[f]}\left(f_{,\sigma}g^{\mu\beta} + f_{,\alpha}\cdot\delta^\mu_\sigma g^{\alpha\beta} - f_{,\gamma}\cdot g^{\gamma\mu}\delta^\beta_\sigma\right)\\[2mm]
+\Gamma^\mu_{\sigma\alpha}\dfrac{F'}{2\cdot F[f]}\left(f_{,\beta}\cdot\delta^\sigma_\mu g^{\alpha\beta} + f_{,\mu}g^{\alpha\sigma} - f_{,\gamma}\cdot g^{\gamma\sigma}\delta^\alpha_\mu\right)\\[2mm]
+\left(\dfrac{F'}{2\cdot F[f]}\right)^2\left[\begin{array}{l}
f_{,\beta}f_{,\alpha} + f_{,\alpha}f_{,\beta} - f_{,\gamma}\cdot g^{\gamma\sigma}g_{\alpha\beta}f_{,\sigma}\\
+f_{,\beta}f_{,\alpha}\cdot n + f_{,\beta}f_{,\alpha} - f_{,\beta}f_{,\alpha}\\
-f_{,\alpha}f_{,\beta} - f_{,\gamma}\cdot g^{\gamma\mu}g_{\alpha\beta}f_{,\mu} + f_{,\alpha}f_{,\beta}
\end{array}\right]g^{\alpha\beta}
\end{array}\right)\\[12mm]
+\begin{array}{l}
\Gamma^\sigma_{\alpha\beta}\Gamma^\mu_{\sigma\mu}g^{\alpha\beta} + \dfrac{\Gamma^\mu_{\sigma\mu}F'}{2\cdot F[f]}\left(f_{,\beta\beta}g^{\sigma\beta} + f_{,\alpha}g^{\alpha\sigma} - f_{,\gamma}\cdot g^{\gamma\sigma}n\right) + \dfrac{F'f_{,\sigma}\cdot n}{2\cdot F[f]}\Gamma^\sigma_{\alpha\beta}g^{\alpha\beta}\\[2mm]
+\left(\dfrac{F'}{2\cdot F[f]}\right)^2 n\left(f_{,\sigma}f_{,\beta\beta}g^{\sigma\beta} + f_{,\sigma}f_{,\alpha}\cdot g^{\alpha\sigma} - f_{,\sigma}f_{,\gamma}\cdot g^{\gamma\sigma}n\right)
\end{array}
\end{array}\right)$$

$$(931)$$

$$
R^* = R^*_{\alpha\beta} G^{\alpha\beta} =
$$

$$
\frac{1}{F}\left\{
\begin{aligned}
&\Gamma^\sigma_{\alpha\beta,\sigma} g^{\alpha\beta} + \frac{FF'' - (F')^2}{2\cdot F[f]^2}\, f_{,\sigma}\left(f_{,\beta}g^{\sigma\beta} + f_{,\alpha}\cdot g^{\alpha\sigma} - f_{,\gamma}\cdot g^{\gamma\sigma} n\right)\\
&+ \frac{F'}{2\cdot F[f]}\left(f_{,\beta\sigma}g^{\sigma\beta} + f_{,\alpha\sigma}g^{\alpha\sigma} - f_{,\gamma\sigma}\cdot g^{\gamma\sigma} n - f_{,\gamma}\cdot\left(g^{\gamma\sigma} g_{\alpha\beta}\right)_{,\sigma} g^{\alpha\beta}\right)\\[4pt]
&-\Gamma^\sigma_{\beta\sigma,\alpha} g^{\alpha\beta} - \frac{(F')^2 f_{,\alpha} f_{,\beta}\cdot n}{2\cdot F[f]^2} g^{\alpha\beta} + \frac{\left(F'' f_{,\alpha} f_{,\beta} + F' f_{,\alpha\beta}\right)\cdot n}{2\cdot F[f]} g^{\alpha\beta}\\[4pt]
&\Gamma^\mu_{\sigma\alpha}\Gamma^\sigma_{\beta\mu} g^{\alpha\beta} + \frac{\Gamma^\sigma_{\beta\mu} F'}{2\cdot F[f]}\left(f_{,\sigma}g^{\mu\beta} + f_{,\alpha}\cdot\delta^\mu_\sigma g^{\alpha\beta} - f_{,\gamma}\cdot g^{\gamma\mu}\delta^\beta_\sigma\right)\\
&+\Gamma^\mu_{\sigma\alpha}\frac{F'}{2\cdot F[f]}\left(f_{,\beta}\cdot\delta^\sigma_\mu g^{\alpha\beta} + f_{,\mu}g^{\alpha\sigma} - f_{,\gamma}\cdot g^{\gamma\sigma}\delta^\alpha_\mu\right)\\
&-\left(\frac{F'}{2\cdot F[f]}\right)^2\left(2f_{,\beta}f_{,\alpha}g^{\alpha\beta} - f_{,\gamma}\cdot g^{\gamma\mu} n f_{,\mu}\right)\\[4pt]
&\Gamma^\sigma_{\alpha\beta}\Gamma^\mu_{\sigma\mu} g^{\alpha\beta} + \frac{\Gamma^\mu_{\sigma\mu} F'}{2\cdot F[f]}\left(f_{,\beta}g^{\sigma\beta} + f_{,\alpha}g^{\alpha\sigma} - f_{,\gamma}\cdot g^{\gamma\sigma} n\right) + \frac{F' f_{,\sigma}\cdot n}{2\cdot F[f]}\Gamma^\sigma_{\alpha\beta}g^{\alpha\beta}\\
&+\left(\frac{F'}{2\cdot F[f]}\right)^2 n\left(f_{,\sigma}f_{,\beta}g^{\sigma\beta} + f_{,\sigma}f_{,\alpha}\cdot g^{\alpha\sigma} - f_{,\sigma}f_{,\gamma}\cdot g^{\gamma\sigma} n\right)
\end{aligned}
\right\}
$$

$$\tag{932}$$

$$
R^* = R^*_{\alpha\beta} G^{\alpha\beta} = \Bigg(\Gamma^\sigma_{\alpha\beta,\sigma} g^{\alpha\beta} - \Gamma^\sigma_{\beta\sigma,\alpha} g^{\alpha\beta} - \Gamma^\mu_{\sigma\alpha} \Gamma^\sigma_{\beta\mu} g^{\alpha\beta} + \Gamma^\sigma_{\alpha\beta} \Gamma^\mu_{\sigma\mu} g^{\alpha\beta}
$$

$$
- \left(\frac{FF'' - (F')^2}{2 \cdot F[f]^2} \left((2-n) f_{,\alpha} f_{,\beta} g^{\alpha\beta} \right) + \frac{F'}{2 \cdot F[f]} \left((2-n) f_{,\beta\alpha} g^{\alpha\beta} - f_{,\gamma} \cdot \left(g^{\gamma\sigma} g_{\alpha\beta} \right)_{,\sigma} g^{\alpha\beta} \right) \right)
$$

$$
- \left(\frac{\left(FF'' - (F')^2 \right) f_{,\alpha} f_{,\beta} \cdot n}{2 \cdot F[f]^2} g^{\alpha\beta} + \frac{\left(F' f_{,\alpha\beta} \right) \cdot n}{2 \cdot F[f]} g^{\alpha\beta} \right)
$$

$$
- \frac{\Gamma^\sigma_{\beta\mu} F'}{2 \cdot F[f]} \left(f_{,\sigma} g^{\mu\beta} + f_{,\alpha} \cdot \delta^\mu_\sigma g^{\alpha\beta} - f_{,\gamma} \cdot g^{\gamma\mu} \delta^\beta_\sigma \right)
$$

$$
+ \Gamma^\mu_{\sigma\alpha} \frac{F'}{2 \cdot F[f]} \left(f_{,\beta} \cdot \delta^\sigma_\mu g^{\alpha\beta} + f_{,\mu} g^{\alpha\sigma} - f_{,\gamma} \cdot g^{\gamma\sigma} \delta^\alpha_\mu \right) + \left(\frac{F'}{2 \cdot F[f]} \right)^2 \left(2 f_{,\beta} f_{,\alpha} g^{\alpha\beta} - f_{,\gamma} \cdot g^{\gamma\mu} n f_{,\mu} \right)
$$

$$
+ \frac{\Gamma^\mu_{\sigma\mu} F'}{2 \cdot F[f]} \left(f_{,\beta} g^{\sigma\beta} + f_{,\alpha} g^{\alpha\sigma} - f_{,\gamma} \cdot g^{\gamma\sigma} n \right) + \frac{F' f_{,\sigma} \cdot n}{2 \cdot F[f]} \Gamma^\sigma_{\alpha\beta} g^{\alpha\beta}
$$

$$
+ \left(\frac{F'}{2 \cdot F[f]} \right)^2 n \left(f_{,\sigma} f_{,\beta} g^{\sigma\beta} + f_{,\sigma} f_{,\alpha} \cdot g^{\alpha\sigma} - f_{,\gamma} \cdot g^{\gamma\sigma} n \right) \Bigg) \frac{1}{F} \tag{933}
$$

$$R^* = R^*_{\alpha\beta} G^{\alpha\beta} =
\left(
\begin{aligned}
& \Gamma^\sigma_{\alpha\beta,\sigma} g^{\alpha\beta} - \Gamma^\sigma_{\beta\sigma,\alpha} g^{\alpha\beta} - \Gamma^\mu_{\sigma\alpha} \Gamma^\sigma_{\beta\mu} g^{\alpha\beta} + \Gamma^\sigma_{\alpha\beta} \Gamma^\mu_{\sigma\mu} g^{\alpha\beta} \\
& + \frac{F'}{2 \cdot F[f]} \left((2-n) f_{,\beta\alpha} g^{\alpha\beta} - f_{,\gamma} \cdot \left(g^{\gamma\sigma} g_{\alpha\beta} \right)_{,\sigma} g^{\alpha\beta} \right) - \left(\frac{F' f_{,\alpha\beta} \cdot n}{2 \cdot F[f]} g^{\alpha\beta} \right) \\
& - \left(\frac{\Gamma^\sigma_{\beta\mu} F'}{2 \cdot F[f]} \left(f_{,\sigma} g^{\mu\beta} + f_{,\alpha} \cdot \delta^\mu_\sigma g^{\alpha\beta} - f_{,\gamma} \cdot g^{\gamma\mu} \delta^\beta_\sigma \right) + \Gamma^\mu_{\sigma\alpha} \frac{F'}{2 \cdot F[f]} \left(f_{,\beta} \cdot \delta^\sigma_\mu g^{\alpha\beta} + f_{,\mu} g^{\alpha\sigma} - f_{,\gamma} \cdot g^{\gamma\sigma} \delta^\alpha_\mu \right) \right) \\
& + \left(\frac{\Gamma^\mu_{\sigma\mu} F'}{2 \cdot F[f]} \left(f_{,\beta} g^{\sigma\beta} + f_{,\alpha} g^{\alpha\sigma} - f_{,\gamma} \cdot g^{\gamma\sigma} n \right) + \frac{F' f_{,\sigma} \cdot n}{2 \cdot F[f]} \Gamma^\sigma_{\alpha\beta} g^{\alpha\beta} \right) \\
& + \frac{FF'' - (F')^2}{2 \cdot F[f]^2} \left((2-n) f_{,\alpha} f_{,\beta} g^{\alpha\beta} \right) - \left(\frac{F'}{2 \cdot F[f]} \right)^2 \left((2-n) f_{,\beta} f_{,\alpha} g^{\alpha\beta} \right) \\
& + \left(\frac{F'}{2 \cdot F[f]} \right)^2 n \left((2-n) f_{,\alpha} f_{,\beta} g^{\sigma\beta} \right) - \frac{\left(FF'' - (F')^2 \right) f_{,\alpha} f_{,\beta} \cdot n}{2 \cdot F[f]^2} g^{\alpha\beta}
\end{aligned}
\right) \frac{1}{F}
\quad (934)$$

$$R^* = R^*_{\alpha\beta} G^{\alpha\beta} = \left(\Gamma^\sigma_{\alpha\beta,\sigma} g^{\alpha\beta} - \Gamma^\sigma_{\beta\sigma,\alpha} g^{\alpha\beta} - \Gamma^\mu_{\sigma\alpha} \Gamma^\sigma_{\beta\mu} g^{\alpha\beta} + \Gamma^\sigma_{\alpha\beta} \Gamma^\mu_{\sigma\mu} g^{\alpha\beta} \right.$$
$$+ \frac{F'}{2 \cdot F[f]} \left((2-n) f_{,\beta\alpha} g^{\alpha\beta} - f_{,\gamma} \cdot \left(g^{\gamma\sigma} g_{\alpha\beta} \right)_{,\sigma} g^{\alpha\beta} \right) - \left(\frac{F' f_{,\alpha\beta} \cdot n}{2 \cdot F[f]} g^{\alpha\beta} \right)$$
$$- \left(\frac{\Gamma^\sigma_{\beta\mu} F'}{2 \cdot F[f]} \left(f_{,\sigma} g^{\mu\beta} + f_{,\alpha} \cdot \delta^\mu_\sigma g^{\alpha\beta} - f_{,\gamma} \cdot g^{\gamma\mu} \delta^\beta_\sigma \right) + \Gamma^\mu_{\sigma\alpha} \frac{F'}{2 \cdot F[f]} \left(f_{,\beta} \cdot \delta^\sigma_\mu g^{\alpha\beta} + f_{,\mu} g^{\alpha\alpha} - f_{,\gamma} \cdot g^{\gamma\sigma} \delta^\alpha_\mu \right) \right) \frac{1}{F}$$
$$+ \left(\frac{\Gamma^\mu_{\sigma\mu} F'}{2 \cdot F[f]} \left(f_{,\beta} g^{\sigma\beta} + f_{,\alpha} g^{\alpha\alpha} - f_{,\gamma} \cdot g^{\gamma\sigma} n \right) + \frac{F' f_{,\sigma} \cdot n}{2 \cdot F[f]} \Gamma^\sigma_{\alpha\beta} g^{\alpha\beta} \right.$$
$$\left. \frac{FF'' - (F')^2}{2 \cdot F[f]^2} \left((2-n) \right) - \left(\frac{F'}{2 \cdot F[f]} \right)^2 \left((2-n) \right) \right.$$
$$\left. + \left(\frac{F'}{2 \cdot F[f]} \right)^2 n \left((2-n) \right) - \frac{(FF'' - (F')^2) \cdot n}{2 \cdot F[f]^2} \right)$$
$$f_{,\alpha} f_{,\beta} g^{\alpha\beta} \qquad (935)$$

$$R^* = R^*_{\alpha\beta}G^{\alpha\beta} = \left(\Gamma^\sigma_{\alpha\beta,\sigma}g^{\alpha\beta} - \Gamma^\sigma_{\beta\sigma,\alpha}g^{\alpha\beta} - \Gamma^\mu_{\sigma\alpha}\Gamma^\sigma_{\beta\mu}g^{\alpha\beta} + \Gamma^\sigma_{\alpha\beta}\Gamma^\mu_{\sigma\mu}g^{\alpha\beta} \right.$$

$$+ \frac{F'}{2\cdot F[f]}\left((2-n)f_{,\beta\alpha}g^{\alpha\beta} - f_{,\gamma}\cdot\left(g^{\gamma\sigma}g_{\alpha\beta}\right)_{,\sigma}g^{\alpha\beta} \right) - \left(\frac{F' f_{,\alpha\beta}\cdot n}{2\cdot F[f]}g^{\alpha\beta} \right)$$

$$- \frac{F'}{2\cdot F[f]}\left(\underbrace{\Gamma^\sigma_{\beta\mu}f_{,\sigma}g^{\mu\beta} + \Gamma^\mu_{\beta\mu}f_{,\alpha}g^{\alpha\beta} - \Gamma^\beta_{\beta\mu}f_{,\gamma}\cdot g^{\gamma\mu}}_{=0} \right) + \frac{F'}{2\cdot F[f]}\left(\underbrace{\Gamma^\mu_{\sigma\alpha}f_{,\mu}g^{\alpha\sigma} + f_{,\beta}\Gamma^\sigma_{\sigma\alpha}g^{\alpha\beta} - f_{,\gamma}\cdot g^{\gamma\sigma}\Gamma^\alpha_{\sigma\alpha}}_{=0} \right)$$

$$+ \frac{F'}{2\cdot F[f]}\left(2\Gamma^\mu_{\sigma\mu}f_{,\beta}g^{\sigma\beta} - \Gamma^\mu_{\sigma\mu}f_{,\gamma}\cdot g^{\gamma\sigma}n \right) + \frac{F' f_{,\sigma}\cdot n}{2\cdot F[f]}\Gamma^\sigma_{\alpha\beta}g^{\alpha\beta}$$

$$\left. + \frac{f_{,\alpha}f_{,\beta}g^{\alpha\beta}(1-n)}{4F^2}\left(4FF'' + (F')^2(n-6) \right) \right) \frac{1}{F} \tag{936}$$

$$R^* = R^*_{\alpha\beta}G^{\alpha\beta} = \left(\Gamma^\sigma_{\alpha\beta,\sigma}g^{\alpha\beta} - \Gamma^\sigma_{\beta\sigma,\alpha}g^{\alpha\beta} - \Gamma^\mu_{\sigma\alpha}\Gamma^\sigma_{\beta\mu}g^{\alpha\beta} + \Gamma^\sigma_{\sigma\alpha}\Gamma^\mu_{\sigma\mu}g^{\alpha\beta} \right.$$

$$+ \frac{F'}{F[f]}\left((1-n)f_{,\beta\alpha}g^{\alpha\beta} \right) + \frac{F'}{2\cdot F[f]}f_{,\sigma}(n-2)\Gamma^\sigma_{\alpha\beta}g^{\alpha\beta}$$

$$- \frac{F'}{2\cdot F[f]}\left(f_{,\gamma}\cdot\left(g^{\gamma\sigma}g_{\alpha\beta}\right)_{,\sigma}g^{\alpha\beta} \right) + \frac{F'}{2\cdot F[f]}\left(2\Gamma^\mu_{\sigma\mu}f_{,\beta}g^{\sigma\beta} - \Gamma^\mu_{\sigma\mu}f_{,\gamma}\cdot g^{\gamma\sigma}n \right)$$

$$\left. + \frac{f_{,\alpha}f_{,\beta}g^{\alpha\beta}(1-n)}{4F^2}\left(4FF'' + (F')^2(n-6) \right) \right) \frac{1}{F} \tag{937}$$

$$
R^{*}=R^{*}_{\alpha\beta}G^{\alpha\beta}=\left(
\begin{array}{c}
\Gamma^{\sigma}_{\alpha\beta,\sigma}g^{\alpha\beta}-\Gamma^{\sigma}_{\beta\sigma,\alpha}g^{\alpha\beta}-\Gamma^{\mu}_{\sigma\alpha}\Gamma^{\sigma}_{\beta\mu}g^{\alpha\beta}+\Gamma^{\sigma}_{\alpha\beta}\Gamma^{\mu}_{\sigma\mu}g^{\alpha\beta}+\dfrac{F'}{F[f]}\left((1-n)f_{,\beta\alpha}g^{\alpha\beta}\right)+\dfrac{F'}{2\cdot F[f]}f_{,\sigma}\,(n-2)\Gamma^{\sigma}_{\alpha\beta}g^{\alpha\beta} \\[2mm]
-\dfrac{F'}{2\cdot F[f]}\left(f_{,\gamma}\cdot\left(g^{\gamma\sigma}g_{\alpha\beta}+g^{\gamma\sigma}g_{\alpha\beta,\sigma}\right)g^{\alpha\beta}\right)+\left(\dfrac{F'}{2\cdot F[f]}g^{\mu\nu}g_{\mu\nu,\sigma}\left(2f_{,\beta}g^{\sigma\beta}-f_{,\gamma}\cdot g^{\gamma\sigma}n\right)\right) \\[2mm]
-\dfrac{f_{,\alpha}f_{,\beta}g^{\alpha\beta}(1-n)}{4F^{2}}\left(4FF''+(F')^{2}(n-6)\right)
\end{array}
\right)\dfrac{1}{F}
$$

$$
=\left(
\begin{array}{c}
\Gamma^{\sigma}_{\alpha\beta,\sigma}g^{\alpha\beta}-\Gamma^{\sigma}_{\beta\sigma,\alpha}g^{\alpha\beta}-\Gamma^{\mu}_{\sigma\alpha}\Gamma^{\sigma}_{\beta\mu}g^{\alpha\beta}+\Gamma^{\sigma}_{\alpha\beta}\Gamma^{\mu}_{\sigma\mu}g^{\alpha\beta}+\dfrac{F'}{F[f]}\left((1-n)f_{,\beta\alpha}g^{\alpha\beta}\right)+\dfrac{F'}{2\cdot F[f]}f_{,\sigma}\,(n-2)\Gamma^{\sigma}_{\alpha\beta}g^{\alpha\beta} \\[2mm]
+\dfrac{F'}{2\cdot F[f]}\left(\left(2f_{,\beta}g^{\mu\nu}g_{\mu\nu,\sigma}g^{\sigma\beta}\right)-f_{,\gamma}\cdot\left(ng^{\gamma\sigma}{}_{,\sigma}+g^{\gamma\sigma}g^{\alpha\beta}g_{\alpha\beta,\sigma}+g^{\gamma\sigma}g^{\mu\nu}g_{\mu\nu,\sigma}n\right)\right) \\[2mm]
-\dfrac{f_{,\alpha}f_{,\beta}g^{\alpha\beta}(1-n)}{4F^{2}}\left(4FF''+(F')^{2}(n-6)\right)
\end{array}
\right)\dfrac{1}{F}
$$

$$
=\left(
\begin{array}{c}
\Gamma^{\sigma}_{\alpha\beta,\sigma}g^{\alpha\beta}-\Gamma^{\sigma}_{\beta\sigma,\alpha}g^{\alpha\beta}-\Gamma^{\mu}_{\sigma\alpha}\Gamma^{\sigma}_{\beta\mu}g^{\alpha\beta}+\Gamma^{\sigma}_{\alpha\beta}\Gamma^{\mu}_{\sigma\mu}g^{\alpha\beta}+\dfrac{F'}{F[f]}\left((1-n)f_{,\beta\alpha}g^{\alpha\beta}\right)+\dfrac{F'}{2\cdot F[f]}f_{,\sigma}\,(n-2)\Gamma^{\sigma}_{\alpha\beta}g^{\alpha\beta} \\[2mm]
+\dfrac{F'}{2\cdot F[f]}\left(\left(2f_{,\gamma}g^{\sigma\gamma}g^{\alpha\beta}g_{\alpha\beta,\sigma}\right)-f_{,\gamma}\cdot\left(ng^{\gamma\sigma}{}_{,\sigma}+g^{\gamma\sigma}g^{\alpha\beta}g_{\alpha\beta,\sigma}(1+n)\right)\right) \\[2mm]
-\dfrac{f_{,\alpha}f_{,\beta}g^{\alpha\beta}(1-n)}{4F^{2}}\left(4FF''+(F')^{2}(n-6)\right)
\end{array}
\right)\dfrac{1}{F}
\qquad (938)
$$

$$
\begin{aligned}
R^* = R^*_{\alpha\beta} G^{\alpha\beta} =\ & \left(\Gamma^\sigma_{\alpha\beta,\sigma} g^{\alpha\beta} - \Gamma^\sigma_{\beta\sigma,\alpha} g^{\alpha\beta} - \Gamma^\mu_{\sigma\alpha} \Gamma^\sigma_{\beta\mu} g^{\alpha\beta} + \Gamma^\sigma_{\alpha\beta} \Gamma^\mu_{\sigma\mu} g^{\alpha\beta} \right. \\
& + \frac{F'}{F[f]}\left((1-n) f_{,\beta\alpha} g^{\alpha\beta}\right) + \frac{F'}{2\cdot F[f]} f_{,\sigma} (n-2)\,\Gamma^\sigma_{\alpha\beta} g^{\alpha\beta} - \frac{f_{,\gamma}\cdot F'}{2\cdot F[f]}\left(n g^{\gamma\sigma}_{,\sigma} + g^{\gamma\sigma} g^{\alpha\beta} g_{\alpha\beta,\sigma} (n-1)\right) \\
& \left. - \frac{f_{,\alpha} f_{,\beta} g^{\alpha\beta} (1-n)}{4F^2}\left(4FF'' + (F')^2 (n-6)\right) \right) \frac{1}{F} \\[8pt]
=\ & \left(\Gamma^\sigma_{\alpha\beta,\sigma} g^{\alpha\beta} - \Gamma^\sigma_{\beta\sigma,\alpha} g^{\alpha\beta} - \Gamma^\mu_{\sigma\alpha} \Gamma^\sigma_{\beta\mu} g^{\alpha\beta} + \Gamma^\sigma_{\alpha\beta} \Gamma^\mu_{\sigma\mu} g^{\alpha\beta} + \frac{F'}{F[f]}\left((1-n) f_{,\beta\alpha} g^{\alpha\beta}\right) \right. \\
& + \frac{F'}{2\cdot F[f]} f_{,\sigma} (n-2)\frac{g^{\sigma\mu}}{2}\left(g_{\alpha\mu,\beta} + g_{\mu\beta,\alpha} - g_{\alpha\beta,\mu}\right) g^{\alpha\beta} - \frac{f_{,\gamma}\cdot F'}{2\cdot F[f]}\left(n g^{\gamma\sigma}_{,\sigma} + g^{\gamma\sigma} g^{\alpha\beta} g_{\alpha\beta,\sigma} (n-1)\right) \\
& \left. - \frac{f_{,\alpha} f_{,\beta} g^{\alpha\beta} (1-n)}{4F^2}\left(4FF'' + (F')^2 (n-6)\right) \right) \frac{1}{F} \\[8pt]
=\ & \left(\Gamma^\sigma_{\alpha\beta,\sigma} g^{\alpha\beta} - \Gamma^\sigma_{\beta\sigma,\alpha} g^{\alpha\beta} - \Gamma^\mu_{\sigma\alpha} \Gamma^\sigma_{\beta\mu} g^{\alpha\beta} + \Gamma^\sigma_{\alpha\beta} \Gamma^\mu_{\sigma\mu} g^{\alpha\beta} + \frac{F'}{F[f]}\left((1-n) f_{,\beta\alpha} g^{\alpha\beta}\right) \right. \\
& + \frac{F'}{4\cdot F[f]} f_{,\gamma} (n-2) g^{\gamma\mu}\left(g_{\alpha\mu,\beta} + g_{\mu\beta,\alpha} - g_{\alpha\beta,\mu}\right) g^{\alpha\beta} - \frac{f_{,\gamma}\cdot F'}{2\cdot F[f]}\left(n g^{\gamma\mu}_{,\mu} + g^{\gamma\mu} g^{\alpha\beta} g_{\alpha\beta,\mu} (n-1)\right) \\
& \left. - \frac{f_{,\alpha} f_{,\beta} g^{\alpha\beta} (1-n)}{4F^2}\left(4FF'' + (F')^2 (n-6)\right) \right) \frac{1}{F}
\end{aligned}
\tag{939}
$$

$$R^* = R^*_{\alpha\beta} G^{\alpha\beta}$$

$$= \begin{pmatrix} \Gamma^{\sigma}_{\alpha\beta,\sigma} g^{\alpha\beta} - \Gamma^{\sigma}_{\beta\sigma,\alpha} g^{\alpha\beta} - \Gamma^{\mu}_{\sigma\alpha} \Gamma^{\sigma}_{\beta\mu} g^{\alpha\beta} + \Gamma^{\sigma}_{\alpha\beta} \Gamma^{\mu}_{\sigma\mu} g^{\alpha\beta} + \dfrac{F'}{F[f]}\left((1-n) f_{,\beta\alpha} g^{\alpha\beta}\right) \\[2ex] + \dfrac{F'}{4 \cdot F[f]} f_{,\gamma} \begin{pmatrix} (n-2)\left(g^{\gamma\mu} g^{\alpha\beta} g_{\alpha\mu,\beta} + g^{\gamma\mu} g^{\alpha\beta} g_{\mu\beta,\alpha}\right) - 2n g^{\gamma\mu}_{,\mu} \\[1ex] -\left((3n-3)\right) g^{\gamma\mu} g^{\alpha\beta} g_{\alpha\beta,\mu} \end{pmatrix} \\[2ex] \dfrac{f_{,\alpha} f_{,\beta} g^{\alpha\beta} (1-n)}{4F^2}\left(4FF'' + (F')^2 (n-6)\right) \end{pmatrix} \dfrac{1}{F} .$$

(940)

For brevity we now simply set $F[f] = f$ (which we can later easily "back-generalize" again) and further evaluate the two lines in the middle of Eq. (940). In order to have an earlier starting point, however, we use Eq. (936) and insert $F = f$:

$$R^* = R^*_{\alpha\beta} G^{\alpha\beta}$$

$$= \begin{pmatrix} \Gamma^{\sigma}_{\alpha\beta,\sigma} g^{\alpha\beta} - \Gamma^{\sigma}_{\beta\sigma,\alpha} g^{\alpha\beta} - \Gamma^{\mu}_{\sigma\alpha} \Gamma^{\sigma}_{\beta\mu} g^{\alpha\beta} + \Gamma^{\sigma}_{\alpha\beta} \Gamma^{\mu}_{\sigma\mu} g^{\alpha\beta} \\[2ex] + \dfrac{1}{f}\left((1-n) f_{,\beta\alpha} g^{\alpha\beta}\right) - \dfrac{1}{2f}\left(f_{,\gamma} \cdot \left(g^{\gamma\sigma} g_{\alpha\beta}\right)_{,\sigma} g^{\alpha\beta}\right) \\[2ex] - \dfrac{1}{2 \cdot f}\left(\Gamma^{\sigma}_{\beta\mu} f_{,\sigma} g^{\mu\beta} + \overbrace{f_{,\alpha} \cdot \Gamma^{\mu}_{\beta\mu} g^{\alpha\beta} - f_{,\gamma} \cdot g^{\gamma\mu} \Gamma^{\sigma}_{\sigma\mu}}^{=0}\right) \\[2ex] + \dfrac{1}{2 \cdot f}\left(f_{,\beta} \cdot \Gamma^{\mu}_{\mu\alpha} g^{\alpha\beta} + f_{,\mu} g^{\alpha\sigma} \Gamma^{\mu}_{\sigma\alpha} - f_{,\gamma} \cdot g^{\gamma\sigma} \Gamma^{\alpha}_{\sigma\alpha}\right) \\[2ex] + \left(\dfrac{\Gamma^{\mu}_{\sigma\mu}}{2 \cdot f}\left((2-n) f_{,\beta} g^{\sigma\beta}\right) + \dfrac{f_{,\sigma} \cdot n}{2 \cdot f} \Gamma^{\sigma}_{\alpha\beta} g^{\alpha\beta}\right) + \dfrac{f_{,\alpha} f_{,\beta} g^{\alpha\beta}}{4 \cdot f^2}(6-n)(n-1) \end{pmatrix} \dfrac{1}{f}$$

(941)

$$R^* = R^*_{\alpha\beta} G^{\alpha\beta}$$

$$= \begin{pmatrix} \Gamma^{\sigma}_{\alpha\beta,\sigma} g^{\alpha\beta} - \Gamma^{\sigma}_{\beta\sigma,\alpha} g^{\alpha\beta} - \Gamma^{\mu}_{\sigma\alpha} \Gamma^{\sigma}_{\beta\mu} g^{\alpha\beta} + \Gamma^{\sigma}_{\alpha\beta} \Gamma^{\mu}_{\sigma\mu} g^{\alpha\beta} \\[2ex] + \dfrac{1}{f}\left((1-n) f_{,\beta\alpha} g^{\alpha\beta}\right) + (n-1)\dfrac{f_{,\sigma}}{2 \cdot f} \Gamma^{\sigma}_{\alpha\beta} g^{\alpha\beta} - \dfrac{1}{2f}\left(f_{,\gamma} \cdot \left(g^{\gamma\sigma} g_{\alpha\beta}\right)_{,\sigma} g^{\alpha\beta}\right) \\[2ex] - \dfrac{1}{2 \cdot f}\left(f_{,\beta} \cdot \Gamma^{\mu}_{\mu\alpha} g^{\alpha\beta} + f_{,\mu} g^{\alpha\sigma} \Gamma^{\mu}_{\sigma\alpha} - f_{,\gamma} \cdot g^{\gamma\sigma} \Gamma^{\alpha}_{\sigma\alpha}\right) \\[2ex] + \left(\dfrac{\Gamma^{\mu}_{\sigma\mu}}{2 \cdot f}\left((2-n) f_{,\beta} g^{\sigma\beta}\right)\right) + \dfrac{f_{,\alpha} f_{,\beta} g^{\alpha\beta}}{4 \cdot f^2}(6-n)(n-1) \end{pmatrix} \dfrac{1}{f}$$

(942)

$$R^* = R^*_{\alpha\beta} G^{\alpha\beta} = \overbrace{\left(\begin{aligned} &\Gamma^\sigma_{\alpha\beta,\sigma} g^{\alpha\beta} - \Gamma^\sigma_{\beta\sigma,\alpha} g^{\alpha\beta} - \Gamma^\mu_{\sigma\alpha}\Gamma^\sigma_{\beta\mu} g^{\alpha\beta} + \Gamma^\sigma_{\alpha\beta}\Gamma^\mu_{\sigma\mu} g^{\alpha\beta} \\ &+\frac{1}{f}\left((1-n) f_{,\beta\alpha} g^{\alpha\beta}\right) + (n-1)\frac{f_{,\sigma}}{2\cdot f}\Gamma^\sigma_{\alpha\beta} g^{\alpha\beta} \\ &= -\frac{1}{2f}\left(f_{,\gamma}\cdot\left(g^{\gamma\sigma}{}_{,\sigma} g_{\alpha\beta} + g^{\gamma\sigma} g_{\alpha\beta,\sigma}\right) g^{\alpha\beta}\right) - \left(\frac{1}{2\cdot f}\left(f_{,\beta}\cdot\Gamma^\mu_{\mu\alpha} g^{\alpha\beta} + f_{,\mu} g^{\alpha\sigma}\Gamma^\mu_{\sigma\alpha} - f_{,\gamma}\cdot g^{\gamma\sigma}\Gamma^\alpha_{\sigma\alpha}\right)\right) \\ &+\left(\frac{\Gamma^\mu_{\sigma\mu}}{2\cdot f}\left((2-n) f_{,\beta} g^{\sigma\beta}\right)\right) + \frac{f_{,\alpha} f_{,\beta} g^{\alpha\beta}}{4\cdot f^2}(6-n)(n-1) \end{aligned} \right)}^{\frac{1}{f}}$$

$$\text{with}: \Gamma^\mu_{\mu\alpha} = \frac{1}{2} g^{\mu\nu} g_{\mu\nu,\alpha}$$

$$= \overbrace{\left(\begin{aligned} &\Gamma^\sigma_{\alpha\beta,\sigma} g^{\alpha\beta} - \Gamma^\sigma_{\beta\sigma,\alpha} g^{\alpha\beta} - \Gamma^\mu_{\sigma\alpha}\Gamma^\sigma_{\beta\mu} g^{\alpha\beta} + \Gamma^\sigma_{\alpha\beta}\Gamma^\mu_{\sigma\mu} g^{\alpha\beta} + \frac{1}{f}\left((1-n) f_{,\beta\alpha} g^{\alpha\beta}\right) \\ &+(n-1)\frac{f_{,\sigma}}{2\cdot f}\Gamma^\sigma_{\alpha\beta} g^{\alpha\beta} - \frac{1}{2\cdot f} f_{,\mu} g^{\alpha\sigma}\Gamma^\mu_{\sigma\alpha} - \frac{1}{2f}\left(f_{,\gamma}\cdot\left(g^{\gamma\sigma}{}_{,\sigma} n + g^{\gamma\sigma} g_{\alpha\beta} g_{\alpha\beta,\sigma}\right)\right) \\ &-\left(\frac{1}{4\cdot f}\left(f_{,\beta} g^{\mu\nu} g_{\mu\nu,\alpha} g^{\alpha\beta} - f_{,\gamma}\cdot g^{\gamma\sigma} g^{\mu\nu} g_{\mu\nu,\sigma}\right)\right) + \left(\frac{g^{\mu\nu} g_{\mu\nu,\sigma}}{4\cdot f}\left((2-n) f_{,\beta} g^{\sigma\beta}\right)\right) + \frac{f_{,\alpha} f_{,\beta} g^{\alpha\beta}}{4\cdot f^2}(6-n)(n-1) \end{aligned} \right)}^{\frac{1}{f}} \tag{943}$$

$$R^* = R^*_{\alpha\beta} G^{\alpha\beta} = \left(\Gamma^\sigma_{\alpha\beta,\sigma} g^{\alpha\beta} - \Gamma^\sigma_{\beta\sigma,\alpha} g^{\alpha\beta} - \Gamma^\mu_{\sigma\alpha} \Gamma^\sigma_{\beta\mu} g^{\alpha\beta} + \Gamma^\sigma_{\alpha\beta} \Gamma^\mu_{\sigma\mu} g^{\alpha\beta} + \frac{1}{f}\left((1-n) f_{,\beta\alpha} g^{\alpha\beta} \right) + (n-1) \frac{f_{,\sigma}}{2\cdot f} \Gamma^\sigma_{\alpha\beta} g^{\alpha\beta} \right.$$
$$- \frac{1}{2\cdot f} f_{,\mu} g^{\alpha\sigma} \Gamma^\mu_{\sigma\alpha} - \frac{1}{2f}\left(f_{,\gamma} \cdot \left(\underbrace{g^{\gamma\sigma}_{,\sigma} n + g^{\gamma\sigma} g^{\alpha\beta} g_{\alpha\beta,\sigma}}_{=0} \right) \right)$$
$$- \frac{1}{4\cdot f}\left(f_{,\beta} g^{\mu\nu} g_{\mu\nu,\alpha} g^{\alpha\beta} - f_{,\gamma} \cdot g^{\gamma\sigma} g^{\mu\nu} g_{\mu\nu,\sigma} \right)$$
$$\left. + \frac{g^{\mu\nu} g_{\mu\nu,\sigma}}{4\cdot f}\left((2-n) f_{,\beta} g^{\sigma\beta} \right) + \frac{f_{,\alpha} f_{,\beta} g^{\alpha\beta}}{4\cdot f^2} (6-n)(n-1) \right) \frac{1}{f} \qquad (944)$$

$$R^* = R^*_{\alpha\beta} G^{\alpha\beta} = \left(\Gamma^\sigma_{\alpha\beta,\sigma} g^{\alpha\beta} - \Gamma^\sigma_{\beta\sigma,\alpha} g^{\alpha\beta} - \Gamma^\mu_{\sigma\alpha} \Gamma^\sigma_{\beta\mu} g^{\alpha\beta} + \Gamma^\sigma_{\alpha\beta} \Gamma^\mu_{\sigma\mu} g^{\alpha\beta} + \frac{1}{f}\left((1-n) f_{,\beta\alpha} g^{\alpha\beta} \right) \right.$$
$$+ (n-1) \frac{f_{,\sigma}}{2\cdot f} \Gamma^\sigma_{\alpha\beta} g^{\alpha\beta} - \frac{1}{2\cdot f} f_{,\mu} g^{\alpha\sigma} \Gamma^\mu_{\sigma\alpha}$$
$$\left. - n\left(\frac{f_{,\gamma} g^{\gamma\sigma}_{,\sigma}}{2f} + \frac{g^{\mu\nu} g_{\mu\nu,\sigma}}{4\cdot f} f_{,\beta} g^{\sigma\beta} \right) + \frac{f_{,\alpha} f_{,\beta} g^{\alpha\beta}}{4\cdot f^2} (6-n)(n-1) \right) \frac{1}{f} \qquad (945)$$

$$R^* = R^*_{\alpha\beta}G^{\alpha\beta} = \left(\Gamma^\sigma_{\alpha\beta,\sigma}g^{\alpha\beta} - \Gamma^\sigma_{\beta\sigma,\alpha}g^{\alpha\beta} - \Gamma^\mu_{\sigma\alpha}\Gamma^\sigma_{\beta\mu}g^{\alpha\beta} + \Gamma^\sigma_{\alpha\beta}\Gamma^\mu_{\sigma\mu}g^{\alpha\beta} \right.$$

$$\left. + \frac{1}{2f}(1-n)\Delta f + \frac{1}{2f}\left((1-n)f_{,\beta\alpha}g^{\alpha\beta} - \underbrace{f_{,\mu}g^{\alpha\alpha}\Gamma^\mu_{\sigma\alpha}}_{=-f_{,\mu}g^{\alpha\alpha}\Gamma^\mu_{\sigma\alpha}} - n\left[f_{,\beta}g^{\beta\sigma}{}_{,\sigma} + \frac{g^{\mu\nu}g^{\sigma\beta}g_{\mu\nu,\sigma}}{2}f_{,\beta} \right] \right) + \frac{f_{,\alpha}f_{,\beta}g^{\alpha\beta}}{4\cdot f^2}(6-n)(n-1) \right)\frac{1}{f} \tag{946}$$

$$= \left(\Gamma^\sigma_{\alpha\beta,\sigma}g^{\alpha\beta} - \Gamma^\sigma_{\beta\sigma,\alpha}g^{\alpha\beta} - \Gamma^\mu_{\sigma\alpha}\Gamma^\sigma_{\beta\mu}g^{\alpha\beta} + \Gamma^\sigma_{\alpha\beta}\Gamma^\mu_{\sigma\mu}g^{\alpha\beta} \right.$$

$$\left. + \frac{1}{2f}(1-n)\Delta f + \frac{1}{2f}\left((1-n)f_{,\beta\alpha}g^{\alpha\beta} - f_{,\mu}g^{\alpha\alpha}\Gamma^\mu_{\sigma\alpha} + nf_{,\mu}g^{\alpha\sigma}\Gamma^\mu_{\sigma\alpha} \right) + \frac{f_{,\alpha}f_{,\beta}g^{\alpha\beta}}{4\cdot f^2}(6-n)(n-1) \right)\frac{1}{f}$$

$$= \left(\Gamma^\sigma_{\alpha\beta,\sigma}g^{\alpha\beta} - \Gamma^\sigma_{\beta\sigma,\alpha}g^{\alpha\beta} - \Gamma^\mu_{\sigma\alpha}\Gamma^\sigma_{\beta\mu}g^{\alpha\beta} + \Gamma^\sigma_{\alpha\beta}\Gamma^\mu_{\sigma\mu}g^{\alpha\beta} \right.$$

$$\left. + \frac{1}{2f}(1-n)\Delta f + \frac{1}{2f}\underbrace{(1-n)\left(f_{,\beta\alpha}g^{\alpha\beta} - f_{,\mu}g^{\alpha\sigma}\Gamma^\mu_{\sigma\alpha}\right)}_{\Delta f} + \frac{f_{,\alpha}f_{,\beta}g^{\alpha\beta}}{4\cdot f^2}(6-n)(n-1) \right)\frac{1}{f}.$$

Finally, we obtain:

$$R^* = R^*_{\alpha\beta}G^{\alpha\beta} = \left(\Gamma^\sigma_{\alpha\beta,\sigma}g^{\alpha\beta} - \Gamma^\sigma_{\beta\sigma,\alpha}g^{\alpha\beta} - \Gamma^\mu_{\sigma\alpha}\Gamma^\sigma_{\beta\mu}g^{\alpha\beta} + \Gamma^\sigma_{\alpha\beta}\Gamma^\mu_{\sigma\mu}g^{\alpha\beta} + \frac{1}{f}(1-n)\Delta f + \frac{f_{,\alpha}f_{,\beta}g^{\alpha\beta}}{4\cdot f^2}(6-n)(n-1) \right)\frac{1}{f} \tag{947}$$

and by going back to Eq. (940) and repeating what we have just done, we get the general result for $F[f]$ as follows:

$$R^* = R^*_{\alpha\beta}G^{\alpha\beta} = \begin{pmatrix} \Gamma^\sigma_{\alpha\beta,\sigma}g^{\alpha\beta} - \Gamma^\sigma_{\beta\sigma,\alpha}g^{\alpha\beta} - \Gamma^\mu_{\sigma\alpha}\Gamma^\sigma_{\beta\mu}g^{\alpha\beta} + \Gamma^\sigma_{\alpha\beta}\Gamma^\mu_{\sigma\mu}g^{\alpha\beta} \\ + \dfrac{F'}{F[f]}(1-n)\Delta f + \dfrac{f_{,\alpha}f_{,\beta}g^{\alpha\beta}(1-n)}{4F^2}\left(4FF'' + (F')^2(n-6)\right) \end{pmatrix}\dfrac{1}{F}.$$

$$(948)$$

8.11   Appendix B: Derivation of the Scaled Ricci Tensor

We start with the result form Appendix A, reading:

$$R^* = R^*_{\alpha\beta}G^{\alpha\beta} = \begin{pmatrix} \Gamma^\sigma_{\alpha\beta,\sigma}g^{\alpha\beta} - \Gamma^\sigma_{\beta\sigma,\alpha}g^{\alpha\beta} - \Gamma^\mu_{\sigma\alpha}\Gamma^\sigma_{\beta\mu}g^{\alpha\beta} + \Gamma^\sigma_{\alpha\beta}\Gamma^\mu_{\sigma\mu}g^{\alpha\beta} \\ + \dfrac{F'}{F[f]}(1-n)\Delta f + \dfrac{f_{,\alpha}f_{,\beta}g^{\alpha\beta}(1-n)}{4F^2}\left(4FF'' + (F')^2(n-6)\right) \end{pmatrix}\dfrac{1}{F}.$$

$$(949)$$

Now we know that:

$$G^{\alpha\beta} = \frac{g^{\alpha\beta}}{F}, \quad \Delta f = g^{\alpha\beta}f_{,\alpha\beta} - g^{\alpha\beta}\Gamma^\gamma_{\alpha\beta}f_{,\gamma} \tag{950}$$

and that:

$$R^*_{\alpha\beta} = \begin{pmatrix} -\dfrac{1}{2}\left(G_{\alpha\beta,ab} + G_{ab,\alpha\beta} - G_{\alpha b,a\beta} - G_{\beta b,a\alpha}\right)G^{ab} \\ +\dfrac{1}{2}\left(\dfrac{1}{2}G_{ac,\alpha}\cdot G_{bd,\beta} + G_{\alpha c,a}\cdot G_{\beta d,b} - G_{\alpha c,a}\cdot G_{\beta b,d}\right)G^{ab}G^{cd} \\ -\dfrac{1}{4}\left(G_{\beta c,\alpha} + G_{\alpha c,\beta} - G_{\alpha\beta,c}\right)\left(2G_{bd,a} - G_{ab,d}\right)G^{ab}G^{cd} \end{pmatrix}$$

$$R_{\alpha\beta} = \begin{pmatrix} -\dfrac{1}{2}\left(g_{\alpha\beta,ab} + g_{ab,\alpha\beta} - g_{\alpha b,a\beta} - g_{\beta b,a\alpha}\right)g^{ab} \\ +\dfrac{1}{2}\left(\dfrac{1}{2}g_{ac,\alpha}\cdot g_{bd,\beta} + g_{\alpha c,a}\cdot g_{\beta d,b} - g_{\alpha c,a}\cdot g_{\beta b,d}\right)g^{ab}g^{cd} \\ -\dfrac{1}{4}\left(g_{\alpha c,\beta} + g_{\beta c,\alpha} - g_{\alpha\beta,c}\right)\left(2g_{bd,a} - g_{ab,d}\right)g^{ab}g^{cd} \end{pmatrix}. \tag{951}$$

From this directly follows that:

$$
R^* = R^*_{\alpha\beta} G^{\alpha\beta} = \left(
\begin{aligned}
&\Gamma^\sigma_{\alpha\beta,\sigma}g^{\alpha\beta} - \Gamma^\sigma_{\beta\sigma,\alpha}g^{\alpha\beta} - \Gamma^\mu_{\sigma\alpha}\Gamma^\sigma_{\beta\mu}g^{\alpha\beta} + \Gamma^\sigma_{\alpha\beta}\Gamma^\mu_{\sigma\mu}g^{\alpha\beta} + \frac{F'}{F[f]}(1-n)\left(g^{\alpha\beta}f_{,\alpha\beta} - g^{\alpha\beta}\Gamma^\gamma_{\alpha\beta}f_{,\gamma}\right)
\end{aligned}
\right)\frac{1}{F}
$$

$$
+\frac{f_{,\alpha}f_{,\beta}g^{\alpha\beta}(1-n)}{4F^2}\left(4FF'' + (F')^2(n-6)\right)
$$

$$
\frac{g^{ab}}{2F}\left(
\begin{aligned}
&F_{,b}g_{\alpha\beta,a} + F_{,a}g_{\alpha\beta,b} + F_{,ab}g_{\alpha\beta} + F_{,\alpha\beta}g_{ab} + F_{,\alpha}g_{ab,\beta} + F_{,\beta}g_{ab,\alpha}\\
&-F_{,\alpha\beta}g_{\alpha b} - F_{,\alpha}g_{\alpha b,\beta} - F_{,\beta}g_{\alpha b,\beta} - F_{,ac}g_{\beta b} - F_{,\beta}g_{\beta b} - F_{,\alpha}g_{\beta b,\alpha}
\end{aligned}
\right)
$$

$$
+\frac{g^{ab}g^{cd}}{2F^2}\left(
\begin{aligned}
&\frac{1}{2}\Big(F\cdot g_{ac,\alpha}F_{,\beta}g_{bd} + F_{,\alpha}g_{ac}\cdot(F\cdot g_{bd,\beta} + F_{,\beta}g_{bd})\Big) + F\cdot g_{ac,\alpha}\cdot\Big(F\cdot g_{bd,\beta} + F_{,\beta}g_{bd}\Big) + F\cdot g_{ac,\alpha}\cdot\Big(F\cdot g_{\beta d,b} + F_{,b}g_{\beta d}\Big)\\
&-\Big(F\cdot g_{ac,\alpha}F_{,d}g_{\beta b} + F_{,\alpha}g_{ac}\cdot(F\cdot g_{\beta b,d} + F_{,d}g_{\beta b})\Big)
\end{aligned}
\right)
$$

$$
= R_{\alpha\beta} -
$$

$$
-\frac{g^{ab}g^{cd}}{4F^2}\left(
\begin{aligned}
&F\cdot g_{ac,\beta}\cdot(2F_{,a}g_{bd} - F_{,d}g_{ab}) + F_{,\beta}g_{ac}\cdot\left(\begin{array}{l}2F\cdot g_{bd,a} - F\cdot g_{ab,d}\\+2F_{,a}g_{bd} - F_{,d}g_{ab}\end{array}\right)\\[4pt]
&+F\cdot g_{\beta c,\alpha}\cdot(2F_{,a}g_{bd} - F_{,d}g_{ab}) + F_{,\alpha}g_{\beta c}\cdot\left(\begin{array}{l}2F\cdot g_{bd,a} - F\cdot g_{ab,d}\\+2F_{,a}g_{bd} - F_{,d}g_{ab}\end{array}\right)\\[4pt]
&-F\cdot g_{\alpha\beta,c}\cdot(2F_{,a}g_{bd} - F_{,d}g_{ab}) - F_{,c}g_{\alpha\beta}\cdot\left(\begin{array}{l}2F\cdot g_{bd,a} - F\cdot g_{ab,d}\\+2F_{,a}g_{bd} - F_{,d}g_{ab}\end{array}\right)
\end{aligned}
\right).
$$

$$(952)$$

This gives us the desired result as follows:

$$\Rightarrow R^*_{\alpha\beta} = R_{\alpha\beta} - \frac{g^{ab}}{2F}\left(\begin{array}{l}
\begin{array}{l}
F_{,b}g_{\alpha\beta,a} + F_{,a}g_{\alpha\beta,b} + F_{,ab}g_{\alpha\beta} + F_{,\alpha\beta}g_{ab}\\
+F_{,\alpha}g_{ab,\beta} + F_{,\beta}g_{ab,\alpha} - F_{,a\beta}g_{\alpha b} - F_{,a}g_{\alpha b,\beta}\\
-F_{,\beta}g_{\alpha b,a} - F_{,a\alpha}g_{\beta b} - F_{,a}g_{\beta b,\alpha} - F_{,\alpha}g_{\beta b,a}
\end{array}\\[12pt]
+\frac{g^{cd}}{F}\left(\begin{array}{l}
\frac{1}{2}\left(\begin{array}{l}
F\cdot g_{ac,\alpha}F_{,\beta}g_{bd} + F_{,\alpha}g_{ac}\cdot\left(F\cdot g_{bd,\beta}+F_{,\beta}g_{bd}\right)\\
+F\cdot g_{\alpha c,a}F_{,b}g_{\beta d} + F_{,a}g_{\alpha c}\cdot\left(F\cdot g_{\beta d,b}+F_{,b}g_{\beta d}\right)\\
-\left(F\cdot g_{\alpha c,a}F_{,d}g_{\beta b}+F_{,a}g_{\alpha c}\cdot\left(F\cdot g_{\beta b,d}+F_{,d}g_{\beta b}\right)\right)
\end{array}\right)
\end{array}\right.\\[24pt]
\left.\begin{array}{l}
-\frac{g^{cd}}{4F}\left(\begin{array}{l}
F\cdot g_{ac,\beta}\cdot\left(2F_{,a}g_{bd}-F_{,d}g_{ab}\right) + F_{,\beta}g_{\alpha c}\cdot\left(\begin{array}{l}2F\cdot g_{bd,a} - F\cdot g_{ab,d}\\+2F_{,a}g_{bd}-F_{,d}g_{ab}\end{array}\right)\\
+F\cdot g_{\beta c,\alpha}\cdot\left(2F_{,a}g_{bd}-F_{,d}g_{ab}\right) + F_{,a}g_{\beta c}\cdot\left(\begin{array}{l}2F\cdot g_{bd,a} - F\cdot g_{ab,d}\\+2F_{,a}g_{bd}-F_{,d}g_{ab}\end{array}\right)\\
-F\cdot g_{\alpha\beta,c}\cdot\left(2F_{,a}g_{bd}-F_{,d}g_{ab}\right) - F_{,c}g_{\alpha\beta}\cdot\left(\begin{array}{l}2F\cdot g_{bd,a} - F\cdot g_{ab,d}\\+2F_{,a}g_{bd}-F_{,d}g_{ab}\end{array}\right)
\end{array}\right)
\end{array}\right)\right) \tag{953}$$

8.12 Appendix C: First Steps for a Tensor-Scaled Metric

$$G_{\delta\gamma} = F[f]_{\delta\gamma}^{\alpha\beta} \cdot g_{\alpha\beta}. \tag{954}$$

And

$$G^{\chi\gamma} = F[f]_{\alpha\beta}^{-1\chi\gamma} \cdot g^{\alpha\beta}. \tag{955}$$

With:

$$G^{\chi\gamma} G_{\delta\gamma} = \delta_{\delta}^{\chi} = F[f]_{\alpha\beta}^{-1\chi\gamma} \cdot g^{\alpha\beta} F[f]_{\delta\gamma}^{k\beta} g_{k\beta} = F[f]_{\alpha\beta}^{-1\chi\gamma} F[f]_{\delta\gamma}^{k\beta} \cdot g^{\alpha\beta} g_{k\beta}$$

$$= F[f]_{\alpha\beta}^{-1\chi\gamma} F[f]_{\delta\gamma}^{k\beta} \delta_{k}^{\alpha} = F[f]_{\alpha\beta}^{-1\chi\gamma} F[f]_{\delta\gamma}^{\alpha\beta}$$

$$\Rightarrow \qquad \delta_{\delta}^{\chi} = F[f]_{\alpha\beta}^{-1\chi\gamma} F[f]_{\delta\gamma}^{\alpha\beta}. \tag{956}$$

$$R = R_{\alpha\beta} g^{\alpha\beta} = \left(\Gamma_{\alpha\beta,\sigma}^{\sigma} - \Gamma_{\beta\sigma,\alpha}^{\sigma} - \Gamma_{\sigma\alpha}^{\mu} \Gamma_{\beta\mu}^{\sigma} + \Gamma_{\alpha\beta}^{\sigma} \Gamma_{\sigma\mu}^{\mu} \right) g^{\alpha\beta}, \tag{957}$$

with:

$$\Gamma_{\alpha\beta}^{\gamma} = \frac{g^{\gamma\sigma}}{2} \left(g_{\sigma\alpha,\beta} + g_{\sigma\beta,\alpha} - g_{\alpha\beta,\sigma} \right). \tag{958}$$

Similarly, we have for R^*:

$$R^* = R^*_{\ \alpha\beta} G^{\alpha\beta} = \left(\Gamma_{\alpha\beta,\sigma}^{*\sigma} - \Gamma_{\beta\sigma,\alpha}^{*\sigma} - \Gamma_{\sigma\alpha}^{*\mu} \Gamma_{\beta\mu}^{*\sigma} + \Gamma_{\alpha\beta}^{*\sigma} \Gamma_{\sigma\mu}^{*\mu} \right) G^{\alpha\beta}$$

$$= \left(\Gamma_{\alpha\beta,\sigma}^{*\sigma} - \Gamma_{\beta\sigma,\alpha}^{*\sigma} - \Gamma_{\sigma\alpha}^{*\mu} \Gamma_{\beta\mu}^{*\sigma} + \Gamma_{\alpha\beta}^{*\sigma} \Gamma_{\sigma\mu}^{*\mu} \right) F[f]_{\xi\zeta}^{-1\alpha\beta} \cdot g^{\xi\zeta}$$

$$= \left(\Gamma_{\alpha\beta,\sigma}^{*\sigma} - \left(\ln\left[\sqrt{-G} \right]_{,\beta} \right)_{,\alpha} - \Gamma_{\sigma\alpha}^{*\mu} \Gamma_{\beta\mu}^{*\sigma} + \Gamma_{\alpha\beta}^{*\sigma} \ln\left[\sqrt{-G} \right]_{,\sigma} \right) F[f]_{\xi\zeta}^{-1\alpha\beta} \cdot g^{\xi\zeta}$$

$$= \left(\Gamma_{\alpha\beta,\sigma}^{*\sigma} - \left(\frac{\sqrt{-G}_{,\beta}}{\sqrt{-G}} \right)_{,\alpha} - \Gamma_{\sigma\alpha}^{*\mu} \Gamma_{\beta\mu}^{*\sigma} + \Gamma_{\alpha\beta}^{*\sigma} \frac{\sqrt{-G}_{,\sigma}}{\sqrt{-G}} \right) F[f]_{\xi\zeta}^{-1\alpha\beta} \cdot g^{\xi\zeta}$$

$$= \left(\Gamma_{\alpha\beta,\sigma}^{*\sigma} - \left(\frac{\sqrt{-G}_{,\beta}}{\sqrt{-G}} \right)_{,\alpha} - \Gamma_{\sigma\alpha}^{*\mu} \Gamma_{\beta\mu}^{*\sigma} + \Gamma_{\alpha\beta}^{*\sigma} \frac{\sqrt{-G}_{,\sigma}}{\sqrt{-G}} \right) F[f]_{\xi\zeta}^{-1\alpha\beta} \cdot g^{\xi\zeta}, \tag{959}$$

with:

$$\Gamma_{\alpha\beta}^{*\gamma} = \frac{g_{\xi\zeta}F[f]^{\xi\zeta\gamma\sigma}}{2}\left(\left[F[f]_{\sigma\alpha}^{\rho\tau}\cdot g_{\rho\tau}\right]_{,\beta} + \left[F[f]_{\sigma\beta}^{\rho\tau}\cdot g_{\rho\tau}\right]_{,\alpha} - \left[F[f]_{\alpha\beta}^{\rho\tau}\cdot g_{\rho\tau}\right]_{,\sigma}\right)$$

$$= \frac{g_{\xi\zeta}}{2}\left(F[f]^{\xi\zeta\gamma\sigma}F[f]_{\sigma\alpha}^{\rho\tau}\cdot g_{\rho\tau,\beta} + F[f]^{\xi\zeta\gamma\sigma}F[f]_{\sigma\beta}^{\rho\tau}\cdot g_{\rho\tau,\alpha} - F[f]^{\xi\zeta\gamma\sigma}F[f]_{\alpha\beta}^{\rho\tau}\cdot g_{\rho\tau,\sigma}\right)$$

$$+ \frac{g_{\xi\zeta}F[f]^{\xi\zeta\gamma\sigma}}{2}\left(F[f]_{\sigma\alpha,\beta}^{\rho\tau}\cdot g_{\rho\tau} + F[f]_{\sigma\beta,\alpha}^{\rho\tau}\cdot g_{\rho\tau} - F[f]_{\alpha\beta,\sigma}^{\rho\tau}\cdot g_{\rho\tau}\right)$$

$$= \frac{g_{\xi\zeta}F[f]^{\xi\zeta\gamma\sigma}}{2}\cdot\left(F[f]_{\sigma\alpha}^{\rho\tau}\cdot g_{\rho\tau,\beta} + F[f]_{\sigma\beta}^{\rho\tau}\cdot g_{\rho\tau,\alpha} - F[f]_{\alpha\beta}^{\rho\tau}\cdot g_{\rho\tau,\sigma}\right)$$

$$+ \frac{g_{\xi\zeta}F[f]^{\xi\zeta\gamma\sigma}}{2}\left(F[f]_{\sigma\alpha,\beta}^{\rho\tau} + F[f]_{\sigma\beta,\alpha}^{\rho\tau} - F[f]_{\alpha\beta,\sigma}^{\rho\tau}\right)\cdot g_{\rho\tau}$$

$$\equiv \Gamma_{*\alpha\beta}^{*\gamma} + \frac{g_{\xi\zeta}F[f]^{\xi\zeta\gamma\sigma}}{2}\left(F[f]_{\sigma\alpha,\beta}^{\rho\tau} + F[f]_{\sigma\beta,\alpha}^{\rho\tau} - F[f]_{\alpha\beta,\sigma}^{\rho\tau}\right)\cdot g_{\rho\tau}$$

$$\equiv \Gamma_{*\alpha\beta}^{*\gamma} + \Gamma_{\alpha\beta}^{**\gamma}, \tag{960}$$

or with:

$$\Gamma_{\alpha\beta}^{*\gamma} = \frac{g^{\xi\zeta}F[f]_{\xi\zeta}^{-1\gamma\sigma}}{2}\left(\left[F[f]_{\sigma\alpha}^{\rho\tau}\cdot g_{\rho\tau}\right]_{,\beta} + \left[F[f]_{\sigma\beta}^{\rho\tau}\cdot g_{\rho\tau}\right]_{,\alpha} - \left[F[f]_{\alpha\beta}^{\rho\tau}\cdot g_{\rho\tau}\right]_{,\sigma}\right)$$

$$= \frac{g^{\xi\zeta}F[f]_{\xi\zeta}^{-1\gamma\sigma}}{2}\cdot\left(F[f]_{\sigma\alpha}^{\rho\tau}\cdot g_{\rho\tau,\beta} + F[f]_{\sigma\beta}^{\rho\tau}\cdot g_{\rho\tau,\alpha} - F[f]_{\alpha\beta}^{\rho\tau}\cdot g_{\rho\tau,\sigma}\right)$$

$$+ \frac{g^{\xi\zeta}F[f]_{\xi\zeta}^{-1\gamma\sigma}}{2}\left(F[f]_{\sigma\alpha,\beta}^{\rho\tau} + F[f]_{\sigma\beta,\alpha}^{\rho\tau} - F[f]_{\alpha\beta,\sigma}^{\rho\tau}\right)\cdot g_{\rho\tau}$$

$$= \frac{g^{\xi\zeta}F[f]_{\xi\zeta}^{-1\gamma\sigma}}{2}\cdot\left(F[f]_{\sigma\alpha}^{\rho\tau}\cdot g_{\rho\tau,\beta} + F[f]_{\sigma\beta}^{\rho\tau}\cdot g_{\rho\tau,\alpha} - F[f]_{\alpha\beta}^{\rho\tau}\cdot g_{\rho\tau,\sigma}\right)$$

$$+ \frac{g^{\xi\zeta}F[f]_{\xi\zeta}^{-1\gamma\sigma}}{2}\left(F'[f]_{\sigma\alpha}^{\rho\tau}f_{,\beta} + F'[f]_{\sigma\beta}^{\rho\tau}f_{,\alpha} - F'[f]_{\alpha\beta}^{\rho\tau}f_{,\sigma}\right)\cdot g_{\rho\tau}$$

$$\equiv \Gamma^{*\gamma}_{*\alpha\beta} + \frac{g^{\xi\zeta}F[f]^{-1\gamma\sigma}_{\ \ \ \ \xi\zeta}}{2}\left(F'[f]^{\rho\tau}_{\sigma\alpha}f_{,\beta} + F'[f]^{\rho\tau}_{\sigma\beta}f_{,\alpha} - F'[f]^{\rho\tau}_{\alpha\beta}f_{,\sigma}\right)\cdot g_{\rho\tau}$$

$$\equiv \Gamma^{*\gamma}_{*\alpha\beta} + \Gamma^{**\gamma}_{\alpha\beta}. \tag{961}$$

Setting this into the second line in Eq. (959) yields:

$$R^* = R^*_{\ \alpha\beta}G^{\alpha\beta} = \left(\Gamma^{*\sigma}_{\alpha\beta,\sigma} - \Gamma^{*\sigma}_{\beta\sigma,\alpha} - \Gamma^{*\mu}_{\sigma\alpha}\Gamma^{*\sigma}_{\beta\mu} + \Gamma^{*\sigma}_{\alpha\beta}\Gamma^{*\mu}_{\sigma\mu}\right)G^{\alpha\beta}$$

$$= \begin{pmatrix} \Gamma^{*\sigma}_{*\alpha\beta,\sigma} - \Gamma^{*\sigma}_{*\beta\sigma,\alpha} - \Gamma^{*\mu}_{*\sigma\alpha}\Gamma^{*\sigma}_{*\beta\mu} + \Gamma^{*\sigma}_{*\alpha\beta}\Gamma^{*\mu}_{*\sigma\mu} \\ -\Gamma^{*\mu}_{*\sigma\alpha}\Gamma^{**\sigma}_{\beta\mu} + \Gamma^{*\sigma}_{*\alpha\beta}\Gamma^{**\mu}_{\sigma\mu} - \Gamma^{**\mu}_{\sigma\alpha}\Gamma^{*\sigma}_{*\beta\mu} + \Gamma^{**\sigma}_{\alpha\beta}\Gamma^{*\mu}_{*\sigma\mu} \\ +\Gamma^{**\sigma}_{\alpha\beta,\sigma} - \Gamma^{**\sigma}_{\beta\sigma,\alpha} - \Gamma^{**\mu}_{\sigma\alpha}\Gamma^{**\sigma}_{\beta\mu} + \Gamma^{**\sigma}_{\alpha\beta}\Gamma^{**\mu}_{\sigma\mu} \end{pmatrix} g_{\xi\zeta}F[f]^{\xi\zeta\alpha\beta}$$

$$\underbrace{\qquad\qquad\qquad\qquad\qquad\qquad\qquad}_{\equiv R^*_*}$$

$$= \left(\Gamma^{*\sigma}_{*\alpha\beta,\sigma} - \Gamma^{*\sigma}_{*\beta\sigma,\alpha} - \Gamma^{*\mu}_{*\sigma\alpha}\Gamma^{*\sigma}_{*\beta\mu} + \Gamma^{*\sigma}_{*\alpha\beta}\Gamma^{*\mu}_{*\sigma\mu}\right)g_{\xi\zeta}F[f]^{\xi\zeta\alpha\beta}$$

$$+ \underbrace{\begin{pmatrix} -\Gamma^{*\mu}_{*\sigma\alpha}\Gamma^{**\sigma}_{\beta\mu} + \Gamma^{*\sigma}_{*\alpha\beta}\Gamma^{**\mu}_{\sigma\mu} - \Gamma^{**\mu}_{\sigma\alpha}\Gamma^{*\sigma}_{*\beta\mu} + \Gamma^{**\sigma}_{\alpha\beta}\Gamma^{*\mu}_{*\sigma\mu} \\ +\Gamma^{**\sigma}_{\alpha\beta,\sigma} - \Gamma^{**\sigma}_{\beta\sigma,\alpha} - \Gamma^{**\mu}_{\sigma\alpha}\Gamma^{**\sigma}_{\beta\mu} + \Gamma^{**\sigma}_{\alpha\beta}\Gamma^{**\mu}_{\sigma\mu} \end{pmatrix}}_{\equiv R^{**}_*} g_{\xi\zeta}F[f]^{\xi\zeta\alpha\beta}$$

$$\equiv R^*_* + R^{**}_*, \tag{962}$$

or

$$R^* = R^*_{\ \alpha\beta}G^{\alpha\beta} = \left(\Gamma^{*\sigma}_{\alpha\beta,\sigma} - \Gamma^{*\sigma}_{\beta\sigma,\alpha} - \Gamma^{*\mu}_{\sigma\alpha}\Gamma^{*\sigma}_{\beta\mu} + \Gamma^{*\sigma}_{\alpha\beta}\Gamma^{*\mu}_{\sigma\mu}\right)G^{\alpha\beta}$$

$$= \begin{pmatrix} \Gamma^{*\sigma}_{*\alpha\beta,\sigma} - \Gamma^{*\sigma}_{*\beta\sigma,\alpha} - \Gamma^{*\mu}_{*\sigma\alpha}\Gamma^{*\sigma}_{*\beta\mu} + \Gamma^{*\sigma}_{*\alpha\beta}\Gamma^{*\mu}_{*\sigma\mu} \\ -\Gamma^{*\mu}_{*\sigma\alpha}\Gamma^{**\sigma}_{\beta\mu} + \Gamma^{*\sigma}_{*\alpha\beta}\Gamma^{**\mu}_{\sigma\mu} - \Gamma^{**\mu}_{\sigma\alpha}\Gamma^{*\sigma}_{*\beta\mu} + \Gamma^{**\sigma}_{\alpha\beta}\Gamma^{*\mu}_{*\sigma\mu} \\ +\Gamma^{**\sigma}_{\alpha\beta,\sigma} - \Gamma^{**\sigma}_{\beta\sigma,\alpha} - \Gamma^{**\mu}_{\sigma\alpha}\Gamma^{**\sigma}_{\beta\mu} + \Gamma^{**\sigma}_{\alpha\beta}\Gamma^{**\mu}_{\sigma\mu} \end{pmatrix} g^{\xi\zeta}F[f]^{\alpha\beta}_{\xi\zeta}$$

$$\underbrace{\qquad\qquad\qquad\qquad\qquad\qquad\qquad}_{\equiv R^*_*}$$

$$= \left(\Gamma^{*\sigma}_{*\alpha\beta,\sigma} - \Gamma^{*\sigma}_{*\beta\sigma,\alpha} - \Gamma^{*\mu}_{*\sigma\alpha}\Gamma^{*\sigma}_{*\beta\mu} + \Gamma^{*\sigma}_{*\alpha\beta}\Gamma^{*\mu}_{*\sigma\mu}\right)g^{\xi\zeta}F[f]^{\alpha\beta}_{\xi\zeta}$$

$$+ \underbrace{\begin{pmatrix} -\Gamma^{*\mu}_{*\sigma\alpha}\Gamma^{**\sigma}_{\beta\mu} + \Gamma^{*\sigma}_{*\alpha\beta}\Gamma^{**\mu}_{\sigma\mu} - \Gamma^{**\mu}_{\sigma\alpha}\Gamma^{*\sigma}_{*\beta\mu} + \Gamma^{**\sigma}_{\alpha\beta}\Gamma^{*\mu}_{*\sigma\mu} \\ +\Gamma^{**\sigma}_{\alpha\beta,\sigma} - \Gamma^{**\sigma}_{\beta\sigma,\alpha} - \Gamma^{**\mu}_{\sigma\alpha}\Gamma^{**\sigma}_{\beta\mu} + \Gamma^{**\sigma}_{\alpha\beta}\Gamma^{**\mu}_{\sigma\mu} \end{pmatrix}}_{\equiv R^{**}_*} g^{\xi\zeta}F[f]^{\alpha\beta}_{\xi\zeta}$$

$$\equiv R^*_* + R^{**}_*. \tag{963}$$

Here the various derivative forms for the adapted affine connection:

$$R^* = R^*_{\ \alpha\beta}G^{\alpha\beta} = \left(\Gamma^{*\sigma}_{\alpha\beta,\sigma} - \Gamma^{*\sigma}_{\beta\sigma,\alpha} - \Gamma^{*\mu}_{\sigma\alpha}\Gamma^{*\sigma}_{\beta\mu} + \Gamma^{*\sigma}_{\alpha\beta}\Gamma^{*\mu}_{\sigma\mu}\right)G^{\alpha\beta}$$

$$\Gamma^{*\sigma}_{\alpha\beta,\sigma} = \Gamma^{*\sigma}_{*\alpha\beta,\sigma} + \left(\frac{g^{\xi\zeta}F[f]^{-1\gamma\sigma}_{\ \ \xi\zeta}}{2}\left(F'[f]^{\rho\tau}_{\gamma\alpha}\,f_{,\beta} + F'[f]^{\rho\tau}_{\gamma\beta}\,f_{,\alpha} - F'[f]^{\rho\tau}_{\alpha\beta}\,f_{,\gamma}\right)\cdot g_{\rho\tau}\right)_{,\sigma}$$

$$= \Gamma^{*\sigma}_{*\alpha\beta,\sigma} + \frac{g^{\xi\zeta}F[f]^{-1\gamma\sigma}_{\ \ \xi\zeta}}{2}\left(F'[f]^{\rho\tau}_{\gamma\alpha}\,f_{,\beta} + F'[f]^{\rho\tau}_{\gamma\beta}\,f_{,\alpha} - F'[f]^{\rho\tau}_{\alpha\beta}\,f_{,\gamma}\right)\cdot g_{\rho\tau,\sigma}$$

$$+ \frac{g^{\xi\zeta}_{\ ,\sigma}F[f]^{-1\gamma\sigma}_{\ \ \xi\zeta}}{2}\left(F'[f]^{\rho\tau}_{\gamma\alpha}\,f_{,\beta} + F'[f]^{\rho\tau}_{\gamma\beta}\,f_{,\alpha} - F'[f]^{\rho\tau}_{\alpha\beta}\,f_{,\gamma}\right)\cdot g_{\rho\tau}$$

$$+ \frac{g^{\xi\zeta}F'[f]^{-1\gamma\sigma}_{\ \ \xi\zeta}\,f_{,\sigma}}{2}\left(F'[f]^{\rho\tau}_{\gamma\alpha}\,f_{,\beta} + F'[f]^{\rho\tau}_{\gamma\beta}\,f_{,\alpha} - F'[f]^{\rho\tau}_{\alpha\beta}\,f_{,\gamma}\right)\cdot g_{\rho\tau}$$

$$\frac{g^{\xi\zeta}F[f]^{-1\gamma\sigma}_{\ \ \xi\zeta}}{2}\left(\begin{array}{l}F'[f]^{\rho\tau}_{\gamma\alpha}\,f_{,\beta\sigma} + F'[f]^{\rho\tau}_{\gamma\beta}\,f_{,\alpha\sigma} - F'[f]^{\rho\tau}_{\alpha\beta}\,f_{,\gamma\sigma}\\ +F''[f]^{\rho\tau}_{\gamma\alpha}\,f_{,\beta}f_{,\sigma} + F''[f]^{\rho\tau}_{\gamma\beta}\,f_{,\alpha}f_{,\sigma} - F''[f]^{\rho\tau}_{\alpha\beta}\,f_{,\gamma}f_{,\sigma}\end{array}\right)\cdot g_{\rho\tau}$$

$$-\Gamma^{*\sigma}_{\beta\sigma,\alpha}$$

$$-\Gamma^{*\mu}_{\sigma\alpha}\Gamma^{*\sigma}_{\beta\mu}$$

$$+\Gamma^{*\sigma}_{\alpha\beta}\Gamma^{*\mu}_{\sigma\mu}$$

$$\Gamma^{*\gamma}_{\alpha\beta} = \frac{g^{\xi\zeta}F[f]^{-1\gamma\sigma}_{\ \ \xi\zeta}}{2}\left(\left[F[f]^{\rho\tau}_{\sigma\alpha}\cdot g_{\rho\tau}\right]_{,\beta} + \left[F[f]^{\rho\tau}_{\sigma\beta}\cdot g_{\rho\tau}\right]_{,\alpha} - \left[F[f]^{\rho\tau}_{\alpha\beta}\cdot g_{\rho\tau}\right]_{,\sigma}\right)$$

$$= \frac{g^{\xi\zeta}F[f]^{-1\gamma\sigma}_{\ \ \xi\zeta}}{2}\cdot\left(F[f]^{\rho\tau}_{\sigma\alpha}\cdot g_{\rho\tau,\beta} + F[f]^{\rho\tau}_{\sigma\beta}\cdot g_{\rho\tau,\alpha} - F[f]^{\rho\tau}_{\alpha\beta}\cdot g_{\rho\tau,\sigma}\right)$$

$$+ \frac{g^{\xi\zeta}F[f]^{-1\gamma\sigma}_{\ \ \xi\zeta}}{2}\left(F[f]^{\rho\tau}_{\sigma\alpha,\beta} + F[f]^{\rho\tau}_{\sigma\beta,\alpha} - F[f]^{\rho\tau}_{\alpha\beta,\sigma}\right)\cdot g_{\rho\tau}$$

$$= \frac{g^{\xi\zeta}F[f]^{-1\gamma\sigma}_{\ \ \xi\zeta}}{2}\cdot\left(F[f]^{\rho\tau}_{\sigma\alpha}\cdot g_{\rho\tau,\beta} + F[f]^{\rho\tau}_{\sigma\beta}\cdot g_{\rho\tau,\alpha} - F[f]^{\rho\tau}_{\alpha\beta}\cdot g_{\rho\tau,\sigma}\right)$$

$$+ \frac{g^{\xi\zeta}F[f]^{-1\gamma\sigma}_{\ \ \xi\zeta}}{2}\left(F'[f]^{\rho\tau}_{\sigma\alpha}\,f_{,\beta} + F'[f]^{\rho\tau}_{\sigma\beta}\,f_{,\alpha} - F'[f]^{\rho\tau}_{\alpha\beta}\,f_{,\sigma}\right)\cdot g_{\rho\tau}$$

$$\equiv \Gamma^{*\gamma}_{*\alpha\beta} + \frac{g^{\xi\zeta}F[f]^{-1\gamma\sigma}_{\xi\zeta}}{2}\left(F'[f]^{\rho\tau}_{\sigma\alpha}f_{,\beta} + F'[f]^{\rho\tau}_{\sigma\beta}f_{,\alpha} - F'[f]^{\rho\tau}_{\alpha\beta}f_{,\sigma}\right)\cdot g_{\rho\tau}$$

$$\equiv \Gamma^{*\gamma}_{*\alpha\beta} + \Gamma^{**\gamma}_{\alpha\beta}. \tag{964}$$

$$R^* = R^*_{\ \alpha\beta}G^{\alpha\beta} = \left(\Gamma^{*\sigma}_{\alpha\beta,\sigma} - \Gamma^{*\sigma}_{\beta\sigma,\alpha} - \Gamma^{*\mu}_{\sigma\alpha}\Gamma^{*\sigma}_{\beta\mu} + \Gamma^{*\sigma}_{\alpha\beta}\Gamma^{*\mu}_{\sigma\mu}\right)G^{\alpha\beta}$$

$$\Gamma^{*\sigma}_{\beta\sigma,\alpha} = \Gamma^{*\sigma}_{*\beta\sigma,\alpha} + \frac{g^{\xi\zeta}F[f]^{-1\gamma\sigma}_{\xi\zeta}}{2}\left(F'[f]^{\rho\tau}_{\gamma\sigma}f_{,\beta} + F'[f]^{\rho\tau}_{\gamma\beta}f_{,\sigma} - F'[f]^{\rho\tau}_{\beta\sigma}f_{,\gamma}\right)\cdot g_{\rho\tau,\alpha}$$

$$+ \frac{g^{\xi\zeta}_{,\alpha}F[f]^{-1\gamma\sigma}_{\xi\zeta}}{2}\left(F'[f]^{\rho\tau}_{\gamma\sigma}f_{,\beta} + F'[f]^{\rho\tau}_{\gamma\beta}f_{,\sigma} - F'[f]^{\rho\tau}_{\beta\sigma}f_{,\gamma}\right)\cdot g_{\rho\tau}$$

$$+ \frac{g^{\xi\zeta}F'[f]^{-1\gamma\sigma}_{\xi\zeta}f_{,\alpha}}{2}\left(F'[f]^{\rho\tau}_{\gamma\sigma}f_{,\beta} + F'[f]^{\rho\tau}_{\gamma\beta}f_{,\sigma} - F'[f]^{\rho\tau}_{\beta\sigma}f_{,\gamma}\right)\cdot g_{\rho\tau}$$

$$+ \frac{g^{\xi\zeta}F[f]^{-1\gamma\sigma}_{\xi\zeta}}{2}\left(\begin{array}{c}F'[f]^{\rho\tau}_{\gamma\sigma}f_{,\beta\alpha} + F'[f]^{\rho\tau}_{\gamma\beta}f_{,\sigma\alpha} - F'[f]^{\rho\tau}_{\beta\sigma}f_{,\gamma\alpha}\\ +F''[f]^{\rho\tau}_{\gamma\sigma}f_{,\beta}f_{,\alpha} + F''[f]^{\rho\tau}_{\gamma\beta}f_{,\sigma}f_{,\alpha} - F''[f]^{\rho\tau}_{\beta\sigma}f_{,\gamma}f_{,\alpha}\end{array}\right)\cdot g_{\rho\tau}$$

$$- \Gamma^{*\mu}_{\sigma\alpha}\Gamma^{*\sigma}_{\beta\mu}$$

$$+ \Gamma^{*\sigma}_{\alpha\beta}\Gamma^{*\mu}_{\sigma\mu}$$

$$\Gamma^{*\gamma}_{\alpha\beta} = \frac{g^{\xi\zeta}F[f]^{-1\gamma\sigma}_{\xi\zeta}}{2}\left(\left[F[f]^{\rho\tau}_{\sigma\alpha}\cdot g_{\rho\tau}\right]_{,\beta} + \left[F[f]^{\rho\tau}_{\sigma\beta}\cdot g_{\rho\tau}\right]_{,\alpha} - \left[F[f]^{\rho\tau}_{\alpha\beta}\cdot g_{\rho\tau}\right]_{,\sigma}\right)$$

$$= \frac{g^{\xi\zeta}F[f]^{-1\gamma\sigma}_{\xi\zeta}}{2}\cdot\left(F[f]^{\rho\tau}_{\sigma\alpha}\cdot g_{\rho\tau,\beta} + F[f]^{\rho\tau}_{\sigma\beta}\cdot g_{\rho\tau,\alpha} - F[f]^{\rho\tau}_{\alpha\beta}\cdot g_{\rho\tau,\sigma}\right)$$

$$+ \frac{g^{\xi\zeta}F[f]^{-1\gamma\sigma}_{\xi\zeta}}{2}\left(F[f]^{\rho\tau}_{\sigma\alpha,\beta} + F[f]^{\rho\tau}_{\sigma\beta,\alpha} - F[f]^{\rho\tau}_{\alpha\beta,\sigma}\right)\cdot g_{\rho\tau}$$

$$= \frac{g^{\xi\zeta}F[f]^{-1\gamma\sigma}_{\xi\zeta}}{2}\cdot\left(F[f]^{\rho\tau}_{\sigma\alpha}\cdot g_{\rho\tau,\beta} + F[f]^{\rho\tau}_{\sigma\beta}\cdot g_{\rho\tau,\alpha} - F[f]^{\rho\tau}_{\alpha\beta}\cdot g_{\rho\tau,\sigma}\right)$$

$$+ \frac{g^{\xi\zeta}F[f]^{-1\gamma\sigma}_{\xi\zeta}}{2}\left(F'[f]^{\rho\tau}_{\sigma\alpha}f_{,\beta} + F'[f]^{\rho\tau}_{\sigma\beta}f_{,\alpha} - F'[f]^{\rho\tau}_{\alpha\beta}f_{,\sigma}\right)\cdot g_{\rho\tau}$$

$$\equiv \Gamma^{*\gamma}_{*\alpha\beta} + \frac{g^{\xi\zeta}F[f]^{-1\gamma\sigma}_{\xi\zeta}}{2}\left(F'[f]^{\rho\tau}_{\sigma\alpha}f_{,\beta} + F'[f]^{\rho\tau}_{\sigma\beta}f_{,\alpha} - F'[f]^{\rho\tau}_{\alpha\beta}f_{,\sigma}\right)\cdot g_{\rho\tau}$$

$$\equiv \Gamma^{*\gamma}_{*\alpha\beta} + \Gamma^{**\gamma}_{\alpha\beta}. \tag{965}$$

8.13 Appendix D: Further Generalization to *R** with Scaling *F^q*

While the Hilbert Lagrange density was just the Ricci scalar *R*, we have no way of knowing the correct exponent for the scale factor *F* times the Ricci scalar of the scaled metric *R**. In the classical approach, the scale factor was *F* = 1 and so it could be any exponent one could put on the term.

Thus, setting Eq. (735) into Eq. (737), setting L_M and Λ to zero and choosing an additional scale factor F^q results in the variation:

$$\delta W = 0 = \delta \int_V d^n x \left(\sqrt{-g \cdot F^n} \times F^q \cdot R^* \right)$$

$$= \delta \int_V d^n x \left[\sqrt{-g \cdot F^n} \times \left(F^q \cdot \left(\begin{array}{c} R + \dfrac{F'}{F}(1-n)\Delta f \\[2mm] + \dfrac{f_{,\alpha} f_{,\beta} g^{\alpha\beta} (1-n)}{4F^2} \left(4FF'' + (F')^2 (n-6) \right) \end{array} \right)^{\frac{1}{F}} \right) \right]$$

$$= \delta \int_V d^n x \left[\sqrt{-g \cdot F^{n+2q}} \times \left(\left(\begin{array}{c} R + \dfrac{F'}{F}(1-n)\Delta f \\[2mm] + \dfrac{f_{,\alpha} f_{,\beta} g^{\alpha\beta} (1-n)}{4F^2} \left(4FF'' + (F')^2 (n-6) \right) \end{array} \right)^{\frac{1}{F}} \right) \right].$$

$$(966)$$

Now we define *p* = *n* + 2*q* and get:

$$\delta_{g_{\alpha\beta}} W = 0 = \left\{ \delta_{g_{\alpha\beta}} \int_V d^n x \frac{\sqrt{-g \cdot F^p}}{F} \times \left(\begin{array}{c} R + \dfrac{F'}{F}(1-n)\Delta f \\[2mm] + \dfrac{f_{,\alpha} f_{,\beta} g^{\alpha\beta} (1-n)}{4F^2} \left(4FF'' + (F')^2 (n-6) \right) \end{array} \right) \right..$$

$$(967)$$

Performing the usual Hilbert variation with respect to the metric tensor $g_{\alpha\beta}$ now leads us to:

$$\delta_{g_{\alpha\beta}} W = 0 = \delta_{g_{\alpha\beta}} \int_V d^n x \left[\sqrt{-g \cdot F^p} \times \left(\left(\begin{array}{c} R + \dfrac{F'}{\sqrt{-g} \cdot F}(1-n)\partial_\beta \sqrt{-g} \cdot g^{\alpha\beta} f_{,\alpha} \\[2mm] + \dfrac{f_{,\alpha} f_{,\beta} g^{\alpha\beta} (1-n)}{4F^2} \left(4FF'' + (F')^2 (n-6) \right) \end{array} \right)^{\frac{1}{F}} \right) \right]$$

$$= \int_V d^n x \frac{\sqrt{-g} \cdot F^p}{F} \times \left(\begin{array}{c} R^{\kappa\lambda} - \dfrac{R}{2} g^{\kappa\lambda} \\[2mm] f_{,\alpha} f_{,\beta} \left(\dfrac{g^{\kappa\lambda} g^{\alpha\beta}}{2} - \dfrac{g^{\kappa\alpha} g^{\lambda\beta} + g^{\lambda\alpha} g^{\kappa\beta}}{2} \right)(1-n) \\[2mm] + \dfrac{}{4F^2} \left(\begin{array}{c} (F')^2 (n-6) \\ +4FF'' \end{array} \right) \end{array} \right) \delta g_{\kappa\lambda}$$

$$+ \delta_{g_{\alpha\beta}} \int_V d^n x \frac{F^{p/2}}{F^2} \cdot \left(F' \cdot (1-n) \partial_\beta \sqrt{-g} \cdot g^{\alpha\beta} f_{,\alpha} \right) + \int_V d^n x \frac{\sqrt{-g} \cdot F^p}{F} \times g^{\alpha\beta} \delta_{g_{\alpha\beta}} R_{\alpha\beta}.$$

(968)

Please note that, as before without the F^q-factor, we separated the Laplace term, because it cannot be varied as easily as the other terms, because we have the metric within an operator. So just like with the Hilbert variation of the Ricci tensor [7] we have to apply caution. We see this, when taking the first term from the last line of Eq. (968) and rewriting the Laplace operator in a different form:

$$\delta_{g_{\alpha\beta}} \int_V d^n x \sqrt{-g} \cdot \frac{F^{p/2}}{F^2} \cdot \left(F' \cdot (1-n) \frac{1}{\sqrt{-g}} \partial_\beta \sqrt{-g} \cdot g^{\alpha\beta} f_{,\alpha} \right)$$

$$= \delta_{g_{\alpha\beta}} \int_V d^n x \sqrt{-g} \cdot \frac{F^{p/2}}{F^2} \cdot \left(F' \cdot (1-n) \left(g^{\alpha\beta} f_{,\alpha\beta} - g^{\alpha\beta} \Gamma^\gamma_{\alpha\beta} f_{,\gamma} \right) \right)$$

$$= \int_V d^n x \sqrt{-g} \cdot \frac{F^{p/2}}{F^2} \cdot \left(F' \cdot (1-n) \left(\frac{g^{\kappa\lambda} g^{\alpha\beta}}{2} - \frac{g^{\kappa\alpha} g^{\lambda\beta} + g^{\lambda\alpha} g^{\kappa\beta}}{2} \right) \left(f_{,\alpha\beta} - \Gamma^\gamma_{\alpha\beta} f_{,\gamma} \right) \right) \delta g_{\kappa\lambda}$$

$$- \int_V d^n x \sqrt{-g} \cdot \frac{F^{p/2}}{F^2} \cdot \left(F' \cdot (1-n) g^{\alpha\beta} \delta_{g_{\alpha\beta}} \Gamma^\gamma_{\alpha\beta} f_{,\gamma} \right). \tag{969}$$

While the first integral makes no problems, we find that the second integral (last line of Eq. (969)) gives us:

$$\int_V d^n x \sqrt{-g} \cdot \frac{F^{p/2}}{F^2} \cdot \left(F' \cdot (1-n) g^{\alpha\beta} \delta_{g_{\alpha\beta}} \Gamma^\gamma_{\alpha\beta} f_{,\gamma} \right)$$

$$= \int_V d^n x \sqrt{-g} \cdot \frac{F^{p/2}}{F^2} \cdot \left(F' \cdot (1-n) g^{\alpha\beta} f_{,\gamma} \frac{1}{2} g^{\gamma\sigma} \left(\delta g_{\sigma\alpha;\beta} + \delta g_{\sigma\beta;\alpha} - \delta g_{\alpha\beta;\sigma} \right) \right).$$

(970)

Thus, we have covariant derivatives of metric variations $\delta g_{\sigma\alpha;\beta} + \delta g_{\sigma\beta;\alpha} - \delta g_{\alpha\beta;\sigma}$ instead of the variation itself $\delta g_{\kappa\lambda}$ as obtained for the other integrals. The total result looks as follows:

$$\delta_{g_{\alpha\beta}} W = 0 = \int_V d^n x \frac{\sqrt{-g \cdot F^p}}{F} \times g^{\alpha\beta} \delta_{g_{\alpha\beta}} R_{\alpha\beta}$$

$$+ \int_V d^n x \frac{\sqrt{-g \cdot F^p}}{F} \times \left(\begin{array}{c} R^{\kappa\lambda} - \dfrac{R}{2} g^{\kappa\lambda} \\ + \dfrac{f_{,\alpha} f_{,\beta} \left(\dfrac{g^{\kappa\lambda} g^{\alpha\beta}}{2} - \dfrac{g^{\kappa\alpha} g^{\lambda\beta} + g^{\lambda\alpha} g^{\kappa\beta}}{2} \right)(1-n)}{4F^2} \left(\begin{array}{c} (F')^2 (n-6) \\ +4FF'' \end{array} \right) \end{array} \right) \delta g_{\kappa\lambda}$$

$$+ \int_V d^n x \sqrt{-g} \cdot \frac{F^{p/2}}{F^2} \cdot \left(F' \cdot (1-n) \left(\frac{g^{\kappa\lambda} g^{\alpha\beta}}{2} - \frac{g^{\kappa\alpha} g^{\lambda\beta} + g^{\lambda\alpha} g^{\kappa\beta}}{2} \right) \left(f_{,\alpha\beta} - \Gamma^{\gamma}_{\alpha\beta} f_{,\gamma} \right) \right) \delta g_{\kappa\lambda}$$

$$- \int_V d^n x \sqrt{-g} \cdot \frac{F^{p/2}}{F^2} \cdot \left(F' \cdot (1-n) g^{\alpha\beta} f_{,\gamma} \frac{1}{2} g^{\gamma\sigma} \left(\delta g_{\sigma\alpha;\beta} + \delta g_{\sigma\beta;\alpha} - \delta g_{\alpha\beta;\sigma} \right) \right)$$

$$= \int_V d^n x \frac{\sqrt{-g \cdot F^p}}{F} \times g^{\alpha\beta} \delta_{g_{\alpha\beta}} R_{\alpha\beta}$$

$$+ \int_V d^n x \frac{\sqrt{-g \cdot F^p}}{F} \times \left(\begin{array}{c} R^{\kappa\lambda} - \dfrac{R}{2} g^{\kappa\lambda} \\ + \dfrac{f_{,\alpha} f_{,\beta} \left(\dfrac{g^{\kappa\lambda} g^{\alpha\beta}}{2} - \dfrac{g^{\kappa\alpha} g^{\lambda\beta} + g^{\lambda\alpha} g^{\kappa\beta}}{2} \right)(1-n)}{4F^2} \left(\begin{array}{c} (F')^2 (n-6) \\ +4FF'' \end{array} \right) \\ + \dfrac{F' \cdot (1-n)}{F} \left(\dfrac{g^{\kappa\lambda} g^{\alpha\beta}}{2} - \dfrac{g^{\kappa\alpha} g^{\lambda\beta} + g^{\lambda\alpha} g^{\kappa\beta}}{2} \right) \left(f_{,\alpha\beta} - \Gamma^{\gamma}_{\alpha\beta} f_{,\gamma} \right) \end{array} \right) \delta g_{\kappa\lambda}$$

$$- \int_V d^n x \frac{\sqrt{-g \cdot F^p}}{F} \cdot \left(\frac{F' \cdot (1-n)}{F} g^{\alpha\beta} f_{,\gamma} \frac{1}{2} g^{\gamma\sigma} \left(\delta g_{\sigma\alpha;\beta} + \delta g_{\sigma\beta;\alpha} - \delta g_{\alpha\beta;\sigma} \right) \right). \quad (971)$$

8.13.1 Variation of the Laplace Operator with R^*-Scaling

Now we have to consider the following term:

$$\delta_{g_{\alpha\beta}} \int_V d^n x \sqrt{-g} \cdot \frac{F^{p/2}}{F^2} \cdot \left(F' \cdot (1-n)\frac{1}{\sqrt{-g}} \partial_\beta \sqrt{-g} \cdot g^{\alpha\beta} f_{,\alpha} \right)$$

$$= \delta_{g_{\alpha\beta}} \int_V d^n x \sqrt{-g} \cdot \frac{F^{p/2}}{F^2} \cdot \left(F' \cdot (1-n)\left(g^{\alpha\beta} f_{,\alpha\beta} - g^{\alpha\beta} \Gamma^\gamma_{\alpha\beta} f_{,\gamma} \right) \right)$$

$$= \int_V d^n x \sqrt{-g} \cdot \frac{F^{p/2}}{F^2} \left(F' \cdot (1-n)\left(\frac{g^{\kappa\lambda} g^{\alpha\beta}}{2} - \frac{g^{\kappa\alpha} g^{\lambda\beta} + g^{\lambda\alpha} g^{\kappa\beta}}{2} \right)\left(f_{,\alpha\beta} - \Gamma^\gamma_{\alpha\beta} f_{,\gamma} \right) \right) \delta g_{\kappa\lambda}$$

$$+ \int_V d^n x \sqrt{-g} \cdot \frac{F^{p/2}}{F^2} \cdot \left(F' \cdot (1-n) g^{\alpha\beta} \delta_{g_{\alpha\beta}} \Gamma^\gamma_{\alpha\beta} f_{,\gamma} \right). \qquad (972)$$

Further evaluation of the last integral yields:

$$\int_V d^n x \sqrt{-g} \cdot \frac{F^{p/2}}{F^2} \cdot \left(F' \cdot (1-n) g^{\alpha\beta} \delta_{g_{\alpha\beta}} \Gamma^\gamma_{\alpha\beta} f_{,\gamma} \right)$$

$$= \int_V d^n x \sqrt{-g} \cdot \frac{F^{p/2}}{F^2} \cdot \left(F' \cdot (1-n) g^{\alpha\beta} \left[\frac{g^{\gamma\sigma}}{2} GX^{\kappa\lambda\rho}_{\alpha\beta\sigma} \delta g_{\kappa\lambda,\rho} - g^{\gamma\lambda} \Gamma^\kappa_{\alpha\beta} \delta g_{\kappa\lambda} \right] f_{,\gamma} \right)$$

with: $\delta_g \left(\Gamma^\gamma_{\alpha\beta} \right) = \frac{1}{2} g^{\gamma\sigma} \left(\delta g_{\sigma\alpha;\beta} + \delta g_{\sigma\beta;\alpha} - \delta g_{\alpha\beta;\sigma} \right) = \frac{g^{\gamma\sigma}}{2} GX^{\kappa\lambda\rho}_{\alpha\beta\sigma} \delta g_{\kappa\lambda,\rho} - g^{\gamma\lambda} \Gamma^\kappa_{\alpha\beta} \delta g_{\kappa\lambda}$

$$\left(g^\kappa_\alpha g^\lambda_\sigma g^\rho_\beta + g^\kappa_\sigma g^\lambda_\beta g^\rho_\alpha - g^\kappa_\alpha g^\lambda_\beta g^\rho_\sigma \right) \equiv GX^{\kappa\lambda\rho}_{\alpha\beta\sigma}. \qquad (973)$$

Thereby:

$$\delta_g \left(\Gamma^\gamma_{\alpha\beta} \right) = \frac{1}{2} g^{\gamma\sigma} \left(\delta g_{\sigma\alpha;\beta} + \delta g_{\sigma\beta;\alpha} - \delta g_{\alpha\beta;\sigma} \right) = \frac{g^{\gamma\sigma}}{2} \left(\delta g_{\sigma\alpha,\beta} + \delta g_{\sigma\beta,\alpha} - \delta g_{\alpha\beta,\sigma} \right)$$

$$- \frac{g^{\gamma\sigma}}{2} \left(\Gamma^\xi_{\beta\sigma} \delta g_{\xi\alpha} + \Gamma^\xi_{\beta\alpha} \delta g_{\sigma\xi} + \Gamma^\xi_{\alpha\sigma} \delta g_{\xi\beta} + \Gamma^\xi_{\alpha\beta} \delta g_{\sigma\xi} - \Gamma^\xi_{\sigma\alpha} \delta g_{\xi\beta} - \Gamma^\xi_{\sigma\beta} \delta g_{\alpha\xi} \right)$$

$$= \frac{g^{\gamma\sigma}}{2} \left(\delta g_{\sigma\alpha,\beta} + \delta g_{\sigma\beta,\alpha} - \delta g_{\alpha\beta,\sigma} \right)$$

$$- \frac{g^{\gamma\sigma}}{2} \left(\Gamma^\kappa_{\beta\sigma} \delta g_{\kappa\alpha} + \Gamma^\kappa_{\beta\alpha} \delta g_{\sigma\kappa} + \Gamma^\kappa_{\alpha\sigma} \delta g_{\kappa\beta} + \Gamma^\kappa_{\alpha\beta} \delta g_{\sigma\kappa} - \Gamma^\kappa_{\sigma\alpha} \delta g_{\kappa\beta} - \Gamma^\kappa_{\sigma\beta} \delta g_{\alpha\kappa} \right)$$

$$\xrightarrow{\delta g_{ij} = \delta g_{ji}} = \frac{g^{\gamma\sigma}}{2} \left(\delta g_{\sigma\alpha,\beta} + \delta g_{\sigma\beta,\alpha} - \delta g_{\alpha\beta,\sigma} \right)$$

$$-\frac{g^{\gamma\sigma}}{2}\left(\Gamma^{\kappa}_{\beta\sigma}g^{\lambda}_{\alpha}+\Gamma^{\kappa}_{\beta\alpha}g^{\lambda}_{\sigma}+\Gamma^{\kappa}_{\alpha\sigma}g^{\lambda}_{\beta}+\Gamma^{\kappa}_{\alpha\beta}g^{\lambda}_{\sigma}-\Gamma^{\kappa}_{\sigma\alpha}g^{\lambda}_{\beta}-\Gamma^{\kappa}_{\sigma\beta}g^{\lambda}_{\alpha}\right)\delta g_{\kappa\lambda}$$

$$\xrightarrow{\Gamma^{\kappa}_{ij}=\Gamma^{\kappa}_{ji}}=\frac{g^{\gamma\sigma}}{2}\left(\delta g_{\sigma\alpha,\beta}+\delta g_{\sigma\beta,\alpha}-\delta g_{\alpha\beta,\sigma}\right)$$

$$-\frac{g^{\gamma\sigma}}{2}\left(\underbrace{\Gamma^{\kappa}_{\beta\sigma}g^{\lambda}_{\alpha}-\Gamma^{\kappa}_{\sigma\beta}g^{\lambda}_{\alpha}}_{=0}+\underbrace{\Gamma^{\kappa}_{\alpha\beta}g^{\lambda}_{\sigma}+\Gamma^{\kappa}_{\beta\alpha}g^{\lambda}_{\sigma}}_{=2\Gamma^{\kappa}_{\alpha\beta}g^{\lambda}_{\sigma}}+\underbrace{\Gamma^{\kappa}_{\alpha\sigma}g^{\lambda}_{\beta}-\Gamma^{\kappa}_{\sigma\alpha}g^{\lambda}_{\beta}}_{=0}\right)\delta g_{\kappa\lambda}$$

$$=\frac{g^{\gamma\sigma}}{2}\left(\delta g_{\sigma\alpha,\beta}+\delta g_{\sigma\beta,\alpha}-\delta g_{\alpha\beta,\sigma}\right)-g^{\gamma\lambda}\Gamma^{\kappa}_{\alpha\beta}\delta g_{\kappa\lambda}. \tag{974}$$

Now we proceed with the result of Eq. (973) (second line) and by applying integration by parts and setting all surface terms equal to zero, we end up with:

$$=\int_{V}d^{n}x\sqrt{-g}\cdot\frac{F^{p/2}}{F^{2}}\cdot\left(F'\cdot(1-n)g^{\alpha\beta}\frac{g^{\gamma\sigma}}{2}GX^{\kappa\lambda\rho}_{\alpha\beta\sigma}f_{,\gamma}\right)\delta g_{\kappa\lambda,\rho}$$

$$=\int_{V}d^{n}x\left[\sqrt{-g}\cdot\frac{F^{p/2}}{F^{2}}\cdot\left(F'\cdot(1-n)g^{\alpha\beta}\frac{g^{\gamma\sigma}}{2}GX^{\kappa\lambda\rho}_{\alpha\beta\sigma}f_{,\gamma}\right)\right]_{,\rho}\delta g_{\kappa\lambda}$$

$$-\int_{V}d^{n}x\left[\sqrt{-g}\cdot\frac{F^{p/2}}{F^{2}}\cdot\left(F'\cdot(1-n)g^{\alpha\beta}\frac{g^{\gamma\sigma}}{2}GX^{\kappa\lambda\rho}_{\alpha\beta\sigma}f_{,\gamma}\right)\delta g_{\kappa\lambda}\right]_{,\rho}$$

$$=\int_{V}d^{n}x\left[\sqrt{-g}\cdot\frac{F^{p/2}}{F^{2}}\cdot\left(F'\cdot(1-n)g^{\alpha\beta}\frac{g^{\gamma\sigma}}{2}GX^{\kappa\lambda\rho}_{\alpha\beta\sigma}f_{,\gamma}\right)\right]_{,\rho}\delta g_{\kappa\lambda}$$

$$-\int_{\partial V}d^{n}x\cdot n_{\rho}\sqrt{-g}\cdot\frac{F^{p/2}}{F^{2}}\cdot\left(F'\cdot(1-n)g^{\alpha\beta}\frac{g^{\gamma\sigma}}{2}GX^{\kappa\lambda\rho}_{\alpha\beta\sigma}f_{,\gamma}\right)\delta g_{\kappa\lambda}$$

$$=\int_{V}d^{n}x\left[\sqrt{-g}\cdot\frac{F^{p/2}}{F^{2}}\cdot\left(F'\cdot(1-n)g^{\alpha\beta}\frac{g^{\gamma\sigma}}{2}GX^{\kappa\lambda\rho}_{\alpha\beta\sigma}f_{,\gamma}\right)\right]_{,\rho}\delta g_{\kappa\lambda}. \tag{975}$$

8.13.2 Back to the Total Variation of Our Scaled Metric Ricci Scalar with Scaling Factor F^{q}

With the help of the results of the last sub-section, we obtain from Eq. (971):

$$\delta_{g_{\alpha\beta}}W=0=\int_{V}d^{n}x\frac{\sqrt{-g\cdot F^{p}}}{F}\times g^{\alpha\beta}\delta_{g_{\alpha\beta}}R_{\alpha\beta}$$

$$+\int_V d^n x \left[\frac{\sqrt{-g}\cdot F^p}{F} \cdot \left(\begin{array}{c} R^{\kappa\lambda} - \dfrac{R}{2}g^{\kappa\lambda} \\[2mm] + \dfrac{f_{,\alpha}f_{,\beta}\left(\dfrac{g^{\kappa\lambda}g^{\alpha\beta}}{2} - \dfrac{g^{\kappa\alpha}g^{\lambda\beta}+g^{\lambda\alpha}g^{\kappa\beta}}{2}\right)(1-n)}{4F^2}\left(\begin{array}{c}(F')^2(n-6)\\+4FF''\end{array}\right) \\[4mm] + \dfrac{F'\cdot(1-n)}{F}\left(\dfrac{g^{\kappa\lambda}g^{\alpha\beta}}{2} - \dfrac{g^{\kappa\alpha}g^{\lambda\beta}+g^{\lambda\alpha}g^{\kappa\beta}}{2}\right)\left(f_{,\alpha\beta}-\Gamma^{\gamma}_{\alpha\beta}f_{,\gamma}\right) \\[4mm] + \dfrac{F'}{F}\cdot(1-n)g^{\alpha\beta}g^{\gamma\lambda}\Gamma^{\kappa}_{\alpha\beta}f_{,\gamma} \end{array}\right) \\[6mm] - \left[\dfrac{\sqrt{-g}\cdot F^p}{F}\cdot\left(\dfrac{F'}{F}\cdot(1-n)g^{\alpha\beta}\dfrac{g^{\gamma\sigma}}{2}GX^{\kappa\lambda\rho}_{\alpha\beta\sigma}f_{,\gamma}\right)\right]_{,\rho} \right]\delta g_{\kappa\lambda}.$$

$$(976)$$

For simplification we further consider the integral above without further treatment of the variation term for the Ricci tensor $\dfrac{\sqrt{-g}\cdot F^p}{F}\cdot\left(g^{\alpha\beta}\delta_{g_{\alpha\beta}}R_{\alpha\beta}\right)$. The rest gives us:

$$\text{Integrand} - \frac{\sqrt{-g}\cdot F^p}{F}\cdot\left(g^{\alpha\beta}\delta_{g_{\alpha\beta}}R_{\alpha\beta}\right)$$

$$\Rightarrow \frac{\sqrt{-g}\cdot F^p}{F}\cdot\left(\begin{array}{c} R^{\kappa\lambda} - \dfrac{R}{2}g^{\kappa\lambda} \\[2mm] + \dfrac{f_{,\alpha}f_{,\beta}\left(\dfrac{g^{\kappa\lambda}g^{\alpha\beta}}{2} - \dfrac{g^{\kappa\alpha}g^{\lambda\beta}+g^{\lambda\alpha}g^{\kappa\beta}}{2}\right)(1-n)}{4F^2}\left(\begin{array}{c}(F')^2(n-6)\\+4FF''\end{array}\right) \\[4mm] + \dfrac{F'\cdot(1-n)}{F}\left(\dfrac{g^{\kappa\lambda}g^{\alpha\beta}}{2} - \dfrac{g^{\kappa\alpha}g^{\lambda\beta}+g^{\lambda\alpha}g^{\kappa\beta}}{2}\right)\left(f_{,\alpha\beta}-\Gamma^{\gamma}_{\alpha\beta}f_{,\gamma}\right) \\[4mm] + \dfrac{F'}{F}\cdot(1-n)g^{\alpha\beta}g^{\gamma\lambda}\Gamma^{\kappa}_{\alpha\beta}f_{,\gamma} \\[4mm] - \left(\dfrac{F'}{F}\cdot(1-n)g^{\alpha\beta}\dfrac{g^{\gamma\sigma}}{2}GX^{\kappa\lambda\rho}_{\alpha\beta\sigma}f_{,\gamma}\right)_{,\rho} \end{array}\right)$$

$$-\left[\frac{\sqrt{-g}\cdot F^p}{F}\right]_{,\rho}\cdot\left(\frac{F'}{F}\cdot(1-n)g^{\alpha\beta}\frac{g^{\gamma\sigma}}{2}GX^{\kappa\lambda\rho}_{\alpha\beta\sigma}f_{,\gamma}\right),\qquad(977)$$

which then gives:

$$\Rightarrow\quad\frac{\sqrt{-g}\cdot F^p}{F}\cdot\left(\begin{array}{c}R^{\kappa\lambda}-\dfrac{R}{2}g^{\kappa\lambda}\\[2mm]+\dfrac{f_{,\alpha}f_{,\beta}\left(\dfrac{g^{\kappa\lambda}g^{\alpha\beta}}{2}-\dfrac{g^{\kappa\alpha}g^{\lambda\beta}+g^{\lambda\alpha}g^{\kappa\beta}}{2}\right)(1-n)}{4F^2}\left(\begin{array}{c}(F')^2(n-6)\\+4FF''\end{array}\right)\\[4mm]+\dfrac{F'\cdot(1-n)}{F}\left(\dfrac{g^{\kappa\lambda}g^{\alpha\beta}}{2}-\dfrac{g^{\kappa\alpha}g^{\lambda\beta}+g^{\lambda\alpha}g^{\kappa\beta}}{2}\right)\left(f_{,\alpha\beta}-\Gamma^{\gamma}_{\alpha\beta}f_{,\gamma}\right)\\[4mm]+\dfrac{F'}{F}\cdot(1-n)g^{\alpha\beta}g^{\gamma\lambda}\Gamma^{\kappa}_{\alpha\beta}f_{,\gamma}\\[4mm]-\dfrac{F'}{F}\cdot(1-n)g^{\alpha\beta}\dfrac{g^{\gamma\sigma}}{2}GX^{\kappa\lambda\rho}_{\alpha\beta\sigma}f_{,\gamma\rho}\\[4mm]-\dfrac{F'}{F}\cdot(1-n)\left(g^{\alpha\beta}\dfrac{g^{\gamma\sigma}}{2}\right)_{,\rho}GX^{\kappa\lambda\rho}_{\alpha\beta\sigma}f_{,\gamma}\\[4mm]-\dfrac{F''}{F}\cdot f_{,\rho}\cdot(1-n)g^{\alpha\beta}\dfrac{g^{\gamma\sigma}}{2}GX^{\kappa\lambda\rho}_{\alpha\beta\sigma}f_{,\gamma}\\[4mm]+\dfrac{(F')^2}{F^2}\cdot f_{,\rho}\cdot(1-n)g^{\alpha\beta}\dfrac{g^{\gamma\sigma}}{2}GX^{\kappa\lambda\rho}_{\alpha\beta\sigma}f_{,\gamma}\end{array}\right)$$

$$+\frac{\sqrt{-g}\cdot F^p}{F}\cdot\left(\frac{g^{\xi\zeta}g_{\xi\zeta,\rho}}{2}-\frac{F'}{F}\cdot f_{,\rho}\left(\frac{p}{2}-1\right)\right)\cdot\left(\frac{F'}{F}\cdot(1-n)g^{\alpha\beta}\frac{g^{\gamma\sigma}}{2}GX^{\kappa\lambda\rho}_{\alpha\beta\sigma}f_{,\gamma}\right).$$

$$(978)$$

And after further simplification and equalizing some of the dummy indices:

$$0 = \left\{ \begin{array}{l} R^{\kappa\lambda} - \dfrac{R}{2} g^{\kappa\lambda} \\[2mm] + \left(g^{\kappa\lambda} g^{\alpha\beta} - g^{\kappa\alpha} g^{\lambda\beta} - g^{\lambda\alpha} g^{\kappa\beta} \right) (1-n) \times \\[2mm] \left[\begin{array}{l} \dfrac{f_{,\alpha} f_{,\beta}}{2 \cdot F^2} \left(\dfrac{1}{4} \left((F')^2 (n-6) + 4FF'' \right) + FF'' - (F')^2 + (F')^2 \left(\dfrac{p}{2} - 1 \right) \right) \\[3mm] + f_{,\alpha\beta} \left(\dfrac{F'}{2 \cdot F} + \dfrac{F'}{2 \cdot F} \right) \end{array} \right] \\[6mm] - \dfrac{F' \cdot (1-n)}{2 \cdot F} \left(g^{\kappa\lambda} g^{\alpha\beta} - g^{\kappa\alpha} g^{\lambda\beta} - g^{\lambda\alpha} g^{\kappa\beta} \right) \Gamma^{\gamma}_{\alpha\beta} f_{,\gamma} \\[3mm] + \dfrac{F'}{F} \cdot (1-n) g^{\alpha\beta} g^{\gamma\lambda} \Gamma^{\kappa}_{\alpha\beta} f_{,\gamma} \\[3mm] + \dfrac{F'}{F} \cdot \dfrac{(1-n)}{2} \left(g^{\kappa\lambda} g^{\alpha\beta} - g^{\kappa\alpha} g^{\lambda\beta} - g^{\lambda\alpha} g^{\kappa\beta} \right)_{,\beta} f_{,\alpha} \\[3mm] + \dfrac{g^{\xi\zeta} g_{\xi\zeta,\beta}}{2} \cdot \left(\dfrac{F'}{F} \cdot \dfrac{(1-n)}{2} \left(g^{\kappa\lambda} g^{\alpha\beta} - g^{\kappa\alpha} g^{\lambda\beta} - g^{\lambda\alpha} g^{\kappa\beta} \right) f_{,\alpha} \right), \end{array} \right. \tag{979}$$

we obtain:

$$0 = \left\{ \begin{array}{l} R^{\kappa\lambda} - \dfrac{R}{2} g^{\kappa\lambda} \\[2mm] + \left(g^{\kappa\lambda} g^{\alpha\beta} - g^{\kappa\alpha} g^{\lambda\beta} - g^{\lambda\alpha} g^{\kappa\beta} \right) (1-n) \\[2mm] \cdot \left[\dfrac{f_{,\alpha} f_{,\beta}}{F^2} \left((F')^2 \dfrac{(2 \cdot p + n - 14)}{8} + FF'' \right) + f_{,\alpha\beta} \dfrac{F'}{F} \right] \\[4mm] - \dfrac{F' \cdot (1-n)}{2 \cdot F} \left(g^{\kappa\lambda} g^{\alpha\beta} - g^{\kappa\alpha} g^{\lambda\beta} - g^{\lambda\alpha} g^{\kappa\beta} \right) \Gamma^{\gamma}_{\alpha\beta} f_{,\gamma} \\[3mm] + \dfrac{F'}{F} \cdot (1-n) g^{\alpha\beta} g^{\gamma\lambda} \Gamma^{\kappa}_{\alpha\beta} f_{,\gamma} \\[3mm] + \dfrac{F'}{F} \cdot \dfrac{(1-n)}{2} \left(g^{\kappa\lambda} g^{\alpha\beta} - g^{\kappa\alpha} g^{\lambda\beta} - g^{\lambda\alpha} g^{\kappa\beta} \right)_{,\beta} f_{,\alpha} \\[3mm] + \dfrac{g^{\xi\zeta} g_{\xi\zeta,\beta}}{2} \cdot \left(\dfrac{F'}{F} \cdot \dfrac{(1-n)}{2} \left(g^{\kappa\lambda} g^{\alpha\beta} - g^{\kappa\alpha} g^{\lambda\beta} - g^{\lambda\alpha} g^{\kappa\beta} \right) f_{,\alpha} \right). \end{array} \right. \tag{980}$$

Now we have to incorporate the variation of the Ricci tensor, respectively its new form, which is now connected with the scale function $F[f]$.

8.13.3 Variation of the Scaled Ricci Tensor Term Together with the New Factor F^q

It becomes immediately clear from the structure of the Ricci tensor integral term from Eq. (968):

$$\int_V d^n x \sqrt{-g} \cdot \frac{F^{p/2}}{F} \cdot g^{\alpha\beta} \delta_{g_{\alpha\beta}} R_{\alpha\beta} = \int_V d^n x \sqrt{-g} \cdot F^{p/2-1} \cdot g^{\alpha\beta} \delta R_{\alpha\beta} \qquad (981)$$

that this cannot give the same surface term (and thus, vanishing [7]) result as the classical form, reading:

$$\int_V d^n x \sqrt{-g} \cdot g^{\alpha\beta} \delta_{g_{\alpha\beta}} R_{\alpha\beta} = \int_V d^n x \sqrt{-g} \cdot g^{\alpha\beta} \delta R_{\alpha\beta}. \qquad (982)$$

The reason is that while Eq. (982) can be made a complete divergence, reading:

$$\int_V d^n x \sqrt{-g} \cdot g^{\alpha\beta} \delta R_{\alpha\beta} = \int_V d^n x \left(\sqrt{-g} \cdot g^{\alpha\beta} \delta \Gamma^{\gamma}_{\alpha\beta} \right)_{,\gamma} - \int_V d^n x \left(\sqrt{-g} \cdot g^{\alpha\beta} \delta \Gamma^{\sigma}_{\beta\sigma} \right)_{,\alpha},$$

$$(983)$$

we will have the following from Eq. (981):

$$\int_V d^n x \sqrt{-g} \cdot F^{p/2-1} \cdot g^{\alpha\beta} \delta R_{\alpha\beta}$$

$$= \int_V d^n x \cdot F^{p/2-1} \cdot \left(\sqrt{-g} \cdot g^{\alpha\beta} \delta \Gamma^{\gamma}_{\alpha\beta} \right)_{,\gamma} - \int_V d^n x \cdot F^{p/2-1} \cdot \left(\sqrt{-g} \cdot g^{\alpha\beta} \delta \Gamma^{\sigma}_{\beta\sigma} \right)_{,\alpha}.$$

$$(984)$$

Only after integration by parts:

$$\int_V d^n x \sqrt{-g} \cdot F^{p/2-1} \cdot g^{\alpha\beta} \delta R_{\alpha\beta}$$

$$= \int_V d^n x \cdot F^{p/2-1} \cdot \left(\sqrt{-g} \cdot g^{\alpha\beta} \delta \Gamma^{\gamma}_{\alpha\beta} \right)_{,\gamma} - \int_V d^n x \cdot F^{p/2-1} \cdot \left(\sqrt{-g} \cdot g^{\alpha\beta} \delta \Gamma^{\sigma}_{\beta\sigma} \right)_{,\alpha}$$

$$= \int_V d^n x \cdot F^{p/2-1}_{,\gamma} \cdot \sqrt{-g} \cdot g^{\alpha\beta} \delta \Gamma^{\gamma}_{\alpha\beta} - \int_V d^n x \cdot \left(F^{p/2-1} \cdot \sqrt{-g} \cdot g^{\alpha\beta} \delta \Gamma^{\gamma}_{\alpha\beta} \right)_{,\gamma}$$

$$- \int_V d^n x \cdot F^{p/2-1}_{,\alpha} \cdot \sqrt{-g} \cdot g^{\alpha\beta} \delta \Gamma^{\sigma}_{\beta\sigma} + \int_V d^n x \cdot \left(F^{p/2-1} \cdot \sqrt{-g} \cdot g^{\alpha\beta} \delta \Gamma^{\sigma}_{\beta\sigma} \right)_{,\alpha}$$

$$= \int_V d^n x \cdot F^{p/2-1}{}_{,\gamma} \cdot \sqrt{-g} \cdot g^{\alpha\beta} \delta\Gamma^{\gamma}_{\alpha\beta} - \int_{\partial V} d^n x \cdot n_\gamma \left(F^{p/2-1} \cdot \sqrt{-g} \cdot g^{\alpha\beta} \delta\Gamma^{\gamma}_{\alpha\beta} \right)$$

$$- \int_V d^n x \cdot F^{p/2-1}{}_{,\alpha} \cdot \sqrt{-g} \cdot g^{\alpha\beta} \delta\Gamma^{\sigma}_{\beta\sigma} + \int_{\partial V} d^n x \cdot n_\alpha \left(F^{p/2-1} \cdot \sqrt{-g} \cdot g^{\alpha\beta} \delta\Gamma^{\sigma}_{\beta\sigma} \right),$$

$$(985)$$

we have the desired surface integrals, which we can assume to vanish (!) using Hilbert's assumption of a boundary free space-time (caution here (!)). The remains are the two integrals:

$$\int_V d^n x \sqrt{-g} \cdot F^{p/2-1} \cdot g^{\alpha\beta} \delta R_{\alpha\beta}$$

$$= \int_V d^n x \cdot F^{p/2-1}{}_{,\gamma} \cdot \sqrt{-g} \cdot g^{\alpha\beta} \delta\Gamma^{\gamma}_{\alpha\beta} - \int_V d^n x \cdot F^{p/2-1}{}_{,\alpha} \cdot \sqrt{-g} \cdot g^{\alpha\beta} \delta\Gamma^{\sigma}_{\beta\sigma}, \quad (986)$$

where we once again find the variated Levi–Civita connections $\delta\Gamma^{\gamma}_{\alpha\beta}, \delta\Gamma^{\sigma}_{\beta\sigma}$, which this time result in:

$$\int_V d^n x \cdot F^{p/2-1}{}_{,\gamma} \cdot \sqrt{-g} \cdot g^{\alpha\beta} \delta\Gamma^{\gamma}_{\alpha\beta} - \int_V d^n x \cdot F^{p/2-1}{}_{,\alpha} \cdot \sqrt{-g} \cdot g^{\alpha\beta} \delta\Gamma^{\sigma}_{\beta\sigma}$$

$$= \int_V d^n x \cdot F^{p/2-1}{}_{,\gamma} \cdot \sqrt{-g} \cdot g^{\alpha\beta} \left(\frac{g^{\gamma\sigma}}{2} \left(\delta g_{\sigma\alpha,\beta} + \delta g_{\sigma\beta,\alpha} - \delta g_{\alpha\beta,\sigma} \right) - g^{\gamma\lambda} \Gamma^{\kappa}_{\alpha\beta} \delta g_{\kappa\lambda} \right)$$

$$- \int_V d^n x \cdot F^{p/2-1}{}_{,\alpha} \cdot \sqrt{-g} \cdot g^{\alpha\beta} \frac{1}{2} \cdot g^{\sigma\lambda} \delta g_{\sigma\lambda;\beta}$$

$$= \int_V d^n x \cdot F^{p/2-1}{}_{,\gamma} \cdot \sqrt{-g} \cdot g^{\alpha\beta} \left(\frac{g^{\gamma\sigma}}{2} G X^{\kappa\lambda\rho}_{\alpha\beta\sigma} \delta g_{\kappa\lambda,\rho} - g^{\gamma\lambda} \Gamma^{\kappa}_{\alpha\beta} \delta g_{\kappa\lambda} \right)$$

$$- \int_V d^n x \cdot F^{p/2-1}{}_{,\alpha} \cdot \sqrt{-g} \cdot g^{\alpha\beta} \frac{1}{2} \cdot g^{\kappa\lambda} \delta g_{\kappa\lambda;\beta}$$

$$= \int_V d^n x \cdot F^{p/2-1}{}_{,\gamma} \cdot \sqrt{-g} \cdot g^{\alpha\beta} \left(\frac{g^{\gamma\sigma}}{2} G X^{\kappa\lambda\rho}_{\alpha\beta\sigma} \delta g_{\kappa\lambda,\rho} - g^{\gamma\lambda} \Gamma^{\kappa}_{\alpha\beta} \delta g_{\kappa\lambda} \right)$$

$$- \int_V d^n x \cdot F^{p/2-1}{}_{,\alpha} \cdot \sqrt{-g} \cdot g^{\alpha\beta} \frac{1}{2} \cdot g^{\kappa\lambda} \left(\delta g_{\kappa\lambda,\beta} - \Gamma^{\xi}_{\beta\kappa} \delta g_{\xi\lambda} - \Gamma^{\xi}_{\beta\lambda} \delta g_{\xi\kappa} \right)$$

$$= \int_V d^n x \cdot F^{p/2-1}_{,\gamma} \cdot \sqrt{-g} \cdot g^{\alpha\beta} \left(\frac{g^{\gamma\sigma}}{2} GX^{\kappa\lambda\rho}_{\alpha\beta\sigma} \delta g_{\kappa\lambda,\rho} - g^{\gamma\lambda} \Gamma^{\kappa}_{\alpha\beta} \delta g_{\kappa\lambda} \right)$$

$$- \int_V d^n x \cdot F^{n/2-1}_{,\alpha} \cdot \sqrt{-g} \cdot g^{\alpha\beta} \frac{1}{2} \left(g^{\kappa\lambda} \delta g_{\kappa\lambda,\beta} - g^{\kappa\lambda} \Gamma^{\xi}_{\beta\kappa} \delta g_{\xi\lambda} - g^{\kappa\lambda} \Gamma^{\xi}_{\beta\lambda} \delta g_{\xi\kappa} \right)$$

$$= \int_V d^n x \cdot F^{n/2-1}_{,\gamma} \cdot \sqrt{-g} \cdot g^{\alpha\beta} \left(\frac{g^{\gamma\sigma}}{2} GX^{\kappa\lambda\rho}_{\alpha\beta\sigma} \delta g_{\kappa\lambda,\rho} - g^{\gamma\lambda} \Gamma^{\kappa}_{\alpha\beta} \delta g_{\kappa\lambda} \right)$$

$$- \int_V d^n x \cdot F^{p/2-1}_{,\alpha} \cdot \sqrt{-g} \cdot g^{\alpha\beta} \frac{1}{2} \left(g^{\kappa\lambda} \delta g_{\kappa\lambda,\beta} - g^{\xi\lambda} \Gamma^{\kappa}_{\beta\xi} \delta g_{\kappa\lambda} - g^{\kappa\xi} \Gamma^{\lambda}_{\beta\xi} \delta g_{\lambda\kappa} \right),$$

$$(987)$$

$$= \int_V d^n x \cdot F^{p/2-1}_{,\gamma} \cdot \sqrt{-g} \cdot g^{\alpha\beta} \left(\frac{g^{\gamma\sigma}}{2} GX^{\kappa\lambda\rho}_{\alpha\beta\sigma} \delta g_{\kappa\lambda,\rho} - g^{\gamma\lambda} \Gamma^{\kappa}_{\alpha\beta} \delta g_{\kappa\lambda} \right)$$

$$- \int_V d^n x \cdot F^{p/2-1}_{,\alpha} \cdot \sqrt{-g} \cdot g^{\alpha\beta} \frac{1}{2} \left(g^{\kappa\lambda} \delta g_{\kappa\lambda,\beta} - g^{\xi\lambda} \Gamma^{\kappa}_{\beta\xi} \delta g_{\kappa\lambda} - g^{\kappa\xi} \Gamma^{\lambda}_{\beta\xi} \delta g_{\lambda\kappa} \right)$$

$$= \int_V d^n x \cdot \sqrt{-g} \cdot g^{\alpha\beta} \left(F^{p/2-1}_{,\gamma} \cdot \frac{g^{\gamma\sigma}}{2} GX^{\kappa\lambda\rho}_{\alpha\beta\sigma} \delta g_{\kappa\lambda,\rho} - F^{p/2-1}_{,\gamma} \cdot g^{\gamma\lambda} \Gamma^{\kappa}_{\alpha\beta} \delta g_{\kappa\lambda} \right)$$

$$- \int_V d^n x \cdot \sqrt{-g} \cdot g^{\alpha\beta} \frac{1}{2} \left(F^{p/2-1}_{,\alpha} \cdot g^{\kappa\lambda} \delta g_{\kappa\lambda,\beta} - F^{p/2-1}_{,\alpha} \cdot g^{\xi\lambda} \Gamma^{\kappa}_{\beta\xi} \delta g_{\kappa\lambda} - F^{p/2-1}_{,\alpha} \cdot g^{\kappa\xi} \Gamma^{\lambda}_{\beta\xi} \delta g_{\lambda\kappa} \right)$$

$$= \int_V d^n x \cdot \sqrt{-g} \cdot g^{\alpha\beta} F^{p/2-1}_{,\gamma} \cdot \frac{g^{\gamma\sigma}}{2} GX^{\kappa\lambda\rho}_{\alpha\beta\sigma} \delta g_{\kappa\lambda,\rho} - \int_V d^n x \cdot \sqrt{-g} \cdot g^{\alpha\beta} \frac{1}{2} F^{p/2-1}_{,\alpha} \cdot g^{\kappa\lambda} \delta g_{\kappa.}$$

$$\int_V d^n x \cdot \sqrt{-g} \cdot g^{\alpha\beta} F^{p/2-1}_{,\gamma} \cdot g^{\gamma\lambda} \Gamma^{\kappa}_{\alpha\beta} \delta g_{\kappa\lambda}$$

$$+ \int_V d^n x \cdot \sqrt{-g} \cdot g^{\alpha\beta} \frac{1}{2} \left(F^{p/2-1}_{,\alpha} \cdot g^{\xi\lambda} \Gamma^{\kappa}_{\beta\xi} + F^{p/2-1}_{,\alpha} \cdot g^{\kappa\xi} \Gamma^{\lambda}_{\beta\xi} \right) \delta g_{\kappa\lambda}. \quad (988)$$

Now we consider the two integrals:

$$\int_V d^n x \cdot \sqrt{-g} \cdot g^{\alpha\beta} F^{p/2-1}_{,\gamma} \cdot \frac{g^{\gamma\sigma}}{2} GX^{\kappa\lambda\rho}_{\alpha\beta\sigma} \delta g_{\kappa\lambda,\rho} - \int_V d^n x \cdot \sqrt{-g} \cdot g^{\alpha\beta} \frac{1}{2} F^{p/2-1}_{,\alpha} \cdot g^{\kappa\lambda} \delta g_{\kappa\lambda,\beta}$$

$$= \int_V d^n x \cdot \left(\sqrt{-g} \cdot g^{\alpha\beta} F^{p/2-1}{}_{,\gamma} \cdot \frac{g^{\gamma\sigma}}{2} GX^{\kappa\lambda\rho}_{\alpha\beta\sigma} \right)_{,\rho} \delta g_{\kappa\lambda}$$

$$- \int_V d^n x \cdot \left(\sqrt{-g} \cdot g^{\alpha\beta} F^{p/2-1}{}_{,\gamma} \cdot \frac{g^{\gamma\sigma}}{2} GX^{\kappa\lambda\rho}_{\alpha\beta\sigma} \delta g_{\kappa\lambda} \right)_{,\rho}$$

$$- \int_V d^n x \cdot \left(\sqrt{-g} \cdot g^{\alpha\beta} \frac{1}{2} F^{p/2-1}{}_{,\alpha} \cdot g^{\kappa\lambda} \right)_{,\beta} \delta g_{\kappa\lambda}$$

$$+ \int_V d^n x \cdot \left(\sqrt{-g} \cdot g^{\alpha\beta} \frac{1}{2} F^{p/2-1}{}_{,\alpha} \cdot g^{\kappa\lambda} \delta g_{\kappa\lambda} \right)_{,\beta}$$

$$= \int_V d^n x \cdot \left(\sqrt{-g} \cdot g^{\alpha\beta} F^{p/2-1}{}_{,\gamma} \cdot \frac{g^{\gamma\sigma}}{2} GX^{\kappa\lambda\rho}_{\alpha\beta\sigma} \right)_{,\rho} \delta g_{\kappa\lambda}$$

$$- \overbrace{\int_{\partial V} d^n x \cdot n_\rho \sqrt{-g} \cdot g^{\alpha\beta} F^{p/2-1}{}_{,\gamma} \cdot \frac{g^{\gamma\sigma}}{2} GX^{\kappa\lambda\rho}_{\alpha\beta\sigma} \delta g_{\kappa\lambda}}^{=0}$$

$$- \int_V d^n x \cdot \left(\sqrt{-g} \cdot g^{\alpha\beta} \frac{1}{2} F^{p/2-1}{}_{,\alpha} \cdot g^{\kappa\lambda} \right)_{,\beta} \delta g_{\kappa\lambda}$$

$$+ \overbrace{\int_{\partial V} d^n x \cdot n_\beta \sqrt{-g} \cdot g^{\alpha\beta} \frac{1}{2} F^{p/2-1}{}_{,\alpha} \cdot g^{\kappa\lambda} \delta g_{\kappa\lambda}}^{=0}. \tag{989}$$

This leads us to:

$$\int_V d^n x \cdot \sqrt{-g} \cdot g^{\alpha\beta} F^{p/2-1}{}_{,\gamma} \cdot \frac{g^{\gamma\sigma}}{2} GX^{\kappa\lambda\rho}_{\alpha\beta\sigma} \delta g_{\kappa\lambda,\rho} - \int_V d^n x \cdot \sqrt{-g} \cdot g^{\alpha\beta} \frac{1}{2} F^{p/2-1}{}_{,\alpha} \cdot g^{\kappa\lambda} \delta g_{\kappa\lambda,\beta}$$

$$= \int_V d^n x \cdot \left[\left(\sqrt{-g} \cdot g^{\alpha\beta} F^{p/2-1}{}_{,\gamma} \cdot \frac{g^{\gamma\sigma}}{2} GX^{\kappa\lambda\rho}_{\alpha\beta\sigma} \right)_{,\rho} - \left(\sqrt{-g} \cdot g^{\alpha\beta} \frac{1}{2} F^{p/2-1}{}_{,\alpha} \cdot g^{\kappa\lambda} \right)_{,\beta} \right] \delta g_{\kappa\lambda}. \tag{990}$$

Thus, in total we have from Eq. (986):

$$\int_V d^n x \sqrt{-g} \cdot F^{p/2-1} \cdot g^{\alpha\beta} \delta R_{\alpha\beta}$$

$$
\begin{aligned}
= \int_V d^n x \cdot \left(\begin{array}{c} \sqrt{-g} \cdot g^{\alpha\beta} \left(\dfrac{F^{n/2-1}_{\ ,\alpha} \cdot g^{\xi\lambda} \Gamma^\kappa_{\beta\xi} + F^{n/2-1}_{\ ,\alpha} \cdot g^{\kappa\xi} \Gamma^\lambda_{\beta\xi}}{2} + F^{n/2-1}_{\ ,\gamma} \cdot g^{\gamma\lambda} \Gamma^\kappa_{\alpha\beta} \right) \\ + \left(\sqrt{-g} \cdot g^{\alpha\beta} F^{p/2-1}_{\ ,\gamma} \cdot \dfrac{g^{\gamma\sigma}}{2} GX^{\kappa\lambda\rho}_{\alpha\beta\sigma} \right)_{,\rho} - \left(\sqrt{-g} \cdot g^{\alpha\beta} \dfrac{1}{2} F^{p/2-1}_{\ ,\alpha} \cdot g^{\kappa\lambda} \right)_{,\beta} \end{array} \right) \delta g_{\kappa\lambda}
\end{aligned}
$$

$$
\begin{aligned}
= \int_V d^n x \cdot \left(\begin{array}{c} \sqrt{-g} \cdot g^{\alpha\beta} \left(\dfrac{F^{p/2-1}_{\ ,\alpha} \cdot g^{\xi\lambda} \Gamma^\kappa_{\beta\xi} + F^{p/2-1}_{\ ,\alpha} \cdot g^{\kappa\xi} \Gamma^\lambda_{\beta\xi}}{2} + F^{p/2-1}_{\ ,\gamma} \cdot g^{\gamma\lambda} \Gamma^\kappa_{\alpha\beta} \right) \\ + \left(\sqrt{-g} \cdot g^{\alpha\beta} F^{p/2-1}_{\ ,\gamma} \dfrac{\left(g^{\kappa\rho} g^{\gamma\lambda} + g^{\rho\lambda} g^{\gamma\kappa} - g^{\kappa\lambda} g^{\gamma\rho} \right)}{2} \right)_{,\rho} \\ - \left(\sqrt{-g} \cdot g^{\alpha\beta} \dfrac{1}{2} F^{p/2-1}_{\ ,\alpha} \cdot g^{\kappa\lambda} \right)_{,\beta} \end{array} \right) \delta g_{\kappa\lambda}.
\end{aligned}
$$

$$(991)$$

8.13.4 Back to the Total Variation of Our Scaled Metric Ricci Scalar with Action Factor F^q

Now incorporating the results for the variated Ricci tensor from section "8.13.3 Variation of the Scaled Ricci Tensor Term Together with the New Factor F^q" gives us:

$$
0 = \frac{\sqrt{-g} \cdot F^p}{F} \cdot \left(\begin{array}{c} R^{\kappa\lambda} - \dfrac{R}{2} g^{\kappa\lambda} \\[2mm] + \left(g^{\kappa\lambda} g^{\alpha\beta} - g^{\kappa\alpha} g^{\lambda\beta} - g^{\lambda\alpha} g^{\kappa\beta} \right)(1-n) \\[2mm] \cdot \left[\dfrac{f_{,\alpha} f_{,\beta}}{F^2} \left(FF'' + \left(F' \right)^2 \cdot \dfrac{2 \cdot p + n - 14}{8} \right) + f_{,\alpha\beta} \dfrac{F'}{F} \right] \\[2mm] - \dfrac{F' \cdot (1-n)}{2 \cdot F} \left(g^{\kappa\lambda} g^{\alpha\beta} - g^{\kappa\alpha} g^{\lambda\beta} - g^{\lambda\alpha} g^{\kappa\beta} \right) \Gamma^\gamma_{\alpha\beta} f_{,\gamma} \\[2mm] + \dfrac{F'}{F} \cdot (1-n) g^{\alpha\beta} g^{\gamma\lambda} \Gamma^\kappa_{\alpha\beta} f_{,\gamma} \\[2mm] + \dfrac{F'}{F} \cdot \dfrac{(1-n)}{2} \left(g^{\kappa\lambda} g^{\alpha\beta} - g^{\kappa\alpha} g^{\lambda\beta} - g^{\lambda\alpha} g^{\kappa\beta} \right)_{,\beta} f_{,\alpha} \\[2mm] + \dfrac{g^{\xi\zeta} g_{\xi\zeta,\beta}}{2} \cdot \left(\dfrac{F'}{F} \cdot \dfrac{(1-n)}{2} \left(g^{\kappa\lambda} g^{\alpha\beta} - g^{\kappa\alpha} g^{\lambda\beta} - g^{\lambda\alpha} g^{\kappa\beta} \right) f_{,\alpha} \right) \end{array} \right)
$$

$$+\sqrt{-g}\cdot g^{\alpha\beta}\left(\frac{F^{p/2-1}{}_{,\alpha}\cdot g^{\xi\lambda}\Gamma^{\kappa}_{\beta\xi}+F^{p/2-1}{}_{,\alpha}\cdot g^{\kappa\xi}\Gamma^{\lambda}_{\beta\xi}}{2}+F^{p/2-1}{}_{,\gamma}\cdot g^{\gamma\lambda}\Gamma^{\kappa}_{\alpha\beta}\right)$$

$$+\left(\sqrt{-g}\cdot g^{\alpha\beta}F^{p/2-1}{}_{,\gamma}\cdot\frac{g^{\gamma\sigma}}{2}GX^{\kappa\lambda\rho}_{\alpha\beta\sigma}\right)_{,\rho}-\left(\sqrt{-g}\cdot\frac{g^{\alpha\beta}}{2}F^{p/2-1}{}_{,\alpha}\cdot g^{\kappa\lambda}\right)_{,\beta}. \qquad (992)$$

Inserting the extension for $GX^{\kappa\lambda\rho}_{\alpha\beta\sigma}$ yields:

$$0=\frac{\sqrt{-g\cdot F^{p}}}{F}\cdot\left(\begin{array}{c} R^{\kappa\lambda}-\dfrac{R}{2}g^{\kappa\lambda} \\[2mm] +\left(g^{\kappa\lambda}g^{\alpha\beta}-g^{\kappa\alpha}g^{\lambda\beta}-g^{\lambda\alpha}g^{\kappa\beta}\right)(1-n) \\[2mm] \cdot\left[\dfrac{f_{,\alpha}f_{,\beta}}{F^{2}}\left(FF''+(F')^{2}\cdot\dfrac{2\cdot p+n-14}{8}\right)+f_{,\alpha\beta}\dfrac{F'}{F}\right] \\[2mm] -\dfrac{F'\cdot(1-n)}{2\cdot F}\left(g^{\kappa\lambda}g^{\alpha\beta}-g^{\kappa\alpha}g^{\lambda\beta}-g^{\lambda\alpha}g^{\kappa\beta}\right)\Gamma^{\gamma}_{\alpha\beta}f_{,\gamma} \\[2mm] +\dfrac{F'}{F}\cdot(1-n)g^{\alpha\beta}g^{\gamma\lambda}\Gamma^{\kappa}_{\alpha\beta}f_{,\gamma} \\[2mm] +\dfrac{F'}{F}\cdot\dfrac{(1-n)}{2}\left(g^{\kappa\lambda}g^{\alpha\beta}-g^{\kappa\alpha}g^{\lambda\beta}-g^{\lambda\alpha}g^{\kappa\beta}\right)_{,\beta}f_{,\alpha} \\[2mm] +\dfrac{g^{\xi\zeta}g_{\xi\zeta,\beta}}{2}\cdot\left(\dfrac{F'}{F}\cdot\dfrac{(1-n)}{2}\left(g^{\kappa\lambda}g^{\alpha\beta}-g^{\kappa\alpha}g^{\lambda\beta}-g^{\lambda\alpha}g^{\kappa\beta}\right)f_{,\alpha}\right) \end{array}\right)$$

$$+\sqrt{-g}\cdot g^{\alpha\beta}\left(\frac{F^{p/2-1}{}_{,\alpha}\cdot g^{\xi\lambda}\Gamma^{\kappa}_{\beta\xi}+F^{p/2-1}{}_{,\alpha}\cdot g^{\kappa\xi}\Gamma^{\lambda}_{\beta\xi}}{2}+F^{p/2-1}{}_{,\gamma}\cdot g^{\gamma\lambda}\Gamma^{\kappa}_{\alpha\beta}\right)$$

$$+\left(\sqrt{-g}\cdot F^{p/2-1}{}_{,\gamma}\cdot\frac{\left(g^{\kappa\rho}g^{\gamma\lambda}+g^{\rho\lambda}g^{\gamma\kappa}-g^{\kappa\lambda}g^{\gamma\rho}\right)}{2}\right)_{,\rho}-\left(\sqrt{-g}\cdot\frac{g^{\alpha\beta}}{2}F^{p/2-1}{}_{,\alpha}\cdot g^{\kappa\lambda}\right)_{,\beta}.$$

$$(993)$$

From there we can evaluate:

$$
0 = \frac{\sqrt{-g} \cdot F^{p}}{F} \cdot \left\{ \left[R^{\kappa\lambda} - \frac{R}{2} g^{\kappa\lambda} + \left(g^{\kappa\lambda} g^{\alpha\beta} - g^{\kappa\alpha} g^{\lambda\beta} - g^{\lambda\alpha} g^{\kappa\beta} \right) (1-n) \right) \cdot \left[\frac{f_{,\alpha} f_{,\beta}}{F^{2}} \left(FF'' + (F')^{2} \cdot \frac{2 \cdot p + n - 14}{8} \right) + f_{,\alpha\beta} \frac{F'}{F} \right] \right. \right.
$$

$$
- \frac{F' \cdot (1-n)}{2 \cdot F} \left(g^{\kappa\lambda} g^{\alpha\beta} - g^{\kappa\alpha} g^{\lambda\beta} - g^{\lambda\alpha} g^{\kappa\beta} \right) \Gamma^{\gamma}_{\alpha\beta} f_{,\gamma} + \frac{F'}{F} \cdot (1-n) g^{\alpha\beta} g^{\gamma\lambda} \Gamma^{\kappa}_{\alpha\beta} f_{,\gamma}
$$

$$
+ \frac{F'}{F} \cdot \frac{(1-n)}{2} \left(g^{\kappa\lambda} g^{\alpha\beta} - g^{\kappa\alpha} g^{\lambda\beta} - g^{\lambda\alpha} g^{\kappa\beta} \right)_{,\beta} f_{,\alpha} + \frac{g^{\xi\zeta} g_{\xi\zeta,\beta}}{2} \cdot \frac{F'}{F} \cdot \frac{(1-n)}{2} \left(g^{\kappa\lambda} g^{\alpha\beta} - g^{\kappa\alpha} g^{\lambda\beta} - g^{\lambda\alpha} g^{\kappa\beta} \right) f_{,\alpha}
$$

$$
+ \sqrt{-g} \cdot g^{\alpha\beta} \left(\frac{p}{2} - 1 \right) F^{p/2-2} F' \left(\frac{f_{,\alpha} \cdot g^{\xi\lambda} \Gamma^{\kappa}_{\beta\xi} + f_{,\alpha} \cdot g^{\kappa\xi} \Gamma^{\lambda}_{\beta\xi}}{2} + f_{,\gamma} \cdot g^{\gamma\lambda} \Gamma^{\kappa}_{\alpha\beta} \right)
$$

$$
+ \sqrt{-g} \cdot F^{p/2-2} F' f_{,\gamma} \cdot \frac{\left(g^{\kappa\rho} g^{\gamma\lambda} + g^{\rho\lambda} g^{\gamma\kappa} - g^{\kappa\lambda} g^{\gamma\rho} \right)_{,\rho}}{2} + \sqrt{-g} \cdot F^{p/2-2} F' f_{,\gamma\rho} \cdot \frac{\left(g^{\kappa\rho} g^{\gamma\lambda} + g^{\rho\lambda} g^{\gamma\kappa} - g^{\kappa\lambda} g^{\gamma\rho} \right)}{2}
$$

$$
+ \sqrt{-g} \cdot F^{p/2-2} F'' f_{,\rho} f_{,\gamma} \cdot \frac{\left(g^{\kappa\rho} g^{\gamma\lambda} + g^{\rho\lambda} g^{\gamma\kappa} - g^{\kappa\lambda} g^{\gamma\rho} \right)}{2} + \sqrt{-g} \cdot \left(\frac{p}{2} - 2 \right) F^{p/2-3} f_{,\rho} F'^{2} f_{,\gamma} \cdot \frac{\left(g^{\kappa\rho} g^{\gamma\lambda} + g^{\rho\lambda} g^{\gamma\kappa} - g^{\kappa\lambda} g^{\gamma\rho} \right)}{2}
$$

$$
+ \sqrt{-g} \cdot \frac{g^{\xi\zeta} g_{\xi\zeta,\rho}}{2} \cdot F^{p/2-2} F' f_{,\gamma} \cdot \frac{\left(g^{\kappa\rho} g^{\gamma\lambda} + g^{\rho\lambda} g^{\gamma\kappa} - g^{\kappa\lambda} g^{\gamma\rho} \right)}{2}
$$

$$
- \left[\sqrt{-g} \cdot \frac{g^{\alpha\beta}}{2} \cdot F^{p/2-2} F'' f_{,\beta} f_{,\alpha} \cdot g^{\kappa\lambda} + \sqrt{-g} \cdot \frac{g^{\alpha\beta}}{2} \cdot F^{p/2-2} F' f_{,\alpha} \cdot g^{\kappa\lambda}_{,\beta} + \sqrt{-g} \cdot \frac{g^{\alpha\beta}}{2} \cdot F^{p/2-2} F' f_{,\alpha\beta} \cdot g^{\kappa\lambda} \right.
$$

$$
+ \sqrt{-g} \cdot \frac{g^{\alpha\beta}_{,\beta}}{2} \cdot F^{p/2-2} F' f_{,\alpha} \cdot g^{\kappa\lambda} + \sqrt{-g} \cdot \frac{g^{\alpha\beta}}{2} \cdot \left(\frac{p}{2} - 2 \right) F^{p/2-3} f_{,\beta} F'^{2} f_{,\alpha} \cdot g^{\kappa\lambda}
$$

$$
\left. \left. + \sqrt{-g} \cdot \frac{g^{\xi\zeta} g_{\xi\zeta,\beta}}{2} \cdot \frac{g^{\alpha\beta}}{2} \cdot F^{p/2-2} F' f_{,\alpha} \cdot g^{\kappa\lambda} \right] \right\}
$$

$$
+ \left(\frac{p}{2} - 1 \right)
$$

This can be simplified:

$$
0 = \frac{\sqrt{-g} \cdot F^p}{F} \cdot
$$

$$
\left\{
\begin{aligned}
& R^{\kappa\lambda} - \frac{R}{2} g^{\kappa\lambda} + \left(g^{\kappa\lambda} g^{\alpha\beta} - g^{\kappa\alpha} g^{\lambda\beta} - g^{\lambda\alpha} g^{\kappa\beta} \right)(1-n) \\
& \cdot \left[\frac{f_{,\alpha} f_{,\beta}}{F^2} \left(F F'' + (F')^2 \cdot \frac{2 \cdot p + n - 14}{8} \right) + f_{,\alpha\beta} \frac{F'}{F} \right] \\
& - \frac{F' \cdot (1-n)}{2 \cdot F} \left(g^{\kappa\lambda} g^{\alpha\beta} - g^{\kappa\alpha} g^{\lambda\beta} - g^{\lambda\alpha} g^{\kappa\beta} \right) \Gamma^\gamma_{\alpha\beta} f_{,\gamma} + \frac{F'}{F} \cdot (1-n) g^{\alpha\beta} g^{\gamma\lambda} \Gamma^\kappa_{\alpha\beta} f_{,\gamma} \\
& + \frac{F'}{F} \cdot \frac{(1-n)}{2} \left(g^{\kappa\lambda} g^{\alpha\beta} - g^{\kappa\alpha} g^{\lambda\beta} - g^{\lambda\alpha} g^{\kappa\beta} \right) f_{,\alpha} \\
& - \frac{F' \cdot (1-n)}{2 \cdot F} \left(g^{\kappa\lambda} g^{\alpha\beta} - g^{\kappa\alpha} g^{\lambda\beta} - g^{\lambda\alpha} g^{\kappa\beta} \right)_{,\beta} f_{,\alpha} + \frac{g^{\xi\xi} g_{\xi\xi,\beta}}{2} \\
& + g^{\alpha\beta} \frac{F'}{F} \left(\frac{f_{,\alpha} \cdot g^{\xi\lambda} \Gamma^\kappa_{\beta\xi} + f_{,\alpha} \cdot g^{\kappa\xi} \Gamma^\lambda_{\beta\xi}}{2} + f_{,\gamma} \cdot g^{\gamma\lambda} \Gamma^\kappa_{\alpha\beta} \right) + \frac{F'}{F} f_{,\gamma} \\
& + \frac{F'}{F} f_{,\gamma\rho} \cdot \frac{\left(g^{\kappa\rho} g^{\gamma\lambda} + g^{\rho\lambda} g^{\gamma\kappa} - g^{\kappa\lambda} g^{\gamma\rho} \right)}{2} + \frac{F''}{F} f_{,\rho} f_{,\gamma} \cdot \frac{\left(g^{\kappa\rho} g^{\gamma\lambda} + g^{\rho\lambda} g^{\gamma\kappa} - g^{\kappa\lambda} g^{\gamma\rho} \right)}{2} \\
& + \frac{F'^2}{F^2} f_{,\rho} f_{,\gamma} \cdot \frac{\left(g^{\kappa\rho} g^{\gamma\lambda} + g^{\rho\lambda} g^{\gamma\kappa} - g^{\kappa\lambda} g^{\gamma\rho} \right)}{2} + \frac{g^{\xi\xi} g_{\xi\xi,\rho}}{2} \frac{F'}{F} f_{,\gamma} \cdot \frac{\left(g^{\kappa\rho} g^{\gamma\lambda} + g^{\rho\lambda} g^{\gamma\kappa} - g^{\kappa\lambda} g^{\gamma\rho} \right)}{2} \\
& + \left(\frac{p}{2} - 1 \right) + \left(\frac{p}{2} - 2 \right) \frac{F'^2}{F^2} f_{,\rho} f_{,\gamma} \\
& - \frac{F'}{F} \frac{g^{\alpha\beta}}{2} \cdot f_{,\alpha} \cdot g^{\kappa\lambda}_{,\beta} - \frac{F'}{F} \frac{g^{\alpha\beta}}{2} \cdot f_{,\alpha\beta} \cdot g^{\kappa\lambda} - \frac{F''}{F} \frac{g^{\alpha\beta}}{2} \cdot f_{,\beta} f_{,\alpha} \cdot g^{\kappa\lambda} \\
& - \frac{g^{\alpha\beta}}{2} \cdot \left(\frac{p}{2} - 2 \right) \frac{F'^2}{F^2} f_{,\beta} f_{,\alpha} \cdot g^{\kappa\lambda} - \frac{F'}{F} \frac{g^{\xi\xi} g_{\xi\xi,\beta}}{2} \frac{g^{\alpha\beta}}{2} \cdot f_{,\alpha} \cdot g^{\kappa\lambda}
\end{aligned}
\right\}
\tag{995}
$$

Now, with the help of the work of other authors (c.f. section "8.6.2 About a More General Kernel within the Einstein–Hilbert Action"), we can derive some further simplifications:

$$\Rightarrow \quad 0 = \left\{ \begin{array}{l} R^{\kappa\lambda} - \dfrac{1}{2}R \cdot g^{\kappa\lambda} + \left(\left(\begin{array}{c} \underbrace{\dfrac{8\pi G T^{\kappa\lambda}}{\kappa} = \text{matter}} \\ 0 = \text{vacuum} \end{array} \right) + \Lambda \cdot g^{\kappa\lambda} \right) \\[2em] -\dfrac{1}{\sqrt{F^p}} \cdot \left(\partial^{;\kappa} \partial^{;\lambda} - g^{\kappa\lambda} \Delta_g \right) F^{\frac{p}{2}-1} \\[1.5em] + \left(g^{\kappa\lambda} g^{\alpha\beta} - g^{\kappa\alpha} g^{\lambda\beta} - g^{\lambda\alpha} g^{\kappa\beta} \right)(1-n) \\[1em] \cdot \left[\dfrac{f_{,\alpha} f_{,\beta}}{F^2} \left(FF'' + (F')^2 \cdot \dfrac{2 \cdot p + n - 14}{8} \right) + f_{,\alpha\beta} \dfrac{F'}{F} \right] \\[1.5em] -\dfrac{F' \cdot (1-n)}{2 \cdot F} \left(g^{\kappa\lambda} g^{\alpha\beta} - g^{\kappa\alpha} g^{\lambda\beta} - g^{\lambda\alpha} g^{\kappa\beta} \right) \Gamma^\gamma_{\alpha\beta} f_{,\gamma} \\[1.5em] +\dfrac{F'}{F} \cdot (1-n) g^{\alpha\beta} g^{\gamma\lambda} \Gamma^\kappa_{\alpha\beta} f_{,\gamma} \\[1.5em] +\dfrac{F'}{F} \cdot \dfrac{(1-n)}{2} \left(g^{\kappa\lambda} g^{\alpha\beta} - g^{\kappa\alpha} g^{\lambda\beta} - g^{\lambda\alpha} g^{\kappa\beta} \right)_{,\beta} f_{,\alpha} \\[1.5em] +\dfrac{g^{\xi\zeta} g_{\xi\zeta,\beta}}{2} \cdot \left(\dfrac{F'}{F} \cdot \dfrac{(1-n)}{2} \left(g^{\kappa\lambda} g^{\alpha\beta} - g^{\kappa\alpha} g^{\lambda\beta} - g^{\lambda\alpha} g^{\kappa\beta} \right) f_{,\alpha} \right). \end{array} \right. \quad (996)$$

However, as said above, we are sure that there are still shorter and simpler forms and we leave it for clever mathematicians to find them.

8.14 Appendix E: Quantum Centers with Scaled Lagrange Density F^q

Now we repeat the variation with respect to the parameters ζ_{ki}, only that this time we incorporate an additional factor F^q:

$$\delta W = 0 = \delta_\zeta \int_V d^n x \left(\sqrt{-G} \cdot F^q \cdot \left(R^* - 2\Lambda + L_M \right) \right)$$

$$= \delta_\zeta \int_V d^n x \left(\sum_{\forall f_k} \sqrt{-G} \cdot F^q \cdot \left(R^* - 2\Lambda + L_M \right) \right) = \sum_{\forall f_k} \delta_\zeta \int_V d^n x \left(\sqrt{-G} \cdot F^q \cdot \left(R^* - 2\Lambda + L_M \right) \right).$$

$$(997)$$

Inserting our result for R^* from the Appendix A, we have:

$$\delta W = 0 = \delta_\zeta \int_V d^n x \left(\sqrt{-G} \cdot F^q \cdot \left(R^* - 2\Lambda + L_M \right) \right)$$

$$= \delta_\zeta \int_V d^n x \left(\sum_{\forall f_k} \sqrt{-G} \cdot F^q \cdot \left(R^* - 2\Lambda + L_M \right) \right) = \sum_{\forall f_k} \delta_\zeta \int_V d^n x \left(\sqrt{-G} \cdot F^q \cdot \left(R^* - 2\Lambda + L_M \right) \right)$$

$$0 = \delta_\zeta \int_V d^n x \left(\sqrt{-G} \cdot F^q \cdot \left(R^* - 2\Lambda + L_M \right) \right)$$

$$= \sum_{\forall f_k} \delta_\zeta \int_V d^n x \left(\sqrt{-G} \cdot F^q \cdot \left(\left(\begin{array}{c} R + \dfrac{F'}{F}(1-n)\Delta f \\[2mm] + \dfrac{f_{,\alpha} f_{,\beta} g^{\alpha\beta}(1-n)}{4F^2}\left(4FF'' + (F')^2 (n-6) \right) \end{array} \right) \dfrac{1}{F} - 2\Lambda + L_M \right) \right)$$

$$= \sum_{\forall f_k} \delta_\zeta \int_V d^n x \left(\sqrt{-G} \cdot F^q \cdot \left(\left(\begin{array}{c} R + \dfrac{F'}{F}(1-n) g^{\alpha\beta}\left(f_{,\alpha\beta} - \Gamma^\gamma_{\alpha\beta} f_{,\gamma} \right) \\[2mm] + \dfrac{f_{,\alpha} f_{,\beta} g^{\alpha\beta}(1-n)}{4F^2}\left(4FF'' + (F')^2 (n-6) \right) \end{array} \right) \dfrac{1}{F} - 2\Lambda + L_M \right) \right)$$

$$= \sum_{\forall f_k} \delta_\zeta \int_V d^n x \left(\sqrt{-g} \cdot F^{\frac{p}{2}} \cdot \left(\left(\begin{array}{c} R + \dfrac{F'}{F}(1-n) g^{\alpha\beta}\left(f_{,\alpha\beta} - \Gamma^\gamma_{\alpha\beta} f_{,\gamma} \right) \\[2mm] + \dfrac{f_{,\alpha} f_{,\beta} g^{\alpha\beta}(1-n)}{4F^2}\left(4FF'' + (F')^2 (n-6) \right) \end{array} \right) \dfrac{1}{F} - 2\Lambda + L_M \right) \right),$$

$$(998)$$

$$= \sum_{\forall f_k} \int_V d^n x \sqrt{-g} \cdot \left(\begin{array}{c} \left(\dfrac{p}{2} - 1 \right) \cdot F' \cdot F^{\frac{p}{2}-2} f_{,\sigma} \\[2mm] \cdot \left(\begin{array}{c} R + \dfrac{F'}{F}(1-n) g^{\alpha\beta}\left(f_{,\alpha\beta} - \Gamma^\gamma_{\alpha\beta} f_{,\gamma} \right) \\[2mm] + \dfrac{f_{,\alpha} f_{,\beta} g^{\alpha\beta}(1-n)}{4F^2}\left(4FF'' + (F')^2 (n-6) \right) \end{array} \right) \\[8mm] + F^{\frac{p}{2}-1}\delta_\zeta \left(\begin{array}{c} \dfrac{F'}{F}(1-n) g^{\alpha\beta}\left(f_{,\alpha\beta} - \Gamma^\gamma_{\alpha\beta} f_{,\gamma} \right) \\[2mm] + \dfrac{f_{,\alpha} f_{,\beta} g^{\alpha\beta}(1-n)}{4F^2}\left(4FF'' + (F')^2 (n-6) \right) \end{array} \right) \\[8mm] + f_{,\sigma} F^{\frac{p}{2}-1} \dfrac{p}{2} F' \cdot \left(L_M - 2\Lambda \right) \end{array} \right) Y^\sigma_{,k} \delta\zeta^k$$

with:
$$Z_{,k} = \frac{\partial Z}{\partial \zeta^k}; \quad Y^\sigma = Y^\sigma - \zeta^{\sigma k},$$

(999)

$$= \sum_{\forall f_k} \int_V d^n x \left| \sqrt{-g} \cdot F^{\frac{p}{2}-1} \left(\left(\frac{\left(\frac{p}{2}-1\right) \cdot F' \frac{f_{,\sigma}}{F}}{\left(R + \frac{F'}{F}(1-n)\Delta f + \frac{f_{,\alpha} f_{,\beta} g^{\alpha\beta}(1-n)}{4F^2}\left(4FF'' + (F')^2(n-6)\right)\right)} \right) \left(\frac{\left(\frac{F'}{F}\right)_{,\sigma}(1-n)\left(f_{,\alpha\beta} - \Gamma^\gamma_{\alpha\beta} f_{,\gamma}\right)}{+\frac{F'}{F}(1-n)\left(f_{,\alpha\beta\sigma} - \Gamma^\gamma_{\alpha\beta} f_{,\gamma\sigma}\right)} \right) + g^{\alpha\beta} \left(+ \left(\frac{f_{,\alpha} f_{,\beta}(1-n)}{4F^2}\left(4FF'' + (F')^2(n-6)\right)\right)_{,\sigma} \right) + f_{,\sigma} \frac{p}{2} F' \cdot (L_M - 2\Lambda) \right) \right| Y^\sigma_{,k} \delta\zeta^k,$$

(1000)

and subsequently obtain the third-order differential equation for f, which in fact becomes f_k now. Please note that during the evaluation it has to be assured that the derivative ∂_σ only acts on the function f. With a bit of simplification:

$$= \sum_{\forall f_k} \int_V d^n x \left| \sqrt{-g} \cdot F^{\frac{p}{2}-1} \left(\left(\frac{\left(\frac{p}{2}-1\right) \cdot F' \frac{f_{,\sigma}}{F}}{\left(R + \frac{F'}{F}(1-n)\Delta f + \frac{f_{,\alpha} f_{,\beta} g^{\alpha\beta}(1-n)}{4F^2}\left(4FF'' + (F')^2(n-6)\right)\right)} \right) + \left(\frac{F'}{F}\right)_{,\sigma}(1-n)g^{\alpha\beta}\overbrace{\left(f_{,\alpha\beta} - \Gamma^\gamma_{\alpha\beta} f_{,\gamma}\right)}^{=\Delta f} + g^{\alpha\beta}\left(+\frac{F'}{F}(1-n)\left(f_{,\alpha\beta\sigma} - \Gamma^\gamma_{\alpha\beta} f_{,\gamma\sigma}\right) + \left(\frac{f_{,\alpha} f_{,\beta}(1-n)}{4F^2}\left(4FF'' + (F')^2(n-6)\right)\right)_{,\sigma} \right) + f_{,\sigma} \frac{p}{2} F' \cdot (L_M - 2\Lambda) \right) \right| Y^\sigma_{,k} \delta\zeta^k,$$

(1001)

$$
= \sum_{\forall f_k} \int_V d^n x \left(\sqrt{-g} \cdot F^{\frac{p}{2}-1} \left(\begin{array}{l} \left(\dfrac{p}{2}-1 \right) \cdot F' \dfrac{f_{,\sigma}}{F} \\[2mm] \cdot \left(\begin{array}{l} R + \dfrac{F'}{F}(1-n)\Delta f \\[2mm] + \dfrac{f_{,\alpha} f_{,\beta} g^{\alpha\beta}(1-n)}{4F^2}\left(4FF'' + (F')^2(n-6)\right) \end{array} \right) \\[6mm] + \left(\dfrac{F''F - (F')^2}{F^2} \right) \cdot f_{,\sigma}(1-n)\Delta f \\[4mm] + g^{\alpha\beta} \left(\begin{array}{l} + \dfrac{F'}{F}(1-n)\left(f_{,\alpha\beta\sigma} - \Gamma^{\gamma}_{\alpha\beta}f_{,\gamma\sigma}\right) \\[2mm] + \left(\dfrac{f_{,\alpha} f_{,\beta}(1-n)}{4F^2}\left(4FF'' + (F')^2(n-6)\right) \right)_{,\sigma} \end{array} \right) \\[6mm] + f_{,\sigma} \dfrac{p}{2} F' \cdot (L_M - 2\Lambda) \end{array} \right) \right) Y^{\sigma}_{,k}\delta\zeta^k,
$$

$$(1002)$$

$$
= \sum_{\forall f_k} \int_V d^n x \left(\sqrt{-g} \cdot F^{\frac{p}{2}-1} \left(\begin{array}{l} \left(\dfrac{p}{2}-1 \right) \cdot F' \dfrac{f_{,\sigma}}{F} \\[2mm] \cdot \left(\begin{array}{l} R + \dfrac{F'}{F}(1-n)\Delta f \\[2mm] + \dfrac{f_{,\alpha} f_{,\beta} g^{\alpha\beta}(1-n)}{4F^2}\left(4FF'' + (F')^2(n-6)\right) \end{array} \right) \\[6mm] + \left(\dfrac{F''F + \left(\dfrac{p}{2}-2\right)(F')^2}{F^2} \right) \cdot f_{,\sigma}(1-n)\Delta f \\[4mm] + g^{\alpha\beta} \left(\begin{array}{l} + \dfrac{F'}{F}(1-n)\left(f_{,\alpha\beta\sigma} - \Gamma^{\gamma}_{\alpha\beta}f_{,\gamma\sigma}\right) \\[2mm] + \left(\dfrac{f_{,\alpha} f_{,\beta}(1-n)}{4F^2}\left(4FF'' + (F')^2(n-6)\right) \right)_{,\sigma} \end{array} \right) \\[6mm] + f_{,\sigma} \dfrac{p}{2} F' \cdot (L_M - 2\Lambda) \end{array} \right) \right) Y^{\sigma}_{,k}\delta\zeta^k,
$$

$$(1003)$$

we realize that even without the sum over the whole k-ensemble, it should be possible to find solutions of the form:

$$0 = \left\{ \begin{array}{l} \left[\left(\frac{p}{2}-1\right)F'\frac{f_{,\sigma}}{F}\cdot\left(R+\frac{f_{,\alpha}f_{,\beta}g^{\alpha\beta}(1-n)}{4F^2}\left(4FF''+(F')^2(n-6)\right)\right)\right] \\[3mm] +\frac{f_{,\sigma}}{F^2}\left(FF''\left(\frac{n}{2}-2\right)(F')^2\right)(1-n)\Delta f \\[3mm] +g^{\alpha\beta}\left(\begin{array}{l} \frac{F'}{F}(1-n)\left(f_{,\alpha\beta\sigma}-\Gamma^\gamma_{\alpha\beta}f_{,\gamma\sigma}\right) \\[2mm] +\left(\frac{f_{,\alpha}f_{,\beta}(1-n)}{4F^2}\left(4FF''+(F')^2(n-6)\right)\right)_{,\sigma} \end{array}\right) \\[6mm] +f_{,\sigma}\frac{p}{2}F'\cdot(L_M-2\Lambda). \end{array} \right. \tag{1004}$$

Such solutions would be intrinsic ones and we here want to investigate some simpler types for certain settings of $F[f]$. Thus, in the case of:

$$0 = 4FF''+(F')^2(n-6) \quad\Rightarrow\quad F = \left(f-C_{f0}\right)^{\frac{4}{n-2}}C_{f1}, \tag{1005}$$

Eq. (1004) simplifies to:

$$0 = \left\{ \begin{array}{l} \left(\frac{p}{2}-1\right)F'\frac{f_{,\sigma}}{F}\cdot R+f_{,\sigma}\frac{p}{2}F'\cdot(L_M-2\Lambda) \\[3mm] +\frac{f_{,\sigma}}{F^2}\left(FF''+\left(\frac{p}{2}-2\right)(F')^2\right)(1-n)\Delta f+g^{\alpha\beta}\left(\frac{F'}{F}(1-n)\left(f_{,\alpha\beta\sigma}-\Gamma^\gamma_{\alpha\beta}f_{,\gamma\sigma}\right)\right) \end{array} \right.$$

$$= \left\{ \begin{array}{l} \left(\frac{p}{2}-1\right)F'\frac{f_{,\sigma}}{F}\cdot R+f_{,\sigma}\frac{p}{2}F'\cdot(L_M-2\Lambda) \\[3mm] +\frac{f_{,\sigma}}{F^2}\left(FF''+\left(\frac{p}{2}-2\right)(F')^2\right)(1-n)\Delta f+g^{\alpha\beta}\left(\frac{F'}{F}(1-n)\left(f_{,\sigma\alpha\beta}-\Gamma^\gamma_{\alpha\beta}f_{,\sigma\gamma}\right)\right) \end{array} \right.$$

$$= \left\{ \begin{array}{l} \left(\frac{p}{2}-1\right)F'\frac{f_{,\sigma}}{F}\cdot R+f_{,\sigma}\frac{p}{2}F'\cdot(L_M-2\Lambda) \\[3mm] +\frac{(1-n)}{F}\left(\frac{f_{,\sigma}}{F}\left(FF''+\left(\frac{p}{2}-2\right)(F')^2\right)\Delta f+F'\Delta\left(f_{,\sigma}\right)\right). \end{array} \right. \tag{1006}$$

Setting, just as an example, $n = 4$, $q = 2$, and inserting $F[f]$ from Eq. (913) (or Eq. (1005)) yields:

$$0 = 4 \cdot f_{,\sigma} \cdot \left(L_M - 2\Lambda\right) \cdot \left(f - C_{f0}\right) \cdot C_{f1}$$

$$+ \frac{3 \cdot \left(\left(f - C_{f0}\right) \cdot \left(f_{,\sigma} \cdot R - 2 \cdot \Delta\left(f_{,\sigma}\right)\right) - 4 \cdot f_{,\sigma} \Delta f\right)}{\left(f - C_{f0}\right)^2}. \tag{1007}$$

References

1. N. Schwarzer, *Societons and Ecotons—The Photons of the Human Society—Control them and Rule the World*, Part 1 of "Medical Socio-Economic Quantum Gravity," Self-published, Amazon Digital Services, 2020, Kindle.

2. N. Schwarzer, *Masses and the Infinity Options Principle: Can We Explain the 3-Generations and the Quantized Mass Problem?*, Part 5 of "Medical Socio-Economic Quantum Gravity," Self-published, Amazon Digital Services, 2020, Kindle.

3. N. Schwarzer, *The World Formula: A Late Recognition of David Hilbert's Stroke of Genius*, 2022, Jenny Stanford Publishing. ISBN: 9789814877206.

4. N. Schwarzer, *Humanitons—The Intrinsic Photons of the Human Body—Understand them and Cure Yourself*, Part 2 of "Medical Socio-Economic Quantum Gravity," Self-published, Amazon Digital Services, 2020, Kindle.

5. N. Schwarzer, *Mastering Human Crises with Quantum-Gravity-based but still Practicable Models—First Measure: SEEING and UNDERSTANDING the WHOLE*, Part 3 of "Medical Socio-Economic Quantum Gravity," Self-published, Amazon Digital Services, 2020, Kindle.

6. N. Schwarzer, *Self-Similar Quantum Gravity: How Infinity Brings Simplicity?*, Part 4 of "Medical Socio-Economic Quantum Gravity," Self-published, Amazon Digital Services, 2020, Kindle.

7. D. Hilbert, Die Grundlagen der Physik, Teil 1, *Göttinger Nachrichten*, 1915, 395–407.

8. N. Schwarzer, *The 3 Generations of Elementary Particles*, Part 6 of "Medical Socio-Economic Quantum Gravity," Self-published, Amazon Digital Services, 2020, Kindle.

9. A. E. Green and W. Zerna, *Theoretical Elasticity*, 1968, Oxford University Press, London.

10. H. Neuber, *Kerspannungslehre*, in German, 3rd edition, Springer-Verlag, Berlin, Heidelberg, New York, Tokyo 1985, ISBN 3-540-13558-8.

11. N. Schwarzer, About Holistic Optimization: Examples from In- and Outside the World of Coatings, *Proceedings of the 58th Annual Technical SVC Conference*, 2015.

12. N. Schwarzer, Scale invariant mechanical surface optimization applying analytical time dependent contact mechanics for layered structures, Chapter 22 in *Applied Nanoindentation in Advanced Materials*, Atul Tiwari (editor), Sridhar Natarajan (co-editor), 2017, ISBN: 978-1-119-08449-5. www.wiley.com/WileyCDA/WileyTitle/productCd-1119084490.html

13. N. Schwarzer, From Interatomic Interaction Potentials Via Einstein Field Equation Techniques To Time Dependent Contact Mechanics, *Materials Research Express, 1, 1*, 2014, IOP Publishing. http://dx.doi.org/10.1088/2053-1591/1/1/015042

14. N. Schwarzer, Completely Analytical Tools for the Next Generation of Surface and Coating Optimization, *Coatings*, 2014, **4**, 263–291. doi:10.3390/coatings4020263

15. N. Schwarzer, Elastic surface deformation due to indenters with arbitrary symmetry of revolution, *J. Phys. D: Appl. Phys.*, 2004, **37**, 2761–2772.

16. N. Schwarzer, Some Basic Equations for the Next Generation of Surface Testers Solving the Problem of Pile-up, Sink-in and Making Area-Function-Calibration Obsolete, JMR Special Focus Issue on "Indentation Methods in Advanced Materials Research," *J. Mater. Res.*, 2009, **24** (3), 1032–1036.

17. V. I. Fabrikant, *Application of Potential Theory in Mechanics: A Selection of New Results*, 1989, Kluver Academic Publishers, Netherlands.

18. V. I. Fabrikant, *Mixed Boundary Value Problems of Potential Theory and their Application in Engineering*, 1991, Kluver Academic Publishers, Netherlands.

19. FilmDoctor, analytical software package for the analysis of complex mechanical problem for inhomogeneous materials, www.siomec.de/filmdoctor

20. P. S. Debnath and B. C. Paul, *Cosmological Models with Variable Gravitational and Cosmological constants in R^2 Gravity*, 2012, arxiv.org/pdf/gr-qc/0508031.pdf

21. N. Schwarzer, *Einstein had it, but he did not see it—Part LXXXI: More Dirac Killing*, Self-published, Amazon Digital Services, 2019, Kindle.

22. A. Einstein, Grundlage der allgemeinen Relativitätstheorie, *Annalen der Physik (ser. 4)*, 1916, **49**, 769–822.

23. N. Schwarzer, *Quantum Gravity Thermodynamics II: Derivation of the Second Law of Thermodynamics and the Metric Driving Force of Evolution*, Self-published, Amazon Digital Services, 2019, Kindle.

24. N. Schwarzer, *Einstein had it, but he did not see it—Part LXXXV: In Conclusion*, Self-published, Amazon Digital Services, 2019, Kindle.

25. N. Schwarzer, *Science Riddles—Riddle No. 20: Second Law of Thermodynamics—Where is its Fundamental Origin?*, Self-published, Amazon Digital Services, 2019, Kindle.

26. C. A. Sporea, *Notes on f(R) Theories of Gravity*, 2014, arxiv.org/pdf/1403.3852.pdf

27. N. Schwarzer, *3rd Epistle to Elementary Particle Physicists—Beyond the Standard Model—Metric Solutions for Neutrino, Electron, Quark*, Self-published, Amazon Digital Services, 2019, Kindle.

28. N. Schwarzer, *Quantum Gravity Thermodynamics: And it May Get Hotter*, Self-published, Amazon Digital Services, 2019, Kindle.

THE FIFTH DAY:
LET THERE BE A DIRAC EQUATION

Chapter 9

The Metric Dirac Equation Revisited and the Geometry of Spinors

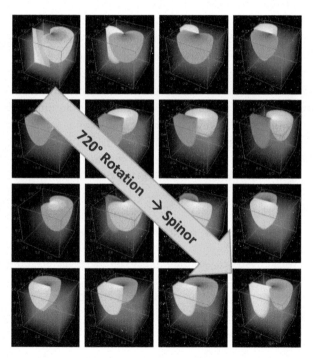

Figure 9.1 A spinor object as evaluated from a scaled 8-dimensional metric here in this chapter.

The Math of Body, Soul, and the Universe
Norbert Schwarzer
Copyright © 2023 Jenny Stanford Publishing Pte. Ltd.
ISBN 978-981-4968-24-9 (Hardcover), 978-1-003-33454-5 (eBook)
www.jennystanford.com

9.1 Abstract: Why Dirac Again?

In our earlier investigations (e.g., [A1] and previous chapters in this book, especially chapters 6 and 7), we saw that important scalar equations of Quantum Theory emerge when we construct a scaled metric by simply multiplying the metric with a factor $F[f]$, thereby assuming the inner part of the scale factor to be an arbitrary function f. f itself then—so it was derived—becomes the wave function. Subjecting this scaled metric in the usual way to the Einstein–Hilbert action [A2] does not only present us with quantum Einstein field equations [A3] and thus, a quantized version of Einstein's General Theory of Relativity [A4], but—along the way (e.g., [A5–A10])—gives us the classical quantum equations. Also, thermodynamics can be derived from, respectively found inside this simple scaled metric approach [A3, A10].

With the Klein–Gordon equation, already falling out of our new approach, which could just be seen as another form of variation, we may just set back and conclude that this way we also have obtained the Dirac equation [A11]. This was classically derived by an operator factorization from the Klein–Gordon equation. Thus, having obtained the latter, automatically gives us the Dirac equation, too. However, when digging deeper (metrically) and also trying to find a completely metric origin and understanding of the Dirac approach (Fig. 9.1), we have to come to the conclusion that the simple scalar metric scaling with $F[f]$ does not suffice. We find that only a vector scaling will probably give us the complete picture [A12].

9.1.1 References for the Abstract

A1. N. Schwarzer, *Masses and the Infinity Options Principle: Can We Explain the 3-Generation and the Quantized Mass Problem?*, Part 5 of "Medical Socio-Economic Quantum Gravity," Self-published, Amazon Digital Services, 2020, Kindle.

A2. D. Hilbert, Die Grundlagen der Physik, Teil 1, *Göttinger Nachrichten*, 1915, 395–407.

A3. N. Schwarzer, *Towards Quantum Einstein Field Equations*, Part 7 of "Medical Socio-Economic Quantum Gravity," Self-published, Amazon Digital Services, 2020, Kindle.

A4. A. Einstein, Grundlage der allgemeinen Relativitätstheorie, *Ann. Phys.*, 1916, **49** (ser. 4), 769–822.

A5. N. Schwarzer, *Brief Proof of Hilbert's World Formula: Dirac, Klein–Gordon, Schrödinger, Einstein, Evolution and the Second Law of*

Thermodynamics all from One Origin, Self-published, Amazon Digital Services, 2020, Kindle.

A6. N. Schwarzer, *Societies and Ecotons—The Photons of the Human Society—Control them and Rule the World*, Part 1 of "Medical Socio-Economic Quantum Gravity," Self-published, Amazon Digital Services, 2020, Kindle.

A7. N. Schwarzer, *Humanitons—The Intrinsic Photons of the Human Body—Understand them and Cure Yourself*, Part 2 of "Medical Socio-Economic Quantum Gravity," Self-published, Amazon Digital Services, 2020, Kindle.

A8. N. Schwarzer, *Mastering Human Crises with Quantum-Gravity-based but still Practicable Models—First Measure: SEEING and UNDERSTANDING the WHOLE*, Part 3 of "Medical Socio-Economic Quantum Gravity," Self-published, Amazon Digital Services, 2020, Kindle.

A9. N. Schwarzer, *Self-Similar Quantum Gravity: How Infinity Brings Simplicity?*, Part 4 of "Medical Socio-Economic Quantum Gravity," Self-published, Amazon Digital Services, 2020, Kindle.

A10. N. Schwarzer, *The World Formula: A Late Recognition of David Hilbert's Stroke of Genius, 2022*, Jenny Stanford Publishing. ISBN: 9789814877206.

A11. P. A. M. Dirac, The Quantum Theory of the Electron, 1928, *Proceedings of the Royal Society A*. DOI: 10.1098/rspa.1928.0023

A12. N. Schwarzer, *My Horcruxes: A Curvy Math to Salvation*, Part 9 of "Medical Socio-Economic Quantum Gravity," Self-published, Amazon Digital Services, 2021, Kindle.

9.2 A Few Basics

9.2.1 Extended (Quantum Gravity) Einstein Field Equations

In earlier papers (e.g., [1]), we already investigated how we could obtain the Dirac equation from a metric origin. In $[2–7]^1$ (and quite some other papers by this author and previous chapters of this book), it was then demonstrated how we can find a direct extraction of the Klein–Gordon, the Schrödinger, and the Dirac equation from the Ricci scalar R^* of a modified metric of the kind $G_{\alpha\beta} = F[f]^* g_{\alpha\beta}.^2$

---

[1] Please note that in most papers we thereby considered metrics g_{ij} with $R = 0$. Here now we need to introduce the general form.

[2] For convenience we here repeated and partially extend these derivations in the appendices of chapter 11.

With a yet arbitrary scalar function $F[f]$, the corresponding modified Ricci scalar R^* reads[3]:

$$R^* = R^*_{\alpha\beta} G^{\alpha\beta} = \left(\begin{array}{c} \Gamma^{\sigma}_{\alpha\beta,\sigma} g^{\alpha\beta} - \Gamma^{\sigma}_{\beta\sigma,\alpha} g^{\alpha\beta} - \Gamma^{\mu}_{\sigma\alpha} \Gamma^{\sigma}_{\beta\mu} g^{\alpha\beta} + \Gamma^{\sigma}_{\alpha\beta} \Gamma^{\mu}_{\sigma\mu} g^{\alpha\beta} \\ + \dfrac{F'}{F}(1-n)\Delta f + \dfrac{f_{,\alpha} f_{,\beta} g^{\alpha\beta}(1-n)}{4F^2} \left(4FF'' + (F')^2 (n-6) \right) \end{array} \right) \dfrac{1}{F}$$

$$= \left(R + \dfrac{F'}{F}(1-n)\Delta f + \dfrac{f_{,\alpha} f_{,\beta} g^{\alpha\beta}(1-n)}{4F^2} \left(4FF'' + (F')^2 (n-6) \right) \right) \dfrac{1}{F}$$

with:
$$F = F[f]; \quad F' = \frac{\partial F[f]}{\partial f}; \quad F'' = \frac{\partial^2 F[f]}{\partial f^2}. \tag{1008}$$

Demanding suitable conditions for the function $F[f]$ and/or f then gives us Dirac or Klein–Gordon equations (see [2–4] and section 6.11). From there, the Schrödinger equation automatically follows by a simple transformation (see section 6.11.4). Only the Dirac equation seems to provide some difficulties, which are less problematic issues rather than interesting (peculiar) options to derive this equation. So, while classically the fact that we already have obtained the Klein–Gordon equation should suffice, because from there Dirac had directly extracted his famous equation, we find that Eq. (1008) offers other paths to end up with the Dirac or Dirac-like equations. For instance, when demanding f to be a Laplace function, which also was the fundament for Dirac's derivation, we obtain from Eq. (1008):

$$R^* = \frac{R}{F[f]} + \frac{(1-n)}{F[f]^3} \cdot \left(\left(\frac{(n-6)}{4} \cdot \left(\frac{\partial F[f]}{\partial f} \right)^2 + F[f] \cdot \frac{\partial^2 F[f]}{\partial f^2} \right) \cdot \overbrace{\left(\tilde{\nabla}_g f \right)^2}^{= f_{,\alpha} g^{\alpha\beta} f_{,\beta}} \right)$$

$$\xrightarrow{n=4; R=0} = \frac{1}{F[f]^3} \cdot \left(\left(3 \cdot F[f] \cdot \frac{\partial^2 F[f]}{\partial f^2} - \frac{3}{2} \cdot \left(\frac{\partial F[f]}{\partial f} \right)^2 \right) \cdot \left(\tilde{\nabla}_g f \right)^2 \right), \tag{1009}$$

which—so it was shown in [2–4]—gives the metric equivalent to the Dirac equation. It was also shown in [2, 3] how this gives the classical Dirac equation in flat space Minkowski metrics. Clearly, we did not extract the Dirac equation from the Klein–Gordon equation, but in a slightly different way. For convenience, the process is been shown below in the section "9.3.2 Repetition: Dirac in the Metric Picture and Its Connection to the Classical Quaternion Form." This is different from the original Dirac approach where

---

[3]The complete evaluation was presented in the appendix A of reference [8] and in chapter 8 in here.

the Klein–Gordon equation is factorized. As said, we are going to investigate this aspect more deeply later in this chapter. We will also try and find out more about the reason of these peculiar redundancies in obtaining the Dirac equation.

Nevertheless, we should state here that Eq. (1008) in connection with certain conditions ($R^* = 0$, mainly) provides us with the most fundamental quantum equations.

Does this scaled metric approach also give us quantum Einstein field equations?

Yes, it does and this was shown in [8, 9] and chapter 8 of this book.

As demonstrated in chapter 8, when starting with the classical Einstein–Hilbert action [10] with the Ricci scalar R^* as kernel or Lagrange density:

$$\delta W = 0 = \delta \int_V d^n x \left(\sqrt{-G} \cdot \left(R^* - 2\Lambda + L_M \right) \right), \tag{1010}$$

we were able to derive new field equations, not only containing the classical part from Einstein's General Theory of Relativity [11], but also a quantum part [8, 9], which is governed by the wave function (and metric scale) f.

The attentive reader will have realized that in Eq. (1010), in contrast to the classical form, we used our somewhat adapted, if not to say scaled metric tensor $G_{\delta\gamma} = F[f] \cdot g_{\delta\gamma}$. Thereby G shall denote the determinant of the metric tensor $G_{\delta\gamma}$, while g will later stand for the corresponding determinant of the metric tensor $g_{\alpha\beta}$. In order to distinguish our new Ricci scalar R^*, being based on $G_{\delta\gamma} = F[f] \cdot g_{\delta\gamma}$ from the usual one R, being based on the metric tensor $g_{\alpha\beta}$, we marked it with the *-superscript. We also have the matter density L_M and the cosmological constant Λ.

Please note that with $G_{\delta\gamma} = F[f] \cdot g_{\delta\gamma}$, we have used the simplest form of metric adaptation, which we could construct as a simplification from a generalization of the typical form of tensor transformations, namely:

$$G_{\alpha\beta} = F\left[f\left[t, x, y, z, \ldots, \xi_k, \ldots, \xi_n \right] \right]_{\alpha\beta}^{ij} g_{ij} \tag{1011}$$

$$\rightarrow G_{\alpha\beta} = F\left[f\left[t, x, y, z, \ldots, \xi_k, \ldots, \xi_n \right] \right] \cdot \delta_\alpha^i \delta_\beta^j g_{ij}$$

$$\rightarrow G_{\alpha\beta} = F\left[f\left[t, x, y, z, \ldots, \xi_k, \ldots, \xi_n \right] \right] \cdot g_{\alpha\beta} = F[f] \cdot g_{\alpha\beta}. \tag{1012}$$

The generalization is been elaborated in [4] and its need (or redundancy) has been discussed in [8]. In chapter 11, we will see how Eq. (1011) could be used to obtain Ricci scalars consisting of systems of Klein–Gordon equations. Setting Eq. (1008) into Eq. (1010) results in the variation:

$$\delta W = 0 = \delta \int_V d^n x \left(\sqrt{-g \cdot F^n} \times \left(\left(\begin{array}{c} R + \dfrac{F'}{F}(1-n)\Delta f \\[2ex] + \dfrac{f_{,\alpha} f_{,\beta} g^{\alpha\beta} (1-n)}{4F^2} \left(4FF'' + (F')^2 (n-6) \right) \\[2ex] -2\Lambda + L_M \end{array} \right)^{\frac{1}{F}} \right) \right)$$

(1013)

Please note that the cosmological constant term, as given in Eq. (1013) in the current form, requires the variation with respect to the metric $G_{\alpha\beta}$. When insisting on the variation with respect to the metric $g_{\alpha\beta}$ instead, we better rewrite Eq. (1013) as follows:

$$\delta_{g_{\alpha\beta}} W = 0 = \left\{ \begin{array}{l} \delta_{g_{\alpha\beta}} \int_V d^n x \dfrac{\sqrt{-g \cdot F^n}}{F} \times \left(\begin{array}{c} R + \dfrac{F'}{F}(1-n)\Delta f \\[2ex] + \dfrac{f_{,\alpha} f_{,\beta} g^{\alpha\beta}(1-n)}{4F^2} \left(4FF'' + (F')^2(n-6) \right) \end{array} \right) \\[4ex] + \delta_{g_{\alpha\beta}} \int_V d^n x \sqrt{-g} \left(L_M - 2\Lambda \right) \end{array} \right.$$

(1014)

in order to make it clear that we do not intend to also scale-adapt the cosmological constant or the Hilbert matter term.

However, when ignoring the cosmological constant and—perhaps—assuming that we would not need any postulated matter terms L_M, simply because our scale-adaptation $F[f]$ and similar "tricks" or "add-ons" (see about inner wrapper functions and intrinsic quantum fields in chapter 11), automatically provides matter we just obtain:

$$\delta W = 0 = \delta \int_V d^n x \left(\sqrt{-g \cdot F^n} \times \left(\left(\begin{array}{c} R + \dfrac{F'}{F}(1-n)\Delta f \\[2ex] + \dfrac{f_{,\alpha} f_{,\beta} g^{\alpha\beta}(1-n)}{4F^2} \left(4FF'' + (F')^2(n-6) \right) \end{array} \right)^{\frac{1}{F}} \right) \right)$$

$$= \delta \int_V d^n x \left(\sqrt{-g \cdot F^n} \times \left(\left(\begin{array}{c} R + \dfrac{F'}{\sqrt{-g} \cdot F}(1-n)\partial_\beta \sqrt{-g} \cdot g^{\alpha\beta} f_{,\alpha} \\[2ex] + \dfrac{f_{,\alpha} f_{,\beta} g^{\alpha\beta}(1-n)}{4F^2} \left(4FF'' + (F')^2(n-6) \right) \end{array} \right)^{\frac{1}{F}} \right) \right).$$

(1015)

Performing the usual Hilbert variation with respect to the metric tensor $g_{\alpha\beta}$ now leads us to (regarding the full evaluation see chapter 8):

$$
\Rightarrow \quad 0 = \left\{ \begin{array}{l} R^{\delta\gamma} - R^* g^{\delta\gamma} + \left(\left(\begin{array}{l} \underbrace{8\pi G T^{\delta\gamma}}_{\kappa} = \text{matter} \\[4pt] 0 = \text{vacuum} \end{array} \right) + \Lambda \cdot g^{\delta\gamma} \right) \\[20pt] - \dfrac{1}{\sqrt{F^n}} \cdot \left(\partial_{;\mu}\partial_{;v} - g_{\mu v}\Delta_g \right) F^{\frac{n}{2}-1} \\[14pt] + \left(g^{\kappa\lambda} g^{\alpha\beta} - g^{\kappa\alpha} g^{\lambda\beta} - g^{\lambda\alpha} g^{\kappa\beta} \right)(1-n) \\[10pt] \cdot \left[\dfrac{f_{,\alpha} f_{,\beta}}{F^2} \left(FF'' + \left(F'\right)^2 \cdot \dfrac{3 \cdot n - 14}{8} \right) + f_{,\alpha\beta} \dfrac{F'}{F} \right] \\[16pt] - \dfrac{F' \cdot (1-n)}{2 \cdot F} \left(g^{\kappa\lambda} g^{\alpha\beta} - g^{\kappa\alpha} g^{\lambda\beta} - g^{\lambda\alpha} g^{\kappa\beta} \right) \Gamma^{\gamma}_{\alpha\beta} f_{,\gamma} \\[16pt] + \dfrac{F'}{F} \cdot (1-n) g^{\alpha\beta} g^{\gamma\lambda} \Gamma^{\kappa}_{\alpha\beta} f_{,\gamma} \\[16pt] + \dfrac{F'}{F} \cdot \dfrac{(1-n)}{2} \left(g^{\kappa\lambda} g^{\alpha\beta} - g^{\kappa\alpha} g^{\lambda\beta} - g^{\lambda\alpha} g^{\kappa\beta} \right)_{,\beta} f_{,\alpha} \\[16pt] + \dfrac{g^{\xi\zeta} g_{\xi\zeta,\beta}}{2} \cdot \left(\dfrac{F'}{F} \cdot \dfrac{(1-n)}{2} \left(g^{\kappa\lambda} g^{\alpha\beta} - g^{\kappa\alpha} g^{\lambda\beta} - g^{\lambda\alpha} g^{\kappa\beta} \right) f_{,\alpha} \right) \end{array} \right. \tag{1016}
$$

Quite some discussions to this new quantum Einstein field equation were presented in the previous chapter.

It has to be pointed out that Hilbert's choice of the Ricci scalar R, acting as the Lagrange density in Eq. (1010), was motivated by his attempt (with success) to reproduce the Einstein field equations in a completely mathematical manner. As his scale factor F was set to one, which is to say as Hilbert did not consider an $F[f]$-scaled metric tensor, we now, when aiming for the Hilbert approach in the $F \to 1$ limit, have the principle degree of freedom of adding a factor F^q to the integrand. This action, reading (when setting L_M and Λ to zero and choosing an additional scale factor F^q):

$$
\delta W = 0 = \delta \int_V d^n x \left(\sqrt{-g \cdot F^n} \times F^q \cdot R^* \right)
$$

$$
= \delta \int_V d^n x \left(\sqrt{-g \cdot F^n} \times \left(F^q \cdot \left(\begin{array}{l} R + \dfrac{F'}{F}(1-n)\Delta f \\[10pt] + \dfrac{f_{,\alpha} f_{,\beta} g^{\alpha\beta} (1-n)}{4F^2} \left(4FF'' + \left(F'\right)^2 (n-6) \right) \end{array} \right)^{\frac{1}{F}} \right) \right)
$$

$$= \delta \int_V d^n x \left(\sqrt{-g \cdot F^{n+2q}} \times \left(\left(R + \frac{F'}{F}(1-n)\Delta f + \frac{f_{,\alpha} f_{,\beta} g^{\alpha\beta}(1-n)}{4F^2} \left(4FF'' + (F')^2 (n-6) \right) \right) \frac{1}{F} \right) \right),$$

(1017)

which would then lead (for the full derivation see previous chapter) to:

$$\Rightarrow \quad 0 = \left\{ \begin{array}{l} R^{\kappa\lambda} - \dfrac{1}{2} R g^{\kappa\lambda} + \left(\left(\begin{array}{c} \underbrace{\dfrac{8\pi G T^{\kappa\lambda}}{\kappa}} = \text{matter} \\ 0 = \text{vacuum} \end{array} \right) + \Lambda \cdot g^{\kappa\lambda} \right) \\[20pt] - \dfrac{1}{\sqrt{F^p}} \cdot \left(\partial^{:\kappa} \partial^{:\lambda} - g^{\kappa\lambda} \Delta_g \right) F^{\frac{p}{2}-1} \\[12pt] + \left(g^{\kappa\lambda} g^{\alpha\beta} - g^{\kappa\alpha} g^{\lambda\beta} - g^{\lambda\alpha} g^{\kappa\beta} \right)(1-n) \\[8pt] \cdot \left[\dfrac{f_{,\alpha} f_{,\beta}}{F^2} \left(FF'' + (F')^2 \cdot \dfrac{2 \cdot p + n - 14}{8} \right) + f_{,\alpha\beta} \dfrac{F'}{F} \right] \\[14pt] - \dfrac{F' \cdot (1-n)}{2 \cdot F} \left(g^{\kappa\lambda} g^{\alpha\beta} - g^{\kappa\alpha} g^{\lambda\beta} - g^{\lambda\alpha} g^{\kappa\beta} \right) \Gamma^\gamma_{\alpha\beta} f_{,\gamma} \\[14pt] + \dfrac{F'}{F} \cdot (1-n) g^{\alpha\beta} g^{\gamma\lambda} \Gamma^\kappa_{\alpha\beta} f_{,\gamma} \\[14pt] + \dfrac{F'}{F} \cdot \dfrac{(1-n)}{2} \left(g^{\kappa\lambda} g^{\alpha\beta} - g^{\kappa\alpha} g^{\lambda\beta} - g^{\lambda\alpha} g^{\kappa\beta} \right)_{,\beta} f_{,\alpha} \\[14pt] + \dfrac{g^{\xi\zeta} g_{\xi\zeta,\beta}}{2} \cdot \left(\dfrac{F'}{F} \cdot \dfrac{(1-n)}{2} \left(g^{\kappa\lambda} g^{\alpha\beta} - g^{\kappa\alpha} g^{\lambda\beta} - g^{\lambda\alpha} g^{\kappa\beta} \right) f_{,\alpha} \right). \end{array} \right. $$

(1018)

Note that we defined $p = n + 2q$.

9.2.2 Example in 6 Dimensions

Assuming Eigenvalue solutions in the case of $n = 6$, Eq. (1008) simplifies as follows:

$$R^* = R^*_{\alpha\beta} G^{\alpha\beta} = \left(R - 5 \cdot \left(\frac{F'}{F} \Delta f + \frac{f_{,\alpha} f_{,\beta} g^{\alpha\beta}}{F} F'' \right) \right) \frac{1}{F}$$

$$\underrightarrow{\Delta f = B \cdot f; \quad f_{,\alpha} f_{,\beta} g^{\alpha\beta} = A \cdot f^2}$$

$$= \left(R - 5 \cdot \left(\frac{F'}{F} B \cdot f + \frac{A \cdot f^2}{F} F'' \right) \right) \frac{1}{F}. \tag{1019}$$

Setting:

$$R = R[f] = R_0 + R_1 \cdot f + R_2 \cdot f^2, \tag{1020}$$

Eq. (1019) gives the following solution:

$$F = e^{-f\sqrt{\frac{R_2}{5 \cdot A}}} \cdot f^{b - \frac{B}{2 \cdot A}} \left(C_{F1} \cdot U \left[-s, 1 + a, 2 \cdot f \sqrt{\frac{R_2}{5 \cdot A}} \right] + C_{F2} \cdot L_s^a \left[2 \cdot f \sqrt{\frac{R_2}{5 \cdot A}} \right] \right)$$

$$s = -b - \frac{\sqrt{5} \cdot R_1}{10\sqrt{A \cdot R_2}}; b = \frac{5 \cdot A + \sqrt{25(A - B)^2 + 20 \cdot A \cdot R_0}}{10 \cdot A}; a = \sqrt{1 + \frac{B^2}{A^2} + \frac{\frac{4R_0}{5} - 2 \cdot B}{A}},$$

$$\tag{1021}$$

with the confluent hypergeometric function $U[\dots, \dots, \dots]$ and the generalized

Laguerre polynomial $L_s^a \left[2 \cdot f \sqrt{\frac{R_2}{5 \cdot A}} \right]$.

We note that F is in danger to vanish or to deliver infinities for $f =$

0 due to the term $f^{b - \frac{B}{2 \cdot A}}$ and we see that the only way to get rid of this

term is by demanding its exponent to vanish. This, however, requires not only the constant curvature term R_0 to be zero, but also to have the two Eigenvalues A and B to be equal, which is to say $R_0 = 0$ and $A = B$. We need to assure this, because otherwise the whole space would have no scale ($F[f = 0] = 0$). It would simply vanish due to our scaled metric assumption (1012) in the case of $f \to 0$.

9.2.3 Can We Now Understand Quantum Theory in a Truly Illustrative Manner?

Yes, but as with many other things it holds that we are not able to truly grasp Quantum Theory without seeing the full picture. Interestingly ... and some may say "unfortunately," only the combination of Quantum Theory with Einstein's General Theory of Relativity will give us this full picture. This means that the task is still unsolved and probably will be so for a long time to come, but taking what we have just presented in this section above, we are of the opinion that the full picture already is at hand.

Here we explain why.

Space and time in Einstein's General Theory of Relativity is explained via objects called metrics.

By allowing these metrics to have a scale factor $F[f]$ as shown in Eqs. (1011) and (1012), and keeping these factors general enough so that they could be functions, we suddenly obtain all those important fundamental quantum equations, like the Dirac, the Klein–Gordon, and the Schrödinger equation (see [3–9], previous chapters of this book and the following main section pages of this chapter). The scaled metrics on the other hand, do not compromise the General Theory of Relativity in any way. On the contrary, now we can derive quantum gravity field equations (1016)–(1018) (see chapter 8 for full derivation).

But there is more and this has to do with the way we can now illustrate quantum effects.

Those scale factors, we added to the metrics, contain what is called wave functions in Quantum Theory and now we can easily (more or less) understand them. A scale to a metric namely changes the volume of the space-time it describes. Thus, our scale functions F, which do contain the wave functions f (being themselves functions of all dimensions of the space-time in question), are just changes in volume of space and time, different at every world point and thereby leading to the permanent jitter of everything that there was, is, and will be.

The distribution of a spinor as shown in Figs. 9.2 and 9.3 in sub-section "9.3.6 A Possible Origin of Spin $l = 1/2$" gives a descriptive example.

Thereby we explicitly choose this example, because it comes with a "complex" problem, namely the occurrence of imaginary portions of results. Usually in Quantum Theory, one simply accepts the presence of such portions, forms the absolute value of the result, calls this a probability density, and everything seems to be in order. We, on the other hand, think that it cannot be so simple. Yes, in section "7.8 Appendix: Taking Care About Complex Curvatures R or R^*" we took care about complex results by forcing the Ricci scalar to give absolute values in the usual quantum theoretical (probability) manner, but can we be sure that this way we do not erase certain information?

After all, we found that all fundamental quantum equations can be extracted from the concept of the scaled metric. Whether in the complicated (tensor transformation-like) form (1011) (see chapter 11) or a simple factor (1012), does not matter here. Thus, the occurrence of complex numbers in connection with the scale factors or scale functions must have a deeper meaning than just the one of the intermediate byproducts. After all, negative numbers can be seen as borrowed length elements, complex numbers as borrowed (or missing) surfaces, and so on. We are not going to completely solve this important riddle of interpretation of the results, we show here in this chapter, but we want to point out, that—obviously—the classical probability function thesis or approach does not seem to be the be-all and end-all of conclusions.

One point, however, seems to have become clear. We already found that the demand for a Ricci-flat space-time $R = 0$ and $R^* = 0$ already gives us the necessary equations for all interactions except gravity. The latter comes in after variation with respect to the metric. In chapter 11 (see also [23]), it will be shown that also the $R = 0$ and the $R^* = 0$ condition originates from the Einstein–Hilbert variation ... if only been done in a complete manner.

9.3   Derivation of the Metric Dirac Equation

In a variety of previous sections of this book and a few papers, we have shown how a metric Dirac equation can be derived from Eq. (1008) (e. g. see [3]). Here now we want to revisit our derivation and show a few more options. Thereby we want to point out that Dirac's original approach [19] started with an extension of the wave function f within the Klein–Gordon equation as follows:

$$0 = \Delta f - B^2 f \xrightarrow{\text{Dirac}} 0 = \Delta f_\lambda - B^2 f_\lambda. \tag{1022}$$

We see that Dirac had made the function **f** a vector and then "proceeded from there"[4] with his famous evaluation, resulting in the classical Dirac equation (further below, e.g., Eq. (1028)). In insisting on a scalar f, we will also investigate the possibilities of intrinsically structured wave functions (c.f. (1061)). This is not strictly the same way as Dirac went, but nevertheless, we will here name it after the great physicist and refer to it as Dirac or Dirac-like approach. In order to work out the similarities (and differences), we will start with Dirac's original work, which is to say the direct introduction of the **f**-vector, and then slip in the intrinsic f-structure and its implications on the Dirac path.

---

[4]Historians will insist that Dirac in fact had first extracted the quaternion root of the Klein–Gordon equation and then introduced the **f**-vector, but this is not of importance here. For the purists, however, and in order to point this fact out, we had set the formulation "proceeded from there" in "...".

9.3.1 Dirac's "Luck" in Connection with the Scaled Ricci Scalar

Applying Eq. (1008) and demanding $F = f + C_f$, $R = R^* = 0$, we obtain:

$$0 = \frac{F'}{F}(1-n)\Delta f + \frac{f_{,\alpha}f_{,\beta}g^{\alpha\beta}(1-n)}{4F^2}\left(4FF'' + (F')^2(n-6)\right)$$

$$= \frac{1}{f+C_f}(1-n)\Delta f + \frac{f_{,\alpha}f_{,\beta}g^{\alpha\beta}(1-n)}{4(f+C_f)^2}(n-6)$$

$$= (f+C_f)\cdot(1-n)\Delta f + \frac{f_{,\alpha}f_{,\beta}g^{\alpha\beta}(1-n)}{4}(n-6)$$

$$\xrightarrow{w=(f+C_f)}$$

$$w\cdot(1-n)\Delta w + \frac{w_{,\alpha}w_{,\beta}g^{\alpha\beta}(1-n)}{4}(n-6) = 0$$

$$\Rightarrow \qquad w\cdot\Delta w + \frac{w_{,\alpha}w_{,\beta}g^{\alpha\beta}}{4}(n-6) = 0. \qquad (1023)$$

Now we assume to have $n = 4$, in Cartesian coordinates and by applying the Dirac matrices as given in Eq. (1082) in the sub-section below, we can rewrite the last line of Eq. (1023) as follows:

$$w\cdot\Delta w - \frac{w_{,\alpha}w_{,\beta}g^{\alpha\beta}}{2} = w^\lambda\cdot\gamma^\beta\partial_\beta\gamma^\alpha\partial_\alpha w_\lambda - \frac{\gamma^\beta w^\lambda_{,\beta}\gamma^\alpha w_{\lambda,\alpha}}{2}$$

$$\Rightarrow \qquad w^\lambda\cdot\gamma^\beta\partial_\beta\gamma^\alpha\partial_\alpha w_\lambda = \frac{\gamma^\beta w^\lambda_{,\beta}\gamma^\alpha w_{\lambda,\alpha}}{2}. \qquad (1024)$$

Thereby—in accordance with the Dirac theory [19]—we had to make the function **w** a vector. Similarly, we may also assume a somehow "innerly" structured **w** with a constant contra-variant "counter"-vector $\mathbf{q}^\lambda$ (for more see sub-section below):

$$w\cdot\Delta w - \frac{w_{,\alpha}w_{,\beta}g^{\alpha\beta}}{2} \xrightarrow{w=q^\lambda h_\lambda} w\cdot q^\lambda\gamma^\beta\partial_\beta\gamma^\alpha\partial_\alpha h_\lambda - \frac{q^\lambda\gamma^\beta h_{\lambda,\beta}q^\lambda\gamma^\alpha h_{\lambda,\alpha}}{2}$$

$$\Rightarrow \qquad w\cdot q^\lambda\gamma^\beta\partial_\beta\gamma^\alpha\partial_\alpha h_\lambda = \frac{q^\lambda\gamma^\beta h_{\lambda,\beta}q^\lambda\gamma^\alpha h_{\lambda,\alpha}}{2}. \qquad (1025)$$

Now we know that the Dirac matrices, combined to the Laplace operator as done in Eqs. (1024) and (1025), assure the outcome of Eigen solutions (because the Dirac equation is something like the operational square root of the Klein–Gordon equation). Thus, we can rewrite Eq. (1025) in the following way ((1024) would just be the same):

$$w \cdot \underbrace{\Delta w}_{=B^2 w} - \frac{w_{,\alpha} w_{,\beta} g^{\alpha\beta}}{2} \xrightarrow{w = q^\lambda h_\lambda} w \cdot q^\lambda \gamma^\beta \partial_\beta \gamma^\alpha \partial_\alpha h_\lambda - \frac{q^\lambda \gamma^\beta h_{\lambda,\beta} q^\lambda \gamma^\alpha h_{\lambda,\alpha}}{2}$$

$$\Rightarrow \qquad w \cdot \underbrace{q^\lambda \gamma^\beta \partial_\beta \gamma^\alpha \partial_\alpha h_\lambda}_{=B^2 w} = B^2 w^2 = \frac{q^\lambda \gamma^\beta h_{\lambda,\beta} q^\lambda \gamma^\alpha h_{\lambda,\alpha}}{2}. \tag{1026}$$

Thereby the Dirac approach gave back the Klein–Gordon result in the following way (c.f. the original work of Dirac [19] or some text books like [13]):

$$0 = q^\lambda \gamma^\beta \partial_\beta \gamma^\alpha \partial_\alpha h_\lambda - B^2 q^\lambda h_\lambda = q^\lambda \left(\gamma^\beta \partial_\beta \gamma^\alpha \partial_\alpha h_\lambda - B^2 h_\lambda \right)$$

$$= q^\lambda \left(\gamma^\beta \partial_\beta + B \right) \left(\gamma^\alpha \partial_\alpha - B \right) h_\lambda. \tag{1027}$$

From there, Dirac gave his equation as follows:

$$0 = q^\lambda \left(\gamma^\alpha \partial_\alpha - B \right) h_\lambda \Rightarrow 0 = \left(\gamma^\alpha \partial_\alpha - B \right) h_\lambda. \tag{1028}$$

Here a little reminder is of need:

It should explicitly be noted that—even though the Dirac equation is often given in the co-variant form of Eq. (1028)—the vector of functions $\mathbf{h}_\lambda$ is not an ordinary vector in the coordinates of our Minkowski space-time, but a vector of solutions to the Klein–Gordon equation. Usually, this is been elaborated with the example of the particle at rest where we have two matter forms (matter and antimatter) and two spins (spin up and spin down). Thus, the summation λ is not the same as the summation over the space-time indices and should therefore better be distinguished via a different choice of symbols for the indices for the Dirac sums and the ordinary coordinate sums. This will be done later in the chapter. Here, for compatibility with the classical theory and its many text-book appearances, we stick to the Greek symbols and just remember that—just here in this sub-section—λ or λ_x ($x = 1, 2, \ldots$) is been used for the Dirac sums in connection with the Dirac matrices.

We also point out that Dirac did not introduce his famous equation in the form of the left-hand side of Eq. (1028), but directly presented the right-hand side without any $\mathbf{q}^\lambda$ counterpart. In fact, Dirac had no such counterpart, because he never considered an intrinsically structured wave function f. This saved him quite some trouble. In order to show what trouble is meant here, we repeat the root extraction process of Eq. (1027) and give a few more details.

Dirac started his work with the Klein–Gordon equation, which—in Cartesian coordinates—we here give as follows[5]:

$$0 = \partial_\beta g^{\alpha\beta} \partial_\alpha f - B^2 f. \tag{1029}$$

Then he applied his gamma matrices and rewrote Eq. (1029) in the following way:

$$0 = \gamma^\beta \partial_\beta \gamma^\alpha \partial_\alpha f - B^2 f. \tag{1030}$$

As, however, the gamma matrices require summation with "something" (here the function f), Dirac concluded that **f** has to be a vector instead of a function and he deduced, respectively postulated the equation(s):

$$0 = \gamma^{\lambda 1 \beta}_{\lambda 3} \partial_\beta \gamma^{\lambda 2 \alpha}_{\lambda 1} \partial_\alpha f_{\lambda 2} - B^2 f_{\lambda 2} = \left(\gamma^{\lambda 1 \beta}_{\lambda 3} \partial_\beta + B \right) \left(\gamma^{\lambda 2 \alpha}_{\lambda 1} \partial_\alpha - B \right) f_{\lambda 2}$$

$$= \left(\gamma^{\lambda 1 \beta}_{\lambda 3} \partial_\beta - B \right) \left(\gamma^{\lambda 2 \alpha}_{\lambda 1} \partial_\alpha + B \right) f_{\lambda 2}$$

$$\Rightarrow \qquad 0 = \left(\gamma^{\lambda 2 \alpha}_{\lambda 1} \partial_\alpha - B \right) f_{\lambda 2}$$

$$\Rightarrow \qquad 0 = \left(\gamma^{\lambda 2 \alpha}_{\lambda 1} \partial_\alpha + B \right) f_{\lambda 2}. \tag{1031}$$

In order—just for formal reasons—to keep the index-situation clean we may add a Kronecker symbol to the equations above as follows:

$$0 = \gamma^{\lambda 1 \beta}_{\lambda 3} \partial_\beta \gamma^{\lambda 2 \alpha}_{\lambda 1} \partial_\alpha f_{\lambda 2} - \delta^{\lambda 2}_{\lambda 3} B^2 f_{\lambda 2} = \left(\gamma^{\lambda 1 \beta}_{\lambda 3} \partial_\beta + \delta^{\lambda 1}_{\lambda 3} B \right) \left(\gamma^{\lambda 2 \alpha}_{\lambda 1} \partial_\alpha - \delta^{\lambda 2}_{\lambda 1} B \right) f_{\lambda 2}$$

$$= \left(\gamma^{\lambda 1 \beta}_{\lambda 3} \partial_\beta - \delta^{\lambda 1}_{\lambda 3} B \right) \left(\gamma^{\lambda 2 \alpha}_{\lambda 1} \partial_\alpha + \delta^{\lambda 2}_{\lambda 1} B \right) f_{\lambda 2}$$

$$\Rightarrow \qquad 0 = \left(\gamma^{\lambda 2 \alpha}_{\lambda 1} \partial_\alpha - \delta^{\lambda 2}_{\lambda 1} B \right) f_{\lambda 2}$$

$$\Rightarrow \qquad 0 = \left(\gamma^{\lambda 2 \alpha}_{\lambda 1} \partial_\alpha + \delta^{\lambda 2}_{\lambda 1} B \right) f_{\lambda 2}. \tag{1032}$$

When insisting on a scalar f starting point and outcome, however, we have to rewrite this as:

$$0 = \Delta f - B^2 f$$

$$\xrightarrow{\quad f = q^\lambda h_\lambda \quad}$$

$$0 = q^{\lambda 3} \left(\gamma^{\lambda 1 \beta}_{\lambda 3} \partial_\beta \gamma^{\lambda 2 \alpha}_{\lambda 1} \partial_\alpha h_{\lambda 2} - \delta^{\lambda 2}_{\lambda 3} B^2 h_{\lambda 2} \right) = q^{\lambda 3} \left(\gamma^{\lambda 1 \beta}_{\lambda 3} \partial_\beta + \delta^{\lambda 1}_{\lambda 3} B \right) \left(\gamma^{\lambda 2 \alpha}_{\lambda 1} \partial_\alpha - \delta^{\lambda 2}_{\lambda 1} B \right) h_{\lambda 2}$$

$$= q^{\lambda 3} \left(\gamma^{\lambda 1 \beta}_{\lambda 3} \partial_\beta - \delta^{\lambda 1}_{\lambda 3} B \right) \left(\gamma^{\lambda 2 \alpha}_{\lambda 1} \partial_\alpha + \delta^{\lambda 2}_{\lambda 1} B \right) h_{\lambda 2}$$

---

[5]In fact, Dirac's goal was to find an equation of first order in the time derivative, equivalent to the Schrödinger equation and so he started more like: $\frac{1}{c^2} \cdot \partial^2_{0 \hat{=} t} f = \gamma^\beta \partial_\beta \gamma^\alpha \partial_\alpha f - \beta^2 \cdot B^2 f; \quad \alpha, \beta = 1,2,3 \hat{=} x, y, z.$ From there he could directly obtain the first-order time-derivative on the left-hand side by a simple root extraction. Naturally this led him to his quaternions on the right-hand side.

$$\Rightarrow \qquad 0 = \left(\gamma_{\lambda 1}^{\lambda 2 \alpha} \partial_\alpha - \delta_{\lambda 1}^{\lambda 2} B \right) h_{\lambda 2}$$

$$\Rightarrow \qquad 0 = \left(\gamma_{\lambda 1}^{\lambda 2 \alpha} \partial_\alpha + \delta_{\lambda 1}^{\lambda 2} B \right) h_{\lambda 2} \tag{1033}$$

and realize that we would need a change of indices, respectively a reordering of the latter, because we had to demand:

$$f = q^\lambda h_\lambda \rightarrow f_{\lambda 2}^{\lambda 3} = q^{\lambda 3} h_{\lambda 2}. \tag{1034}$$

We note that Dirac had not decomposed f (or w in our case as introduced in Eq. (1023)) in the way we did here, but directly introduced the vector function he needed as a collection of solutions (solutions to the Klein–Gordon equation). As one can deduce from the first line of Eq. (1031), this collection of solutions still shows up in the end, when the two Dirac operators are being combined to form the Klein–Gordon equation again. Thus, in principle, Dirac already assumed a Klein–Gordon equation acting not on a scalar, but on a function vector.

Now we shall proceed with our investigation of Eq. (1023) in connection with the classical Dirac theory. Thereby we will—for the time being—ignore the index issues just elaborated.

Extraction of the square root in Eq. (1023) (last line), however, leads us to:

$$\sqrt{2} \cdot B \cdot w = \sqrt{2} \cdot B \cdot q^\lambda h_\lambda = q^\lambda \gamma^\alpha h_{\lambda,\alpha}$$

$$\Rightarrow \qquad q^\lambda \left(\gamma^\alpha h_{\lambda,\alpha} - \sqrt{2} \cdot B \cdot h_\lambda \right) = 0$$

$$\Rightarrow \qquad \gamma^\alpha h_{\lambda,\alpha} - \sqrt{2} \cdot B \cdot h_\lambda = 0 \quad ?, \tag{1035}$$

where we also recognize the Dirac equation (c.f. (1079) further below), but with a different outcome as Eigenvalue than Eq. (1028). While the Laplace operator part in Eq. (1023) (last line) gives us B^2, the $f_{,\alpha} f_{,\beta} g^{\alpha\beta}$ operator results in $-2 * B^2$. This will not fulfill the last line of Eq. (1023) as it should when we demand $R = 0$.

So, something is not quite right here! In order to make the Dirac approach solve both operator equations (Laplace and $f_{,\alpha} f_{,\beta} g^{\alpha\beta}$) in the same way, which is to say with the same Eigenvalue as outcome and then sum up to zero in Eq. (1023), we have to find a suitable function $F[f]$ first. In order to find this function, we have to take the first line of Eq. (1023) and extract the following differential equation for $F[f]$:

$$0 = F' \Delta f + \frac{f_{,\alpha} f_{,\beta} g^{\alpha\beta}}{4F} \left(4FF'' + \left(F' \right)^2 (n-6) \right)$$

$$0 = F' B^2 f + f^2 \frac{B^2}{4F} \left(4FF'' + \left(F' \right)^2 (n-6) \right)$$

$$\Rightarrow \qquad 0 = F'f + \frac{f^2}{4F}\left(4FF'' + \left(F'\right)^2 (n-6)\right)$$

$$\xrightarrow{\quad n=4 \quad}$$

$$\Rightarrow \qquad 0 = 2 \cdot F \cdot \left(F' + f \cdot F''\right) - f \cdot \left(F'\right)^2. \qquad (1036)$$

The last line gives us the solution:

$$F[f] = C_{f1}\left(C_{f0} \pm \ln[f]\right)^2. \qquad (1037)$$

Interestingly, when assuming positive real C_{f0} and taking the "+," this solution has a minimum in its real part for non-zero f (for $f > 0$ the imaginary part vanishes anyway). So, if there were not the pole at $f = 0$, we'd be reminded of the structure of the Higgs field.

On the other hand, we may assume the situation where—for whatever reason—the Laplace operator delivers the Eigenvalue B^2 while $f_{,\alpha}f_{,\beta}g^{\alpha\beta}$ operator gives $-B^2$.

In this case a few things in Eq. (1023) (respectively (1008)) change and Eq. (1036) has to be reevaluated as follows:

$$0 = F'\Delta f + \frac{f_{,\alpha}f_{,\beta}g^{\alpha\beta}}{4F}\left(4FF'' + \left(F'\right)^2 (n-6)\right)$$

$$0 = -F'B^2 f + f^2 \frac{B^2}{4F}\left(4FF'' + \left(F'\right)^2 (n-6)\right)$$

$$\Rightarrow \qquad 0 = -F'f + \frac{f^2}{4F}\left(4FF'' + \left(F'\right)^2 (n-6)\right)$$

$$\xrightarrow{\quad n=4 \quad}$$

$$\Rightarrow \qquad 0 = 2 \cdot F \cdot \left(-F' + f \cdot F''\right) - f \cdot \left(F'\right)^2 \qquad (1038)$$

and the corresponding solution for $F[f]$ reads:

$$\Rightarrow \qquad F[f] = \left(C_{f0}^2 \pm f^2\right)^2 \cdot C_{f1}. \qquad (1039)$$

Here now, we exactly have the typical so-called Mexican hat Higgs field structure (for more see section "9.5 Higgs with $F[f]$" below in this chapter). The only question would be how one could construct the opposite signs for the Eigenvalues of the Laplace and the $f_{,\alpha}f_{,\beta}g^{\alpha\beta}$ operator. But even without the demand of Eigenvalue solutions and sticking to $n = 4$, $R = 0$, and Eq. (1039), we'd end up with:

$$0 = f \cdot \Delta f + f_{,\alpha}f_{,\beta}g^{\alpha\beta}. \qquad (1040)$$

Even though we are not considering this equation any further here, we deduce from the particle at rest case of Eq. (1040), which would read:

$$0 = f \cdot f_{,tt} + f_{,t}f_{,t} \tag{1041}$$

that we will probably not find any suitable solutions for Eq. (1040) (not in simple cases with $f[t]$ or $f[r]$-dependencies anyway).

Nevertheless, we found that with the setting Eq. (1037), the Dirac equation solves (1008) with the conditions $R^* = R = 0$ and $n = 4$.

We conclude that, due to the fact that the classical Dirac equation just assures the outcome of Eigen solutions to both, the Laplace and the $f_{,\alpha}f_{,\beta}g^{\alpha\beta}$ operator in Eq. (1023) (last line), automatically also gives solutions to the flat space as long as we also demand Eq. (1037) for the function wrapper $F[f]$.

So, no wonder that the Dirac theory works so well, as its ingenuous structure already provides some metric compatibility. In the case of massless particles, this equation solves both the Laplace operator and the $f_{,\alpha}f_{,\beta}g^{\alpha\beta}$ operator. With masses, this equation is a solution for the flat space condition $R^* = R = 0$ in connection with the condition (1037). Furthermore, in 4 dimensions, the Dirac equation gives a total solution to Eq. (1008) with the condition $R^* = 0$:[6]

$$0 = FF'' - 2 \cdot \left(F'\right)^2 \quad \Rightarrow \quad F[f] = C_{f1} \cdot \left(C_{f0} \pm f\right)^2, \tag{1042}$$

leading to:

$$0 = R \cdot \left(C_{f0} \pm f\right) \mp 6 \cdot \Delta f, \tag{1043}$$

where the Ricci scalar R takes on the position of the Eigenvalue for the Klein–Gordon equation. Please note that—due to condition $R^* = 0$—the space-time is still flat even though $R \neq 0$, because the scaling factor to the metric effectively compensates for the R-appearances.

What about the situation with non-vanishing Ricci curvature R in those cases where we do not also demand the condition (1042)?

Well, concentrating first on the intrinsically structured function f, this changes Eq. (1026) to:

$$3 \cdot R \cdot w^2 = w \cdot q^\lambda \gamma^\beta \partial_\beta \gamma^\alpha \partial_\alpha h_\lambda - \frac{q^\lambda \gamma^\beta h_{\lambda,\beta} q^\lambda \gamma^\alpha h_{\lambda,\alpha}}{2}, \tag{1044}$$

which we might also write as:

$$3 \cdot R \cdot w^2 = \left[w \cdot q^\lambda \gamma^\beta \partial_\beta - \frac{q^\lambda \gamma^\beta h_{\lambda,\beta}}{2}\right] q^\lambda \gamma^\alpha h_{\lambda,\alpha}. \tag{1045}$$

---

[6]It has to be pointed out that Eq. (1042) could also be interpreted as a Higgs field as long as f has an internal structure like $f = h^2$. This will be considered around Eq. (1059) and further below in this chapter.

We realize that we have to take care about the co- and contra-variant form of the constant vectors $\mathbf{q}^\lambda$, because inserting a general Dirac solution from Eq. (1028) via:

$$\gamma^\alpha h_{\lambda,\alpha} - B \cdot h_\lambda = 0, \tag{1046}$$

which gives:

$$3 \cdot R \cdot w^2 = w \cdot B^2 w - \frac{q^\lambda \gamma^\beta h_{\lambda,\beta} q^\lambda \gamma^\alpha h_{\lambda,\alpha}}{2} = B^2 \cdot w^2 - \overbrace{\frac{q^\lambda B q^\lambda B}{2}}^{?} \cdot w^2. \tag{1047}$$

So, we better try the following for the Dirac equation:

$$\gamma^\alpha h_{\lambda,\alpha} - B \cdot h_\lambda = 0$$

$$\gamma^\alpha h^\lambda{}_{,\alpha} - B \cdot h^\lambda = 0 \tag{1048}$$

and

$$3 \cdot R \cdot w^2 = w \cdot q^\lambda \gamma^\beta \partial_\beta \gamma^\alpha \partial_\alpha h_\lambda - \frac{q_\lambda \gamma^\beta h^\lambda{}_{,\beta} q^\lambda \gamma^\alpha h_{\lambda,\alpha}}{2}, \tag{1049}$$

instead of Eq. (1044), which yields:

$$3 \cdot R \cdot w^2 = \left(B^2 - \overbrace{\frac{q_\lambda B q^\lambda B}{2}}^{!} \right) \cdot w^2 = \left(B^2 - \frac{B^2}{2} \right) \cdot w^2 = \frac{B^2}{2} \cdot w^2$$

$$\Rightarrow \qquad\qquad B^2 = 6 \cdot R. \tag{1050}$$

This, so it seems, fixes the curvature of space-time to an object with Dirac Eigenvalue B to:

$$R = \frac{B^2}{6}. \tag{1051}$$

Apart from the fact that we should see a constant Ricci curvature kind of problematic, which we will discuss below again, we also find other issues in our simple direct application of the Dirac equation within the metric picture when trying for intrinsically structured function f. For once, we find that there needs to be an interchange of the order of summation, because Eq. (1049) should read:

$$3 \cdot R \cdot w^2 = w \cdot q^\lambda \gamma^\beta \partial_\beta \gamma^\alpha \partial_\alpha h_\lambda - \frac{q_{\lambda 1} \gamma^\beta h^{\lambda 1}{}_{,\beta} q^{\lambda 2} \gamma^\alpha h_{\lambda 2,\alpha}}{2}, \tag{1052}$$

ending up in:

$$3 \cdot R \cdot w^2 = \left(B^2 - \overbrace{\frac{q_{\lambda1} B q^{\lambda2} B}{2}}^{!} \right) \cdot w^2 = B^2 \left(1 - \overbrace{\frac{q_{\lambda1} q^{\lambda2}}{2}}^{!} \right) \cdot w^2. \quad (1053)$$

This is similar to the index-problem already mentioned in connection with the classical Dirac theory and Eqs. (1033) and (1034).

We conclude that the quaternion structure of the Dirac approach and the metric theory cannot be brought together as simply as we have tried in this sub-section, at least not in the case of intrinsic structures for the wave function f.

Do things improve for the $R \neq 0$-situation when using sets of "external" Klein–Gordon solutions $\mathbf{f}_\lambda$ as Dirac did?

This time Eq. (1026) changes to:

$$3 \cdot R \cdot \left(w_\lambda \right)^2 = w_\lambda \cdot \gamma^\beta \partial_\beta \gamma^\alpha \partial_\alpha w_\lambda - \frac{\gamma^\beta w_{\lambda,\beta} \gamma^\alpha w_{\lambda,\alpha}}{2}, \quad (1054)$$

respectively:

$$3 \cdot R \cdot \left(w_\lambda \right)^2 = \left[w_\lambda \cdot \gamma^\beta \partial_\beta - \frac{\gamma^\beta w_{\lambda,\beta}}{2} \right] \gamma^\alpha w_{\lambda,\alpha}. \quad (1055)$$

This time we have no problem with the constant vectors $\mathbf{q}^\lambda$ and inserting a general Dirac solution as from Eq. (1028) via:

$$\gamma^\alpha w_{\lambda,\alpha} - B \cdot w_\lambda = 0, \quad (1056)$$

gives:

$$3 \cdot R \cdot \left(w_\lambda \right)^2 = w_\lambda \cdot B^2 w_\lambda - \frac{\gamma^\beta w_{\lambda,\beta} \gamma^\alpha w_{\lambda,\alpha}}{2} = B^2 \cdot \left(w_\lambda \right)^2 - \frac{B^2}{2} \cdot \left(w_\lambda \right)^2$$

$$\Rightarrow \qquad 3 \cdot R = \frac{B^2}{2}. \quad (1057)$$

We directly end up with our condition (1051), which connects the Eigenvalues B with the Ricci scalar R.

Just out of interest, we repeat the last evaluation with the interesting condition (1039) (interesting because this $F[f]$ has the shape of the Higgs field). With $R \neq 0$ and in the style of Eq. (1026), Eq. (1040) reads:

$$3 \cdot R \cdot \left(w_\lambda \right)^2 = w_\lambda \cdot \gamma^\beta \partial_\beta \gamma^\alpha \partial_\alpha w_\lambda + \gamma^\beta w_{\lambda,\beta} \gamma^\alpha w_{\lambda,\alpha}$$

$$\xrightarrow{\gamma^\alpha w_{\lambda,\alpha} - B \cdot w_\lambda = 0}$$

$$3 \cdot R \cdot \left(w_\lambda \right)^2 = w_\lambda \cdot B^2 \cdot w_\lambda + B \cdot w_\lambda B \cdot w_\lambda$$

$$\Rightarrow \qquad 3 \cdot R = 2 \cdot B^2 \quad \Rightarrow \quad R = \frac{2}{3} \cdot B^2. \qquad (1058)$$

Only using a simple direct introduction of the Dirac approach (in a classical and an intrinsic **f**-vector form), we saw that in connection with a metric derivation and interpretation of the Dirac theory, we have issues with an intrinsically structured function f, but no bigger problems in connection with the classical Dirac approach, where a set of Klein–Gordon solutions forms a vector $\mathbf{f}_\lambda$ and then is subjected to the classical Dirac treatment. Non-zero R may even provide suitable Higgs field-structures for the scalar function $F[f]$. Thereby it should be pointed out that Higgs or Higgs-like fields may also be seen in the options for the functional wrapper $F[f]$ even without non-zero R (c.f. the various settings for $F[f]$ in Eqs. (1037), (1039) or even the simple case (1042) when assuming an intrinsically structured f with $f = h^a \cdot h_a$ or simply just $f \rightarrow h^2$). Especially the simple case (1042) in $n = 4$ is quite interesting when we apply the substitution $f \rightarrow h^2$. This gives us:

$$0 = FF'' - 2 \cdot (F')^2 \quad \Rightarrow \quad F[f] = C_{f1} \cdot (C_{f0} \pm f)^2$$

$$\Rightarrow \qquad F[f = h^2] = C_{f1} \cdot (C_{f0} \pm h^2)^2, \qquad (1059)$$

and makes Eq. (1008) to:

$$0 = R \cdot (C_{f0} \pm h^2) \mp 6 \cdot \Delta h^2 = R \cdot (C_{f0} \pm h^2) \mp \frac{6}{\sqrt{g}} \cdot \partial_\beta \left(\sqrt{g} \cdot g^{\alpha\beta} \partial_\alpha h^2 \right)$$

$$= R \cdot (C_{f0} \pm h^2) \mp \frac{12}{\sqrt{g}} \cdot \partial_\beta \left(\sqrt{g} \cdot g^{\alpha\beta} h \cdot h_{,\alpha} \right)$$

$$= R \cdot (C_{f0} \pm h^2) \mp 12 \cdot \left(h_{,\beta} g^{\alpha\beta} h_{,\alpha} + \frac{h}{\sqrt{g}} \cdot \partial_\beta \left(\sqrt{g} \cdot g^{\alpha\beta} h_{,\alpha} \right) \right)$$

$$= R \cdot (C_{f0} \pm h^2) \mp 12 \cdot \left(h_{,\beta} g^{\alpha\beta} h_{,\alpha} + h \cdot \Delta h \right). \qquad (1060)$$

We see that the wrapper $F[f]$ in Eq. (1059) clearly is of Higgs character and once again we have a combination of the Laplace and the squared Nabla operator in connection with a non-zero Ricci scalar. All terms can be summed up (due to our scaling wrapper $F[f]$, which here gives the impression and action of a Higgs field) to a vanishing total curvature R^*, leaving us with a perfectly flat space-time. The path forward—so it seems—is to use a list of Klein–Gordon solutions, which is to say Eigenvalue solutions to the Laplace operator and then try to solve Eq. (1060) or other versions of Eq. (1008) with it. This, however, is exactly the principal recipe of the Dirac theory.

Thus, taking the fact that Dirac never intended his theory to be compatible with a metric origin like it was worked out here from the kernel of the Einstein–Hilbert action (which is the Ricci scalar R^* of a scaled metric), but purely based it upon the postulation of the existence of a square root to the classical Klein–Gordon equation in 4-dimensional space-times[7] with Minkowski character, we may well consider it an extremely wise or lucky approach he had made there. It is for this reason that we choose our peculiar title with the word "lucky" to this sub-section.

In the following, we are going to reconsider the structure of our starting Eq. (1008) with respect to its connection to the classical Dirac theory with intrinsic and external **f**-vectors.

9.3.2 Repetition: Dirac in the Metric Picture and Its Connection to the Classical Quaternion Form

In connection with our scaling or quantum wrapping approach $F[f]$, it should be pointed out that with the condition for f to be a scalar, there are still many more options to have vector and tensor functions included, namely, e.g., via:

$$f = h + h_\alpha q^\alpha + h_{\alpha\beta} q^{\alpha\beta} + \dots. \tag{1061}$$

It needs to be pointed out that also arbitrary spinor forms can be incorporated into Eq. (1061), which may be written as follows:

$$f = h + h_{\alpha A} q^{\alpha A} + h_{\alpha\beta A} q^{\alpha\beta A} + h_{\alpha\beta AB} q^{\alpha\beta AB} \dots. \tag{1062}$$

Going back to Eq. (1008), we now assume the Laplace operator to have Eigen solutions, which give us:

$$R^* = R^*{}_{\alpha\beta} G^{\alpha\beta} = \left(R + \frac{F'}{F}(1-n)\Delta f + \frac{f_{,\alpha} f_{,\beta} g^{\alpha\beta}(1-n)}{4F^2}\left(4FF'' + \left(F'\right)^2(n-6)\right) \right) \frac{1}{F}$$

with: $$F = F[f]; \quad F' = \frac{\partial F[f]}{\partial f}; \quad F'' = \frac{\partial^2 F[f]}{\partial f^2}$$

$$\xrightarrow{\Delta f = B^2 f}$$

$$= \left(R + \frac{F'}{F}(1-n)B^2 f + \frac{f_{,\alpha} f_{,\beta} g^{\alpha\beta}(1-n)}{4F^2}\left(4FF'' + \left(F'\right)^2(n-6)\right) \right) \frac{1}{F}. \tag{1063}$$

---

[7]Mainly in order to obtain a relativistic (Special Theory of Relativity) quantum equation with a first-order derivative in time.

As F[f] could still be freely chosen, we assume (just as within the derivation of the Schrödinger equation in section 6.11 in this book) F being linear dependent on f and set:

$$F = f - C_f.\qquad(1064)$$

This simplifies Eq. (1063) to:

$$R^* = R^*_{\alpha\beta}G^{\alpha\beta} = \left(R + \frac{1}{f - C_f}(1-n)\Delta f + \frac{f_{,\alpha}f_{,\beta}g^{\alpha\beta}(1-n)}{4(f - C_f)^2}(n-6)\right)\frac{1}{f - C_f}.$$

$$(1065)$$

Now we demand:

$$R^* = R^*_{\alpha\beta}G^{\alpha\beta} = 0\qquad(1066)$$

and obtain:

$$0 = \frac{4 \cdot R \cdot (f - C_f)^2}{(1-n)\cdot(n-6)} + \frac{4 \cdot (f - C_f)}{n-6} \cdot B^2 \cdot f + f_{,\alpha}f_{,\beta}g^{\alpha\beta}$$

$$= \frac{4}{(1-n)\cdot(n-6)}\left(R \cdot \left(f^2 - 2\cdot C_f \cdot f + C_f^2\right) + (1-n)\cdot\left(f^2 - C_f \cdot f\right)\cdot B^2\right) + f_{,\alpha}f_{,\beta}g^{\alpha\beta}$$

$$= \frac{4}{(1-n)\cdot(n-6)}\left(\begin{array}{c} f^2 \cdot \left(R + (1-n)\cdot B^2\right) \\ -2\cdot f \cdot \left(C_f \cdot \left(R + \frac{(1-n)\cdot B^2}{2}\right)\right) + C_f^2 \cdot R \end{array}\right) + f_{,\alpha}f_{,\beta}g^{\alpha\beta}$$

$$= \frac{4 \cdot \left(R + (1-n)\cdot B^2\right)}{(1-n)\cdot(n-6)}\left(f^2 - 2\cdot f \cdot \frac{C_f \cdot \left(R + \frac{(1-n)\cdot B^2}{2}\right)}{R + (1-n)\cdot B^2} + \frac{C_f^2 \cdot R}{R + (1-n)\cdot B^2}\right) + f_{,\alpha}f_{,\beta}g^{\alpha\beta}$$

$$= \frac{4 \cdot \left(R + (1-n)\cdot B^2\right)}{(1-n)\cdot(n-6)}\left(\begin{array}{c} \left(f - \dfrac{C_f \cdot \left(R + \frac{(1-n)\cdot B^2}{2}\right)}{R + (1-n)\cdot B^2}\right)^2 \\ -\left(\dfrac{C_f \cdot \left(R + \frac{(1-n)\cdot B^2}{2}\right)}{R + (1-n)\cdot B^2}\right)^2 + \dfrac{C_f^2 \cdot R}{R + (1-n)\cdot B^2} \end{array}\right) + f_{,\alpha}f_{,\beta}g^{\alpha\beta}.$$

$$(1067)$$

We see that we find no reasonable setting for either B or C_f for which we could make the term in big parentheses in the last line of the equation above a complete square. We, therefore, assume that there may exist a function $F[f]$ for which we find a development for Eq. (1063) resulting in:

$$R^* = 0 = \left(R + \frac{F'}{F}(1-n)B^2 f + \frac{f_{,\alpha}f_{,\beta}g^{\alpha\beta}(1-n)}{4F^2}\left(4FF'' + \left(F'\right)^2(n-6)\right)\right)\frac{1}{F}$$

$$\Rightarrow \quad 0 = C_{f1}^2 \cdot \left(f - C_{f0}\right)^2 + f_{,\alpha}f_{,\beta}g^{\alpha\beta}; \quad C_{f0} = C_{f0}[n,B,R]; \quad C_{f1} = C_{f1}[n,B,R].$$

$$(1068)$$

Alternatively, as finding a suitable $F[f]$ seems to be an extremely difficult task, one may also just assume to have an R-dependency of the kind:

$$R = R[f] = \left(R_0 + R_1 \cdot f + R_2 \cdot f^2\right) \cdot C_{f0}^2 \cdot (n-2) \cdot (n-1). \qquad (1069)$$

We could also choose $F = e^{f \cdot C_{f0}} \cdot C_{f1}$. This makes Eq. (1063) with Eq. (1066) to:

$$0 = \frac{R}{(1-n)} + \frac{F'}{F}B^2 f + \frac{f_{,\alpha}f_{,\beta}g^{\alpha\beta}}{4F^2}\left(4FF'' + \left(F'\right)^2(n-6)\right)$$

$$\Rightarrow \qquad 0 = \frac{4F^2}{4FF'' + \left(F'\right)^2(n-6)}\left(\frac{R}{(1-n)} + \frac{F'}{F}B^2 f\right) + f_{,\alpha}f_{,\beta}g^{\alpha\beta}$$

$$0 = \frac{B^2 \cdot C_{f0} \cdot f \cdot (n-1) - R}{C_{f0}^2 \cdot (n-2) \cdot (n-1)} + f_{,\alpha}f_{,\beta}g^{\alpha\beta}$$

$$= f \cdot \frac{B^2}{C_{f0} \cdot (n-2)} - \left(R_0 + R_1 \cdot f + R_2 \cdot f^2\right) + f_{,\alpha}f_{,\beta}g^{\alpha\beta}$$

$$\Rightarrow \qquad R_0 + R_1 \cdot f + R_2 \cdot f^2 - f \cdot \frac{B^2}{C_{f0} \cdot (n-2)} = f_{,\alpha}f_{,\beta}g^{\alpha\beta}$$

$$= R_2 \cdot \left(f^2 + \frac{f}{R_2}\left(R_1 - \frac{B^2}{R_2 \cdot C_{f0} \cdot (n-2)}\right) + \frac{R_0}{R_2}\right) = f_{,\alpha}f_{,\beta}g^{\alpha\beta}. \qquad (1070)$$

Now we demand:

$$\frac{1}{R_2}\left(R_1 - \frac{B^2}{R_2 \cdot C_{f0} \cdot (n-2)}\right) = 2 \cdot \sqrt{\frac{R_0}{R_2}}, \qquad (1071)$$

thereby fixing the constant:

$$C_{f0} = \frac{B^2}{R_2 \cdot \left(R_1 - 2 \cdot \sqrt{R_0 \cdot R_2}\right) \cdot (n-2)} \tag{1072}$$

and subsequently result in:

$$R_2 \cdot \left(f + \sqrt{\frac{R_0}{R_2}}\right)^2 = f_{,\alpha} f_{,\beta} g^{\alpha\beta}. \tag{1073}$$

Now we apply our intrinsic structural setting for f as given in Eq. (1061), restrict it to its simplest form, which is to say:

$$f = h_\lambda q^\lambda \tag{1074}$$

and obtain:

$$R_2 \cdot \left(f + \sqrt{\frac{R_0}{R_2}}\right)^2 = f_{,\alpha} f_{,\beta} g^{\alpha\beta}$$

$$\xrightarrow{\quad f = h_\lambda q^\lambda \quad}$$

$$R_2 \cdot \left(f + \sqrt{\frac{R_0}{R_2}}\right)^2 = f_{,\alpha} f_{,\beta} g^{\alpha\beta} = -h_{\lambda,\alpha} q^\lambda g^{\alpha\beta} h_{\lambda,\beta} q^\lambda \xrightarrow{\quad g^{\alpha\beta} = e^\alpha \cdot e^\beta \quad}$$

$$= -h_{\lambda,\alpha} q^\lambda e^\alpha \cdot e^\beta h_{\lambda,\beta} q^\lambda \xrightarrow{\quad q^\lambda e^\alpha = Q^{\lambda\alpha} \quad} = -h_{\lambda,\alpha} Q^{\lambda\alpha} \cdot Q^{\lambda\beta} h_{\lambda,\beta}. \tag{1075}$$

Please note that in principle the evaluation should have been:

$$R_2 \cdot \left(f + \sqrt{\frac{R_0}{R_2}}\right)^2 = f_{,\alpha} f_{,\beta} g^{\alpha\beta} = -h_{\lambda,\alpha} q^\lambda g^{\alpha\beta} h_{\lambda,\beta} q^\lambda$$

$$= -\left(h_\lambda q^\lambda{}_{,\alpha} + h_{\lambda,\alpha} q^\lambda\right) g^{\alpha\beta} \left(h_\lambda q^\lambda{}_{,\beta} + h_{\lambda,\beta} q^\lambda\right)$$

$$= -h_\lambda q^\lambda{}_{,\alpha} g^{\alpha\beta} h_\lambda q^\lambda{}_{,\beta} - h_{\lambda,\alpha} q^\lambda g^{\alpha\beta} h_\lambda q^\lambda{}_{,\beta} - h_\lambda q^\lambda{}_{,\alpha} g^{\alpha\beta} h_{\lambda,\beta} q^\lambda - h_{\lambda,\alpha} q^\lambda g^{\alpha\beta} h_{\lambda,\beta} q^\lambda. \tag{1076}$$

However, as we have no need to fix both vectors h_λ or q^λ, yet, we assume that one is a vector of constants and by making q^λ this constant vector we obtain Eq. (1075) instead of the complicated result in Eq. (1076).

Now, only in order to obtain the classical Dirac equation, we approximate the curvature term on the left-hand side of Eq. (1075) as a square of the linear function of the scalar product f and therefore we can write:

$$R_2 \cdot \left(f + \sqrt{\frac{R_0}{R_2}} \right)^2 = h_\lambda q^\lambda \mathbf{M} \cdot \mathbf{M} h_\lambda q^\lambda = -h_{\lambda,\alpha} q^\lambda g^{\alpha\beta} h_{\lambda,\beta} q^\lambda$$

$$\xrightarrow{g^{\alpha\beta} = \mathbf{e}^\alpha \cdot \mathbf{e}^\beta} = -h_{\lambda,\alpha} q^\lambda \mathbf{e}^\alpha \cdot \mathbf{e}^\beta h_{\lambda,\beta} q^\lambda$$

$$\Rightarrow \qquad \mp h_\lambda q^\lambda \mathbf{M} = i \cdot h_{\lambda,\alpha} q^\lambda \mathbf{e}^\alpha$$

$$\Rightarrow \qquad 0 = i \cdot h_{\lambda,\alpha} \mathbf{e}^\alpha \pm h_\lambda \mathbf{M}, \tag{1077}$$

$$\Rightarrow \qquad 0 = i \cdot h_{\lambda,\alpha} \mathbf{e}^\alpha \pm h_\lambda \mathbf{M}. \tag{1078}$$

Please note that Eq. (1078) represents a generalized "Dirac" equation in an arbitrary number of dimensions. We say this and simultaneously apply the "..."-signs to emphasize the fact that Eq. (1078) is of Dirac character but does not fully mirror the classical form, yet. However, by reducing this general and arbitrarily n-dimensional equation to 4 dimensions, we almost directly recognize the Dirac equation, only that in our metric case the mass-terms have to be at least a vector, while the Dirac matrices for the derivatives of the function vector $\mathbf{h}_\lambda$ (c.f. [19]) are becoming the metric base vector components. In order to obtain the classical Dirac equation in the natural units form:

$$0 = i \cdot \gamma^\alpha h_{\lambda,\alpha} \pm h_\lambda \cdot m, \tag{1079}$$

which we can also write as:

$$0 = i \cdot \gamma^\alpha h_{\lambda,\alpha} \pm I \cdot h_\lambda \cdot m, \tag{1080}$$

or (taking the "negative masses" into account)

$$0 = i \cdot \gamma^\alpha h_{\lambda,\alpha} + \beta \cdot h_\lambda \cdot m; \quad \beta = \begin{pmatrix} 1 & & & \\ & 1 & & \\ & & -1 & \\ & & & -1 \end{pmatrix}, \tag{1081}$$

and with the Dirac Gamma matrices γ^α and the unit matrix I given as follows:

$$\gamma^0 = \begin{pmatrix} 1 & & & \\ & 1 & & \\ & & -1 & \\ & & & -1 \end{pmatrix}; \quad \gamma^1 = \begin{pmatrix} & & & 1 \\ & & 1 & \\ & -1 & & \\ -1 & & & \end{pmatrix}$$

$$\gamma^2 = \begin{pmatrix} & & & -i \\ & & i & \\ & i & & \\ -i & & & \end{pmatrix}; \quad \gamma^3 = \begin{pmatrix} & & 1 & \\ & & & -1 \\ -1 & & & \\ & 1 & & \end{pmatrix}; \quad I = \begin{pmatrix} 1 & & & \\ & 1 & & \\ & & 1 & \\ & & & 1 \end{pmatrix}, \tag{1082}$$

we simply need to multiply Eq. (1078) with a suitable object **C**. This gives us the identities[8]:

$$\mathbf{C} \cdot \mathbf{e}^\alpha = \gamma^\alpha; \quad \mathbf{C} \cdot \mathbf{M} = I \cdot m, \tag{1083}$$

and provides a connection between the classical quaternion and the metric Dirac equation. As the classical Dirac equation was given in Minkowski coordinates, with base vectors of the kind:

$$\mathbf{e}^0 = \begin{pmatrix} i \\ 0 \\ 0 \\ 0 \end{pmatrix}; \quad \mathbf{e}^1 = \begin{pmatrix} 0 \\ 1 \\ 0 \\ 0 \end{pmatrix}; \quad \mathbf{e}^2 = \begin{pmatrix} 0 \\ 0 \\ 1 \\ 0 \end{pmatrix}; \quad \mathbf{e}^3 = \begin{pmatrix} 0 \\ 0 \\ 0 \\ 1 \end{pmatrix}, \tag{1084}$$

we could find the corresponding structure of the object **C** via the first equation in Eq. (1083) and it becomes clear that we need something of the kind:

$$C^\alpha_\beta \cdot \mathbf{e}^\beta = \gamma^\alpha, \tag{1085}$$

$$\Rightarrow \quad \left\{ \begin{array}{l} C^0_\alpha = \begin{pmatrix} & -i & & \\ & & 1 & \\ & & & -1 \\ & & & & -1 \end{pmatrix}; \quad C^1_\alpha = \begin{pmatrix} & & & 1 \\ & & 1 & \\ & & -1 & \\ & i & & \end{pmatrix} \\[6mm] C^2_\alpha = \begin{pmatrix} & & -i & \\ & i & & \\ & i & & \\ -1 & & & \end{pmatrix}; \quad C^3_\alpha = \begin{pmatrix} & & & 1 \\ & & -1 & \\ & i & & \\ 1 & & & \end{pmatrix}. \end{array} \right. \tag{1086}$$

From here we deduce that **M** must be something more complex than just a vector as only the following approach gives us what we need:

$$C^\alpha_\beta \cdot \mathbf{M} = I \cdot m \quad \Rightarrow \quad C^\alpha_\beta \cdot M^\beta_\lambda = \delta^\alpha_\lambda \cdot m = I \cdot m. \tag{1087}$$

Thereby we need to point out that $h_\lambda q^\lambda \mathbf{M} \cdot \mathbf{M} h_\lambda q^\lambda$ in Eq. (1077) only is a linear approximation of the complex expression $R_2 \cdot \left(f + \sqrt{\dfrac{R_0}{R_2}} \right)^2$ and that

---

[8]It should be noted that classically one does not use the identity matrix for the second equation in Eq. (1083), but the so called β-matrix instead. It reads $\beta = \begin{pmatrix} 1 & & & \\ & 1 & & \\ & & -1 & \\ & & & -1 \end{pmatrix}$, but will not be considered here, as it makes no difference for the results of this chapter whether we use I or β in Eq. (1083).

the classical Dirac equation was just postulated [19], while our new Dirac equation (1073) has a fundamental metric and extremal-principle origin. The result after solving Eq. (1087) with respect to the object M_λ^β reads:

$$\Rightarrow \quad \left\{ \begin{array}{ll} M_0^\alpha = \begin{pmatrix} i & & & \\ & 1 & & \\ & & -1 & \\ & & & -1 \end{pmatrix}; & M_1^\alpha = \begin{pmatrix} & & & -i \\ & & -1 & \\ & 1 & & \\ 1 & & & \end{pmatrix} \\[30pt] M_2^\alpha = \begin{pmatrix} & & & -1 \\ & & -i & \\ & -i & & \\ i & & & \end{pmatrix}; & M_3^\alpha = \begin{pmatrix} & & & -i \\ & & 1 & \\ & 1 & & \\ -1 & & & \end{pmatrix} \end{array} \right. \tag{1088}$$

and gives us the connection of the classical Dirac equation (1079) from [19] with our new metric Dirac equation (1073) in its linear approximated form in Eq. (1077). We realize that in the metric picture the corresponding inertia parameters are not just scalars. With this connection found, we automatically also assure that f solves the Klein–Gordon equation and thus, the condition $\Delta f = B^2 \cdot f$.

It needs to be pointed out here that the summation over β in Eq. (1085) is not the only option we have in order to obtain the appearance of the Dirac gamma matrices. Similarly, we may just perform the evaluation as follows:

$$C_a^{cb} \cdot e^\beta = C_a^{cb} \cdot E_b{}^\beta \equiv D_a^{c\beta} = \gamma^\beta, \tag{1089}$$

$$\Rightarrow \quad \left\{ \begin{array}{ll} C_a^{c0} = \begin{pmatrix} -i & & & \\ & 1 & & \\ & & -1 & \\ & & & -1 \end{pmatrix}; & C_a^{c1} = \begin{pmatrix} & & & 1 \\ & & 1 & \\ & -1 & & \\ i & & & \end{pmatrix} \\[30pt] C_a^{c2} = \begin{pmatrix} & & & -i \\ & & i & \\ & i & & \\ -1 & & & \end{pmatrix}; & C_a^{c3} = \begin{pmatrix} & & & 1 \\ & & -1 & \\ & i & & \\ 1 & & & \end{pmatrix}. \end{array} \right. \tag{1090}$$

Consequently, we obtain for **M**:

$$C_a^{cb} \cdot \mathbf{M} = I \cdot m \quad \Rightarrow \quad C_a^{cb} \cdot M_{b\lambda}^a = \delta_\lambda^c \cdot m = I \cdot m. \tag{1091}$$

The result after solving Eq. (1091) with respect to the object $M^a_{b\lambda}$ reads:

$$\Rightarrow \quad \begin{cases} M^a_{0\lambda} = \begin{pmatrix} & i & & \\ 1 & & & \\ & & -1 & \\ & & & -1 \end{pmatrix}; & M^a_{1\lambda} = \begin{pmatrix} & & & -i \\ & & -1 & \\ & 1 & & \\ 1 & & & \end{pmatrix} \\[30pt] M^a_{2\lambda} = \begin{pmatrix} & & -1 & \\ & -i & & \\ -i & & & \\ i & & & \end{pmatrix}; & M^a_{3\lambda} = \begin{pmatrix} & & & -i \\ & & 1 & \\ 1 & & & \\ & & & -1 \end{pmatrix}. \end{cases} \quad (1092)$$

Here it has to be pointed out, however, that our complex forms of **M** are a result of the approximation from Eq. (1077) (first line) and not necessarily a genuine equivalence to real inertia. More discussion will be of need here.

Insisting on the outcome of the β-matrix with $\beta = \begin{pmatrix} 1 & & & \\ & 1 & & \\ & & -1 & \\ & & & -1 \end{pmatrix}$ in

Eq. (1091), leading to:

$$C^{cb}_a \cdot \mathbf{M} = \beta \cdot m \quad \Rightarrow \quad C^{cb}_a \cdot M^a_{b\lambda} = \beta \cdot m \qquad (1093)$$

does give the following solution with respect to the object $M^a_{b\lambda}$:

$$\Rightarrow \quad \begin{cases} M^a_{0\lambda} = \begin{pmatrix} & i & & \\ 1 & & & \\ & & 1 & \\ & & & 1 \end{pmatrix}; & M^a_{1\lambda} = \begin{pmatrix} & & & i \\ & & 1 & \\ & 1 & & \\ 1 & & & \end{pmatrix} \\[30pt] M^a_{2\lambda} = \begin{pmatrix} & & 1 & \\ & i & & \\ & -i & & \\ i & & & \end{pmatrix}; & M^a_{3\lambda} = \begin{pmatrix} & & & i \\ & & -1 & \\ 1 & & & \\ & & & -1 \end{pmatrix}. \end{cases} \quad (1094)$$

We saw that with a simple matrix multiplication, we had come from our metric Dirac equation (1078) to the classical one (1081):

$$0 = i \cdot h_{\lambda,\alpha} e^\alpha \pm h_\lambda \mathbf{M} / \cdot \mathbf{C}$$

$$\Rightarrow \qquad 0 = i \cdot \gamma^\alpha h_{\lambda,\alpha} + \beta \cdot h_\lambda \cdot m. \qquad (1095)$$

Thereby, in the case of 4-dimensional Minkowski coordinates, the objects C and M are given via Eqs. (1090) and (1094). Our metric Dirac equation (1078), on the other hand, just came from a simple condition about the Ricci scalar. In chapters 10 and 11 (also see [23]), we will show that this is actually just a by-product of a more complete Einstein–Hilbert variation.

The well-informed reader may ask at this point why we explicitly tried to obtain the Dirac equation from the $f_{,\alpha}f_{,\beta}g^{\alpha\beta}$ operator, while Dirac in [19] only extracted the square root (operational factorization) from the Klein–Gordon equation via the introduction of quaternions. The reason for this excursion lies in the fact that the factorization of the $f_{,\alpha}f_{,\beta}g^{\alpha\beta}$ operator is straight forward and works for all kinds of symmetric metrics, while this is not the case for the Laplace operator in the Klein–Gordon equation. In sub-section "9.3.4 Dirac's 'Luck' and the Flat Space Condition," we will show that for certain (scale-Ricci-flat) space-times there is a direct connection between the two operators, which makes our evaluation, even though it does not use Dirac's starting point (the Klein–Gordon equation), above absolutely worthwhile ... if not to say essential!

9.3.3 Intermediate Sum-Up: The Classical and the Intrinsic Dirac Approach

We saw that the classical Dirac theory—in principle but not necessarily in the original work of Dirac—started with the introduction of a vector of functions subjected to the Klein–Gordon equation, reading in 4-dimensional Minkowski coordinates (c.f. Eq. (1022)):

$$0 = \partial_\beta g^{\alpha\beta} \partial_\alpha f - B^2 f \xrightarrow{\text{Dirac}} 0 = \partial_\beta g^{\alpha\beta} \partial_\alpha f_\lambda - B^2 f_\lambda. \qquad (1096)$$

Then Dirac introduced his gamma matrices and obtained a factorized version of Eq. (1096), where just one factor suffices to satisfy the whole (see Eqs. (1027) and (1028)). In formal form, his evaluation should read:

$$0 = \partial_\beta g^{\alpha\beta} \partial_\alpha f_\lambda - B^2 f_\lambda$$

$$= \partial_\beta E_c^{\ \alpha} E^{c\beta} \partial_\alpha f_\lambda - B^2 f_\lambda$$

$$= \begin{cases} \left(\partial_\beta E^{c\beta} + B \right) \left(E_c^{\ \alpha} \partial_\alpha - B \right) f_\lambda \\ \left(\partial_\beta E^{c\beta} - B \right) \left(E_c^{\ \alpha} \partial_\alpha + B \right) f_\lambda \end{cases}$$

$$\Rightarrow \qquad 0 = \left(E_c{}^\alpha \partial_\alpha - B \right) f_\lambda \quad \& \quad 0 = \left(E_c{}^\alpha \partial_\alpha + B \right) f_\lambda \xrightarrow{\;B=?\;}$$

$$\Rightarrow \qquad 0 = \left(E_c{}^\alpha \partial_\alpha - \overset{?}{B} \right) f_\lambda \quad \& \quad 0 = \left(E_c{}^\alpha \partial_\alpha + \overset{?}{B} \right) f_\lambda. \qquad (1097)$$

Already here, we realize that B cannot be a scalar, but has to be a vector like the base-vectors $E_c{}^\alpha, E^{c\beta}$. We marked this for the moment by putting a "?"-sign over the symbol. Now we multiply the last line of Eq. (1097) with an object $Q_a^{\lambda c}$:

$$\Rightarrow \qquad 0 = Q_a^{\lambda c}\left(E_c{}^\alpha \partial_\alpha - \overset{?}{B} \right) f_\lambda \quad \& \quad 0 = Q_a^{\lambda c}\left(E_c{}^\alpha \partial_\alpha + \overset{?}{B} \right) f_\lambda \qquad (1098)$$

and it is clear that B should become B_c, leading us to:

$$\Rightarrow \qquad 0 = Q_a^{\lambda c}\left(E_c{}^\alpha \partial_\alpha - B_c \right) f_\lambda \quad \& \quad 0 = Q_a^{\lambda c}\left(E_c{}^\alpha \partial_\alpha + B_c \right) f_\lambda. \qquad (1099)$$

Still there is the problem of the scalar product B^2 being properly formed out of the vectors $B_c{}^* B^c$, but this will be discussed below at the end of this sub-section.

The resulting products:

$$Q_a^{\lambda c} E_c{}^\alpha = \gamma^\alpha; \quad Q_a^{\lambda c} B_c = \beta_a^\lambda \qquad (1100)$$

are the Dirac gamma matrices and another object β which has no expression in the classical Dirac theory and the postulated Dirac equation.

Now we perform the evaluation above for an intrinsically structured function f. As we will deepen this investigation later in this chapter, we here only consider the simplest case assuming a scalar product $f = q^c h_c$, with a vector of constants q^c, changing Eq. (1096) as follows:

$$0 = \partial_\beta g^{\alpha\beta} \partial_\alpha f - B^2 f \xrightarrow{\;f = q^b h_b\;} 0 = q^b \left(\partial_\beta g^{\alpha\beta} \partial_\alpha h_b - B^2 h_b \right). \qquad (1101)$$

Now the equivalent evaluation to Eq. (1097) looks as follows:

$$0 = q^b \left(\partial_\beta g^{\alpha\beta} \partial_\alpha h_b - B^2 h_b \right)$$

$$= q^b \left(\partial_\beta E_c{}^\alpha E^{c\beta} \partial_\alpha h_b - B^2 h_b \right)$$

$$= \begin{cases} q^b \left(\partial_\beta E^{c\beta} + B \right)\left(E_c{}^\alpha \partial_\alpha - B \right) h_b \\ q^b \left(\partial_\beta E^{c\beta} - B \right)\left(E_c{}^\alpha \partial_\alpha + B \right) h_b \end{cases}$$

$$\Rightarrow \qquad 0 = q^b\left(E_c{}^\alpha \partial_\alpha - B\right)h_b \quad \& \quad 0 = q^b\left(E_c{}^\alpha \partial_\alpha + B\right)h_b \xrightarrow{\;B=?\;}$$

$$0 = q^b\left(E_c{}^\alpha \partial_\alpha - \overset{?}{B}\right)h_b \quad \& \quad 0 = q^b\left(E_c{}^\alpha \partial_\alpha + \overset{?}{B}\right)h_b. \tag{1102}$$

As before with the Dirac path, we have marked the problematic B-terms with a "?"-sign. Again we multiply the last line of Eq. (1102) with an object Q_{ab}^c (we note the switch of order for the co-variant indices in comparison to Eq. (1098)):

$$\Rightarrow \qquad 0 = Q_a^c q^b\left(E_c{}^\alpha \partial_\alpha - \overset{?}{B}\right)h_b \quad \& \quad 0 = Q_a^c q^b\left(E_c{}^\alpha \partial_\alpha + \overset{?}{B}\right)h_b \tag{1103}$$

and—as before—making B to B_c, is leading us to:

$$\Rightarrow \qquad 0 = Q_a^c q^b\left(E_c{}^\alpha \partial_\alpha - B_c\right)h_b \quad \& \quad 0 = Q_a^c q^b\left(E_c{}^\alpha \partial_\alpha + B_c\right)h_b. \tag{1104}$$

This time, the Dirac gamma matrices and the new object β should be presented by the following products:

$$Q_a^c q^b E_c{}^\alpha = \gamma^\alpha; \quad Q_a^c q^b B_c = \beta_a^b. \tag{1105}$$

It is important to note that the necessity of the introduction of the Eigenvalue vector B_c in Eqs. (1099) and (1104) in principle requires its scalar product in Eq. (1096) in the form $B^2 = B_c {}^* B^c$. Thus, the Dirac trick in applying the third binomial formula while operational factorizing the Klein–Gordon equation in Eqs. (1097) (Classical Dirac) and (1102) (Dirac with intrinsically structured f) is only valid in cases where the a-b-c space can be considered Cartesian like and we would have $B_c = B^c$. Alternatively, we have to assure that Eq. (1097) should be adapted as follows:

$$0 = \partial_\beta g^{\alpha\beta} \partial_\alpha f_\lambda - B^2 f_\lambda$$

$$= \partial_\beta E_c{}^\alpha E^{c\beta} \partial_\alpha f_\lambda - B^c B_c f_\lambda \Rightarrow \partial_\beta E^{c\beta} E_c{}^\alpha \partial_\alpha f_\lambda = B^c B_c f_\lambda$$

$$\Rightarrow \qquad E^{c\beta} E_c{}^\alpha \partial_\beta \partial_\alpha f_\lambda = B^c B_c f_\lambda \Rightarrow \begin{cases} E_c{}^\alpha \partial_\alpha f_\lambda = B_c f_\lambda \\ E^{c\beta} \partial_\beta f_\lambda = B^c f_\lambda \end{cases}$$

$$\Rightarrow \qquad 0 = \left(E_c{}^\alpha \partial_\alpha - B_c\right)f_\lambda \quad \& \quad 0 = \left(E^{c\beta} \partial_\beta - B^c\right)f_\lambda, \tag{1106}$$

which—as the third line clearly shows—can only work in those cases where the base vectors consist of constants. As we here explicitly only considered the Minkowski space-time, this condition already is fulfilled. The rest of the evaluation then is equivalent to Eq. (1099), only that we have two forms for our Q-terms, namely co- and contra-variantly dominated as follows:

$$\Rightarrow \qquad 0 = Q_a^{\lambda c}\left(E_c^{\;\alpha}\partial_\alpha - B_c\right)f_\lambda \quad \& \quad 0 = Q_{ac}^{\lambda}\left(E^{c\beta}\partial_\beta - B^c\right)f_\lambda. \qquad (1107)$$

Now the different Eigenvalues for the mass are not coming from the third binomial formula, but the co- and contra-variant B. The evaluation for Eqs. (1102), (1104), and (1105) should be adapted similarly.

9.3.4 Dirac's "Luck" and the Flat Space Condition

We already saw that we can obtain both the Klein–Gordon and the Dirac equation out of Eq. (1008) with the simple assumption of the flat space, which is to say a vanishing Ricci scalar $R^* = 0$. Taking for instance the condition:

$$4FF'' + (F')^2(n - 6) = 0, \qquad (1108)$$

Eq. (1008) results in:

$$0 = R^* = R^*_{\;\alpha\beta}G^{\alpha\beta} = \left(R + \frac{F'}{F}(1-n)\Delta f\right)\frac{1}{F}$$

$$0 = R + \frac{F'}{F}(1-n)\Delta f$$

$$\xrightarrow{\Delta f = B^2 f}$$

$$0 = R + \frac{F'}{F}(1-n)B^2 f. \qquad (1109)$$

We may see Eq. (1109) (second line) as a generalized n-dimensional $R \neq 0$ Klein–Gordon equation and could now apply the usual Dirac procedures to extract the operational square roots and obtain the subsequent Dirac equations. However, as this is a textbook material, we refrain from such a presentation. Instead, we reconsider Eq. (1008) with the assumption of the flat space for the scaled metric (1012), which is to say a vanishing Ricci scalar $R^* = 0$ and realize that there is a direct connection between the Laplace operator and the $f_{,\alpha}f_{,\beta}g^{\alpha\beta}$ operator, namely:

$$R^* = R^*_{\;\alpha\beta}G^{\alpha\beta} = 0$$

$$= \left(R + \frac{F'}{F}(1-n)\Delta f + \frac{f_{,\alpha}f_{,\beta}g^{\alpha\beta}(1-n)}{4F^2}\left(4FF'' + (F')^2(n-6)\right)\right)\frac{1}{F}$$

$$\Rightarrow \qquad -\Delta f = \frac{F}{F'\cdot(1-n)}R + \frac{f_{,\alpha}f_{,\beta}g^{\alpha\beta}}{4F'\cdot F}\left(4FF'' + (F')^2(n-6)\right). \qquad (1110)$$

For cases where also the unscaled metric is flat, which is to say where we also have $R = 0$, we even obtain:

$$\Delta f = -f_{,\alpha} f_{,\beta} g^{\alpha\beta} \left(\frac{F''}{F'} + \frac{F'}{F} \cdot \frac{n-6}{4} \right) \xrightarrow{R=0} = f_{,\alpha} f_{,\beta} g^{\alpha\beta} \left(\frac{F'}{F} \cdot \frac{6-n}{4} - \frac{F''}{F'} \right). \quad (1111)$$

This extremely interesting connection between the linear Laplace operator and the apparently nonlinear $f_{,\alpha} f_{,\beta} g^{\alpha\beta}$ operator via a yet arbitrary wrapping function $F[f]$ guarantees that in certain space-times (space-times of scale-Ricci-flatness $R^* = 0$ namely), we can always factorize the Laplace operator without the help of Dirac's quaternions. We illustrate this via the following equation:

$$\Delta f \xrightarrow{R^*=0} = \overbrace{\frac{F}{F' \cdot (n-1)}}^{A^2} R - f_{,\alpha} f_{,\beta} \overbrace{g^{\alpha\beta}}^{e^\alpha \cdot e^\beta} \overbrace{\left(\frac{F''}{F'} + \frac{F'}{F} \cdot \frac{n-6}{4} \right)}^{B^2}$$

$$\Rightarrow \quad \text{``}\sqrt{\Delta f}^{+\alpha}\text{''} * \text{``}\sqrt{\Delta f}^{-\beta}\text{''} = \left(\mathbf{A} + f_{,\alpha} e^\alpha B \right) * \left(\mathbf{A} - f_{,\beta} e^\beta B \right)$$

$$\text{with}: \quad \mathbf{A} \cdot f_{,\alpha} e^\alpha B - \mathbf{A} \cdot f_{,\beta} e^\beta B = 0, \quad (1112)$$

$$\Delta f \xrightarrow{R^*=R=0} = -f_{,\alpha} f_{,\beta} \overbrace{g^{\alpha\beta}}^{e^\alpha \cdot e^\beta} \overbrace{\left(\frac{F''}{F'} + \frac{F'}{F} \cdot \frac{n-6}{4} \right)}^{B^2} = f_{,\alpha} f_{,\beta} g^{\alpha\beta} \left(\frac{F'}{F} \cdot \frac{6-n}{4} - \frac{F''}{F'} \right)$$

$$\Rightarrow \quad \text{``}\sqrt{\Delta f}^{\alpha}\text{''} * \text{``}\sqrt{\Delta f}^{\beta}\text{''} = \left(i \cdot f_{,\alpha} e^\alpha B \right) * \left(i \cdot f_{,\beta} e^\beta B \right). \quad (1113)$$

Even if it seems from here that the path is clear and we should just further investigate the possible factors of the $f_{,\alpha} f_{,\beta} g^{\alpha\beta}$ operator, we are nevertheless looking for other options and possibilities before actually coming back to the findings Eqs. (1110) and (1111), respectively Eqs. (1112) and (1113).

We see, however, that in the flat space Eq. (1113) gives us a direct option for the factorization of the Laplace operator. In 4 dimensions this gives us automatically the possibility to obtain Dirac equations in spaces or space-times free of mass and potentials. Then, namely, we can write the factorized Eq. (1113) as follows:

$$\text{``}\sqrt{\Delta f}^{\,\alpha}\text{''} * \text{``}\sqrt{\Delta f}^{\,\beta}\text{''} = 0 = \left(i \cdot f_{,\alpha} \mathbf{e}^\alpha B\right) * \left(i \cdot f_{,\beta} \mathbf{e}^\beta B\right)$$

$$\Rightarrow \qquad\qquad i \cdot f_{,\alpha} \mathbf{e}^\alpha B = 0 \qquad\qquad\qquad (1114)$$

and by applying the recipe from Eq. (1089) in the form:

$$i \cdot f_{,\alpha} \mathbf{e}^\alpha B = 0 \,/\, C_a^{cb}$$

$$\Rightarrow C_a^{cb} \cdot i \cdot f_{,\alpha} \mathbf{e}^\alpha B = \xrightarrow{\ f \to f_c\ } = i \cdot f_{c,\alpha} C_a^{cb} \cdot E_b^{\ \alpha} B = i \cdot f_{c,\alpha} D_a^{c\alpha} B = i \cdot f_{c,\alpha} \gamma^\alpha B = 0,$$

$$(1115)$$

immediately results in the Dirac equation.

If we could only also have the possibility to code inertia via additional dimensions and thus, have everything (inertia and potentials) not in the curvature terms R, but given by dimensions and consequently effectively resulting in the Laplace operator, we would have an extremely straight forward method to create Dirac equations.

Therefore, in the following, we will investigate the dimensional origin of masses and potentials. In the sub-section "9.3.9.2 To Dirac from 6 dimensions in a simpler (luckier) way" later on, we will then give an example how this works.

9.3.5   Mass and Its Metric Dimensional Origin

It was shown in [3] how mass can just occur due to the existence of additional dimensions. We then, by observation, register these dimensions as certain physical properties (like mass).

At first, we intend to give one example in 8 dimensions:

$$g_{\alpha\beta}^8 = \begin{pmatrix} -c^2 & 0 & 0 & 0 & \cdots & 0 \\ 0 & 1 & 0 & 0 & \cdots & 0 \\ 0 & 0 & r^2 & 0 & \cdots & 0 \\ 0 & 0 & 0 & r^2 \cdot \sin^2 \varphi_1 & \cdots & 0 \\ \cdots & \cdots & \cdots & \cdots & \ddots & 0 \\ 0 & 0 & 0 & 0 & 0 & g_{77} \end{pmatrix}$$

$$\times F\left[f\left[t, r, \vartheta, \varphi_1, \varphi_2, \varphi_3, \varphi_4, \varphi_5\right]\right]; \quad F[f] = \left(C_f + f\right)^{\frac{2}{3}}$$

$$f\big[t,r,\vartheta,\varphi_1,\varphi_2,\varphi_3,\varphi_4,\varphi_5\big]=f_t[t]\cdot f_r[r]\cdot f_\vartheta[\vartheta]\cdot\prod_{i=1}^{5}f_{\varphi_i}\big[\varphi_i\big];$$

$$g_{44}=\Big(g_{\varphi_2}\big[\varphi_2\big]'\Big)^2\cdot r^0;$$

$$g_{55}=\Big(g_{\varphi_3}\big[\varphi_3\big]'\Big)^2\cdot r^1;$$

$$g_{66}=\Big(g_{\varphi_4}\big[\varphi_4\big]'\Big)^2\cdot r^2;$$

$$g_{77}=\Big(g_{\varphi_5}\big[\varphi_5\big]'\Big)^2\cdot r^3. \tag{1116}$$

Thereby, our choice for the function $F[f]$ assures the vanishing of the Nabla-terms in Eq. (1008), because it guarantees:

$$4FF''+\big(F'\big)^2\big(n-6\big)=0. \tag{1117}$$

Unfortunately, in order to obtain a suitable equation, we have to demand $C_f = 0$, which does not exactly provide us with the desired asymptotic flat space behavior for bigger r. But for here and now we want to ignore this (alternatives will be discussed in chapter 11). Thereby—for justification—we may simply assume an object which indeed vanishes with big r and so does the space-time it represents. Other objects with different asymptotic behavior or even a surrounding overall space-time might guarantee the flat space on bigger distances to the object described in Eq. (1116).

We partially "fix" some of the functions as follows (for i = 2, 3, 4, 5):

$$f_{\varphi_i}\big[\varphi_i\big]=f_{\varphi_i}\big[g_{\varphi_i}\big[\varphi_i\big]\big]\equiv f_{\varphi_i}\big[g_{\varphi_i}\big];\quad f_{\varphi_i}\big[g_{\varphi_i}\big]=C_{-i}\cdot e^{-A_i\cdot g_{\varphi_i}}+C_{+i}\cdot e^{+A_i\cdot g_{\varphi_i}},$$

$$\tag{1118}$$

$$f_t[t]=C_{t1}\cdot\cos[c\cdot E\cdot t]+C_{t2}\cdot\sin[c\cdot E\cdot t]. \tag{1119}$$

This gives us the following differential equation:

$$R^*=0=\left(\frac{3}{2}\cdot\frac{\partial}{r\cdot\partial r}+\left(\frac{A_6^2}{r^3}+\frac{A_5^2}{r^2}+\frac{A_4^2}{r}+A_3^2\right)+\Delta_{3\text{D-sphere}}+E^2+\frac{111}{28\cdot r^2}\right)\Psi$$

$$\Psi=\Psi[r,\vartheta,\varphi]. \tag{1120}$$

It should be pointed out that negative exponents can be constructed, which are leading to infinitely growing potential terms (as known, respectively expected for the quarks). So, for example:

$$g^8_{\alpha\beta} = \begin{pmatrix} -c^2 & 0 & 0 & 0 & \cdots & 0 \\ 0 & 1 & 0 & 0 & \cdots & 0 \\ 0 & 0 & r^2 & 0 & \cdots & 0 \\ 0 & 0 & 0 & r^2 \cdot \sin^2\varphi_1 & \cdots & 0 \\ \cdots & \cdots & \cdots & \cdots & \ddots & 0 \\ 0 & 0 & 0 & 0 & 0 & g_{77} \end{pmatrix}$$

$$\times F\left[f\left[t,r,\vartheta,\varphi_1,\varphi_2,\varphi_3,\varphi_4,\varphi_5\right]\right]; \quad F[f] = \left(C_f + f\right)^{\frac{2}{3}}$$

$$f\left[t,r,\vartheta,\varphi_1,\varphi_2,\varphi_3,\varphi_4,\varphi_5\right] = f_t[t] \cdot f_r[r] \cdot f_\vartheta[\vartheta] \cdot \prod_{i=1}^{5} f_{\varphi_i}\left[\varphi_i\right];$$

$$g_{44} = \left(g_{\varphi_2}\left[\varphi_2\right]'\right)^2 \cdot r^0;$$

$$g_{55} = \left(g_{\varphi_3}\left[\varphi_3\right]'\right)^2 \cdot r^1;$$

$$g_{66} = \left(g_{\varphi_4}\left[\varphi_4\right]'\right)^2 \cdot r^2;$$

$$g_{77} = \left(g_{\varphi_5}\left[\varphi_5\right]'\right)^2 \cdot r^{-1}. \tag{1121}$$

This gives us the following differential equation for the spatial coordinates:

$$R^* = 0 = \left(\frac{1}{2} \cdot \frac{\partial}{r \cdot \partial r} + \left(r \cdot A_6^2 + \frac{A_5^2}{r^2} + \frac{A_4^2}{r} + A_3^2\right) + \Delta_{\text{3D-sphere}} + E^2 + \frac{27}{28 \cdot r^2}\right)\Psi$$

$$\Psi = \Psi[r,\vartheta,\varphi]. \tag{1122}$$

Note: Further generalization of potentials with respect to other coordinates is easily possible via structural "playing" with the additional—compactified—coordinate components. For instance, the metrics already considered in chapter 6, section "6.9 More Metric Options and Subsequent Quantum Equations Plus Their Solutions" are giving angular dependencies.

9.3.6 A Possible Origin of Spin *I* = 1/2

The metric we would be interested in here reads:

$$g_{\alpha\beta}^{8} = \begin{pmatrix} -c^2 & 0 & 0 & 0 & \cdots & 0 \\ 0 & 1 & 0 & 0 & \cdots & 0 \\ 0 & 0 & r^2 & 0 & \cdots & 0 \\ 0 & 0 & 0 & r^2 \cdot \sin^2\varphi_1 & \cdots & 0 \\ \cdots & \cdots & \cdots & \cdots & \ddots & 0 \\ 0 & 0 & 0 & 0 & 0 & g_{77} \end{pmatrix}$$

$$\times F\Big[f\big[t,r,\vartheta,\varphi_1,\varphi_2,\varphi_3,\varphi_4,\varphi_5\big]\Big]; \quad F[f] = \big(C_f + f\big)^{\frac{2}{3}}$$

$$f\big[t,r,\vartheta,\varphi_1,\varphi_2,\varphi_3,\varphi_4,\varphi_5\big] = f_t[t] \cdot f_r[r] \cdot f_\vartheta[\vartheta] \cdot \prod_{i=1}^{5} f_{\varphi_i}\big[\varphi_i\big];$$

$$g_{44} = \Big(g_{\varphi_2}\big[\varphi_2\big]'\Big)^2 \cdot r^1;$$

$$g_{55} = \Big(g_{\varphi_3}\big[\varphi_3\big]'\Big)^2 \cdot r^1;$$

$$g_{66} = \Big(g_{\varphi_4}\big[\varphi_4\big]'\Big)^2 \cdot r^0;$$

$$g_{77} = \Big(g_{\varphi_5}\big[\varphi_5\big]'\Big)^2 \cdot r^0. \tag{1123}$$

Now we partially "fix" some of the functions as follows (for $i = 2, 3, 4, 5^9$):

$$f_{\varphi_i}\big[\varphi_i\big] = f_{\varphi_i}\Big[g_{\varphi_i}\big[\varphi_i\big]\Big] \equiv f_{\varphi_i}\big[g_{\varphi_i}\big]; \quad f_{\varphi_i}\big[g_{\varphi_i}\big] = C_{-i} \cdot e^{-A_i \cdot g_{\varphi_i}} + C_{+i} \cdot e^{+A_i \cdot g_{\varphi_i}}$$

$$f_{\varphi_2}\big[g_{\varphi_2}\big] = C_{-2} \cdot e^{-i \cdot A_2 \cdot g_{\varphi_2}} + C_{+2} \cdot e^{+i \cdot A_2 \cdot g_{\varphi_2}}, \tag{1124}$$

$$f_t[t] = C_{t1} \cdot \cos[c \cdot E \cdot t] + C_{t2} \cdot \sin[c \cdot E \cdot t], \tag{1125}$$

$$f_\vartheta[\vartheta] = C_{P\vartheta} \cdot P_L^{A_2}\big[\cos[\vartheta]\big] + C_{Q\vartheta} \cdot Q_L^{A_2}\big[\cos[\vartheta]\big], \tag{1126}$$

with the associated Legendre polynomials $P_L^{A_2}, Q_L^{A_2}$. This gives us the following differential equation for the r-dependency of $f[\ldots]$:

$$R^* = 0 = \left(\frac{3 - 4 \cdot L \cdot (L+1)}{4 \cdot r^2} + \frac{A_3^2 + A_4^2}{r} + A_5^2 + A_6^2 + \frac{\partial}{r \cdot \partial r} + \overbrace{\Delta_{\text{3D-sphere}}[r]}^{\frac{\partial}{r \cdot \partial r} + \frac{\partial^2}{\partial r^2}} - E^2 \right)\Psi$$

$$\Psi = \Psi[r,\vartheta,\varphi] = f_r[r] \cdot f_\vartheta[\vartheta] \cdot f_{\varphi_1}\big[\varphi_1\big]$$

---

^{9}If not used as index, "i" stands for the imaginary number $i = \sqrt{-1}$.

$$\Rightarrow 0 = f_\vartheta\left[\vartheta\right]\cdot f_{\varphi_1}\left[\varphi_1\right]\cdot\left(\frac{3-4\cdot L\cdot(L+1)}{4\cdot r^2}+\frac{A_3^2+A_4^2}{r}+A_5^2+A_6^2+\frac{3\cdot\partial}{r\cdot\partial r}+\frac{\partial^2}{\partial r^2}-E^2\right)f_r[r].$$

$$(1127)$$

We find that for $L = 1/2$ the terms with r^{-2} vanish, resulting in the differential equation for the remaining coordinate r:

$$0 = \left(\frac{A_3^2+A_4^2}{r}+A_5^2+A_6^2+\frac{3\cdot\partial}{r\cdot\partial r}+\frac{\partial^2}{\partial r^2}-E^2\right)f_r[r]. \qquad (1128)$$

This is extremely interesting, because we have obtained a condition, namely:

$$\frac{3-4\cdot L\cdot(L+1)}{4\cdot r^2}=0 \quad\Rightarrow\quad L=\frac{1}{2} \qquad (1129)$$

forcing us to except the existence of spinors with a geometry given due to the metric (1123). With respect to the cases of half spins regarding the Legendre functions, we refer to the section 6.16 and a variety of previous papers by this author (mainly within the series of "Einstein had it … ").

Now we can also solve Eqs. (1127) and (1128) via:

$$f_r[r] = e^{-r\cdot\sqrt{E^2-\left(A5^2+A6^2\right)}}\cdot r^{L-\frac{1}{2}}\cdot\left(\begin{array}{c}C_U U\left[-nn, L\cdot(1+L), 2\cdot r\cdot\sqrt{E^2-\left(A5^2+A6^2\right)}\right]\\[2mm]+C_L L_{nn}^{1+2\cdot L}\left[2\cdot r\cdot\sqrt{E^2-\left(A5^2+A6^2\right)}\right]\end{array}\right)$$

$$nn = \frac{A_3^2+A_4^2}{2\cdot\sqrt{E^2-\left(A5^2+A6^2\right)}}-1-L, \qquad (1130)$$

$$f_r[r] = e^{-r\cdot\sqrt{E^2-\left(A5^2+A6^2\right)}}\left(\begin{array}{c}C_U U\left[-nn, 3, 2\cdot r\cdot\sqrt{E^2-\left(A5^2+A6^2\right)}\right]\\[2mm]+C_L L_{nn}^2\left[2\cdot r\cdot\sqrt{E^2-\left(A5^2+A6^2\right)}\right]\end{array}\right)$$

$$nn = \frac{A_3^2+A_4^2-3\cdot\sqrt{E^2-\left(A5^2+A6^2\right)}}{2\cdot\sqrt{E^2-\left(A5^2+A6^2\right)}}, \qquad (1131)$$

respectively.

With the hypergeometric function $U[a,b,z]$ and the Laguerre polynomials $L_{nn}^{1+2\cdot L}$ in the solutions above, we recognize the similarity to the Schrödinger hydrogen solution (e.g., [13]). Only considering the ordinary spatial coordinates r, ϑ, φ, we can illustrate the spatial distribution, our solution would produce in ordinary space-time (Figs. 9.2 and 9.3).

The pictures in the graphics-grid below (Fig. 9.2) shows $\pi/4$-rotations in a clock-wise manner (top left to bottom right). If this was a classical object, then the 8th picture ($8*\pi/4 = 2\,\pi = 360°$) should be exactly equal to the first one, but obviously it is not. On the contrary, it has changed its sign (Fig. 9.3) and thus, is a spinor.

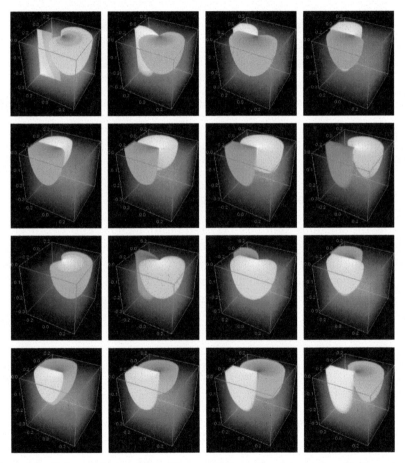

Figure 9.2 Spinor-object as derived from an 8-dimensional space-time (4 dimensions appear compactified to the human "eye" / sensors). Rotation angle from picture to picture is $\pi/4$ clockwise from left to right and π clockwise from up to down.

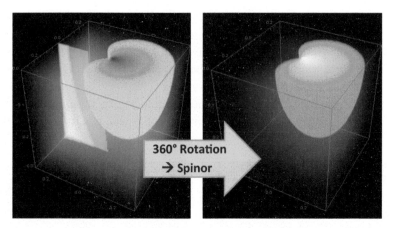

Figure 9.3 Spinor-object as derived from an 8-dimensional space-time (4 dimensions appear compactified to the human "eye" / sensors). Rotation angles 0 and $2*\pi$. Please note the changed color coding, which is clearly showing the sign-switch of the metric object after one full rotation ($360° = 2*\pi$). The corresponding 8-dimensional solution, being used for the evaluation here, is been presented in Eqs. (1123)–(1131).

9.3.7 Connection to the Classical Picture

Taking the solutions from Eqs. (1124)–(1126), Eq. (1127) is clearly giving us the connection to the classical picture with potential, mass, and so on. Thereby the angular part is completely been sucked into the L-term, which—as shown above—vanishes in the setting $L = 1/2$.

In this case, we can extract from Eq. (1128) the connection to the classical picture with potential $V[r]$, operator, energy E, and mass terms M as follows:

$$0 = \left(\overbrace{\frac{\partial}{r \cdot \partial r} + \frac{A_3^2 + A_4^2}{r}}^{V[r]} + \overbrace{A_5^2 + A_6^2}^{M} + \overbrace{\frac{2 \cdot \partial}{r \cdot \partial r} + \frac{\partial^2}{\partial r^2}}^{\Delta_r} - \overbrace{E^2}^{\text{energy}} \right) f_r[r]. \quad (1132)$$

9.3.8 Seeds for Particles

Evaluation of the Ricci scalar R of the metric (1123) from Eq. (1008) via the setting $F = 1$ gives:

$$R = -\frac{3}{2 \cdot r^2}. \quad (1133)$$

Thus, we have a metric with a r^{-2} asymptotically vanishing curvature, which apparently acts as a seed for spin-1/2-particles or at least particle-like objects when adding (and plying with) a scalar factor $F[f]$ as done in Eq. (1123). For $r = 0$, this curvature would become infinity and the scalar factor F prevents this as it simply makes R^* (the Ricci scalar of the scaled metric) to zero.

9.3.9 Several Time-Like Dimensions and More Origins of Mass

Assuming curled up or compactified time-like dimensions, one might take a metric as follows:

$$g_{\alpha\beta}^6 = \begin{pmatrix} -c^2 & 0 & 0 & 0 & 0 & 0 \\ 0 & 1 & 0 & 0 & 0 & 0 \\ 0 & 0 & r^2 & 0 & 0 & 0 \\ 0 & 0 & 0 & r^2 \cdot \sin[\vartheta]^2 & 0 & 0 \\ 0 & 0 & 0 & 0 & -c_u^2 & 0 \\ 0 & 0 & 0 & 0 & 0 & -c_v^2 \end{pmatrix} \cdot \left(C_f + g[u,v] \cdot f[t,r,\vartheta,\varphi]\right).$$

$$(1134)$$

Further assuming that t, r, ϑ, φ are the usual 4D space-time coordinates, while u and v are compactified, and applying the approach $g[u,v] = Gu[u]*Gv[v]$ plus the usual separation approach for the function $f[\ ...\ ] = f_t[t] \cdot f_r[r] \cdot f_\vartheta[\vartheta] \cdot f_\varphi[\varphi]$, we obtain the following solution for the case $R^* = 0$:

$$Gv[v] = C_{v1} \cdot \cos\left[c_v \cdot E_v \cdot v\right] + C_{v2} \cdot \sin\left[c_v \cdot E_v \cdot v\right], \qquad (1135)$$

$$Gu[u] = C_{u1} \cdot \cos\left[c_u \cdot E_u \cdot u\right] + C_{u2} \cdot \sin\left[c_u \cdot E_u \cdot u\right], \qquad (1136)$$

$$f_t[t] = C_{t1} \cdot \cos[c \cdot E \cdot t] + C_{t2} \cdot \sin[c \cdot E \cdot t], \qquad (1137)$$

$$f_\varphi[\varphi] = C_{\varphi1} \cdot \cos[m \cdot \varphi] + C_{\varphi2} \cdot \sin[m \cdot \varphi], \qquad (1138)$$

$$f_\vartheta[\vartheta] = C_{P\vartheta} \cdot P_\ell^m\left[\cos[\vartheta]\right] + C_{Q\vartheta} \cdot Q_\ell^m\left[\cos[\vartheta]\right], \qquad (1139)$$

with the associated Legendre polynomials P_ℓ^m (Remember: In the sub-section above, we have also discussed the option for half spin solutions. Also see 6.16.)

The intermediate result above makes us end up with the following differential equation for $f_r[r]$:

$$R^* = \frac{1}{F[g \cdot f]^3} \cdot \left(-5 \cdot F[g \cdot f] \cdot \frac{\partial F[g \cdot f]}{\partial (g \cdot f)} \cdot \Delta_g f \right); \quad F[g \cdot f] = C_f + f[t,r,\vartheta,\varphi] \cdot g[u,v]$$

$$C_f \neq 0$$

$$\xrightarrow{6D} R^* = 0$$

$$\Rightarrow \quad 0 = \left(\ell(1+\ell) - \left(E^2 + E_u^2 + E_v^2 \right) r^2 \right) f_r[r] - r \left(2 f_r[r]' + r \cdot f_r[r]'' \right). \quad (1140)$$

This time and in contrast to our metric choice in 8 dimensions above (c.f. Eq. (1116)), we do not have to demand $C_f = 0$ and thus, it can assure the asymptotic flat space-time behavior. The solution for Eq. (1140) (last line) consists of the spherical Bessel functions j_n and y_n:

$$f_r[r] = C_{r1} \cdot j_\ell \left[\sqrt{E^2 + E_u^2 + E_v^2} \cdot r \right] + C_{r2} \cdot y_\ell [\sqrt{E^2 + E_u^2 + E_v^2} \cdot r]. \quad (1141)$$

Changing the metric (1134) to space-like coordinates u and v:

$$g_{\alpha\beta}^6 = \begin{pmatrix} -c^2 & 0 & 0 & 0 & 0 & 0 \\ 0 & 1 & 0 & 0 & 0 & 0 \\ 0 & 0 & r^2 & 0 & 0 & 0 \\ 0 & 0 & 0 & r^2 \cdot \sin[\vartheta]^2 & 0 & 0 \\ 0 & 0 & 0 & 0 & c_u^2 & 0 \\ 0 & 0 & 0 & 0 & 0 & c_v^2 \end{pmatrix} \cdot \left(C_f + g[u,v] \cdot f[t,r,\vartheta,\varphi] \right),$$

$$(1142)$$

but leaving the approach in Eqs. (1135) to (1139) results in the following differential equation for f_r:

$$0 = \left(\ell(1+\ell) - \left(E^2 - E_u^2 - E_v^2 \right) r^2 \right) f_r[r] - r \left(2 f_r[r]' + r \cdot f_r[r]'' \right), \quad (1143)$$

which results in:

$$f_r[r] = C_{r1} \cdot j_{-1-\ell} \left[\sqrt{E^2 - E_u^2 - E_v^2} \cdot r \right] + C_{r2} \cdot y_{-1-\ell} [\sqrt{E^2 - E_u^2 - E_v^2} \cdot r].$$

$$(1144)$$

While with the solution in Eq. (1141), the function j should be favored as it does not lead to a singularity for $r = 0$, we have to choose the function y in the case of Eq. (1144) for the same reason, which is to say avoiding singularities.

However, when also changing the approach for the function Gu and Gv to:

$$Gv[v] = C_{v1} \cdot e^{-c_v \cdot E_v \cdot v} + C_{v2} \cdot e^{c_v \cdot E_v \cdot v}, \tag{1145}$$

$$Gu[u] = C_{u1} \cdot e^{-c_u \cdot E_u \cdot u} + C_{u2} \cdot e^{c_u \cdot E_u \cdot u}, \tag{1146}$$

we are back with Eq. (1140) and the corresponding solution (1141) for f_r. Now, together with the choice $C_{u2} = C_{v2} = 0$ for positive u and v and $C_{u1} = C_{v1} = 0$ for negative, we can even assure asymptotically vanishing (thus, compactified) add-on coordinates u and v.

More such dimensions ω_i could be added easily in the same way and seeing the parameters $E_{\omega i}$ either as energies or masses $m_{\omega i}$ via $E_{\omega i} = c^{2*} m_{\omega i}$, we do not only have the equivalency of mass and energy, but also a great variety of options to add mass and energy as such entangled (entangled via the function $g[u, v, \dots] = g[\omega_0, \omega_1, \omega_2, \dots]$) dimensions.

Thus, this way we can add arbitrary masses and energies just as additional dimensions either time- or space-like. Together with the add-on coordinates, respectively dimensions the additional masses and energies will automatically come as quanta.

9.3.9.1 To Dirac in 6 dimensions

Now we want to consider the example above with respect to the Dirac equation. In order to avoid—just for the time being—the complications coming in due to the spherical coordinates, we change the metrics in Eqs. (1134) and (1142) to their Cartesian form:

$$g_{\alpha\beta}^6 = \begin{pmatrix} -c^2 & 0 & 0 & 0 & 0 & 0 \\ 0 & 1 & 0 & 0 & 0 & 0 \\ 0 & 0 & 1 & 0 & 0 & 0 \\ 0 & 0 & 0 & 1 & 0 & 0 \\ 0 & 0 & 0 & 0 & \pm c_u^2 & 0 \\ 0 & 0 & 0 & 0 & 0 & \pm c_v^2 \end{pmatrix} \cdot \left(C_f + g[u,v] \cdot f[t,x,y,z] \right). \tag{1147}$$

We keep the separation approach and reuse the following functions from the section above:

$$Gv[v] = C_{v1} \cdot \cos\left[c_v \cdot E_v \cdot v \right] + C_{v2} \cdot \sin\left[c_v \cdot E_v \cdot v \right], \tag{1148}$$

$$Gu[u] = C_{u1} \cdot \cos\left[c_u \cdot E_u \cdot u \right] + C_{u2} \cdot \sin\left[c_u \cdot E_u \cdot u \right], \tag{1149}$$

$$f_t[t] = C_{t1} \cdot \cos[c \cdot E \cdot t] + C_{t2} \cdot \sin[c \cdot E \cdot t]. \tag{1150}$$

Now Eq. (1008) gives us:

$$R^* = \frac{1}{F[g \cdot f]^3} \left(-5 \cdot F[g \cdot f] \cdot \frac{\partial F[g \cdot f]}{\partial (g \cdot f)} \cdot \Delta_g f \right)$$

$$F[g \cdot f] = C_f + \overbrace{f_t[t] \cdot f_{xyz}[x,y,z]}^{f=f[t,x,y,z]} \cdot g[u,v]; \quad C_f \neq 0$$

$$\xrightarrow{6D} R^* = 0$$

$$\Rightarrow \quad \begin{cases} 0 = \Delta_g f \cdot g = \left(\pm \left(E_u^2 + E_v^2 \right) + \Delta_{4D} \right) f[t,x,y,z] \\ = \left(E^2 \pm \left(E_u^2 + E_v^2 \right) + \Delta_{3D} \right) f_{xyz}[x,y,z] \end{cases}. \tag{1151}$$

We realize that with:

$$0 = \Delta_g f \cdot g = \left(\pm \left(E_u^2 + E_v^2 \right) + \Delta_{4D} \right) f[t,x,y,z], \tag{1152}$$

we have just obtained the classical Klein–Gordon equation with quite some flexibility to add mass or other forms of inertia. For reasons of recognition, we set:

$$B^2 = E_u^2 + E_v^2 \tag{1153}$$

and can now directly repeat our evaluation from sub-section "9.3.3 Intermediate Sum-Up: The Classical and the Intrinsic Dirac Approach" and Eq. (1096). There we had concluded that in connection with a formalized Dirac development of the Klein–Gordon equation B has to become a vector $\mathbf{B_c}$ and in connection with our metrically derived mass equivalents (via additional dimensions and the corresponding Eigenvalues E_u, E_v), we are now in the position to better understand this.

In order to keep the discussion simple, we choose $F[f]$ in accordance with the following condition:

$$4FF'' + (F')^2 (n - 6) = 0 \tag{1154}$$

and Minkowski-like coordinates in an arbitrary number of dimensions. While a certain number of dimensions provide masses in accordance with our derivation above, we consider all other dimensions as space-time dimensions. For simplicity, we name the dimensions leading to masses "mass-dimensions." Taking our example in 6 dimensions from this sub-section, we find $F[f]$ to be linear in f and, applying suitable settings for the mass-dimensions (see our choice for $g[u,v]$, respectively $Gu[u]$ and $Gv[v]$ in Eqs. (1148) and (1149)), end up with the last line of Eq. (1151). We realize that, even though our metric choice in Eq. (1147) does not result in any curvature $R = 0$, we still obtain a classical Klein–Gordon equation in 4 dimensions and completely with mass (as the examples in the previous sub-sections starting with sub-section "9.3.5 Mass and Its Metric Dimensional Origin" have shown, we could also have other forms of physical properties and potentials). We see that our completely Ricci-curvature-free space-time still gives particle equations when just suitably adjusting solutions to additional dimensions. So from Eqs. (1151) and (1152), we take the classical Klein–Gordon equation:

$$0 = \left(\pm \overbrace{\left(E_u^2 + E_v^2 \right)}^{B^2} + \Delta_{4D} \right) f[t, x, y, z] \tag{1155}$$

and by developing this into a Dirac equation, thereby using our results from sub-section "9.3.3 Intermediate Sum-Up: The Classical and the Intrinsic Dirac Approach," we obtain the following (for simplicity, we only take the time-like solutions for the dimensions u and v):

$$0 = \partial_\beta g^{\alpha\beta} \partial_\alpha f_\lambda - B^2 f_\lambda = \partial_\beta g^{\alpha\beta} \partial_\alpha f_\lambda - \left(E_u^2 + E_v^2 \right) f_\lambda$$

$$= \partial_\beta E_c{}^\alpha E^{c\beta} \partial_\alpha f_\lambda - \sqrt{\left(E_u^2 + E_v^2 \right)}^c \sqrt{\left(E_u^2 + E_v^2 \right)}_c f_\lambda$$

$$\Rightarrow \qquad \partial_\beta E^{c\beta} E_c{}^\alpha \partial_\alpha f_\lambda = \sqrt{\left(E_u^2 + E_v^2 \right)}^c \sqrt{\left(E_u^2 + E_v^2 \right)}_c f_\lambda$$

$$\Rightarrow \qquad E^{c\beta} E_c{}^\alpha \partial_\beta \partial_\alpha f_\lambda = \sqrt{\left(E_u^2 + E_v^2 \right)}^c \sqrt{\left(E_u^2 + E_v^2 \right)}_c f_\lambda$$

$$\Rightarrow \qquad \begin{bmatrix} E_c{}^\alpha \partial_\alpha f_\lambda = \sqrt{\left(E_u^2 + E_v^2 \right)}_c f_\lambda \\ E^{c\beta} \partial_\beta f_\lambda = \sqrt{\left(E_u^2 + E_v^2 \right)}^c f_\lambda \end{bmatrix}$$

$$\Rightarrow \quad 0 = \left(E_c{}^\alpha \partial_\alpha - \sqrt{\left(E_u^2 + E_v^2 \right)}_c \right) f_\lambda \quad \& \quad 0 = \left(E^{c\beta} \partial_\beta f_\lambda - \sqrt{\left(E_u^2 + E_v^2 \right)}^c \right) f_\lambda. \tag{1156}$$

Dirac-like equations can now be created from the last line of Eq. (1156) via:

$$\Rightarrow 0 = Q_a^{\lambda c} \left(E_c{}^\alpha \partial_\alpha - \sqrt{\left(E_u^2 + E_v^2 \right)}_c \right) f_\lambda \quad \& \quad 0 = Q_{ac}^\lambda \left(E^{c\beta} \partial_\beta - \sqrt{\left(E_u^2 + E_v^2 \right)}^c \right) f_\lambda \tag{1157}$$

and the Q-terms as given in Eq. (1090). As elaborated at the end of sub-section "9.3.2 Repetition: Dirac in the Metric Picture and Its Connection to the

Classical Quaternion Form," this also forces us to assume certain structures for the mass-terms $\sqrt{\left(E_u^2 + E_v^2\right)}_c$ and $\sqrt{\left(E_u^2 + E_v^2\right)}^c$. Namely, when demanding to obtain the perfect mirroring of the original Dirac equation, we would have to obtain (with the Kronecker symbols δ_a^λ):

$$Q_a^{\lambda c}\sqrt{\left(E_u^2 + E_v^2\right)}_c \overset{!}{=} M \cdot \delta_a^\lambda \quad \& \quad Q_{ac}^{\lambda}\sqrt{\left(E_u^2 + E_v^2\right)}^c \overset{!}{=} M \cdot \delta_a^\lambda, \tag{1158}$$

which is not possible with a simple vector $\sqrt{\left(E_u^2 + E_v^2\right)}_c$ or $\sqrt{\left(E_u^2 + E_v^2\right)}^c$. We would either (in accordance with Eq. (1092) and only for the co-variant mass-term) require something more like:

$$\frac{\sqrt{\left(E_u^2 + E_v^2\right)}^a}{M}_{0c} = \begin{pmatrix} i & & \\ & 1 & \\ & & -1 \\ & & & -1 \end{pmatrix}; \quad \frac{\sqrt{\left(E_u^2 + E_v^2\right)}^a}{M}_{1c} = \begin{pmatrix} & & & -i \\ & & -1 & \\ & 1 & & \\ 1 & & & \end{pmatrix}$$

$$\frac{\sqrt{\left(E_u^2 + E_v^2\right)}^a}{M}_{2c} = \begin{pmatrix} & & & -1 \\ & & -i & \\ & -i & & \\ i & & & \end{pmatrix}; \quad \frac{\sqrt{\left(E_u^2 + E_v^2\right)}^a}{M}_{3c} = \begin{pmatrix} & & & -i \\ & & 1 & \\ 1 & & & \\ & & & -1 \end{pmatrix},$$

$$\tag{1159}$$

including a suitable switch of summation indices or accept that the terms $Q_a^{\lambda c}\sqrt{\left(E_u^2 + E_v^2\right)}_c, Q_{ac}^{\lambda}\sqrt{\left(E_u^2 + E_v^2\right)}^c$ are not summing up to the simple scalar masses as given (desired) in Eq. (1158) and as assumed in the classical Dirac theory. In other words, the vector of Eigenvalues to the dimensions u and v clearly provides masses for the Klein–Gordon equation in its classical shape, but this does not result in the classical Dirac equation with respect to the same scalar masses. Instead, we here have to deal with mass or inertia vectors and when transforming the whole to the corresponding Dirac structure (1157) we end up with complex inertia matrices as follows:

$$Q_a^{\lambda c}\sqrt{\left(E_u^2 + E_v^2\right)}_c = \beta_a^\lambda \quad \& \quad Q_{ac}^{\lambda}\sqrt{\left(E_u^2 + E_v^2\right)}^c = \beta_a^\lambda. \tag{1160}$$

We realize that the transition from the metric to the classical form requires some more discussion. For this we need to consider more options than the $R^* = R = 0$ in connection with Eq. (1008) does provide. We will come back to this example at the end of this chapter in the sub-section "9.4.9 Example in $n = 6$, $n = 8$ and Higher Dimensions."

9.3.9.2 To Dirac from 6 dimensions in a simpler (luckier) way

On the other hand, we see that our example metric in Eq. (1147) fulfills the conditions $R^* = R = 0$ and thus, we can directly apply Eq. (1113), combine it with our intermediate solutions from Eq. (1152) for the Laplace operator, leading us to:

$$\xrightarrow{R^* = R = 0}$$

$$\Delta f_{\text{whole}} = f_{\text{whole},\alpha} f_{\text{whole},\beta} g^{\alpha\beta} \left(\frac{F'}{F} \cdot \frac{6-n}{4} - \frac{F''}{F'} \right)$$

$$\Rightarrow \quad \begin{cases} \Delta f_{\text{whole}} = \Delta_g (f \cdot g) = \left(\pm \left(E_u^2 + E_v^2 \right) + \Delta_{4D} \right) \overbrace{f[t,x,y,z]}^{=f} \\[2mm] = (f \cdot g)_{,\alpha} (f \cdot g)_{,\beta} \overbrace{g^{\alpha\beta}}^{\mathbf{e}^\alpha \cdot \mathbf{e}^\beta} \left(\frac{F'}{F} \cdot \frac{6-n}{4} - \frac{F''}{F'} \right) \end{cases}$$

$$\Rightarrow \quad \begin{cases} \overbrace{\sqrt{\Delta(f \cdot g)}^\alpha}^{``} {}^{``} * {}^{``} \overbrace{\sqrt{\Delta(f \cdot g)}^\beta}^{``} \\[2mm] = \left((f \cdot g)_{,\alpha} \mathbf{e}^\alpha \left(\frac{F'}{F} \cdot \frac{6-n}{4} - \frac{F''}{F'} \right) \right) * \left((f \cdot g)_{,\beta} \mathbf{e}^\beta \left(\frac{F'}{F} \cdot \frac{6-n}{4} - \frac{F''}{F'} \right) \right). \end{cases}$$

$$(1161)$$

Thereby we have assumed the function F to be general again. Now, taking our solutions for the function g from Eqs. (1148) and (1149), results the last line of Eq. (1161) to:

$$\xrightarrow{a,b=\{t,x,y,z\}}$$

$$\Rightarrow \quad = \begin{cases} \left(\left[(f \cdot g)_{,a} \mathbf{e}^a + (f \cdot g)_{,u} \mathbf{e}^u + (f \cdot g)_{,v} \mathbf{e}^v \right] \left(\frac{F'}{F} \cdot \frac{6-n}{4} - \frac{F''}{F'} \right) \right) \\[2mm] * \left(\left[(f \cdot g)_{,b} \mathbf{e}^b + (f \cdot g)_{,u} \mathbf{e}^u + (f \cdot g)_{,v} \mathbf{e}^v \right] \left(\frac{F'}{F} \cdot \frac{6-n}{4} - \frac{F''}{F'} \right) \right) \end{cases}$$

$$= \begin{cases} \left(\left[f_{,a} \mathbf{e}^a \cdot g + f \cdot \left(g_{,u} \mathbf{e}^u + g_{,v} \mathbf{e}^v \right) \right] \left(\frac{F'}{F} \cdot \frac{6-n}{4} - \frac{F''}{F'} \right) \right) \\[2mm] * \left(\left[f_{,b} \mathbf{e}^b \cdot g + f \cdot \left(g_{,u} \mathbf{e}^u + g_{,v} \mathbf{e}^v \right) \right] \left(\frac{F'}{F} \cdot \frac{6-n}{4} - \frac{F''}{F'} \right) \right). \end{cases}$$

$$(1162)$$

For the reason of simplicity, we set the function $Gv[v] = 0$ and thus, only have to deal with $Gu[u]$. Here we set $C_{u1} = C_u$ and $C_{u2} = i^*C_u$, giving us:

$$Gu[u] = C_u \cdot e^{i \cdot c_u \cdot E_u \cdot u} \tag{1163}$$

and making Eq. (1162) to:

$$\Delta f_{\text{whole}} = f_{\text{whole},\alpha} f_{\text{whole},\beta} g^{\alpha\beta} \left(\frac{F'}{F} \cdot \frac{6-n}{4} - \frac{F''}{F'} \right)$$

$$= \left\{ \begin{array}{l} \left(\left[f_{,a} \mathbf{e}^a \cdot g + i \cdot c_u \cdot E_u \cdot f \cdot g \cdot \mathbf{e}^u \right] \left(\frac{F'}{F} \cdot \frac{6-n}{4} - \frac{F''}{F'} \right) \right) \\ * \left(\left[f_{,b} \mathbf{e}^b \cdot g + i \cdot c_u \cdot E_u \cdot f \cdot g \cdot \mathbf{e}^u \right] \left(\frac{F'}{F} \cdot \frac{6-n}{4} - \frac{F''}{F'} \right) \right) \end{array} \right.$$

$$= \left[f_{,a} \mathbf{e}^a + i \cdot c_u \cdot E_u \cdot f \cdot \mathbf{e}^u \right] * \left[f_{,b} \mathbf{e}^b + i \cdot c_u \cdot E_u \cdot f \cdot \mathbf{e}^u \right] \cdot g^2 \left(\frac{F'}{F} \cdot \frac{6-n}{4} - \frac{F''}{F'} \right)^2.$$

$$\tag{1164}$$

In principle, these are three different factors (in fact it is four, but we here ignore the trivial case of $g = 0$) of which only one needs to vanish to fulfill the condition $\Delta f_{\text{whole}} = 0$. One, namely $\left(\frac{F'}{F} \cdot \frac{6-n}{4} - \frac{F''}{F'} \right)^2$ and the fixing of $F[f]$ via condition (1154), we are already acquainted with. The other two factors $[f_{,a} \mathbf{e}^a + i \cdot c_u \cdot E_u \cdot f \cdot \mathbf{e}^u]$, $[f_{,b} \mathbf{e}^b + i \cdot c_u \cdot E_u \cdot f \cdot \mathbf{e}^u]$ are vectors, being combined in a scalar product and therefore require great caution.

Applying the Dirac way, thereby using a set (vector) of functions f, all satisfying the Klein–Gordon equation, respectively—as shown here—just satisfying the Laplace equation with add-on dimensions providing inertia, and assuring the right operational outcome via quaternions may be one way to properly perform the factorization. But it is not the only one and that is why we need a bit more discussion.

For here and now we simply conclude that assuring the outcome of $\Delta f_{\text{whole}} = 0$, also on the right-hand side of Eq. (1164), gives us the additional condition:

$$\Delta f_{\text{whole}} = f_{\text{whole},\alpha} f_{\text{whole},\beta} g^{\alpha\beta} \overbrace{\left(\frac{F'}{F} \cdot \frac{6-n}{4} - \frac{F''}{F'} \right)}^{=B^2} = 0. \tag{1165}$$

Thereby it does not matter in which way we guarantee the correct appearance of the $f_{whole,\alpha} f_{whole,\beta} g^{\alpha\beta}$ operator and thus, could easily apply Dirac's recipe plus his gamma matrices and write[10]:

$$\underrightarrow{f_{whole} \to f_\lambda}$$

$$\Delta f_\lambda = f_{\lambda,\alpha} \gamma^\alpha f_{\lambda,\beta} \gamma^\beta \left(\frac{F'}{F} \cdot \frac{6-n}{4} - \frac{F''}{F'} \right) = 0. \tag{1166}$$

Now, however, the gamma matrices would be matrices in 6 and not in 4 dimensions. The Dirac equation would now have to read:

$$f_{\lambda,\alpha} \gamma^\alpha = 0; \quad \alpha = 0,1,2,\ldots,5. \tag{1167}$$

Using the simplifications for Gv and Gu as introduced above ($Gv = 0$ and $Gu[u]$ from Eq. (1163)), we immediately realize that we could write Eq. (1166) as follows:

$$\Delta f_\lambda = f_{\lambda,\alpha} \gamma^\alpha f_{\lambda,\beta} \gamma^\beta \left(\frac{F'}{F} \cdot \frac{6-n}{4} - \frac{F''}{F'} \right) =$$

$$\left(f_{\lambda,a} \gamma^a - E_u \cdot f_\lambda \right)\left(f_{\lambda,b} \gamma^b + E_u \cdot f_\lambda \right) \cdot g^2 \left(\frac{F'}{F} \cdot \frac{6-n}{4} - \frac{F''}{F'} \right) = 0. \tag{1168}$$

Here now we have the classical 4-dimensional Dirac gamma matrices (reminder $a,b = \{t,x,y,z\}$; c.f. Eq. (1162), first line) and with

$$f_{\lambda,a} \gamma^a - E_u \cdot f = \left(\gamma^a \partial_{,a} - E_u \right) f_\lambda = 0$$

$$f_{\lambda,b} \gamma^b + E_u \cdot f_\lambda = \left(\gamma^b \partial_{,b} + E_u \right) f_\lambda = 0, \tag{1169}$$

we have obtained the classical Dirac equations.

We conclude that whenever we have the condition $R^* = R = 0$, Eq. (1113) immediately and directly gives us a connection between the Laplace operator and the $f_{whole,\alpha} f_{whole,\beta} g^{\alpha\beta}$ operator. As we can introduce inertia via additional dimensions, the condition $\Delta f_{whole} = 0$ then automatically also allows us to create Dirac equations, not as operational (the classical Dirac way starting from the Klein–Gordon equation) but as true factorized terms from Eq. (1166).

---

[10]Please note that it was shown in sub-section "9.3.4 Dirac's "Luck" and the Flat Space Condition" via the recipe in Eq. (1089), which is to say:

$$f_{whole,\alpha} e^\alpha B = 0 / \cdot C_a^{\lambda b}$$

$$\Rightarrow \quad C_a^{\lambda b} \cdot f_{whole,\alpha} e^\alpha B = \underrightarrow{f_{whole} \to f_\lambda} = f_{\lambda,\alpha} C_a^{\lambda b} \cdot E_b{}^\alpha B = f_{\lambda,\alpha} D_a^{\lambda\alpha} B = f_{\lambda,\alpha} \gamma^\alpha B = 0,$$

to come from the metric operator form Eq. (1165) to the Dirac quaternion form Eq. (1166).

9.3.9.3 The infinity options principle

It should be pointed out that by applying the infinity option principle (c.f. chapter 6 and [3, 5]), which is to say space-times of infinitely many dimensions, Eq. (1008) would evolve:

$$\lim_{n\to\infty} \frac{\overset{*}{R}}{n^2} = \lim_{n\to\infty} \frac{\overset{*}{R}_{\alpha\beta} G^{\alpha\beta}}{n^2}$$

$$= \lim_{n\to\infty} \left(\begin{array}{c} \Gamma^{\sigma}_{\alpha\beta,\sigma} g^{\alpha\beta} - \Gamma^{\sigma}_{\beta\sigma,\alpha} g^{\alpha\beta} - \Gamma^{\mu}_{\sigma\alpha}\Gamma^{\sigma}_{\beta\mu} g^{\alpha\beta} + \Gamma^{\sigma}_{\alpha\beta}\Gamma^{\mu}_{\sigma\mu} g^{\alpha\beta} \\ + \dfrac{F'}{F}(1-n)\Delta f + \dfrac{f_{,\alpha} f_{,\beta} g^{\alpha\beta}(1-n)}{4F^2}\left(4FF'' + (F')^2(n-6)\right) \end{array} \right) \frac{1}{n^2 F}$$

$$\lim_{n\to\infty} \frac{\overset{*}{R}}{n^2} = \lim_{n\to\infty} \frac{R}{n^2 F} - \frac{f_{,\alpha} f_{,\beta} g^{\alpha\beta}}{4F^3}(F')^2$$

$$\Rightarrow \quad \underbrace{\lim_{n\to\infty} \frac{\overset{*}{R}}{n^2} - \lim_{n\to\infty} \frac{R}{n^2 F}}_{\equiv -R_n^{\infty}} = -\frac{f_{,\alpha} f_{,\beta} g^{\alpha\beta}}{4F^3}(F')^2 \quad \Rightarrow \quad R_n^{\infty} = \frac{f_{,\alpha} f_{,\beta} g^{\alpha\beta}}{4F^3}(F')^2$$

$$\Rightarrow \qquad \boxed{\frac{4F^3 R_n^{\infty}}{(F')^2} = f_{,\alpha} f_{,\beta} g^{\alpha\beta}}. \tag{1170}$$

Assuming vanishing curvature for infinitely dimensional space-times, which is to say:

$$\lim_{n\to\infty} \frac{\overset{*}{R}}{n^2} - \lim_{n\to\infty} \frac{R}{n^2 F} = \lim_{n\to\infty} \frac{\overset{*}{R}_{\alpha\beta} G^{\alpha\beta}}{n^2} - \lim_{n\to\infty} \frac{R}{n^2 F} \equiv -R_n^{\infty} = 0, \tag{1171}$$

results in the simple equation:

$$f_{,\alpha} f_{,\beta} g^{\alpha\beta} = 0. \tag{1172}$$

It was shown in the sub-sections "9.3.4 Dirac's 'Luck' and the Flat Space Condition" and "9.3.9.2 To Dirac from 6 dimensions in a simpler (luckier) way," how we can obtain Dirac equations from there. For convenience, we here repeat the evaluation using the procedure from sub-section 9.3.4 and the recipe in Eq. (1089), which is to say:

$$f_{,\alpha} f_{,\beta} g^{\alpha\beta} = 0 = f_{,\alpha} f_{,\beta} \mathbf{e}^{\alpha} \cdot \mathbf{e}^{\beta} \quad \Rightarrow \quad f_{,\alpha} \mathbf{e}^{\alpha} = 0 \, / \cdot C_a^{\lambda b}$$

$$\Rightarrow \quad C_a^{\lambda b} \cdot f_{,\alpha} \mathbf{e}^{\alpha} = \xrightarrow{f\to f_{\lambda}} = f_{\lambda,\alpha} C_a^{\lambda b} \cdot E_b^{\ \alpha} = f_{\lambda,\alpha} D_a^{\lambda \alpha} = f_{\lambda,\alpha} \gamma^{\alpha} = 0. \tag{1173}$$

9.3.10 More Discussion and a Bit Redundancy

As the corresponding setting of conditions requires some discussion, we here want to consider Eq. (1008) again. Here now with the explicit demand of the following:

$$R^* = R^*_{\alpha\beta} G^{\alpha\beta} = 0$$

$$\Rightarrow \quad 0 = \left(\begin{array}{c} \Gamma^\sigma_{\alpha\beta,\sigma} g^{\alpha\beta} - \Gamma^\sigma_{\beta\sigma,\alpha} g^{\alpha\beta} - \Gamma^\mu_{\sigma\alpha} \Gamma^\sigma_{\beta\mu} g^{\alpha\beta} + \Gamma^\sigma_{\alpha\beta} \Gamma^\mu_{\sigma\mu} g^{\alpha\beta} \\ + \dfrac{F'}{F}(1-n)\Delta f + \dfrac{f_{,\alpha} f_{,\beta} g^{\alpha\beta} (1-n)}{4F^2} \left(4FF'' + (F')^2 (n-6) \right) \end{array} \right) \dfrac{1}{F}$$

$$= \left(R + \dfrac{F'}{F}(1-n)\Delta f + \dfrac{f_{,\alpha} f_{,\beta} g^{\alpha\beta} (1-n)}{4F^2} \left(4FF' + (F')^2 (n-6) \right) \right) \dfrac{1}{F}$$

$$\Rightarrow \quad 0 = R \cdot F^2 + FF'(1-n)\Delta f + f_{,\alpha} f_{,\beta} g^{\alpha\beta} (1-n) \left(FF'' + (F')^2 \cdot \dfrac{(n-6)}{4} \right),$$

(1174)

we can always find a solution for $F[f]$ assuring us:

$$FF'' + (F')^2 \cdot \dfrac{(n-6)}{4} = 0,$$ (1175)

which makes the last line of Eq. (1174) to:

$$0 = R \cdot F + F'(1-n)\Delta f.$$ (1176)

Unfortunately, the general solution to Eq. (1175) is[11]:

$$F[f] = C_2 \cdot \left(f \cdot (n-2) + C_1 \right)^{\frac{4}{n-2}}$$ (1177)

and thus, Eq. (1176) is only going to be linear in f in either the case $R = 0$ or the dimensional setting $n = 6$. In all other cases, we obtain non-linear differential equations.

It is for this reason that we ask whether there might be another way to approach the last line of Eq. (1174) and land with the Dirac approach by demanding:

$$\Delta f = 0,$$ (1178)

---

[11]Except in the case $n = 2$ where we have $F[f] = C_F \cdot e^{C_f \cdot f}$.

thereby giving us:

$$0 = R \cdot F^2 + f_{,\alpha} f_{,\beta} g^{\alpha\beta} (1-n) \left(FF'' + (F')^2 \cdot \frac{(n-6)}{4} \right). \tag{1179}$$

By also setting:

$$F = f - C_f, \tag{1180}$$

we end up with:

$$0 = R \cdot (f - C_f)^2 + f_{,\alpha} f_{,\beta} g^{\alpha\beta} (1-n) \cdot \frac{(n-6)}{4}$$

$$\Rightarrow \qquad R \cdot (f - C_f)^2 = f_{,\alpha} f_{,\beta} g^{\alpha\beta} (n-1) \cdot \frac{(n-6)}{4}. \tag{1181}$$

Now we could once again (c.f. sub-section "9.3.2 Repetition: Dirac in the Metric Picture and Its Connection to the Classical Quaternion Form") aim for a root extraction and derive suitable Dirac equations in *n* dimensions. Thereby we have to make sure that a solution to Eq. (1181) also solves condition (1178). As such a condition may be unsuitable (as too harsh), we repeat our considerations from sub-section 9.3.2 in a slightly different way and see what we could simplify. Going back to our starting point at the beginning of the section and demanding a kind of Eigenvalue solution with a set-up to the Laplace operator:

$$\Delta f = B^2 \cdot (f - C_f) \tag{1182}$$

and still sticking with the setting (1180), we would obtain:

$$0 = R \cdot (f - C_f)^2 + (f - C_f)^2 \cdot (1-n) \cdot B^2 + f_{,\alpha} f_{,\beta} g^{\alpha\beta} (1-n) \cdot \frac{(n-6)}{4}$$

$$\Rightarrow \qquad 0 = (f - C_f)^2 \cdot ((1-n) \cdot B^2 + R) + f_{,\alpha} f_{,\beta} g^{\alpha\beta} (1-n) \cdot \frac{(n-6)}{4}. \tag{1183}$$

We substitute $w = f - C_f$ and obtain:

$$0 = w^2 \cdot ((1-n) \cdot B^2 + R) + w_{,\alpha} w_{,\beta} g^{\alpha\beta} (1-n) \cdot \frac{(n-6)}{4}$$

$$\Rightarrow \qquad w^2 \cdot ((1-n) \cdot B^2 + R) = w_{,\alpha} w_{,\beta} g^{\alpha\beta} (n-1) \cdot \frac{(n-6)}{4}. \tag{1184}$$

Now we can either use Dirac's quaternion approach (with respect to the results from the beginning of this section we would rather need octonions), perform the root extraction via the base-vector separation or apply the vectorial root extraction from [15].

9.3.10.1 Vectorial root extraction

For the reason of simplicity, we choose the latter option.

The technique of the "vectorial root extraction" was already introduced in chapter 6 of this book and thus readers can refer to section 6.12.2.

9.3.10.2 Decomposition of the metric Eq. (1184) → generalized metric Dirac

Now we want to apply the new decomposition method on our Eq. (1184) (last line). We assume a Minkowski space-time with $n = 4$ in spatial spherical coordinates and abbreviate the terms in Eq. (1184) as follows:

$$w^2 \cdot \left((1-n) \cdot B^2 + R\right) = w_{,\alpha} w_{,\beta} g^{\alpha\beta} (n-1) \cdot \frac{(n-6)}{4}$$

$$\Rightarrow \quad w^2 \cdot \left(\frac{\overbrace{(1-n) \cdot B^2 + R}^{M^2 = \mathbf{M} \cdot \mathbf{M}}}{(n-1) \cdot \frac{(n-6)}{4}} \right) = w_{,\alpha} w_{,\beta} g^{\alpha\beta}. \qquad (1185)$$

Using the decomposition-method from the last sub-section, we can write:

$$0 = \left\{ \begin{array}{l} a_0 \cdot \partial_0 + b_0 \cdot \partial_1 + c_0 \cdot \partial_2 + d_0 \cdot \partial_3 \\[4pt] b_1 \cdot \partial_1 + c_1 \cdot \partial_2 + d_1 \cdot \partial_3 \\[4pt] c_2 \cdot \partial_2 + d_2 \cdot \partial_3 \\[4pt] d_3 \cdot \partial_3 \end{array} \right\} w \pm w \cdot \mathbf{M}$$

$$= \left\{ \begin{array}{l} a_0 \cdot \dfrac{i}{c} \cdot \partial_t + b_0 \cdot \partial_r + c_0 \cdot \dfrac{1}{r} \cdot \partial_\vartheta + d_0 \cdot \dfrac{1}{r \cdot \sin(\vartheta)} \cdot \partial_\varphi \\[10pt] b_1 \cdot \partial_r + c_1 \cdot \dfrac{1}{r} \cdot \partial_\vartheta + d_1 \cdot \dfrac{1}{r \cdot \sin(\vartheta)} \cdot \partial_\varphi \\[10pt] c_2 \cdot \dfrac{1}{r} \cdot \partial_\vartheta + d_2 \cdot \dfrac{1}{r \cdot \sin(\vartheta)} \cdot \partial_\varphi \\[10pt] d_3 \cdot \dfrac{1}{r \cdot \sin(\vartheta)} \cdot \partial_\varphi \end{array} \right\} w \pm w \cdot \mathbf{M} \quad (1186)$$

with the constants a_i to d_i to be obtained from section 6.12.2. Once again using the fact that the function w can have a sub-structure, respectively any arbitrary intrinsic structure like:

$$w = h + h_\alpha q^\alpha + h_{\alpha\beta} q^{\alpha\beta} + \ldots \xrightarrow{\text{here only}} w = h_\alpha q^\alpha, \tag{1187}$$

or with spinors:

$$w = h + h_{\alpha A} q^{\alpha A} + h_{\alpha\beta A} q^{\alpha\beta A} + h_{\alpha\beta AB} q^{\alpha\beta AB} \ldots, \tag{1188}$$

we obtain the $\mathbf{V}_\Omega$-vector equivalent to Eq. (1184), respectively the last line of Eq. (1186) as follows:

$$0 = \left[\left\{ \begin{array}{c} a_0 \cdot \dfrac{i}{c} \cdot \partial_t + b_0 \cdot \partial_r + c_0 \cdot \dfrac{1}{r} \partial_\vartheta + d_0 \cdot \dfrac{1}{r \cdot \sin(\vartheta)} \cdot \partial_\varphi \\[2mm] b_1 \cdot \partial_r + c_1 \cdot \dfrac{1}{r} \partial_\vartheta + d_1 \cdot \dfrac{1}{r \cdot \sin(\vartheta)} \cdot \partial_\varphi \\[2mm] c_2 \cdot \dfrac{1}{r} \partial_\vartheta + d_2 \cdot \dfrac{1}{r \cdot \sin(\vartheta)} \cdot \partial_\varphi \\[2mm] d_3 \cdot \dfrac{1}{r \cdot \sin(\vartheta)} \cdot \partial_\varphi \end{array} \right\} \pm \left\{ \begin{array}{c} M \\ M \\ M \\ M \end{array} \right\} q^\lambda h_\lambda \right]$$

$$\Rightarrow \quad 0 = \left[q^\lambda \left\{ \begin{array}{c} a_0 \cdot \dfrac{i}{c} \cdot \partial_t + b_0 \cdot \partial_r + c_0 \cdot \dfrac{1}{r} \partial_\vartheta + d_0 \cdot \dfrac{1}{r \cdot \sin(\vartheta)} \cdot \partial_\varphi \\[2mm] b_1 \cdot \partial_r + c_1 \cdot \dfrac{1}{r} \partial_\vartheta + d_1 \cdot \dfrac{1}{r \cdot \sin(\vartheta)} \cdot \partial_\varphi \\[2mm] c_2 \cdot \dfrac{1}{r} \partial_\vartheta + d_2 \cdot \dfrac{1}{r \cdot \sin(\vartheta)} \cdot \partial_\varphi \\[2mm] d_3 \cdot \dfrac{1}{r \cdot \sin(\vartheta)} \cdot \partial_\varphi \end{array} \right\} \pm q^\lambda \left\{ \begin{array}{c} M \\ M \\ M \\ M \end{array} \right\} h_\lambda \right].$$

$$\tag{1189}$$

Generalization to any number of dimensions or coordinates is straight forward.

Still, we are not happy with the "outsourced" condition (1182), potentially compromising the whole approach. But remembering the identities in Eqs. (1112) and (1113), we immediately see that the generalized form of Eq. (1185), without any dimensional, F or Eigenvalue-setting or assumption for the Laplace operator would read:

$$\xrightarrow{R^* = 0}$$

$$\Delta w = \overbrace{\dfrac{F}{F' \cdot (n-1)}}^{A^2} R - w_{,\alpha} w_{,\beta} \overbrace{g^{\alpha\beta}}^{e^\alpha \cdot e^\beta} \overbrace{\left(\dfrac{F''}{F'} + \dfrac{F'}{F} \cdot \dfrac{n-6}{4} \right)}^{C^2}$$

$$\Rightarrow \qquad ``\sqrt{\Delta w}^{+\alpha}" * ``\sqrt{\Delta w}^{-\beta}" = \left(\mathbf{A} + w_{,\alpha}\mathbf{e}^{\alpha}C\right) * \left(\mathbf{A} - w_{,\beta}\mathbf{e}^{\beta}C\right)$$

with: $\qquad \mathbf{A} \cdot w_{,\alpha}\mathbf{e}^{\alpha}C - \mathbf{A} \cdot w_{,\beta}\mathbf{e}^{\beta}C = 0.$ (1190)

Now we combine this with the Eigenvalue assumption $\Delta w = w \cdot B^2$ as already used in Eq. (1185) and obtain:

$$w^2 \cdot B^2 = A^2 - w_{,\alpha}w_{,\beta} \overbrace{g^{\alpha\beta}}^{\mathbf{e}^{\alpha}\cdot\mathbf{e}^{\beta}} C^2$$

$$\Rightarrow \qquad ``\sqrt{\Delta w}^{+\alpha}" * ``\sqrt{\Delta w}^{-\beta}" = ``\sqrt{w^2 \cdot B^2}^{+\alpha}" * ``\sqrt{w^2 \cdot B^2}^{-\beta}"$$

$$= \left(\mathbf{A} + w_{,\alpha}\mathbf{e}^{\alpha}C\right) * \left(\mathbf{A} - w_{,\beta}\mathbf{e}^{\beta}C\right)$$

with: $\qquad \mathbf{A} \cdot w_{,\alpha}\mathbf{e}^{\alpha}C - \mathbf{A} \cdot w_{,\beta}\mathbf{e}^{\beta}C = 0.$ (1191)

We see that this requires a vector structure for the Eigenvalues $\mathbf{B}$ as follows:

$$``\sqrt{w^2 \cdot B^2}^{+\alpha}" * ``\sqrt{w^2 \cdot B^2}^{-\beta}" = \left(\mathbf{A} + w_{,\alpha}\mathbf{e}^{\alpha}C\right) * \left(\mathbf{A} - w_{,\beta}\mathbf{e}^{\beta}C\right)$$

$$[w \cdot \mathbf{B}]^{+\alpha} = \mathbf{A} + w_{,\alpha}\mathbf{e}^{\alpha}C$$

$$[w \cdot \mathbf{B}]^{-\beta} = \mathbf{A} - w_{,\beta}\mathbf{e}^{\beta}C$$

with: $\qquad \mathbf{A} \cdot w_{,\alpha}\mathbf{e}^{\alpha}C - \mathbf{A} \cdot w_{,\beta}\mathbf{e}^{\beta}C = 0.$ (1192)

It needs to be pointed out that $\mathbf{A}$ depends on the Function F, its first derivative and the Ricci scalar of the unscaled metric $g_{\alpha\beta}$. Thus, $\mathbf{A}$ is a function of w. The assumption of a linear dependency $\mathbf{A}*w$ gives us the usual Dirac forms. But even without this approximation we can just reset $\mathbf{A}$ in Eq. (1190) as follows:

$$\Delta w = \overbrace{\underbrace{\frac{F}{F' \cdot (n-1)}}_{A^2 \cdot w} R}^{R^*=0} - w_{,\alpha}w_{,\beta} \overbrace{g^{\alpha\beta}}^{\mathbf{e}^{\alpha}\cdot\mathbf{e}^{\beta}} \overbrace{\left(\frac{F''}{F'} + \frac{F'}{F} \cdot \frac{n-6}{4}\right)}^{C^2}$$

$$\Rightarrow \qquad ``\sqrt{\Delta w}^{+\alpha}" * ``\sqrt{\Delta w}^{-\beta}" = \left(\mathbf{A} \cdot w + w_{,\alpha}\mathbf{e}^{\alpha}C\right) * \left(\mathbf{A} \cdot w - w_{,\beta}\mathbf{e}^{\beta}C\right)$$

with: $\qquad \mathbf{A} \cdot w \cdot w_{,\alpha}\mathbf{e}^{\alpha}C - \mathbf{A} \cdot w \cdot w_{,\beta}\mathbf{e}^{\beta}C = 0$ (1193)

and immediately recognize the familiar structure for the Dirac equations when reevaluating Eq. (1192), because subsequently we get:

$$\text{“}\sqrt{w^2 \cdot B^2}^{+\alpha}\text{”} * \text{“}\sqrt{w^2 \cdot B^2}^{-\beta}\text{”} = \left(w \cdot \mathbf{A} + w_{,\alpha} \mathbf{e}^{\alpha} C\right) * \left(w \cdot \mathbf{A} - w_{,\beta} \mathbf{e}^{\beta} C\right)$$

$$[w \cdot \mathbf{B}]^{+\alpha} = w \cdot \mathbf{A} + w_{,\alpha} \mathbf{e}^{\alpha} C$$

$$[w \cdot \mathbf{B}]^{-\beta} = w \cdot \mathbf{A} - w_{,\beta} \mathbf{e}^{\beta} C$$

with:
$$w \cdot \mathbf{A} \cdot w_{,\alpha} \mathbf{e}^{\alpha} C - w \cdot \mathbf{A} \cdot w_{,\beta} \mathbf{e}^{\beta} C = 0. \qquad (1194)$$

9.3.11 Further Factorization

9.3.11.1 Insertion: The classical way to obtain curved Dirac equations

Here we want to briefly cover the classical (and partially problematic) way to factorize the Klein–Gordon equation, respectively the Laplace operator, which in principle uses the Einstein–Cartan theory (e.g., [21, 22], see also https://www.youtube.com/watch?v=gwpWIYBJmq4). We start with the so-called curved Dirac equation:

$$\left(i\hbar \gamma^{\mu} \nabla_{\mu} - m \cdot c\right)\Psi = 0, \qquad (1195)$$

where we have the following curved "quadratic Nabla" operator:

$$\nabla_{\mu}\Psi = \left(\partial_{\mu} - \frac{i}{4} \cdot \omega_{\mu}^{ab} \sigma_{ab}\right)\Psi. \qquad (1196)$$

Thereby the spin connection ω_{μ}^{ab} can be derived as follows:

$$\omega_{\mu}^{ab} = \begin{pmatrix} \frac{1}{2}e^{av}\left(\partial_{\mu}e_{v}^{b} - \partial_{v}e_{\mu}^{b}\right) + \frac{1}{4}e^{a\rho}e^{b\sigma}\left(\partial_{\sigma}e_{\rho}^{c} - \partial_{\rho}e_{\sigma}^{c}\right)e_{c\mu} \\ -\frac{1}{2}e^{bv}\left(\partial_{\mu}e_{v}^{a} - \partial_{v}e_{\mu}^{a}\right) + \frac{1}{4}e^{b\rho}e^{a\sigma}\left(\partial_{\sigma}e_{\rho}^{c} - \partial_{\rho}e_{\sigma}^{c}\right)e_{c\mu} \end{pmatrix}$$

$$\eta_{ab}e_{\mu}^{a}e_{v}^{b} = g_{\mu v}; \quad e^{av} = g^{\mu v}e_{\mu}^{a}; \quad e^{av} = \eta^{ab}e_{b}^{v}; \quad \sigma^{ab} = \frac{i}{2}\left(\gamma^{a}\gamma^{b} + \gamma^{b}\gamma^{a}\right).$$

$$(1197)$$

Here the Minkowski metric is given via η_{ab}, η^{ab} and we have the symbol "e" for the various forms of the tetrad or base-vector. We find the curved Clifford algebra for the curved gamma matrices with the important connection:

$$\left(\gamma^{\mu}\gamma^{v} + \gamma^{v}\gamma^{\mu}\right) = 2 \cdot g^{\mu v}. \qquad (1198)$$

Using the tetrad, one can express the curved gamma matrices by the usual flat ones applying the formula:

$$\gamma^\mu = e_a^\mu \gamma^a. \tag{1199}$$

An example for the situation in a Minkowski space-time in spatial spherical coordinates is given below in Eq. (1221).

9.3.11.2 The connection to the metric from

The connection to the metric form can—once again—easily be shown via our result in Eq. (1112). Let Eq. (1190) be our general starting point. While Eq. (1195) should operate as follows:

$$\left(i\hbar\gamma^\mu\nabla_\mu + m\cdot c\right)\left(i\hbar\gamma^\mu\nabla_\mu - m\cdot c\right)\Psi = \left(\hbar^2\cdot\Delta - m^2\cdot c^2\right)\Psi = 0, \tag{1200}$$

we directly have from Eq. (1112):

$$\Delta\Psi = \overbrace{\frac{F}{F'\cdot(n-1)}}^{A^2}R - \Psi_{,\alpha}\Psi_{,\beta}\overbrace{g^{\alpha\beta}}^{e^\alpha\cdot e^\beta}\overbrace{\left(\frac{F''}{F'}+\frac{F'}{F}\cdot\frac{n-6}{4}\right)}^{c^2} \xrightarrow{\;R^*=0\;}$$

$$\Rightarrow \qquad \text{``}\sqrt{\Delta\Psi}^{+\alpha}\text{''}*\text{``}\sqrt{\Delta\Psi}^{-\beta}\text{''} = \left(\mathbf{A}+\Psi_{,\alpha}\mathbf{e}^\alpha C\right)*\left(\mathbf{A}-\Psi_{,\beta}\mathbf{e}^\beta C\right)$$

with: $$\mathbf{A}\cdot\Psi_{,\alpha}\mathbf{e}^\alpha C - \mathbf{A}\cdot\Psi_{,\beta}\mathbf{e}^\beta C = 0. \tag{1201}$$

At first, we realize that this is not the same situation as given in the classical Eq. (1200), because we would only have a nice relation between the Laplace and the $\Psi_{,\alpha}\Psi_{,\beta}g^{\alpha\beta} = \Psi_{,\alpha}\Psi_{,\beta}e^\alpha\cdot e^\beta$ operator (and could factorize the whole as shown above), but then we'd miss the mass-term.

Then, however, we remember that additional dimensions can provide us with the mass and that thus, we may just use the A as such mass in the Dirac form and adapt A in Eq. (1201) as follows:

$$\Delta\Psi = \overbrace{\frac{F}{F'\cdot(n-1)}}^{A^2\Psi}R - \Psi_{,\alpha}\Psi_{,\beta}\overbrace{g^{\alpha\beta}}^{e^\alpha\cdot e^\beta}\overbrace{\left(\frac{F''}{F'}+\frac{F'}{F}\cdot\frac{n-6}{4}\right)}^{c^2} \xrightarrow{\;R^*=0\;}$$

$$\Rightarrow \qquad \text{``}\sqrt{\Delta\Psi}^{+\alpha}\text{''}*\text{``}\sqrt{\Delta\Psi}^{-\beta}\text{''} = \left(\mathbf{A}\Psi+\Psi_{,\alpha}\mathbf{e}^\alpha C\right)*\left(\mathbf{A}\Psi-\Psi_{,\beta}\mathbf{e}^\beta C\right)$$

with: $$\Psi\mathbf{A}\cdot\Psi_{,\alpha}\mathbf{e}^\alpha C - \Psi\mathbf{A}\cdot\Psi_{,\beta}\mathbf{e}^\beta C = 0. \tag{1202}$$

Now A clearly stands for the masses and may even code what we classically see as potentials.

Even the case $A = 0$ does provide us with the usual Dirac equation (see sub-section "9.3.9.2 To Dirac from six dimensions in a simpler (luckier) way"). For convenience, we repeat the evaluation here with the classical symbols. We take Eq. (1201) and set $A = 0$ (de facto this is just the same as applying Eq. (1113) directly):

$$\Delta\Psi = \overbrace{\frac{F}{F'\cdot(n-1)}}^{A^2=0}R - \Psi_{,\alpha}\Psi_{,\beta}\,\overbrace{g^{\alpha\beta}}^{e^\alpha\cdot e^\beta}\overbrace{\left(\frac{F''}{F'}+\frac{F'}{F}\cdot\frac{n-6}{4}\right)}^{C^2}$$

$$= -\Psi_{,\alpha}\Psi_{,\beta}\,\overbrace{g^{\alpha\beta}}^{e^\alpha\cdot e^\beta}\overbrace{\left(\frac{F''}{F'}+\frac{F'}{F}\cdot\frac{n-6}{4}\right)}^{C^2} = \Psi_{,\alpha}\Psi_{,\beta}\,\overbrace{g^{\alpha\beta}}^{e^\alpha\cdot e^\beta}\overbrace{\left(\frac{F'}{F}\cdot\frac{6-n}{4}-\frac{F''}{F'}\right)}^{B^2} = 0.$$

$$(1203)$$

As already stated above, it does not matter in which way we guarantee the correct appearance of the $\Psi_{,\alpha}\Psi_{,\beta}g^{\alpha\beta}$ operator and thus, we could easily apply Dirac's recipe plus his gamma matrices and write[12]:

$$\Delta\Psi = \Psi_{,\alpha}\gamma^\alpha\Psi_{,\beta}\gamma^\beta\left(\frac{F'}{F}\cdot\frac{6-n}{4}-\frac{F''}{F'}\right) = 0. \qquad (1204)$$

Now, however, the gamma matrices would be matrices in a higher number of dimensions. The Dirac equation would now have to read:

$$\Psi_{,\alpha}\gamma^\alpha = 0; \quad \alpha = 0,1,2,\ldots,n-1; \quad n > 4. \qquad (1205)$$

Using our example from above in 6 dimensions and once again applying the simplifications for Gv and Gu as introduced above ($Gv = 0$ and $Gu[u]$ from Eq. (1163)), we immediately realize that we could write Eq. (1204) as follows:

$$\Delta\Psi = \Psi_{,\alpha}\gamma^\alpha\Psi_{,\beta}\gamma^\beta\left(\frac{F'}{F}\cdot\frac{6-n}{4}-\frac{F''}{F'}\right) = 0$$

$$\left(\Psi_{,a}\gamma^a - E_u\cdot\Psi\right)\left(\Psi_{,b}\gamma^b + E_u\cdot\Psi\right)\cdot g^2\left(\frac{F'}{F}\cdot\frac{6-n}{4}-\frac{F''}{F'}\right) = 0$$

with: $\qquad a,b = 0,1,2,3 \;\text{ (classical 4 dimensions)}. \qquad (1206)$

---

[12]Please note that it was shown in sub-section "9.3.4 Dirac's "Luck" and the Flat Space Condition" via the recipe in Eq. (1089), which is to say:

$$f_{\text{whole},\alpha}e^\alpha B = 0 / \cdot C_a^{\lambda b}$$

$$\Rightarrow\; C_a^{\lambda b}\cdot f_{\text{whole},\alpha}e^\alpha B = \xrightarrow{f_{\text{whole}}\to f_\lambda} = f_{\lambda,\alpha}C_a^{\lambda b}\cdot E_b^{\,\alpha}B = f_{\lambda,\alpha}D_a^{\lambda\alpha}B = f_{\lambda,\alpha}\gamma^\alpha B = 0,$$

to come from the metric operator form Eq. (1165) to the Dirac quaternion form Eq. (1166).

Here now we have the 4-dimensional **curved** Dirac gamma matrices (c.f. Eqs. (1198) and (1199)) and with

$$\Psi_{,a}\gamma^a - E_u \cdot \Psi = \left(\gamma^a \partial_{,a} - E_u\right)\Psi = 0$$

$$\Psi_{,b}\gamma^b + E_u \cdot \Psi = \left(\gamma^b \partial_{,b} + E_u\right)\Psi = 0, \qquad (1207)$$

we have obtained the classical "curved" (meaning Ricci flat but curvilinear) Dirac equations.

However, there is also another option. Moving back to Eq. (1203), we see that it does not matter how we split up the Ricci-term in the case where we do not have $R^* = R = 0$. Thus, we are free to set the following:

$$\Delta\Psi = \overbrace{\frac{F}{F' \cdot (n-1)}\left(R - F \cdot R^*\right)}^{A^2 = A_\Delta^2 \cdot \Psi - A_\nabla^2 \cdot \Psi^2} + \Psi_{,\alpha}\Psi_{,\beta}\overbrace{g^{\alpha\beta}}^{e^\alpha \cdot e^\beta}\overbrace{\left(\frac{F'}{F} \cdot \frac{6-n}{4} - \frac{F''}{F'}\right)}^{B^2}$$

$$= A_\Delta^2 \cdot \Psi - A_\nabla^2 \cdot \Psi^2 + \Psi_{,\alpha}\Psi_{,\beta}\overbrace{g^{\alpha\beta}}^{e^\alpha \cdot e^\beta} B^2. \qquad (1208)$$

Naturally, the A-terms are functions of F, R, and R^*, and so—strictly speaking— when treating them as constants, we have to consider the second line in Eq. (1208) an approximation, but we will see below what nice connection to the classical picture this can provide.

Please note that we also have not—not yet—demanded $\Delta\Psi = 0$ as we had done it before in Eq. (1204). This gives us:

$$\Delta\Psi - A_\Delta^2 \cdot \Psi = \Psi_{,\alpha}\Psi_{,\beta}\overbrace{g^{\alpha\beta}}^{e^\alpha \cdot e^\beta} B^2 - A_\nabla^2 \cdot \Psi^2. \qquad (1209)$$

We see that we can now factorize the right-hand side in the usual way, thereby giving us:

$$\xrightarrow{R^* \neq 0; \quad R \neq 0}$$

$$\Delta\Psi - A_\Delta^2 \cdot \Psi = \Psi_{,\alpha}\Psi_{,\beta}\overbrace{g^{\alpha\beta}}^{e^\alpha \cdot e^\beta} B^2 - A_\nabla^2 \cdot \Psi^2 = \left(\mathbf{A}_\nabla\Psi + \Psi_{,\alpha}e^\alpha B\right) * \left(\Psi_{,\beta}e^\beta B - \mathbf{A}_\nabla\Psi\right)$$

with: $$\mathbf{A}_\nabla\Psi \cdot \Psi_{,\alpha}e^\alpha B - \mathbf{A}_\nabla\Psi \cdot \Psi_{,\beta}e^\beta B = 0. \qquad (1210)$$

Not only do we recognize the typical Dirac structures in the two factors on the right-hand side, but we also realize that the A-terms could code all sorts of what classically is been seen as masses and potentials and that

with Eq. (1210) we always have a straight forward—truly algebraic—connection between the Laplace or Klein–Gordon operator form and the $\Psi_{,\alpha}\Psi_{,\beta}g^{\alpha\beta}$-one, we can directly factorize (in contrast to the Klein–Gordon operator, which factorization—in general—is problematic). In order to get back to the metric terms, we only have to remember the definitions of our A and B, making the connection (1210) to:

$$\xrightarrow{\quad R^* \neq 0; \quad R \neq 0 \quad}$$

$$\Delta\Psi - \left(\frac{F}{F'} \cdot \frac{\left(R - F \cdot R^*\right)}{(n-1)\cdot\Psi} + A_\nabla^2 \cdot \Psi \right)\cdot\Psi$$

$$= \left(A_\nabla\Psi + \Psi_{,\alpha}e^\alpha \sqrt{\left(\frac{F'}{F}\cdot\frac{6-n}{4} - \frac{F''}{F'}\right)} \right) * \left(\Psi_{,\beta}e^\beta \sqrt{\left(\frac{F'}{F}\cdot\frac{6-n}{4} - \frac{F''}{F'}\right)} - A_\nabla\Psi \right)$$

$$\text{with}:\quad A_\nabla\Psi\cdot\Psi_{,\alpha}e^\alpha \sqrt{\left(\frac{F'}{F}\cdot\frac{6-n}{4} - \frac{F''}{F'}\right)} - A_\nabla\Psi\cdot\Psi_{,\beta}e^\beta \sqrt{\left(\frac{F'}{F}\cdot\frac{6-n}{4} - \frac{F''}{F'}\right)} = 0.$$

$$(1211)$$

As the $A^2 = A_\Delta^2 \cdot \Psi - A_\nabla^2 \cdot \Psi^2$-split-up in Eq. (1208) was totally arbitrary, we could always set one of the two terms to zero, which is giving us back the complete metric form:

$$\Delta\Psi - \frac{F}{F'}\cdot\frac{\left(R - F \cdot R^*\right)}{(n-1)} = \Psi_{,\alpha}e^\alpha\Psi_{,\beta}e^\beta\left(\frac{F'}{F}\cdot\frac{6-n}{4} - \frac{F''}{F'} \right)$$

or

$$\Delta\Psi = \frac{F}{F'}\cdot\frac{\left(R - F \cdot R^*\right)}{(n-1)} + \Psi_{,\alpha}e^\alpha\Psi_{,\beta}e^\beta\left(\frac{F'}{F}\cdot\frac{6-n}{4} - \frac{F''}{F'} \right). \qquad (1212)$$

So far, we had always concentrated on at least $R^* = 0$, which is to say an effectively Ricci-flat space-time.

Thus, our relations in Eq. (1212) may better look as follows:

$$\Delta\Psi - \frac{F}{F'}\cdot\frac{R}{(n-1)} = \Psi_{,\alpha}e^\alpha\Psi_{,\beta}e^\beta\left(\frac{F'}{F}\cdot\frac{6-n}{4} - \frac{F''}{F'} \right)$$

or

$$\Delta\Psi = \frac{F}{F'}\cdot\frac{R}{(n-1)} + \Psi_{,\alpha}e^\alpha\Psi_{,\beta}e^\beta\left(\frac{F'}{F}\cdot\frac{6-n}{4} - \frac{F''}{F'} \right), \qquad (1213)$$

which simplifies in the case of the totally flat space-time with also $R = 0$ to:

$$\Delta\Psi = \Psi_{,\alpha}e^{\alpha}\Psi_{,\beta}e^{\beta}\left(\frac{F'}{F}\cdot\frac{6-n}{4}-\frac{F''}{F'}\right). \tag{1214}$$

As we have seen above that all sorts of inertia and potentials can just pop out from the Laplace operator via suitable additional dimensions (e.g., c.f. subsection "9.3.9.2 To Dirac from 6 dimensions in a simpler (luckier) way"), we still have everything we'd need to construct the classical Dirac equation in curvilinear coordinates (of Ricci-flat space-times) without the need to resort to the Einstein–Cartan theory [21, 22].

9.3.11.3 An asymmetry

Yes, we have found a way to construct Dirac equations in arbitrary space-times directly out of the Ricci scalar of a scaled metric. However, the fact that our current approach relies on certain condition with respect to the Laplace operator in Eq. (1008) does not exactly make us truly happy. We, therefore, intend to find a more generalized Dirac approach.

Once again, just to have something to start with, using $R^* = 0$ and:

$$F = f - C_f, \tag{1215}$$

in Eq. (1008) we end up with:

$$0 = R\cdot\left(f - C_f\right)^2 + \left(1-n\right)\cdot\left(f - C_f\right)\cdot\Delta f + f_{,\alpha}f_{,\beta}g^{\alpha\beta}\left(1-n\right)\cdot\frac{\left(n-6\right)}{4}$$

$$\xrightarrow{w = f - C_f}$$

$$\Rightarrow \qquad 0 = R\cdot w^2 + \left(1-n\right)\cdot w\cdot\Delta w + w_{,\alpha}w_{,\beta}g^{\alpha\beta}\left(1-n\right)\cdot\frac{\left(n-6\right)}{4}. \tag{1216}$$

Partial expansion of the Laplace operator in the last line brings us to:

$$0 = R\cdot w^2 + \left(1-n\right)\cdot w\cdot\frac{1}{\sqrt{g}}\partial_{\alpha}\sqrt{g}\cdot g^{\alpha\beta}\partial_{\beta}w + w_{,\alpha}w_{,\beta}g^{\alpha\beta}\left(1-n\right)\cdot\frac{\left(n-6\right)}{4}$$

$$\Rightarrow \qquad \frac{4\cdot R}{\left(n-1\right)\cdot\left(n-6\right)}\cdot w^2 = \frac{4}{n-6}\cdot w\cdot\frac{1}{\sqrt{g}}\partial_{\alpha}\sqrt{g}\cdot g^{\alpha\beta}\partial_{\beta}w + w_{,\alpha}w_{,\beta}g^{\alpha\beta}$$

$$\Rightarrow \qquad \overbrace{\frac{4\cdot R}{\left(n-1\right)\cdot\left(n-6\right)}}^{M^+\cdot M^-}\cdot w^2 = \frac{4}{n-6}\cdot w\cdot\frac{1}{\sqrt{g}}\partial_{\alpha}\sqrt{g}\cdot g^{\alpha\beta}\partial_{\beta}w + w_{,\alpha}w_{,\beta}g^{\alpha\beta}.$$

$$\tag{1217}$$

We realize that in general, due to the Laplace operator, only an asymmetric factorization of Eq. (1217) is possible. In order to have a simple example to start with, we take the Minkowski metric from our last sub-section above and try to approach such a factorization of the right-hand side of Eq. (1217) via the two operator vectors:

$$V_{\Omega}^+ \cdot V_{\Omega}^- = ... = \overbrace{\frac{4}{n-6} \cdot w}^{\equiv w^*} \cdot \frac{1}{\sqrt{g}} \partial_\alpha \sqrt{g} \cdot g^{\alpha\beta} \partial_\beta w + w_{,\alpha} w_{,\beta} g^{\alpha\beta}. \quad (1218)$$

With eight addends (remember: Minkowski with spherical spatial coordinates), we may try to set the vectors as follows:

$$V_{\Omega}^+ = \begin{cases} \left[w^* \cdot \left(a_{\Delta 0} \cdot \dfrac{i}{c} \cdot \partial_t + b_{\Delta 0} \cdot \dfrac{1}{r^2} \cdot \partial_r + c_{\Delta 0} \cdot \dfrac{1}{r \cdot \sin(\vartheta)} \cdot \partial_\vartheta + d_{\Delta 0} \cdot \dfrac{1}{r \cdot \sin(\vartheta)} \cdot \partial_\varphi \right) \right. \\ \left. + a_0 \cdot \dfrac{i}{c} \cdot w_{,t} + b_0 \cdot w_{,r} + c_0 \cdot \dfrac{1}{r} \cdot w_{,\vartheta} + d_0 \cdot \dfrac{1}{r \cdot \sin(\vartheta)} \cdot w_{,\varphi} \right] \\[2mm] \left[w^* \cdot \left(b_{\Delta 1} \cdot \dfrac{1}{r^2} \cdot \partial_r + c_{\Delta 1} \cdot \dfrac{1}{r \cdot \sin(\vartheta)} \cdot \partial_\vartheta + d_{\Delta 1} \cdot \dfrac{1}{r \cdot \sin(\vartheta)} \cdot \partial_\varphi \right) \right. \\ \left. + a_1 \cdot \dfrac{i}{c} \cdot w_{,t} + b_1 \cdot w_{,r} + c_1 \cdot \dfrac{1}{r} \cdot w_{,\vartheta} + d_1 \cdot \dfrac{1}{r \cdot \sin(\vartheta)} \cdot w_{,\varphi} \right] \\[2mm] \left[w^* \cdot \left(c_{\Delta 2} \cdot \dfrac{1}{r \cdot \sin(\vartheta)} \cdot \partial_\vartheta + d_{\Delta 2} \cdot \dfrac{1}{r \cdot \sin(\vartheta)} \cdot \partial_\varphi \right) \right. \\ \left. + a_2 \cdot \dfrac{i}{c} \cdot w_{,t} + b_2 \cdot w_{,r} + c_2 \cdot \dfrac{1}{r} \cdot w_{,\vartheta} + d_2 \cdot \dfrac{1}{r \cdot \sin(\vartheta)} \cdot w_{,\varphi} \right] \\[2mm] \vdots \\[2mm] c_6 \cdot \dfrac{1}{r} \cdot w_{,\vartheta} + d_6 \cdot \dfrac{1}{r \cdot \sin(\vartheta)} \cdot w_{,\varphi} \\[2mm] d_7 \cdot \dfrac{1}{r \cdot \sin(\vartheta)} \cdot w_{,\varphi} \end{cases}$$

$$(1219)$$

$$V_\Omega^- = \left\{ \begin{array}{c} \left\{ \begin{array}{l} a_{\Delta 0} \cdot \dfrac{i}{c} \cdot w_{,t} + b_{\Delta 0} \cdot r^2 \cdot w_{,r} + c_{\Delta 0} \cdot \dfrac{\sin(\vartheta)}{r} \cdot w_{,\vartheta} + d_{\Delta 0} \cdot \dfrac{1}{r \cdot \sin(\vartheta)} \cdot w_{,\varphi} \\[2mm] + a_0 \cdot \dfrac{i}{c} \cdot w_{,t} + b_0 \cdot w_{,r} + c_0 \cdot \dfrac{1}{r} \cdot w_{,\vartheta} + d_0 \cdot \dfrac{1}{r \cdot \sin(\vartheta)} \cdot w_{,\varphi} \end{array} \right\} \\[6mm] \left[\begin{array}{l} b_{\Delta 1} \cdot r^2 \cdot w_{,r} + c_{\Delta 1} \cdot \dfrac{\sin(\vartheta)}{r} \cdot w_{,\vartheta} + d_{\Delta 1} \cdot \dfrac{1}{r \cdot \sin(\vartheta)} \cdot w_{,\varphi} \\[2mm] + a_1 \cdot \dfrac{i}{c} \cdot w_{,t} + b_1 \cdot w_{,r} + c_1 \cdot \dfrac{1}{r} \cdot w_{,\vartheta} + d_1 \cdot \dfrac{1}{r \cdot \sin(\vartheta)} \cdot w_{,\varphi} \end{array} \right] \\[6mm] \left\{ \begin{array}{l} c_{\Delta 2} \cdot \dfrac{\sin(\vartheta)}{r} \cdot w_{,\vartheta} + d_{\Delta 2} \cdot \dfrac{1}{r \cdot \sin(\vartheta)} \cdot w_{,\varphi} \\[2mm] + a_2 \cdot \dfrac{i}{c} \cdot w_{,t} + b_2 \cdot w_{,r} + c_2 \cdot \dfrac{1}{r} \cdot w_{,\vartheta} + d_2 \cdot \dfrac{1}{r \cdot \sin(\vartheta)} \cdot w_{,\varphi} \end{array} \right\} \\[4mm] \vdots \\[2mm] c_6 \cdot \dfrac{1}{r} \cdot w_{,\vartheta} + d_6 \cdot \dfrac{1}{r \cdot \sin(\vartheta)} \cdot w_{,\varphi} \\[4mm] d_7 \cdot \dfrac{1}{r \cdot \sin(\vartheta)} \cdot w_{,\varphi} \end{array} \right\}.$$

$$(1220)$$

Such an approach would be equivalent to the octonion technique, which is to say an operational math with 8-dimensional matrix objects.

We realize, however, that this system would require an 8-dimensional **w**-function vector (or spinor or whatever).

For comparison, we here remind the classical Dirac equation in spherical coordinates as it was given by Dietterich in https://www.youtube.com/watch?v=gwpWIYBJmq4 (see also references of the next chapter):

$$0 = \begin{pmatrix} \dfrac{i\hbar}{c}\partial_t & 0 & \dfrac{i\hbar}{r\sin\vartheta}\partial_\varphi & i\hbar\left(-\dfrac{i}{r}\Theta + R\right) \\[4mm] 0 & \dfrac{i\hbar}{c}\partial_t & i\hbar\left(\dfrac{i}{r}\Theta + R\right) & -\dfrac{i\hbar}{r\sin\vartheta}\partial_\varphi \\[4mm] -\dfrac{i\hbar}{r\sin\vartheta}\partial_\varphi & i\hbar\left(\dfrac{i}{r}\Theta - R\right) & -\dfrac{i\hbar}{c}\partial_t & 0 \\[4mm] i\hbar\left(-\dfrac{i}{r}\Theta - R\right) & \dfrac{i\hbar}{r\sin\vartheta}\partial_\varphi & 0 & -\dfrac{i\hbar}{c}\partial_t \end{pmatrix}$$

$$\times \begin{pmatrix} \Psi_1 \\ \Psi_2 \\ \Psi_3 \\ \Psi_4 \end{pmatrix} - mc \begin{pmatrix} \Psi_1 \\ \Psi_2 \\ \Psi_3 \\ \Psi_4 \end{pmatrix}$$

$$\Theta = \frac{1}{\sqrt{\sin\vartheta}} \partial_\vartheta \left(\sqrt{\sin\vartheta} (\dots) \right); \quad R = \frac{1}{r} \partial_r \left(r(\dots) \right). \tag{1221}$$

Thus, our 8-dimensional **w**-function vector in just 4 dimensions appears strange and even though there were quite some discussions about the apparent success and insight due to the introduction of octonions, this author is of the opinion that the approach above is not an irreducible one. The reason for this assumption is to be found in the fact that—so far—the degree of freedom, the wrapper function $F[f]$ does provide, has not been used.

For instance, a significant simplification to Eqs. (1219) and (1220) can be obtained when demanding the following condition for this wrapper function $F[f]$:

$$\frac{4FF'' + (F')^2(n-6)}{4F} = \frac{F'}{f} \quad \Rightarrow \quad F = \begin{cases} \left(C_{f0}^2 \pm f^2 \right)^{\frac{4}{n-2}} \cdot C_{f1} & n > 2 \\ e^{\pm C_{f0}^2 \cdot f^2} \cdot C_{f1} & n = 2. \end{cases} \tag{1222}$$

This makes Eq. (1008) to:

$$R^* = R^*_{\alpha\beta} G^{\alpha\beta} = \left(R + (1-n) \cdot \frac{F'}{F} \left(\Delta f + \frac{1}{f} \cdot f_{,\alpha} f_{,\beta} g^{\alpha\beta} \right) \right) \frac{1}{F}$$

$$= \left(R + (1-n) \cdot \frac{F'}{f \cdot F} \left(\frac{f}{\sqrt{g}} \partial_\alpha \sqrt{g} \cdot g^{\alpha\beta} \partial_\beta f + f_{,\alpha} f_{,\beta} g^{\alpha\beta} \right) \right) \frac{1}{F}$$

$$= \left(R + (1-n) \cdot \frac{F'}{f \cdot F} \left(\frac{1}{\sqrt{g}} \partial_\alpha f \cdot \sqrt{g} \cdot g^{\alpha\beta} \partial_\beta f \right) \right) \frac{1}{F} \xrightarrow{R^* = R^*_{\alpha\beta} G^{\alpha\beta} = 0}$$

$$\Rightarrow \quad R \cdot \frac{f \cdot F}{(n-1) \cdot F'} = \frac{1}{\sqrt{g}} \partial_\alpha f \cdot \sqrt{g} \cdot g^{\alpha\beta} \partial_\beta f \xrightarrow{\frac{4FF'' + (F')^2(n-6)}{4F} = \frac{F'}{f}}$$

$$\Rightarrow \quad \frac{1}{\sqrt{g}} \partial_\alpha f \cdot \sqrt{g} \cdot g^{\alpha\beta} \partial_\beta f = R \cdot \begin{cases} \frac{1}{8} \cdot \frac{2-n}{n-1} \cdot \left(C_{f0}^2 \pm f^2 \right) & n > 2 \\ \pm \frac{1}{2} \cdot C_{f0}^2 & n = 2. \end{cases} \tag{1223}$$

We recognize the new (summed up due to the chain rule) operator, which could be seen as some kind of Ricci or curvature operator to the scale (wave) function f:

$$\frac{1}{\sqrt{g}}\partial_\alpha f \cdot \sqrt{g} \cdot g^{\alpha\beta}\partial_\beta f \equiv \Omega f^2. \tag{1224}$$

Factorization could now be approached via:

$$\frac{1}{\sqrt{g}}\partial_\alpha f \cdot \sqrt{g} \cdot g^{\alpha\beta}\partial_\beta f = V_\Omega^+ \cdot V_\Omega^- = R \cdot \begin{cases} \dfrac{1}{8}\cdot\dfrac{2-n}{n-1}\cdot\left(C_{f0}+f\right)\cdot\left(C_{f0}-f\right) & n>2 \\[2ex] \pm\dfrac{1}{2}\cdot C_{f0}\cdot C_{f0} & n=2. \end{cases} \tag{1225}$$

As before, we take the example of spherical spatial symmetry in a 4-dimensional Minkowski space-time. Our approach reads:

$$V_\Omega^+ = \begin{cases} \left\{a_0\cdot\dfrac{i}{c}\partial_t + b_0\cdot\dfrac{1}{r^2}\cdot\partial_r + c_0\cdot\dfrac{1}{r\cdot\sin(\vartheta)}\cdot\partial_\vartheta + d_0\cdot\dfrac{1}{r\cdot\sin(\vartheta)}\cdot\partial_\varphi\right\} \\[2ex] \left\{b_1\cdot\dfrac{1}{r^2}\cdot\partial_r + c_1\cdot\dfrac{1}{r\cdot\sin(\vartheta)}\cdot\partial_\vartheta + d_1\cdot\dfrac{1}{r\cdot\sin(\vartheta)}\cdot\partial_\varphi\right\} \\[2ex] \left\{c_2\cdot\dfrac{1}{r\cdot\sin(\vartheta)}\cdot\partial_\vartheta + d_2\cdot\dfrac{1}{r\cdot\sin(\vartheta)}\cdot\partial_\varphi\right\} \\[2ex] \left\{d_3\cdot\dfrac{1}{r\cdot\sin(\vartheta)}\cdot\partial_\varphi\right\} \end{cases} f \tag{1226}$$

$$V_\Omega^- = \begin{cases} \left\{a_0\cdot\dfrac{i}{c}f_{,t} + b_0\cdot r^2\cdot f_{,r} + c_0\cdot\dfrac{\sin(\vartheta)}{r}\cdot f_{,\vartheta} + d_0\cdot\dfrac{1}{r\cdot\sin(\vartheta)}\cdot f_{,\varphi}\right\} \\[2ex] \left\{b_1\cdot r^2\cdot f_{,r} + c_1\cdot\dfrac{\sin(\vartheta)}{r}\cdot f_{,\vartheta} + d_1\cdot\dfrac{1}{r\cdot\sin(\vartheta)}\cdot f_{,\varphi}\right\} \\[2ex] \left\{c_2\cdot\dfrac{\sin(\vartheta)}{r}\cdot f_{,\vartheta} + d_2\cdot\dfrac{1}{r\cdot\sin(\vartheta)}\cdot f_{,\varphi}\right\} \\[2ex] \left\{d_3\cdot\dfrac{1}{r\cdot\sin(\vartheta)}\cdot f_{,\varphi}\right\} \end{cases}. \tag{1227}$$

With the metric of our example known to be perfectly diagonal:

$$G_{\alpha\beta}^4 = F[f] \cdot \begin{pmatrix} -c^2 & 0 & 0 & 0 \\ 0 & 1 & 0 & 0 \\ 0 & 0 & r^2 & 0 \\ 0 & 0 & 0 & r^2 \cdot \sin^2 \vartheta \end{pmatrix}, \qquad (1228)$$

we see that we can just use the results from sub-section 6.12.2 in order to have all constants a_j, b_j, c_j, and d_j determined.

Does this also give us reasonable results?

In order to obtain a first and quick answer to this question, there is no need to actually use the new and asymmetric Dirac equations, but just look at the starting point Eq. (1224). Assuming a function f solely depending on the radius coordinate, and demanding R to be zero (flat space, which is clear due to our choice of the Minkowski coordinates anyway), we result in the following differential equation and its solution:

$$\frac{1}{\sqrt{g}} \partial_\alpha f \cdot \sqrt{g} \cdot g^{\alpha\beta} \partial_\beta f = 0 \xrightarrow{f=f[r]} 0 = (f')^2 + f \cdot \left(\frac{2 \cdot f'}{r} + f'' \right)$$

$$\Rightarrow \qquad f[r] = C_{r1} \cdot \sqrt{1 \pm \frac{C_{r0}}{r}}. \qquad (1229)$$

We recognize the structure of the metric components time and radius from the Schwarzschild metric [16]. However, with the singularity at $r = 0$, we are anything but happy and therefore see how to set another, perhaps more suitable, condition for the functional wrapper $F[r]$. We see that with:

$$\frac{4FF'' + (F')^2 (n-6)}{4F} = -\frac{F'}{f} \quad \Rightarrow \quad F = \begin{cases} \left(C_{f0} \pm \ln f \right)^{\frac{4}{n-2}} \cdot C_{f1} & n > 2 \\ f^{\pm C_{f0}^2} \cdot C_{f1} & n = 2 \end{cases}$$

$$(1230)$$

and subsequently Eq. (1008) changing to:

$$R^* = R^*_{\alpha\beta} G^{\alpha\beta} = \left(R + (1-n) \cdot \frac{F'}{F} \left(\Delta f - \frac{1}{f} f_{,\alpha} f_{,\beta} g^{\alpha\beta} \right) \right) \frac{1}{F}$$

$$= \left(R + (1-n) \cdot \frac{F'}{f \cdot F} \left(\frac{f}{\sqrt{g}} \partial_\alpha \sqrt{g} \cdot g^{\alpha\beta} \partial_\beta f - f_{,\alpha} f_{,\beta} g^{\alpha\beta} \right) \right) \frac{1}{F}$$

$$\xrightarrow{R^* = R^*_{\alpha\beta} G^{\alpha\beta} = 0}$$

$$\Rightarrow \qquad R \cdot \frac{f \cdot F}{(n-1) \cdot F'} = \frac{f}{\sqrt{g}} \partial_\alpha \sqrt{g} \cdot g^{\alpha\beta} \partial_\beta f - f_{,\alpha} f_{,\beta} g^{\alpha\beta}$$

$$\xrightarrow{\quad \frac{4FF'' + (F')^2(n-6)}{4F} = -\frac{F'}{f} \quad}$$

$$\Rightarrow \qquad \frac{1}{\sqrt{g}} \partial_\alpha f \cdot \sqrt{g} \cdot g^{\alpha\beta} \partial_\beta f = R \cdot \begin{cases} \pm \dfrac{1}{4} \cdot \dfrac{n-2}{n-1} \cdot (C_{f0} \pm \ln f) \cdot f^2 & n>2 \\ \pm f^2 \cdot C_{f0}^2 & n=2, \end{cases} \qquad (1231)$$

the corresponding calculation to our $f[r]$-problem from above changes to:

$$\frac{f}{\sqrt{g}} \partial_\alpha \sqrt{g} \cdot g^{\alpha\beta} \partial_\beta f - f_{,\alpha} f_{,\beta} g^{\alpha\beta} = 0 \xrightarrow{\quad f=f[r] \quad} 0 = f \cdot \left(\frac{2 \cdot f'}{r} + f'' \right) - (f')^2$$

$$\Rightarrow \qquad f[r] = C_{r1} \cdot e^{-\frac{C_{r0}}{r}} . \qquad (1232)$$

This time the singularity has disappeared from the function f when assuming $C_{r0} > 0$ (and real), but this does not help us, as it requires its return (the singularity's return, we mean), when inserting f into F, thereby having to use Eq. (1230).

We simply realize that—similar to the ordinary Laplace equation—a pure dependence on r does not give us suitable results. When, however, assuming a non-vanishing (perhaps constant) curvature term R, then things become more interesting, because we have (here only considering condition from Eq. (1222)):

$$\xrightarrow{\quad R^* = R^*_{\alpha\beta} G^{\alpha\beta} = 0 \quad} \& \xrightarrow{\quad \frac{4FF'' + (F')^2(n-6)}{4F} = \frac{F'}{f} \quad} \& \xrightarrow{\quad n=4 \quad}$$

$$\frac{1}{\sqrt{g}} \partial_\alpha f \cdot \sqrt{g} \cdot g^{\alpha\beta} \partial_\beta f + R \cdot \frac{1}{12} \cdot \left(C_{f0}^2 \pm f^2 \right) = 0$$

$$\xrightarrow{\quad f=f[r] \quad}$$

$$0 = (f')^2 + f \cdot \left(\frac{2 \cdot f'}{r} + f'' \right) + R \cdot \frac{1}{12} \cdot \left(C_{f0}^2 \pm f^2 \right)$$

$$\xrightarrow{\quad C_{f0}^2 = 0; R \cdot \frac{1}{12} = E^2 \quad}$$

$$\Rightarrow \qquad \text{for} \quad \pm f^2 \to f[r] = r^{-\frac{1}{2}} \cdot C_{r1} \cdot \cos\left[\sqrt{2} \cdot \frac{E^2 \cdot r}{\sqrt{\pm E^2}} - C_{r0} \right]^{\frac{1}{2}} . \qquad (1233)$$

Now we might not like to imagine an overall constant Ricci curvature in order to assure an outcome as obtained above and so we want to look for a way to achieve the same result via a suitable situation with a function f not only depending on the radius r.

Assuming a function $f[t,r]$ with the separation approach $f[t, r] = f_t[t] \cdot f_r[r]$ and the following time dependency:

$$f_t[t] = \begin{cases} C_{tc} \cdot \cos[c \cdot E \cdot t] \\ C_{ts} \cdot \sin[c \cdot E \cdot t], \end{cases} \tag{1234}$$

always apply only either of the two functions, we obtain:

$$\underrightarrow{R^* = R^*{}_{\alpha\beta} G^{\alpha\beta} = 0} \,\&\, \underrightarrow{\frac{4FF'' + (F')^2 (n-6)}{4F} = \frac{F'}{f}} \,\&\, \underrightarrow{n=4}$$

$$\frac{1}{\sqrt{g}} \partial_\alpha f \cdot \sqrt{g} \cdot g^{\alpha\beta} \partial_\beta f + R \cdot \frac{1}{12} \cdot \left(C_{f0}^2 \pm f^2 \right) = 0$$

$$\underrightarrow{f = f[t,r] = f_t[t] \cdot f_r[r]; C_{f0}^2 = 0; R = 0}$$

$$0 = \left(f_r' \right)^2 + f_r \cdot \left(\frac{2 \cdot f_r'}{r} + f_r'' \right) + 2 \cdot E \cdot f_r^2$$

$$\Rightarrow \qquad f[r] = C_{r1} \cdot \sqrt{\frac{\cos[E \cdot r - C_{r0}]}{r}}. \tag{1235}$$

Please note that in order to avoid the singularity at $r = 0$, we have to set $C_{r0} = \pm \pi/2$.

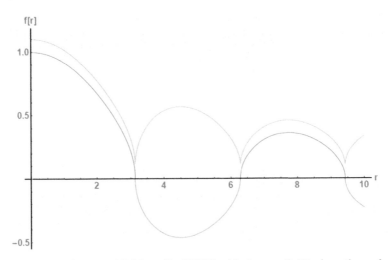

Figure 9.4 Distribution of $f[r]$ from Eq. (1235) with $C_{r0} = -\pi/2$. We show the real part (blue), the imaginary part (time-1 in yellow), and the absolute value (green). To the latter we added a constant (0.1) in order to make it distinguishable from the real part.

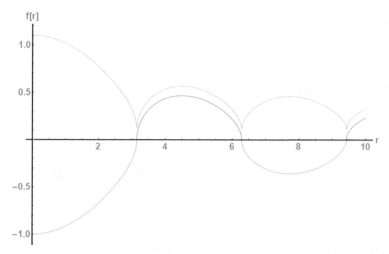

Figure 9.5   Distribution of $f[r]$ from Eq. (1235) with $C_{r0} = +\pi/2$. We show the real part (blue), the imaginary part (time-1 in yellow), and the absolute value (green). To the latter we added a constant (0.1) in order to make it distinguishable from the real part.

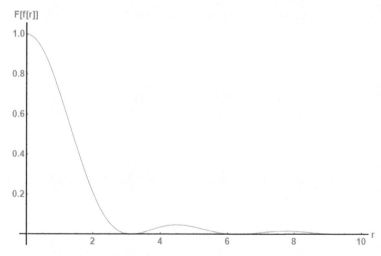

Figure 9.6   Distribution of $F[f[r]]$ from Eqs. (1235) and (1222) with $C_{r0} = \pm\pi/2$.

We see that we can also obtain an almost suitable wave or scale function, only that we do not obtain the asymptotic behavior of the flat space for infinite r. This is caused by the fact that we have not been able to find a solution to the task in Eq. (1235) with a setting $C_{f0} \neq 0$.

Things are getting much easier when applying the condition from Eq. (1175) and also having the flat space with $R = 0$:

$$\frac{4FF'' + (F')^2 (n-6)}{4F} = 0 \quad \Rightarrow \quad F = \begin{cases} \left(C_{f0} \pm f\right)^{\frac{4}{n-2}} \cdot C_{f1} & n > 2 \\ e^{\pm f \cdot C_{f0}} \cdot C_{f1} & n = 2. \end{cases} \quad (1236)$$

Then we obtain from Eq. (1008):

$$R^* = R^*_{\alpha\beta} G^{\alpha\beta} = \left(R + (1-n) \cdot \frac{F'}{F} \left(\Delta f - \frac{1}{f} \cdot f_{,\alpha} f_{,\beta} g^{\alpha\beta} \right) \right) \frac{1}{F}$$

$$= \left(R + (1-n) \cdot \frac{F'}{f \cdot F} \left(\frac{f}{\sqrt{g}} \partial_\alpha \sqrt{g} \cdot g^{\alpha\beta} \partial_\beta f - f_{,\alpha} f_{,\beta} g^{\alpha\beta} \right) \right) \frac{1}{F}$$

$$\xrightarrow{R^* = R^*_{\alpha\beta} G^{\alpha\beta} = 0}$$

$$\Rightarrow \qquad R \cdot \frac{F}{(n-1) \cdot F'} = \frac{f}{\sqrt{g}} \partial_\alpha \sqrt{g} \cdot g^{\alpha\beta} \partial_\beta f - f_{,\alpha} f_{,\beta} g^{\alpha\beta}$$

$$\xrightarrow{\frac{4FF'' + (F')^2 (n-6)}{4F} = 0}$$

$$\Rightarrow \quad \frac{1}{\sqrt{g}} \partial_\alpha f \cdot \sqrt{g} \cdot g^{\alpha\beta} \partial_\beta f = R \cdot \begin{cases} \mp \frac{1}{4} \cdot \frac{n-2}{n-1} \cdot \left(C_{f0} \pm f \right) & n > 2 \\ \pm C_{f0}^{-1} & n = 2. \end{cases} \xrightarrow{R=0} \Delta f = 0$$

$$(1237)$$

The corresponding result for our spherical problem reads:

$$\xrightarrow{R^* = R^*_{\alpha\beta} G^{\alpha\beta} = 0} \& \xrightarrow{\frac{4FF'' + (F')^2 (n-6)}{4F} = 0} \& \xrightarrow{n=4}$$

$$\frac{1}{\sqrt{g}} \partial_\alpha f \cdot \sqrt{g} \cdot g^{\alpha\beta} \partial_\beta f + R \cdot \frac{1}{6} \cdot \left(C_{f0} + f \right) = 0$$

$$\xrightarrow{f = f[t,r] = f_t[t] \cdot f_r[r]; C_{f0} = 1; R = 0}$$

$$0 = \frac{2 \cdot f_r'}{r} + f_r'' + E^2 \cdot f_r$$

$$\Rightarrow \qquad f[r] = C_{r1} \cdot \frac{e^{i \cdot E \cdot r} - e^{-i \cdot E \cdot r}}{r}. \qquad (1238)$$

Thereby we used the following function for $f_t[t]$:

$$f_t[t] = C_{tc} \cdot \cos[c \cdot E \cdot t] + C_{ts} \cdot \sin[c \cdot E \cdot t] \qquad (1239)$$

and demanded the solution for the radius to be free of singularities (especially at $r = 0$). As we see in Fig. 9.7, this gives us the right asymptotic behavior for r to infinity.

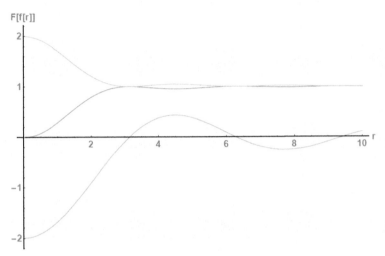

Figure 9.7 Distribution of $F[f[r]]$ according to condition in Eq. (1175) with solution Eq. (1238) (last line). Blue = real part, Yellow = imaginary part, Green = absolute value. We see that we obtain the correct asymptotic behavior for bigger r.

9.3.11.4 Generalization

Now we intend to investigate some generalizations of the options to set the F-factor in connection with the Nabla operators. We start with a simple factor fac:

$$\xrightarrow{R^* = R^*_{\alpha\beta} G^{\alpha\beta} = 0} \& \xrightarrow{\dfrac{4FF'' + (F')^2(n-6)}{4F} = fac \cdot \dfrac{F'}{f}} \& \xrightarrow{n=4}$$

$$F[f] = \left(C_{f0}^2 \pm f^{1+fac} \right)^{\frac{4}{n-2}} C_{f1}$$

$$\xrightarrow{f = f[r]} \xrightarrow{C_{f0}^2 = 0; R \cdot \frac{1}{12} = E^2}$$

$$\Rightarrow \quad \text{for} \quad \pm f^2 \rightarrow f[r] = r^{-\frac{1}{1+fac}} \cdot C_{r1} \cdot \cos\left[\frac{E^2 \cdot r}{\sqrt{\pm\frac{E^2}{1+fac}}} - C_{r0}\right]^{\frac{1}{1+fac}}. \quad (1240)$$

Further generalization is possible via an arbitrary function $H[f]$ with the following condition:

$$\frac{4FF'' + (F')^2(n-6)}{4F} = F' \cdot \frac{H'}{H} \Rightarrow F = \begin{cases} \left(C_{f0}^2 - \int\limits_1^f H[\phi] \cdot \frac{2-n}{4} \cdot d\phi\right)^{\frac{4}{n-2}} \cdot C_{f1} & n > 2 \\ e^{C_{f0}^{-2} \cdot \int\limits_1^f H[\phi] \cdot d\phi} \cdot C_{f1} & n = 2. \end{cases}$$

$$(1241)$$

This changes Eq. (1008) to:

$$R^* = R^*{}_{\alpha\beta} G^{\alpha\beta} = \left(R + (1-n) \cdot \frac{F'}{F}\left(\Delta f + \frac{H'}{H} \cdot f_{,\alpha} f_{,\beta} g^{\alpha\beta}\right)\right)\frac{1}{F}$$

$$= \left(R + (1-n) \cdot \frac{F'}{H \cdot F}\left(\frac{H}{\sqrt{g}}\partial_\alpha\left(\sqrt{g} \cdot g^{\alpha\beta}\partial_\beta f\right) + H' \cdot f_{,\alpha} f_{,\beta} g^{\alpha\beta}\right)\right)\frac{1}{F}$$

$$= \left(R + (1-n) \cdot \frac{F'}{H \cdot F}\left(\frac{1}{\sqrt{g}}\partial_\alpha\left(H \cdot \sqrt{g} \cdot g^{\alpha\beta}\partial_\beta f\right)\right)\right)\frac{1}{F}$$

$$\xrightarrow{R^* = R^*{}_{\alpha\beta} G^{\alpha\beta} = 0}$$

$$\Rightarrow \quad R \cdot \frac{H \cdot F}{(n-1) \cdot F'} = \frac{1}{\sqrt{g}}\partial_\alpha\left(H \cdot \sqrt{g} \cdot g^{\alpha\beta}\partial_\beta f\right) \equiv \Omega\{H[f]f\}$$

$$\xrightarrow{\frac{4FF'' + (F')^2(n-6)}{4F} = \frac{H'}{H} \cdot F'}$$

$$\Rightarrow \quad \frac{1}{\sqrt{g}}\partial_\alpha H \cdot \sqrt{g} \cdot g^{\alpha\beta}\partial_\beta f = R \cdot \frac{1}{n-1} \cdot \begin{cases} \left(C_{f0}^2 - \frac{2-n}{4} \cdot \int\limits_1^f H[\phi] \cdot d\phi\right) & n > 2 \\ C_{f0}^2 & n = 2. \end{cases}$$

$$(1242)$$

9.3.12 A Multitude of Options and the Collapse of the Wave Function

When comparing the various solutions to the spherical problem with the various settings for the term:

$$\frac{4FF'' + \left(F'\right)^2 (n-6)}{4F} = ?, \tag{1243}$$

as been obtained and illustrated in the section "9.3.11 Further Factorization" above, we find that some options do provide reasonable results but do not show the "right"[13] asymptotic behavior. Now we ask whether only the solutions with the assumed-to-be-correct asymptotic behavior for bigger r should be allowed, but we might also just assume that in fact all solutions are allowed, respectively exist parallel, because we could just split up Eq. (1008) into a sum of such equations:

$$R^* = \sum_{k=1}^{\Omega} R_k^* {}_{\alpha\beta} G_k^{\alpha\beta} = \left(R_k + (1-n) \cdot \frac{F_k'}{F_k} \left(\Delta f_k + \frac{1}{f_k} \cdot f_{k,\alpha} f_{k,\beta} g_k^{\alpha\beta} \right) \right) \frac{1}{F_k}$$

$$\xrightarrow{\quad R^* = R^*_{\alpha\beta} G^{\alpha\beta} = 0 \quad}$$

$$\Rightarrow \qquad R_k + (1-n) \cdot \frac{F_k'}{F_k} \left(\Delta f_k + \frac{1}{f_k} \cdot f_{k,\alpha} f_{k,\beta} g_k^{\alpha\beta} \right) = 0. \tag{1244}$$

Now, depending on our choice of the condition (1243) for a certain ensemble of k-equations, we can have a great variety of equations of the types:

$$\frac{4FF'' + \left(F'\right)^2 (n-6)}{4F} = \frac{H'}{H} \cdot F'$$

$$\xrightarrow{\hspace{3cm}}$$

$$\Rightarrow \qquad \frac{1}{\sqrt{g}} \partial_\alpha H \cdot \sqrt{g} \cdot g^{\alpha\beta} \partial_\beta f = R \cdot \frac{1}{n-1} \cdot \begin{cases} \left(C_{f0}^2 - \dfrac{2-n}{4} \cdot \displaystyle\int_1^f H[\phi] \cdot d\phi \right) & n > 2 \\[4mm] C_{f0}^2 & n = 2, \end{cases}$$

$$\tag{1245}$$

---

[13]Here we applied the quotation marks, because the asymptotic behavior is purely postulated. It could well be that the objects we consider do vanish for infinite r but others (space-time objects or space-time portions) prevent us from seeing this "disappearance" as they (the "others") "take over" or "come into sight" when moving away from the very one object with the "wrong" (vanishing) asymptotic behavior. This coincides with the idea of a sub-structured space-time as elaborated in [4] and the stories of T. Bodan given there in the introduction part of this book.

$$\xrightarrow{\frac{4FF''+\left(F'\right)^2(n-6)}{4F}=0} F[f] = \begin{cases} C_{f1} \cdot \left(f + C_{f0}\right)^{\frac{4}{n-2}} & n > 2 \\ C_{f1} \cdot e^{C_{f0} \cdot f} & n = 2 \end{cases}$$

$$\Rightarrow 0 = R \cdot F + F'\left(1-n\right)\Delta f \Rightarrow \frac{1}{\sqrt{g}}\partial_\alpha\sqrt{g} \cdot g^{\alpha\beta}\partial_\beta f = R \cdot \frac{1}{n-1} \cdot \begin{cases} \left(f + C_{f0}\right) & n > 2 \\ C_{f0}^{-1} & n = 2, \end{cases}$$

$$(1246)$$

and—obviously—also many other possibilities.

In principle, the choice of F decides on the result for f and as there are so many options (namely principally infinitely many) for our wrapper $F[f]$, there are also infinitely many possibilities for the wave function f.

The question is: Which one will be seen in a measurement?

... and:

Is it possible that the multitude of options has something to do with the "sudden selection of certain solutions" when a measurement is been performed, which became classically known as the "collapse of the wave-function"?

... and:

Will the variation of the kernel Lagrangian R^* or function$[R^*]$ help us to find the answers to the questions above?

9.3.13 Application to the Schwarzschild Metric → Getting Rid of the Singularity

Instead of metric Eq. (1228), we now take the Schwarzschild metric [16]:

$$G_{\alpha\beta}^4 = F[f] \cdot \begin{pmatrix} -c^2 \cdot \left(1 - \dfrac{r_s}{r}\right) & 0 & 0 & 0 \\ 0 & \left(1 - \dfrac{r_s}{r}\right)^{-1} & 0 & 0 \\ 0 & 0 & r^2 & 0 \\ 0 & 0 & 0 & r^2 \cdot \sin^2\vartheta \end{pmatrix} \quad (1247)$$

and repeat our evaluations form the previous sub-section.

Once again assuming the function f only to depend on the radius and setting the wrapper $F[f]$ in accordance with the conditions in Eqs. (1222) and (1230), we obtain the following tasks and their subsequent solutions:

$$\frac{1}{\sqrt{g}}\partial_\alpha f\cdot\sqrt{g}\cdot g^{\alpha\beta}\partial_\beta f=0\xrightarrow{f=f[r]}0=\left(f'\right)^2+f\cdot\left(\frac{2\cdot f'}{r}+f''-\frac{f'}{r}\cdot\frac{r_s}{r}\cdot\left(\frac{1}{1-\frac{r_s}{r}}\right)\right)$$

$$\Rightarrow\qquad f[r]=C_{r1}\cdot\sqrt{\frac{r_s}{r^2}-\frac{2}{r}}+C_{r0}, \qquad (1248)$$

$$\frac{f}{\sqrt{g}}\partial_\alpha\sqrt{g}\cdot g^{\alpha\beta}\partial_\beta f-f_{,\alpha}f_{,\beta}g^{\alpha\beta}=0\xrightarrow{f=f[r]}0=f\cdot\left(\frac{2\cdot f'}{r}+f''-\frac{f'}{r}\cdot\frac{r_s}{r}\cdot\left(\frac{1}{1-\frac{r_s}{r}}\right)\right)-\left(f'\right)^2$$

$$\Rightarrow\qquad f[r]=C_{r1}\cdot e^{-\frac{C_{r0}}{r^2}\cdot(r_s-2\cdot r)}, \qquad (1249)$$

for the cases in Eqs. (1222) and (1230), respectively. In Eq. (1248), we recognize the typical structures for the metric of charged massive spherical objects [17, 18].

Applying the condition from (1175), we can directly use the results already published in [20]. Applying the separation approach $f[t,r,\vartheta,\varphi] = g[t]*h[r]*Y[\vartheta]*Z[\varphi]$ (for the reason of recognition in connection with [20], we use $f_r[r] = h[r]$) with the separation parameters E and m, we obtained the following immediate solution for $g[t]$ and $Z[\varphi]$:

$$g[t]=C_{t1}\cdot\cos[c\cdot E\cdot t]+C_{t2}\cdot\sin[c\cdot E\cdot t], \qquad (1250)$$

$$Z_m[\varphi]=C_{\varphi1}\cdot\cos[m\cdot\varphi]+C_{\varphi2}\cdot\sin[m\cdot\varphi]. \qquad (1251)$$

The two remaining differential equations can be constructed as follows:

$$-\frac{m^2}{\sin[\vartheta]^2}+\frac{\cot[\vartheta]Y'[\vartheta]+Y''[\vartheta]}{B^2Y[\vartheta]}=l(1+l), \qquad (1252)$$

$$\frac{l(1+l)}{r^2}(r_s-r)-\frac{\left((r_s-2\cdot r)\cdot h'[r]+(r_s-r)\cdot r\cdot h''[r]\right)}{r^2\cdot h[r]}=\frac{r\cdot E^2}{c^2}-\frac{R^*}{6}$$

$$\xrightarrow{R^*=0}$$

$$\frac{l(1+l)}{r^2}(r_s-r)-\frac{\left((r_s-2\cdot r)\cdot h'[r]+(r_s-r)\cdot r\cdot h''[r]\right)}{r^2\cdot h[r]}-\frac{r\cdot E^2}{c^2}=0. \qquad (1253)$$

While the first equation of the two gives the usual solution with the associate Legendre Polynomials P_l^m and Q_l^m, as also known from the Schrödinger hydrogen solution:

$$Y_l^m[\vartheta] = C_{P\vartheta} \cdot P_l^m\left[\cos[\vartheta]\right] + C_{Q\vartheta} \cdot Q_l^m\left[\cos[\vartheta]\right], \tag{1254}$$

we have a rather non-Schrödinger-like solution with respect to the radius coordinate in the case of $R^* = E = 0$, namely:

$$h[r] = C_{\mathrm{Pr}} \cdot P_l\left[2 \cdot \frac{r}{r_s} - 1\right] + C_{Qr} \cdot Q_l\left[2 \cdot \frac{r}{r_s} - 1\right]. \tag{1255}$$

This time we have the simple Legendre Polynomials P_1 and Q_1.

Unfortunately, we were not able to find a solution for the interesting case of $f = f[t,r]$, but had to set $E = 0$ and thus, we have to resort to numerical integration for $E \neq 0$. Still, we can find physically meaningful parameter settings with the right asymptotic behavior (Figs. 9.8–9.10) and—so it seems—even parameter combinations, where we might avoid the singularity at $r = 0$. But the latter is only a guess.

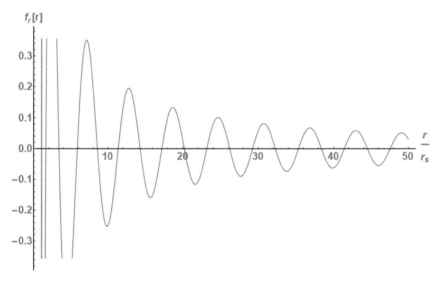

Figure 9.8 Distribution of $f[r]$ for the Schwarzschild metric with the parameter setting $E = r_s$.

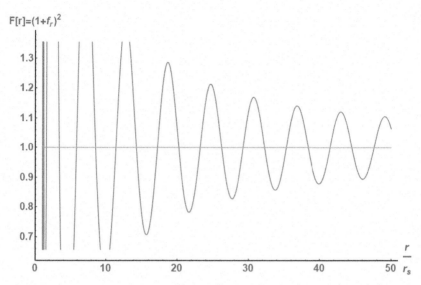

Figure 9.9 Distribution of $F[f[r]]$ according to condition in Eq. (1175) for the Schwarzschild metric with the parameter setting $E = r_s$.

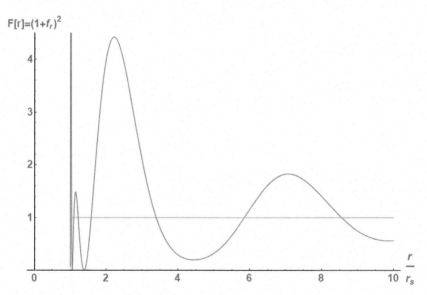

Figure 9.10 Distribution of $F[f[r]]$ according to condition in Eq. (1175) for the Schwarzschild metric with the parameter setting $E = r_s$.

9.4 Adaptation of the Dirac Approach

From all the discussion above, we extract the conclusion that the classical Dirac approach is not necessarily the best way to solve Eq. (1008), when—for whatever purpose—one does not want to apply the condition from Eq. (1175). Some may not see this as a true adaptation of Dirac's original approach but a mere re-interpretation. Well, this author actually does not care about such verbal trivia as he sees the scaled metric ansatz as a new concept anyway, which consequently also requires some new interpretations of the classical "probability" theory including the collapse of the wave function, the measurement process, and the wave function itself. Before actually starting with this interpretation work, however, the author intends to put together some more basics in order to better see through all the "mess" of options the **curvature originated Quantum Theory**, as presented here so far, already does provide.

9.4.1 More Tedious Work

So, our starting point shall be an F being linear dependent on f and we set:

$$F = C_f + f. \tag{1256}$$

This simplifies Eq. (1008) or Eq. (1063) to:

$$R^* = R^*_{\alpha\beta} G^{\alpha\beta} = \left(R + \frac{1}{f + C_f}(1-n)\Delta f + \frac{f_{,\alpha} f_{,\beta} g^{\alpha\beta}(1-n)}{4(f + C_f)^2}(n-6) \right) \frac{1}{f + C_f}. \tag{1257}$$

Again, we also demand to have an effective Ricci flat space-time via:

$$R^* = R^*_{\alpha\beta} G^{\alpha\beta} = 0 \tag{1258}$$

resulting in:

$$R + \frac{1}{f + C_f}(1-n)\Delta f + \frac{f_{,\alpha} f_{,\beta} g^{\alpha\beta}(1-n)}{4(f + C_f)^2}(n-6) = 0. \tag{1259}$$

Now we reactivate the intrinsic structure approach from section "9.3 Derivation of the Metric Dirac Equation" and assume f to be built out of an ensemble of functions like:

$$f = h + h_\alpha q^\alpha + h_{\alpha\beta} q^{\alpha\beta} + \dots \tag{1260}$$

or even:

$$f = h + h_{\alpha A} q^{\alpha A} + h_{\alpha \beta A} q^{\alpha \beta A} + h_{\alpha \beta AB} q^{\alpha \beta AB} \dots, \tag{1261}$$

if interested in the incorporation of spinors or similar funny objects, not necessarily summing up to the n dimensions of our ordinary space-time (or any space-time). Thus, the internal functions can be quite general and so could be their sum, which we here want to point out via Latin indices in Eq. (1260):

$$f = h + h_a q^a + h_{ab} q^{ab} + \dots. \tag{1262}$$

We also demand that f shall always be a Laplace function:

$$\Delta f = \Delta \left(h + h_a q^a + h_{ab} q^{ab} + \dots \right) = 0. \tag{1263}$$

From Eq. (1259), this would give us the following equation:

$$f_{,\alpha} f_{,\beta} g^{\alpha \beta} = 4 \cdot \left(C_f + f \right)^2 \cdot R \cdot \frac{1}{(n-6) \cdot (n-1)}. \tag{1264}$$

Now, just as before in section "9.3 Derivation of the Metric Dirac Equation" (only that things are simpler), we can factorize the metric tensor:

$$f_{,\alpha} f_{,\beta} g^{\alpha \beta} = 4 \cdot \left(C_f + f \right)^2 \cdot R \cdot \frac{1}{(n-6) \cdot (n-1)}$$

$$\left(C_f + f \right)^2 \cdot 4 \cdot R \cdot \overbrace{\frac{1}{(n-6) \cdot (n-1)}}^{M^2 = M \cdot M} = \left(C_f + f \right) \cdot M \cdot M \cdot \left(C_f + f \right) = f_{,\alpha} f_{,\beta} g^{\alpha \beta}$$

$$\xrightarrow{w = C_f + f = h_a q^a}$$

$$h_a q^a \cdot M \cdot M \cdot h_a q^a = h_{a,\alpha} q^a g^{\alpha \beta} h_{b,\beta} q^b \xrightarrow{g^{\alpha \beta} = e^{\alpha} \cdot e^{\beta}} = h_{a,\alpha} q^a e^{\alpha} \cdot e^{\beta} h_{b,\beta} q^b$$

$$\Rightarrow \qquad\qquad \mp h_a q^a M = h_{a,\alpha} q^a e^{\alpha}$$

$$\Rightarrow \qquad\qquad 0 = q^a \left(h_{a,\alpha} e^{\alpha} \pm h_a M \right). \tag{1265}$$

For simplicity, we assumed $\mathbf{q^a}$ to be a vector of constants, but it needs to be pointed out that this is not a necessary condition. We could also—just to give an example—assume a symmetric decomposition (f-structuring) as follows:

$$f_{,\alpha} f_{,\beta} g^{\alpha \beta} = 4 \cdot \left(C_f + f \right)^2 \cdot R \cdot \frac{1}{(n-6) \cdot (n-1)}$$

$$\left(C_f + f\right)^2 \cdot 4 \cdot R \cdot \overbrace{\frac{1}{(n-6)\cdot(n-1)}}^{M^2 = \mathbf{M}\cdot\mathbf{M}} = \left(C_f + f\right)\cdot \mathbf{M}\cdot\mathbf{M}\cdot\left(C_f + f\right) = f_{,\alpha}f_{,\beta}g^{\alpha\beta}$$

$$\xrightarrow{\;w = C_f + f = h_a h^a\;}$$

$$h_a h^a \cdot \mathbf{M}\cdot\mathbf{M}\cdot h_a h^a = \left(h_{a,\alpha}h^a + h_a h^a{}_{,\alpha}\right)g^{\alpha\beta}\left(h_{b,\beta}h^b + h_b h^b{}_{,\beta}\right)$$

$$\xrightarrow{\;g^{\alpha\beta} = \mathbf{e}^\alpha \cdot \mathbf{e}^\beta\;} = \left(h_{a,\alpha}h^a + h_a h^a{}_{,\alpha}\right)\mathbf{e}^\alpha \cdot \mathbf{e}^\beta\left(h_{b,\beta}h^b + h_b h^b{}_{,\beta}\right)$$

$$\Rightarrow \qquad \mp h_a h^a \mathbf{M} = \left(h_{a,\alpha}h^a + h_a h^a{}_{,\alpha}\right)\mathbf{e}^\alpha$$

$$\Rightarrow \qquad 0 = \left(h_{a,\alpha}h^a + h_a h^a{}_{,\alpha}\right)\mathbf{e}^\alpha \pm h_a h^a \mathbf{M}. \tag{1266}$$

We boldly take it that we may also be able to find a Cartesian-like system for the a-b-coordinates, allowing us to change the last line of Eq. (1266) to:

$$0 = h_a\left(2\cdot h^a{}_{,\alpha}\cdot \mathbf{e}^\alpha \pm h^a \mathbf{M}\right). \tag{1267}$$

However, we may also just adapt the intermediate transformations in Eq. (1266) as follows:

$$h_a h^a \cdot \mathbf{M}\cdot\mathbf{M}\cdot h_b h^b = \left(h_{a,\alpha}h^a + h_a h^a{}_{,\alpha}\right)g^{\alpha\beta}\left(h_{b,\beta}h^b + h_b h^b{}_{,\beta}\right)$$

$$\xrightarrow{\;g^{\alpha\beta} = \mathbf{e}^\alpha \cdot \mathbf{e}^\beta\;} = \left(h_{a,\alpha}h^a + h_a h^a{}_{,\alpha}\right)\mathbf{e}^\alpha \cdot \mathbf{e}^\beta\left(h_{b,\beta}h^b + h_b h^b{}_{,\beta}\right)$$

$$\Rightarrow \qquad \mp \frac{h_a h^a + h_a h^a}{2}\mathbf{M} = \left(h_{a,\alpha}h^a + h_a h^a{}_{,\alpha}\right)\mathbf{e}^\alpha$$

$$\Rightarrow \qquad 0 = \left(h_{a,\alpha}h^a + h_a h^a{}_{,\alpha}\right)\mathbf{e}^\alpha \pm \frac{h_a h^a + h_a h^a}{2}\mathbf{M}$$

$$\Rightarrow \qquad 0 = \begin{cases} h_a\left(h^a{}_{,\alpha}\mathbf{e}^\alpha \pm \dfrac{h^a}{2}\mathbf{M}\right) \\[2ex] h^a\left(h_{a,\alpha}\mathbf{e}^\alpha \pm \dfrac{h_a}{2}\mathbf{M}\right). \end{cases} \tag{1268}$$

This can also be done in an asymmetric case where $\mathbf{q^a}$ is no vector of constants, leading to:

$$\xrightarrow{\;w = C_f + f = h_a q^a\;}$$

$$h_a q^a \cdot \mathbf{M}\cdot\mathbf{M}\cdot h_b q^b = \left(h_{a,\alpha}q^a + h_a q^a{}_{,\alpha}\right)g^{\alpha\beta}\left(h_{b,\beta}q^b + h_b q^b{}_{,\beta}\right)$$

$$\xrightarrow{\;g^{\alpha\beta} = \mathbf{e}^\alpha \cdot \mathbf{e}^\beta\;} = \left(h_{a,\alpha}q^a + h_a q^a{}_{,\alpha}\right)\mathbf{e}^\alpha \cdot \mathbf{e}^\beta\left(h_{b,\beta}q^b + h_b q^b{}_{,\beta}\right)$$

$$\Rightarrow \qquad \mp \frac{h_a q^a + h_a q^a}{2} \mathbf{M} = \left(h_{a,\alpha} q^a + h_a q^a{}_{,\alpha} \right) \mathbf{e}^\alpha$$

$$\Rightarrow \qquad 0 = \begin{cases} q^a \left(h_{a,\alpha} \mathbf{e}^\alpha \pm h_a \mathbf{M} \right) \\ h_a \left(q^a{}_{,\alpha} \mathbf{e}^\alpha \pm q^a \mathbf{M} \right). \end{cases} \qquad (1269)$$

Now we apply a few degrees of freedom as they exist in the construction of base vectors as discussed in [4] (section 4) and assume a matrix form as follows:

$$\mathbf{e}^\alpha \cdot \mathbf{e}^\beta = E^\alpha{}_c \cdot E^{\beta c}; \quad \mathbf{M} \cdot \mathbf{M} = M_c \cdot M^c. \qquad (1270)$$

The constant vector assumption for $\mathbf{q^a}$ then yields:

$$f_{,\alpha} f_{,\beta} g^{\alpha\beta} = 4 \cdot \left(C_f + f \right)^2 \cdot R \cdot \frac{1}{(n-6)\cdot(n-1)}$$

$$\left(C_f + f \right)^2 \cdot 4 \cdot R \cdot \overbrace{\frac{1}{(n-6)\cdot(n-1)}}^{M^2 = \mathbf{M}\cdot\mathbf{M} = M_c \cdot M^c} = \left(C_f + f \right) \cdot M_c \cdot M^c \cdot \left(C_f + f \right) = f_{,\alpha} f_{,\beta} g^{\alpha\beta}$$

$$\xrightarrow{\;w = C_f + f = h_a q^a\;}$$

$$h_a q^a \cdot M_c \cdot M^c \cdot h_b q^b = h_{a,\alpha} q^a g^{\alpha\beta} h_{b,\beta} q^b \xrightarrow{\;g^{\alpha\beta} = E^\alpha{}_c \cdot E^{\beta c}\;} = h_{a,\alpha} q^a E^\alpha{}_c \cdot E^{\beta c} h_{b,\beta} q^b$$

$$\Rightarrow \qquad \mp h_a q^a M_c = h_{a,\alpha} q^a E^\alpha{}_c$$

$$\Rightarrow \qquad 0 = q^a \left(h_{a,\alpha} E^\alpha{}_c \pm h_a M_c \right). \qquad (1271)$$

The symmetric decomposition for f on the other hand gives us:

$$f_{,\alpha} f_{,\beta} g^{\alpha\beta} = 4 \cdot \left(C_f + f \right)^2 \cdot R \cdot \frac{1}{(n-6)\cdot(n-1)}$$

$$\left(C_f + f \right)^2 \cdot 4 \cdot R \cdot \overbrace{\frac{1}{(n-6)\cdot(n-1)}}^{M^2 = \mathbf{M}\cdot\mathbf{M} = M_c \cdot M^c} = \left(C_f + f \right) \cdot M_c \cdot M^c \cdot \left(C_f + f \right) = f_{,\alpha} f_{,\beta} g^{\alpha\beta}$$

$$\xrightarrow{\;w = C_f + f = h_a h^a\;}$$

$$h_a h^a \cdot M_c \cdot M^c \cdot h_b h^b = \left(h_{a,\alpha} h^a + h_a h^a{}_{,\alpha} \right) g^{\alpha\beta} \left(h_{b,\beta} h^b + h_b h^b{}_{,\beta} \right)$$

$$\xrightarrow{\;g^{\alpha\beta} = E^\alpha{}_c \cdot E^{\beta c}\;} = \left(h_{a,\alpha} h^a + h_a h^a{}_{,\alpha} \right) E^\alpha{}_c \cdot E^{\beta c} \left(h_{b,\beta} h^b + h_b h^b{}_{,\beta} \right)$$

$$\Rightarrow \qquad \mp h_a h^a M_c = 2 \cdot \begin{cases} h_{a,\alpha} h^a E^\alpha{}_c \\ h_a h^a{}_{,\alpha} E^\alpha{}_c \end{cases}$$

$$\Rightarrow \qquad 0 = \begin{cases} h^a \left(2 \cdot h_{a,\alpha} E^\alpha{}_c \pm h_a M_c \right) \\ h_a \left(2 \cdot h^a{}_{,\alpha} E^\alpha{}_c \pm h^a M_c \right). \end{cases} \qquad (1272)$$

In completely general or asymmetric cases (factorization) we obtain:

$$f_{,\alpha} f_{,\beta} g^{\alpha\beta} = 4 \cdot \left(C_f + f \right)^2 \cdot R \cdot \frac{1}{(n-6) \cdot (n-1)}$$

$$\left(C_f + f \right)^2 \cdot 4 \cdot R \cdot \frac{1}{(n-6) \cdot (n-1)} \overset{M^2 = \mathbf{M} \cdot \mathbf{M} = M_c \cdot M^c}{=} \left(C_f + f \right) \cdot M_c \cdot M^c \cdot \left(C_f + f \right) = f_{,\alpha} f_{,\beta} g^{\alpha\beta}$$

$$\overset{w = C_f + f = h_a q^a}{\longrightarrow}$$

$$h_a q^a \cdot M_c \cdot M^c \cdot h_b q^b = \left(h_{a,\alpha} q^a + h_a q^a{}_{,\alpha} \right) g^{\alpha\beta} \left(h_{b,\beta} q^b + h_b q^b{}_{,\beta} \right)$$

$$\overset{g^{\alpha\beta} = E^\alpha{}_c \cdot E^{\beta c}}{\longrightarrow} = \left(h_{a,\alpha} q^a + h_a q^a{}_{,\alpha} \right) E^\alpha{}_c \cdot E^{\beta c} \left(h_{b,\beta} q^b + h_b q^b{}_{,\beta} \right)$$

$$\Rightarrow \qquad \mp \frac{h_a q^a + h_a q^a}{2} M_c = \left(h_{a,\alpha} q^a + h_a q^a{}_{,\alpha} \right) E^\alpha{}_c$$

$$\Rightarrow \qquad 0 = \left(h_{a,\alpha} q^a + h_a q^a{}_{,\alpha} \right) E^\alpha{}_c \pm \frac{h_a q^a + h_a q^a}{2} M_c$$

$$\Rightarrow \qquad 0 = \begin{cases} h_a \left(q^a{}_{,\alpha} E^\alpha{}_c \pm \dfrac{q^a}{2} M_c \right) \\ q^a \left(h_{a,\alpha} E^\alpha{}_c \pm \dfrac{h_a}{2} M_c \right). \end{cases} \qquad (1273)$$

Please note that any splitting is possible in our setting in Eq. (1273) for the mass-vectors, which could also generalize to:

$$\Rightarrow \qquad \mp \frac{C_1 \cdot h_a q^a + C_2 \cdot h_a q^a}{2} M_c = \left(h_{a,\alpha} q^a + h_a q^a{}_{,\alpha} \right) E^\alpha{}_c; \quad C_1 + C_2 = 2$$

$$\Rightarrow \qquad 0 = \left(h_{a,\alpha} q^a + h_a q^a{}_{,\alpha} \right) E^\alpha{}_c \pm \frac{C_1 \cdot h_a q^a + C_2 \cdot h_a q^a}{2} M_c$$

$$\Rightarrow \qquad 0 = \begin{cases} h_a \left(q^a{}_{,\alpha} E^\alpha{}_c \pm C_2 \cdot \dfrac{q^a}{2} M_c \right) \\ q^a \left(h_{a,\alpha} E^\alpha{}_c \pm C_1 \cdot \dfrac{h_a}{2} M_c \right). \end{cases} \qquad (1274)$$

Hereby we could also aim for the following symmetrization in Eq. (1273):

$$f_{,\alpha} f_{,\beta} g^{\alpha\beta} = 4 \cdot \left(C_f + f \right)^2 \cdot R \cdot \frac{1}{(n-6) \cdot (n-1)}$$

$$\left(C_f + f \right)^2 \cdot 4 \cdot R \cdot \frac{1}{(n-6) \cdot (n-1)} = \overbrace{\left(C_f + f \right) \cdot M_c \cdot M^c}^{M^2 = \mathbf{M} \cdot \mathbf{M} = M_c \cdot M^c} \cdot \left(C_f + f \right) = f_{,\alpha} f_{,\beta} g^{\alpha\beta}$$

$$\xrightarrow{\quad w = C_f + f = \frac{h_a q^a + q_a h^a}{2} \quad}$$

$$\left(h_a q^a + q_a h^a \right) \cdot M_c \cdot M^c \cdot \left(h_b q^b + q_b h^b \right)$$

$$= \left(h_{a,\alpha} q^a + q_{a,\alpha} h^a + h_a q^a_{,\alpha} + q_a h^a_{,\alpha} \right) g^{\alpha\beta} \left(h_{b,\beta} q^b + q_{b,\beta} h^b + h_b q^b_{,\beta} + q_b h^b_{,\beta} \right)$$

$$\xrightarrow{\quad g^{\alpha\beta} = E^\alpha_c \cdot E^{\beta c} \quad}$$

$$= \left(h_{a,\alpha} q^a + q_{a,\alpha} h^a + h_a q^a_{,\alpha} + q_a h^a_{,\alpha} \right) E^\alpha_c \cdot E^{\beta c} \left(h_{b,\beta} q^b + q_{b,\beta} h^b + h_b q^b_{,\beta} + q_b h^b_{,\beta} \right)$$

$$\Rightarrow \mp \frac{h_a q^a + q_a h^a + h_a q^a + q_a h^a}{2} \cdot M_c = \left(h_{a,\alpha} q^a + q_{a,\alpha} h^a + h_a q^a_{,\alpha} + q_a h^a_{,\alpha} \right) E^\alpha_c$$

$$\Rightarrow 0 = \left(h_{a,\alpha} q^a + q_{a,\alpha} h^a + h_a q^a_{,\alpha} + q_a h^a_{,\alpha} \right) E^\alpha_c \pm \frac{h_a q^a + q_a h^a + h_a q^a + q_a h^a}{2} \cdot M_c. \tag{1275}$$

Subsequently, we can now separate the last line of Eq. (1275) as we already did above:

$$0 = \left(h_{a,\alpha} q^a + q_{a,\alpha} h^a + h_a q^a_{,\alpha} + q_a h^a_{,\alpha} \right) E^\alpha_c \pm \frac{h_a q^a + q_a h^a + h_a q^a + q_a h^a}{2} \cdot M_c$$

$$0 = \begin{cases} q^a \left(h_{a,\alpha} E^\alpha_c \pm \dfrac{h_a}{2} \cdot M_c \right) \\[2mm] h^a \left(q_{a,\alpha} E^\alpha_c \pm \dfrac{q_a}{2} \cdot M_c \right) \\[2mm] q_a^{} \left(h^a_{,\alpha} E^\alpha_c \pm \dfrac{h^a}{2} \cdot M_c \right) \\[2mm] h_a \left(q^a_{,\alpha} E^\alpha_c \pm \dfrac{q^a}{2} \cdot M_c \right) \end{cases} \tag{1276}$$

including the option for a more general splitting of the mass vector terms:

$$0 = \begin{cases} \left(h_{a,\alpha} q^a + q_{a,\alpha} h^a + h_a q^a_{,\alpha} + q_a h^a_{,\alpha} \right) E^\alpha_c \\[2mm] \pm \dfrac{C_1 \cdot \left(h_a q^a + q_a h^a \right) + C_2 \cdot \left(h_a q^a + q_a h^a \right)}{2} \cdot M_c \end{cases}$$

$$0 = \begin{cases} q^a \left(h_{a,\alpha} E^\alpha{}_c \pm C_1 \cdot \dfrac{h_a}{2} \cdot M_c \right) \\[2mm] h^a \left(q_{a,\alpha} E^\alpha{}_c \pm C_2 \cdot \dfrac{q_a}{2} \cdot M_c \right) \\[2mm] q_a{}^a \left(h^a{}_{,\alpha} E^\alpha{}_c \pm C_1 \cdot \dfrac{h^a}{2} \cdot M_c \right) \\[2mm] h_a \left(q^a{}_{,\alpha} E^\alpha{}_c \pm C_2 \cdot \dfrac{q^a}{2} \cdot M_c \right) \end{cases} ; \quad C_1 + C_2 = 2. \qquad (1277)$$

Taking the simplest form in Eq. (1271), with the **q**-vector assumed to be a vector of constants, we can achieve the Dirac form via an exchange of summation as follows:

$$q_a h^a \cdot M_c \cdot M^c \cdot q_b h^b = q_a h^a{}_{,\alpha} g^{\alpha\beta} q_b h^b{}_{,\beta} \xrightarrow{\ w = C_f + f = q_a h^a\ } = q_a h^a{}_{,\alpha} g^{\alpha\beta} q_b h^b{}_{,\beta} \xrightarrow{\ g^{\alpha\beta} = E^\alpha{}_c \cdot E^{\beta c}\ } = q_a h^a{}_{,\alpha} E^\alpha{}_c \cdot E^{\beta c} q_b h^b{}_{,\beta}$$

$$\Rightarrow \qquad \mp q_a h^a M_c = q_a h^a{}_{,\alpha} E^\alpha{}_c \xrightarrow{\ \text{interchange a,c}\ } \mp q_c h^a M_a = q_c h^a{}_{,\alpha} E^\alpha{}_a$$

$$\Rightarrow \qquad 0 = q_c \left(h^a{}_{,\alpha} E^\alpha{}_a \pm h^a M_a \right). \qquad (1278)$$

The recipe in the sub-section "9.3.2 Repetition: Dirac in the Metric Picture and Its Connection to the Classical Quaternion Form" tells us how to evaluate the classical Dirac equation from the last line of Eq. (1278), namely by finding an object **C**, which satisfies Eq. (1083). Along the way, we had realized that the classical mass-objects are no scalars, but have to be something more complex.

In addition, we may also just see the path forward, when combining the *E*-, *M*-, and *q*-forms in Eq. (1278) in the following way:

$$\Rightarrow \qquad \mp q_a h^a M_c = q_a h^a{}_{,\alpha} E^\alpha{}_c \xrightarrow{\ \text{interchange a,c}\ } \mp q_c h^a M_a = q_c h^a{}_{,\alpha} E^\alpha{}_a$$

$$\Rightarrow \qquad 0 = q_c \left(h^a{}_{,\alpha} E^\alpha{}_a \pm h^a M_a \right)$$

$$0 = \begin{cases} q_a h^a{}_{,\alpha} E^\alpha{}_c \pm q_a h^a M_c \\[2mm] q_c h^a{}_{,\alpha} E^\alpha{}_a \pm q_c h^a M_a \end{cases} \xrightarrow[\ q_c E^\alpha{}_a = Q^\alpha_{ca}\, ;\, q_c h^a M_a = h^a M_{ca}\ ]{\ q_a E^\alpha{}_c = Q^\alpha_{ac}\, ;\, q_a h^a M_c = h^a M_{ac}\ } 0 = \begin{cases} h^a{}_{,\alpha} Q^\alpha{}_{ac} \pm h^a M_{ac} \\[2mm] h^a{}_{,\alpha} Q^\alpha{}_{ca} \pm h^a M_{ca} \end{cases}.$$

$$(1279)$$

Now we would not even need the interchange-process of the *a-b-c*-indices, but just a suitable combination of the base-vectors and the *M*-term with the constant **q**-vectors to some kind of matrices, which—similar to what was shown in sub-section "9.3.2 Repetition: Dirac in the Metric Picture and Its Connection to the Classical Quaternion Form"—result in the classical Dirac matrices. For convenience, we give the result for Eq. (1278) without the discussion on interchanging inside the *a-b*-system:

$$\xrightarrow{\quad w=C_f+f=q_ah^a \quad}$$

$$q_ah^a \cdot M_c \cdot M^c \cdot q_bh^b = q_ah^a{}_{,\alpha}g^{\alpha\beta}q_bh^b{}_{,\beta} \xrightarrow{\quad g^{\alpha\beta}=E^\alpha{}_c\cdot E^{\beta c} \quad} = q_ah^a{}_{,\alpha}E^\alpha{}_c\cdot E^{\beta c}q_bh^b{}_{,\beta}$$

$$\Rightarrow \qquad \mp q_ah^a M_c = q_ah^a{}_{,\alpha}E^\alpha{}_c \Rightarrow 0 = q_ah^a{}_{,\alpha}E^\alpha{}_c \pm q_ah^a M_c$$

$$\xrightarrow{\quad q_aE^\alpha{}_c=Q^\alpha_{ac}\,;\,q_ah^aM_c=h^aM_{ac} \quad}$$

$$\Rightarrow \qquad 0 = h^a{}_{,\alpha}Q^\alpha{}_{ac} \pm h^a M_{ac}. \qquad (1280)$$

The task remaining is to find an object **C**, being similar to Eq. (1083), but this time satisfying the equation:

$$\mathbf{C}\cdot Q^\alpha{}_{ac} = \mathbf{C}\cdot q_aE^\alpha{}_c = \gamma^\alpha; \quad \mathbf{C}\cdot M_{ac} = \mathbf{C}\cdot q_aM_c = I\cdot m. \qquad (1281)$$

We realize that this is not just a task with respect to finding **C,** but may also give us the necessary conditions for the extraction of the vector of constants **q**_a. We shall leave the solution for later and first consider the symmetric and the general asymmetric decomposition of the function *f*.

In the case of the symmetric decomposition in Eq. (1272), we first try to find a way to achieve the Dirac form via an exchange of summation as follows:

$$\xrightarrow{\quad w=C_f+f=h_ah^a \quad}$$

$$h_ah^a \cdot M_c \cdot M^c \cdot h_bh^b = \left(h_{a,\alpha}h^a + h_ah^a{}_{,\alpha}\right)g^{\alpha\beta}\left(h_{b,\beta}h^b + h_bh^b{}_{,\beta}\right)$$

$$\xrightarrow{\quad g^{\alpha\beta}=E^\alpha{}_c\cdot E^{\beta c} \quad} = \left(h_{a,\alpha}h^a + h_ah^a{}_{,\alpha}\right)E^\alpha{}_c\cdot E^{\beta c}\left(h_{b,\beta}h^b + h_bh^b{}_{,\beta}\right)$$

$$= \begin{pmatrix} h_{a,\alpha}h^a \\ h_{a,\alpha}h^a \\ h_ah^a{}_{,\alpha} \\ h_ah^a{}_{,\alpha} \end{pmatrix} E^\alpha{}_c \cdot E^{\beta c} \begin{pmatrix} h_{b,\beta}h^b \\ h_bh^b{}_{,\beta} \\ h_{b,\beta}h^b \\ h_bh^b{}_{,\beta} \end{pmatrix} = \begin{pmatrix} h_{a,\alpha}h^aE^\alpha{}_c \\ h_{a,\alpha}h^aE^\alpha{}_c \\ h_ah^a{}_{,\alpha}E^\alpha{}_c \\ h_ah^a{}_{,\alpha}E^\alpha{}_c \end{pmatrix} \cdot \begin{pmatrix} E^{\beta c}h_{b,\beta}h^b \\ E^{\beta c}h_bh^b{}_{,\beta} \\ E^{\beta c}h_{b,\beta}h^b \\ E^{\beta c}h_bh^b{}_{,\beta} \end{pmatrix}$$

$$\xrightarrow{\quad \text{interchange }\{a,b\},c \quad}$$

$$= \begin{pmatrix} h_{a,\alpha}h^a E^\alpha{}_c \\ h_{a,\alpha}h^a E^\alpha{}_c \\ \boxed{h_c h^a{}_{,\alpha} E^\alpha{}_a} \\ \boxed{h_c h^a{}_{,\alpha} E^\alpha{}_a} \end{pmatrix} \cdot \begin{pmatrix} \boxed{E^{\beta b} h_{b,\beta} h^c} \\ E^{\beta c} h_b h^b{}_{,\beta} \\ \boxed{E^{\beta b} h_{b,\beta} h^c} \\ E^{\beta c} h_b h^b{}_{,\beta} \end{pmatrix}. \tag{1282}$$

While for the equation in the last line of Eq. (1278), we can easily find the transition to the classical Dirac form via the recipe described in sub-section "9.3.2 Repetition: Dirac in the Metric Picture and Its Connection to the Classical Quaternion Form," we see that we face some difficulties for the symmetric case in Eq. (1284). In the constant **q**-vector-situation, we directly have obtained the starting point for the application of our recipe to end up with the classical Dirac equation. However, we have a bit more work in the symmetric case. We see that we always have one factor of the scalar product in the last line (terms in boxes) of Eq. (1282) in the classical Dirac form when performing yet another process of index-interchanging:

$$h_{a,\alpha}h^a E^\alpha{}_c E^{\beta c} h_b h^b{}_{,\beta} \xrightarrow{\text{interchange } E^\alpha{}_c E^{\beta c} \to E^\beta{}_c E^{\alpha c}} h_{a,\alpha}h^a E^\beta{}_c E^{\alpha c} h_b h^b{}_{,\beta}$$

$$\Rightarrow \begin{pmatrix} h_{a,\alpha}h^a E^\alpha{}_c \\ h_{a,\alpha}h^c E^{\alpha c} \\ \boxed{h_c h^a{}_{,\alpha} E^\alpha{}_a} \\ \boxed{h_c h^a{}_{,\alpha} E^\alpha{}_a} \end{pmatrix} \cdot \begin{pmatrix} \boxed{E^{\beta b} h_{b,\beta} h^c} \\ E^\beta{}_c h_b h^b{}_{,\beta} \\ \boxed{E^{\beta b} h_{b,\beta} h^c} \\ E^{\beta c} h_b h^b{}_{,\beta} \end{pmatrix} = \begin{pmatrix} h_{a,\alpha}h^a E^\alpha{}_c \\ h_{a,\alpha}h^c E^{\alpha a} \\ \boxed{h_c h^a{}_{,\alpha} E^\alpha{}_a} \\ \boxed{h_c h^a{}_{,\alpha} E^\alpha{}_a} \end{pmatrix} \cdot \begin{pmatrix} \boxed{E^{\beta b} h_{b,\beta} h^c} \\ E^\beta{}_b h_c h^b{}_{,\beta} \\ \boxed{E^{\beta b} h_{b,\beta} h^c} \\ E^{\beta c} h_b h^b{}_{,\beta} \end{pmatrix} \tag{1283}$$

and can further simplify:

$$h_c h^a \cdot M_a \cdot M^b \cdot h_b h^c = \begin{pmatrix} h_{a,\alpha}h^a E^\alpha{}_c \\ \boxed{h_{a,\alpha}h^c E^{\alpha a}} \\ \boxed{h_c h^a{}_{,\alpha} E^\alpha{}_a} \\ \boxed{h_c h^a{}_{,\alpha} E^\alpha{}_a} \end{pmatrix} \cdot \begin{pmatrix} \boxed{E^{\beta b} h_{b,\beta} h^c} \\ \boxed{E^\beta{}_b h_c h^b{}_{,\beta}} \\ \boxed{E^{\beta b} h_{b,\beta} h^c} \\ E^{\beta c} h_b h^b{}_{,\beta} \end{pmatrix}$$

$$= \begin{pmatrix} h_{a,\alpha}h^a E^\alpha{}_c + \boxed{h_c h^a{}_{,\alpha} E^\alpha{}_a} \\ \boxed{h_c h^a{}_{,\alpha} E^\alpha{}_a} \end{pmatrix} \begin{pmatrix} \boxed{E^{\beta b} h_{b,\beta} h^c} \\ E^{\beta c} h_b h^b{}_{,\beta} + \boxed{E^\beta{}_b h_c h^b{}_{,\beta}} \end{pmatrix}$$

$$\Rightarrow \quad \begin{pmatrix} \dfrac{1}{2} \cdot h_c h^a \cdot M_a \cdot M^b \cdot h_b h^c = \left(h_{a,\alpha} h^a E^\alpha{}_c + \boxed{h_c h^a{}_{,\alpha} E^\alpha{}_a} \right) \cdot \boxed{E^{\beta b} h_{b,\beta} h^c} \\ \dfrac{1}{2} \cdot h_c h^a \cdot M_a \cdot M^b \cdot h_b h^c = \boxed{h_c h^a{}_{,\alpha} E^\alpha{}_a} \cdot \left(E^{\beta c} h_b h^b{}_{,\beta} + \boxed{E^\beta{}_b h_c h^b{}_{,\beta}} \right) \end{pmatrix}$$

$$\Rightarrow \quad \begin{cases} \mp M^b \cdot h_b h^c = 2 \cdot E^{\beta b} h_{b,\beta} h^c \\ \mp h_c h^a \cdot M_a = 2 \cdot h_c h^a{}_{,\alpha} E^\alpha{}_a \end{cases}$$

$$\Rightarrow \quad 0 = \begin{cases} h^c \left(2 \cdot E^{\beta b} h_{b,\beta} \pm M^b \cdot h_b \right) \\ h_c \left(2 \cdot h^a{}_{,\alpha} E^\alpha{}_a \pm h^a \cdot M_a \right). \end{cases} \tag{1284}$$

It can easily be seen from the equation above and its rather artificial character, where only in systems of Cartesian-like *a-b-c*-spaces things are simple, that we have every reason to be critical about the result. Nevertheless, the factorization at least seems to work somehow even without the need to resort to Cartesian *a-b*-systems.

We realize, however, that obviously such a Dirac separation is not possible with the general form $w = C_f + f = q_a h^a$ (e.g., Eq. (1273)), where both vectors are general function-vectors unless the *a-b*-system is of completely Cartesian character. The reason is that we will always be left with at least one sum, where we cannot find a way to make it Dirac equivalent by the means of the interchange of indices as shown above. For instance, from Eq. (1273) we would obtain the following four equations:

$$f_{,\alpha} f_{,\beta} g^{\alpha\beta} = 4 \cdot \left(C_f + f \right)^2 \cdot R \cdot \frac{1}{(n-6) \cdot (n-1)}$$

$$\overbrace{M^2 = \mathbf{M} \cdot \mathbf{M} = M_c \cdot M^c}$$
$$\left(C_f + f \right)^2 \cdot 4 \cdot R \cdot \frac{1}{(n-6) \cdot (n-1)} = \left(C_f + f \right) \cdot M_c \cdot M^c \cdot \left(C_f + f \right) = f_{,\alpha} f_{,\beta} g^{\alpha\beta}$$

$$\xrightarrow{w = C_f + f = h_a q^a}$$
$$h_a q^a \cdot M_c \cdot M^c \cdot h_b q^b = \left(h_{a,\alpha} q^a + h_a q^a{}_{,\alpha} \right) g^{\alpha\beta} \left(h_{b,\beta} q^b + h_b q^b{}_{,\beta} \right)$$

$$\xrightarrow{g^{\alpha\beta} = E^\alpha{}_c \cdot E^{\beta c}} = \left(h_{a,\alpha} q^a + h_a q^a{}_{,\alpha} \right) E^\alpha{}_c \cdot E^{\beta c} \left(h_{b,\beta} q^b + h_b q^b{}_{,\beta} \right)$$

$$\xrightarrow{\text{interchange } \{a,b\},c}$$

$$
= \begin{pmatrix} h_{a,\alpha}q^a E^\alpha{}_c \\ h_{a,\alpha}q^a E^\alpha{}_c \\ \boxed{h_c q^a{}_{,\alpha} E^\alpha{}_a} \\ \boxed{h_c q^a{}_{,\alpha} E^\alpha{}_a} \end{pmatrix} \cdot \begin{pmatrix} \boxed{E^{\beta b}h_{b,\beta}q^c} \\ \boxed{E^{\beta b}h_{b,\beta}h^c} \\ E^{\beta c}h_b h^b{}_{,\beta} \\ E^{\beta c}h_b h^b{}_{,\beta} \end{pmatrix} = \begin{pmatrix} h_{a,\alpha}h^a E^\alpha{}_c \\ \boxed{2\cdot h_c h^a{}_{,\alpha} E^\alpha{}_a} \end{pmatrix} \cdot \begin{pmatrix} \boxed{2\cdot E^{\beta b}h_{b,\beta}h^c} \\ E^{\beta c}h_b h^b{}_{,\beta} \end{pmatrix}
$$

$$\Rightarrow \qquad \mp \frac{h_a q^a + h_a q^a}{2} M_c = \left(h_{a,\alpha}q^a + h_a q^a{}_{,\alpha} \right) E^\alpha{}_c$$

$$\Rightarrow \qquad 0 = \left(h_{a,\alpha}q^a + h_a q^a{}_{,\alpha} \right) E^\alpha{}_c \pm \frac{h_a q^a + h_a q^a}{2} M_c$$

$$\Rightarrow \qquad 0 = \begin{cases} h_a \left(q^a{}_{,\alpha} E^\alpha{}_c \pm \dfrac{q^a}{2} M_c \right) \\ q^a \left(h_{a,\alpha} E^\alpha{}_c \pm \dfrac{h_a}{2} M_c \right). \end{cases} \qquad (1285)$$

Consequently, we have to conclude that the approach with the intrinsically structured function f is anything but satisfying.

9.4.2 The Klein–Gordon Option

We easily see that the evaluations above can also be performed with the assumption of the Laplace operator producing Eigenvalues. This changes Eq. (1259) to:

$$R + \frac{1}{f + C_f}(1-n)\Delta f + \frac{f_{,\alpha}f_{,\beta}g^{\alpha\beta}(1-n)}{4(f+C_f)^2}(n-6) = 0$$

$$\xrightarrow{\Delta f = B^2 \cdot f}$$

$$R + \frac{1}{f + C_f}(1-n)\cdot B^2 \cdot f + \frac{f_{,\alpha}f_{,\beta}g^{\alpha\beta}(1-n)}{4(f+C_f)^2}(n-6) = 0 \qquad (1286)$$

and subsequently with:

$$\Delta f = \Delta\left(h + h_a q^a + h_{ab}q^{ab} + \ldots \right) = B^2 \cdot f \qquad (1287)$$

leads us via:

$$f_{,\alpha}f_{,\beta}g^{\alpha\beta} = 4 \cdot \left(\left(C_f + f\right)^2 \cdot R \cdot \frac{1}{(n-6)\cdot(n-1)} + \frac{\Delta f \cdot \left(C_f + f\right)}{n-6} \right)$$

$$\xrightarrow{\quad w \equiv C_f + f \quad}$$

$$w_{,\alpha}w_{,\beta}g^{\alpha\beta} = 4 \cdot w^2 \cdot \left(R \cdot \frac{1}{(n-6)\cdot(n-1)} + \frac{B^2}{n-6} \right) \qquad (1288)$$

to:

$$w_{,\alpha}w_{,\beta}g^{\alpha\beta} = w^2 \cdot 4 \cdot \left(R \cdot \frac{1}{(n-6)\cdot(n-1)} + \frac{B^2}{n-6} \right)$$

$$\overbrace{w^2 \cdot 4 \cdot \left(R \cdot \frac{1}{(n-6)\cdot(n-1)} + \frac{B^2}{n-6} \right)}^{M^2 = \mathbf{M} \cdot \mathbf{M}} = \left(C_f + f\right) \cdot \mathbf{M} \cdot \mathbf{M} \cdot \left(C_f + f\right) = f_{,\alpha}f_{,\beta}g^{\alpha\beta}$$

$$\xrightarrow{\quad w = C_f + f = h_a q^a \quad}$$

$$h_a q^a \cdot \mathbf{M} \cdot \mathbf{M} \cdot h_a q^a = h_{a,\alpha}q^a g^{\alpha\beta}h_{b,\beta}q^b \xrightarrow{\quad g^{\alpha\beta} = \mathbf{e}^\alpha \cdot \mathbf{e}^\beta \quad} = h_{a,\alpha}q^a \mathbf{e}^\alpha \cdot \mathbf{e}^\beta h_{b,\beta}q^b$$

$$\Rightarrow \qquad \mp h_a q^a \mathbf{M} = h_{a,\alpha}q^a \mathbf{e}^\alpha$$

$$\Rightarrow \qquad 0 = q^a \left(h_{a,\alpha}\mathbf{e}^\alpha \pm h_a \mathbf{M} \right). \qquad (1289)$$

Applying the procedure from the section(s) above, we could now bring Eq. (1289) into a suitable Dirac form, but here refrain from this task as it only is formal and does not add to the understanding.

9.4.3 About Higher Order Quaternions

We already saw that our fundamental Eq. (1008) together with the assumption of the effectively Ricci-flat space-time or $R^* = 0$ can give us all fundamental classical quantum equations. However, in order to separate Klein–Gordon and Dirac equations, we needed special conditions for either the function $F[f]$ or the function f itself. So, for instance, the condition in Eq. (1175) for $F[f]$ gave us the metric Klein–Gordon equation in the form of Eq. (1176), while demanding the satisfaction of the Laplace equation for the function f resulted in the metric Dirac equation (see sub-section "9.3.9.2 To Dirac from 6 dimensions in a simpler (luckier) way").

Can we also obtain a suitable quantum equation without the condition in Eq. (1175) or $\Delta f = 0$? The dimension of the space-time the Laplace operator is working on (or in) is higher than four because we also want to have inertia and potentials coded this way, see sub-section "9.3.9.2 To Dirac from 6 dimensions in a simpler (luckier) way").

Sticking with the effective flat-space condition, we directly have from Eq. (1112):

$$\xrightarrow{R^{*}=0}$$

$$\Delta\Psi = \overbrace{\frac{F}{F'\cdot(n-1)}}^{A^2} R - \Psi_{,\alpha}\Psi_{,\beta}\overbrace{g^{\alpha\beta}}^{e^{\alpha}\cdot e^{\beta}}\overbrace{\left(\frac{F''}{F'}+\frac{F'}{F}\cdot\frac{n-6}{4}\right)}^{C^2}$$

$$=\left(\mathbf{A}+\Psi_{,\alpha}\mathbf{e}^{\alpha}C\right)*\left(\mathbf{A}-\Psi_{,\beta}\mathbf{e}^{\beta}C\right)$$

$$\Rightarrow \text{``}\sqrt{\Delta\Psi}^{+\alpha}\text{''}*\text{``}\sqrt{\Delta\Psi}^{-\beta}\text{''}=\left(i\hbar\gamma^{\mu}\nabla_{\mu}\right)\left(i\hbar\gamma^{\mu}\nabla_{\mu}\right)\Psi=\left(\mathbf{A}+\Psi_{,\alpha}\gamma^{\alpha}C\right)*\left(\mathbf{A}-\Psi_{,\beta}\gamma^{\beta}C\right)$$

with:

$$\Psi_{,\alpha}\gamma^{\alpha}C\cdot\mathbf{A}-\mathbf{A}\cdot\Psi_{,\beta}\gamma^{\beta}C=0. \tag{1290}$$

Please note that this time we took care to take the direction of multiplication for the factors in the last line in Eq. (1290) into account. It will become clear later in this section. Above we also used the definitions for the curved Nabla operator ∇_{μ} and the curved gamma matrices from sub-section "9.3.11.1 Insertion: The classical way to obtain curved Dirac equations," with:

$$\nabla_{\mu}\Psi = \left(\partial_{\mu}-\frac{i}{4}\cdot\omega_{\mu}^{ab}\sigma_{ab}\right)\Psi, \tag{1291}$$

$$\omega_{\mu}^{ab} = \begin{pmatrix}\frac{1}{2}e^{av}\left(\partial_{\mu}e_{v}^{b}-\partial_{v}e_{\mu}^{b}\right)+\frac{1}{4}e^{a\rho}e^{b\sigma}\left(\partial_{\sigma}e_{\rho}^{c}-\partial_{\rho}e_{\sigma}^{c}\right)e_{c\mu}\\ -\frac{1}{2}e^{bv}\left(\partial_{\mu}e_{v}^{a}-\partial_{v}e_{\mu}^{a}\right)+\frac{1}{4}e^{b\rho}e^{a\sigma}\left(\partial_{\sigma}e_{\rho}^{c}-\partial_{\rho}e_{\sigma}^{c}\right)e_{c\mu}\end{pmatrix}$$

$$\eta_{ab}e_{\mu}^{a}e_{v}^{b}=g_{\mu v};\quad e^{av}=g^{\mu v}e_{\mu}^{a};\quad e^{av}=\eta^{ab}e_{b}^{v};\quad \sigma^{ab}=\frac{i}{2}\left(\gamma^{a}\gamma^{b}+\gamma^{b}\gamma^{a}\right), \tag{1292}$$

$$\left(\gamma^{\mu}\gamma^{v}+\gamma^{v}\gamma^{\mu}\right)=2\cdot g^{\mu v}, \tag{1293}$$

and:

$$\gamma^{\mu}=e_{a}^{\mu}\gamma^{a}. \tag{1294}$$

We see that in Eq. (1290) (third line), we have an operational and a purely ordinary factorization on the left-hand and on the right-hand side, respectively. We also see that we can apply the gamma matrices concept also on the right-hand side and consequently obtain:

$$\left(i\hbar\gamma^{\mu}\nabla_{\mu}\right)\left(i\hbar\gamma^{\nu}\nabla_{\nu}\right)\Psi = \left(\mathbf{A} + C\cdot\gamma^{\mu}\partial_{\mu}\Psi\right) * \left(\mathbf{A} - C\cdot\gamma^{\nu}\partial_{\nu}\Psi\right)$$

with:
$$\mathbf{A}\cdot C\cdot\gamma^{\mu}\partial_{\mu}\Psi - \mathbf{A}\cdot C\cdot\gamma^{\nu}\partial_{\nu}\Psi = 0. \tag{1295}$$

This gives us:

$$\left(i\hbar\gamma^{\mu}\nabla_{\mu}\right)\left(i\hbar\gamma^{\nu}\nabla_{\nu}\right)\Psi - \left(\mathbf{A} + C\cdot\gamma^{\mu}\partial_{\mu}\Psi\right) * \left(\mathbf{A} - C\cdot\gamma^{\nu}\partial_{\nu}\Psi\right) = 0. \tag{1296}$$

By assuming that each of the factors in Eq. (1296), respectively Eq. (1295) could be used to fulfill the equation, we might be allowed to set:

$$\left(i\hbar\gamma^{\mu}\nabla_{\mu}\right) - \left(\mathbf{A} + C\cdot\gamma^{\mu}\partial_{\mu}\Psi\right) = 0$$

$$\left(i\hbar\gamma^{\nu}\nabla_{\nu}\right)\Psi - \left(\mathbf{A} - C\cdot\gamma^{\nu}\partial_{\nu}\Psi\right) = 0. \tag{1297}$$

We find that the asymmetry looks rather strange and so apply the split-up technique we already introduced in sub-section "9.3.11.2 The connection to the metric form." Thereby, in contrast to our earlier usage of this "trick" above, we at first refrain from fixing the A-terms to any kind of Ψ-dependency and obtain:

$$\Delta\Psi = \left(i\hbar\gamma^{\mu}\nabla_{\mu}\right)\left(i\hbar\gamma^{\nu}\nabla_{\nu}\right)\Psi = \overbrace{\frac{F}{F'\cdot(n-1)}}^{A^2 = A_{\Delta}^2 + A_{\partial}^2}R - \Psi_{,\alpha}\Psi_{,\beta}\,\overbrace{g^{\alpha\beta}}^{e^{\alpha}\cdot e^{\beta}}\overbrace{\left(\frac{F'}{F}\cdot\frac{n-6}{4} + \frac{F''}{F'}\right)}^{C^2}$$

$$= A_{\Delta}^2 + A_{\partial}^2 - C\cdot\gamma^{\mu}\partial_{\mu}\Psi C\cdot\gamma^{\nu}\partial_{\nu}\Psi$$

$$\Rightarrow \left(i\hbar\gamma^{\mu}\nabla_{\mu} + \frac{\mathbf{A}_{\Delta}}{\Psi}\right)\left(i\hbar\gamma^{\nu}\nabla_{\nu} - \frac{\mathbf{A}_{\Delta}}{\Psi}\right)\Psi = \left(\mathbf{A}_{\partial} + C\cdot\gamma^{\mu}\partial_{\mu}\Psi\right) * \left(\mathbf{A}_{\partial} - C\cdot\gamma^{\nu}\partial_{\nu}\Psi\right)$$

with: $\mathbf{A}_{\nabla}\cdot C\cdot\gamma^{\mu}\partial_{\mu}\Psi - \mathbf{A}_{\nabla}\cdot C\cdot\gamma^{\nu}\partial_{\nu}\Psi = 0;$ $\left(\frac{\mathbf{A}_{\Delta}}{\Psi}\gamma^{\mu}\nabla_{\mu} - \frac{\mathbf{A}_{\Delta}}{\Psi}\cdot\gamma^{\nu}\nabla_{\nu}\right)\Psi = 0.$

$$\tag{1298}$$

It should be noted that the simplicity of the condition in the last line with respect to the right-hand side is only valid if we assume that the curved Nabla operators do not act on the A-terms. We realize that with the originally assumed Ψ-dependency things become easier, which we will therefore consider below. Taking these earlier applied Ψ-dependency for the A-terms we obtain, namely:

$$\Delta\Psi = \left(i\hbar\gamma^{\mu}\nabla_{\mu}\right)\left(i\hbar\gamma^{\nu}\nabla_{\nu}\right)\Psi = \overbrace{\frac{F}{F'\cdot(n-1)}}^{A^2 = A_{\Delta}^2\cdot\Psi + A_{\nabla}^2\cdot\Psi^2}R - \Psi_{,\alpha}\Psi_{,\beta}\overbrace{g^{\alpha\beta}}^{e^{\alpha}\cdot e^{\beta}}\overbrace{\left(\frac{F'}{F}\cdot\frac{n-6}{4}+\frac{F''}{F'}\right)}^{C^2}$$

$$= A_{\Delta}^2\cdot\Psi + A_{\partial}^2\cdot\Psi^2 - C\cdot\gamma^{\mu}\partial_{\mu}\Psi C\cdot\gamma^{\nu}\partial_{\nu}\Psi$$

$$\Rightarrow\left(i\hbar\gamma^{\mu}\nabla_{\mu}+\mathbf{A}_{\Delta}\right)\left(i\hbar\gamma^{\nu}\nabla_{\nu}-\mathbf{A}_{\Delta}\right)\Psi = \left(\mathbf{A}_{\partial}\Psi + C\cdot\gamma^{\mu}\partial_{\mu}\Psi\right)*\left(\mathbf{A}_{\partial}\Psi - C\cdot\gamma^{\nu}\partial_{\nu}\Psi\right)$$

with: $\quad\Psi\mathbf{A}_{\nabla}\cdot C\cdot\gamma^{\mu}\partial_{\mu}\Psi - \Psi\mathbf{A}_{\nabla}\cdot C\cdot\gamma^{\nu}\partial_{\nu}\Psi = 0;\quad\left(\gamma^{\mu}\nabla_{\mu}\mathbf{A}_{\Delta}-\mathbf{A}_{\Delta}\cdot\gamma^{\nu}\nabla_{\nu}\right)\Psi = 0.$

$$(1299)$$

This time, we can have a relatively symmetric factorization and write:

$$\left(i\hbar\gamma^{\mu}\nabla_{\mu}+\mathbf{A}_{\Delta}\right)\Psi = \left(\mathbf{A}_{\nabla}+C\cdot\gamma^{\mu}\partial_{\mu}\right)\Psi$$

$$\left(i\hbar\gamma^{\nu}\nabla_{\nu}-\mathbf{A}_{\Delta}\right)\Psi = \left(\mathbf{A}_{\nabla}-C\cdot\gamma^{\nu}\partial_{\nu}\right)\Psi$$

or

$$\left(i\hbar\gamma^{\mu}\nabla_{\mu}+\mathbf{A}_{\Delta}\right)\Psi = \left(\mathbf{A}_{\nabla}-C\cdot\gamma^{\nu}\partial_{\nu}\right)\Psi$$

$$\left(i\hbar\gamma^{\nu}\nabla_{\nu}-\mathbf{A}_{\Delta}\right)\Psi = \left(\mathbf{A}_{\nabla}+C\cdot\gamma^{\mu}\partial_{\mu}\right)\Psi.$$

$$(1300)$$

Apparently, which is to say on first sight, we have not obtained more than a rather trivial identity, but before discussing this issue, we want to consider the whole approach without resorting to the gamma matrices.

Naturally, the A-terms are functions of F, R, and R^* and so—strictly speaking—when treating them as constants, we have to consider the second line in Eq. (1208) an approximation, but we will see below what nice connection to the classical picture this can provide. At first, we rewrite the first line of Eq. (1299) as follows:

$$\Delta\Psi = \overbrace{\frac{F}{F'\cdot(n-1)}}^{A^2 = A_{\Delta}^2\cdot\Psi - A_{\nabla}^2\cdot\Psi^2}(R-R^*F)+\Psi_{,\alpha}\Psi_{,\beta}\overbrace{g^{\alpha\beta}}^{e^{\alpha}\cdot e^{\beta}}\overbrace{\left(\frac{F'}{F}\cdot\frac{6-n}{4}-\frac{F''}{F'}\right)}^{B^2}.$$

$$(1301)$$

Please note that we here also have not—not yet—demanded $\Delta\Psi = 0$ as we had it before in Eq. (1204). This gives us:

$$\Delta\Psi - A_{\Delta}^2\cdot\Psi = \Psi_{,\alpha}\Psi_{,\beta}\overbrace{g^{\alpha\beta}}^{e^{\alpha}\cdot e^{\beta}}B^2 - A_{\nabla}^2\cdot\Psi^2.$$

$$(1302)$$

We see that we can now factorize the right-hand side in the usual way, thereby giving us:

$$\xrightarrow{R^* \neq 0; \ R \neq 0}$$

$$\Delta\Psi - A_\Delta^2 \cdot \Psi = \Psi_{,\alpha}\Psi_{,\beta}\overbrace{g^{\alpha\beta}}^{e^\alpha \cdot e^\beta} B^2 - A_\nabla^2 \cdot \Psi^2 = \left(\mathbf{A}_\nabla\Psi + \Psi_{,\alpha}e^\alpha B\right)*\left(\Psi_{,\beta}e^\beta B - \mathbf{A}_\nabla\Psi\right)$$

with: $\qquad\qquad \Psi\cdot\Psi_{,\alpha}e^\alpha B\cdot\mathbf{A}_\nabla - \mathbf{A}_\nabla\Psi\cdot\Psi_{,\beta}e^\beta B = 0.$ $\qquad$ (1303)

Now we want to take care about the apparent triviality, we made out when applying the gamma matrices above. We also do not like the fact that the last line of Eq. (1299) requires the curved Nabla operator not to act on the $\mathbf{A}_\Delta$-terms. Thus, we move back to Eq. (1290) and reshape it as follows:

$$\xrightarrow{R^* \neq 0}$$

$$\Delta\Psi = \overbrace{\frac{F}{F'\cdot(n-1)}(R - R^*F)}^{A^2 = \mathbf{A}\cdot\mathbf{A}} + \Psi_{,\alpha}\Psi_{,\beta}\overbrace{g^{\alpha\beta}}^{e^\alpha \cdot e^\beta}\overbrace{\left(\frac{F'}{F}\cdot\frac{6-n}{4} - \frac{F''}{F'}\right)}^{B^2}$$

$$= \left(\mathbf{A} + \Psi_{,\alpha}e^\alpha B\right)*\left(\mathbf{A} + \Psi_{,\beta}e^\beta B\right)$$

$$\Rightarrow \text{``}\sqrt{\Delta\Psi}^{+\alpha}\text{''}*\text{``}\sqrt{\Delta\Psi}^{-\beta}\text{''} = \left(i\hbar\gamma^\mu\nabla_\mu\right)\left(i\hbar\gamma^\mu\nabla_\mu\right)\Psi = \left(\mathbf{A} + \Psi_{,\alpha}\gamma^\alpha B\right)\left(\mathbf{A} + \Psi_{,\beta}\gamma^\beta B\right)$$

with: $\quad \Psi_{,\alpha}\gamma^\alpha B\cdot\mathbf{A} + \mathbf{A}\cdot\Psi_{,\beta}\gamma^\beta B = 0; \quad \Psi_{,\alpha}\gamma^\alpha B\Psi_{,\beta}\gamma^\beta B \triangleq \Psi_{,\alpha}\Psi_{,\beta}g^{\alpha\beta}.$

$$(1304)$$

This time we can have a symmetric factorization:

$$i\hbar\gamma^\mu\nabla_\mu\Psi \overset{!}{=} \left(\mathbf{A} + \Psi_{,\alpha}\gamma^\alpha B\right), \qquad (1305)$$

but the necessary condition in the last line of Eq. (1304) forces us to apply the Pauli matrices σ_k ($k = 1, 2, 3$) in the following way:

$$\sigma_1 \cdot i\hbar\gamma^\mu\nabla_\mu\Psi = \sigma_2\cdot\mathbf{A} + \sigma_3\cdot\Psi_{,\alpha}\gamma^\alpha B$$

with: $\quad \sigma_k = \begin{pmatrix} \delta_k^3 & \delta_k^1 - i\cdot\delta_k^2 \\ \delta_k^1 + i\cdot\delta_k^2 & -\delta_k^3 \end{pmatrix}; \quad \sigma_j\cdot\sigma_k = \begin{cases} 0 \ \text{ for } \ j \neq k \\ I = \begin{pmatrix} 1 & 0 \\ 0 & 1 \end{pmatrix} \ \text{ for } \ j = k. \end{cases}$

$$(1306)$$

Thereby we need to remind that $\mathbf{A}$ can be an arbitrary function (also of Ψ). As with the classical Dirac theory we also have Ψ to be a vector of functions. However, there is even more. Namely, in order to make the first line of Eq.

(1306) a reasonable equation, the terms $i\hbar\gamma^\mu\nabla_\mu\Psi$, $\mathbf{A}$, $\Psi_{,\alpha}\gamma^\alpha B$ need to be spinors. Thus, we have the n-dimensional Ψ-vectors for the curved Dirac gamma matrices in n dimensions plus the 2-dimensional spinors for the terms $i\hbar\gamma^\mu\nabla_\mu\Psi$, $\mathbf{A}$, $\Psi_{,\alpha}\gamma^\alpha B$ for the additional Pauli matrices.

9.4.4 A Few Words about the Metric Form(s) and Its Connection to the Dirac Approach

As our general intention of this chapter is the consideration of a metric Dirac equation, we should just work ourselves through the options to factorize Eq. (1008) under the condition $R^* = 0$. Obviously, as already shown here and in our previous papers [2–9], this gives us all necessary equations for all forms of interactions except gravity. The latter just requires the variation under the Einstein–Hilbert action integral as given in Eq. (1010).

We saw that the transition from the metric to the Dirac form does not only require the extension of the wave function f from a scalar to a vector, but also needs a switch of summation. In the following we will now try to work out ways of factorization where we could potentially avoid such awkward "tricks."

9.4.5 Decomposition of the Ricci Scalar and Tensor

In the case of an effectively flat space, which is to say $R^* = 0$, we are able to further decompose the first line in Eq. (1304) as follows:

$$\xrightarrow{R^*=0}$$

$$\Delta\Psi = \overbrace{\frac{F}{F'\cdot(n-1)}}^{A^2} R - \Psi_{,\alpha}\Psi_{,\beta} \overbrace{g^{\alpha\beta}}^{\mathbf{e}^\alpha\cdot\mathbf{e}^\beta} \overbrace{\left(\frac{F'}{F}\cdot\frac{n-6}{4}+\frac{F''}{F'}\right)}^{C^2}$$

$$= \left(A^2 \cdot R_{\alpha\beta} - \Psi_{,\alpha}\Psi_{,\beta}\cdot C^2\right)\overbrace{g^{\alpha\beta}}^{\mathbf{e}^\alpha\cdot\mathbf{e}^\beta}. \tag{1307}$$

As we know that the Ricci tensor is symmetric, we can decompose it into two forms like we can decompose the metric tensor into tetrads or base vectors:

$$\Delta\Psi = \left(A^2\cdot R_{\alpha\beta} - \Psi_{,\alpha}\Psi_{,\beta}\cdot C^2\right)\overbrace{g^{\alpha\beta}}^{\mathbf{e}^\alpha\cdot\mathbf{e}^\beta} = \left(A^2\cdot\mathbf{R}_\alpha\cdot\mathbf{R}_\beta - \Psi_{,\alpha}\Psi_{,\beta}\cdot C^2\right)\mathbf{e}^\alpha\cdot\mathbf{e}^\beta. \tag{1308}$$

For the moment we ignore the differences in the vector-status for the A and C terms and we write:

$$\Delta\Psi = \left(A^2 \cdot \mathbf{R}_\alpha \cdot \mathbf{R}_\beta - \Psi_{,\alpha}\Psi_{,\beta} \cdot C^2\right)\mathbf{e}^\alpha \cdot \mathbf{e}^\beta = \left(A \cdot \mathbf{R}_\alpha + \Psi_{,\alpha} \cdot C\right)\left(A \cdot \mathbf{R}_\beta - \Psi_{,\beta} \cdot C\right)\mathbf{e}^\alpha \cdot \mathbf{e}^\beta$$

with: $\qquad\left(\Psi_{,\alpha} \cdot C \cdot A \cdot \mathbf{R}_\beta - A \cdot \mathbf{R}_\alpha \cdot \Psi_{,\beta} \cdot C\right)\mathbf{e}^\alpha \cdot \mathbf{e}^\beta = 0.$ \hfill (1309)

In order to take care about the vector-states, we may refine our last evaluation or separation as follows:

$$\Delta\Psi = \left(A^2 \cdot \mathbf{R}_\alpha \cdot \mathbf{R}_\beta - \Psi_{,\alpha}\Psi_{,\beta} \cdot C^2\right)\mathbf{e}^\alpha \cdot \mathbf{e}^\beta$$

$$= \left(A \cdot \mathbf{R}_\alpha + \mathbf{X} \cdot \Psi_{,\alpha} \cdot C\right)\left(A \cdot \mathbf{R}_\beta - \mathbf{X} \cdot \Psi_{,\beta} \cdot C\right)\mathbf{e}^\alpha \cdot \mathbf{e}^\beta$$

with: $\qquad\left(\mathbf{X} \cdot \Psi_{,\alpha} \cdot C \cdot A \cdot \mathbf{R}_\beta - A \cdot \mathbf{R}_\alpha \cdot \mathbf{X} \cdot \Psi_{,\beta} \cdot C\right)\mathbf{e}^\alpha \cdot \mathbf{e}^\beta = 0.$ \hfill (1310)

Thereby the new object $\mathbf{X}$ could be kind of arbitrary. It would only be important that it has the same dimension as the various $\mathbf{R}$-"tetrad-parts." As we do not like the asymmetry, we may once more opt for the following evaluation:

$$\xrightarrow{R^* = 0}$$

$$\Delta\Psi = \overbrace{\frac{F}{F' \cdot (n-1)}}^{A^2}R + \Psi_{,\alpha}\Psi_{,\beta}\,\overbrace{g^{\alpha\beta}}^{\mathbf{e}^\alpha \cdot \mathbf{e}^\beta}\overbrace{\left(\frac{F'}{F} \cdot \frac{6-n}{4} - \frac{F''}{F'}\right)}^{B^2}$$

$$= \left(A^2 \cdot R_{\alpha\beta} + \Psi_{,\alpha}\Psi_{,\beta} \cdot B^2\right)\overbrace{g^{\alpha\beta}}^{\mathbf{e}^\alpha \cdot \mathbf{e}^\beta},$$ \hfill (1311)

$$\Rightarrow \quad \Delta\Psi = \left(A^2 \cdot R_{\alpha\beta} + \Psi_{,\alpha}\Psi_{,\beta} \cdot B^2\right)\overbrace{g^{\alpha\beta}}^{\mathbf{e}^\alpha \cdot \mathbf{e}^\beta} = \left(A^2 \cdot \mathbf{R}_\alpha \cdot \mathbf{R}_\beta + \Psi_{,\alpha}\Psi_{,\beta} \cdot B^2\right)\mathbf{e}^\alpha \cdot \mathbf{e}^\beta,$$ \hfill (1312)

$$\Delta\Psi = \left(A^2 \cdot \mathbf{R}_\alpha \cdot \mathbf{R}_\beta + \Psi_{,\alpha}\Psi_{,\beta} \cdot B^2\right)\mathbf{e}^\alpha \cdot \mathbf{e}^\beta$$

$$= \left(A \cdot \mathbf{R}_\alpha + \mathbf{X} \cdot \Psi_{,\alpha} \cdot B\right)\left(A \cdot \mathbf{R}_\beta + \mathbf{X} \cdot \Psi_{,\beta} \cdot B\right)\mathbf{e}^\alpha \cdot \mathbf{e}^\beta$$

with: $\left(\mathbf{X} \cdot \Psi_{,\alpha} \cdot B \cdot A \cdot \mathbf{R}_\beta + A \cdot \mathbf{R}_\alpha \cdot \mathbf{X} \cdot \Psi_{,\beta} \cdot B\right)\mathbf{e}^\alpha \cdot \mathbf{e}^\beta = 0; \quad \mathbf{X} \cdot \mathbf{X} = 1.$ \hfill (1313)

We see that the demands for $\mathbf{X}$ become more complicated and require deeper consideration.

On the other hand, we may just accept the asymmetry in Eq. (1310) and try for a total factorization by the means of the Dirac gamma matrices as follows:

$$\Delta\Psi = \left(A\cdot\mathbf{R}_\alpha + \mathbf{X}\cdot\Psi_{,\alpha}\cdot C\right)\left(A\cdot\mathbf{R}_\beta - \mathbf{X}\cdot\Psi_{,\beta}\cdot C\right)e^\alpha\cdot e^\beta$$

$$= \left(A\cdot\mathbf{R}_\alpha + \mathbf{X}\cdot\Psi_{,\alpha}\cdot C\right)\left(A\cdot\mathbf{R}_\beta - \mathbf{X}\cdot\Psi_{,\beta}\cdot C\right)\frac{\left(\gamma^\alpha\gamma^\beta + \gamma^\beta\gamma^\alpha\right)}{2}$$

$$\Rightarrow \left(\gamma^\alpha\nabla_\alpha\right)\left(\gamma^\beta\nabla_\beta\right)\Psi = \left(A\cdot\mathbf{R}_\alpha + \mathbf{X}\cdot\Psi_{,\alpha}\cdot C\right)\left(A\cdot\mathbf{R}_\beta - \mathbf{X}\cdot\Psi_{,\beta}\cdot C\right)\frac{\left(\gamma^\alpha\gamma^\beta + \gamma^\beta\gamma^\alpha\right)}{2}$$

$$\left(\gamma^\alpha\nabla_\alpha\right)\left(\gamma^\beta\nabla_\beta\right)\Psi = \frac{1}{2}\left(\begin{array}{l}\gamma^\alpha\left(A\cdot\mathbf{R}_\alpha + \mathbf{X}\cdot\Psi_{,\alpha}\cdot C\right)\left(A\cdot\mathbf{R}_\beta - \mathbf{X}\cdot\Psi_{,\beta}\cdot C\right)\gamma^\beta \\ +\gamma^\beta\left(A\cdot\mathbf{R}_\alpha + \mathbf{X}\cdot\Psi_{,\alpha}\cdot C\right)\left(A\cdot\mathbf{R}_\beta - \mathbf{X}\cdot\Psi_{,\beta}\cdot C\right)\gamma^\alpha\end{array}\right)$$

$$= \frac{1}{2}\left(\begin{array}{l}\gamma^\alpha\left(A\cdot\mathbf{R}_\alpha + \mathbf{X}\cdot\Psi_{,\alpha}\cdot C\right)\left(A\cdot\mathbf{R}_\beta - \mathbf{X}\cdot\Psi_{,\beta}\cdot C\right)\gamma^\beta \\ +\gamma^\beta\left(A\cdot\mathbf{R}_\beta - \mathbf{X}\cdot\Psi_{,\beta}\cdot C\right)\left(A\cdot\mathbf{R}_\alpha + \mathbf{X}\cdot\Psi_{,\alpha}\cdot C\right)\gamma^\alpha\end{array}\right). \tag{1314}$$

Now we boldly take it that we can split up the Laplace operator as follows:

$$\Delta\Psi = \frac{\left(\gamma^\alpha\nabla_\alpha\right)\left(\gamma^\beta\nabla_\beta\right) + \left(\gamma^\beta\nabla_\beta\right)\left(\gamma^\alpha\nabla_\alpha\right)}{2}\Psi \tag{1315}$$

and subsequently evaluate the following:

$$\Delta\Psi = \frac{\left(\gamma^\alpha\nabla_\alpha\right)\left(\gamma^\beta\nabla_\beta\right) + \left(\gamma^\beta\nabla_\beta\right)\left(\gamma^\alpha\nabla_\alpha\right)}{2}\Psi$$

$$= \frac{1}{2}\left(\begin{array}{l}\gamma^\alpha\left(A\cdot\mathbf{R}_\alpha + \mathbf{X}\cdot\Psi_{,\alpha}\cdot C\right)\left(A\cdot\mathbf{R}_\beta - \mathbf{X}\cdot\Psi_{,\beta}\cdot C\right)\gamma^\beta \\ +\gamma^\beta\left(A\cdot\mathbf{R}_\beta - \mathbf{X}\cdot\Psi_{,\beta}\cdot C\right)\left(A\cdot\mathbf{R}_\alpha + \mathbf{X}\cdot\Psi_{,\alpha}\cdot C\right)\gamma^\alpha\end{array}\right)$$

$$\Rightarrow \left(\gamma^\alpha\nabla_\alpha\right)\left(\gamma^\beta\nabla_\beta\right)\Psi + \left(\gamma^\beta\nabla_\beta\right)\left(\gamma^\alpha\nabla_\alpha\right)\Psi$$

$$= \left(\begin{array}{l}\gamma^\alpha\left(A\cdot\mathbf{R}_\alpha + \mathbf{X}\cdot\Psi_{,\alpha}\cdot C\right)\left(A\cdot\mathbf{R}_\beta - \mathbf{X}\cdot\Psi_{,\beta}\cdot C\right)\gamma^\beta \\ +\gamma^\beta\left(A\cdot\mathbf{R}_\beta - \mathbf{X}\cdot\Psi_{,\beta}\cdot C\right)\left(A\cdot\mathbf{R}_\alpha + \mathbf{X}\cdot\Psi_{,\alpha}\cdot C\right)\gamma^\alpha\end{array}\right)$$

$$\Rightarrow \left|\begin{array}{l}\left(\gamma^\alpha\nabla_\alpha\right)\left(\gamma^\beta\nabla_\beta\right)\Psi = \gamma^\alpha\left(A\cdot\mathbf{R}_\alpha + \mathbf{X}\cdot\Psi_{,\alpha}\cdot C\right)\left(A\cdot\mathbf{R}_\beta - \mathbf{X}\cdot\Psi_{,\beta}\cdot C\right)\gamma^\beta \\ \qquad\Rightarrow \left\{\begin{array}{l}\left(\gamma^\alpha\nabla_\alpha\right)\Psi = \gamma^\alpha\left(A\cdot\mathbf{R}_\alpha + \mathbf{X}\cdot\Psi_{,\alpha}\cdot C\right) \\ \left(\gamma^\beta\nabla_\beta\right)\Psi = \left(A\cdot\mathbf{R}_\beta - \mathbf{X}\cdot\Psi_{,\beta}\cdot C\right)\gamma^\beta\end{array}\right. \\ \left(\gamma^\beta\nabla_\beta\right)\left(\gamma^\alpha\nabla_\alpha\right)\Psi = \gamma^\beta\left(A\cdot\mathbf{R}_\beta - \mathbf{X}\cdot\Psi_{,\beta}\cdot C\right)\left(A\cdot\mathbf{R}_\alpha + \mathbf{X}\cdot\Psi_{,\alpha}\cdot C\right)\gamma^\alpha \\ \qquad\Rightarrow \left\{\begin{array}{l}\left(\gamma^\beta\nabla_\beta\right)\Psi = \left(A\cdot\mathbf{R}_\beta - \mathbf{X}\cdot\Psi_{,\beta}\cdot C\right)\gamma^\beta \\ \left(\gamma^\alpha\nabla_\alpha\right)\Psi = \gamma^\alpha\left(A\cdot\mathbf{R}_\alpha + \mathbf{X}\cdot\Psi_{,\alpha}\cdot C\right).\end{array}\right.\end{array}\right.$$

$$\tag{1316}$$

We have obtained the two Dirac equations:

$$\left(\gamma^\beta \nabla_\beta\right)\Psi = \left(A \cdot \mathbf{R}_\beta - \mathbf{X} \cdot \Psi_{,\beta} \cdot C\right)\gamma^\beta$$

$$\left(\gamma^\alpha \nabla_\alpha\right)\Psi = \gamma^\alpha \left(A \cdot \mathbf{R}_\alpha + \mathbf{X} \cdot \Psi_{,\alpha} \cdot C\right). \tag{1317}$$

As usual, we require a function vector for Ψ in order to obtain reasonable equations. We see the discrepancy in rank on both sides of the two equations above and conclude that the factorization of the Laplace operator in Eq. (1316) requires the introduction of yet another auxiliary vector $\mathbf{Y}$ as follows:

$$\Delta\Psi = \left(\gamma^\alpha \nabla_\alpha\right)\mathbf{Y} \cdot \mathbf{Y}\left(\gamma^\beta \nabla_\beta\right)\Psi = \gamma^\alpha \left(A \cdot \mathbf{R}_\alpha + \mathbf{X} \cdot \Psi_{,\alpha} \cdot C\right)\left(A \cdot \mathbf{R}_\beta - \mathbf{X} \cdot \Psi_{,\beta} \cdot C\right)\gamma^\beta$$

$$\Rightarrow \quad \begin{cases} \mathbf{Y}\left(\gamma^\alpha \nabla_\alpha\right)\Psi = \gamma^\alpha \left(A \cdot \mathbf{R}_\alpha + \mathbf{X} \cdot \Psi_{,\alpha} \cdot C\right) \\ \mathbf{Y}\left(\gamma^\beta \nabla_\beta\right)\Psi = \left(A \cdot \mathbf{R}_\beta - \mathbf{X} \cdot \Psi_{,\beta} \cdot C\right)\gamma^\beta \end{cases}$$

$$\left(\gamma^\beta \nabla_\beta\right)\left(\gamma^\alpha \nabla_\alpha\right)\Psi = \gamma^\beta \left(A \cdot \mathbf{R}_\beta - \mathbf{X} \cdot \Psi_{,\beta} \cdot C\right)\left(A \cdot \mathbf{R}_\alpha + \mathbf{X} \cdot \Psi_{,\alpha} \cdot C\right)\gamma^\alpha$$

$$\Rightarrow \quad \begin{cases} \mathbf{Y}\left(\gamma^\beta \nabla_\beta\right)\Psi = \left(A \cdot \mathbf{R}_\beta - \mathbf{X} \cdot \Psi_{,\beta} \cdot C\right)\gamma^\beta \\ \mathbf{Y}\left(\gamma^\alpha \nabla_\alpha\right)\Psi = \gamma^\alpha \left(A \cdot \mathbf{R}_\alpha + \mathbf{X} \cdot \Psi_{,\alpha} \cdot C\right). \end{cases} \tag{1318}$$

9.4.6 A Potentially Problematic Issue

When assuming a flat space-time, we should be able to simplify the results from Eq. (1318) as follows:

$$\left(\gamma^\beta \nabla_\beta\right)\Psi = \Psi_{,\beta} \cdot C\gamma^\beta$$

$$\left(\gamma^\alpha \nabla_\alpha\right)\Psi = \gamma^\alpha \Psi_{,\alpha} \cdot C. \tag{1319}$$

While in Cartesian coordinates, this seems to be a simple identity, because we have:

$$\left(\gamma^\beta \partial_\beta\right)\Psi = \Psi_{,\beta} \cdot C\gamma^\beta$$

$$\left(\gamma^\alpha \partial_\alpha\right)\Psi = \gamma^\alpha \Psi_{,\alpha} \cdot C, \tag{1320}$$

which we can easily satisfy with $C = 1$, we find that in curvilinear coordinates something seems to be wrong. So, for instance, we have in spherical spatial coordinates (c.f. Eq. (1221)):

$$\left(\gamma^{\alpha}\nabla_{\alpha}\right)\Psi = \left(\gamma^{\alpha}\nabla_{\alpha}\right)\begin{pmatrix}\Psi_1\\\Psi_2\\\Psi_3\\\Psi_4\end{pmatrix}$$

$$= \begin{pmatrix}\dfrac{1}{c}\partial_t & 0 & \dfrac{\partial_\varphi}{r\sin\vartheta} & \left(R-\dfrac{i}{r}\Theta\right)\\[2ex] 0 & \dfrac{1}{c}\partial_t & \left(\dfrac{i}{r}\Theta+R\right) & -\dfrac{\partial_\varphi}{r\sin\vartheta}\\[2ex] -\dfrac{\partial_\varphi}{r\sin\vartheta}\partial_\varphi & \left(\dfrac{i}{r}\Theta-R\right) & -\dfrac{1}{c}\partial_t & 0\\[2ex] \left(-\dfrac{i}{r}\Theta-R\right) & \dfrac{\partial_\varphi}{r\sin\vartheta} & 0 & -\dfrac{1}{c}\partial_t\end{pmatrix}\begin{pmatrix}\Psi_1\\\Psi_2\\\Psi_3\\\Psi_4\end{pmatrix}$$

$$\Theta = \frac{1}{\sqrt{\sin\vartheta}}\partial_\vartheta\left(\sqrt{\sin\vartheta}(...)\right); \quad R = \frac{1}{r}\partial_r\left(r(...)\right). \tag{1321}$$

Apparently, with a constant or only Ψ-dependent C, we cannot always fulfill Eq. (1319). The way out seems to be the introduction of more complex C when moving from the purely metric to the gamma-matrix form in Eq. (1314). Now, of course, we know that the classical Dirac equation was postulated while our equations are derived from a scaled metric with the assumption of a vanishing Ricci scalar. In addition, we have the kind of "artificial introduction" of the spinor into the curved space-times. We also already pointed out some problematic issues regarding a switch of summation when moving from the metric to the gamma matrix forms (c.f. sub-section "9.4.1 More Tedious Work"), but nevertheless (especially as the Dirac equation obviously is so successful in describing quantum reality) we realize the need for quite some discussion. This, however, shall be part of the next chapters of this book.

Here we want to finish our chapter with the re-consideration of the metric form with intrinsically structured wave functions Ψ.

9.4.7 Back to Intrinsically Structured Wave Functions

As in the sub-section before, we once again assume an effectively flat space, which is to say $R^* = 0$ and further decompose the first line in Eq. (1304) with an intrinsically structured function Ψ as follows:

$$\Delta\Psi = \Delta\left(q^a h_a\right) = 0 = \overbrace{\frac{F}{F' \cdot (n-1)}}^{A^2} R - \Psi_{,\alpha}\Psi_{,\beta}\, \overbrace{g^{\alpha\beta}}^{e^\alpha \cdot e^\beta} \overbrace{\left(\frac{F'}{F} \cdot \frac{n-6}{4} + \frac{F''}{F'}\right)}^{C^2}$$

$$= \left(A^2 \cdot R_{\alpha\beta} - q^a q^b h_{a,\alpha} h_{b,\beta} \cdot C^2\right)\overbrace{g^{\alpha\beta}}^{e^\alpha \cdot e^\beta}$$

with:
$$q^a = \text{const}^a;\quad \Delta h_a = 0,\tag{1322}$$

$$\Delta\Psi = 0 = \left(A^2 \cdot R_{\alpha\beta} - q^a q^b h_{a,\alpha} h_{b,\beta} \cdot C^2\right)\overbrace{g^{\alpha\beta}}^{e^\alpha \cdot e^\beta} = \left(A^2 \cdot \mathbf{R}_\alpha \cdot \mathbf{R}_\beta - q^a q^b h_{a,\alpha} h_{b,\beta} \cdot C^2\right)e^\alpha \cdot e^\beta.$$
$$\tag{1323}$$

The reader notes that we have demanded the Ψ-function to be a Laplace function. This does not mean that we have excluded any masses or potentials, because these can easily be constructed by suitable ensembles of additional dimensions, as we have already shown in the sections above (especially in "9.3.9.2 To Dirac from 6 dimensions in a simpler (luckier) way"). Ignoring the differences in the vector-status for the A and C terms, we write:

$$0 = \left(A^2 \cdot \mathbf{R}_\alpha \cdot \mathbf{R}_\beta - q^a q^b h_{a,\alpha} h_{b,\beta} \cdot C^2\right)e^\alpha \cdot e^\beta$$

$$= \left(A \cdot \mathbf{R}_\alpha + q^a h_{a,\alpha} \cdot C\right)\left(A \cdot \mathbf{R}_\beta - q^b h_{b,\beta} \cdot C\right)e^\alpha \cdot e^\beta$$

with:
$$\left(q^a h_{a,\alpha} \cdot C \cdot A \cdot \mathbf{R}_\beta - A \cdot \mathbf{R}_\alpha \cdot q^b h_{b,\beta} \cdot C\right)e^\alpha \cdot e^\beta = 0.\tag{1324}$$

In order to take care about the vector-states, we may refine our last evaluation or separation as follows:

$$\Delta\Psi = \left(A^2 \cdot \mathbf{R}_\alpha \cdot \mathbf{R}_\beta - q^a q^b h_{a,\alpha} h_{b,\beta} \cdot C^2\right)e^\alpha \cdot e^\beta$$

$$= \left(A \cdot \mathbf{R}_\alpha + \mathbf{X} \cdot q^a h_{a,\alpha} \cdot C\right)\left(A \cdot \mathbf{R}_\beta - \mathbf{X} \cdot q^b h_{b,\beta} \cdot C\right)e^\alpha \cdot e^\beta = 0$$

with:
$$\left(\mathbf{X} \cdot q^a h_{a,\alpha} \cdot C \cdot A \cdot \mathbf{R}_\beta - A \cdot \mathbf{R}_\alpha \cdot \mathbf{X} \cdot q^b h_{b,\beta} \cdot C\right)e^\alpha \cdot e^\beta = 0.\tag{1325}$$

As before, the new object $\mathbf{X}$ could be kind of arbitrary. It would only be important that it has the same dimension as the various $\mathbf{R}$-"tetrad-parts."

This results in the following two first-order differential equations:

$$\left(A \cdot \mathbf{R}_\alpha + \mathbf{X} \cdot q^a h_{a,\alpha} \cdot C\right)e^\alpha = 0$$

$$\left(A \cdot \mathbf{R}_\beta - \mathbf{X} \cdot q^b h_{b,\beta} \cdot C\right)e^\beta = 0$$

with: $\left(\mathbf{X} \cdot q^a h_{a,\alpha} \cdot C \cdot A \cdot \mathbf{R}_\beta - A \cdot \mathbf{R}_\alpha \cdot \mathbf{X} \cdot q^b h_{b,\beta} \cdot C\right)\mathbf{e}^\alpha \cdot \mathbf{e}^\beta = 0.$ (1326)

In the power-law case of $F[\Psi] = (C_f + \Psi)^p$, the equations (1326) simplify to systems of linear differential equations (as long as the Ricci-vectors do not depend on Ψ):

$$\left(\sqrt{\frac{C_f + q^a h_a}{p \cdot (n-1)}} \cdot \mathbf{R}_\alpha + \mathbf{X} \cdot q^a h_{a,\alpha} \cdot \sqrt{\frac{p \cdot (n-2) - 4}{4\left(C_f + q^a h_a\right)}}\right) \mathbf{e}^\alpha = 0$$

$$\left(\sqrt{\frac{C_f + q^a h_a}{p \cdot (n-1)}} \cdot \mathbf{R}_\beta - \mathbf{X} \cdot q^b h_{b,\beta} \cdot \sqrt{\frac{p \cdot (n-2) - 4}{4\left(C_f + q^a h_a\right)}}\right) \mathbf{e}^\beta = 0$$

$$\Rightarrow \begin{cases} \left(2 \cdot \left(C_f + q^a h_a\right)\sqrt{\dfrac{1}{p \cdot (n-1)\left(p \cdot (n-2) - 4\right)}} \cdot \mathbf{R}_\alpha + \mathbf{X} \cdot q^a h_{a,\alpha}\right) \mathbf{e}^\alpha = 0 \\[4mm] \left(2 \cdot \left(C_f + q^a h_a\right)\sqrt{\dfrac{1}{p \cdot (n-1)\left(p \cdot (n-2) - 4\right)}} \cdot \mathbf{R}_\beta - \mathbf{X} \cdot q^b h_{b,\beta}\right) \mathbf{e}^\beta = 0. \end{cases}$$

(1327)

Likewise, one could perform the factorization as follows:

$$\Delta \Psi = 0 = \left(A^2 \cdot R_{\alpha\beta} - q^a q^b h_{a,\alpha} h_{b,\beta} \cdot C^2\right) \overbrace{g^{\alpha\beta}}^{\mathbf{e}^\alpha \cdot \mathbf{e}^\beta} = A^2 \cdot R - q^a q^b h_{a,\alpha} h_{b,\beta} \cdot C^2 \mathbf{e}^\alpha \cdot \mathbf{e}^\beta$$

$$\Rightarrow A^2 \cdot R \equiv A^2 \cdot \mathbf{R}_\mathbf{X} \cdot \mathbf{R}_\mathbf{X} = q^a q^b h_{a,\alpha} h_{b,\beta} \cdot C^2 \mathbf{e}^\alpha \cdot \mathbf{e}^\beta \Rightarrow \begin{cases} A \cdot \mathbf{R}_\mathbf{X} = q^a h_{a,\alpha} \cdot C \cdot \mathbf{e}^\alpha \\ A \cdot \mathbf{R}_\mathbf{X} = q^b h_{b,\beta} \cdot C \cdot \mathbf{e}^\beta \end{cases}$$

$$\xrightarrow{F[\Psi]=(C_f+\Psi)^p}$$

$$\Rightarrow \begin{cases} 2 \cdot \left(C_f + q^a h_a\right)\sqrt{\dfrac{1}{p \cdot (n-1)\left(p \cdot (n-2) - 4\right)}} \cdot \mathbf{R}_\mathbf{X} = q^a h_{a,\alpha} \cdot \mathbf{e}^\alpha \\[4mm] 2 \cdot \left(C_f + q^a h_a\right)\sqrt{\dfrac{1}{p \cdot (n-1)\left(p \cdot (n-2) - 4\right)}} \cdot \mathbf{R}_\mathbf{X} = q^b h_{b,\beta} \cdot \mathbf{e}^\beta. \end{cases}$$

(1328)

As we already pointed out earlier in this chapter that the base vector separation is not the only option, we refrain from discussing these new metric Dirac equations here, but intend to use a similar form, one can extract via the vectorial factorization. This we will do in the next sub-section.

9.4.8 From Tetrads or Base Vectors to Real Vectors

When applying the vectorial factorization as introduced in [15] and demonstrated here in sub-section "9.3.10 More Discussion and a Bit Redundancy" and sub-section "9.3.10.2 Decomposition of the metric Eq. (1184) $\rightarrow$ generalized metric Dirac," we also considered an example in 4 dimensions. We shall use this here and by adjusting our vector as follows:

$$\mathbf{V}_\Omega = \left\{ \begin{array}{c} a_0 \cdot a + b_0 \cdot b + c_0 \cdot c + d_0 \cdot d, \\ b_1 \cdot b + c_1 \cdot c + d_1 \cdot d, \\ c_2 \cdot c + d_2 \cdot d, \\ d_3 \cdot d \end{array} \right\}$$

$$\equiv \left[a_j \cdot a + b_j \cdot b + c_j \cdot c + d_j \cdot d \right]_\Omega; \quad a = \Psi_{,0}; \quad b = \Psi_{,1}; \quad a = \Psi_{,2}; \quad a = \Psi_{,3},$$

(1329)

we only need to find suitable coefficients a_j, b_j, c_j, and d_j so that the scalar product:

$$\mathbf{V}_\Omega \cdot \mathbf{V}_\Omega = \left\{ \left(\begin{array}{c} a_0 \cdot a + b_0 \cdot b + c_0 \cdot c + d_0 \cdot d, \\ b_1 \cdot b + c_1 \cdot c + d_1 \cdot d, \\ c_2 \cdot c + d_2 \cdot d, \\ d_3 \cdot d \end{array} \right) \cdot \left(\begin{array}{c} a_0 \cdot a + b_0 \cdot b + c_0 \cdot c + d_0 \cdot d, \\ b_1 \cdot b + c_1 \cdot c + d_1 \cdot d, \\ c_2 \cdot c + d_2 \cdot d, \\ d_3 \cdot d \end{array} \right) \right\}$$

(1330)

gives:

$$\mathbf{V}_\Omega \cdot \mathbf{V}_\Omega = \Psi_{,\alpha} \Psi_{,\beta} g^{\alpha\beta}.$$

(1331)

In order to emphasize the vector character, we this time wrote the symbol $\mathbf{V}_\Omega$ in bold letter. The subsequent equations for the determination of the constants a_j, b_j, c_j, and d_j can be written as:

$$a_0^2 = g^{00}, \quad b_0^2 + b_1^2 = g^{11}, \quad c_0^2 + c_1^2 + c_2^2 = g^{22}, \quad d_0^2 + d_1^2 + d_2^2 + d_3^2 = g^{33},$$

$$a_0 \cdot b_0 = g^{01}, \quad a_0 \cdot c_0 = g^{02}, \quad a_0 \cdot d_0 = g^{03},$$

$$b_0 \cdot c_0 + b_1 \cdot c_1 = g^{12}, \quad b_0 \cdot d_0 + b_1 \cdot d_1 = g^{13}, \quad c_0 \cdot d_0 + c_1 \cdot d_1 + c_2 \cdot d_2 = g^{23}.$$

(1332)

We obtain 16 solutions to the system of linear equations (1332) with respect to our a_j, b_j, c_j, and d_j and thus, 16 vectors $\mathbf{V}_\Omega$.

Inserting this into Eq. (1322), we find:

$$\Delta\Psi = \Delta\left(q^a h_a\right) = 0 = \overbrace{\frac{F}{F'\cdot(n-1)}}^{A^2} \xrightarrow{R^\cdot = 0} R - \mathbf{V}_\Omega \cdot \mathbf{V}_\Omega \overbrace{\left(\frac{F'}{F}\cdot\frac{n-6}{4}+\frac{F''}{F'}\right)}^{C^2}$$

$$= A^2 \cdot R - \mathbf{V}_\Omega \cdot \mathbf{V}_\Omega \cdot C^2. \tag{1333}$$

From the last line of the equation above, we extract the square root as follows:

$$\Delta\Psi = 0 = A^2 \cdot R - \mathbf{V}_\Omega \cdot \mathbf{V}_\Omega \cdot C^2$$

$$\Rightarrow \qquad A^2 \cdot R \equiv A^2 \cdot \mathbf{R}_X \cdot \mathbf{R}_X = \mathbf{V}_\Omega \cdot \mathbf{V}_\Omega \cdot C^2 \Rightarrow \begin{cases} A \cdot \mathbf{R}_X = C \cdot \mathbf{V}_\Omega \\ A \cdot \mathbf{R}_X = C \cdot \mathbf{V}_\Omega \end{cases}$$

$$\Rightarrow \qquad \left\{ 2 \cdot \left(C_f + q^a h_a\right) \right\} \xrightarrow{F[\Psi] = \left(C_f + \Psi\right)^p} \sqrt{\frac{1}{p \cdot (n-1)\left(p \cdot (n-2)-4\right)}} \cdot \mathbf{R}_X = C \cdot \mathbf{V}_\Omega. \tag{1334}$$

Thereby, the object $\mathbf{R}_X$ has to have the same rank as the $\mathbf{V}_\Omega$-vector.

In the case of an example based on metrics, being purely diagonal, we find that one of the solutions to Eq. (1332) would simply be:

$$\mathbf{V}_\Omega = \left\{ \begin{array}{c} \sqrt{g^{00}} \cdot \Psi_{,0} \\ \sqrt{g^{11}} \cdot \Psi_{,1} \\ \sqrt{g^{22}} \cdot \Psi_{,2} \\ \sqrt{g^{33}} \cdot \Psi_{,3} \end{array} \right\}. \tag{1335}$$

We immediately see that this cannot generally lead to a proper system of equations in Eq. (1334), at least not as long as the curvature vector $\mathbf{R}_X$ is been considered a constant vector, meaning, of course, a vector of constants.

However, as we have seen in connection with the vectorial root factorization (or "Vectorial Root Extraction" in the corresponding sub-section 6.12.2), we could even have virtual Ricci vectors $\mathbf{R}_X$ summing up to vanishing total Ricci R curvature in the scalar product $\mathbf{R}_X \cdot \mathbf{R}_X$. Please see the example in sub-section 6.12.2, where we have introduced "virtual" parameters, which can be incorporated into the $\mathbf{V}_\Omega$-vector as follows:

$$V_\Omega = \left\{ a+b+c, a+b-c, a-b+i\cdot c, a-b-i\cdot c \right\} \equiv \left[a \pm b \pm I \cdot c\right]_\Omega. \tag{1336}$$

Here the parameter c is been considered virtual, because, even though it clearly shows up in the vector, it vanishes in subsequent scalar product:

$$V_\Omega \cdot V_\Omega = a^2 + b^2. \tag{1337}$$

Thus, with this fundamental degree of freedom, of course, we can construct reasonable systems of differential equations out of the metric Dirac equations as obtained in the second line of Eq. (1334). We even see the connection to the classical form (see Eq. (1338) below). There, namely, the gamma matrices organize the derivatives of Ψ into a matrix system and the function comes in connection with the mass-term. In our metric form, both function and derivatives are structured in vectors.

$$\text{classical:} \quad 0 = \left(\gamma^\alpha \partial_\alpha - B \right) \Psi_\lambda$$

$$\text{metrical:} \quad A[F[\Psi]] \cdot R_X[\Psi] = C[F[\Psi]] \cdot V_\Omega[\Psi]. \tag{1338}$$

9.4.9 Example in $n = 6$, $n = 8$ and Higher Dimensions

Taking the example from Eq. (1147) and generalizing it to:

$$g_{\alpha\beta}^6 = \begin{pmatrix} g_{00} & 0 & 0 & 0 & 0 & 0 \\ 0 & g_{11} & 0 & 0 & 0 & 0 \\ 0 & 0 & g_{22} & 0 & 0 & 0 \\ 0 & 0 & 0 & g_{33} & 0 & 0 \\ 0 & 0 & 0 & 0 & \pm c_u^2 & 0 \\ 0 & 0 & 0 & 0 & 0 & \pm c_v^2 \end{pmatrix} \cdot \left(C_f + g[u,v] \cdot f[x_0, x_1, x_2, x_3] \right) \tag{1339}$$

plus using the solutions for $g[u,v]$ from Eqs. (1148) and (1149), we immediately have a classical Klein–Gordon situation, where Eq. (1333) (now in 6 dimensions) results in:

$$\Delta_{4D} \Psi + \left(\mp E_u^2 \mp E_v^2 \right) \Psi = A^2 \cdot R - V_{6\Omega} \cdot V_{6\Omega} \cdot C^2. \tag{1340}$$

Using the simplifications for Gu as introduced above (see $Gu[u]$ from Eq. (1163)) and setting $C_{v1} = C_u$ and $C_{v2} = i^* C_v$, giving us:

$$Gv[v] = C_v \cdot e^{i \cdot c_v \cdot E_v \cdot v}, \tag{1341}$$

we immediately realize that we could further simplify Eq. (1340) as follows:

$$\Delta_{4D} \Psi \mp E_u^2 \Psi \mp E_v^2 \Psi = A^2 \cdot R - \left(V_{4\Omega} \cdot V_{4\Omega} + \left(\mp E_u^2 \mp E_v^2 \right) \Psi^2 \right) \cdot C^2. \tag{1342}$$

We note that we are back with our 4-dimensional $\mathbf{V}_\Omega$-vectors, while in Eq. (1340) they and the $\mathbf{R_X}$ would have to be 6-dimensional. The corresponding solutions to Eq. (1332) in the case of Eq. (1342) would be simply all 16 sign-combinations for the vector:

$$
\mathbf{V}_\Omega = \left\{ \begin{array}{c} \pm\sqrt{g^{00}} \cdot \Psi_{,0} \\ \pm\sqrt{g^{11}} \cdot \Psi_{,1} \\ \pm\sqrt{g^{22}} \cdot \Psi_{,2} \\ \pm\sqrt{g^{33}} \cdot \Psi_{,3} \end{array} \right\}
$$

$$
= \left\{ \begin{bmatrix} \sqrt{g^{00}} \cdot \Psi_{,0} \\ \sqrt{g^{11}} \cdot \Psi_{,1} \\ \sqrt{g^{22}} \cdot \Psi_{,2} \\ \sqrt{g^{33}} \cdot \Psi_{,3} \end{bmatrix}, \begin{bmatrix} -\sqrt{g^{00}} \cdot \Psi_{,0} \\ \sqrt{g^{11}} \cdot \Psi_{,1} \\ \sqrt{g^{22}} \cdot \Psi_{,2} \\ \sqrt{g^{33}} \cdot \Psi_{,3} \end{bmatrix}, \ldots, \begin{bmatrix} -\sqrt{g^{00}} \cdot \Psi_{,0} \\ -\sqrt{g^{11}} \cdot \Psi_{,1} \\ -\sqrt{g^{22}} \cdot \Psi_{,2} \\ \sqrt{g^{33}} \cdot \Psi_{,3} \end{bmatrix}, \begin{bmatrix} -\sqrt{g^{00}} \cdot \Psi_{,0} \\ -\sqrt{g^{11}} \cdot \Psi_{,1} \\ -\sqrt{g^{22}} \cdot \Psi_{,2} \\ -\sqrt{g^{33}} \cdot \Psi_{,3} \end{bmatrix} \right\}.
$$

$$(1343)$$

We realize that any number of dimensions could lead us to a similar structure with more parameters. For instance, in 8 dimensions with the metric:

$$
g^8_{\alpha\beta} = \begin{pmatrix} g_{00} & 0 & 0 & 0 & \cdots & 0 \\ 0 & g_{11} & 0 & 0 & \cdots & 0 \\ 0 & 0 & g_{22} & 0 & \cdots & 0 \\ 0 & 0 & 0 & g_{33} & \cdots & 0 \\ \cdots & \cdots & \cdots & \cdots & \ddots & 0 \\ 0 & 0 & 0 & 0 & 0 & g_{77} \end{pmatrix}
$$

$$
g_{44} = \pm c_4^2; \quad g_{55} = \pm c_5^2; \quad g_{66} = \pm c_6^2; \quad g_{77} = \pm c_7^2. \tag{1344}
$$

This extends Eq. (1342) to:

$$
\Delta_{4D}\Psi - \left(\pm E_4^2 \pm E_5^2 \pm E_6^2 \pm E_7^2 \ldots \right)\Psi
$$

$$
= A^2 \cdot R - \left(\mathbf{V}_{4\Omega} \cdot \mathbf{V}_{4\Omega} - \left(\pm E_4^2 \pm E_5^2 \pm E_6^2 \pm E_7^2 \ldots \right)\Psi^2 \right) \cdot C^2, \tag{1345}
$$

where we have assumed a separation approach

$$
\Psi = f\left[x_0, x_1, x_2, x_3\right] \cdot \prod_{k=4}^{n} G_k\left[x_k\right]:
$$

$$G_k\left[x_k\right] = C_k \cdot e^{i \cdot c_k \cdot E_k \cdot x_k} . \tag{1346}$$

We already hinted via the "..."-signs in Eq. (1345) that we can arbitrarily extend the number of parameters E_k with additional dimensions, diagonal metric structures as in Eq. (1344) and functions of the type Eq. (1346).

Obviously, we get into trouble when simultaneously demanding to have Laplace functions in the n-dimensional space-time, leading to:

$$\Delta_{4D}\Psi - \left(\pm E_4^2 \pm E_5^2 \pm E_6^2 \pm E_7^2 ...\right)\Psi = 0 \tag{1347}$$

and thus, Klein–Gordon solutions in the subsequent (remaining) 4 dimensions, and $R = 0$. This namely leaves us with:

$$0 = \left(\mathbf{V}_{4\Omega} \cdot \mathbf{V}_{4\Omega} - \left(\pm E_4^2 \pm E_5^2 \pm E_6^2 \pm E_7^2 ...\right)\Psi^2\right) \cdot C^2 \tag{1348}$$

and it does not occur as an easy task to satisfy this equation with just Klein–Gordon solutions (namely the solutions from Eq. (1347) at hand. Of course, we could just demand $C^2 = 0$, which can be achieved via Eq. (1175) and its subsequent solution in Eq. (1176) or simply:

$$F[\Psi = f] = C_2 \cdot \left(f + C_1\right)^{\frac{4}{n-2}} , \tag{1349}$$

but this is not the generality we are aiming for in the chapter here.

It looks like that, in the case of effectively, which is to say R^*-, flat space-times, we have to give up the Dirac idea of sticking to Klein–Gordon solutions in connection with also R-flat space-times. In fact, we may just see the Klein–Gordon solutions as one option allowing us the extraction of curvature situations for various $F[f]$-settings via:

$$\Delta_{4D}\Psi - \left(\pm E_4^2 \pm E_5^2 \pm E_6^2 \pm E_7^2 ...\right)\Psi = 0$$

$$\Rightarrow \qquad A^2 \cdot R = \left(\frac{\Psi_{,\alpha}\Psi_{,\beta}g^{\alpha\beta}}{\mathbf{V}_{4\Omega} \cdot \mathbf{V}_{4\Omega} - \left(\pm E_4^2 \pm E_5^2 \pm E_6^2 \pm E_7^2 ...\right)\Psi^2}\right) \cdot C^2 . \tag{1350}$$

For instance, when assuming a flat space-time and spherical spatial symmetry with $\Psi = \Psi[r] = f[r]$ and arbitrary $F[\Psi]$, we would have the first line of Eq. (1350) fulfilled via:

$$0 = \frac{2 \cdot f_r{}'}{r} + f_r{}'' + \overbrace{\left(\mp E_4^2 \mp E_5^2 \mp E_6^2 \mp E_7^2\right)}^{E^2} \cdot f_r \Rightarrow f[r] = \frac{C_{r1} \cdot e^{i \cdot E \cdot r} + C_{r2} \cdot e^{-i \cdot E \cdot r}}{r} .$$

$$\tag{1351}$$

In order to avoid singularities at $r = 0$, we set:

$$C_{r2} = -C_{r1} \Rightarrow f[r] = C_{r1} \cdot \frac{e^{i \cdot E \cdot r} - e^{-i \cdot E \cdot r}}{r} = C_r \cdot \frac{\sin[E \cdot r]}{r}. \quad (1352)$$

Due to the flat-space-condition, we automatically have $R = 0$ and thus should demand:

$$0 = \left(\frac{\overbrace{\Psi_{,\alpha} \Psi_{,\beta} g^{\alpha\beta}}}{V_{4\Omega} \cdot V_{4\Omega}} - \left(\pm E_4^2 \pm E_5^2 \pm E_6^2 \pm E_7^2 \ldots \right) \Psi^2 \right) \cdot c^2. \quad (1353)$$

Not willing to achieve this via an $F[\Psi]$ in accordance with condition in Eq. (1175), we require the satisfaction of:

$$0 = \overbrace{V_{4\Omega} \cdot V_{4\Omega}}^{\Psi_{,\alpha} \Psi_{,\beta} g^{\alpha\beta}} - \left(\pm E_4^2 \pm E_5^2 \pm E_6^2 \pm E_7^2 \ldots \right) \Psi^2. \quad (1354)$$

We see that this is not possible. One way out would be to demand a non-zero total Ricci curvature R^* leading to:

$$A^2 \cdot R^* \cdot F = \left(\frac{\overbrace{\Psi_{,\alpha} \Psi_{,\beta} g^{\alpha\beta}}}{V_{4\Omega} \cdot V_{4\Omega}} - \left(\pm E_4^2 \pm E_5^2 \pm E_6^2 \pm E_7^2 \ldots \right) \Psi^2 \right) \cdot c^2 \quad (1355)$$

or to try and satisfy Eq. (1354) by the means of a series of functions $f[r]$ as obtained in Eq. (1352). Because we intend to avoid the evaluation with the squares in Eq. (1354), we factorize as follows:

$$0 = \overbrace{V_{4\Omega} \cdot V_{4\Omega}}^{\Psi_{,\alpha} \Psi_{,\beta} g^{\alpha\beta}} - \overbrace{\left(\mp E_4^2 \mp E_5^2 \mp E_6^2 \mp E_7^2 \right)}^{E^2} \Psi^2 \Rightarrow V_{4\Omega} \cdot V_{4\Omega} = E^2 \Psi^2$$

$$\Rightarrow \qquad V_{4\Omega} = E \cdot \Psi. \quad (1356)$$

Please note that, for the reason of continuity and recognition, we kept the symbol of the 4-dimensional $V_{4\Omega}$-vector even though in this reduced case here, it would only be $V_{r\Omega}$-vector. We found that the constant E needs to become a vector $\mathbf{E}$. Plugging the solution for the Klein–Gordon part from Eq. (1352) into Eq. (1354) or directly into the last line of Eq. (1356), we find that we result in an impossible "equation" as follows:

$$\frac{\sin[E \cdot r]}{E \cdot r} \rightarrow \overset{!?!}{=} \leftarrow \cos[E \cdot r] + \sin[E \cdot r]. \quad (1357)$$

So, we ask, is there another way to find general solutions to generally Ricci-flat $(R^* = R = 0)$ space-times of spatial spherical symmetry with wave functions only depending on the radius r?

Applying the results from the sub-section "9.3.11.3 An asymmetry," we try to find a function F as follows:

$$\frac{4FF'' + \left(F'\right)^2 (n-6)}{4F} = \frac{F'}{\Psi} \quad \Rightarrow \quad F = \begin{cases} \left(C_{f0}^2 \pm \Psi^2\right)^{\frac{4}{n-2}} \cdot C_{f1} & n > 2 \\ e^{\pm C_{f0}^2 \cdot \Psi^2} \cdot C_{f1} & n = 2. \end{cases} \tag{1358}$$

From Eq. (1008), we then obtain:

$$R^* = R^*_{\alpha\beta} G^{\alpha\beta} = \left(R + (1-n) \cdot \frac{F'}{F} \left(\Delta\Psi + \frac{1}{\Psi} \cdot \Psi_{,\alpha} \Psi_{,\beta} g^{\alpha\beta} \right) \right) \frac{1}{F}$$

$$= \left(R + (1-n) \cdot \frac{F'}{\Psi \cdot F} \left(\frac{\Psi}{\sqrt{g}} \partial_\alpha \sqrt{g} \cdot g^{\alpha\beta} \partial_\beta \Psi + \Psi_{,\alpha} \Psi_{,\beta} g^{\alpha\beta} \right) \right) \frac{1}{F}$$

$$= \left(R + (1-n) \cdot \frac{F'}{\Psi \cdot F} \left(\frac{1}{\sqrt{g}} \partial_\alpha \Psi \cdot \sqrt{g} \cdot g^{\alpha\beta} \partial_\beta \Psi \right) \right) \frac{1}{F}$$

$$\xrightarrow{\quad R^* = R^*_{\alpha\beta} G^{\alpha\beta} = 0 \quad}$$

$$\Rightarrow \qquad R \cdot \frac{\Psi \cdot F}{(n-1) \cdot F'} = \frac{1}{\sqrt{g}} \partial_\alpha \Psi \cdot \sqrt{g} \cdot g^{\alpha\beta} \partial_\beta \Psi$$

$$\xrightarrow{\quad \frac{4FF'' + \left(F'\right)^2 (n-6)}{4F} = \frac{F'}{\Psi} \quad}$$

$$\Rightarrow \qquad \frac{1}{\sqrt{g}} \partial_\alpha \Psi \cdot \sqrt{g} \cdot g^{\alpha\beta} \partial_\beta \Psi = R \cdot \begin{cases} \frac{1}{8} \cdot \frac{2-n}{n-1} \cdot \left(C_{f0}^2 \pm \Psi^2 \right) & n > 2 \\ \pm \frac{1}{2} \cdot C_{f0}^2 & n = 2. \end{cases} \tag{1359}$$

Now we substitute as follows:

$$\partial_\beta \Psi = \frac{g_{\beta\gamma} h^\gamma}{\Psi \cdot \sqrt{g}}. \tag{1360}$$

Assuming $R = 0$ and only r-dependency as defined in our problem considered above, we would have to solve the simple differential equation:

$$\Rightarrow \qquad \frac{1}{\sqrt{g}} \partial_\alpha \Psi \cdot \sqrt{g} \cdot g^{\alpha\beta} \partial_\beta \Psi = 0$$

$$\Rightarrow \quad \frac{1}{\sqrt{g}} \partial_\alpha \Psi \cdot \sqrt{g} \cdot g^{\alpha\beta} \frac{g_{\beta\gamma} h^\gamma}{\Psi \cdot \sqrt{g}} = \frac{1}{\sqrt{g}} \partial_\alpha g^{\alpha\beta} g_{\beta\gamma} h^\gamma = \frac{1}{\sqrt{g}} \partial_\alpha \delta^\alpha_\gamma h^\gamma = \frac{1}{\sqrt{g}} \partial_\alpha h^\alpha = 0$$

$$\Rightarrow \qquad \partial_\alpha h^\alpha = 0 \quad \Rightarrow \quad \partial_r h^r = \partial_r h[r] = 0 \quad \Rightarrow \quad h[r] = \text{const} = C_r. \qquad (1361)$$

The corresponding wave function can now be obtained from Eq. (1360) via:

$$\int \Psi \cdot \partial_\beta \Psi \, dx^\beta = \frac{1}{2} \cdot \Psi^2 = \int \frac{g_{\beta\gamma} h^\gamma}{\sqrt{g}} \, dx^\beta = \int \frac{h^r}{r^2} \, dr = \int \frac{C_r}{r^2} \, dr = C_r \left(C_{rr} - \frac{1}{r} \right)$$

$$\Rightarrow \qquad \Psi = C_{r0} \cdot \sqrt{C_{r1} - \frac{1}{r}}. \qquad (1362)$$

Things are getting a bit more interesting in the case of a radius-time dependency. Then Eq. (1361) results in:

$$\partial_\alpha h^\alpha = 0 \quad \Rightarrow \quad \partial_t h^t + \partial_r h^r = 0. \qquad (1363)$$

Choosing the approach:

$$h^t = -p[t,r]_{,r} \, ; \quad h^r = p[t,r]_{,t} \qquad (1364)$$

with yet arbitrary functions $p[t,r]$ gives us the desired result. However, when plugging this into the first line of Eq. (1362) we see that it results in:

$$\frac{1}{2} \cdot \Psi^2 = \int \frac{g_{\beta\gamma} h^\gamma}{\sqrt{g}} \, dx^\beta = \int \frac{h^r \, dr - c^2 \cdot h^t \, dt}{r^2} = \int \frac{p[t,r]_{,t} \, dr + c^2 \cdot p[t,r]_{,r} \, dt}{r^2}$$

$$(1365)$$

and leaves us with functions $p_t[t]$ and $p_r[r]$ as "integrational constants." Thus, the substitution (1360) does not truly simplify our life. Nevertheless, we want to generalize the approach.

9.4.10 The Substitution Approach

Now we intend to generalize the technique, we introduced at the end of the last sub-section.

We see that such a generalization is possible via an arbitrary function

$$H[f = \Psi] = \frac{\partial \Phi'[f]}{\partial f} = \Phi'[f] = \Phi' \text{ with the following condition:}$$

$$\frac{4FF'' + (F')^2 (n-6)}{4F} = F' \cdot \frac{\Phi''}{H = \Phi'}$$

$$\Rightarrow \qquad F = \begin{cases} \left(C_{f0}^2 - \int\limits_1^f \Phi'[\phi] \cdot \frac{2-n}{4} \cdot d\phi \right)^{\frac{4}{n-2}} \cdot C_{f1} & n > 2 \\ e^{C_{f0}^{-2} \cdot \int\limits_1^f \Phi'[\phi] \cdot d\phi} \cdot C_{f1} & n = 2. \end{cases} \qquad (1366)$$

This changes Eq. (1008) to:

$$R^* = R^*_{\alpha\beta} G^{\alpha\beta} = \left(R + (1-n) \cdot \frac{F'}{F} \left(\Delta f + \frac{\Phi''}{\Phi'} \cdot f_{,\alpha} f_{,\beta} g^{\alpha\beta} \right) \right) \frac{1}{F}$$

$$= \left(R + (1-n) \cdot \frac{F'}{\Phi' \cdot F} \left(\frac{\Phi'}{\sqrt{g}} \partial_\alpha \left(\sqrt{g} \cdot g^{\alpha\beta} \partial_\beta f \right) + \Phi'' \cdot f_{,\alpha} f_{,\beta} g^{\alpha\beta} \right) \right) \frac{1}{F}$$

$$= \left(R + (1-n) \cdot \frac{F'}{\Phi' F} \left(\frac{1}{\sqrt{g}} \partial_\alpha \left(\Phi' \cdot \sqrt{g} \cdot g^{\alpha\beta} \partial_\beta f \right) \right) \right) \frac{1}{F}$$

$$\xrightarrow{\quad R^* = R^*_{\alpha\beta} G^{\alpha\beta} = 0 \quad}$$

$$\Rightarrow \qquad R \cdot \frac{\Phi' \cdot F}{(n-1) \cdot F'} = \frac{1}{\sqrt{g}} \partial_\alpha \left(\Phi' \cdot \sqrt{g} \cdot g^{\alpha\beta} \partial_\beta f \right) \equiv \Omega\{\Phi'[f]f\}$$

$$\xrightarrow{\quad \frac{4FF'' + (F')^2 (n-6)}{4F} = \frac{\Phi''}{\Phi'} \cdot F' \quad}$$

$$\Rightarrow \qquad \frac{1}{\sqrt{g}} \partial_\alpha \Phi' \cdot \sqrt{g} \cdot g^{\alpha\beta} \partial_\beta f = R \cdot \frac{1}{n-1} \cdot \begin{cases} \left(C_{f0}^2 - \frac{2-n}{4} \int\limits_1^f \Phi'[\phi] \cdot d\phi \right) & n > 2 \\ C_{f0}^2 & n = 2. \end{cases}$$

$$(1367)$$

Now we can try to solve this equation via the following substitution:

$$R \cdot \frac{1}{n-1} \cdot \left\{ \begin{array}{l} \left(C_{f0}^2 - \frac{2-n}{4} \cdot \int\limits_1^f \Phi'[\phi] \cdot d\phi \right) \quad n > 2 \\[3mm] C_{f0}^2 \quad n = 2 \end{array} \right\} = \frac{1}{\sqrt{g}} \partial_\alpha \Phi' \cdot \sqrt{g} \cdot g^{\alpha\beta} \partial_\beta f$$

$$\xrightarrow{\partial_\beta f = \frac{g_{\beta\gamma} h^\gamma}{\Phi' \cdot \sqrt{g}}}$$

$$= \frac{1}{\sqrt{g}} \partial_\alpha g_{\beta\gamma} g^{\alpha\beta} h^\gamma = \frac{1}{\sqrt{g}} \partial_\alpha \delta_\gamma^\alpha h^\gamma = \frac{1}{\sqrt{g}} \partial_\alpha h^\alpha$$

$$= R \cdot \frac{1}{n-1} \cdot \left\{ \begin{array}{l} \left(C_{f0}^2 - \frac{2-n}{4} \cdot \int\limits_1^{f=\int \frac{g_{\beta\gamma} h^\gamma}{\Phi' \cdot \sqrt{g}} dx^\beta} \Phi'[\phi] \cdot d\phi \right) \quad n > 2 \\[3mm] C_{f0}^2 \quad n = 2 \end{array} \right\}. \tag{1368}$$

We see that in the case of $R = 0$, we obtain the following significant simplification:

$$\frac{1}{\sqrt{g}} \partial_\alpha h^\alpha = 0 \quad \Rightarrow \quad \partial_\alpha h^\alpha = 0 \tag{1369}$$

and can obtain the function f via the integral(s):

$$\int \Phi' \cdot f_\beta \, dx^\beta = \int \frac{g_{\beta\gamma} h^\gamma}{\sqrt{g}} dx^\beta \quad \Rightarrow \quad \Phi[f] = \int \frac{g_{\beta\gamma} h^\gamma}{\sqrt{g}} dx^\beta$$

$$\Rightarrow \qquad f = \Phi^{\{-1\}}[f] = \left[\int \frac{g_{\beta\gamma} h^\gamma}{\sqrt{g}} dx^\beta \right]^{\{-1\}}. \tag{1370}$$

9.4.11 Metric Dirac Equation—An Outlook

We assume that the application of vectors $\mathbf{f_a}$ instead of scalar functions f does not change the principal structure of Eq. (1008) with respect to f. We adapt the metric scale variation as follows from Eq. (1011):

$$G_{\alpha\beta} = q_\alpha f^a \cdot g_{\alpha\beta} \tag{1371}$$

and

$$G^{\alpha\beta} = q^a f_a g^{\alpha\beta}. \tag{1372}$$

With the condition:

$$G^{\chi\gamma}G_{\delta\gamma} = \delta_\delta^\chi \quad \Rightarrow \quad f_a f^a = 1; \quad q^\chi q_\delta = \delta_\delta^\chi \tag{1373}$$

we see that the function f-vectors and the q-vectors have to behave like tetrads. Nevertheless, we do not explicitly use bold letters here to emphasize this fact, but just try to remember it throughout the rest of the chapter (and later on; c.f. [23]). A more thorough investigation will follow elsewhere [23] (part of this will also be presented in chapter 11 of this book). Our goal shall be to find *q-f*-combinations where the $f_{a,c}f_{a,\beta}g^{\alpha\beta}$ operator directly vanishes due to a suitable summation of derivatives to the **f**-vector components (similar to the Galerkin-ansatz in linear elasticity). Thereby demanding all *f*-components to be Laplace equations would immediately solve Eq. (1008) in the case $R^* = R$ = 0. Otherwise, we are back at the factorization problem as discussed with respect to the "9.4.5 Decomposition of the Ricci Scalar and Tensor."

One may see the scale-adaptation of the metric simply as some other kind of variation. That this new type of variation now led us to fundamental quantum equations via a simple effective Ricci-flat space-time condition $R^* = 0$ prompts the conclusion that the latter condition itself is a rather fundamental one and that the universe works always very hard—which is to say via a multitude of variational forms—to reach the $R^* = 0$-goal. We immediately see that there are many options. For instance, one may also set:

$$G_{\alpha\beta} = f_\alpha^a \cdot g_{\alpha\beta} \tag{1374}$$

and

$$G^{\alpha\beta} = \overline{f}_a^\alpha g^{\alpha\beta}, \tag{1375}$$

with the condition:

$$G^{\chi\gamma}G_{\delta\gamma} = \delta_\delta^\chi \quad \Rightarrow \quad f_\delta^a \overline{f}_a^\chi = \delta_\delta^\chi. \tag{1376}$$

One obvious choice in connection with metrically producing Dirac equations would probably consist of a function vector and an object of rank three as follows:

$$G_{\alpha\beta} = \zeta_{\alpha\beta}^a f^b \cdot g_{ab} \tag{1377}$$

and

$$G^{\alpha\beta} = \zeta_a^{\alpha\beta} f_b g^{ab}. \tag{1378}$$

The subsequent condition would then be:

$$G^{\alpha\gamma}G_{\gamma\beta} = \zeta_a^{\alpha\gamma} f_b g^{ab} \zeta_{\gamma\beta}^c f^d \cdot g_{cd} = \zeta_a^{\alpha\gamma} f^a \zeta_{\gamma\beta}^c f_c = \delta_\beta^\alpha. \tag{1379}$$

Similarly, one may also take:

$$G_{\alpha\beta} = \zeta_{\alpha b}^{a} f^{b} \cdot g_{\alpha\beta} \tag{1380}$$

and

$$G^{\alpha\beta} = \zeta_{a}^{\alpha b} f_{b} g^{\alpha\beta}, \tag{1381}$$

and the condition would then be:

$$G^{\alpha\gamma} G_{\gamma\beta} = \zeta_{a}^{\alpha b} f_{b} g^{\alpha\gamma} \zeta_{\gamma d}^{c} f^{d} \cdot g_{c\beta} = \zeta^{\alpha b\gamma} f_{b} \zeta_{\gamma d\beta} f^{d} = \delta_{\beta}^{\alpha}. \tag{1382}$$

Only giving the co-variant form, one may also insist on completely symmetric scaled (adapted) metrics and sums up Eqs. (1377), (1378), (1380), and (1381) and all other options, which are reading:

$$G_{\alpha\beta} = \begin{Bmatrix} \zeta_{\alpha a}^{a} f^{b} g_{b\beta}, \zeta_{\beta a}^{a} f^{b} g_{\alpha b}, \zeta_{\alpha b}^{a} f^{b} g_{\alpha\beta}, \zeta_{\beta b}^{a} f^{b} g_{\alpha a}, \\ \zeta_{\alpha\beta}^{a} f^{b} g_{ab}, \zeta_{\beta\alpha}^{a} f^{b} g_{ab}, \zeta_{\alpha b}^{a} f^{b} g_{\alpha\beta}, \dots \end{Bmatrix}, \tag{1383}$$

and so on, as follows:

$$G_{\alpha\beta} = \frac{1}{24} \begin{pmatrix} \zeta_{\alpha a}^{a} f^{b} g_{b\beta} + \zeta_{\beta a}^{a} f^{b} g_{\alpha b} + \zeta_{\alpha b}^{a} f^{b} g_{\alpha\beta} + \zeta_{\beta b}^{a} f^{b} g_{\alpha a} \\ + \zeta_{\alpha\beta}^{a} f^{b} g_{ab} + \zeta_{\beta\alpha}^{a} f^{b} g_{ab} + \zeta_{\alpha b}^{a} f^{b} g_{\alpha\beta}, \dots \end{pmatrix}. \tag{1384}$$

We are going to consider and investigate these options in [23].

9.5 Higgs with $F[f]$

Out of curiosity, we also check the situation in the case of $F[\Psi] = (C_f + \Psi^2)^p$. In this case, the first line of Eq. (1325) would read:

$$\Delta\Psi = \left(\frac{(C_f + \Psi^2)}{2p\Psi(n-1)} \cdot \mathbf{R}_{\alpha} \cdot \mathbf{R}_{\beta} - q^a q^b h_{a,\alpha} h_{b,\beta} \cdot \frac{(2C_f + \Psi^2(p(n-2)-2))}{2\Psi(C_f + \Psi^2)} \right) \mathbf{e}^{\alpha} \cdot \mathbf{e}^{\beta}. \tag{1385}$$

Here now we see, that the assumption of a const Ψ would force a Higgs field F with the solution:

$$C_f + \Psi^2 = 0 \tag{1386}$$

to come into existence in order to satisfy Eq. (1385). Thereby only an exponent p with:

$$p = \frac{4}{n-2} \tag{1387}$$

assures non-singularities at the Nabla terms in the case that there might be small (infinitesimal small) residuals of the first derivatives of the function Ψ. Especially in 4 dimensions, we obtain the expected Mexican hat shape as shown in Fig. 9.11.

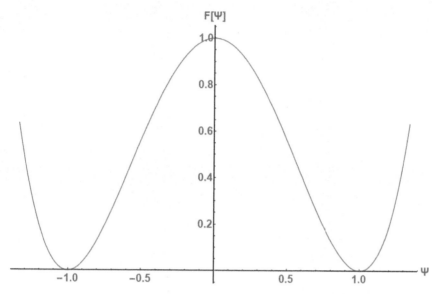

Figure 9.11   Higgs field as resulting from Eq. (1385) for const Ψ in the case of non-vanishing overall Ricci curvature R under the condition of $R^* = 0$ (regarding the definition and evaluation of R^* see Eq. (1008) and chapter 8).

We have found that a condition which demands the effective curvature R^* to vanish produces a Higgs field $F[\Psi]$ for constant (or almost constant) functions Ψ, the moment the fundamental (unscaled) metric $g_{\alpha\beta}$ is of such character that the ordinary (unscaled) Ricci scalar R does not vanish.

9.6   Conclusions Regarding Our "Dirac Trials" Here So Far

We have shown that the Klein–Gordon equation can directly be extracted from a scaled metric via the assumption of an effectively Ricci-flat space-time in n dimensions. Thereby we have assumed the scaling to be of scalar character.

With the Klein–Gordon equation, already falling out of our new approach—which, interpreted generously, could just be seen as another form of variation, by the way—we may just sit comfortably back and conclude that this way we also have obtained the Dirac equation. This namely was classically derived by an operator factorization from the Klein–Gordon equation. Thus, having obtained the latter, automatically gives us the Dirac equation, too. However, when digging deeper (metrically) and also trying to find a completely metric origin and illustrative understanding of the Dirac approach, we have to come to the conclusion that the simple scalar metric scaling with a function $F[f]$ multiplied with the metric does not suffice. We have found that only a vector scaling will probably give us the complete picture. In the next section, we will see how to resolve this entanglement.

9.7 Appendix A: The Derivation of the Scaled Ricci Scalar R^*

In order to evaluate the Ricci scalar of the scaled metric $G_{\alpha\beta} = F[f_a]^* g_{\alpha\beta} = F^* g_{\alpha\beta}$ with a function vector $\mathbf{f_a}$, we need a variety of identities. At first, we require some definitions:

$$G_{\alpha\gamma} G^{\alpha\beta} = \delta_\gamma^\beta; \quad G_{\alpha\beta} = F \cdot g_{\alpha\beta} = F\Big[f_a\big[x_k\big]\Big] \cdot g_{\alpha\beta};$$

$$G^{\alpha\beta} = \frac{g^{\alpha\beta}}{F} = \frac{g^{\alpha\beta}}{F\Big[f_a\big[x_k\big]\Big]}, \tag{1388}$$

$$\Gamma_{\alpha\beta}^{*\gamma} = \frac{g^{\gamma\sigma}}{2 \cdot F[\mathbf{f}]}\left(\Big[F[\mathbf{f}] \cdot g_{\sigma\alpha}\Big]_{,\beta} + \Big[F[\mathbf{f}] \cdot g_{\sigma\beta}\Big]_{,\alpha} - \Big[F[\mathbf{f}] \cdot g_{\alpha\beta}\Big]_{,\sigma}\right)$$

$$= \frac{g^{\gamma\sigma}}{2 \cdot F[\mathbf{f}]}\left(\Big[F[\mathbf{f}] \cdot g_{\sigma\alpha}\Big]_{,\beta} + \Big[F[\mathbf{f}] \cdot g_{\sigma\beta}\Big]_{,\alpha} - \Big[F[\mathbf{f}] \cdot g_{\alpha\beta}\Big]_{,\sigma}\right)$$

$$= \frac{g^{\gamma\sigma}}{2}\left(g_{\sigma\alpha,\beta} + g_{\sigma\beta,\alpha} - g_{\alpha\beta,\sigma}\right) + \frac{g^{\gamma\sigma}\mathbf{F'}}{2 \cdot F[\mathbf{f}]}\left(\mathbf{f}_{,\beta} \cdot g_{\sigma\alpha} + \mathbf{f}_{,\alpha} \cdot g_{\sigma\beta} - \mathbf{f}_{,\sigma} \cdot g_{\alpha\beta}\right)$$

$$= \Gamma_{\alpha\beta}^\gamma + \frac{g^{\gamma\sigma}\mathbf{F'}}{2 \cdot F[\mathbf{f}]}\left(\mathbf{f}_{,\beta} \cdot \delta_\alpha^\gamma + \mathbf{f}_{,\alpha} \cdot \delta_\beta^\gamma - \mathbf{f}_{,\sigma} \cdot g_{\alpha\beta}\right)$$

$$\equiv \Gamma_{\alpha\beta}^\gamma + \Gamma_{\alpha\beta}^{**\gamma}$$

$$\mathbf{f} = \left\{ f_0, f_1, f_2, \ldots \right\}; \quad \mathbf{F}' = \frac{\partial F}{\partial \mathbf{f}} = \left\{ \frac{\partial F}{\partial f_0}, \frac{\partial F}{\partial f_1}, \frac{\partial F}{\partial f_2}, \ldots \right\}^T = \left\{ \begin{array}{c} \dfrac{\partial F}{\partial f_0} \\[2mm] \dfrac{\partial F}{\partial f_1} \\[2mm] \dfrac{\partial F}{\partial f_2} \\[2mm] \ldots \end{array} \right\}, \quad (1389)$$

$$\mathbf{F}'' = \left\{ \begin{array}{c} \dfrac{\partial^2 F}{\partial f_0^2}, \dfrac{\partial^2 F}{\partial f_0 \partial f_1}, \dfrac{\partial^2 F}{\partial f_0 \partial f_2}, \ldots \\[3mm] \dfrac{\partial^2 F}{\partial f_1 \partial f_0}, \dfrac{\partial^2 F}{\partial f_1^2}, \dfrac{\partial^2 F}{\partial f_1 \partial f_2}, \ldots \\[3mm] \dfrac{\partial^2 F}{\partial f_2 \partial f_0}, \dfrac{\partial^2 F}{\partial f_2 \partial f_1}, \dfrac{\partial^2 F}{\partial f_2^2}, \ldots \\[3mm] \ldots \end{array} \right\} \quad \left(\mathbf{F}' \right)^2 = \left\{ \begin{array}{c} \left(\dfrac{\partial F}{\partial f_0} \right)^2, \dfrac{\partial F}{\partial f_0} \dfrac{\partial F}{\partial f_1}, \dfrac{\partial F}{\partial f_0} \dfrac{\partial F}{\partial f_2}, \ldots \\[3mm] \dfrac{\partial F}{\partial f_1} \dfrac{\partial F}{\partial f_0}, \left(\dfrac{\partial F}{\partial f_1} \right)^2, \dfrac{\partial F}{\partial f_1} \dfrac{\partial F}{\partial f_2}, \ldots \\[3mm] \dfrac{\partial F}{\partial f_2} \dfrac{\partial F}{\partial f_0}, \dfrac{\partial F}{\partial f_2} \dfrac{\partial F}{\partial f_1}, \left(\dfrac{\partial F}{\partial f_2} \right)^2, \ldots \\[3mm] \ldots \end{array} \right\}, \quad (1390)$$

and then the necessary terms for the evaluation for the Ricci scalar:

$$\Gamma^{*\gamma}_{\alpha\beta} = \Gamma^{\gamma}_{\alpha\beta} + \frac{g^{\gamma\sigma}\mathbf{F}'}{2 \cdot F[\mathbf{f}]} \left(\mathbf{f}_{,\beta} \cdot \delta^{\gamma}_{\alpha} + \mathbf{f}_{,\alpha} \cdot \delta^{\gamma}_{\beta} - \mathbf{f}_{,\sigma} \cdot g_{\alpha\beta} \right)$$

$$R^* = R^*_{\ \alpha\beta} G^{\alpha\beta} = \left(\Gamma^{*\sigma}_{\alpha\beta,\sigma} - \Gamma^{*\sigma}_{\beta\sigma,\alpha} - \Gamma^{*\mu}_{\sigma\alpha} \Gamma^{*\sigma}_{\beta\mu} + \Gamma^{*\sigma}_{\alpha\beta} \Gamma^{*\mu}_{\sigma\mu} \right) G^{\alpha\beta}, \quad (1391)$$

$$\Gamma^{*\sigma}_{\alpha\beta} = \Gamma^{\sigma}_{\alpha\beta} + \frac{\mathbf{F}'}{2 \cdot F[\mathbf{f}]} \left(\mathbf{f}_{,\beta} \cdot \delta^{\sigma}_{\alpha} + \mathbf{f}_{,\alpha} \cdot \delta^{\sigma}_{\beta} - \mathbf{f}_{,\gamma} \cdot g^{\gamma\sigma} g_{\alpha\beta} \right)$$

$$\Gamma^{*\mu}_{\sigma\mu} = \Gamma^{\mu}_{\sigma\mu} + \frac{\mathbf{F}'}{2 \cdot F[\mathbf{f}]} \left(\mathbf{f}_{,\sigma} \cdot \delta^{\mu}_{\mu} + \mathbf{f}_{,\mu} \cdot \delta^{\mu}_{\sigma} - \mathbf{f}_{,\gamma} \cdot g^{\gamma\mu} g_{\sigma\mu} \right)$$

$$= \Gamma^{\mu}_{\sigma\mu} + \frac{\mathbf{F}'}{2 \cdot F[\mathbf{f}]} \left(\mathbf{f}_{,\sigma} \cdot n + \mathbf{f}_{,\sigma} - \mathbf{f}_{,\sigma} \right) = \Gamma^{\mu}_{\sigma\mu} + \frac{\mathbf{F}' \mathbf{f}_{,\sigma} \cdot n}{2 \cdot F[\mathbf{f}]}. \quad (1392)$$

$$\Gamma^{*\sigma}_{\beta\mu} = \Gamma^{\sigma}_{\beta\mu} + \frac{1}{2 \cdot F[\mathbf{f}]} \left(F[\mathbf{f}]_{,\beta} \cdot \delta^{\sigma}_{\mu} + F[\mathbf{f}]_{,\mu} \cdot \delta^{\sigma}_{\beta} - F[\mathbf{f}]_{,\gamma} \cdot g^{\gamma\sigma} g_{\mu\beta} \right)$$

$$\Gamma^{*\mu}_{\sigma\alpha} = \Gamma^{\mu}_{\sigma\alpha} + \frac{1}{2 \cdot F[\mathbf{f}]} \left(F[\mathbf{f}]_{,\sigma} \cdot \delta^{\mu}_{\alpha} + F[\mathbf{f}]_{,\alpha} \cdot \delta^{\mu}_{\sigma} - F[\mathbf{f}]_{,\gamma} \cdot g^{\gamma\mu} g_{\alpha\sigma} \right). \quad (1393)$$

$$\Gamma^{*\sigma}_{\beta\sigma} = \Gamma^{\sigma}_{\beta\sigma} + \frac{1}{2 \cdot F[\mathbf{f}]} \left(F[\mathbf{f}]_{,\sigma} \cdot \delta^{\sigma}_{\beta} + F[\mathbf{f}]_{,\beta} \cdot n - F[\mathbf{f}]_{,\beta} \right)$$

$$= \Gamma^{\sigma}_{\beta\sigma} + \frac{1}{2 \cdot F[\mathbf{f}]} \left(F[\mathbf{f}]_{,\beta} + F[\mathbf{f}]_{,\beta} \cdot n - F[\mathbf{f}]_{,\beta} \right) = \Gamma^{\sigma}_{\beta\sigma} + \frac{\mathbf{F}' \mathbf{f}_{,\beta} \cdot n}{2 \cdot F[\mathbf{f}]}. \quad (1394)$$

Without much fuss we now evaluate R^* as follows:

$$R^* = R^*_{\alpha\beta} G^{\alpha\beta}$$

$$= \left(\begin{array}{l} \left(\Gamma^\sigma_{\alpha\beta} + \dfrac{\mathbf{F}'}{2\cdot F[\mathbf{f}]}\left(\mathbf{f}_{,\beta}\cdot\delta^\sigma_\alpha + \mathbf{f}_{,\alpha}\cdot\delta^\sigma_\beta - \mathbf{f}_{,\gamma}\cdot g^{\gamma\sigma}g_{\alpha\beta}\right)\right)_{,\sigma} \\[2mm] -\left(\Gamma^\sigma_{\beta\sigma} + \dfrac{\mathbf{F}'\mathbf{f}_{,\beta}\cdot n}{2\cdot F[\mathbf{f}]}\right)_{,\alpha} \\[2mm] -\left(\Gamma^\mu_{\sigma\alpha} + \dfrac{1}{2\cdot F[\mathbf{f}]}\left(F[\mathbf{f}]_{,\sigma}\cdot\delta^\mu_\alpha + F[\mathbf{f}]_{,\alpha}\cdot\delta^\mu_\sigma - F[\mathbf{f}]_{,\gamma}\cdot g^{\gamma\mu}g_{\alpha\sigma}\right)\right) \\[2mm] \cdot\left(\Gamma^\sigma_{\beta\mu} + \dfrac{1}{2\cdot F[\mathbf{f}]}\left(F[\mathbf{f}]_{,\beta}\cdot\delta^\sigma_\mu + F[\mathbf{f}]_{,\mu}\cdot\delta^\sigma_\beta - F[\mathbf{f}]_{,\gamma}\cdot g^{\gamma\sigma}g_{\mu\beta}\right)\right) \\[2mm] +\left(\Gamma^\sigma_{\alpha\beta} + \dfrac{\mathbf{F}'}{2\cdot F[\mathbf{f}]}\left(\mathbf{f}_{,\beta}\cdot\delta^\sigma_\alpha + \mathbf{f}_{,\alpha}\cdot\delta^\sigma_\beta - \mathbf{f}_{,\gamma}\cdot g^{\gamma\sigma}g_{\alpha\beta}\right)\right)\left(\Gamma^\mu_{\sigma\mu} + \dfrac{\mathbf{F}'\mathbf{f}_{,\beta}\cdot n}{2\cdot F[\mathbf{f}]}\right) \end{array}\right)\dfrac{g^{\alpha\beta}}{F}$$

$$(1395)$$

$$R^* = R^*_{\alpha\beta} G^{\alpha\beta}$$

$$= \left(\begin{array}{l} \left(\Gamma^\sigma_{\alpha\beta,\sigma} + \dfrac{\mathbf{F}\mathbf{F}'' - (\mathbf{F}')^2}{2\cdot F[\mathbf{f}]^2}\mathbf{f}_{,\sigma}\left(\mathbf{f}_{,\beta}\cdot\delta^\sigma_\alpha + \mathbf{f}_{,\alpha}\cdot\delta^\sigma_\beta - \mathbf{f}_{,\gamma}\cdot g^{\gamma\sigma}g_{\alpha\beta}\right)\right. \\[2mm] \left.+\dfrac{\mathbf{F}'}{2\cdot F[\mathbf{f}]}\left(\mathbf{f}_{,\beta\sigma}\cdot\delta^\sigma_\alpha + \mathbf{f}_{,\alpha\sigma}\cdot\delta^\sigma_\beta - \mathbf{f}_{,\gamma\sigma}\cdot g^{\gamma\sigma}g_{\alpha\beta} - \mathbf{f}_{,\gamma}\cdot\left(g^{\gamma\sigma}g_{\alpha\beta}\right)_{,\sigma}\right)\right) \\[2mm] -\left(\Gamma^\sigma_{\beta\sigma,\alpha} - \dfrac{(\mathbf{F}')^2\mathbf{f}_{,\alpha}\mathbf{f}_{,\beta}\cdot n}{2\cdot F[\mathbf{f}]^2} + \dfrac{\left(\mathbf{F}''\mathbf{f}_{,\alpha}\mathbf{f}_{,\beta} + \mathbf{F}'\mathbf{f}_{,\alpha\beta}\right)\cdot n}{2\cdot F[\mathbf{f}]}\right) \\[2mm] -\left(\Gamma^\mu_{\sigma\alpha} + \dfrac{\mathbf{F}'}{2\cdot F[\mathbf{f}]}\left(\mathbf{f}_{,\sigma}\cdot\delta^\mu_\alpha + \mathbf{f}_{,\alpha}\cdot\delta^\mu_\sigma - \mathbf{f}_{,\gamma}\cdot g^{\gamma\mu}g_{\alpha\sigma}\right)\right) \\[2mm] \cdot\left(\Gamma^\sigma_{\beta\mu} + \dfrac{\mathbf{F}'}{2\cdot F[\mathbf{f}]}\left(\mathbf{f}_{,\beta}\cdot\delta^\sigma_\mu + \mathbf{f}_{,\mu}\cdot\delta^\sigma_\beta - \mathbf{f}_{,\gamma}\cdot g^{\gamma\sigma}g_{\mu\beta}\right)\right) \\[2mm] +\left(\Gamma^\sigma_{\alpha\beta} + \dfrac{\mathbf{F}'}{2\cdot F[\mathbf{f}]}\left(\mathbf{f}_{,\beta}\cdot\delta^\sigma_\alpha + \mathbf{f}_{,\alpha}\cdot\delta^\sigma_\beta - \mathbf{f}_{,\gamma}\cdot g^{\gamma\sigma}g_{\alpha\beta}\right)\right)\left(\Gamma^\mu_{\sigma\mu} + \dfrac{\mathbf{F}'\mathbf{f}_{,\beta}\cdot n}{2\cdot F[\mathbf{f}]}\right) \end{array}\right)\dfrac{g^{\alpha\beta}}{F}$$

$$(1396)$$

$$R^* = R^*_{\alpha\beta} G^{\alpha\beta}$$

$$
= \left(
\begin{array}{l}
\left(
\begin{array}{l}
\Gamma^\sigma_{\alpha\beta,\sigma} + \dfrac{FF'' - \left(\mathbf{F}'\right)^2}{2 \cdot \underbrace{F[\mathbf{f}]}_{=F}{}^2} \mathbf{f}_{,\sigma}\left(\mathbf{f}_{,\beta}\cdot\delta^\sigma_\alpha + \mathbf{f}_{,\alpha}\cdot\delta^\sigma_\beta - \mathbf{f}_{,\gamma}\cdot g^{\gamma\sigma}g_{\alpha\beta}\right) \\[3mm]
+ \dfrac{\mathbf{F}'}{2\cdot F}\left(\mathbf{f}_{,\beta\sigma}\cdot\delta^\sigma_\alpha + \mathbf{f}_{,\alpha\sigma}\cdot\delta^\sigma_\beta - \mathbf{f}_{,\gamma\sigma}\cdot g^{\gamma\sigma}g_{\alpha\beta} - \mathbf{f}_{,\gamma}\cdot\left(g^{\gamma\sigma}g_{\alpha\beta}\right)_{,\sigma}\right)
\end{array}
\right) \\[10mm]
- \left(\Gamma^\sigma_{\beta\sigma,\alpha} - \dfrac{\left(\mathbf{F}'\right)^2\mathbf{f}_{,\alpha}\mathbf{f}_{,\beta}\cdot n}{2\cdot F^2} + \dfrac{\left(\mathbf{F}''f_{,\alpha}f_{,\beta} + \mathbf{F}'\mathbf{f}_{,\alpha\beta}\right)\cdot n}{2\cdot F}\right) \\[10mm]
= \left(
\begin{array}{l}
\Gamma^\mu_{\sigma\alpha}\Gamma^\sigma_{\beta\mu} + \dfrac{\Gamma^\sigma_{\beta\mu}\mathbf{F}'}{2\cdot F}\left(\mathbf{f}_{,\sigma}\cdot\delta^\mu_\alpha + \mathbf{f}_{,\alpha}\cdot\delta^\mu_\sigma - \mathbf{f}_{,\gamma}\cdot g^{\gamma\mu}g_{\alpha\sigma}\right) \\[3mm]
+ \Gamma^\mu_{\sigma\alpha}\dfrac{\mathbf{F}'}{2\cdot F}\left(\mathbf{f}_{,\beta}\cdot\delta^\sigma_\mu + \mathbf{f}_{,\mu}\cdot\delta^\sigma_\beta - \mathbf{f}_{,\gamma}\cdot g^{\gamma\sigma}g_{\mu\beta}\right) \\[3mm]
+ \left(\dfrac{\mathbf{F}'}{2\cdot F}\right)^2
\left(
\begin{array}{l}
\mathbf{f}_{,\beta}\cdot\delta^\sigma_\mu f_{,\sigma}\cdot\delta^\mu_\alpha + \mathbf{f}_{,\mu}\cdot\delta^\sigma_\beta f_{,\sigma}\cdot\delta^\mu_\alpha - \mathbf{f}_{,\gamma}\cdot g^{\gamma\sigma}g_{\mu\beta}\mathbf{f}_{,\sigma}\cdot\delta^\mu_\alpha \\[2mm]
+ \mathbf{f}_{,\beta}\cdot\delta^\sigma_\mu\mathbf{f}_{,\alpha}\cdot\delta^\mu_\sigma + \mathbf{f}_{,\mu}\cdot\delta^\sigma_\beta f_{,\alpha}\cdot\delta^\mu_\sigma - \mathbf{f}_{,\gamma}\cdot g^{\gamma\sigma}g_{\mu\beta}\mathbf{f}_{,\alpha}\cdot\delta^\mu_\sigma \\[2mm]
- \mathbf{f}_{,\gamma}\cdot g^{\gamma\mu}g_{\alpha\sigma}\mathbf{f}_{,\beta}\cdot\delta^\sigma_\mu - \mathbf{f}_{,\gamma}\cdot g^{\gamma\mu}g_{\alpha\sigma}\mathbf{f}_{,\mu}\cdot\delta^\sigma_\beta \\[2mm]
+ \mathbf{f}_{,\gamma}\cdot g^{\gamma\mu}g_{\alpha\sigma}\mathbf{f}_{,\gamma}\cdot g^{\gamma\sigma}g_{\mu\beta}
\end{array}
\right)
\end{array}
\right) \\[20mm]
+ \left(
\begin{array}{l}
\Gamma^\sigma_{\alpha\beta}\Gamma^\mu_{\sigma\mu} + \dfrac{\Gamma^\mu_{\sigma\mu}\mathbf{F}'}{2\cdot F}\left(\mathbf{f}_{,\beta}\cdot\delta^\sigma_\alpha + \mathbf{f}_{,\alpha}\cdot\delta^\sigma_\beta - \mathbf{f}_{,\gamma}\cdot g^{\gamma\sigma}g_{\alpha\beta}\right) \\[3mm]
+ \dfrac{\mathbf{F}'f_{,\sigma}\cdot n}{2\cdot F}\Gamma^\sigma_{\alpha\beta} + \dfrac{\left(\mathbf{F}'\right)^2}{4\cdot F^2}n\left(\mathbf{f}_{,\sigma}\mathbf{f}_{,\beta}\cdot\delta^\sigma_\alpha + \mathbf{f}_{,\sigma}\mathbf{f}_{,\alpha}\cdot\delta^\sigma_\beta - \mathbf{f}_{,\sigma}\mathbf{f}_{,\gamma}\cdot g^{\gamma\sigma}g_{\alpha\beta}\right)
\end{array}
\right)
\end{array}
\right) \dfrac{g^{\alpha\beta}}{F}.
$$

(1397)

The rest of the evaluation is tedious and so we only present the most important steps:

$$
R^* = R^*_{\alpha\beta}G^{\alpha\beta} = \left(
\begin{array}{l}
\Gamma^\sigma_{\alpha\beta,\sigma}g^{\alpha\beta} - \Gamma^\sigma_{\beta\sigma,\alpha}g^{\alpha\beta} - \Gamma^\mu_{\sigma\alpha}\Gamma^\sigma_{\beta\mu}g^{\alpha\beta} + \Gamma^\sigma_{\alpha\beta}\Gamma^\mu_{\sigma\mu}g^{\alpha\beta} \\[3mm]
+ \dfrac{\mathbf{F}'}{F}\left((1-n)\mathbf{f}_{,\beta\alpha}g^{\alpha\beta}\right) \\[3mm]
+ \dfrac{\mathbf{F}'}{4\cdot F}\mathbf{f}_{,\gamma}\left(
\begin{array}{l}
(n-2)\left(g^{\gamma\mu}g^{\alpha\beta}g_{\alpha\mu,\beta} + g^{\gamma\mu}g^{\alpha\beta}g_{\mu\beta,\alpha}\right) - 2ng^{\gamma\mu}{}_{,\mu} \\[2mm]
-\left((3n-3)\right)g^{\gamma\mu}g^{\alpha\beta}g_{\alpha\beta,\mu}
\end{array}
\right) \\[5mm]
\dfrac{\mathbf{f}_{,\alpha}\mathbf{f}_{,\beta}g^{\alpha\beta}(1-n)}{4F^2}\left(4FF'' + \left(\mathbf{F}'\right)^2(n-6)\right)
\end{array}
\right)\dfrac{1}{F}.
$$

(1398)

This can be further simplified to:

$$R^* = R^*{}_{\alpha\beta}G^{\alpha\beta} = \begin{pmatrix} \Gamma^\sigma_{\alpha\beta,\sigma}g^{\alpha\beta} - \Gamma^\sigma_{\beta\sigma,\alpha}g^{\alpha\beta} - \Gamma^\mu_{\sigma\alpha}\Gamma^\sigma_{\beta\mu}g^{\alpha\beta} + \Gamma^\sigma_{\alpha\beta}\Gamma^\mu_{\sigma\mu}g^{\alpha\beta} \\ +\dfrac{\mathbf{F}'}{F}(1-n)\Delta\mathbf{f} + \dfrac{\mathbf{f}_{,\alpha}\mathbf{f}_{,\beta}g^{\alpha\beta}(1-n)}{4F^2}\left(4\mathbf{FF}'' + \left(\mathbf{F}'\right)^2(n-6)\right) \end{pmatrix}\frac{1}{F}.$$

$$(1399)$$

In order to make it clear that this is still a scalar, we rewrite the expression Eq. (1399) by using Latin indices in co- and contra-variant form:

$$R^* = R^*{}_{\alpha\beta}G^{\alpha\beta} = \begin{pmatrix} \Gamma^\sigma_{\alpha\beta,\sigma}g^{\alpha\beta} - \Gamma^\sigma_{\beta\sigma,\alpha}g^{\alpha\beta} - \Gamma^\mu_{\sigma\alpha}\Gamma^\sigma_{\beta\mu}g^{\alpha\beta} + \Gamma^\sigma_{\alpha\beta}\Gamma^\mu_{\sigma\mu}g^{\alpha\beta} \\ +\dfrac{F'^a}{F}(1-n)\Delta f_a + \dfrac{f_{a,\alpha}f_{b,\beta}g^{\alpha\beta}(1-n)}{4F^2}\left(4FF''^{ab} + F'^a F'^b(n-6)\right) \end{pmatrix}\frac{1}{F}.$$

$$(1400)$$

Applying the definition from the appendix E of [26], we can abbreviate the last equation as follows:

$$R^* = R^*{}_{\alpha\beta}G^{\alpha\beta} = \left(R + (1-n)\left(\Theta^a\Delta f_a + f_{a,\alpha}f_{b,\beta}g^{\alpha\beta}\Xi^{ab}\right)\right)\frac{1}{F}$$

$$= \frac{R + (1-n)\left(\Theta^a\Delta + f_{b,\beta}g^{\alpha\beta}\Xi^{ab}\partial_\alpha\right)f_a}{F}.$$

$$(1401)$$

9.7.1 For Illustration

We start with a function F being linear in f-vec, which is to say $F[f] = q^a f_a$ and assume $R = 0$:

$$R^* = \left(R + \frac{F'^a}{F}(1-n)\Delta f_a + \frac{f_{a,\alpha}f_{b,\beta}g^{\alpha\beta}(1-n)}{4F^2}\left(\overbrace{4FF''^{ab}}^{=0} + F'^a F'^b(n-6)\right)\right)\frac{1}{F}$$

$$= (1-n)F'^a\left(\Delta f_a + \frac{f_{a,\alpha}f_{b,\beta}g^{\alpha\beta}F'^b(n-6)}{4F}\right)\frac{1}{F^2}.$$

$$(1402)$$

Also demanding $n>1$ and $R^* = 0$, we get:

$$0 = \Delta f_a + \frac{f_{a,\alpha}f_{b,\beta}g^{\alpha\beta}F'^b(n-6)}{4F}.$$

$$(1403)$$

Each of the functions $\mathbf{f_a}$ shall have an Eigenvalue solution to the Laplace equation, meaning:

$$0 = B^2 \cdot f_a + \frac{f_{a,\alpha} f_{b,\beta} g^{\alpha\beta} F'^b (n-6)}{4F}. \tag{1404}$$

This gives the following new Eigenvalue equation:

$$0 = B^2 \cdot f_a + f_{a,\alpha} \cdot \frac{f_b' F'^b (n-6)}{4F} \tag{1405}$$

and as—because of:

$$\Delta f_a = B^2 \cdot f_a \tag{1406}$$

the $\mathbf{f_a}$ are already fixed, we now look for a suitable vector $\mathbf{f_a}$ in $F[\mathbf{f_a}]$, which satisfies Eq. (1405). Because of our choice of $F[\mathbf{f_a}]$, this de facto results in a search for suitable coefficients, respectively a vector of constant coefficients $\mathbf{q^a}$, which satisfies Eq. (1405). This can be written as follows:

$$0 = B^2 \cdot f_a + f_{a,\alpha} \cdot \frac{f_b'^\alpha q^b (n-6)}{4F}. \tag{1407}$$

Thus, our linear approach for $F[f_a]$ gives us:

$$0 = B \cdot f_a + f_{a,\alpha} \cdot \frac{f_b'^\alpha q^b (n-6)}{4 f_b q^b}. \tag{1408}$$

Multiplication with $\mathbf{q^a}$ and a bit of rearrangement yields:

$$0 = B^2 \cdot f_a q^a + q^a f_{a,\alpha} \cdot \frac{f_b'^\alpha q^b (n-6)}{4 f_b q^b} \Rightarrow \frac{4B^2}{(6-n)} \cdot f_a q^a f_b q^b = q^a f_{a,\alpha} \cdot f_b'^\alpha q^b$$

$$\frac{4B^2}{(6-n)} \cdot f_a q^a f_b q^b = q^a f_{a,\alpha} g^{\alpha\beta} f_{b,\beta} q^b = q^a f_{a,\alpha} \mathbf{e}^\alpha \cdot \mathbf{e}^\beta f_{b,\beta} q^b$$

$$\Rightarrow \quad \left\{ \begin{array}{c} \dfrac{\pm 2\mathbf{B}}{\sqrt{(6-n)}} \cdot f_a q^a = q^a f_{a,\alpha} \mathbf{e}^\alpha \\[2ex] \dfrac{\pm 2\mathbf{B}}{\sqrt{(6-n)}} \cdot f_b q^b = \mathbf{e}^\beta f_{b,\beta} q^b \end{array} \right\}. \tag{1409}$$

As in the main section of this paper, we realize that $\mathbf{B}$ has to be an object with the characteristics of a base vector. Interestingly, we see that the terms $\mathbf{e}^\beta q^b$, $q^a \mathbf{e}^\alpha$ already have the ranks and structure of the Dirac gamma matrices. As these objects have been extracted via a factorization of the $q^a f_{a,\alpha} g^{\alpha\beta} f_{b,\beta} q^b$ operator, they automatically also guarantee the following outcome in Minkowski coordinates:

$$q^a \mathbf{e}^\beta \partial_\beta \mathbf{e}^\alpha f_{a,\alpha} = q^a \partial_\beta \mathbf{e}^\beta \mathbf{e}^\alpha f_{a,\alpha} = q^a \partial_\beta g^{\beta\alpha} f_{a,\alpha} = q^a \Delta_{\text{Minkowski}} f_a. \tag{1410}$$

But this was, per definition from Eq. (1406), our starting point, which is to say:

$$q^a e^\beta \partial_\beta e^\alpha f_{a,\alpha} = \ldots = q^a \Delta_{\text{Minkowski}} f_a = B^2 \cdot f_a q^a \qquad (1411)$$

and is exactly equivalent to the functionality of the Dirac equation, only that the Dirac theory is based on quaternions and not on base vectors. Thus, decompositions of the kind $e^\beta q^b$, $q^a e^\alpha$ and sets of functions f_a, satisfying the outcome in Eq. (1406) (solving the Klein–Gordon equation) and the two last lines in Eq. (1409) are Dirac equivalent solutions to the "hard" flat space-time condition $R^* = R = 0$. The question is whether such solutions exist or whether we'd rather have a soft flat space-time condition with $R^* = 0$ only. The latter was already discussed at the end of sub-section "9.3.1 Dirac's "Luck" in Connection with the Scaled Ricci Scalar." What we did not discuss there, however, was the degree of freedom the function F does provide in connection with our Dirac-like approach. Thus, we move back to Eq. (1402) and by leaving F general, but sticking to the condition $R^* = R = 0$, we obtain:

$$R^* = 0 = \left(R + \frac{F'^a}{F}(1-n)\Delta f_a + \frac{f_{a,\alpha} f_{b,\beta} g^{\alpha\beta}(1-n)}{4F^2}\left(4FF''^{ab} + F'^a F'^b(n-6) \right) \right)\frac{1}{F}$$

$$= (1-n)\left(F'^a \Delta f_a + \frac{f_{a,\alpha} f_{b,\beta} g^{\alpha\beta}}{4F}\left(4FF''^{ab} + F'^a F'^b(n-6) \right) \right)\frac{1}{F^2}$$

$$0 = F'^a \Delta f_a + \frac{f_{a,\alpha} f_{b,\beta} g^{\alpha\beta}}{4F}\left(4FF''^{ab} + F'^a F'^b(n-6) \right). \qquad (1412)$$

Also sticking to the Eigenvalue solution to the Laplace operator in Eq. (1406) yields:

$$R^* = 0 = F'^a B^2 f_a + \frac{f_{a,\alpha} f_{b,\beta} g^{\alpha\beta}}{4F}\left(4FF''^{ab} + F'^a F'^b(n-6) \right). \qquad (1413)$$

From there we may still aim for Dirac equivalent equations by demanding something like:

$$\frac{4FF''^{ac} + F'^a F'^c(n-6)}{4F} = -\frac{\left[A^2 \right]^{ac}}{F'^b f_b}. \qquad (1414)$$

Now we result in:

$$R^* = 0 = F'^a B^2 f_a - f_{a,\alpha} f_{c,\beta} g^{\alpha\beta} \cdot \frac{\left[A^2 \right]^{ac}}{F'^b f_b}$$

$$\Rightarrow \qquad B^2 F'^a f_a F'^b f_b = f_{a,\alpha} f_{b,\beta} g^{\alpha\beta} \cdot \left[A^2 \right]^{ab}$$

$\Rightarrow$ $$B^2 F'^a f_a F'^b f_b = f_{a,\alpha} f_{b,\beta} \mathbf{e}^\beta \cdot \mathbf{e}^\alpha \left[A^2\right]^{ab}$$ (1415)

and can directly factorize that last line, when assuming that:

$$\left[A^2\right]^{ab} = A^a A^b$$ (1416)

which is giving us:

$\Rightarrow$ $$\begin{cases} \pm \mathbf{B} \cdot f_a F'^a = A^a \mathbf{e}^\alpha f_{a,\alpha} \\ \pm \mathbf{B} \cdot f_b F'^b = A^b \mathbf{e}^\beta f_{b,\beta} \end{cases}.$$ (1417)

We realize that the objects $A^a \mathbf{e}^\alpha$, $\mathbf{e}^\beta A^b$ are the general metric equivalents to the Dirac gamma matrices and knowing the latter, we might extract the corresponding solutions to the wrapping function F from Eq. (1414).

In the case of functional dependencies of the kind $\mathbf{q^a f_a}$ with both vectors $\mathbf{q^a}$ and $\mathbf{f_a}$ being function-vectors, Eq. (1412) needs to be extended as follows:

$$R^* = 0 = \left(R + \frac{(1-n)}{F} \left(\begin{array}{l} F'^a \Delta f_a + \dfrac{f_{a,\alpha} f_{b,\beta} g^{\alpha\beta}}{4F}\left(4FF''^{ab} + F'^a F'^b (n-6)\right) \\ + F'_a \Delta q^a + \dfrac{h^a{}_{,\alpha} f_{b,\beta} g^{\alpha\beta}}{4F}\left(4FF''^b_a + F'_a F'^b (n-6)\right) \\ + \dfrac{h^a{}_{,\alpha} h^b{}_{,\beta} g^{\alpha\beta}}{4F}\left(4FF''_{ab} + F'_a F'_b (n-6)\right) \end{array} \right) \right) \frac{1}{F}$$

$$= (1-n)\left(\begin{array}{l} F''^a \Delta f_a + \dfrac{f_{a,\alpha} f_{b,\beta} g^{\alpha\beta}}{4F}\left(4FF''^{ab} + F'^a F'^b (n-6)\right) \\ + F'_a \Delta q^a + \dfrac{h^a{}_{,\alpha} f_{b,\beta} g^{\alpha\beta}}{4F}\left(4FF''^b_a + F'_a F'^b (n-6)\right) \\ + \dfrac{h^a{}_{,\alpha} h^b{}_{,\beta} g^{\alpha\beta}}{4F}\left(4FF''_{ab} + F'_a F'_b (n-6)\right) \end{array} \right) \frac{1}{F^2}$$

$$0 = \begin{cases} F'^a \Delta f_a + \dfrac{f_{a,\alpha} f_{b,\beta} g^{\alpha\beta}}{4F}\left(4FF''^{ab} + F'^a F'^b (n-6)\right) \\ + F'_a \Delta q^a + \dfrac{h^a{}_{,\alpha} f_{b,\beta} g^{\alpha\beta}}{4F}\left(4FF''^b_a + F'_a F'^b (n-6)\right) \\ + \dfrac{h^a{}_{,\alpha} h^b{}_{,\beta} g^{\alpha\beta}}{4F}\left(4FF''_{ab} + F'_a F'_b (n-6)\right). \end{cases}$$ (1418)

9.8 Appendix B: Derivation of the Scaled Ricci Tensor—With an Extension

We start with the result from Appendix A, reading:

$$R^* = R^*_{\alpha\beta} G^{\alpha\beta} = \left(\begin{array}{c} \Gamma^\sigma_{\alpha\beta,\sigma} g^{\alpha\beta} - \Gamma^\sigma_{\beta\sigma,\alpha} g^{\alpha\beta} - \Gamma^\mu_{\sigma\alpha} \Gamma^\sigma_{\beta\mu} g^{\alpha\beta} + \Gamma^\sigma_{\alpha\beta} \Gamma^\mu_{\sigma\mu} g^{\alpha\beta} \\ + \dfrac{F'^a}{F}(1-n)\Delta f_a + \dfrac{f_{a,\alpha} f_{b,\beta} g^{\alpha\beta}(1-n)}{4F^2}\left(4FF''^{ab} + F'^a F'^b (n-6)\right) \end{array} \right) \dfrac{1}{F}. \tag{1419}$$

Now we know that:

$$G^{\alpha\beta} = \frac{g^{\alpha\beta}}{F} \tag{1420}$$

and that in general we have the following identity:

$$\Delta f_a = \left\{ \begin{array}{c} g^{\alpha\beta} f_{a,\alpha\beta} - g^{\alpha\beta} \dfrac{g_{\alpha\beta}}{n} \Gamma^\gamma_{\kappa\lambda} g^{\kappa\lambda} f_{a,\gamma} \\ g^{\alpha\beta} f_{a,\alpha\beta} - g^{\alpha\beta} \Gamma^\gamma_{\alpha\beta} f_{a,\gamma} \end{array} \right\} = g^{\alpha\beta} \left\{ \begin{array}{c} f_{a,\alpha\beta} - \dfrac{g_{\alpha\beta}}{n} \Gamma^\gamma_{\kappa\lambda} g^{\kappa\lambda} f_{a,\gamma} \\ f_{a,\alpha\beta} - \Gamma^\gamma_{\alpha\beta} f_{a,\gamma} \end{array} \right\}. \tag{1421}$$

From this directly follows that:

$$R^* = R^*_{\alpha\beta} G^{\alpha\beta} = \left(\begin{array}{c} \Gamma^\sigma_{\alpha\beta,\sigma} g^{\alpha\beta} - \Gamma^\sigma_{\beta\sigma,\alpha} g^{\alpha\beta} - \Gamma^\mu_{\sigma\alpha} \Gamma^\sigma_{\beta\mu} g^{\alpha\beta} + \Gamma^\sigma_{\alpha\beta} \Gamma^\mu_{\sigma\mu} g^{\alpha\beta} \\ + \dfrac{F'^a}{F}(1-n)\left(g^{\alpha\beta} f_{a,\alpha\beta} - g^{\alpha\beta} \left\{ \begin{array}{c} \dfrac{g_{\alpha\beta}}{n} \Gamma^\gamma_{\kappa\lambda} g^{\kappa\lambda} f_{a,\gamma} \\ \Gamma^\gamma_{\alpha\beta} f_{a,\gamma} \end{array} \right\} \right) \\ + \dfrac{f_{a,\alpha} f_{b,\beta} g^{\alpha\beta}(1-n)}{4F^2}\left(4FF''^{ab} + F'^a F'^b (n-6)\right) \end{array} \right) \dfrac{1}{F}$$

$$= \left(\begin{array}{c} \left(\Gamma^\sigma_{\alpha\beta,\sigma} - \Gamma^\sigma_{\beta\sigma,\alpha} - \Gamma^\mu_{\sigma\alpha} \Gamma^\sigma_{\beta\mu} + \Gamma^\sigma_{\alpha\beta} \Gamma^\mu_{\sigma\mu} \right) \\ + \dfrac{F'^a}{F}(1-n)\left(f_{a,\alpha\beta} - \left\{ \begin{array}{c} \dfrac{g_{\alpha\beta}}{n} \Gamma^\gamma_{\kappa\lambda} g^{\kappa\lambda} f_{a,\gamma} \\ \Gamma^\gamma_{\alpha\beta} f_{a,\gamma} \end{array} \right\} \right) \dfrac{g^{\alpha\beta}}{F} \\ + \dfrac{f_{a,\alpha} f_{b,\beta} (1-n)}{4F^2}\left(4FF''^{ab} + F'^a F'^b (n-6)\right) \end{array} \right). \tag{1422}$$

With the help of the equations presented in section 8.11, this gives us the desired result as follows:

$$\Rightarrow$$

$$
R^*_{\alpha\beta} = R_{\alpha\beta} - \frac{g^{ab}}{2F}
\left(
\begin{array}{c}
\begin{pmatrix} F_{,b}g_{\alpha\beta,a} + F_{,a}g_{\alpha\beta,b} + F_{,ab}g_{\alpha\beta} + F_{,\alpha\beta}g_{ab} + F_{,\alpha}g_{ab,\beta} + F_{,\beta}g_{ab,\alpha} \\ -F_{,\alpha\beta}g_{ab} - F_{,a}g_{\alpha b,\beta} - F_{,\beta}g_{\alpha b,a} - F_{,\alpha\alpha}g_{\beta b} - F_{,a}g_{\beta b,\alpha} - F_{,\alpha}g_{\beta b,a} \end{pmatrix} \\[2em]
+\dfrac{g^{cd}}{F}
\begin{pmatrix}
\frac{1}{2}\Big(F\cdot g_{ac,\alpha}F_{,\beta}g_{bd} + F_{,\alpha}g_{ac}\cdot\big(F\cdot g_{bd,\beta} + F_{,\beta}g_{bd}\big)\Big) \\
+F\cdot g_{\alpha c,a}F_{,b}g_{\beta d} + F_{,a}g_{\alpha c}\cdot\big(F\cdot g_{\beta d,b} + F_{,b}g_{\beta d}\big) \\
-\Big(F\cdot g_{\alpha c,a}F_{,d}g_{\beta b} + F_{,a}g_{\alpha c}\cdot\big(F\cdot g_{\beta b,d} + F_{,d}g_{\beta b}\big)\Big)
\end{pmatrix} \\[3em]
-\dfrac{g^{cd}}{4F}
\begin{pmatrix}
F\cdot g_{\alpha c,\beta}\cdot\big(2F_{,a}g_{bd} - F_{,d}g_{ab}\big) + F_{,\beta}g_{\alpha c}\cdot\begin{pmatrix} 2F\cdot g_{bd,a} - F\cdot g_{ab,d} \\ +2F_{,a}g_{bd} - F_{,d}g_{ab}\end{pmatrix} \\
+F\cdot g_{\beta c,\alpha}\cdot\big(2F_{,a}g_{bd} - F_{,d}g_{ab}\big) + F_{,\alpha}g_{\beta c}\cdot\begin{pmatrix} 2F\cdot g_{bd,a} - F\cdot g_{ab,d} \\ +2F_{,a}g_{bd} - F_{,d}g_{ab}\end{pmatrix} \\
-F\cdot g_{\alpha\beta,c}\cdot\big(2F_{,a}g_{bd} - F_{,d}g_{ab}\big) - F_{,c}g_{\alpha\beta}\cdot\begin{pmatrix} 2F\cdot g_{bd,a} - F\cdot g_{ab,d} \\ +2F_{,a}g_{bd} - F_{,d}g_{ab}\end{pmatrix}
\end{pmatrix}
\end{array}
\right)
$$

with
$$\frac{\partial F}{\partial f} = F' \rightarrow F'^{a} \tag{1423}$$

References

1. N. Schwarzer, *Einstein had it, but he did not see it: Part LXXXI: More Dirac Killing*, Self-published, Amazon Digital Services, 2019, Kindle.

2. N. Schwarzer, *Societies and Ecotons—The Photons of the Human Society— Control them and Rule the World*, Part 1 of "Medical Socio-Economic Quantum Gravity," Self-published, Amazon Digital Services, 2020, Kindle.

3. N. Schwarzer, *Masses and the Infinity Options Principle? Can We Explain the 3-Generation and the Quantized Mass Problem?*, Part 5 of "Medical Socio-Economic Quantum Gravity," Self-published, Amazon Digital Services, 2020, Kindle.

4. N. Schwarzer, *The World Formula: A Late Recognition of David Hilbert's Stroke of Genius*, 2022, Jenny Stanford Publishing. ISBN: 9789814877206.

5. N. Schwarzer, *Humanitons—The Intrinsic Photons of the Human Body— Understand them and Cure Yourself*, Part 2 of "Medical Socio-Economic Quantum Gravity," Self-published, Amazon Digital Services, 2020, Kindle.

6. N. Schwarzer, *Mastering Human Crises with Quantum-Gravity-based but still Practicable Models—First Measure: SEEING and UNDERSTANDING the WHOLE: Part 3 of Medical Socio-Economic Quantum Gravity*, Self-published, Amazon Digital Services, 2020, Kindle.

7. N. Schwarzer, *Self-Similar Quantum Gravity: How Infinity Brings Simplicity?*, Part 4 of "Medical Socio-Economic Quantum Gravity," Self-published, Amazon Digital Services, 2020, Kindle.

8. N. Schwarzer, *Towards Quantum Einstein Field Equations*, Part 7 of "Medical Socio-Economic Quantum Gravity," Self-published, Amazon Digital Services, 2020, Kindle.

9. N. Schwarzer, *The 3 Generation of Elementary Particles*, Part 6 of "Medical Socio-Economic Quantum Gravity," Self-published, Amazon Digital Services, 2020, Kindle.

10. D. Hilbert, Die Grundlagen der Physik, Teil 1, *Göttinger Nachrichten*, 1915, 395–407.

11. A. Einstein, Grundlage der allgemeinen Relativitätstheorie, 1916, *Ann. Phys.*, 1916, **49** (ser. 4), 769–822.

12. C. A. Sporea, *Notes on f(R) Theories of Gravity*, arxiv.org/pdf/1403.3852.pdf.

13. H. Haken and H. Chr. Wolf, *Atom- und Quantenphysik* (in German), 4th Ed., 1990, Springer Heidelberg, ISBN 0-387-52198-4.

14. N. Schwarzer, *Einstein had it, but he did not see it: Part LXXIX: Dark Matter Options*, Self-published, Amazon Digital Services, 2016, Kindle.

15. N. Schwarzer, *The Theory of Everything—Quantum and Relativity is Everywhere—A Fermat Universe*, 2020, Pan Stanford Publishing, ISBN-10: 9814774472.

16. K. Schwarzschild, Über das Gravitationsfeld einer Kugel aus inkompressibler Flüssigkeit nach der Einsteinschen Theorie *[On the Gravitational Field of a Ball of Incompressible Fluid following Einstein's Theory]*, Sitzungsberichte der Königlich-Preussischen Akademie der Wissenschaften (in German), Berlin, 1916, 424–434.

17. G. Nordström, On the Energy of the Gravitation Field in Einstein's Theory, *Koninklijke Nederlandse Akademie van Weteschappen Proceedings Series B Physical Sciences*, 20:1238(1245), 1918.

18. H. Reissner, Über die Eigengravitation des elektrischen Feldes nach der Einsteinschen Theorie, *Annalen der Physik*, 355(9):106(120), 1916.

19. P. A. M. Dirac, *The Quantum Theory of the Electron*, 1928. DOI: 10.1098/rspa.1928.0023

20. N. Schwarzer, *Science Riddles—Riddle No. 8: Could the Schwarzschild Metric Contain its own Quantum Solution?*, Self-published, Amazon Digital Services, 2019, Kindle.

21. É. Cartan, Sur une généralisation de la notion de courbure de Riemann et les espaces à torsión, *C. R. Acad. Sci. (Paris)*, 1922, **174**, 593–595.

22. É. Cartan, Sur les variétés à connection affine et la théorie de la relativité généralisée, *Part I: Ann. Éc. Norm*, 1923, **40**, 325–412 and *ibid*, **41**, 1924, 1–25; *Part II: ibid*, 1925, **42**, 17–88.

23. N. Schwarzer, *My Horcruxes: A Curvy Math to Salvation*, Part 9 of "Medical Socio-Economic Quantum Gravity," Self-published, Amazon Digital Services, 2021, Kindle.

24. W. Pauli, Über den Zusammenhang des Abschlusses der Elektronengruppen im Atom mit der Komplexstruktur der Spektren, *Zeitschrift für Physik*, 1925, **31**, 765–783. Bibcode:1925ZPhy...31..765P. doi:10.1007/BF02980631

25. N. Schwarzer, *General Quantum Relativity*, Self-published, Amazon Digital Services, 2016, Kindle.

26. N. Schwarzer, *The Metric Dirac Equation Revisited and the Geometry of Spinors*, Part 8 of "Medical Socio-Economic Quantum Gravity," Self-published, Amazon Digital Services, 2020, Kindle.

Chapter 10

The Dirac Miracle

$$0 = \left(\overbrace{\left(V^{\xi} \cdot \delta_{\beta}^{\zeta} - \frac{1}{2} g^{\xi\zeta} \cdot V_{\beta} \right)}^{\rightarrow \hbar \cdot \gamma^{\zeta}} \partial_{\zeta} + \underbrace{V_{\beta} \left(\frac{1}{2} - \frac{1}{n} \right) \Gamma_{\kappa\lambda}^{\xi} g^{\kappa\lambda}}_{\rightarrow m \cdot c} \right) \Phi_{\xi}$$

Figure 10.1 The other gravitational "Dirac equation."

10.1 Abstract: Toward a Variety of Dirac Paths

After deriving quantum Einstein field equations (see previous chapters and [A1, A2]) directly out of the usual Einstein–Hilbert action [A3, A4] only that we used a scaled metric, which directly brought us the Klein–Gordon and the Schrödinger equations [A5–A8], but realizing some problems in deriving the Dirac equation [A9, A10] (see also [A11] and chapter 9), we now intend to sort these problems out. We will find that it only requires a more general reading/application of the variation process of the Einstein–Hilbert action to solve these problems (Fig. 10.1).

The Math of Body, Soul, and the Universe
Norbert Schwarzer
Copyright © 2023 Jenny Stanford Publishing Pte. Ltd.
ISBN 978-981-4968-24-9 (Hardcover), 978-1-003-33454-5 (eBook)
www.jennystanford.com

Along the way there also appears the second path for the derivation of the Dirac, respectively a Dirac-like equation, which is free from quaternions and only contains metric objects. Still the Dirac quaternion approach appears to be the more elegant path.

We will also see that there is a possibility for a gravitational Dirac equation, similar to the Einstein field equations and quantum Einstein field equations, which we are going to derive in this chapter (also directly out of the Einstein–Hilbert action with a scaled metric tensor).

10.1.1 References for the Abstract

A1. N. Schwarzer, *Towards Quantum Einstein Field Equations*, Part 7 of "Medical Socio-Economic Quantum Gravity," Self-published, Amazon Digital Services, 2020, Kindle.

A2. N. Schwarzer, *The Quantum Einstein Field Equations in their Simplest Form*, Part 7a of "Medical Socio-Economic Quantum Gravity," Self-published, Amazon Digital Services, 2021, Kindle.

A3. D. Hilbert, Die Grundlagen der Physik, Teil 1, *Göttinger Nachrichten*, 1915, 395–407.

A4. A. Einstein, Grundlage der allgemeinen Relativitätstheorie, A. Einstein, Grundlage der allgemeinen Relativitätstheorie, *Ann. Phys.*, 1916, **49** (ser. 4), 769–822

A5. N. Schwarzer, *Societies and Ecotons—The Photons of the Human Society—Control them and Rule the World*, Part 1 of "Medical Socio-Economic Quantum Gravity," Self-published, Amazon Digital Services, 2020, Kindle.

A6. N. Schwarzer, *Masses and the Infinity Options Principle: Can We Explain the 3-Generation and the Quantized Mass Problem?*, Part 5 of "Medical Socio-Economic Quantum Gravity," Self-published, Amazon Digital Services, 2020, Kindle.

A7. N. Schwarzer, *The World Formula: A Late Recognition of David Hilbert's Stroke of Genius*, 2022, Jenny Stanford Publishing. ISBN: 9789814877206.

A8. N. Schwarzer, *The 3-Generation of Elementary Particles*, Part 6 of "Medical Socio-Economic Quantum Gravity," Self-published, Amazon Digital Services, 2020, Kindle.

A9. N. Schwarzer, *The Metric Dirac Equation Revisited and the Geometry of Spinors*, Part 8 of "Medical Socio-Economic Quantum Gravity," Self-published, Amazon Digital Services, 2020, Kindle.

A10. P. A. M. Dirac, *The Quantum Theory of the Electron*, 1928. *Proceedings of the Royal Society A.* DOI: 10.1098/rspa.1928.0023.

A11. N. Schwarzer, *My Horcruxes: A Curvy Math to Salvation*, Part 9 of "Medical Socio-Economic Quantum Gravity," Self-published, Amazon Digital Services, 2021, Kindle.

10.2   Quantum Einstein Field Equations

As demonstrated in [1, 2] and previous chapters, when starting with the classical Einstein–Hilbert action [3]:

$$\delta W = 0 = \delta \int_V d^n x \left(\sqrt{-G} \cdot \left(R^* - 2\Lambda + L_M \right) \right), \tag{1424}$$

with the Ricci scalar R^*:

$$R^* = R^*{}_{\alpha\beta} G^{\alpha\beta} = \left(\begin{array}{c} \Gamma^{\sigma}_{\alpha\beta,\sigma} g^{\alpha\beta} - \Gamma^{\sigma}_{\beta\sigma,\alpha} g^{\alpha\beta} - \Gamma^{\mu}_{\sigma\alpha} \Gamma^{\sigma}_{\beta\mu} g^{\alpha\beta} + \Gamma^{\sigma}_{\alpha\beta} \Gamma^{\mu}_{\sigma\mu} g^{\alpha\beta} \\[2mm] + \dfrac{F'}{F}(1-n)\Delta f + \dfrac{f_{,\alpha} f_{,\beta} g^{\alpha\beta} (1-n)}{4F^2} \left(4FF'' + \left(F' \right)^2 (n-6) \right) \end{array} \right) \dfrac{1}{F}$$

$$= \left(R + \frac{F'}{F}(1-n)\Delta f + \frac{f_{,\alpha} f_{,\beta} g^{\alpha\beta} (1-n)}{4F^2} \left(4FF'' + \left(F' \right)^2 (n-6) \right) \right) \frac{1}{F}$$

with:

$$F = F[f]; \quad F' = \frac{\partial F[f]}{\partial f}; \quad F'' = \frac{\partial^2 F[f]}{\partial f^2}, \tag{1425}$$

as kernel or Lagrange density, we were able to derive new field equations, not only containing the classical part from Einstein's General Theory of Relativity [4], but also a quantum part [5–9], which is governed by the wave-function f and the metric scale $F[f]$.

It needs to be pointed out that in Eq. (1010), we used a scaled metric tensor $G_{\delta\gamma} = F[f] \cdot g_{\delta\gamma}$. Thereby G shall denote the determinant of the metric tensor $G_{\delta\gamma}$, while g will later stand for the corresponding determinant of the metric tensor $g_{\alpha\beta}$. In order to distinguish our new or scaled Ricci scalar R^*, being based on $G_{\delta\gamma} = F[f] \cdot g_{\delta\gamma}$ from the usual one R, being based on the metric tensor $g_{\alpha\beta}$, we marked it with the *-superscript. We also have the matter density L_M and the cosmological constant Λ.

Please note that with $G_{\delta\gamma} = F[f] \cdot g_{\delta\gamma}$ we have used the simplest form of metric adaptation, which we could construct as a simplification from a generalization of the typical form of tensor transformations, namely:

$$G_{\alpha\beta} = F\Big[f\big[t,x,y,z,\ldots,\xi_k,\ldots,\xi_n\big]\Big]^{ij}_{\alpha\beta} g_{ij} \tag{1426}$$

$$\to G_{\alpha\beta} = F\Big[f\big[t,x,y,z,\ldots,\xi_k,\ldots,\xi_n\big]\Big] \cdot \delta^i_\alpha \delta^j_\beta g_{ij}$$

$$\to G_{\alpha\beta} = F\Big[f\big[t,x,y,z,\ldots,\xi_k,\ldots,\xi_n\big]\Big] \cdot g_{\alpha\beta} = F[f]\cdot g_{\alpha\beta}. \tag{1427}$$

The generalization is been elaborated in [7] and its need (or rather redundancy) has been discussed in [1].

Setting Eq. (1008) into Eq. (1010) results in the variational task:

$$\delta W = 0 = \delta \int_V d^n x \ \sqrt{-g \cdot F^n} \times \left(\left(\begin{array}{c} R + \dfrac{F'}{F}(1-n)\Delta f \\[2mm] + \dfrac{f_{,\alpha} f_{,\beta} g^{\alpha\beta}(1-n)}{4F^2}\left(4FF'' + (F')^2(n-6)\right) \\[2mm] -2\Lambda + L_M \end{array} \right)^{\frac{1}{F}} \right). \tag{1428}$$

Performing the Hilbert variation with respect to the scale adapted metric tensor $G_{\alpha\beta}$ and applying the chain rule, gives us the following task:

$$\delta W = 0 = \delta_{G_{\alpha\beta}} \int_V d^n x \ \sqrt{-g \cdot F^n} \times \left(\left(\begin{array}{c} R + \dfrac{F'}{F}(1-n)\Delta f \\[2mm] + \dfrac{f_{,\alpha} f_{,\beta} g^{\alpha\beta}(1-n)}{4F^2}\left(4FF'' + (F')^2(n-6)\right) \\[2mm] -2\Lambda + L_M \end{array} \right)^{\frac{1}{F}} \right). \tag{1429}$$

$$0 = \int_V d^n x \left(\sqrt{-G} \times \left(R^{*\kappa\lambda} - \frac{1}{2}R^* G^{\kappa\lambda} + \kappa\cdot T^{\kappa\lambda} + \Lambda\cdot G^{\kappa\lambda} \right) \delta G_{\kappa\lambda} \right)$$

$$= \int_V d^n x \left(\sqrt{-g\cdot F^n} \times \underbrace{\left(R^*_{\ \alpha\beta} - \frac{1}{2}R^* G_{\alpha\beta} + \kappa\cdot T_{\alpha\beta} + \Lambda\cdot G_{\alpha\beta} \right)}_{\text{INT}} \delta G^{\alpha\beta} \right). \tag{1430}$$

Inserting the results from [2] into Eq. (1430), gives us the following variation:

$$\delta W = 0 = \int_V d^n x \ \sqrt{-g \cdot F^n} \left(\begin{array}{c} R_{\alpha\beta} - R \cdot \dfrac{g_{\alpha\beta}}{2} \\[2mm] + \dfrac{F'}{F}(1-n)\left(f_{,\alpha\beta} - \dfrac{g_{\alpha\beta}}{n}\Gamma^\gamma_{\kappa\lambda}g^{\kappa\lambda}f_{,\gamma} - \dfrac{g_{\alpha\beta}}{2}\Delta f \right) \\[2mm] + \dfrac{(1-n)}{4F^2}\left(\begin{array}{c} 4FF'' \\ +F'F'(n-6) \end{array}\right)\left(\begin{array}{c} f_{,\alpha}f_{,\beta} \\ -f_{,\mu}f_{,\nu}g^{\mu\nu}\dfrac{g_{\alpha\beta}}{2} \end{array}\right) \\[2mm] +\kappa T_{\alpha\beta} + \Lambda \cdot F \cdot g_{\alpha\beta} \end{array}\right) \delta G^{\alpha\beta}.$$

$$(1431)$$

10.2.1 A Side-Note

It has to be pointed out (hinted in [8], appendix A or appendix E) that there are options in splitting up the Laplace operator in the scaled Ricci scalar. We know that:

$$G^{\alpha\beta} = \frac{g^{\alpha\beta}}{F} \tag{1432}$$

and that in general, we have the following identities:

$$\Delta f = \left\{ \begin{array}{c} g^{\alpha\beta}f_{,\alpha\beta} - g^{\alpha\beta}\Gamma^\gamma_{\alpha\beta}f_{,\gamma} \\[2mm] g^{\alpha\beta}f_{,\alpha\beta} - g^{\alpha\beta}\dfrac{g_{\alpha\beta}}{n}\Gamma^\gamma_{\kappa\lambda}g^{\kappa\lambda}f_{,\gamma} \\[2mm] g^{\alpha\beta}\dfrac{g_{\alpha\beta}}{n}g^{\kappa\lambda}f_{,\kappa\lambda} - g^{\alpha\beta}\Gamma^\gamma_{\alpha\beta}f_{,\gamma} \\[2mm] g^{\alpha\beta}\dfrac{g_{\alpha\beta}}{n}g^{\kappa\lambda}f_{,\kappa\lambda} - g^{\alpha\beta}\dfrac{g_{\alpha\beta}}{n}\Gamma^\gamma_{\kappa\lambda}g^{\kappa\lambda}f_{,\gamma} \end{array}\right\}$$

$$= g^{\alpha\beta}\left\{ \begin{array}{c} f_{,\alpha\beta} - \Gamma^\gamma_{\alpha\beta}f_{,\gamma} \\[2mm] f_{,\alpha\beta} - \dfrac{g_{\alpha\beta}}{n}\Gamma^\gamma_{\kappa\lambda}g^{\kappa\lambda}f_{,\gamma} \\[2mm] \dfrac{g_{\alpha\beta}}{n}g^{\kappa\lambda}f_{,\kappa\lambda} - \Gamma^\gamma_{\alpha\beta}f_{,\gamma} \\[2mm] \dfrac{g_{\alpha\beta}}{n}g^{\kappa\lambda}f_{,\kappa\lambda} - \dfrac{g_{\alpha\beta}}{n}\Gamma^\gamma_{\kappa\lambda}g^{\kappa\lambda}f_{,\gamma} \end{array}\right\}. \tag{1433}$$

This leaves quite a few options for the treatment of the $R^{*}_{\alpha\beta}$-term in Eq. (1430) and its subsequent appearances in Eq. (1431). Thus, the generalized version of the latter equation should rather be:

$$\delta W = 0$$

$$= \int_V d^n x \, \sqrt{-g \cdot F^n} \left(\left(R_{\alpha\beta} - R \cdot \frac{g_{\alpha\beta}}{2} \right. \right.$$
$$\left. + \frac{F'}{F}(1-n) \left\{ \begin{array}{c} f_{,\alpha\beta} - \Gamma^\gamma_{\alpha\beta} f_{,\gamma} \\[2mm] f_{,\alpha\beta} - \dfrac{g_{\alpha\beta}}{n} \Gamma^\gamma_{\kappa\lambda} g^{\kappa\lambda} f_{,\gamma} \\[2mm] \dfrac{g_{\alpha\beta}}{n} g^{\kappa\lambda} f_{,\kappa\lambda} - \Gamma^\gamma_{\alpha\beta} f_{,\gamma} \\[2mm] \dfrac{g_{\alpha\beta}}{n} g^{\kappa\lambda} f_{,\kappa\lambda} - \dfrac{g_{\alpha\beta}}{n} \Gamma^\gamma_{\kappa\lambda} g^{\kappa\lambda} f_{,\gamma} \end{array} \right\} - \frac{g_{\alpha\beta}}{2} \Delta f$$
$$+ \frac{(1-n)}{4F^2} \left(\begin{array}{c} 4FF'' \\ +F'F'(n-6) \end{array} \right) \left(\begin{array}{c} f_{,\alpha} f_{,\beta} \\ -f_{,\mu} f_{,\nu} g^{\mu\nu} \dfrac{g_{\alpha\beta}}{2} \end{array} \right)$$
$$\left. \left. + \kappa T_{\alpha\beta} + \Lambda \cdot F \cdot g_{\alpha\beta} \right) \right) \delta G^{\alpha\beta} .$$

$$(1434)$$

For the time being, however, we will only apply the simple version from Eq. (1431).

10.2.2 Back to the Main Section and Eq. (1431)

We know that the variation of the scaled metric $\delta G_a^{\alpha\beta}$ can be split up into the variation of the function F and the variation of the unscaled metric and subsequently delivers the following:

$$
\delta W = 0 = \int_V d^n x \left\{ \sqrt{-g_a \cdot F^{an}} \left[\left(R_{a\alpha\beta} - R_a \cdot \frac{g_{a\alpha\beta}}{2} + \frac{F^{a\prime}}{F^a}(1-n)\left(f^a_{,\alpha\beta} - \frac{g_{\alpha\beta}}{n}\Gamma^\gamma_{\kappa\lambda}g^{\kappa\lambda}f_{,\gamma} - \frac{g_{a\alpha\beta}}{2}\Delta f^a \right) \right. \right. \right.
$$
$$
\left. + \frac{(1-n_a)}{4F^{a2}}\left(\begin{array}{c} 4F^aF^{a\prime\prime} \\ +F^{a\prime}F^{a\prime}(n_a-6) \end{array} \right) f^a_{,\alpha}f^a_{,\beta} - f^a_{,\mu}f^a_{,\nu}g^{\mu\nu}\frac{g_{\alpha\beta}}{2} \right)
$$
$$
\left. \left. + \kappa T_{a\alpha\beta} + \Lambda \cdot F^a \cdot g_{a\alpha\beta} \right] g_a^{\alpha\beta}\,\delta\frac{1}{F^a} \right\}. \tag{1435}
$$

The total variation now takes the form (c.f. [2] for details):

$$
0 = \int_V d^n x \left\{ \sqrt{-g_a \cdot F^{an_a}} \left(R_a + \frac{F^{a\prime}}{F^a}(1-n_a)\Delta_a f^a \right)\left(1-\frac{n_a}{2} \right) \right.
$$
$$
+ \frac{f^a_{,\alpha}f^a_{,\beta}(1-n_a)}{4F^{a2}}g_a^{\alpha\beta}\left(4F^aF^{a\prime\prime} + F^{a\prime}F^{a\prime}(n_a-6) \right)\times\left(1-\frac{n_a}{2} \right) + \kappa T_{a\alpha\beta}g_a^{\alpha\beta} + \Lambda\cdot F^a\cdot n_a \right)\, \delta\frac{1}{F^a}
$$
$$
+ \sqrt{-g_a\cdot(F^a)^{n_a-2}} \left[\left(R_{a\alpha\beta} - R_a\cdot\frac{g_{a\alpha\beta}}{2} + \frac{F^{a\prime}}{F^a}(1-n_a)\left(f^a_{,\alpha\beta} - \frac{g_{\alpha\beta}}{n}\Gamma^\gamma_{\kappa\lambda}g^{\kappa\lambda}f_{,\gamma} - \frac{g_{a\alpha\beta}}{2}\Delta f^a \right) \right. \right.
$$
$$
+ \frac{(1-n_a)}{4F^{a2}}\left(\begin{array}{c} 4F^aF^{a\prime\prime} \\ +F^{a\prime}F^{a\prime}(n_a-6) \end{array} \right) f^a_{,\alpha}f^a_{,\beta} - f^a_{,\mu}f^a_{,\nu}g^{\mu\nu}\frac{g_{\alpha\beta}}{2} \right)
$$
$$
\left. \left. \times\left(1-\frac{n_a}{2} \right) + \kappa T_{a\alpha\beta} + \Lambda\cdot F^a\cdot g_{a\alpha\beta} \right]\,\delta g_a^{\alpha\beta} \right\}. \tag{1436}
$$

Subsequently, demanding the vanishing of the whole integral via the integrand being zero, we obtain two equations, namely:

$$\textbf{1.} \quad 0 = \begin{pmatrix} \left(R_a + \dfrac{F^{a\prime}}{F^a}(1-n_a)\Delta_a f^a \right)\left(1 - \dfrac{n_a}{2} \right) \\[2em] + \dfrac{f^a{}_{,\alpha} f^a{}_{,\beta}(1-n_a)}{4F^{a2}} g_a{}^{\alpha\beta}\left(4F^a F^{a\prime\prime} + F^{a\prime} F^{a\prime}(n_a - 6) \right)\left(1 - \dfrac{n_a}{2} \right) \\[2em] + \kappa T_{a\alpha\beta} g_a{}^{\alpha\beta} + \Lambda \cdot F^a \cdot n_a \end{pmatrix}$$

$$\textbf{2.} \quad 0 = \begin{pmatrix} R_{a\alpha\beta} - R_a \cdot \dfrac{g_{a\alpha\beta}}{2} \\[1.5em] + \dfrac{F^{a\prime}}{F^a}(1-n_a)\left(f^a{}_{,\alpha\beta} - \dfrac{g_{\alpha\beta}}{n}\Gamma^{\gamma}_{\kappa\lambda} g^{\kappa\lambda} f_{,\gamma} - \dfrac{g_{a\alpha\beta}}{2}\Delta f^a \right) \\[1.5em] + \dfrac{(1-n_a)}{4F^{a2}}\begin{pmatrix} 4F^a F^{a\prime\prime} \\ +F^{a\prime} F^{a\prime}(n_a - 6) \end{pmatrix}\begin{pmatrix} f^a{}_{,\alpha} f^a{}_{,\beta} \\ -f^a{}_{,\mu} f^a{}_{,\nu} g^{\mu\nu}\dfrac{g_{\alpha\beta}}{2} \end{pmatrix} \\[1.5em] + \kappa T_{a\alpha\beta} + \Lambda \cdot F^a \cdot g_{a\alpha\beta} \end{pmatrix} . \qquad (1437)$$

One of the equations (the second one) **may** be seen as set of the quantum Einstein field equations, which determines the metric and the other would fix the wave functions F and f, whereby only one of the two functions (F or f) needs to be determined, while the other can be chosen in an arbitrary manner (probably so that one results in the simplest equations).

The reader finds quite some discussion about these results in [2] and later in here in chapter 11.

10.3 Quantum Einstein Field Equations in Their Simplest Form

Assuming that no extra matter is of need, but that the matter simply is materialized via certain forms or appearances of the wave function, we then find the simplest possible form of quantum Einstein field equations [2]. For convenience, we here repeat the evaluation.

Avoiding any vector character for the function f, the general quantum Einstein field equations are given via:

1. $\quad 0 = \left(\begin{array}{c} \left(R + \dfrac{F'}{F}(1-n)\Delta f \right)\left(1 - \dfrac{n}{2} \right) \\[2mm] + \dfrac{f_{,\alpha}f_{,\beta}(1-n)}{4F^2} g^{\alpha\beta}\left(4FF'' + F'F'(n-6) \right)\left(1 - \dfrac{n}{2} \right) \\[2mm] + \kappa T_{\alpha\beta}g^{\alpha\beta} + \Lambda \cdot F \cdot n \end{array} \right)$

$\Rightarrow \quad -R = \left(\begin{array}{c} \left(\dfrac{F'}{F}(1-n)\Delta f \right) \\[2mm] + \dfrac{f_{,\alpha}f_{,\beta}(1-n)}{4F^2} g^{\alpha\beta}\left(4FF'' + F'F'(n-6) \right) \\[2mm] + \dfrac{2}{2-n}\left(\kappa T_{\alpha\beta}g^{\alpha\beta} + \Lambda \cdot F \cdot n \right) \end{array} \right)$

$\Rightarrow \quad$ **2.** $\quad 0 = \left(\begin{array}{c} \left(R_{\alpha\beta} + \left(\begin{array}{c} \left(\dfrac{F'}{F}(1-n)\Delta f \right) \\[2mm] + \dfrac{f_{,\kappa}f_{,\lambda}(1-n)}{4F^2} g^{\kappa\lambda}\left(4FF'' + F'F'(n-6) \right) \\[2mm] + \dfrac{2}{2-n}\left(\kappa T_{\kappa\lambda}g^{\kappa\lambda} + \Lambda \cdot F \cdot n \right) \end{array} \right) \cdot \dfrac{g_{\alpha\beta}}{2} \right) \\[4mm] + \dfrac{F'}{F}(1-n)\left(f_{,\alpha\beta} - \dfrac{g_{\alpha\beta}}{n}\Gamma^{\gamma}_{\kappa\lambda}g^{\kappa\lambda}f_{,\gamma} - \dfrac{g_{\alpha\beta}}{2}\Delta f \right) \\[4mm] + \dfrac{(1-n)}{4F^2}\left(\begin{array}{c} 4FF'' \\ + F'F'(n-6) \end{array} \right)\left(\begin{array}{c} f_{,\alpha}f_{,\beta} \\ - f_{,\mu}f_{,\nu}g^{\mu\nu}\dfrac{g_{\alpha\beta}}{2} \end{array} \right) \\[4mm] + \kappa T_{\alpha\beta} + \Lambda \cdot F \cdot g_{\alpha\beta} \end{array} \right).$ (1438)

Sticking to the assumption that no extra matter is of need this may be simplified as follows:

1. $\quad 0 = \left(\begin{array}{c} \left(R + \dfrac{F'}{F}(1-n)\Delta f \right)\left(1 - \dfrac{n}{2} \right) \\[2mm] + \dfrac{f_{,\alpha}f_{,\beta}(1-n)}{4F^2} g^{\alpha\beta}\left(4FF'' + F'F'(n-6) \right)\left(1 - \dfrac{n}{2} \right) \end{array} \right)$

$$\Rightarrow \qquad -R = \left(\begin{array}{c} \left(\dfrac{F'}{F}(1-n)\Delta f \right) \\[2mm] + \dfrac{f_{,\alpha} f_{,\beta}(1-n)}{4F^2} g^{\alpha\beta}\left(4FF'' + F'F'(n-6)\right) \end{array} \right)$$

$$\Rightarrow \qquad \textbf{2.} \ \ 0 = \left(\begin{array}{c} \left(R_{\alpha\beta} + \left(\begin{array}{c} \left(\dfrac{F'}{F}(1-n)\Delta f \right) \\[2mm] + \dfrac{f_{,\kappa} f_{,\lambda}(1-n)}{4F^2} g^{\kappa\lambda}\left(4FF'' + F'F'(n-6)\right) \end{array} \right) \cdot \dfrac{g_{\alpha\beta}}{2} \right) \\[6mm] + \dfrac{F'}{F}(1-n)\left(f_{,\alpha\beta} - \dfrac{g_{\alpha\beta}}{n}\Gamma^\gamma_{\kappa\lambda} g^{\kappa\lambda} f_{,\gamma} - \dfrac{g_{\alpha\beta}}{2}\Delta f \right) \\[6mm] + \dfrac{(1-n)}{4F^2}\left(\begin{array}{c} 4FF'' \\ +F'F'(n-6) \end{array} \right)\left(\begin{array}{c} f_{,\alpha} f_{,\beta} \\ -f_{,\mu} f_{,\nu} g^{\mu\nu} \dfrac{g_{\alpha\beta}}{2} \end{array} \right) \end{array} \right) . \quad (1439)$$

With the condition $4FF'' + F'F'(n-6) = 0$, we obtain the very brief form of quantum Einstein field equations as:

$$\textbf{1.} \quad -R = \frac{F'}{F}(1-n)\Delta f$$

$$\Rightarrow \qquad \textbf{2.} \quad R_{\alpha\beta} = \frac{F'}{F}(n-1)\left(f_{,\alpha\beta} - \frac{g_{\alpha\beta}}{n}\Gamma^\gamma_{\kappa\lambda} g^{\kappa\lambda} f_{,\gamma} \right)$$

$$\Rightarrow \qquad R_{\alpha\beta} = \frac{F'}{F}(n-1) f_{:\alpha\beta}. \qquad (1440)$$

With F given as:

$$4FF'' + F'F'(n-6) = 0 \ \Rightarrow \ F = \left(C_f \pm f\right)^{\frac{4}{n-2}}; \quad n > 2, \qquad (1441)$$

this results in:

$$\Rightarrow \qquad \boxed{R_{\alpha\beta} = \pm \frac{4\cdot(n-1)}{(n-2)\left(C_f \pm f\right)} f_{:\alpha\beta}}. \qquad (1442)$$

This also was discussed in [2] and further investigation will be published elsewhere [10] and here in chapter "11 A Curvy Math to Salvation."

We conclude that the simplest form of quantum Einstein field equations can be given via Eq. (1442), but this is not the only option. We

will discuss other possibilities in chapter "11 A Curvy Math to Salvation." (see also [10]). It should also be pointed out that the fundamental law leading to our field equations is the minimum principle in Eq. (1013). This, however, is a scalar, which is formed out of an integral sum, which stretches over a certain volume, while the volume and the dimension of the integral are not fixed. And so, it does not necessarily truly require the integrand to be zero in order to make Eq. (1013) being fulfilled. Other options, which are not requiring the quantum Einstein field equations, we have derived elsewhere [10], but which could also be extracted from Eq. (1013), should be considered, too. For instance, there is still the question about the surface terms, centers of gravity, centers of wave functions [1, 6, 7], or other variational possibilities. Some of these will be discussed further below in this chapter and chapter 11.

Nevertheless, here we have found that the simplest and shortest form of quantum Einstein field equations evolves from condition in Eq. (1441) and, bringing it into the form similar to the one shown on some earlier publications (for recognition), reads:

$$\boxed{R_{\alpha\beta} \cdot \left(C_f \pm f\right) = \pm 4 \cdot \frac{(n-1)}{(n-2)} f_{;\alpha\beta}}. \tag{1443}$$

Thus, it may even be seen as a kind of Eigen equation, becoming a true Eigen equation in the case of non-f-dependent Ricci tensors.

Here we chose the $-$sign in order to emphasize the other possibilities for the decomposition of the Laplace operator as shown in Eq. (1433). By using Eq. (1434) rather than Eq. (1431), Eq. (1443) would read:

$$R_{\alpha\beta} \cdot \left(C_f \pm f\right) = \pm 4 \cdot \frac{(n-1)}{(n-2)} \left\{ \begin{array}{c} f_{,\alpha\beta} - \Gamma^\gamma_{\alpha\beta} f_{,\gamma} \\[2mm] f_{,\alpha\beta} - \dfrac{g_{\alpha\beta}}{n} \Gamma^\gamma_{\kappa\lambda} g^{\kappa\lambda} f_{,\gamma} \\[2mm] \dfrac{g_{\alpha\beta}}{n} g^{\kappa\lambda} f_{,\kappa\lambda} - \Gamma^\gamma_{\alpha\beta} f_{,\gamma} \\[2mm] \dfrac{g_{\alpha\beta}}{n} g^{\kappa\lambda} f_{,\kappa\lambda} - \dfrac{g_{\alpha\beta}}{n} \Gamma^\gamma_{\kappa\lambda} g^{\kappa\lambda} f_{,\gamma} \end{array} \right\}. \tag{1444}$$

Alternatively, one may also seek for a solution of the whole integrand of Eq. (1431), leading us to:

$$0 = R_{\alpha\beta} - \frac{1}{2}R \cdot g_{\alpha\beta} + \frac{F'}{F}(1-n)\left\{\left[\begin{array}{c} f_{,\alpha\beta} - \Gamma^{\gamma}_{\alpha\beta}f_{,\gamma} \\[2mm] f_{,\alpha\beta} - \dfrac{g_{\alpha\beta}}{n}\Gamma^{\gamma}_{\kappa\lambda}g^{\kappa\lambda}f_{,\gamma} \\[2mm] \dfrac{g_{\alpha\beta}}{n}g^{\kappa\lambda}f_{,\kappa\lambda} - \Gamma^{\gamma}_{\alpha\beta}f_{,\gamma} \\[2mm] \dfrac{g_{\alpha\beta}}{n}g^{\kappa\lambda}f_{,\kappa\lambda} - \dfrac{g_{\alpha\beta}}{n}\Gamma^{\gamma}_{\kappa\lambda}g^{\kappa\lambda}f_{,\gamma} \end{array}\right] - \frac{1}{2}\Delta f \cdot g_{\alpha\beta}\right\},$$

$$(1445)$$

respectively the simplified form:

$$0 = R_{\alpha\beta} - \frac{1}{2}R \cdot g_{\alpha\beta} + \frac{F'}{F}(1-n)\left(f_{,\alpha\beta} - \frac{g_{\alpha\beta}}{n}\Gamma^{\gamma}_{\kappa\lambda}g^{\kappa\lambda}f_{,\gamma} - \frac{1}{2}\Delta f \cdot g_{\alpha\beta}\right), \quad (1446)$$

when applying the condition in Eq. (1441) and assuming no matter and a zero cosmological constant. One may now ask, why then at all looking for a solution just to Eq. (1443) if this only solves "half" of the whole (meaning Eq. (1445))?

Well, one answer might be:

"As we think that the universe has means to take care about Eq. (1443) and the rest, which is to say:

$$0 = -\frac{1}{2}R \cdot g_{\alpha\beta} + \frac{F'}{F}(1-n)\left(-\frac{1}{2}\Delta f \cdot g_{\alpha\beta}\right) \quad \Rightarrow \quad 0 = R + \frac{F'}{F}(1-n)\Delta f, \quad (1447)$$

in different ways." We think that there is a reason, why the first line of Eq. (1440), being "the rest of Eq. (1445)" and thus, Eq. (1447) appears to us as a quantum equation, while the gravity equations containing the Ricci tensor act on bigger scales. We even think that different volumes have to or could be applied to both parts of the integrand within the variation. This may mathematically be done as follows:

- Starting with the result from Eq. (1431) and setting the matter term and the cosmological constant to zero:

$$\delta W = 0 = \int_V d^n x \sqrt{-g} \cdot F^n \left(\begin{array}{c} R_{,\alpha\beta} - R \cdot \dfrac{g_{\alpha\beta}}{2} \\[3mm] +\dfrac{F'}{F}(1-n)\left(f_{,\alpha\beta} - \dfrac{g_{\alpha\beta}}{n}\Gamma^{\gamma}_{\kappa\lambda}g^{\kappa\lambda}f_{,\gamma} - \dfrac{g_{\alpha\beta}}{2}\Delta f\right) \\[3mm] +\dfrac{f_{,\alpha}f_{,\beta}(1-n)}{4F^2}\left(4FF'' + F'F'(n-6)\right)\left(1-\dfrac{n}{2}\right) \end{array}\right)\delta G^{\alpha\beta},$$

$$(1448)$$

we can simplify and split up to:

$$0 = \int\limits_{V} d^n x \left(\sqrt{-g \cdot F^n} \left(\xrightarrow{\; 4FF'' + F'F'(n-6)=0 \;} \left(R_{\alpha\beta} - R \cdot \frac{g_{\alpha\beta}}{2} + \frac{F'}{F}(1-n)\left(f_{,\alpha\beta} - \frac{g_{\alpha\beta}}{n}\Gamma^{\gamma}_{\kappa\lambda}g^{\kappa\lambda}f_{,\gamma} - \frac{g_{\alpha\beta}}{2}\Delta f \right) \right) \right) \delta G^{\alpha\beta} \right)$$

$$\Rightarrow \quad 0 = \left(\begin{array}{c} \int\limits_{V_G} d^n x \left(\sqrt{-g \cdot F^n} \left(R_{\alpha\beta} + \frac{F'}{F}(1-n)\left(f_{,\alpha\beta} - \frac{g_{\alpha\beta}}{n}\Gamma^{\gamma}_{\kappa\lambda}g^{\kappa\lambda}f_{,\gamma} \right) \right) \delta G^{\alpha\beta} \right) \\ -\frac{1}{2} \cdot \int\limits_{V_Q} d^n x \left(\sqrt{-g \cdot F^n} \left(R + \frac{F'}{F}(1-n)\Delta f \right) g_{\alpha\beta}\delta G^{\alpha\beta} \right) \end{array} \right).$$

(1449)

Thereby the volumes of the two variations do not necessarily need to be the same.

For symmetry reasons, where we want to have equal total volumes for both terms in the last equation above, one may just extend the last line as follows:

$$0 = \left(\begin{array}{c} \int\limits_{V_G} d^n x \left(\sqrt{-g \cdot F^n} \left(R_{\alpha\beta} + \frac{F'}{F}(1-n)\left(f_{,\alpha\beta} - \frac{g_{\alpha\beta}}{n}\Gamma^{\gamma}_{\kappa\lambda}g^{\kappa\lambda}f_{,\gamma} \right) \right) \delta G^{\alpha\beta} \right) \\ -\frac{1}{2} \cdot \sum_{i=1}^{N} \int\limits_{V_{Qi}} d^n x \left(\sqrt{-g \cdot F^n} \left(R + \frac{F'}{F}(1-n)\Delta f \right) g_{\alpha\beta}\delta G^{\alpha\beta} \right) \end{array} \right)$$

$$V_G = \sum_{i=1}^{N} V_{Qi}. \qquad (1450)$$

Thus, it appears totally reasonable to assume that, for whatever reasons, the scales of the two variations could be significantly different and while the volume V_Q (Q = Quantum) might be extremely small, forcing us to watch the variation:

$$0 = \int\limits_{V_Q} d^n x \left(\sqrt{-g \cdot F^n} \left(R + \frac{F'}{F}(1-n)\Delta f \right) g_{\alpha\beta}\delta G^{\alpha\beta} \right) \qquad (1451)$$

at smaller, which is to say quantum scales, we would have to take care about the tensorial or gravitational variation:

$$0 = \int_{V_G} d^n x \left(\sqrt{-g \cdot F^n} \left(R_{\alpha\beta} + \frac{F'}{F}(1-n)\left(f_{,\alpha\beta} - \frac{g_{\alpha\beta}}{n} \Gamma^{\gamma}_{\kappa\lambda} g^{\kappa\lambda} f_{,\gamma} \right) \right) \delta G^{\alpha\beta} \right)$$

(1452)

at much bigger scales, respectively bigger volumes V_G. Thus, our reasoning for the consideration of solutions to Eq. (1443) (and other split-ups) with respect to the inner structure of any metric object (perhaps even black holes as considered in [13]) would be based on the observation of the different scales of quantum effects and gravity. Of course, this requires more discussion about the variational, respectively scale-dependent split-up in Eq. (1449) and the possible substructure of space or space-time with respect to the subsequent small-scale solutions to Eq. (1451) (or Eq. (1447) or the first line of Eq. (1440)). This discussion was presented elsewhere [10] and will partially be repeated in the next chapter of this book. Assuming, to pick an example, that the apparent matter (meaning, what we see as matter) inside a black hole (where we actually cannot see the matter) is solely made of scaled quantum states of our scaling functions F and f, we could also split-up our variation as follows:

$$0 = \left(\begin{array}{c} \int_{V_G} d^n x \sqrt{-g \cdot F^n} \left(\left(R_{\alpha\beta} - \frac{1}{2} \cdot R \cdot g_{\alpha\beta} \right) \delta G^{\alpha\beta} \right) \\ + \frac{1}{2} \cdot \sum_{i=1}^{N} \int_{V_{Qi}} d^n x \sqrt{-g \cdot F^n} \frac{F'}{F}(n-1)\left(\Delta f \cdot g_{\alpha\beta} - 2 \cdot \left(f_{,\alpha\beta} - \frac{g_{\alpha\beta}}{n} \Gamma^{\gamma}_{\kappa\lambda} g^{\kappa\lambda} f_{,\gamma} \right) \right) \delta G^{\alpha\beta} \end{array} \right).$$

(1453)

Now we ask, could perhaps the sum of V_{Qi} variations in our ensemble of scale split-ups as given above (c.f. Eqs.(1449)–(1453)) act as the matter, which was introduced (by postulation) by Hilbert [3]?

10.4 Where Does Hilbert's Matter Come From?

Discovering the classical Field-Equation structure (c.f. Eq. (1453)) among our ensemble of options (see [10, 14] and previous chapters), one might want to demonstrate the connection to the Einstein–Hilbert matter approach. We start with the assumption of one global metric $g_{\alpha\beta}$, but an ensemble of F_i and f_i, leading us to:

$$0 = \begin{pmatrix} \int_{V_G} d^n x \sqrt{-G} \left(\left(R_{\alpha\beta} - \frac{1}{2} \cdot R \cdot g_{\alpha\beta} \right) \delta G^{\alpha\beta} \right) \\ + \frac{1}{2} \cdot \sum_{i=1}^{N} \int_{V_{Qi}} d^n x \sqrt{-g} \cdot F^n \frac{F'}{F} (n-1) \left(\Delta f \cdot g_{\alpha\beta} - 2 \cdot \overbrace{\left(f_{,\alpha\beta} - \frac{g_{\alpha\beta}}{n} \Gamma^{\gamma}_{\kappa\lambda} g^{\kappa\lambda} f_{,\gamma} \right)}^{\equiv f_{\alpha\beta}} \right) \delta G^{\alpha\beta} \end{pmatrix}$$

$$\Rightarrow = \int_{V_G} d^n x \sqrt{-g} \left(\sum_{i=1}^{N} \sqrt{F_i^n} \cdot \left(R_{\alpha\beta} - \frac{1}{2} \cdot R \cdot g_{\alpha\beta} + \frac{(n-1)}{2} \cdot \frac{F_i'}{F_i} \left(\Delta f_i \cdot g_{\alpha\beta} - 2 \cdot f_{i;\alpha\beta} \right) \right) \delta G^{\alpha\beta} \right).$$

$$(1454)$$

This gives us the classical non-vacuum field equations via:

$$0 = \sum_{i=1}^{N} \sqrt{F_i^n} \cdot \left(R_{\alpha\beta} - \frac{1}{2} \cdot R \cdot g_{\alpha\beta} \right) + \sum_{i=1}^{N} \sqrt{F_i^n} \cdot \left(\frac{(n-1)}{2} \cdot \frac{F_i'}{F_i} \left(\Delta f_i \cdot g_{\alpha\beta} - 2 \cdot f_{i;\alpha\beta} \right) \right)$$

$$\Rightarrow \quad 0 = R_{\alpha\beta} - \frac{1}{2} \cdot R \cdot g_{\alpha\beta} + \frac{\sum_{i=1}^{N} \sqrt{F_i^n} \cdot \left(\frac{(n-1)}{2} \cdot \frac{F_i'}{F_i} \left(\Delta f_i \cdot g_{\alpha\beta} - 2 \cdot f_{i;\alpha\beta} \right) \right)}{\sum_{i=1}^{N} \sqrt{F_i^n}}$$

$$= R_{\alpha\beta} - \frac{1}{2} \cdot R \cdot g_{\alpha\beta} + \frac{\sum_{i=1}^{N} \sqrt{F_i^n} \cdot \left(\frac{(n-1)}{2} \cdot \frac{F_i'}{F_i} \left(\Delta f_i \cdot g_{\alpha\beta} - 2 \cdot \left(f_{,\alpha\beta} - \frac{g_{\alpha\beta}}{n} \Gamma^{\gamma}_{\kappa\lambda} g^{\kappa\lambda} f_{,\gamma} \right) \right) \right)}{\sum_{i=1}^{N} \sqrt{F_i^n}}$$

$$(1455)$$

and the last term in the last line above is handing us the quantum or scale equivalent of the Einstein–Hilbert matter term, respectively the energy-momentum tensor:

$$\Rightarrow \quad \kappa T_{\alpha\beta} = \frac{\sum_{i=1}^{N} \sqrt{F_i^n} \cdot \left(\frac{(n-1)}{2} \cdot \frac{F_i'}{F_i} \left(\Delta f_i \cdot g_{\alpha\beta} - 2 \cdot f_{i;\alpha\beta} \right) \right)}{\sum_{i=1}^{N} \sqrt{F_i^n}}$$

$$\frac{\sum_{i=1}^{N}\sqrt{F_i^n}\cdot\left(\frac{(n-1)}{2}\frac{F_i'}{F_i}\left(\Delta f_i\cdot g_{\alpha\beta}-2\cdot\left(f_{,\alpha\beta}-\frac{g_{\alpha\beta}}{n}\Gamma_{\kappa\lambda}^{\gamma}g^{\kappa\lambda}f_{,\gamma}\right)\right)\right)}{\sum_{i=1}^{N}\sqrt{F_i^n}}. \quad (1456)$$

10.5 Solving an Inconsistency Problem with the Quantum Einstein Field Equations

In [2], we had briefly discussed an interesting inconsistency in connection with the classical Einstein field equations [3, 4], which we now are able to resolve.

Let us briefly repeat the problematic issue here.

We take the Einstein field equations in vacuum [3, 4], where summation (contraction) with the metric tensor provides us with a peculiar simplification, namely:

$$0=\left(R_{\alpha\beta}-R\cdot\frac{g_{\alpha\beta}}{2}\right)g^{\alpha\beta}=R-R\cdot\frac{n}{2}=R\cdot\left(1-\frac{n}{2}\right). \quad (1457)$$

Here R gives the Ricci scalar, $R_{\alpha\beta}$ the Ricci tensor and $g_{\alpha\beta}$ the metric tensor. We find the result in Eq. (1457) peculiar, because for non-vanishing R or general n unequal to 2, the equation cannot be fulfilled. Only in the non-vacuum case:

$$0=\left(R_{\alpha\beta}-R\cdot\frac{g_{\alpha\beta}}{2}+\kappa T_{\alpha\beta}+\Lambda\cdot g_{\alpha\beta}\right)g^{\alpha\beta}=R\cdot\left(1-\frac{n}{2}\right)+\kappa T_{\alpha\beta}g^{\alpha\beta}+\Lambda\cdot n \quad (1458)$$

the problem vanishes.

However, in the classical theory, the matter term was artificially introduced by Einstein and Hilbert and thus, the classical General Theory of Relativity is somehow incomplete and inconsistent or both.

We saw (e.g. [1, 2]) that in order to sort this out, we simply have to allow the metric to be scaled by a functional wrapper $F[f]$. We also saw that this also—rather automatically—gives us a Quantum General Theory of Relativity (in its simplest form, c.f. [10] and next chapter).

Thereby the corresponding process of derivation is rather simple.

We here start with the generalized Einstein–Hilbert action:

$$\delta W=0=\delta_{G_{\alpha\beta}}\int_V d^n x\left(\sqrt{-G}\times R^*\right)=\delta_{G_{\alpha\beta}}\int_V d^n x\left(\sqrt{-g\cdot F^n}\times R^*\right)$$

with:

$$G_{\alpha\beta}=F[f]\cdot g_{\alpha\beta}, \quad (1459)$$

which provides us with the matter-free quantum equivalent of the Einstein field equations in n dimensions. The symbols $R^*_{\alpha\beta}, R^*$ are giving the Ricci tensor and the Ricci scalar, respectively, only that these are built upon the scaled metric $G_{\alpha\beta}$. As our wrapper function $F[f]$ is totally arbitrary, we can easily apply conditions, where the resulting field equations take on the simplest forms in any number of dimensions n. As shown in [2], we result in two equations for all cases $n > 2$, namely:

$$\textbf{1.} \quad R = 4 \cdot \frac{n-1}{n-2} \cdot \frac{1}{C_f + f} \cdot \Delta f$$

$$\textbf{2.} \quad R_{\alpha\beta} = 4 \cdot \frac{n-1}{n-2} \cdot \frac{1}{C_f + f} \cdot f_{:\alpha\beta}. \tag{1460}$$

Thus, different to the classical Einstein field equations, where there is only a set of tensor equations, we have a scalar and a tensor equation in the new, more general case. We also find that the inconsistency of the classical theory is resolved without the need of a cosmological constant or matter terms. The only thing of need was a scale factor $F[f]$ of the metric.

One might argue now that the classical Hilbert variation should also work and give consistent results, simply because we could always assume a scaled metric where the scale factor is sucked into the various components like, for instance, here demonstrated on a scaled Schwarzschild solution [11]:

$$G_{\alpha\beta} = F[f] \cdot \begin{pmatrix} -c^2 \cdot \left(1 - \dfrac{r_s}{r}\right) & 0 & 0 & 0 \\ 0 & \dfrac{1}{\left(1 - \dfrac{r_s}{r}\right)} & 0 & 0 \\ 0 & 0 & r^2 & 0 \\ 0 & 0 & 0 & r^2 \cdot \sin(\vartheta)^2 \end{pmatrix}$$

$$= \begin{pmatrix} -F[f] \cdot c^2 \cdot \left(1 - \dfrac{r_s}{r}\right) & 0 & 0 & 0 \\ 0 & \dfrac{F[f]}{\left(1 - \dfrac{r_s}{r}\right)} & 0 & 0 \\ 0 & 0 & F[f] \cdot r^2 & 0 \\ 0 & 0 & 0 & F[f] \cdot r^2 \cdot \sin(\vartheta)^2 \end{pmatrix}. \tag{1461}$$

Then variation with respect to such a metric leads to the classical field equations only that instead of the ordinary R- and $R_{\alpha\beta}$-terms we have:

$$\delta W = 0 = \delta_{G_{\alpha\beta}} \int_V d^n x \left(\sqrt{-G} \times R^*\right) = \delta_{G_{\alpha\beta}} \int_V d^n x \left(\sqrt{-g \cdot F^n} \times R^*\right)$$

$$= \int_V d^n x \left(\sqrt{-G} \times \left(R^*_{\alpha\beta} - \frac{1}{2} R^* G_{\alpha\beta} \right) \delta G^{\alpha\beta} \right). \tag{1462}$$

And only considering the scaled metric and its derivatives, one would indeed be left with the task of having to solve the classical Einstein field equations with just:

$$R^*_{\alpha\beta} - \frac{1}{2} R^* G_{\alpha\beta} = 0. \tag{1463}$$

Well, this argument seems to have its justification, on first shallow sight, but one should not forget that the complete integrand of the variation in Eq. (1462) is not just given by Eq. (1463), but actually reads:

$$\sqrt{-G} \times \left(R^*_{\alpha\beta} - \frac{1}{2} R^* G_{\alpha\beta} \right) \delta G^{\alpha\beta}. \tag{1464}$$

Thus, we have to fulfill the equation:

$$\sqrt{-G} \times \left(R^*_{\alpha\beta} - \frac{1}{2} R^* G_{\alpha\beta} \right) \delta G^{\alpha\beta} = 0. \tag{1465}$$

Ruling out the determinant G of the metric to give zero (which is also a possibility we had discussed earlier, by the way), we still have, in addition to the $\boxed{R^*_{\alpha\beta} - \frac{1}{2} R^* G_{\alpha\beta}}$ -term, the variation $\delta G^{\alpha\beta}$ and this, as it was shown earlier (c.f. [2]), can be split up into:

$$\delta G^{\alpha\beta} = g^{\alpha\beta} \cdot \delta \frac{1}{F} + \frac{1}{F} \cdot \delta g^{\alpha\beta}. \tag{1466}$$

In consequence the total variation gives:

$$\sqrt{-G} \times \left(R^*_{\alpha\beta} - \frac{1}{2} R^* G_{\alpha\beta} \right) \left(g^{\alpha\beta} \cdot \delta \frac{1}{F} + \frac{1}{F} \cdot \delta g^{\alpha\beta} \right) = 0. \tag{1467}$$

This, however, results in the two field equations (1460), which we consider quantum Einstein field equations.

In other words, Hilbert in principle had it [3], but in not recognizing and exploiting the possible inner, respectively scaled structure of the metric, his variation was not complete. The completed variation, taking this inner structure of the variated metric into account (1466), results in two equations, a scalar and a tensor.

10.6 The Classical Dirac Equation, Derived from the Einstein–Hilbert Action

The classical Dirac approach for the derivation of the Dirac equation [12] is an operational factorization of the Klein–Gordon equation. As the latter was directly derived from the Einstein–Hilbert action with a scaled metric (see [5–10] or previous chapters, especially 6 and 8, in here), leading to the Ricci scalar Eq. (1008), we might find ourselves in the understanding of also having—automatically—solved the metric derivation or origination of the Dirac equation. This assumption, however, is wrong, as we have shown in chapter 8. Indeed, the fact that the scaled Ricci scalar does not necessarily need to give an equivalence to the mass-term M^2 (with: $M^2 = \dfrac{m^2 \cdot c^2}{\hbar^2}$, thereby m = rest mass, $\hbar$ = reduced Planck's constant, c = speed of light in vacuum), because this could be achieved via additional dimensions and suitable entanglement to our ordinary space-time [6], saved us quite some problems, but there is still the issue of the wave-function f suddenly becoming a vector in the Dirac picture, while it was a scalar in the Klein–Gordon realm, and also has to be a scalar in our derivation of the Klein–Gordon equation from the Einstein–Hilbert action (c.f. section 6.11 and papers [1, 2, 5–9]). While in the Dirac approach this was not a problem at all, because both, the Klein–Gordon and the Dirac equations were postulations anyway, the question how and why f should be a vector comes of importance.

Here, we now want to sort this out.

At first, we apply the following split-up option (c.f. section "10.3 Quantum Einstein Field Equations in their Simplest Form"):

$$0 = \left(\begin{array}{l} \displaystyle\int_{V_G} d^n x \sum_{i=1}^{N} \sqrt{-g} \cdot F_i^n \left(\left(R_{\alpha\beta} - \frac{F_i'}{F_i}(n-1)\left(f_{i,\alpha\beta} - \frac{g_{\alpha\beta}}{n}\Gamma^\gamma_{\kappa\lambda} g^{\kappa\lambda} f_{i,\gamma} \right) \right) \delta G^{\alpha\beta} \right) \\[4mm] \displaystyle + \frac{1}{2} \cdot \sum_{i=1}^{N} \int_{V_{Qi}} d^n x \sqrt{-g} \cdot F_i^n \left(\frac{F_i'}{F_i}(n-1)\Delta f_i - R \right) \cdot g_{\alpha\beta} \delta G^{\alpha\beta} \end{array} \right).$$

$$(1468)$$

It should be noted that, as shown for instance in [6], the whole could also be derived with the scaling split-up:

$$
0 = \left(\begin{array}{l} \int\limits_{V_G} d^n x \sum_{i=1}^{N} \sqrt{-g \cdot F_i^{\,n}} \left(\left(R_{\alpha\beta} - \frac{1}{2} \cdot R \cdot g_{\alpha\beta} - \frac{F_i'}{F_i}(n-1) \left(f_{i,\alpha\beta} - \frac{g_{\alpha\beta}}{n} \Gamma_{\kappa\lambda}^{\gamma} g^{\kappa\lambda} f_{i,\gamma} \right) \right) \delta G^{\alpha\beta} \right) \\ + \frac{1}{2} \cdot \sum_{i=1}^{N} \int\limits_{V_{Qi}} d^n x \sqrt{-g \cdot F_i^{\,n}} \left(\frac{F_i'}{F_i}(n-1) \Delta f_i \right) \cdot g_{\alpha\beta} \delta G^{\alpha\beta} \end{array}\right)
$$

$$(1469)$$

and the introduction of masses via additional dimensions and their entanglement with our ordinary 4-dimensional space-time. Quite some examples, also containing spin-1/2-solutions, are also given in chapters 9, 11, and [10]. Here, however, we will apply Eq. (1468), as this coincides structurally with the Klein–Gordon equation from section 6.11, $F = F[f]$ is always fixed via the condition in Eq. (1441). We demand both lines in Eq. (1468) to give zero separately as follows:

$$
0 = \int\limits_{V_G} d^n x \sum_{i=1}^{N} \sqrt{-g \cdot F_i^{\,n}} \left(\left(R_{\alpha\beta} - \frac{F_i'}{F_i}(n-1) \left(f_{i,\alpha\beta} - \frac{g_{\alpha\beta}}{n} \Gamma_{\kappa\lambda}^{\gamma} g^{\kappa\lambda} f_{i,\gamma} \right) \right) \delta G^{\alpha\beta} \right)
$$

$$
0 = \sum_{i=1}^{N} \int\limits_{V_{Qi}} d^n x \sqrt{-g \cdot F_i^{\,n}} \left(\frac{F_i'}{F_i}(n-1) \Delta f_i - R \right) \cdot g_{\alpha\beta} \delta G^{\alpha\beta}. \qquad (1470)
$$

10.6.1 Insertion: The Classical Way to Obtain Curved Dirac Equations

We briefly here want to cover the classical way to factorize the Klein–Gordon equation, respectively the Laplace operator, which already was presented in [8] and the previous chapter. We start with the so-called curved Dirac equation:

$$
\left(i\hbar \gamma^\mu \nabla_\mu - m \cdot c \right) \Psi = 0, \qquad (1471)
$$

where we have the following curved "quadratic Nabla" operator:

$$
\nabla_\mu \Psi = \left(\partial_\mu - \frac{i}{4} \cdot \omega_\mu^{ab} \sigma_{ab} \right) \Psi. \qquad (1472)
$$

Thereby the function Ψ has to be considered a vector $\mathbf{f_i}$ and the spin connection ω_μ^{ab} can be derived as follows:

$$\omega_{\mu}^{ab} = \begin{pmatrix} \frac{1}{2}e^{av}\left(\partial_{\mu}e_{v}^{b} - \partial_{v}e_{\mu}^{b}\right) + \frac{1}{4}e^{ap}e^{b\sigma}\left(\partial_{\sigma}e_{\rho}^{c} - \partial_{\rho}e_{\sigma}^{c}\right)e_{c\mu} \\ -\frac{1}{2}e^{bv}\left(\partial_{\mu}e_{v}^{a} - \partial_{v}e_{\mu}^{a}\right) + \frac{1}{4}e^{bp}e^{a\sigma}\left(\partial_{\sigma}e_{\rho}^{c} - \partial_{\rho}e_{\sigma}^{c}\right)e_{c\mu} \end{pmatrix}$$

$$\eta_{ab}e_{\mu}^{a}e_{v}^{b} = g_{\mu v}; \quad e^{av} = g^{\mu v}e_{\mu}^{a}; \quad e^{av} = \eta^{ab}e_{b}^{v}; \quad \sigma^{ab} = \frac{i}{2}\left(\gamma^{a}\gamma^{b} + \gamma^{b}\gamma^{a}\right). \quad (1473)$$

Here the Minkowski metric is given via η_{ab}, η^{ab} and we have the symbol e for the various forms of the tetrad or base-vector. We find the curved Clifford algebra for the curved gamma matrices with the important connection:

$$\left(\gamma^{\mu}\gamma^{v} + \gamma^{v}\gamma^{\mu}\right) = 2 \cdot g^{\mu v}. \quad (1474)$$

Using the tetrad one can express the curved gamma matrices by the usual flat ones applying the formula:

$$\gamma^{\mu} = e_{a}^{\mu}\gamma^{a}. \quad (1475)$$

An example for the situation in a Minkowski space-time in spatial spherical coordinates was given in chapter 9.

10.6.2 Back to the Main Section

While Eq. (1195) should operate as follows:

$$\left(i\hbar\gamma^{\mu}\nabla_{\mu} + m \cdot c\right)\left(i\hbar\gamma^{\mu}\nabla_{\mu} - m \cdot c\right)\Psi = \left(\hbar^{2} \cdot \Delta - m^{2} \cdot c^{2}\right)\Psi \quad (1476)$$

and thus, provides us with the desired factorization of the Klein–Gordon equation (on the right-hand side of the first line of Eqs. (1200)), we directly have from the second line of Eq. (1468) or (1470):

$$0 = \sum_{i=1}^{N}\int_{V_{Qi}} d^{n}x\sqrt{-g} \cdot F_{i}^{n}\left(\frac{F_{i}'}{F_{i}}(n-1)\Delta f_{i} - R\right) \cdot g_{\alpha\beta}\delta G^{\alpha\beta}$$

$$\Rightarrow \qquad 0 = \sum_{i=1}^{N}\sqrt{F_{i}^{n}}\left(\frac{F_{i}'}{F_{i}}(n-1)\Delta f_{i} - R\right). \quad (1477)$$

Together with the condition in Eq. (1441), this results in the following:

$$\Rightarrow \quad 0 = \sum_{i=1}^{N}\sqrt{f_{i}^{n\cdot\frac{4}{n-2}-2}}\left(\frac{4\cdot(n-1)}{(n-2)}\Delta f_{i} - f_{i} \cdot R\right) = \sum_{i=1}^{N}f_{i}^{\frac{n+2}{n-2}} \cdot \left(\frac{4\cdot(n-1)}{(n-2)}\Delta f_{i} - f_{i} \cdot R\right).$$

$$(1478)$$

This, however, does not give us the desired structure of the right-hand side of Eq. (1200), unless one could assume that each of the function f_i fulfills the Klein–Gordon equation:

$$0 = \frac{4 \cdot (n-1)}{(n-2)} \Delta f_i - f_i \cdot R. \tag{1479}$$

For completeness, we remember that the variational term $\delta G^{\alpha\beta}$ should be written as given in Eq. (1466), which means $\delta G^{\alpha\beta} = g^{\alpha\beta} \cdot \delta \frac{1}{F} + \frac{1}{F} \cdot \delta g^{\alpha\beta}$.

Ignoring the variation of the F^{-1}-term (by considering it negligible) we have to change Eq. (1477) as follows:

$$0 = \sum_{i=1}^{N} \int_{V_{Qi}} d^n x \sqrt{-g} \cdot F_i^n \left(\frac{F_i'}{F_i}(n-1)\Delta f_i - R \right) \cdot \frac{g_{\alpha\beta}}{F_i} \delta g^{\alpha\beta}$$

$$\Rightarrow \quad \left\{ 0 = \sum_{i=1}^{N} \sqrt{F_i^{n-2}} \left(\frac{F_i'}{F_i}(n-1)\Delta f_i - R \right) = \sum_{i=1}^{N} \left(\sqrt{F_i^{n-4}} \cdot F_i' \cdot (n-1)\Delta f_i - \sqrt{F_i^{n-2}} \cdot R \right) \right.$$
$$\tag{1480}$$

Unless demanding that each f_i fulfills the Klein–Gordon equation, we realize— once again—that this cannot lead to the right-hand side of Eq. (1200) when still sticking to the condition in Eq. (1441). But by changing Eq. (1468) to:

$$0 = \begin{pmatrix} \int_{V_G} d^n x \sum_{i=1}^{N} \sqrt{-g} \cdot F_i^n \begin{pmatrix} \left(R_{\alpha\beta} - \frac{R}{2} \cdot g_{\alpha\beta} - \frac{F_i'}{F_i}(n-1)\left(f_{i,\alpha\beta} - \frac{g_{\alpha\beta}}{n}\Gamma_{\kappa\lambda}^{\gamma} g^{\kappa\lambda} f_{i,\gamma} \right) \right. \\ + \frac{(1-n)}{4F_i^2}\begin{pmatrix} F_i F_i'' \\ +F_i'F_i'(n-6) \end{pmatrix} \begin{pmatrix} f_{i,\alpha}f_{i,\beta} \\ -f_{i,\mu}f_{i,\nu}g^{\mu\nu}\frac{g_{\alpha\beta}}{2} \end{pmatrix} \end{pmatrix} \delta G^{\alpha\beta} \\ + \frac{1}{2} \cdot \sum_{i=1}^{N} \int_{V_{Qi}} d^n x \sqrt{-g} \cdot F_i^n \frac{F_i'}{F_i}(n-1)\Delta f_i \cdot g_{\alpha\beta}\delta G^{\alpha\beta} \end{pmatrix}$$

$$= \begin{pmatrix} \int_{V_G} d^n x \sum_{i=1}^{N} \sqrt{-g} \cdot F_i^n \begin{pmatrix} \left(R_{\alpha\beta} - \frac{R}{2} \cdot g_{\alpha\beta} - \frac{F_i'}{F_i}(n-1)\left(f_{i,\alpha\beta} - \frac{g_{\alpha\beta}}{n}\Gamma_{\kappa\lambda}^{\gamma} g^{\kappa\lambda} f_{i,\gamma} \right) \right. \\ + \frac{(1-n)}{4F_i^2}\begin{pmatrix} F_i F_i'' \\ +F_i'F_i'(n-6) \end{pmatrix} \begin{pmatrix} f_{i,\alpha}f_{i,\beta} \\ -f_{i,\mu}f_{i,\nu}g^{\mu\nu}\frac{g_{\alpha\beta}}{2} \end{pmatrix} \end{pmatrix} \delta G^{\alpha\beta} \\ + \frac{1}{2} \cdot \sum_{i=1}^{N} \int_{V_{Qi}} d^n x \sqrt{-g} \cdot F_i^n \frac{F_i'}{F_i^2}(n-1)\Delta f_i \cdot g_{\alpha\beta}\delta g^{\alpha\beta} \end{pmatrix}$$
$$\tag{1481}$$

assuming the Einstein field equations classically fulfilled, which is to say $R_{\alpha\beta} - \dfrac{R}{2} \cdot g_{\alpha\beta} = 0$ choosing $n = 4$, we end up with:

$$
0 = \left(
\begin{array}{l}
\displaystyle\int_{V_G} d^n x \sum_{i=1}^{N} \sqrt{-g} \cdot F_i^{n-4} \left(\left(
\begin{array}{l}
-F_i'(n-1)\left(f_{i,\alpha\beta} - \dfrac{g_{\alpha\beta}}{n} \Gamma_{\kappa\lambda}^{\gamma} g^{\kappa\lambda} f_{i,\gamma} \right) \\[2ex]
+\dfrac{(1-n)}{4F_i^2}\left(\begin{array}{l} F_i F_i'' \\ +F_i'F_i'(n-6) \end{array} \right)\left(\begin{array}{l} f_{i,\alpha} f_{i,\beta} \\ -f_{i,\mu} f_{i,\nu} g^{\mu\nu} \dfrac{g_{\alpha\beta}}{2} \end{array} \right)
\end{array}
\right) \delta g^{\alpha\beta}\right) \\[6ex]
+\dfrac{1}{2} \cdot \displaystyle\sum_{i=1}^{N} \int_{V_{Qi}} d^n x \sqrt{-g} \cdot F_i^{n} \dfrac{F_i'}{F_i^2}(n-1)\Delta f_i \cdot g_{\alpha\beta}\delta g^{\alpha\beta}
\end{array}
\right)
$$

$$\xrightarrow{\;n=4\;}$$

$$
= 3 \cdot \left(
\begin{array}{l}
\displaystyle\int_{V_G} d^n x \sum_{i=1}^{N} \sqrt{-g} \left(\left(
\begin{array}{l}
-\dfrac{1}{F_i}\left(2F_i F_i'' - F_i'F_i'\right)\left(\begin{array}{l} f_{i,\alpha} f_{i,\beta} \\ -f_{i,\mu} f_{i,\nu} g^{\mu\nu} \dfrac{g_{\alpha\beta}}{2} \end{array} \right) \\[2ex]
-F_i'\left(f_{i,\alpha\beta} - \dfrac{g_{\alpha\beta}}{4} \Gamma_{\kappa\lambda}^{\gamma} g^{\kappa\lambda} f_{i,\gamma} \right)
\end{array}
\right) \delta g^{\alpha\beta}\right) \\[6ex]
+\dfrac{1}{2} \cdot \displaystyle\sum_{i=1}^{N} \int_{V_{Qi}} d^n x \sqrt{-g} \cdot F_i'\Delta f_i \cdot g_{\alpha\beta}\delta g^{\alpha\beta}
\end{array}
\right)
$$

$$
= 3 \cdot \left(
\begin{array}{l}
\displaystyle\sum_{i=1}^{N} \int_{V_{Qi}} d^n x \sqrt{-g} \cdot \left(
\begin{array}{l}
-\dfrac{1}{F_i}\left(2F_i F_i'' - F_i'F_i'\right)\left(\begin{array}{l} f_{i,\alpha} f_{i,\beta} \\ -f_{i,\mu} f_{i,\nu} g^{\mu\nu} \dfrac{g_{\alpha\beta}}{2} \end{array} \right) \\[2ex]
-F_i'\left(f_{i,\alpha\beta} - \dfrac{g_{\alpha\beta}}{4} \Gamma_{\kappa\lambda}^{\gamma} g^{\kappa\lambda} f_{i,\gamma} \right)
\end{array}
\right) \delta g^{\alpha\beta} \\[6ex]
+\dfrac{1}{2} \cdot \displaystyle\sum_{i=1}^{N} \int_{V_{Qi}} d^n x \sqrt{-g} \cdot F_i'\Delta f_i \cdot g_{\alpha\beta}\delta g^{\alpha\beta}
\end{array}
\right).
$$

$$(1482)$$

Adjusting the integration according to:

$$0 = 3 \cdot \sum_{i=1}^{N} \int_{V_{Qi}} d^n x \sqrt{-g} \cdot \left(\begin{pmatrix} -\dfrac{1}{F_i}\left(2F_i F_i{}'' - F_i{}'F_i{}'\right) \begin{pmatrix} f_{i,\alpha} f_{i,\beta} \\ -f_{i,\mu} f_{i,\nu} g^{\mu\nu} \dfrac{g_{\alpha\beta}}{2} \end{pmatrix} \\ -F_i{}'\left(f_{i,\alpha\beta} - \dfrac{g_{\alpha\beta}}{4}\Gamma^{\gamma}_{\kappa\lambda} g^{\kappa\lambda} f_{i,\gamma}\right) \\ +\dfrac{1}{2}F_i{}'\Delta f_i \cdot g_{\alpha\beta} \end{pmatrix} \delta g^{\alpha\beta} \right),$$

$$(1483)$$

would allow us to demand the following:

$$0 = \begin{pmatrix} -\dfrac{1}{F_i}\left(2F_i F_i{}'' - F_i{}'F_i{}'\right) \begin{pmatrix} f_{i,\alpha} f_{i,\beta} \\ -f_{i,\mu} f_{i,\nu} g^{\mu\nu} \dfrac{g_{\alpha\beta}}{2} \end{pmatrix} \\ -F_i{}'\left(f_{i,\alpha\beta} - \dfrac{g_{\alpha\beta}}{4}\Gamma^{\gamma}_{\kappa\lambda} g^{\kappa\lambda} f_{i,\gamma}\right) + \dfrac{1}{2}F_i{}'\Delta f_i \cdot g_{\alpha\beta} \end{pmatrix}$$

$$0 = -\dfrac{1}{F_i}\left(2\dfrac{F_i F_i{}''}{F_i{}'} - F_i{}'\right) \begin{pmatrix} f_{i,\alpha} f_{i,\beta} \\ -f_{i,\mu} f_{i,\nu} g^{\mu\nu} \dfrac{g_{\alpha\beta}}{2} \end{pmatrix} - \left(f_{i,\alpha\beta} - \dfrac{g_{\alpha\beta}}{4}\Gamma^{\gamma}_{\kappa\lambda} g^{\kappa\lambda} f_{i,\gamma}\right) + \dfrac{1}{2}\Delta f_i \cdot g_{\alpha\beta}.$$

$$(1484)$$

Summing over with the contra-variant metric tensor gives:

$$0 = -g^{\alpha\beta}\dfrac{1}{F_i}\left(2\dfrac{F_i F_i{}''}{F_i{}'} - F_i{}'\right) \begin{pmatrix} f_{i,\alpha} f_{i,\beta} \\ -f_{i,\mu} f_{i,\nu} g^{\mu\nu} \dfrac{g_{\alpha\beta}}{2} \end{pmatrix} - \left(g^{\alpha\beta} f_{i,\alpha\beta} - \Gamma^{\gamma}_{\kappa\lambda} g^{\kappa\lambda} f_{i,\gamma}\right) + 2\Delta f_i$$

$$= \underbrace{-g^{\alpha\beta}\dfrac{1}{F_i}\left(2\dfrac{F_i F_i{}''}{F_i{}'} - F_i{}'\right) \begin{pmatrix} f_{i,\alpha} f_{i,\beta} \\ -f_{i,\mu} f_{i,\nu} g^{\mu\nu} \dfrac{g_{\alpha\beta}}{2} \end{pmatrix}}_{=-f_i \cdot M^2} + \Delta f_i = \Delta f_i - f_i \cdot M^2. \quad (1485)$$

Thereby we have chosen to define the mass M via the expression:

$$f_i \cdot M^2 = f_i \cdot M\left[f_i, g^{\alpha\beta}\right]^2 \equiv g^{\alpha\beta} \frac{1}{F_i}\left(2\frac{F_i F_i''}{F_i'} - F_i'\right)\left(\begin{array}{c} f_{i,\alpha} f_{i,\beta} \\ -f_{i,\mu} f_{i,\nu} g^{\mu\nu} \dfrac{g_{\alpha\beta}}{2} \end{array}\right).$$

(1486)

Now as we see that our result at the end of Eq. (1485) perfectly mirrors the right-hand side of Eq. (1200), which could also be given as follows:

$$\left(\hbar^2 \cdot \Delta - m^2 \cdot c^2\right)\Psi = \sum_{i=1}^{N}\left(\hbar^2 \cdot \Delta - m^2 \cdot c^2\right)f_i = 0 \qquad (1487)$$

and already provides the necessary vector character of f_i, we could consider the task of deriving the Dirac equation from the Einstein–Hilbert action completed.

However, as we will see below, there is a peculiar second possibility to obtain a Dirac or Dirac-like equation and this fact alone probably merits quite some discussion.

Before we come to that, however, we intend to generalize our result to arbitrary n and will also demonstrate that there is a much simpler and more direct way to obtain the classical Dirac equation out of the Einstein–Hilbert action.

10.6.3 Extension to Arbitrary n and Non-Vacuum Einstein Compatible Space-Times

We omit the condition $R_{\alpha\beta} - \dfrac{R}{2}\cdot g_{\alpha\beta} = 0$, go back to Eq. (1481) and end up with:

$$0 = \left(\begin{array}{c} \displaystyle\int_{V_G} d^n x \sum_{i=1}^{N} \sqrt{-g}\cdot F_i^{n-4}\left(\left(\left(R_{\alpha\beta} - \frac{R}{2}\cdot g_{\alpha\beta}\right)F_i - F_i'(n-1)\left(f_{i,\alpha\beta} - \frac{g_{\alpha\beta}}{n}\Gamma^{\gamma}_{\kappa\lambda} g^{\kappa\lambda} f_{i,\gamma}\right)\right) \\ + \frac{(1-n)}{4F_i^2}\left(\begin{array}{c} F_i F_i'' \\ +F_i' F_i'(n-6) \end{array}\right)\left(\begin{array}{c} f_{i,\alpha} f_{i,\beta} \\ -f_{i,\mu} f_{i,\nu} g^{\mu\nu} \dfrac{g_{\alpha\beta}}{2} \end{array}\right)\right)\delta g^{\alpha\beta}\right) \\ + \dfrac{1}{2}\cdot \displaystyle\sum_{i=1}^{N}\int_{V_{Qi}} d^n x \sqrt{-g}\cdot F_i^n \dfrac{F_i'}{F_i^2}(n-1)\Delta f_i \cdot g_{\alpha\beta}\delta g^{\alpha\beta} \end{array}\right)$$

$$
= \sum_{i=1}^{N} \int_{V_{Qi}} d^n x \sqrt{-g} \cdot F_i^{n-4} \left(\frac{\left(R_{\alpha\beta} - \frac{R}{2} \cdot g_{\alpha\beta} \right) F_i}{n-1} - F_i' \left(f_{i,\alpha\beta} - \frac{g_{\alpha\beta}}{n} \Gamma_{\kappa\lambda}^{\gamma} g^{\kappa\lambda} f_{i,\gamma} - \frac{\Delta f_i}{2} \cdot g_{\alpha\beta} \right) \\ - \frac{1}{4F_i^2} \left(\begin{matrix} F_i F_i'' \\ + F_i' F_i' (n-6) \end{matrix} \right) \left(\begin{matrix} f_{i,\alpha} f_{i,\beta} \\ -f_{i,\mu} f_{i,\nu} g^{\mu\nu} \frac{g_{\alpha\beta}}{2} \end{matrix} \right) \right) \delta g^{\alpha\beta}
$$

$$(1488)$$

and repeating the procedure from above, we result in:

$$
0 = \left(\frac{R}{(n-1)} \cdot \left(1 - \frac{n}{2} \right) F_i - F_i' \left(g^{\alpha\beta} f_{i,\alpha\beta} - \Gamma_{\kappa\lambda}^{\gamma} g^{\kappa\lambda} f_{i,\gamma} - \frac{\Delta f_i}{2} \cdot n \right) \\ - \frac{1}{4F_i} \left(\begin{matrix} F_i F_i'' \\ + F_i' F_i' (n-6) \end{matrix} \right) \left(\begin{matrix} f_{i,\alpha} f_{i,\beta} \\ -f_{i,\mu} f_{i,\nu} g^{\mu\nu} \frac{g_{\alpha\beta}}{2} \end{matrix} \right) \right)
$$

$$
= \left(\left(\frac{R \cdot F_i}{(n-1)} - \frac{f_{i,\alpha} f_{i,\beta}}{4F_i} \left(F_i F_i'' + F_i' F_i' (n-6) \right) \right) \cdot \left(1 - \frac{n}{2} \right) - F_i' \left(1 - \frac{n}{2} \right) \Delta f_i \right)
$$

$$
\Rightarrow \qquad 0 = \underbrace{\frac{f_{i,\alpha} f_{i,\beta}}{4F_i F_i'} \left(F_i F_i'' + F_i' F_i' (n-6) \right) - \frac{R \cdot F_i}{F_i'(n-1)} + \Delta f_i}_{\equiv -f_i \cdot M^2}. \qquad (1489)
$$

As before, we have chosen to define the mass M such that the Laplace operator term remains via the expression:

$$
f_i \cdot M^2 = f_i \cdot M \left[f_i, g^{\alpha\beta} \right]^2 \equiv \frac{R \cdot F_i}{F_i'(n-1)} - \frac{f_{i,\alpha} f_{i,\beta}}{4F_i} g^{\alpha\beta} \left(\frac{4F_i F_i''}{F_i'} + F_i'(n-6) \right).
$$

$$(1490)$$

Again we see that our result at the end of Eq. (1489) perfectly mirrors the right-hand side of Eq. (1200) and already provides the necessary vector character of f_i. Thus, we could consider the task of deriving the Dirac equation from the Einstein–Hilbert action in the case of arbitrary n and curved space-times completed. The remaining task consist of the quaternion factorization of the Klein–Gordon equation at the end of Eq. (1489), respectively:

$$
0 = \Delta f_i - f_i \cdot M^2. \qquad (1491)
$$

However, as already hinted above and as we will show below, there is a peculiar second possibility to obtain a Dirac or Dirac-like equation.

Please note that while in the simplified cases in Eqs. (1485) and (1486), we can always transform away the mass via the condition in Eq. (1441) and thus, a suitable choice for $F_i[f_i]$, this is not any longer possible in the general cases Eqs. (1489) and (1490). In the latter case, the application of the condition in Eq. (1441) just leads to $(n \neq 2)$:

$$f_i \cdot M^2 = f_i \cdot M\left[f_i, g^{\alpha\beta}\right]^2 = \frac{R \cdot F_i}{F_i'(n-1)} = \frac{R}{n-1} \cdot \frac{n-2}{4} f_i \quad \Rightarrow \quad M^2 = \frac{R}{n-1} \cdot \frac{n-2}{4}.$$

(1492)

An exception we have in the case $n = 2$, which results in (C_f = arbitrary constant):

$$f_i \cdot M^2 = f_i \cdot M\left[f_i, g^{\alpha\beta}\right]^2 = \frac{R \cdot F_i}{F_i'(n-1)} = \frac{R}{n-1} \cdot \frac{1}{C_f} f_i \quad \Rightarrow \quad M^2 = \frac{R}{n-1} \cdot \frac{1}{C_f}.$$

(1493)

Now we can directly apply the classical Dirac approach, thereby ending up with the typical quaternion equations.

10.6.3.1 The Ricci-flat space-time

Thereby, which is to say, when having transformed away the mass M either directly via:

$$M^2 = M\left[f_i, g^{\alpha\beta}\right]^2 \equiv \frac{\dfrac{R \cdot F_i}{F_i'(n-1)} - \dfrac{f_{i,\alpha} f_{i,\beta}}{4F_i} g^{\alpha\beta}\left(\dfrac{4F_i F_i''}{F_i'} + F_i'(n-6)\right)}{f_i} = 0$$

(1494)

or by the means of the condition in Eq. (1441) in connection with a Ricci-flat space-time, which is to say $R = 0$, we could still have a mass. This was demonstrated in our previous papers, especially [6–8] and requires the entanglement of the ordinary 4D-space-time with additional dimensions.

Let us assume to have the following metric in 6 dimensions:

$$G_{\alpha\beta}^6 = \begin{pmatrix} -c^2 & 0 & 0 & 0 & 0 & 0 \\ 0 & 1 & 0 & 0 & 0 & 0 \\ 0 & 0 & 1 & 0 & 0 & 0 \\ 0 & 0 & 0 & 1 & 0 & 0 \\ 0 & 0 & 0 & 0 & \pm c_u^2 & 0 \\ 0 & 0 & 0 & 0 & 0 & \pm c_v^2 \end{pmatrix} \cdot F\left[g[u,v] \cdot f[t,x,y,z]\right]. \quad (1495)$$

Even though it does not matter here, we also give—for completeness—the solution for the function $F[g*f]$ from our condition in Eq. (1441):

$$4F_iF_i'' + (F_i')^2(n-6) = 0 \xrightarrow{n=6} F_i = C_{f1} \cdot f_i + C_{f0}. \qquad (1496)$$

We apply the separation approach for $g[u,v]$ via $g[u,v] = g_v[v]*g_u[u]$ and use the following functions:

$$g_v[v] = C_{v1} \cdot \cos[c_v \cdot E_v \cdot v] + C_{v2} \cdot \sin[c_v \cdot E_v \cdot v], \qquad (1497)$$

$$g_u[u] = C_{u1} \cdot \cos[c_u \cdot E_u \cdot u] + C_{u2} \cdot \sin[c_u \cdot E_u \cdot u]. \qquad (1498)$$

Now Eq. (1491) gives us:

$$0 = \Delta_G(f_i \cdot g) = \left(\pm(E_u^2 + E_v^2) + \Delta_{4D}\right)f_i[t,x,y,z]. \qquad (1499)$$

We realize that with Eq. (1499) we have just obtained the classical Klein–Gordon equation with quite some flexibility to add mass or other forms of inertia. For the reasons of recognition, we set:

$$-M^2 = -\frac{m^2 \cdot c^2}{\hbar^2} = \pm(E_u^2 + E_v^2), \qquad (1500)$$

obtain from Eq. (1499):

$$0 = \Delta_G(f_i \cdot g) = \left(\Delta_{4D} - \frac{m^2 \cdot c^2}{\hbar^2}\right)f_i[t,x,y,z], \qquad (1501)$$

and could now directly apply the Dirac factorization of the Klein–Gordon equation, we have obtained here.

It should be pointed out at this point that we only chose the case $n = 6$, because this leads to a very simple solution for the wrapper function $F[f]$ with respect to our condition in Eq. (1532). Otherwise, there is no restriction regarding the choice of n.

10.6.4 Instead of the Quaternions—The Stamler-Approach [15–18]

Reusing the result from Eq. (1489) and instead of applying the condition in Eq. (1441), we intend to try the Klein–Gordon condition:

$$0 = \frac{f_{i,\alpha}f_{i,\beta}}{4F_i}g^{\alpha\beta}\left(\frac{4F_iF_i''}{F_i'} + F_i'(n-6)\right) - \frac{R \cdot F_i}{F_i'(n-1)} + \Delta f_i$$

$$\Delta f_i = M^2 \cdot f_i$$

$$\Rightarrow \qquad 0 = \frac{f_{i,\alpha} f_{i,\beta}}{4F_i} g^{\alpha\beta} \left(\frac{4F_i F_i''}{F_i'} + F_i'(n-6) \right) - \frac{R \cdot F_i}{F_i'(n-1)} + M^2 \cdot f_i. \qquad (1502)$$

With the setting of $F[f] = f$ for all i, we obtain:

$$0 = f_{i,\alpha} f_{i,\beta} g^{\alpha\beta} + \left(M^2 - \frac{R}{(n-1)} \right) \cdot \frac{4}{(n-6)} \cdot f_i^2. \qquad (1503)$$

Using the base-vector decomposition gives us:

$$-f_{i,\alpha} f_{i,\beta} g^{\alpha\beta} = \overbrace{\left(M^2 - \frac{R}{(n-1)} \right) \cdot \frac{4}{n-6} \cdot f_i^2}^{\equiv \mu^2 = \mu \cdot \mu}$$

$$\xrightarrow{g^{\alpha\beta} = \mathbf{e}^\alpha \cdot \mathbf{e}^\beta}$$

$$-f_{i,\alpha} \mathbf{e}^\alpha \cdot \mathbf{e}^\beta f_{i,\beta} = \mu^2 \cdot f_i^2$$

$$\Rightarrow \qquad \mp f_i \mu = i \cdot f_{i,\alpha} \mathbf{e}^\alpha. \qquad (1504)$$

$$\Rightarrow \qquad 0 = i \cdot f_{i,\alpha} \mathbf{e}^\alpha \pm f_i \mu. \qquad (1505)$$

It has to be pointed out that our way of decomposition is slightly incorrect or at least requires some discussion as we have chosen to use a simple root extraction onto both sides of the second last line of Eq. (1504). This may be ok for the right-hand side, but could be seen as improper for the product $f_{i,\alpha} \mathbf{e}^\alpha \cdot \mathbf{e}^\beta f_{i,\beta}$. For here and now, we could just assume to make use of the principle degree of freedom in naming the dummy indices and considering $f_{i,\alpha} \mathbf{e}^\alpha \cdot \mathbf{e}^\alpha f_{i,\alpha} = (f_{i,\alpha} \mathbf{e}^\alpha)^2$. This, too, however, requires some discussion, because we are talking about a scalar product of vectors and not just a simple square of scalar objects. We will take care about these potential issues later on in this sub-section.

Please note that Eq. (1505) represents a generalized "Dirac" equation in an arbitrary number of dimensions. We say this and simultaneously apply the "..."-signs to emphasize the fact that Eq. (1505) is of Dirac character but does not fully mirror the classical form, yet. However, by reducing this general and arbitrarily n-dimensional equation to 4 dimensions, we almost directly recognize the Dirac equation, only that in our metric case the mass terms have to be at least a vector, while the Dirac matrices for the derivatives of the function vector $\mathbf{h}_\lambda$ (c.f. [19]) are becoming the metric base vector components. In order to obtain the classical Dirac equation in the natural units form:

$$0 = i \cdot \gamma^\alpha f_{i,\alpha} \pm f_i \cdot m, \tag{1506}$$

which we can also write as:

$$0 = i \cdot \gamma^\alpha f_{i,\alpha} \pm I \cdot f_i \cdot m, \tag{1507}$$

and with the Dirac Gamma matrices γ^α and the unit matrix I given as follows:

$$\gamma^0 = \begin{pmatrix} 1 & & & \\ & 1 & & \\ & & -1 & \\ & & & -1 \end{pmatrix}; \; \gamma^1 = \begin{pmatrix} & & & 1 \\ & & 1 & \\ & -1 & & \\ -1 & & & \end{pmatrix}$$

$$\gamma^2 = \begin{pmatrix} & & & -i \\ & & i & \\ & i & & \\ -i & & & \end{pmatrix}; \; \gamma^3 = \begin{pmatrix} & & 1 & \\ & & & -1 \\ -1 & & & \\ & 1 & & \end{pmatrix}; \; I = \begin{pmatrix} 1 & & & \\ & 1 & & \\ & & 1 & \\ & & & 1 \end{pmatrix},$$

$$\tag{1508}$$

we simply need to multiply Eq. (1505) with a suitable object $\mathbf{C}$. This gives us the identities[1]:

$$\mathbf{C} \cdot \mathbf{e}^\alpha = \gamma^\alpha; \quad \mathbf{C} \cdot \mathbf{\mu} = I \cdot m, \tag{1509}$$

and provides a connection between the classical quaternion and the metric Dirac equation. As the classical Dirac equation was given in Minkowski coordinates, with base vectors of the kind:

$$\mathbf{e}^0 = \begin{pmatrix} i \\ 0 \\ 0 \\ 0 \end{pmatrix}; \; \mathbf{e}^1 = \begin{pmatrix} 0 \\ 1 \\ 0 \\ 0 \end{pmatrix}; \; \mathbf{e}^2 = \begin{pmatrix} 0 \\ 0 \\ 1 \\ 0 \end{pmatrix}; \; \mathbf{e}^3 = \begin{pmatrix} 0 \\ 0 \\ 0 \\ 1 \end{pmatrix}, \tag{1510}$$

we could find the corresponding structure of the object $\mathbf{C}$ via the first equation in Eq. (1509) and it becomes clear that we need something of the kind:

$$C_\beta^\alpha \cdot \mathbf{e}^\beta = \gamma^\alpha, \tag{1511}$$

---

[1]It should be noted that classically one does not use the identity matrix for the second equation in Eq. (1509), but the so called β-matrix instead. It reads $\beta = \begin{pmatrix} 1 & & & \\ & 1 & & \\ & & -1 & \\ & & & -1 \end{pmatrix}$ and will be considered further below. For here and now it makes no difference for the results of this paper whether we use I or β in Eq. (1509).

$$\Rightarrow \quad \left\{ \begin{array}{l} C_\alpha^0 = \begin{pmatrix} -i & & & \\ & 1 & & \\ & & -1 & \\ & & & -1 \end{pmatrix} ; \quad C_\alpha^1 = \begin{pmatrix} & & & 1 \\ & & 1 & \\ & -1 & & \\ i & & & \end{pmatrix} \\[40pt] C_\alpha^2 = \begin{pmatrix} & & -i & \\ & i & & \\ i & & & \\ & & & -1 \end{pmatrix} ; \quad C_\alpha^3 = \begin{pmatrix} & 1 & & \\ & & & -1 \\ i & & & \\ & & 1 & \end{pmatrix} . \end{array} \right. \tag{1512}$$

From here we deduce that μ must be something more complex than just a vector as only the following approach gives us what we need:

$$C_\beta^\alpha \cdot \quad = I \cdot m \quad \Rightarrow \quad C_\beta^\alpha \cdot \mu_\lambda^\beta = \delta_\lambda^\alpha \cdot m = I \cdot m. \tag{1513}$$

Thereby we need to point out that the scalar character of m in the classical Dirac equation (1506) was postulated [12], while our new Dirac-like equation (1505) has a fundamental metric and extremal-principle origin. The result after solving Eq. (1513) with respect to the object μ_λ^β reads:

$$\Rightarrow \quad \left\{ \begin{array}{l} \mu_0^\alpha = \begin{pmatrix} i & & & \\ & 1 & & \\ & & -1 & \\ & & & -1 \end{pmatrix} ; \quad \mu_1^\alpha = \begin{pmatrix} & & & -i \\ & & -1 & \\ & 1 & & \\ 1 & & & \end{pmatrix} \\[40pt] \mu_2^\alpha = \begin{pmatrix} & & -1 & \\ & -i & & \\ -i & & & \\ i & & & \end{pmatrix} ; \quad \mu_3^\alpha = \begin{pmatrix} & & -i & \\ & & & 1 \\ 1 & & & \\ & -1 & & \end{pmatrix} \end{array} \right. \tag{1514}$$

and gives us the connection of the classical Dirac equation (1506) from [12] with our new metric Dirac equation (1505). We realize that in the metric picture, the corresponding inertia parameters are not just scalars. With this connection found, we automatically also assure that f solves the Klein–Gordon equation and thus, the condition $\Delta f_i = M^2 \cdot f_i$.

As the transformation of the metric to the classical form, if performed via the connections just worked out, could be seen as problematic (see [8]), it needs to be pointed out that the summation over the index β in Eq. (1511) is not the only option we have in order to obtain the appearance of the Dirac gamma matrices. Similarly, we may just perform the evaluation as follows:

$$C_a^{cb} \cdot e^\beta = C_a^{cb} \cdot E_b^{\ \beta} \equiv D_a^{c\beta} = \gamma^\beta, \tag{1515}$$

$$\Rightarrow \quad \begin{cases} C_a^{c0} = \begin{pmatrix} -i & & & \\ & 1 & & \\ & & -1 & \\ & & & -1 \end{pmatrix}; & C_a^{c1} = \begin{pmatrix} & & & 1 \\ & & 1 & \\ & -1 & & \\ i & & & \end{pmatrix} \\[3em] C_a^{c2} = \begin{pmatrix} & & & -i \\ & & i & \\ & i & & \\ -1 & & & \end{pmatrix}; & C_a^{c3} = \begin{pmatrix} & & 1 & \\ & & & -1 \\ i & & & \\ & 1 & & \end{pmatrix}. \end{cases} \tag{1516}$$

Consequently, we obtain for **μ**:

$$C_a^{cb} \cdot \mu = I \cdot m \quad \Rightarrow \quad C_a^{cb} \cdot \mu_{b\lambda}^a = \delta_\lambda^c \cdot m = I \cdot m. \tag{1517}$$

The result after solving Eq. (1517) with respect to the object $\mu_{b\lambda}^a$ reads:

$$\Rightarrow \quad \begin{cases} \mu_{0\lambda}^a = \begin{pmatrix} i & & & \\ & 1 & & \\ & & -1 & \\ & & & -1 \end{pmatrix}; & \mu_{1\lambda}^a = \begin{pmatrix} & & & -i \\ & & -1 & \\ & 1 & & \\ 1 & & & \end{pmatrix} \\[3em] \mu_{2\lambda}^a = \begin{pmatrix} & & & -1 \\ & & -i & \\ & -i & & \\ i & & & \end{pmatrix}; & \mu_{3\lambda}^a = \begin{pmatrix} & & & -i \\ & & 1 & \\ 1 & & & \\ & -1 & & \end{pmatrix}. \end{cases} \tag{1518}$$

Here it has to be pointed out, however, that our complex forms of **μ** are a result of the approximation scalarization in Eq. (1517), which we need to reproduce the classical Dirac form in Eq. (1507) and not necessarily a genuine equivalence to real inertia. More discussion will be of need here.

We also need to point out that there is still the task to make the Klein–Gordon-setting for the Laplace term (c.f. our evaluation step from Eq. (1502)) and the resulting metric Dirac equation (1505), consistent with each other regarding the mass-terms. We shall leave this task for interested readers.

Insisting on the outcome of the β-matrix with $\beta = \begin{pmatrix} 1 & & & \\ & 1 & & \\ & & -1 & \\ & & & -1 \end{pmatrix}$ in

Eq. (1517), leading to the equation:

$$C_a^{cb} \cdot \mathbf{\mu} = \beta \cdot m \quad \Rightarrow \quad C_a^{cb} \cdot \mu_{b\lambda}^a = \beta \cdot m, \tag{1519}$$

does give the following solution with respect to the object $\mu_{b\lambda}^a$:

$$\Rightarrow \begin{cases} \mu_{0\lambda}^a = \begin{pmatrix} i & & & \\ & 1 & & \\ & & 1 & \\ & & & 1 \end{pmatrix}; \quad \mu_{1\lambda}^a = \begin{pmatrix} & & & i \\ & & 1 & \\ & 1 & & \\ 1 & & & \end{pmatrix} \\[3em] \mu_{2\lambda}^a = \begin{pmatrix} & & & 1 \\ & & i & \\ & -i & & \\ i & & & \end{pmatrix}; \quad \mu_{3\lambda}^a = \begin{pmatrix} & & i & \\ & & & -1 \\ 1 & & & \\ & -1 & & \end{pmatrix}. \end{cases} \tag{1520}$$

We saw that with a simple matrix multiplication we had come from our metric Dirac equation (1505) to the classical one Eq. (1507) (whether with the β-matrix or I does not matter):

$$0 = i \cdot f_{i,\alpha} e^\alpha \pm f_i \mu / \cdot C$$

$$\Rightarrow$$

$$0 = i \cdot \gamma^\alpha f_{i,\alpha} + \beta \cdot f_i \cdot m. \tag{1521}$$

Thereby, in the case of 4-dimensional Minkowski coordinates, the objects C and M are given via Eqs. (1516) and (1518). Our metric Dirac equation (1505) on the other hand, just came from a simple condition about the Ricci scalar. In chapter "11 A Curvy Math to Salvation," we will show that this is actually just a by-product of a more complete Einstein–Hilbert variation.

10.6.4.1 The factorization problem and simple derivation of the Dirac equation in its classical form

In connection with our way of decomposing the metric tensor into base vectors, we already hinted the problematic issue of using a simple root extraction onto both sides of the second last line of Eq. (1504). Here now we intend to avoid such a simplification by going back to Eq. (1503), once again defining the mass-inertia term μ^2 via:

$$0 = f_{i,\alpha} f_{i,\beta} g^{\alpha\beta} + \overbrace{\left(M^2 - \frac{R}{(n-1)} \right)}^{\equiv \mu^2 = \mu \cdot \mu} \frac{4}{(n-6)} \cdot f_i^2 \tag{1522}$$

and looking for a factorization of the kind:

$$0 = f_{i,\alpha} f_{i,\beta} g^{\alpha\beta} + \boldsymbol{\mu} \cdot \boldsymbol{\mu} \cdot f_i^2 = f_{i,\alpha} \mathbf{e}^\alpha \cdot \mathbf{e}^\beta f_{i,\beta} + \boldsymbol{\mu} \cdot \boldsymbol{\mu} \cdot f_i^2$$

$$= \left(f_{i,\alpha} \mathbf{e}^\alpha - i \cdot \boldsymbol{\mu} \cdot f_i \right) \left(\mathbf{e}^\beta f_{i,\beta} + i \cdot \boldsymbol{\mu} \cdot f_i \right)$$

$$\Rightarrow \qquad\qquad 0 = i \cdot f_{i,\alpha} \mathbf{e}^\alpha \pm f_i \boldsymbol{\mu}. \qquad (1523)$$

We see that this gives us our result from Eq. (1505) for a metric "Dirac equation."

Taking the result from the transformation to the original Dirac equation in Eq. (1521), we see that Eq. (1505) should be taken as:

$$0 = i \cdot f_{i,\alpha} \mathbf{e}^\alpha \pm f_i \boldsymbol{\mu} \;\Rightarrow\; 0 = i \cdot f_{i,\alpha} E_b{}^\alpha \pm f_i \mu_{b\lambda}^a. \qquad (1524)$$

The only way to make this equation reasonable is by setting the indexes in the equation on the right as follows:

$$0 = i \cdot f_{i,\alpha} E_b{}^\alpha \pm f_q \mu_{bi}^q. \qquad (1525)$$

Consequently, we face some problems in obtaining the classical Dirac equation (1521) (second line). While everything is still ok with the reproduction of the Dirac gamma matrices via multiplication and subsequent contraction with the **C**-objects from Eq. (1516):

$$C_a^{cb} \cdot \mathbf{e}^\alpha f_{i,\alpha} = C_a^{cb} \cdot E_b{}^\alpha f_{i,\alpha} \equiv D_a^{c\alpha} f_{i,\alpha} \Rightarrow D_a^{i\alpha} f_{i,\alpha} = \gamma^\alpha f_{i,\alpha}$$

$$\Rightarrow \qquad\qquad C_a^{ib} \cdot E_b{}^\alpha f_{i,\alpha} = D_a^{i\alpha} f_{i,\alpha} = \gamma^\alpha f_{i,\alpha}, \qquad (1526)$$

we see that things are compromised with respect to the mass-inertia term μ, where the transition has to be performed as follows:

$$C_a^{ib} \cdot f_q \mu_{bi}^q \;\Rightarrow\; C_a^{ib} \cdot \mu_{bi}^q = \beta_a^q \cdot m. \qquad (1527)$$

Here the problematic issue lies in the fact that summation over the index a is required in order to make Eq. (1521) (first line) or Eq. (1525) meaningful, while this does not allow to find an easy transition via our C-objects to the classical Dirac form in Eq. (1521) (second line) if not also demanding that the C- and the μ-objects are behaving Cartesian-like, which is to say:

$$C_a^{ib} = C_{ib}^a \;\Rightarrow\; \gamma^\alpha = \gamma_\alpha; \; \mu_q^{bi} = \mu_{bi}^q \;\Rightarrow\; \mu^\alpha = \mu_\alpha. \qquad (1528)$$

This is of need in order to allow for the factorization as given in Eq. (1523). As this only concerns the Latin letters in our metric Dirac equation (1525), we are faced with the task of finding "Cartesian" decompositions of the metric tensor and the mass-inertia terms.

A more comfortable way to extract the classical Dirac equation from Eq. (1522) would be as follows:

$$0 = f_{i,\alpha} f_{i,\beta} g^{\alpha\beta} + \mu^2 \cdot f_i^2 \Rightarrow \quad = \quad \Rightarrow f_{i,\alpha} f_{i,\beta} \left(\frac{\gamma^\alpha \gamma^\beta + \gamma^\beta \gamma^\alpha}{2} \right)$$

$$+ \frac{\beta^2 + \beta^2}{2} \cdot m^2 \cdot f_i^2 = 0$$

$$\Rightarrow \qquad f_{i,\alpha} f_{i,\beta} \gamma^\alpha \gamma^\beta + \beta^2 \cdot m^2 \cdot f_i^2 = 0$$

$$\Rightarrow \qquad \left(f_{i,\alpha} \gamma^\alpha + i \cdot \beta \cdot m \cdot f_i \right) \left(f_{i,\beta} \gamma^\beta - i \cdot \beta \cdot m \cdot f_i \right) = 0$$

$$\Rightarrow \qquad f_{i,\alpha} f_{i,\beta} \gamma^\beta \gamma^\alpha + \beta^2 \cdot m^2 \cdot f_i^2 = 0$$

$$\Rightarrow \qquad \left(f_{i,\alpha} \gamma^\alpha + i \cdot \beta \cdot m \cdot f_i \right) \left(f_{i,\beta} \gamma^\beta - i \cdot \beta \cdot m \cdot f_i \right) = 0. \tag{1529}$$

Comparing the results from Eq. (1525) (metric form) with the classical Dirac equation directly above, we see that in the metric form, the mass-inertia or curvature terms μ are of complex matrix character, while it is the derivative terms in the classical equation.

We, therefore, conclude that curvature, masses, and inertia are the true reason for the occurrence of spinors in space-time. As we have already shown in earlier papers that masses can be produced via entanglement with additional dimensions [6, 7, 8], we also suspect hidden (compactified) dimensions as the source for the occurrence of spinors. This we will further elaborate in our next chapter (see also [10]).

10.6.4.2 The Laplace condition

Note that instead of the Klein–Gordon condition, we may just also apply $\Delta f_i = 0$ which does not change Eq. (1505) but the parameter μ^2 in there as follows (with the setting of $F[f] = f$ for all i):

$$-f_{i,\alpha} f_{i,\beta} g^{\alpha\beta} = \overbrace{\left(\frac{R}{(n-1)} \right) \cdot \frac{4}{6-n}}^{\equiv \mu^2 = \mu \cdot \mu} \cdot f_i^2$$

$$\xrightarrow{g^{\alpha\beta} = e^\alpha \cdot e^\beta}$$

$$-f_{i,\alpha} e^\alpha \cdot e^\beta f_{i,\beta} = \mu^2 \cdot f_i^2$$

$$\Rightarrow \qquad 0 = i \cdot f_{i,\alpha} e^\alpha \pm f_i \mu. \tag{1530}$$

10.6.4.2.1 *Ricci-flat space-time and the metric Dirac equation*

As before with the classical Dirac approach and the corresponding Klein–Gordon starting point in the previous main section, we now want to consider the case of a Ricci-flat space-time. As we see from Eq. (1530), this would automatically give a massless equation due to the fact that we have $R = 0$. On the other hand, we know that mass can also be obtained via entanglement of dimensions (chapter "6 Masses and the Infinity Options Principle" and [6–8]) and thus, we want to investigate this possibility here for the otherwise flat space-time situation (here meaning Ricci-flat in the sense of $R = 0$ and not necessarily $R^* = 0$).

Let us assume to have the following metric in 6 dimensions:

$$G_{\alpha\beta}^6 = \begin{pmatrix} -c^2 & 0 & 0 & 0 & 0 & 0 \\ 0 & 1 & 0 & 0 & 0 & 0 \\ 0 & 0 & 1 & 0 & 0 & 0 \\ 0 & 0 & 0 & 1 & 0 & 0 \\ 0 & 0 & 0 & 0 & \pm c_u^2 & 0 \\ 0 & 0 & 0 & 0 & 0 & \pm c_v^2 \end{pmatrix} \cdot F\big[g[u,v] \cdot f[t,x,y,z]\big]. \quad (1531)$$

We know the solution for the function $F[g*f]$ from our condition (1441), which reads:

$$4F_iF_i'' + \left(F_i'\right)^2 (n-6) = 0 \xrightarrow{\ n=6\ } F_i = C_{f1} \cdot f_i + C_{f0}. \quad (1532)$$

As we do not want to have the first derivative terms to vanish, however, we apply $F[f] = f^2/2$ and obtain from the last line of Eq. (1489) for a Ricci-flat space-time $R = 0$ the following equation:

$$0 = f_{i,\alpha} f_{i,\beta} g^{\alpha\beta} \left(\frac{1}{2 \cdot f_i}\right) + \Delta f_i. \quad (1533)$$

We apply the separation approach for $g[u,v]$ via $g[u,v] = g_v[v]*g_u[u]$ and use the following functions:

$$g_v[v] = C_{v\pm} \cdot \exp\big[\pm c_v \cdot E_v \cdot v\big], \quad (1534)$$

$$g_u[u] = C_{u\pm} \cdot \exp\big[\pm c_u \cdot E_u \cdot u\big]. \quad (1535)$$

Now Eq. (1489) gives us:

$$0 = f_{i,\alpha} f_{i,\beta} g^{\alpha\beta} + 2 \cdot f_i \Delta f_i = f_{i,\alpha} f_{i,\beta} g^{\alpha\beta} + 2 \cdot f_i \left(\left(E_u^2 + E_v^2\right) + \Delta_{4D}\right) f_i[t,x,y,z].$$

$$(1536)$$

Sticking to the Laplace condition $\Delta f_i = 0$, this equation simplifies to:

$$0 = f_{i,\alpha}f_{i,\beta}g^{\alpha\beta} + \overbrace{2 \cdot f_i \Delta f_i}^{=0} = \left(f_{i,a}f_{i,b}g^{ab} \pm \left(E_u^2 + E_v^2\right) \cdot f_i[t,x,y,z]^2\right)g_u^2[u]g_v^2[v]$$

$$0 = f_{i,a}f_{i,b}g^{ab} \pm \left(E_u^2 + E_v^2\right) \cdot f_i[t,x,y,z]^2 ; \quad a,b = 0,1,2,3$$

$$\Rightarrow \qquad -f_{i,a}f_{i,b}g^{ab} = \pm\left(E_u^2 + E_v^2\right) \cdot f_i[t,x,y,z]^2. \qquad (1537)$$

Applying the base vector decomposition from Eq. (1530), we obtain a first-order differential equation as follows:

$$-f_i[t,x,y,z]_{,a}\, f_i[t,x,y,z]_{,b}\, g^{ab} = \pm\overbrace{\left(E_u^2 + E_v^2\right)}^{\equiv \mu^2 = \mu\cdot\mu} \cdot f_i[t,x,y,z]^2$$

$$\xrightarrow{\;g^{\alpha\beta}=\mathbf{e}^\alpha\cdot\mathbf{e}^\beta\;}$$

$$-f_{i,\alpha}\mathbf{e}^\alpha \cdot \mathbf{e}^\beta f_{i,\beta} = \mu^2 \cdot f_i^2$$

$$\Rightarrow \qquad 0 = i \cdot f_{i,\alpha}\mathbf{e}^\alpha \pm f_i\mu. \qquad (1538)$$

We realize that with Eq. (1538) we can still have a metric Dirac equation with mass even when $R = 0$ and $\Delta f_i = 0$. Thereby the latter condition forces us to only apply Laplace solutions in $n = 6$, which are, due to the entanglement of the t-x-y-z-space-time with the u-v-coordinates, just Klein–Gordon solutions in 4D to the following equation:

$$0 = \left(\pm\left(E_u^2 + E_v^2\right) + \Delta_{4D}\right)f_i[t,x,y,z]. \qquad (1539)$$

As we know that also the solutions to the classical Dirac equation are all solutions to the Klein–Gordon equation, the situation is not new to us.

10.6.4.3 The $F' = 0$-condition

Similarly to the Laplace condition, one might also demand $F' = 0$, which gives us from the last line of Eq. (1489):

$$0 = \frac{R \cdot F_i}{(n-1)} - f_{i,\alpha}f_{i,\beta}g^{\alpha\beta}F_i''. \qquad (1540)$$

As with the Laplace condition above, this does not change Eq. (1505) but we have to adjust the parameter μ^2 in there as follows:

$$f_{i,\alpha}f_{i,\beta}g^{\alpha\beta} = \overbrace{\left(\frac{R}{(n-1)} \cdot \frac{F_i}{F_i''}\right)}^{\equiv \mu^2 = \mu\cdot\mu}. \qquad (1541)$$

Thus, μ becomes f-dependent and we result in:

$$f_{i,\alpha} f_{i,\beta} g^{\alpha\beta} = \overbrace{\left(\frac{R}{(n-1)} \cdot \frac{F_i}{F_i''} \right)}^{=\mu[f]^2 = \mu[f] \cdot \mu[f]}$$

$$\xrightarrow{g^{\alpha\beta} = e^\alpha \cdot e^\beta}$$

$$f_{i,\alpha} e^\alpha \cdot e^\beta f_{i,\beta} = \mu[f]^2$$

$$\Rightarrow \qquad 0 = f_{i,\alpha} e^\alpha \pm \mu[f]. \qquad (1542)$$

Please note that this equation always requires $F[f]$ under the condition $F'[f] = 0$.

10.6.4.4 The $H[f]$-approach from chapter 9, section 9.4.10

Further generalization is possible via an arbitrary function $H[f]$ with the following condition:

$$\frac{4F_i F_i'' + \left(F_i'\right)^2 (n-6)}{4F_i} = F_i' \cdot \frac{H_i'}{H_i} \quad \Rightarrow \quad F_i = \begin{cases} \left(C_{f0}^2 - \displaystyle\int_1^{f_i} H_i[\phi] \cdot \frac{2-n}{4} \cdot d\phi \right)^{\frac{4}{n-2}} \cdot C_{f1} & n > 2 \\ \\ e^{\displaystyle C_{f0}^{-2} \int_1^{f_i} H_i[\phi] \cdot d\phi} \cdot C_{f1} & n = 2. \end{cases}$$

$$(1543)$$

This changes Eq. (1489) to:

$$0 = \left(\left(\frac{R \cdot F_i}{(n-1)} - \frac{f_{i,\alpha} f_{i,\beta}}{4F_i} g^{\alpha\beta} \left(4F_i F_i'' + F_i' F_i'' (n-6) \right) \right) \cdot \left(1 - \frac{n}{2} \right) - F_i' \left(1 - \frac{n}{2} \right) \Delta f_i \right)$$

$$\Rightarrow \qquad 0 = \frac{R \cdot F_i}{(n-1)} - \frac{F_i'}{H_i} \cdot \left(f_{i,\alpha} f_{i,\beta} g^{\alpha\beta} + H_i \Delta f_i \right)$$

$$= \frac{R \cdot F_i}{(n-1)} - \frac{F_i'}{H_i} \cdot \left(f_{i,\alpha} f_{i,\beta} g^{\alpha\beta} H_i' + \frac{H_i}{\sqrt{g}} \partial_\alpha \left(\sqrt{g} \cdot g^{\alpha\beta} \partial_\beta f_i \right) \right)$$

$$= \frac{R \cdot F_i}{(n-1)} - \frac{F_i'}{H_i} \cdot \frac{1}{\sqrt{g}} \partial_\alpha \left(H_i \sqrt{g} \cdot g^{\alpha\beta} \partial_\beta f_i \right)$$

$$\Rightarrow \qquad \frac{R \cdot F_i}{(n-1)} \cdot \frac{H_i}{F_i'} = \frac{1}{\sqrt{g}} \partial_\alpha \left(H_i \sqrt{g} \cdot g^{\alpha\beta} \partial_\beta f_i \right). \qquad (1544)$$

Now we intend to find $H[f]$ in such a way that the operator on the right-hand side of the last line of Eq. (1544) becomes an ordinary Laplace operator on the basis of the scaled metric $G_{\alpha\beta}$. In order to achieve this, we need to demand:

$$H_i \sqrt{g} \cdot g^{\alpha\beta} = \sqrt{G_i} \cdot G_i^{\alpha\beta} \Rightarrow H_i = F_i^{n/2-1} \Rightarrow \frac{4F_i F_i'' + \left(F_i'\right)^2 \left(n-6\right)}{4F_i} = 1$$

(1545)

and the last line of Eq. (1544) evolves to:

$$\frac{R}{(n-1)} \cdot \frac{1}{F_i'} = \frac{1}{\sqrt{G_i}} \partial_\alpha \left(\sqrt{G_i} \cdot G_i^{\alpha\beta} \partial_\beta f_i\right) = \Delta_G f_i.$$

(1546)

As before we can now apply the classical Dirac approach (see sub-section "10.6.1 Insertion: The Classical Way to Obtain Curved Dirac Equations") and subsequently obtain Dirac equations for our curved and scaled metric. Interestingly, when demanding a Ricci flat space-time ($R = 0$), we cannot have any mass terms in Eq. (1546) without additional dimensions, because we would have:

$$0 = \frac{1}{\sqrt{G_i}} \partial_\alpha \left(\sqrt{G_i} \cdot G_i^{\alpha\beta} \partial_\beta f_i\right) = \Delta_G f_i.$$

(1547)

However, as shown in chapter 6 and [6–8], additional (potentially curved up or compactified) dimensions could bring mass even in the Ricci flat space. In order to demonstrate this, we chose an example with $n = 6$ and the following metric:

$$G_{\alpha\beta}^6 = \begin{pmatrix} -c^2 & 0 & 0 & 0 & 0 & 0 \\ 0 & 1 & 0 & 0 & 0 & 0 \\ 0 & 0 & 1 & 0 & 0 & 0 \\ 0 & 0 & 0 & 1 & 0 & 0 \\ 0 & 0 & 0 & 0 & \pm c_u^2 & 0 \\ 0 & 0 & 0 & 0 & 0 & \pm c_v^2 \end{pmatrix} \cdot F\left[\frac{=f_i=f_i[u,v,t,x,y,z]}{g[u,v]\cdot f[t,x,y,z]}\right].$$

(1548)

Even though it does not matter here, because we want to simplify it anyway, we also give—for completeness—the general solution for the function $F[g*f]$ from our condition in Eq. (1545) in the case of $n = 6$:

$$\frac{4F_i F_i'' + \left(F_i'\right)^2 \left(n-6\right)}{4F_i} = 1 \xrightarrow{n=6} F_i = \frac{1}{2} \cdot f_i^2 + C_{f1} \cdot f_i + C_{f0}$$

(1549)

of which we only use the first term by setting $C_{f1} = 0$ and $C_{f0} = 0$. Now, Eq. (1547) gives us the following differential equation:

$$0 = g^{\alpha\beta}\left(f_{i,\alpha\beta} + f_{i,\alpha}f_{i,\beta} \cdot f_i^{-2}\right) \quad \Rightarrow \quad 0 = g^{\alpha\beta}\left(f_{i,\alpha\beta} \cdot f_i^2 + f_{i,\alpha}f_{i,\beta}\right). \quad (1550)$$

We apply the separation approach for $g[u,v]$ via $g[u,v] = g_v[v]*g_u[u]$ and use the following functions:

$$g_v[v] = C_{v1} \cdot \cos\left[c_v \cdot E_v \cdot v\right] + C_{v2} \cdot \sin\left[c_v \cdot E_v \cdot v\right], \quad (1551)$$

$$g_u[u] = C_{u1} \cdot \cos\left[c_u \cdot E_u \cdot u\right] + C_{u2} \cdot \sin\left[c_u \cdot E_u \cdot u\right]. \quad (1552)$$

Now Eq. (1547) gives us:

$$0 = \left(\pm\left(E_u^2 + E_v^2\right) + \Delta_{4D}\right)f_i[t,x,y,z] \cdot (f \cdot g)_i^2 \cdot g_i^2 + g^{\alpha\beta}(f \cdot g)_{i,\alpha}(f \cdot g)_{i,\beta}. \quad (1553)$$

We realize that with the first addend in Eq. (1151), we have just obtained the classical Klein–Gordon equation with quite some flexibility to add mass or other forms of inertia. For the reasons of recognition, we set:

$$-M^2 = -\frac{m^2 \cdot c^2}{\hbar^2} = \pm\left(E_u^2 + E_v^2\right), \quad (1554)$$

obtain from Eq. (1151):

$$0 = \left(\Delta_{4D} - \frac{m^2 \cdot c^2}{\hbar^2}\right)f_i[t,x,y,z] \cdot (f \cdot g)_i^2 \cdot g_i^2 + g^{\alpha\beta}(f \cdot g)_{i,\alpha}(f \cdot g)_{i,\beta}$$

$$\Rightarrow \quad 0 = \left(\Delta_{4D} - \frac{m^2 \cdot c^2}{\hbar^2}\right)f_i[t,x,y,z] + \frac{g^{\alpha\beta}(f \cdot g)_{i,\alpha}(f \cdot g)_{i,\beta}}{(f \cdot g)_i^2 \cdot g_i^2} \quad (1555)$$

and could now interpret the second addend as some kind of potential. A direct application of the Dirac factorization of the Klein–Gordon equation, we have obtained here, is not possible at least we do not see how, but when writing the whole as an approximation of the from:

$$0 = \left(\Delta_{4D} - \frac{m^2 \cdot c^2}{\hbar^2}\right)f_i[t,x,y,z] + V \cdot f_i[t,x,y,z], \quad (1556)$$

we might still be able to find some use for the $H[f]$-approach.

It should be pointed out at this point that we only chose the case $n = 6$, because this leads to a very simple solution for the wrapper function $F[f]$ with respect to our condition in Eq. (1549).

10.6.5 A Note

One may ask at this point why we explicitly tried to obtain the Dirac equation from the $f_{,\alpha}f_{,\beta}g^{\alpha\beta}$ operator, while Dirac in [12] only extracted the square root (operational factorization) from the Klein–Gordon equation via the introduction of quaternions. The reason for this excursion lies in the fact that the factorization of the $f_{,\alpha}f_{,\beta}g^{\alpha\beta}$ operator is straightforward and works for all kinds of symmetric metrics, while this is not the case for the Laplace operator in the Klein–Gordon equation. In our previous chapter about this matter, in sub-section "9.3.4 Dirac's 'Luck' and the Flat Space Condition," we have shown that for certain (scale-Ricci-flat) space-times there is a direct connection between the two operators, which make our evaluation above, even though it does not use Dirac's starting point (the Klein–Gordon equation), absolutely worthwhile ... if no to say essential!

10.6.6 Excursion: The $n = 1$ – Case

As in the case $n = 1$ the Ricci scalar is supposed to give zero, because when we have only 1 dimension, which is a curve, the intrinsic curvature of a curve does not exist and thus, R vanishes. Closely observing Eq. (1492), however, we find that there could still be a mass, because with the limit $n\to 0$ we obtain:

$$\lim_{n\to 1} f_i \cdot M^2 = \lim_{n\to 1} \frac{R \cdot F_i}{F_i{}'(n-1)} \to \frac{0 \cdot F_i}{F_i{}' \cdot 0}. \tag{1557}$$

By also inserting Eq. (1441), we even get:

$$\lim_{n\to 1} f_i \cdot M^2 = \lim_{n\to 1} \frac{R \cdot F_i}{F_i{}'(n-1)} = \frac{\lim\limits_{n\to 1} R \cdot F_i}{F_i{}' \lim\limits_{n\to 1}(n-1)} \to \frac{0}{0}\left(-\frac{1}{4}\cdot f_i\right)$$

$$\Rightarrow \qquad \lim_{n\to 1} M^2 = \frac{\lim\limits_{n\to 1} R}{\lim\limits_{n\to 1}(n-1)}\cdot\left(-\frac{1}{4}\right) \to \frac{0}{0}\cdot\left(-\frac{1}{4}\right) \to ? \to M_{1D}^2. \tag{1558}$$

We state that, even though there is no curvature in a 1-dimensional space, there could still be a mass.

10.7 Brief Sum-Up: "The Dirac Miracle"

For convenience, we here repeat the derivation of the classical Dirac equation directly from the Einstein–Hilbert action. The latter reads as follows (see also Eq. (1010)):

$$\delta W = 0 = \delta \int_V d^n x \left(\sqrt{-G} \cdot \left(R^* - 2\Lambda + L_M \right) \right). \tag{1559}$$

Thereby we apply the scaled Ricci scalar R^*:

$$R^* = R^*_{\alpha\beta} G^{\alpha\beta} = \left(\begin{array}{c} \Gamma^\sigma_{\alpha\beta,\sigma} g^{\alpha\beta} - \Gamma^\sigma_{\beta\sigma,\alpha} g^{\alpha\beta} - \Gamma^\mu_{\sigma\alpha} \Gamma^\sigma_{\beta\mu} g^{\alpha\beta} + \Gamma^\sigma_{\alpha\beta} \Gamma^\mu_{\sigma\mu} g^{\alpha\beta} \\ + \dfrac{F'}{F}(1-n)\Delta f + \dfrac{f_{,\alpha} f_{,\beta} g^{\alpha\beta}(1-n)}{4F^2} \left(4FF'' + \left(F'\right)^2(n-6) \right) \end{array} \right) \dfrac{1}{F}$$

$$= \left(R + \frac{F'}{F}(1-n)\Delta f + \frac{f_{,\alpha} f_{,\beta} g^{\alpha\beta}(1-n)}{4F^2} \left(4FF'' + \left(F'\right)^2(n-6) \right) \right) \frac{1}{F}$$

with:

$$F = F[f]; \quad F' = \frac{\partial F[f]}{\partial f}; \quad F'' = \frac{\partial^2 F[f]}{\partial f^2}, \tag{1560}$$

being based upon the scaled metric $G_{\delta\gamma} = F[f] \cdot g_{\delta\gamma}$. Assuming the Lagrange matter density L_M and the cosmological constant Λ to vanish, the variation results in:

$$0 = \left[\int_{V_G} d^n x \sum_{i=1}^N \sqrt{-g} \cdot F_i^n \left(\begin{array}{c} \left(R_{\alpha\beta} - \dfrac{1}{2} \cdot R \cdot g_{\alpha\beta} - \dfrac{F_i'}{F_i}(n-1) \right) \\ \times \left(f_{i,\alpha\beta} - \Gamma^\gamma_{\alpha\beta} f_{i,\gamma} - \dfrac{g_{\alpha\beta}}{2} \Delta f_i \right) \\ + \dfrac{(1-n)}{4F_i^2} \left(\begin{array}{c} F_i F_i'' \\ + F_i' F_i'(n-6) \end{array} \right) \left(\begin{array}{c} f_{i,\alpha} f_{i,\beta} \\ - f_{i,\mu} f_{i,\nu} g^{\mu\nu} \dfrac{g_{\alpha\beta}}{2} \end{array} \right) \end{array} \right) \delta G^{\alpha\beta} \right]. \tag{1561}$$

Thereby we have also assumed to have many scaling factors and volumes to perform the integration over (for more discussion about this see [10]). By splitting up the variation of the scaled metric:

$$0 = \left(\int_{V_G} d^n x \sum_{i=1}^{N} \sqrt{-g} \cdot F_i^n \middle| \left(\left(\begin{array}{c} R_{\alpha\beta} - \dfrac{1}{2} \cdot R \cdot g_{\alpha\beta} - \dfrac{F_i{}'}{F_i}(n-1) \\[2mm] \times \left(f_{i,\alpha\beta} - \Gamma_{\alpha\beta}^{\gamma} f_{i,\gamma} - \dfrac{g_{\alpha\beta}}{2} \Delta f_i \right) \\[2mm] + \dfrac{(1-n)}{4F_i^2} \left(\begin{array}{c} F_i F_i{}'' \\ + F_i{}' F_i{}' (n-6) \end{array} \right) \left(\begin{array}{c} f_{i,\alpha} f_{i,\beta} \\ - f_{i,\mu} f_{i,v} g^{\mu v} \dfrac{g_{\alpha\beta}}{2} \end{array} \right) \end{array} \right) \left(\begin{array}{c} g^{\alpha\beta} \delta \dfrac{1}{F_i} \\[2mm] + \dfrac{\delta g^{\alpha\beta}}{F_i} \end{array} \right) \right) \right),$$

$$(1562)$$

we result in two equations:

$$0 = \left(\int_{V_G} d^n x \sum_{i=1}^{N} \sqrt{-g} \cdot F_i^n \middle| \left(\left(1 - \dfrac{n}{2} \right) \left(\begin{array}{c} R - \dfrac{F_i{}'}{F_i}(n-1)\Delta f_i \\[2mm] + \dfrac{f_{i,\alpha} f_{i,\beta}(1-n)}{4F_i^2} g^{\alpha\beta} \left(\begin{array}{c} 4F_i F_i{}'' \\ + F_i{}' F_i{}' (n-6) \end{array} \right) \end{array} \right) \delta \dfrac{1}{F_i} \right) \right),$$

$$(1563)$$

$$0 = \left(\int_{V_G} d^n x \sum_{i=1}^{N} \sqrt{-g} \cdot F_i^n \middle| \left(\left(\begin{array}{c} R_{\alpha\beta} - \dfrac{1}{2} \cdot R \cdot g_{\alpha\beta} - \dfrac{F_i{}'}{F_i}(n-1) \\[2mm] \times \left(f_{i,\alpha\beta} - \Gamma_{\alpha\beta}^{\gamma} f_{i,\gamma} - \dfrac{g_{\alpha\beta}}{2} \Delta f_i \right) \\[2mm] + \dfrac{(1-n)}{4F_i^2} \left(\begin{array}{c} F_i F_i{}'' \\ + F_i{}' F_i{}' (n-6) \end{array} \right) \left(\begin{array}{c} f_{i,\alpha} f_{i,\beta} \\ - f_{i,\mu} f_{i,v} g^{\mu v} \dfrac{g_{\alpha\beta}}{2} \end{array} \right) \end{array} \right) \dfrac{\delta g^{\alpha\beta}}{F_i} \right) \right).$$

$$(1564)$$

While we consider the second equation a generalized or quantum Einstein field equation, we use the first one to derive the classical Dirac equation [12]. Setting the integrand equal to zero we obtain the following differential equation:

$$0 = R - \frac{F_i'}{F_i}(n-1)\Delta f_i + \frac{f_{i,\alpha} f_{i,\beta}(1-n)}{4F_i^2} g^{\alpha\beta}\left(4F_i F_i'' + F_i' F_i'(n-6)\right). \quad (1565)$$

Now we add in the Dirac starting point, who assumed that:

(a) The space-time can be assumed Ricci flat and thus, $R = 0$.

(b) The Laplace operator term in Eq. (1565) has an Eigenvalue M^2.

Also setting $F[f] = f$ gives us:

$$0 = f_{i,\alpha} f_{i,\beta} g^{\alpha\beta} + M^2 \cdot \overbrace{\frac{4}{(n-6)}}^{\equiv \mu^2 = \mu \cdot \mu} \cdot f_i^2. \quad (1566)$$

Now we apply the known relation between the Dirac matrices in Eq. (1082) and the metric tensor, reading:

$$g^{\alpha\beta} \cdot I = \frac{\gamma^\alpha \gamma^\beta + \gamma^\beta \gamma^\alpha}{2}, \quad (1567)$$

and put Eq. (1566) into the following form:

$$0 = f_{i,\alpha} f_{i,\beta} g^{\alpha\beta} + \mu^2 \cdot f_i^2 \Rightarrow \quad = \quad \Rightarrow f_{i,\alpha} f_{i,\beta}\left(\frac{\gamma^\alpha \gamma^\beta + \gamma^\beta \gamma^\alpha}{2}\right)$$

$$+ \frac{\beta^2 + \beta^2}{2} \cdot m^2 \cdot f_i^2 = 0$$

$$\Rightarrow \quad f_{i,\alpha} f_{i,\beta} \gamma^\alpha \gamma^\beta + \beta^2 \cdot m^2 \cdot f_i^2 = 0$$

$$\Rightarrow \quad \left(f_{i,\alpha} \gamma^\alpha + i \cdot \beta \cdot m \cdot f_i\right)\left(f_{i,\beta} \gamma^\beta - i \cdot \beta \cdot m \cdot f_i\right) = 0$$

$$\Rightarrow \quad f_{i,\alpha} f_{i,\beta} \gamma^\beta \gamma^\alpha + \beta^2 \cdot m^2 \cdot f_i^2 = 0$$

$$\Rightarrow \quad \left(f_{i,\alpha} \gamma^\alpha + i \cdot \beta \cdot m \cdot f_i\right)\left(f_{i,\beta} \gamma^\beta - i \cdot \beta \cdot m \cdot f_i\right) = 0. \quad (1568)$$

Thereby we have also used the β-matrix with $\beta = \begin{pmatrix} 1 & & & \\ & 1 & & \\ & & -1 & \\ & & & -1 \end{pmatrix}$.

This gives us the Dirac equation in its classical form:

$$0 = f_{i,\beta} \gamma^\beta - i \cdot \beta \cdot m \cdot f_i. \quad (1569)$$

The advantage of the Dirac approach clearly is that here the structure of the equation with the quaternions already assures the satisfaction of the Eigenvalue equation for the Laplace operator term, which it to say the Klein–Gordon condition:

$$\Delta f_i = -M^2 \cdot f_i. \tag{1570}$$

Seeing the simplicity of the derivation of the Dirac equation with the Einstein–Hilbert action as starting point, as just performed here and comparing this with Dirac's path, one can only wonder "how on earth this genius came up with this idea of introducing the quaternions."

It is—we have no other word for it—a miracle.

Now we have obtained the Klein–Gordon equation (sub-sections 6.11.1–6.11.3), the Schrödinger equation (section 6.11.4) and the Dirac equation (see above) directly out of the Einstein–Hilbert action. We also saw that there is enough freedom via various options for the wrapper function $F[f]$ plus the split-up of the variational integral to also obtain other quantum equations. Thus, so we conclude, the classical Quantum Theory is to be found in the scalar part of the variation in Eq. (1559), which is to say Eq. (1563) ... with quite many options for the split-up of the integration and the variation itself.

Gravity, however, requires the consideration of the tensorial integrand of Eq. (1559), being Eq. (1564) (also with many variational degrees of freedom it seems), and quantum gravity is it all just put together.

In the next sections, we are going to consider this gravity part.

10.8 Is There a Gravitational Dirac Equation?

So far, we only considered the scalar equation resulting from the Einstein–Hilbert action. In section 6.11 and the sections above in this chapter we have shown that this scalar part in Eq. (1563) gives the most important classical quantum equations.

Repeating our evaluation from above with just taking into account more options for the decomposition of the Laplace operator, we have:

$$
0 = \int_{V_G} d^n x \sum_{i=1}^{N} \sqrt{-g} \cdot F_i^n \left[\underbrace{\left(R_{\alpha\beta} - \frac{1}{2} \cdot R \cdot g_{\alpha\beta} - \frac{F_i'}{F_i}(n-1) \times \underbrace{\left(\begin{array}{c} f_{i,\alpha\beta} - \Gamma_{\alpha\beta}^{\gamma} f_{i,\gamma} \\ -\dfrac{g_{\alpha\beta}}{n}\Gamma_{\kappa\lambda}^{\gamma} g^{\kappa\lambda} f_{i,\gamma} \\ +\dfrac{g_{\alpha\beta}}{n} g^{\kappa\lambda} f_{i,\kappa\lambda} - \dfrac{g_{\alpha\beta}}{n}\Gamma_{\kappa\lambda}^{\gamma} g^{\kappa\lambda} f_{i,\gamma} \end{array}\right)} \right. \\ \left. + \frac{(1-n)}{4F_i^2}\underbrace{\left(\begin{array}{c} F_i F_i'' \\ +F_i' F_i'(n-6)\end{array}\right)}\underbrace{\left(\begin{array}{c} f_{i,\alpha} f_{i,\beta} \\ -f_{i,\mu} f_{i,\nu} g^{\mu\nu}\dfrac{g_{\alpha\beta}}{2}\end{array}\right)} \right) - \underbrace{\frac{g_{\alpha\beta}}{2}\Delta f_i} \right] \underbrace{\delta G^{\alpha\beta}} \,. \tag{1571}
$$

By splitting up the variation of the scaled metric:

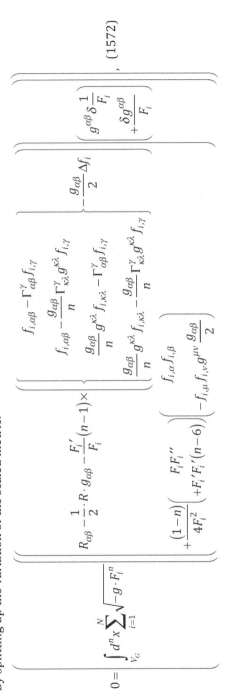

$$
0 = \int_{V_G} d^n x \sum_{i=1}^{N} \sqrt{-g} \cdot F_i^n \left[\underbrace{\left(R_{\alpha\beta} - \frac{1}{2} \cdot R \cdot g_{\alpha\beta} - \frac{F_i'}{F_i}(n-1) \times \underbrace{\left(\begin{array}{c} f_{i,\alpha\beta} - \Gamma_{\alpha\beta}^{\gamma} f_{i,\gamma} \\ -\dfrac{g_{\alpha\beta}}{n}\Gamma_{\kappa\lambda}^{\gamma} g^{\kappa\lambda} f_{i,\gamma} \\ +\dfrac{g_{\alpha\beta}}{n} g^{\kappa\lambda} f_{i,\kappa\lambda} - \dfrac{g_{\alpha\beta}}{n}\Gamma_{\kappa\lambda}^{\gamma} g^{\kappa\lambda} f_{i,\gamma} \end{array}\right)} \right. \\ \left. + \frac{(1-n)}{4F_i^2}\underbrace{\left(\begin{array}{c} F_i F_i'' \\ +F_i' F_i'(n-6)\end{array}\right)}\underbrace{\left(\begin{array}{c} f_{i,\alpha} f_{i,\beta} \\ -f_{i,\mu} f_{i,\nu} g^{\mu\nu}\dfrac{g_{\alpha\beta}}{2}\end{array}\right)} \right) - \underbrace{\frac{g_{\alpha\beta}}{2}\Delta f_i} \right] \underbrace{\left(g^{\alpha\beta}\delta\frac{1}{F_i} + \frac{\delta g^{\alpha\beta}}{F_i}\right)} \,, \tag{1572}
$$

we result in two equations:

$$
0 = \int_{V_G} d^n x \sum_{i=1}^{N} \sqrt{-g} \cdot F_i^n \left(1 - \frac{n}{2}\right) \left(R - \frac{F_i'}{F_i}(n-1)\Delta f_i + \frac{f_{i,\alpha}f_{i,\beta}(1-n)}{4F_i^2} g^{\alpha\beta}\left(\begin{array}{c} 4F_i F_i'' \\ +F_i' F_i'(n-6) \end{array} \right) \right) \delta \frac{1}{F_i} ,
\tag{1573}
$$

$$
0 = \int_{V_G} d^n x \sum_{i=1}^{N} \sqrt{-g} \cdot F_i^n \left(R_{\alpha\beta} - \frac{1}{2}\cdot R \cdot g_{\alpha\beta} - \frac{F_i'}{F_i}(n-1)\times \left(\begin{array}{c} f_{i,\alpha\beta} - \Gamma_{\alpha\beta}^\gamma f_{i,\gamma} \\ f_{i,\alpha\beta} - \dfrac{g_{\alpha\beta}}{n} \Gamma_{\kappa\lambda}^\gamma g^{\kappa\lambda} f_{i,\gamma} \\ \dfrac{g_{\alpha\beta}}{n} g^{\kappa\lambda} f_{i,\kappa\lambda} - \Gamma_{\alpha\beta}^\gamma f_{i,\gamma} \\ \dfrac{g_{\alpha\beta}}{n} g^{\kappa\lambda} f_{i,\kappa\lambda} \end{array} \right) - \frac{g_{\alpha\beta}}{2}\Delta f_i \right.
$$
$$
\left. + \frac{(1-n)}{4F_i^2}\left(\begin{array}{c} F_i F_i'' \\ +F_i' F_i'(n-6) \end{array} \right) \left(\begin{array}{c} f_{i,\alpha}f_{i,\beta} \\ -f_{i,\mu}f_{i,\nu}g^{\mu\nu}\dfrac{g_{\alpha\beta}}{2} \end{array} \right) \right) \frac{\delta g^{\alpha\beta}}{F_i} .
\tag{1574}
$$

Demanding the integrand to give zero for the second equation, we end up with the following:

$$
0 = \begin{pmatrix} R_{\alpha\beta} - \dfrac{1}{2}\cdot R\cdot g_{\alpha\beta} - \dfrac{F_i'}{F_i}(n-1) \\[2ex] \times \left\{ \begin{pmatrix} f_{i,\alpha\beta} - \Gamma^\gamma_{\alpha\beta} f_{i,\gamma} \\[1.5ex] f_{i,\alpha\beta} - \dfrac{g_{\alpha\beta}}{n}\Gamma^\gamma_{\kappa\lambda}g^{\kappa\lambda}f_{i,\gamma} \\[1.5ex] \dfrac{g_{\alpha\beta}}{n}g^{\kappa\lambda}f_{i,\kappa\lambda} - \Gamma^\gamma_{\alpha\beta}f_{i,\gamma} \\[1.5ex] \dfrac{g_{\alpha\beta}}{n}g^{\kappa\lambda}f_{i,\kappa\lambda} - \dfrac{g_{\alpha\beta}}{n}\Gamma^\gamma_{\kappa\lambda}g^{\kappa\lambda}f_{i,\gamma} \end{pmatrix} - \dfrac{g_{\alpha\beta}}{2}\Delta f_i \right\} \\[6ex] + \dfrac{(1-n)}{4F_i^2}\begin{pmatrix} F_iF_i'' \\ +F_i'F_i'(n-6) \end{pmatrix}\begin{pmatrix} f_{i,\alpha}f_{i,\beta} \\ -f_{i,\mu}f_{i,\nu}g^{\mu\nu}\dfrac{g_{\alpha\beta}}{2} \end{pmatrix} \end{pmatrix}.
\tag{1575}
$$

From here, we are now going to try to extract a tensorial or gravitational Dirac equation. In order to have something to start with, however, we will concentrate on just one option of the Laplace operator decomposition.

10.9 The Other "Dirac Equation"

After rather intensive investigations (previous chapters and e.g., [8]) of many possibilities to derive the Dirac equation [12], more or less directly from the Einstein–Hilbert action and partially also succeeding, but not without special assumptions and—in some cases—approximations, we now just want to consider the option of obtaining an equation of first-order derivatives out of Eq. (1010) (with the cosmological constant and the artificial matter set to zero). Perhaps this gives us a better insight about the occurrence of the Dirac equation.

10.9.1 The Other Origin of a "Dirac Equation"

That being said, we shall start with Eq. (1445) and extend it as follows:

$$
0 = R_{\alpha\beta} - \frac{1}{2}R\cdot g_{\alpha\beta} + \frac{F'}{F}(1-n)\left(f_{,\alpha\beta} - \frac{g_{\alpha\beta}}{n}\Gamma^\gamma_{\kappa\lambda}g^{\kappa\lambda}f_{,\gamma} - \frac{1}{2}\Delta f\cdot g_{\alpha\beta} \right)
$$

$$= R_{\alpha\beta} - \frac{1}{2} R \cdot g_{\alpha\beta} + \frac{F'}{F}(1-n)\left(f_{,\alpha\beta} - \frac{g_{\alpha\beta}}{n} \Gamma^\gamma_{\kappa\lambda} g^{\kappa\lambda} f_{,\gamma} - \frac{1}{2} g^{\xi\zeta}\left(f_{,\xi\zeta} - \frac{g_{\xi\zeta}}{n} \Gamma^\gamma_{\kappa\lambda} g^{\kappa\lambda} f_{,\gamma} \right) \cdot g_{\alpha\beta} \right).$$

$$(1576)$$

Now we assume the classical Einstein part to be $0 = R_{\alpha\beta} - \frac{1}{2} R \cdot g_{\alpha\beta}$ and reshape Eq. (1576) a bit:

$$0 = f_{,\alpha\beta} - \frac{1}{2} g^{\xi\zeta} f_{,\xi\zeta} \cdot g_{\alpha\beta} - \frac{g_{\alpha\beta}}{n} \Gamma^\gamma_{\kappa\lambda} g^{\kappa\lambda} f_{,\gamma} + \frac{1}{2} \Gamma^\gamma_{\kappa\lambda} g^{\kappa\lambda} f_{,\gamma} \cdot g_{\alpha\beta}$$

$$= \left(\delta^\xi_\alpha \delta^\zeta_\beta - \frac{1}{2} g^{\xi\zeta} \cdot g_{\alpha\beta} \right) f_{,\xi\zeta} + \frac{1}{2} \Gamma^\gamma_{\kappa\lambda} g^{\kappa\lambda} f_{,\gamma} \cdot g_{\alpha\beta} - \frac{g_{\alpha\beta}}{n} \Gamma^\gamma_{\kappa\lambda} g^{\kappa\lambda} f_{,\gamma}$$

$$= \left(\delta^\xi_\alpha \delta^\zeta_\beta - \frac{1}{2} g^{\xi\zeta} \cdot g_{\alpha\beta} \right) f_{,\xi\zeta} + g_{\alpha\beta} \left(\frac{1}{2} - \frac{1}{n} \right) \Gamma^\gamma_{\kappa\lambda} g^{\kappa\lambda} f_{,\gamma}$$

$$= \left(\delta^\xi_\alpha \delta^\zeta_\beta - \frac{1}{2} g^{\xi\zeta} \cdot g_{\alpha\beta} \right) f_{,\xi\zeta} + g_{\alpha\beta} \left(\frac{1}{2} - \frac{1}{n} \right) \Gamma^\xi_{\kappa\lambda} g^{\kappa\lambda} f_{,\xi}$$

$$= \left(\left(\delta^\xi_\alpha \delta^\zeta_\beta - \frac{1}{2} g^{\xi\zeta} \cdot g_{\alpha\beta} \right) \partial_\zeta + g_{\alpha\beta} \left(\frac{1}{2} - \frac{1}{n} \right) \Gamma^\xi_{\kappa\lambda} g^{\kappa\lambda} \right) f_{,\xi}. \qquad (1577)$$

We realize that we could substitute the first derivative $f_{,\xi}$ by an arbitrary vector Φ_ξ and already recognize the typical Dirac structure, only that we have "matrices" with the derivative term $\left(\delta^\xi_\alpha \delta^\zeta_\beta - \frac{1}{2} g^{\xi\zeta} \cdot g_{\alpha\beta} \right) \partial_\zeta$ of rank 4 rather than 3:

$$0 = \left(\left(\delta^\xi_\alpha \delta^\zeta_\beta - \frac{1}{2} g^{\xi\zeta} \cdot g_{\alpha\beta} \right) \partial_\zeta + g_{\alpha\beta} \left(\frac{1}{2} - \frac{1}{n} \right) \Gamma^\xi_{\kappa\lambda} g^{\kappa\lambda} \right) \Phi_\xi. \qquad (1578)$$

This could easily be remediated with an arbitrary vector V^α as follows:

$$0 = V^\alpha \cdot \left(\left(\delta^\xi_\alpha \delta^\zeta_\beta - \frac{1}{2} g^{\xi\zeta} \cdot g_{\alpha\beta} \right) \partial_\zeta + g_{\alpha\beta} \left(\frac{1}{2} - \frac{1}{n} \right) \Gamma^\xi_{\kappa\lambda} g^{\kappa\lambda} \right) \Phi_\xi$$

$$= \left(\left(V^\xi \cdot \delta^\zeta_\beta - \frac{1}{2} g^{\xi\zeta} \cdot V_\beta \right) \partial_\zeta + V_\beta \left(\frac{1}{2} - \frac{1}{n} \right) \Gamma^\xi_{\kappa\lambda} g^{\kappa\lambda} \right) \Phi_\xi. \qquad (1579)$$

This clearly now has the structure of a rather general Dirac equation. Generally we say, because it is automatically given in n dimensions and arbitrary—potentially also curved—space-times under the condition of:

$$0 = R_{\alpha\beta} - \frac{1}{2} R \cdot g_{\alpha\beta} + \frac{(1-n)}{4F^2}\left(\begin{matrix} FF'' \\ +F'F'(n-6) \end{matrix} \right)\left(\begin{matrix} f_{,\alpha} f_{,\beta} \\ -f_{,\mu} f_{,\nu} g^{\mu\nu} \frac{g_{\alpha\beta}}{2} \end{matrix} \right) + \kappa T_{\alpha\beta} + \Lambda \cdot F \cdot g_{\alpha\beta}.$$

$$(1580)$$

Knowing that we can always apply Eq. (1441), resulting in:

$$0 = R_{\alpha\beta} - \frac{1}{2} R \cdot g_{\alpha\beta} + \kappa T_{\alpha\beta} + \Lambda \cdot F \cdot g_{\alpha\beta}, \tag{1581}$$

and avoiding the postulated matter term and cosmological constant anyway, we end up with the vacuum Einstein Field equations:

$$0 = R_{\alpha\beta} - \frac{1}{2} R \cdot g_{\alpha\beta} \tag{1582}$$

as an additional condition in connection with the Dirac-like equation Eq. (1578) or:

$$0 = \left(\left(V^{\xi} \cdot \delta_{\beta}^{\zeta} - \frac{1}{2} g^{\xi\zeta} \cdot V_{\beta} \right) \partial_{\zeta} + V_{\beta} \left(\frac{1}{2} - \frac{1}{n} \right) \Gamma_{\kappa\lambda}^{\xi} g^{\kappa\lambda} \right) \Phi_{\xi}. \tag{1583}$$

Now we may ask where the **V**-vector could come from and remind ourselves that Eq. (1576) originated from the variational task in Eq. (1010) and therefore our starting point Eq. (1445) should always be seen as follows:

$$\delta W = 0 = \int_V d^n x \left| \sqrt{-g} \cdot F^n \left(\begin{array}{c} R_{\alpha\beta} - R \cdot \dfrac{g_{\alpha\beta}}{2} \\[2mm] + \dfrac{F'}{F}(1-n)\left(f_{,\alpha\beta} - \dfrac{g_{\alpha\beta}}{n} \Gamma_{\kappa\lambda}^{\gamma} g^{\kappa\lambda} f_{,\gamma} - \dfrac{g_{\alpha\beta}}{2} \Delta f \right) \\[2mm] + \dfrac{(1-n)}{4F^2}\left(\begin{array}{c} FF'' \\ +F'F'(n-6) \end{array} \right) \left(\begin{array}{c} f_{,\alpha} f_{,\beta} \\ -f_{,\mu} f_{,\nu} g^{\mu\nu} \dfrac{g_{\alpha\beta}}{2} \end{array} \right) \\[2mm] + \kappa T_{\alpha\beta} + \Lambda \cdot F \cdot g_{\alpha\beta} \end{array} \right) \delta G^{\alpha\beta} \right|. \tag{1584}$$

Using all the assumptions we also applied to Eq. (1576), we end up with the following variational integral:

$$\delta W = 0 = \int_V d^n x \left| \sqrt{-g} \cdot F^n \left(\begin{array}{c} R_{\alpha\beta} - R \cdot \dfrac{g_{\alpha\beta}}{2} \\[2mm] + \dfrac{F'}{F}(1-n)\left(f_{,\alpha\beta} - \dfrac{g_{\alpha\beta}}{n} \Gamma_{\kappa\lambda}^{\gamma} g^{\kappa\lambda} f_{,\gamma} - \dfrac{g_{\alpha\beta}}{2} \Delta f \right) \end{array} \right) \delta G^{\alpha\beta} \right|. \tag{1585}$$

Also, as we did above, demanding the classical Einstein field equations to be fulfilled via:

$$R_{\alpha\beta} - R \cdot \frac{g_{\alpha\beta}}{2} = 0, \qquad (1586)$$

results in:

$$0 = \int_V d^n x \left(\sqrt{-g} \cdot F^n \left(\frac{F'}{F}(1-n) \left(f_{,\alpha\beta} - \frac{g_{\alpha\beta}}{n} \Gamma^\gamma_{\kappa\lambda} g^{\kappa\lambda} f_{,\gamma} - \frac{g_{\alpha\beta}}{2} \Delta f \right) \right) \delta G^{\alpha\beta} \right) \qquad (1587)$$

and following our derivations from Eqs. (1576)–(1578) yield:

$$0 = \int_V d^n x \left(\sqrt{-g} \cdot F^n \left(\frac{F'}{F}(1-n) \left(\left(\left(\delta^\xi_\alpha \delta^\zeta_\beta - \frac{1}{2} g^{\xi\zeta} \cdot g_{\alpha\beta} \right) \partial_\zeta \right. \right. \right. \right.$$
$$\left. \left. \left. \left. + g_{\alpha\beta} \left(\frac{1}{2} - \frac{1}{n} \right) \Gamma^\xi_{\kappa\lambda} g^{\kappa\lambda} \right) \Phi_\xi \right) \right) \delta G^{\alpha\beta} \right). \qquad (1588)$$

Now simply imagine to decompose the variational term $\delta G^{\alpha\beta}$ into the **V**-vector and something else, which structure here is not of importance, is leading us to:

$$0 = \int_V d^n x \left(\sqrt{-g} \cdot F^n \left(\frac{F'}{F}(1-n) \left(\left(\left(\delta^\xi_\alpha \delta^\zeta_\beta - \frac{1}{2} g^{\xi\zeta} \cdot g_{\alpha\beta} \right) \partial_\zeta \right. \right. \right. \right.$$
$$\left. \left. \left. \left. + g_{\alpha\beta} \left(\frac{1}{2} - \frac{1}{n} \right) \Gamma^\xi_{\kappa\lambda} g^{\kappa\lambda} \right) \Phi_\xi \right) \right) V^\alpha \delta?^\beta \right) \qquad (1589)$$

and thus:

$$0 = \int_V d^n x \left(\sqrt{-g} \cdot F^n \left(\frac{F'}{F}(1-n) \left(\left(\left(V^\xi \cdot \delta^\zeta_\beta - \frac{1}{2} g^{\xi\zeta} \cdot V_\beta \right) \partial_\zeta \right. \right. \right. \right.$$
$$\left. \left. \left. \left. + V_\beta \left(\frac{1}{2} - \frac{1}{n} \right) \Gamma^\xi_{\kappa\lambda} g^{\kappa\lambda} \right) \Phi_\xi \right) \right) \delta?^\beta \right). \qquad (1590)$$

As the core of the integrand of Eq. (1590) is just our result from Eq. (1583), we now know where we could get the **V**-vector from in order to obtain our additional Dirac-like, but completely quaternion free and n-dimensional equation.

10.9.2 For Completeness

10.9.2.1 A consistency-check

We realize that we could also multiply Eq. (1578) (from the left-hand side) with the contra-variant metric tensor and sum, respectively contract as follows:

$$
\begin{aligned}
0 &= g^{\alpha\beta}\left(\left(\delta_\alpha^\xi \delta_\beta^\zeta - \frac{1}{2}g^{\xi\zeta}\cdot g_{\alpha\beta}\right)\partial_\zeta + g_{\alpha\beta}\left(\frac{1}{2}-\frac{1}{n}\right)\Gamma_{\kappa\lambda}^\xi g^{\kappa\lambda}\right)\Phi_\xi \\
&= \left(\left(1-\frac{n}{2}\right)g^{\xi\zeta}\partial_\zeta + \left(\frac{n}{2}-1\right)\Gamma_{\kappa\lambda}^\xi g^{\kappa\lambda}\right)\Phi_\xi \\
&= \left(1-\frac{n}{2}\right)\left(g^{\xi\zeta}\partial_\zeta - g^{\kappa\lambda}\Gamma_{\kappa\lambda}^\xi\right)\Phi_\xi \\
&= \left(1-\frac{n}{2}\right)\left(g^{\xi\zeta}\partial_\zeta - g^{\kappa\lambda}\Gamma_{\kappa\lambda}^\xi\right)f_{,\xi} = \left(1-\frac{n}{2}\right)\Delta f,
\end{aligned}
\tag{1591}
$$

which, of course, is consistent with the following derivation, where we start directly with the first line of Eq. (1576) and evaluate:

$$
\begin{aligned}
0 &= R_{\alpha\beta} - \frac{1}{2}R\cdot g_{\alpha\beta} + \overbrace{\frac{F'}{F}(1-n)}^{=0}g^{\alpha\beta}\left(f_{,\alpha\beta} - \frac{g_{\alpha\beta}}{n}\Gamma_{\kappa\lambda}^\gamma g^{\kappa\lambda}f_{,\gamma} - \frac{1}{2}\Delta f\cdot g_{\alpha\beta}\right) \\
&= \left(g^{\alpha\beta}f_{,\alpha\beta} - \Gamma_{\kappa\lambda}^\gamma g^{\kappa\lambda}f_{,\gamma} - \frac{1}{2}\Delta f\cdot n\right) \\
&= \left(1-\frac{1}{2}n\right)\Delta f.
\end{aligned}
\tag{1592}
$$

10.9.2.2 Another derivation

One might feel tempted to just take the classical Dirac assumption of *f* being a Klein–Gordon function (meaning a function that fulfills the Klein–Gordon equation) and assuming the mass term to occur, as shown in our previous papers, within the Laplace operator due to higher dimensions and their entanglement with our ordinary space-time (e.g., [5–10]). Then we'd have the condition $\Delta f = 0$ and the evaluation in Eq. (1577) could be simplified as follows:

$$
0 = f_{,\alpha\beta} - \frac{g_{\alpha\beta}}{n}\Gamma_{\kappa\lambda}^\gamma g^{\kappa\lambda}f_{,\gamma}
$$

$$= \left(\delta_\alpha^\xi \delta_\beta^\zeta \right) f_{,\xi\zeta} - \frac{g_{\alpha\beta}}{n} \Gamma_{\kappa\lambda}^\gamma g^{\kappa\lambda} f_{,\gamma}$$

$$= \left(\delta_\alpha^\xi \delta_\beta^\zeta \right) f_{,\xi\zeta} - \frac{g_{\alpha\beta}}{n} \Gamma_{\kappa\lambda}^\xi g^{\kappa\lambda} f_{,\xi}$$

$$= \left(\left(\delta_\alpha^\xi \delta_\beta^\zeta \right) \partial_\zeta - \frac{g_{\alpha\beta}}{n} \Gamma_{\kappa\lambda}^\xi g^{\kappa\lambda} \right) f_{,\xi}. \tag{1593}$$

This apparently simpler equation, however, is—in contrast to our earlier result from Eq. (1578)—strongly restricted, because it requires the additional condition of $\Delta f = 0$.

10.9.2.3 Brief example: Flat space—Minkowski-like

In order to give an example, we try to find a simple solution to Eq. (1578) in the case of a 4-dimensional Minkowski space-time in Cartesian coordinates. Equation (1578) then simplifies as follows:

$$0 = \left(\delta_\alpha^\xi \delta_\beta^\zeta - \frac{1}{2} g^{\xi\zeta} \cdot g_{\alpha\beta} \right) \partial_\zeta \Phi_\xi = \partial_\beta \Phi_\alpha - \frac{1}{2} \cdot g_{\alpha\beta} \cdot \partial^\xi \Phi_\xi$$

$$= \partial_\beta f_{,\alpha} - \frac{1}{2} \cdot g_{\alpha\beta} \cdot \partial^\xi f_{,\xi}. \tag{1594}$$

We might seek for a solution of f-components being linear in the coordinates, which is a huge restriction and may be seen as useless anyway.

With the $\mathbf{V}$-vector from Eq. (1583) the Minkowski simplification reads:

$$0 = \left(V^\xi \cdot \delta_\beta^\zeta - \frac{1}{2} g^{\xi\zeta} \cdot V_\beta \right) \partial_\zeta \Phi_\xi = V^\xi \cdot \partial_\beta \Phi_\xi - \frac{1}{2} \cdot V_\beta \cdot \partial^\xi \Phi_\xi$$

$$= V^\xi \cdot \partial_\beta f_{,\xi} - \frac{1}{2} \cdot V_\beta \cdot \partial^\xi f_{,\xi} \tag{1595}$$

and we obtain more options for reasonable solutions. The wave-approach with:

$$f = f[\pm c \cdot t + x] \tag{1596}$$

would then require a $\mathbf{V}$-vector:

$$V^\xi = \begin{pmatrix} \pm V_x \\ V_x \\ V_y \\ V_z \end{pmatrix}. \tag{1597}$$

10.9.3 Incorporation of the Volume Split-Up Option

In section "10.3 Quantum Einstein Field Equations in their Simplest Form," we discussed the possibility of using different scales within the variation of the Einstein–Hilbert action. One of these options was given in Eq. (1453) and reads (here now with different, summable scale factor F_i and wave function f_i)

$$
0 = \left(\begin{array}{c}
\int_{V_G} d^n x \sqrt{-G} \left(\left(R_{\alpha\beta} - \frac{1}{2} \cdot R \cdot g_{\alpha\beta} \right) \delta G^{\alpha\beta} \right) \\[2ex]
+ \frac{1}{2} \cdot \sum_{i=1}^N \int_{V_{Qi}} d^n x \sqrt{-g} \cdot F^n \frac{F'}{F} (n-1) \left(\Delta f \cdot g_{\alpha\beta} - 2 \cdot \left(\overbrace{f_{,\alpha\beta} - \frac{g_{\alpha\beta}}{n} \Gamma^\gamma_{\kappa\lambda} g^{\kappa\lambda} f_{,\gamma}}^{\equiv f_{:\alpha\beta}} \right) \right) \delta G^{\alpha\beta}
\end{array} \right)
$$

$$
\Rightarrow \quad = \int_{V_G} d^n x \sqrt{-g} \left(\sum_{i=1}^N \sqrt{F_i^n} \cdot \left(R_{\alpha\beta} - \frac{1}{2} \cdot R \cdot g_{\alpha\beta} + \frac{(n-1)}{2} \cdot \frac{F_i'}{F_i} \left(\Delta f_i \cdot g_{\alpha\beta} - 2 \cdot f_{i:\alpha\beta} \right) \right) \delta G^{\alpha\beta} \right).
$$

$$(1598)$$

Demanding the integrand to give zero and thus, to fulfill the equation above, plus distinguishing between the classical field equation part and the "rest" leads to:

$$
0 = \sum_{i=1}^N \sqrt{F_i^n} \cdot \left(R_{\alpha\beta} - \frac{1}{2} \cdot R \cdot g_{\alpha\beta} + \frac{(n-1)}{2} \cdot \frac{F_i'}{F_i} \left(\Delta f_i \cdot g_{\alpha\beta} - 2 \cdot f_{i:\alpha\beta} \right) \right) \delta G^{\alpha\beta}
$$

$$
\Rightarrow \quad \left\{ \begin{array}{c}
0 = R_{\alpha\beta} - \frac{1}{2} \cdot R \cdot g_{\alpha\beta} \\[2ex]
0 = \sum_{i=1}^N \sqrt{F_i^n} \cdot \left(\frac{(n-1)}{2} \cdot \frac{F_i'}{F_i} \left(\Delta f_i \cdot g_{\alpha\beta} - 2 \cdot f_{i:\alpha\beta} \right) \right) \delta G^{\alpha\beta}.
\end{array} \right.
$$

$$(1599)$$

Inserting the result from section "10.9.1 The Other Origin of A "Dirac Equation"" allows us to reshape the last equation as follows:

$$
0 = \sum_{i=1}^N \sqrt{F_i^n} \cdot \left((n-1) \cdot \frac{F_i'}{F_i} \left(\left(\delta_\alpha^\xi \delta_\beta^\zeta - \frac{1}{2} g^{\xi\zeta} \cdot g_{\alpha\beta} \right) \partial_\zeta + g_{\alpha\beta} \left(\frac{1}{2} - \frac{1}{n} \right) \Gamma^\xi_{\kappa\lambda} g^{\kappa\lambda} \right) f_{i,\xi} \right) \delta G^{\alpha\beta}
$$

$$
0 = \sum_{i=1}^N \sqrt{F_i^n} \cdot \left(\frac{F_i'}{F_i} \left(\left(\delta_\alpha^\xi \delta_\beta^\zeta - \frac{1}{2} g^{\xi\zeta} \cdot g_{\alpha\beta} \right) \partial_\zeta + g_{\alpha\beta} \left(\frac{1}{2} - \frac{1}{n} \right) \Gamma^\xi_{\kappa\lambda} g^{\kappa\lambda} \right) f_{i,\xi} \right) \delta G^{\alpha\beta}.
$$

$$(1600)$$

With the **V**-vector approach, we can have a sum of Dirac-like form:

$$0 = \sum_{i=1}^{N} \sqrt{F_i^n} \cdot \left(\frac{F_i'}{F_i} \left(\left(\delta_\alpha^\xi \delta_\beta^\zeta - \frac{1}{2} g^{\xi\zeta} \cdot g_{\alpha\beta} \right) \partial_\zeta + g_{\alpha\beta} \left(\frac{1}{2} - \frac{1}{n} \right) \Gamma_{\kappa\lambda}^\xi g^{\kappa\lambda} \right) f_{i,\xi} \right) V^\alpha \delta?^\beta ,$$

(1601)

and thus:

$$0 = \sum_{i=1}^{N} \sqrt{F_i^n} \cdot \frac{F_i'}{F_i} \left(\left(V^\xi \cdot \delta_\beta^\zeta - \frac{1}{2} g^{\xi\zeta} \cdot V_\beta \right) \partial_\zeta + V_\beta \left(\frac{1}{2} - \frac{1}{n} \right) \Gamma_{\kappa\lambda}^\xi g^{\kappa\lambda} \right) f_{i,\xi} .$$

(1602)

Thereby all scale-functions $F_i = F_i[f]$ have to be chosen in accordance with the condition in Eq. (1441). This finally gives us:

$$0 = \frac{4}{n-2} \cdot \sum_{i=1}^{N} \frac{\sqrt{f_i^{\frac{4}{n-2} \cdot n}}}{f_i} \left(\left(V^\xi \cdot \delta_\beta^\zeta - \frac{1}{2} g^{\xi\zeta} \cdot V_\beta \right) \partial_\zeta + V_\beta \left(\frac{1}{2} - \frac{1}{n} \right) \Gamma_{\kappa\lambda}^\xi g^{\kappa\lambda} \right) f_{i,\xi}$$

$$\Rightarrow \begin{cases} 0 = \sum_{i=1}^{N} \sqrt{f_i^{2+\frac{8}{n-2}}} \left(\left(V^\xi \cdot \delta_\beta^\zeta - \frac{1}{2} g^{\xi\zeta} \cdot V_\beta \right) \partial_\zeta + V_\beta \left(\frac{1}{2} - \frac{1}{n} \right) \Gamma_{\kappa\lambda}^\xi g^{\kappa\lambda} \right) f_{i,\xi} \\ = \sum_{i=1}^{N} f_i^{\frac{n+2}{n-2}} \left(\left(V^\xi \cdot \delta_\beta^\zeta - \frac{1}{2} g^{\xi\zeta} \cdot V_\beta \right) \partial_\zeta + V_\beta \left(\frac{1}{2} - \frac{1}{n} \right) \Gamma_{\kappa\lambda}^\xi g^{\kappa\lambda} \right) f_{i,\xi} . \end{cases}$$

(1603)

As before in the section "10.6 The Classical Dirac Equation, Derived from the Einstein–Hilbert Action," we now demand each addend to give zero and obtain:

$$0 = \left(\left(V^\xi \cdot \delta_\beta^\zeta - \frac{1}{2} g^{\xi\zeta} \cdot V_\beta \right) \partial_\zeta + V_\beta \left(\frac{1}{2} - \frac{1}{n} \right) \Gamma_{\kappa\lambda}^\xi g^{\kappa\lambda} \right) f_{i,\xi} . \qquad (1604)$$

10.9.4 Extension of Eq. (1604) to Arbitrary n and non-Vacuum-Einstein-Compatible Space-Times

We omit the condition $R_{\alpha\beta} - \frac{R}{2} \cdot g_{\alpha\beta} = 0$, go back to the first line of Eq. (1599) and by inserting the results from section "10.9.1 The Other Origin of A "Dirac Equation"" we end up with:

$$0 = \sum_{i=1}^{N} \sqrt{F_i^n} \cdot \left(R_{\alpha\beta} - \frac{1}{2} \cdot R \cdot g_{\alpha\beta} - \frac{(n-1)}{2} \cdot \frac{F_i'}{F_i} \left(\Delta f_i \cdot g_{\alpha\beta} - 2 \cdot f_{i:\alpha\beta} \right) \right) \delta G^{\alpha\beta}$$

$$\Rightarrow \left\{ \begin{array}{l} 0 = \sum_{i=1}^{N} \sqrt{F_i^n} \cdot \left(R_{\alpha\beta} - \frac{1}{2} \cdot R \cdot g_{\alpha\beta} - (n-1) \cdot \frac{F_i'}{F_i} \left(\begin{array}{c} \left(\delta_\alpha^\xi \delta_\beta^\zeta - \frac{1}{2} g^{\xi\zeta} \cdot g_{\alpha\beta} \right) \partial_\zeta \\ + g_{\alpha\beta} \left(\frac{1}{2} - \frac{1}{n} \right) \Gamma_{\kappa\lambda}^\xi g^{\kappa\lambda} \end{array} \right) f_{i,\xi} \right) \delta G^{\alpha\beta} \\[3em] = \sum_{i=1}^{N} \sqrt{F_i^n} \cdot \left(\frac{R_{\alpha\beta} - \frac{1}{2} \cdot R \cdot g_{\alpha\beta}}{n-1} - \frac{F_i'}{F_i} \left(\begin{array}{c} \left(\delta_\alpha^\xi \delta_\beta^\zeta - \frac{1}{2} g^{\xi\zeta} \cdot g_{\alpha\beta} \right) \partial_\zeta \\ + g_{\alpha\beta} \left(\frac{1}{2} - \frac{1}{n} \right) \Gamma_{\kappa\lambda}^\xi g^{\kappa\lambda} \end{array} \right) f_{i,\xi} \right) \delta G^{\alpha\beta}. \end{array} \right. \tag{1605}$$

With the **V**-vector approach, we can have a sum of Dirac-like form:

$$0 = \sum_{i=1}^{N} \sqrt{F_i^n} \cdot \left(\frac{R_{\alpha\beta} - \frac{1}{2} \cdot R \cdot g_{\alpha\beta}}{n-1} - \frac{F_i'}{F_i} \left(\begin{array}{c} \left(\delta_\alpha^\xi \delta_\beta^\zeta - \frac{1}{2} g^{\xi\zeta} \cdot g_{\alpha\beta} \right) \partial_\zeta \\ + g_{\alpha\beta} \left(\frac{1}{2} - \frac{1}{n} \right) \Gamma_{\kappa\lambda}^\xi g^{\kappa\lambda} \end{array} \right) f_{i,\xi} \right) V^\alpha \delta?^\beta, \tag{1606}$$

and thus:

$$0 = \sum_{i=1}^{N} \sqrt{F_i^n} \cdot \left(\frac{V^\alpha R_{\alpha\beta} - \frac{1}{2} \cdot R \cdot V_\beta}{n-1} - \frac{F_i'}{F_i} \left(\begin{array}{c} \left(V^\xi \cdot \delta_\beta^\zeta - \frac{1}{2} g^{\xi\zeta} \cdot V_\beta \right) \partial_\zeta \\ + V_\beta \left(\frac{1}{2} - \frac{1}{n} \right) \Gamma_{\kappa\lambda}^\xi g^{\kappa\lambda} \end{array} \right) f_{i,\xi} \right). \tag{1607}$$

Thereby all scale-functions $F_i = F_i[f]$ have to be chosen in accordance with condition in Eq. (1441) and we demand all F_i and f_i to fulfill the equation above separately. This finally gives us for Eq. (1605):

$$0 = \left(R_{\alpha\beta} - \frac{1}{2} \cdot R \cdot g_{\alpha\beta} \right) \cdot \frac{(n-2)}{4 \cdot (n-1)} \cdot f_i - \left(\begin{array}{c} \left(\delta_\alpha^\xi \delta_\beta^\zeta - \frac{1}{2} g^{\xi\zeta} \cdot g_{\alpha\beta} \right) \partial_\zeta \\ + g_{\alpha\beta} \left(\frac{1}{2} - \frac{1}{n} \right) \Gamma_{\kappa\lambda}^\xi g^{\kappa\lambda} \end{array} \right) f_{i,\xi} \tag{1608}$$

and for the **V**-vector form:

$$0 = \frac{V^\alpha R_{\alpha\beta} - \frac{1}{2} \cdot R \cdot V_\beta}{4 \cdot (n-1)} \cdot (n-2) \cdot f_i - \left(\begin{array}{c} \left(V^\xi \cdot \delta_\beta^\zeta - \frac{1}{2} g^{\xi\zeta} \cdot V_\beta \right) \partial_\zeta \\ + V_\beta \left(\frac{1}{2} - \frac{1}{n} \right) \Gamma_{\kappa\lambda}^\xi g^{\kappa\lambda} \end{array} \right) f_{i,\xi}. \tag{1609}$$

Naturally, when performing the summation with the contra-variant metric tensor in Eq. (1608), we should obtain the same result as we got in the last line of Eq. (1489), when also applying the condition from Eq. (1441). This, in fact, we do:

$$0 = R \cdot \left(1 - \frac{n}{2}\right) \cdot \frac{(n-2)}{4 \cdot (n-1)} \cdot f_i - \left(\left(1 - \frac{n}{2}\right) g^{\xi\zeta} \partial_\zeta + \left(\frac{n}{2} - 1\right) \Gamma^{\xi}_{\kappa\lambda} g^{\kappa\lambda}\right) f_{i,\xi}$$

$$= \left(1 - \frac{n}{2}\right) \cdot \left(R \cdot \frac{(n-2)}{4 \cdot (n-1)} \cdot f_i - \left(g^{\xi\zeta} \partial_\zeta - \Gamma^{\xi}_{\kappa\lambda} g^{\kappa\lambda}\right) f_{i,\xi}\right)$$

$$\Rightarrow \qquad 0 = R \cdot \frac{(n-2)}{4 \cdot (n-1)} \cdot f_i - \left(g^{\xi\zeta} \partial_\zeta - \Gamma^{\xi}_{\kappa\lambda} g^{\kappa\lambda}\right) f_{i,\xi}$$

$$\Rightarrow \qquad 0 = R \cdot \frac{(n-2)}{4 \cdot (n-1)} \cdot f_i - \Delta f_i. \tag{1610}$$

10.9.5 A Note about Potential Simplifications and Evaluation Techniques

It should be pointed out that Eq. (1608) can be brought a bit more into the typical Dirac structure when demanding the condition:

$$\left(\delta^{\xi}_\alpha \delta^{\zeta}_\beta - \frac{1}{2} g^{\xi\zeta} \cdot g_{\alpha\beta}\right) \partial_\zeta f_{i,\xi} = O_{\alpha\beta} \cdot f_i, \tag{1611}$$

which leads to:

$$0 = \left(\left(R_{\alpha\beta} - \frac{1}{2} \cdot R \cdot g_{\alpha\beta}\right) \cdot \frac{(n-2)}{4 \cdot (n-1)} - O_{\alpha\beta}\right) \cdot f_i - g_{\alpha\beta} \left(\frac{1}{2} - \frac{1}{n}\right) \Gamma^{\xi}_{\kappa\lambda} g^{\kappa\lambda} f_{i,\xi}. \tag{1612}$$

Besides, summing up the latter equation as follows:

$$0 = \left(R \cdot \left(1 - \frac{1}{2} \cdot n\right) \cdot \frac{(n-2)}{4 \cdot (n-1)} - g^{\alpha\beta} O_{\alpha\beta}\right) \cdot f_i - \left(\frac{n}{2} - 1\right) \Gamma^{\xi}_{\kappa\lambda} g^{\kappa\lambda} f_{i,\xi}$$

$$\Rightarrow \qquad 0 = \left(R \cdot \frac{(n-2)}{4 \cdot (n-1)} + 2 \cdot \frac{g^{\alpha\beta} O_{\alpha\beta}}{(n-2)}\right) \cdot f_i + \Gamma^{\xi}_{\kappa\lambda} g^{\kappa\lambda} f_{i,\xi}$$

$$\Rightarrow \qquad \left(R \cdot \frac{(n-2)}{4 \cdot (n-1)} + 2 \cdot \frac{g^{\alpha\beta} O_{\alpha\beta}}{(n-2)}\right) \cdot f_i = -\Gamma^{\xi}_{\kappa\lambda} g^{\kappa\lambda} f_{i,\xi} \tag{1613}$$

results in a scalar differential equation of first order for the function f_i.

Another possibility for the simplified treatment of the first line of Eq. (1599) might be obtained via:

$$0 = \int_{V_G} d^n x \sqrt{-g} \left(\sum_{i=1}^{N} \sqrt{F_i^n} \cdot \left(\frac{F_i'}{F_i} \left(\Delta f_i \cdot g_{\alpha\beta} - 2 \cdot \overbrace{\left(f_{i,\alpha\beta} - \frac{g_{\alpha\beta}}{n} \Gamma_{\kappa\lambda}^\gamma g^{\kappa\lambda} f_{i,\gamma} \right)}^{\equiv X_{\alpha\beta}\Delta\Phi_i} \right) \right) \right) \delta G^{\alpha\beta} \right)$$

$$\Rightarrow \quad X_{\alpha\beta}\Delta\Phi_i = f_{i,\alpha\beta} - \frac{g_{\alpha\beta}}{n} \Gamma_{\kappa\lambda}^\gamma g^{\kappa\lambda} f_{i,\gamma} \rightarrow X_{\alpha\beta} \left(g^{\mu\nu}\Phi_{i,\mu\nu} - \Gamma_{\kappa\lambda}^\gamma g^{\kappa\lambda}\Phi_{i,\gamma} \right)$$

$$= f_{i,\alpha\beta} - \frac{g_{\alpha\beta}}{n} \Gamma_{\kappa\lambda}^\gamma g^{\kappa\lambda} f_{i,\gamma}$$

$$\rightarrow X_{\alpha\beta} g^{\mu\nu}\Phi_{i,\mu\nu} = f_{i,\alpha\beta}; \quad X_{\alpha\beta}\Gamma_{\kappa\lambda}^\gamma g^{\kappa\lambda}\Phi_{i,\gamma} = \frac{g_{\alpha\beta}}{n} \Gamma_{\kappa\lambda}^\gamma g^{\kappa\lambda} f_{i,\gamma}. \quad (1614)$$

We see no inconsistency when choosing:

$$\Phi_i = f_i + C_{\Phi i}; \quad X_{\alpha\beta} = \frac{g_{\alpha\beta}}{n}. \quad (1615)$$

Then the necessary conditions for our little transformational choice in Eq. (1614):

$$\rightarrow \frac{g_{\alpha\beta}}{n} g^{\mu\nu}\Phi_{i,\mu\nu} = f_{i,\alpha\beta} = \frac{g_{\alpha\beta}}{n} g^{\mu\nu} f_{i,\mu\nu}; \quad \Phi_{i,\gamma} = f_{i,\gamma}, \quad (1616)$$

leads to:

$$0 = \int_{V_G} d^n x \sqrt{-g} \left(\sum_{i=1}^{N} \sqrt{F_i^n} \cdot \left(R_{\alpha\beta} - \frac{1}{2} \cdot R \cdot g_{\alpha\beta} + \frac{(n-1)}{2} \cdot \frac{F_i'}{F_i} \left(1 - \frac{2}{n} \right) \Delta\Phi_i \cdot g_{\alpha\beta} \right) \delta G^{\alpha\beta} \right).$$

$$(1617)$$

Please note that this approach simplifies Eq. (1599) but also dramatically restricts the possibilities for solutions to arbitrary classical non-vacuum cases $R_{\alpha\beta} - \dfrac{1}{2} \cdot R \cdot g_{\alpha\beta} \neq 0$. We will therefore not consider this approach any further in this chapter.

10.9.6 Generalization

Now we intend to repeat our evaluation from above with our more general decomposition of the Laplace operator from Eq. (1433):

$$0 = R_{\alpha\beta} - \frac{1}{2} R \cdot g_{\alpha\beta} + \frac{F'}{F}(1-n) \left(\left\{ \begin{array}{c} f_{,\alpha\beta} - \Gamma_{\alpha\beta}^{\gamma} f_{,\gamma} \\[2mm] f_{,\alpha\beta} - \dfrac{g_{\alpha\beta}}{n} \Gamma_{\kappa\lambda}^{\gamma} g^{\kappa\lambda} f_{,\gamma} \\[3mm] \dfrac{g_{\alpha\beta}}{n} g^{\kappa\lambda} f_{,\kappa\lambda} - \Gamma_{\alpha\beta}^{\gamma} f_{,\gamma} \\[3mm] \dfrac{g_{\alpha\beta}}{n} g^{\kappa\lambda} f_{,\kappa\lambda} - \dfrac{g_{\alpha\beta}}{n} \Gamma_{\kappa\lambda}^{\gamma} g^{\kappa\lambda} f_{,\gamma} \end{array} \right\} - \frac{1}{2} \Delta f \cdot g_{\alpha\beta} \right)$$

$$= R_{\alpha\beta} - \frac{1}{2} R \cdot g_{\alpha\beta} + \frac{F'}{F}(1-n) \left(\left\{ \begin{array}{c} f_{,\alpha\beta} - \Gamma_{\alpha\beta}^{\gamma} f_{,\gamma} \\[2mm] f_{,\alpha\beta} - \dfrac{g_{\alpha\beta}}{n} \Gamma_{\kappa\lambda}^{\gamma} g^{\kappa\lambda} f_{,\gamma} \\[3mm] \dfrac{g_{\alpha\beta}}{n} g^{\kappa\lambda} f_{,\kappa\lambda} - \Gamma_{\alpha\beta}^{\gamma} f_{,\gamma} \\[3mm] \dfrac{g_{\alpha\beta}}{n} g^{\kappa\lambda} f_{,\kappa\lambda} - \dfrac{g_{\alpha\beta}}{n} \Gamma_{\kappa\lambda}^{\gamma} g^{\kappa\lambda} f_{,\gamma} \\[3mm] -\dfrac{1}{2} g^{\xi\zeta} \left(f_{,\xi\zeta} - \dfrac{g_{\xi\zeta}}{n} \Gamma_{\kappa\lambda}^{\gamma} g^{\kappa\lambda} f_{,\gamma} \right) \cdot g_{\alpha\beta} \end{array} \right\} \right). \quad (1618)$$

Now we assume the classical Einstein part to be $0 = R_{\alpha\beta} - \dfrac{1}{2} R \cdot g_{\alpha\beta}$ and reshape Eq. (1618):

$$0 = \left\{ f_{,\alpha\beta}, \frac{g_{\alpha\beta}}{n} g^{\kappa\lambda} f_{,\kappa\lambda} \right\} - \frac{1}{2} g^{\xi\zeta} f_{,\xi\zeta} \cdot g_{\alpha\beta} - \left\{ \Gamma_{\alpha\beta}^{\gamma} f_{,\gamma}, \frac{g_{\alpha\beta}}{n} \Gamma_{\kappa\lambda}^{\gamma} g^{\kappa\lambda} f_{,\gamma} \right\}$$

$$+ \frac{1}{2} \Gamma_{\kappa\lambda}^{\gamma} g^{\kappa\lambda} f_{,\gamma} \cdot g_{\alpha\beta}$$

$$= \left(\left\{ \delta_{\alpha}^{\xi} \delta_{\beta}^{\zeta}, \frac{g_{\alpha\beta}}{n} g^{\xi\zeta} \right\} - \frac{1}{2} g^{\xi\zeta} \cdot g_{\alpha\beta} \right) f_{,\xi\zeta} + \frac{1}{2} \Gamma_{\kappa\lambda}^{\gamma} g^{\kappa\lambda} \cdot g_{\alpha\beta} - \left\{ \Gamma_{\alpha\beta}^{\gamma}, \frac{g_{\alpha\beta}}{n} \Gamma_{\kappa\lambda}^{\gamma} g^{\kappa\lambda} \right\} f_{,\gamma}$$

$$= \left(\left\{ \delta_{\alpha}^{\xi} \delta_{\beta}^{\zeta}, \frac{g_{\alpha\beta}}{n} g^{\xi\zeta} \right\} - \frac{1}{2} g^{\xi\zeta} \cdot g_{\alpha\beta} \right) f_{,\xi\zeta} + \left(\frac{1}{2} \Gamma_{\kappa\lambda}^{\gamma} g^{\kappa\lambda} \cdot g_{\alpha\beta} - \left\{ \Gamma_{\alpha\beta}^{\gamma}, \frac{g_{\alpha\beta}}{n} \Gamma_{\kappa\lambda}^{\gamma} g^{\kappa\lambda} \right\} \right) f_{,\gamma}$$

$$= \left(\left\{ \delta_{\alpha}^{\xi} \delta_{\beta}^{\zeta}, \frac{g_{\alpha\beta}}{n} g^{\xi\zeta} \right\} - \frac{1}{2} g^{\xi\zeta} \cdot g_{\alpha\beta} \right) f_{,\xi\zeta} + \left(\frac{1}{2} \Gamma_{\kappa\lambda}^{\xi} g^{\kappa\lambda} \cdot g_{\alpha\beta} - \left\{ \Gamma_{\alpha\beta}^{\xi}, \frac{g_{\alpha\beta}}{n} \Gamma_{\kappa\lambda}^{\xi} g^{\kappa\lambda} \right\} \right) f_{,\xi}$$

$$= \left(\left(\left\{ \delta_{\alpha}^{\xi} \delta_{\beta}^{\zeta}, \frac{g_{\alpha\beta}}{n} g^{\xi\zeta} \right\} - \frac{1}{2} g^{\xi\zeta} \cdot g_{\alpha\beta} \right) \partial_{\zeta} + \frac{1}{2} \Gamma_{\kappa\lambda}^{\xi} g^{\kappa\lambda} \cdot g_{\alpha\beta} - \left\{ \Gamma_{\alpha\beta}^{\xi}, \frac{g_{\alpha\beta}}{n} \Gamma_{\kappa\lambda}^{\xi} g^{\kappa\lambda} \right\} \right) f_{,\xi}.$$

$$(1619)$$

As before, we realize that we could substitute the first derivative $f_{,\xi}$ by an arbitrary vector Φ_ξ and already recognize the typical Dirac structure, only that we have "matrices" with the derivative term $\left(\left\{\delta_\alpha^\xi \delta_\beta^\zeta, \dfrac{g_{\alpha\beta}}{n} g^{\xi\zeta}\right\} - \dfrac{1}{2} g^{\xi\zeta} \cdot g_{\alpha\beta}\right)\partial_\zeta$ of rank 4 rather than 3:

$$0 = \left(\begin{array}{l}\left(\left\{\delta_\alpha^\xi \delta_\beta^\zeta, \dfrac{g_{\alpha\beta}}{n} g^{\xi\zeta}\right\} - \dfrac{1}{2} g^{\xi\zeta} \cdot g_{\alpha\beta}\right)\partial_\zeta \\ + \dfrac{1}{2}\Gamma_{\kappa\lambda}^\xi g^{\kappa\lambda} \cdot g_{\alpha\beta} - \left\{\Gamma_{\alpha\beta}^\xi, \dfrac{g_{\alpha\beta}}{n}\Gamma_{\kappa\lambda}^\xi g^{\kappa\lambda}\right\}\end{array}\right)\Phi_\xi. \tag{1620}$$

This could easily be remediated with an arbitrary vector $\mathbf{V}^\alpha$ as follows:

$$0 = V^\alpha \cdot \left(\begin{array}{l}\left(\left\{\delta_\alpha^\xi \delta_\beta^\zeta, \dfrac{g_{\alpha\beta}}{n} g^{\xi\zeta}\right\} - \dfrac{1}{2} g^{\xi\zeta} \cdot g_{\alpha\beta}\right)\partial_\zeta \\ + \dfrac{1}{2}\Gamma_{\kappa\lambda}^\xi g^{\kappa\lambda} \cdot g_{\alpha\beta} - \left\{\Gamma_{\alpha\beta}^\xi, \dfrac{g_{\alpha\beta}}{n}\Gamma_{\kappa\lambda}^\xi g^{\kappa\lambda}\right\}\end{array}\right)\Phi_\xi$$

$$= \left(\begin{array}{l}\left(\left\{V^\xi \cdot \delta_\beta^\zeta, \dfrac{V_\beta}{n} g^{\xi\zeta}\right\} - \dfrac{1}{2} g^{\xi\zeta} \cdot g_{\alpha\beta}\right)\partial_\zeta \\ + \dfrac{1}{2}\Gamma_{\kappa\lambda}^\xi g^{\kappa\lambda} \cdot V_\beta - \left\{V^\alpha \cdot \Gamma_{\alpha\beta}^\xi, \dfrac{V_\beta}{n}\Gamma_{\kappa\lambda}^\xi g^{\kappa\lambda}\right\}\end{array}\right)\Phi_\xi. \tag{1621}$$

This clearly now has the structure of a rather general Dirac equation. Generally we say, because it is automatically given in n dimensions and arbitrary—potentially also curved—space-times under the condition of:

$$0 = R_{\alpha\beta} - \frac{1}{2}R \cdot g_{\alpha\beta} + \frac{(1-n)}{4F^2}\left(\begin{array}{c}FF'' \\ +F'F'(n-6)\end{array}\right)\left(\begin{array}{c}f_{,\alpha}f_{,\beta} \\ -f_{,\mu}f_{,\nu}g^{\mu\nu}\dfrac{g_{\alpha\beta}}{2}\end{array}\right)$$

$$+ \kappa T_{\alpha\beta} + \Lambda \cdot F \cdot g_{\alpha\beta}. \tag{1622}$$

Knowing that we can always apply Eq. (1441), resulting in:

$$0 = R_{\alpha\beta} - \frac{1}{2}R \cdot g_{\alpha\beta} + \kappa T_{\alpha\beta} + \Lambda \cdot F \cdot g_{\alpha\beta} \tag{1623}$$

and avoiding the postulated matter term and cosmological constant anyway, we end up with the vacuum Einstein field equations:

$$0 = R_{\alpha\beta} - \frac{1}{2}R \cdot g_{\alpha\beta} \tag{1624}$$

as an additional condition in connection with the Dirac-like equation (1620) or:

$$0 = \begin{pmatrix} \left(\left\{ V^\xi \cdot \delta^\zeta_\beta , \frac{V_\beta}{n} g^{\xi\zeta} \right\} - \frac{1}{2} g^{\xi\zeta} \cdot V_\beta \right) \partial_\zeta \\ + \frac{1}{2} \Gamma^\xi_{\kappa\lambda} g^{\kappa\lambda} \cdot V_\beta - \left\{ V^\alpha \cdot \Gamma^\xi_{\alpha\beta} , \frac{V_\beta}{n} \Gamma^\xi_{\kappa\lambda} g^{\kappa\lambda} \right\} \end{pmatrix} \Phi_\xi. \tag{1625}$$

As the terms in { ... } are arbitrary in the way we combine them, we can here also give the versions with the highest degree of freedom, reading:

$$0 = \left(\left(\delta^\xi_\alpha \delta^\zeta_\beta - \frac{1}{2} g^{\xi\zeta} \cdot g_{\alpha\beta} \right) \partial_\zeta + \frac{1}{2} \Gamma^\xi_{\kappa\lambda} g^{\kappa\lambda} \cdot g_{\alpha\beta} - \Gamma^\xi_{\alpha\beta} \right) \Phi_\xi \tag{1626}$$

and

$$0 = \left(\left(V^\xi \cdot \delta^\zeta_\beta - \frac{1}{2} g^{\xi\zeta} \cdot V_\beta \right) \partial_\zeta + \frac{1}{2} \Gamma^\xi_{\kappa\lambda} g^{\kappa\lambda} \cdot V_\beta - V^\alpha \cdot \Gamma^\xi_{\alpha\beta} \right) \Phi_\xi. \tag{1627}$$

We might consider these equations as gravitational or gravity Dirac equations.

The potential origination for the **V**-vector was already discussed in connection with Eq. (1583).

Once again avoiding the Einstein vacuum condition $R_{\alpha\beta} - \frac{1}{2} \cdot R \cdot g_{\alpha\beta} = 0$ leads to:

$$0 = \left(R_{\alpha\beta} - \frac{1}{2} \cdot R \cdot g_{\alpha\beta} \right) \cdot \frac{(n-2)}{4 \cdot (n-1)} \cdot f_i - \begin{pmatrix} \left(\delta^\xi_\alpha \delta^\zeta_\beta - \frac{1}{2} g^{\xi\zeta} \cdot g_{\alpha\beta} \right) \partial_\zeta \\ + \frac{1}{2} \Gamma^\xi_{\kappa\lambda} g^{\kappa\lambda} \cdot g_{\alpha\beta} - \Gamma^\xi_{\alpha\beta} \end{pmatrix} f_{i,\xi} \tag{1628}$$

and

$$0 = \frac{V^\alpha R_{\alpha\beta} - \frac{1}{2} \cdot R \cdot V_\beta}{4 \cdot (n-1)} \cdot (n-2) \cdot f_i - \begin{pmatrix} \left(V^\xi \cdot \delta^\zeta_\beta - \frac{1}{2} g^{\xi\zeta} \cdot V_\beta \right) \partial_\zeta \\ + \frac{1}{2} \Gamma^\xi_{\kappa\lambda} g^{\kappa\lambda} \cdot V_\beta - V^\alpha \cdot \Gamma^\xi_{\alpha\beta} \end{pmatrix} f_{i,\xi}. \tag{1629}$$

10.10 The Gravity Dirac Equation

We use Eq. (1618) in the following form:

$$
0 = \begin{pmatrix} R_{\alpha\beta} - \dfrac{1}{2} \cdot R \cdot g_{\alpha\beta} - \dfrac{F_i'}{F_i}(n-1)\left(f_{i,\alpha\beta} - \Gamma_{\alpha\beta}^{\gamma} f_{i,\gamma} - \dfrac{g_{\alpha\beta}}{2} \Delta f_i \right) \\[3mm] + \dfrac{(1-n)}{4F_i^2}\left(\begin{array}{c} F_i F_i'' \\ +F_i'F_i'(n-6) \end{array} \right)\left(\begin{array}{c} f_{i,\alpha} f_{i,\beta} \\ -f_{i,\mu}f_{i,\nu}g^{\mu\nu}\dfrac{g_{\alpha\beta}}{2} \end{array} \right) \end{pmatrix} \qquad (1630)
$$

and set $F[f] = f$, which gives us:

$$
0 = \begin{pmatrix} R_{\alpha\beta} - \dfrac{1}{2} \cdot R \cdot g_{\alpha\beta} - \dfrac{1}{f_i}(n-1)\left(f_{i,\alpha\beta} - \Gamma_{\alpha\beta}^{\gamma} f_{i,\gamma} - \dfrac{g_{\alpha\beta}}{2} \Delta f_i \right) \\[3mm] + \dfrac{(1-n)}{4f_i^2}(n-6)\left(f_{i,\alpha} f_{i,\beta} - f_{i,\mu}f_{i,\nu}g^{\mu\nu}\dfrac{g_{\alpha\beta}}{2} \right) \end{pmatrix}. \qquad (1631)
$$

Now we assume the possibility for a decomposition of the following form:

$$
\beta_\alpha \cdot \beta_\beta = \begin{pmatrix} R_{\alpha\beta} - \dfrac{1}{2} \cdot R \cdot g_{\alpha\beta} - \dfrac{1}{f_i}(n-1)\left(f_{i,\alpha\beta} - \Gamma_{\alpha\beta}^{\gamma} f_{i,\gamma} - \dfrac{g_{\alpha\beta}}{2} \Delta f_i \right) \\[3mm] - \dfrac{(1-n)}{8f_i^2}(n-6) f_{i,\mu}f_{i,\nu}g^{\mu\nu}g_{\alpha\beta} \end{pmatrix} \qquad (1632)
$$

with some yet unknown objects β. This leads us to the following equation:

$$
0 = \beta_\alpha \cdot \beta_\beta \cdot \underbrace{\dfrac{4}{(1-n)(n-6)\left(1-\dfrac{n}{2}\right)}}_{\equiv \chi^2} \cdot f_i^2 + f_{i,\alpha} f_{i,\beta}
$$

$$
= \beta_\alpha \cdot \beta_\beta \cdot \chi^2 \cdot f_i^2 + f_{i,\alpha} f_{i,\beta}. \qquad (1633)
$$

We realize that Eq. (1633) could just be obtained as scalar product of the following vectors:

$$
\frac{1}{\sqrt{2}} \begin{bmatrix} (\beta_\alpha \cdot \chi \cdot f_i + f_{i,\alpha}) \\ (\beta_\alpha \cdot \chi \cdot f_i - f_{i,\alpha}) \end{bmatrix} ; \frac{1}{\sqrt{2}} \begin{bmatrix} (\beta_\beta \cdot \chi \cdot f_i + f_{i,\beta}) \\ (\beta_\beta \cdot \chi \cdot f_i - f_{i,\beta}) \end{bmatrix}
$$

$$\Rightarrow \quad \frac{1}{\sqrt{2}}\begin{Bmatrix}\left(\beta_\alpha\cdot\chi\cdot f_i+f_{i,\alpha}\right)\\\left(\beta_\alpha\cdot\chi\cdot f_i-f_{i,\alpha}\right)\end{Bmatrix}\cdot\frac{1}{\sqrt{2}}\begin{Bmatrix}\left(\beta_\beta\cdot\chi\cdot f_i+f_{i,\beta}\right)\\\left(\beta_\beta\cdot\chi\cdot f_i-f_{i,\beta}\right)\end{Bmatrix}$$

$$=\beta_\alpha\cdot\beta_\beta\cdot\chi^2\cdot f_i^2+f_{i,\alpha}f_{i,\beta}. \tag{1634}$$

Even though this simple vector form is not a suitable decomposition for us, because it gives two completely different equations for the same wave function, we conclude that decomposition should be possible when finding the right objects to result in the connection of the terms in Eq. (1633). Our Dirac-like approach shall be:

$$\left(a\cdot\beta_\alpha\cdot\chi\cdot f_i+A\cdot f_{i,\alpha}\right)\left(b\cdot\beta_\beta\cdot\chi\cdot f_i+B\cdot f_{i,\beta}\right)=0. \tag{1635}$$

With the objects a, A, b, and B being general 2×2 matrices, we find a variety of solutions of which we here only present a very simple one, namely:

$$A=\begin{pmatrix}1&1\\1&1\end{pmatrix};\quad a=\begin{pmatrix}-1&1\\-1&1\end{pmatrix};\quad B=\begin{pmatrix}1&-1\\1&-1\end{pmatrix};\quad b=\begin{pmatrix}-1&1\\1&-1\end{pmatrix}. \tag{1636}$$

This makes Eq. (1635) to result in:

$$\left(a\cdot\beta_\alpha\cdot\chi\cdot f_i+A\cdot f_{i,\alpha}\right)\left(b\cdot\beta_\beta\cdot\chi\cdot f_i+B\cdot f_{i,\beta}\right)$$

$$=2\cdot\begin{pmatrix}1&-1\\1&-1\end{pmatrix}\cdot\left(\beta_\alpha\cdot\beta_\beta\cdot\chi^2\cdot f_i^2+f_{i,\alpha}f_{i,\beta}\right) \tag{1637}$$

and gives us the gravity "Dirac equations":

$$a\cdot\beta_\alpha\cdot\chi\cdot f_i+A\cdot f_{i,\alpha}=\begin{pmatrix}-1&1\\-1&1\end{pmatrix}\cdot\beta_\alpha\cdot\chi\cdot f_i+\begin{pmatrix}1&1\\1&1\end{pmatrix}\cdot f_{i,\alpha}=0$$

$$b\cdot\beta_\beta\cdot\chi\cdot f_i+B\cdot f_{i,\beta}=\begin{pmatrix}-1&1\\1&-1\end{pmatrix}\cdot\beta_\beta\cdot\chi\cdot f_i+\begin{pmatrix}1&-1\\1&-1\end{pmatrix}\cdot f_{i,\beta}=0. \tag{1638}$$

Please note that the decomposition in Eq. (1632) may also be performed as follows:

$$\beta_\alpha\cdot\beta_\beta=R_{\alpha\beta}-\frac{1}{2}\cdot R\cdot g_{\alpha\beta}-\frac{1}{f_i}(n-1)\left(f_{i,\alpha\beta}-\Gamma_{\alpha\beta}^\gamma f_{i,\gamma}-\frac{g_{\alpha\beta}}{2}\Delta f_i\right). \tag{1639}$$

This leads us to the following equation:

$$0=\beta_\alpha\cdot\beta_\beta\cdot\underbrace{\frac{4}{(1-n)(n-6)\left(1-\dfrac{n}{2}\right)}}_{\equiv\chi^2}\cdot f_i^2+f_{i,\alpha}f_{i,\beta}-\frac{f_{i,\mu}f_{i,\nu}g^{\mu\nu}g_{\alpha\beta}}{2}$$

$$= \beta_\alpha \cdot \beta_\beta \cdot \chi^2 \cdot f_i^2 + f_{i,\alpha} f_{i,\beta} - \frac{f_{i,\mu} f_{i,v} g^{\mu v} g_{\alpha\beta}}{2}. \tag{1640}$$

Now our Dirac-like approach shall be:

$$\begin{pmatrix} a \cdot \beta_\alpha \cdot \chi \cdot f_i + A \cdot f_{i,\alpha} \\ +k \cdot \dfrac{i}{\sqrt{2}} \cdot f_{i,\mu} \cdot \mathbf{e}^\mu \cdot \mathbf{e}_\alpha \end{pmatrix} \times \begin{pmatrix} b \cdot \beta_\beta \cdot \chi \cdot f_i + B \cdot f_{i,\beta} \\ +K \cdot \dfrac{i}{\sqrt{2}} \cdot f_{i,v} \cdot \mathbf{e}^v \cdot \mathbf{e}_\beta \end{pmatrix} = 0. \tag{1641}$$

A more general solution and some discussion will be presented elsewhere.

10.11 Conclusions

We think that we finally found a way to directly extract the Dirac equation from the Einstein–Hilbert action. Thereby, we only required a scaled metric with a scaling function $F[f]$ and an extension of the variation allowing us for a split-up of the integral volume into smaller volume pieces with a variety of $F[f]$.

Along the way we also discovered that there should be a Gravity Dirac Equation, which we derived.

References

1. N. Schwarzer, *Towards Quantum Einstein Field Equations*, Part 7 of "Medical Socio-Economic Quantum Gravity," Self-published, Amazon Digital Services, December 2020, Kindle.

2. N. Schwarzer, *The Quantum Einstein Field Equations in their Simplest Form*, Part 7a of "Medical Socio-Economic Quantum Gravity," Self-published, Amazon Digital Services, 2021, Kindle.

3. D. Hilbert, Die Grundlagen der Physik, Teil 1, *Göttinger Nachrichten*, 1915, 395–407.

4. A. Einstein, Grundlage der allgemeinen Relativitätstheorie, *Ann. Phys.*, 1916, **49** (ser. 4), 769–822.

5. N. Schwarzer, *Societies and Ecotons—The Photons of the Human Society—Control them and Rule the World*, Part 1 of "Medical Socio-Economic Quantum Gravity," Self-published, Amazon Digital Services, 2020, Kindle.

6. N. Schwarzer, *Masses and the Infinity Options Principle: Can We Explain the 3-Generation and the Quantized Mass Problem?*, Part 5 of "Medical Socio-

Economic Quantum Gravity," Self-published, Amazon Digital Services, 2020, Kindle.

7. N. Schwarzer, *The World Formula: A Late Recognition of David Hilbert's Stroke of Genius*, 2022, Jenny Stanford Publishing. ISBN: 9789814877206.

8. N. Schwarzer, *The Metric Dirac Equation Revisited and the Geometry of Spinors*, Part 8 of "Medical Socio-Economic Quantum Gravity," Self-published, Amazon Digital Services, 2020, Kindle.

9. N. Schwarzer, *The 3-Generation of Elementary Particles*, Part 6 of "Medical Socio-Economic Quantum Gravity," Self-published, Amazon Digital Services, 2020, Kindle.

10. N. Schwarzer, *My Horcruxes: A Curvy Math to Salvation*, Part 9 of "Medical Socio-Economic Quantum Gravity," Self-published, Amazon Digital Services, 2021, Kindle.

11. K. Schwarzschild, Über das Gravitationsfeld eines Massenpunktes nach der Einsteinschen Theorie, *Sitzungsberichte der Königlich-Preussischen Akademie der Wissenschaften*, Reimer, Berlin. 1916, 189–196.

12. P. A. M. Dirac, *The Quantum Theory of the Electron*, 1928. *Proceedings of the Royal Society A.* DOI: 10.1098/rspa.1928.00231.

13. N. Schwarzer, *The Quantum Black Hole*, Part 7b of "Medical Socio-Economic Quantum Gravity," Self-published, Amazon Digital Services, 2021, Kindle.

14. N. Schwarzer, *Quantum Cosmology: A Simple Answer to the Flatness Riddle*, Part 7c of "Medical Socio-Economic Quantum Gravity," Self-published, Amazon Digital Services, 2021, Kindle.

15. T. Bodan, *7 Days—How to Explain the World to My Dying Child*, Self-published, 2021, BoD Classic, ISBN 978-3-7526-3972-8.

16. T. Bodan, *The Eighth Day—Two Jews against the Third Reich*, illustrated version, Self-published, 2021, BoD Classic, ISBN 978-3-7534-1725-7.

17. N. Schwarzer, *The Theory of Everything—Quantum and Relativity is Everywhere—A Fermat Universe*, 2020, Pan Stanford Publishing, ISBN-10: 9814774472.

18. N. Schwarzer, *General Quantum Relativity*, Self-published, Amazon Digital Services, 2016, Kindle.

19. N. Schwarzer, *The Theory of Everything—Quantum and Relativity is Everywhere: A Fermat Universe*, 2020, Pan Stanford Publishing, ISBN-10: 9814774472.

THE SIXTH DAY:
A MATH FOR BODY, SOUL, AND UNIVERSE

Chapter 11

A Curvy Math to Salvation

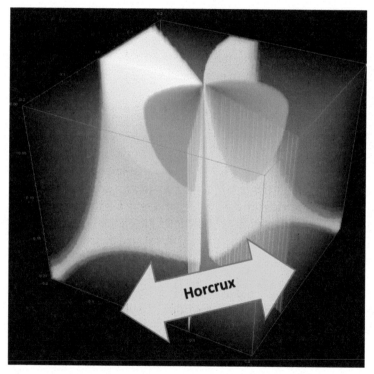

Figure 11.1 A spinor object during split-up.

The Math of Body, Soul, and the Universe
Norbert Schwarzer
Copyright © 2023 Jenny Stanford Publishing Pte. Ltd.
ISBN 978-981-4968-24-9 (Hardcover), 978-1-003-33454-5 (eBook)
www.jennystanford.com

11.1 Personal Note and Motivation

"Do you fear death?"

Most people would probably answer "yes" and explain this by the simple fact that—apart from the fear of physical pain in connection with the process of dying—their fear is in principal made out and fed by the fact that they do not know what comes after death. Clear though it is that death destroys the body of the individual and gives all of its components back to the circle of life (recycles it), most human beings still think that there is more to them than the human eye sees. In other words, they think that there is also something, which is connected with the body, while being alive, but which does not die with it at the end.

The goal of this book was and still is to work out the mathematical fundamentals for the description of what often unscrupulously is called the "soul" (Fig. 11.1). Now, with so many basics brought together in the previous chapters, we think that we are finally able to begin with this task.

11.2 Abstract

In the previous chapters (also [A1]) we saw that important scalar equations of Quantum Theory emerge when we construct a scaled metric by simply multiplying the metric with a factor $F[f]$, thereby assuming the inner part of the scale factor to be an arbitrary function f. f itself then—so it was derived—becomes the wave function.[1] Subjecting this scaled metric in the usual way to the Einstein–Hilbert action [A2] not only presents us with quantum Einstein field equations [A3] and thus, a quantized version of Einstein's General Theory of Relativity [A4], but also—along the way (e.g., [A5–A10])—gives us the classical quantum equations. Also, thermodynamics can be derived from, respectively found inside this simple scaled metric approach [A3, A10]. This was all also presented in the previous chapters of this book.

In the present chapter, it will be shown how various extensions and generalizations lead to a great variety of metric options of which all are treatable with sums of Laplace, respectively Klein–Gordon equations. This multitude of degrees of freedom and the corresponding duality of metric and wave functions gives the impression of body-soul-structures.

---

[1]For convenience, we will repeat these derivations for the Klein–Gordon and the Schrödinger equation in the appendices of this chapter. We will also repeat the derivation of the Dirac equation in its briefest form.

References for the Abstract

A1. N. Schwarzer, *Masses and the Infinity Options Principle—Can We Explain the 3-Generations and the Quantized Mass Problem?, Part 5 of Medical Socio-Economic Quantum Gravity*, Self-published, Amazon Digital Services, 2020, Kindle.

A2. D. Hilbert, Die Grundlagen der Physik, Teil 1, *Göttinger Nachrichten*, 1915, 395–407.

A3. N. Schwarzer, *Towards Quantum Einstein Field Equations, Part 7 of Medical Socio-Economic Quantum Gravity*, Self-published, Amazon Digital Services, December 2020, Kindle.

A4. A. Einstein, Grundlage der allgemeinen Relativitätstheorie, *Ann. Phys.*, 1916, **49** (ser. 4), 769–822.

A5. N. Schwarzer, *Brief Proof of Hilbert's World Formula—Dirac, Klein–Gordon, Schrödinger, Einstein, Evolution and the Second Law of Thermodynamics all from one origin*, Self-published, Amazon Digital Services, 2020, Kindle.

A6. N. Schwarzer, *Societons and Ecotons—The Photons of the Human Society: Control Them and Rule the World, Part 1 of Medical Socio-Economic Quantum Gravity*, Self-published, Amazon Digital Services, 2020, Kindle.

A7. N. Schwarzer, *Humanitons –The Intrinsic Photons of the Human Body—Understand Them and Cure Yourself, Part 2 of Medical Socio-Economic Quantum Gravity*, Self-published, Amazon Digital Services, 2020, Kindle.

A8. N. Schwarzer, *Mastering Human Crises with Quantum-Gravity-Based but Still Practicable Models—First Measure: SEEING and UNDERSTANDING the WHOLE, Part 3 of Medical Socio-Economic Quantum Gravity*, Self-published, Amazon Digital Services, 2020, Kindle.

A9. N. Schwarzer, *Self-Similar Quantum Gravity—How Infinity Brings Simplicity, Part 4 of Medical Socio-Economic Quantum Gravity*, Self-published, Amazon Digital Services, 2020, Kindle.

A10. N. Schwarzer, *The World Formula: A Late Recognition of David Hilbert's Stroke of Genius*, 2022, Jenny Stanford Publishing. ISBN: 9789814877206.

11.3 What We Will Need

11.3.1 Repetition: Extended (Quantum Gravity) Einstein Field Equations

It was already shown in this book and earlier papers [1–9][2] how we can find a direct extraction of the Klein–Gordon, the Schrödinger, and the Dirac equation from the Ricci scalar R^* of a modified metric of the kind $G_{\alpha\beta} = F[f]^* g_{\alpha\beta}$.[3] This just means that, in essence, we had derived the fundamental quantum equation from the kernel of the Einstein–Hilbert action [10], from which the Einstein field equation and thus, the General Theory of Relativity [11] can be obtained, too. It was also shown how this approach provides us with new—and intrinsically consistent (see section "10.5 Solving an Inconsistency Problem with the Quantum Einstein Field Equations" in the previous chapter) Einstein field equations clearly showing quantum properties. From there, we think that we can derive mathematical structures, allowing us the description of metric wave function–wrapped-objects, which is to say "body–soul systems."

In order to work this out, we need to repeat a few essentials.

Taking the modified metric from above $G_{\alpha\beta} = F[f]^* g_{\alpha\beta}$ with a yet arbitrary scalar function $F[f]$ and a general function of all coordinates $f[x_0, x_1, \ldots]$, the corresponding modified Ricci scalar R^* reads[4]:

$$R^* = R^*_{\alpha\beta} G^{\alpha\beta}$$

$$= \left(\begin{array}{c} \Gamma^{\sigma}_{\alpha\beta,\sigma} g^{\alpha\beta} - \Gamma^{\sigma}_{\beta\sigma,\alpha} g^{\alpha\beta} - \Gamma^{\mu}_{\sigma\alpha}\Gamma^{\sigma}_{\beta\mu} g^{\alpha\beta} + \Gamma^{\sigma}_{\alpha\beta}\Gamma^{\mu}_{\sigma\mu} g^{\alpha\beta} \\ -\dfrac{1}{2F}\left(\begin{array}{c} 2F_{,ij}(n-1)g^{ij} + 2\Gamma^{a}_{ij}F_{,a}g^{ij} \\ -F_{,i}g^{ab}g_{jb,a}g^{ij} - F_{,j}g^{ab}g_{ib,a}g^{ij} \\ -n\Gamma^{d}_{ij}F_{,d}g^{ij} + \dfrac{n}{2}F_{,d}g^{cd}g_{ab,c}g^{ab} \end{array} \right) - \dfrac{F_{,i}\cdot F_{,j}}{4F^2}g^{ij}\left((n-6)(n-1)\right) \end{array} \right) \dfrac{1}{F}$$

$$= \left(R - \dfrac{1}{2F}\left(\begin{array}{c} 2F_{,ij}(n-1)g^{ij} + 2\Gamma^{a}_{ij}F_{,a}g^{ij} \\ -F_{,i}g^{ab}g_{jb,a}g^{ij} - F_{,j}g^{ab}g_{ib,a}g^{ij} \\ -n\Gamma^{d}_{ij}F_{,d}g^{ij} + \dfrac{n}{2}F_{,d}g^{cd}g_{ab,c}g^{ab} \end{array} \right) - \dfrac{F_{,i}\cdot F_{,j}}{4F^2}g^{ij}\left((n-6)(n-1)\right) \right) \dfrac{1}{F}$$

---

[2]Please note that in most papers we thereby considered metrics g_{ij} with $R = 0$. Here now we need to introduce the general form.

[3]For convenience we here repeat and partially extend these derivations in the appendices of this paper.

[4]The complete evaluation was presented in the appendix A of reference [8]. An extension to cases of f being a vector is been given in [13] and the two previous chapters.

$$
\begin{aligned}
&= \left(R - \frac{F'}{2F} \left(\begin{array}{l} 2f_{,ij}(n-1)g^{ij} + 2\Gamma_{ij}^{a}f_{,a}g^{ij} \\ -f_{,i}g^{ab}g_{jb,a}g^{ij} - f_{,j}g^{ab}g_{ib,a}g^{ij} \\ -n\Gamma_{ij}^{d}f_{,d}g^{ij} + \frac{n}{2}f_{,d}g^{cd}g_{ab,c}g^{ab} \end{array} \right) \right. \\
&\qquad \left. -(n-1)\frac{f_{,i}\cdot f_{,j}}{4F^2}g^{ij}\left(4F''F + F'F'(n-6) \right) \right) \frac{1}{F}
\end{aligned}
$$

with: $\quad F = F[f]; \quad F' = \dfrac{\partial F[f]}{\partial f}; \quad F'' = \dfrac{\partial^2 F[f]}{\partial f^2}.$ \hfill (1642)

This simplifies dramatically as follows:

$$
\begin{aligned}
R^* = R^*{}_{\alpha\beta}G^{\alpha\beta} &= \left(\overbrace{\Gamma^{\sigma}_{\alpha\beta,\sigma}g^{\alpha\beta} - \Gamma^{\sigma}_{\beta\sigma,\alpha}g^{\alpha\beta} - \Gamma^{\mu}_{\sigma\alpha}\Gamma^{\sigma}_{\beta\mu}g^{\alpha\beta} + \Gamma^{\sigma}_{\alpha\beta}\Gamma^{\mu}_{\sigma\mu}g^{\alpha\beta}}^{R} \right. \\
&\qquad \left. + \frac{F'}{F}(1-n)\Delta f + \frac{f_{,\alpha}f_{,\beta}g^{\alpha\beta}(1-n)}{4F^2}\left(4FF'' + (F')^2(n-6) \right) \right) \frac{1}{F} \\
&= \left(R + \frac{F'}{F}(1-n)\Delta f + \frac{f_{,\alpha}f_{,\beta}g^{\alpha\beta}(1-n)}{4F^2}\left(4FF' + (F')^2(n-6) \right) \right) \frac{1}{F}
\end{aligned}
$$

with: $\quad F = F[f]; \quad F' = \dfrac{\partial F[f]}{\partial f}; \quad F'' = \dfrac{\partial^2 F[f]}{\partial f^2},$ \hfill (1643)

when restricting to typical standard metrics with the following properties:

$$
g_{ij} = \begin{pmatrix} g_{00} & \cdots & 0 \\ \vdots & \ddots & \vdots \\ 0 & \cdots & g_{n-1n-1} \end{pmatrix}; \quad g_{jj,j} = 0. \tag{1644}
$$

Demanding certain conditions for the function $F[f]$ and/or f then gives us Dirac or Klein–Gordon equations (e.g., [2–4] and appendix A). From there the Schrödinger equation automatically follows a simple transformation (see appendix B). Only the Dirac equation seems to provide some "difficulties," which are lesser problematic issues rather than interesting options to derive this equation (chapters 9 and 10 and [12, 13]). So, while classically the fact that we already have obtained the Klein–Gordon equation should suffice, because from there Dirac had directly extracted his famous equation, we find that Eq. (1643) offers other paths to end up with the Dirac or Dirac-like equations. For instance, when demanding f to be a Laplace function, we obtain from Eq. (1643):

$$R^* = \frac{R}{F[f]} + \frac{(1-n)}{F[f]^3} \cdot \left(\left(\frac{(n-6)}{4} \cdot \left(\frac{\partial F[f]}{\partial f} \right)^2 + F[f] \cdot \frac{\partial^2 F[f]}{\partial f^2} \right) \cdot \overbrace{\left(\tilde{\nabla}_g f \right)^2}^{= f_{,\alpha} g^{\alpha\beta} f_{,\beta}} \right)$$

$$\xrightarrow{n=4;R=0} = \frac{1}{F[f]^3} \cdot \left(\left(3 \cdot F[f] \cdot \frac{\partial^2 F[f]}{\partial f^2} - \frac{3}{2} \cdot \left(\frac{\partial F[f]}{\partial f} \right)^2 \right) \cdot \left(\tilde{\nabla}_g f \right)^2 \right), \quad (1645)$$

which—so it was shown in [2–4]—gives the metric equivalent to the Dirac equation. Thereby, a very simple way to directly end up with the Dirac equation in the classical form was presented in [32] and chapter 10 (see section "10.7 Brief Sum-Up: 'The Dirac Miracle'"). For convenience we are going to repeat this derivation from an equation of the type (1645) in the section "11.4.2 The Simplest Way to Derive a Metric Dirac Equation from the Scaled Ricci Scalar R^*." It was also shown in [2, 3] how this gives the classical Dirac equation in flat space Minkowski metrics. Clearly, we did not extract the Dirac equation from the Klein–Gordon equation as Dirac himself had done, but in a slightly different way. The process and its differences from the classical path were presented in a comprehensive manner in [13] in the section "Dirac in the Metric Picture and Its Connection to the Classical Quaternion Form" and the two previous chapters. This path is different from the original Dirac approach where the Klein–Gordon equation is factorized (operationally factorized). When trying to find out more about the reason of these peculiar redundancies in obtaining the Dirac equation, we had to conclude that—obviously—there is more behind the whole Dirac apparatus than classically assumed [13, 32]. In [32] (here repeated in chapter 10), we therefore had started to work on a tensorial, respectively gravity Dirac equation, which automatically forced us to bring in matrix objects like Dirac did with his quaternions (see [32] and section "10.10 The Gravity Dirac Equation"). We are going to repeat the essentials of this work here (see section "11.11 The Gravity Dirac Equation Revisited").

Never the less, we should state here that Eq. (1643) in connection with certain conditions ($R^* = 0$, mainly) provides us with the most fundamental quantum equations.

But does this scaled metric approach also give us quantum Einstein field equations?

Yes, it does and this was first shown in [8, 9]!

As demonstrated in [8], when starting with the classical Einstein–Hilbert action [10] with the Ricci scalar R^* as kernel or Lagrange density:

$$\delta W = 0 = \delta \int_V d^n x \left(\sqrt{-G} \cdot \left(R^* - 2\Lambda + L_M \right) \right), \tag{1646}$$

we were able to derive new field equations, containing not only the classical part from Einstein's General Theory of Relativity [11] but also a quantum part [8, 9], which is governed by the wave function (and metric scale) f.

The attentive reader will have realized that in (1646), in contrast to the classical form, we used our somewhat adapted, if not to say scaled metric tensor $G_{\delta\gamma} = F[f] \cdot g_{\delta\gamma}$. Thereby G shall denote the determinant of the metric tensor $G_{\delta\gamma}$, while g will later stand for the corresponding determinant of the metric tensor $g_{\alpha\beta}$. In order to distinguish our new Ricci scalar R^*, being based on $G_{\delta\gamma} = F[f] \cdot g_{\delta\gamma}$ from the usual one R, being based on the metric tensor $g_{\alpha\beta}$, we marked it with the *-superscript. We also have the matter density L_M and the cosmological constant Λ.

Please note that with $G_{\delta\gamma} = F[f] \cdot g_{\delta\gamma}$ we have used the simplest form of a scale metric adaptation, which we could construct as a simplification from a generalization of the typical form of tensor transformations, namely:

$$G_{\alpha\beta} = F\left[f\left[t,x,y,z,\ldots,\xi_k,\ldots,\xi_n \right] \right]_{\alpha\beta}^{ij} g_{ij} \tag{1647}$$

$$\rightarrow G_{\alpha\beta} = F\left[f\left[t,x,y,z,\ldots,\xi_k,\ldots,\xi_n \right] \right] \cdot \delta_\alpha^i \delta_\beta^j g_{ij}$$

$$\rightarrow G_{\alpha\beta} = F\left[f\left[t,x,y,z,\ldots,\xi_k,\ldots,\xi_n \right] \right] \cdot g_{\alpha\beta} = F[f] \cdot g_{\alpha\beta}. \tag{1648}$$

The generalization has been elaborated in [4] and its need (or redundancy) has been discussed in [8]. In this chapter now we will work out that the general form (1647) can be seen as a mathematical recipe for a complete linearization of the Ricci scalar in n-dimensions (see section "11.12 The Math for Body, Soul and Universe?")

Setting (1643) into (1646) results in the variation:

$$\delta W = 0 = \delta \int_V d^n x \left(\sqrt{-g \cdot F^n} \times \left(\left(R - \frac{1}{2F} \left(\begin{array}{c} 2F_{,\alpha\beta}(n-1)g^{\alpha\beta} + 2\Gamma^a_{\alpha\beta}F_{,a}g^{\alpha\beta} \\ -F_{,\alpha}g^{ab}g_{\beta b,a}g^{\alpha\beta} - F_{,\beta}g^{ab}g_{\alpha b,a}g^{\alpha\beta} \\ -n\Gamma^d_{\alpha\beta}F_{,d}g^{\alpha\beta} + \frac{n}{2}F_{,d}g^{cd}g_{ab,c}g^{ab} \end{array} \right) \right. \right. \right.$$
$$\left. \left. \left. - \frac{F_{,\alpha} \cdot F_{,\beta}}{4F^2} g^{\alpha\beta} \left((n-6)(n-1)\right) \right) \frac{1}{F} \right) \right) . \\ -2\Lambda + L_M$$

$$\tag{1649}$$

Thereby, in order to keep things simple and still general enough, we here often restrict ourselves to diagonal metrics with the following properties:

$$g_{ij} = \begin{pmatrix} g_{00} & \cdots & 0 \\ \vdots & \ddots & \vdots \\ 0 & \cdots & g_{n-1\,n-1} \end{pmatrix}; \quad g_{jj,j} = 0, \tag{1650}$$

which gives the following Hilbert integral:

$$\delta W = 0 = \delta \int_V d^n x \left(\sqrt{-g \cdot F^n} \times \left(\left(\begin{array}{c} R + \dfrac{F'}{F}(1-n)\Delta f \\[2mm] + \dfrac{f_{,\alpha} f_{,\beta} g^{\alpha\beta}(1-n)}{4F^2}\left(4FF'' + (F')^2(n-6)\right) \\[2mm] -2\Lambda + L_M \end{array} \right)^{\frac{1}{F}} \right) \right). \tag{1651}$$

Please note that the cosmological constant term as given in (1651) in the current form, requires the variation with respect to the metric $G_{\alpha\beta}$. When insisting on the variation with respect to the metric $g_{\alpha\beta}$ instead, we better rewrite (1651) as follows:

$$\delta_{g_{\alpha\beta}} W = 0 = \left\{ \begin{array}{c} \delta_{g_{\alpha\beta}} \int_V d^n x \dfrac{\sqrt{-g \cdot F^n}}{F} \times \left(\begin{array}{c} R + \dfrac{F'}{F}(1-n)\Delta f \\[2mm] + \dfrac{f_{,\alpha} f_{,\beta} g^{\alpha\beta}(1-n)}{4F^2}\left(4FF'' + (F')^2(n-6)\right) \end{array} \right) \\[6mm] + \delta_{g_{\alpha\beta}} \int_V d^n x \sqrt{-g}\left(L_M - 2\Lambda\right) \end{array} \right. \tag{1652}$$

in order to make it clear that we do not intend to also scale-adapt the cosmological constant or the Hilbert matter term.

However, when ignoring the cosmological constant and—**PERHAPS—assuming that we would not need any postulated matter terms L_M, simply because our scale adaptation $F[f]$, dimensional entanglement, and similar "tricks" or "add-ons" automatically provide matter**, we just obtain:

$$\delta W = 0 = \delta \int_V d^n x \left(\sqrt{-g \cdot F^n} \times \left(\left(\begin{array}{c} R + \dfrac{F'}{F}(1-n)\Delta f \\[2mm] + \dfrac{f_{,\alpha} f_{,\beta} g^{\alpha\beta}(1-n)}{4F^2}\left(4FF'' + (F')^2(n-6)\right) \end{array} \right)^{\frac{1}{F}} \right) \right)$$

$$= \delta \int_V d^n x \left(\sqrt{-g \cdot F^n} \times \left(\left(\begin{array}{c} R + \dfrac{F'}{\sqrt{-g} \cdot F}(1-n)\partial_\beta \sqrt{-g} \cdot g^{\alpha\beta} f_{,\alpha} \\[3mm] + \dfrac{f_{,\alpha} f_{,\beta} g^{\alpha\beta}(1-n)}{4F^2}\left(4FF'' + \left(F'\right)^2(n-6)\right) \end{array} \right) \dfrac{1}{F} \right) \right). \quad (1653)$$

Performing the usual Hilbert variation with respect to the metric tensor $g_{\alpha\beta}$ now leads us to (regarding the full evaluation see [8]):

$$\Rightarrow \quad 0 = \left\{ \begin{array}{l} R^{\delta\gamma} - Rg^{\delta\gamma} + \left(\left(\begin{array}{c} \underbrace{\dfrac{8\pi G}{\kappa}T^{\delta\gamma}}_{} = \text{matter} \\[3mm] 0 = \text{vacuum} \end{array} \right) + \Lambda \cdot g^{\delta\gamma} \right) \\[8mm] -\dfrac{1}{\sqrt{F^n}} \cdot \left(\partial_{;\mu}\partial_{;\nu} - g_{\mu\nu}\Delta_g\right)F^{\frac{n}{2}-1} \\[5mm] + \left(g^{\kappa\lambda}g^{\alpha\beta} - g^{\kappa\alpha}g^{\lambda\beta} - g^{\lambda\alpha}g^{\kappa\beta}\right)(1-n) \\[3mm] \cdot \left[\dfrac{f_{,\alpha} f_{,\beta}}{F^2}\left(FF'' + \left(F'\right)^2 \cdot \dfrac{3 \cdot n - 14}{8}\right) + f_{,\alpha\beta}\dfrac{F'}{F} \right] \\[5mm] - \dfrac{F' \cdot (1-n)}{2 \cdot F}\left(g^{\kappa\lambda}g^{\alpha\beta} - g^{\kappa\alpha}g^{\lambda\beta} - g^{\lambda\alpha}g^{\kappa\beta}\right)\Gamma^\gamma_{\alpha\beta}f_{,\gamma} \\[5mm] + \dfrac{F'}{F} \cdot (1-n)g^{\alpha\beta}g^{\gamma\lambda}\Gamma^\kappa_{\alpha\beta}f_{,\gamma} \\[5mm] + \dfrac{F'}{F} \cdot \dfrac{(1-n)}{2}\left(g^{\kappa\lambda}g^{\alpha\beta} - g^{\kappa\alpha}g^{\lambda\beta} - g^{\lambda\alpha}g^{\kappa\beta}\right)_{,\beta}f_{,\alpha} \\[5mm] + \dfrac{g^{\xi\zeta}g_{\xi\zeta,\beta}}{2} \cdot \left(\dfrac{F'}{F} \cdot \dfrac{(1-n)}{2}\left(g^{\kappa\lambda}g^{\alpha\beta} - g^{\kappa\alpha}g^{\lambda\beta} - g^{\lambda\alpha}g^{\kappa\beta}\right)f_{,\alpha}\right). \end{array} \right. \quad (1654)$$

Quite some discussion to this new quantum Einstein field equation is presented in [8] and chapter 8 of this book.

It has to be pointed out that Hilbert's choice of the Ricci scalar R, acting as the Lagrange density in (1646), was motivated by his attempt (with success) to reproduce the Einstein field equations in a completely mathematical manner from a minimum principle. As his scale factor F was set to one, which is to say as Hilbert did not consider any $F[f]$-scaled metric tensor, we now, when aiming for the Hilbert approach in the $F \rightarrow 1$ limit, have the principal degree of freedom of adding a factor F^q to the integrand. This action, reading (when setting L_M and Λ to zero and choosing an additional scale factor F^q):

$$\delta W = 0 = \delta \int_V d^n x \left(\sqrt{-g \cdot F^n} \times F^q \cdot R^* \right)$$

$$= \delta \int_V d^n x \left(\sqrt{-g \cdot F^n} \times \left(F^q \cdot \left(\left(R + \frac{F'}{F}(1-n)\Delta f + \frac{f_{,\alpha} f_{,\beta} g^{\alpha\beta}(1-n)}{4F^2} \left(4FF'' + (F')^2(n-6) \right) \right) \frac{1}{F} \right) \right) \right)$$

$$= \delta \int_V d^n x \left(\sqrt{-g \cdot F^{n+2q}} \times \left(\left(R + \frac{F'}{F}(1-n)\Delta f + \frac{f_{,\alpha} f_{,\beta} g^{\alpha\beta}(1-n)}{4F^2} \left(4FF'' + (F')^2(n-6) \right) \right) \frac{1}{F} \right) \right)$$

$$(1655)$$

would then lead (for the full derivation see [8]) to:

$$\Rightarrow \quad 0 = \begin{cases} R^{\kappa\lambda} - \frac{1}{2} R g^{\kappa\lambda} + \left(\left(\underbrace{8\pi G}_{\kappa} T^{\kappa\lambda} = \text{matter} \atop 0 = \text{vacuum} \right) + \Lambda \cdot g^{\kappa\lambda} \right) \\[2em] - \frac{1}{\sqrt{F^p}} \cdot \left(\partial^{:\kappa} \partial^{:\lambda} - g^{\kappa\lambda} \Delta_g \right) F^{\frac{p}{2}-1} \\[1em] + \left(g^{\kappa\lambda} g^{\alpha\beta} - g^{\kappa\alpha} g^{\lambda\beta} - g^{\lambda\alpha} g^{\kappa\beta} \right)(1-n) \\[0.5em] \cdot \left[\frac{f_{,\alpha} f_{,\beta}}{F^2} \left(FF'' + (F')^2 \cdot \frac{2 \cdot p + n - 14}{8} \right) + f_{,\alpha\beta} \frac{F'}{F} \right] \\[1em] - \frac{F' \cdot (1-n)}{2 \cdot F} \left(g^{\kappa\lambda} g^{\alpha\beta} - g^{\kappa\alpha} g^{\lambda\beta} - g^{\lambda\alpha} g^{\kappa\beta} \right) \Gamma^{\gamma}_{\alpha\beta} f_{,\gamma} \\[1em] + \frac{F'}{F} \cdot (1-n) g^{\alpha\beta} g^{\gamma\lambda} \Gamma^{\kappa}_{\alpha\beta} f_{,\gamma} \\[1em] + \frac{F'}{F} \cdot \frac{(1-n)}{2} \left(g^{\kappa\lambda} g^{\alpha\beta} - g^{\kappa\alpha} g^{\lambda\beta} - g^{\lambda\alpha} g^{\kappa\beta} \right)_{,\beta} f_{,\alpha} \\[1em] + \frac{g^{\xi\zeta} g_{\xi\zeta,\beta}}{2} \cdot \left(\frac{F'}{F} \cdot \frac{(1-n)}{2} \left(g^{\kappa\lambda} g^{\alpha\beta} - g^{\kappa\alpha} g^{\lambda\beta} - g^{\lambda\alpha} g^{\kappa\beta} \right) f_{,\alpha} \right). \end{cases} \quad (1656)$$

Note that we defined $p = n + 2q$.

11.3.2 Repetition: Can We Now Understand Quantum Theory in a Truly Illustrative Manner? Part I

Yes, but as with many other things it holds that we are not able to truly grasp Quantum Theory without seeing the full picture. Interestingly ... and some may say "unfortunately," only the combination of Quantum Theory with Einstein's General Theory of Relativity will give us this full picture. To some this means that the task is still unsolved and probably will be so for a long time to come, but taking what we have just presented in this section above, we are of the opinion that the full picture already is at hand.

Here we explain why:

Space and time in Einstein's General Theory of Relativity are explained via objects called metrics.

By allowing these metrics to have a scale factor $F[f]$ (or factors ... c.f. section "11.12 The Math for Body, Soul and Universe?") as shown in (1647) and (1648) and keeping these factors general enough so that they could be functions, we suddenly obtain all those important fundamental quantum equations, such as the Dirac, the Klein–Gordon and the Schrödinger equation (see [3–9], appendices A and B, and the following main section pages of this paper). The scaled metrics on the other hand, do not compromise the General Theory of Relativity in any way. On the contrary, now we can derive quantum gravity field equations (1654) to (1656) (see [8], chapter 8 for full derivation ... or further below in this chapter).

But there is more and this has to do with the way we can now illustrate quantum effects.

Those scale factors that we added to the metrics contain what is called wave functions in Quantum Theory and now we can easily (more or less) understand them. A scale to a metric namely changes the volume of the space-time it describes. Thus, our scale functions F, which do contain the wave functions f (being themselves functions of all dimensions of the space-time in question), are just changes in volume of space and time, different at every world point and thereby leading to the permanent jitter of everything that there was, is, and will be.

The distribution of a spinor as shown in Figs. 11.3 and 11.4 in sub-section "11.5.3 A Possible Origin of Spin $l = 1/2$" gives a descriptive example.

Thereby we explicitly choose this example, because it comes with a "complex" problem, namely, the occurrence of imaginary portions of results. Usually in Quantum Theory one simply accepts the presence of such portions, forms the absolute value of the result, calls this a probability density, and everything seems to be in order. We, on the other hand, think that it cannot be so simple. Yes, in chapter 7, section "7.8 Appendix: Taking Care about Complex Curvatures R or R^*" we took care about complex results by forcing the Ricci scalar to give absolute values in the usual quantum theoretical (probability) manner, but can we be sure that this way we do not erase certain information?

After all, we found that fundamental quantum equations can be extracted from the concept of the scaled metric. Whether in the complicated (tensor transformation-like) form (1647) or a simple factor (1648) does not matter here. Thus, the occurrence of complex numbers in connection with the scale factors or scale functions must have a deeper meaning than just the one of an intermediate by-product. After all, negative numbers can be seen as borrowed length elements, complex numbers as borrowed (or missing) surfaces, and so on. We are not going to completely solve this important riddle of interpretation of the results we show here in this book, but we want to point out that—obviously—the classical probability function thesis or approach does not seem to be the be-all and end-all of conclusions.

For example, we will show in section "11.10 Metric Observables" that the assumption of purely positive real "observable results" automatically leads to the funny possibility of fermions changing into bosons ... well, or rather Dirac equations changing into Klein–Gordon ones.

One point, however, *seems* to have become clear. We found that the demand for a Ricci-flat space-time $R = 0$ and $R^* = 0$ already gives us the necessary equations for all interactions except gravity. The latter comes in after variation with respect to the metric.

But can this be?

After all, the extremal principle is a crucial one to all physics and suddenly there would be just this $R^* = 0$ condition without any need for a variation to obtain fundamental laws. Something seems to be wrong here and that is why we underlined the word "seems" above. We will demonstrate in section "11.6 The Simpler Route" further below how all clears up and the minimum principle can be easily saved.

11.4 About the Derivation of the Metric Dirac Equation

In the previous chapter of this book, we have already shown how to directly extract the Dirac equation from the Einstein–Hilbert action. However, the derivation alone does not truly explain why such a peculiar equation has to exist. In this chapter now we want to discuss this point and we will try to—also illustratively—elaborate the deeper structural or geometrical reasoning for something so horribly theoretical like quaternions to have some bearing on the real world. In order to do so, we need to repeat a few earlier evaluations.

In the appendix A of this chapter, we show that the Klein–Gordon equation can directly be extracted from a scaled metric via the assumption of an effectively Ricci-flat space-time in n dimensions. Thereby we have assumed the scaling, which is to say the functional multiplicator $F[f]$, to be of scalar character.

With the Klein–Gordon equation already falling out of our new approach, we may just lean comfortably back and conclude that this way we also have obtained the Dirac equation [12]. This namely was classically derived by an operator factorization from the Klein–Gordon equation. Thus, having obtained the latter automatically gives us the Dirac equation, too. However, when digging deeper (metrically) and also trying to find a completely metric origin and illustrative understanding of the Dirac approach, we have to come to the conclusion that the simple scalar metric scaling with a function $F[f]$ multiplied with the metric does not suffice [13]. We have found that only a vector scaling will probably give us the complete picture [13]. This was especially worked out in chapters 9 and 10. Here in this chapter, with a little bit of repetition, we want to briefly investigate this option.

Before we come to that, however, we want to point out that Dirac's original approach [12] started with an extension of the wave function f within the Klein–Gordon equation as follows:

$$0 = \Delta f - B^2 f \xrightarrow{\text{Dirac}} 0 = \Delta f_\lambda - B^2 f_\lambda. \tag{1657}$$

We see that Dirac had made the function f a vector and then "proceeded from there"[5] with his famous evaluation, resulting in the classical Dirac equation (see [13] for more details). In insisting on a scalar f for Eq. (1643), we have also investigated the possibilities of intrinsically structured f (see chapter 9 and [13]). This is not strictly the same way as Dirac went, but nevertheless, we named it after the great physicist and refer to it as Dirac or Dirac-like approaches. It appears as a peculiar accident that even though Dirac did never intend to make his approach metrically originated, his derivation

---

[5]As admitted before, we have cheated here a bit. Historians will insist that Dirac in fact had first extracted the quaternion root of the Klein-Gordon equation and then introduced the **f** vector, but this is not of importance here. For the purists, however, and in order to point this fact out, we had set the formulation "proceeded from there" in "..."-signs.

and its success in describing quantum reality prompted us to investigate also non-scalar scaling factors of the metric in order to have a consistent transition to his classical theory. In [13], where we have tried to work out the similarities (and differences) of the metric and the classical Dirac approach, we had therefore started with Dirac's original work, which is to say the direct introduction of the **f** vector, and then let slip in the intrinsic *f* structure and its implications on the Dirac path. For convenience we repeated this approach in chapter 9 maybe write the full title of the chapter?. On this way, we had to realize that—apparently—only non-scalar metric scales provide us with the right structures to find the true metric origin of the Dirac theory. Later then, however, we also saw that certain degrees of freedom regarding the scale split-up of the integration of the Einstein–Hilbert action (1646) also can give us what we need in order to obtain the Dirac equation without fuzz ([32] and chapter 10). Because of its importance for the discussion in this chapter, we will repeat the necessary derivations further below in this section.

11.4.1 Toward Metric Dirac Equation: About the Starting Point

We assume that the application of vectors $\mathbf{f}_a$ instead of scalar functions f does not change the principal structure of (1643) with respect to f. We adapt the metric scale variation as follows from (1647):

$$G_{\alpha\beta} = \mathbf{q}_\alpha \mathbf{f}^a \cdot g_{a\beta} \tag{1658}$$

and

$$G^{\alpha\beta} = \mathbf{q}^\alpha \mathbf{f}_a g^{a\beta}. \tag{1659}$$

Thereby we used bold letters for $\mathbf{q}$ and $\mathbf{f}$ in order to point out that we don't know yet what objects these are, respectively what rank is needed for them. With the condition:

$$G^{\chi\gamma} G_{\delta\gamma} = \delta^\chi_\delta \quad \Rightarrow \quad \mathbf{f}_a \mathbf{f}^a = 1; \quad \mathbf{q}^\chi \mathbf{q}_\delta = \delta^\chi_\delta \tag{1660}$$

we see that the function **f** vectors and the **q** vectors have to behave like tetrads. A more thorough investigation will follow elsewhere, respectively, should—for the moment—be left to the interested and mathematically skilled reader. Our conclusion here would rather go into the following direction:

With the number of dimensions being open anyway, the introduction of a vector function $\mathbf{f}_a$ instead of a scalar function just increases the number of degrees of freedom and thus, the number of dimensions. So, we ask, why can't we just take a space of higher dimensions instead of offering the space at hand a higher number of degrees of freedom?

The two options should be equivalent and when seeing the need for the extension of the scalar f to a vector f_a or any other object of higher rank, we may just check whether we could also apply an increase in the number of dimensions of our system.

At any rate, our goal with respect to (1658) and (1659) shall be to find $q\,f$ combinations where the $f_{a,\alpha}f_{a,\beta}g^{\alpha\beta}$-operator directly vanishes due to a suitable summation of derivatives to the **f** vector components (potentially similar to the Galerkin ansatz in linear elasticity). Thereby demanding all f components to satisfy Laplace equations would immediately solve Eq. (1643) in the case $R^* = R = 0$. Otherwise, we are back at the factorization problem as discussed with respect to the "Decomposition of the Ricci Scalar and Tensor" in [13] and chapter 9.

One may see the scale adaptation of the metric simply as some other kind of variation. That this new type of variation now led us to fundamental quantum equations via a simple effective Ricci-flat space-time condition $R^* = 0$ prompts the conclusion that the latter condition itself is a rather fundamental one and that the universe works always very hard—which is to say via a multitude of variational forms—to reach the $R^* = 0$ goal. Further below in this chapter, however, we will work out that this assumption is a bit flawed, because—as we will see then—the $R^* = 0$ condition also is only a direct result of the Hamilton minimum principle realized via the Einstein–Hilbert action.

We immediately see that there are many options for our vector-scaling constructions. For instance, one may also set:

$$G_{\alpha\beta} = \mathbf{f}_\alpha^a \cdot g_{\alpha\beta} \tag{1661}$$

and

$$G^{\alpha\beta} = \overline{\mathbf{f}}_a^\alpha\, g^{\alpha\beta}, \tag{1662}$$

with the condition:

$$G^{\chi\gamma}G_{\delta\gamma} = \delta_\delta^\chi \quad \Rightarrow \quad \mathbf{f}_\delta^a\overline{\mathbf{f}}_a^\chi = \delta_\delta^\chi. \tag{1663}$$

One obvious, because rather Dirac-matrix-like, choice in connection with metrically producing Dirac equations would probably consist of a function vector and an object of rank three as follows:

$$G_{\alpha\beta} = \zeta_{\alpha\beta}^a f^b \cdot g_{ab} \tag{1664}$$

and

$$G^{\alpha\beta} = \zeta_a^{\alpha\beta} f_b g^{ab}. \tag{1665}$$

The subsequent condition would then be:

$$G^{\alpha\gamma} G_{\gamma\beta} = \zeta_a^{\alpha\gamma} f_b g^{ab} \zeta_{\gamma\beta}^c f^d \cdot g_{cd} = \zeta_a^{\alpha\gamma} f^a \zeta_{\gamma\beta}^c f_c = \delta_\beta^\alpha. \tag{1666}$$

Similarly, one may also take:

$$G_{\alpha\beta} = \zeta_{\alpha b}^a f^b \cdot g_{\alpha\beta} \tag{1667}$$

and

$$G^{\alpha\beta} = \zeta_a^{\alpha b} f_b g^{\alpha\beta}, \tag{1668}$$

and the condition would then be:

$$G^{\alpha\gamma} G_{\gamma\beta} = \zeta_a^{\alpha b} f_b g^{\alpha\gamma} \zeta_{\gamma d}^c f^d \cdot g_{c\beta} = \zeta^{\alpha b\gamma} f_b \zeta_{\gamma d\beta} f^d = \delta_\beta^\alpha. \tag{1669}$$

Only giving the covariant form, one may also insist on completely symmetric scaled (adapted) metrics and sum up (1664), (1665), (1667), (1668) and all other options, which are reading:

$$G_{\alpha\beta} = \begin{Bmatrix} \zeta_{\alpha a}^a f^b g_{b\beta}, \zeta_{\beta a}^a f^b g_{\alpha b}, \zeta_{\alpha b}^a f^b g_{\alpha\beta}, \zeta_{\beta b}^a f^b g_{\alpha\alpha}, \\ \zeta_{\alpha\beta}^a f^b g_{ab}, \zeta_{\beta\alpha}^a f^b g_{ab}, \zeta_{\alpha b}^a f^b g_{\alpha\beta}, \cdots \end{Bmatrix}, \tag{1670}$$

and so on, as follows:

$$G_{\alpha\beta} = \frac{1}{24} \begin{pmatrix} \zeta_{\alpha a}^a f^b g_{b\beta} + \zeta_{\beta a}^a f^b g_{\alpha b} + \zeta_{\alpha b}^a f^b g_{\alpha\beta} + \zeta_{\beta b}^a f^b g_{\alpha\alpha} \\ + \zeta_{\alpha\beta}^a f^b g_{ab} + \zeta_{\beta\alpha}^a f^b g_{ab} + \zeta_{\alpha b}^a f^b g_{\alpha\beta}, \cdots \end{pmatrix}. \tag{1671}$$

Closer investigation of these options shows us that an approach of the form:

$$G_{\alpha\beta} = \zeta_{\alpha b}^c f^a \cdot g_{\alpha\beta} \tag{1672}$$

and

$$G^{\alpha\beta} = \zeta_c^{\alpha b} f_a g^{\alpha\beta} \tag{1673}$$

evolves as the most promising and the condition would then be:

$$G^{\alpha\gamma} G_{\gamma\beta} = \zeta_c^{\alpha b} f_a g^{\alpha\gamma} \zeta_{\beta f}^e f^d \cdot g_{\gamma d} = \zeta_c^{\alpha b} f_\gamma \zeta_{\beta f}^e f^\gamma = \delta_\beta^\alpha. \tag{1674}$$

Finding solutions to this condition seems to be impossible, but when changing (1674) to:

$$G^{\alpha\gamma} G_{\gamma\beta} + G^{\beta\gamma} G_{\gamma\alpha} = \delta_\beta^\alpha + \delta_\alpha^\beta, \tag{1675}$$

which does lead us to:

$$\zeta_c^{\alpha b} \zeta_{\beta f}^e + \zeta_c^{\beta b} \zeta_{\alpha f}^e = \delta_\beta^\alpha + \delta_\alpha^\beta; \quad f_\gamma f^\gamma = 1, \tag{1676}$$

we can directly apply the Dirac gamma matrices γ^α. Please note, however, that—as we are considering n-dimensional space-times—these are Dirac matrices in n dimensions.

In 4 dimensions, these and the unit matrix read as follows:

$$\gamma^0 = \begin{pmatrix} 1 & & & \\ & 1 & & \\ & & -1 & \\ & & & -1 \end{pmatrix}; \quad \gamma^1 = \begin{pmatrix} & & & 1 \\ & & 1 & \\ & -1 & & \\ -1 & & & \end{pmatrix}$$

$$\gamma^2 = \begin{pmatrix} & & & -i \\ & & i & \\ & i & & \\ -i & & & \end{pmatrix}; \quad \gamma^3 = \begin{pmatrix} & & 1 & \\ & & & -1 \\ -1 & & & \\ & 1 & & \end{pmatrix}; \quad I = \begin{pmatrix} 1 & & & \\ & 1 & & \\ & & 1 & \\ & & & 1 \end{pmatrix}. \quad (1677)$$

Knowing that the Dirac matrices follow the rule:

$$\left(\gamma^\mu \gamma^\nu + \gamma^\nu \gamma^\mu \right) = 2 \cdot g_{\text{Cartesian}}^{\ \mu\nu} \cdot I = 2 \cdot I, \quad (1678)$$

we suddenly see that we can just use the Dirac matrices with the setting:

$$\zeta_c^{ab} \zeta_{\beta f}^e + \zeta_c^{\beta b} \zeta_{\alpha f}^e = \gamma^\alpha \gamma^\beta + \gamma^\beta \gamma^\alpha = 2 \cdot I = \delta_\beta^\alpha + \delta_\alpha^\beta$$

$$\zeta_c^{ab} = \gamma^\alpha; \quad \zeta_{\beta f}^e = \gamma^\beta. \quad (1679)$$

11.4.2 The Simplest Way to Derive a Metric Dirac Equation from the Scaled Ricci Scalar R^*

A very simple way to directly end up with the Dirac equation in classical form was first presented in [32].

For convenience, here we repeat this derivation of the classical Dirac equation [12] directly from the Einstein–Hilbert action. The latter reads (1646):

$$\delta W = 0 = \delta \int_V d^n x \left(\sqrt{-G} \cdot \left(R^* - 2\Lambda + L_M \right) \right).$$

$$(1680)$$

Thereby we apply the scaled Ricci scalar R^*:

$$R^* = R^*_{\alpha\beta} G^{\alpha\beta} = \begin{pmatrix} \Gamma^\sigma_{\alpha\beta,\sigma} g^{\alpha\beta} - \Gamma^\sigma_{\beta\sigma,\alpha} g^{\alpha\beta} - \Gamma^\mu_{\sigma\alpha} \Gamma^\sigma_{\beta\mu} g^{\alpha\beta} + \Gamma^\sigma_{\alpha\beta} \Gamma^\mu_{\sigma\mu} g^{\alpha\beta} \\ + \dfrac{F'}{F}(1-n)\Delta f + \dfrac{f_{,\alpha} f_{,\beta} g^{\alpha\beta} (1-n)}{4F^2} \left(4FF'' + \left(F' \right)^2 (n-6) \right) \end{pmatrix} \dfrac{1}{F}$$

$$= \left(R + \dfrac{F'}{F}(1-n)\Delta f + \dfrac{f_{,\alpha} f_{,\beta} g^{\alpha\beta} (1-n)}{4F^2} \left(4FF'' + \left(F' \right)^2 (n-6) \right) \right) \dfrac{1}{F}$$

with:
$$F = F[f]; \quad F' = \frac{\partial F[f]}{\partial f}; \quad F'' = \frac{\partial^2 F[f]}{\partial f^2}, \tag{1681}$$

being based upon the scaled metric $G_{\delta\gamma} = F[f] \cdot g_{\delta\gamma}$. Assuming the Lagrange matter density L_M and the cosmological constant Λ to vanish, the variation results in:

$$0 = \int_{V_G} d^n x \sum_{i=1}^{N} \sqrt{-g} \cdot F_i^n \left(\left(\left(R_{\alpha\beta} + \overset{**}{R}_{\alpha\beta} \right) - \frac{g_{\alpha\beta}}{2} \cdot \left(R - (n-1) \left(\frac{F_i'}{F_i} \Delta f_i + \frac{1}{4F_i^2} \left(\frac{4F_i F_i''}{+F_i' F_i'(n-6)} \right) f_{i,\mu} f_{i,\nu} g^{\mu\nu} \right) \right) \right) \right) \delta G^{\alpha\beta}$$

$$\overset{**}{R}_{\alpha\beta} \equiv \frac{1}{2F_i} \left(\begin{matrix} (2-n)F_{i,\alpha\beta} + \frac{1}{2F}\left((3n-4)F_{i,\alpha}F_{i,\beta} + (2-n)F_{i,c}g_{\alpha\beta}F_{i,a}g^{ac} \right) \\ + \left(\begin{matrix} \frac{1}{2}\left(\begin{matrix} F_{i,\alpha}g_{bd,\beta}g^{bd} + g_{ac,\alpha}F_{i,\beta}g^{ac} - g_{ac,\beta}F_{i,a}g^{cd} \\ -\left(F_{i,a}g^{ac}(2-n)\left(g_{\alpha c,\beta} + g_{\beta c,\alpha} - g_{\alpha\beta,c} \right) \right) \end{matrix} \right) \\ -\left(\begin{matrix} 2F_{i,b}g_{\alpha\beta,a} + F_{i,a}g_{\alpha\beta,b} + F_{i,ab}g_{\alpha\beta} \\ +\frac{1}{2}\left(F_{i,\alpha}g_{ab,\beta} + F_{i,\beta}g_{ab,\alpha} \right) - F_{i,a}g_{\alpha b,\beta} - F_{i,a}g_{\beta b,\alpha} \\ +\frac{1}{2}F_{i,c}g_{\alpha\beta}\left(2g_{bd,a} - g_{ab,d} \right)g^{cd} \end{matrix} \right)g^{ab} \end{matrix} \right) \end{matrix} \right).$$

$$\tag{1682}$$

Thereby we have also assumed to have many scaling factors and volumes to perform the integration over (for more discussion about this see further below in this chapter). By splitting up the variation of the scaled metric:

$$0 = \int_{V_G} d^n x \sum_{i=1}^{N} \sqrt{-g} \cdot F_i^n \left(\left(\left(R_{\alpha\beta} + \overset{**}{R}_{\alpha\beta} \right) - \frac{g_{\alpha\beta}}{2} \cdot \left(R - (n-1) \left(\frac{F_i'}{F_i} + \frac{1}{4F_i^2} \left(\frac{4F_i F_i''}{+F_i' F_i'(n-6)} \right) f_{i,\mu} f_{i,\nu} g^{\mu\nu} \right) \right) \right) \left(g^{\alpha\beta} \delta \frac{1}{F_i} + \frac{\delta g^{\alpha\beta}}{F_i} \right) \right),$$

$$\tag{1683}$$

we obtain two equations:

$$0 = \int_{V_G} d^n x \sum_{i=1}^{N} \sqrt{-g} \cdot F_i^n \left(\left(-\frac{g_{\alpha\beta}}{2} \cdot \left(R - (n-1) \left(\frac{\frac{F_i'}{F_i}}{\overset{R_{\alpha\beta} + R_{\alpha\beta}^{**}}{}} + \frac{1}{4F_i^2} \left(\frac{4F_i F_i''}{+F_i' F_i'(n-6)} \right) f_{i,\mu} f_{i,\nu} g^{\mu\nu} \right) \right) g^{\alpha\beta} \delta \frac{1}{F_i} \right) \right)$$

$$= \int_{V_G} d^n x \sum_{i=1}^{N} \sqrt{-g} \cdot F_i^n \left(1 - \frac{n}{2} \right) \left(R - \frac{F_i'}{F_i}(n-1)\Delta f_i + \frac{f_{i,\alpha} f_{i,\beta}(1-n)}{4F_i^2} g^{\alpha\beta} \left(\frac{4F_i F_i''}{+F_i' F_i'(n-6)} \right) \right) \delta \frac{1}{F_i},$$

$$(1684)$$

$$0 = \int_{V_G} d^n x \sum_{i=1}^{N} \sqrt{-g} \cdot F_i^n \left(\left(-\frac{g_{\alpha\beta}}{2} \cdot \left(R - (n-1) \left(\frac{\frac{F_i'}{F_i}}{\overset{R_{\alpha\beta} + R_{\alpha\beta}^{**}}{}} + \frac{1}{4F_i^2} \left(\frac{4F_i F_i''}{+F_i' F_i'(n-6)} \right) f_{i,\mu} f_{i,\nu} g^{\mu\nu} \right) \right) \frac{\delta g^{\alpha\beta}}{F_i} \right) \right).$$

$$(1685)$$

While we consider the second equation a generalized or quantum Einstein field equation, we are going to use the first one to derive the classical Dirac equation [12]. Setting the integrand equal to zero, we obtain the following differential equation:

$$0 = R - \frac{F_i'}{F_i}(n-1)\Delta f_i + \frac{f_{i,\alpha} f_{i,\beta}(1-n)}{4F_i^2} g^{\alpha\beta} \left(4F_i F_i'' + F_i' F_i'(n-6) \right). \quad (1686)$$

Now we add in the Dirac starting point, who took for granted that:

(a) The space-time can be assumed Ricci-flat and thus, $R = 0$.

(b) The Laplace operator term in (1686) has an eigenvalue M^2.

Also setting $F[f] = f$ gives us:

$$0 = f_{i,\alpha} f_{i,\beta} g^{\alpha\beta} + M^2 \cdot \overset{\equiv \mu^2 = \mu \cdot \mu}{\frac{4}{(n-6)}} \cdot f_i^2. \quad (1687)$$

Now we apply the known relation between the Dirac matrices

$$\gamma^0 = \begin{pmatrix} 1 & & & \\ & 1 & & \\ & & -1 & \\ & & & -1 \end{pmatrix}; \quad \gamma^1 = \begin{pmatrix} & & & 1 \\ & & 1 & \\ & -1 & & \\ -1 & & & \end{pmatrix}$$

$$\gamma^2 = \begin{pmatrix} & & & -i \\ & & i & \\ & i & & \\ -i & & & \end{pmatrix}; \quad \gamma^3 = \begin{pmatrix} & & 1 & \\ & & & -1 \\ -1 & & & \\ & 1 & & \end{pmatrix}; \quad I = \begin{pmatrix} 1 & & & \\ & 1 & & \\ & & 1 & \\ & & & 1 \end{pmatrix}, \quad (1688)$$

and the metric tensor, reading:

$$g^{\alpha\beta} \cdot I = \frac{\gamma^\alpha \gamma^\beta + \gamma^\beta \gamma^\alpha}{2}, \quad (1689)$$

and put (1687) into the following form:

$$0 = \left(f_{i,\alpha} f_{i,\beta} g^{\alpha\beta} + \mu^2 \cdot f_i^2 \right) \cdot I$$

$$\Rightarrow = \Rightarrow \qquad f_{i,\alpha} f_{i,\beta} \left(\frac{\gamma^\alpha \gamma^\beta + \gamma^\beta \gamma^\alpha}{2} \right) + \frac{\beta^2 + \beta^2}{2} \cdot m^2 \cdot f_i^2 = 0$$

$$\Rightarrow \qquad f_{i,\alpha} f_{i,\beta} \gamma^\alpha \gamma^\beta + \beta^2 \cdot m^2 \cdot f_i^2 = 0$$

$$\Rightarrow \qquad \left(f_{i,\alpha} \gamma^\alpha + i \cdot \beta \cdot m \cdot f_i \right)\left(f_{i,\beta} \gamma^\beta - i \cdot \beta \cdot m \cdot f_i \right) = 0$$

$$\Rightarrow \qquad f_{i,\alpha} f_{i,\beta} \gamma^\beta \gamma^\alpha + \beta^2 \cdot m^2 \cdot f_i^2 = 0$$

$$\Rightarrow \qquad \left(f_{i,\alpha} \gamma^\alpha + i \cdot \beta \cdot m \cdot f_i \right)\left(f_{i,\beta} \gamma^\beta - i \cdot \beta \cdot m \cdot f_i \right) = 0. \quad (1690)$$

Thereby we have also used the β matrix with $\beta = \begin{pmatrix} 1 & & & \\ & 1 & & \\ & & -1 & \\ & & & -1 \end{pmatrix}$.

This gives us the Dirac equation in its classical form [12]:

$$0 = f_{i,\beta} \gamma^\beta - i \cdot \beta \cdot m \cdot f_i. \quad (1691)$$

The advantage of the Dirac approach clearly is that here the structure of the equation with the quaternions already assures the satisfaction of the eigenvalue equation for the Laplace operator term, which is to say the Klein–Gordon condition:

$$\Delta f_i = -M^2 \cdot f_i. \quad (1692)$$

Seeing the simplicity of the derivation of the Dirac equation with the Einstein–Hilbert action as starting point, as just performed here and comparing this with Dirac's path, we can only repeat our amazement about "how on earth this genius came up with this idea of introducing the quaternions."

It is—we have no other word for it—a miracle [32], which becomes more obvious as the more one considers all the possible other options as we have considered in the two previous chapters.

However, as we can see from the metric origin of the classical Dirac equation [12] and the possible alternatives being derived in chapters 9 and 10, we conclude that Dirac's ingenious introduction of quaternions is not the only option, nor is it necessarily the best approach.

This becomes obvious in curvilinear coordinates [33, 34] or—even more so—in potentially non-Ricci-flat space-times.

11.4.3 What Does the Occurrence of Spinors in Quantum Gravity Mean?

But what does it mean that metric variations such as (1672) and (1673) are realized with the Dirac matrices and—consequently—also with spinors f_a?

This question shall be discussed in the next section.

11.5 The Meaning of the Occurrence of Spinors: A Suggestion

Well, we already hinted it above: Seeing the need for the introduction of vector functions f_a or any other functional object of rank > 1 should be equivalent to an increase in the number of dimensions of the system in question. The example of the occurrence of spinors in connection with the Dirac equation will be treated here in order to find a truly geometrical meaning and extension.

Let us observe Fig. 11.2. We directly cite from [14] (p. 1135):

"Take a cube (Fig. 11.2). Rotate it about one axis through 90°. Then pick another axis at right angles to the first. About it rotate the cube again through 90°. In this way the cube is carried from the orientation marked "Initial" to that marked "Final." How can one make this net transformation in a single step, with a single rotation? In other words, what is the law for the combination of rotations?

...

To any rotation is associated a quantity (Hamilton's 'quaternions'; today's 'spin matrix' or 'spinor transformation' or 'rotation operator'):

$$\text{Rot} = \cos\left[\frac{\theta}{2}\right] - i \cdot \sin\left[\frac{\theta}{2}\right]\left(\sigma_x \cos[\alpha] + \sigma_y \cos[\beta] + \sigma_z \cos[\gamma]\right)$$

with:
$$\sigma_x = \begin{Vmatrix} 0 & 1 \\ 1 & 0 \end{Vmatrix}; \quad \sigma_y = \begin{Vmatrix} 0 & -i \\ i & 0 \end{Vmatrix}; \quad \sigma_z = \begin{Vmatrix} 1 & 0 \\ 0 & -1 \end{Vmatrix},$$
(1693)

where θ is the angle of rotation and α, β, γ are the angles between the axis of rotation and the coordinate axes. A rotation described be Rot_1 followed by a rotation described by Rot_2 gives a net change in orientation described by the single rotation:

$$\text{Rot}_3 = \text{Rot}_1\,\text{Rot}_2.\text{"}$$
(1694)

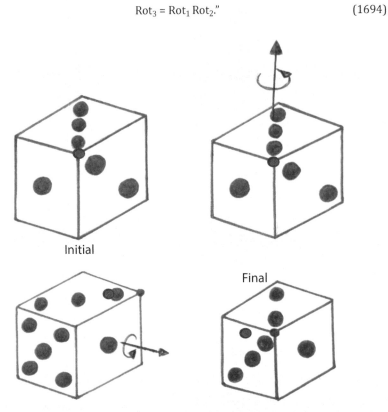

Figure 11.2 Cited from [14] (p. 1136), "Rotation about the vertical axis through 90°, followed by rotation about the horizontal axis through 90°, gives a net change in orientation that can be achieved by a single rotation through 120° about an axis emergent from the center through the corner" marked here with the red dot.

As we here will not further discuss these spinors and their math, the reader is referred to chapter 41 of [14]. The aspect we are interested in here lies in the fact that the object in Fig. 11.2 really needs to be an object with a certain volume, shape, and so on, which is to say, a mathematical point would not need such a bother. It requires an object having a surface (just like the cube in Fig. 11.2) to show such a complexity when being rotated. We therefore conclude that the need to apply spinors and Dirac gamma matrices as scaling factors to metrics in order to obtain suitable descriptions of the physical reality directly leads us to non-point-like objects and thus, consequently, to systems of higher dimensions than the actual metric, which has originally spawned the need for a spinor, describes.

The same also holds, as already said above, in connection with the metric adaptations (1664), (1665), (1667), (1668), where we do not explicitly have used the Dirac gamma matrices but only saw the need for the introduction of wave functions f with ranks higher than 0.

Example: Let us consider a metric in 4 dimensions. We know that in order to obtain Dirac equations in this 4-dimensional space-time we need to scale-factor the metric as given in (1672) and (1673) (that is at least one possibility). **To give this operation a metric meaning, we conclude that this spinor scaling is necessary, because points in our 4-dimensional space-time are in fact objects of a finite number of dimensions itself. In other words, the "points" in classical space-time cannot have 0 dimension ... at least not when needing to be treated with the Dirac equation.**

11.5.1 A Spherical Object in Cartesian Coordinates Regarding Its Position

We start with a spherical object, having its own time coordinate and "living" in our ordinary space-time with t, x, y, and z.

$$g_{\alpha\beta}^8 = \begin{pmatrix} -c^2 & 0 & 0 & 0 & \cdots & 0 \\ 0 & 1 & 0 & 0 & \cdots & 0 \\ 0 & 0 & 1 & 0 & \cdots & 0 \\ 0 & 0 & 0 & 1 & \cdots & 0 \\ \cdots & \cdots & \cdots & \cdots & \ddots & 0 \\ 0 & 0 & 0 & 0 & 0 & g_{77} \end{pmatrix}$$

$$\times F\Big[f[t,x,y,z,\tau,r,\vartheta,\varphi]\Big]; \quad F[f] = \big(C_f + f\big)^{\frac{2}{3}}$$

$$f[t,x,y,z,\tau,r,\vartheta,\varphi] = f_\tau[\tau] \cdot f_r[r] \cdot f_\vartheta[\vartheta] \cdot f_\varphi[\varphi] \cdot h[t,x,y,z];$$

$$g_{44} = -c_4^2;$$

$$g_{55} = 1;$$

$$g_{66} = r^2;$$

$$g_{77} = r^2 \cdot \sin^2 \varphi. \tag{1695}$$

Now we partially "fix" some of the functions as follows:

$$f_\varphi[\varphi] = C_{\varphi-} \cdot e^{-i \cdot A \cdot \varphi} + C_{\varphi+} \cdot e^{+i \cdot A \cdot \varphi}, \tag{1696}$$

$$f_\tau[\tau] = C_{\tau 1} \cdot \cos[c_4 \cdot E \cdot \tau] + C_{\tau 2} \cdot \sin[c_4 \cdot E \cdot \tau], \tag{1697}$$

$$f_\vartheta[\vartheta] = C_{P\vartheta} \cdot P_L^A[\cos[\vartheta]] + C_{Q\vartheta} \cdot Q_L^A[\cos[\vartheta]], \tag{1698}$$

$$f_r[r] = C_j \cdot j_l[B \cdot r] + C_y \cdot y_l[B \cdot r], \tag{1699}$$

with the associated Legendre polynomials $P_L^{A_2}, Q_L^{A_2}$ and the spherical Bessel functions j_l and y_l. This gives us the following differential equation for the t, x, y, z dependency of $h[...]$:

$$0 = (B^2 + E^2) h[t,x,y,z] - \Delta_{\{t,x,y,z\}} h[t,x,y,z]. \tag{1700}$$

We see that in fact we have obtained the classical Klein–Gordon equation of a 3-dimensional object with its own time τ and even its very own speed of light c_4 existing on the ordinary space-time t, x, y, and z. Understanding the coordinates x, y, z as the positions of the spherical object, we may take the quantum numbers resulting from the solutions (1696) to (1699) as inner properties of the object. Interestingly, among such properties will also be spin $1/2$ or spin $k/2$ ($k = 1,3,5, ...$) solutions as it will be shown in appendix C.

11.5.2 Spherical Symmetry around the Sphere and an *r*-Connection

Things are getting more interesting when assuming an outer spherical symmetry and a connection between the inner and outer spheres via the radius r.

The metric we would be interested in here reads:

$$g^8_{\alpha\beta} = \begin{pmatrix} -c^2 & 0 & 0 & 0 & \cdots & 0 \\ 0 & 1 & 0 & 0 & \cdots & 0 \\ 0 & 0 & r^2 & 0 & \cdots & 0 \\ 0 & 0 & 0 & r^2 \cdot \sin^2 \varphi_1 & \cdots & 0 \\ \cdots & \cdots & \cdots & \cdots & \ddots & 0 \\ 0 & 0 & 0 & 0 & 0 & g_{77} \end{pmatrix}$$

$$\times F\big[f[t,r,\vartheta,\varphi,\tau,\rho,\theta,\phi]\big]; \quad F[f] = f^{\frac{2}{3}}$$

$$f[t,r,\vartheta,\varphi,\tau,\rho,\theta,\phi] = f_t[t] \cdot f_r[r] \cdot f_\vartheta[\vartheta] \cdot f_\varphi[\varphi] \cdot f_\tau[\tau] \cdot f_\rho[\rho] \cdot f_\theta[\theta] \cdot f_\phi[\phi];$$

$$g_{44} = -c_4^2 \cdot r;$$

$$g_{55} = r;$$

$$g_{66} = \rho^2 \cdot r;$$

$$g_{77} = \rho^2 \cdot \sin[\theta]^2 \cdot r. \tag{1701}$$

This time we "fix" some of the functions as follows:

$$f_\varphi[\varphi] = C_{\varphi-} \cdot e^{-i \cdot A \cdot \varphi} + C_{\varphi+} \cdot e^{+i \cdot A \cdot \varphi}, \tag{1702}$$

$$f_\tau[\tau] = C_{\tau 1} \cdot \cos[c_4 \cdot E \cdot \tau] + C_{\tau 2} \cdot \sin[c_4 \cdot E \cdot \tau], \tag{1703}$$

$$f_\vartheta[\vartheta] = C_{P\vartheta} \cdot P_L^A[\cos[\vartheta]] + C_{Q\vartheta} \cdot Q_L^A[\cos[\vartheta]], \tag{1704}$$

$$f_\phi[\phi] = C_{\phi-} \cdot e^{-i \cdot B \cdot \phi} + C_{\phi+} \cdot e^{+i \cdot B \cdot \phi}, \tag{1705}$$

$$f_t[t] = C_{t1} \cdot \cos[c \cdot E_t \cdot t] + C_{t2} \cdot \sin[c \cdot E_t \cdot t], \tag{1706}$$

$$f_\theta[\theta] = C_{P\theta} \cdot P_\ell^B[\cos[\theta]] + C_{Q\theta} \cdot Q_\ell^B[\cos[\theta]]. \tag{1707}$$

This gives us the following differential equation for the r-ρ dependency of $f[\ldots]$:

$$R^* = 0 = 14 \cdot \left(\begin{matrix} \dfrac{27}{14 \cdot r^2} - \dfrac{\ell \cdot (\ell+1)}{\rho^2 \cdot r} - \dfrac{L \cdot (L+1)}{\rho \cdot r^2} + \dfrac{1}{r}\left(\dfrac{2 \cdot \partial}{\rho \cdot \partial\rho} + \dfrac{\partial^2}{\partial\rho^2} \right) \\ +2 \cdot \dfrac{\partial}{r \cdot \partial r} + \overbrace{\Delta_{\text{3D-sphere}}[r]}^{\frac{2}{r \cdot \partial r}\frac{\partial}{} + \frac{\partial^2}{\partial r^2}} - \dfrac{E^2}{r} - E_t^2 \end{matrix} \right) f. \tag{1708}$$

We find that we cannot solve this equation with the separation approach for the two radius coordinates simultaneously in its general form. However, we may look for an approximation by treating the two radii as constant to each other. This results in a variety of options. For one we could just demand:

$$0 = \left(\frac{2 \cdot \partial}{\rho \cdot \partial \rho} + \frac{\partial^2}{\partial \rho^2} - \frac{\ell \cdot (\ell + 1)}{\rho^2} - E_\rho^2 \right) f_\rho, \tag{1709}$$

leading us to:

$$f_\rho[\rho] = C_j \cdot j_\ell \left[E_\rho \cdot \rho \right] + C_y \cdot y_\ell \left[E_\rho \cdot \rho \right] \tag{1710}$$

and subsequently:

$$0 = \left(\frac{27}{14 \cdot r^2} - \frac{L \cdot (L+1)}{\rho \cdot r^2} + \frac{E_\rho^2}{r} + 2 \cdot \frac{\partial}{r \cdot \partial r} + \overbrace{\Delta_{\text{3D-sphere}}[r]}^{2\frac{\partial}{r \cdot \partial r} + \frac{\partial^2}{\partial r^2}} - \frac{E^2}{r} - E_t^2 \right) f_r. \tag{1711}$$

Similarly, we may choose:

$$0 = \left(\frac{2 \cdot \partial}{\rho \cdot \partial \rho} + \frac{\partial^2}{\partial \rho^2} - \frac{\ell \cdot (\ell + 1)}{\rho^2} - E^2 - E_\rho^2 \right) f_\rho, \tag{1712}$$

giving us:

$$f_\rho[\rho] = C_j \cdot j_\ell \left[\sqrt{E^2 + E_\rho^2} \cdot \rho \right] + C_y \cdot y_\ell \left[\sqrt{E^2 + E_\rho^2} \cdot \rho \right] \tag{1713}$$

and:

$$0 = \left(\frac{27}{14 \cdot r^2} - \frac{L \cdot (L+1)}{\rho \cdot r^2} + \frac{E_\rho^2}{r} + 2 \cdot \frac{\partial}{r \cdot \partial r} + \overbrace{\Delta_{\text{3D-sphere}}[r]}^{2\frac{\partial}{r \cdot \partial r} + \frac{\partial^2}{\partial r^2}} - E_t^2 \right) f_r. \tag{1714}$$

Both differential equations for r ((1711) and (1714)) are leading to the Laguerre polynomials as known from the Schrödinger hydrogen atom [15].

Another set of options arises with the following setting for the ρ coordinates

$$0 = \left(2 \frac{\partial}{\rho \cdot \partial \rho} + \frac{\partial^2}{\partial \rho^2} - \frac{\ell \cdot (\ell + 1)}{\rho^2} - \frac{L \cdot (L+1)}{\rho \cdot r} - E_\rho^2 - \frac{E_{\rho r}^2}{r} \right) f_\rho, \tag{1715}$$

leading us to Laguerre polynomials for the function f_ρ and subsequently with respect to the function f_r via:

$$0 = \left(\frac{27}{14 \cdot r^2} - \frac{E_{\rho r}^2}{r^2} + \frac{E_\rho^2}{r} + 2 \cdot \frac{\partial}{r \cdot \partial r} + \overbrace{\Delta_{\text{3D-sphere}}[r]}^{2\frac{\partial}{r \cdot \partial r} + \frac{\partial^2}{\partial r^2}} - \frac{E^2}{r} - E_t^2 \right) f_r. \quad (1716)$$

Similarly, we may choose the condition just as above:

$$0 = \left(2 \cdot \frac{\partial}{\rho \cdot \partial \rho} + \frac{\partial^2}{\partial \rho^2} - \frac{\ell \cdot (\ell+1)}{\rho^2} - \frac{L \cdot (L+1)}{\rho \cdot r} - E^2 - E_\rho^2 - \frac{E_{\rho r}^2}{r} \right) f_\rho, \quad (1717)$$

also leading to Laguerre polynomials when treating r as constant in this equation. The corresponding differential equation for the function f_r now reads

$$0 = \left(\frac{27}{14 \cdot r^2} - \frac{L \cdot (L+1)}{\rho \cdot r^2} + \frac{E_\rho^2}{r} + 2 \cdot \frac{\partial}{r \cdot \partial r} + \overbrace{\Delta_{\text{3D-sphere}}[r]}^{2\frac{\partial}{r \cdot \partial r} + \frac{\partial^2}{\partial r^2}} - E_t^2 \right) f_r. \quad (1718)$$

As before the solution to (1717) and (1718) are Laguerre polynomials.

A non-approximated solution to (1708) can be found in the following way:

$$R^* = 0 = 14 \cdot \left(\begin{array}{c} \dfrac{27}{14 \cdot r^2} - \dfrac{\ell \cdot (\ell+1)}{\rho^2 \cdot r} - \overbrace{\dfrac{L \cdot (L+1)}{\rho \cdot r^2}}^{=0 \rightarrow L=0} + \dfrac{1}{r}\left(\dfrac{2 \cdot \partial}{\rho \cdot \partial \rho} + \dfrac{\partial^2}{\partial \rho^2} \right) \\[4mm] +2 \cdot \dfrac{\partial}{r \cdot \partial r} + 2\dfrac{\partial}{r \cdot \partial r} + \dfrac{\partial^2}{\partial r^2} - \dfrac{E^2}{r} - E_t^2 \end{array} \right) f. \quad (1719)$$

This leads us to the following two well-separated equations for the two radii:

$$0 = \frac{1}{r}\left(\left(\frac{2 \cdot \partial}{\rho \cdot \partial \rho} + \frac{\partial^2}{\partial \rho^2} \right) - \frac{\ell \cdot (\ell+1)}{\rho^2} - E_\rho^2 \right) f_\rho, \quad (1720)$$

$$0 = 14 \cdot \left(\frac{27}{14 \cdot r^2} + 4 \cdot \frac{\partial}{r \cdot \partial r} + \frac{\partial^2}{\partial r^2} + \frac{E_\rho^2}{r} - \frac{E^2}{r} - E_t^2 \right) f_r. \quad (1721)$$

The solutions are:

$$f_\rho[\rho] = C_j \cdot j_\ell[E_\rho \cdot \rho] + C_y \cdot y_\ell[E_\rho \cdot \rho], \quad (1722)$$

$$f_r[r] = e^{-r \cdot E_t} \cdot r^{\frac{3}{14}(\sqrt{7}-7)} \cdot \left(C_U U\left[nn, 1 + \frac{3}{\sqrt{7}}, 2 \cdot i \cdot r \cdot E_t \right] + C_L L_{-nn}^{\frac{3}{\sqrt{7}}} \left[2 \cdot i \cdot r \cdot E_t \right] \right)$$

$$nn = \frac{7 + 3 \cdot \sqrt{7}}{14} + i \cdot \frac{E^2 - E_\rho^2}{2 \cdot E_t}. \tag{1723}$$

While the ρ solution is rather ordinary, the r one appears very strange and this is not just due to the fact that we have a negative exponent for the r, but also the mixed imaginary constellation in the special functions.

We find a similar complex situation in the case of a metric with the following r behavior:

$$g_{\alpha\beta}^8 = \begin{pmatrix} -c^2 & 0 & 0 & 0 & \cdots & 0 \\ 0 & 1 & 0 & 0 & \cdots & 0 \\ 0 & 0 & r^2 & 0 & \cdots & 0 \\ 0 & 0 & 0 & r^2 \cdot \sin^2 \varphi_1 & \cdots & 0 \\ \cdots & \cdots & \cdots & \cdots & \ddots & 0 \\ 0 & 0 & 0 & 0 & 0 & g_{77} \end{pmatrix}$$

$$\times F\left[f[t,r,\vartheta,\varphi,\tau,\rho,\theta,\phi] \right]; \quad F[f] = f^{\frac{2}{3}}$$

$$f[t,r,\vartheta,\varphi,\tau,\rho,\theta,\phi] = f_t[t] \cdot f_r[r] \cdot f_\vartheta[\vartheta] \cdot f_\varphi[\varphi] \cdot f_\tau[\tau] \cdot f_\rho[\rho] \cdot f_\theta[\theta] \cdot f_\phi[\phi];$$

$$g_{44} = -c_4^2;$$

$$g_{55} = 1;$$

$$g_{66} = \rho^2 \cdot r;$$

$$g_{77} = \rho^2 \cdot \sin[\theta]^2 \cdot r. \tag{1724}$$

Here a complete separation can only be achieved when we apply the solutions (1702) to (1707) and demand:

$$\frac{3}{7} + \ell + \ell^2 = 0 \quad \rightarrow \quad \ell_{1,2} = \frac{1}{14}\left(\pm i \cdot \sqrt{35} - 7 \right), \tag{1725}$$

leading us to:

$$0 = \frac{1}{r}\left(\left(\frac{2 \cdot \partial}{\rho \cdot \partial\rho} + \frac{\partial^2}{\partial\rho^2} \right) + \frac{3}{7 \cdot \rho^2} + E_\rho^2 \right) f_\rho, \tag{1726}$$

$$0 = \left(\frac{3}{4 \cdot r^2} - \frac{L \cdot (L+1)}{r^2} + 3 \cdot \frac{\partial}{r \cdot \partial r} + \frac{\partial^2}{\partial r^2} + E_\rho^2 - E^2 - E_t^2 \right) f_r. \quad (1727)$$

The solutions are:

$$f_\rho[\rho] = C_j \cdot j_p \left[E_\rho \cdot \rho \right] + C_y \cdot y_p \left[E_\rho \cdot \rho \right]; \quad p = \frac{1}{14} \left(i \cdot \sqrt{35} - 7 \right), \quad (1728)$$

$$f_r[r] = \frac{1}{r} \cdot \left(C_J \cdot J_{L+\frac{1}{2}} \left[E_r \cdot r \right] + C_Y \cdot Y_{L+\frac{1}{2}} \left[E_r \cdot r \right] \right); \quad E_r = \sqrt{E^2 - E_\rho^2 - E_t^2}, \quad (1729)$$

where the latter consists of the ordinary Bessel functions J and Y of the first and the second kind, respectively.

A much simpler, because directly separational, set of ρ-r dependencies occurs for the following metric setting:

$$g_{\alpha\beta}^8 = \begin{pmatrix} -c^2 & 0 & 0 & 0 & \cdots & 0 \\ 0 & 1 & 0 & 0 & \cdots & 0 \\ 0 & 0 & r^2 & 0 & \cdots & 0 \\ 0 & 0 & 0 & r^2 \cdot \sin^2 \varphi_1 & \cdots & 0 \\ \cdots & \cdots & \cdots & \cdots & \ddots & 0 \\ 0 & 0 & 0 & 0 & 0 & g_{77} \end{pmatrix}$$

$$\times F \left[f[t, r, \vartheta, \varphi, \tau, \rho, \theta, \phi] \right]; \quad F[f] = f^{\frac{2}{3}}$$

$$f[t, r, \vartheta, \varphi, \tau, \rho, \theta, \phi] = f_t[t] \cdot f_r[r] \cdot f_\vartheta[\vartheta] \cdot f_\varphi[\varphi] \cdot f_\tau[\tau] \cdot f_\rho[\rho] \cdot f_\theta[\theta] \cdot f_\phi[\phi];$$

$$g_{44} = -c_4^2;$$
$$g_{55} = r;$$
$$g_{66} = \rho^2 \cdot r;$$
$$g_{77} = \rho^2 \cdot \sin[\theta]^2 \cdot r. \quad (1730)$$

Applying the solutions as given in (1702) to (1707), we have for the scaled Ricci scalar R^*:

$$R^* = 0 = \frac{7}{6} \cdot \left(\frac{9}{7 \cdot r^2} - \frac{\ell \cdot (\ell+1)}{\rho^2} - \frac{L \cdot (L+1)}{r^2} + \frac{2 \cdot \partial}{\rho \cdot \partial \rho} + \frac{\partial^2}{\partial \rho^2} \right.$$
$$\left. + \frac{3}{2} \cdot \frac{\partial}{r \cdot \partial r} + \overbrace{\Delta_{\text{3D-sphere}}}^{2\frac{\partial}{r \cdot \partial r} + \frac{\partial^2}{\partial r^2}} [r] - E^2 - E_t^2 \right) f. \quad (1731)$$

This time we can separate the two radii easily and obtain the two differential equations:

$$0 = \left(-\frac{\ell \cdot (\ell+1)}{\rho^2} + \frac{2 \cdot \partial}{\rho \cdot \partial \rho} + \frac{\partial^2}{\partial \rho^2} - E_\rho^2 \right) f_\rho, \tag{1732}$$

$$0 = \left(\frac{9}{7 \cdot r^2} - \frac{L \cdot (L+1)}{r^2} + \frac{3}{2} \cdot \frac{\partial}{r \cdot \partial r} + \overbrace{\Delta_{\text{3D-sphere}}[r]}^{2\frac{\partial}{r \cdot \partial r} + \frac{\partial^2}{\partial r^2}} - E^2 - E_t^2 + E_\rho^2 \right) f_r. \tag{1733}$$

While (1732) gives us the spherical Bessel functions again via:

$$f_\rho[\rho] = C_j \cdot j_\ell \left[E_\rho \cdot \rho \right] + C_y \cdot y_\ell \left[E_\rho \cdot \rho \right], \tag{1734}$$

we obtain the following solution for f_r:

$$f_r[r] = e^{-r \cdot \sqrt{-E_t^2 - E^2}} \cdot r^{-\frac{5}{4} + \sqrt{\frac{31}{112} + L(1+L)}} \cdot \left(\begin{array}{c} C_U U\left[nn, 1 + P, 2 \cdot r \cdot \sqrt{-E_t^2 - E^2} \right] \\ + C_L L_{-nn}^P \left[2 \cdot r \cdot \sqrt{-E_t^2 - E^2} \right] \end{array} \right)$$

$$nn = \frac{1}{2} \cdot \left(1 + P + \frac{E_\rho^2}{\sqrt{-E_t^2 - E^2}} \right); \quad P = \frac{1}{2} \sqrt{\frac{31}{7} + 16 \cdot L(1+L)}. \tag{1735}$$

With the hypergeometric function $U[a, b, z]$ and the Laguerre polynomials L_{-nn}^P in the solutions above, we recognize the similarity to the Schrödinger hydrogen solution (e.g., [15]). We require quite some discussion with respect to the meaning of our result, but this we will omit at this point here, because there are other simpler-to-discuss solutions and we are going to consider those in the next sub-section. Thus, we leave the illustration, interpretation, and discussion of the solutions (1702) to (1707), (1734), and (1735) for later.

Again, we point out that there are also half spin solutions to the Legendre polynomial solutions with the coordinates θ and ϑ (see appendix C for details). In the next sub-section, we investigate a metric setting which kind of even forces such half spins to come forth.

11.5.3   A Possible Origin of Spin I = 1/2

The metric we would be interested in here reads:

$$
g_{\alpha\beta}^8 = \begin{pmatrix}
-c^2 & 0 & 0 & 0 & \cdots & 0 \\
0 & 1 & 0 & 0 & \cdots & 0 \\
0 & 0 & r^2 & 0 & \cdots & 0 \\
0 & 0 & 0 & r^2 \cdot \sin^2 \varphi_1 & \cdots & 0 \\
\cdots & \cdots & \cdots & \cdots & \ddots & 0 \\
0 & 0 & 0 & 0 & 0 & g_{77}
\end{pmatrix}
$$

$$
\times F\Big[f\big[t,r,\vartheta,\varphi_1,\varphi_2,\varphi_3,\varphi_4,\varphi_5 \big] \Big]; \quad F[f] = f^{\tfrac{2}{3}}
$$

$$
f\big[t,r,\vartheta,\varphi_1,\varphi_2,\varphi_3,\varphi_4,\varphi_5 \big] = f_t[t] \cdot f_r[r] \cdot f_\vartheta[\vartheta] \cdot \prod_{i=1}^{5} f_{\varphi_i}[\varphi_i];
$$

$$
g_{44} = \left(g_{\varphi_2}\big[\varphi_2 \big]' \right)^2 \cdot r^1;
$$

$$
g_{55} = \left(g_{\varphi_3}\big[\varphi_3 \big]' \right)^2 \cdot r^1;
$$

$$
g_{66} = \left(g_{\varphi_4}\big[\varphi_4 \big]' \right)^2 \cdot r^0;
$$

$$
g_{77} = \left(g_{\varphi_5}\big[\varphi_5 \big]' \right)^2 \cdot r^0. \tag{1736}
$$

Now we "fix" some of the functions as follows (for $i = 3, 4, 5^6$):

$$
f_{\varphi_i}\big[\varphi_i \big] = f_{\varphi_i}\Big[g_{\varphi_i}\big[\varphi_i \big] \Big] \equiv f_{\varphi_i}\big[g_{\varphi_i} \big]; \quad f_{\varphi_i}\big[g_{\varphi_i} \big] = C_{-i} \cdot e^{-A_i \cdot g_{\varphi_i}} + C_{+i} \cdot e^{+A_i \cdot g_{\varphi_i}}
$$

$$
f_{\varphi_2}\big[g_{\varphi_2} \big] = C_{-2} \cdot e^{-i \cdot A_2 \cdot g_{\varphi_2}} + C_{+2} \cdot e^{+i \cdot A_2 \cdot g_{\varphi_2}}, \tag{1737}
$$

$$
f_t[t] = C_{t1} \cdot \cos[c \cdot E \cdot t] + C_{t2} \cdot \sin[c \cdot E \cdot t], \tag{1738}
$$

$$
f_\vartheta\big[\vartheta = \varphi_1 \big] = C_{P\vartheta} \cdot P_L^{A_2}\Big[\cos\big[\vartheta = \varphi_1 \big] \Big] + C_{Q\vartheta} \cdot Q_L^{A_2}\Big[\cos\big[\vartheta = \varphi_1 \big] \Big], \tag{1739}
$$

with the associated Legendre polynomials $P_L^{A_2}, Q_L^{A_2}$. This gives us the following differential equation for the r dependency of $f[\ldots]$:

$$
R^* = 0 = \left(\frac{3 - 4 \cdot L \cdot (L+1)}{4 \cdot r^2} + \frac{A_2^2 + A_3^2}{r} + A_4^2 + A_5^2 + \frac{\partial}{r \cdot \partial r} + \underbrace{\frac{2 \frac{\partial}{r \cdot \partial r} + \frac{\partial^2}{\partial r^2}}{\Delta_{\text{3D-sphere}}[r]}} - E^2 \right) \Psi
$$

$$\Psi = \Psi\left[r,\vartheta = \varphi_1,\varphi\right] = f_r\left[r\right]\cdot f_\vartheta\left[\vartheta = \varphi_1\right]\cdot f_{\varphi_2}\left[\varphi_2\right]$$

$$\Rightarrow$$

$$0 = f_\vartheta\left[\vartheta\right]\cdot f_{\varphi_2}\left[\varphi_2\right]\cdot\left(\frac{3 - 4\cdot L\cdot(L+1)}{4\cdot r^2} + \frac{A_2^2 + A_3^2}{r} + A_4^2 + A_5^2 + \frac{3\cdot\partial}{r\cdot\partial r} + \frac{\partial^2}{\partial r^2} - E^2\right)f_r\left[r\right].$$

$$(1740)$$

We find that for $L = 1/2$ the terms with r^{-2} vanish, resulting in the differential equation for the remaining coordinate r:

$$0 = \left(\frac{A_2^2 + A_3^2}{r} + A_4^2 + A_5^2 + \frac{3\cdot\partial}{r\cdot\partial r} + \frac{\partial^2}{\partial r^2} - E^2\right)f_r\left[r\right].\qquad(1741)$$

This is extremely interesting, because now we have obtained a condition, namely:

$$\frac{3 - 4\cdot L\cdot(L+1)}{4\cdot r^2} = 0 \quad\Rightarrow\quad L = \frac{1}{2}\qquad(1742)$$

forcing us to except the existence of spinors with a geometry given due to the metric (1736). With respect to the cases of half spins regarding the Legendre functions, we refer to appendix C and a variety of previous papers by this author (mainly within the series of "Einstein had it …").

Now we can also solve (1740) and (1741) via:

$$f_r\left[r\right] = e^{-r\cdot\sqrt{E^2 - \left(A_4^2 + A_5^2\right)}}\cdot r^{L - \frac{1}{2}}\cdot\left(\begin{array}{c} C_U U\left[-nn, L\cdot(1+L), 2\cdot r\cdot\sqrt{E^2 - \left(A_4^2 + A_5^2\right)}\right] \\ + C_L L_{nn}^{1+2\cdot L}\left[2\cdot r\cdot\sqrt{E^2 - \left(A_4^2 + A_5^2\right)}\right] \end{array}\right)$$

$$nn = \frac{A_2^2 + A_3^2}{2\cdot\sqrt{E^2 - \left(A_4^2 + A_5^2\right)}} - 1 - L,\qquad(1743)$$

$$f_r\left[r\right] = e^{-r\cdot\sqrt{E^2 - \left(A_4^2 + A_5^2\right)}}\left(\begin{array}{c} C_U U\left[-nn, 3, 2\cdot r\cdot\sqrt{E^2 - \left(A_4^2 + A_5^2\right)}\right] \\ + C_L L_{nn}^2\left[2\cdot r\cdot\sqrt{E^2 - \left(A_4^2 + A_5^2\right)}\right] \end{array}\right)$$

$$nn = \frac{A_2^2 + A_3^2 - 3\cdot\sqrt{E^2 - \left(A_4^2 + A_5^2\right)}}{2\cdot\sqrt{E^2 - \left(A_4^2 + A_5^2\right)}},\qquad(1744)$$

respectively.

With the hypergeometric function $U[a, b, z]$ and the Laguerre polynomials $L_{nn}^{1+2 \cdot L}$ in the solutions above, we recognize the similarity to the Schrödinger hydrogen solution (e.g., [15]). Only considering the ordinary spatial coordinates r, ϑ, φ, we can illustrate the spatial distribution our solution would produce in ordinary space-time (Figs. 11.3 and 11.4).

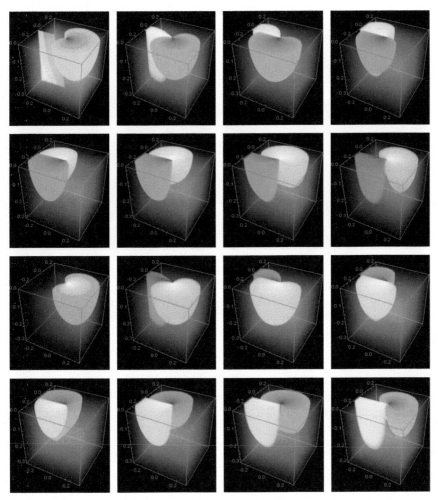

Figure 11.3 Spinor-object as derived from an 8-dimensional space-time (4 dimensions appear compactified to the human "eye"/sensors). Rotation angle from picture to picture is $\pi/4$ clockwise from left to right and π clockwise from up to down. We evaluated the wave function f.

The pictures in the graphics grid above (Fig. 11.3) show $\pi/4$ rotations in a clockwise manner (top left to bottom right). If this was a classical object, then the ninth picture ($8*\pi/4 = 2\pi = 360°$) should be exactly equal to the first one, but obviously it is not. On the contrary, it has changed its sign (Fig. 11.4) and thus is a spinor.

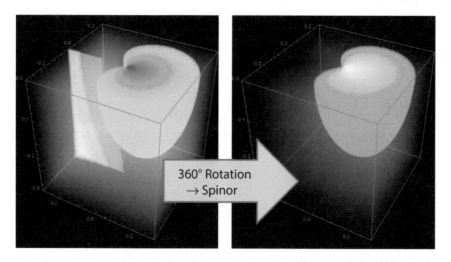

Figure 11.4 Spinor object as derived from an 8-dimensional space-time (4 dimensions appear compactified to the human "eye"/sensors). Rotation angles 0 and $2*\pi$. Please note the changed color coding, which is clearly showing the sign switch of the metric object after one full rotation ($360° = 2*\pi$). The corresponding 8-dimensional solution, being used for the evaluation here, has been presented in Eqs. (1736) to (1744). We evaluated the wave function f.

The following set of pictures (Figs. 11.5 to 11.36) shows the resulting metric fields of such a spinor for the scale function $F[f]$, which scales the metric via $G_{\alpha\beta} = F[f]*g_{\alpha\beta}$.

Figure 11.5 Real part of $F[f]$ for spinor object as derived from an 8-dimensional space-time (4 dimensions appear compactified to the human "eye"/sensors). Rotation angle 0. The corresponding 8-dimensional solution, being used for the evaluation here, has been presented in Eqs. (1736) to (1744).

Figure 11.6 Real part of $F[f]$ for spinor object as derived from an 8-dimensional space-time (4 dimensions appear compactified to the human "eye"/sensors). Rotation angle $\pi/4$. The corresponding 8-dimensional solution, being used for the evaluation here, has been presented in Eqs. (1736) to (1744).

Figure 11.7 Real part of $F[f]$ for spinor object as derived from an 8-dimensional space-time (4 dimensions appear compactified to the human "eye"/sensors). Rotation angle $2*\pi/4$. The corresponding 8-dimensional solution, being used for the evaluation here, has been presented in Eqs. (1736) to (1744).

Figure 11.8 Real part of $F[f]$ for spinor object as derived from an 8-dimensional space-time (4 dimensions appear compactified to the human "eye"/sensors). Rotation angle $3*\pi/4$. The corresponding 8-dimensional solution, being used for the evaluation here, has been presented in Eqs. (1736) to (1744).

Figure 11.9 Real part of $F[f]$ for spinor object as derived from an 8-dimensional space-time (4 dimensions appear compactified to the human "eye"/sensors). Rotation angle $4*\pi/4$. The corresponding 8-dimensional solution, being used for the evaluation here, has been presented in Eqs. (1736) to (1744).

Figure 11.10 Real part of $F[f]$ for spinor object as derived from an 8-dimensional space-time (4 dimensions appear compactified to the human "eye"/sensors). Rotation angle $5*\pi/4$. The corresponding 8-dimensional solution, being used for the evaluation here, has been presented in Eqs. (1736) to (1744).

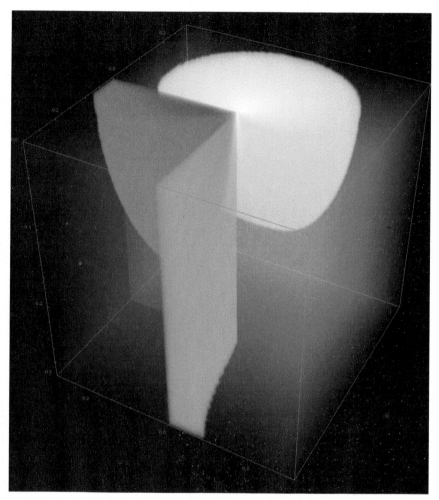

Figure 11.11 Real part of $F[f]$ for spinor object as derived from an 8-dimensional space-time (4 dimensions appear compactified to the human "eye"/sensors). Rotation angle $6*\pi/4$. The corresponding 8-dimensional solution, being used for the evaluation here, has been presented in Eqs. (1736) to (1744).

Figure 11.12   Real part of $F[f]$ for spinor object as derived from an 8-dimensional space-time (4 dimensions appear compactified to the human "eye"/sensors). Rotation angle $7*\pi/4$. The corresponding 8-dimensional solution, being used for the evaluation here, has been presented in Eqs. (1736) to (1744).

Figure 11.13 Real part of $F[f]$ for spinor object as derived from an 8-dimensional space-time (4 dimensions appear compactified to the human "eye"/sensors). Rotation angle $8*\pi/4$. The corresponding 8-dimensional solution, being used for the evaluation here, has been presented in Eqs. (1736) to (1744).

Figure 11.14 Real part of $F[f]$ for spinor object as derived from an 8-dimensional space-time (4 dimensions appear compactified to the human "eye"/sensors). Rotation angle $9*\pi/4$. The corresponding 8-dimensional solution, being used for the evaluation here, has been presented in Eqs. (1736) to (1744).

Figure 11.15 Real part of $F[f]$ for spinor object as derived from an 8-dimensional space-time (4 dimensions appear compactified to the human "eye"/sensors). Rotation angle $10*\pi/4$. The corresponding 8-dimensional solution, being used for the evaluation here, has been presented in Eqs. (1736) to (1744).

Figure 11.16 Real part of $F[f]$ for spinor object as derived from an 8-dimensional space-time (4 dimensions appear compactified to the human "eye"/sensors). Rotation angle $11*\pi/4$. The corresponding 8-dimensional solution, being used for the evaluation here, has been presented in Eqs. (1736) to (1744).

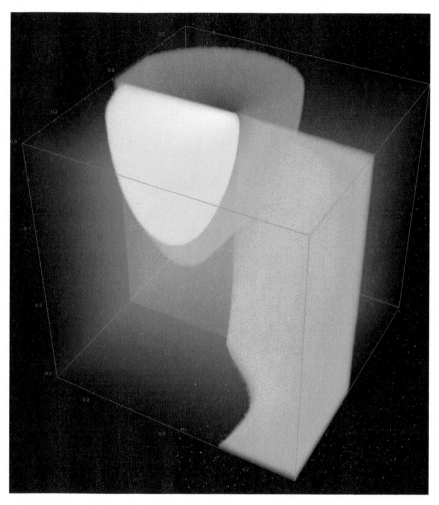

Figure 11.17 Real part of $F[f]$ for spinor object as derived from an 8-dimensional space-time (4 dimensions appear compactified to the human "eye"/sensors). Rotation angle $12*\pi/4$. The corresponding 8-dimensional solution, being used for the evaluation here, has been presented in Eqs. (1736) to (1744).

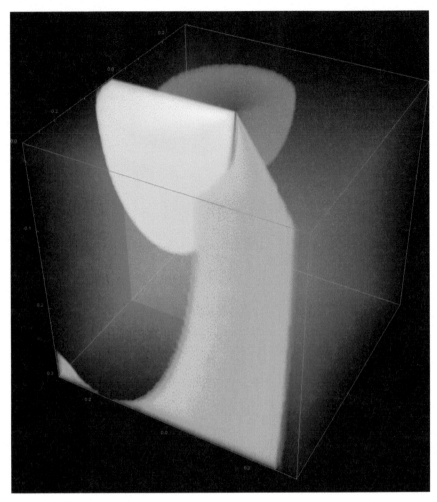

Figure 11.18 Real part of $F[f]$ for spinor object as derived from an 8-dimensional space-time (4 dimensions appear compactified to the human "eye"/sensors). Rotation angle $13*\pi/4$. The corresponding 8-dimensional solution, being used for the evaluation here, has been presented in Eqs. (1736) to (1744).

Figure 11.19 Real part of $F[f]$ for spinor object as derived from an 8-dimensional space-time (4 dimensions appear compactified to the human "eye"/sensors). Rotation angle $14*\pi/4$. The corresponding 8-dimensional solution, being used for the evaluation here, has been presented in Eqs. (1736) to (1744).

Figure 11.20 Real part of *F*[*f*] for spinor object as derived from an 8-dimensional space-time (4 dimensions appear compactified to the human "eye"/sensors). Rotation angle 15*π/4. The corresponding 8-dimensional solution, being used for the evaluation here, has been presented in Eqs. (1736) to (1744).

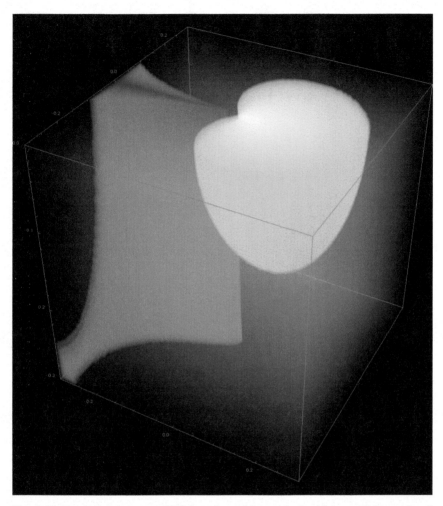

Figure 11.21 Imaginary part of $F[f]$ for spinor object as derived from an 8-dimensional space-time (4 dimensions appear compactified to the human "eye"/sensors). Rotation angle 0. The corresponding 8-dimensional solution, being used for the evaluation here, has been presented in Eqs. (1736) to (1744).

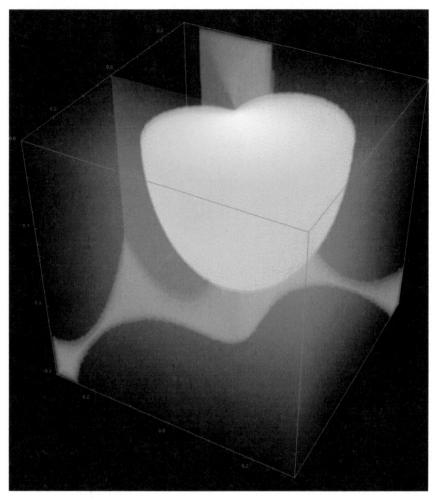

Figure 11.22 Imaginary part of $F[f]$ for spinor object as derived from an 8-dimensional space-time (4 dimensions appear compactified to the human "eye"/sensors). Rotation angle $\pi/4$. The corresponding 8-dimensional solution, being used for the evaluation here, has been presented in Eqs. (1736) to (1744).

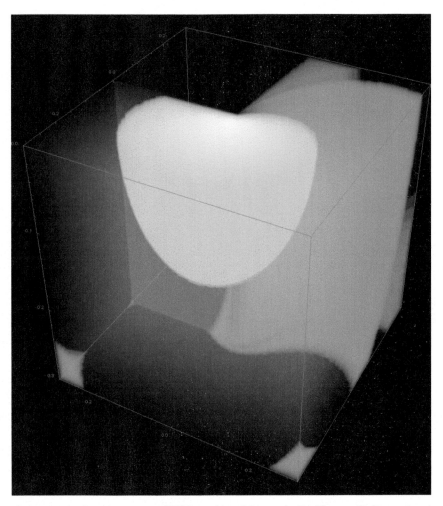

Figure 11.23 Imaginary part of $F[f]$ for spinor object as derived from an 8-dimensional space-time (4 dimensions appear compactified to the human "eye"/sensors). Rotation angle $2*\pi/4$. The corresponding 8-dimensional solution, being used for the evaluation here, has been presented in Eqs. (1736) to (1744).

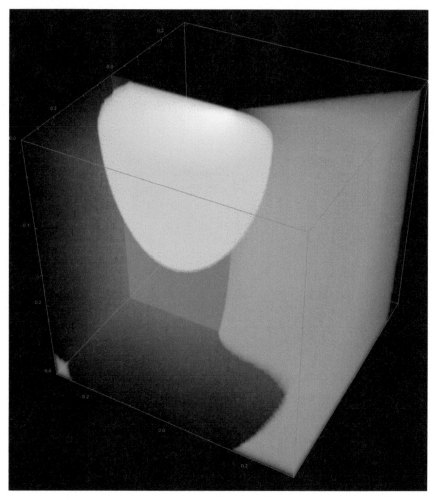

Figure 11.24 Imaginary part of $F[f]$ for spinor object as derived from an 8-dimensional space-time (4 dimensions appear compactified to the human "eye"/sensors). Rotation angle $3*\pi/4$. The corresponding 8-dimensional solution, being used for the evaluation here, has been presented in Eqs. (1736) to (1744).

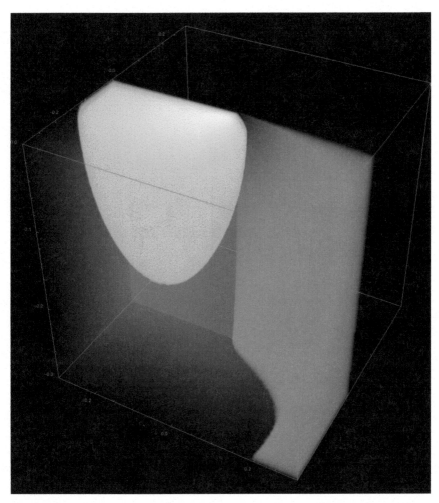

Figure 11.25 Imaginary part of $F[f]$ for spinor object as derived from an 8-dimensional space-time (4 dimensions appear compactified to the human "eye"/sensors). Rotation angle $4*\pi/4$. The corresponding 8-dimensional solution, being used for the evaluation here, has been presented in Eqs. (1736) to (1744).

Figure 11.26 Imaginary part of $F[f]$ for spinor object as derived from an 8-dimensional space-time (4 dimensions appear compactified to the human "eye"/sensors). Rotation angle $5*\pi/4$. The corresponding 8-dimensional solution, being used for the evaluation here, has been presented in Eqs. (1736) to (1744).

Figure 11.27   Imaginary part of $F[f]$ for spinor object as derived from an 8-dimensional space-time (4 dimensions appear compactified to the human "eye"/sensors). Rotation angle $6*\pi/4$. The corresponding 8-dimensional solution, being used for the evaluation here, has been presented in Eqs. (1736) to (1744).

Figure 11.28 Imaginary part of $F[f]$ for spinor object as derived from an 8-dimensional space-time (4 dimensions appear compactified to the human "eye"/sensors). Rotation angle $7*\pi/4$. The corresponding 8-dimensional solution, being used for the evaluation here, has been presented in Eqs. (1736) to (1744).

Figure 11.29 Imaginary part of $F[f]$ for spinor object as derived from an 8-dimensional space-time (4 dimensions appear compactified to the human "eye"/sensors). Rotation angle $8*\pi/4$. The corresponding 8-dimensional solution, being used for the evaluation here, has been presented in Eqs. (1736) to (1744).

Figure 11.30 Imaginary part of $F[f]$ for spinor object as derived from an 8-dimensional space-time (4 dimensions appear compactified to the human "eye"/sensors). Rotation angle $9*\pi/4$. The corresponding 8-dimensional solution, being used for the evaluation here, has been presented in Eqs. (1736) to (1744).

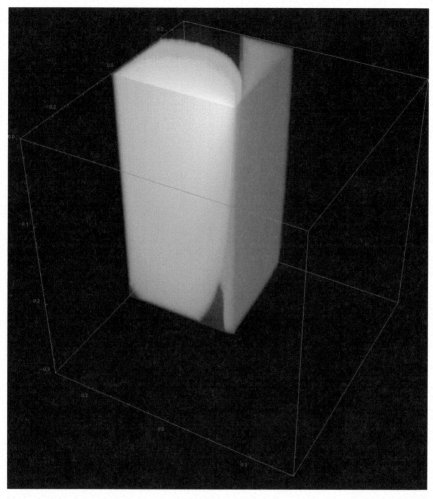

Figure 11.31 Imaginary part of $F[f]$ for spinor object as derived from an 8-dimensional space-time (4 dimensions appear compactified to the human "eye"/ sensors). Rotation angle $10*\pi/4$. The corresponding 8-dimensional solution, being used for the evaluation here, has been presented in Eqs. (1736) to (1744).

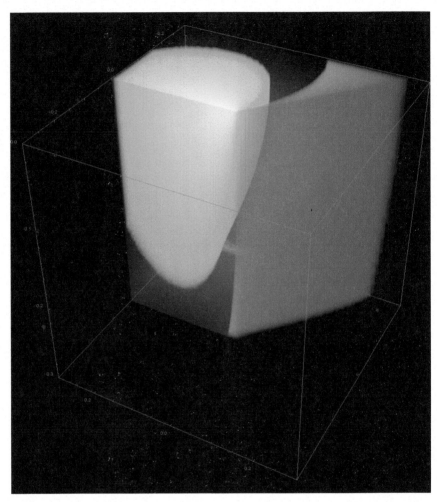

Figure 11.32 Imaginary part of $F[f]$ for spinor object as derived from an 8-dimensional space-time (4 dimensions appear compactified to the human "eye"/ sensors). Rotation angle $11*\pi/4$. The corresponding 8-dimensional solution, being used for the evaluation here, has been presented in Eqs. (1736) to (1744).

Figure 11.33 Imaginary part of $F[f]$ for spinor object as derived from an 8-dimensional space-time (4 dimensions appear compactified to the human "eye"/ sensors). Rotation angle $12*\pi/4$. The corresponding 8-dimensional solution, being used for the evaluation here, has been presented in Eqs. (1736) to (1744).

Figure 11.34 Imaginary part of *F*[*f*] for spinor object as derived from an 8-dimensional space-time (4 dimensions appear compactified to the human "eye"/ sensors). Rotation angle 13*π/4. The corresponding 8-dimensional solution, being used for the evaluation here, has been presented in Eqs. (1736) to (1744).

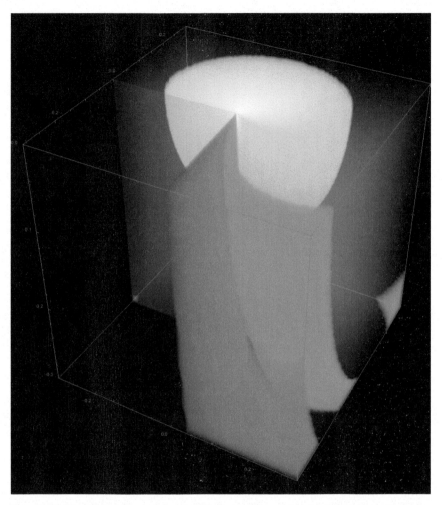

Figure 11.35 Imaginary part of $F[f]$ for spinor object as derived from an 8-dimensional space-time (4 dimensions appear compactified to the human "eye"/ sensors). Rotation angle $14*\pi/4$. The corresponding 8-dimensional solution, being used for the evaluation here, has been presented in Eqs. (1736) to (1744).

Figure 11.36 Imaginary part of $F[f]$ for spinor object as derived from an 8-dimensional space-time (4 dimensions appear compactified to the human "eye"/ sensors). Rotation angle $15*\pi/4$. The corresponding 8-dimensional solution, being used for the evaluation here, has been presented in Eqs. (1736) to (1744).

11.5.4 An Adapted Half Spin Metric and a Bit of Interpretation Work

We change the metric (1736) as follows:

$$g_{\alpha\beta}^{8} = \begin{pmatrix} -c^2 & 0 & 0 & 0 & \cdots & 0 \\ 0 & 1 & 0 & 0 & \cdots & 0 \\ 0 & 0 & r^2 & 0 & \cdots & 0 \\ 0 & 0 & 0 & r^2 \cdot \sin^2 \varphi_1 & \cdots & 0 \\ \cdots & \cdots & \cdots & \cdots & \ddots & 0 \\ 0 & 0 & 0 & 0 & 0 & g_{77} \end{pmatrix}$$

$$\times F\Big[f\big[t,r,\vartheta,\varphi,\tau,\rho,\theta,\phi\big]\Big]; \quad F[f] = f^{\frac{2}{3}}$$

$$f\big[t,r,\vartheta,\varphi,\tau,\rho,\theta,\phi\big] = f_t[t] \cdot f_r[r] \cdot f_{\vartheta}[\vartheta] \cdot f_{\varphi}[\varphi] \cdot f_{\tau}[\tau] \cdot f_{\rho}[\rho] \cdot f_{\theta}[\theta] \cdot f_{\phi}[\phi];$$

$$g_{44} = -c_4^2 \cdot r;$$

$$g_{55} = r;$$

$$g_{66} = \rho^2;$$

$$g_{77} = \rho^2 \cdot \sin[\theta]^2. \tag{1745}$$

Thereby we have used the coordinates from (1730) and we will also use the solutions (1702) to (1707), giving us the following scaled Ricci scalar R^*:

$$R^* = 0 = \frac{7}{3} \cdot \left(\begin{array}{c} \dfrac{3}{4 \cdot r^2} - \dfrac{L \cdot (L+1)}{r^2} - \dfrac{\ell \cdot (\ell+1)}{\rho^2} - \dfrac{3}{7 \cdot \rho^2} + \dfrac{2 \cdot \partial}{r \cdot \rho \cdot \partial \rho} + \dfrac{\partial^2}{r \cdot \partial \rho^2} \\ + \dfrac{3}{7 \cdot r \cdot \rho^2} + \dfrac{\partial}{r \cdot \partial r} + \dfrac{2 \cdot \partial}{r \cdot \partial r} + \dfrac{\partial^2}{\partial r^2} - E^2 - \dfrac{E_t^2}{r} \end{array} \right) f. \tag{1746}$$

As before we cannot separate the two radii with the approach from (1745) and thus try for the solution of the following condition for ℓ and the two differential equations:

$$0 = -\frac{\ell \cdot (\ell+1)}{\rho^2} - \frac{3}{7 \cdot \rho^2} \quad \Rightarrow \quad \ell_{1,2} = \pm i \cdot \frac{\sqrt{35}}{14} - \frac{1}{2}$$

$$0 = \left(\frac{2 \cdot \partial}{r \cdot \rho \cdot \partial \rho} + \frac{\partial^2}{r \cdot \partial \rho^2} + \frac{3}{7 \cdot r \cdot \rho^2} \mp \frac{E_{\rho}^2}{r} \right) f_{\rho}, \tag{1747}$$

$$0 = \left(\frac{3}{4 \cdot r^2} - \frac{L \cdot (L+1)}{r^2} + \frac{E_{\rho}^2}{r} + \frac{\partial}{r \cdot \partial r} + \frac{2 \cdot \partial}{r \cdot \partial r} + \frac{\partial^2}{\partial r^2} - E^2 - \frac{E_t^2}{r} \right) f_r. \tag{1748}$$

While (1747) gives us the spherical Bessel functions again via:

$$f_\rho[\rho] = C_j \cdot j_{\ell_+}[E_\rho \cdot \rho] + C_y \cdot y_{\ell_+}[E_\rho \cdot \rho]; \quad \ell_+ = i \cdot \frac{\sqrt{35}}{14} - \frac{1}{2}$$

$$f_\rho[\rho] = C_j \cdot j_{\ell_-}[-i \cdot E_\rho \cdot \rho] + C_y \cdot y_{\ell_-}[-i \cdot E_\rho \cdot \rho]; \quad \ell_- = -i \cdot \frac{\sqrt{35}}{14} - \frac{1}{2}, \quad (1749)$$

we obtain the following solution for f_r:

$$f_r[r] = e^{-r \cdot E} \cdot r^{L - \frac{1}{2}} \cdot \left(C_U \cdot U[nn, 2 \cdot (L+1), 2 \cdot r \cdot E] + C_L \cdot L_{-nn}^{1+2 \cdot L}[2 \cdot r \cdot E] \right)$$

$$nn = \frac{1}{2} \cdot \left(2 \cdot (L+1) + \frac{E_t^2 \pm E_\rho^2}{E} \right). \quad (1750)$$

Again we recognize the hypergeometric function $U[a, b, z]$ and the Laguerre polynomials $L_{-nn}^{1+2 \cdot L}$ in the solutions above, we know from the Schrödinger hydrogen solution (e.g., [15]).

In the interesting case of $L = 1/2$ (1750) simplifies to:

$$f_r[r] = e^{-r \cdot E} \cdot \left(C_U \cdot U[nn, 3, 2 \cdot r \cdot E] + C_L \cdot L_{-nn}^2[2 \cdot r \cdot E] \right)$$

$$nn = \frac{1}{2} \cdot \left(3 + \frac{E_t^2 \pm E_\rho^2}{E} \right). \quad (1751)$$

11.5.5 Connection to the Classical Picture

Taking the solutions from (1737) to (1739), Eq. (1740) is clearly giving us the connection to the classical picture with potential, mass and so on. Thereby the angular part is completely sucked into the L term, which—as shown above—vanishes in the setting $L = 1/2$.

In this case, we can extract the connection to the classical picture with potential $V[r]$, operator, energy E and mass terms M from (1741) as follows:

$$0 = \left(\overbrace{\frac{\partial}{r \cdot \partial r}}^{?} + \overbrace{\frac{A_3^2 + A_4^2}{r}}^{V[r]} + \overbrace{A_5^2 + A_6^2}^{M} + \overbrace{\frac{2 \cdot \partial}{r \cdot \partial r} + \frac{\partial^2}{\partial r^2}}^{\Delta_{r3D\text{-sphere}}} - \overbrace{E^2}^{energy} \right) f_r[r]. \quad (1752)$$

Similarly, we can allocate $V[r]$, M, and energy to the terms in (1748) as follows:

$$0 = \left(\overbrace{\frac{\partial}{r \cdot \partial r}}^{?} + \overbrace{\frac{3}{4 \cdot r^2} - \frac{L \cdot (L+1)}{r^2}}^{\text{momentum}} + \overbrace{\frac{E_\rho^2}{r} - \frac{E_t^2}{r}}^{V[r]} + \overbrace{\frac{2 \cdot \partial}{r \cdot \partial r} + \frac{\partial^2}{\partial r^2}}^{\Delta_{r\text{3D-sphere}}} - \overbrace{E^2}^{\text{energy}} \right) f_r. \tag{1753}$$

This time we miss the masses, but as we know that we could always add those via additional dimensions (c.f. previous chapters, especially chapter 6) like:

$$g_{\alpha\beta}^{10} = \begin{pmatrix} -c^2 & 0 & 0 & 0 & \cdots & 0 \\ 0 & 1 & 0 & 0 & \cdots & 0 \\ 0 & 0 & r^2 & 0 & \cdots & 0 \\ 0 & 0 & 0 & r^2 \cdot \sin^2 \varphi_1 & \cdots & 0 \\ \cdots & \cdots & \cdots & \cdots & \ddots & 0 \\ 0 & 0 & 0 & 0 & 0 & g_{99} \end{pmatrix}$$

$$\times F\left[f[t,r,\vartheta,\varphi,\tau,\rho,\theta,\phi] \right]; \quad F[f] = f^{\frac{1}{2}}$$

$$f[t,r,\vartheta,\varphi,\tau,\rho,\theta,\phi,x,y] = \begin{pmatrix} f_t[t] \cdot f_r[r] \cdot f_\vartheta[\vartheta] \cdot f_\varphi[\varphi] \cdot f_\tau[\tau] \\ \cdot f_\rho[\rho] \cdot f_\theta[\theta] \cdot f_\phi[\phi] \cdot f_x[x] \cdot f_y[y] \end{pmatrix};$$

$$g_{44} = -c_4^2 \cdot r; \quad g_{55} = r; \quad g_{66} = \rho^2; \quad g_{77} = \rho^2 \cdot \sin[\theta]^2; \quad g_{88} = 1; \quad g_{99} = 1$$

$$f_x[x] = C_{-x} \cdot e^{-A_x \cdot x} + C_{+x} \cdot e^{+A_x \cdot x};$$

$$f_y[y] = C_{-y} \cdot e^{-A_y \cdot y} + C_{+y} \cdot e^{+A_y \cdot y}$$

$$f_{\varphi_i}\left[g_{\varphi_i} \right] = C_{-i} \cdot e^{-A_i \cdot g_{\varphi_i}} + C_{+i} \cdot e^{+A_i \cdot g_{\varphi_i}} \tag{1754}$$

resulting in the r dependency for the corresponding solutions to (1748), respectively to (1753) as follows:

$$0 = \left(\overbrace{\frac{\partial}{r \cdot \partial r}}^{?} + \overbrace{\frac{3}{4 \cdot r^2} - \frac{L \cdot (L+1)}{r^2}}^{\text{momentum}} + \overbrace{A_x^2 + A_y^2}^{M} + \overbrace{\frac{E_\rho^2}{r} - \frac{E_t^2}{r}}^{V[r]} + \overbrace{\frac{2 \cdot \partial}{r \cdot \partial r} + \frac{\partial^2}{\partial r^2}}^{\Delta_{r\text{3D-sphere}}} - \overbrace{E^2}^{\text{energy}} \right) f_r, \tag{1755}$$

we see no reason why this should bother us.

11.5.6 Seeds for Particles

Evaluation of the Ricci scalar R of the metric (1736) from (1643) via the setting $F = 1$ gives:

$$R = -\frac{3}{2 \cdot r^2}.$$

(1756)

Thus, we have a metric with a r^{-2} asymptotically vanishing curvature, which apparently acts as a seed for spin 1/2 particles or at least particle-like objects when adding (and plying with) a scalar factor $F[f]$ as done in (1736). For $r = 0$, this curvature would become infinite and the scalar factor F prevents this singularity to occur as it simply makes R^* (the Ricci scalar of the scaled metric) to zero.

11.6 The Simpler Route

As discussed in sub-section 11.3.2 "Can We Now Understand Quantum Theory in a Truly Illustrative Manner? Part I," we need to find out the reason why—all of a sudden—it just suffices to demand something as simple as the flatness of space via $R^* = 0$, which is to say, without any/obvious/minimum principle put into action (quite literally) to obtain principal natural laws. In this section, we will work out that things are in fact a little bit more complicated and that the apparent result of $(R =)R^* = 0$ containing it all, indeed is only an apparent, respectively an accidental one.

Once again, our starting point shall be the Einstein–Hilbert action (1646). Performing the Hilbert variation with respect to the scale adapted metric tensor $G_{\alpha\beta}$ and applying the chain rule, gives us the following task:

$$\delta W = 0 = \delta_{G_{\alpha\beta}} \int_V d^n x \left(\sqrt{-g} \cdot F^n \times \left(\left(\begin{array}{c} R + \dfrac{F'}{F}(1-n)\Delta f \\[2mm] + \dfrac{f_{,\alpha} f_{,\beta} g^{\alpha\beta}(1-n)}{4F^2}\left(4FF'' + (F')^2(n-6)\right) \\[2mm] -2\Lambda + L_M \end{array} \right)^{\frac{1}{F}} \right) \right).$$

(1757)

$$0 = \int_V d^n x \left(\sqrt{-G} \times \left(R^{*\kappa\lambda} - \frac{1}{2} R^* G^{\kappa\lambda} + \kappa \cdot T^{\kappa\lambda} + \Lambda \cdot G^{\kappa\lambda} \right) \delta G_{\kappa\lambda} \right)$$

$$= \int_V d^n x \left(\sqrt{-g \cdot F^n} \times \underbrace{\left(R^*_{\alpha\beta} - \frac{1}{2} R^* G_{\alpha\beta} + \kappa \cdot T_{\alpha\beta} + \Lambda \cdot G_{\alpha\beta} \right) \delta G^{\alpha\beta}}_{\text{INT}} \right). \quad (1758)$$

For the moment, just concentrating on the integrand INT, we may simplify the last line above as follows:

$$\text{INT} = \left(- \left(\begin{array}{c} \overbrace{\Gamma^\sigma_{\alpha\beta,\sigma} - \Gamma^\sigma_{\beta\sigma,\alpha} - \Gamma^\mu_{\sigma\alpha}\Gamma^\sigma_{\beta\mu} + \Gamma^\sigma_{\alpha\beta}\Gamma^\mu_{\sigma\mu} + R^{**}_{\alpha\beta}}^{=R_{\alpha\beta}} \\[4mm] \left(\begin{array}{c} \overbrace{\Gamma^\sigma_{\kappa\lambda,\sigma}g^{\kappa\lambda} - \Gamma^\sigma_{\lambda\sigma,\kappa}g^{\kappa\lambda} - \Gamma^\mu_{\sigma\kappa}\Gamma^\sigma_{\lambda\mu}g^{\kappa\lambda} + \Gamma^\sigma_{\kappa\lambda}\Gamma^\mu_{\sigma\mu}g^{\kappa\lambda}}^{=R} \\[3mm] + \dfrac{F'^a}{F}(1-n)\Delta f_a + \dfrac{f_{a,\mu}f_{b,\nu}g^{\mu\nu}(1-n)}{4F^2}\left(4FF''^{ab} + F'^a F'^b (n-6)\right) \end{array} \right) \dfrac{G_{\alpha\beta}}{2F} \\[6mm] + \left(\left(\begin{array}{c} \underbrace{8\pi G T_{\alpha\beta}}_{\kappa} = \text{matter} \\[2mm] 0 = \text{vacuum} \end{array} \right) + \Lambda \cdot G_{\alpha\beta} \right) \end{array} \right) \right).$$

$$(1759)$$

$$\Rightarrow$$

$$\text{INT} = \left(\begin{array}{c} \left(\begin{array}{c} R_{\alpha\beta} + \dfrac{F'^a}{F}(1-n)\left(f_{a,\alpha\beta} - \Gamma^\gamma_{\alpha\beta}f_{a,\gamma}\right) \\[3mm] + \dfrac{f_{a,\alpha}f_{b,\beta}(1-n)}{4F^2}\left(4FF''^{ab} + F'^a F'^b (n-6)\right) \end{array} \right) \\[8mm] - \left(R + \dfrac{F'^a}{F}(1-n)\Delta f_a + \dfrac{f_{a,\mu}f_{b,\nu}g^{\mu\nu}(1-n)}{4F^2}\left(4FF''^{ab} + F'^a F'^b (n-6)\right) \right)\dfrac{G_{\alpha\beta}}{2 \cdot F} \\[6mm] + \left(\left(\begin{array}{c} \underbrace{8\pi G T_{\alpha\beta}}_{\kappa} = \text{matter} \\[2mm] 0 = \text{vacuum} \end{array} \right) + \Lambda \cdot G_{\alpha\beta} \right) \end{array} \right)$$

$$= \left(\begin{array}{c} R_{\alpha\beta} - R \cdot \dfrac{G_{\alpha\beta}}{2 \cdot F} + \dfrac{F'^a}{F}(1-n)\left(f_{a,\alpha\beta} - \Gamma^\gamma_{\alpha\beta}f_{a,\gamma} - \dfrac{G_{\alpha\beta}}{2 \cdot F}\Delta f_a\right) \\[4mm] + \dfrac{(1-n)}{4F^2}\left(4FF''^{ab} + F'^a F'^b (n-6)\right)\left(f_{a,\alpha}f_{b,\beta} - f_{a,\mu}f_{b,\nu}g^{\mu\nu}\dfrac{G_{\alpha\beta}}{2 \cdot F}\right) \\[6mm] + \left(\left(\begin{array}{c} \underbrace{8\pi G T_{\alpha\beta}}_{\kappa} = \text{matter} \\[2mm] 0 = \text{vacuum} \end{array} \right) + \Lambda \cdot G_{\alpha\beta} \right) \end{array} \right)$$

$$
\begin{pmatrix}
R_{\alpha\beta} - R \cdot \dfrac{G_{\alpha\beta}}{2 \cdot F} + \dfrac{F'^{a}}{F}(1-n)\left(f_{a;\alpha\beta} - \dfrac{g_{\alpha\beta}}{2}\Delta f_a\right) \\[2mm]
= \begin{array}{l} + \dfrac{(1-n)}{4F^2}\left(4FF''^{ab} + F'^{a}F'^{b}(n-6)\right)\left(f_{a,\alpha}f_{b,\beta} - f_{a,\mu}f_{b,\nu}g^{\mu\nu}\dfrac{g_{\alpha\beta}}{2}\right) \end{array} \\[2mm]
+ \left(\begin{pmatrix}\underbrace{\dfrac{8\pi G T_{\alpha\beta}}{\kappa}}_{} = \text{matter} \\[1mm] 0 = \text{vacuum}\end{pmatrix} + \Lambda \cdot G_{\alpha\beta}\right)
\end{pmatrix}. \quad (1760)
$$

Only for completeness, we thereby also considered the option of f being a vector of arbitrary functions as it was of need in deriving the classical Dirac equation (c.f. subsection "11.4.2 The Simplest Way to Derive a Metric Dirac Equation from the Scaled Ricci Scalar R^{*}" in this chapter and [32]). Inserting this result into (1758), gives us the variation:

$$
\delta W = 0 = \int_V d^n x \sqrt{-g \cdot F^n}
\begin{pmatrix}
R_{\alpha\beta} + R_{\alpha\beta}^{**} - R \cdot \dfrac{g_{\alpha\beta}}{2} + \dfrac{F'^{a}}{F}(1-n)\dfrac{g_{\alpha\beta}}{2}\Delta f_a \\[2mm]
- \dfrac{(1-n)}{4F^2}\left(\begin{array}{l}4FF''^{ab}\\ +F'^{a}F'^{b}(n-6)\end{array}\right)f_{a,\mu}f_{b,\nu}g^{\mu\nu}\dfrac{g_{\alpha\beta}}{2} \\[2mm]
+ \kappa T_{\alpha\beta} + \Lambda \cdot F \cdot g_{\alpha\beta}
\end{pmatrix} \delta G^{\alpha\beta}.
$$

$$(1761)$$

In the case $F[f_a] = F[f] \sim f$, we obtain:

$$
\delta W = 0 = \int_V d^n x \sqrt{-g \cdot F^n}
\begin{pmatrix}
R_{\alpha\beta} + R_{\alpha\beta}^{**} - R \cdot \dfrac{g_{\alpha\beta}}{2} + \dfrac{F'^{a}}{F}(1-n)\dfrac{g_{\alpha\beta}}{2}\Delta f_a \\[2mm]
- \dfrac{(1-n)}{4F^2}\left(\begin{array}{l}4FF''^{ab}\\ +F'^{a}F'^{b}(n-6)\end{array}\right)f_{a,\mu}f_{b,\nu}g^{\mu\nu}\dfrac{g_{\alpha\beta}}{2} \\[2mm]
+ \kappa T_{\alpha\beta} + \Lambda \cdot F \cdot g_{\alpha\beta}
\end{pmatrix} \delta G^{\alpha\beta}.
$$

$$(1762)$$

$$
0 =
\begin{pmatrix}
R_{\alpha\beta} + R_{\alpha\beta}^{**} - R \cdot \dfrac{g_{\alpha\beta}}{2} + \dfrac{F'^{a}}{F}(1-n)\dfrac{g_{\alpha\beta}}{2}\Delta f_a \\[2mm]
- \dfrac{(1-n)}{4F^2}\left(\begin{array}{l}4FF''^{ab}\\ +F'^{a}F'^{b}(n-6)\end{array}\right)f_{a,\mu}f_{b,\nu}g^{\mu\nu}\dfrac{g_{\alpha\beta}}{2} \\[2mm]
+ \kappa T_{\alpha\beta} + \Lambda \cdot F \cdot g_{\alpha\beta}
\end{pmatrix}. \quad (1763)
$$

The latter can be considered quantum Einstein field equations in the special case of $F[f]\sim f$. For illustration only we here consider it via options, like the following:

$$0 = f^2\overbrace{\left(R_{\alpha\beta} - R\cdot\frac{g_{\alpha\beta}}{2}\right)}^{=0} + f^2 R_{\alpha\beta}^{**} - (1-n)\left(f\frac{g_{\alpha\beta}}{2}\Delta f + \frac{n-6}{4}f_{a,\mu}f_{b,\nu}g^{\mu\nu}\frac{g_{\alpha\beta}}{2}\right)$$

$$n=6:\quad \frac{f}{1-n}\cdot R_{\alpha\beta}^{**} = \frac{g_{\alpha\beta}}{2}\Delta f \xrightarrow{\Delta f = M^2 f} \frac{f}{1-n}\cdot R_{\alpha\beta}^{**} = M^2\frac{g_{\alpha\beta}}{2}f$$

trial: $$\Rightarrow R_{\alpha\beta}^{**} = \frac{g_{\alpha\beta}}{2}, \tag{1764}$$

or potential combinations of f and the metrics:

$$0 = \left\{ \begin{array}{l} f^{a2}\overbrace{\left(R_{a\alpha\beta} - R_a\cdot\frac{g_{a\alpha\beta}}{2}\right)}^{GE_{a\alpha\beta}} + f^{a2}R_{\alpha\beta}^{**} \\[3mm] -(1-n)\left(f^a\frac{g_{a\alpha\beta}}{2}\Delta f^a + \frac{n-6}{4}f_{a,\mu}f_{b,\nu}g^{\mu\nu}\frac{g_{\alpha\beta}}{2}\right) \end{array} \right\}$$

$$n=6, GE_{a\alpha\beta}=0:\quad \frac{f^{a2}R_{\alpha\beta}^{**}}{(1-n)} = \frac{g_{a\alpha\beta}}{2}\Delta f^a \xrightarrow{\Delta f^a = M^2} \frac{f^a R_{\alpha\beta}^{**}}{(1-n)} = \frac{g_{a\alpha\beta}}{2}M^2 f^a$$

trial: $$\Rightarrow R_{\alpha\beta}^{**} = \frac{g_{a\alpha\beta}}{2}, \tag{1765}$$

where in the latter case we once again assumed the existence of various scaled metrices and marked those with the Latin index "a" (no summation convention, but only a coding for various scaled metrics and the corresponding R, F, and so on). Thereby we should remember that under the variational integral we would find the following:

$$\delta W = 0$$

$$= \int_V d^n x\ \sqrt{-g_a}\cdot F^{an}\left(\begin{array}{l} \left(\begin{array}{l} R_{a\alpha\beta} + R_{a\alpha\beta}^{***} - R_a\cdot\dfrac{g_{a\alpha\beta}}{2} \\[2mm] +\dfrac{F^{a\prime}}{F^a}(1-n)\left(f^a_{,\alpha\beta} - \Gamma^\gamma_{\alpha\beta}f^a_{,\gamma} - \dfrac{g_{a\alpha\beta}}{2}\Delta f^a\right) \\[3mm] +\dfrac{(1-n)}{4F^{a2}}\left(\begin{array}{l} 4FF''^{ab} \\ +F'^a F'^b(n-6) \end{array} \right)\left(\begin{array}{l} f_{a,\alpha}f_{b,\beta} \\ -f_{a,\mu}f_{b,\nu}g^{\mu\nu}\dfrac{g_{\alpha\beta}}{2} \end{array} \right) \\[4mm] +\kappa T_{a\alpha\beta} + \Lambda\cdot F^a\cdot g_{a\alpha\beta} \end{array} \right)\delta G_a^{\ \alpha\beta} \right).$$

$$\tag{1766}$$

In order to keep things shorter and always illustrate the special case of 6 dimensions, we will use the the following definition throughout the book:

$$
\overset{***}{R}_{a\alpha\beta} \equiv \overset{**}{R}_{a\alpha\beta} + (n-1)\left(\frac{F^{a\,\prime}}{F^a}\left(f^a_{\ ,\alpha\beta} - \Gamma^\gamma_{\alpha\beta}f^a_{\ ,\gamma} - \frac{g_{a\alpha\beta}}{2}\Delta f^a\right) + \frac{1}{4F^{a2}}\left(\begin{array}{c}4FF''^{ab}\\ +F'^a F'^b (n-6)\end{array}\right)\left(\begin{array}{c}f_{a,\alpha}f_{b,\beta}\\ -f_{a,\mu}f_{b,\nu}g^{\mu\nu}\frac{g_{\alpha\beta}}{2}\end{array}\right)\right). \qquad (1767)
$$

We know that the variation of the scaled metric $\delta G_a^{\ \alpha\beta}$ can be split up into the variation of the function F and the variation of the unscaled metric and subsequently delivers the following:

$$\delta W = 0 =$$

$$
\int_V d^n x \left(\begin{array}{c}
\sqrt{-g_a \cdot F^{an}}\left(\begin{array}{c}
R_{a\alpha\beta} + \overset{***}{R}_{a\alpha\beta} - R_a \cdot \dfrac{g_{a\alpha\beta}}{2}\\[2mm]
+\dfrac{F^{a\,\prime}}{F^a}(1-n)\left(f^a_{\ ,\alpha\beta} - \Gamma^\gamma_{\alpha\beta}f^a_{\ ,\gamma} - \dfrac{g_{a\alpha\beta}}{2}\Delta f^a\right)\\[2mm]
+\dfrac{(1-n)}{4F^{a2}}\left(\begin{array}{c}4FF''^{ab}\\ +F'^a F'^b (n-6)\end{array}\right)\left(\begin{array}{c}f_{a,\alpha}f_{b,\beta}\\ -f_{a,\mu}f_{b,\nu}g^{\mu\nu}\dfrac{g_{\alpha\beta}}{2}\end{array}\right)\\[2mm]
+\kappa T_{a\alpha\beta} + \Lambda \cdot F^a \cdot g_{a\alpha\beta}
\end{array}\right)g_a^{\ \alpha\beta}\delta\dfrac{1}{F^a}\\[20mm]
+\sqrt{-g_a \cdot F^{an}}\left(\begin{array}{c}
R_{a\alpha\beta} + \overset{***}{R}_{a\alpha\beta} - R_a \cdot \dfrac{g_{a\alpha\beta}}{2}\\[2mm]
+\dfrac{F^{a\,\prime}}{F^a}(1-n)\left(f^a_{\ ,\alpha\beta} - \Gamma^\gamma_{\alpha\beta}f^a_{\ ,\gamma} - \dfrac{g_{a\alpha\beta}}{2}\Delta f^a\right)\\[2mm]
+\dfrac{(1-n)}{4F^{a2}}\left(\begin{array}{c}4FF''^{ab}\\ +F'^a F'^b (n-6)\end{array}\right)\left(\begin{array}{c}f_{a,\alpha}f_{b,\beta}\\ -f_{a,\mu}f_{b,\nu}g^{\mu\nu}\dfrac{g_{\alpha\beta}}{2}\end{array}\right)\\[2mm]
+\kappa T_{a\alpha\beta} + \Lambda \cdot F^a \cdot g_{a\alpha\beta}
\end{array}\right)\dfrac{1}{F^a}\delta g_a^{\ \alpha\beta}
\end{array}\right),
$$

$$ (1768) $$

$$ n \to n_a $$

$$
\begin{aligned}
= \int_V d^{n_a}x \Bigg\{ & \sqrt{-g_a \cdot F^{an_a}} \left. \left(R_a - R_a\cdot\frac{n_a}{2} + \frac{F^{a\prime}}{F^a}(1-n_a)\left(\Delta_a f^a - \frac{n_a}{2}\Delta_a f^a\right) + \frac{f^a_{,\alpha}f^a_{,\beta}(1-n_a)}{4F^{a2}}g_a^{\alpha\beta}\left(4F^aF^{a\prime\prime}+F^{a\prime}F^{a\prime}(n_a-6)\right)\right) \right. \\
& \times\left(1-\frac{n_a}{2}\right) + \kappa T_{a\alpha\beta}g_a^{\alpha\beta} + \Lambda\cdot F^a\cdot n_a \Bigg\rvert \delta\frac{1}{F^a} \\[1em]
+ & \sqrt{-g_a\cdot(F^a)^{n_a-2}} \left(R_{a\alpha\beta}+R^{****}_{a\alpha\beta} - R_a\cdot\frac{g_{a\alpha\beta}}{2} + \frac{F^{a\prime}}{F^a}(1-n)\left(f^a_{,\alpha\beta} - \Gamma^\gamma_{\alpha\beta}f^a_{,\gamma} - \frac{g_{a\alpha\beta}}{2}\Delta f^a\right) \right. \\
& \left. \left. + \frac{(1-n_a)}{4F^{a2}}\left(\begin{array}{c}4FF^{\prime\prime ab}\\+F^{\prime a}F^{\prime b}(n_a-6)\end{array}\right)\left(\begin{array}{c}f_{a,\alpha}f_{b,\beta}\\-f_{a,\mu}f_{b,\nu}g^{\mu\nu}\dfrac{g_{\alpha\beta}}{2}\end{array}\right) + \kappa T_{a\alpha\beta} + \Lambda\cdot F^a\cdot g_{a\alpha\beta}\right) \delta g_a^{\alpha\beta} \Bigg\}
\end{aligned}
$$

$$
\begin{aligned}
= \int_V d^{n}x \Bigg\{ & \sqrt{-g_a \cdot F^{an_a}} \left. \left(\left(R_a + \frac{F^{a\prime}}{F^a}(1-n_a)\Delta_a f^a\right)\left(1-\frac{n_a}{2}\right) + \frac{f^a_{,\alpha}f^a_{,\beta}(1-n_a)}{4F^{a2}}g_a^{\alpha\beta}\left(4F^aF^{a\prime\prime}+F^{a\prime}F^{a\prime}(n_a-6)\right)\right) \right. \\
& \times\left(1-\frac{n_a}{2}\right) + \kappa T_{a\alpha\beta}g_a^{\alpha\beta} + \Lambda\cdot F^a\cdot n_a \Bigg\rvert \delta\frac{1}{F^a} \\[1em]
+ & \sqrt{-g_a\cdot(F^a)^{n_a-2}} \left(R_{a\alpha\beta}+R^{****}_{a\alpha\beta} - R_a\cdot\frac{g_{a\alpha\beta}}{2} + \frac{F^{a\prime}}{F^a}(1-n_a)\left(f^a_{,\alpha\beta} - \Gamma^\gamma_{\alpha\beta}f^a_{,\gamma} - \frac{g_{a\alpha\beta}}{2}\Delta f^a\right) \right. \\
& \left. \left. + \frac{(1-n_a)}{4F^{a2}}\left(\begin{array}{c}4FF^{\prime\prime ab}\\+F^{\prime a}F^{\prime b}(n_a-6)\end{array}\right)\left(\begin{array}{c}f_{a,\alpha}f_{b,\beta}\\-f_{a,\mu}f_{b,\nu}g^{\mu\nu}\dfrac{g_{\alpha\beta}}{2}\end{array}\right) + \kappa T_{a\alpha\beta} + \Lambda\cdot F^a\cdot g_{a\alpha\beta}\right) \delta g_a^{\alpha\beta} \Bigg\}. \quad (1769)
\end{aligned}
$$

The total variation now takes the form:

$$0 = \int_V d^n x \left(\sqrt{-g_a \cdot F^{a n_a}} \left(\begin{array}{c} \left(R_a + \dfrac{F^{a\,\prime}}{F^a}(1-n_a)\Delta_a f^a \right)\left(1 - \dfrac{n_a}{2}\right) \\[2mm] + \dfrac{f^a{}_{,\alpha} f^a{}_{,\beta}(1-n_a)}{4F^{a2}} g_a{}^{\alpha\beta}\left(4F^a F^{a\,\prime\prime} + F^{a\,\prime}F^{a\,\prime}(n_a - 6)\right) \\[2mm] \times \left(1 - \dfrac{n_a}{2}\right) + \kappa T_{a\alpha\beta} g_a{}^{\alpha\beta} + \Lambda \cdot F^a \cdot n_a \end{array} \right) \delta \dfrac{1}{F^a} \right.$$

$$\left. + \sqrt{-g_a \cdot (F^a)^{n_a - 2}} \left(\begin{array}{c} R_{a\alpha\beta} + R^{***}_{a\alpha\beta} - R_a \cdot \dfrac{g_{a\alpha\beta}}{2} \\[2mm] + \dfrac{F^{a\,\prime}}{F^a}(1-n_a)\left(f^a{}_{,\alpha\beta} - \Gamma^{\gamma}_{\alpha\beta} f^a{}_{,\gamma} - \dfrac{g_{a\alpha\beta}}{2}\Delta f^a \right) \\[2mm] + \dfrac{(1-n_a)}{4F^{a2}}\left(\begin{array}{c} 4FF''^{ab} \\ +F'^a F'^b (n-6) \end{array} \right)\left(\begin{array}{c} f_{a,\alpha} f_{b,\beta} \\ -f_{a,\mu} f_{b,\nu} g^{\mu\nu} \dfrac{g_{\alpha\beta}}{2} \end{array} \right) \\[2mm] + \kappa T_{a\alpha\beta} + \Lambda \cdot F^a \cdot g_{a\alpha\beta} \end{array} \right) \delta g_a{}^{\alpha\beta} \right) .$$

$$(1770)$$

Subsequently, demanding the vanishing of the whole integral via the integrand being zero, we obtain two equations, namely:

1. $$0 = \left(\begin{array}{c} \left(R_a + \dfrac{F^{a\,\prime}}{F^a}(1-n_a)\Delta_a f^a \right)\left(1 - \dfrac{n_a}{2}\right) \\[2mm] + \dfrac{f^a{}_{,\alpha} f^a{}_{,\beta}(1-n_a)}{4F^{a2}} g_a{}^{\alpha\beta}\left(4F^a F^{a\,\prime\prime} + F^{a\,\prime}F^{a\,\prime}(n_a - 6)\right)\left(1 - \dfrac{n_a}{2}\right) \\[2mm] + \kappa T_{a\alpha\beta} g_a{}^{\alpha\beta} + \Lambda \cdot F^a \cdot n_a \end{array} \right)$$

2. $$0 = \left(\begin{array}{c} R_{a\alpha\beta} + R^{***}_{a\alpha\beta} - R_a \cdot \dfrac{g_{a\alpha\beta}}{2} \\[2mm] + \dfrac{F^{a\,\prime}}{F^a}(1-n_a)\left(f^a{}_{,\alpha\beta} - \Gamma^{\gamma}_{\alpha\beta} f^a{}_{,\gamma} - \dfrac{g_{a\alpha\beta}}{2}\Delta f^a \right) \\[2mm] + \dfrac{(1-n_a)}{4F^{a2}}\left(\begin{array}{c} 4FF''^{ab} \\ +F'^a F'^b (n-6) \end{array} \right)\left(\begin{array}{c} f_{a,\alpha} f_{b,\beta} \\ -f_{a,\mu} f_{b,\nu} g^{\mu\nu} \dfrac{g_{\alpha\beta}}{2} \end{array} \right) \\[2mm] + \kappa T_{a\alpha\beta} + \Lambda \cdot F^a \cdot g_{a\alpha\beta} \end{array} \right) .$$

$$(1771)$$

One of the equations (the second one) "**may**" be seen as set of the quantum Einstein field equations, which determines the metric and the other would fix the wave functions F and f, whereby only one of the two functions (f or F) needs to be determined, while the other can be chosen in an arbitrary manner (probably so that one obtains the simplest equations).

Differences to the equations we published earlier in [8] and represented in chapter 8 of this book are caused by the fact that we have—per definition— different boundary conditions. While in [8] and chapter 8 our approach started with the variation with respect to the unscaled metric and it was assumed that this variation gives zero for complete boundary terms, we here have the same boundary condition for the scaled metric. Further below we will take this fact into account.

Now we see that the condition $R^* = 0$, we discussed above, which allows us the derivation of all essential scalar quantum equations, including the Dirac equation, is identical with the first equation of (1771) and, thus, also just is a—scalar—result of the variation of the Einstein–Hilbert action. In other words, it also comes from the Hamilton minimum principle.

However, it "disturbs" us that the two equations are not truly independent ones, but that the first one could be obtained out of the second one by multiplication with the covariant metric tensor and forming an inner product or contraction. One might argue that this situation is not different from what we also find with the Einstein-Field equations, where summation with the metric tensor also provides a peculiar simplification, namely (here the vacuum case):

$$0 = \left(R_{\alpha\beta} - R \cdot \frac{g_{\alpha\beta}}{2} \right) g^{\alpha\beta} = R - R \cdot \frac{n}{2} = R \cdot \left(1 - \frac{n}{2} \right), \qquad (1772)$$

(here the non-vacuum case):

$$0 = \left(R_{\alpha\beta} - R \cdot \frac{g_{\alpha\beta}}{2} + \kappa T_{\alpha\beta} + \Lambda \cdot g_{\alpha\beta} \right) g^{\alpha\beta} = R \cdot \left(1 - \frac{n}{2} \right) + \kappa T_{\alpha\beta} g^{\alpha\beta} + \Lambda \cdot n.$$

$$(1773)$$

While we have no problems in the situation $n = 2$ also in the classical case, we see that there can be no matter-free solution to the problem for general n.

We will easily see what this means for our case (1771) when—for a start—investigating the vacuum case.

11.6.1   Checking Ricci-Flat and Vacuum Conditions for (1771)

Using the assumption of a matter-free, Ricci-flat space-time, we may simplify (1771) as follows:

$$\xrightarrow{R_{a\alpha\beta}=R_a=T_{a\alpha\beta}g_a{}^{\alpha\beta}=\Lambda=0}$$

1. $0=\left(\dfrac{F^{a\prime}}{F^a}(1-n_a)\Delta_a f^a+\dfrac{f^a{}_{,\alpha}f^a{}_{,\beta}(1-n_a)}{4F^{a2}}g_a{}^{\alpha\beta}\left(4F^a F^{a\prime\prime}+F^{a\prime}F^{a\prime}(n_a-6)\right)\right)\left(1-\dfrac{n_a}{2}\right)$

2. $0=\begin{pmatrix}\left(\overset{***}{R_{a\alpha\beta}}+\dfrac{F^{a\prime}}{F^a}(1-n_a)\left(f^a{}_{,\alpha\beta}-\Gamma^\gamma_{\alpha\beta}f^a{}_{,\gamma}-\dfrac{g_{a\alpha\beta}}{2}\Delta f^a\right)\right)\\[2em]+\dfrac{(1-n_a)}{4F^{a2}}\begin{pmatrix}4FF^{\prime\prime ab}\\+F^{\prime a}F^{\prime b}(n-6)\end{pmatrix}\begin{pmatrix}f_{a,\alpha}f_{b,\beta}\\-f_{a,\mu}f_{b,\nu}g^{\mu\nu}\dfrac{g_{\alpha\beta}}{2}\end{pmatrix}\end{pmatrix}$

$$\xrightarrow{n_a\neq 1,2}$$

1. $0=F^{a\prime}\Delta_a f^a+\dfrac{f^a{}_{,\alpha}f^a{}_{,\beta}}{4F^a}g_a{}^{\alpha\beta}\left(4F^a F^{a\prime\prime}+F^{a\prime}F^{a\prime}(n_a-6)\right)$

2. $0=\begin{pmatrix}F^a\dfrac{\overset{***}{R_{a\alpha\beta}}}{1-n_a}+F^{a\prime}\left(f^a{}_{,\alpha\beta}-\Gamma^\gamma_{\alpha\beta}f^a{}_{,\gamma}-\dfrac{g_{a\alpha\beta}}{2}\Delta f^a\right)\\[2em]+\dfrac{1}{4F^a}\begin{pmatrix}4FF^{\prime\prime ab}\\+F^{\prime a}F^{\prime b}(n-6)\end{pmatrix}\begin{pmatrix}f_{a,\alpha}f_{b,\beta}\\-f_{a,\mu}f_{b,\nu}g^{\mu\nu}\dfrac{g_{\alpha\beta}}{2}\end{pmatrix}\end{pmatrix}.$ (1774)

Knowing that we could always demand (see appendix A for the case $n=2$):

$$4F^a F^{a\prime\prime}+F^{a\prime}F^{a\prime}(n_a-6)=0 \quad\Rightarrow\quad F^a=\left(C_f\pm f^a\right)^{\frac{4}{n-2}};\quad n>2,\quad (1775)$$

we would obtain:

1. $0=F^{a\prime}\Delta_a f^a$

2. $0=F^a\dfrac{\overset{***}{R_{a\alpha\beta}}}{1-n_a}+F^{a\prime}\left(f^a{}_{,\alpha\beta}-\Gamma^\gamma_{\alpha\beta}f^a{}_{,\gamma}-\dfrac{g_{a\alpha\beta}}{2}\Delta f^a\right).$ (1776)

Forming the following sum with the second equation above:

$$0=F^a\dfrac{\overset{***}{R_{a\alpha\beta}}}{1-n_a}g_a{}^{\alpha\beta}+F^{a\prime}\left(f^a{}_{,\alpha\beta}-\Gamma^\gamma_{\alpha\beta}f^a{}_{,\gamma}-\dfrac{g_{a\alpha\beta}}{2}\Delta f^a\right)g_a{}^{\alpha\beta}$$

$$=F^a\dfrac{\overset{***}{R_{a\alpha\beta}}}{1-n_a}g_a{}^{\alpha\beta}+F^{a\prime}\Delta f^a\cdot\left(1-\dfrac{n}{2}\right) \qquad (1777)$$

reproduces the first equation of (1776) only in the case of $F^a \dfrac{\overset{***}{R}_{a\alpha\beta}}{1-n_a} = 0.$

11.6.2 Checking Non-Ricci-Flat Conditions for (1771)

We may repeat this with the non-Ricci flat space-times and obtain:

1. $$0 = \left(\begin{array}{c} \underbrace{\left(R_a + \dfrac{F^{a\,\prime}}{F^a}(1-n_a)\Delta_a f^a \right)\left(1 - \dfrac{n_a}{2} \right)}_{=0} \\[2mm] + \dfrac{(1-n_a)}{4F^{a2}} g_a{}^{\alpha\beta} \overbrace{\left(4F^a F^{a\,\prime\prime} + F^{a\,\prime}F^{a\,\prime}(n_a - 6) \right)}(...) \\[2mm] \underbrace{+\kappa T_{a\alpha\beta}g_a{}^{\alpha\beta} + \Lambda \cdot F^a \cdot n_a}_{=0} \end{array} \right)$$

$$(...) \equiv \left(f_{a,\alpha}f_{b,\beta} - f_{a,\mu}f_{b,\nu}g^{\mu\nu}\dfrac{g_{\alpha\beta}}{2} \right)$$

$\Rightarrow \qquad\qquad R_a = \dfrac{F^{a\,\prime}}{F^a}(n_a - 1)\Delta_a f^a$

$\Rightarrow \qquad$ 2. $0 = \overset{***}{R}_{a\alpha\beta} + R_{a\alpha\beta} - \dfrac{F^{a\,\prime}}{F^a}(n_a - 1)\Delta_a f^a \cdot \dfrac{g_{a\alpha\beta}}{2}$

$$+ \dfrac{F^{a\,\prime}}{F^a}(1-n_a)\left(f^a{}_{,\alpha\beta} - \Gamma^\gamma_{\alpha\beta}f^a{}_{,\gamma} - \dfrac{g_{a\alpha\beta}}{2}\Delta f^a \right)$$

$$= \overset{***}{R}_{a\alpha\beta} + R_{a\alpha\beta} + \dfrac{F^{a\,\prime}}{F^a}(1-n_a)\left(f^a{}_{,\alpha\beta} - \Gamma^\gamma_{\alpha\beta}f^a{}_{,\gamma} \right), \qquad (1778)$$

with the latter line giving a rather simple (but still general) form of the vacuum quantum Einstein field equations.

As before we multiply the second equation of (1778) with the contravariant metric tensor and obtain:

$$0 = \left(\overset{***}{R}_{a\alpha\beta} + R_{a\alpha\beta} + \dfrac{F^{a\,\prime}}{F^a}(1-n_a)\left(f^a{}_{,\alpha\beta} - \Gamma^\gamma_{\alpha\beta}f^a{}_{,\gamma} \right) \right)g_a^{\alpha\beta}$$

$$= \overset{***}{R}_{a\alpha\beta}g_a^{\alpha\beta} + R_a + \dfrac{F^{a\,\prime}}{F^a}(1-n_a)\Delta f^a, \qquad (1779)$$

which is to say, the first equation of (1778) when $\overset{***}{R}_{a\alpha\beta}g_a^{\alpha\beta} = 0$. Thus, the two equations are not truly independent ones. However, solving the equation in the last line of (1778) also solves (1779) and thus the pair in (1778).

11.6.3 Checking Non-Ricci-Flat-Non-Vacuum Conditions for (1771)

We may also repeat this with the non-Ricci-flat-non-vacuum space-times and obtain:

$$
\textbf{1.} \quad 0 = \begin{pmatrix} \left(R_a + \dfrac{F^{a\,\prime}}{F^a}(1-n_a)\Delta_a f^a\right)\left(1-\dfrac{n_a}{2}\right) \\[2mm] + \dfrac{(1-n_a)}{4F^{a2}} g_a{}^{\alpha\beta} \overbrace{\left(4F^a F^{a\,\prime\prime} + F^{a\,\prime} F^{a\,\prime}(n_a - 6)\right)}^{=0}(\ldots) \\[2mm] + \kappa T_{a\alpha\beta} g_a{}^{\alpha\beta} + \Lambda \cdot F^a \cdot n_a \end{pmatrix}
$$

$$
\Rightarrow \quad R_a = \dfrac{F^{a\,\prime}}{F^a}(n_a - 1)\Delta_a f^a + \dfrac{\kappa T_{a\alpha\beta} g_a{}^{\alpha\beta} + \Lambda \cdot F^a \cdot n_a}{\dfrac{n_a}{2} - 1}
$$

$$
\Rightarrow \textbf{2.} \quad 0 = \begin{pmatrix} R_{a\alpha\beta}^{***} + R_{a\alpha\beta} - \left(\dfrac{F^{a\,\prime}}{F^a}(n_a - 1)\Delta_a f^a + \dfrac{\kappa T_{a\alpha\beta} g_a{}^{\alpha\beta} + \Lambda \cdot F^a \cdot n_a}{\dfrac{n_a}{2} - 1}\right) \cdot \dfrac{g_{a\alpha\beta}}{2} \\[3mm] + \dfrac{F^{a\,\prime}}{F^a}(1-n_a)\left(f^a{}_{,\alpha\beta} - \Gamma^{\gamma}_{\alpha\beta} f^a{}_{,\gamma} - \dfrac{g_{a\alpha\beta}}{2}\Delta f^a\right) \\[3mm] + \kappa T_{a\alpha\beta} + \Lambda \cdot F^a \cdot g_{a\alpha\beta} \end{pmatrix},
$$

$$
(1780)
$$

with the latter line once again giving the most general form of quantum Einstein field equations.

As before we multiply the second equation of (1780) with the contravariant metric tensor and after summation we obtain:

$$
0 = \begin{pmatrix} R_{a\alpha\beta}^{***} + R_{a\alpha\beta} - \left(\dfrac{F^{a\,\prime}}{F^a}(n_a - 1)\Delta_a f^a + \dfrac{\kappa T_{a\alpha\beta} g_a{}^{\alpha\beta} + \Lambda \cdot F^a \cdot n_a}{\dfrac{n_a}{2} - 1}\right) \cdot \dfrac{g_{a\alpha\beta}}{2} \\[3mm] + \dfrac{F^{a\,\prime}}{F^a}(1-n_a)\left(f^a{}_{,\alpha\beta} - \Gamma^{\gamma}_{\alpha\beta} f^a{}_{,\gamma} - \dfrac{g_{a\alpha\beta}}{2}\Delta f^a\right) \\[3mm] + \kappa T_{a\alpha\beta} + \Lambda \cdot F^a \cdot g_{a\alpha\beta} \end{pmatrix} g_a{}^{\alpha\beta}
$$

$$
\begin{aligned}
&= R_{a\alpha\beta}^{***} g_a^{\alpha\beta} + \left(\begin{array}{c} \left(R_a - \left(\dfrac{F^{a\,\prime}}{F^a}(n_a - 1)\Delta_a f^a\, g_a^{\alpha\beta} + \dfrac{\kappa T_{a\alpha\beta}g_a^{\alpha\beta} + \Lambda \cdot F^a \cdot n_a}{\dfrac{n_a}{2} - 1} \right) \right) \cdot \dfrac{n_a}{2} \\[2ex] + \dfrac{F^{a\,\prime}}{F^a}(1 - n_a)\left(1 - \dfrac{n_a}{2}\right)\Delta_a f^a + \kappa T_{a\alpha\beta}g_a^{\alpha\beta} + \Lambda \cdot F^a \cdot n_a \end{array} \right)
\end{aligned}
$$

$$
= R_{a\alpha\beta}^{***} g_a^{\alpha\beta} + R_a + \frac{F^{a\,\prime}}{F^a}(1 - n_a)^2 \Delta_a f^a - \frac{2}{n_a - 2}\left(\kappa T_{a\alpha\beta}g_a^{\alpha\beta} + \Lambda \cdot F^a \cdot n_a \right),
$$

$$
\tag{1781}
$$

which is to say, the first equation of (1780) when $R_{a\alpha\beta}^{***} g_a^{\alpha\beta} = 0$. Then, the two equations are not truly independent ones.

11.6.4 Generalizing the Langrangian of the Einstein–Hilbert Action, but Keeping the R-Linearity

Before concluding, however, that this should have been clear to us right from the start, we remind that the Einstein–Hilbert action of a simply scaled metric as applied in (1648) has one additional degree of freedom. Knowing that the generalized Lagrangian of the Einstein–Hilbert action could be a function $\varphi(R)$ instead of just R, we obtain the following variational integrand with the help of [16]:

$$
\varphi'(R)R_{\mu\nu} - \frac{1}{2}\varphi(R)g_{\mu\nu} + \Lambda g_{\mu\nu} - \left(\partial_{;\mu}\partial_{;\nu} - g_{\mu\nu}\Delta_g\right)\varphi'(R)
$$

$$
= \begin{cases} 0 & \dots\text{``vacuum''} \\ -8\pi G T_{\mu\nu} & \dots\text{postulated matter.} \end{cases}
\tag{1782}
$$

For the reason of simplicity and brevity we avoid to consider sets of f^a functions, metric, and so on in the following.

Now we also know that non-linear R do not result in the Newton laws and thus have been ruled out. However, we could still ask for an arbitrary factor F^q, making the Lagrangian to $\varphi(R) = F[f]^{q*}R$, because for the limit of $F \to 1$ this always results in the classical action. Taking such a factor into account, however, the variation with respect to the scaled metric reads:

$$0 = \int_V d^n x \left\{ \sqrt{-g}\cdot F^n \left[\left(R + \frac{F'}{F}(1-n)\Delta f \right)\left(1-\frac{n}{2}\right) + \frac{f_{,\alpha}f_{,\beta}(1-n)}{4F^2}\, g^{\alpha\beta}\left(4FF'' + F'F'(n-6)\right) \right] \delta\frac{1}{F} \right.$$

$$\times \left(1-\frac{n}{2}\right) + \kappa T_{\alpha\beta}g^{\alpha\beta} + \Lambda\cdot F\cdot n$$

$$-\sqrt{-g}\cdot F^n\, g^{\alpha\beta}\left(\partial_{;\alpha}\partial_{;\beta} - F\cdot G_{\alpha\beta}\Delta_G\right)F^q\cdot \delta\frac{1}{F}$$

$$+ F^q \sqrt{-g}\cdot F^{n-2}\left[R^{***}_{\alpha\beta} + R_{\alpha\beta} - R\cdot\frac{g_{\alpha\beta}}{2} + \frac{F'}{F}(1-n)\left(f_{,\alpha\beta} - \Gamma^\gamma_{\alpha\beta}f_{,\gamma} - \frac{g_{\alpha\beta}}{2}\Delta f\right)\right.$$

$$\left. + \frac{(1-n)}{4F^2}\left(4FF'' + F'F'(n-6)\right)\left(f_{,\alpha}f_{,\beta} - f_{,\mu}f_{,\nu}g^{\mu\nu}\frac{g_{\alpha\beta}}{2}\right) + \kappa T_{\alpha\beta} + \Lambda\cdot F\cdot g_{\alpha\beta}\right.$$

$$\left.\left. -\sqrt{-g}\cdot F^{n-2}\left(\partial_{;\alpha}\partial_{;\beta} - F\cdot G_{\alpha\beta}\Delta_G\right)F^q\cdot \delta g^{\alpha\beta}\right]\right\}, \tag{1783}$$

$$0 = \int_V d^n x \left\{ F^q \sqrt{-g}\cdot F^n \left[\left(R + \frac{F'}{F}(1-n)\Delta f \right)\left(1-\frac{n}{2}\right) + \frac{f_{,\alpha}f_{,\beta}(1-n)}{4F^2}\, g^{\alpha\beta}\left(4FF'' + F'F'(n-6)\right) \right] \delta\frac{1}{F} \right.$$

$$\times \left(1-\frac{n}{2}\right) + \kappa T_{\alpha\beta}g^{\alpha\beta} + \Lambda\cdot F\cdot n - g^{\alpha\beta}\frac{1}{F^q}\left(\partial_{;\alpha}\partial_{;\beta} - F\cdot g_{\alpha\beta}\Delta_G\right)F^q$$

$$+ F^q \sqrt{-g}\cdot F^{n-2}\left[R^{***}_{\alpha\beta} + R_{\alpha\beta} - R\cdot\frac{g_{\alpha\beta}}{2} + \frac{F'}{F}(1-n)\left(f_{,\alpha\beta} - \Gamma^\gamma_{\alpha\beta}f_{,\gamma} - \frac{g_{\alpha\beta}}{2}\Delta f\right)\right.$$

$$\left. + \frac{(1-n)}{4F^2}\left(4FF'' + F'F'(n-6)\right)\left(f_{,\alpha}f_{,\beta} - f_{,\mu}f_{,\nu}g^{\mu\nu}\frac{g_{\alpha\beta}}{2}\right) + \kappa T_{\alpha\beta} + \Lambda\cdot F\cdot g_{\alpha\beta}\right.$$

$$\left.\left. -\frac{1}{F^q}\left(\partial_{;\alpha}\partial_{;\beta} - F\cdot g_{\alpha\beta}\Delta_G\right)F^q\right] \delta g^{\alpha\beta}\right\}. \tag{1784}$$

Thereby we have to note that the Laplace operator with respect to the scaled metric reads:

$$\Delta_G f = \frac{1}{\sqrt{g \cdot F^n}} \partial_\kappa \sqrt{g \cdot F^n} \cdot F^{-1} \cdot g^{\kappa\lambda} \partial_\lambda f = \frac{1}{\sqrt{g \cdot F^n}} \partial_\kappa \sqrt{g \cdot F^2}^{\frac{n-1}{2}} \cdot g^{\kappa\lambda} \partial_\lambda f = F^{-1} \cdot \Delta f + \left(\frac{n}{2}-1\right) F^{-2} F' g^{\kappa\lambda} f_{,\kappa} f_{,\lambda}. \qquad (1785)$$

Thus, the operator contains the scaling function $F[f]$. The same holds for the second covariant derivative $f_{;\alpha\beta}$, which we will consider later. For the time being, we simply mark it with a box $\boxed{f_{;\alpha\beta}}$.

$$
\begin{aligned}
0 = \int_V d^n x \Bigg\{ \quad & F^q \sqrt{-g \cdot F^n} \left(
\begin{aligned}
&\left(R+\frac{F'}{F}(1-n)\Delta f\right)\left(1-\frac{n}{2}\right) + \frac{f_{,\alpha} f_{,\beta}(1-n)}{4F^2} g^{\alpha\beta}\left(4FF''+F'F'(n-6)\right) \\
&\times\left(1-\frac{n}{2}\right) + \kappa T_{\alpha\beta} g^{\alpha\beta} + \Lambda\cdot F\cdot n \\
&-g^{\alpha\beta}\frac{1}{F^q}\left(
\begin{aligned}
&q\cdot F^{q-1}\cdot F'\cdot\left(\boxed{f_{;\alpha\beta}} - g_{\alpha\beta}\left(\Delta f + \left(\frac{n}{2}-1\right)\times F^{-1}F' g^{\kappa\lambda} f_{,\kappa} f_{,\lambda}\right)\right) \\
&+\left(q\cdot(q-1)\cdot F^{q-2}\cdot(F')^2 + q\cdot F^{q-1}\cdot F''\right)\times\left(f_{,\alpha} f_{,\beta} - g_{\alpha\beta} f_{,\sigma} f^{,\sigma}\right)
\end{aligned}\right)
\end{aligned}\right) \delta\frac{1}{F} \\[2ex]
+ & F^q \sqrt{-g\cdot F^{n-2}}\left(
\begin{aligned}
&R^{***}_{\alpha\beta}+R_{\alpha\beta}-R\cdot\frac{g_{\alpha\beta}}{2}+\frac{g_{\alpha\beta}}{2}\frac{F'}{F}(1-n)\left(f_{,\alpha\beta}-\Gamma^\gamma_{\alpha\beta} f_{,\gamma}-\frac{g_{\alpha\beta}}{2}\Delta f\right) \\
&+\frac{(1-n)}{4F^2}\left(4FF''+F'F'(n-6)\right)\left(f_{,\alpha} f_{,\beta}-f_{,\mu} f_{,\nu} g^{\mu\nu}\frac{g_{\alpha\beta}}{2}\right) + \kappa T_{\alpha\beta}+\Lambda\cdot F\cdot g_{\alpha\beta} \\
&-\frac{1}{F^q}\left(
\begin{aligned}
&q\cdot F^{q-1}\cdot F'\cdot\left(\boxed{f_{;\alpha\beta}}-g_{\alpha\beta}\left(\Delta f+\left(\frac{n}{2}-1\right)\times F^{-1}F' g^{\kappa\lambda} f_{,\kappa} f_{,\lambda}\right)\right) \\
&+\left(q\cdot(q-1)\cdot F^{q-2}\cdot(F')^2 + q\cdot F^{q-1}\cdot F''\right)\times\left(f_{,\alpha} f_{,\beta}-g_{\alpha\beta} f_{,\sigma} f^{,\sigma}\right)
\end{aligned}\right)
\end{aligned}\right) \delta g^{\alpha\beta}
\Bigg\}.
\end{aligned}
\qquad (1786)
$$

$$
\begin{aligned}
0 = \int_V d^n x \Bigg[\; & F^q \sqrt{-g}\cdot F^n \Bigg\{ \left(R + \frac{F'}{F}(1-n)\Delta f\right)\left(1-\frac{n}{2}\right) + \frac{f_{,\alpha}f_{,\beta}(1-n)}{4F^2}g^{\alpha\beta}\big(4FF'' + F'F'(n-6)\big) \\
& \times\left(1-\frac{n}{2}\right) + \kappa T_{\alpha\beta}g^{\alpha\beta} + \Lambda\cdot F\cdot n \\
& -\frac{1}{F^q}\Bigg[\, q\cdot F^{q-1}\cdot F'\cdot\left(g^{\alpha\beta}\boxed{f_{;\alpha\beta}} - n\left(\Delta f + \left(\frac{n}{2}-1\right)\times F^{-1}F'g^{\kappa\lambda}f_{,\kappa}f_{,\lambda}\right)\right) \\
& + \Big(q\cdot(q-1)\cdot F^{q-2}\cdot(F')^2 + q\cdot F^{q-1}\cdot F''\Big)\times\Big(g^{\alpha\beta}f_{;\alpha}f_{;\beta} - n\cdot f_{;\sigma}f^{;\sigma}\Big)\Bigg]\Bigg\}\,\delta\frac{1}{F} \\[6pt]
+\; & F^q \sqrt{-g}\cdot F^{n-2} \Bigg\{ R^{***}_{\alpha\beta} + R_{\alpha\beta} - R\cdot\frac{g_{\alpha\beta}}{2} + \frac{F'}{F}(1-n)\left(f_{,\alpha\beta} - \Gamma^{\gamma}_{\alpha\beta}f_{,\gamma} - \frac{g_{\alpha\beta}}{2}\Delta f\right) \\
& + \frac{(1-n)}{4F^2}\big(4FF'' + F'F'(n-6)\big)\left(f_{,\alpha}f_{,\beta} - f_{,\mu}f_{,\nu}g^{\mu\nu}\frac{g_{\alpha\beta}}{2}\right) + \kappa T_{\alpha\beta} + \Lambda\cdot F\cdot g_{\alpha\beta} \\
& -\frac{1}{F^q}\Bigg(q\cdot F^{q-1}\cdot F'\cdot\left(\boxed{f_{;\alpha\beta}} - g_{\alpha\beta}\left(\Delta f + \left(\frac{n}{2}-1\right)\times F^{-1}F'g^{\kappa\lambda}f_{,\kappa}f_{,\lambda}\right)\right) \\
& + \Big(q\cdot(q-1)\cdot F^{q-2}\cdot(F')^2 + q\cdot F^{q-1}\cdot F''\Big)\times\Big(f_{;\alpha}f_{;\beta} - g_{\alpha\beta}f_{;\sigma}f^{;\sigma}\Big)\Bigg)\Bigg\}\,\delta g^{\alpha\beta} \Bigg].
\end{aligned}
\tag{1787}
$$

$$
\begin{aligned}
0 = \int_V d^n x \Bigg[\; & F^q \sqrt{-g} \cdot F^n \Bigg(\left(\left(R + \frac{F'}{F}(1-n)\Delta f\right)\left(1-\frac{n}{2}\right) + \frac{f_{,\alpha}f_{,\beta}(1-n)}{4F^2}\, g^{\alpha\beta}\left(4FF'' + F'F'(n-6)\right)\right) \\[4pt]
& \times \left(1-\frac{n}{2}\right) + \kappa T_{\alpha\beta}\, g^{\alpha\beta} + \Lambda \cdot F \cdot n \\[4pt]
& - q \cdot \Bigg[F^{-1} \cdot F' \cdot \left(g^{\alpha\beta}\,\boxed{f_{,\alpha\beta}} - n\left(\Delta f + \left(\frac{n}{2}-1\right)\times F^{-1}F'\, g^{\kappa\lambda} f_{,\kappa} f_{,\lambda}\right)\right) \\[4pt]
& \qquad + \left((q-1)\cdot F^{-2}\cdot (F')^2 + F^{-1}\cdot F''\right)\times \left(g^{\alpha\beta} f_{,\alpha} f_{;\beta} - n\cdot f_{;\sigma} f^{;\sigma}\right) \Bigg] \Bigg)\, \delta\frac{1}{F} \\[10pt]
+ \; & F^q \sqrt{-g}\cdot F^{n-2} \Bigg(R^{***}_{\alpha\beta} + R_{\alpha\beta} - R\cdot\frac{g_{\alpha\beta}}{2} + \frac{F'}{F}(1-n)\left[f_{,\alpha\beta} - \Gamma^\gamma_{\alpha\beta} f_{,\gamma} - \frac{g_{\alpha\beta}}{2}\Delta f\right] \\[4pt]
& + \frac{(1-n)}{4F^2}\left(4FF'' + F'F'(n-6)\right)\left[f_{,\alpha}f_{,\beta} - f_{,\mu}f_{,\nu}g^{\mu\nu}\frac{g_{\alpha\beta}}{2}\right] \\[4pt]
& + \kappa T_{\alpha\beta} + \Lambda\cdot F\cdot g_{\alpha\beta} \\[4pt]
& - q\cdot \Bigg[F^{-1}\cdot F'\cdot \left(\boxed{f_{,\alpha\beta}} - g_{\alpha\beta}\left(\Delta f + \left(\frac{n}{2}-1\right)\times F^{-1}F'\, g^{\kappa\lambda} f_{,\kappa} f_{,\lambda}\right)\right) \\[4pt]
& \qquad + \left((q-1)\cdot F^{-2}\cdot (F')^2 + F^{-1}\cdot F''\right)\times \left(f_{;\alpha} f_{;\beta} - g_{\alpha\beta} f_{;\sigma} f^{;\sigma}\right) \Bigg] \Bigg)\, \delta g^{\alpha\beta} \Bigg]
\end{aligned}
$$

$$(1788)$$

Because f is a scalar, we can further simplify:

$$
\begin{aligned}
0 = \int_V d^n x \Bigg\{ & F^q \sqrt{-g \cdot F^n}\left[\left(R + \frac{F'}{F}(1-n)\Delta f\right)\left(1 - \frac{n}{2}\right) + \frac{f_{,\alpha}f_{,\beta}(1-n)}{4F^2}g^{\alpha\beta}\left(4FF'' + F'F'(n-6)\right)\right. \\
& \times \left(1 - \frac{n}{2}\right) + \kappa T_{\alpha\beta}g^{\alpha\beta} + \Lambda \cdot F \cdot n \\
& -q\cdot\left(F^{-1}\cdot F'\cdot\left(g^{\alpha\beta}\boxed{f_{;\alpha\beta}} - n\left(\Delta f + \left(\frac{n}{2}-1\right)\times F^{-1}F'g^{\kappa\lambda}f_{,\kappa}f_{,\lambda}\right)\right)\right. \\
& \left.\left. + \left((q-1)\cdot F^{-2}\cdot(F')^2 + F^{-1}\cdot F''\right)\times\left(g^{\alpha\beta}f_{,\alpha}f_{,\beta} - n\cdot g^{\kappa\lambda}f_{,\kappa}f_{,\lambda}\right)\right)\right]\,\delta\frac{1}{F} \\[6pt]
& + F^q\sqrt{-g}\cdot F^{n-2}\left[\overset{***}{R_{\alpha\beta}} + R_{\alpha\beta} - R\cdot\frac{g_{\alpha\beta}}{2}\right. \\
& + \frac{F'}{F}(1-n)\left(f_{,\alpha\beta} - \Gamma^\gamma_{\alpha\beta}f_{,\gamma} - \frac{g_{\alpha\beta}}{2}\Delta f\right) \\
& + \frac{(1-n)}{4F^2}\left(4FF'' + F'F'(n-6)\right)\left(f_{,\alpha}f_{,\beta} - f_{,\mu}f_{,\nu}g^{\mu\nu}\frac{g_{\alpha\beta}}{2}\right) \\
& + \kappa T_{\alpha\beta} + \Lambda \cdot F \cdot g_{\alpha\beta} \\
& -q\cdot\left(F^{-1}\cdot F'\cdot\left(\boxed{f_{;\alpha\beta}} - g_{\alpha\beta}\left(\Delta f + \left(\frac{n}{2}-1\right)\times F^{-1}F'g^{\kappa\lambda}f_{,\kappa}f_{,\lambda}\right)\right)\right. \\
& \left.\left.\left. + \left((q-1)\cdot F^{-2}\cdot(F')^2 + F^{-1}\cdot F''\right)\times\left(f_{,\alpha}f_{,\beta} - g_{\alpha\beta}g^{\kappa\lambda}f_{,\kappa}f_{,\lambda}\right)\right)\right]\,\delta g^{\alpha\beta}\right\}
\end{aligned}
\tag{1789}
$$

Now we only need to take special care about the second covariant derivative $\boxed{f_{;\alpha\beta}}$ because it still is with respect to the scaled metric and not the unscaled one. We can evaluate this term as follows (c.f. appendix A of [8]):

$$\Rightarrow$$

$$\boxed{f_{;\alpha\beta} = f_{,\alpha\beta} - \Gamma^{*\gamma}_{\alpha\beta}f_{,\gamma}} \; ; \quad \Gamma^{*\gamma}_{\alpha\beta} = \Gamma^{\gamma}_{\alpha\beta} + \frac{F'}{2\cdot F}\left(f_{,\beta}\cdot\delta^{\gamma}_{\alpha} + f_{,\alpha}\cdot\delta^{\gamma}_{\beta} - f_{,\sigma}\cdot g^{\gamma\sigma}g_{\alpha\beta}\right)$$

$$\boxed{f_{;\alpha\beta} = f_{,\alpha\beta} - \Gamma^{\gamma}_{\alpha\beta}f_{,\gamma} - \frac{F'}{2\cdot F}\left(f_{,\beta}\cdot\delta^{\gamma}_{\alpha} + f_{,\alpha}\cdot\delta^{\gamma}_{\beta} - f_{,\sigma}\cdot g^{\gamma\sigma}g_{\alpha\beta}\right)f_{,\gamma}\cdot}$$

(1790)

Inserting this into (1789) and a bit of simplification yields:

$$0 = \int_V d^n x$$

$$F^q\sqrt{-g}\cdot F^n \left\{ \begin{array}{l} \left(R + \frac{F'}{F}(1-n)\Delta f\right)\left(1-\frac{n}{2}\right) + \frac{f_{,\alpha}f_{,\beta}(1-n)}{4F^2}g^{\alpha\beta}\left(4FF'' + F'F'(n-6)\right)\times\left(1-\frac{n}{2}\right) + \kappa T_{\alpha\beta}g^{\alpha\beta} + \Lambda\cdot F\cdot n \\[2mm] -q\cdot\frac{F'}{F}\cdot\left[\underbrace{g^{\alpha\beta}f_{,\alpha\beta} - g^{\alpha\beta}\Gamma^{\gamma}_{\alpha\beta}f_{,\gamma}}_{=\Delta f} - \frac{F'}{2\cdot F}\times(2-n)g^{\alpha\beta}f_{,\alpha}\cdot f_{,\beta}\right] \\[3mm] -n\left(\Delta f + \left(\frac{n}{2}-1\right)\times F^{-1}F'g^{\kappa\lambda}f_{,\kappa}f_{,\lambda}\right) \\[2mm] +\left((q-1)\cdot F^{-2}\cdot(F')^2 + F^{-1}\cdot F''\right)\times(1-n)g^{\alpha\beta}f_{,\alpha}f_{,\beta} \end{array}\right\}\delta\frac{1}{F}$$

$$+F^q\sqrt{-g}\cdot F^{n-2}\left\{ \begin{array}{l} R^{***}_{\alpha\beta} + R_{\alpha\beta} - R\cdot\frac{g_{\alpha\beta}}{2} + \frac{F'}{F}(1-n)\left(f_{,\alpha\beta} - \Gamma^{\gamma}_{\alpha\beta}f_{,\gamma} - \frac{g_{\alpha\beta}}{2}\Delta f\right) \\[2mm] +\frac{(1-n)}{4F^2}\left(4FF'' + F'F'(n-6)\right)\left(f_{,\alpha}f_{,\beta} - f_{,\mu}f_{,\nu}g^{\mu\nu}\frac{g_{\alpha\beta}}{2}\right) + \kappa T_{\alpha\beta} + \Lambda\cdot F\cdot g_{\alpha\beta} \\[3mm] -q\cdot\frac{F'}{F}\cdot\left[f_{,\alpha\beta} - \Gamma^{\gamma}_{\alpha\beta}f_{,\gamma} - \frac{F'}{2\cdot F}\times\left(2\cdot f_{,\beta}f_{,\alpha} - f_{,\sigma}f_{,\gamma}g^{\gamma\sigma}g_{\alpha\beta}\right)\right] \\[2mm] -g_{\alpha\beta}\left(\Delta f + \left(\frac{n}{2}-1\right)\times F^{-1}F'g^{\kappa\lambda}f_{,\kappa}f_{,\lambda}\right) \\[2mm] +\left((q-1)\cdot F^{-2}\cdot(F')^2 + F^{-1}\cdot F''\right)\times\left(f_{,\alpha}f_{,\beta} - g_{\alpha\beta}g^{\kappa\lambda}f_{,\kappa}f_{,\lambda}\right) \end{array}\right\}\delta g^{\alpha\beta}$$

(1791)

Further simplification gives us:

$$0 = \int_V d^n x \left(F^q \sqrt{-g \cdot F^n} \left(\begin{array}{c} \left(R + \dfrac{F'}{F}(1-n)\Delta f\right)\left(1 - \dfrac{n}{2}\right) \\[2mm] + \dfrac{f_{,\alpha} f_{,\beta}(1-n)}{4F^2} g^{\alpha\beta}\left(4FF'' + F'F'(n-6)\right) \\[2mm] \times\left(1 - \dfrac{n}{2}\right) + \kappa T_{\alpha\beta} g^{\alpha\beta} + \Lambda \cdot F \cdot n \\[2mm] -q \cdot \dfrac{1}{F^2} \cdot \left(\begin{array}{c} FF'\Delta f(1-n) - \dfrac{F'F'}{2}(2-n)g^{\alpha\beta} f_{,\alpha} \cdot f_{,\beta} \\[2mm] -n\left(\dfrac{n}{2}-1\right)F'F' g^{\alpha\beta} f_{,\alpha} f_{,\beta} \\[2mm] + \left((q-1)\cdot (F')^2 + F \cdot F''\right)(1-n) g^{\alpha\beta} f_{,\alpha} f_{,\beta} \end{array}\right) \end{array}\right) \delta \dfrac{1}{F} \right.$$

$$\left. + F^q \sqrt{-g \cdot F^{n-2}} \left(\begin{array}{c} R^{***}_{\alpha\beta} + R_{\alpha\beta} - R \cdot \dfrac{g_{\alpha\beta}}{2} \\[2mm] + \dfrac{F'}{F}(1-n)\left(f_{,\alpha\beta} - \Gamma^\gamma_{\alpha\beta} f_{,\gamma} - \dfrac{g_{\alpha\beta}}{2}\Delta f\right) \\[2mm] + \dfrac{(1-n)}{4F^2}\left(\begin{array}{c} 4FF'' \\ +F'F'(n-6) \end{array}\right)\left(\begin{array}{c} f_{,\alpha} f_{,\beta} \\ -f_{,\mu} f_{,\nu} g^{\mu\nu} \dfrac{g_{\alpha\beta}}{2} \end{array}\right) \\[2mm] + \kappa T_{\alpha\beta} + \Lambda \cdot F \cdot g_{\alpha\beta} \\[2mm] -q \cdot \dfrac{F'}{F} \cdot \left(\begin{array}{c} \boxed{\begin{array}{c} f_{,\alpha\beta} - \Gamma^\gamma_{\alpha\beta} f_{,\gamma} - \dfrac{F'}{2 \cdot F} \\ \times\left(2\cdot f_{,\beta} f_{,\alpha} - f_{,\sigma} f_{,\gamma} g^{\gamma\sigma} g_{\alpha\beta}\right) \end{array}} \\[4mm] -g_{\alpha\beta}\left(\begin{array}{c} \Delta f + \left(\dfrac{n}{2}-1\right) \\ \times F^{-1} F' g^{\kappa\lambda} f_{,\kappa} f_{,\lambda} \end{array}\right) \end{array}\right) \\[2mm] + \left((q-1)\cdot F^{-2}\cdot(F')^2 + F^{-1}\cdot F''\right) \\[2mm] \times\left(f_{,\alpha} f_{,\beta} - g_{\alpha\beta} g^{\kappa\lambda} f_{,\kappa} f_{,\lambda}\right) \end{array}\right) \delta g^{\alpha\beta} \right)$$

$$\tag{1792}$$

Summing up yields:

$$
0 = \int_V d^n x \left(
\begin{array}{l}
F^q \sqrt{-g \cdot F^n} \left(
\begin{array}{l}
\left(\left(R\left(1 - \dfrac{n}{2}\right) + \dfrac{F'}{F}(1-n)\Delta f\left(\left(1 - \dfrac{n}{2}\right) - q\right) \right) \right) \\[2mm]
+ \dfrac{f_{,\alpha} f_{,\beta}(1-n)}{4F^2} g^{\alpha\beta}\left(4FF'' + F'F'(n-6)\right) \\[2mm]
\times\left(1 - \dfrac{n}{2}\right) + \kappa T_{\alpha\beta} g^{\alpha\beta} + \Lambda \cdot F \cdot n \\[2mm]
-q \cdot \left(\dfrac{1}{F^2} \cdot \left(\left(\dfrac{2q-4+n}{2}\cdot(F')^2 + F\cdot F'' \right) \right. \right. \\[2mm]
\left. \left. \times(1-n)g^{\alpha\beta} f_{,\alpha}\cdot f_{,\beta} \right) \right)
\end{array}
\right) \delta\dfrac{1}{F} \\[20mm]
+ F^q \sqrt{-g\cdot F^{n-2}} \left(
\begin{array}{l}
R_{\alpha\beta}^{***} + R_{\alpha\beta} - R\cdot\dfrac{g_{\alpha\beta}}{2} \\[2mm]
+ \dfrac{F'}{F}(1-n)\left(f_{,\alpha\beta} - \Gamma_{\alpha\beta}^{\gamma}f_{,\gamma} - \dfrac{g_{\alpha\beta}}{2}\Delta f\right) \\[2mm]
+ \dfrac{(1-n)}{4F^2}\left(\begin{array}{l}4FF'' \\ +F'F'(n-6)\end{array}\right)\left(\begin{array}{l}f_{,\alpha}f_{,\beta} \\ -f_{,\mu}f_{,\nu}g^{\mu\nu}\dfrac{g_{\alpha\beta}}{2}\end{array}\right) \\[2mm]
+\kappa T_{\alpha\beta} + \Lambda\cdot F\cdot g_{\alpha\beta} \\[2mm]
-q\cdot\left(
\begin{array}{l}
\dfrac{F'}{F}\cdot\left(
\begin{array}{l}
\boxed{\begin{array}{l} f_{,\alpha\beta} - \Gamma_{\alpha\beta}^{\gamma}f_{,\gamma} - \dfrac{F'}{2\cdot F} \\ \times\left(2\cdot f_{,\beta}f_{,\alpha} - f_{,\sigma}f_{,\gamma}g^{\gamma\sigma}g_{\alpha\beta}\right)\end{array}} \\[2mm]
-g_{\alpha\beta}\left(\begin{array}{l}\Delta f + \left(\dfrac{n}{2}-1\right) \\ \times F^{-1}F'g^{\kappa\lambda}f_{,\kappa}f_{,\lambda}\end{array}\right)
\end{array}
\right) \\[2mm]
+\left((q-1)\cdot F^{-2}\cdot(F')^2 + F^{-1}\cdot F''\right) \\[2mm]
\times\left(f_{,\alpha}f_{,\beta} - g_{\alpha\beta}g^{\kappa\lambda}f_{,\kappa}f_{,\lambda}\right)
\end{array}
\right)
\end{array}
\right) \delta g^{\alpha\beta}
\end{array}
\right),
$$

$$(1793)$$

$$0 = \int_V d^n x \left(F^q \sqrt{-g \cdot F^n} \left(\begin{array}{c} \left(R\left(1-\dfrac{n}{2}\right) + \dfrac{F'}{F}(1-n)\Delta f\left(\left(1-\dfrac{n}{2}\right)-q\right)\right) \\[2ex] + \dfrac{f_{,\alpha} f_{,\beta}(1-n)}{4F^2} g^{\alpha\beta} \left(\begin{array}{c} 4FF'' \cdot \left(1-\dfrac{n}{2}-q\right) \\[2ex] + F'F'\left(\begin{array}{c} (n-6)\left(1-\dfrac{n}{2}\right) \\[1ex] -q \cdot (4q-8+2n) \end{array}\right) \end{array}\right) \\[6ex] + \kappa T_{\alpha\beta} g^{\alpha\beta} + \Lambda \cdot F \cdot n \end{array}\right) \delta\dfrac{1}{F} \right.$$

$$\left. + F^q \sqrt{-g \cdot F^{n-2}} \left(\begin{array}{c} R_{\alpha\beta}^{***} + R_{\alpha\beta} - R \cdot \dfrac{g_{\alpha\beta}}{2} \\[2ex] + \dfrac{F'}{F}\left(\begin{array}{c} \left(f_{,\alpha\beta} - \Gamma_{\alpha\beta}^{\gamma} f_{,\gamma}\right)(1-n-q) \\[2ex] - \dfrac{g_{\alpha\beta}}{2}\Delta f(1-2q) \end{array}\right) \\[4ex] + \dfrac{(1-n)}{4F^2}\left(\begin{array}{c} 4FF'' \\ + F'F'(n-6) \end{array}\right)\left(\begin{array}{c} f_{,\alpha} f_{,\beta} \\ -f_{,\mu} f_{,\nu} g^{\mu\nu} \dfrac{g_{\alpha\beta}}{2} \end{array}\right) \\[4ex] + \kappa T_{\alpha\beta} + \Lambda \cdot F \cdot g_{\alpha\beta} \\[2ex] -q \cdot \left(\begin{array}{c} + g^{\kappa\lambda} f_{,\kappa} f_{,\lambda} g_{\alpha\beta}\left(\dfrac{(F')^2}{2F^2}(5-n-2q) - \dfrac{F''}{F}\right) \\[2ex] + f_{,\beta} f_{,\alpha}\left(\left((q-2)\cdot\dfrac{(F')^2}{F^2} + \dfrac{F''}{F}\right)\right) \end{array}\right) \end{array}\right) \delta g^{\alpha\beta} \right)$$

$$(1794)$$

Demanding the integrand to be zero gives us the following two equations:

1.
$$0 = \begin{pmatrix} \left(R\left(1-\dfrac{n}{2}\right) + \dfrac{F'}{F}(1-n)\Delta f\left(\left(1-\dfrac{n}{2}\right)-q\right) \right) \\ + \dfrac{f_{,\alpha}f_{,\beta}(1-n)}{4F^2}g^{\alpha\beta}\begin{pmatrix} 4FF''\cdot\left(1-\dfrac{n}{2}-q\right) \\ +(F')^2\left((n-6)\left(1-\dfrac{n}{2}\right)-q\cdot(4q-8+2n)\right) \end{pmatrix} \\ +\kappa T_{\alpha\beta}g^{\alpha\beta} + \Lambda\cdot F\cdot n \end{pmatrix}$$

2.
$$0 = \begin{pmatrix} R^{***}_{\alpha\beta} + R_{\alpha\beta} - R\cdot\dfrac{g_{\alpha\beta}}{2} \\ +\dfrac{F'}{F}\left(\left(f_{,\alpha\beta}-\Gamma^{\gamma}_{\alpha\beta}f_{,\gamma}\right)(1-n-q)-\dfrac{g_{\alpha\beta}}{2}\Delta f(1-2q)\right) \\ +\dfrac{(1-n)}{4F^2}\left(4FF''+F'F'(n-6)\right)\left(f_{,\alpha}f_{,\beta}-f_{,\mu}f_{,\nu}g^{\mu\nu}\dfrac{g_{\alpha\beta}}{2}\right) \\ +\kappa T_{\alpha\beta} + \Lambda\cdot F\cdot g_{\alpha\beta} \\ -q\cdot\begin{pmatrix} +g^{\kappa\lambda}f_{,\kappa}f_{,\lambda}g_{\alpha\beta}\left(\dfrac{(F')^2}{2F^2}(5-n-2q)-\dfrac{F''}{F}\right) \\ +f_{,\beta}f_{,\alpha}\left(\left((q-2)\cdot\dfrac{(F')^2}{F^2}+\dfrac{F''}{F}\right)\right) \end{pmatrix} \end{pmatrix}. \qquad (1795)$$

With this result at hand we can also derive the corresponding variation for the above-mentioned case of the direct variation with respect to the unscaled metric, including the assumption that the surface term is zero only with respect to the unscaled metric and not as taken above (see elaboration underneath Eq. (1771)) with respect to the scaled metric. The resulting variation should then be defined as follows (demanding the surface term to be zero also with respect to the unscaled metric):

$$\delta W = 0 = \delta_{g_{\alpha\beta}}\int_V d^n x \left(\sqrt{-g\cdot F^n}\cdot\left(\left(\begin{matrix} R + \dfrac{F'}{F}(1-n)\Delta f \\ +\dfrac{f_{,\alpha}f_{,\beta}g^{\alpha\beta}(1-n)}{4F^2}\left(4FF''+(F')^2(n-6)\right) \\ -2\Lambda + L_M \end{matrix}\right)^{\frac{1}{F}}\right)\right).$$

$$(1796)$$

In contrast to the task (1757), it results in:

$$
0 = \int_V d^n x \left(
\sqrt{-g \cdot F^n} \left(
\begin{array}{c}
\left(R\left(1 - \dfrac{n}{2}\right) + \dfrac{F'}{F}(1-n)\Delta f\left(\left(1 - \dfrac{n}{2}\right) - \dfrac{n}{2}\right)\right) \\[2ex]
+ \dfrac{f_{,\alpha} f_{,\beta}(1-n)}{4F^2} g^{\alpha\beta}
\left(
\begin{array}{c}
4FF'' \cdot \left(1 - \dfrac{n}{2} - \dfrac{n}{2}\right) \\[1ex]
+F'F'\left(
\begin{array}{c}
(n-6)\left(1 - \dfrac{n}{2}\right) \\[1ex]
-\dfrac{n}{2}\cdot\left(4\dfrac{n}{2} - 8 + 2n\right)
\end{array}
\right)
\end{array}
\right) \\[4ex]
+ \kappa T_{\alpha\beta} g^{\alpha\beta} + \Lambda \cdot F \cdot n
\end{array}
\right) \delta \dfrac{1}{F}
\right.
$$

$$
\left.
+ \sqrt{-g \cdot F^{n-2}}\left(
\begin{array}{c}
R^{***}_{\alpha\beta} + R_{\alpha\beta} - R \cdot \dfrac{g_{\alpha\beta}}{2} \\[2ex]
+ \dfrac{F'}{F}\left(
\begin{array}{c}
\left(f_{,\alpha\beta} - \Gamma^{\gamma}_{\alpha\beta} f_{,\gamma}\right)\left(1 - n - \dfrac{n}{2}\right) \\[1ex]
- \dfrac{g_{\alpha\beta}}{2}\Delta f\left(1 - 2\dfrac{n}{2}\right)
\end{array}
\right) \\[3ex]
+ \dfrac{(1-n)}{4F^2}\left(
\begin{array}{c}
4FF'' \\ +F'F'(n-6)
\end{array}
\right)\left(
\begin{array}{c}
f_{,\alpha} f_{,\beta} \\ -f_{,\mu} f_{,\nu} g^{\mu\nu} \dfrac{g_{\alpha\beta}}{2}
\end{array}
\right) \\[3ex]
+ \kappa T_{\alpha\beta} + \Lambda \cdot F \cdot g_{\alpha\beta} \\[2ex]
-\dfrac{n}{2}\cdot\left(
\begin{array}{c}
+ g^{\kappa\lambda} f_{,\kappa} f_{,\lambda} g_{\alpha\beta}\left(\dfrac{(F')^2}{2F^2}\left(5 - n - 2\dfrac{n}{2}\right) - \dfrac{F''}{F}\right) \\[1ex]
+ f_{,\beta} f_{,\alpha}\left(\left(\dfrac{n}{2} - 2\right)\cdot\dfrac{(F')^2}{F^2} + \dfrac{F''}{F}\right)
\end{array}
\right)
\end{array}
\right) \delta g^{\alpha\beta}
\right),
$$

$$(1797)$$

$$
0 = \int_V d^n x \left(
\begin{array}{l}
\sqrt{-g \cdot F^n}\left[
\begin{array}{l}
\left(R\left(1-\dfrac{n}{2}\right)+\dfrac{F'}{F}(1-n)^2 \Delta f\right) \\[2ex]
+\dfrac{f_{,\alpha} f_{,\beta}(1-n)}{4F^2} g^{\alpha\beta}
\left(
\begin{array}{l}
4FF'' \cdot (1-n) \\[1ex]
+F'F'\left(
\begin{array}{l}
(n-6)\left(1-\dfrac{n}{2}\right) \\[1ex]
-\dfrac{n}{2}\cdot(6n-8)
\end{array}
\right)
\end{array}
\right) \\[4ex]
+\kappa T_{\alpha\beta} g^{\alpha\beta} + \Lambda \cdot F \cdot n
\end{array}
\right]\delta\dfrac{1}{F} \\[10ex]
+\sqrt{-g \cdot F^{n-2}}\left[
\begin{array}{l}
\left(
\begin{array}{l}
\overset{***}{R}_{\alpha\beta} + R_{\alpha\beta} - R\cdot\dfrac{g_{\alpha\beta}}{2} \\[2ex]
+\dfrac{F'}{F}\left(
\begin{array}{l}
\left(f_{,\alpha\beta}-\Gamma^\gamma_{\alpha\beta} f_{,\gamma}\right)\left(1-3\dfrac{n}{2}\right) \\[1ex]
-\dfrac{g_{\alpha\beta}}{2}\Delta f(1-n)
\end{array}
\right)
\end{array}
\right. \\[5ex]
+\dfrac{(1-n)}{4F^2}\left(
\begin{array}{l}
4FF'' \\[1ex]
+F'F'(n-6)
\end{array}
\right)\left(
\begin{array}{l}
f_{,\alpha} f_{,\beta} \\[1ex]
-f_{,\mu} f_{,\nu} g^{\mu\nu}\dfrac{g_{\alpha\beta}}{2}
\end{array}
\right) \\[4ex]
\left.+\kappa T_{\alpha\beta} + \Lambda \cdot F \cdot g_{\alpha\beta}\right. \\[2ex]
-\dfrac{n}{2}\cdot\left(
\begin{array}{l}
+g^{\kappa\lambda} f_{,\kappa} f_{,\lambda} g_{\alpha\beta}\left(\dfrac{(F')^2}{2F^2}(5-2n)-\dfrac{F''}{F}\right) \\[2ex]
+f_{,\beta} f_{,\alpha}\left(\left(\dfrac{n}{2}-2\right)\cdot\dfrac{(F')^2}{F^2}+\dfrac{F''}{F}\right)
\end{array}
\right)
\end{array}
\right]\delta g^{\alpha\beta}
\end{array}
\right)
$$

$$(1798)$$

and could just be obtained by using (1795) and setting $q = n/2$ withing the corresponding parts of the integrand.

For completeness we also give the situation of the variation with respect to the unscaled metric (including the corresponding boundary condition as Hilbert has used it, which is to say zero of the variation along a boundary with respect to the unscaled metric) plus an arbitrary factor F^q within the Lagrangian, which makes the total variation to:

$$\delta W = 0 = \delta_{g_{\alpha\beta}} \int_V d^n x \left(\sqrt{-g} \cdot F^n \cdot F^q \cdot \left(\left(R + \frac{F'}{F}(1-n)\Delta f + \frac{f_{,\alpha}f_{,\beta}g^{\alpha\beta}(1-n)}{4F^2}\left(4FF'' + (F')^2(n-6)\right)\right)\frac{1}{F} - 2\Lambda + L_M \right)\right) \tag{1799}$$

and gives us:

$$0 = \int_V d^n x \left(\underbrace{ F^q \sqrt{-g} \cdot F^n \left(\frac{f_{,\alpha}f_{,\beta}(1-n)}{4F^2} g^{\alpha\beta} \left(4FF'' \cdot (1-n-q) + F'F'\left(n-6\right)\left(1-\frac{n}{2}\right) - \left(q+\frac{n}{2}\right)\cdot(4q-8+4n)\right) \right.}_{} \right.$$

$$\left. + \left(R\left(1-\frac{n}{2}\right) + \frac{F'}{F}(1-n)\Delta f(1-n-q)\right) + \kappa T_{\alpha\beta}g^{\alpha\beta} + \Lambda \cdot F \cdot n \right) \delta\frac{1}{F}$$

$$+ F^q \sqrt{-g} \cdot F^{n-2} \underbrace{\left(\overset{***}{R_{\alpha\beta}} + R_{\alpha\beta} - R \cdot \frac{g_{\alpha\beta}}{2} \right.}_{}$$

$$+ \frac{F'}{F}\left(\left(f_{,\alpha\beta} - \Gamma^\gamma_{\alpha\beta} f_{,\gamma}\right)\left(1 - n - \left(q+\frac{n}{2}\right)\right) - \frac{g_{\alpha\beta}}{2}\Delta f\left(1 - 2\left(q+\frac{n}{2}\right)\right)\right)$$

$$+ \frac{(1-n)}{4F^2}\left(4FF'' + F'F'(n-6) \right)\left(f_{,\alpha}f_{,\beta} - f_{,\mu}f_{,\nu}g^{\mu\nu}\frac{g_{\alpha\beta}}{2} \right) + \kappa T_{\alpha\beta} + \Lambda \cdot F \cdot g_{\alpha\beta}$$

$$- \left(q+\frac{n}{2}\right)\cdot\left(f_{,\alpha}f_{,\beta} - \frac{F''}{F} + g^{\kappa\lambda}f_{,\kappa}f_{,\lambda}g_{\alpha\beta}\left(\frac{(F')^2}{2F^2}(5-2q) - \frac{F''}{F}\right) + f_{,\beta}f_{,\alpha}\left(\left(q+\frac{n}{2}-2\right)\cdot\left(\frac{(F')^2}{F^2} + \frac{F''}{F}\right)\right) \right) \right) \delta g^{\alpha\beta} \tag{1800}$$

11.6.5 The Hilbert Dimension

All the considerations above bring us to yet another degree of freedom. We may assume that there exists a dimension which is independent on the true dimension for which the Hilbert surface term assumption truly holds. Denoting this dimension with the letter d and naming it the Hilbert dimension, we have the following variational task:

$$
\delta W = 0 = \delta_{g^d_{\alpha\beta}} \int_V d^n x \left(\sqrt{-g \cdot F^{d+(n-d)} \cdot F^q} \cdot \left(\left(\frac{R + \dfrac{F'}{F}(1-n)\Delta f}{4F^2} + \frac{f_{,\alpha} f_{,\beta} g^{\alpha\beta}(1-n)}{4F^2} \times \left(4FF'' + (F')^2(n-6)\right) \right) \left| \frac{1}{F} - 2\Lambda + L_M \right. \right)
$$

$$
= \delta_{g^d_{\alpha\beta}} \int_V d^n x \left(\sqrt{-g \cdot F^d \cdot F^{q+\frac{n-d}{2}}} \cdot \left(\left(\frac{R + \dfrac{F'}{F}(1-n)\Delta f}{4F^2} + \frac{f_{,\alpha} f_{,\beta} g^{\alpha\beta}(1-n)}{4F^2} \left(4FF'' + (F')^2(n-6)\right) \right) \left| \frac{1}{F} - 2\Lambda + L_M \right. \right).
$$

$$(1801)$$

With the help of (1795) we can directly obtain:

$$
0 = \int_V d^n x \left(
\begin{array}{l}
F^{\left(q+\frac{n-d}{2}\right)}\sqrt{-g \cdot F^d} \times \left(
\begin{array}{l}
\left(R\left(1-\frac{n}{2}\right)+\frac{F'}{F}(1-n)\Delta f\left(\begin{array}{l}\left(1-\frac{n}{2}\right)\\-\left(q+\frac{n-d}{2}\right)\end{array}\right)\right)\\
+\dfrac{f_{,\alpha}f_{,\beta}(1-n)}{4F^2}g^{\alpha\beta}\\
\left(\begin{array}{l}4FF''\cdot\left(1-\frac{n}{2}-\left(q+\frac{n-d}{2}\right)\right)\\
+F'F'\left(\begin{array}{l}(n-6)\left(1-\frac{n}{2}\right)-\left(q+\frac{n-d}{2}\right)\\
\cdot\left(4\left(q+\frac{n-d}{2}\right)-8+2n\right)\end{array}\right)\end{array}\right)\\
+\kappa T_{\alpha\beta}g^{\alpha\beta}+\Lambda\cdot F\cdot n
\end{array}
\right)\delta\frac{1}{F}\\[1cm]
+F^{\left(q+\frac{n-d}{2}\right)}\sqrt{-g\cdot F^{d-2}}\left(
\begin{array}{l}
R^{***}_{\alpha\beta}+R_{\alpha\beta}-R\cdot\dfrac{g_{\alpha\beta}}{2}\\
+\dfrac{F'}{F}\left(\begin{array}{l}\left(f_{,\alpha\beta}-\Gamma^{\gamma}_{\alpha\beta}f_{,\gamma}\right)\left(1-n-\left(q+\frac{n-d}{2}\right)\right)\\
-\dfrac{g_{\alpha\beta}}{2}\Delta f\left(1-2\left(q+\frac{n-d}{2}\right)\right)\end{array}\right)\\
+\dfrac{(1-n)}{4F^2}\left(\begin{array}{l}4FF''\\+F'F'(n-6)\end{array}\right)\left(\begin{array}{l}f_{,\alpha}f_{,\beta}\\-f_{,\mu}f_{,\nu}g^{\mu\nu}\dfrac{g_{\alpha\beta}}{2}\end{array}\right)\\
+\kappa T_{\alpha\beta}+\Lambda\cdot F\cdot g_{\alpha\beta}\\
-\left(q+\dfrac{n-d}{2}\right)\cdot\\
\left(\begin{array}{l}+g^{\kappa\lambda}f_{,\kappa}f_{,\lambda}g_{\alpha\beta}\left(\dfrac{(F')^2}{2F^2}\left(\begin{array}{l}5-n\\-2\left(q+\frac{n-d}{2}\right)\end{array}\right)-\dfrac{F''}{F}\right)\\
+f_{,\beta}f_{,\alpha}\left(\left(\left(q+\frac{n-d}{2}\right)-2\right)\cdot\dfrac{(F')^2}{F^2}+\dfrac{F''}{F}\right)\end{array}\right)
\end{array}
\right)\delta g^{\alpha\beta}
\end{array}
\right),
$$

$$(1802)$$

$$
0 = \int_V d^n x
\begin{pmatrix}
F^{\left(q+\frac{n-d}{2}\right)} \sqrt{-g \cdot F^d}
\begin{pmatrix}
\left(R\left(1-\frac{n}{2}\right)+\frac{F'}{F}(1-n)\Delta f\left(1-n-q-\frac{d}{2}\right)\right) \\[6pt]
+\dfrac{f_{,\alpha}f_{,\beta}(1-n)}{4F^2}g^{\alpha\beta} \\[6pt]
\times
\begin{pmatrix}
4FF'' \cdot \left(1-n-q-\frac{d}{2}\right) \\[6pt]
+F'F' \cdot \begin{pmatrix}(n-6)\left(1-\frac{n}{2}\right)-\left(q+\frac{n-d}{2}\right)\\[4pt] \cdot\left(4\left(q+\frac{n-d}{2}\right)-8+2n\right)\end{pmatrix}
\end{pmatrix} \\[6pt]
+\kappa T_{\alpha\beta}g^{\alpha\beta}+\Lambda\cdot F\cdot n
\end{pmatrix}\delta\frac{1}{F} \\[20pt]
+F^{\left(q+\frac{n-d}{2}\right)}\sqrt{-g\cdot F^{d-2}}
\begin{pmatrix}
R_{\alpha\beta}^{***}+R_{\alpha\beta}-R\cdot\dfrac{g_{\alpha\beta}}{2} \\[6pt]
+\dfrac{F'}{F}\begin{pmatrix}\left(f_{,\alpha\beta}-\Gamma^\gamma_{\alpha\beta}f_{,\gamma}\right)\left(1-n-\left(q+\frac{n-d}{2}\right)\right)\\[4pt]-\dfrac{g_{\alpha\beta}}{2}\Delta f(1-2q-n+d)\end{pmatrix} \\[6pt]
+\dfrac{(1-n)}{4F^2}\begin{pmatrix}4FF''\\+F'F'(n-6)\end{pmatrix}\begin{pmatrix}f_{,\alpha}f_{,\beta}\\-f_{,\mu}f_{,\nu}g^{\mu\nu}\dfrac{g_{\alpha\beta}}{2}\end{pmatrix} \\[6pt]
+\kappa T_{\alpha\beta}+\Lambda\cdot F\cdot g_{\alpha\beta}-\left(q+\dfrac{n-d}{2}\right)\cdot \\[6pt]
\begin{pmatrix}+g^{\kappa\lambda}f_{,\kappa}f_{,\lambda}g_{\alpha\beta}\left(\dfrac{(F')^2}{2F^2}(5-2n+d-2q)-\dfrac{F''}{F}\right)\\[6pt]+f_{,\beta}f_{,\alpha}\left(\left(\left(q+\dfrac{n-d}{2}-2\right)\cdot\dfrac{(F')^2}{F^2}+\dfrac{F''}{F}\right)\right)\end{pmatrix}
\end{pmatrix}\delta g^{\alpha\beta}
\end{pmatrix}.
$$

$$(1803)$$

11.6.6 Looking for Suitable Solutions: The Quantum Einstein Field Equations

In order to obtain suitable solutions to the equations derived above, we find that simultaneously demanding the $f_{,\alpha}f_{,\beta}$ operator terms to vanish (or at least to be equal) for the metric variation part and the $1/F$ variation results in the condition $n = 2$ (if insisted in integer dimensions) for both cases (1800) and

(1803). This result supports the foam-theory as fundament for "everything" as other authors have discussed and as we have already derived in [9] and represented in chapter 7 of this book.

However, what can be done in a general number of dimensions?

Well, here we already saw that only in the case of our variation result (1771) things seem to be consistent in an arbitrary number of dimensions. The result (1771), however, was obtained under the assumption:

(a) that the variation has been performed with respect to the scaled metric $G_{\alpha\beta}$,

(b) that the Hilbert surface term is zero also with respect to the scaled metric,

(c) that the generally free to choose factor F^q of a generalize Einstein–Hilbert Lagrangian F^q*R is set to be 1, which is to say $q = 0$,

(d) that the "Hilbert dimension" $d = 0$.

Assuming that no extra matter is of need, but that the matter simply is presented by certain forms or appearances of the wave function(s)[7], we then find that the simplest possible forms of quantum Einstein field equations are given via (1778) (last line). Thereby we had made use of condition (1775). Without this condition, the general quantum Einstein field equations are given via:

$$
0 = \begin{pmatrix}
R_{\alpha\beta} - \dfrac{g_{\alpha\beta}}{2}R + \dfrac{g_{\alpha\beta}}{2}\dfrac{1}{2F}\begin{pmatrix} 2F_{,ij}(n-1)g^{ij} + 2\Gamma^a_{ij}F_{,a}g^{ij} \\ -F_{,i}g^{ab}g_{jb,a}g^{ij} - F_{,j}g^{ab}g_{ib,a}g^{ij} \\ -n\Gamma^d_{ij}F_{,d}g^{ij} + \dfrac{n}{2}F_{,d}g^{cd}g_{ab,c}g^{ab} \end{pmatrix} \\ + \dfrac{F_{,i}\cdot F_{,j}}{4F^2}g^{ij}\left((n-6)(n-1)\right) \\[2ex]
-\dfrac{1}{2F}\begin{pmatrix} F_{,\alpha\beta}(n-2) + F_{,ab}g_{\alpha\beta}g^{ab} \\ +F_{,a}g^{ab}\left(g_{\beta b,\alpha} - g_{\beta\alpha,b}\right) - F_{,\alpha}g^{ab}g_{\beta b,a} - F_{,\beta}g^{ab}g_{\alpha b,a} \\ +F_{,d}g^{cd}\dfrac{1}{2}n\left(\dfrac{2}{n}g_{\alpha c,\beta} - g_{\alpha c,\beta} - g_{\beta c,\alpha} + g_{\alpha\beta,c} + \dfrac{1}{n}g_{\alpha\beta}g_{ab,c}g^{ab}\right) \\ -\dfrac{1}{2F}\left(F_{,\alpha}\cdot F_{,\beta}(3n-6) + g_{\alpha\beta}F_{,c}F_{,d}g^{cd}(4-n)\right) \\ +\kappa T_{\alpha\beta} + \Lambda\cdot F\cdot g_{\alpha\beta} \end{pmatrix}
\end{pmatrix},
$$

(1804)

---

[7]In fact, it is the plural we require here, which will be elaborated in section "11.12 The Math for Body, Soul and Universe?".

which for metrics of the form (1650) simplify to:

1. $0 = \begin{pmatrix} \left(R + \dfrac{F'}{F}(1-n)\Delta f\right)\left(1-\dfrac{n}{2}\right) \\[2mm] + \dfrac{f_{,\alpha}f_{,\beta}(1-n)}{4F^2}g^{\alpha\beta}\left(4FF'' + F'F'(n-6)\right)\left(1-\dfrac{n}{2}\right) \\[2mm] + \kappa T_{\alpha\beta}g^{\alpha\beta} + \Lambda \cdot F \cdot n \end{pmatrix}$

$\Rightarrow \qquad -R = \begin{pmatrix} \left(\dfrac{F'}{F}(1-n)\Delta f\right) \\[2mm] + \dfrac{f_{,\alpha}f_{,\beta}(1-n)}{4F^2}g^{\alpha\beta}\left(4FF'' + F'F'(n-6)\right) \\[2mm] + \dfrac{2}{2-n}\left(\kappa T_{\alpha\beta}g^{\alpha\beta} + \Lambda \cdot F \cdot n\right) \end{pmatrix}$

$\Rightarrow \quad$ **2.** $0 = \begin{pmatrix} R_{\alpha\beta}^{***} + R_{\alpha\beta} + \begin{pmatrix} \left(\dfrac{F'}{F}(1-n)\Delta f\right) \\[2mm] + \dfrac{f_{,\kappa}f_{,\lambda}(1-n)}{4F^2}g^{\kappa\lambda}\left(4FF'' + F'F'(n-6)\right) \\[2mm] + \dfrac{2}{2-n}\left(\kappa T_{\kappa\lambda}g^{\kappa\lambda} + \Lambda \cdot F \cdot n\right) \end{pmatrix} \cdot \dfrac{g_{\alpha\beta}}{2} \\[4mm] + \dfrac{F'}{F}(1-n)\left(f_{,\alpha\beta} - \Gamma_{\alpha\beta}^{\gamma}f_{,\gamma} - \dfrac{g_{\alpha\beta}}{2}\Delta f\right) \\[4mm] + \dfrac{(1-n)}{4F^2}\left(4FF'' + F'F'(n-6)\right)\left(f_{,\alpha}f_{,\beta} - f_{,\mu}f_{,\nu}g^{\mu\nu}\dfrac{g_{\alpha\beta}}{2}\right) \\[4mm] + \kappa T_{\alpha\beta} + \Lambda \cdot F \cdot g_{\alpha\beta} \end{pmatrix}.$

$$(1805)$$

Sticking to the assumption that no extra matter is of need, this may be simplified as follows:

1. $0 = \begin{pmatrix} \left(R + \dfrac{F'}{F}(1-n)\Delta f\right)\left(1-\dfrac{n}{2}\right) \\[2mm] + \dfrac{f_{,\alpha}f_{,\beta}(1-n)}{4F^2}g^{\alpha\beta}\left(4FF'' + F'F'(n-6)\right)\left(1-\dfrac{n}{2}\right) \end{pmatrix}$

$\Rightarrow \qquad -R = \left(\left(\dfrac{F'}{F}(1-n)\Delta f\right) + \dfrac{f_{,\alpha}f_{,\beta}(1-n)}{4F^2}g^{\alpha\beta}\left(4FF'' + F'F'(n-6)\right)\right)$

$$\Rightarrow \quad 2. \quad 0 = \left(R_{\alpha\beta}^{***} + R_{\alpha\beta} + \left(\begin{array}{c} \left(\dfrac{F'}{F}(1-n)\Delta f \right) \\ + \dfrac{f_{,\kappa} f_{,\lambda}(1-n)}{4F^2} g^{\kappa\lambda} \left(4FF'' + F'F'(n-6) \right) \end{array} \right) \cdot \dfrac{g_{\alpha\beta}}{2} \\ + \dfrac{F'}{F}(1-n) \left(f_{,\alpha\beta} - \Gamma_{\alpha\beta}^{\gamma} f_{,\gamma} - \dfrac{g_{\alpha\beta}}{2}\Delta f \right) \\ + \dfrac{(1-n)}{4F^2} \left(4FF'' + F'F'(n-6) \right) \left(f_{,\alpha} f_{,\beta} - f_{,\mu} f_{,\nu} g^{\mu\nu} \dfrac{g_{\alpha\beta}}{2} \right) \right).$$

(1806)

With condition (1775), we obtain the very brief form of quantum Einstein field equations as:

$$1. \quad -R = \frac{F'}{F}(1-n)\Delta f$$

$$\Rightarrow \qquad 2. \quad R_{\alpha\beta}^{***} + R_{\alpha\beta} = \frac{F'}{F}(n-1)\left(f_{,\alpha\beta} - \Gamma_{\alpha\beta}^{\gamma} f_{,\gamma} \right)$$

$$\Rightarrow \qquad R_{\alpha\beta}^{***} + R_{\alpha\beta} = \frac{F'}{F}(n-1) f_{;\alpha\beta}.$$

(1807)

With F given as:

$$4FF'' + F'F'(n-6) = 0 \quad \Rightarrow \quad F = \left(C_f \pm f \right)^{\frac{4}{n-2}}; \quad n > 2,$$

(1808)

this results in:

$$\Rightarrow \qquad \boxed{R_{\alpha\beta}^{***} + R_{\alpha\beta} = \pm \frac{4 \cdot (n-1)}{(n-2)\left(C_f \pm f \right)} f_{;\alpha\beta}}.$$

(1809)

Even though this is a nice compact set of equations, we should not ignore other possibilities, which, as it was shown in our previous paper of the series [13] and has been represented in chapters 9 and 10 in this book, can be obtained via the degrees of freedom offering themselves in form of the wrapper function $F[f]$.

For instance, knowing that, instead of condition (1808), we could also have any other condition in (1806) such as (with an arbitrary function $H'[f] = \dfrac{\partial H[f]}{\partial f} = H'$):

$$\frac{4FF'' + \left(F' \right)^2 (n-6)}{4F} = F' \cdot H'$$

$$\Rightarrow \qquad F = \begin{cases} \left(C_{f0} + \int\limits_1^f e^{H[\phi]} \cdot \dfrac{n-2}{4} \cdot d\phi \right)^{\frac{4}{n-2}} \cdot C_{f1} & n > 2 \\[2em] C_{f1} \cdot e^{C_{f0} \cdot \int\limits_1^f e^{H|\phi|} d\phi} & n = 2, \end{cases} \qquad (1810)$$

leading us to:

1. $\quad 0 = \left(1 - \dfrac{n}{2}\right)\left(R + \dfrac{F'}{F}(1-n)\Delta f + \dfrac{f_{,\alpha} f_{,\beta}(1-n)}{F} g^{\alpha\beta} H' F' \right)$

$$\Rightarrow \qquad -R = \dfrac{F'}{F}(1-n)\left(\Delta f + f_{,\alpha} f_{,\beta} g^{\alpha\beta} H' \right)$$

$$\Rightarrow \qquad \textbf{2.} \quad 0 = \left(\begin{array}{l} \overset{***}{R}_{\alpha\beta} + R_{\alpha\beta} + \dfrac{F'}{F}(1-n)\left(\Delta f + f_{,\kappa} f_{,\lambda} g^{\kappa\lambda} H' \right) \cdot \dfrac{g_{\alpha\beta}}{2} \\[1.5em] + \dfrac{F'}{F}(1-n)\left(f_{,\alpha\beta} - \Gamma^\gamma_{\alpha\beta} f_{,\gamma} - \dfrac{g_{\alpha\beta}}{2}\Delta f + \left(\begin{array}{c} f_{,\alpha} f_{,\beta} \\[0.5em] -f_{,\mu} f_{,\nu} g^{\mu\nu} \dfrac{g_{\alpha\beta}}{2} \end{array} \right) H' \right) \end{array} \right)$$

$$= \overset{***}{R}_{\alpha\beta} + R_{\alpha\beta} + \dfrac{F'}{F}(1-n)\cdot\left(\dfrac{f_{,\kappa} f_{,\lambda} g^{\kappa\lambda} H' g_{\alpha\beta}}{2} + f_{,\alpha\beta} - \Gamma^\gamma_{\alpha\beta} f_{,\gamma} + \left(\begin{array}{c} f_{,\alpha} f_{,\beta} \\[0.5em] -f_{,\mu} f_{,\nu} g^{\mu\nu} \dfrac{g_{\alpha\beta}}{2} \end{array} \right) H' \right)$$

$$= \overset{***}{R}_{\alpha\beta} + R_{\alpha\beta} + \dfrac{F'}{F}(1-n)\cdot\left(\dfrac{f_{,\kappa} f_{,\lambda} g^{\kappa\lambda} H' g_{\alpha\beta}}{2} + f_{:\alpha\beta} + \left(f_{,\alpha} f_{,\beta} - f_{,\mu} f_{,\nu} g^{\mu\nu} \dfrac{g_{\alpha\beta}}{2} \right) H' \right).$$

$$(1811)$$

With the introduction of this function $H = H[f]$, we can develop the two differential operators in the first equation in (1811) into one.

$$0 = R - \dfrac{n-1}{F}\left(\dfrac{F'}{\sqrt{g}}\left(\sqrt{g} g^{\mu\nu} f_{,\mu} \right)_{,\nu} + \overbrace{\dfrac{1}{4F}\left(4FF'' - (6-n)F'F' \right)}^{F'H'} f_{,\mu} f_{,\nu} g^{\mu\nu} \right)$$

$$= R - (n-1)\dfrac{F'}{F}\left(\dfrac{1}{\sqrt{g}}\left(\sqrt{g} g^{\mu\nu} f_{,\mu} \right)_{,\nu} + H' f_{,\mu} f_{,\nu} g^{\mu\nu} \right)$$

$$= R - (n-1)\dfrac{F'}{F}\left(\dfrac{1}{\sqrt{g}}\left(\sqrt{g} g^{\mu\nu} f_{,\mu} \right)_{,\nu} + \dfrac{\sqrt{g}}{\sqrt{g}} H' f_{,\mu} f_{,\nu} g^{\mu\nu} \right)$$

$$= R - (n-1)\frac{F'}{F}\left(\frac{1}{\sqrt{g}}\left(\sqrt{g} \cdot g^{\mu\nu} H \cdot f_{,\mu}\right)_{,\nu}\right) \tag{1812}$$

By defining a new metric $h^{\mu\nu}$ with the following properties:

$$\sqrt{g} \cdot g^{\mu\nu} H = \sqrt{h} \cdot h^{\mu\nu}, \tag{1813}$$

we end up with an equation containing a simple Laplace operator as follows:

$$0 = R - (n-1)\frac{F'}{F}\left(\frac{1}{\sqrt{g}}\left(\sqrt{h} \cdot h^{\mu\nu} \cdot f_{,\mu}\right)_{,\nu}\right)$$

$$\Rightarrow \quad 0 = \frac{\sqrt{g}}{\sqrt{h}} \cdot \frac{F}{F'} \cdot R_g - \frac{(n-1)}{\sqrt{h}}\left(\sqrt{h} \cdot h^{\mu\nu} \cdot f_{,\mu}\right)_{,\nu} = \frac{\sqrt{g}}{\sqrt{h}} \cdot \frac{F}{F'} \cdot R_g - (n-1)\Delta_h f.$$

$$\tag{1814}$$

Thereby we need to distinguish between the metric for the Laplace operator applied on f, being $h^{\mu\nu}$, and the one being the basis for the Ricci scalar R, which is $g^{\mu\nu}$.

11.6.7 Can We Now Understand Quantum Theory in a Truly Illustrative Manner? Part II

We conclude that the simplest form of quantum Einstein field equations can be given via (1809), but this is not the only option. We see this when changing condition (1808) to (1810), for instance, we end up with a set of completely different equations, namely (1811). And here we only used one degree of freedom, which determines the functional wrapper F. There are more.

It should also be pointed out that the fundamental law leading to our field equations is the minimum principle (1651). This, however, is a scalar, which is formed out of an integral sum, which stretches over a certain volume, while the volume and the dimension of the integral are not fixed. And so, it does not necessarily truly require the integrand to be zero in order to make (1651) being fulfilled. Other options, which is to say options not requiring the quantum Einstein field equations, we have derived, but which could also be extracted from (1651), should be considered, too. For instance, there is still the question about the surface terms, centers of gravity, centers of wave functions ([4, 8], chapter 8), or other variational possibilities.

Furthermore, in the subsection ("11.6.8 Scaling Options"), we will also consider possibilities of different scale split-ups of the Hilbert variation, which is leading to even more options for the appearance of quantum Einstein field equations.

Nevertheless, here we have found that the simplest and shortest form of quantum Einstein field equations evolves from condition (1808) and, bringing it into the form similar to the one shown on the cover of [17], reads:

$$\left(R_{\alpha\beta}^{***} + R_{\alpha\beta} \right) \cdot \left(C_f \pm f \right) = \pm 4 \cdot \frac{(n-1)}{(n-2)} f_{;\alpha\beta}. \qquad (1815)$$

Thus, it may even be seen as a kind of eigen equation, becoming a true eigen equation in the case of non-f-dependent Ricci tensors.

Just as a side note we should state that the Schwarzschild solution [18] fulfills Eq. (1815) without the need of a wave function f (thus, f = const $\rightarrow$ F = const). Thus, also black holes (non-charged) [18] are—in principle—perfectly well described by the classical Schwarzschild solution and—at least for the outside region—do not sport an own quantum field (not as long as one only considers the 4-D Schwarzschild object without any surrounding fluctuations anyway). This author thinks, however, that this only holds for the outside region of the massive object, while the inside solution requires some more sophisticated approaches with respect to Eq. (1815) most likely including a non-zero, respectively non-constant wave function (see section "11.9 Quantum Black Hole Solutions $\rightarrow$ Inner Black Hole Metrics without Singularities" and [25]). This wave function, respectively its second covariant derivative then acts as the classical matter. Thereby, it should be pointed out, however, that the outside and inside regions do not necessarily have to have the same dimension, nor do they both have to follow the simple form (1815). There may be a lot more possibilities behind Hamilton's minimum principle or Hilbert's variation problem (1651). A simple and potentially inner solution to (1815) will be presented further below in this chapter.

We see that what classically is called vacuum solutions also provides solutions to (1815) with constant or linear wave functions, leading to vanishing second covariant derivatives for f when $R_{\alpha\beta}^{***} = 0$ as follows:

$$f_{;\alpha\beta} = 0. \qquad (1816)$$

As there are quite many such solutions (c.f. [19], for instance), we temporarily conclude that there have to be objects without quantum fields at least with respect to their surroundings. This, however, only holds for the object and its metric field and not for the surrounding area around it. The "vacuum" around the object is still free to potentially produce its own fluctuational quantum field. Also, the interactions (entanglements) with the object in question can (or most likely also do) sport quantum fields.

To put it illustratively, one may simply say that, for some metrics, the universe has the option to not produce scaling or quantum fields. In fact, the universe can't even produce such fields without simultaneously introducing "something" else into the total equation. Whether this "something" is matter (meaning it appears as matter to us) or simply confines a great volume to the satisfaction of the governing minimum principle (1651) does not matter. Apparently, there are many options and the universe uses them, respectively it variates among them. Our measurement process, however, only seems to see (detect) the results of the variation with just one selected possibility out of the many and it does not observe the variation itself (not directly anyway).

The interested reader finds some more discussion to (1815) in [22].

11.6.8 Scaling Options

It has to be pointed out that a solution to (1815) does not suffice to solve the whole variation problem as this would also require a solution to the first line of (1807). Alternatively, one may also seek for a solution of the whole integrand of (1651), leading us to:

$$0 = R^{***}_{\alpha\beta} + R_{\alpha\beta} - \frac{1}{2}R \cdot g_{\alpha\beta} + \frac{F'}{F}(1-n)\left(f_{,\alpha\beta} - \Gamma^{\gamma}_{\alpha\beta} f_{,\gamma} - \frac{1}{2}\Delta f \cdot g_{\alpha\beta} \right), \quad (1817)$$

when applying condition (1808) and assuming no matter and a zero cosmological constant.

So, one might ask, why then at all bothering about solutions consisting of solutions to (1815) without also being solutions to the first line of (1807)?

We think that there is a reason why the first line of (1807), being "the rest of (1817)," appears to us as a quantum equation, while the gravity equations containing the Ricci tensor act on bigger scales. We even think that different volumes have to or could be applied to both parts of the integrand within the same variation. This may mathematically be done as follows (partial repetition from chapters 9 and 10):

Starting with (1651), respectively the result from (1766), and setting the matter term and the cosmological constant to zero:

$$\delta W = 0 = \int_V d^n x \left(\sqrt{-g \cdot F^n} \left(\begin{array}{c} R_{\alpha\beta}^{***} + R_{\alpha\beta} - R \cdot \dfrac{g_{\alpha\beta}}{2} \\[2mm] + \dfrac{F'}{F}(1-n)\left(f_{,\alpha\beta} - \Gamma_{\alpha\beta}^{\gamma} f_{,\gamma} - \dfrac{g_{\alpha\beta}}{2}\Delta f \right) \\[2mm] + \dfrac{(1-n)}{4F^2}\left(\begin{array}{c} 4FF'' \\ +F'F'(n-6) \end{array} \right)\left(\begin{array}{c} f_{,\alpha}f_{,\beta} \\ -f_{,\mu}f_{,v}g^{\mu v}\dfrac{g_{\alpha\beta}}{2} \end{array} \right) \end{array} \right) \delta G^{\alpha\beta} \right)$$

(1818)

and also using $4FF'' + F'F'(n-6) = 0$ again, we can simplify and split up to:

$$\xrightarrow{\;4FF''+F'F'(n-6)=0\;}$$

$$0 = \int_V d^n x \left(\sqrt{-g \cdot F^n} \left(\begin{array}{c} R_{\alpha\beta}^{***} + R_{\alpha\beta} - R \cdot \dfrac{g_{\alpha\beta}}{2} \\[2mm] + \dfrac{F'}{F}(1-n)\left(f_{,\alpha\beta} - \Gamma_{\alpha\beta}^{\gamma} f_{,\gamma} - \dfrac{g_{\alpha\beta}}{2}\Delta f \right) \end{array} \right) \delta G^{\alpha\beta} \right)$$

$$\Rightarrow \quad 0 = \left(\begin{array}{c} \int_{V_G} d^n x \left(\sqrt{-g \cdot F^n}\left(R_{\alpha\beta}^{***} + R_{\alpha\beta} + \dfrac{F'}{F}(1-n)\left(f_{,\alpha\beta} - \Gamma_{\alpha\beta}^{\gamma} f_{,\gamma} \right) \right) \delta G^{\alpha\beta} \right) \\[4mm] - \dfrac{1}{2} \cdot \int_{V_Q} d^n x \left(\sqrt{-g \cdot F^n}\left(R + \dfrac{F'}{F}(1-n)\Delta f \right) g_{\alpha\beta}\delta G^{\alpha\beta} \right) \end{array} \right).$$

(1819)

Thereby the volumes of the two variations do not necessarily need to be the same.

For symmetry reasons, where we want to have equal total volumes for both terms in the last equation above, one may just extend the last line as follows:

$$0 = \left(\begin{array}{c} \int_{V_G} d^n x \left(\sqrt{-g \cdot F^n}\left(R_{\alpha\beta}^{***} + R_{\alpha\beta} + \dfrac{F'}{F}(1-n)\left(f_{,\alpha\beta} - \Gamma_{\alpha\beta}^{\gamma} f_{,\gamma} \right) \right) \delta G^{\alpha\beta} \right) \\[4mm] - \dfrac{1}{2} \cdot \sum_{i=1}^{N} \int_{V_{Qi}} d^n x \left(\sqrt{-g \cdot F^n}\left(R + \dfrac{F'}{F}(1-n)\Delta f \right) g_{\alpha\beta}\delta G^{\alpha\beta} \right) \end{array} \right)$$

$$V_G = \sum_{i=1}^{N} V_{Qi}. \tag{1820}$$

Thus, it appears totally reasonable to assume that, for whatever reasons, the scales of the two variations could be extremely different and while the volume V_Q (Q denotes quantum) might be extremely small, forcing us to watch the variation:

$$0 = \int_{V_Q} d^n x \left(\sqrt{-g} \cdot F^n \left(R + \frac{F'}{F}(1-n)\Delta f \right) g_{\alpha\beta} \delta G^{\alpha\beta} \right) \tag{1821}$$

at smaller, which is to say—perhaps—quantum scales, we would have to take care about the tensorial or gravitational variation:

$$0 = \int_{V_G} d^n x \left(\sqrt{-g} \cdot F^n \left(\overset{***}{R_{\alpha\beta}} + R_{\alpha\beta} + \frac{F'}{F}(1-n)\left(f_{,\alpha\beta} - \Gamma^\gamma_{\alpha\beta} f_{,\gamma} \right) \right) \delta G^{\alpha\beta} \right) \tag{1822}$$

at much bigger scales, respectively bigger volumes V_G. Thus, our reasoning for the consideration of solutions to (1815) with respect to certain scales would be based on the observation of the different scales of quantum effects and gravity. Of course, this requires more discussion about the variational, respectively scale-dependent split-up in (1819), and the possible substructure of space or space-time with respect to the subsequent small-scale solutions to (1821). For instance, with respect to the split-up, we should take into account all possibilities resulting from our $F[f]^* g_{\alpha\beta}$ scaled metric approach. So, in addition to (1820) or the complete integrand of the variation (1646) (matter term and cosmological constant set to zero) with either:

$$0 = \int_V d^n x \left(\sqrt{-g} \cdot F^n \left(\begin{array}{c} \overset{***}{R_{\alpha\beta}} + R_{\alpha\beta} - R \cdot \dfrac{g_{\alpha\beta}}{2} \\ + \dfrac{F'}{F}(1-n)\left(f_{,\alpha\beta} - \Gamma^\gamma_{\alpha\beta} f_{,\gamma} - \dfrac{g_{\alpha\beta}}{2}\Delta f \right) \end{array} \right) \delta G^{\alpha\beta} \right), \tag{1823}$$

or:

$$0 = \sum_{i=1}^{N} \int_{V_{Qi}} d^n x \sqrt{-g} \cdot F^n \left(\begin{array}{c} \overset{***}{R_{\alpha\beta}} + R_{\alpha\beta} - R \cdot \dfrac{g_{\alpha\beta}}{2} \\ + \dfrac{F'}{F}(1-n)\left(f_{,\alpha\beta} - \Gamma^\gamma_{\alpha\beta} f_{,\gamma} - \dfrac{g_{\alpha\beta}}{2}\Delta f \right) \end{array} \right) \delta G^{\alpha\beta}, \tag{1824}$$

we can also have:

$$0 = \left[\begin{array}{c} \int\limits_{V_G} d^n x \sqrt{-g} \cdot F^n \left(\left(\overset{***}{R}_{\alpha\beta} + R_{\alpha\beta} - \frac{1}{2} \cdot R \cdot g_{\alpha\beta} + \frac{F'}{F}(1-n)\left(f_{,\alpha\beta} - \Gamma^\gamma_{\alpha\beta} f_{,\gamma} \right) \right) \delta G^{\alpha\beta} \right) \\ -\frac{1}{2} \cdot \sum_{i=1}^{N} \int\limits_{V_{Qi}} d^n x \sqrt{-g} \cdot F^n \left(\frac{F'}{F}(1-n)\Delta f \cdot g_{\alpha\beta} \delta G^{\alpha\beta} \right) \end{array} \right],$$

$$(1825)$$

$$0 = \left[\begin{array}{c} \int\limits_{V_G} d^n x \sqrt{-g} \cdot F^n \left(\left(\overset{***}{R}_{\alpha\beta} + R_{\alpha\beta} - \frac{1}{2} \cdot R \cdot g_{\alpha\beta} \right) \delta G^{\alpha\beta} \right) \\ +\frac{1}{2} \cdot \sum_{i=1}^{N} \int\limits_{V_{Qi}} d^n x \sqrt{-g} \cdot F^n \frac{F'}{F}(n-1)\left(\Delta f \cdot g_{\alpha\beta} - 2 \cdot \left(f_{,\alpha\beta} - \Gamma^\gamma_{\alpha\beta} f_{,\gamma} \right) \right) \delta G^{\alpha\beta} \end{array} \right],$$

$$(1826)$$

$$0 = \left[\begin{array}{c} \int\limits_{V_G} d^n x \sqrt{-g} \cdot F^n \left(\overset{***}{R}_{\alpha\beta} + R_{\alpha\beta} \right) \delta G^{\alpha\beta} \\ +\frac{1}{2} \cdot \sum_{i=1}^{N} \int\limits_{V_{Qi}} d^n x \sqrt{-g} \cdot F^n \left(-R \cdot g_{\alpha\beta} + \frac{F'}{F}(n-1)\left(\Delta f \cdot g_{\alpha\beta} - 2 \cdot \left(f_{,\alpha\beta} - \Gamma^\gamma_{\alpha\beta} f_{,\gamma} \right) \right) \right) \delta G^{\alpha\beta} \end{array} \right],$$

$$(1827)$$

$$0 = \left[\begin{array}{c} -\frac{1}{2} \cdot \int\limits_{V_G} d^n x \sqrt{-g} \cdot F^n \left(R \cdot g_{\alpha\beta} \delta G^{\alpha\beta} \right) \\ +\frac{1}{2} \cdot \sum_{i=1}^{N} \int\limits_{V_{Qi}} d^n x \sqrt{-g} \cdot F^n \left(2 \cdot \left(\overset{***}{R}_{\alpha\beta} + R_{\alpha\beta} \right) + \frac{F'}{F}(n-1) \left(\begin{array}{c} \Delta f \cdot g_{\alpha\beta} \\ -2 \cdot \left(f_{,\alpha\beta} - \Gamma^\gamma_{\alpha\beta} f_{,\gamma} \right) \end{array} \right) \right) \delta G^{\alpha\beta} \end{array} \right],$$

$$(1828)$$

$$\vdots, \qquad (1829)$$

and so on.

Naturally, we can have all these options with interchanged integral boundaries like illustrated with the complete integrand in (1823) and (1824).

One should also discuss the possibility of surface terms in connection with the quantum-variational integrals. Realizing that one of the options given above (namely (1826)) already mirrors the classical Einstein field equations we may try to obtain an explanation for the classical matter terms, which were—so far—always artificially introduced into the General Theory of Relativity.

11.6.9 Origin of the Energy–Momentum Tensor in the General Theory of Relativity

Discovering the classical Field-Equation structure (c.f. (1826)) among our ensemble of scaling and scaling split-up options, one might want to demonstrate the connection to the Einstein–Hilbert matter approach. We start, for the reason of simplicity and brevity, with condition (1808), which is to say $4FF'' + F'F'(n-6) = 0$ and the assumption of one global metric $g_{\alpha\beta}$, but an ensemble of F_i and f_i, leading us to:

$$
0 = \begin{pmatrix} \int_{V_G} d^n x \sqrt{-G}\left(\left(\overset{***}{R}_{\alpha\beta} + R_{\alpha\beta} - \frac{1}{2}\cdot R\cdot g_{\alpha\beta}\right)\delta G^{\alpha\beta}\right) \\ + \frac{1}{2}\cdot\sum_{i=1}^{N}\int_{V_{Qi}} d^n x\sqrt{-g}\cdot F^n\,\frac{F'}{F}(n-1)\left(\Delta f\cdot g_{\alpha\beta} - 2\cdot\left(f_{,\alpha\beta} - \Gamma_{\alpha\beta}^{\gamma} f_{,\gamma}\right)\right)\delta G^{\alpha\beta} \end{pmatrix}
$$

$$
\Rightarrow \quad = \int_{V_G} d^n x\sqrt{-g}\left(\sum_{i=1}^{N}\sqrt{F_i^n}\cdot\begin{pmatrix} \overset{***}{R}_{\alpha\beta} + R_{\alpha\beta} - \frac{1}{2}\cdot R\cdot g_{\alpha\beta} \\ + \frac{(n-1)}{2}\cdot\frac{F_i'}{F_i}\left(\Delta f_i\cdot g_{\alpha\beta} - 2\cdot f_{i;\alpha\beta}\right) \end{pmatrix}\delta G^{\alpha\beta}\right). \quad (1830)
$$

This gives us the classical non-vacuum field equations via:

$$
0 = \sum_{i=1}^{N}\sqrt{F_i^n}\cdot\left(\overset{***}{R}_{\alpha\beta} + R_{\alpha\beta} - \frac{1}{2}\cdot R\cdot g_{\alpha\beta}\right) + \sum_{i=1}^{N}\sqrt{F_i^n}\cdot\left(\frac{(n-1)}{2}\cdot\frac{F_i'}{F_i}\left(\Delta f_i\cdot g_{\alpha\beta} - 2\cdot f_{i;\alpha\beta}\right)\right)
$$

$$
\Rightarrow \quad 0 = \overset{***}{R}_{\alpha\beta} + R_{\alpha\beta} - \frac{1}{2}\cdot R\cdot g_{\alpha\beta} + \frac{\sum_{i=1}^{N}\sqrt{F_i^n}\cdot\left(\frac{(n-1)}{2}\cdot\frac{F_i'}{F_i}\left(\Delta f_i\cdot g_{\alpha\beta} - 2\cdot f_{i;\alpha\beta}\right)\right)}{\sum_{i=1}^{N}\sqrt{F_i^n}}
$$

$$(1831)$$

and the last term in the last line above is handing us the quantum or scale equivalent of the Einstein–Hilbert matter term, respectively the energy–momentum tensor:

$$\Rightarrow \quad \kappa T_{\alpha\beta} = \frac{R_{\alpha\beta}^{***} + \sum_{i=1}^{N} \sqrt{F_i^n} \cdot \left(\frac{(n-1)}{2} \cdot \frac{F_i'}{F_i} \left(\Delta f_i \cdot g_{\alpha\beta} - 2 \cdot f_{i;\alpha\beta} \right) \right)}{\sum_{i=1}^{N} \sqrt{F_i^n}}. \qquad (1832)$$

For completeness, we also want to present the result without the simplification $4FF'' + F'F'(n-6) = 0$. This renders (1830) as follows:

$$0 = \left(\begin{array}{c} \int_{V_G} d^n x \sqrt{-G} \left(\left(R_{\alpha\beta}^{***} + R_{\alpha\beta} - \frac{1}{2} \cdot R \cdot g_{\alpha\beta} \right) \delta G^{\alpha\beta} \right) \\[2em] + \frac{1}{2} \cdot \sum_{i=1}^{N} \int_{V_{Qi}} d^n x \sqrt{-g} \cdot F^n \left(\begin{array}{c} \frac{F'}{F}(n-1)\left(\Delta f \cdot g_{\alpha\beta} - 2 \cdot \left(f_{,\alpha\beta} - \Gamma_{\alpha\beta}^{\gamma} f_{,\gamma} \right) \right) \\[1em] + \frac{(1-n)}{2F_i^2} \left(\begin{array}{c} F_i F_i'' \\ + F_i' F_i'(n-6) \end{array} \right) \left(\begin{array}{c} f_{i,\alpha} f_{i,\beta} \\ - f_{i,\mu} f_{i,\nu} g^{\mu\nu} \frac{g_{\alpha\beta}}{2} \end{array} \right) \end{array} \right) \delta G^{\alpha\beta} \end{array} \right)$$

$$\Rightarrow = \int_{V_G} d^n x \sqrt{-g} \left| \sum_{i=1}^{N} \sqrt{F_i^n} \cdot \left| \times \left(\begin{array}{c} R_{\alpha\beta}^{***} + R_{\alpha\beta} - \frac{1}{2} \cdot R \cdot g_{\alpha\beta} + \frac{(n-1)}{2} \\[1em] \frac{F_i'}{F_i} \left(\Delta f_i \cdot g_{\alpha\beta} - 2 \cdot f_{i;\alpha\beta} \right) \\[1em] + \frac{1}{2F_i^2} \left(\begin{array}{c} F_i F_i'' \\ + F_i' F_i'(n-6) \end{array} \right) \left(\begin{array}{c} f_{i,\alpha} f_{i,\beta} \\ - f_{i,\mu} f_{i,\nu} g^{\mu\nu} \frac{g_{\alpha\beta}}{2} \end{array} \right) \end{array} \right) \right| \delta G^{\alpha\beta}$$

$$(1833)$$

and gives us the matter in the form:

$$T_{\alpha\beta} = \frac{R_{\alpha\beta}^{***} + \sum_{i=1}^{N} \sqrt{F_i^n} \cdot \left(\frac{(n-1)}{2} \cdot \left(\begin{array}{c} \frac{F_i'}{F_i} \left(\Delta f_i \cdot g_{\alpha\beta} - 2 \cdot f_{i;\alpha\beta} \right) \\[1em] + \frac{1}{2F_i^2} \left(\begin{array}{c} F_i F_i'' \\ + F_i' F_i'(n-6) \end{array} \right) \left(\begin{array}{c} f_{i,\alpha} f_{i,\beta} \\ - f_{i,\mu} f_{i,\nu} g^{\mu\nu} \frac{g_{\alpha\beta}}{2} \end{array} \right) \end{array} \right) \right)}{\kappa \cdot \sum_{i=1}^{N} \sqrt{F_i^n}}.$$

$$(1834)$$

11.6.10 The Quantum Origin of the Energy–Momentum Tensor to the Scale-Split-Up (1820)

Out of curiosity and for completeness, we also want to investigate the possibility of the appearance of matter terms in connection with our scale split-up (1820).

Taking the variational scale split-up (1820), we interchange the summation and reshape as follows:

$$
0 =
\begin{pmatrix}
\int\limits_{V_G} d^n x \left(\sqrt{-g} \cdot F^n \left(\overset{***}{R}_{\alpha\beta} + R_{\alpha\beta} + \dfrac{F'}{F}(1-n)\left(f_{,\alpha\beta} - \Gamma^\gamma_{\alpha\beta} f_{,\gamma}\right) \right) \delta G^{\alpha\beta} \right) \\[2ex]
-\dfrac{1}{2} \cdot \sum\limits_{i=1}^{N} \int\limits_{V_{Qi}} d^n x \left(\sqrt{-g} \cdot F^n \left(R + \dfrac{F'}{F}(1-n)\Delta f \right) g_{\alpha\beta} \delta G^{\alpha\beta} \right)
\end{pmatrix}
$$

$$
=
\begin{pmatrix}
\sum\limits_{i=1}^{N} \int\limits_{V_{Qi}} d^n x \left(\sqrt{-g} \cdot F^n \left(\overset{***}{R}_{\alpha\beta} + R_{\alpha\beta} + \dfrac{F'}{F}(1-n)\left(f_{,\alpha\beta} - \Gamma^\gamma_{\alpha\beta} f_{,\gamma}\right) \right) \delta G^{\alpha\beta} \right) \\[2ex]
-\dfrac{1}{2} \cdot \sum\limits_{i=1}^{N} \int\limits_{V_{Qi}} d^n x \left(\sqrt{-g} \cdot F^n \left(R + \dfrac{F'}{F}(1-n)\Delta f \right) g_{\alpha\beta} \delta G^{\alpha\beta} \right)
\end{pmatrix}
$$

$$\Rightarrow$$

$$
= \int\limits_{V_G} d^n x \sum\limits_{i=1}^{N} \sqrt{-g_i} \sqrt{F_i^n} \cdot \left(\overset{***}{R}_{i\alpha\beta} + R_{i\alpha\beta} - \dfrac{1}{2} \cdot R_i \cdot g_{i\alpha\beta} + \dfrac{(n-1)}{2} \cdot \dfrac{F_i'}{F_i}\left(\Delta f_i \cdot g_{i\alpha\beta} - 2 \cdot f_{i;\alpha\beta}\right) \right) \delta G_i^{\alpha\beta}.
$$

$$(1835)$$

Assuming that we could globally sum up (average) the wave function, metric, and the curvature terms, one may rewrite the last line of (1835) as follows:

$$
0 = \int\limits_{V_G} d^n x \sum\limits_{i=1}^{N} \sqrt{-g_i} \sqrt{F_i^n} \cdot
\begin{pmatrix}
\overset{***}{R}_{i\alpha\beta} + R_{i\alpha\beta} - (n-1)\cdot \dfrac{F_i'}{F_i} f_{i;\alpha\beta} \\[2ex]
-\dfrac{1}{2}\cdot\left(R_i \cdot g_{i\alpha\beta} - (n-1)\cdot \dfrac{F_i'}{F_i}\Delta f_i \cdot g_{i\alpha\beta} \right)
\end{pmatrix}
\delta G_i^{\alpha\beta}
$$

$$
\Rightarrow \qquad 0 = \sum\limits_{i=1}^{N} \sqrt{-g_i} \cdot F_i^n \cdot
\begin{pmatrix}
\overset{***}{R}_{i\alpha\beta} + R_{i\alpha\beta} - (n-1)\cdot \dfrac{F_i'}{F_i} f_{i;\alpha\beta} \\[2ex]
-\dfrac{1}{2}\cdot\left(R_i \cdot g_{i\alpha\beta} - (n-1)\cdot \dfrac{F_i'}{F_i}\Delta f_i \cdot g_{i\alpha\beta} \right)
\end{pmatrix}
$$

$$0 = R_{i\alpha\beta}^{***} + R_{\alpha\beta} - (n-1) \cdot \frac{F'}{F} f_{;\alpha\beta} - \frac{1}{2} \cdot \frac{\overbrace{\sum\limits_{i=1}^{N} \sqrt{-g_i \cdot F_i^n} \cdot \left(R_i \cdot g_{i\alpha\beta} - (n-1) \cdot \frac{F_i'}{F_i} \Delta f_i \cdot g_{i\alpha\beta} \right)}^{\text{matter}}}{\sum\limits_{i=1}^{N} \sqrt{-g_i \cdot F_i^n}}.$$

(1836)

Even though we marked the third term in (1836) as "matter," it has to be pointed out that due to the fact that we do not have the classical field equation structure with dominant (global) metric terms $R_{\alpha\beta} - \frac{1}{2} \cdot R \cdot g_{\alpha\beta}$, our "matter" here is not the same as it is the energy momentum tensor in the classical theory (not as we derived it for the last subsection via (1834)). But this does not mean that our version of interpretation or our structural reading of the scale options for the various variational forms could not also be realized by the universe on a certain scale.

Please note that in (1836) the function F is already fixed via the condition $4FF'' + F'F'(n - 6) = 0$.

As we have found that taking into account a scaled metric not only leads to a quantum gravity field theory but also provides options for the structuring of space-time, we may think about not adjusting the new possibilities to the classical equations, but rather to the experimental observations. Obviously, there are a variety of different ways for the universe to satisfy the fundamental variational task (1646) and it only seems logical to assume that the universe has not only picked one of these options for all scales.

11.7 The Special Situation in 2 Dimensions

We start with our previous result of the Einstein–Hilbert action:

$$\delta W = 0 = \int\limits_V d^n x \, \sqrt{-g \cdot F^n} \left(\begin{array}{l} R_{\alpha\beta}^{***} + R_{\alpha\beta} - R \cdot \dfrac{g_{\alpha\beta}}{2} \\[2mm] + \dfrac{F'}{F}(1-n)\left(f_{,\alpha\beta} - \Gamma_{\alpha\beta}^{\gamma} f_{,\gamma} - \dfrac{g_{\alpha\beta}}{2} \Delta f \right) \\[2mm] + \dfrac{(1-n)}{4F^2}\left(\begin{array}{c} 4FF'' \\ +F'F'(n-6) \end{array} \right)\left(f_{,\alpha} f_{,\beta} - f_{,\mu} f_{,\nu} g^{\mu\nu} \dfrac{g_{\alpha\beta}}{2} \right) \\[2mm] + \kappa T_{\alpha\beta} + \Lambda \cdot F \cdot g_{\alpha\beta} \end{array} \right) \delta G^{\alpha\beta}.$$

(1837)

In 2D, the kernel of (1837) always gives zero when assuming the cosmological constant and the energy momentum tensor to vanish. Thereby $F[f]$ is not arbitrary, because we want the term:

$$\frac{(1-n)}{4F^2}\left(4FF'' + F'F'(n-6)\right)\left(f_{,\alpha}f_{,\beta} - f_{,\mu}f_{,\nu}g^{\mu\nu}\frac{g_{\alpha\beta}}{2}\right)\xrightarrow{n=2} = 0 \qquad (1838)$$

to vanish by choosing $F[f]$ as follows:

$$F = C_{f1}\cdot e^{C_{f0}\cdot f}. \qquad (1839)$$

This gives us the total integrand:

$$0 = \overset{***}{R}_{\alpha\beta} + R_{\alpha\beta} - R\cdot\frac{g_{\alpha\beta}}{2} + \frac{F'}{F}(1-n)\left(f_{,\alpha\beta} - \Gamma^{\gamma}_{\alpha\beta}f_{,\gamma} - \frac{g_{\alpha\beta}}{2}\Delta f\right). \qquad (1840)$$

But as we also always have:

$$0 = R_{\alpha\beta} - R\cdot\frac{g_{\alpha\beta}}{2}, \qquad (1841)$$

it follows that:

$$0 = \overset{***}{R}_{\alpha\beta} + f_{,\alpha\beta} - \Gamma^{\gamma}_{\alpha\beta}f_{,\gamma} - \frac{g_{\alpha\beta}}{2}\Delta f \qquad (1842)$$

must always be automatically fulfilled. This, however, does not mean that we could not seek for solutions to the equations:

$$0 = \overset{***}{R}_{\alpha\beta} + R_{\alpha\beta} + \frac{F'}{F}(1-n)\left(f_{,\alpha\beta} - \Gamma^{\gamma}_{\alpha\beta}f_{,\gamma}\right), \qquad (1843)$$

$$0 = \left(R + \frac{F'}{F}(1-n)\Delta f\right)\cdot g_{\alpha\beta} = R + \frac{F'}{F}(1-n)\Delta f. \qquad (1844)$$

We simply applied some kind of comparison of coefficients with respect to the metric tensor.

With the help of (1841) and (1842), the first equation above, (1843), just becomes the second one (1844) when $\overset{***}{R}_{\alpha\beta} = 0$ and this is leaving us with only one equation, instead of two:

$$0 = \overbrace{R_{\alpha\beta}}^{=R\cdot\frac{g_{\alpha\beta}}{2}} + \frac{F'}{F}(1-n)\overbrace{\left(f_{,\alpha\beta} - \Gamma^{\gamma}_{\alpha\beta}f_{,\gamma}\right)}^{\frac{g_{\alpha\beta}}{2}\Delta f}$$

$$\Rightarrow \qquad 0 = R\cdot\frac{g_{\alpha\beta}}{2} + \frac{F'}{F}(1-n)\frac{g_{\alpha\beta}}{2}\Delta f = R\cdot\frac{g_{\alpha\beta}}{2} - \frac{F'}{F}\frac{g_{\alpha\beta}}{2}\Delta f$$

$$\Rightarrow \qquad 0 = R - \frac{F'}{F} \cdot \Delta f. \tag{1845}$$

Incorporating condition (1838) gives:

$$0 = R - C_{f0} \cdot \Delta f. \tag{1846}$$

Avoiding condition (1838) for $F[f]$ results in the following equation for the total integrand:

$$0 = \left(\begin{array}{c} \overset{***}{R}_{\alpha\beta} + R_{\alpha\beta} - R \cdot \dfrac{g_{\alpha\beta}}{2} + \kappa T_{\alpha\beta} + \Lambda \cdot F \cdot g_{\alpha\beta} \\[2mm] + \dfrac{F'}{F}(1-n)\left(f_{,\alpha\beta} - \Gamma^{\gamma}_{\alpha\beta} f_{,\gamma} - \dfrac{g_{\alpha\beta}}{2}\Delta f \right) \\[2mm] + \dfrac{(1-n)}{4F^2}\left(4FF'' + F'F'(n-6)\right)\left(f_{,\alpha} f_{,\beta} - f_{,\mu} f_{,\nu} g^{\mu\nu} \dfrac{g_{\alpha\beta}}{2} \right) \end{array} \right). \tag{1847}$$

Setting the matter term and the cosmological constant equal to zero, this simplifies to:

$$0 = \left(\begin{array}{c} \overset{***}{R}_{\alpha\beta} + R_{\alpha\beta} - R \cdot \dfrac{g_{\alpha\beta}}{2} \\[2mm] + \dfrac{F'}{F}(1-n)\left(f_{,\alpha\beta} - \Gamma^{\gamma}_{\alpha\beta} f_{,\gamma} - \dfrac{g_{\alpha\beta}}{2}\Delta f \right) \\[2mm] + \dfrac{(1-n)}{4F^2}\left(4FF'' + F'F'(n-6)\right)\left(f_{,\alpha} f_{,\beta} - f_{,\mu} f_{,\nu} g^{\mu\nu} \dfrac{g_{\alpha\beta}}{2} \right) \end{array} \right)$$

$$\Rightarrow \qquad 0 = R_{\alpha\beta} - R \cdot \dfrac{g_{\alpha\beta}}{2}$$

$$0 = \left(\begin{array}{c} \overset{***}{R}_{\alpha\beta} + \dfrac{F'}{F}(1-n)\left(f_{,\alpha\beta} - \Gamma^{\gamma}_{\alpha\beta} f_{,\gamma} - \dfrac{g_{\alpha\beta}}{2}\Delta f \right) \\[2mm] + \dfrac{(1-n)}{4F^2}\left(4FF'' + F'F'(n-6)\right)\left(f_{,\alpha} f_{,\beta} - f_{,\mu} f_{,\nu} g^{\mu\nu} \dfrac{g_{\alpha\beta}}{2} \right) \end{array} \right). \tag{1848}$$

As before, we can construct intrinsic conditions via "comparison of coefficients" with respect to the metric tensor as follows:

$$\textbf{1.} \quad 0 = \left(\begin{array}{c} \overset{***}{R}_{\alpha\beta} + R_{\alpha\beta} + \dfrac{F'}{F}(1-n)\left(f_{,\alpha\beta} - \Gamma^{\gamma}_{\alpha\beta} f_{,\gamma} \right) \\[2mm] + \dfrac{(1-n)}{4F^2}\left(4FF'' + F'F'(n-6)\right)\left(f_{,\alpha} f_{,\beta} \right) \end{array} \right)$$

2. $\quad 0 = \left(R + \dfrac{F'}{F}(1-n)\Delta f + \dfrac{(1-n)}{4F^2}\left(4FF'' + F'F'(n-6)\right)f_{,\mu}f_{,\nu}g^{\mu\nu} \right) \cdot \dfrac{g_{\alpha\beta}}{2}$

$\Rightarrow \quad$ **1.** $\quad 0 = \overset{***}{R_{\alpha\beta}} + R_{\alpha\beta} - \dfrac{F'}{F}\left(f_{,\alpha\beta} - \Gamma^{\gamma}_{\alpha\beta}f_{,\gamma}\right) - \dfrac{f_{,\alpha}f_{,\beta}}{F^2}\left(FF'' - F'F'\right)$

$\quad$ **2.** $\quad 0 = \left(R - \dfrac{F'}{F}\Delta f - \dfrac{1}{F^2}\left(FF'' - F'F'\right)f_{,\mu}f_{,\nu}g^{\mu\nu} \right) \cdot \dfrac{g_{\alpha\beta}}{2}, \qquad (1849)$

which also gives us an interesting equation for the Ricci tensor:

$$R_{\alpha\beta} = R \cdot \frac{g_{\alpha\beta}}{2} - \overset{***}{R_{\alpha\beta}}$$

$$= (n-1)\left(\frac{F'}{F}\Delta f + \frac{1}{4F^2}\left(4FF'' + F'F'(n-6)\right)f_{,\mu}f_{,\nu}g^{\mu\nu} \right) \cdot \frac{g_{\alpha\beta}}{2} - \overset{***}{R_{\alpha\beta}}$$

$$= \left(\frac{F'}{F}\Delta f + \frac{1}{F^2}\left(FF'' - F'F'\right)f_{,\mu}f_{,\nu}g^{\mu\nu} \right) \cdot \frac{g_{\alpha\beta}}{2} - \overset{***}{R_{\alpha\beta}}. \qquad (1850)$$

This time, we need to point out that the scaling function $F[f]$ is completely arbitrary! Thus, in 2 dimensions, where the vacuum Einstein field equations (with scaled or un-scaled metric does not matter) are always automatically fulfilled, we could still have the following—rather general—intrinsic condition:

$$0 = R - \frac{F'}{F}\Delta f - \frac{1}{F^2}\left(FF'' - F'F'\right)f_{,\mu}f_{,\nu}g^{\mu\nu}. \qquad (1851)$$

Introducing a function $H = H[f]$, we can develop the two differential operators in (1851) into one:

$$0 = R - \frac{F'}{F}\left(f_{,\mu\nu} - \Gamma^{\gamma}_{\mu\nu}f_{,\gamma}\right)g^{\mu\nu} - \frac{1}{F^2}\left(FF'' - F'F'\right)f_{,\mu}f_{,\nu}g^{\mu\nu}$$

$$= R - \frac{1}{F}\left(F'\left(f_{,\mu\nu} - \Gamma^{\gamma}_{\mu\nu}f_{,\gamma}\right) + \frac{1}{F}\left(FF'' - F'F'\right)f_{,\mu}f_{,\nu} \right)g^{\mu\nu}$$

$$= R - \frac{1}{F}\left(\frac{F'}{\sqrt{g}}\left(\sqrt{g}\,g^{\mu\nu}f_{,\mu}\right)_{,\nu} + \overbrace{\frac{1}{F}\left(FF'' - F'F'\right)}^{F'H'}f_{,\mu}f_{,\nu}g^{\mu\nu} \right)$$

$$= R - \frac{F'}{F}\left(\frac{1}{\sqrt{g}}\left(\sqrt{g}\,g^{\mu\nu}f_{,\mu}\right)_{,\nu} + H'f_{,\mu}f_{,\nu}g^{\mu\nu} \right)$$

$$= R - \frac{F'}{F}\left(\frac{1}{\sqrt{g}}\left(\sqrt{g}\,g^{\mu\nu}f_{,\mu}\right)_{,\nu} + \frac{\sqrt{g}}{\sqrt{g}}H'f_{,\mu}f_{,\nu}g^{\mu\nu} \right)$$

$$= R - \frac{F'}{F} \left(\frac{1}{\sqrt{g}} \left(\sqrt{g} \cdot g^{\mu\nu} H \cdot f_{,\mu} \right)_{,\nu} \right) \tag{1852}$$

and by defining a new metric $h^{\mu\nu}$ with the following properties:

$$\sqrt{g} \cdot g^{\mu\nu} H = \sqrt{h} \cdot h^{\mu\nu}, \tag{1853}$$

we end up with an equation containing a simple Laplace operator as follows:

$$0 = R - \frac{F'}{F} \left(\frac{1}{\sqrt{g}} \left(\sqrt{h} \cdot h^{\mu\nu} \cdot f_{,\mu} \right)_{,\nu} \right)$$

$$\Rightarrow \qquad 0 = \frac{\sqrt{g}}{\sqrt{h}} \cdot \frac{F}{F'} \cdot R_g - \frac{1}{\sqrt{h}} \left(\sqrt{h} \cdot h^{\mu\nu} \cdot f_{,\mu} \right)_{,\nu} = \frac{\sqrt{g}}{\sqrt{h}} \cdot \frac{F}{F'} \cdot R_g - \Delta_h f. \tag{1854}$$

There by it needs to be distinguished between the metric for the Laplace operator applied on f, being $h^{\mu\nu}$, and the one being the basis for the Ricci scalar R, which is $g^{\mu\nu}$. The evaluation of the function $F[f]$ via the condition for $H[f]$ as given in the third line of (1852) was already presented in section "11.6 The Simpler Route" and results in the following solution:

$$\frac{4FF'' + \left(F' \right)^2 (n-6)}{4F} = F' \cdot H' \xrightarrow{n=2} F = C_{f1} \cdot e^{C_{f0} \cdot \int_1^f e^{H|\phi|} d\phi}. \tag{1855}$$

11.8 Repetition: A Funny Inconsistency Resolved

Within the section "11.6 The Simpler Route," we had briefly discussed an interesting inconsistency in connection with the classical Einstein field equations (see also [22] and chapter 10), which we now are able to resolve.

Let us briefly repeat the problematic issue here.

We take the Einstein field equations in vacuum [10, 11], where summation with the metric tensor provides us with a peculiar simplification, namely:

$$0 = \left(R_{\alpha\beta} - R \cdot \frac{g_{\alpha\beta}}{2} \right) g^{\alpha\beta} = R - R \cdot \frac{n}{2} = R \cdot \left(1 - \frac{n}{2} \right). \tag{1856}$$

Here R gives the Ricci scalar, $R_{\alpha\beta}$ the Ricci tensor, and $g_{\alpha\beta}$ the metric tensor. We find the result in (1856) peculiar, because for non-vanishing R or general n unequal to 2, the equation cannot be fulfilled. Only in the non-vacuum case:

$$0 = \left(R_{\alpha\beta} - R \cdot \frac{g_{\alpha\beta}}{2} + \kappa T_{\alpha\beta} + \Lambda \cdot g_{\alpha\beta} \right) g^{\alpha\beta} = R \cdot \left(1 - \frac{n}{2} \right) + \kappa T_{\alpha\beta} g^{\alpha\beta} + \Lambda \cdot n \tag{1857}$$

there would be no problem.

However, the matter term was artificially introduced by Einstein and Hilbert, and thus we are of the opinion that the classical General Theory of Relativity has to be incomplete or inconsistent or both.

We saw that in order to sort this out we simply have to allow the metric to be scaled by a functional wrapper $F[f]$. We also saw, that this also—rather automatically—gives us a Quantum General Theory of Relativity.

Thereby the corresponding process of derivation is rather simple. We here start with (1806), which is our matter-free equivalent of the Einstein field equations in n dimensions. As our wrapper function F is totally arbitrary, we can easily apply condition (1775) resulting in the two equations (1807) (as just one option of many, c.f. subsection "11.6.8 Scaling Options"), which we here give again:

$$\textbf{1.}\quad R = \frac{F'}{F}(n-1)\Delta f$$

$$\textbf{2.}\quad R_{\alpha\beta} = \frac{F'}{F}(n-1)f_{;\alpha\beta}. \tag{1858}$$

Thus, different from the classical Einstein field equations, where there is only a tensor equation, we have a scalar and a tensor equation in the new, more general case. We also find that the inconsistency of the classical theory is resolved without the need of a cosmological constant or matter terms. The only thing of need was a scale factor of the metric.

We should also repeat the counterargument to the following possible point of critic:

Taking it for granted that the classical Hilbert variation should also work and give consistent results, simply because we could always assume a scaled metric where the scale factor is sucked into the various components like, for instance, here demonstrated on a scaled hyperspherical Schwarzschild solution[8]:

$$G_{\alpha\beta} = F[f] \cdot \begin{pmatrix} -c^2 \cdot \left(1 - \frac{r_s}{r}\right) & 0 & 0 & 0 \\ 0 & \dfrac{1}{\left(1 - \dfrac{r_s}{r}\right)} & 0 & 0 \\ 0 & 0 & r^2 \cdot \cosh(\vartheta)^{-2} & 0 \\ 0 & 0 & 0 & r^2 \cdot \cosh(\vartheta)^{-2} \end{pmatrix}$$

---

[8]The reader may prove that for $F[f] = 1$ the metric gives $R_{ij} = 0$.

$$
= \begin{pmatrix} -F[f] \cdot c^2 \cdot \left(1 - \dfrac{r_s}{r}\right) & 0 & 0 & 0 \\[2ex] 0 & \dfrac{F[f]}{\left(1 - \dfrac{r_s}{r}\right)} & 0 & 0 \\[2ex] 0 & 0 & F[f] \cdot r^2 \cdot \cosh(\vartheta)^{-2} & 0 \\[2ex] 0 & 0 & 0 & F[f] \cdot r^2 \cdot \cosh(\vartheta)^{-2} \end{pmatrix}.
$$

$$(1859)$$

Then variation with respect to this metric leads to the classical field equations only that instead of the ordinary R and $R_{\alpha\beta}$ terms we have:

$$
\delta W = 0 = \delta_{G_{\alpha\beta}} \int_V d^n x \left(\sqrt{-G} \times R^*\right) = \delta_{G_{\alpha\beta}} \int_V d^n x \left(\sqrt{-g \cdot F^n} \times R^*\right)
$$

$$
= \int_V d^n x \left(\sqrt{-G} \times \left(R^*{}_{\alpha\beta} - \frac{1}{2} R^* G_{\alpha\beta}\right) \delta G^{\alpha\beta}\right).
$$

$$(1860)$$

And only considering the scaled metric and its derivatives, one would indeed be left with the task of having to solve the classical Einstein field equations with just:

$$
R^*{}_{\alpha\beta} - \frac{1}{2} R^* G_{\alpha\beta} = 0.
$$

$$(1861)$$

Thus, so it seems, no problem was solved by the new approach?

However, as already shown in the previous chapter, one should not forget that the complete integrand of the variation in (1860) is not just given by (1861), but actually reads:

$$
\sqrt{-G} \times \left(R^*{}_{\alpha\beta} - \frac{1}{2} R^* G_{\alpha\beta}\right) \delta G^{\alpha\beta}.
$$

$$(1862)$$

Ruling out the determinant of the metric G to give zero (which is also a possibility we had discussed earlier, by the way), we still have, in addition to the $\boxed{R^*{}_{\alpha\beta} - \dfrac{1}{2} R^* G_{\alpha\beta}}$ term, the variation $\delta G^{\alpha\beta}$ and this, as it was shown above, can be split up into:

$$
\delta G^{\alpha\beta} = g^{\alpha\beta} \cdot \delta \frac{1}{F} + \frac{1}{F} \cdot \delta g^{\alpha\beta}.
$$

$$(1863)$$

This, however, results in the new TWO field equations (1858), which we consider quantum Einstein field equations.

In other words, Hilbert in principle had it [10], but in not recognizing—and exploiting—the possible inner, respectively scaled structure of the metric, his variation was not complete. The completed variation, taking this inner structure of the variated metric into account (1863), results in two equations, a scalar and a tensor one.

11.9 Quantum Black Hole Solutions → Inner Black Hole Metrics without Singularities

Classically the inner Schwarzschild solution [20] has been assumed to have a matter-core of a perfect fluid. Therefore, Schwarzschild had to introduce certain terms for the energy momentum tensor into the classical Einstein field equations [10, 11].

Intending to avoid any artificial introduction of matter but only to look for additional solutions to Eqs. (1815) or (1817) we want to see whether these can be used as inner solutions. In fact, such solutions were already found [25]. Here we will only briefly introduce them, while the reader is referred to [25] with respect to their application and discussion.

One can directly prove that the classical Schwarzschild solution [18] also solves (1815) with a wave function or scale term F = const. This solution can be given as:

$$g_{\alpha\beta} = \begin{pmatrix} -c^2 \cdot \left(1 - \dfrac{r_s}{r}\right) & 0 & 0 & 0 \\ 0 & \dfrac{1}{\left(1 - \dfrac{r_s}{r}\right)} & 0 & 0 \\ 0 & 0 & r^2 & 0 \\ 0 & 0 & 0 & r^2 \cdot \sin(\vartheta)^2 \end{pmatrix}. \tag{1864}$$

Here r_s denotes the Schwarzschild radius.

In addition, however, the following solution to (1817) can be found, too:

$$g_{\alpha\beta} = C_1 \cdot f[t]^2 \cdot \begin{pmatrix} -c^2 & 0 & 0 & 0 \\ 0 & t^2 & 0 & 0 \\ 0 & 0 & t^2 \cdot \sin(r)^2 & 0 \\ 0 & 0 & 0 & t^2 \cdot \sin(r)^2 \cdot \sin(\vartheta)^2 \end{pmatrix}$$

$$f[t] = t^{-1\pm i\cdot c} \cdot C_f. \tag{1865}$$

We realize that r has become an angle while t took over the position of a radius. One could even assume the object to just "live" with a fixed time-shell $t = \rho$ and obtain a solution as follows:

$$g_{\alpha\beta} = C_1 \cdot f[t]^2 \cdot \begin{pmatrix} -c^2 & 0 & 0 & 0 \\ 0 & \rho^2 & 0 & 0 \\ 0 & 0 & \rho^2 \cdot \sin(r)^2 & 0 \\ 0 & 0 & 0 & \rho^2 \cdot \sin(r)^2 \cdot \sin(\vartheta)^2 \end{pmatrix}$$

$$f[t] = e^{\pm \frac{i\cdot c}{\rho} \cdot t} \cdot C_f. \tag{1866}$$

The same can be found in 3 dimensions as follows:

$$g_{\alpha\beta} = C_1 \cdot f[t]^4 \cdot \begin{pmatrix} -c^2 & 0 & 0 \\ 0 & t^2 & 0 \\ 0 & 0 & t^2 \cdot \sin(r)^2 \end{pmatrix}.$$

$$f[t] = \sqrt{t^{\pm i\cdot c - 1}} \cdot C_f. \tag{1867}$$

As before in 4 dimensions we see that r has become an angle while t took over the position of the radius. The shell-like object similar to (1866) would be given as follows:

$$g_{\alpha\beta} = C_1 \cdot f[t]^4 \cdot \begin{pmatrix} -c^2 & 0 & 0 \\ 0 & \rho^2 & 0 \\ 0 & 0 & \rho^2 \cdot \sin(r)^2 \end{pmatrix}$$

$$f[t] = e^{\pm \frac{i\cdot c}{2\cdot\rho} \cdot t} \cdot C_f. \tag{1868}$$

We assume that there is a change of dimension at the black hole objects boundary, which prevents us from the demand of smooth transition conditions for all coordinates. Discontinuities might be acceptable for the coordinates, which vanish, interchange, or merge. In the case of the Schwarzschild metric, this is clearly t and r, which we can easily see, when observing the metric (1864) near the event horizon, which is to say around $r = r_s$. Assuming a transition from 4 to 3 dimensions, we could move from (1864) to (1868) and set the following conditions at a certain $r = r_b$:

$$\begin{pmatrix} r^2 & 0 \\ 0 & r^2 \cdot \sin(\vartheta)^2 \end{pmatrix}_{r=r_b} = ? = \left[\left(e^{\pm \frac{i \cdot c}{2 \cdot \rho} \cdot t} \cdot C_f \right)^4 \begin{pmatrix} \rho^2 & 0 \\ 0 & \rho^2 \cdot \sin(r)^2 \end{pmatrix} \right]_{\rho = r_b \ \& \ r = \vartheta} .$$

(1869)

We realize that even this is not truly possible due to the periodic scale function $f[t]$, but we think that by using the degree of freedom the complete variation of the Einstein–Hilbert action does provide, we can also fix this problem (see subsection "A Robertson–Walker Approach" in [25]).

Now, as we see that ρ clearly is a length, we want to derive its properties. From basic Quantum Theory, we know that a particle at rest has the time dependency:

$$f[t] = e^{\pm i \cdot \frac{m \cdot c^2}{\hbar} \cdot t} \cdot C_f,$$

(1870)

with m giving the rest mass of the particle and $\hbar$ denoting the reduced Planck constant. Comparing with the $f[t]$ function from the metric solution (1868), we find:

$$\frac{m \cdot c}{\hbar} = \frac{1}{2 \cdot \rho}.$$

(1871)

Inserting the Schwarzschild radius $r_s = \dfrac{2 \cdot m \cdot G}{c^2}$ (G is Newton's constant and c the speed of light in vacuum), thereby substituting the rest mass m, leaves us with:

$$\frac{r_s \cdot c^3}{2 \cdot \hbar \cdot G} = \frac{1}{2 \cdot \rho} \quad \Rightarrow \quad \rho = \frac{1}{r_s} \cdot \left(\frac{c^3}{\hbar \cdot G} \right)^{-1} = \frac{\ell_P^2}{r_s}.$$

(1872)

Here ℓ_P denotes the Planck length.

Thus, with increasing mass of the object (increasing Schwarzschild radius), the shell radius parameter ρ of the object decreases.

11.9.1 Discussion of the Missing Smoothness Conditions

We already discussed in [25] that the inner and outer solutions, with (1868) and (1864) for the inner and the outer part, respectively, can only be adjusted via boundary conditions, which are non-smooth. The same holds for other inner solutions being introduced in [25]. Smooth conditions would require more complex metric solutions with more free parameters. Such free parameters, however, may also just come from matter fields like electromagnetic ones or spin fields, and we wonder whether the fact that elementary particles rarely come without such fields also is connected with the question of the boundary conditions. We should also remind ourselves about the many additional options our variational split-up form (1819) could provide.

However, this author rather takes the metric discontinuities at $r = r_b$ (Fig. 11.37) of the new solutions, which is to say the little kinks there, than living on with a singularity at $r = 0$ (Fig. 11.38) in the classical Schwarzschild metric.

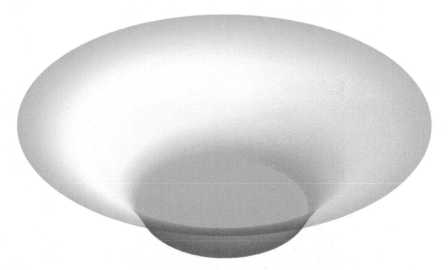

Figure 11.37 Flamm's [21] cylinder for the solution with inner and outer regions for the Schwarzschild object according to a reduced dimension according to [18] and (1864), respectively.

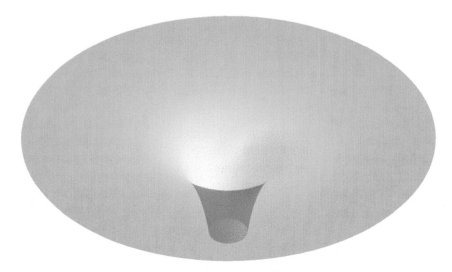

Figure 11.38 Flamm's [13] cylinder for the classical (here outer) Schwarzschild solution according to (1864).

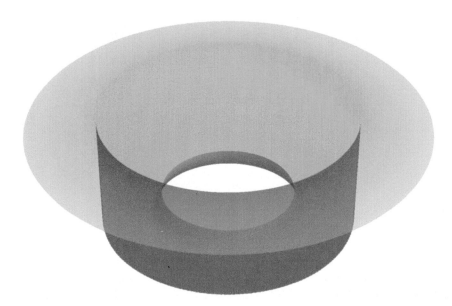

Figure 11.39 Illustration of the metric radius component for an inner solution similar to (1868), but based on the Robertson–Walker metric, the classical solution (1864) for the outer region, and the boundary conditions given in [25] (our setting for r_s was $r_s = 1$ and the natural constants were chosen to be $c = 1$ and Planck length = 0.2).

It should be pointed out, however, that by superposing a variety of inner-outer solutions of the type given above, respectively in [4], section 3.8 and (1864) (Fig. 11.37) could be used for the construction of an arbitrarily smoothed out boundary. Mathematically, this could be done similar to the method this author had introduced to construct gradient coatings for layered half-spaces in linear elastic theory (see [25], appendix C). We think that such an approach (smoothening the boundary at r_b via super-positioning of solutions of the inner-outer type) is appropriate, because quantum effects should not allow for absolutely fixed Schwarzschild radii or black hole positions. This automatically smooths the boundary anyway. The same holds for the combination of the Robertson–Walker and the Schwarzschild metric (Figs. 11.37 and 11.39). In order to achieve this, however, we'd require an additional variation with respect to any of the parameters defining the black hole. The one parameter not compromising our solution to reach this goal would be the position of the black hole. The corresponding evaluation was already introduced in [8] and given here in chapter 8.

Yet another possibility occurs when incorporating the complete solution (1868), which we have investigated further in [25]. Here one could apply the superposition of many (1868)-objects in order to achieve an ever better smoothness.

Regarding the Robertson–Walker approach as shown in Fig. 11.38, the boundary conditions give us solutions for the boundary $r = r_b > r_s$. Thus, here, signals from collapsing heavy stars should give different results than the classical Schwarzschild theory predicts when the star's surface reaches the region $r < r_b$ (see [25] with respect to the metric and boundary conditions).

That is not necessarily the only option, meaning that we can also have true black holes with $R < r_s$, which can be seen when incorporating the Bekenstein thought experiment [23, 24] and combine it with the new solutions to the Schwarzschild object, which we did in [25].

11.10 Metric Observables

In Quantum Theory, observables are assumed to be based on positive reals. Metrics cannot be put under such a stringent treatment, simply because we know that there have to be certain signatures in order to have a possibility to distinguish between time- and space-like dimensions, for instance. However, our approach with the scaled metric as introduced in section "11.3 What We Will Need" and reading:

$$G_{\alpha\beta} = F\left[f\left[t,x,y,z,\ldots,\xi_k,\ldots,\xi_n\right]\right]\cdot\delta_\alpha^i\delta_\beta^j g_{ij}$$

$$= F\left[f\left[t,x,y,z,\ldots,\xi_k,\ldots,\xi_n\right]\right]\cdot g_{\alpha\beta} = F[f]\cdot g_{\alpha\beta}, \tag{1873}$$

where the scale becomes the wave functions, should not disrupt the principal signature of a metric. We might therefore like to resort to a slightly different scaling approach, given via:

$$G_{\alpha\beta} = F\cdot\bar{F}\cdot g_{\alpha\beta}, \tag{1874}$$

where $\bar{F}$ denotes the conjugate complex of F.

In 4 dimensions this changes R^* in difference to (1643) as follows:

$$R^* = \text{sgn}\left[F[f]\right]^2 \cdot \left(\begin{array}{c} \dfrac{R}{F[f]^2} - \dfrac{3}{2\cdot F[f]^4} \\ \times \left(\begin{array}{c} 2\cdot\left(1+\text{sgn}\left[F[f]\right]^2\right)\cdot F[f]\cdot\dfrac{\partial F[f]}{\partial f}\cdot\Delta f \\ + \left(\begin{array}{c} 2\cdot\left(1+\text{sgn}\left[F[f]\right]^2\right)\cdot F[f]\cdot\dfrac{\partial^2 F[f]}{\partial f^2} \\ -\left(\text{sgn}\left[F[f]\right]^2 - 1\right)^2\cdot\left(\dfrac{\partial F[f]}{\partial f}\right)^2 \end{array}\right)\cdot f_{,\alpha}g^{\alpha\beta}f_{,\beta} \end{array}\right) \end{array}\right). \tag{1875}$$

The corresponding Ricci tensor reads:

$$R^* = R^*{}_{\alpha\beta}G^{\alpha\beta} = R^*{}_{\alpha\beta}\left(F\cdot\bar{F}\right)^{-1}\cdot g_{\alpha\beta}$$

$$\Rightarrow \qquad R^*{}_{\alpha\beta}g_{\alpha\beta} =$$

$$\left(\begin{array}{c} R_{\alpha\beta}^{***} + R_{\alpha\beta} - \dfrac{3}{2\cdot F[f]^2} \\ \left(\begin{array}{c} 2\cdot\left(1+\text{sgn}\left[F[f]\right]^2\right)\cdot F[f]\cdot\dfrac{\partial F[f]}{\partial f}\cdot\left(f_{,\alpha\beta} - \Gamma^\gamma_{\alpha\beta}f_{,\gamma}\right) \\ \times \left(\begin{array}{c} 2\cdot\left(1+\text{sgn}\left[F[f]\right]^2\right)\cdot F[f]\cdot\dfrac{\partial^2 F[f]}{\partial f^2} \\ + -\left(\text{sgn}\left[F[f]\right]^2 - 1\right)^2\cdot\left(\dfrac{\partial F[f]}{\partial f}\right)^2 \end{array}\right)\cdot f_{,\alpha}f_{,\beta} \end{array}\right) \end{array}\right)g^{\alpha\beta}, \tag{1876}$$

$\Rightarrow$

$$R^*_{\alpha\beta} = \left(R^{***}_{\alpha\beta} + R_{\alpha\beta} - \frac{3}{2\cdot F[f]^2} \times \left(\begin{pmatrix} 2\cdot\left(1+\mathrm{sgn}\left[F[f]\right]^2\right)\cdot F[f]\cdot\dfrac{\partial F[f]}{\partial f}\cdot\left(f_{,\alpha\beta}-\Gamma^\gamma_{\alpha\beta}f_{,\gamma}\right) \\ + \begin{pmatrix} 2\cdot\left(1+\mathrm{sgn}\left[F[f]\right]^2\right)\cdot F[f]\cdot\dfrac{\partial^2 F[f]}{\partial f^2} \\ -\left(\mathrm{sgn}\left[F[f]\right]^2-1\right)^2\cdot\left(\dfrac{\partial F[f]}{\partial f}\right)^2 \end{pmatrix}\cdot f_{,\alpha}f_{,\beta} \end{pmatrix} \right) \right),$$

(1877)

After variation under the Einstein–Hilbert action integral with respect to the metric, we obtain:

$$\delta W = 0 = \int_V d^n x\sqrt{-g\cdot(F\cdot\bar{F})^4}\,\delta G_a^{\ \alpha\beta}$$

$$\left(\begin{array}{l} R^{***}_{\alpha\beta} + R_{\alpha\beta} - R\cdot\dfrac{g_{\alpha\beta}}{2} \\[2mm] +\dfrac{3\cdot g_{\alpha\beta}}{4\cdot F[f]^2}\cdot\left(\begin{matrix} 2\cdot\left(1+\mathrm{sgn}\left[F[f]\right]^2\right)\cdot F[f]\cdot\dfrac{\partial F[f]}{\partial f}\cdot\Delta f \\ + \begin{pmatrix} 2\cdot\left(1+\mathrm{sgn}\left[F[f]\right]^2\right)\cdot F[f]\cdot\dfrac{\partial^2 F[f]}{\partial f^2} \\ -\left(\mathrm{sgn}\left[F[f]\right]^2-1\right)^2\cdot\left(\dfrac{\partial F[f]}{\partial f}\right)^2 \end{pmatrix}\cdot f_{,\mu}g^{\mu\nu}f_{,\nu} \end{matrix} \right) \\[6mm] -\dfrac{3}{2\cdot F[f]^2}\cdot\left(\begin{matrix} 2\cdot\left(1+\mathrm{sgn}\left[F[f]\right]^2\right)\cdot F[f]\cdot\dfrac{\partial F[f]}{\partial f}\cdot\left(f_{,\alpha\beta}-\Gamma^\gamma_{\alpha\beta}f_{,\gamma}\right) \\ + \begin{pmatrix} 2\cdot\left(1+\mathrm{sgn}\left[F[f]\right]^2\right)\cdot F[f]\cdot\dfrac{\partial^2 F[f]}{\partial f^2} \\ -\left(\mathrm{sgn}\left[F[f]\right]^2-1\right)^2\cdot\left(\dfrac{\partial F[f]}{\partial f}\right)^2 \end{pmatrix}\cdot f_{,\alpha}f_{,\beta} \end{matrix} \right) \\[6mm] +\kappa T_{a\alpha\beta}+\Lambda\cdot F^a\cdot g_{a\alpha\beta} \end{array} \right).$$

(1878)

This now allows for sudden changes of the equations simply due to changes of the signature function sgn[F] and can even lead to switches from Klein–

Gordon $\left(\left(1 - \mathrm{sgn}\left[F[f] \right]^2 \right) = 0 \ \& \ \dfrac{\partial^2 F[f]}{\partial f^2} = 0 \right)$ to Dirac equations (1 +

sgn[F[f]]^2 = 0). This, however, would be equivalent to a switch of spinless particles (or composite particles, like the pion) to a fermion.

11.11   The Gravity Dirac Equation Revisited

We use (1833) in the following form:

$$0 = \left(\begin{array}{c} \overset{***}{R}_{\alpha\beta} + R_{\alpha\beta} - \dfrac{1}{2} \cdot R \cdot g_{\alpha\beta} - \dfrac{F_i{}'}{F_i}(n-1)\left(f_{i,\alpha\beta} - \Gamma^{\gamma}_{\alpha\beta} f_{i,\gamma} - \dfrac{g_{\alpha\beta}}{2}\Delta f_i \right) \\[2ex] + \dfrac{(1-n)}{4F_i^2}\left(F_i F_i{}'' + F_i{}' F_i{}'(n-6) \right)\left(f_{i,\alpha} f_{i,\beta} - f_{i,\mu} f_{i,\nu} g^{\mu\nu} \dfrac{g_{\alpha\beta}}{2} \right) \end{array} \right) \quad (1879)$$

and set $F[f] = f$, which gives us:

$$0 = \left(\begin{array}{c} \overset{***}{R}_{\alpha\beta} + R_{\alpha\beta} - \dfrac{1}{2} \cdot R \cdot g_{\alpha\beta} - \dfrac{1}{f_i}(n-1)\left(f_{i,\alpha\beta} - \Gamma^{\gamma}_{\alpha\beta} f_{i,\gamma} - \dfrac{g_{\alpha\beta}}{2}\Delta f_i \right) \\[2ex] + \dfrac{(1-n)}{4f_i^2}(n-6)\left(f_{i,\alpha} f_{i,\beta} - f_{i,\mu} f_{i,\nu} g^{\mu\nu} \dfrac{g_{\alpha\beta}}{2} \right) \end{array} \right). \quad (1880)$$

Now we assume the possibility for a decomposition of the following form:

$$\beta_{\alpha} \cdot \beta_{\beta} = \left(\begin{array}{c} \overset{***}{R}_{\alpha\beta} + R_{\alpha\beta} - \dfrac{1}{2} \cdot R \cdot g_{\alpha\beta} - \dfrac{1}{f_i}(n-1)\left(f_{i,\alpha\beta} - \Gamma^{\gamma}_{\alpha\beta} f_{i,\gamma} - \dfrac{g_{\alpha\beta}}{2}\Delta f_i \right) \\[2ex] - \dfrac{(1-n)}{8f_i^2}(n-6) f_{i,\mu} f_{i,\nu} g^{\mu\nu} g_{\alpha\beta} \end{array} \right) \quad (1881)$$

with some yet unknown objects β. This leads us to the following equation:

$$0 = \beta_{\alpha} \cdot \beta_{\beta} \cdot \underbrace{\dfrac{4}{(1-n)(n-6)\left(1 - \dfrac{n}{2}\right)}}_{\equiv \chi^2} \cdot f_i^2 + f_{i,\alpha} f_{i,\beta} = \beta_{\alpha} \cdot \beta_{\beta} \cdot \chi^2 \cdot f_i^2 + f_{i,\alpha} f_{i,\beta}.$$

$$(1882)$$

We realize that (1882) could just be obtained as scalar product of the following vectors:

$$\frac{1}{\sqrt{2}}\left\{\begin{matrix}(\beta_\alpha \cdot \chi \cdot f_i + f_{i,\alpha})\\ (\beta_\alpha \cdot \chi \cdot f_i - f_{i,\alpha})\end{matrix}\right\} \cdot \frac{1}{\sqrt{2}}\left\{\begin{matrix}(\beta_\beta \cdot \chi \cdot f_i + f_{i,\beta})\\ (\beta_\beta \cdot \chi \cdot f_i - f_{i,\beta})\end{matrix}\right\}$$

$$\Rightarrow \frac{1}{\sqrt{2}}\left\{\begin{matrix}(\beta_\alpha \cdot \chi \cdot f_i + f_{i,\alpha})\\ (\beta_\alpha \cdot \chi \cdot f_i - f_{i,\alpha})\end{matrix}\right\} \cdot \frac{1}{\sqrt{2}}\left\{\begin{matrix}(\beta_\beta \cdot \chi \cdot f_i + f_{i,\beta})\\ (\beta_\beta \cdot \chi \cdot f_i - f_{i,\beta})\end{matrix}\right\} = \beta_\alpha \cdot \beta_\beta \cdot \chi^2 \cdot f_i^2 + f_{i,\alpha} f_{i,\beta}.$$

(1883)

Even though this simple vector form is not a suitable decomposition for us, because it gives two completely different equations for the same wave function, we conclude that decomposition should be possible when finding the right objects to result in the connection of the terms in (1882). Our Dirac-like approach shall be:

$$\left(a \cdot \beta_\alpha \cdot \chi \cdot f_i + A \cdot f_{i,\alpha}\right)\left(b \cdot \beta_\beta \cdot \chi \cdot f_i + B \cdot f_{i,\beta}\right) = 0. \qquad (1884)$$

With the objects a, A, b and B being general 2×2 matrices, we find a variety of solutions of which we here only present a very simple one, namely:

$$A = \begin{pmatrix} 1 & 1 \\ 1 & 1 \end{pmatrix}; \quad a = \begin{pmatrix} -1 & 1 \\ -1 & 1 \end{pmatrix}; \quad B = \begin{pmatrix} 1 & -1 \\ 1 & -1 \end{pmatrix}; \quad b = \begin{pmatrix} -1 & 1 \\ 1 & -1 \end{pmatrix}. \qquad (1885)$$

This makes (1884) to result in:

$$\left(a \cdot \beta_\alpha \cdot \chi \cdot f_i + A \cdot f_{i,\alpha}\right)\left(b \cdot \beta_\beta \cdot \chi \cdot f_i + B \cdot f_{i,\beta}\right)$$

$$= 2 \cdot \begin{pmatrix} 1 & -1 \\ 1 & -1 \end{pmatrix} \cdot \left(\beta_\alpha \cdot \beta_\beta \cdot \chi^2 \cdot f_i^2 + f_{i,\alpha} f_{i,\beta}\right) \qquad (1886)$$

and gives us the gravity "Dirac equations":

$$a \cdot \beta_\alpha \cdot \chi \cdot f_i + A \cdot f_{i,\alpha} = \begin{pmatrix} -1 & 1 \\ -1 & 1 \end{pmatrix} \cdot \beta_\alpha \cdot \chi \cdot f_i + \begin{pmatrix} 1 & 1 \\ 1 & 1 \end{pmatrix} \cdot f_{i,\alpha} = 0$$

$$b \cdot \beta_\beta \cdot \chi \cdot f_i + B \cdot f_{i,\beta} = \begin{pmatrix} -1 & 1 \\ 1 & -1 \end{pmatrix} \cdot \beta_\beta \cdot \chi \cdot f_i + \begin{pmatrix} 1 & -1 \\ 1 & -1 \end{pmatrix} \cdot f_{i,\beta} = 0. \qquad (1887)$$

Please note that the decomposition (1881) may also be performed as follows:

$$\beta_\alpha \cdot \beta_\beta = R_{\alpha\beta}^{***} + R_{\alpha\beta} - \frac{1}{2} \cdot R \cdot g_{\alpha\beta} - \frac{1}{f_i}(n-1)\left(f_{i,\alpha\beta} - \Gamma_{\alpha\beta}^\gamma f_{i,\gamma} - \frac{g_{\alpha\beta}}{2}\Delta f_i\right). \qquad (1888)$$

This leads us to the following equation:

$$0 = \beta_\alpha \cdot \beta_\beta \cdot \underbrace{\frac{4}{(1-n)(n-6)\left(1-\dfrac{n}{2}\right)}}_{\equiv \chi^2} \cdot f_i^2 + f_{i,\alpha} f_{i,\beta} - \frac{f_{i,\mu} f_{i,\nu} g^{\mu\nu} g_{\alpha\beta}}{2}$$

$$= \beta_\alpha \cdot \beta_\beta \cdot \chi^2 \cdot f_i^2 + f_{i,\alpha} f_{i,\beta} - \frac{f_{i,\mu} f_{i,\nu} g^{\mu\nu} g_{\alpha\beta}}{2}. \tag{1889}$$

Now our Dirac-like approach shall be:

$$\left(\begin{array}{c} a \cdot \beta_\alpha \cdot \chi \cdot f_i + A \cdot f_{i,\alpha} \\ + k \cdot \dfrac{i}{\sqrt{2}} \cdot f_{i,\mu} \cdot \mathbf{e}^\mu \cdot \mathbf{e}_\alpha \end{array} \right) \times \left(\begin{array}{c} b \cdot \beta_\beta \cdot \chi \cdot f_i + B \cdot f_{i,\beta} \\ + K \cdot \dfrac{i}{\sqrt{2}} \cdot f_{i,\nu} \cdot \mathbf{e}^\nu \cdot \mathbf{e}_\beta \end{array} \right) = 0. \tag{1890}$$

A more general solution and some discussion will be presented elsewhere.

11.12 The Math for Body, Soul, and Universe?

From the material presented in the previous sections, one might feel tempted to conclude that the scale factors to every metric may be interpreted as the soul of objects. That things are not so simple can be demonstrated by the fact that in our 4-dimensional space-time the corresponding scale factor provides the electromagnetic field as it was shown in section 2 of [4] and has been repeated here in chapter 2 of this book. So, our question here, how can we come to a gravity quantum description which embeds more dimensions with metric and quantum fields? The solution to this problem can be found via the following approach:

$$G_{\alpha\beta} = F_n\left[f_n\right] \begin{pmatrix} F_{n-1}\cdots F_2 \cdot g_{00} & F_{n-1}\cdots F_2 \cdot g_{10} & \cdots & \cdots & \cdots & g_{n0} \\ F_{n-1}\cdots F_2 \cdot g_{10} & F_{n-1}\cdots F_2 \cdot g_{11} & \cdots & \cdots & \cdots & \vdots \\ F_{n-1}\cdots F_3 \cdot g_{20} & F_{n-1}\cdots F_3 \cdot g_{21} & \ddots & \cdots & \cdots & \vdots \\ F_{n-1}\cdots F_4 \cdot g_{30} & F_{n-1}\cdots F_4 \cdot g_{31} & \cdots & \ddots & \cdots & \vdots \\ \cdots & \cdots & & \cdots & \cdots & \ddots & \vdots \\ g_{n0} & \cdots & & \cdots & \cdots & \cdots & g_{nn} \end{pmatrix}$$

$$F_i = F_i\left[f_i\right]. \tag{1891}$$

Using the definition:

$$^2G_{\alpha\beta} = F_2 \cdot {}^2g_{\alpha\beta}$$

$$^3G_{\alpha\beta} = F_3 \cdot {}^3g_{\alpha\beta}$$

$$\vdots$$

$$^nG_{\alpha\beta} = F_n \cdot {}^ng_{\alpha\beta},$$ (1892)

we can write (1891) as follows:

$$^2G_{\alpha\beta} = \left(F_2 \cdot \left({}^2g_{\alpha\beta}\right)\right) = F_2 \cdot \begin{pmatrix} {}^2g_{00} & {}^2g_{01} \\ {}^2g_{01} & {}^2g_{11} \end{pmatrix}$$

$$^3G_{\alpha\beta} = F_3 \cdot {}^3g_{\alpha\beta} = F_3 \cdot \begin{pmatrix} {}^3g_{00} & {}^3g_{01} & {}^3g_{02} \\ {}^3g_{01} & {}^3g_{11} & {}^3g_{12} \\ {}^3g_{02} & {}^3g_{12} & {}^3g_{22} \end{pmatrix} = F_3 \cdot \begin{pmatrix} F_2 \cdot \begin{pmatrix} {}^2g_{00} & {}^2g_{01} \\ {}^2g_{01} & {}^2g_{11} \end{pmatrix} & {}^3g_{02} \\ & {}^3g_{12} \\ {}^3g_{02} & {}^3g_{12} & {}^3g_{22} \end{pmatrix}$$

$$\vdots$$

$$^nG_{\alpha\beta} = F_n \cdot \begin{pmatrix} \cdots F_3 \cdot \begin{pmatrix} F_2 \cdot \begin{pmatrix} {}^2g_{00} & {}^2g_{01} & {}^3g_{02} \\ {}^2g_{01} & {}^2g_{11} & {}^3g_{12} \\ {}^3g_{02} & {}^3g_{12} & {}^3g_{22} \end{pmatrix} \cdots \end{pmatrix} & {}^ng_{0m} \\ & \vdots \\ & \vdots \\ {}^ng_{0m} \quad \cdots \quad \cdots \quad \cdots \quad \cdots \quad {}^ng_{mm} \end{pmatrix} ; \quad m = n-1 \quad (1893)$$

and obtain the corresponding multi-scaled Ricci scalar to be:

$$R^* = \sum_{m=1}^{n-1}\left(\sum_{\alpha=0}^{m} 2 \cdot {}^{m+1}R_{m\alpha} \, {}^{m+1}g^{m\alpha} - {}^{m+1}R_{mm} \, {}^{m+1}g^{mm}\right) + \sum_{m=2}^{n} \frac{F_m'}{F_m} \cdot \Delta_m f_m$$

$$\Delta_m = \frac{1}{\sqrt{{}^mg}} \sum_{\alpha,\beta=0}^{m-1} \partial_\beta \left(\sqrt{{}^mg} \cdot {}^mg^{\alpha\beta} \cdot \partial_\alpha f\right).$$ (1894)

Thereby the ${}^jR_{\alpha\beta}$ are denoting the Ricci tensor to the ${}^jg_{\alpha\beta}$ metric and all functions F_i have to fulfill the condition:

$$4 \cdot F_i F_i'' + F_i' F_i' (n-6) = 0 \quad \Rightarrow \quad F_i = \left(C_f \pm f^a\right)^{\frac{4}{n-2}}; \quad n > 2$$

$$4 \cdot F_2 F_2'' + F_2' F_2' (n-6) = 0 \quad \Rightarrow \quad F_2 = C_F \cdot e^{C_f \cdot f}; \quad n = 2.$$ (1895)

Most interestingly, we obtain a sum of Laplace operations in various dimensions on the scale factors (wave functions) f plus curvature terms. Thus, what originally (on first sight) has been a typical GTR task, namely the evaluation of the Ricci tensor and the subsequent Ricci scalar, evolves to a complex ensemble of quantum equations if only be brought into the right form. It should explicitly be pointed out that we can also construct string, brane, multi-verse, and similar structures. Here we just give a few examples:

$$G_{\alpha\beta} = {}_nF\left[f_n\right] \begin{pmatrix} {}_2F_1\left[f_1\right]\cdot{}_1^2g_{\alpha\beta} & {}^2 0 & \cdots & \cdots & \cdots & {}^2 0 \\ {}^2 0 & {}_2F_2\left[f_2\right]\cdot{}_2^2g_{\alpha\beta} & \cdots & \cdots & \cdots & \vdots \\ {}^2 0 & {}^2 0 & \ddots & \cdots & \cdots & \vdots \\ {}^2 0 & {}^2 0 & \cdots & \ddots & \cdots & \vdots \\ \cdots & \cdots & \cdots & \cdots & \ddots & \vdots \\ {}^2 0 & \cdots & \cdots & \cdots & \cdots & {}_{n/2}^2 g_{\alpha\beta} \end{pmatrix}$$

$$_i^2 g_{\alpha\beta} = \begin{pmatrix} {}_ig_{00} & {}_ig_{01} \\ {}_ig_{01} & {}_ig_{11} \end{pmatrix}; \quad {}_2F_i\left[f_i\right] = C_{f1i}\cdot e^{C_{fi}\cdot f_i} = C_{f1i}\cdot e^{C_{fi}\cdot f_i\left[x_i,y_i\right]}; \quad {}^2 0 = \begin{pmatrix} 0 & 0 \\ 0 & 0 \end{pmatrix},$$

$$(1896)$$

$$G_{\alpha\beta} = {}_nF\left[f_n\right] \begin{pmatrix} {}_3F_1\left[{}_3f_1\right]\cdot{}_1^3g_{\alpha\beta} & {}^2 0 & \cdots & \cdots & \cdots & {}^3 0 \\ {}^3 0 & {}_3F_2\left[{}_3f_2\right]\cdot{}_2^3g_{\alpha\beta} & \cdots & \cdots & \cdots & \vdots \\ {}^3 0 & {}^3 0 & \ddots & \cdots & \cdots & \vdots \\ {}^3 0 & {}^3 0 & \cdots & \ddots & \cdots & \vdots \\ \cdots & \cdots & \cdots & \cdots & \ddots & \vdots \\ {}^3 0 & \cdots & \cdots & \cdots & \cdots & {}_{n/3}^3 g_{\alpha\beta} \end{pmatrix}$$

$$_i^3 g_{\alpha\beta} = \begin{pmatrix} {}_2F_i\left[{}_2f_i\right]\cdot\begin{pmatrix} {}_ig_{00} & {}_ig_{01} \\ {}_ig_{01} & {}_ig_{11} \end{pmatrix} & {}_ig_{02} \\ & {}_ig_{11} \\ {}_ig_{02} & {}_ig_{21} & {}_ig_{22} \end{pmatrix};$$

$$_2F_i\left[{}_2f_i\right] = C_{f1i}\cdot e^{C_{fi}\cdot{}_2f_i} = C_{f1i}\cdot e^{C_{fi}\cdot{}_2f_i\left[x_i,y_i\right]}$$

$$_3F_i\left[{}_3f_i\right] = C_{f1i}\cdot\left(C_{fi} + {}_3f_i\left[t_i,x_i,y_i,z_i\right]\right)^4; \quad {}^3 0 = \begin{pmatrix} 0 & 0 & 0 \\ 0 & 0 & 0 \\ 0 & 0 & 0 \end{pmatrix}, \quad (1897)$$

$$G_{\alpha\beta} = {}_nF\big[f_n\big]\begin{pmatrix} {}_4F_1\big[f_1\big]\cdot{}_1^4g_{\alpha\beta} & {}^40 & \cdots & \cdots & \cdots & {}^40 \\ {}^40 & {}_4F_2\big[f_2\big]\cdot{}_2^4g_{\alpha\beta} & \cdots & \cdots & \cdots & \vdots \\ {}^40 & {}^40 & \ddots & \cdots & \cdots & \vdots \\ {}^40 & {}^40 & & \ddots & \cdots & \vdots \\ \cdots & \cdots & & \cdots & \ddots & \vdots \\ {}^40 & \cdots & & \cdots & \cdots & {}_{n/4}^4g_{\alpha\beta} \end{pmatrix}$$

$$_i^4g_{\alpha\beta} = \begin{pmatrix} {}_2F_{i1}\big[{}_2f_{i1}\big]\cdot\begin{pmatrix} {}_ig_{00} & {}_ig_{01} \\ {}_ig_{01} & {}_ig_{11} \end{pmatrix} & {}^20 \\ {}^20 & {}_2F_{i2}\big[{}_2f_{i2}\big]\cdot\begin{pmatrix} {}_ig_{00} & {}_ig_{01} \\ {}_ig_{01} & {}_ig_{11} \end{pmatrix} \end{pmatrix};$$

$$_4F_i\big[f_i\big] = C_{f1i}\cdot\big(C_{fi} + f_i\big[t_i,x_i,y_i,z_i\big]\big)^2; \quad {}^40 = \begin{pmatrix} 0 & 0 & 0 & 0 \\ 0 & 0 & 0 & 0 \\ 0 & 0 & 0 & 0 \\ 0 & 0 & 0 & 0 \end{pmatrix}. \tag{1898}$$

When evaluating the Ricci scalar, we obtain—just as shown above with the Matryoshka structure from (1893)—sums of Laplace operators applied on the various wave functions appearing in (1896) to (1898). Almost arbitrary combinations of scaled sub-metrics in accordance with the examples given above are possible.

Examples and discussion will be presented elsewhere [27].

With this new multitude of options, however, the number of degrees of freedom does not end. As elaborated in [4] and [8], there is also the option of centers of gravity and centers of wave functions, which simply means that we can seek for a satisfaction of (1646) via variation with respect to the position of such centers. Seeing that then the last line of (1893) would become generalized as follows:

$$^nG_{\alpha\beta} = F_n\big[X_n\big]\cdot\begin{pmatrix} \Big(\cdots F_3\big[X_3\big]\cdot\begin{pmatrix} F_2\big[X_2\big]\cdot\begin{pmatrix} {}^2g_{00} & {}^2g_{01} & {}^3g_{02} \\ {}^2g_{01} & {}^2g_{11} & {}^3g_{12} \\ {}^3g_{02} & {}^3g_{12} & {}^3g_{22} \end{pmatrix} \cdots\Big)^n g_{0m} & \vdots \\ \cdots & \vdots \\ {}^ng_{0m} & \cdots \quad \cdots \quad \cdots \quad \cdots & {}^ng_{mm} \end{pmatrix}$$

$$X_i = x_i - \xi_i; \quad {}^ig_{\alpha\beta} = {}^ig_{\alpha\beta}\big[Y_i\big] = {}^ig_{\alpha\beta}\big[y_i - \zeta_i\big], \tag{1899}$$

we realize a huge number of possible combinations, all fulfilling (1646) and many having the possibility to code not just matter, dark matter, energy, but also objects the common language might better express with the word "soul."

11.13 Appendix

Nearly having reached the end of our journey, respectively the book, we here want once more to give the most important connections between certain classical quantum equations and the scaled metric as used repeatedly throughout the book. Thereby we are extending our partially presented previous evaluations with respect to what was learned within the course.

11.13.1 Appendix A1: Derivation of the Metric Klein–Gordon Equation from the Scaled Ricci Scalar R^*

In order to obtain the metric Klein–Gordon equation, we take the scaled Ricci scalar from (1643) and demand:

$$0 = 4FF'' + \left(F'\right)^2 \left(n-6\right) \quad \Rightarrow \quad F = \begin{cases} \left(f - C_{f0}\right)^{\frac{4}{n-2}} \cdot C_{f1} & n > 2 \\ e^{C_{f0} \cdot f} \cdot C_{f1} & n = 2, \end{cases} \quad (1900)$$

which immediately gives us:

$$R^* = R^*_{\alpha\beta} G^{\alpha\beta} = \left(\begin{array}{l} \Gamma^\sigma_{\alpha\beta,\sigma} g^{\alpha\beta} - \Gamma^\sigma_{\beta\sigma,\alpha} g^{\alpha\beta} - \Gamma^\mu_{\sigma\alpha} \Gamma^\sigma_{\beta\mu} g^{\alpha\beta} + \Gamma^\sigma_{\alpha\beta} \Gamma^\mu_{\sigma\mu} g^{\alpha\beta} \\ + \dfrac{F'}{F}\left(1-n\right)\Delta f + \dfrac{f_{,\alpha} f_{,\beta} g^{\alpha\beta} \left(1-n\right)}{4F^2}\left(4FF'' + \left(F'\right)^2 \left(n-6\right)\right) \end{array} \right) \dfrac{1}{F}$$

$$= \left(R + \dfrac{F'}{F}\left(1-n\right)\Delta f + \dfrac{f_{,\alpha} f_{,\beta} g^{\alpha\beta} \left(1-n\right)}{4F^2}\left(4FF'' + \left(F'\right)^2 \left(n-6\right)\right) \right)\dfrac{1}{F}$$

$$= \left(R + \dfrac{F'}{F}\left(1-n\right)\Delta f \right)\dfrac{1}{F}$$

with: $\qquad F = F[f]; \quad F' = \dfrac{\partial F[f]}{\partial f}; \quad F'' = \dfrac{\partial^2 F[f]}{\partial f^2}.$ $\qquad (1901)$

Now we assume the total curvature R^* to vanish (where this condition comes from has been elaborated above, mainly in section "11.6 The Simpler Route") and result in:

$$R^* = 0 \quad \rightarrow \quad 0 = F \cdot R + F' \cdot (1-n) \cdot \Delta f$$

$$\Rightarrow \qquad 0 = \begin{cases} \left(f - C_{f0} \right)^{\frac{4}{n-2}} \cdot C_{f1} \left(R + \dfrac{(1-n)}{\left(f - C_{f0} \right)} \cdot \Delta f \right) & n > 2 \\[3mm] e^{C_{f0} \cdot f} \cdot C_{f1} \left(R + C_{f0} \cdot (1-n) \cdot \Delta f \right) & n = 2. \end{cases} \qquad (1902)$$

Thus, in the case of $n > 2$, we always also have the option for a constant (broken symmetry) solution of the kind:

$$0 = f - C_{f0} \quad \Rightarrow \quad f = C_{f0}. \qquad (1903)$$

We may interpret this as a Higgs solution.

Otherwise, we have the simple equations:

$$0 = \begin{cases} \left(f - C_{f0} \right) \cdot R + (1-n) \cdot \Delta f & n > 2 \\[2mm] R + C_{f0} \cdot (1-n) \cdot \Delta f & n = 2. \end{cases} \qquad (1904)$$

Depending on the curvature term R, which substitutes the classical terms for mass and potential, we can have the classical Klein–Gordon equation or more complex versions of the latter. We also need to point out that potential and mass (or other forms of inertia and fields) can also be obtained via certain (potentially compactified) dimensions when simultaneously choosing suitable structures for the functions f with respect to these add-on dimensions. This was shown in the previous chapters of this book as well as our book "World Formula ..." [4], but especially and with quite many "practical" examples in [3] and chapter 6 of this book.

Meanwhile we have also learned how we can answer the question about the origin of the condition $R^* = 0$ (section 11.6).

Assuming the Lagrange matter density L_M and the cosmological constant Λ to vanish, the variation of the Einstein–Hilbert action with a scaled metric $G_{\delta\gamma} = F[f] \cdot g_{\delta\gamma}$ results in:

$$0 = \left(\int_{V_G} d^n x \sum_{i=1}^{N} \sqrt{-g} \cdot F_i^n \left(\begin{pmatrix} R_{\alpha\beta}^{***} + R_{\alpha\beta} - \dfrac{1}{2} \cdot R \cdot g_{\alpha\beta} - \dfrac{F_i'}{F_i} (n-1) \\[3mm] \times \left(f_{i,\alpha\beta} - \Gamma_{\alpha\beta}^{\gamma} f_{i,\gamma} - \dfrac{g_{\alpha\beta}}{2} \Delta f_i \right) \\[3mm] + \dfrac{(1-n)}{4F_i^2} \begin{pmatrix} 4 F_i F_i'' \\ + F_i' F_i' (n-6) \end{pmatrix} \left(f_{i,\alpha} f_{i,\beta} - f_{i,\mu} f_{i,\nu} g^{\mu\nu} \dfrac{g_{\alpha\beta}}{2} \right) \end{pmatrix} \delta G^{\alpha\beta} \right).$$

$$(1905)$$

Thereby we have also assumed to have many scaling factors and volumes to perform the integration over (see section 11.4 and following for motivation). This way we will also be able to directly extract a metric Dirac equation as it was shown in section 11.4.2.

By splitting up the variation of the scaled metric:

$$0 = \left(\int_{V_G} d^n x \sum_{i=1}^{N} \sqrt{-g} \cdot F_i^n \left(\begin{pmatrix} R_{\alpha\beta}^{***} + R_{\alpha\beta} - \dfrac{1}{2} \cdot R \cdot g_{\alpha\beta} - \dfrac{F_i'}{F_i}(n-1) \\ \times \left(f_{i,\alpha\beta} - \Gamma_{\alpha\beta}^{\gamma} f_{i,\gamma} - \dfrac{g_{\alpha\beta}}{2} \Delta f_i \right) \\ + \dfrac{(1-n)}{4F_i^2}\left(4F_i F_i'' + F_i' F_i'(n-6) \right) \\ \times \left(f_{i,\alpha} f_{i,\beta} - f_{i,\mu} f_{i,\nu} g^{\mu\nu} \dfrac{g_{\alpha\beta}}{2} \right) \end{pmatrix} \begin{pmatrix} g^{\alpha\beta}\delta\dfrac{1}{F_i} \\ + \dfrac{\delta g^{\alpha\beta}}{F_i} \end{pmatrix} \right) \right),$$

$$(1906)$$

we obtain two equations:

$$0 = \int_{V_G} d^n x \sum_{i=1}^{N} \sqrt{-g} \cdot F_i^n \left(\begin{pmatrix} R_{\alpha\beta}^{***} + R_{\alpha\beta} - \dfrac{1}{2} \cdot R \cdot g_{\alpha\beta} - \dfrac{F_i'}{F_i}(n-1) \\ \times \left(f_{i,\alpha\beta} - \Gamma_{\alpha\beta}^{\gamma} f_{i,\gamma} - \dfrac{g_{\alpha\beta}}{2} \Delta f_i \right) \\ + \dfrac{(1-n)}{4F_i^2}\left(4F_i F_i'' + F_i' F_i'(n-6) \right) \\ \times \left(f_{i,\alpha} f_{i,\beta} - f_{i,\mu} f_{i,\nu} g^{\mu\nu} \dfrac{g_{\alpha\beta}}{2} \right) \end{pmatrix} g^{\alpha\beta}\delta\dfrac{1}{F_i} \right)$$

$$= \int_{V_G} d^n x \sum_{i=1}^{N} \sqrt{-g} \cdot F_i^n \left(\begin{pmatrix} R\left(1 - \dfrac{n}{2}\right) - \dfrac{F_i'}{F_i}(n-1)\left(1 - \dfrac{n}{2}\right)\Delta f_i \\ + \dfrac{(1-n)}{4F_i^2}\left(4F_i F_i'' + F_i' F_i'(n-6) \right) \\ \times \left(f_{i,\alpha} f_{i,\beta} - f_{i,\mu} f_{i,\nu} g^{\mu\nu} \dfrac{n}{2} \right) \end{pmatrix} \delta\dfrac{1}{F_i} \right)$$

$$= \int_{V_G} d^n x \sum_{i=1}^{N} \sqrt{-g} \cdot F_i^n \left(\left(1 - \frac{n}{2}\right) \left(\begin{array}{c} R - \dfrac{F_i'}{F_i}(n-1)\Delta f_i \\ + \dfrac{f_{i,\alpha} f_{i,\beta}(1-n)}{4F_i^2} g^{\alpha\beta} \left(\begin{array}{c} 4F_i F_i'' \\ +F_i' F_i'(n-6) \end{array} \right) \end{array} \right) \right) \delta \frac{1}{F_i} ,$$

$$(1907)$$

$$0 = \int_{V_G} d^n x \sum_{i=1}^{N} \sqrt{-g} \cdot F_i^n \left(\left(\left(\begin{array}{c} R_{\alpha\beta}^{***} + R_{\alpha\beta} - \dfrac{1}{2} \cdot R \cdot g_{\alpha\beta} - \dfrac{F_i'}{F_i}(n-1) \\ \times \left(f_{i,\alpha\beta} - \Gamma_{\alpha\beta}^{\gamma} f_{i,\gamma} - \dfrac{g_{\alpha\beta}}{2}\Delta f_i \right) \\ + \dfrac{(1-n)}{4F_i^2} \left(\begin{array}{c} 4F_i F_i'' \\ +F_i' F_i'(n-6) \end{array} \right) \left(f_{i,\alpha} f_{i,\beta} - f_{i,\mu} f_{i,\nu} g^{\mu\nu} \dfrac{g_{\alpha\beta}}{2} \right) \end{array} \right) \right) \frac{\delta g^{\alpha\beta}}{F_i} \right) .$$

$$(1908)$$

While we consider the second equation a generalized or quantum Einstein field equation, we are going to use the first one to extract the classical Klein–Gordon equation simply by demanding condition (1900) for the functional wrapper $F[f]$. Now, setting the integrand equal to zero, we obtain the following differential equation:

$$0 = R - \frac{F_i'}{F_i}(n-1)\Delta f_i + \frac{f_{i,\alpha} f_{i,\beta}(1-n)}{4F_i^2} g^{\alpha\beta} \left(4F_i F_i'' + F_i' F_i'(n-6) \right). \quad (1909)$$

Incorporation of condition (1900) finally gives us:

$$0 = R - \frac{F_i'}{F_i}(n-1)\Delta f_i \quad (1910)$$

and thus (1904). The introduction of mass and potential via dimensional entanglement was extensively discussed in chapter 6 of this book.

11.13.2 Appendix A2: Derivation of the Metric Dirac Equation [12] from the Scaled Ricci Scalar R^*

Using (1909), we now add in the Dirac starting point, who took for granted that:

(a) The space-time can be assumed Ricci flat and thus, $R = 0$.

(b) The Laplace operator term in (1909) has an eigenvalue M^2, which is just the Klein–Gordon case.

Also setting $F[f] = f$ gives us:

$$0 = f_{i,\alpha} f_{i,\beta} g^{\alpha\beta} + M^2 \cdot \overbrace{\frac{4}{(n-6)}}^{\equiv \mu^2 = \mu \cdot \mu} \cdot f_i^2. \tag{1911}$$

Now we apply the known relation between the Dirac matrices

$$\gamma^0 = \begin{pmatrix} 1 & & & \\ & 1 & & \\ & & -1 & \\ & & & -1 \end{pmatrix}; \quad \gamma^1 = \begin{pmatrix} & & & 1 \\ & & 1 & \\ & -1 & & \\ -1 & & & \end{pmatrix}$$

$$\gamma^2 = \begin{pmatrix} & & & -i \\ & & i & \\ & i & & \\ -i & & & \end{pmatrix}; \quad \gamma^3 = \begin{pmatrix} & & 1 & \\ & & & -1 \\ -1 & & & \\ & 1 & & \end{pmatrix}; \quad I = \begin{pmatrix} 1 & & & \\ & 1 & & \\ & & 1 & \\ & & & 1 \end{pmatrix}, \tag{1912}$$

and the metric tensor, reading:

$$g^{\alpha\beta} \cdot I = \frac{\gamma^\alpha \gamma^\beta + \gamma^\beta \gamma^\alpha}{2}, \tag{1913}$$

and put (1911) into the following form:

$$0 = \left(f_{i,\alpha} f_{i,\beta} g^{\alpha\beta} + \mu^2 \cdot f_i^2 \right) \cdot I$$

$$\Rightarrow \; = \; \Rightarrow \qquad f_{i,\alpha} f_{i,\beta} \left(\frac{\gamma^\alpha \gamma^\beta + \gamma^\beta \gamma^\alpha}{2} \right) + \frac{\beta^2 + \beta^2}{2} \cdot m^2 \cdot f_i^2 = 0$$

$$\Rightarrow \qquad f_{i,\alpha} f_{i,\beta} \gamma^\alpha \gamma^\beta + \beta^2 \cdot m^2 \cdot f_i^2 = 0$$

$$\Rightarrow \qquad \left(f_{i,\alpha} \gamma^\alpha + i \cdot \beta \cdot m \cdot f_i \right) \left(f_{i,\beta} \gamma^\beta - i \cdot \beta \cdot m \cdot f_i \right) = 0$$

$$\Rightarrow \qquad f_{i,\alpha} f_{i,\beta} \gamma^\beta \gamma^\alpha + \beta^2 \cdot m^2 \cdot f_i^2 = 0$$

$$\Rightarrow \qquad \left(f_{i,\alpha} \gamma^\alpha + i \cdot \beta \cdot m \cdot f_i \right) \left(f_{i,\beta} \gamma^\beta - i \cdot \beta \cdot m \cdot f_i \right) = 0. \tag{1914}$$

Thereby we have also used the β matrix with $\beta = \begin{pmatrix} 1 & & & \\ & 1 & & \\ & & -1 & \\ & & & -1 \end{pmatrix}$.

This gives us the Dirac equation in its classical form [12]:

$$0 = f_{i,\beta}\gamma^\beta - i \cdot \beta \cdot m \cdot f_i. \tag{1915}$$

The advantage of the Dirac approach clearly is that here the structure of the equation with the quaternions already assures the satisfaction of the eigenvalue equation for the Laplace operator term, which is to say the Klein–Gordon condition:

$$\Delta f_i = -M^2 \cdot f_i. \tag{1916}$$

For further discussion, the reader is referred to section 11.4.2.

11.13.3 Appendix B: Derivation of the Metric Schrödinger Equation from the Scaled Ricci Scalar R^*

The Schrödinger equation, in principle, is only a kind of approximated offspring of the Klein–Gordon equation under the assumption of non-reletivistic conditions. However, with its first-order time-drivative and second-order spatial derivatives, it could also be seen as something between the Klein–Gordon and the Dirac equation with just "summed-up" or scalarized quaternions. That being said, we should be able to derive "Schrödinger" equations from both, the Klein–Gordon and the Dirac equation.

We start with the Klein–Gordon equation and use our result of the appendix A1, which is to say Eq. (1904), only that, for the reason of recognition and better comparison with the classical form, instead of the symbol f we apply the classical Greek symbol Ψ. We also, in order to be close enough to the classical Schrödinger case, assume a Minkowski-like metric time component. We write (1904) as follows:

$$0 = \begin{cases} \left(\Psi - C_{f0}\right) \cdot R + \left(1-n\right) \cdot \left(\Delta_{n-1}\Psi - \dfrac{\partial^2 \Psi}{c^2 \cdot \partial t^2}\right) & n > 2 \\[4mm] R + C_{f0} \cdot \left(1-n\right) \cdot \left(\Delta_{n-1}\Psi - \dfrac{\partial^2 \Psi}{c^2 \cdot \partial t^2}\right) & n = 2. \end{cases} \tag{1917}$$

We ignore the $n = 2$-case and take it that R can be assumed being linear to f, which is to say that the curvature of space-time is caused by and proportional to the quantum effects residing in it. This gives us:

$$0 = \left(\Psi - C_{f0}\right) \cdot R_c \cdot \Psi + \left(1-n\right) \cdot \left(\Delta_{n-1}\Psi - \dfrac{\partial^2 \Psi}{c^2 \cdot \partial t^2}\right). \tag{1918}$$

Assuming the quantum function to be small against the classical physics (an important aspect in Schrödinger's work anyway and also agrees with the practical observation that the space-time curvature is small as we experience our surrounding space as being mainly flat), we discard of the quadratic term, resulting in:

$$0 = -C_{f0} \cdot R_c \cdot \Psi + (1-n) \cdot \left(\Delta_{n-1} \Psi - \frac{\partial^2 \Psi}{c^2 \cdot \partial t^2} \right)$$

$$= \left[(1-n) \cdot \left(\Delta_{n-1} - \frac{\partial^2}{c^2 \cdot \partial t^2} \right) - C_R \right] \Psi. \tag{1919}$$

Then we can separate the time derivative and—for the reason of recognition in connection with Schrödinger's classical approach—we reformulate everything a bit as follows:

$$0 = -C_{f0} \cdot R_c \cdot \Psi + (1-n) \cdot \left(\Delta_{n-1} \Psi - \frac{\partial^2 \Psi}{c^2 \cdot \partial t^2} \right) = \left[(1-n) \cdot \left(\Delta_{n-1} - \frac{\partial^2}{c^2 \cdot \partial t^2} \right) - C_R \right] \Psi$$

$$\underrightarrow{\left(\Delta_{n-1} - \frac{\partial^2}{c^2 \cdot \partial t^2} \right) - \frac{C_R}{(1-n)} = C1 \cdot \frac{\partial_t^2}{c^2} + \Delta_{n-1} - M^2}$$

$$\Rightarrow \qquad 0 = \left[\Delta - M^2 \right] \Psi = \left[\left(C1 \cdot \frac{\partial_t^2}{c^2} + \underbrace{\frac{1}{\sqrt{g}} \partial_\alpha \sqrt{g} \cdot g^{\alpha\beta} \partial_\beta}_{\text{3D-}\Delta\text{-Operator}} \right) - M^2 \right] \Psi. \tag{1920}$$

Please note that our Greek indices α and β are running only from 1 to n and not—as usual—from 0 to n, because the time or 0-component has been separated. Now we introduce a function $\Psi = \Phi + X$ and demand the following additional condition:

$$\partial_t \Psi = c^2 \cdot C2 \cdot (\Phi - X). \tag{1921}$$

Together with (1920), we obtain:

$$0 = -M^2 (\Phi + X) + \left(C1 \cdot C2 \cdot \partial_t (\Phi - X) + \frac{1}{\sqrt{g}} \partial_\alpha \sqrt{g} \cdot g^{\alpha\beta} \partial_\beta (\Phi + X) \right). \tag{1922}$$

The following two equations summed up would result in (1922):

$$0 = -M^2 \Phi + \left(C1 \cdot C2 \cdot \partial_t \Phi + \frac{1}{\sqrt{g}} \partial_\alpha \sqrt{g} \cdot g^{\alpha\beta} \partial_\beta \Phi \right)$$

$$0 = -M^2 X + \left(\frac{1}{\sqrt{g}} \partial_\alpha \sqrt{g} \cdot g^{\alpha\beta} \partial_\beta X - C1 \cdot C2 \cdot \partial_t X \right). \tag{1923}$$

Comparing with the original Schrödinger equation as given in the form below:

$$\left[-i \cdot \hbar \cdot \partial_t - \frac{\hbar^2}{2m} \Delta_{\text{Schrödinger}} + V_{\text{Schrödinger}} \right] \Psi = 0 \qquad (1924)$$

not only gives us the matter and antimatter solutions (c.f. [4, 28]) but also shows us—as seen and discussed before [29]—that mass m and the potential $V_{\text{Schrödinger}}$ are now taken on by the metric and the results in hidden (potentially compactified) dimensions (especially chapter 6 of this book), potentially due to entanglement (see also [3] for more). We also realize that the scalar curvature plays an important role in coding the mass and potential terms of the classical Schrödinger equation. Thus, a distorted metric and additional dimensions act like an effective potential and/or mass in the Schrödinger approximation and vice versa, which is to say: what in classical physics has been described as a potential would now become a scalar curvature or (derivation see [3]) a set of potentially compactified dimensions, which is providing the necessary interaction. Similarly, we have to formulate for the mass M that what appears as (rest) mass to us is just permanently and locally curved space (showing up in our equations above within the Ricci scalar), whereby the curvature can also be (c.f. [3]) the one of additional dimensions, being potentially entangled ... either among each other, with other "hidden" or compactified dimensions, or with our "visible" dimensions. Disregarding the antimatter solution here, setting the constants $C1$, $C2$, and reshaping the classical Schrödinger equation as:

$$C1 \cdot C2 = \frac{i \cdot 2 \cdot m}{\hbar}; \quad X = 0; \quad 0 = \left[\frac{i \cdot 2 \cdot m}{\hbar} \cdot \partial_t + \Delta_{\text{Schrödinger}} - 2 \cdot V_{\text{Schrödinger}} \right] \Psi,$$

$$(1925)$$

gives us the proportionality:

$$\Rightarrow \qquad \left[\Delta_{3D} - 2 \cdot V_{\text{Schrödinger}} \right] \Psi \overset{\triangle}{=} -M^2 \Phi + \frac{1}{\sqrt{g}} \partial_\alpha \sqrt{g} \cdot g^{\alpha\beta} \partial_\beta \Phi$$

$$\Rightarrow \qquad \Delta_{3D} \Psi \overset{\triangle}{=} \Delta_{3D} \Phi; \quad \Rightarrow \quad V_{\text{Schrödinger}} \overset{\triangle}{=} \frac{M^2 \Phi}{2 \cdot \Psi} \overset{\triangle}{=} \frac{C_R}{(1-n)} \frac{\Phi}{2 \cdot \Psi}. \qquad (1926)$$

This way we can now link classical Schrödinger solutions and the corresponding Schrödinger potentials with their metric analog.

Again, we introduce a function $\Psi = \Phi + X$ and demand the following additional condition:

$$\partial_t \Psi = i \cdot c \cdot C2 \cdot (\Phi - X). \qquad (1927)$$

This results in the following after inserting the new function into (1911):

$$0 = f_{i,0} f_{i,0} g^{00} + f_{i,\alpha} f_{i,\beta} g^{\alpha\beta} + M^2 \cdot \overbrace{\frac{4}{(n-6)}}^{\equiv \mu^2 = \mu \cdot \mu} \cdot f_i^2; \quad \alpha, \beta = \{\cancel{0}, 1, 2, 3, ...\}$$

with:

$$\partial_t f_i = i \cdot c \cdot C2_i \cdot (\Phi_i - X_i)$$

$$0 = -c^2 \cdot C2_i^2 \cdot (\Phi_i - X_i)^2 + (\Phi_i + X_i)_{,\alpha} (\Phi_i + X_i)_{,\beta} g^{\alpha\beta} + \mu \cdot \mu \cdot (\Phi_i + X_i)^2$$

$$\text{(1928)}$$

$$\Rightarrow \quad \begin{cases} 0 = -c^2 \cdot C2_i^2 \cdot (\Phi_i)^2 + (\Phi_i)_{,\alpha} (\Phi_i)_{,\beta} g^{\alpha\beta} + \mu \cdot \mu \cdot (\Phi_i)^2 \\ 0 = -c^2 \cdot C2_i^2 \cdot (-X_i)^2 + (X_i)_{,\alpha} (X_i)_{,\beta} g^{\alpha\beta} + \mu \cdot \mu \cdot (X_i)^2. \end{cases} \quad \text{(1929)}$$

Now we apply the known relation between the Dirac matrices and the metric tensor, reading:

$$g^{\alpha\beta} \cdot I = \frac{\gamma^\alpha \gamma^\beta + \gamma^\beta \gamma^\alpha}{2}, \quad \text{(1930)}$$

and put (1929) into the following form:

$$\Rightarrow \quad \begin{cases} 0 = -c^2 \cdot C2_i^2 \cdot (\Phi_i)^2 + (\Phi_i)_{,\alpha} (\Phi_i)_{,\beta} g^{\alpha\beta} + \mu \cdot \mu \cdot (\Phi_i)^2 \\ 0 = -c^2 \cdot C2_i^2 \cdot (-X_i)^2 + (X_i)_{,\alpha} (X_i)_{,\beta} g^{\alpha\beta} + \mu \cdot \mu \cdot (X_i)^2 \end{cases}$$

$$0 = \left((\mu \cdot \mu - c^2 \cdot C2_i^2) \cdot (\Phi_i)^2 + (\Phi_i)_{,\alpha} (\Phi_i)_{,\beta} g^{\alpha\beta} \right) \cdot I$$

$$0 = \left(\mu \cdot \mu - c^2 \cdot C2_i^2 \right) \cdot \overbrace{I \cdot (\Phi_i)^2}^{\equiv \frac{\beta^2 + \beta^2}{2} \cdot m^2} + (\Phi_i)_{,\alpha} (\Phi_i)_{,\beta} g^{\alpha\beta} \cdot I$$

$$\Rightarrow = \Rightarrow \quad (\Phi_i)_{,\alpha} (\Phi_i)_{,\beta} \left(\frac{\gamma^\alpha \gamma^\beta + \gamma^\beta \gamma^\alpha}{2} \right) + \frac{\beta^2 + \beta^2}{2} \cdot m^2 \cdot (\Phi_i)^2 = 0$$

$$\Rightarrow \quad (\Phi_i)_{,\alpha} (\Phi_i)_{,\beta} \gamma^\alpha \gamma^\beta + \beta^2 \cdot m^2 \cdot (\Phi_i)^2 = 0$$

$$\Rightarrow \quad \left(\Phi_{i,\alpha} \gamma^\alpha + i \cdot \beta \cdot m \cdot \Phi_i \right)\left(\Phi_{i,\beta} \gamma^\beta - i \cdot \beta \cdot m \cdot \Phi_i \right) = 0$$

$$\Rightarrow \quad \Phi_{i,\alpha} \Phi_{i,\beta} \gamma^\beta \gamma^\alpha + \beta^2 \cdot m^2 \cdot \Phi_i^2 = 0$$

$$\Rightarrow \quad \left(\Phi_{i,\alpha} \gamma^\alpha + i \cdot \beta \cdot m \cdot \Phi_i \right)\left(\Phi_{i,\beta} \gamma^\beta - i \cdot \beta \cdot m \cdot \Phi_i \right) = 0, \quad \text{(1931)}$$

$$0 = -c^2 \cdot C2_i^{\ 2} \cdot \left(-X_i\right)^2 + \left(X_i\right)_{,\alpha}\left(X_i\right)_{,\beta} g^{\alpha\beta} + \mu \cdot \mu \cdot \left(X_i\right)^2$$

$$\Rightarrow \qquad 0 = -\left(-c \cdot C2_i\right)^2 \cdot \left(X_i\right)^2 + \left(X_i\right)_{,\alpha}\left(X_i\right)_{,\beta} g^{\alpha\beta} + \mu \cdot \mu \cdot \left(X_i\right)^2$$

$$0 = \left(\left(\mu \cdot \mu - \left(-c \cdot C2_i\right)^2\right) \cdot \left(X_i\right)^2 + \left(X_i\right)_{,\alpha}\left(X_i\right)_{,\beta} g^{\alpha\beta}\right) \cdot I$$

$$0 = \underbrace{\left(\mu \cdot \mu - \left(-c \cdot C2_i\right)^2\right) \cdot I}_{\equiv \frac{\beta^2 + \beta^2}{2} \cdot M^2} \cdot \left(X_i\right)^2 + \left(X_i\right)_{,\alpha}\left(X_i\right)_{,\beta} g^{\alpha\beta} \cdot I$$

$$\Rightarrow = \Rightarrow \qquad \left(X_i\right)_{,\alpha}\left(X_i\right)_{,\beta}\left(\frac{\gamma^\alpha\gamma^\beta + \gamma^\beta\gamma^\alpha}{2}\right) + \frac{\beta^2 + \beta^2}{2} \cdot M^2 \cdot \left(X_i\right)^2 = 0$$

$$\Rightarrow \qquad \left(X_i\right)_{,\alpha}\left(X_i\right)_{,\beta} \gamma^\alpha\gamma^\beta + \beta^2 \cdot M^2 \cdot \left(X_i\right)^2 = 0$$

$$\Rightarrow \qquad \left(X_{i,\alpha}\gamma^\alpha + i \cdot \beta \cdot M \cdot X_i\right)\left(X_{i,\beta}\gamma^\beta - i \cdot \beta \cdot M \cdot X_i\right) = 0$$

$$\Rightarrow \qquad X_{i,\alpha}X_{i,\beta}\gamma^\beta\gamma^\alpha + \beta^2 \cdot M^2 \cdot X_i^{\ 2} = 0$$

$$\Rightarrow \qquad \left(X_{i,\alpha}\gamma^\alpha + i \cdot \beta \cdot M \cdot X_i\right)\left(X_{i,\beta}\gamma^\beta - i \cdot \beta \cdot M \cdot X_i\right) = 0. \tag{1932}$$

Please note that the matrix β is now defined via:

$$\left(\mu \cdot \mu - c^2 \cdot C2_i^{\ 2}\right) \cdot I \equiv \frac{\beta^2 + \beta^2}{2} \cdot m^2 \tag{1933}$$

and:

$$\left(\mu \cdot \mu - \left(-c \cdot C2_i\right)^2\right) \cdot I \equiv \frac{\beta^2 + \beta^2}{2} \cdot M^2. \tag{1934}$$

This gives us two Dirac equations in almost classical form [12] for the two functions Φ and X:

$$0 = \Phi_{i,\beta}\gamma^\beta - i \cdot \beta \cdot m \cdot \Phi_i, \tag{1935}$$

$$0 = X_{i,\beta}\gamma^\beta - i \cdot \beta \cdot M \cdot X_i, \tag{1936}$$

only that this is not a classical Dirac equation, but a Schrödinger-Dirac equation, where in contrast to the classical form, the Dirac matrices have dimension $(n - 1) \times (n - 1)$ and the indices are only running from 1 to $n - 1$ and not from 0 to $n - 1$.

So, in the Minkowski flat space with dimension $n = 5$, we could directly apply the classical Dirac matrices with just a new indexing:

$$\gamma^1 = \begin{pmatrix} 1 & & & \\ & 1 & & \\ & & -1 & \\ & & & -1 \end{pmatrix}; \quad \gamma^2 = \begin{pmatrix} & & & 1 \\ & & 1 & \\ & -1 & & \\ -1 & & & \end{pmatrix}$$

$$\gamma^3 = \begin{pmatrix} & & & -i \\ & & i & \\ & i & & \\ -i & & & \end{pmatrix}; \quad \gamma^4 = \begin{pmatrix} & & 1 & \\ & & & -1 \\ -1 & & & \\ & 1 & & \end{pmatrix}; \quad I = \begin{pmatrix} 1 & & & \\ & 1 & & \\ & & 1 & \\ & & & 1 \end{pmatrix}, \quad (1937)$$

and the index β in (1935) and (1936) would run from 1 to 4.

11.13.4   Appendix C: A Few Words about Spin 1/2, 3/2, 5/2 and So On

In the following, we want to investigate potential half spin solutions for the associated Legendre polynomials, which we needed in various forms within this chapter.

Here, as an example, we intend to consider the simple form:

$$Y_l^m[v] = C_{Pv} \cdot P_l^m[\cos[v]] + C_{Qv} \cdot Q_l^m[\cos[v]]. \quad (1938)$$

Observing the solution $Y[v]$ more closely, we find that there exist singularity-free spin $l = n/2$-solutions for $m = -n/2$ with $n = 1, 3, 5, 7, \ldots$ in the case of $C_{Pv} \neq 0$, $C_{Qv} = 0$ and for $m = +n/2$ in the case of $C_{Pv} = 0$, $C_{Qv} \neq 0$. Figures 11.40 and 11.41 illustrate the corresponding distributions within the definition range for the angle v from 0 to π.

The general solution to the partial differential equation resulting from the Schrödinger hydrogen problem [13] can be given as follows:

$$f_{n,l,m}[t,r,\vartheta,\varphi] = g[t] \cdot \Psi_{n,l,m}[r,\vartheta,\varphi] = g[t] \cdot e^{i \cdot m \cdot \varphi} \cdot P_l^m[\cos \vartheta] \cdot R_{n,l}[r]$$

$$= \left(C_1 \cdot \cos[c \cdot C_{tt} \cdot t] + C_2 \cdot \sin[c \cdot C_{tt} \cdot t] \right) \cdot N \cdot e^{-\rho/2} \cdot \rho^l \cdot L_{n-l-1}^{2l+1}[\rho] \cdot Y_l^m[\vartheta,\varphi]$$

$$\rho = \frac{2 \cdot r}{n \cdot a_0}; \quad N = \sqrt{\left(\frac{2}{n \cdot a_0} \right)^3 \frac{(n-l-1)!}{2 \cdot n \cdot (n+l)!}}. \quad (1939)$$

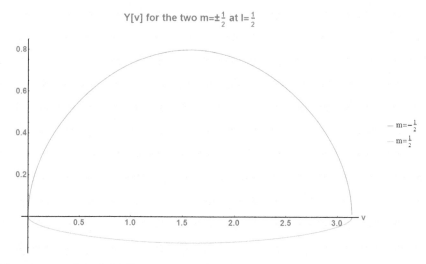

Figure 11.40 Spin $l = 1/2$ situation with the two possible spin states $m = \pm 1/2$ according to our angular quantum gravity solution (1938). For better illustration and comparability, we divided the "Q Legendres" by 10.

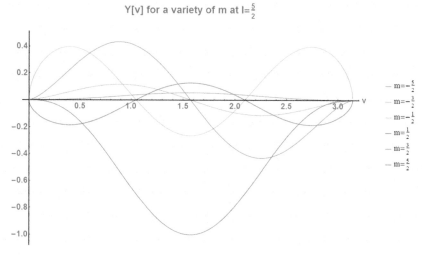

Figure 11.41 Spin $l = 5/2$ situation with the five possible spin states $m = \pm\{1/2, 3/2, 5/2\}$ according to our angular quantum gravity solution (1938). Regarding the evaluation, see text. For better illustration, we divided the "Q Legendres" by 10.

It totally suffices here to consider the classical problems in order to also discuss the angular aspects of (1740) as the results for these coordinates in both cases are the same functional dependencies.

For nostalgic reasons, we used the symbol "n" (here not the dimension) as the so-called main quantum number. All quantum numbers n, l, and the parameter a_0 depend on the constants E, A_i in (1740) and have to satisfy certain quantum conditions in order to result in singularity-free solutions for $f[...]$.

Regarding the conditions for the quantum numbers n, l, and m, we not only have the usual:

$$\{n,l,m\} \in \mathbb{Z}; \quad n \geq 0; \quad l < n; \quad -l \leq m \leq +l, \tag{1940}$$

but also found the suitable solutions for the half spin forms as discussed above and derived in the Figs. 11.40 and 11.41. The corresponding main quantum numbers for half spin l numbers with $l = 1/2, 3/2, ...$ are simply (just as before with the integers) $n = l + 1 = 3/2, 5/2, 7/2, ...$

It should explicitly be noted, however, that the usual spherical harmonics are inapplicable in cases of half spin. For $\{n, l, m\} = \{1/2, 3/2, 5/2, 7/2, ...\}$ the wave function (1939) has to be adapted as follows:

$$f_{n,l,m}[t,r,\vartheta,\varphi] = e^{\pm i \cdot c \cdot C_t \cdot t} \cdot N \cdot e^{-\rho/2} \cdot \rho^l \cdot L_{n-l-1}^{2l+1}[\rho] \cdot Z_m[\varphi] \cdot \begin{Bmatrix} P_l^{m<0}[\cos\vartheta] \\ Q_l^{m>0}[\cos\vartheta] \end{Bmatrix}$$

$$= e^{\pm i \cdot c \cdot C_t \cdot t} \cdot N \cdot e^{-\rho/2} \cdot \rho^l \cdot L_{n-l-1}^{2l+1}[\rho] \cdot \begin{Bmatrix} \cos[m \cdot \varphi] \\ \sin[m \cdot \varphi] \end{Bmatrix} \cdot \begin{Bmatrix} P_l^{m<0}[\cos\vartheta] \\ Q_l^{m>0}[\cos\vartheta] \end{Bmatrix}. \tag{1941}$$

Thereby it was elaborated elsewhere [30] that in fact the sin and the cos functions seem to make the Pauli exclusion [31] and not the "+" and "–" of the m. However, in order to have the usual Fermionic statistic, we can simply define as follows:

$$f_{n,l,m}[t,r,\vartheta,\varphi] = e^{\pm i \cdot c \cdot C_t \cdot t} \cdot N \cdot e^{-\rho/2} \cdot \rho^l \cdot L_{n-l-1}^{2l+1}[\rho] \cdot \begin{Bmatrix} \sin[m \cdot \varphi]_{m<0} \\ \cos[m \cdot \varphi]_{m>0} \end{Bmatrix} \cdot \begin{Bmatrix} P_l^{m<0}[\cos\vartheta] \\ Q_l^{m>0}[\cos\vartheta] \end{Bmatrix}. \tag{1942}$$

As derived in [30], the resolution of the degeneration with respect to half spin requires a break of the symmetry, which we achieved by introducing elliptical geometry instead of the spherical one.

Thus, we have metrically derived a fairly general "hydrogen atom." In addition to the Schrödinger structure, our form also sports a time-dependent factor clearly showing the options for matter and antimatter via the ± sign in the $g[t]$ function. We also found the half spin states and were able to resolve the spin degeneration via a simple symmetry break by switching from spherical to elliptical coordinates (see [30] regarding the evaluation). A few illustrations of the half spin solutions are given in section 11.5.

References

1. N. Schwarzer, *Einstein Had It, but He Did Not See It, Part LXXXI: More Dirac Killing*, Self-published, Amazon Digital Services, 2019, Kindle.

2. N. Schwarzer, *Societons and Ecotons—The Photons of the Human Society: Control Them and Rule the World, Part 1 of Medical Socio-Economic Quantum Gravity*, Self-published, Amazon Digital Services, 2020, Kindle.

3. N. Schwarzer, *Masses and the Infinity Options Principle: Can We Explain the 3-Generations and the Quantized Mass Problem?, Part 5 of Medical Socio-Economic Quantum Gravity*, Self-published, Amazon Digital Services, 2020, Kindle.

4. N. Schwarzer, *The World Formula: A Late Recognition of David Hilbert's Stroke of Genius*, 2022, Jenny Stanford Publishing. ISBN: 9789814877206.

5. N. Schwarzer, *Humanitons—The Intrinsic Photons of the Human Body—Understand Them and Cure Yourself, Part 2 of Medical Socio-Economic Quantum Gravity*, Self-published, Amazon Digital Services, 2020, Kindle.

6. N. Schwarzer, *Mastering Human Crises with Quantum-Gravity-Based but Still Practicable Models—First Measure: SEEING and UNDERSTANDING the WHOLE, Part 3 of Medical Socio-Economic Quantum Gravity*, Self-published, Amazon Digital Services, 2020, Kindle.

7. N. Schwarzer, *Self-Similar Quantum Gravity—How Infinity Brings Simplicity, Part 4 of Medical Socio-Economic Quantum Gravity*, Self-published, Amazon Digital Services, 2020, Kindle.

8. N. Schwarzer, *Towards Quantum Einstein Field Equations, Part 7 of Medical Socio-Economic Quantum Gravity*, Self-published, Amazon Digital Services, December 2020, Kindle.

9. N. Schwarzer, *The 3 Generations of Elementary Particles, Part 6 of Medical Socio-Economic Quantum Gravity*, Self-published, Amazon Digital Services, December 2020, Kindle.

10. D. Hilbert, Die Grundlagen der Physik, Teil 1, *Göttinger Nachrichten*, 1915, 395–407.

11. A. Einstein, Grundlage der allgemeinen Relativitätstheorie, *Ann. Phys.*, 1916, **49** (ser. 4), 769–822.

12. P. A. M. Dirac, The quantum theory of the electron, *Proc. R. Soc. A*, 1928, **117**(778), doi: 10.1098/rspa.1928.0023.

13. N. Schwarzer, *The Metric Dirac Equation Revisited and the Geometry of Spinors, Part 8 of Medical Socio-Economic Quantum Gravity*, Self-published, Amazon Digital Services, December 2020, Kindle.

14. Ch. W. Misner, K. S. Thorne, J. A. Wheeler, *Gravitation*, 20th ed., W. H. Freeman and Company, New York, 1997, ISBN 0-7167-0344.

15. H. Haken, H. Chr. Wolf, *Atom- und Quantenphysik*, 4th ed. (in German), Springer Heidelberg, 1990, ISBN 0-387-52198-4.

16. C. A. Sporea, Notes on $f(R)$ Theories of Gravity, arxiv.org/pdf/1403.3852.pdf.

17. N. Schwarzer, *The Quantum Einstein Field Equations in Their Simplest Form, Part 7a of Medical Socio-Economic Quantum Gravity*, Self-published, Amazon Digital Services, December 2021, Kindle.

18. K. Schwarzschild, Über das Gravitationsfeld eines Massenpunktes nach der Einsteinschen Theorie, Sitzungsberichte der Königlich-Preussischen Akademie der Wissenschaften, Reimer, Berlin 1916, pp. 189–196.

19. H. Stephani, D. Kramer, M. MacCallum, C. Hoenselaers, E. Herlt, *Exact Solutions of Einstein's Field Equations*, Cambridge University Press, 2009, ISBN 978-0-521-46702-5.

20. K. Schwarzschild, Über das Gravitationsfeld einer Kugel aus inkompressibler Flüssigkeit nach der Einsteinschen Theorie [On the gravitational field of a ball of incompressible fluid following Einstein's theory], Sitzungsberichte der Königlich-Preussischen Akademie der Wissenschaften (in German), Berlin, 1916, 424–434.

21. L. Flamm, Beiträge zur Einsteinschen Gravitationstheorie, *Phys. Z.*, 1916, **17**, 448.

22. N. Schwarzer, *My Horcruxes – A Curvy Math to Salvation, Part 9 of "Medical Socio-Economic Quantum Gravity*, Self-published, Amazon Digital Services, 2021, Kindle, ASIN: B096SPB5MW.

23. J. D. Bekenstein, Black holes and entropy, *Phys. Rev. D*, 1973, **7**, 2333–2346.

24. J. D. Bekenstein, Information in the holographic universe, *Sci. Am.*, 2003, **289**(2), 61.

25. N. Schwarzer, *The Quantum Black Hole, Part 7b of Medical Socio-Economic Quantum Gravity*, Self-published, Amazon Digital Services, April 2021, Kindle.

26. N. Schwarzer, *Quantum Cosmology—A Simple Answer to the Flatness Riddle, Part 7c of Medical Socio-Economic Quantum Gravity*, Self-published, Amazon Digital Services, April 2021, Kindle.

27. N. Schwarzer, *Strings, Branes, Friedmanns and the Matryoshka Universe, Part 7d of Medical Socio-Economic Quantum Gravity*, Self-published, Amazon Digital Services, December 2021, Kindle.

28. N. Schwarzer, *The Theory of Everything: Quantum and Relativity Is Everywhere—A Fermat Universe*, Jenny Stanford Publishing, 2020, ISBN-10: 9814774472.

29. N. Schwarzer, *General Quantum Relativity*, Self-published, Amazon Digital Services, 2016, Kindle.

30. N. Schwarzer, *Einstein Had It, but He Did Not See It, Part LXXIX: Dark Matter Options*, Self-published, Amazon Digital Services, 2016, Kindle.

31. W. Pauli, Über den Zusammenhang des Abschlusses der Elektronengruppen im Atom mit der Komplexstruktur der Spektren, *Z. Phys.*, 1925, **31**, 765–783, Bibcode:1925ZPhy...31..765P, doi:10.1007/BF02980631.

32. N. Schwarzer, *The Dirac Miracle, Part 8a of Medical Socio-Economic Quantum Gravity*, Self-published, Amazon Digital Services, May 2021, Kindle.

33. E. Ley-Koo, R. C. Wang, Dirac equation in ortogonal curvilinear coordinates, *Rev. Mex. de Fis.*, 1988, **34**(2), 206–919.

34. T. Brennan, General Relativity and the Dirac Equation, 2020, https://quantumthom.com/GenRelAndDiracEqH2.pdf.

Chapter 12

About the Flatness Problem in Cosmology

In [26], based on the scaling extensions [1, 2] of the works of Hilbert and Einstein [3, 4] and our earlier publications [5–10], we were able to show that the new quantum Einstein field equations (1817) automatically demand the Robertson–Walker metric to be flat (Fig. 12.1). In the following we repeat the essentials of our derivations.

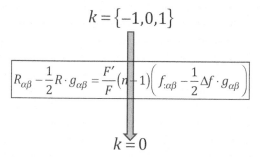

$$k = \{-1, 0, 1\}$$

$$R_{\alpha\beta} - \frac{1}{2} R \cdot g_{\alpha\beta} = \frac{F'}{F}(n-1)\left(f_{:\alpha\beta} - \frac{1}{2}\Delta f \cdot g_{\alpha\beta} \right)$$

$$k = 0$$

Figure 12.1 Quantum Einstein field equations and the flatness problem.

12.1 The Flatness Problem in Cosmology

It has long been a mystery why our universe seems to be so extremely close to the flat case, meaning close to an almost vanishing curvature [11, 12]. The

The Math of Body, Soul, and the Universe
Norbert Schwarzer
Copyright © 2023 Jenny Stanford Publishing Pte. Ltd.
ISBN 978-981-4968-24-9 (Hardcover), 978-1-003-33454-5 (eBook)
www.jennystanford.com

standard cosmological model, mainly based on the Friedmann universe [13, 14], predicts that there should either have been a perfectly flat universe or an extreme increase or a decrease of the parameter of the density ρ since the beginning of time (assumed to be the big bang). The equation containing the matter density and comparing it with its critical value guaranteeing a perfectly flat universe ρ_c in classical cosmology reads:

$$\left(\frac{\rho_c}{\rho} - 1\right) \cdot \rho \cdot a[t]^2 = \text{const.} \tag{1943}$$

Thereby a[t] gives the scale parameter of the universe, which—as the universe expands—is time-dependent. For simplicity, we cite from the corresponding Wikipedia page [12]:

"The right-hand side of ... (1943) contains constants only and therefore the left hand side must remain constant throughout the evolution of the universe.

As the universe expands the scale factor a[t] increases, but the density ρ decreases as matter (or energy) becomes spread out. For the standard model of the universe which contains mainly matter and radiation for most of its history, ρ decreases more quickly than $a[t]^2$ increases, and so the factor $\rho*a[t]^2$ will decrease. Since the time of the Planck era, shortly after the Big Bang, this term has decreased by a factor of around 10^{60} (according to ref. *Peter Coles; Francesco Lucchin (1997). Cosmology. Chichester: Wiley. ISBN 978-0-471-95473-6*) and so $\left(\frac{\rho_c}{\rho} - 1\right)$ must have increased by a similar amount to retain the constant value of their product."

Most interestingly and in contrast to this theoretical prediction, however, the density was found to still be very near the critical values of a flat universe.

The most prominent attempt to cure the situation is seen in the so-called inflation theory [15, 16, 17], but there are also other attempts using anisotropic gravity approaches or non-constant speed-of-light assumptions (e.g., [18, 19]).

In [23], we also investigated the possibility of additional dimensions leading to suitable mass, spin, and Higgs fields (see also [6, 7]) and providing the necessary conditions for a balanced universe with nearly flat characteristics for principally infinite time scales.

All these attempts to solve the flatness problem were based on the classical Einstein field equations. Here we will show that the problem appears to be extremely simple when applying the quantum Einstein field equations.

12.2   Robertson–Walker Metric

Using the assumption of homogeneity and isotropy of the distribution of matter within the observable universe on big enough scales, it can be derived that the corresponding metric in a 4-dimensional space-time has been given as follows [20] (please note that the author of [20] used a different sign convention as we apply here):

$$
g_{\alpha\beta} =
\begin{pmatrix}
-c^2 & 0 & 0 & 0 \\[2mm]
0 & \dfrac{f[t]^2}{\left(1+\dfrac{k}{4}\cdot\left(\dfrac{r}{R}\right)^2\right)^2} & 0 & 0 \\[4mm]
0 & 0 & \dfrac{f[t]^2\cdot r^2}{\left(1+\dfrac{k}{4}\cdot\left(\dfrac{r}{R}\right)^2\right)^2} & 0 \\[4mm]
0 & 0 & 0 & \dfrac{f[t]^2\cdot r^2}{\left(1+\dfrac{k}{4}\cdot\left(\dfrac{r}{R}\right)^2\right)^2}\cdot\sin(\vartheta)^2
\end{pmatrix}
$$

$$k = \{-1,0,1\}. \tag{1944}$$

The three cases of k code sphere ($k = 1$), flat space ($k = 0$) and hypersphere ($k = -1$). The function $f[t]$ here acts as the cosmological spatial scale.

12.3   Toward a Simple Quantum Cosmology

As already stated earlier in this book, it has to be pointed out that a solution to (1815) does not suffice to solve the whole variation problem as this would also require a solution to the first line of (1807). Alternatively, one may also seek for a solution of the whole integrand of (1837), leading us to:

$$
0 = \overset{***}{R_{\alpha\beta}} + R_{\alpha\beta} - \frac{1}{2}R\cdot g_{\alpha\beta} + \frac{F'}{F}(1-n)\left(f_{,\alpha\beta} - \Gamma^{\gamma}_{\alpha\beta}f_{,\gamma} - \frac{1}{2}\Delta f\cdot g_{\alpha\beta}\right), \tag{1945}
$$

when applying condition (1808) and assuming no matter and a zero cosmological constant.

However, when bringing (1944) into the following form:

$$G_{\alpha\beta} = f[t]^2 \cdot g_{\alpha\beta} =$$

$$f[t]^2 \cdot \begin{pmatrix} \dfrac{-c^2}{f[t]^2} & 0 & 0 & 0 \\[3mm] 0 & \dfrac{1}{\left(1+\dfrac{k}{4}\cdot\left(\dfrac{r}{R}\right)^2\right)^2} & 0 & 0 \\[3mm] 0 & 0 & \dfrac{r^2}{\left(1+\dfrac{k}{4}\cdot\left(\dfrac{r}{R}\right)^2\right)^2} & 0 \\[3mm] 0 & 0 & 0 & \dfrac{r^2\cdot\sin(\vartheta)^2}{\left(1+\dfrac{k}{4}\cdot\left(\dfrac{r}{R}\right)^2\right)^2} \end{pmatrix}$$

$$k = \{-1,0,1\}, \tag{1946}$$

we directly obtain the required scaled form of the metric tensor with:

$$F[f[t]] = f[t]^2. \tag{1947}$$

Thereby $F[f]$ is satisfying condition (1808) and we can directly plug (1946) into our quantum gravity equation (1945) (already given in (1817)) (with $C_f = 0$). We obtain:

$$\overset{***}{R_{\alpha\beta}} + R_{\alpha\beta} - \frac{1}{2}R\cdot g_{\alpha\beta} + \frac{F'}{F}(1-n)\left(f_{,\alpha\beta} - \frac{g_{\alpha\beta}}{n}\Gamma^{\gamma}_{\kappa\lambda}g^{\kappa\lambda}f_{,\gamma} - \frac{1}{2}\Delta f\cdot g_{\alpha\beta}\right)$$

$$= \begin{pmatrix} A & 0 & 0 & 0 \\ 0 & B & 0 & 0 \\ 0 & 0 & r^2\cdot B & 0 \\ 0 & 0 & 0 & r^2\cdot\sin(\vartheta)^2\cdot B \end{pmatrix}$$

$$A = \frac{3\left(c^2 k + f'[t]^2 + f[t]f''[t]\right)}{f[t]^2} - \frac{3f''[t]}{f[t]}$$

$$B = \frac{16\left(2c^2 k + 2f'[t]^2 + f[t]f''[t]\right)}{c^2\left(4+kr^2\right)^2} - \frac{3\left(c^2 k + f'[t]^2 + f[t]f''[t]\right)}{c^2\left(1+\dfrac{kr^2}{4}\right)^2} \tag{1948}$$

and find the only possible solution via:

$$f[t] = i \cdot c \cdot \sqrt{k} \cdot t + C_{t1}. \tag{1949}$$

As this is not a very satisfying solution for our universe, we try to find more suitable solutions via the following generalization of our metric approach:

$$g_{\alpha\beta} = f[t]^2 \cdot \begin{pmatrix} \dfrac{-c^2}{f[t]^2} \cdot h[t]^2 & 0 & 0 & 0 \\[3mm] 0 & \dfrac{1}{\left(1 + \dfrac{k}{4} \cdot \left(\dfrac{r}{R}\right)^2\right)^2} & 0 & 0 \\[3mm] 0 & 0 & \dfrac{r^2}{\left(1 + \dfrac{k}{4} \cdot \left(\dfrac{r}{R}\right)^2\right)^2} & 0 \\[3mm] 0 & 0 & 0 & \dfrac{r^2 \cdot \sin(\vartheta)^2}{\left(1 + \dfrac{k}{4} \cdot \left(\dfrac{r}{R}\right)^2\right)^2} \end{pmatrix}$$

$$k = \{-1, 0, 1\}. \tag{1950}$$

The subsequent solution for the function $f[t]$ reads:

$$f[t] = C_{t1} \pm C_{t0} \cdot c \cdot \sqrt{k} \cdot \int_1^t h[\tau] d\tau. \tag{1951}$$

Interestingly, we can obtain quite a variety of cosmic expansion scenarios, but in the more exciting case of the flat universe, which is to say $k = 0$, we would be left with no expansion at all, because $f[t]$ reduces to a constant.

On bigger scales, however, our universe appears flat (almost perfectly so) and it is for this reason that we intend to find out whether the complete variational integrand being set to zero really is the only option. In other words, it is for this reason that we also want to have a look at Eq. (1815) and other—similar—options.

One may now ask why then at all looking for a solution just to (1815) if this only solves "half" of the whole (meaning (1817), (1945))? Well, the first and somewhat sloppy answer could be

"just for fun"

and the second and more elaborate one might be:

"because we think that the universe has means to take care about (1815) and the rest, which is to say:

$$0 = -\frac{1}{2}R \cdot g_{\alpha\beta} + \frac{F'}{F}(1-n)\left(-\frac{1}{2}\Delta f \cdot g_{\alpha\beta}\right) \quad \Rightarrow \quad 0 = R + \frac{F'}{F}(1-n)\Delta f,$$

(1952)

in different ways." We think that there is a reason why the first line of (1807), being "the rest of (1945)" and thus (1952), appears to us as a quantum equation, while the gravity equations containing the Ricci tensor act on bigger scales. We even think that different volumes have to or could be applied to both parts of the integrand within the variation. We already have considered this possibility in connection with the **f** vectors being of need for the Dirac approach (c.f. chapters 9 to 11). Here, now our focus is more about the degrees of freedom regarding the variational (and quite free) choice of volume[1] for the split-up of the Einstein–Hilbert action integral. This may mathematically be done as follows (repetition):

$\rightarrow$ Starting with the result from (1837) and setting the matter term and the cosmological constant to zero:

$$\delta W = 0 = \int_V d^n x \sqrt{-g \cdot F^n} \left(\begin{pmatrix} R_{\alpha\beta}^{***} + R_{\alpha\beta} - R \cdot \dfrac{g_{\alpha\beta}}{2} \\[2mm] + \dfrac{F'}{F}(1-n)\left(f_{,\alpha\beta} - \Gamma_{\alpha\beta}^{\gamma}f_{,\gamma} - \dfrac{g_{\alpha\beta}}{2}\Delta f\right) \\[2mm] + \dfrac{(1-n)}{4F^2}\begin{pmatrix} 4FF'' \\ +F'F'(n-6) \end{pmatrix}\begin{pmatrix} f_{,\alpha}f_{,\beta} \\ -f_{,\mu}f_{,\nu}g^{\mu\nu}\dfrac{g_{\alpha\beta}}{2}\end{pmatrix} \end{pmatrix} \delta G^{\alpha\beta} \right),$$

(1953)

we can simplify and split up to:

$$\xrightarrow{\quad 4FF''+F'F'(n-6)=0 \quad}$$

$$0 = \int_V d^n x \sqrt{-g \cdot F^n} \left(\begin{pmatrix} R_{\alpha\beta}^{***} + R_{\alpha\beta} - R \cdot \dfrac{g_{\alpha\beta}}{2} \\[2mm] + \dfrac{F'}{F}(1-n)\left(f_{,\alpha\beta} - \Gamma_{\alpha\beta}^{\gamma}f_{,\gamma} - \dfrac{g_{\alpha\beta}}{2}\Delta f\right) \end{pmatrix} \delta G^{\alpha\beta} \right)$$

---

[1]It should be pointed out that also the number of dimensions, surfaces on which the variation vanishes and other things could be considered as degrees of freedom with respect to the Einstein-Hilbert variation.

$$\Rightarrow \quad 0 = \left| \begin{array}{c} \int\limits_{V_G} d^n x \left(\sqrt{-g} \cdot F^n \left(R_{\alpha\beta} + \frac{F'}{F}(1-n)\left(f_{,\alpha\beta} - \Gamma^\gamma_{\alpha\beta} f_{,\gamma}\right) \right) \delta G^{\alpha\beta} \right) \\ -\frac{1}{2} \cdot \int\limits_{V_Q} d^n x \left(\sqrt{-g} \cdot F^n \left(R + \frac{F'}{F}(1-n)\Delta f \right) g_{\alpha\beta} \delta G^{\alpha\beta} \right) \end{array} \right| . \quad (1954)$$

Thereby the volumes of the two variations do not necessarily need to be the same.

Still, for symmetry reasons, where we want to have at least equal total volumes for both terms in the last equation above, one may just extend the last line as follows:

$$0 = \left(\begin{array}{c} \int\limits_{V_G} d^n x \left(\sqrt{-g} \cdot F^n \left(\overset{***}{R}_{\alpha\beta} + R_{\alpha\beta} + \frac{F'}{F}(1-n)\left(f_{,\alpha\beta} - \Gamma^\gamma_{\alpha\beta} f_{,\gamma}\right) \right) \delta G^{\alpha\beta} \right) \\ -\frac{1}{2} \cdot \sum\limits_{i=1}^{N} \int\limits_{V_{Qi}} d^n x \left(\sqrt{-g} \cdot F^n \left(R + \frac{F'}{F}(1-n)\Delta f \right) g_{\alpha\beta} \delta G^{\alpha\beta} \right) \end{array} \right)$$

$$V_G = \sum_{i=1}^{N} V_{Qi}. \quad (1955)$$

Thus, it appears totally reasonable to assume that, for whatever reasons, the scales of the two variations could be extremely different and while the volume V_Q (Q denotes quantum) might be extremely small, forcing us to watch the variation:

$$0 = \int\limits_{V_Q} d^n x \left(\sqrt{-g} \cdot F^n \left(R + \frac{F'}{F}(1-n)\Delta f \right) g_{\alpha\beta} \delta G^{\alpha\beta} \right) \quad (1956)$$

at smaller, which is to say quantum scales, we would have to take care about the tensorial or gravitational variation:

$$0 = \int\limits_{V_G} d^n x \left(\sqrt{-g} \cdot F^n \left(\overset{***}{R}_{\alpha\beta} + R_{\alpha\beta} + \frac{F'}{F}(1-n)\left(f_{,\alpha\beta} - \Gamma^\gamma_{\alpha\beta} f_{,\gamma}\right) \right) \delta G^{\alpha\beta} \right) \quad (1957)$$

at much bigger scales, respectively bigger volumes V_G. Thus, our reasoning for the consideration of solutions to (1815) or other split-up options with respect to the universe on a cosmological scale would be based on the observation of the different scales of quantum effects and gravity. Of course, this requires more discussion about the variational, respectively scale-dependent split-up in (1955) and the possible substructure of space or space-time with respect to the subsequent small-scale solutions to (1952) (or (1821) or the first line of (1807)). For instance, with respect to the split-up, we should take into account all possibilities resulting from our $F[f]^* g_{\alpha\beta}$ scaled metric approach. So, in addition to (1954) or the complete integrand with either:

$$0 = \int_V d^n x \left(\sqrt{-g \cdot F^n} \left(\begin{array}{c} R_{\alpha\beta}^{***} + R_{\alpha\beta} - R \cdot \dfrac{g_{\alpha\beta}}{2} \\[2mm] + \dfrac{F'}{F}(1-n)\left(f_{,\alpha\beta} - \Gamma_{\alpha\beta}^{\gamma} f_{,\gamma} - \dfrac{g_{\alpha\beta}}{2}\Delta f \right) \end{array} \right) \delta G^{\alpha\beta} \right),$$

(1958)

or:

$$0 = \sum_{i=1}^{N} \int_{V_{Qi}} d^n x \sqrt{-g \cdot F^n} \left(\begin{array}{c} R_{\alpha\beta}^{***} + R_{\alpha\beta} - R \cdot \dfrac{g_{\alpha\beta}}{2} \\[2mm] + \dfrac{F'}{F}(1-n)\left(f_{,\alpha\beta} - \Gamma_{\alpha\beta}^{\gamma} f_{,\gamma} - \dfrac{g_{\alpha\beta}}{2}\Delta f \right) \end{array} \right) \delta G^{\alpha\beta},$$

(1959)

we can also have:

$$0 = \left(\begin{array}{c} \displaystyle\int_{V_G} d^n x \sqrt{-g \cdot F^n} \left(\left(R_{\alpha\beta}^{***} + R_{\alpha\beta} - \dfrac{1}{2} \cdot R \cdot g_{\alpha\beta} + \dfrac{F'}{F}(1-n)\left(f_{,\alpha\beta} - \Gamma_{\alpha\beta}^{\gamma} f_{,\gamma} \right) \right) \delta G^{\alpha\beta} \right) \\[3mm] -\dfrac{1}{2} \cdot \displaystyle\sum_{i=1}^{N} \int_{V_{Qi}} d^n x \sqrt{-g \cdot F^n} \left(\dfrac{F'}{F}(1-n)\Delta f \cdot g_{\alpha\beta} \delta G^{\alpha\beta} \right) \end{array} \right),$$

(1960)

$$0 = \left(\begin{array}{c} \displaystyle\int_{V_G} d^n x \sqrt{-g \cdot F^n} \left(\left(R_{\alpha\beta}^{***} + R_{\alpha\beta} - \dfrac{1}{2} \cdot R \cdot g_{\alpha\beta} \right) \delta G^{\alpha\beta} \right) \\[3mm] +\dfrac{1}{2} \cdot \displaystyle\sum_{i=1}^{N} \int_{V_{Qi}} d^n x \sqrt{-g \cdot F^n} \dfrac{F'}{F}(n-1)\left(\Delta f \cdot g_{\alpha\beta} - 2 \cdot \left(f_{,\alpha\beta} - \Gamma_{\alpha\beta}^{\gamma} f_{,\gamma} \right) \right) \delta G^{\alpha\beta} \end{array} \right),$$

(1961)

$$0 = \begin{pmatrix} \int\limits_{V_G} d^n x \sqrt{-g} \cdot F^n \left(R^{***}_{\alpha\beta} + R_{\alpha\beta} \right) \delta G^{\alpha\beta} \\ -\frac{1}{2} \cdot \sum_{i=1}^{N} \int\limits_{V_{Qi}} d^n x \sqrt{-g} \cdot F^n \left(R \cdot g_{\alpha\beta} - \frac{F'}{F}(n-1) \begin{pmatrix} \Delta f \cdot g_{\alpha\beta} \\ -2 \cdot \left(f_{,\alpha\beta} - \Gamma^{\gamma}_{\alpha\beta} f_{,\gamma} \right) \end{pmatrix} \right) \delta G^{\alpha\beta} \end{pmatrix},$$

$$(1962)$$

$$0 = \begin{pmatrix} -\frac{1}{2} \cdot \int\limits_{V_G} d^n x \sqrt{-g} \cdot F^n \left(R \cdot g_{\alpha\beta} \delta G^{\alpha\beta} \right) \\ +\frac{1}{2} \cdot \sum_{i=1}^{N} \int\limits_{V_{Qi}} d^n x \sqrt{-g} \cdot F^n \left(2 \cdot \left(R^{***}_{\alpha\beta} + R_{\alpha\beta} \right) + \frac{F'}{F}(n-1) \begin{pmatrix} \Delta f \cdot g_{\alpha\beta} \\ -2 \cdot \left(f_{,\alpha\beta} - \Gamma^{\gamma}_{\alpha\beta} f_{,\gamma} \right) \end{pmatrix} \right) \delta G^{\alpha\beta} \end{pmatrix},$$

$$(1963)$$

$$\vdots,$$

$$(1964)$$

and so on.

Naturally, we can have all these options with interchanged integral boundaries, like we have illustrated in the previous chapter with the complete integrand in (1823) and (1824).

One should also discuss the possibility of surface terms in connection with the quantum-variational integrals. Parts of this discussion will be presented elsewhere [10].

Realizing that one of the options given above (namely (1826)) already mirrors the classical Einstein field equations and thus, the Friedmann approach if it comes to cosmology and knowing that the classical, which is to say unscaled, Friedmann equations [13, 14] give no satisfying answer to the flatness problem, we conclude that other options, like the scale split-up might give better results. We simply assume that in our universe a scale split-up of the kind (1955) is "active," meaning that on bigger scales we have to watch the variational integral:

$$0 = \int\limits_{V_G} d^n x \sqrt{-g} \cdot F^n \left(\left(R^{***}_{\alpha\beta} + R_{\alpha\beta} - \frac{1}{2} \cdot R \cdot g_{\alpha\beta} \right) \delta G^{\alpha\beta} \right) \qquad (1965)$$

and on smaller scales:

$$0 = \sum_{i=1}^{N} \int\limits_{V_{Qi}} d^n x \sqrt{-g} \cdot F^n \frac{F'}{F}(n-1) \left(\Delta f \cdot g_{\alpha\beta} - 2 \cdot \left(f_{,\alpha\beta} - \Gamma^{\gamma}_{\alpha\beta} f_{,\gamma} \right) \right) \delta G^{\alpha\beta}. \qquad (1966)$$

Thus, when aiming for a solution to (1965), we use (1946) and can plug it into our resulting big scale or gravity equation:

$$0 = \left(\overset{***}{R_{\alpha\beta}} + R_{\alpha\beta} - \frac{1}{2} \cdot R \cdot g_{\alpha\beta} \right) \delta G^{\alpha\beta} \quad \Rightarrow \quad \overset{***}{R_{\alpha\beta}} + R_{\alpha\beta} - \frac{1}{2} \cdot R \cdot g_{\alpha\beta} = 0 \quad (1967)$$

when demanding the integrand of (1965) to vanish. As these are just the classical Einstein field equations in vacuum, the result is well known. It reads:

$$\overset{***}{R_{\alpha\beta}} + R_{\alpha\beta} - \frac{R}{2} g_{\alpha\beta} =$$

$$= \begin{pmatrix} -\dfrac{3 \cdot c^2 \cdot k}{f[t]^2} & 0 & 0 & 0 \\[4mm] 0 & \dfrac{k}{\left(1 + \dfrac{k}{4} \cdot \left(\dfrac{r}{R} \right)^2 \right)^2} & 0 & 0 \\[8mm] 0 & 0 & \dfrac{k \cdot r^2}{\left(1 + \dfrac{k}{4} \cdot \left(\dfrac{r}{R} \right)^2 \right)^2} & 0 \\[8mm] 0 & 0 & 0 & \dfrac{k \cdot r^2 \cdot \sin(\vartheta)^2}{\left(1 + \dfrac{k}{4} \cdot \left(\dfrac{r}{R} \right)^2 \right)^2} \end{pmatrix} \quad (1968)$$

and we find the only possible solution via:

$$k = 0. \quad (1969)$$

This means that the space had to be flat in order to allow for a homogeneous and isotropic universe on a cosmological scale. Things would be different, of course, when allowing for a cosmological constant or matter terms, but we are not going to consider these options here. In fact, we consider the artificially added matter and also the cosmological constant unnecessary on a cosmic scale. Admittedly, we have shown that additional dimensions show up as matter (e.g., [6, 7], previous chapters, and section "12.5 Discussion" further below in this chapter), but here we are not considering such additional dimensions, and thus we should not—artificially—leave the path of pure geometry and incorporate terms we do not know where they come from (geometrically come from, we mean). This may be clear for the matter term, while for the cosmological constant, we simply state that—until now—there is no experimental evidence for this constant to be anything but zero. Ok, we have the observation of an accelerated expanding universe, but such an outcome can easily be constructed within our theory. Simply by changing (1946) to:

$$G_{\alpha\beta} = f[t]^2 \cdot \begin{pmatrix} -c^2 \cdot \dfrac{h[t]^2}{f[t]^2} \cdot & 0 & 0 & 0 \\[2em] 0 & \dfrac{1}{\left(1+\dfrac{k}{4}\cdot\left(\dfrac{r}{R}\right)^2\right)^2} & 0 & 0 \\[2em] 0 & 0 & \dfrac{r^2}{\left(1+\dfrac{k}{4}\cdot\left(\dfrac{r}{R}\right)^2\right)^2} & 0 \\[2em] 0 & 0 & 0 & \dfrac{r^2\cdot\sin(\vartheta)^2}{\left(1+\dfrac{k}{4}\cdot\left(\dfrac{r}{R}\right)^2\right)^2} \end{pmatrix}$$

$$k = \{-1,0,1\}, \tag{1970}$$

we obtain:

$$k = 0 \tag{1971}$$

again, with the function $h[t]$ being totally arbitrary.

12.4 The Sub-Space Idea and the Global Time

Assuming that the space we see consists of many smaller n spherical objects, we want to investigate their t dependency. In [24], we found the following solution to (1817) in the case of $n = 4$:

$$G_{\alpha\beta} = C_1 \cdot f[t]^2 \cdot \begin{pmatrix} -c^2 & 0 & 0 & 0 \\ 0 & t^2 & 0 & 0 \\ 0 & 0 & t^2\cdot\sin(r)^2 & 0 \\ 0 & 0 & 0 & t^2\cdot\sin(r)^2\cdot\sin(\vartheta)^2 \end{pmatrix}$$

$$f[t] = t^{-1\pm i\cdot c} \cdot C_f. \tag{1972}$$

It was shown in [24] that this is just the Robertson–Walker metric with r being treated like an angle. The following transformation:

$$\sin(r) = r_{RW} \cdot \left(1 + \frac{r_{RW}^2}{4}\right)^{-1} \tag{1973}$$

would give us back a $F[f[t]]$-scaled Robertson–Walker like metric with the scale being applied on all metric components and not just on the spatial ones. We realize that while r has become an angle, t took over the position of a radius. One could even assume the object to just "live" with a fixed time-shell $t = \rho$ and obtain a solution as follows (c.f. section 11.9):

$$G_{\alpha\beta} = C_1 \cdot f[t]^2 \cdot \begin{pmatrix} -c^2 & 0 & 0 & 0 \\ 0 & \rho^2 & 0 & 0 \\ 0 & 0 & \rho^2 \cdot \sin(r)^2 & 0 \\ 0 & 0 & 0 & \rho^2 \cdot \sin(r)^2 \cdot \sin(\vartheta)^2 \end{pmatrix}$$

$$f[t] = e^{\pm \frac{i \cdot c}{\rho} \cdot t} \cdot C_f. \tag{1974}$$

The same can be found in 3 dimensions as follows:

$$G_{\alpha\beta} = C_1 \cdot f[t]^4 \cdot \begin{pmatrix} -c^2 & 0 & 0 \\ 0 & t^2 & 0 \\ 0 & 0 & t^2 \cdot \sin(r)^2 \end{pmatrix}$$

$$f[t] = \sqrt{t^{\pm i \cdot c - 1}} \cdot C_f. \tag{1975}$$

As before in 4 dimensions, we see that r has become an angle while t took over the position of the radius. The shell-like object would be given as follows:

$$G_{\alpha\beta} = C_1 \cdot f[t]^4 \cdot \begin{pmatrix} -c^2 & 0 & 0 \\ 0 & \rho^2 & 0 \\ 0 & 0 & \rho^2 \cdot \sin(r)^2 \end{pmatrix}$$

$$f[t] = e^{\pm \frac{i \cdot c}{2 \cdot \rho} \cdot t} \cdot C_f. \tag{1976}$$

Now, as we see that ρ clearly is a length, we want to derive its properties. This we already did in section 11.9, but as the result is quite surprising, because it connects Dirac's particle at rest without cosmological investigation, we are here going to repeat it. From basic Quantum Theory, we know that a particle at rest has the time dependency:

$$f[t] = e^{\pm i \cdot \frac{m \cdot c^2}{\hbar} \cdot t} \cdot C_f, \qquad (1977)$$

with m giving the rest mass of the particle and $\hbar$ denoting the reduced Planck constant. Comparing with the $f[t]$ function from the metric solution (1976), we find:

$$\frac{m \cdot c}{\hbar} = \frac{1}{2 \cdot \rho}. \qquad (1978)$$

Inserting the Schwarzschild radius $r_s = \dfrac{2 \cdot m \cdot G}{c^2}$ (G is Newton's constant and c the speed of light in vacuum), thereby substituting the rest mass m, leaves us with:

$$\frac{r_s \cdot c^3}{2 \cdot \hbar \cdot G} = \frac{1}{2 \cdot \rho} \quad \Rightarrow \quad \rho = \frac{1}{r_s} \cdot \left(\frac{c^3}{\hbar \cdot G} \right)^{-1} = \frac{\ell_p^2}{r_s}. \qquad (1979)$$

Here ℓ_p denotes the Planck length.

By the way, in order to have the right units for the t spheres in (1972) and (1975), we should introduce spatial "speed of light" velocities c_x and obtain the following dimensionally improved solutions:

$$G_{\alpha\beta} = C_1 \cdot f[t]^2 \cdot \begin{pmatrix} -c^2 & 0 & 0 & 0 \\ 0 & c_x^2 \cdot t^2 & 0 & 0 \\ 0 & 0 & c_x^2 \cdot t^2 \cdot \sin(r)^2 & 0 \\ 0 & 0 & 0 & c_x^2 \cdot t^2 \cdot \sin(r)^2 \cdot \sin(\vartheta)^2 \end{pmatrix}$$

$$f[t] = t^{-1 \pm i \cdot c / c_x} \cdot C_f, \qquad (1980)$$

$$G_{\alpha\beta} = C_1 \cdot f[t]^4 \cdot \begin{pmatrix} -c^2 & 0 & 0 \\ 0 & c_x^2 \cdot t^2 & 0 \\ 0 & 0 & c_x^2 \cdot t^2 \cdot \sin(r)^2 \end{pmatrix}$$

$$f[t] = \sqrt{t^{\pm i \cdot c / c_x - 1}} \cdot C_f. \qquad (1981)$$

With (1974) and (1976) we find pulsating "universes" and are strongly reminded of the suggestion of the "Friedmanns" in Bodan's adventurous stories [25, 26][2], where Bodan suggested little, potentially pulsating universes forming our space-time. And what, so it was asked in Bodan's stories, forms those little "Friedmann-universe?" The answer simply was: "Even smaller Friedmanns" and the attentive reader might already guess what the answer to the question "And what do the little Friedmanns of second order consist of?" would be. Thus, we have the suggestion of a self-similar structure of universes forming universes on different scales, thereby the smaller scale providing the building blocks for the next and so on.

The solutions (1980) and (1981) do also oscillate, but with a significantly t-dependent amplitude (Fig. 12.2).

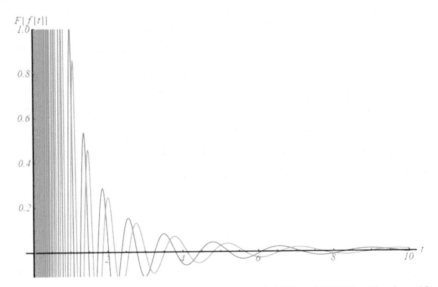

Figure 12.2 t-Dependency of the scale-solutions (1980) and (1981) with $c/c_x = 10$ (real part = blue, imaginary part = yellow).

As we see that the scale dependency in our "Friedmanns" is made by a t, acting as a radius coordinate, we might just assume a permanent jitter of all space points caused by these oscillations, which in consequence gives the global time. In other words, the space formed out of the oscillating and jittering Friedmann objects (1972), (1975), (1980), and (1981) (or corresponding

---

[2]One of the stories [25] is been reprinted in this book in the chapter "Seven Days Or how to Explain the World to my Dying Child."

objects in higher dimensions) has its own time due to the jittering of its building blocks. Thereby a great variety of vibrational options is at hand. For instance, one may use a simple generalization of (1980) and (1981) as follows:

$$G_{\alpha\beta} = C_1 \cdot f[t]^2 \cdot \begin{pmatrix} -c^2 & 0 & 0 & 0 \\ 0 & T^2 & 0 & 0 \\ 0 & 0 & T^2 \cdot \sin(r)^2 & 0 \\ 0 & 0 & 0 & T^2 \cdot \sin(r)^2 \cdot \sin(\vartheta)^2 \end{pmatrix}$$

$$T^2 = c_x^2 \cdot t^2 + \rho^2; \quad f[t] = \frac{\left(c_x \left(c_x \cdot t + |T|\right)\right)^{-\frac{i \cdot c}{cx}}}{|T|^{\frac{1}{2}}} \cdot C_f, \tag{1982}$$

$$G_{\alpha\beta} = C_1 \cdot f[t]^4 \cdot \begin{pmatrix} -c^2 & 0 & 0 \\ 0 & T^2 & 0 \\ 0 & 0 & T^2 \cdot \sin(r)^2 \end{pmatrix}$$

$$T^2 = c_x^2 \cdot t^2 + \rho^2; \quad f[t] = \frac{\left(c_x \left(c_x \cdot t + |T|\right)\right)^{-\frac{i \cdot c}{2 \cdot cx}}}{|T|^{\frac{1}{4}}} \cdot C_f. \tag{1983}$$

Obviously this is some kind of mix between the shell-like solutions (1972) and (1975) and the t-spheres (1980) and (1981). Figure 12.3 illustrates the outcome.

Please note that (1982) and (1983) are not just simple transformations with $T = T[t]$ from (1980) and (1981) as such would have to read:

$$G_{\alpha\beta} = \begin{pmatrix} C_1 \cdot f[T[t]]^2 & & & \\ & \begin{pmatrix} -c^2 \cdot T'[t]^2 & 0 & 0 & 0 \\ 0 & c_x^2 \cdot T[t]^2 & 0 & 0 \\ 0 & 0 & c_x^2 \cdot T[t]^2 \cdot \sin(r)^2 & 0 \\ 0 & 0 & 0 & c_x^2 \cdot T[t]^2 \cdot \sin(r)^2 \cdot \sin(\vartheta)^2 \end{pmatrix} \end{pmatrix}$$

$$f[t] = T[t]^{-1 \pm i \cdot c / c_x} \cdot C_f, \tag{1984}$$

$$G_{\alpha\beta} = C_1 \cdot f[T[t]]^4 \cdot \begin{pmatrix} -c^2 \cdot T'[t]^2 & 0 & 0 \\ 0 & c_x^2 \cdot T[t]^2 & 0 \\ 0 & 0 & c_x^2 \cdot T[t]^2 \cdot \sin(r)^2 \end{pmatrix}$$

$$f[t] = \sqrt{T[t]}^{\pm i \cdot c / c_x - 1} \cdot C_f. \tag{1985}$$

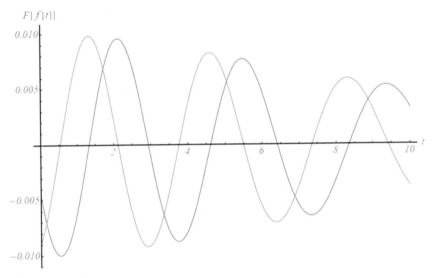

Figure 12.3 t-Dependency of the scale solutions (1982) and (1983) with $c = 10$, $c_x = 1$ and $\rho = 10$ (real part = blue, imaginary part = yellow).

12.5 Discussion

Unless we do not artificially incorporate (per postulate) matter or demand a non-zero cosmological constant, the newly found quantum Einstein field equations demand a flat space when we assume the classical part, which is to say the classical Einstein field equations, to be dominant on bigger scales. This holds in a spacetime of 4 dimensions under the assumption of cosmological homogeneity and isotropy, thus just as the cosmological principle requires.

We saw, however, that matter could be added without artificial matter terms, as introduced by Hilbert and Einstein [3, 4], simply via adding additional degrees of freedom to the system, which is to say dimensions (e.g., [6, 7]) or scaling functions. Thus, when considering such add-on dimensions into our system, we not only obtain matter but also—by entanglement of the additional dimensions with the classical spatial ones—obtain a certain distribution of the latter. This compromises the spatial homogeneity and isotropy, but only locally. On cosmological scales, we may just claim that the add-on dimensions are not important or smoothed out and by omitting them in the model, we have the Robertson–Walker metric as our starting approach. By inserting this metric into the new quantum Einstein field equations and select for suitable scale split-up options (here (1955)), we can automatically obtain the flat space condition.

Consequently, in order to explain the little inhomogeneities to be observed in the cosmological background radiation [21], we need to consider such add-on dimensions and their entanglement with the cosmological spatial and potentially also the cosmological time dimension. However, with these inhomogeneities to be found extremely small [21, 22], we are forced to conclude that the corrections coming from such add-on dimensions in comparison to the simple Robertson–Walker metric are relatively small, too. Nevertheless, further investigation is of need and we now know in which direction to go.

In addition, we found that the cosmological Robertson–Walker model might just consist of many small such "universes" forming the building blocks of our universe. Here we only considered two very simple forms of such mini-universes, but already found that due to the fact that they would show a permanent oscillation and thus jitter, we might have an explanation for our notion of time.

12.5.1 Repetition: A Quantum Origin of the Classical Energy– Momentum Tensor

Could perhaps the sum of V_{Qi} variations in our ensemble of scale split-up as given above (c.f. Eqs. (1955) to (1964)) act as the matter, which was introduced by Hilbert [3]?

Discovering the classical field equation structure (c.f. (1961)) among our ensemble of options, one might want to demonstrate the connection to the Einstein–Hilbert matter approach. We start with the assumption of one global metric $g_{\alpha\beta}$, but an ensemble of F_i and f_i, leading us to:

$$0 = \begin{pmatrix} \int_{V_G} d^n x \sqrt{-G} \left(\left(\overset{***}{R_{\alpha\beta}} + R_{\alpha\beta} - \frac{1}{2} \cdot R \cdot g_{\alpha\beta} \right) \delta G^{\alpha\beta} \right) \\ + \frac{1}{2} \cdot \sum_{i=1}^{N} \int_{V_{Qi}} d^n x \sqrt{-g} \cdot F^n \frac{F'}{F} (n-1) \left(\Delta f \cdot g_{\alpha\beta} - 2 \cdot \left(f_{,\alpha\beta} - \Gamma^\gamma_{\alpha\beta} f_{,\gamma} \right) \right) \delta G^{\alpha\beta} \end{pmatrix}$$

$$\Rightarrow$$

$$= \int_{V_G} d^n x \sqrt{-g} \left(\sum_{i=1}^{N} \sqrt{F_i^n} \cdot \left(\overset{***}{R_{\alpha\beta}} + R_{\alpha\beta} - \frac{1}{2} \cdot R \cdot g_{\alpha\beta} + \frac{(n-1)}{2} \cdot \frac{F_i'}{F_i} \left(\Delta f_i \cdot g_{\alpha\beta} - 2 \cdot f_{i;\alpha\beta} \right) \right) \delta G^{\alpha\beta} \right).$$

$$(1986)$$

This gives us the classical non-vacuum field equations via:

$$0 = \sum_{i=1}^{N} \sqrt{F_i^n} \cdot \left(R_{\alpha\beta} - \frac{1}{2} \cdot R \cdot g_{\alpha\beta} \right) + \sum_{i=1}^{N} \sqrt{F_i^n} \cdot \left(\frac{(n-1)}{2} \cdot \frac{F_i'}{F_i} \left(\Delta f_i \cdot g_{\alpha\beta} - 2 \cdot f_{i;\alpha\beta} \right) \right)$$

$$\Rightarrow$$

$$0 = \overset{***}{R_{\alpha\beta}} + R_{\alpha\beta} - \frac{1}{2} \cdot R \cdot g_{\alpha\beta} + \frac{\sum_{i=1}^{N} \sqrt{F_i^n} \cdot \left(\frac{(n-1)}{2} \cdot \frac{F_i'}{F_i} \left(\Delta f_i \cdot g_{\alpha\beta} - 2 \cdot f_{i;\alpha\beta} \right) \right)}{\sum_{i=1}^{N} \sqrt{F_i^n}} \quad (1987)$$

and the last term in the last line above is handing us the quantum or scale equivalent of the Einstein–Hilbert matter term, respectively the energy–momentum tensor:

$$\Rightarrow \quad \kappa T_{\alpha\beta} = \frac{\overset{***}{R_{\alpha\beta}} + \sum_{i=1}^{N} \sqrt{F_i^n} \cdot \left(\frac{(n-1)}{2} \cdot \frac{F_i'}{F_i} \left(\Delta f_i \cdot g_{\alpha\beta} - 2 \cdot f_{i;\alpha\beta} \right) \right)}{\sum_{i=1}^{N} \sqrt{F_i^n}}. \quad (1988)$$

For entertainment and illustration, we have also derived resulting "effective matter" for an alternative scale split-up. This was presented in section 11.6.10 and will not be repeated here.

References

1. N. Schwarzer, *Towards Quantum Einstein Field Equations, Part 7 of Medical Socio-Economic Quantum Gravity*, Self-published, Amazon Digital Services, December 2020, Kindle.

2. N. Schwarzer, *The Quantum Einstein Field Equations in Their Simplest Form, Part 7a of Medical Socio-Economic Quantum Gravity*, Self-published, Amazon Digital Services, March 2021, Kindle.

3. D. Hilbert, Die Grundlagen der Physik, Teil 1, *Göttinger Nachrichten*, 1915, 395–407.

4. A. Einstein, Grundlage der allgemeinen Relativitätstheorie, *Ann. Phys. (ser. 4)*, 1916, **49**, 769–822.

5. N. Schwarzer, *Societons and Ecotons—The Photons of the Human Society: Control them and Rule the World, Part 1 of Medical Socio-Economic Quantum Gravity*, Self-published, Amazon Digital Services, 2020, Kindle.

6. N. Schwarzer, *Masses and the Infinity Options Principle: Can We Explain the 3-Generations and the Quantized Mass Problem?, Part 5 of Medical Socio-Economic Quantum Gravity*, Self-published, Amazon Digital Services, 2020, Kindle.

7. N. Schwarzer, *The World Formula: A Late Recognition of David Hilbert's Stroke of Genius*, 2022, Jenny Stanford Publishing. ISBN: 9789814877206.

8. N. Schwarzer, *The Metric Dirac Equation Revisited and the Geometry of Spinors, Part 8 of Medical Socio-Economic Quantum Gravity*, Self-published, Amazon Digital Services, December 2021, Kindle.

9. N. Schwarzer, *The 3 Generations of Elementary Particles, Part 6 of Medical Socio-Economic Quantum Gravity*, Self-published, Amazon Digital Services, December 2020, Kindle.

10. N. Schwarzer, *My Horcruxes—A Curvy Math to Salvation, Part 9 of Medical Socio-Economic Quantum Gravity*, Self-published, Amazon Digital Services, 2021, Kindle.

11. www.astro.umd.edu/~richard/ASTRO340/class23_RM_2015.pdf.

12. en.wikipedia.org/wiki/Flatness_problem.

13. A. Friedman, Über die Krümmung des Raumes, *Z. Phys.* (in German), 1922, **10**(1), 377–386, Bibcode:1922ZPhy...10..377F, doi: 10.1007/BF01332580 (English translation: A. Friedman, On the curvature of space, *Gen. Relativ. Gravitation*, 1999, **31**(12), 1991–2000, Bibcode:1999GReGr..31.1991F, doi:10.1023/A:1026751225741).

14. A. Friedmann, Über die Möglichkeit einer Welt mit konstanter negativer Krümmung des Raumes, *Z. Phys.* (in German), 1924, **21**(1), 326–332, Bibcode:1924ZPhy...21..326F, doi:10.1007/BF01328280 (English translation: A. Friedmann, On the possibility of a world with constant negative curvature of space, *Gen. Relativ. Gravitation*, 1999, **31**(12): 2001–2008, Bibcode:1999GReGr..31.2001F, doi:10.1023/A:1026755309811).

15. A. H. Guth, Fluctuations in the new inflationary universe, *Phys. Rev. Lett.*, 1982, **49**(15), 1110–1113, Bibcode:1982PhRvL..49.1110G, doi: 10.1103/PhysRevLett.49.1110.

16. A. Linde, A new inflationary universe scenario: a possible solution of the horizon, flatness, homogeneity, isotropy and primordial monopole problems, *Phys. Lett. B*, 1982, **108**(6), 389–393, Bibcode:1982PhLB..108..389L, doi: 10.1016/0370-2693(82)91219-9.

17. St. Hawking, Th. Hertog, A smooth exit from eternal inflation, *J. High Energy Phys.*, 2018, arXiv:1707.07702, Bibcode:2018JHEP...04..147H, doi: 10.1007/JHEP04(2018)147.

18. S. F. Bramberger, A. Coates, J. Magueijo, S. Mukohyama, R. Namba, Y. Watanabe, Solving the flatness problem with an anisotropic instanton in Hořava-Lifshitz gravity, *Phys. Rev. D*, 2018, **97**, 043512, arXiv:1709.07084.

19. N. Schwarzer, *Einstein Had It, but He Did Not See It, Part LXXVI: Quantum Universes—We Don't Need No ... an Inflation*, Self-published, Amazon Digital Services, 2019, Kindle.

20. H. Goenner, *Einführung in die spezielle und allgemeine Relativitätstheorie* (in German), Spektrum Akad. Verlag, Heidelberg, Berlin, Oxford, 1996, ISBN: 3-86025-333-6.

21. en.wikipedia.org/wiki/Cosmic_microwave_background.

22. Y. Akrami, et al. (Planck Collaboration), Planck 2018 results. I. Overview, and the cosmological legacy of Planck, *Astron. Astrophys.*, 2020, **641**, A1. arXiv:1807.06205. Bibcode:2020A&A...641A...1P. doi:10.1051/0004-6361/201833880. S2CID 119185252.

23. N. Schwarzer, *Science Riddles—Riddle No. 13: How to Solve the Flatness Problem?*, Self-published, Amazon Digital Services, 2019, Kindle.

24. N. Schwarzer, *The Quantum Black Hole, Part 7b of Medical Socio-Economic Quantum Gravity*, Self-published, Amazon Digital Services, March 2021, Kindle.

25. T. Bodan, *7 Days—How to Explain the World to My Dying Child, in German—7 Tage: Wie erkläre ich meinem sterbenden Kind die Welt*, Self-published, BoD Classic, ISBN 9783752639728.

26. T. Bodan, *The Eighth Day—Two Jews against the Third Reich*, illustrated version, Self-published, BoD Classic, 2021, ISBN 9783753417257.

27. N. Schwarzer, *The Theory of Everything: Quantum and Relativity Is Everywhere—A Fermat Universe*, Jenny Stanford Publishing, 2020, ISBN-10: 9814774472.

28. N. Schwarzer, *General Quantum Relativity*, Self-published, Amazon Digital Services, 2016, Kindle.

29. N. Schwarzer, *The Dirac Miracle, Part 8a of Medical Socio-Economic Quantum Gravity*, Self-published, Amazon Digital Services, May 2021, Kindle.

Chapter 13

Mathematical Tools for Socioeconomic and Psychological Simulations

13.1 Mass Formation Psychosis as an Example

13.1.1 About Math as a Tool in Psychology and Socioeconomics

Many people are of the opinion that it is impossible to mathematically describe human behavior in a satisfactory manner. Thereby they accept millions of artificial intelligence solutions doing exactly this in every second and with respect to almost every aspect of their life.

It is widely agreed that human beings can be characterized by a list of attributes, which partially entangle with each other ... thereby we mean, that the attributes entangle, not the humans. Even though this list may be extremely long and the interaction (or entanglement) of the various attributes might be very complex, but in essence we are characterizable by such a list. After all, nothing else is it when we try to—more or less comprehensively and fairly—describe another person or—usually with much less vigor, enthusiasm and rigorousness—ourselves.

Accepting the fact that humans can be described by a list of entangled attributes, however, still does leave us with the question how to mathematically handle this list in order to describe processes of action, human interaction, individuality and other characteristics. Most surprisingly, this work was

The Math of Body, Soul, and the Universe
Norbert Schwarzer
Copyright © 2023 Jenny Stanford Publishing Pte. Ltd.
ISBN 978-981-4968-24-9 (Hardcover), 978-1-003-33454-5 (eBook)
www.jennystanford.com

already done. Over 100 years ago, it was the great mathematician David Hilbert who found a way to put a set of properties, attributes, degrees of freedom or—as all these other terms are just coding—**dimensions** into a mathematical formalism, which was already in his time known for years as the one fundamental principle for the determination of the laws of our universe, the Hamilton extremal principle. All truly fundamental natural laws are originating from this principle. The only problem is to find the right ingredients to "feed the monster of the Hamilton principle." The essential ingredient namely, the Lagrange density function, seems to differ from problem to problem and this cannot be when assuming that the Hamilton mechanism should be truly fundamental. Here now came the stroke of genius of Hilbert [1], who realized that a given set of attributes is nothing else but a set of dimensions and thus, forms a mathematical space or space-time. The corresponding Lagrange density function is then always the same, namely the so-called Ricci scalar of the very system or space-time. Hilbert applied his calculus on our ordinary 4-dimensional space-time and derived the Einstein field equations and thus, in just one fluid and conclusive derivation, the complete mathematical basis for the General Theory of Relativity of Albert Einstein [2]. Admittedly, Hilbert never extended his calculus to the space or space-time of attributes characterizing human beings and/or their societies, but the fact that the title of his work [1], in which he derived Einstein's great theory, translates "The Fundaments of Physics" clearly shows us that he saw his approach in a much more general way than just the minimum-principle-origin of Einstein's field equations. Thus, no matter the system we are considering, the moment we have its basic attributes and know their entanglement (various interactions), we also immediately know how to place all this knowledge into Hilbert's calculus to come up with a most fundamental and unbiased model for this very system.

Hilbert's creation, de facto, was—no, is—a Theory of Everything [3] or World Formula [4–19], even though he himself had little chance of fully realizing this. Even in physics, where we now can show that Hilbert's fundamental equation covers both great theories, the General Theory of Relativity and the Quantum Theory, the time was not ripe for such a discovery, simply because the mathematical apparatus of the Quantum Theory was not fully developed then. While Hilbert brought out his great work in 1915 and knew about the Einstein field equations at the time, the basic quantum equations such as the Schrödinger, the Klein–Gordon and the Dirac equation would not follow before the second half of the 1920s.

Even though we will only illustratively discuss the process of mass formation psychosis in this chapter, for convenience, falsifiability and rigorousness, we are pointing out that all necessary equations are given in this book. The reader finds the corresponding extended Hilbert approach in chapter 11, especially Eqs. (1643), (1649), and (1804). From these equations, suitable solutions for 5- to 10-dimensional space-times, fit for simple psychological or socioeconomic simulations, are presented in section 11.5.

13.1.2 Mass Formation Psychosis

In this section, where we want to concentrate on the illustrative discussion of the application of math to potentially and effectively describe this psychological process of the masses, we are only going to give a very brief overview about the process of mass formation and mass formation psychosis. With respect to more information, we refer to Prof. Mattias Desmet's book about "The Psychology of Totalitarianism" [20].

Contrary to dictatorships, where obedience is created from a basic fear of the dictator, totalitarianism produces obedience via mass hypnosis. In psychology, this mass hypnosis is called mass formation.

Such a process requires four conditions:

(1) The masses must feel alone and isolated.
(2) Their lives must feel pointless and meaningless.
    These conditions have been growing for years with social media (i.e., narcissism, sociopathy, addiction and mental illness) and an education system which generally suppresses holistic analysis and broader, if not say lateral, thinking. In other and very simple words, people are mentally squeezed into boxes and these boxes are not just becoming smaller and smaller, they are also having lesser and lesser dimensions or attributes. It is therefore not surprising that psychological studies [20] and the daily experience show us a greater susceptibility to totalitarian control mechanisms with higher grades of education. The education system creates box-thinkers and Pavlov dogs instead of holistic thinkers and independent true individuals.
(3) The masses then must experience constant free-floating anxiety.
(4) They must experience free-floating frustration and aggression.

Free-floating means that there is no discernible source for anxiety or aggression. And so, the person begins to irrationally crave a remedy, no matter how absurd or destructive it may be.

The moment these conditions are met, people are ripe for hypnosis. It suffices to present a "solution" to them.

No matter how stupid, senseless or even harmful the "solution" is, people are made to feel solidarity, to contribute in some free-floating, undiscernible way, when following the proposed "solution." This validates the whole thing for them and they are becoming blind towards the damage the remedy might cause to them, to the society and yes, even their own children.

They are lesser than Pavlov's famous dogs, but even degenerated ones.

They are now changed, no longer rational.

People become intolerant and cruel.

By the way, when trying to understand this process, some experts try to apply game theory. However, the creation of irrationality as one goal of the process automatically excludes mathematical approaches like Nash's game theory, where all agents are assumed to act completely rationally.

13.1.3 Illustrative "Mathematical," Quantum Gravity Discussion of the Process of Mass Formation Psychosis

In this discussion, we will now aim for an effective mathematical quantum gravity description of the process of mass formation psychosis. We will do so without explicitly introducing the math again, but only refer to the corresponding mathematical parts of this book.

The important aspect for our goal of a mathematical description of the process of mass formation requires the understanding and falsifiable presentation of dimensional mass creation and its entanglement with the individual fields of all agents or players in a given socioeconomic space-time.

In appendix A1 of chapter 11, we concentrated on the metric derivation of the Klein–Gordon equation and mass as one of the essential parameters in this very equation. From there, it is simple to also extract the metric appearance of mass for Dirac (appendix A2 of chapter 11), Schrödinger (appendix B of chapter 11) and any other quantum equation also for arbitrarily big systems. The reader finds an extensive introduction to the process of dimensional mass creation in chapter 6 of his book. An example for the dimensional mass creation in a 10-dimensional space-time is given in section 11.5.5.

We could have a simple model, which does not aim for a "perfect" mirroring of a society of individuals, because this would not be possible even on the best supercomputers, but only for the illustration of the method on a small scale. In such an approach, the individuality is illustrated as wave functions (which actually it is anyway). The more dynamic, healthy and active the individual is, the more complex the multidimensional orbital appearance of the individual space-time wave density. It is also important to note that the individuality requires the wave function to be fermionic, which means it follows the Pauli exclusion principle and wants to keep its own individuality. Suitable solutions can be obtained directly from our derivation in section 11.5 of this book. There is nothing artificial about this picture, because, as explained in the introduction part to this chapter, all our oh-so-human individuality summed up is just a huge set of attributes (partially entangled) and each attribute contributes to the complexity and appearance of the individual's very own halo (psi-field some call it) in the space-time of society. Of course, we cannot show the full dimensionality of the wave function, because we are unable to visualize more than 3 dimensions, but we can sum up a few dimensions and illustrate the result in the 3 dimensions (c.f. Figs. 11.3 to 11.36). This still gives us at least a feeling about the appearance of a person's individual field inside a "socioeconomic space-time" (just another word for society). So, to give an example, the wave function figures given in chapter 11 were actually calculated from individual fields of up to 10 dimensions of which only the 3 spatial ones are being shown. In this chapter, we used some of the solutions given in chapter 11 to evaluate examples of such wave fields (Fig. 13.1).

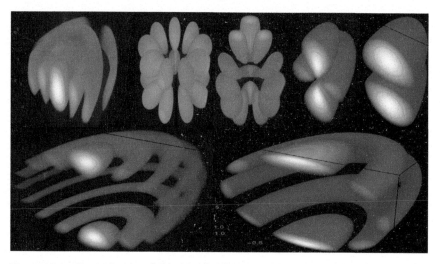

Figure 13.1 Example wave fields from half spin solutions given in chapter 11.

Please note that none of the fields has any sharp structures, but always comes as a slightly fuzzy density distribution. This is in perfect congruence with its natural equivalent, namely the field of individuality. This, too, can never be a sharp object but always only is a nebular something. It even needs to be pointed out here that the sharper such an orbital object becomes, the lower its creative content, its innovative potential, its individuality. The sharpest fields or orbitals are to be found near the ground states. Thus, when aiming for low individuality in order to herd the masses into a formation psychosis, the goal has to be to load the individual agents within a socioeconomic system with so much mass that they are forced into their ground states, where they not only lose their mobility, but also their potential to individually excite anything or bind to other individuals. In other words, they stop thinking, communicating and their individual acting (seen from the physical meaning of action) and thus, their true human being.

In a mass formation psychosis, these individual fields need to be squashed, distorted, made less complex, made less fermionic ($\rightarrow$ bosonic then) in order to switch off the Pauli principle and make the individual compliant and ready for compactification ... in other words, we need a process of deindividualization. This is done by external—free floating—fields and one already sees that anxiety is just one such field. In both, socioeconomic reality and the model, such a field requires the introduction or activation of an additional degree of freedom (could be a fear factor against a certain omnipresent entity like a virus). But this does not suffice, because there can always be threats around and nobody gives a damn about them. In order to make the process work, there also needs to be a strong entanglement with dimensions defining the individuals (again it is the same in reality and the model, which is to say the quantum gravity model already is quite realistic). And here the media, politicians and false experts come in with their fear mongering. This entanglement then automatically creates what in physics is called inertia and as inertia always distorts the space-time they are residing in, we have the desired distortion and squashing of the individual wave fields.

And there is another important effect.

We know that the increase in inertia due to an additional (free floating) field of fear degenerates individuality depending on the strength of entanglement (coupling of the fear-field with the individual wave functions or metrics). More inertia, however, leads to a lower probability of excited states (ideas, innovation, evolutionary progressive freedom, readiness to assume a productive risk, seeking new opportunities, ...) and a higher probability of ground states of mainly bosonic character (compliance).

With respect to the current COVID-19 crisis, we think that the statistical data of vaccine compliance on the one hand and the innovative potential, activity or even job-market developments, on the other, would directly fit into the model of inertia and the multitudes of states of the individual wave-fields. Stagnation or even decline here (in reality) would be directly correlated to a lower number of exited states (or a shift of their local distribution), more bosonic wave-fields and much more confined orbitals in the model. As we know that this reduction of individuality in the model comes from the inertia produced by the fear-field, we can extract some measurable/quantifiable effect from the fear-mongering and fear-entangling taking place in the real world.

In other words, we would have a coupling constant and could make your model more robust (mathematically we mean) and thus, also falsifiable in a Karl Popper–like sense.

All these smaller, distorted and bosonized (de-fermionized) residuals of those former healthy individuals are then ready for herding into whatever small barn and way of life they (whoever they is) want the masses of former healthy individuals in (Fig. 13.2).

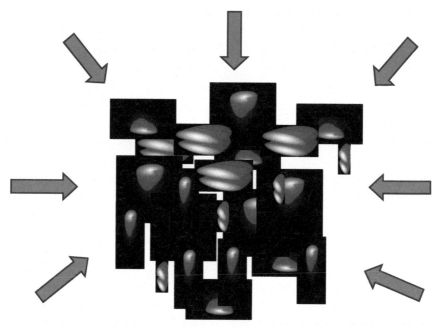

Figure 13.2 Herding: Distorted and deindividualized wave fields are ready for compactification (see the text).

Not just in the model, but also in reality, the individual with such a degenerated individuality will accept very little for itself ("having nothing but being happy all the same," which, in fact, is rubbish as it cannot work for true human beings ... at least not for elongated time spans).

But there are severe problems involved.

Problem A with all this: Because the add-on dimensions of fear (after all it is a free-floating and thus non-graspable anxiety) produces entanglement not only with the target group but in general (per definition it cannot be contained), the whole will also affect the source. It is bound to backfire onto the group who created the fear-field, and this makes it very likely that the whole system collapses in the end (black hole formation).

Problem B with all this: Humans are humans because they have individuality. They are defined by it and they define themselves throughout their individuality. Striping them off it makes each and every individual degenerated wave function unstable and renders it ready for all sorts of phase transitions ... in other words, violence is one of the more likely outcomes.

As we saw that the process of the creation of inertia (mass) is the fundament for both, the practical process and the model of the mass formation psychosis, we again refer to chapter 6 and section 11.5.5, where the process of dimensional mass creation is elaborated in quite some detail. After all, this aspect is essential for the understanding of the process of deindividualization (and thus, perhaps also for seeing possibilities to hold off this process or even reverse it). The math is not too complicated and we even think that the fact that we all have this inner picture of ourselves losing individuality and psychologically crumbling when being loaded with too much "mass" or "mess" (whatever type of mass or mess it is) helps in understanding why such a Hilbert approach is anything but far off. As this process is somehow the key, we probably have to endure the math in here, because this is the tool to truly and fundamentally understand the whole process.

References

1. D. Hilbert, *Die Grundlagen der Physik*, Teil 1, Göttinger Nachrichten, 395–407 (1915).

2. A. Einstein, Grundlage der allgemeinen Relativitätstheorie, *Ann. Phys.*, 1916, **49** (ser. 4), 769–822.

3. N. Schwarzer, *The Theory of Everything—Quantum and Relativity Is Everywhere—A Fermat Universe*," Pan Stanford Publishing, (2020), ISBN-10: 9814774472.

4. N. Schwarzer, *The World Formula: A Late Recognition of David Hilbert's Stroke of Genius*, 2022, Jenny Stanford Publishing. ISBN: 9789814877206.

5. N. Schwarzer, *Societies and Ecotons—The Photons of the Human Society—Control Them and Rule the World*, Part 1 of "Medical Socio-Economic Quantum Gravity," Self-published, Amazon Digital Services, 2020, Kindle.

6. N. Schwarzer, *Humanitons—The Intrinsic Photons of the Human Body—Understand Them and Cure Yourself*, Part 2 of "Medical Socio-Economic Quantum Gravity," Self-published, Amazon Digital Services, 2020, Kindle.

7. N. Schwarzer, *Mastering Human Crises with Quantum-Gravity-Based but Still Practicable Models—First Measure: SEEING and UNDERSTANDING the WHOLE*, Part 3 of "Medical Socio-Economic Quantum Gravity," Self-published, Amazon Digital Services, 2020, Kindle.

8. N. Schwarzer, *Self-Similar Quantum Gravity—How Infinity Brings Simplicity*, Part 4 of "Medical Socio-Economic Quantum Gravity," Self-published, Amazon Digital Services, 2020, Kindle.

9. N. Schwarzer, *Masses and the Infinity Options Principle—Can We Explain the 3-Generations and the Quantized Mass Problem?*, Part 5 of "Medical Socio-Economic Quantum Gravity," Self-published, Amazon Digital Services, 2020, Kindle.

10. N. Schwarzer, *The 3 Generations of Elementary Particles*, Part 6 of "Medical Socio-Economic Quantum Gravity," Self-published, Amazon Digital Services, December 2020, Kindle.

11. N. Schwarzer, *Towards Quantum Einstein Field Equations*," Part 7 of "Medical Socio-Economic Quantum Gravity," Self-published, Amazon Digital Services, December 2020, Kindle.

12. N. Schwarzer, *The Quantum Einstein Field Equations in Their Simplest Form*," Part 7a of "Medical Socio-Economic Quantum Gravity," Self-published, Amazon Digital Services, March 2021, Kindle.

13. N. Schwarzer, *The Quantum Black Hole*, Part 7b of "Medical Socio-Economic Quantum Gravity," Self-published, Amazon Digital Services, March 2021, Kindle.

14. N. Schwarzer, *Quantum Cosmology—A Simple Answer to the Flatness Riddle*, Part 7c of "Medical Socio-Economic Quantum Gravity," Self-published, Amazon Digital Services, April 2021, Kindle.

15. N. Schwarzer, *Strings, Branes, Friedmanns and the Matryoshka Universe*, Part 7d of "Medical Socio-Economic Quantum Gravity," Self-published, Amazon Digital Services, December 2021, Kindle.

16. N. Schwarzer, *The Metric Dirac Equation Revisited and the Geometry of Spinors*, Part 8 of "Medical Socio-Economic Quantum Gravity," Self-published, Amazon Digital Services, December 2020, Kindle.

17. N. Schwarzer, *The Dirac Miracle*, Part 8a of "Medical Socio-Economic Quantum Gravity," Self-published, Amazon Digital Services, May 2021, Kindle.

18. N. Schwarzer, *My Horcruxes—A Horrible Math to Salvation*, Part 9 of "Medical Socio-Economic Quantum Gravity," Self-published, Amazon Digital Services, 2021, Kindle.

19. N. Schwarzer, *How Will the Nearby Unification of Physics Change Future Warfare—A Brief Study*, Part 10 of "Medical Socio-Economic Quantum Gravity," Self-published, Amazon Digital Services, June 2021, Kindle.

20. M. Desmet, *The Psychology of Totalitarianism*, 2022, Chelsea Green Publishing, ISBN: 9781645021728.

THE SEVENTH DAY: GIVE IT SOME REST

Chapter 14

About the Origin of the Minimum Principle

14.1 Brief Story about the Minimum Principle—Part II

I was once again with my Schabracke on wanderings. Our destination was the dike, where we wanted to look over the Bodden and have a little picnic.

In the meantime I was feeling much better. After all, I was about to finish my book and the burden of so many pages still to be written started to disappear. Schabracke, as I always called my 9-year-old daughter, marched along beside me and told me about a pair of butterflies she had observed the other day.

"One of the two butterflies," she explained, "certainly the female, was dead and the other, which was then the male, was always fluttering around his lifeless mate. It was kind of very sad."

"Hmm!" was all I could think of to say.

"The butterfly didn't flee either when I got really close, bent over the two of them, and could have touched him at any time," she continued to report.

"It was as if it also wanted to die where its mate lay."

"Hmm!"

We walked side by side in silence for quite a while. Then we reached a piece of meadow where just then many butterflies were fluttering around seemingly full of life.

"You see, Daddy," my daughter suddenly said, "life goes on."

The Math of Body, Soul, and the Universe
Norbert Schwarzer
Copyright © 2023 Jenny Stanford Publishing Pte. Ltd.
ISBN 978-981-4968-24-9 (Hardcover), 978-1-003-33454-5 (eBook)
www.jennystanford.com

And although I had never said anything to the contrary, at least I couldn't remember it at the moment, I immediately agreed with her:

"Yes, you're absolutely right, my little one!"

"You know, Daddy," Schabracke then began, "I've been thinking about nothingness again, too."

I perked up my ears, but said nothing.

"You know, the nothing that created the world for itself because it wanted to know why it was there in the first place, that nothing."

"Yes, of course I remember, my little one."

"So if you want a total of zero, nothingness, you need something like +1 and −1, right?"

"Yes!" I agreed with her.

"Fine, let's assume that +1 is the world, our world, then there would have to be a counterworld, that is, the −1 world, right?"

My daughter looked at me eagerly from the side. I, however, just went: "Hmm!"

In the meantime, we had reached the dike and sat down next to each other on a piece of wall edge. I took the backpack off my shoulder, poured us both some tea sweetened with honey, and also whipped out some chocolate chip cookies.

Schabracke nibbled with pleasure from the cookies and slurped audibly the still quite hot tea. Looking pensively across the Bodden, she suddenly said:

"Of course, one 1 could just be a 1, while the other makes up a whole world."

"What do you mean?" asked I in amazement.

"Well, if you have −1 and +1 and both together make nothingness, it's not a foregone conclusion that the +1 must always remain just a +1, is it?"

I just looked wide-eyed and said nothing. So Schabracke answered her question, which was obviously meant to be rhetorical anyway, herself and continued:

"Just as well the +1 could be 2 half or 4 quarters or something else, something much more split-up. Then one would see what would come out of it as something terribly complicated, thus a world for example, and somewhere there would still be a quite simple −1 which makes the world to nothing. Yet the −1 doesn't stand out at all because it seems so incredibly simple against the complicated world, and yet it's what makes the world possible in the first place."

'Wow!' is all I think, now digesting much more than a sip of tea and some cookie.

We are silent for a while, enjoying the view, the singing of the birds, the play of the butterflies that also exist here, and the wonderful fresh sea air.

"But," my little girl asks after having gulped down her last tea, "when all this just works so nicely on its own, what's God to do then?"

I'm about to say my "hmm?" again, but even before I can do so, Schabracke answers her question, which again—obviously—was more or less just a rhetorical one.

"Well," she says, "I think he just enjoys to have a little bit of spare time once in a while."

I'm about to chuckle, but can just catch myself just in time when I realize that Schabracke actually meant what she had said, because she looks at me so openly and honestly that I can only shrug and say:

"Probably?!?"

She raises an eyebrow in suspicion, because I may not have sounded very convincing, but then she just looks away and gets interested in a couple of swans landing on the water right in front of us.

Again, we are silent for quite some time.

"Tell me Dad, what actually comes out when you divide a quarter again?"

My daughter shows me with her little hands the angles of pieces of cake to make clear what she means.

I laugh and am so happy that she finally has a question that I am able to answer.

Index